T5-BCJ-392

Kuhlenbeck: The Central Nervous System of Vertebrates

Hartwig Kuhlenbeck

The Central Nervous System of Vertebrates

A General Survey of its Comparative Anatomy with an Introduction to the Pertinent Fundamental Biologic and Logical Concepts

S. Karger · Basel · München · Paris · London · New York · Sydney

Volume 4

Spinal Cord and Deuterencephalon

With 430 figures

19 75

S. Karger · Basel · München · Paris · London · New York · Sydney

S. Karger · Basel · München · Paris · London · New York · Sydney
Arnold-Böcklin-Strasse 25, CH-4011 Basel (Switzerland)

Printed in Switzerland by Pochon-Jent AG, Berne
ISBN 3-8055-1732-7

«Die vergleichende Anatomie, Histologie, Architektonik und Embryologie des Zentralnervensystems bildet ferner einen umfangreichen Zweig und zugleich eine unentbehrliche Methode der neurobiologischen Forschung. Sie verrät die zahlreichen Wege, durch welche die Evolution der Nervensysteme der verschiedene Tiersorten im phylogenetischen Zusammenhang ihre heutige Verschiedenartigkeit zustande gebracht hat. Vertieft man sich dabei genügend in den Zusammenhang von Form und Funktion, so gelangt man in eine wunderbare Welt der Harmonie zwischen Geist und lebendem Nervensystem...

Wer vergleichende Anatomie des Nervensystems sagt, sagt also auch vergleichende Physiologie – Psychologie und – Biologie, und das ist ein Gebiet, aus welchem die künftige Forschung mit vollen Zügen schöpfen kann...

Dass der Mensch für den Menschen sich zunächst interessiert, ist verzeihlich und naheliegend. Hat er aber einmal erkannt, dass er nur ein Glied in der Tierreihe bildet und dass sein Hirn, das Organ seiner Seele, aus dem Tiergehirn und somit aus der Tierseele stammt, so muss er doch zur Erkenntnis gelangen, dass das Studium der Neurobiologie dieser seiner Verwandten das grösste Licht auf sein eigenes Nerven- und Seelenleben werfen muss.»

August Forel
(«Die Aufgaben der Neurobiologie»)

Preface

The lectures on the central nervous system of vertebrates, given by the author during his first sojourn in Japan, 1924–1927 (Taishô 13 to Shôwa 2), intended to foster the interest in comparative neurologic studies based upon the morphologic principles established by the *Gegenbaur* or *Jena-Heidelberg School of Comparative Anatomy*. Notwithstanding their introductory and elementary nature, these lectures, published by Gustav Fischer, Jena, in 1927, included a number of advanced as well as independent concepts, and represented, as it were, the outline of a further program.

Despite various vicissitudes, and although I found the prevailing intellectual climate in the realm of biologic sciences rather unfavorable to the pursuit of investigations related to the domain of classical morphology, I have, *tant bien que mal*, carried on with my studies as originally planned, and propose to summarize my viewpoints in the present series, designed to represent a general survey, and projected to comprise five separate volumes, of which the first two are now completed. It can easily be seen that the present series follows closely the outline of my old 'Vorlesungen', meant to stress 'die grossen Hauptlinien der Hirnarchitektur und die allgemeinen Gesetzmässigkeiten, welche in Bau und Funktion des Nervensystems erkannt werden können'.

Comparative anatomy of the vertebrate central nervous system requires a very broad and comprehensive background of biological data, evaluated by means of a rational, consistent, and appropriate logical procedure. Without the relevant unifying concepts, comparative neurology becomes no more than a trivial description of apparently unrelated miscellaneous and bewildering configurational varieties, loosely held together by a string of hazy 'functional' notions. A perusal of the multitudinous literature dealing with matters involving the morphologic aspects of neurobiology reveals, to the critical observer, considerable confusion as regards many fundamental questions.

For this reason, the present attempt at an integrated overall presentation includes a somewhat detailed scrutiny of problems concerning the significance of configuration and configurational variety with respect to evolution and to correlated reasonably 'natural' taxonomic classifications. Because comparative anatomy of the central nervous system embodies the morphological clues required to infer the presumable phylogenetic evolution of the brain, a number of general questions referring to ontogenetic evolution are critically considered: it is evident that both the inferred phylogenetic sequences and the observable ontogenetic sequences represent evolutionary processes suitable for a comparison outlining the obtaining invariants.

Moreover, the comparison of organic forms involves procedures closely related to *analysis situs*. Thus, a simplified and elementary discussion of the here relevant principles of *topology* was deemed necessary.

Finally, since vertebrate comparative anatomy and vertebrate evolution, including the origin of vertebrates, cannot be properly assessed in default of an at least moderately adequate familiarity with the vast array of invertebrate organic forms, a general and elementary survey of invertebrate comparative neurology from the vertebrate neurobiologist's viewpoint, that is as seen by an 'outsider' with a modicum of first-hand acquaintance, has been included as volume two of this series. The approximately 20 pages and 12 figures dealing with this matter in my 1927 'Vorlesungen' have thus, of necessity, become rather expanded.

US N.I.H. Grant NB 4999, which is acknowledged with due appreciation, made possible the completion of Volumes 1 and 2 of this series, and, for the time being, the continuation of these studies, by supporting a 'Research Professorship' established to that effect, following my superannuation, at the Woman's Medical College of Pennsylvania.

Concluding this preamble to the present series, I may state with CICERO (*De oratore*, III, 61, 228): '*Edidi quae potui, non ut volui sed ut me temporis angustiae coegerunt; scitum est enim causam conferre in tempus, cum afferre plura si cupias non queas.*'

H. K.

Foreword to Volume 4

The present volume, containing chapters VIII to XI, deals with the spinal cord and the derivatives of the deuterencephalon. These latter are medulla oblongata *sensu latiori* (i.e. including the Mammalian pons), cerebellum, and mesencephalon. In accordance with the viewpoints elaborated in the preceding volumes of the work's *General Subdivision*, an attempt is made to bring a critical review of the adult Vertebrate central nervous system's regional configurational features. The fundamental morphologic respectively *topologic* components of the neuraxis, such as alar and basal plate with their further subdivisions, are stressed, and an account of the main communication channels interconnecting the relevant grisea is given. Although the topologic relationships are thereby emphasized, in contradistinction to a presentation arranged on the basis of 'functional systems', the significant functional aspects are pertinently taken into consideration.

With regard to *receptor structures*, an account of the otic apparatus, not dealt with in the section on peripheral nerve endings of chapter VII of the preceding volume, has been included in chapter IX. The optic receptors, derived from the diencephalon, and the olfactory organ, closely related to the telencephalon, shall be discussed in the apposite chapters XII and XIII of volume 5, now in preparation.

Because of the manifold increase in available data recorded subsequently to the publication of the latest extant standard treatises, by KAPPERS (1920, 1921, 1947; KAPPERS, HUBER and CROSBY, 1936) and by BECCARI (1943), severe limitations, *qua* inclusion of details, are now imposed on a single author endeavouring to survey the wide domain of comparative neurology and overall neurobiology. Yet, the unity and comprehensiveness of such attempt might counterbalance some of the disadvantages inherent in a more detailed multi-authored presentation which, if actually accomplished for the entire subject matter under consideration, would fill substantially more than a score of large volumes.

The bibliographies appended to the chapters of this series are therefore meant to be selective, but should easily enable those interested in further particulars to find the required additional references.

As in the preceding volumes, numerous duly credited illustrations were taken from the public domain of published scientific literature. Illustrations without credit reference are previously unpublished originals from my own studies.

As before, I am obliged to the *Medical College of Pennsylvania* for the facilities of my '*Laboratory of Morphologic Brain Research*', and particularly grateful to my many former students, who, through the *Alumnae Association* of the whilom *Woman's Medical College of Pennsylvania*, have again generously contributed to the funds necessary for my work.

Table of Contents of the Present Volume

Volume 4 Spinal Cord and Deuterencephalon

Table of Contents of the Complete Work

Volume 1 Propaedeutics to Comparative Neurology

Volume 5 Derivatives of the Prosencephalon; The Neuraxis as a Whole

VIII. The Spinal Cord

1. General Pattern

The *medulla spinalis* of Vertebrates is an extensive caudal subdivision of the neural tube. In the course of ontogenesis, it becomes demarcated from the cerebral 'vesicles' by an ill-defined boundary neighborhood representing a zone of more or less gradual transition. Enclosed in the *vertebral canal*, the spinal cord of the adult neuraxis displays, in all *Craniota*, a lesser degree of complexity than the brain, remaining, with regard to many aspects, at a simpler stage of the primitive tubular anlage's differentiation.

The paired lateral wall portions or *lateral plates*, providing the dorsal *alar* and the ventral *basal plates*, undergo a substantial thickening, whereby the original lumen, generally lined by ependyma, becomes relatively reduced to the narrow *central canal*[1] of the adult stage (Fig. 1). In addition to this relative reduction, a dorsal portion of the embryonic central canal of many Vertebrates tends to obliterate, resulting in a posterior glial or ependymal septum (Figs. 1 C, D, 2).

The unpaired median dorsal and ventral wall portions *(roof plate* and *floor plate)* apparently do not participate to a substantial or significant degree in the production of functional neuronal elements.[2] Their

[1] The central canal in adult representatives of apparently all major Vertebrate groups (including the Acraniote Amphioxus) may contain an extension of *Reissner's fiber*, which is not ordinarily found in the adult human neuraxis. In various adult lower Vertebrates, however, it may reach as far as the canal's caudal end. In Craniota, *Reissner's fiber*, a non-neuronal structure, generally seems to begin at the subcommissural organ, being perhaps, at least in part, produced by that structure. Cf. vol. 3, part I, pp. 366–367 of this series, and the papers by REISSNER (1860), DENDY (1907), JORDAN (1919), NICHOLLS (1912, 1917), EBERL-ROTHE (1951), WISLOCKI and LEDUC (1952, 1954), GILBERT (1956) and OLSSON (1955, 1966), as well as the comments in the Treatises by BECCARI (1943) and KAPPERS (1947).

[2] Among the exceptions to this essentially fairly valid general rule, internuncial, commissural neurons seem to originate in the floor plate of Amblystoma and have been designated as *floor-plate cells* (COGHILL, 1929). These perhaps transitory elements are of particular significance for certain early larval swimming reflexes discussed further below. In addition, the floor plate appears to develop, in certain Amphibian forms such as

ependymal cells, however, may, by their long processes, provide septal fibers in ventral and dorsal raphé, of which the latter one, as just mentioned, can also to some extent result from a process of partial lumen obliteration.

The *sulcus limitans*, whose significance was repeatedly discussed in several sections of chapter VI, is, as a rule, clearly recognizable during at least some stages of ontogenesis in most or all forms, but becomes effaced by further developmental changes, being finally included into a dorsal region of the central canal's lining.

As regards the functional neural elements, the nerve cell bodies (perikarya), with usually most of their neuropil and with their appurtenant glia cells, remain inwardly located, within the surroundings of the central canal, while the long fiber systems, running parallel to the rostro-caudal axis, assume a peripheral respectively superficial location. This configuration corresponds to a generalized overall pattern characteristic, at early ontogenetic stages, for the entire developing neuraxis.

In the Acraniote Amphioxus and in Craniote Cyclostomes all nerve fibers are essentially non-medullated.[3] In Selachians and in the other

---

Amblystoma, from a ventral midline ridge of the neural tube, resting directly upon the notochord, and described as '*neural keel*' (BAKER, 1927). On the basis of the available data, it remains uncertain whether a 'keel' of this type is of widespread occurrence in Vertebrate ontogenesis.

[3] On the basis of electron microscopy it has been shown that true medullated (myelinated) fibers, present in all Craniote Vertebrates except Cyclostomes, are characterized by tightly wound spiral ultrastructural double membranes located externally to the fiber's own surface membrane. These lipid containing extraneous membranes are provided by the 'unit membranes' of glia respectively lemnoblast cells. In both central and peripheral nerve fibers, this 'myelin sheath' is interrupted by '*nodes of Ranvier*'. Within the nervous system of some Invertebrates, certain large nerve fibers are surrounded by 'loosely wrapped' lipid sheaths of glial provenance differing in various details from true Vertebrate myelin sheaths and lacking '*nodes of Ranvier*'. In some Crustaceans, however, 'tightly wound' myelin sheats with *nodes of Ranvier* of Vertebrate type have been observed around certain central axons (cf. also pp. 255–256 in vol. 2). BECCARI (1943, p. 39) assumed that the large *fibers of Müller* in the neuraxis of Cyclostomes were provided '*con guaina mielinica*'. Because their appearance in preparations with routine light-microscopic techniques may indeed give that impression, I considered the occurrence of some sort of myelin sheath around these fibers not altogether improbable (cf. chapter V, p. 201 in vol. 3, part I). On the other hand, SCHULTZ *et al.* (1956) obtained electron microscopic pictures showing *Müller's fibers* devoid of a myelin sheath. If it granted that the findings of these authors are of general validity for Cyclostomes, the exceptional occurrence of myelin sheats in this Vertebrate class, namely as restricted to *Müller's fibers*, must be

groups of gnathostome Vertebrates, most long fiber tracts are medullated and appear whitish in the fresh condition. The neuronal cell masses, with their neuropil and with fewer medullated fibers, present, in contrast, a grayish aspect. In cross-sections, the spinal cord thus displays two more or less distinctive components, namely the *central gray matter* and the surrounding *peripheral white substance*. This distinction, however, does not necessarily imply well defined or linear boundaries, although, particularly in Man and other Mammals, a rather conspicuous demarcation is generally seen. In lower as well as in higher Vertebrates, the central gray substance is commonly penetrated by discrete medullated fiber bundles, while ramifications of dendrites and even nerve cell bodies may be located in the 'white substance'. The differences in vascularization between gray and white matter were discussed and depicted in chapter V, section 5, pp. 320–326 of volume 3, part I. Other significant anatomical aspects of blood supply are dealt with in chapter VI, section 7 of the preceding volume 3, part II.

The neuronal cell populations with their neuropil, forming the relevant texture of the Vertebrate spinal cord's gray core, have the tendency to become arranged in more or less protruding bilateral dorsal and ventral aggregates, known as *dorsal or posterior and ventral or anterior horns*, respectively *columns*.[4] Together with a narrower midline bridge of gray containing the central canal, these configurations, protruding into the white substance, display an H-like or butterfly-like outline in cross-sections. The white substance, in turn, is thereby roughly subdivided into *funiculi* or *white 'columns'*. The anterior or ventral funiculus is located medially and ventrally to the anterior gray column, being approximately bounded from the lateral funiculus by the ventral root fiber fascicles. The dorsal or posterior funiculus lies between dorsal gray column with its roots, and the midline. The lateral funiculus occupies the space between posterior and anterior white columns. Fibers crossing the midline in close dorsal or ventral vicinity of the central canal within either white or gray substance form the variable *posterior* or *anterior gray* respectively *white commissures*.

---

discounted. Concerning the somewhat similar 'giant fibers' of Amphioxus, whose nervous system is considered entirely to lack medullated fibers, no conclusive ultrastructural data have so far come to my attention.

[4] An arrangement of this type is not displayed by the Acraniote Amphioxus. Again, in Cyclostomes, particularly in Petromyzonts, the typical protrusion of dorsal and ventral horns is mostly only barely suggested and may even be missing.

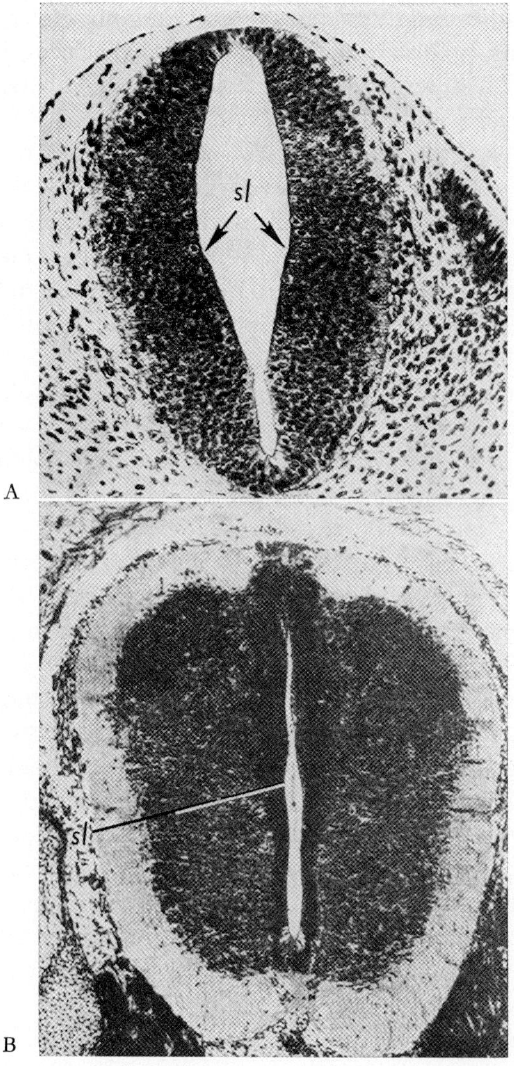

Figure 1. Four stages in the transformation of the human spinal cord's central canal as seen in cross-sections (after KOLMER and MARBURG, 1929, with added labels). A At about 4 weeks (4.6 mm); B About 9 weeks (36 mm), lumbar region. C About 9 weeks (37 mm), showing beginning dorsal septum formation. D About 9 weeks (36 mm), cervical region, showing considerable reduction. sl: sulcus limitans. The differentiation of the spinal cord and the relative reduction of the central canal proceed in accordance with a cranio-caudal gradient. The histogenetically important stage depicted in Figure 21, p. 23, of volume 3/I lies between A and B of the present Figure.

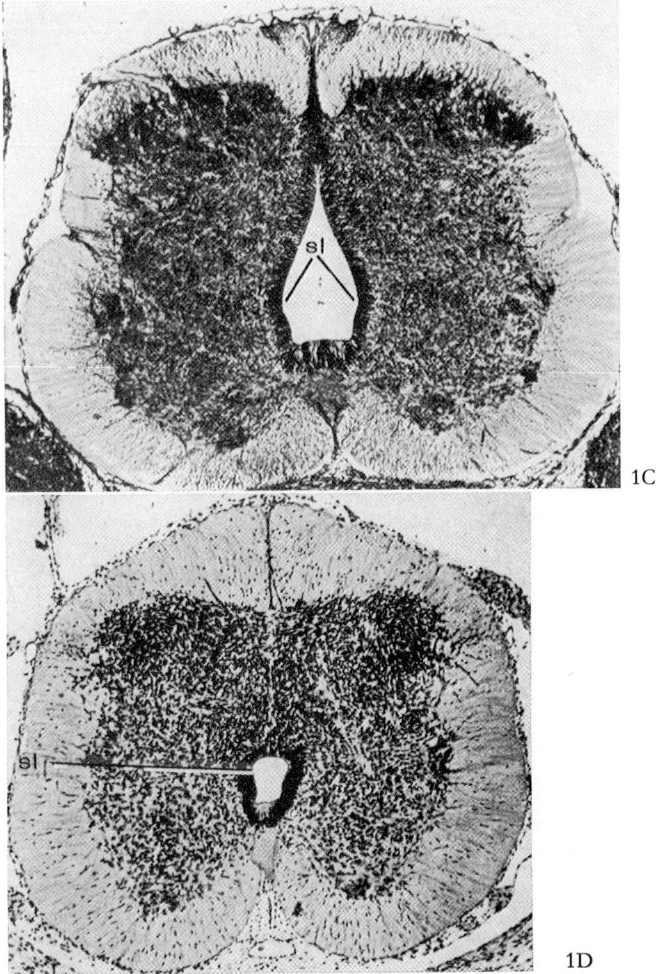

1C

1D

On the surface of the spinal cord, a *posterior (dorsal)* and an *anterior (ventral) longitudinal median sulcus* are commonly displayed, the anterior one being frequently more pronounced and even becoming a rather deep *anterior median fissure*. Depending on taxonomic forms and spinal cord regions, additional external longitudinal sulci may be present, which shall be pointed out, where relevant, in the subsequent descriptions.[5]

---

[5] The *meninges* habe been dealt with in chapter VI, section 7 of the preceding volume.

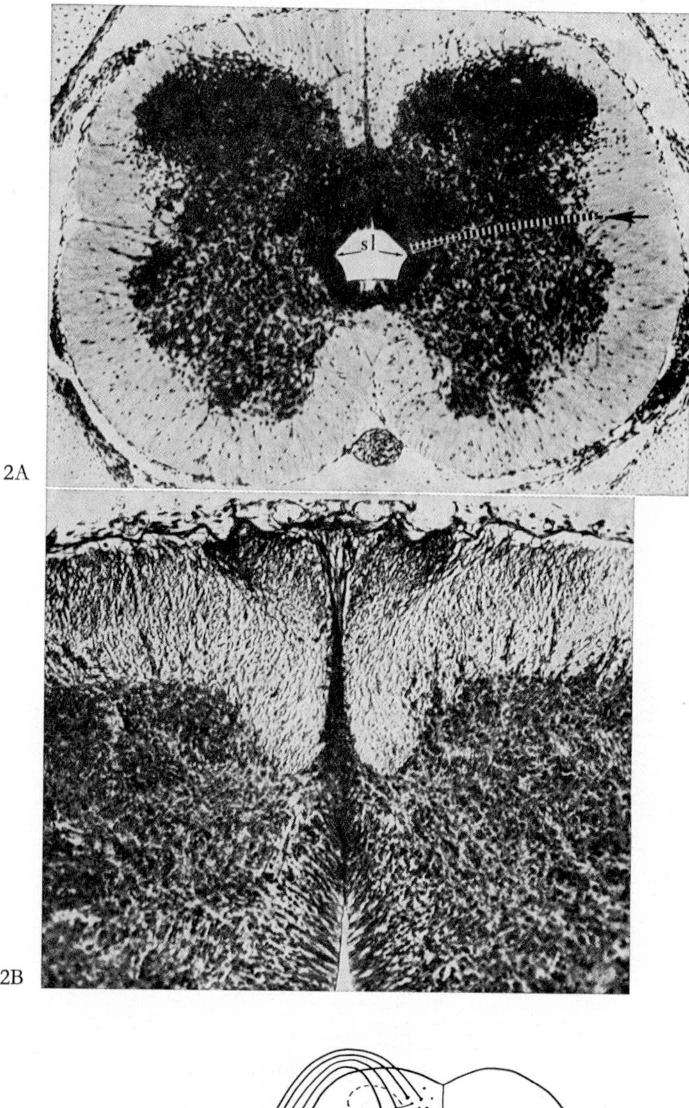

2A

2B

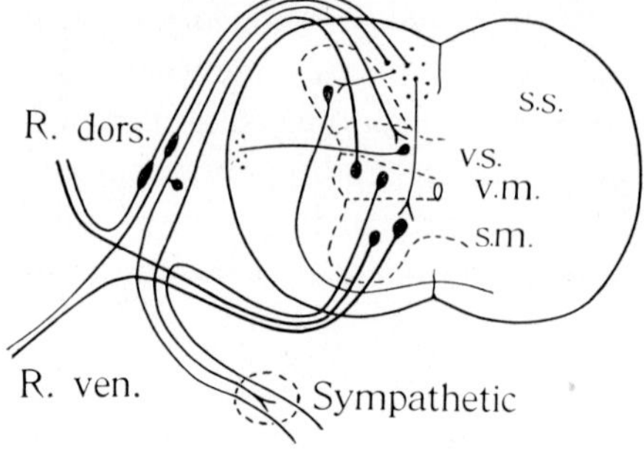

3

The grisea of the dorsal columns are closely related to the dorsal roots[6] and derive from the alar plate. The grisea of the ventral columns originate from the basal plate. The bodies of efferent nerve cells, whose neurites provide spinal output to muscles or glands, seem thus to originate exclusively in the basal plate and to remain located in its derivative. The further subdivision of the fundamental longitudinal zones represented by alar and basal plates in deuterencephalon and spinal cord was reviewed in section 4 of chapter VI as well as in sections 2 and 4 of chapter VII in the preceding volume. It will be recalled that this secondary zonal system is less differentiated in spinal cord than in deuterencephalon. In agreement with the doctrine of functional nerve components, the presence of at most four longitudinal zones in medulla spinalis can be assumed, namely, from dorsally ventralward, '*somatic sensory*' and *visceral sensory*' zones of alar plate, and '*visceral motor*' and '*somatic motor*' zones of basal plate, as illustrated in Figure 3 (to be compared with Figure 155 of the preceding volume). From a purely morphologic viewpoint, these zones correspond to *dorsal*, *intermediodorsal*, *intermedioventral*, and *ventral zones*, respectively, of formanalytic terminology.

There is indeed substantial evidence for a distinction of intermedioventral zone ('visceral motor') and ventral zone ('somatic motor') as intrinsic components of the ventral horn. The latter zone contains the motoneurons for striated musculature, and the former the somewhat smaller preganglionic cells of the spinal autonomic system. In various Vertebrate forms, preganglionic elements may even aggregate at certain levels to form a protruding *intermediolateral column* or '*lateral horn*' at the dorsal base of the ventral horn. Some levels of such 'lateral horn' however, can also include other elements than preganglionics. It is, on the other hand, rather dubious whether a definite and delimitable subdivision of the dorsal horn into 'somatic sensory' and 'visceral sen-

[6] Origin and general significance of *spinal ganglia* were dealt with in chapters V (section 1) and VII (section 2) of the preceding volumes.

Figure 2. Outline of central canal (A), and (B) detail of dorsal septum formation in embryonic human spinal cord (56 mm) at about 12 weeks (after Kolmer and Marburg, 1929). Added label and line in A roughly indicates boundary of alar and basal plates.

Figure 3. Diagram of trunk segment in a generalized Vertebrate spinal cord showing the 4 'functional divisions' in Johnston's interpretation (from Johnston, 1906). R. dors.: radix dorsalis; R. ven.: radix ventralis; s.m.: somatic motor; s.s.: somatic sensory; v.m.: visceral motor; v.s.: visceral sensory.

sory' zones[7] as postulated and depicted by JOHNSTON (1906) and HERRICK (1931) can be considered compatible with the actual structural and functional organization of the spinal cord.

In addition to the dorso-ventral zonal pattern based on morphologic subdivisions, such as alar and basal plates, or on the concept of functional components such as 'somatic' and 'visceral', a dorso-ventral *cytoarchitectural lamination pattern* of the spinal cord's gray substance can be recognized in at least some Vertebrate groups. OBERSTEINER (1912), on the basis of reports by SANO, depicted a laminar stratification *(Schichtung der Nervenzellen)* in three distinct layers as obtaining for the Mammalian gray at the tip of the posterior horn and including the substantia gelatinosa. About 40 years later, apparently quite independently and almost simultaneously, REXED (1952, 1964) and HIRASAWA (1955) emphasized an arrangement characterized, in cross-sections, by a dorso-ventral succession of fairly distinctive cell layers. REXED distinguishes nine or ten laminae (Fig. 4) and HIRASAWA five or six. FUKUYAMA (1955), who, at my suggestion, investigated this lamination in the rat, favored, in accordance with our joint observations, a subdivision into seven laminae, which might, however, be reduced to six. This lamination, moreover, is not in all instances adequately sharp. Also, depending on the regions, gradients and overlaps exist. Thus, all numerical designations of layers, and lines drawn on paper for the separation of such laminae, are to a large extent arbitrary. The stratification in the dorsal horn is more pronounced than that in the ventral horn. In this latter, on account of a tendency toward cell grouping or clustering, the laminar pattern is commonly barely recognizable and rather questionable. The intermediolateral column, corresponding to the 'visceral motor' zone of JOHNSTON and HERRICK, is located in lamina VII of REXED's terminology, which can be regarded as a dorsal derivative of the basal plate. A moderate gradient seems to indicate a zone of transition between the probable derivatives of alar and of basal plate.

It should be added that the cytoarchitectural lamination pointed out by REXED and HIRASAWA, although frequently rather distinct in the spinal cord of Mammals, is much more faintly shown in Sauropsi-

---

[7] In addition to JOHNSTON's diagram depicted above as Figure 3 in the text, cf. Figures 155 A–D of the preceding volume, which show HERRICK's concept both in that author's view and in a slightly modified version tentatively adopted by myself in 1927.

dan Amniota and barely suggested in Anamnia. In these submammalian forms, there obtains, of course, an evident difference in cytoarchitecture between the ventral cell populations of larger motoneurons, and those of more dorsal smaller elements, including, in some Anamnia, characteristic arrays of large dorsal cells. Attempts to delimit specific layers, even in a reduced number, would doubtless be much more difficult and arbitrary in the submammalian groups than in Mammals.

Roughly speaking, and discounting conditions in Amphioxus as well as secondary details of boundaries and of overlap, the brain is located in the *cranial cavity*, while the spinal cord extends to a variable length through the lumen of the *vertebral canal*.

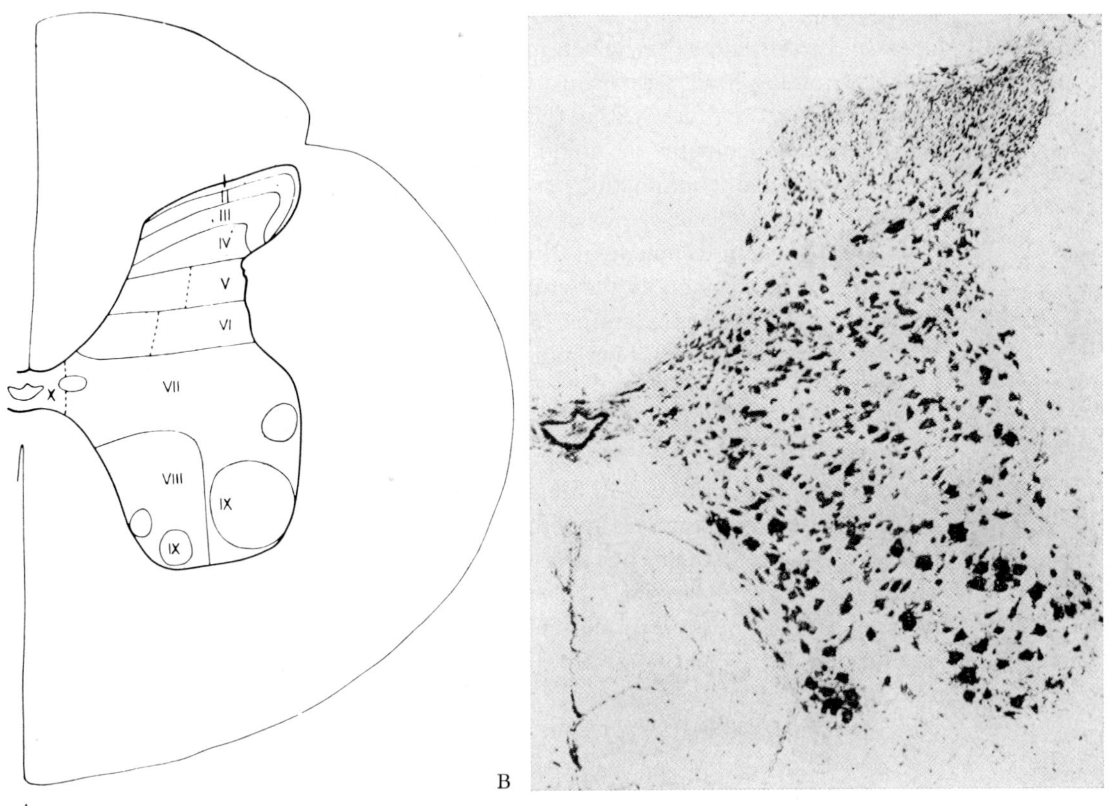

Figure 4. Lamination of the gray matter in the fifth lumbar segment of the adult cat's spinal cord (after REXED, 1964). A REXED's interpretation of laminar sequence. B Unlabelled cross-section (*Nissl-stain*, approx. ×37) shown by REXED for comparison.

From a fairly rigorous morphologic viewpoint, the spinal cord can be said to have its rostral beginning at the level containing the motoneurons of the first 'true' spinal nerve segment, and the exit respectively entrance of that segment's ventral and dorsal roots. This statement, if generalized as applying to the entire Craniote Vertebrate series, nevertheless sidesteps the not quite satisfactorily solved questions pertaining to the diversified manifestations of spino-occipital nerves in Anamnia. These nerves, together with related problems concerning the evolution of the hypoglossal (n. XII) and accessory nerve (n. XI) were discussed in chapter VII, section 3 of the preceding volume 3, part II.

In Anamnia as well as in Amniote Sauropsidans, the spinal cord generally reaches caudalward approximately as far as the canal's caudal end.[8] Particularly conspicuous among diverse exceptions of that rule is the short spinal cord of the Teleosts *Lophius piscatorius* and *Orthagoriscus mola*. In this latter fish, which has been characterized as 'an enormous swimming head with a short trunk and without tail', the spinal cord remains more or less within the cranial cavity, while the vertebral canal essentially contains the spinal nerve roots and the rudimentary string-like caudal continuation of spinal cord designated as *filum terminale*. The bundle of caudolateralward diverging nerve roots surrounding the filum terminale in those Vertebrates whose spinal cord does not extend throughout the length of the vertebral canal is known as the *cauda equina*. Again, among Amphibians, the Anurans likewise possess a commonly very short spinal cord, which can end at the level of third or fourth vertebra.

Unlike the spinal cord of Sauropsida, that of adult Mammals seldom extends through the entire length of the vertebral canal, whose caudal portion merely contains filum terminale and cauda equina as in some Teleosts or in Anuran Amphibia. At early mammalian embryonic stages, however, spinal cord and vertebral canal are co-extensive. In adult Ungulates, moreover, the spinal cord generally extends as far as the sacral portion of the vertebral canal. An interesting difference obtains in the two extant groups of Monotreme Prototheria. The spinal cord of Echidna only reaches about midway within the vertebral canal, while it extends as far as the canal's sacral portion in

[8] In Plagiostome Selachians a 'postchordal' spinal cord portion may even extend beyond the vertebral canal's end into the tail fin.

Ornithorhynchus. Several inconclusive hypotheses concerning the factors influencing discrepancy in length between spinal cord and vertebral canal have been suggested (cf. KAPPERS *et al.*, 1936; KAPPERS 1947). Some additional comments on spinal cord length relations are included further below, where required, in dealing with the configurational features displayed by the different Vertebrate groups.

With regard to the spinal cord's caudal extremity, a vesicular enlargement either of the reduced cord or of the still more rudimentary filum terminale has been recorded in at least some Vertebrates. Moreover, the caudal region of the spinal cord in various Fishes may contain large and commonly polynuclear cells of glandular type. In some, but not all Teleosts, these cells form part of a peculiar terminal apparatus known as '*renflement caudal*' of VERNE (1914), hypophysis caudalis of FAVARO (1925), or 'caudal neurosecretory system' (urohypophysis, neurohypophysis spinalis caudalis) of ENAMI and IMAI (1955), ENAMI (1959), and SANO *et al.* (1962).

In a morphological as well as structural aspect related to the rostrocaudal metameric sequence of spinal nerve roots, and with respect to some of its functional activities, a segmental organization of the spinal cord doubtless obtains. However, discounting the peripheral spinal nerve arrangement and some reports on early but transitory embryonic metamery (e.g. BOLK, 1906), such segmentation is not manifested by the gross configuration of the spinal cord. On the other hand, the development of the extremities in Tetrapod Vertebrates (Amphibia, Sauropsida, Mammals) ist correlated with corresponding more or less pronounced enlargements of the spinal cord representing *intumescentia cervicalis* and *intumescentia lumbalis*.

The output (efferent) and the input (afferent) channels linking the spinal cord with the body's periphery, i.e. with the various effector and receptor structures, are provided by the paired ventral and dorsal spinal nerve roots. The former represent predominantly output channels, while the latter predominantly mediate input, in overall agreement with the formulation of *Magendie-Bell's* '*law*', whose merely relative validity was discussed in section 2, chapter VII of the preceding volume. Although, as shall be pointed out further below in section 2 of the present chapter, the efferent root fibers of the spinal cord generally precede the afferent ones in the course of ontogeny, the *number* of these latter (i.e. of peripheral input channels) greatly exceeds that of the efferent fibers not only with respect to the spinal cord but also to the entire neuraxis.

Fairly detailed estimates concerning the number of spinal root fibers have been recorded for some Vertebrates, including Man, a few Mammals, as well as a few submammalian forms (cf. e.g. v. PODHRADZKY, 1933). The available data, with various pertinent discussions, are contained in the treatise of BLINKOV and GLEZER (1968). In Man, the total number of medullated and non-medullated posterior root fibers on one side may have an order of magnitude of about one million, the overall ration of myelinated to unmyelinating fibers being approximately 7:3, with considerable segmental and individual variations. The ventral root fibers on one side may reach a total of about 200,000, with an approximate overall ratio of 7:1.3 respective to medullated and non-medullated fibers. Considering the obtaining uncertainties, our own estimate of the number of large motoneurons in one side of the human spinal cord as no less than 80,000 and probably no more than 160,000 (SIRKEN and K., 1966) appears quite compatible with the just cited data.

As regards the number of both dorsal and ventral root fibers, there are evidently substantial differences between those segments related to brachial or lumbosacral plexuses and those thoracic segments merely related to intercostal nerves. Thus, considering only medullated fibers, approximately 25,000 or more may be found in a posterior root at segment C VII, and somewhat more than 5,000 in the corresponding anterior one. At segment T IX, however, these figures become reduced to about 9,000 and 3,500 respectively. Moreover, considerable variations are reported with respect to one and the same root in different individuals. Some of these variations may indeed be actual, while others could be attributed to the difficulties in obtaining accurate counts, to the technical procedures involved, and to various additional sources of error inherent in studies of this sort by different investigators. Another category of in some cases substantial numerical variations is that obtaining between right and left side of the body, i.e. between antimeric structures, although, as a general rule, the total number of neuroectodermal elements in both sides of the body seems to remain within roughly equal orders of magnitude. The variable differences in cell number between right and left sides of the midsagittal plane are related to but not in all respects coincident with, diverse additional asymmetries displayed by the essentially bilateral-symmetric Vertebrate body.

In a 'lower' Mammalian form such as the mouse, a total of about 48,000 fibers (medullated and non-medullated) was found in the posterior roots of one side, while the corresponding total unilateral number of ventral root fibers amounted to approximately 24,000 (AGDUHR, 1934). In an anuran Amphibian (Toad), the respective figures were given, by the just cited author, as about 8,200 (dorsal roots) and 5,700 (ventral roots). Generally speaking, and evaluating, despite some exceptions to that rule, the dorsal roots as essentially afferent and the ventral ones as essentially efferent, one could perhaps infer from the available scattered data that, with regard to the spinal cord, the numerical difference between input and output fibers is probably more pronounced in 'higher' than in 'lower' Vertebrate forms. However, it is claimed by CUNNINGHAM (1877) and HATSCHEK (1896), that, in Cetacea, the ventral roots are larger and contain more fibers than the dorsal ones. According to KAPPERS *et al.* (1936) this is 'due to the poor development of skin sensibility, including pain'.

A much discussed problem has been the *numerical relation of dorsal root fibers* to spinal ganglion cells. GAULE and LEWIN (1897) reported, for a particular spinal nerve segment in the rabbit, 4,270 fibers in the posterior root, 27,614 spinal ganglion cells, 2,997 fibers in the ventral root, and 9,022 fibers in the common spinal nerve. The considerable discrepancy between dorsal root fibers and spinal ganglion cells was subsequently easily explained by the fact that only the medullated fibers had been counted. The excess of fibers in the common spinal nerve (namely 1,755 more than 4,270 + 2,997 = 7,267) can likewise be explained on the basis of presumptive fiber bifurcations near the origin of that common nerve. The data obtained by subsequent authors, taking into consideration the presence of non-medullated fibers, disclosed a much lesser excess of cells over fibers in posterior roots and spinal ganglia of various Vertebrates. The investigations of DUNCAN and KEYSER (1936, 1938), moreover, seemed to indicate a 1:1 ratio in the cat, with even an occasional slight excess of fibers over cells. BARNES and DAVENPORT (1937), who examined cat and cow material, remained convinced that at least some dorsal roots have more cells in their ganglion than fibers directed centrally, while other such roots might indeed have the same number in accordance with the ratio 1:1.

BARNES and DAVENPORT (1937) discuss the various sources of error inhering in technical procedures as well as in the methods of counting. It is of special interest that the cited authors exchanged preparations with DUNCAN and KEYSER. While the cell counts on identical material by different observers was thereby shown fairly well to agree, considerable differences were noted in the counting of fibers. This disagreement in the results reported by different workers indicated that, in this respect, 'the psychological element is not negligible' (BARNES and DAVENPORT, 1937).

Should a numerical excess of spinal ganglion cells over posterior root fibers indeed obtain, a number of different explanations are possible. Thus, the adult ganglion might include some incompletely differentiated nerve cells who did not develop neurites. Also, since vegetative postganglionics seem to originate from the neural crest, a few abortive postganglionics could remain within the spinal ganglion. Again, if, as postulated by some authors, the spinal ganglion might include a number of functionally active postganglionics, several of these could receive terminals from a single efferent dorsal root fiber. A number of additional relationships between spinal ganglia and sympathetic nervous system were postulated by HIRT (1928) but need not be considered in this context. Also, a number of internuncials of *Golgi II type* might conceivably be present within a spinal ganglion. It is well established that a variety of different types of spinal ganglion cells occurs, including larger and smaller elements as well as irregular forms, displaying paraphytes (cf. section 2, chapter V, vol. 3, part I) or a 'multipolar' aspect. As a rule, however, these latter seem to be the result of fixation artifacts. SCHARF (1958), in dealing with this topic, enumerates no less than 8 sorts of neuronal elements in the spinal ganglia and depicts various hypothetically postulated patterns of fiber course and synaptic connections.

On the whole, nevertheless, except for the possibility of isolated retained postganglionics, and, depending on Vertebrate forms, a few 'visceral efferents' which, moreover, would rather tend to increase the number of fibers in relation to spinal ganglion cells, since they seem merely to pass through the ganglion on their way to more peripheral vegetative plexuses, the various hypotheses on complex structural arrangements in the spinal ganglia of Vertebrates have remained rather unconvincing.

For practical purposes, and with the necessary *reservatio mentalis* required in view of the rather shaky and in part flimsy foundation of many neurobiologic concepts, it seems reasonable to retain, in essence, the following two time-honored concepts. (1) Except for the transit of a few efferent fibers originating in the spinal cord, the Vertebrate spinal ganglion contains predominantly if not exclusively bipolar or pseudounipolar cells with a central and a peripheral neurite. In some instances (cf. chapter VII, volume 3, part II) this latter might mediate vasodilator effects. (2) The ratio of spinal ganglion cells to posterior root fibers is roughly 1:1 with minor fluctuations in favor of cells (abortive elements, retained postganglionics exceeding the number of preganglionic neurites in posterior root) or of fibers ('visceral efferents').

As regards the *ventral root*, and discounting the peculiarities displayed by Amphioxus, 'recurrent' posterior root fibers entering the radix anterior have been reliably reported in Man, some other Mammals, and in the Frog, by several authors (cf. SCHARF, 1958). This accounts in part for pain sensations upon electrical stimulation of intact or transected human ventral roots, as recorded by certain neurosurgeons. There is, moreover, some evidence for the exit of afferent, especially proprioceptive fibers through the ventral roots in at least various vertebrates. That some 'pain fibers' might be included in such afferents cannot be entirely excluded. No reliable information is available on the location of the pericarya pertaining to the presumed afferent fibers of ventral spinal roots. Another unsettled problem is the conduction of 'sensory' or 'pain impulses' through the sympathetic trunk cranialward over several segments, and the entrance of the corresponding fibers into the neuraxis by way of ventral roots. The location of the cell bodies pertaining to assumed fibers of this type remains highly uncertain. Various authors, particularly neurosurgeons, have propounded diverse hypotheses concerning this topic.

Again, considering only the well substantiated data, and for practical purposes, one may tentatively assume that, except for the exit of some afferent proprioceptive fibers in at least several Vertebrate forms, the ventral roots are essentially provided by efferent fibers originating from motoneurons or from γ-cells, and innervating the 'somatic' musculature, as well as by efferent preganglionic fibers from visceral-efferent neurons located at certain levels.

2. Intrinsic and Extrinsic Mechanisms (L. EDINGER's *Eigenapparat* and *Verbindungsapparat*); Some Comments on the Reflex Concept and on Myelogeny

From a functional viewpoint, L. EDINGER (1912 *et passim*) distinguished in the Vertebrate spinal cord two different, but intimately associated and co-operating systems, which he designated as *Eigenapparat (intrisic mechanism)* and *Verbindungsapparat (connecting or extrinsic mechanism)*.

The *intrinsic system*, which, in turn, consists of a *segmental* subdivision and of an *intersegmental* correlating mechanism, receives its input essentially through the afferent fibers of the dorsal roots reaching the gray substance. This latter, again, provides output to the periphery through efferent fibers which take their course mainly by way of the ventral roots but occasionally also through the dorsal ones.[9]

In the metamerically organized segmental subdivision of the *Eigenapparat*, the afferent fibers effect, at the level of their entrance, connections with efferent neurons *(root cells)* whose neurites leave the spinal cord at that same segmental level. These connections may be either direct, by synaptic endings of the primary afferent terminal branches on efferent neurons, or indirect, through the mediation of short internuncials, e.g. of *Golgi II-type (Binnenzellen)* within the griseum of the segment concerned.

The intersegmental correlating subdivision of the *Eigenapparat* includes the afferent fibers respectively their branches or collaterals which, from their level of entrance, run in ascending or descending direction and provide synaptic connections at other more rostral or caudal segmental levels. Additional intersegmental correlations are provided by internuncial neurons (*Strangzellen*, funicular neurons) whose neurites ascend or descend homolaterally or contralaterally in the white matter, interconnecting the different segments of the spinal cord and forming the so-called *fasciculi proprii*. This intersegmental mechanism allows for the coordinated function of several segments during complex activities, such, e.g. as compound or combined movements of limbs and body.

The intrinsic mechanism of the spinal cord is independently capable to perform numerous nervous reactions *(reflexes)* of 'somatic' or 'visceral' type without any participation by the brain. The behavioral activities of decapitated frogs or chickens are well-known examples of that ability.[10] Normally, however, the Vertebrate spinal cord operates under the control of various grisea of higher order located in different regions of the brain. Long descending fiber systems originating in these grisea reach the spinal cord and, by excitatory or inhibitory im-

---

[9] Cf. the comments on '*Bell-Magendie's law*' in section 2, p. 826 ff., of chapter VII in the preceding volume of this series.

[10] STEINER (1900), citing the authority of H. KRONECKER, mentions that the unsavory Roman emperor COMMODUS (180–192 AD) particularly enjoyed, during circus games, the decapitation of running ostriches by means of arrows with sickle-shaped heads, in order to watch the headless animals carrying on their race.

pulses, regulate, initiate or suppress the activities of the intrinsic spinal mechanisms. Conversely, long ascending fiber systems of the spinal cord convey to the cerebral grisea relevant signals from the periphery and from the intrinsic spinal apparatus. Such fibers are long neurites of tract cells within the spinal cord's gray substance or, to some extent, also afferent root fibers whose cell bodies are located in the spinal ganglia. The totality of these connections between spinal cord and brain represents EDINGER's *Verbindungsapparat (extrinsic mechanism)* whose fiber bundles form homolateral and contralateral tracts. Some of these latter decussate already at spinal levels in the above-mentioned commissures which also include fibers of the intrinsic system.

Generally speaking, the descending and ascending systems of the spinal cord's *Verbindungsapparat* form more or less unified although overlapping and not always quite homogeneous bundles, traditionally designated as *tracts* in accordance with their assumed origin and (main) termination. Thus, there are, depending on taxonomic forms, various systems such as cortico-spinal, tecto-spinal, vestibulo-spinal, reticulo-spinal or bulbo-spinal tracts, as well as spino-bulbar, spino-cerebellar, spino-tectal and spino-thalamic ones. Although, in diverse instances and in a number of forms many tracts can be regarded as fairly well understood and located, the features of these connections are far less clarified with respect to their details than is seemingly believed by numerous authors. Also, the collateral connections of such tracts introduce additional complications.

The inferences upon which the systematization of these tracts is based are drawn from evidence provided by a wide variety of histological or experimental techniques as well as pathological and clinical observations. The histological methods comprise myelin stains, *Golgi* or other metallic impregnations of normal material, and the study of myelogenesis pioneered by FLECHSIG. The combined histologic and experimental methods make use of tigrolysis, of the *Marchi technique* and other procedures involving the study of degeneration effects. Among these latter procedures, the *Glees* and *Nauta-Gygax methods* and the still more recent *Fink-Heimer technique* (1967) have provided further refinements. Essentially experimental methods include the registration of evoked potentials and of recordings with extra- and intracellular micro-electrodes in addition to other approaches such e.g. as the investigation of reflexes.

All these methods, however, being subject to greater or lesser imperfections and intrinsic weaknesses, require considerable caution and experience for the interpretation of their results.[11] Thus, although considerable progress in the tracing of fiber tracts, and in the understanding of their connections as well as functions has been made in the course of the last hundred years, numerous uncertainties remain. In many Vertebrate forms only a very generalized and vague knowledge restricted to some representative fiber systems, constituting main communication channels, has been attained.

Prima facie, it would be plausible to assume that in Vertebrates with a highly developed connecting apparatus related to dominant grisea of a correspondingly differentiated brain, the intrinsic mechanism of the spinal cord might become less independent than in taxonomic forms with simpler cerebral control and thus be unable to perform, by itself, complex activities such as the coordinated locomotion mentioned above in the case of frogs and birds. This assumption seems to be corroborated by observations in mammals and especially man. In this latter, clinical experiences disclose that even all reflex activities of the spinal cord become at first, and for variable but rather long periods, abolished below the level of transections. Similar, but shorter periods of '*spinal shock*' or '*diaschisis*' also occur in lower forms, including submammalian Vertebrates.

Disconnected from the brain, the spinal cord has been a classic experimental material for the investigation of *reflexes*, particularly in the frog and in various mammals. Together with expressions such as '*neuron*' and '*synapse*', the designation '*reflex*' has become a 'household-word' of neurobiology. Yet, despite its evident usefulness and verbal as well as experimental or clinical validity, it includes, in accordance with the intrinsic limitations and the fictional nature of language, certain more or less hidden semantic traps.

The term '*reflex*' in analogy with the backward-directed 'reflection' obtaining in physical optics, was perhaps first used about 200 years

---

[11] Concerning the interpretation of degenerative processes and of the results obtained by the diverse technical procedures, pertinent critical comments are included in the preceding volume 3, part I of this series. Even with regard to the recently much praised *Fink-Heimer-technique*, it still seems to me likely that an investigator postulating certain fiber connections will always be able, *per fas et nefas*, to provide a semblance of a 'proof' by means of this method (cf. vol. 3, part I, chapter V, sect. 8, p. 667).

A historical survey of trends in neuroanatomy, with particular reference to the relevant technical procedures up to the 5th decade of this century was undertaken by A.T. Rasmussen (1947).

ago[12] by UNZER (1771) for the reaction of a Vertebrate organism upon an external stimulus resulting in what he already called 'afferent' and 'efferent' nerve conduction. STEPHEN HALES, about 1730, had previously stated that animal reactions of this type (in the frog) were dependent on the spinal cord's integrity.[13]

Subsequently, the reflex concept was further elaborated by MARSHALL HALL (1833). In this author's opinion, a stimulus affecting e.g. the skin becomes transmitted to the spinal cord and is, by the latter, '*reflected*' to the musculature, which then is set in motion. Moreover, as particularly stressed by HALL, this entire nervous transmission event is said to occur without 'sensation' *qua* 'reception', and without 'will' *qua* 'reaction', or, in other words, without involving 'consciousness' respectively any 'psychic activity'. HALL accordingly postulated two categories of cerebrospinal nerves, of which the first was called *sensorimotor*, including the 'sensitive' nerves mediating conscious sensations, and the *spontaneous-motor* nerves for 'willed' movements. The second, or '*excito-motor*' category, whose activities were presumed to be unconscious, comprised 'exciting' (afferent) nerves, and *reflecto-motor* (efferent) ones for 'involuntary' movements. As an additional third category HALL recognized 'ganglionic nerves' which represented components of what is now designated as vegetative, sympathetic, respectively visceral nervous system.

Consciousness in general and the poorly definable 'faculty' called '*will*' are strictly private phenomena exclusively recognizable by self-observation (introspection) and cannot be directly observed or recorded by instruments 'in' other living beings. At best, the presumed occur-

---

[12] Cf. also BLASIUS (1962). A detailed historical study of the reflex-concept in the wider sense, involving various aspects of neuromuscular action, was undertaken by FEARING (1939). The treatise of this author (a psychologist) contains references to the views expressed by classical Greek and Roman writers as well as by VESALIUS, DESCARTES, SWAMMERDAM and others who preceded the development of experimental studies initiated by the 18th century investigators. DESCARTES (1596–1650), who did not expound the actual structures of the nervous system, about which no relevant knowledge was available at the time, nevertheless attempted to describe mechanisms capable of performing its functions. Thus, he propounded a rather ingenious model for the coordinated reciprocal innervation of the eye muscles (cf. also BAYLISS, 1924). Some authors (e.g. BRAIN, 1969) therefore regard DESCARTES as the originator of the reflex action concept.

[13] STEPHEN HALES (1677–1761) an Anglican cleric with pronounced interest in physiology, graduated from *Cambridge*, and parish minister at *Teddington*, ranks, although second to HARVEY, as a pioneer in hemodynamics. He is author of '*Statical Essays*' in 2 volumes (London 1731–1733). Cf. also C.C. and F.A. METTLER, *History of Medicine* (1947).

rence of consciousness or '*will*' 'in' an organism other than the observer is an inference from verifiable observed behavior. Such inference has a variable, greater or lesser degree of non-verifiable probability based on analogy with self-experience. The behavior of another organic being is both directly observable and recordable by instruments, but the presumed states of consciousness 'in' such organism are neither observable nor verifiable.

The vague term '*will*', in this respect comparable to the conscious experience of the color red, is a strictly private '*qualité pure*' of consciousness in POINCARÉ's impeccable formulation. '*Will*' subsumes (1) experienced apparently 'free' ('intended') simple motor actions or comparable more complex physical output activities of the '*self*', moreover (2) experienced '*affective*' or '*emotional*' states conditioning judgments, decisions, and mental attitudes toward events (wish, desire, command, purpose, choice, intention, inclination, pleasure, displeasure, suffering). The circularity in attempting to define will by intention (*i.e.* will by will) is here evident.

Thus, from the viewpoint of 'objective science', a definition of reflex-activities (that is, of a certain type of recordable behavioral manifestations), in terms of 'consciousness' or 'will' evidently becomes, *ab initio*, encumbered with considerable weaknesses. A few years after MARSHALL HALL's elaborations, VOLKMANN (1838) put forth a very interesting and still readable critical review of HALL's reflex concept. Although VOLKMANN failed to recognize the essential epistemologic questions involved in that topic, and somewhat oddly discussed the possibility of psychic activities *('Mitwirkung der Seele')* in the decapitated frog's spinal cord, he nevertheless regarded consciousness as a function of the brain mechanisms and not of the conducting fiber tracts.[14]

VOLKMANN, moreover, on the basis of his own careful experiments with decapitated frogs,[15] stressed the apparently 'purposeful'

---

[14] '*Bekannte Erfahrungen beweisen, dass die Empfindung, in sofern als sie auf Bewusstsein beruht, nicht im Innern der Nervenstränge, sondern im Gehirn zu Stande kommt. Daher wird das Gehirn als Sensorium, und die Nerven werden als Leiter betrachtet. Diese Ansicht ist unantastbar, nur ist die Betrachtung noch nicht am Ziele*'. VOLKMANN (1838, p. 42).

[15] With reference to earlier experiments on non-decapitated frogs reported by JOHANNES MÜLLER in his important 'Textbook of Physiology' (vol. 1, 1834), VOLKMANN stresses the differences in the reflex behavior of the decapitated and the non-decapitated animal. Since in this latter some reflexes, which are easily elicited in the former, commonly fail to occur, VOLKMANN assumed that the brain could not only *initiate* but also *inhibit* the activities of the spinal cord *('weil der Wille das Vermögen hat, die Reflexbewegungen zu beschränken')*.

(*'zweckmässig'*), i.e. properly *coordinated* and *directed* nature of many spinal reflexes, whose different segmental and intersegmental, homolateral and contralateral patterns he described.[16] This author also recorded various entirely 'spontaneous' movements of the spinal animal, apparently made to 'improve' its 'sitting position'. However, after the frog had taken up the appropriate '*sitzende Stellung*', 'spontaneous motion' was only rarely observed. In accordance with the view that the dorsal roots represented the afferent (*'excitirende'*) and the ventral roots the efferent (*'reflectirende'*) pathway of the reflex arc, VOLKMANN performed numerous relevant experiments with different transections of roots and peripheral nerves.

Substantial advances in the experimental and semantic development of the reflex concept resulted, since about 1899, from the fundamental studies of SHERRINGTON (1906, 1947). This author defined reflexes as manifestations of behavior 'in which there follows on an initiating reaction an end-effect reached through the mediation of a conductor, itself incapable either of the end-effect, or, under natural conditions, of the inception of the reaction'. It will be noted that SHERRINGTON omitted here, quite appropriately, any reference to 'consciousness' or 'will', although, in the last chapter of his classic treatise (1906, 1947, p. 385) he referred to '*pure reflexes*', which, so far as introspection can discover, are 'devoid of psychical accompaniment'.[17] Three separable structures were pointed out, namely 'an *initiating organ* or *receptor* whence the reaction starts', 'a *conducting nervous path* or *conductor*', and an '*effector*' (e.g. muscle or gland), to which the conductor path leads. 'The receptor is best included as a part of the nervous system, and so it is convenient to speak of the whole chain of structures – receptor, conductor, and effector – as a *reflex arc*. All the part of the chain which

---

[16] In performing longitudinal transections of the spinal cord, VOLKMANN noted with respect to contralateral and bilateral reflexes upon unilateral stimulation: '*dass Längstheilung des Rückenmarkes die Ausdehnung der Reflexbewegungen über alle Muskeln beider Körperhälften nicht hindere, so lange nur irgend ein Theil des eigentlichen Rückenmarkes in der Mittellinie verbunden bleibt*'.

[17] SHERRINGTON justly remarks, with respect to 'volitional control', that 'reflexes ordinarily outside its pale can by training brought within it'. He adds that 'volitional movement can certainly become involuntary, and, conversely, involuntary movements can sometimes be brought under subjection to the will' (l.c., p. 386). Quite evidently, such 'training' is related both to 'conditioning' and to practices of *Yoga* and similar disciplines, which I have discussed in the monograph 'Brain and Consciousness' (1957, pp. 288–301).

leads up to but does not include the effector and the nerve-cell attached to this latter, is conveniently distinguished as the afferent-arc'. The remainder of the chain is then, accordingly, the efferent arc. To these structures correspond three processes, *initiation*, *conduction*, and *end-effect*.

A *simple reflex*, justly regarded by SHERRINGTON as a convenient fiction,[18] occurs when an effector organ responds to excitement of a receptor, all other parts of the organism being supposed indifferent to and indifferent for that reaction. Such reflex arc can be considered to represent the unit mechanism of the nervous system, the reflex action of this mechanism being the unit reaction in nervous integration. In contradistinction to the 'simple reflex', more complex reflexes are compounded of simpler ones. The compounding of simultaneously or successively proceeding reflexes with orderliness *qua* coadjustment and *qua* sequence constitutes coordination, or, in other words, a manifestation of the 'integrative action of the nervous system'. Events of this sort can spread over a wide range of nervous arcs or circuits provided by complex interconnected chains of internuncial conductors.[19]

Now, as regards SHERRINGTON's above-mentioned definition of reflexes, said to be triggered by an '*initiating reaction*', should or should not the locus of this reaction be restricted to an external or internal 'receptor' reacting upon a non-neural stimulus? In other words, should a reflex-like manifestation of behavior originated by 'spontaneous' (although presumably determinate) activity of central grisea be excluded from or included in the logical class 'reflexes'?

In a definition adopted by HERRICK (1931), a reflex act 'is an invariable, mechanically determined adaptive response to the stimulation of a sense organ, involving the use of a center of adjustment and the conductor necessary to connect this center with the appropriate receptor and effector apparatus. The act is not voluntarily performed, though one may become aware of the reaction during and after its perform ance'. HERRICK,[19a] moreover, designates reflexes as unspecified 'sim-

[18] 'A simple reflex is probably a purely abstract conception, because all parts of the nervous system are connected together and no part of it is probably ever capable of reaction without affecting and being affected by various other parts, and it is a system never absolutely at rest' (SHERRINGTON, l.c., p. 7).

[19] Cf. the introductory chapter I in the first volume of this series.

[19a] Referring to a distinction made by LANDACRE, HERRICK (1931) also stresses a difference between '*correlation*' and '*coordination*' in neural activities: 'The term *correlation*

pler reactions to stimulation' of this sort and elaborates on their mechanism which consists (1) of a receiving (receptor or sensitive) organ, (2) a conductor (afferent or sensory nerve) transmitting the nervous impulse inward from the receptor, (3) a correlation center or adjustor, generally located within the nervous system, (4) a second conductor (efferent or motor nerve) transmitting the nervous impulse outward from the center to (5) the effector apparatus, consisting of the organs of response (muscles, glands) and the terminals of the efferent nerves upon them.

Discounting the vagueness of the term '*simpler reactions*', the weaknesses in HERRICK's fairly expedient elaboration are here the use of the words 'invariable' and 'adaptive', as well as the reference to voluntary performance and awareness (consciousness). Evidently, even rather 'simple' reflexes are not always invariable but can be modified by various factors, particularly those related to 'storage' of information (engraphy), such as PAVLOV's *conditioned reflexes* (cf. vol. 1, chapter I, pp. 31–33 of this series). The semantic traps inherent in the here inappropriate reference to 'adaptation', 'will', and consciousness have been repeatedly discussed and need not again be pointed out in this context.

More recently, BULLOCK *et al.* (1965) have formulated the following definition in their 'glossary': Reflexes are, 'in neurophysiology, a relatively simple action' (cf. above HERRICK's 'simpler reactions') 'produced by an afferent influx to a nerve center and its reflection as an efferent discharge back to the periphery to appropriate effectors, independently of volition'.[20] 'Reflexes are simpler than instincts in number of muscles, of successive movements, and of specification of stimuli,

---

is applied to those combinations of the afferent impulses within the sensory centers which provide for the integration of these impulses into appropriate or adaptive responses'. To the term *coordination* is given 'a restricted significance, applying it only to those processes employing anatomically fixed arrangements of the motor apparatus which provide for the co-working of particular muscle groups (or other effectors) for the performance of definite adaptively useful responses'. One may find this distinction including its basic definitions not very enlightening, and HERRICK himself (J. comp. Neurol., 1930) is compelled to admit that these essentially physiological concepts lack a clear application to the relevant anatomical mechanisms (or 'apparatus' of his definition, p. 27, 1930).

[20] One may well wonder how the cited authors, who deal with Invertebrates, and pertain to a highly sophisticated neurobiological group of the *Establishment*, emphasizing 'exact' technical procedures, are able to register 'volition' in the CNS of Invertebrates.

they are more readily evoked repeatedly'. Nevertheless, a gleam of diffidence seems to cross the cited authors' minds, prompting them to add: 'There is considerable overlap between the ethological element called a fixed action pattern and the reflex'.

Similarly, the vague terms '*simple*' and '*will*' play a role in a definition by BLASIUS (1962): '*Von Reflexen wird nur gesprochen, wenn es sich um einfache zentralnervöse Abläufe auf dem Wege etwa des einfachen Funktionskreises*' (referring to a diagram of a monosynaptic reflex) '*handelt, die den Character des Selbsttätigen oder Automatischen besitzen, bei denen also der Reflexerfolg bei gleichbleibendem Reiz regelmässig auftritt und innerhalb gewisser Grenzen ein gleiches Ausmass hat. Je nach dem sichtbaren Erfolg kann man motorische, sekretorische und Hemmungsreflexe unterscheiden. Wenn ein Reflex Vorgänge betrifft, die auch durch Einwirkungen des Willens beeinflusst werden können, spricht man von animalen, bei denen dies nicht der Fall ist, von vegetativen Reflexen*'. The cited author furthermore remarks: '*Alle natürlichen, instinktiven oder willkürlichen Bewegungen verlaufen im Gegensatz zu den Reflexen niemals automatisch, sondern werden dauernd nach den jeweiligen Bedürfnissen abgewandelt. Es ist daher nicht möglich, die natürlichen Bewegungen als reine Reflexbewegungen aufzufassen. Der Reflex ist lediglich ein "Fragment" der natürlichen Bewegung. Periphere und zentrale Anteile des Nervensystems sind zusammen mit den Ausführungsorganen durch eine Anzahl von Funktionskreisen oder Regelkreisen zu höheren Funktionssystemen zusammengeschlossen*'.

'*Bei der "Regelung" wird im Gegensatz zur "Steuerung", die von einer übergeordneten, unabhängigen Stelle erfolgt, die einzuhaltende Grösse dauernd überwacht und jede Abweichung von der Konstanz durch Gegenmassnahmen ausgeglichen*'.

One might here question the possibility of a semantically rigorous distinction between '*Regelung*' (control) and '*Steuerung*' (guidance) as postulated by BLASIUS. The concept of '*Funktionskreise*' or '*Regelkreise*' evidently refers to a wide variety of circuits whose activities involve SHERRINGTON's 'compounding' of reflexes, thereby providing '*höhere Funktionssysteme*'.

Clearly, *qua* 'natural' or 'normal' behavior, the term '*reflex*' applies to arbitrarily circumscribed patterns of neural events presumed to be 'units' or 'components' of which the overall neural activity resulting in the behavior is 'compounded'. *Qua* experimental, i.e. essentially artificial or 'abnormal' situation, the term '*reflex*' may subsume either the behavioral responses elicited and registered by the observer, or the corresponding conceptualized neural events.

Finally, from the clinician's viewpoint, R. BRAIN *(Lord Brain)*, in the posthumous 7th edition (1969) of his very creditable text *Diseases of the Nervous System*, designates the reflex as 'the simplest form of *involuntary response* to a stimulus'. Here, of course, the reference to '*will*' is fully justified. Obviously, in clinical medicine, the unobservable but inferred states of a patient's consciousness (which include the phenomena of 'will' as well as of 'pain' etc.) play a most significant role.

Assuming that the term '*reflex*' should be restricted to neural reactions upon an '*extraneural*' (e.g. sensory) stimulus, there are thus, depending upon the adopted viewpoint, at least three aspects of the neurobiologically highly significant reflex concept.[21]

(1) *Introspectively (erlebnispsychologisch)* 'reflexes' represent a class of motor or secretory reactions, which are unconscious or involuntary or both, upon a stimulus and its (excitatory or inhibitory) effect. It is here irrelevant whether the stimulus and its reactions are (secondarily) experienced or not experienced in consciousness.

(2) *Behavioristically*, reflex designates a more or less definable class of observable positive (active) or negative (inhibitory) reactions upon generally non-neural *events* describable as *stimuli*, the observed living organism being not that of the observer.

(3) *In terms of conceptual neural mechanisms*, a reflex is the transmission of events affecting a receptor, by means of transduced neural signals carried over more or less definable neuronal channels, upon a then positively or negatively reacting effector.

An oversimplified but nevertheless quite useful and instructive diagram of the mammalian central nervous system was drawn up by BAYLISS (1924) and adopted, in a slightly modified form, by RANSON (1943) as here shown in Figure 5. This elementary diagram, although discounting the complexities of the relevant grisea, particularly of the cerebral cortex, graphically illustrate the incontestable fact that the neuronal events resulting in muscular response to a sensory stimulus are in essence *(*or '*quiddity*'*)* basically of one and the same character *('nature')* regardless whether the 'impulse' transmission proceeds over a short pathway with few synapses, or a long one by way of the cerebral cortex. Since patterned *control of compound reflexes*, as well as '*chain reflexes*', *de-*

---

[21] Cf. K., 1966 *(Weitere Bemerkungen zur Maschinentheorie des Gehirns.* Confin. neurol. *27:* 295–328).

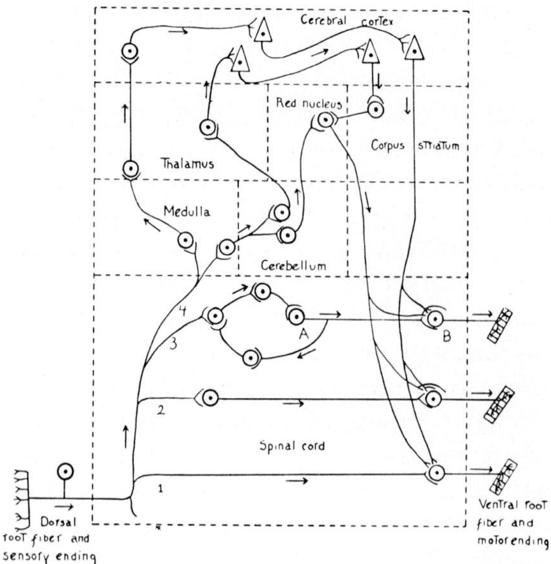

Figure 5. Diagram displaying essential circuit arrangements of the mammalian central nervous system, as interpreted by BAYLISS (1924) on the basis of the then available data, and slightly modified by RANSON (1943). Thus, in the upper part of 'spinal cord', RANSON has added a self-re-exciting circuit of the type defined by myself in the '*Vorlesungen*' of 1927, but here alleged to represent an entirely new concept introduced by RANSON and HINSEY in 1930 (cf. also vol. 1, chapter I, p. 11–12 of this series).

layed reflexes and *conditioned reflexes*[21a] are generally accepted forms of *'reflex' activity*, one might indeed wonder where a rigorously definable upper limit, *qua* sequence and number of circuits, could possibly be applied to the behavioristic and conceptual mechanistic formulations of

[21a] A brief discussion on the overall import of conditioned reflexes as investigated by PAVLOV and his school was included on pp. 31–32 of the introductory chapter in volume 1 (1967). Some present-day psychologists are wont to insist on a fundamental distinction between the original response techniques elaborated by PAVLOV and subsequent developments such as '*instrumental*' or '*operant conditioning*' as e.g. undertaken by SKINNER, in which the behavior becomes affected by its consequences. Although, in the cited introductory remarks, I pointed out the significance of such behavioral activities with emotional feedback effect, I do not believe that this distinction is sufficiently relevant to preclude the inclusion of '*operant conditioning*' into the general concept of '*Pavlovian conditioning*'.

the reflex concept. The stock reply, asserting that a reflex is either *unconscious* or *involuntary* or both, has no validity whatsoever in the behavioristic and in the mechanistic semantic models, in which consciousness is devoid of logical existence.[22] In the *introspective (erlebnispsychologischen)* or the *parallelistic model*, moreover, the presupposition that numerous *cortical neuronal activities* remain *unconscious* seems reasonably well substantiated and can be upheld. A pathway involving the cerebral cortex is, therefore, not a valid criterion for any semantic limitation applicable to neural circuits whose activities may be subsumed under the reflex-concept.

Thus, in stopping one's car at a red traffic light, in picking up a ringing telephone, in complying at table with the request 'please pass the salt', or even in a conversation with another person, the describable and instrumentally registrable neural events in no way differ from those in a *complex reflex*.

Although a reflex *sensu strictiori* is understood to be initiated by a receptor or 'sensory' organ generaly located *externally to the neuraxis*, Sherrington's definition, postulating an '*initiating reaction*' causing the 'end-effect', is not incompatible with the inclusion of 'spontaneous' behavioral activities[23] in the concept of reflexes *sensu latiori*. Accordingly, the neuronal events occurring while writing a letter and thereby combined with thought-activities are no more and no less than internally initiated complex compounded chain reflexes in which storage, delay, and modulating circuits play a significant role.

Hence, within the consistent and valid structural formalism of the behavioristic and of the mechanistic semantic models, we are no more than *reflex automata* or, from the viewpoint of postulational psycho-

---

[22] Behavioristically, conscious events *per se* 'in' an organism can neither be observed by an extraneous observer, nor be registered by means of any instrument or detecting device. In the mechanistic model, no path in a given time involving a definable biochemical respectively biophysical energy transfer or transformation can be described as connecting physical with mental events (consciousness).

[23] 'Spontaneous' or 'intrinsic neural activities' which can be regarded as strictly causal and therefore determined biologic events, were discussed in the introductory chapter I (section 3, on pp. 17–18), volume 1 (1967) of this series. Moreover, although the timing of Sherrington's 'initiating reaction' would here depend on the intrinsic activity of certain grisea within the central neuraxis, an undefined number of additional extrinsic necessary (contingent) conditions must obtain not only within the central nervous system but also within the body in general and within its environment.

physical parallelism, *conscious automata*[24] whose thoughts and actions can be conceived as strictly determined, thereby excluding 'free will'. Freedom of decision or choice refers here merely to the restricted capability of acting or thinking in accordance with the outcome of intrinsic neural events related to self-programming mechanisms.[25]

Introspectively, of course, we experience what is called 'free will'.[26]

---

[24] The following quatrains from FITZGERALD's *Omar* are here appropriate:

"For in and out, above, about, below
'Tis nothing but a Magic Shadow-show
Play'd in a Box whose Candle is the Sun
Round which we Phantom Figures come and go."
(1859, XLVI)

"We are no other than a moving row
Of Magic Shadow-shapes that come and go
Round with the Sun-illumin'd Lantern held
In Midnight by the Master of the Show;"
(1872, LXVIII)

" 'Tis all a Chequer-board of Nights and Days
Where Destiny with Men for Pieces plays:
Hither and thither moves, and mates and slays,
And one by one back in the Closet lays."
(1859, XLIX)

"Impotent Pieces of the Game It plays
Upon this Chequer-board of Night and Days;
Hither and thither moves, and checks, and slays,
And one by one back in the Closet lays."
(*sic secus*, 1872, LXIX)

"The Ball no Question makes of Ayes and Noes,
But Right or Left as strikes the Player goes;
And It that toss'd you down into the Field,
It knows about it all – IT knows – IT knows."
(*sic secus*, 1868, LXXV)

[25] It should be recalled that non-living systems such as computers and control devices can be provided with self-programming mechanisms. There obtains, in this respect, no essential difference between living organisms and artefacts.

[26] We can indeed maintain, as SCHOPENHAUER justly stated *(ed. Grisebach*, vol. 3, *'Über die Freiheit des menschlichen Willens'): Ich bin frei, denn 'ich kann thun was ich will'*. SCHOPENHAUER then asks: *'kannst du auch wollen, was du willst?'* and points ont the further sequence: *'kannst du auch wollen, was du wollen willst?'*. SCHOPENHAUER, moreover, fully agrees with SPINOZA, who remarked in his *Epistola 62: Atque haec humana illa libertas est, quam omnes habere jactant, et quae in hoc solo consistit, quod homines sui appetitus sint conscii, et causarum, a quibus determinatur, ignari'*. In *Ethices Pars II, propos*. 48, SPINOZA remarks: *'In mente nulla est absoluta sive libera voluntas, sed mens ad hoc vel illud volendum determinatur a causa, quae etiam ab alia determinata est, et haec iterum ab alia, et sic in infinitum.'*

The concept of *liberum arbitrium* bearing upon this conscious experience may be considered, approximately in KANT's sense, a 'postulate of practical reason'. Freedom of decision and action, as well as responsibility and moral values, based on relevant fictions in VAIHINGER's sense, are therefore, from a practical viewpoint, fully consistent with the strictest determinism, as was well expounded by HUME, SCHOPENHAUER and other deterministic philosophers.[27]

It seems quite evident that an undefined combination of structural and functional features may be arbitrarily conceived or 'reified' as representing a more or less adjustable '*basic reflex pattern*' which can be triggered in the case of motor behavior induced by higher centers, including 'voluntary' respectively cortical motor activity. EASTON (1972) has recently elaborated on this concept which is already clearly implied in SHERRINGTON's views concerning the integrative action of the nervous system. In EASTON's phraseology, the neuraxis is 'designed' to 'respond automatically' to certain stimuli with certain 'basic reflexes or coordinative structures' (CS), such that 'reflexes' form the 'basic language' of the 'motor program'.

Reverting, after this '*excursus*', to an appropriate predominantly behavioristic and mechanistic classification of reflexes, with particular emphasis on those in the spinal cord,[28] the following criteria may be applied: (1) number of successive neuronal links respectively synapses involved in the pathway for the transmission of the 'impulse'; (2) relation of input segment to output segment; (3) homolateral, contralateral or bilateral response to unilateral input; (4) number of output segments in relation to input level or levels; this includes, e.g. a distinction of

---

[27] The *quantum theories* of physics, which require a statistical approach, quite logically and consistently excluding determinism from their semantic models based on a fictional concept of physical space-time, cannot be used as valid arguments against the concept of determinism. This was fully understood by MAX PLANCK, who originated the quantum concept, as well as by ALBERT EINSTEIN. Both outstanding physicists, although hardly very profound philosophers and epistemologists, were, nevertheless, endowed with unusally sound philosophical 'common-sense' and repeatedly expressed their preference for the concept of causality despite its apparent unsuitability for the mathematical formulation of fictional small-scale physical events. PLANCK's and EINSTEIN's insight are conspicuously missing in a number of eminent and even canonized quantum physicists with considerable mathematical proficiency.

[28] Reflexes essentially or primarily involving cerebral grisea shall be discussed in the subsequent chapters concerning the subdivisions of the brain, thus omitting here the pupillary reflexes which are in part mediated by cervical cord grisea.

'unimuscular' and 'multimuscular' reflex responses; (5) location of receptor in respect to effector, i.e. within the same or in different organs *('Eigenreflexe'* and *'Fremdreflexe');* (6) classification in accordance with the doctrine of functional nerve components or in accordance with SHERRINGTON's terminology *(somatic* and *visceral*, *exteroceptive*, *proprioceptive* and *interoceptive* reflexes); (7) temporal characteristics such as short contractions *('phasische Reflexe')*, continuous contraction *(tonic reflexes)*, rhythmic activity *(clonic reflexes)*, delayed reflexes; (8) degree of coordination, *'geordnete Reflexe'*, or lack of coordination *('convulsions', 'Reflexkrämpfe');* (9) excitatory or inhibitory effects or both (e.g. in reflexes manifesting *'reciprocal innervation);* (10) predominantly *clinical* criteria, such as e.g. tendon reflexes, cutaneous reflexes, and mucosal reflexes; (11) in physiologic experiments, *postural (attitudinal, righting)*

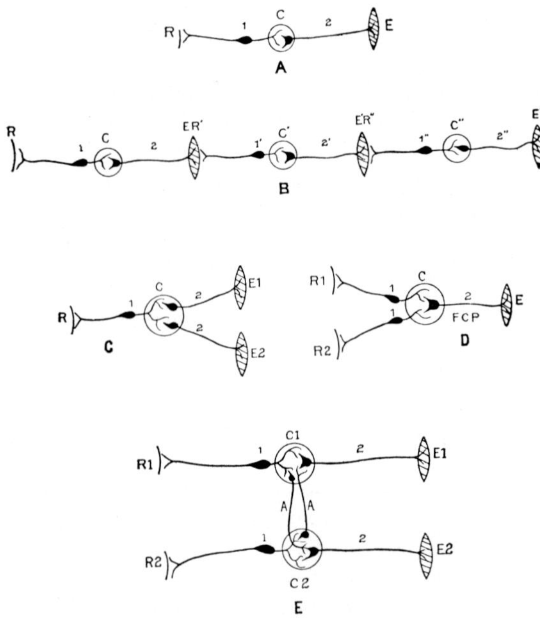

Figure 6. Diagrams representing the relations of neurons in five types of reflex arcs (after HERRICK, 1931). A: simple reflex arc; B: chain reflex; C: a complex system illustrating allied and antagonistic reflexes; D: complex system with common final path; E: complex system with 'association'. AA: association neurons; C, C′, C″, C1, C2: 'centers' or 'adjustors'; E, E′, E″, E1, E2: 'effectors'; FCP: final common path; R, R′, R″, R1, R2: receptors; 1: afferent neuron; 2: efferent neuron.

reflexes of static or statokinetic type have been described, many of which involve not only spinal cord but also brain-stem, particularly vestibular (labyrinthine) circuits. Some of these, e.g. essentially spinal tonic neck reflexes, are also clinically significant; (12) the so-called *psychogalvanic reflex* and allied phenomena, in part related to 'reflectory' innervation of the sweat glands.

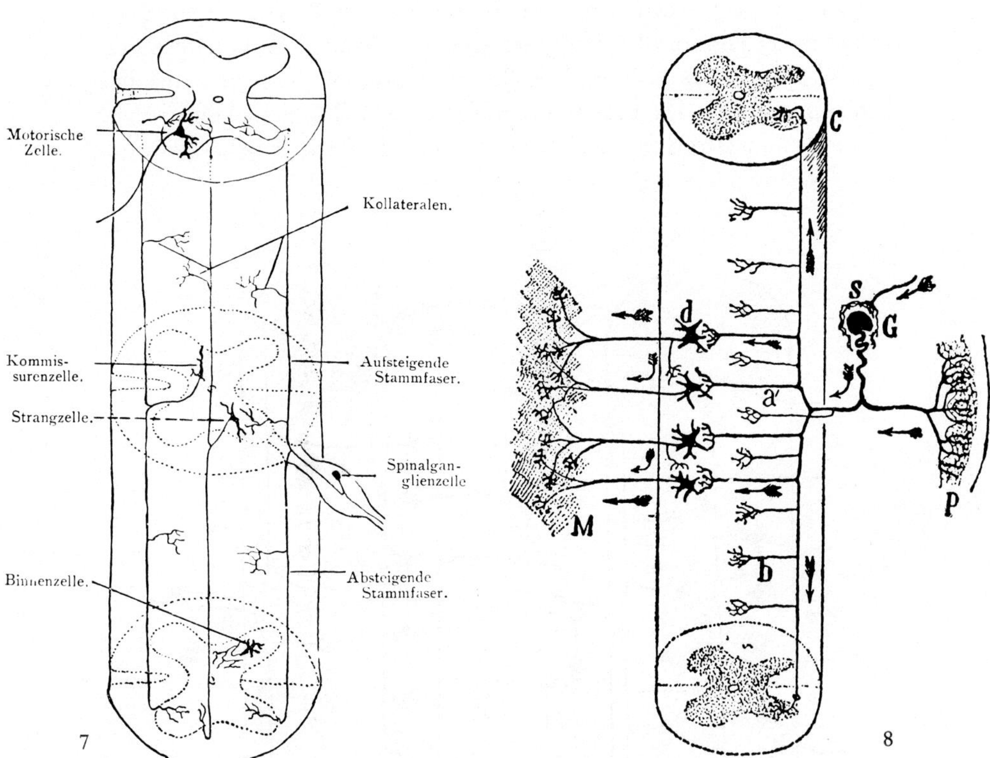

Figure 7. Diagram of Vertebrate (Mammalian) spinal cord showing neuronal connections of the *Eigenapparat* (after Stöhr-Möllendorff from K., 1927).

Figure 8. Cajal's diagram of unilateral 'circumscribed' inter- or plurisegmental reflex connections (from Cajal, 1909, 1952). G: spinal ganglion cell; M: striated musculature; P: skin; S: 'arborisation sympathique péricellulaire' (this sort of connection, suspected by Cajal, and which, in 1927, I likewise presumed to exist as a visceral-afferent mechanism, has not been corroborated); a′: monosynaptic reflex-collaterals; b: short collaterals; c: afferent ending in dorsal horn; d: motoneuron; '*les flèches indiquent le sens des courants*'.

Since a number of these reflex categories overlap, a given reflex response can usually be classified as pertaining to more than one of such categories, whose foregoing enumeration, moreover, is by no means intended to be exhaustive. As regards, e.g., semantic overlap, it is evident that a central 'vegetative reflex', defined as a reflex whose efferent arc pertains to the 'vegetative' ('autonomic') nervous system and thus involves a preganglionic and a postganglionic neuron, may have an afferent arc of undefined neuronal sequence pertaining to any type of somatic or visceral category.

With respect to category (1), a '*simple reflex arc*' is traditionally illustrated by a diagram of spinal cord showing dorsal root collaterals effecting synapses on a homolateral motoneuron at the root's entrance level (cf. e.g. Fig. 1A, p. 7, vol. 1 of this series). The actual occurrence of such connections was amply corroborated by the *Golgi preparations*

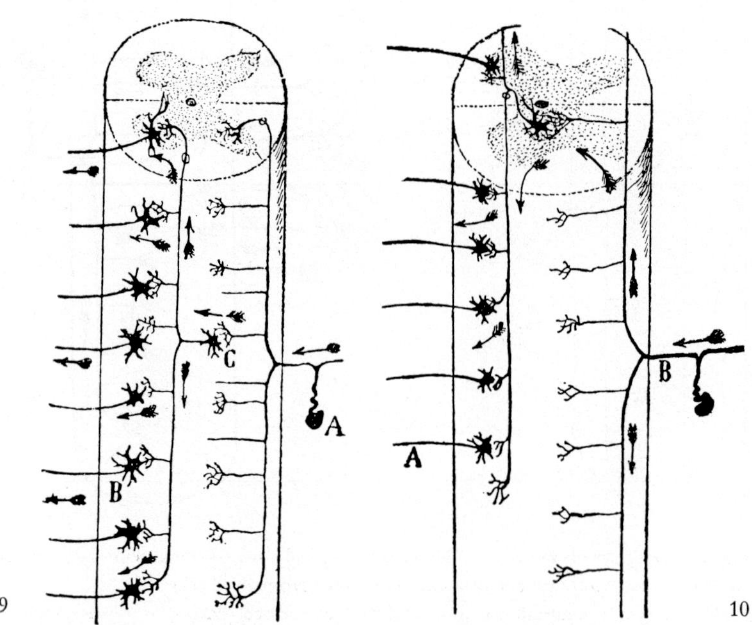

Figure 9. CAJAL's diagram of unilateral 'diffuse' inter- or plurisegmental reflex connections (from CAJAL 1909, 1952). A: spinal ganglion cell; B: motoneuron; C: internuncial neuron (*'cellule funiculaire'*, *'Strangzelle'* of funiculus proprius.)

Figure 10. CAJAL's diagram of contralateral inter- or plurisegmental reflex connections (from CAJAL 1909, 1952). A: ventral root; B: dorsal root.

which CAJAL and others obtained, as well as by the subsequent studies with refined neurophysiologic techniques. Reflexes of this type are monosynaptic,[29] segmental, and homolateral. Such monosynaptic arcs seem to be characteristic for proprioceptive reflexes mediated by large afferent A-fibers from muscle spindles.[30] To which extent various other proprioceptive or exteroceptive reflexes may likewise be monosynaptic, remains, despite numerous investigations and discussions of that topic, a poorly elucidated question.

Reflexes not involving a synapse within the neuronal connecting pathway are the so-called *axon reflexes* assumed to obtain in peripheral ganglia of the vegetative nervous system and in vasodilation (cf. chapter VII in the preceding volume of this series). If preganglionic elements should be capable of axon reflexes, then this sort of neural transmission could still involve a synapse between pre- and postganglionic neuron. Again, because of the presence of this latter synapse, all typical vegetative reflexes are thus presumed to comprise *at least* a chain of three neurons with two intervening synapses.

The diagram of Figures 6–10 illustrate diverse aspects of monosynaptic, multisynaptic, segmental, intersegmental, homolateral and contralateral reflex mechanisms obtaining in the spinal cord's *eigenapparat*. With regard to HOFFMANN's (1934, 1952) definition of '*Eigenreflexe*' and '*Fremdreflexe*', of which the former originate within the responding organ itself and are thereby 'proprioceptive', reference to the somewhat different interpretations of the term 'proprioceptive' as discussed in section 1 (pp. 805–806, footnote 26) of the preceding chapter VII (vol. 3, part II) will here be sufficient. Concerning reaction-time, delayed responses and the phenomena related to 'after-discharge', Figure 11 illustrates a number of different neuronal connections which can be conceived, on the basis of inferences derived from a

---

[29] *Monosynaptic* with respect to the neural conduction pathway. Including the neuromuscular junctions and the peripheral input locus, three connections of synaptic type could be enumerated in such monosynaptic reflex arcs. These latter are also occasionally called '*direct reflexes*' in contradistinction to '*indirect reflexes*' including one or more internuncial elements.

[30] It will be recalled (cf. chapter VII, section 1, p. 806 in the preceding volume 3, part II of this series) that the occurrence of muscle spindles is apparently restricted to tetrapod Vertebrates (Amphibia, Sauropsida, Mammals). The structural and functional details concerning this sort of proprioceptive innervation have been particularly investigated in Mammals.

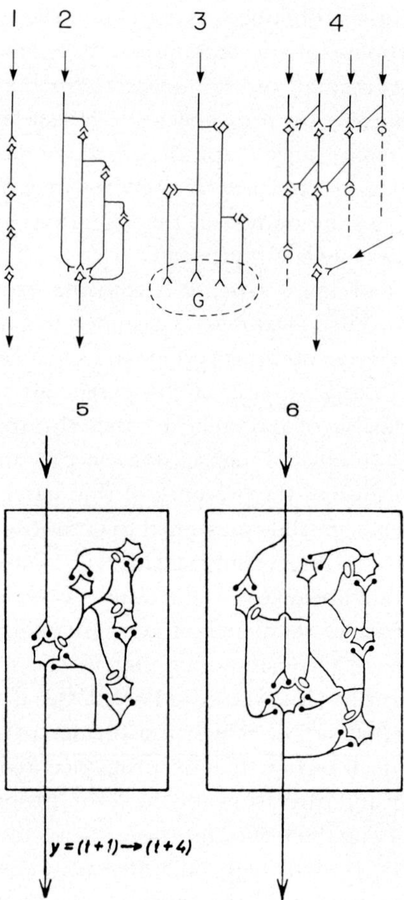

Figure 11. Diagrams of neuronal models of delay lines, after-discharge, and self-reexciting circuits. 1 and 2 simplified after K., 1957. 3 and 4 adapted after BLASIUS, 1962. 5 and 6 from K., 1957. 1: delay through internuncial chain; 2: after-discharge resulting from collateral channels without spread; 3: collateral delay combined with spread into griseum G; 4: delay with collateral spread limited by extinction (dotted lines) based on insufficient spatial summation; 5, 6: self-reexciting (reverberating) circuits with automatic cut-off through *Dale-type* inhibiting neuron (light, oval synapses signify inhibition). In accordance with the adopted theoretical postulates such as validity or non-validity of *Dale's principle*, necessity or non-necessity of internuncials for self-reexcitation, and assumed requirements for spatial and temporal summation, a wide variety of different models illustrating identical sorts of end-effects can be construed.

variety of data, as providing the relevant structural arrangements. Among these latter, 'reverberating' or 'self-reexciting' respectively 'self-inhibiting' circuits may play an important role. Other factors bearing upon reflex-time, such as *synaptic delay* and *conduction velocities*, were dealt with in section 8 (p. 578f., 618, 625, and 627) of volume 3, part I.

With regard to coordinated (*'geordnete'*) and apparently 'purposeful' spinal reflexes involving several segments, the well-known *'wiping reflex'* in the spinal frog and the *'scratch reflex'* investigated in the spinal dog by SHERRINGTON (cf. Fig. 12) are typical examples. As an instance of bilateral respectively contralateral response, SHERRINGTON's *crossed extensor reflex* may be mentioned (Fig. 13). The contralateral component of this reflex becomes inhibited or modified upon simultaneous bilateral stimulation.

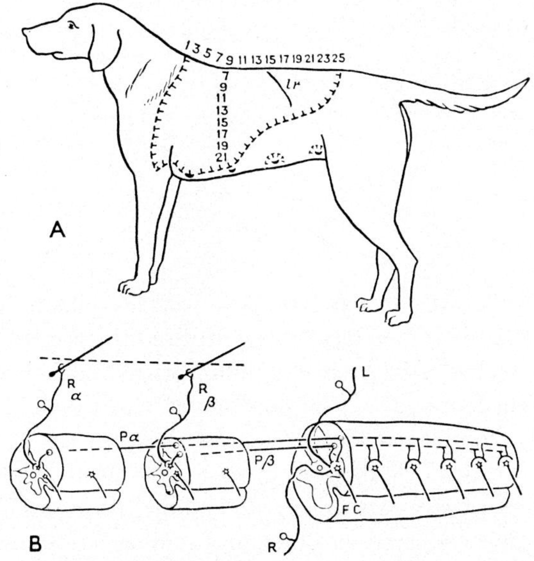

Figure 12. Diagrams illustrating 'receptive field' and spinal arcs involved in the dog's 'scratch reflex' (from SHERRINGTON, 1906, 1947). A Saddle-shaped area of dorsal skin whence the scratch-reflex of the left hindlimb can be elicited after low cervical transection of spinal cord. lr: indicates position of the last rib. B Diagram of the neural connections mediating the reflex; L: afferent pathway from left foot; R: afferent pathways from opposite foot; Rα, Rβ: afferent path from hairs in dorsal skin of left side; FC: final common path from motoneuron to flexor muscle of hip; Pα, Pβ: 'proprio-spinal neurons' (i.e. internuncial pathway in fasciculi proprii of intersegmental aubdivision of eigenapparat).

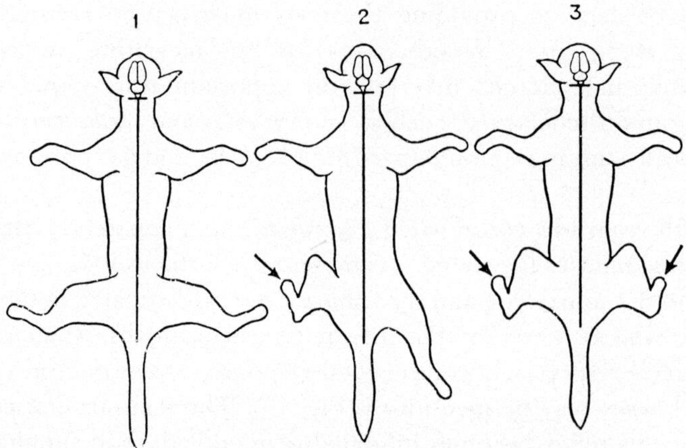

Figure 13. Diagrams illustrating the predominant uncrossed flexor-reflex of the hind-limb inhibiting the crossed extensor-reflex otherwise obtainable by stimulation of the opposite limb in the 'spinal cat' (from SHERRINGTON, 1906, 1947).

1. The initial pose of the spinal animal (the site of transection is indicated by the line caudal to the brain).

2. The pose assumed after stimulation of the left hind-foot; the flexors of the left hip, knee, and ankle, and the extensors of the right hip, knee, and ankle are in active contraction.

3. The pose assumed after simultaneous stimulation of both hind feet. The extensor action of the hip, knee and ankle that would appear from either side as a crossed reflex is bilaterally inhibited and the antagonistic flexor-reflexes bilaterally prevail.

Although the early investigations of reflexes and the various diagrams depicting their neural pathways stressed the transmission of excitatory impulses, the occurrence of inhibitory neural activities was observed and reported by the older authors (cf. e.g. above, footnote 15). Subsequently, inhibitory effects, such as those of the vagus on the heart, were recorded by numerous investigators. With respect to spinal reflexes, SHERRINGTON particularly studied the reciprocal inhibition of reflexes and the manifestations of reciprocal innervation combining excitatory and inhibitory[31] effects required for innervation of antagonists

[31] The equivalent importance of inhibitory and excitatory synapses became particularly evident in connection with the development of circuit algebra based on *Boolean algebra* and, in principle, applicable to neural networks (cf. K.: Brain and Consciousness, 1957; Gehirn und Intelligenz, 1965). Biochemical and electrical aspects of excitatory and inhibitory synaptic activity were briefly discussed in chapter V, section 8 of volume 3, part I of this series. A recent publication by ECCLES (The inhibitory pathways of the central nervous system, 1969) deals with the details of that topic. This author upholds

and synergists in coordinated movements (Fig. 14, cf. also the simplified Fig. 13A on p. 14 in volume 1 of this series). The pertinent inhibitory mechanism, as interpreted by ECCLES (1969), is shown in the diagram of Figure 15 based on the postulate of a specific inhibiting internuncial in accordance with *Dale's principle*. LLOYD (1946, 1960), how-

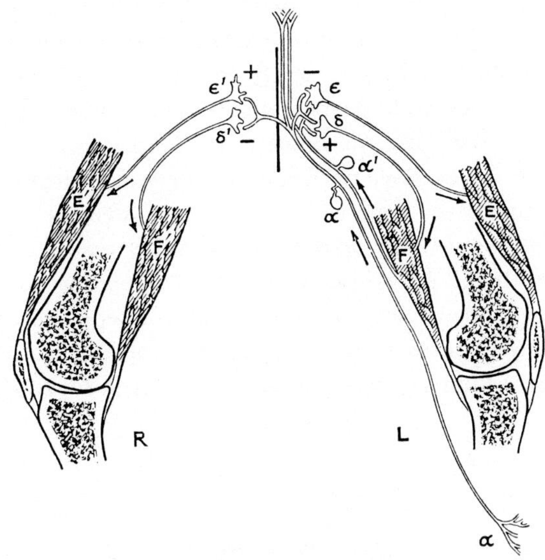

Figure 14. SHERRINGTON's diagram of reciprocal innervation, 'indicating connexions and actions of two afferent spinal root cells, α and α^1, in regard to their reflex influence on the extensor and flexor muscles of the two knees (from SHERRINGTON, 1906, 1947). α: Root-cell afferent from skin below knee; α^1: root-cell afferent from flexor muscle of knee (e.g. hamstring); δ, δ^1 efferent neurones to flexor muscles of knees; ε, ε': efferent neurones to extensor muscles; E, E': extensor muscles; F, F': flexor muscles; +, —: excitatory and inhibitory synapses. SHERRINGTON adds here: 'the effect of strychnine and of tetanus toxin is to convert the minus sign into plus sign'. It will be seen that *Dale's principle* is not considered in this diagram. The postulated validity of said principle does not significantly affect the soundness of the diagram, but would require additional complications depicting presumed details of synaptology.

Dale's principle postulating that at all synaptic output terminals of a neuron the same transmitter substance is liberated. Hence, a given nerve cell would have either excitatory or inhibitory function but not both. In the spinal cord glycine or a related substance could be the inhibitory transmitter. At higher levels of the neuraxis, GABA might have that function. Although *Dale's principle* appears reasonably well substantiated in the case of various neuronal elements of Vertebrates, its universal cogency seems questionable, especially since exceptions to said principle were reported in the nervous system of Invertebrates.

ever, maintained the presumed occurrence of a monosynaptic inhibition inconsistent with said principle. Despite ECCLES' emphatic claims, supported by his evaluation of intracellular recordings, this question may be regarded as not entirely clarified.

The intraspinal recurrent collaterals of ventral horn motoneuron axons, displayed in *Golgi preparations*, and recorded by CAJAL as well as other authors, are believed to bring about the inhibition of other motoneurons. This type of inhibition caused by the firing of motoneurons, and described by LLOYD (1941) and RENSHAW (1941), is nowadays generally presumed to be mediated by special inhibitory internuncial neurons of small size, designated as *Renshaw cells* (Figs. 16, 17). These latter, however, have not been definitely identified in histologic preparations up to the time of this writing. Nevertheless, since the socalled '*Renshaw-effect or phenomenon*' can be easily explained, in accordance with the prevailing views, by postulating specific inhibiting neuronal elements, it is not unjustified to intepret relatively small nerve cells in the ventral portion of the anterior horn as representing '*Renshaw cells*'. Other neuronal elements of smaller size than typical

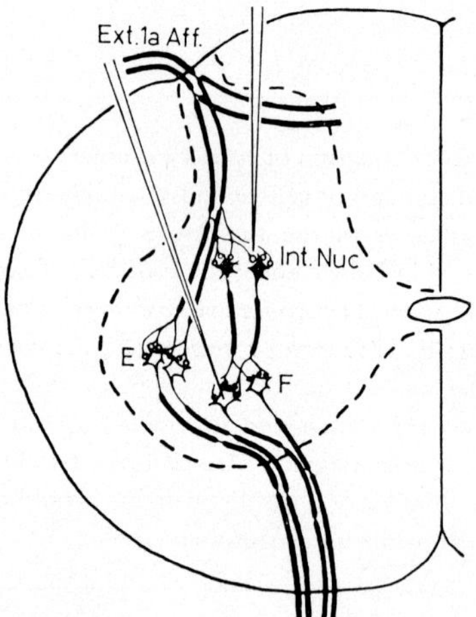

Figure 15. Diagram of transverse section of a mammalian spinal cord in the lower lumbar region, illustrating the inhibitory action on motoneurons of antagonistic muscles in accordance with ECCLES' views (from ECCLES, 1969). E: extensor motoneurons; Ext. 1a Aff.: large primary afferents (of so-called group Ia) from extensor muscle spindles; F: flexor motoneurons; Int. Nuc.: postulated 'intermediate nucleus' with inhibiting interneurons for flexors. The diagram shows microelectrodes in intermediate nucleus and F-group of motoneurons.

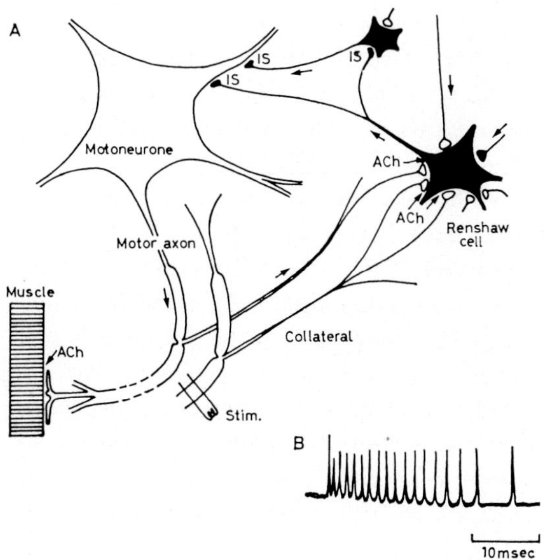

Figure 16. Diagram purporting to show, in A, the inhibitory pathways to spinal motoneurons by their axon collaterals and '*Renshaw cells*'. B shows what is interpreted as the 'extracellular recording of a *Renshaw cell* being excited by an antidromic volley in the motor fibres of lateral gastrocnemius muscle' (from ECCLES, 1969). Ach: acetylcholine; IS: inhibitory synapses.

motoneurons, and located within the ventral gray column of Amniota, can be presumed to represent the γ-cells for the motor innervation of muscle spindles.[32] The assumed connections pertaining to the system of motoneurons, γ-cells and *Renshaw cells*, including some typical registrations of impulses from some peripheral root fibers are shown in Figure 18.

---

[32] The nerve cells making up the Mammalian ventral column have been arbitrarily subdivised into three groups: large, medium and small (cf. e.g. ROMANES, 1964). Some of the medium or small cells at the ventral border of the gray substance about the bundles of anterior root fibers may be *Renshaw cells* but were also tentatively identified as γ-cells. These latter could also be represented by other, less superficially located medium or small cells. It is, moreover, most probable that typical motoneurons are of different sizes, varying at least between 'large' and 'medium'. In our preliminary computations of the number of motoneurons in the human spinal cord (SIRKEN and K., 1966), we stressed the difficulties concerning the reasonably certain identification, in cell stain preparations, of ventral horn perikarya as those of motoneurons. We arbitrarily distinguished on the basis of size, shape, as well as location, 'very probable', 'probable', and 'dubious' motoneurons.

With respect to reflexes, and particularly also for a proper interpretation or evaluation of simplified neuronal circuit diagrams such as shown in Figures 5 to 11, and 14 to 18, it is appropriate to keep in mind the great multiformity of interneuronal couplings or junctions, which are still, in part, incompletely understood and can be classified in accordance with a diversity of viewpoints as elaborated in chapter V, volume 3, part I of this series. Thus, stressing the direction of the impulse, respectively the presence or absence of a '*rectifier-effect*' at the interneuronal connection, unpolarized, facultatively polarized and permanently polarized junctions can be distinguished. If SHERRINGTON's term '*synapse*' *sensu strictiori* is used for the latter type, then the two other ones can be designated as asynaptic respectively protosynaptic. Considering the effect of the impulse transmission, such couplings may be excitatory or inhibitory, or possibly both (cf. above p. 36 and footnote 31). As regards the transmission process, couplings may be

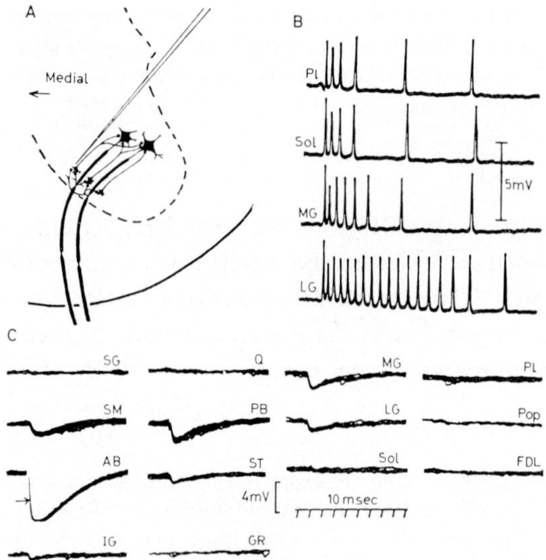

Figure 17. Recording 'from a *Renshaw cell*' (A), said to show (B) 'that it fires repetitively to single volleys in the motor fibres to four different muscles'. C is said to show by intracellular recording 'that IPSPs of various sizes are produced in an anterior biceps motoneurone by single volleys in the motor fibres supplying seven different muscles of the same hindlimb' (from ECCLES, 1969). AB: anterior biceps; FDL: flexor digitorum longus; GR: gracilis; IG: inferior gluteal; LG: lateral gastrocnemius; MG: medial gastrocnemius; PB: posterior biceps; PL: plantaris; Pop: popliteus; Q: quadriceps; SG: superior gluteal; SM: semimembranosus; Sol: soleus; ST: semitendinosus.

electrical, biochemical, or both. Biochemical couplings, again, can be classified according to the relevant transmitter substances.

In anatomical classifications, one might consider junctions involving typical, easily definable conventional nerve cell processes such as dendrites and neurites, or, again, junctions involving cell processes that cannot be unambiguously defined, such as the 'apotiles' discussed in chapter V, volume 3, part I of this treatise. Axo-dendritic, axo-somatic, and axo-axonic couplings refer to 'typical' neuronal processes.

Most interneuronal couplings are junctions by contiguity, while a few well-documented instances of continuity (without interposition of cell membranes) are known to occur. Structural differences observable by light microscopy have led to the description of numerous types of synaptic junctions, and electron microscopy has added a multitude of ultrastructural details concerning pre- and postsynaptic membranes, desmosomic junctions', 'gap-junctions', 'mixed junctions', 'reciprocal

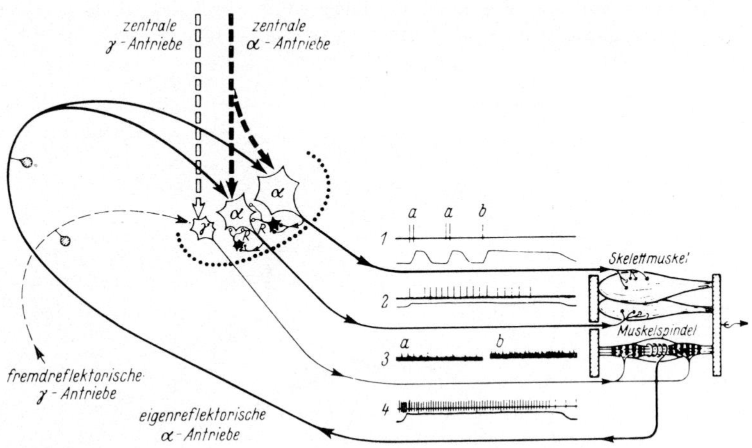

Figure 18. Diagram purporting to illustrate relationships of the Mammalian α and γ motoneuron system (after BLASIUS, 1962). The larger of the two α-elements is supposed to have a 'phasic' discharge, that of the smaller one being 'tonic'. The former is believed related to 'rapid', and the latter to 'slow' components of the 'extrafusal' muscle fibers. 'Typical' examples of registration are depicted as follows. 1: phasic α-stretch-reflex upon short (a) and longer (b) duration of stretch; 2: tonic α-stretch-reflex upon continuous stretch; 3: 'spontaneous' activity of a γ-fiber before (a) and after (b) central activitation; 4: muscle-spindle discharges upon continuous stretch of the musculature; R: *Renshaw-cells.*

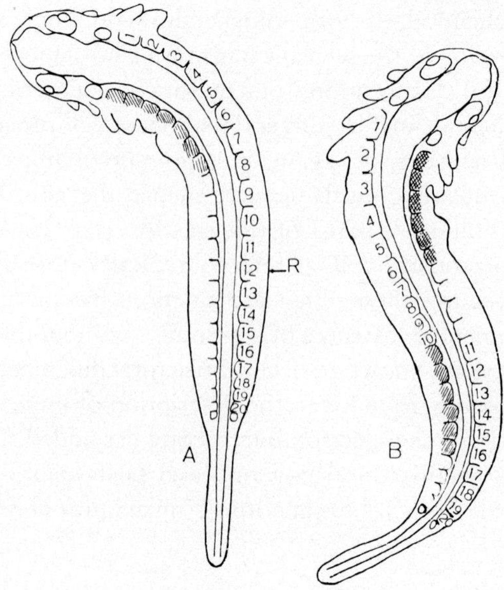

Figure 19. Initial stages during the swimming reflex of an Amblystoma larva (after HERRICK and COGHILL, 1915, from K., 1927). R: stimulated body region.

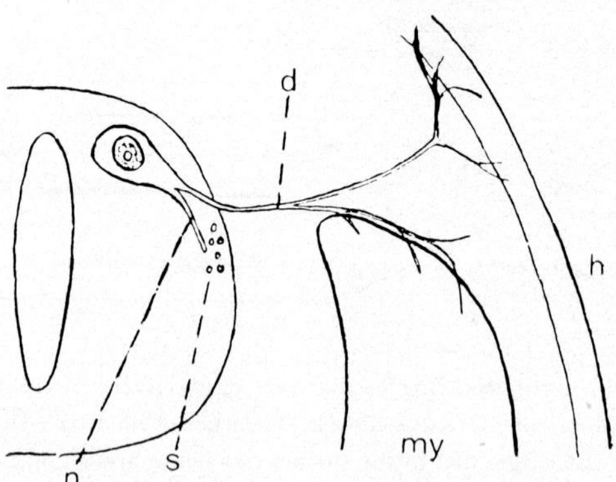

Figure 20. Peripheral connections of a *Rohon-Beard cell* in the spinal cord of larval Amblystoma (after HERRICK and COGHILL, 1915, from K., 1927). d: peripheral, afferent neurite (so-called 'dendrite'); h: skin; my: myotome; n: central (cellulifugal) neurite; s: dorsolateral 'sensory tract'.

junctions', types of synaptic vesicles, and 'serial synapses' (i.e. synapses upon synapses). Moreover, 'dendro-dendritic', 'somato-somatic' and 'somato-axonic' couplings have been claimed to occur. It is evident that, quite apart from the multitudinous poorly elucidated details whose significance still remains unknown, a convenient single and overall valid classification of interneuronal couplings or 'synapses' cannot be given. Many interpretations remain rather unconvincing. A not very critical and somewhat overenthusiastic attempt was recently propounded by BODIAN (1972).

A peculiar type of reflex activity is displayed by the spinal cord mechanisms of larval Amphibia, which are capable to swim and to react upon environmental stimuli at relatively very early ontogenetic stages. In *Amblystoma*, the behavior of such larvae and the organization of their spinal cord were investigated by COGHILL (1913, 1914, 1929) in cooperation with HERRICK (HERRICK and COGHILL, 1915).

During the so-called non-motile stage, defined by COGHILL (1929) as that stage when the animal can contract its muscles but cannot do so in response to the stimulus of of light touch upon the skin, there are

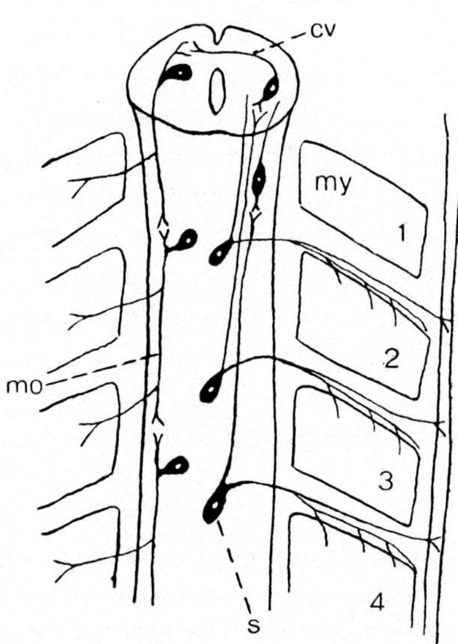

Figure 21. Neuronal connections of the 'eigenapparat' in larval Amblystoma (simplified after HERRICK and COGHILL, 1915, from K., 1927). cv: ventral commissure; mo: 'motor tract'; my: myotome; s: afferent dorsal *(Rohon-Beard)* cell.

already motor nerve roots reaching about twelve of the most anterior myotomes. Subsequently, the youngest larvae reacting upon an external (touch) stimulus bend their rostral body-end away from the side of contact. At a closely following, only slightly more developed stage, any such touch-stimulus elicits a characteristic swimming motion (Fig. 19).

As regards the organization of the spinal cord corresponding to this latter stage, the primary afferent neurons are here represented by segmentally distributed large transitory dorsal or *Rohon-Beard cells*, whose peripheral axon-process (frequently referred to as 'dendrite') bifurcates into an exteroceptive branch supplying the skin, and into a proprioceptive branch with endings in the myotome (Fig. 20). The central axon-processes of these cells ascend rostrad within the spinal cord, forming a dorsolaterally located fiber tract. At the level of transition from spinal cord to medulla oblongata, this tract ends and connects with commissural internuncials[33] whose neurites reach the contralateral side through a ventral decussation developed at that level. The impulses from the *Rohon-Beard cells* are thereby transmitted to a synaptically connected series of peculiar primitive motoneurons. These latter do not display a strictly segmental arrangement, but are distributed as a chain whose descending neurites form a ventrolateral longitudinal 'motor tract' of which the ventral roots, innervating the myotomes, are collateral branches (cf. Fig. 21). At this stage, no effective functional connections other than those of the just described system seem to be present, since all other neuronal elements can here be interpreted as being functionally still immature neuroblasts.

Thus, if e.g. a touch stimulus affects the skin of the right caudal trunk region, the corresponding nervous impulse, originating at the exteroceptive input locus of the dorsal cell's or first neuron's peripheral axon, is conveyed by way of the ascending central axon to a commissural cell (2nd neuron), proceeds over the rostral commissure to the contralateral side, and is then transmitted to the rostralmost motor cell

---

[33] Although, as pointed out in chapter VI, p. 291 of the preceding volume, and again mentioned above in the introductory remarks on the spinal cord's general pattern, the floor plate does not play a significant role in the production of neuronal elements, the commissural cells, representing very relevant links in the early swimming reflex mechanism of Amblystoma, have been identified as floor plate derivatives and were therefore also designated as '*floor plate cells*' (COGHILL, 1929). In Amblystoma and perhaps some other species, the floor plate forms a so-called '*neural keel*'.

(3rd neuron). This latter discharges through its collaterals into the appurtenant myotomes and, moreover, through the longitudinal 'primitive motor tract', upon more caudally located other motor elements. Hence, a lateral flexion of the left rostral body region results (cf. Fig. 19A). While the myotomic contraction wave progresses caudalward, the contraction of the rostral left myotomes has provided a proprioceptive stimulus for the myotomic input loci of the corresponding dorsal cells' peripheral neurites, and the resulting nervous impulses discharge, by way of the commissure, into the rostral motor elements on the right side. These events produce an S-shaped contraction of the body (cf. Fig. 19B) which, in turn, by a wave-like rostro-caudal progression, results in an undulating swimming motion, while the nervous impulses oscillate backward and forward as well as from one side to the other. If no additional stimulus occurs, the swimming motion subsides because of still poorly understood decremental factors such as 'inner resistance' or 'inertia' of the system.

At a subsequent developmental stage this simple reflex path becomes substituted by a more complex organization. At first, the transitory dorsal cells seem to disappear and slightly later the permanent motor root fibers arise as neurites of segmentally arranged typical ventral horn motoneurons.

A noteworthy peculiarity of this early larval reflex mechanism elucidated by COGHILL and HERRICK, consists, as recognized by these authors, and also stressed by KAPPERS (1920), in the fact that exteroceptive and proprioceptive input is mediated by afferent terminals pertaining to the peripheral neurite of one and the same nerve cell (dorsal or *Rohon-Beard cell*). Both sorts of input, moreover, become channeled not only through a common central tract, but also through one and the same neurite within that tract. In addition, the reflex path at these early ontogenetic stages is relatively very long and passes, especially on the motor side, through many synapses. However, since this reflex mechanism may be evaluated as being highly 'adapted' to the life mode of aquatic amphibian larvae, its peculiarities could be considered a secondary 'acquisition' whose detailed features do not provide significant arguments for particular phylogenetic speculations.

Concerning the *ontogenetic development of reflex activities*, it has been noted by several investigators that the appearance of motor performances commonly antedates sensory control. In the case of Amblystoma, this is clearly related to the fact that the ventral root fibers described above as collaterals of the primitive longitudinal motor tract are apparently here the first nerves growing, as non-medullated threads, from neurons in the spinal cord to their endings on peripheral structures. Since said outgrowth seems to precede that of dorsal root fibers from *Rohon-Beard cells*, it is not possible to derive in this and perhaps in various additional cases, the properties of the output of the nervous system directly from the pattern of the sensory input. Yet, a close relationship of input to output appears significant for other aspects of neurophysiological and behavioristic analysis.

WEISS (1955) pointed out that the central nervous system ontogenetically develops a finite repertory of behavioral performances which are pre-functional in origin and ready to be exhibited as soon as the proper effector apparatus becomes available. HIS (1904 *et passim)* as well as FLECHSIG (1927) noted that, in human fetuses, the outgrowth of motor roots from the spinal cord precedes, during the 4th week of intrauterine development (ca. 4.4 mm length), the ingrowth of dorsal roots. At this time, the spinal ganglion anlage contains only bipolar neuroblasts devoid of long processes, and afferent dorsal root fibers are thus lacking. Within one week, all ventral spinal roots are laid down, their establishment being followed by that of the dorsal roots. Subsequently, the anterior commissure of the spinal cord becomes one of the earliest fiber systems developing in this subdivision of the neural tube.

According to WINDLE (1940), the very first neurons that can be recognized in mammalian embryos are primary motor elements of spinal and cranial nerves, present e.g. in rat embryos of 3 mm length. Early local internuncials are second, and primary afferent fibers appear third. On the basis of his observations concerning the sequential development of reflex mechanisms in Mammalian embryos, the cited author therefore concludes that the component neural channels of these systems are laid down from efferent to afferent side, and that the simplest reflex pathways are formed before those mediating 'higher activities' of integration or 'perception' (perhaps better: 'registration').

Despite general agreement on the commonly obtaining temporal precedence of efferent connections with respect to afferent ones,[34] two

[34] SCHOPENHAUER, who was greatly interested in problems of brain anatomy and neurobiology, would presumably have interpreted the precedence of motor output with respect to 'sensory' input as a confirmation of his views concerning the primacy of 'will' with respect to 'intellect'. In both volumes of his great work '*Die Welt als Wille und Vorstellung*' this thesis of 'primacy' is elaborated with considerable detail. Although I consider myself in various regards a disciple of SCHOPENHAUER, I maintain a somewhat more

apparently contradictory views on the correlated development of neural structures and behavior have been expressed. COGHILL (1929) and supporters of his opinion believe that the essential conduction pathways as well as the primary neural mechanisms emerge from a preneural dynamic pattern. The resulting total neural pattern is considered to arise as a perfectly integrated unit. This totally integrated pattern expands through the organism, and local reflexes emerge as a 'quality upon a ground', that is as special features within a more diffuse but dominant mechanism of integration of the whole organism. In other words, a gradual individuation arising from a background of mass reactions is assumed.

However, according to WINDLE (1950) and others, certain types of activity are territorially localized from the beginning, and individuation from mass action does not apply to such performances (cf. also WEISS, 1955). These conflicting views may be reconciled if the diversity of sample species and techniques is duly considered. It is not unlikely that both principles obtain, and that each is valid for a certain aspect of behavioral development.

Besides COGHILL's investigations primarily dealing with the urodele Amphibian Amblystoma, a number of studies concerning ontogenetic development of reflexes in other Vertebrates, including Fishes (ARMSTRONG and HIGGINS, 1971; COGHILL, 1933; PATON, 1905, 1911; TRACY, 1926, 1959; WINTREBERT, 1920), Reptiles (SMITH and DANIEL, 1946), Birds (HAMBURGER *et al.*, 1966; KUO, 1932–1938; ORR and WINDLE, 1934; TUGE, 1937), Mammals and Man (ANGULO, 1927,

---

sceptical attitude and have remained unconvinced by his attempts to equate the vague term '*will*', which subsumes various aspects of conscious experience, with KANT's '*Ding an sich*' (cf. K., 1957, 1961a, 1961b, 1966). Nevertheless, as interpreted by SCHOPENHAUER, 'will' is doubtless related to a significant universal '*action principle*' and represents a useful fiction in VAIHINGER's sense. Thus, seen from this viewpoint, many of SCHOPENHAUER's elaborations on that topic appear not unjustified and are still of considerable interest.

It should, moreover, be added that, in accordance with such hypostatization or reification of the concept '*will*', all physical interactions whatsoever, involving mass or energy or both, and manifested by either non-living or living matter, thus also including all reflexes, become '*objectivations of will*'. Since, however, a restriction of the term 'will' to certain phenomena of consciousness seems, in my opinion, mandatory, said term should be entirely eliminated from the materialistic and behavioristic 'universe of discourse'. Evidently, 'will' is neither an objectively observable nor measurable phenomenon which could be registered by instruments, but can merely be experienced by (conscious) introspection.

1929, 1932; HOOKER, 1952, 1954; MINKOWSKI, 1928; WINDLE, 1940, 1950) have been undertaken by various authors. An early treatise on the physiology of the embryo was published by PREYER (1885). Many of the available data can be interpreted in accordance with differing theoretical viewpoints and thus remain, in this respect, inconclusive despite their value *qua* actual records of the temporal sequence of behavioral manifestations displayed during ontogenetic evolution in a small number of lower and higher Vertebrate forms. It should, moreover, be mentioned that the first embryonic movements in many or perhaps all Vertebrates are '*myogenic*', i.e. not caused by the yet lacking motor innervation but occurring either through 'spontaneous' activity of the developing musculature or upon its direct external (e.g. mechanical) stimulation. Concerning the development of motility during ontogenic evolution of the rat, ANGULO (1932) distinguished four stages: (1) the nonmotile, (2) the myogenic, (3) the neurogenic, and (4) the reflex stage. As regards embryonic behavioral development in a Teleost Fish, ARMSTRONG and HIGGINS (1971) distinguish (1) a spinal phase, (2) a hindbrain phase, and (3) an optic-midbrain phase.

Numerous subsequent studies on what is now called 'behavioral embryology' have been recently summarized in a publication edited by GOTTLIEB (1973). The confused and hazy state of this new 'specialty', overlapping with aspects of so-called 'ethology' will here become evident to the critical reader.

It seems obvious that the extent to which embryonic neural development and activities are dependent on (a) genetically determined, and (b) on environmental factors, is extremely difficult to ascertain. The precise formulation of these problems is encumbered by numerous semantic traps.

In a chapter of the cited publication, HAMBURGER (1973) stresses an 'incongruity between neurogenesis and overt motility in amniote embryos'. The early unorganized movements in the Chick embryo do not 'reflect the neurogenetic events which go on in the meantime, so to speak below the surface'. Therefore the notion that neurogenesis fully 'explains' or 'determines' embryonic behavior development is said to be not valid as a generalization. Even 'the most detailed knowledge of neural organization, including all significant synapses, in chick and rat embryos at a given stage, would permit no prediction' of the actual movements performed at that stage. 'Nor would a progression in synaptogenesis from one stage to the other be reflected in details of motility'. HAMBURGER (loc. cit.), however, concedes that in Amblystoma embryos the correspondence of progression in neurogenesis and behavior is indeed very close. With regard to this activity, investigated by COGHILL and HERRICK, and dealt with further above (cf. Figs. 19–21), HAMBURGER nevertheless adds that, in the 'early flexure stage', the head can move either to the left or to the right, though there is a high probability that it will move away from a unilateral stimulus. He concludes by remarking: 'All one can say is that the state of differentiation of the nervous system at a given stage delimits the range of behavioral potentialities'.

One could here reply that unpredictability, based on incomplete knowledge of the numerous multifactorial variables, should not be equated with 'indeterminacy'. Thus, spreading excitation may be inhibited by a variety of poorly understood inhibitory processes. Concerning the just mentioned 'probability' of the 'early flexure stage' response in Amblystoma, either direct myotomic stimulation, or random variations in the synaptic pattern of the established early pathways may play a role, since the description of the relevant connections by COGHILL and HERRICK (1915), although doubtless valid, may not be complete *qua* various additional details.

It is therefore quite justifiable to maintain that a strictly determinate neural mechanism evolves ontogenetically and thus 'gradually' within the neuraxis, resulting in a finite repertory of behavioral performances which, to a significant degree, are pre-functional if referred to their developmental (genetic) origin.

Following the preceding general discussion of reflexes, some additional remarks on a few basic functional concepts related to reflex activities are perhaps appropriate. Still more fundamental neural events involving stimulation, threshold, excitation, inhibition, adaptation, impulse conduction and synaptic transmission were dealt with in section 8, chapter V, of volume 3, part I. SHERRINGTON's concepts of common final path and convergence were discussed in the introductory chapter of volume 1 (p. 19). Another important concept pertaining to relationships between the central neuraxis and peripheral effectors (striated musculature) is the principle of modulation established by the investigations of P. WEISS, and involving 'homologous response' as well as 'myotopic function' recorded by means of experimental studies in Amphibians. It seems likely that a retrograde (cellulipetal) neuronal flow may play a role in this phenomenon as briefly discussed on pp. 641–642 in section 8, chapter V of volume 3/I and on pp. 120 f of the monograph 'Brain and Consciousness' (K., 1957).

The relevant performances of the nervous system can be subsumed under two more or less opposed categories, namely '*activation*' and '*inhibition*', or, since both phenomena imply transmission of signals concomitant with transformations of energy, *positive* and *negative action*. An important component of such actions is the either '*spontaneous*' or '*triggered*' '*discharge*' of neuronal elements.[34a] Discounting here the spontaneous discharges, 'essentially' or primarily caused by metabolic events within the nerve cell's body, triggered discharges imply the neuron's excitation by supraliminal extrinsic impulses overcoming the obtaining 'threshold'. Several impulses concomitantly affecting a number of different synapses on a given neuron, or separate successive impulses reaching a synapse can manifest an *additive effect* designated as *summation* or *facilitation*,[35] briefly mentioned on p. 576 and 619 of chapter V (volume 3, part I). *Spatial summation* subsumes the effect of different simultaneous impulses at separate terminals, and *temporal summation* the effect of successive impulses through one and the same channel. It is evident that spatial and temporal summation, being not mutually exclusive, can be combined in an undefined variety of patterns. With further analysis of the complex events involved in impulse conduction

---

[34a] With respect to the triggered discharge of neuronal elements, the impulse conduction through nerve fibers representing 'neurites' occurs in accordance with the *all-or-none law* (discussed on pp. 583–585, section 8, chapter V in volume 3, part I) and displays characteristic *spike potentials*. On the other hand, oscillating local excitatory or inhibitory events in regions of the neuronal soma may take place as '*graded responses*' *qua* amplitude and strength. Similar local fluctuations could also obtain in synaptic endings. Such events involving perikarya and synapses are presumably related to the recordable '*synaptic noise*', of which they may be a significant component not excluding additional causes. VERWORN (1922) designated the all-or-none type of behavior as *isobolic* (or *holobolic*), and that characterized by graded responses as *heterobolic*. He suggested that in contradistinction to the heterobolic 'nerve cell' (perikaryon), the nerve fiber (neurite) was holobolic (cf. K., 1927, p. 57). Through a *lapsus calami*, I absent-mindedly reversed that statement on p. 125 of 'Brain and Consciousness' (1957) and failed to detect this error upon proof reading. Said *lapsus* was then subsequently retained on p. 157 of the 1973 German edition. This latter contains a few additional minor mistakes of translation for which I must assume full responsibility since my load of work permitted me to advise the able translators (Prof. GERLACH and Dr. PROTZER) only on various points of primary importance in this difficult text.

[35] LUCAS (1917) stated that a first stimulus traveling down a nerve but failing to pass a neuromuscular junction facilitates the passage of the junction by a second impulse arriving a little later. SHERRINGTON (1906, 1947) used the term 'facilitating influence' *(bahnung)*' as essentially synonymous with (temporal) summation.

and transmission, it became advisable to distinguish 'facilitation' *(Bahnung)* from summation in the narrower sense. Thus, this latter has been defined as the additive effect of subliminal impulses and stimuli, and 'facilitation' as such effect by supraliminal ones (e.g. BLASIUS, 1962).[36] Again, according to BULLOCK *et al.* (1965) facilitation is characterized by the additional effect of a second stimulus over and above the summed affects of the first and second stimuli if they had been separate'.[37]

Other effects of additive type are designated as 'recruitment', 'reinforcement', 'irradiation', and 'rebound'. Certain reflexes gradually increase to a maximum when a stimulus of unaltered intensity is merely prolonged. This phenomenon, caused by the activation of a progressively greater number of elements is called *recruitment* in the wider sense, which subsumes a number of special cases, including some findings obtained by experiments involving evoked potentials and repetitive stimuli at particular frequencies. '*Reinforcement*' was defined by SHERRINGTON (1906, 1947) as the overflow of reflex action into channels belonging primarily to other reflex-arcs than those under stimulation, and leading 'to the production by the single stimulus of a wide, compound reflex which is tantamount in effect to a simultaneous combination of several allied reflexes'. Reinforcement has been regarded as a form of 'summation'.

---

[36] BLASIUS adds the following comment: '*Insofern ist der Begriff Summation irreführend, als nervöse Erregungen nicht einfach additiv in gleicher Wertigkeit zusammengesetzt werden können. Auch für die Summation gilt, dass das Ganze mehr ist als die Summe der Teile. Je komplexer die Neurenketten, desto länger können die zeitlichen Intervalle für die Summations- und Bahnungserscheinungen werden, z.B. sind im Grosshirn mit seinen ausgeprägten Gedächtnisleistungen Bahnungen noch nach ausserordentlich langer Zeit wirksam*'. Although an approximate algebraic summation of effects may indeed obtain in some sorts of neural processes, the fact that such summation does not hold for numerous significant types of central reflex interactions was already stressed by SHERRINGTON (1906, 1947).

[37] If, instead of the additive effect, a diminished one upon successive stimuli occurs, '*defacilitation*' or '*antifacilitation*' is said to occur. Experimental investigations of the still very incompletely understood phenomena pertaining to neural signal transmission have resulted in a highly sophisticated terminology based on a combination of inconclusive data with in part unconvincing interpretations, and including questionable concepts such as '*facilitation of facilitation*'. A less sophisticated, but still quite useful discussion of reflex actions, with definitions of relevant concepts, and ignoring the term 'facilitation' is included in BEST's and TAYLOR's text of applied physiology (1950).

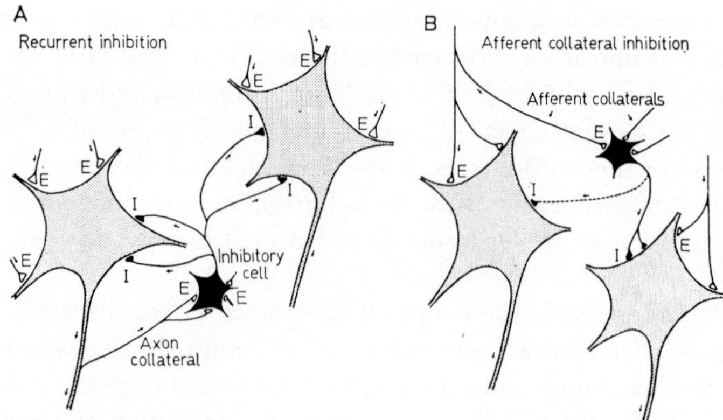

Figure 22. Diagram purporting to illustrate two types of inhibiting pathways (from Eccles, 1969). Inhibitory cells respectively inhibiting synapses are shown in black.

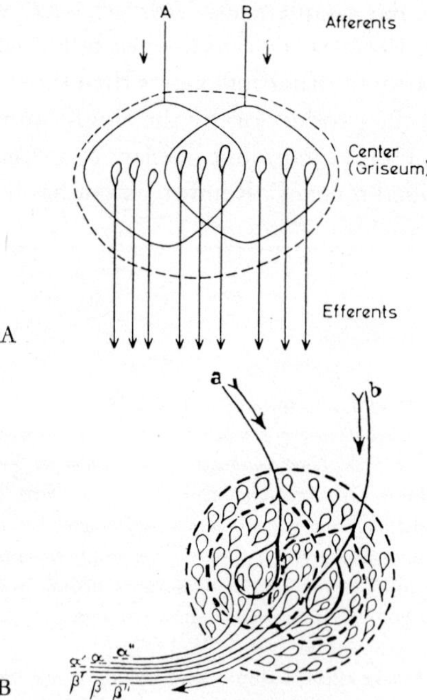

Figure 23. Diagrams purporting to illustrate subliminal fringe and occlusion (A from K., 1961, B from Best and Taylor, 1950). In B, relatively weak supraliminal stimulation of afferent a or b restricts the respective fields as shown by the continuous line limit, and each afferent activates only 1 motoneuron (α and β). Concurrently, however, owing to summation in the overlap of the subliminal fields indicated by the broken outlines, 4 units may be activated (α', α, β', β).

If the strength of a stimulus is gradually increased, the central excitatory process, spreading to a wider population of motoneurons, causes the participation of additional muscle groups in the reflex response. This spread has been called '*irradiation*', also known as *Pflüger's* '*laws of spread*'.[38] Irradiation commonly extends *per saltum* rather than *gradatim*. When, during the elicitation of a reflex, this latter is suppressed by stimulation of an inhibitory pathway, a conspicuous augmentation of the reflex-response may commonly occur when the inhibitory stimulus is withdrawn. This is known as the '*rebound*' *phenomenon*.

Again, the discharge of impulses from a stimulated or activated griseum, respectively a given reflex response, may continue after the initiating stimulus has ceased.[39] Such activity is known as *after-discharge*. It corresponds to a persistence of the 'central excitatory state' and was also figuratively likened by SHERRINGTON to a reflex-arc 'momentum', in contradistinction to the short delay in reflex response upon stimulation, comparable to 'inertia'. There is little doubt that after-discharge may be the result of a variety of different factors, among which asynchronous successive arrival of impulses transmitted through 'parallel delay lines, and various self-reexciting or reverberating circuits, as depicted above in Figure 11 could be mentioned. In some instances, reexcitation of a 'center' by proprioceptive impulses, as e.g. in the larval swimming reflexes of Amblystoma, maintains a central excitatory state.

As regards '*central inhibitory states*', a number of diverse mechanisms can be inferred in accordance with possible neuronal connections for which *Dale's principle* may or may not be postulated to hold. Through the activity of reverberating circuits, inhibitory neurons could be kept firing in a variety of ways. Figure 22 shows two types of inhibitory mechanisms conjectured by ECCLES (1969) and presumed to explain, in agreement with *Dale's principle*, the phenomena of recurrent respectively afferent collateral inhibition.

The fact that, in central reflex interaction and at the common final path, an *algebraic summation* of afferent arc effects does not commonly obtain, is illustrated by the principles of *convergence*, *occlusion* and *subliminal fringe* elucidated by SHERRINGTON and his school (Fig. 23). Thus,

---

[38] PFLÜGER, about 1853, formulated four specific '*laws of spread*' which, however, could not be upheld as regards their particular details (cf. SHERRINGTON, 1906, 1947).

[39] SHERRINGTON quotes the saying '*sublata causa non tollitur effectus*'.

two afferents A and B converge upon a griseum in which some of their respective terminals overlap. Upon a supraliminal impulse through channel A, 6 motoneurons (motor units) will be activated. A comparable impulse through channel B will likewise activate 6 units. Such simultaneous impulses through A and B, however, will activate 9 but not 12 (6+6) units, thereby displaying occlusion, since synaptic fields A and B are not disjoined, having a non-void intersection. This implies a spatial overlap of central excitation in a reflex activity, which can be inferred where the reflex effects going on together amount to less than the algebraic sum of their separate effects. Conversely, subliminal impulses through either A or B will not be effective, but, if simultaneous, or close succession, may activate 3 motoneurons of the subliminal fringe by spatial or temporal summation or both. Again, in Figure 23B, a relatively weak supraliminal stimulus through either channel a or b but not both may activate only one motoneuron within the continuous line limit, being inefficient within the subliminal fields indicated by the broken lines. Conjointly, however, they can activate 4 (instead of 1+1=2) motoneurons owing to the summation of their effects in the overlap of their subliminal fields.

With respect to the various patterns of connections provided by the neuronal network, such as, e.g. depicted above in the few particular examples of Figure 11, a brief mention should be made concerning the type of transmission which Cajal (1909) has designated as '*avalanche conduction*' (*avalanche de conduction*, *Lawinenleitung*) and described as follows: '*Chaque neurone est relié, par les innombrables divisions de ses appareils protoplasmiques et axiles, à une quantité souvent considérable d'autres neurones. Aussi, l'impression reçue à la périphérie par une expansion dendritique se propage-t-elle, en éventail, en cône, embrassant à chacun de ses passages une multitude de plus en plus croissante de neurones. Elle avance donc, c'est notre comparaison, comme l'avalanche, qui, à mesure de sa chute, entraîne une masse de plus en plus accrue de matériaux. Par suite, aucune chaîne réflexe de neurones n'est isolée; elle a, dans les centres avec ses voisines et même avec des chaînes éloignées, des neurones communs en plus ou moins grand nombre.*'

It is evident that this principle represents merely a particular formulation of the more generalized *principle of one-many transformations*, which includes, e.g. dispersion, divergence, and irradiation. Its opposite is the likewise obtaining *principle of many-one transformations*, which include, e.g. various instances of convergence and common final path. Cajal (1909) justly added that the significance of his avalanche conduction concept should not be exaggerated, and added the comment:

'*nous verrons qu'il ne faut pas trop en exagérer l'étendue, car jamais la diffusion de l'unité d'impression n'est telle, qu'elle rende impossible la localisation des images sensitives et sensorielles en des foyers déterminés de l'écorce cérébrale.*'

Avalanche conduction is obviously kept in bounds, respectively becomes suppressed, by the multiple inhibiting synaptic activities which, although already well known at the time of CAJAL's cited writings, were not properly recognized, as regards their significance, by this author. If the effect of inhibiting synaptic activities is suppressed, as, e.g., according to some contemporary views, in strychnine or picrotoxin poisoning or in tetanus infection (cf. vol. 3, part I, pp. 635–636), the resulting widespread convulsions can be conceived as manifestations of *Cajal's 'avalanche de conduction'*. An additional *constraint*, counteracting said conduction, is provided by the obtaining '*thresholds*', or, crudely speaking, 'synaptic resistances'.

As regards the particular types of neural networks implying avalanche conduction, CAJAL (1909, 1911) especially enumerates, on the basis of his fundamental investigations, the following structures: '*bulbe olfactif*', '*écorce cérébrale*', '*voies acoustiques*', '*voies olfactives*', and '*voies optiques*'. Strangely enough, CAJAL does not specifically refer to the cerebellar cortex, which, as manifestly shown by his own remarkable studies, displays one of the most clear-cut instances of avalanche conduction. This latter principle, moreover, is closely combined with that of *feedback*, whose significance CAJAL failed to realize, although many of his own actual findings clearly indicated the substantial importance of that principle which he apparently merely subsumed under his avalanche conduction concept (cf. also the comments on pp. 8–14 in volume 1 of this series).

Since about 1872, PAUL FLECHSIG (1876, 1896, 1920, 1927), amplifying studies on secondary degeneration by TÜRCK and BOUCHARD, as well as incidental observations by others, developed the concept of *myelogeny* and initiated a long series of epochal investigations based on this particular approach which disclosed an orderly sequence in the ontogenetic development of medullary sheaths *(zeitlich streng geordnete Reihenfolge der Markscheidenbildung)*. Three fundamental principles of myelogenesis *(myelogenetische Grundgesetze)* have been formulated and repeatedly emphasized by this author (1927):

'1. *Gleichwertige, d. h. in gleicher Weise eingeschaltete Nervenfasern erhalten ihr Mark annähernd gleichzeitig, verschiedenartige Systeme in gesetzmässiger Reihenfolge, unter Einhaltung bestimmter Altersstufen.*

2. *Die Bildung der Markscheiden wiederholt zeitlich ganz allgemein die erste Anlage der Achsenfasern durch die Neuroblasten* (FLECHSIG, HIS).

3. *Die Myelogenese wiederholt auch die phylogenetische Entwicklungsreihe des gesamten Nervensystems, entsprechend dem biogenetischen Grundgesetz* E. HAECKELS.'

Despite attempts by several authors, especially O. and C. VOGT, to disprove FLECHSIG's conclusions, and to minimize the significance of myelogenesis, the basic validity of FLECHSIG's rules 1 and 2 can, in my opinion, be fully upheld. Rule 3 may be evaluated as a first approximation reasonably serviceable for phylogenetic speculations but evidently, because of manifold cenogenetic parameters, subject to numerous exceptions.[40]

Concerning the *human spinal cord*, and also that of at least some other Vertebrate forms, the ventral roots, which are the first nerve fibers to develop (cf. above, footnote 34) become subsequently also the first with respect to myelinization, namely in the course of the fourth month of human intrauterine development. The myelinization of the dorsal roots begins considerably later. With regard to the individual neuron, myelogeny seems to begin in the vicinity of the perikaryon, and, in the case of long neurites, to progress gradually toward the periphery. Thus, in the human telencephalon, the cortico-spinal tract is already conspicuously medullated toward the end of intrauterine development, being still essentially non-medullated in the spinal cord at, and for some time after, birth.

Again, on the basis of its myelinization, investigated about hundred years ago by FLECHSIG, the *dorsal spino-cerebellar tract*, originating in the *nucleus dorsalis of Clarke*, became the first tract of the human neuraxis which could be accurately traced from origin to end.

---

[40] ERNST HAECKEL's outstanding merits as one of the two main founders of present-day phylogenetic theory (the other being, of course, DARWIN) do not consist in having outlined a definite and detailed outline of phylogenetic evolution, but in having conclusively demonstrated, *on morphologic grounds*, the valid overall features of evolutionary biology and the general course of phylogenesis. His 'biogenetic law' doubtless plays here a relevant although not yet sufficiently well understood role and expresses at least some important aspects of transformism. Despite one's well-founded conviction that 'something of this sort must have taken place', one can indeed regard with considerable scepticism not only the actual, specific details of HAECKEL's early 'phylogenetic trees' *(Stammbäume)* but also the detailed conclusions illustrated by the much more sophisticated present-day '*adaptive radiations*'. The multiplicity of the factors pertaining to phylogenesis has, so far, prevented a satisfactory analysis.

This bundle, also known as *Flechsig's tract*, begins to acquire myelin sheaths at about 7 months of intrauterine development.

FLECHSIG came to the conclusion that complete functional capability of a tract normally medullated in the adult is reached only when its fibers have become myelinated. With the necessary emphasis on the term '*complete*', this view can still be considered essentially correct.[41] In a general way, correlated investigations[42] on myelination and behavioral development in pouch-young opossums, cat embryos and kittens, as well as in human fetuses and infants have shown an evident relationship between maturation of behavior and the myelination of various fiber tracts. On the other hand, it became not possible to draw specific correlations in all cases. Quite apart from the fact that, in the adult Vertebrate neuraxis, a number of significant and functionally efficient fiber systems are permanently non-medullated, there can be a great deal of well organized activity in the central nervous system before any nerve fibers become medullated.[43] According to WINDLE (1940), 'about all that one can say is that the first tracts to develop in the embryo are the first to begin to be myelinated and the last to form in the late fetus are the last to receive sheaths. Some tracts never develop significant numbers of myelin sheaths. It is quite probable that conduction of impulses may be improved with the acquisition of myelin, but myelination is certainly not an essential corollary of function. With increasing fetal size, distances between points in the nervous system become greater. Perhaps myelin is laid down to compensate by increasing the conduction speed of the fibers'.

Be that as it may, and discounting a teleologic interpretation of WINDLE's foregoing remark on conduction velocity, one could main-

---

[41] As e.g. evidenced by a normally occurring *Babinski sign* in the human newborn and early infant. The loss of function in various demyelinating disorders in which, despite loss of myelin, the axons as well as the perikarya are initially preserved, could likewise be interpreted in accordance with this view. Nevertheless, because of poorly understood additional factors obtaining in this group of diseases, functional impairment may here be related to a variety of different parameters.

[42] Fairly recent general discussions on this topic can be found in the publications by ANGULO (1929), LANGWORTHY (1933), and WINDLE (1940). An anatomic study of the myelinization of the central nervous system in the snake Natrix sipedon was undertaken by WARNER (1952).

[43] Thus, ANGULO (1929), in summarizing his conclusions, remarks that 'myelination is not a criterion of functional insulation of conducting paths' nor 'an absolute index of behavioral capability'.

tain (1) that myelogenesis is correlated with the functional capability of those fiber systems which are normally medullated in the adult, and (2) that myelogenesis, as investigated in accordance with FLECHSIG's concepts, provides important data not only concerning the anatomy of fiber tracts but also of significance with respect to an analysis of numerous functional systems.

3. Amphioxus

The neuraxis of *Amphioxus (Branchiostoma)* consists of a spinal cord which displays anteriorly an epichordal rostral or apical vesicle (Figs. 24, 25). This latter, on rather convincing morphologic grounds, can be regarded as an *archencephalon* homologous to the *embryonic prosencephalon* of craniote Vertebrates. Since, nevertheless, Amphioxus seems to represent an essentially *'spinal' animal*, lacking brain subdivisions comparable in detail to Craniote medulla oblongata, cerebellum, mesencephalon, diencephalon and telencephalon, the entire neuraxis of Branchiostoma shall briefly be considered in the present chapter dealing with the spinal cord. Additional general comments on the nervous system of Amphioxus were included in section 1, chapter VI, and sections 1 and 2, chapter VII of the preceding volume.

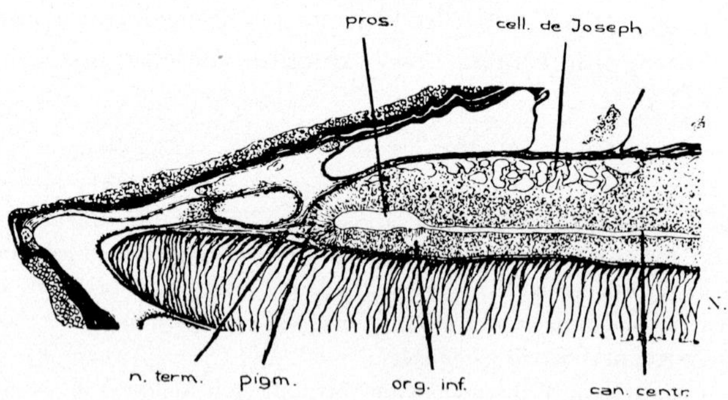

Figure 24. Approximately midsagittal section through the rostral end of the neuraxis in adult Amphioxus, showing archencephalic vesicle, and location of *cells of Joseph* in postapical region of spinal cord (from KAPPERS, 1947). can. centr.: central canal of spinal cord; n. term.: nervus terminalis seu apicis; org. inf.: infundibular organ; pigm.: pigmented spot in neuroporic neighborhood; pros.: archencephalic vesicle; N: notochord.

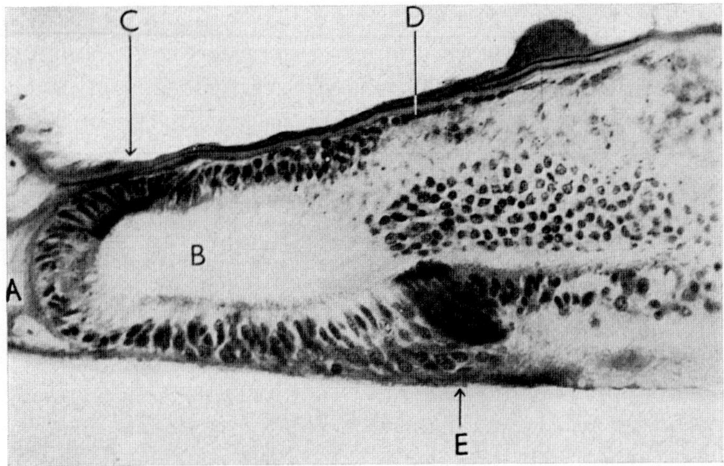

Figure 25A. Midsagittal section through rostral end of neuraxis in adult Amphioxus (from OLSSON and WINGSTRAND, 1954). A: decolorized pigment cells; B: lumen of archencephalic vesicle; C: granulated ependyma in the recessus neuroporicus; D: granulated ependyma in the roof of the so-called 'dorsal central canal'; E: infundibular organ. Bouin, 6 μ paraffin section, *Gomori-stain,* orange filter, original magnification ×600, roughly reduced to ×450 in the present reproduction.

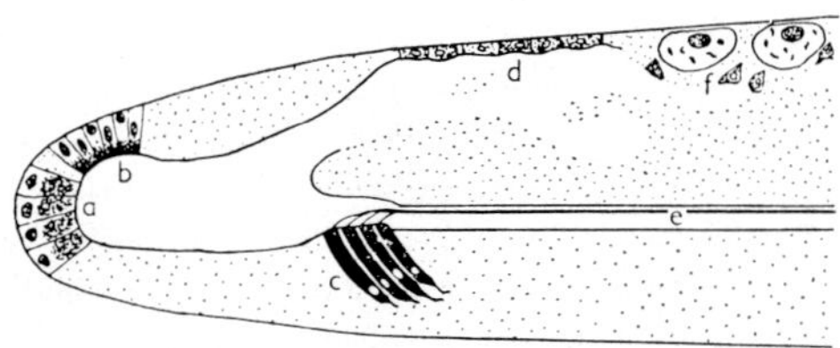

Figure 25B. Diagram of the rostral end of the neuraxis in Amphioxus, showing attachment of *Reissner's fiber* to infundibular organ (from OLSSON and WINGSTRAND, 1954). a: pigment spot; b: secretory ependyma in recessus neuroporicus; c: infundibular organ; d: granular ependyma in wall of 'dorsal central canal; e: *Reissner's fiber* in the central canal; f: granulated ependymal cells and large *cells of Joseph.*

Nothing certain is known concerning functional significance or neuronal structures respectively connections of the *cerebral vesicle,* whose cavity is lined by ciliated cells of ependymal type. In the region of the neuroporic recess, a pigment spot and a somewhat more ventral granulated ependyma[44] have been described. Basally to the neuroporic region, the paired *nervus apicis* or *nervus terminalis*[45] is connected with the 'brain'. It seems to be an afferent nerve carrying impulses from the sense organs of the oral hood and its tentacles, as well as from the unpaired '*pit of Kölliker*' or '*olfactory pit*' usually displaced toward the left side. This pit (Fig. 26), apparently a remnant of the external neuropore, might indeed be a chemoreceptor organ, but no conclusive evidence as to its function is available. Again basally, and more caudally, *the infundibular organ* is formed by distinctive tall columnar cells whose ventricular surfaces carry flagella (one or two per cell), and whose abventricular extremities each give origin to a nerve fiber (neurite). This 'infundibular' cell aggregate would, accordingly, include neuronal elements of epithelial type, and therefore seems to represent a ventricular sense organ. Since it is located in a topologic neighborhood corresponding to the mammillo-infundibular or posterior hypothalamic region of Vertebrates, the infundibular organ, regardless of its dubious function, might be evaluated as kathomologous to the Craniote Vertebrate neurohypophysis or to the saccus vasculosus of Fishes, or to a to-

---

[44] OLSSON and WINGSTRAND (1955) interpret such granulation of neuroectodermal cells, demonstrable by means of the *Gomori method*, as an indication of secretory activities. These authors also attempt to derive the hypothalamic-hypophysial neurosecretory system of Vertebrates from the infundibular organ of Amphioxus. However, nondescript granular material suggestive of 'secretory' activity can be found elsewhere in the central nervous system of Amphioxus, and similar evidence of 'secretory' activity by neuroectodermal structures is displayed in diverse regions of the Vertebrate neuraxis. Moreover, the apparently 'sensory' infundibular organ of Amphioxus and the Vertebrate hypothalamo-hypophysial complex have little in common *qua* structural details. Nevertheless, both saccus vasculosus and neurohypophysial complex are two separate derivatives of a larger common topologic neighborhood represented by the Vertebrate mammillo-infundibular region and corresponding to the neighborhood which includes the 'infundibular organ' of Branchiostoma. There is also, within the ectodermal oral cavity of Amphioxus, a ciliated '*wheel organ*' of which a dorsal extension, known as '*Hatschek's pit*', has been considered homologous to the adenohypophysial anlage, i.e. to *Rathkes pouch* of Craniota. *Hatschek's pit*, however, is located basally to the notochord, being entirely separated by this latter from the neuraxis. Amphioxus does thus not display any configuration structurally comparable to the hypophysial complex of Vertebrates, although, in a very general way, the topologic neighborhoods corresponding to that complex are, of course, included in the overall bauplan obtaining for Branchiostoma.

[45] The nervus terminalis of Amphioxus has been considered, on reasonably defensible morphologic grounds, which are likewise topologically fairly sound, as kathomologous with the nervous terminalis of Vertebrates (cf. KAPPERS, 1947).

pologic neighborhood providing a common configurational origin for both. A preoral pit in the roof of the oral cavity *(Hatschek's pit)*, located to the right of the notochord, could represent a kathomologon of the Craniote Vertebrate adenohypophysis.

Reverting to the *infundibular organ* it should be added that, according to BOEKE (1908, 1913) and FRANZ (1923) some of the fibers originating from this structure decussate just ventrocaudally of their origin, but further course and destination of the 'infundibular' fiber system is unknown. In addition, OLSSON and WINGSTRAND (1955) have shown that *Reissner's fiber* originates from the infundibular organ, which, in this respect, seems analogous but certainly not homologous to the 'subcommissural organ' of Vertebrates.

The spinal cord of Amphioxus can be said to have its rostral boundary at a neighborhood-level just caudal to the infundibular organ. The here beginning anterior, postapical portion of the spinal cord is related to the first two pairs of dorsal roots, but does not display ventral roots and differs also in another conspicuous aspect from the more caudal or typical main extent of the spinal cord, namely by the presence of *large dorsal cells (dorsal cells of Rohde*, 1888, *cells of Joseph*, 1904). These large cells have been identified as photoreceptor cells by WELSCH (1968) who investigated their structure with the electron mi-

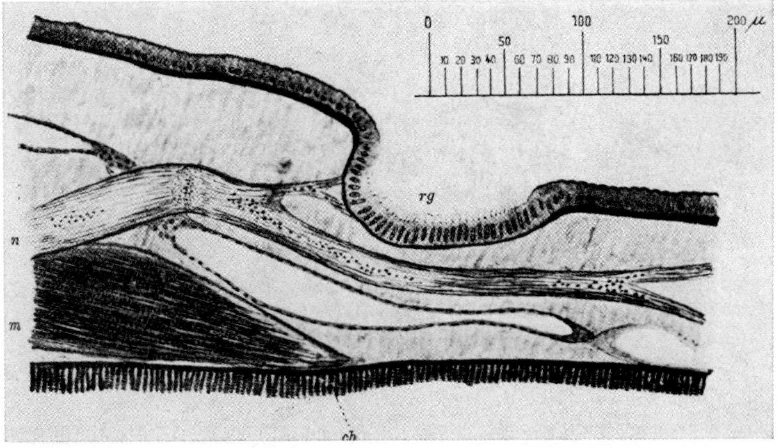

Figure 26. Sagittal section through 'nasal pit' of Amphioxus, showing its innervation by a branch of the n. apicis (from KRAUSE, 1923). ch: notochord; m: myomere; n: stem of n. apicis; rg: 'nasal pit' *(Riechgrube')*. Rostral end at right.

croscope (cf. Fig. 27), but their synaptic connections still remain unknown. KAPPERS (1947) cites findings by PARKER indicating that these cells were not demonstrably photosensitive, but these data do not exclude their relationship to a highly specialized (e.g. humoral or 'circadian') photosensory system.

Being a region of transition between medulla spinalis and archencephalon, the modified rostral or post-apical portion of the spinal cord can be evaluated as homologous to the Craniote *deuterencephalon*. Since it is not altogether unjustified to regard the first dorsal root as a ka-

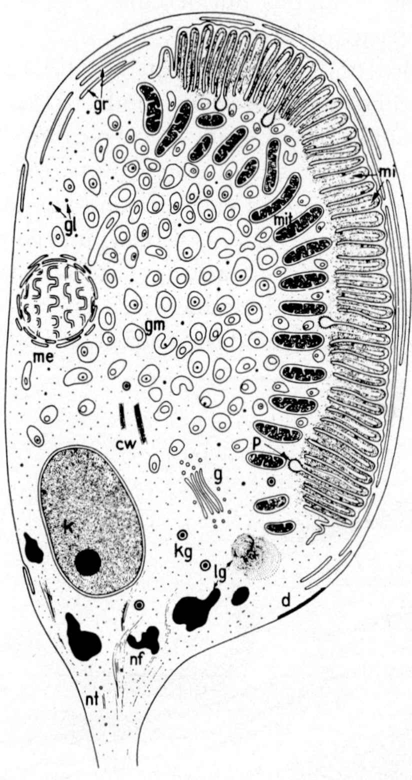

Figure 27. Ultrastructural diagram of a *cell of Joseph*, based on electron photomicrographs (from WELSCH, 1968). d: desmosome; cw: roots of cilia; g: *Golgi-apparatus;* gl: glycogen granules; gm: smooth, tubular membrane system with internal structures; gr: smooth reticulum at cellular periphery; k: nucleus; kg: granula of katecholamine type; lg: granula of lipofuscin appearance; me: 'meandering tubular system'; mi: microvilli of the rhabdome; mit: mitochondria; nf: neurofilaments; nt: neurotubules; p: pinocytic vesicle (marginal vesicle, *Saumbläschen)* at base of microvilli.

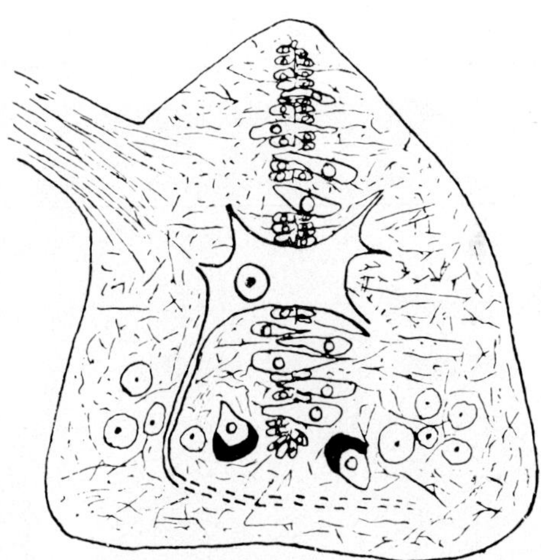

Figure 28. Diagrammatic cross-section through spinal cord of Amphioxus (partly after FRANZ, 1923, from K., 1927). A dorsal root, a *Rohde cell* with decussating neurite, some giant fibers, and two spinal eyes are shown. The unpaired median giant fiber (cf. Fig. 36) was inadvertently omitted.

thomologon of Craniote ophthalmicus profundus, and the second one as the kathomologon of maxillo-mandibularis, the post-apical portion might be said to include a rostral 'mesencephalic' and a caudal 'rhombencephalic' neighborhood, but such interpretation is, at best, valid in a highly generalized way, which disregards not only the complete absence of a cerebellar configuration but also a variety of other substantial differences.[46]

The *spinal cord proper* of Amphioxus is roughly isosceles triangular in outline (Fig. 28), displaying a rounded dorsal apex and a slightly concave base closely adjacent to the notochord sheath. Because of shrinkage effects, the contour of a cross-section may also assume an approximately trapezoid shape (Fig. 29). Peculiar bilateral cellular connections, piercing the notochordal sheath at more or less regular intervals, and apparently anchoring the spinal cord to the notochord, were described by BONE (1960) and others.

The *central canal* of the medulla spinalis is a very narrow dorsoventral slit, lined by ependymal cells and bridged at numerous places by

[46] A concise discussion of this question can be found in KAPPERS' posthumous treatise 'Anatomie du Système Nerveux' (1947, pp. 53–55).

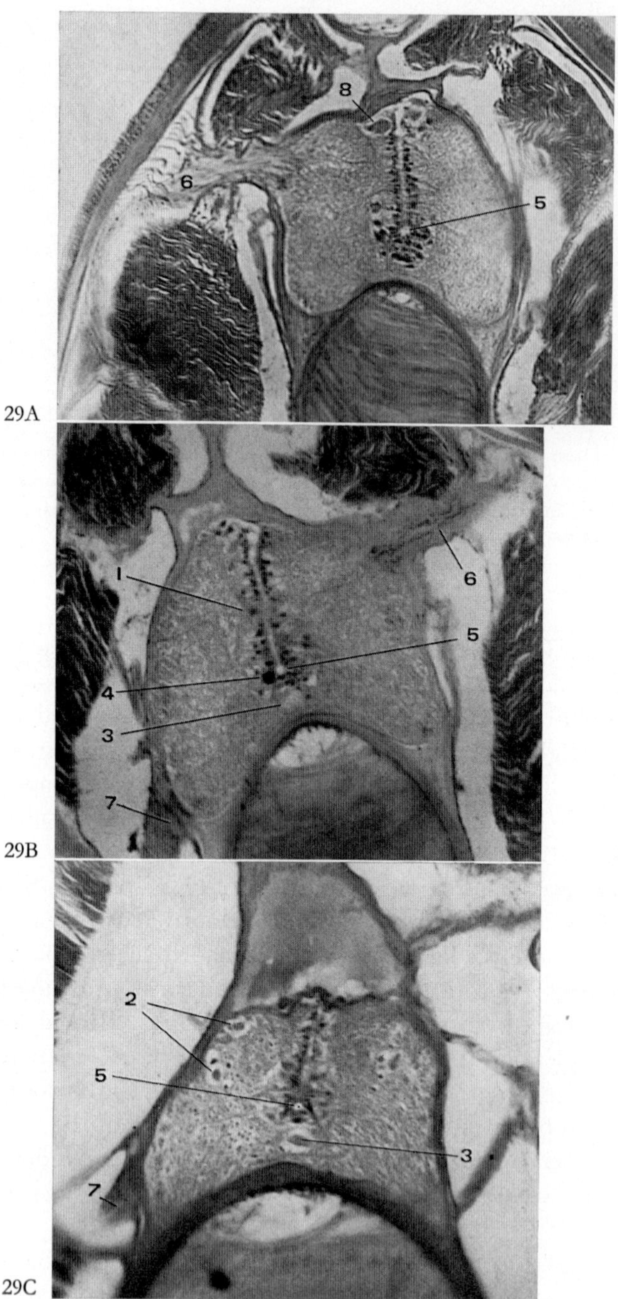
8
6
5
29A
1
6
5
4
3
7
29B
2
5
3
7
29C

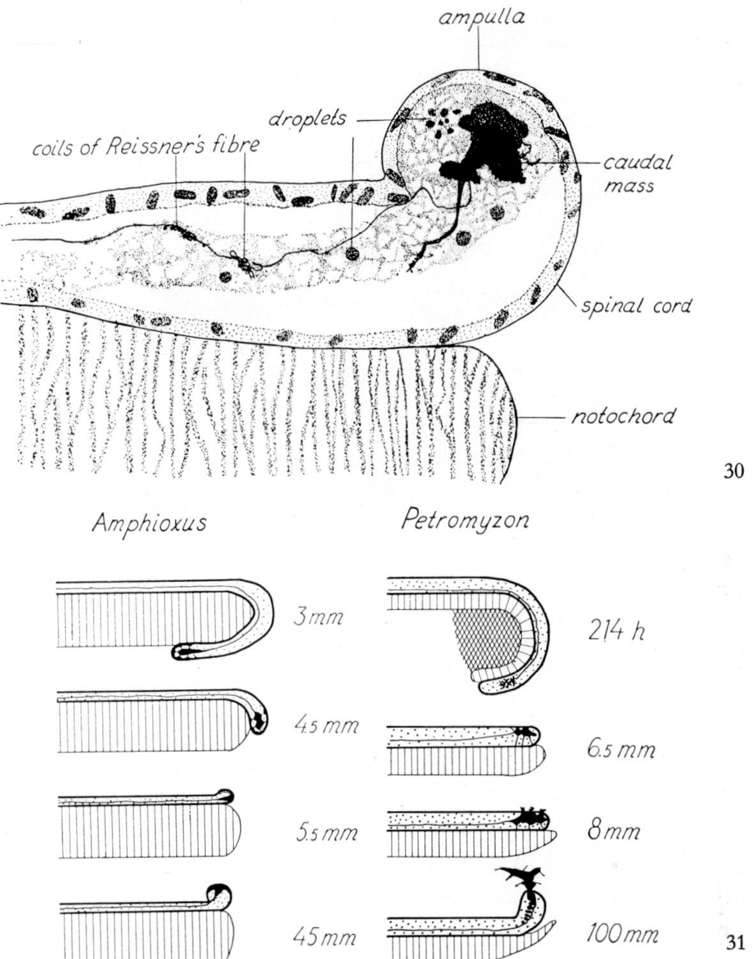

Figure 30. Semi-diagrammatic median reconstruction of the spinal cord's caudal end in adult Amphioxus, showing terminal mass of *Reissner's fiber* (from Olsson, 1955). Magnification in this reproduction approximately ×700.

Figure 31. Diagrammatic reconstructions of stages in the ontogenetic changes involving the caudal end of spinal cord in Amphioxus and Petromyzon (from Olsson, 1955). Vertical lines: notochord; crossed lines (in Petromyzon, right top): yolk; dotted structure: spinal cord; black line in spinal cord: *Reissner's fiber* ending with caudal mass. The diagrams are not drawn to scale.

Figure 29 A–C. Cross-sections through the spinal cord of a young adult Amphioxus as seen in routine hematoxylin-eosin preparations (×400, red. $^{4}/_{5}$). 1: portion of Rohde cell; 2: giant *(Rohde)* fibers; 3: median giant fiber; 4: spinal eye; 5: *Reissner's fiber;* 6: dorsal root; 7: ventral root; 8: caudalmost *cells of Joseph*.

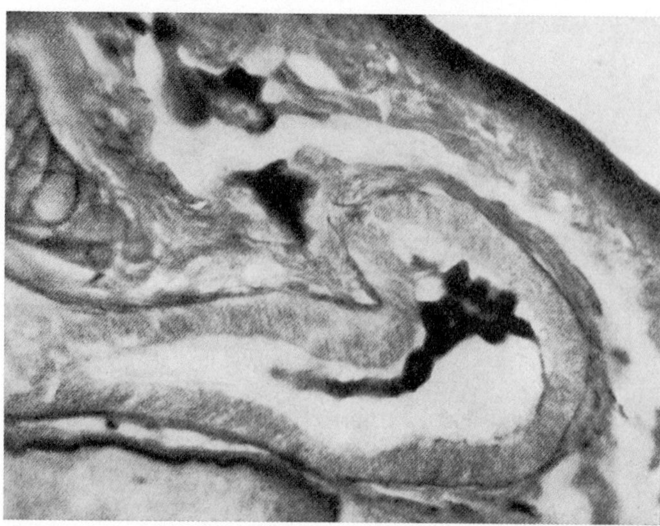

Figure 32. Sagittal section through caudal end of spinal cord in Petromyzon, showing terminal mass of *Reissner's fiber*, for comparison with Figures 30 and 31. A portion of terminal mass substance can be seen in connective tissue dorsal to cord (from OLSSON, 1955; *Gomori's hematoxylin*, 6 μ section, magnification in reproduction approximately ×530).

the bodies of neuronal elements which have assumed a median location. The basal portion of the central canal commonly displays a somewhat more distinct rounded lumen which contains the rather thin *fiber of Reissner* (Figs. 25B, 29).[47] The caudal end-portion of the spinal cord becomes a sort of filum terminale with only epithelial i.e. ependymal wall. At the posterior end, this hollow filum runs out as a somewhat variable ampulla in which *Reissner's fiber* terminates with a 'caudal mass' (Figs. 30, 31). OLSSON (1955), who investigated the caudal end of *Reissner's fiber* in various Vertebrates, did not find a posterior neuropore in the larval and adult Amphioxus material which he examined. In *Petromyzon*, however, the terminal 'sinus' of the spinal cord opens through a neural pore, through which the 'caudal mass' of *Reissner's fiber* protrudes into the surrounding tissues (Figs. 31, 32).

[47] In Amphioxus, *Reissner's fiber* appears rather inconspicuous in routine preparation, but was reported and depicted by KRAUSE (1923) as well as particularly investigated by OLSSON (1955) and OLSSON and WINGSTRAND (1955). These latter authors point out that, while WOLFF (1907) had already recorded that structure in Amphioxus, AGDUHR (1922) denied its presence in the neuraxis of this animal.

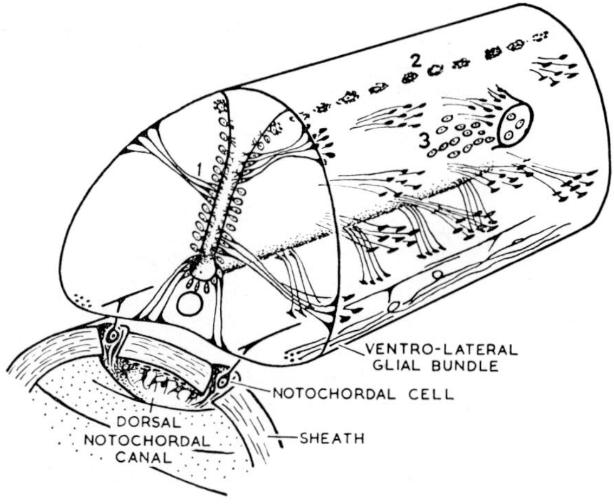

Figure 33. Arrangement of ependymal cells, 'neuroglia', and 'notochord' cells in the spinal cord of Amphioxus (from Bone, 1960). 1: ependyma; 2: cells interpreted by Bone as '*Schneider's type* of glial cells'; 3: cells interpreted as '*Schwann cell* analogues'. The dorsal notochordal canal is also clearly recognizable in Figures 29A–C.

As regards the *structural components* of the spinal cord of Amphioxus, the *supporting elements* seem to be mainly provided by ependymal cells, whose radiating peripheral processes form more or less regularly arranged distinctive fascicles and terminate with end feet at the surface of the neuraxis. They may be related to a sort of external limiting membrane. Some ependymal fibers appear also to form longitudinal bundles. In addition, several dubious types of neuroglia cells have been described (Fig. 33). There are, however, various difficulties in the identification of diverse cellular elements of Amphioxus as either glial or neuronal. The neuraxis does not seem to be penetrated by bloodvessels, but, at its base, abutting on the notochord, processes of so-called '*notochordal cells*' can be seen to enter the spinal cord as components of the 'anchoring' structure mentioned above. Bone (1960) considers the possibility that these cells might be neuronal elements pertaining to a bizarre notochordal proprioceptive system.

Besides the *dorsal cells of Rohde* or *Joseph*, presumed to be photoreceptors, and dealt with above in the discussion of the spinal cord's post-apical (deuterencephalic) region, rather typical photoreceptor cells are

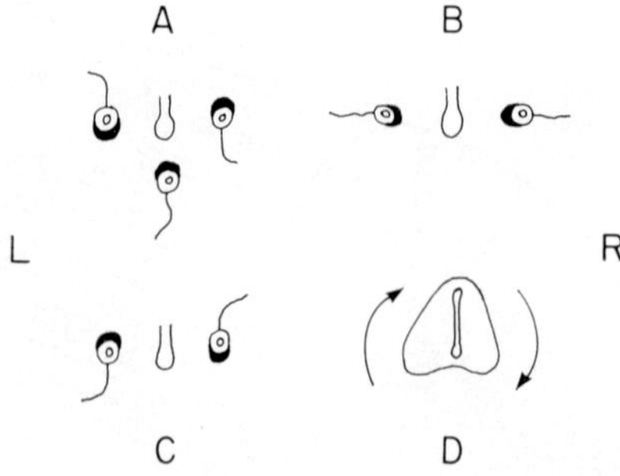

Figure 34. Diagrams illustrating the orientation of the spinal eyes in Amphioxus and the direction of the animal's spiralling when swimming (adapted after FRANZ, 1923, and YOUNG, 1955). A: anterior group; B: middle group; C: posterior group; D: direction of spiralling motion. The nuclei of the pigment cells are not indicated. L: left; R: right.

present in the ventral part of the spinal cord proper, being scattered throughout most of its length (Figs. 28, 29, 34). These cells are of neuronal epithelial type. Their neurite arises at one pole, while the photosensory striated opposite pole is covered by a pigmented cup provided by a pigment cell. Eeach 'spinal cord eye' (*'Rückenmarksauge'*, *'Becherauge'*) consists thus of a neuroepithelial element screened by a pigment cell.[48]

At the level of third or fourth myotome, one or two 'eyes' may be found on each side and gradually increase in number caudalward as far as midlength of the body, where they may become scarce or lacking. More caudalward, their number increases again; the cord's tail region, however, is generally devoid of eyes. Slight variations of this overall grouping have been reported.

By means of their pigment cups, the 'spinal eyes' are oriental toward light coming from a particular direction. Rostrally, the right and ventromedian eyes 'look' ventralward, the left ones dorsalward; at

[48] The spinal cord eyes of Amphioxus thus represent a much more simplified, *bicellular* version of the otherwise in some respect comparable *multicellular* cerebral eyes of Turbellaria, discussed in chapter IV, volume 2 of this series (pp. 32–33, and Figs. 18, 22 and 25, l.c.). It can also be noted that the 'spinal eye' photoreceptors are *inverted*, i.e. directed away from the light, as in the retina of Vertebrates. The photoreceptors of Invertebrates, on the other hand, are frequently oriented toward the light (cf. vol. 2, p. 268 f. and Figs. 162, 185, 186, 190). The somewhat irregular orientation of the *cells of Joseph* in the postapical spinal cord seems, on the whole, to be directed toward the periphery, i.e. toward incident light.

mid-body levels, the right eyes look toward the right and the left ones toward the left. More caudally, the orientation of the eyes is opposite to that obtaining at rostral levels (cf. Fig. 34). Nothing is known concerning the neuronal connections effected by the neurites of the spinal photoreceptor cells. A thin beam of light directed to the head ('brain') region is said to be ineffective, but elicits movements when shining upon the body with its 'spinal eyes'. It is not improbable that the clockwise (as seen from behind) spiraling rotation of the body during the swimming movements of Amphioxus, briefly considered further below, is related to the peculiar orientation of the spinal eyes.

Because of the difficulties in obtaining suitable preparations disclosing the connections and other relevant features of the nerve cells in this Chordate form, the neuronal organization of its spinal cord remains poorly elucidated, despite numerous detailed and careful studies by different and very competent investigators, whose interpretations frequently show considerable disagreements.

As regards the *dorsal roots*,[49] BONE (1960), comparing his own findings with those recorded in the literature, describes several types[50] of dorsally located presumably afferent elements, whose peripheral processes leave the spinal cord as dorsal root fibers (Fig. 35). In addition, some afferent dorsal root fibers presumably arise peripherally from primary (neuronal) sensory cells. Efferent dorsal root fibers are believed to be neurites of 'viscero-motor' cells. Two types (large and small) of these cells are described as located in the most basal part of the spinal cord, including the midline ventral to the central canal (Fig. 36). Some of these cells may directly innervate the 'visceral' (i.e. non-somatic) musculature, while others could perhaps represent preganglionic elements, since, in the region of the so-called atrium, plexuses with elements interpreted as multipolar cells of postganglionic type have been noted by several authors. Others, however, believe that all visceral efferent ('autonomic') fibers in Amphioxus terminate in the effectors without the intercalation of an additional postganglionic neuron, which is characteristic for the 'typical' autonomic outflow of the Vertebrate neuraxis.

---

[49] It will be recalled (cf. vol. 3, part II, chapter VII, section 2) that the dorsal roots of Amphioxus do not have spinal ganglia.

[50] These include two sorts of '*Retzius bipolar cells*', and three additional sorts of dorsal root cells. The afferent processes of some of these intramedullary cells are said to pass through the contralateral dorsal roots.

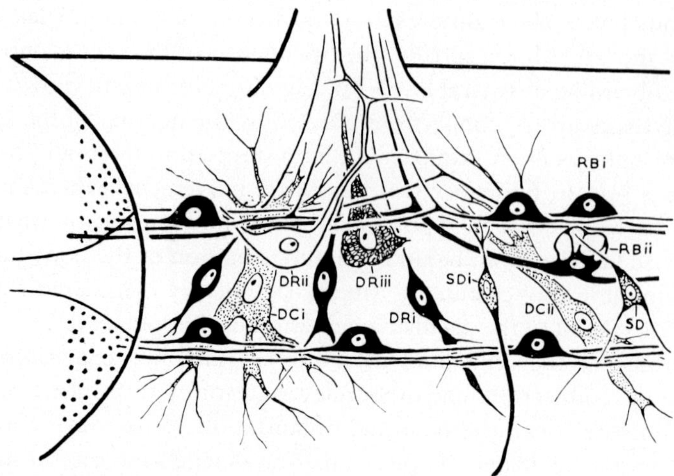

Figure 35. The organization of different neuronal types related to a dorsal root of Amphioxus, as interpreted by BONE. The semidiagrammatic figure shows 'only a small number of some of the cell types' (from BONE, 1960). DCi, ii: dorsal commissural cell of 'type I' and 'type II'; DRi, ii, iii: dorsal root cells of 'type I, II, III'; RBi, ii: '*Retzius bipolar cells*, type I and II'; SD: small dorsal cell; SDi: small dorsal cell, 'special type'.

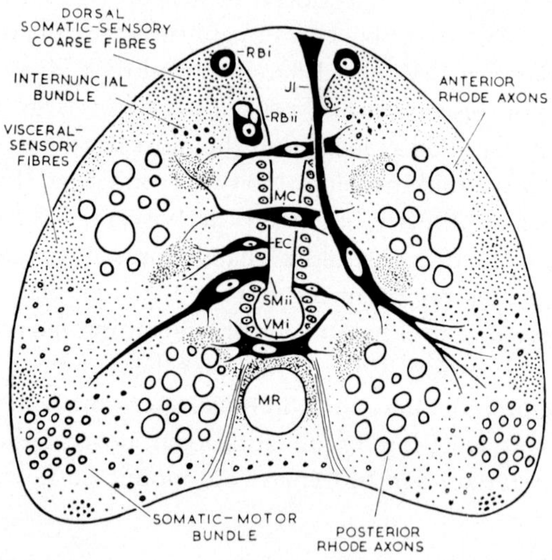

Figure 36. Semidiagrammatic transverse section through spinal cord of Amphioxus in mid-region of body (from BONE, 1960). EC: '*Edinger cell*'; Jl: '*Johnston internuncial*' ('vertical cell'); MC: mid-dorsal cell; MR: median giant fiber; RBi, ii: '*Retzius bipolar cells*'; SMii: ('small') 'somatic motor cell'; VMi: ('large') 'visceromotor cell'. In this figure which embodies BONE's interpretation of cell types, the *Rohde cell* (cf. Figs. 28, 29 and 38) is not included.

An electron-microscopic study of the dorsal roots by PETERS (1963) indicated that the cellular elements scattered between the fiber bundles can be regarded as sheath cells, presumably corresponding to the Vertebrate *Schwann cells* of non-medullated fibers. The cited author made some fiber counts and reported that one dorsal root of the branchial region contained 1988 nerve fibers. In summary, the dorsal roots of Amphioxus can be said to include 'somatic afferent', 'visceral afferent' and 'visceral efferent' fibers in agreement with BONE's (1960) conclusions, but, with regard to further details, much remains rather uncertain.

Concerning the 'somatic-motor' system and the *ventral roots*, BONE (1960) distinguishes several types (large and small) of 'somatic-motor' cells whose bodies are situated in the ventral portion of the spinal cord, but dorsally to the location of the 'viscero-motor' elements. He emphasizes, however, that 'little or nothing is known of the central course or connections of the somatic motor axons, or of the somatic motor cells giving rise to them'. BONE, nevertheless, described a longitudinal ventrolateral 'somatic-motor bundle', formed by these axons, and running fairly close to the surface of the spinal cord. According to the cited author, the efferent ventral root fibers innervating the somatic musculature were provided by collaterals arising from said longitudinal bundle, as well as by some axons of somatic-motor cells which he depicted and described to 'pass directly to the motor root, branch, and issue to the muscle fibers'.

AYERS (1921) had described four branches of the 'ventral roots', a rostral one to the 'dorsal fork' of the preceding myotome, a caudal one to the 'ventral fork' of the succeeding myotomes, and two branches to the myotome of the root's segment, one supplying the dorsal and the other the ventral 'fork'.

As already pointed out and discussed in the preceding volume of this series, electron microscopic studies by FLOOD (1966) have rather convincingly shown that the ventral roots of Amphioxus do not contain neurites of 'somatic-motor' neurons, but are essentially formed by processes of muscle fibers[51] establishing neuromuscular junctions at the surface of the spinal cord (Fig. 37). The region in which these junctions occur corresponds more or less to the longitudinal somatic-motor bundle described by BONE (1960). The mesodermal myotomic ven-

[51] As also mentioned by FLOOD (1960), this was already suspected by SCHNEIDER (1879) and ROHDE (1888, 1892). Cf. chapter VII, sections 1 and 2 of the preceding volume 3, part II).

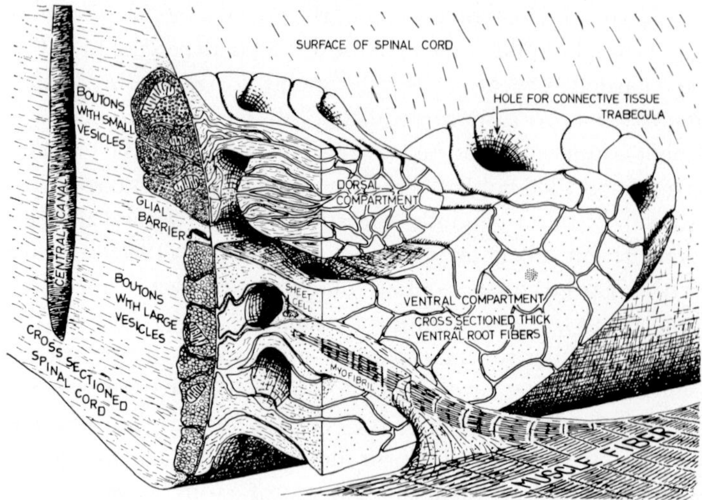

Figure 37. Diagram illustrating the ultrastructure of the 'central motor endplate' on the surface of the spinal cord in Amphioxus as disclosed by electron microscopy (from FLOOD, 1966).

tral root fibers end with conical expansions, separated 'from an extensive layer of axon endings or boutons by an extracellular cleft containing a basement membrane'. Two compartments (dorsal and ventral) of these muscular ventral roots can be distinguished (cf. Fig. 37).

It remains possible, however, that, in addition to the (mesodermal) muscular ventral root fibers, a few genuine (neuroectodermal) nerve fibers are present in the ventral roots, representing afferent (proprioceptive) components. BONE (1960) describes a 'fine-fiber component', originating from cells of the 'mid-dorsal group' and which is said to be 'almost certainly somatic afferent'.[52]

A variety of neuronal elements within the spinal cord provides the internuncials effecting the still unknown connections mediating the

[52] According to BONE, the fine-fiber bundle in question 'lies above the major part of the root, in a small groove upon its dorsal surface'. BONE's description, however, fits FLOOD's dorsal compartment of the muscular ventral root (cf. Fig. 37). This, nevertheless, does not entirely exclude the presence of some proprioceptive ventral root nerve fibers. On the other hand, such fibers could perhaps be included in the dorsal roots or, again, in view of the peculiarities displayed by Amphioxus, one could even assume that its somitic musculature might function entirely without proprioceptive innervation. This latter, moreover, might be substituted by an unusual proprioceptive system related to the notochord, as suggested by BONE (1960).

nervous system's integrative activities. Several types of 'small dorsal cells', and of 'dorsal commissural cells' have been described, in addition to 'middorsal cells' (including '*Edinger cells*' related to the central canal),[53] vertical *cells of Johnston*, and 'ventral longitudinal cells'.

The most conspicuous internuncials, however, are the *Rohde cells* (cf. Figs. 28, 38) which represent the largest neurons in the central nervous system of Amphioxus.[54] These multipolar cells commonly lie across the central canal. Their dendrites extend bilaterally into the dorsal and dorsolateral spinal cord neighborhoods, while their axon passes ventrad, decussating below the central canal and then joins a dorsal or a ventral bundle of giant fibers, similar to the neurochords of Invertebrates and to the *Müller fibers* of Cyclostomes. A rostral and a caudal group of giant cells are separated by an interspace devoid of these elements and corresponding to the body's mid-region. The anterior cells send their neurites caudalward in a more dorsal bundle, and the posterior ones rostrad in a ventral bundle. The most anterior giant cell sends its axon, the median giant fiber, caudad in the ventral midline below the central canal. It is the largest *Rohde cell* with the thickest giant fiber (cf. Fig. 36). As in the case of *Müller's giant cells* in Cyclostomes,[55] numerical cell constancy obtains for the giant cells of Amphioxus. The rostral group regularly comprises 12 and the caudal group 14 cells.[56]

The axons of the *Rohde cells* of successive segments alternate by originating from the cell body on either right or left side of the spinal cord, such that one giant fiber then decussates toward the left and the next one toward the right (cf. Fig. 38). Various uncertainties obtain concerning the manner in which the giant fibers end, as well as with regard to their collaterals. RETZIUS and BONE have noted some of these on the side of the giant fiber's origin, seemingly related to the lateral 'somatic-motor' longitudinal tract, but none were, so far, demonstrat-

---

[53] The '*Edinger cells*' have been suspected to be 'intraependymal neurosensory cells', but this interpretation is not generally accepted (cf. also the comments by KAPPERS, 1947).

[54] Although described in detail by ROHDE (1888), they were apparently discovered by STIEDA (1873; Mém. Acad. Imp. Sc. St. Petersbg. ser. 7, 19, no. 7, cited after BONE, 1960).

[55] In Entosphenus japonicus, 12 pairs of *Müller's cells* were recorded and classified by my associate T. SAITO (1928, cf. the discussion on constant cell numbers on p. 708 f. vol. 3, part I of this series).

[56] In Amphioxus larvae, BONE (1960) found 5 large cells of *Rohde-type* rostral to the most anterior adult giant cells. No trace of these cells could be found in the adult. Any phylogenetic speculation based upon this finding must remain rather uncertain.

ed along the contralateral course of the giant fibers, yet it appears most likely that these fibers are also functionally related to the 'somatic-motor' bundle of the contralateral side.

With respect to the distinction between an intrinsic and an extrinsic mechanism of the spinal cord *(eigenapparat* and *verbindungsapparat)*, which is doubtless semantically valid in dealing with Vertebrates, little can be said concerning the essentially spinal animal Amphioxus. The only actually recorded system that might be interpreted as pertaining to a *verbindungsapparat* is an inconspicuous, partly crossed and partly uncrossed fiber tract, apparently originating from the neuroepithelial cells of the infundibular region, and reaching the spinal cord, where its connections could not yet be ascertained.

Adult Amphioxus may reach a length of approximately 5 cm or less, and is capable of free swimming, although being essentially a burrowing marine animal, living in the sand at small depths and found in all seas of the world. Relatively few species of Cephalochorda (Leptocardii) are extant. According to YOUNG (1955), some eight species of Amphioxus are recognized, and there is also a group of about six species subsumed under the genus Asymmetron. In addition, a still somewhat unclarified genus Amphioxides has been described. Amphioxus commonly seems to remain in a sedentary position, collecting its food by means of the current originated through its cilia. All stimuli, such as touch or illumination (of the body, but not of the head) appear to elicit movements of 'flight'.

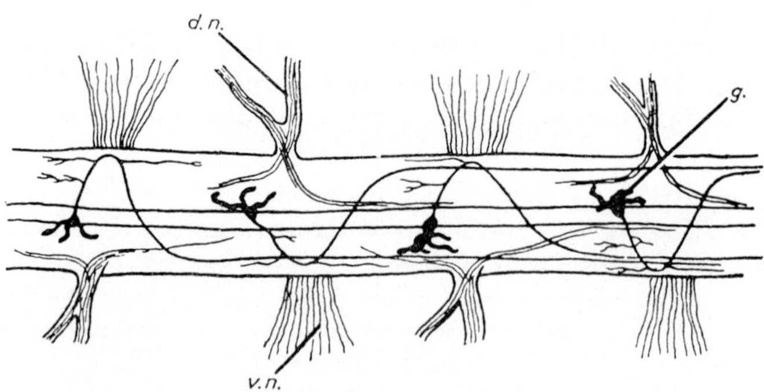

Figure 38. Diagram depicting longitudinal (horizontal) aspect of spinal cord in Amphioxus, and indicating the contralateral course taken by the giant fibers (simplified after RETZIUS, from YOUNG, 1955). d.n.: dorsal root; g.: giant cell of Rohde; v.n.: ventral root. Assuming that the here depicted cells pertain to the anterior group, the neurites, after crossing, run in a caudal direction (toward the right of the diagram).

According to TEN CATE (1938), localized reflexes are entirely lacking in Amphioxus, the responses to external stimuli being always sinuous *('schlängelnde')* movements of the body. Swimming requires waves of myotomal contractions progressing along the body as in the swimming reflex of Amblystoma discussed further above in section 2. It is most likely that the mechanism provided by the *giant cells and fibers of Rohde* plays a major role in the swimming movements of Amphioxus,[57] which commonly display two further characteristics, namely a posture with the rostral end pointing downward, and a rotation around the longitudinal axis, resulting in a spiraling movement (clockwise as seen from the tail). YOUNG (1955) is inclined to assume the presence of some sort of gravitational receptor mechanism. The synaptic arrangements in the spinal cord might, according to the cited author, favor burrowing into the sand when the anterior part of the body, normally exposed while feeding, is touched, but elicit reverse movements of emergence and escape, when the hind part becomes stimulated. Again, the peculiar orientation of the spinal eyes (cf. Fig. 33) could play a role in the just mentioned spiraling swimming motion.

Because the notochord represents an incompressible elastic rod, the lateral bending of the body by the contraction of the longitudinally arranged muscle fibers takes place without significant change in actual body length, in contradistinction to such changes during the motions of e.g. the earthworm Lumbricus.[58] As the animal swims, the notochord nevertheless bends. BONE (1960) tentatively suggests that the peculiar 'notochord cells' connected with the spinal cord (mentioned above on p. 67) might thus become stimulated and 'provide information used in regulating the swimming movements'. However, since BONE assumes the presence of true proprioceptive 'muscle afferents' in

---

[57] This view, already suggested by FRANZ (1923) was also expressed in my Vorlesungen (K., 1927).

[58] YOUNG (1955), in dealing with this topic, refers to the interpretation of the notochord as a '*supporting structure*' and adds: 'but, of course, an animal such as a fish in water needs no "support".' Be that as it may, one can evidently subsume relatively rigid '*bracing structures*', providing a skeletal 'framework' and including the notochord, under the logical class 'supporting structures'. Gnathostome Craniote fishes, in addition to their cranium, display a cartilaginous respectively bony vertebral column of substantial solidity and, despite their habitat in a buoyant medium, of considerable significance for the maintenance of their body shape.

Amphioxus, he expresses the following doubts about this interpretation: 'it therefore hardly seems probable that the bizarre notochordal cells form a second system concerned with the reflex regulation of swimming'.

While certain similarities between the swimming reflexes of larval Amblystoma as analyzed by HERRICK and COGHILL (1915) and the behavior of adult Amphioxus can easily be recognized, both sorts of behavior display conspicuous differences. The neuronal mechanisms in larval Amblystoma, moreover, are rather simple and have been reasonably well elucidated. In contradistinction, the corresponding structural arrangements in the neuraxis of Amphioxus are highly complex and have, until now, defied a comparable and sufficiently relevant analysis.

Amphioxus is doubtless, seen both from a morphological and from a phylogenetic viewpoint, an exceedingly interesting animal, whose overall bauplan displays conspicuous homologies with craniote Vertebrate features. There is thus substantial justification for the 'old saw' stating that, if this organism had not been actually found to exist, it would have to be invented. YOUNG (1955) assumes 'that Amphioxus shows us approximately the condition of the early fish-like chordates living in the Silurian some 400 million years ago, and that it has undergone relatively little change in all the time since'.

On the other hand, so many discrepancies obtain, in detail, between morphologic and structural configurations of Amphioxus and of true Vertebrates, that all specific phylogenetic speculations based on the anatomy of Branchiostoma remain encumbered by considerable difficulties and uncertainties, and are therefore rather unconvincing. Depending on the viewpoint[59] one is justified to say that there is, in Amphioxus, both much more and much less 'than meets the eye'.

4. Cyclostomes

The spinal cord of adult *Cyclostomes*, both in *Petromyzonts* and *Myxinoids*, assumes a more or less dorsoventrally flattened aspect. Figures 39, 40 and 42 illustrate the general changes in outline occurring during the course of ontogenesis. ALLEN (1916) believes that this 'secondary flattening' is due to a pressure effect by the sizeable notochord

---

[59] Cf. the comments on p. 334 f. in volume 2 of this series.

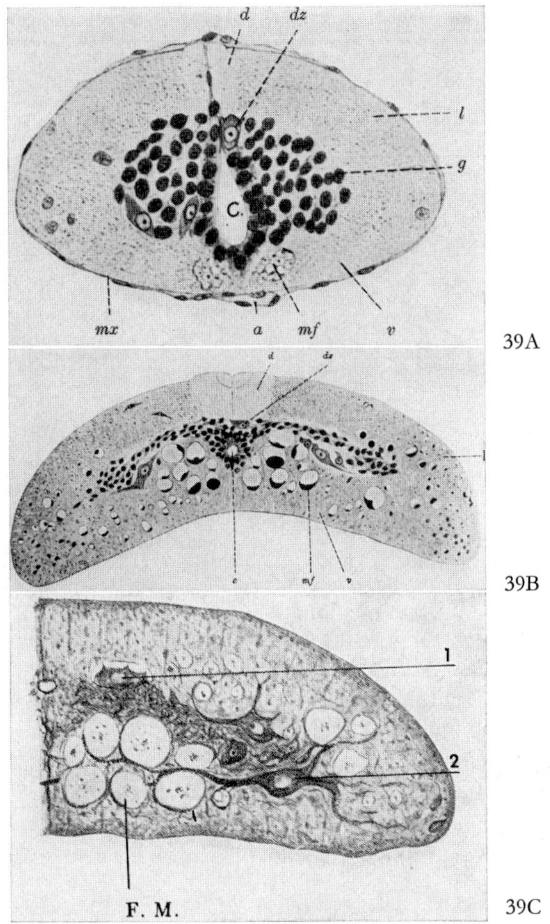

Figure 39A. Spinal cord in a Petromyzon (Ammocoetes) larva of 19 mm length (from KUPFFER, 1906). a: 'tractus arteriosus; d: 'dorsal funiculus; dz: large dorsal cell; g: 'gray substance'; l: 'lateral funiculus; mf: *Müller's fibers;* mx: endomeninx; v: 'ventral funiculus'; C: central canal.

Figure 39B. Spinal cord in a 12 cm Ammocoetes (from KUPFFER, 1906). Designations as in A. BECCARI (1943) characterizes this outline of the Cyclostome spinal cord as '*aspetto nastriforme, depresso in senso dorsoventrale, leggermente incavato sul mezzo dal lato ventrale*'.

Figure 39C. Section of spinal cord in Petromyzon fluviatilis drawn by means of the camera lucida (from STEFANELLI, 1933). 1: dorsal cell; 2: motoneuron; F.M.: *Müller fiber*. This drawing indicates the relationship of motoneuron dendrites to *Müller fibers*. One of the lateral giant fibers presumably represents ROVAINEN's '*Mauthner fiber*'. In addition, it will be seen that the accurately drawn pattern of large fibers in this specimen of Petromyzon fluviatilis differs in various details from that in the advanced larva of Petromyzon marinus depicted by Figure 39D.

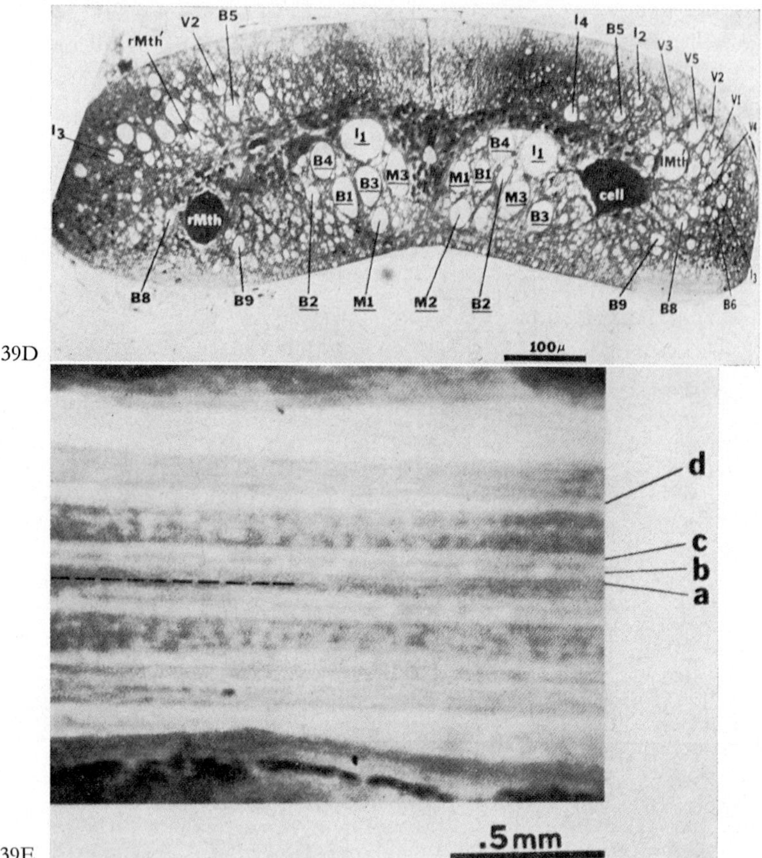

39D

39E

Figure 39D. Cross-section of the spinal cord of a larva of Petromyzon marinus 'in the final states of transformation', 3 mm beyond the end of the fourth ventricle (from Rovainen, 1967). The letters and numbers indicate axons of specific 'reticular cells'. B: bulbar *Müller cells;* I: isthmic *Müller cells;* M: mesencephalic *Müller cells;* Mth, Mth′ (l: left, r: right): *Mauthner cells;* V: cells of 'vagal region reticular group'. The dense staining of the rMth axon is attributed by Rovainen 'to injury during dissection'.

Figure 39E. 'Living spinal cord' in the gill region of a 16 mm (Petromyzon marinus) ammocoetes, as seen by means of transillumination (from Rovainen, 1967). a: position of relatively thin 'medial axon', next to midline (indicated by dashes); b, c: 'inner' and 'outer medial giant axons'; d: 'giant lateral axon'. The light spots over b and c are interpreted as 'dorsal cells', the mottled area between c and d as a 'mixture of cells and axons', and the gray area lateral to d as 'composed of small axons'.

which remains here the basic structure of the axial skeleton.[60] With regard to Petromyzon, TRETJAKOFF (1909) suggests that the flattening, which reduces the distance of internal elements from the cord's meningeal surface, is related to the lack of intraspinal blood vessels.[61] Concerning this explanation, however, KAPPERS *et al.* (1936) justly point out that the likewise flat spinal cord of Myxinoids displays some degree of intrinsic vascularization. While ALLEN's explanation might perhaps be partially valid, it seems probable that the flattening of the spinal cord in Cyclostomes results from far more complex multifactorial developmental events.

ALLEN (1916) justly stresses the fact that the shape of a 'typical' embryonic spinal cord in the vertebrate series manifests considerable variations with respect to its tubular contour. This author, moreover, recorded an interesting abnormality in the spinal cord of one specimen of a young adult Myxinoid Polistotrema (Bdellostoma) which displayed an expansion of the roof plate forming a dorsal recess of the central canal (Fig. 41A). He interpreted this expansion, which somewhat resembles the neuroepithelial roof of the fourth ventricle, as a 'mutation' related to the presumptive genetic factors originating the formation of choroid plexuses in ancestral Vertebrates. In a still more general way, one might indeed regard this abnormality as a manifestation of activated latent histogenetic potencies inherent in a particular morphologic zone (in this instance, the roof plate of the neuraxis).[62]

Reissner's fiber was investigated in Petromyzon by OLSSON (1955) who particularly described the aboral end of that structure and its relationship to the spinal cord's caudal extremity discussed above in connection with the corresponding structures of Amphioxus (Figs. 31, 32).

---

[60] Cyclostomes are, in fact, not 'true' Vertebrates, if this latter term is interpreted in a very literal sense. Myxinoids entirely lack vertebral elements, although some cartilages participate in the formation of posterior fin supports. Petromyzonts display two pairs of small dorsal cartilaginous elements alongside the spinal cord in each segment, but lack ventral vertebral components. In all Cyclostomes, the axial skeleton is essentially represented by a large notochord with thick fibrous sheath.

[61] Cf. vol. 3, part II, chapter VI, section 7, p. 703 of this series.

[62] With regard to other structures, cf. K. and FRIEDMAN: The manifestation of certain germ layer potencies in epithelial neoplasms of the urinary bladder. Anat. Rec. *103:* 479 (1949); also: FRIEDMAN and K.: Adenomatoid tumors of the bladder reproducing renal structures (nephrogenic adenoma). J. Urol. *64:* 657–670 (1950).

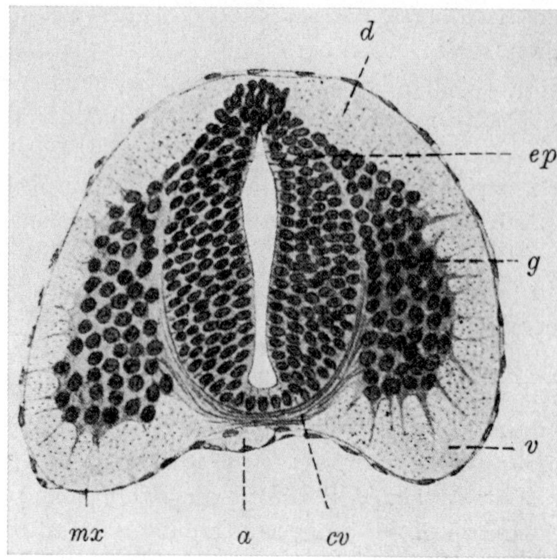

Figure 40A. Spinal cord in a fairly advanced *('älterer Embryo')* of Bdellostoma (from KUPFFER, 1906). a: 'arteria spinalis ventralis'; cv: ventral commissure; ep: ependymal layer. Other abbreviations as in preceding figures from that author.

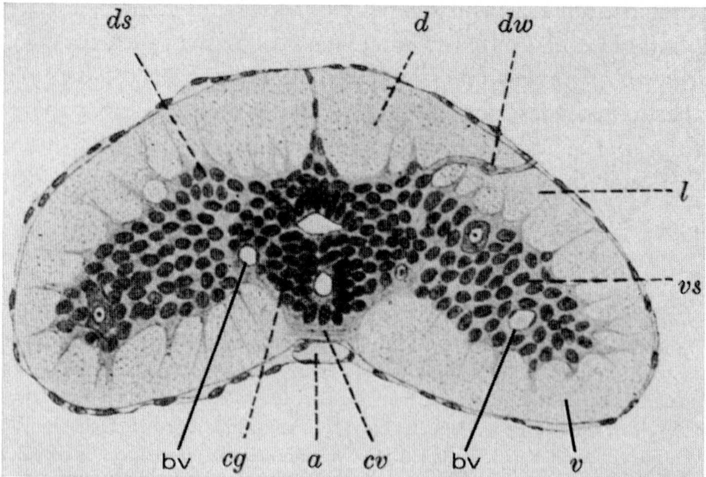

Figure 40B. Spinal cord in an embryo of Bdellostoma at an advanced stage just preceding hatching (after KUPFFER, 1906). bv: intraspinal blood vessels; cg: 'central gray substance; ds: suggestion of 'dorsal column'; dw: dorsal root; vs: suggestion of 'ventral column'. Other abbreviations as in preceding figures. It can be seen that the ventral portion of central canal is partially obliterated.

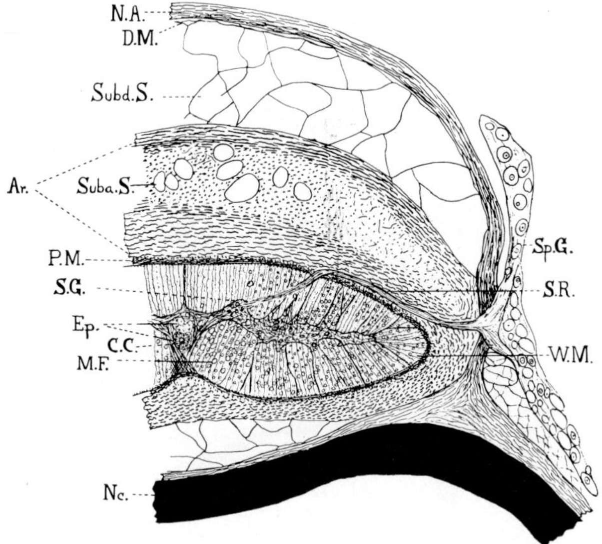

Figure 40C. Transverse section through anterior portion of spinal cord and adjacent structures in the adult Myxinoid Polistotrema (from ALLEN, 1916). Ar: 'arachnoid'; C.C.: central canal; D.M.: 'dura mater'; Ep.: ependyma; M.F.: *Müllerian fibers;* N.A.: membranous neural arch; Nc.: notochord; P.M.: 'pia mater'; S.G.: 'substantia gelatinosa' (suggestion of 'dorsal column'); Sp.G.: spinal ganglion; S.R.: dorsal root; Suba.S.: 'subarachnoid cavities'; Subd.S.: 'subdural spaces'; W.M.: 'white matter'.

In *Petromyzonts*, dorsal and ventral roots alternate and do not join,[63] thereby manifesting an arrangement quite similar to that in Amphioxus. In contradistinction to the ventral roots in this latter, however, those of Petromyzon are formed, as in all Craniote Vertebrates, by true nerve fibers (neurites) originating from intraspinal motoneurons and visceral-efferent cells. Whether some afferent proprioceptive fibers are included in the ventral roots remains an open question. Again, the dorsal roots of Petromyzon display spinal ganglia with mostly bipolar afferent nerve cells, but a few pseudounipolar cells, similar to those prevailing in higher Vertebrates such as Mammals, have been recorded. In addition, some of the afferent dorsal root fibers appear to originate from intraspinal dorsal cells. Visceral-efferent fibers likewise aris-

[63] Cf. Figure 417 in chapter VII, dealing with the peripheral nervous system, of the preceding volume 3, part II.

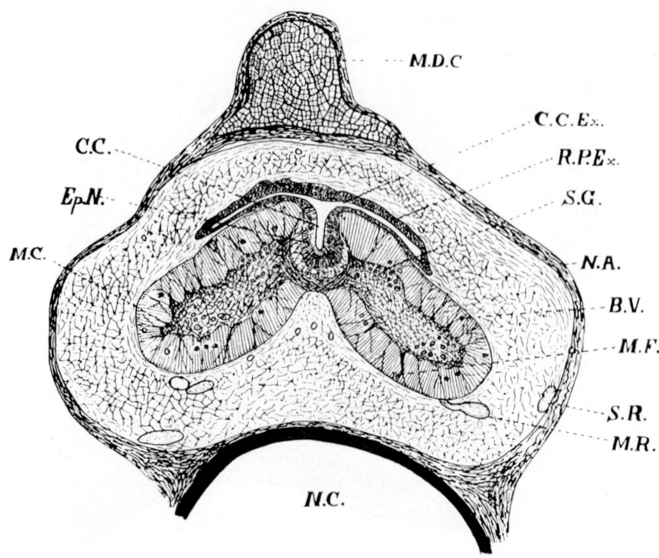

Figure 41 A. Transverse section through rostral portion of spinal cord in a 20 cm Polistotrema (Bdellostoma) with some surrounding structures. A peculiar roof plate extension covers a large dorsal area of spinal cord, its cavity being in communication with the central canal (from ALLEN, 1916). B.V.: blood vessel; C.C.: central canal; C.C.Ex.: central canal extension into roof plate expansion; Ep.N.: 'layer of ependymal nuclei'; M.C.: 'motor cells'; M.D.C.: median dorsal cartilaginous bar; M.F.: Müllerian fiber; M.R.: ventral root; N.A.: membranous neural arch; N.C.: notochord; R.P. Ex.: roof plate expansion; S.G.: 'substantia gelatinosa' (suggestion of 'dorsal horn'); S.R.: portion of dorsal root.

ing in the spinal cord and joining the dorsal roots were also described.[64]

In *Myxinoids*, on the other hand, a variable degree of union between dorsal and ventral roots obtains, concomitantly with other peculiarities, such that, at certain levels, one dorsal and two ventral roots are given off. As in Petromyzonts, the dorsal roots carry spinal ganglia.

The *intraspinal neuronal elements* of Cyclostomes (Figs. 39, 40, 42–44) show an arrangement whose pattern is intermediate between that ob-

[64] JOHNELS (1956, 1958) has undertaken extensive investigations of the peripheral autonomic nervous system in the trunk region of Lampetra, and of the dorsal ganglion cells in the spinal cord of Lampreys. Despite the careful studies of this author many details must be considered poorly elucidated because of inherent technical difficulties and uncertainties of interpretation.

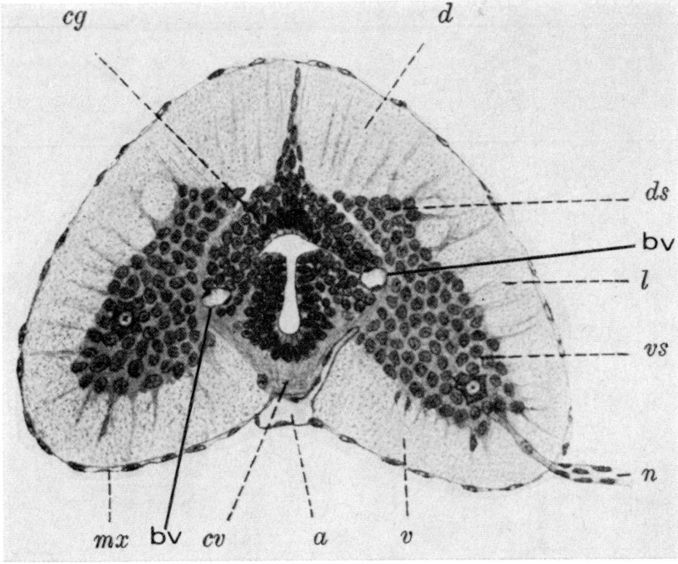

Figure 41 B. Cross-section through the spinal cord in an advanced embryo of Bdellostoma at a stage between those of Figures 40A and B (from KUPFFER, 1906). This figure shows a slight dorsal expansion of the central canal which, if becoming exaggerated, might result in the abnormality depicted by ALLEN in figure 41A. n: ventral root. Other abbreviations as in Figures 39 A to 40B.

taining on one hand in Amphioxus and, on the other hand, in Gnathostome Vertebrates. As in these latter, the central canal has a narrow, tubular (cylindrical) lumen, and the surrounding cell populations extend lateralward, forming a wing-like expansion on both sides. However, a distinctive outline of ventral and dorsal horns is not displayed by Petromyzonts, but may, to a variable degree, become only very faintly suggested in Myxinoids (Fig. 40).

Concerning the cellular structure of the spinal cord in Petromyzonts, relevant, but still to a large extent inconclusive details were elucidated by the investigations of OWSYANNIKOW (1903), TRETJAKOFF (1909), SCHULTZ *et al.* (1956), JOHNELS (1958) and others. Roughly speaking, the smaller dorsal neuronal elements of that cell population represent internuncials, many of which appear related to afferent root fibers. The larger ventral cells are presumably motoneurons respectively visceral-efferent cells. Additional ventral cells of intermediate size may be internuncials and perhaps also include some visceral-efferents. The motor neurites emerging through the ventral roots seem, as a rule, to

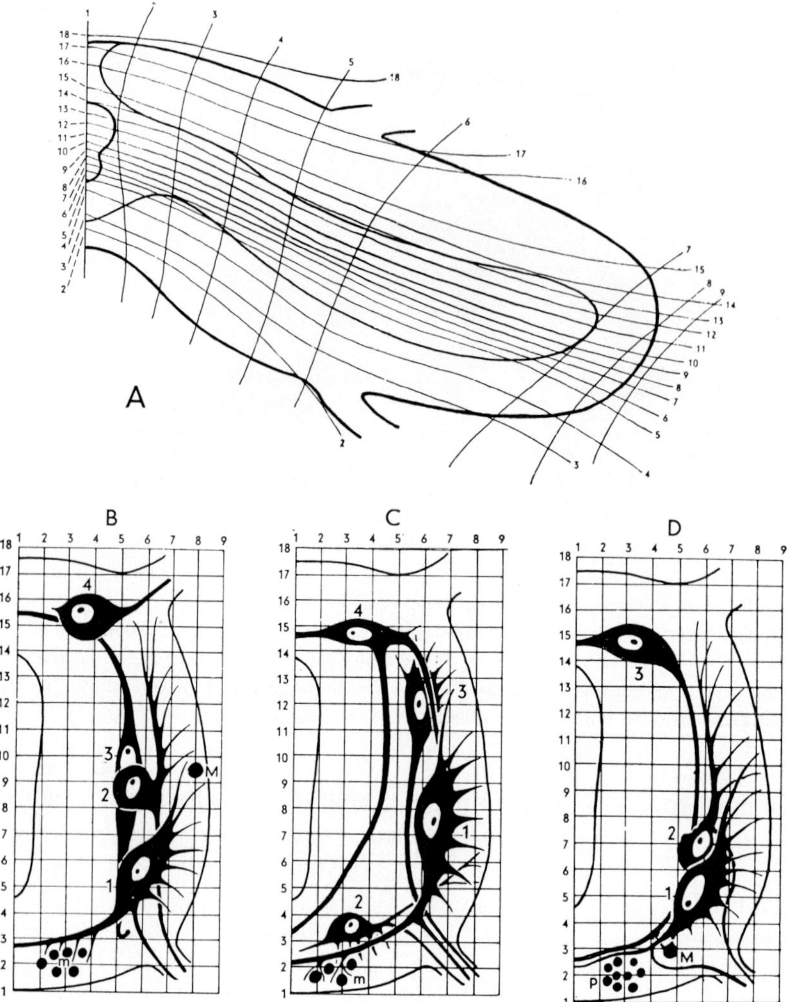

Figure 42. Diagrams comparing outline of the adult spinal cord in Myxine (A) with the configuration of the spinal cord at certain ontogenetic stages in Petromyzon, in a 'generalized Cyclostome' and in Larval Amblystoma (from BONE, 1963). BONE adapts here THOMPSON's (1942) method of transformation by means of curvilinear coordinates derived from a *Cartesian grid* (cf. my discussion on p. 186–192 in volume 1 of this series). A: adult Myxine; B: Petromyzont pro-ammocoete larva with neuronal elements in the interpretation of WHITING, 1948); C: Some of the types of Neurons of adult Myxine, as interpreted by BONE, 'transferred to their relative positions' within a generalized embryonic Cyclostome medulla; D: larval spinal cord of Amblystoma with neuronal elements in YOUNGSTROM's (1940) interpretation. 1: 'primary somatic-motor neuron' (B), 'first type of somatic motor neuron' (C), 'primary motor neuron (D); 2: 'secondary somatic-motor

take, centrally to their exit, an at first longitudinal course, directed either rostrad or caudad.

Another peculiarity of motoneurons and other nerve cells is the expansion of their ramified dendrites toward the surface of the spinal cord, where these processes appear to reach the limiting membrane, providing rather dense plexuses (Fig. 44). Neuropil formations with synaptic connections are thus not restricted to the spinal cord's central 'griseum'. A similar, but perhaps less pronounced peripheral expansion of dendrites into the cord's white substance is, however, also manifested in numerous Gnathostome Anamnia and even occurs in Amniota.

Among the internuncials or 'interneurons', ROVAINEN (1974a) distinguishes *'edge cells' (Randzellen* of older authors), located within the lateral fiber tracts of the spinal cord, and *'lateral cells'* within the more central main cell population. The neurites of the former are said to extend rostrad on either homo- or contralateral side. Some 'edge cells' seem to have an inhibitory function. The 'lateral cells' occurring at branchial and trunk levels are reported to extend their long ipsilateral neurites as far caudad as tail levels, and to produce weak IPSPs 'in unidentified neurons' The *'giant interneurons'* (cf. also further below) are interpreted as a third type of internuncials, whose long ascending contralateral neurites provide part of 'a convergent, multispecific sensory system reaching the brain' (ROVAINEN, 1974a).

By means of electron microscopy SCHULTZ *et al.* (1956) recorded, in 'many regions', synaptic structures characterized by 'synaptic membranes' and vesicles. These latter were sometimes seen close to the membrane in two adjacent axons[65] or even found centrally placed in

[65] This could possibly be interpreted as indicating the presence of some *'protosynaptic connections'* in addition to permanently polarized synaptic ones (cf. vol. 3, part I, chapter V, section 9, p. 639 of this series).

---

neuron (B), 'second type of somatic-motor neuron' (C); 'secondary motor neuron' (D); 3: 'vertical internuncial neuron' (B), 'viscero-motor (?) neuron' (C), 'dorsal intercalated neuron' (D); 4: '*Rohon-Beard neuron,* possibly equivalent to the adult petromyzont *Hinterzelle* (B), 'large dorsal internuncial neuron (C); M: *Mauthner axon* (B_1 'with collaterals to 1 and 3 in D); m: '*Müller axons*' (B), '*Müller-type axons* in fasciculus longitudinalis medialis' (C); p: 'primitive motor tract'. BONE adds that these diagrams 'are not to the same scale, and include various types of neuron which are not found in the same section of the spinal medulla'. I would, moreover, remark that, except for some *Rohon-Beard cells* and efferent elements of the basal plate, the proper identification of neuronal elements is, in my opinion, much less certain than most authors seem to assume.

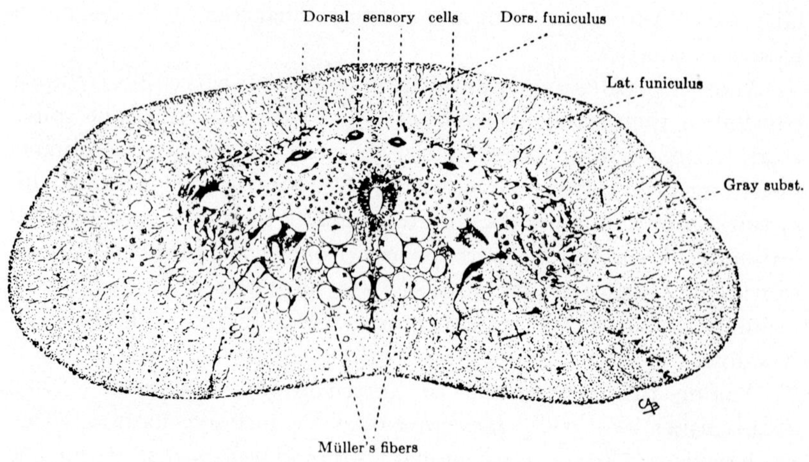

Figure 43. Cross-section through the spinal cord of Petromyzon (from KAPPERS *et al.*, 1936). It can be seen that in such sections, *Müller's fibers* show a coagulated dot-like axon surrounded by a large shrinkage space which resembles a dissolved myelin sheath (cf. also Fig. 39B and the discussion of giant fiber structure in the text).

some axons. Differentiated 'synaptic knobs' were apparently absent and the vesicles seemed to be 'rare in parts of the central gray and in most terminals around he cell membrane'.[66] The cell bodies of motoneurons appear generally fusiform with transverse long axis in 'horizontal' longitudinal sections, which also disclose a fusiform shape of the large dorsal cells, whose long axis, however, shows a rostro-caudal orientation.

As regards the *large dorsal cells*, whose peripheral process emerges through the dorsal roots, several types can be distinguished on the basis of rather dubious criteria. The assumption, by KAPPERS (1947) and several previous authors that these cells are intramedullary afferents (perhaps essentially 'somatic') seems rather plausible, at least for many of these cells, and, in this respect, their comparison with *Rohon-Beard cells* would appear justified. BECCARI (1943) found that, in adult Petro-

[66] The apparent lack of differentiated 'synaptic knobs' in Petromyzon should not be interpreted to mean that such knobs are present only in Tetrapod Vertebrates. Well developed 'boutons', clearly demonstrable by high-resolution light microscopy, can be found in the neuraxis of numerous Gnathostome Fishes, and are e.g. conspicuous for the synaptic connections of *Mauthner cells*. Cf. also the detailed investigation of synaptic structures in Teleosts by KIRSCHE (1967).

myzonts, these cells have the tendency to assume a 'fenestrated' or irregular 'multipolar' appearance, presumably related to the growth of paraphytes as discussed in chapter V, pp. 79–82 of volume 3, part I of this series. BECCARI also recorded rather dorsal multipolar cells whose neurite decussates in the ventral commissure, bifurcating into ascending and descending branches within the anterior funiculus. He regards these cells as comparable to the *Rohde cells* of Amphioxus: '*queste cellule*

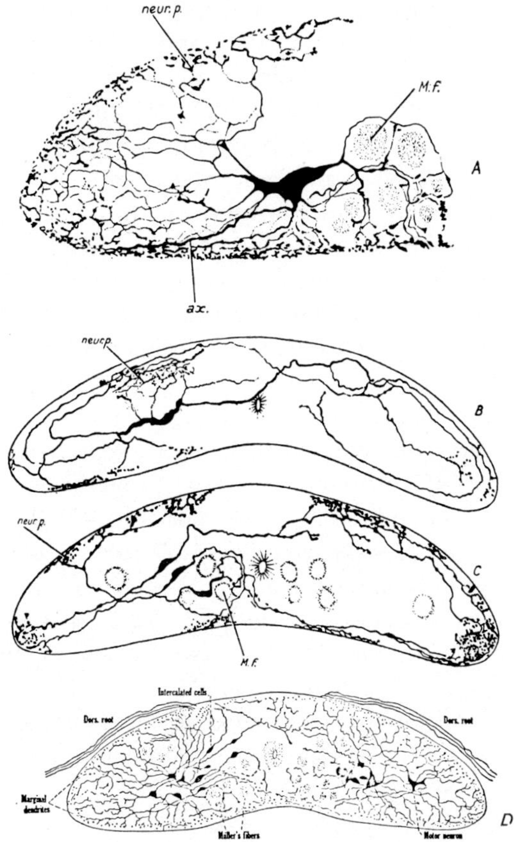

Figure 44. Cross-sections through the spinal cord of larval Petromyzon (Ammocoetes), showing the dendritic ramifications of neuronal elements (after TRETJAKOFF, 1909, A–C from YOUNG, 1955, D from KAPPERS *et al.*, 1936). In A and B large 'motor' cells, with dendrites reaching the opposite side. In C some small cells with widespread 'dendrites' but no recognizable 'axon'. In D an additional overall picture. ax.: 'axon'; neur. p.: peripheral neuropil; M.f.: Müllers' fibers.

commessurali sarebbero omologhe delle cellule colossali dell'Amphioxus, ed il ramo ascendente delle loro fibre costituirebbe la più primitiva via sensitiva centrale (Ariëns Kappers,' 20). La maggior parte delle fibre termina nel midollo spinale, ma non è escluso che alcune raggiungano la oblongata, portando eccitamenti sensitivi a grosse cellule multipolare situate nel suo tegmento.' Johnels (1958), however, claims that some large dorsal cells of a particular type represent visceral efferent elements. While this interpretation cannot be summarily dismissed, the observations and arguments in favor of that possibility do not appear very convincing. Likewise, the results of experimental investigations with the use of microelectrodes undertaken and reported by Rovainen (1967) appear rather inconclusive but are not inconsistent with the interpretation of the large dorsal cells as '*sensory neurons*'. In addition, the cited investigation seems to confirm the identification, by Edinger, Beccari, and others, of large internuncials, which Rovainen calls '*giant interneurons*'.

The central dorsal root fibers originating from spinal ganglion cells generally bifurcate into an ascending rostral branch and a descending caudal one, both of which take their course through the homolateral dorsal region of the spinal cord. Kappers *et al.* (1936) mention also 'finer fibers which turn in a longitudinal direction' without bifurcating.

Besides the ciliated ependymal cells surrounding the central canal, the intramedullary non-neuronal elements are represented by neuroglia cells of astrocytic type, whose processes, together with ependymal fibers, join the external limiting membrane (Fig. 45). The presence of oligodendroglia and *Hortega cells* comparable to those of Gnathostome Vertebrates has not been recorded.[67]

The synaptic connections of the spinal cord's *eigenapparat* and *verbindungsapparat ('voies secondaires de la moelle'*, Kappers, 1947) are very incompletely understood. Kappers (1947) believes that they consist predominantly of short intersegmental pathways mediating reflexes. He assumes, however, that some ascending fibers reach the vestibular

---

[67] *Prima facie*, a progressive differentiation of 'supporting elements' in the series Amphioxus, Cyclostomes and Gnathostomes might be assumed, corresponding to a phylogenetic sequence characterized by only ependymal cells in Amphioxus, ependymal cells and 'primitive astrocytes' in Cyclostomes, and finally ependymal cells, astrocytes, oligodendroglia and *Hortega cells* in Gnathostomes. Although the presence of 'neuroglia cells' in Amphioxus is somewhat dubious, it can nevertheless be tentatively maintained (cf. above, p. 67), and would exclude Amphioxus from said hypothetical series.

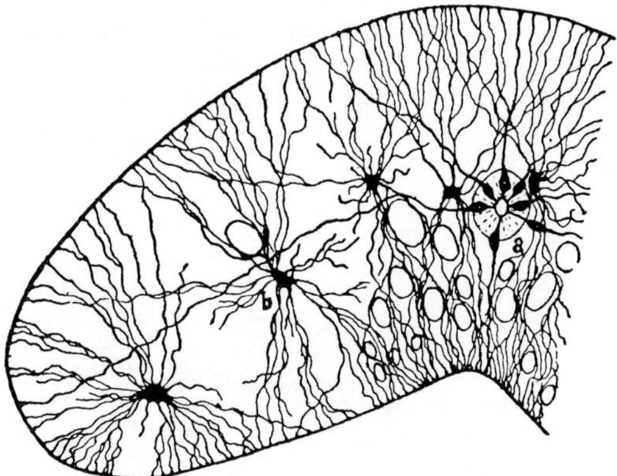

Figure 45. Cross-section through the spinal cord of Petromyzon, showing ependyma and astrocytic elements in *Golgi-impregnation* (after LENHOSSÉK, from CAJAL, 1909).

and optic regions of the brain stem, thereby providing spino-bulbar and spino-tectal pathways within unspecified locations of the cord's longitudinal fiber bundles. Crossing fibers pertaining to either intrinsic or extrinsic systems *(eigenapparat* or *verbindungsapparat)* can be seen to form an anterior decussation ventral to the central canal. In addition, it seems quite likely that some ascending components of the bifurcating afferent dorsal root fibers located within the 'dorsal funiculus' may reach the medulla oblongata.

As regards descending components of the cord's extrinsic apparatus, the most conspicuous system is that provided by the thick *fibers of Müller.*[68] There are about 8 such fibers on each side ventrally to the central cell aggregate (i.e. in the 'ventral funiculus') and a lesser, but apparently variable number can be found in lateral and even dorsal

[68] In his studies on the comparative anatomy of Myxinoids, published as Abhandlungen der Königl. Akad. d. Wiss., Berlin 1834–1843, JOHANNES MÜLLER, had described relatively large, pale '*Bänder*' or '*bandartige durchaus platte Fäden*' in the spinal cord. LEYDIG, in his *Lehrbuch der Histologie des Menschen und der Thiere* (1854) again calls attention to the '*sehr breiten, von* JOH. MÜLLER *zuerst beschriebenen Nervenfasern*'. He adds the following details concerning the spinal cord of Ammocoetes and Petromyzon: '*Die Nervenfasern des Rückenmarks sind einmal feine, welche auf beiden Seiten der Medulla die äusserste Schicht bilden, und zweitens die s.g. Müller'schen Fasern, jene kolossal breiten, welche sich zu beiden Seiten des Centralkanales finden. Sie haben nicht im ganzen Rückenmark denselben Durchmesser, sind gegen den Kopftheil am breitesten, verschmälern sich aber in der Schwanzgegend so beträchtlich, dass sie endlich nur vom Breitendurchmesser der übrigen Längsfasern sind. Im verlängerten Mark gehen sie in grosse runde Ganglienzellen über.*'

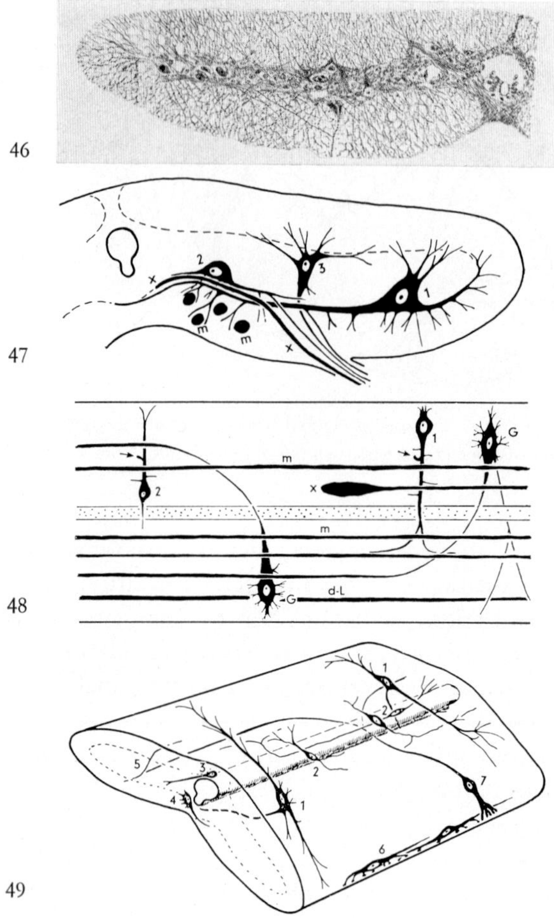

Figure 46. Cross-section through the spinal cord of Myxine as seen with *Weigert's glia stain* (after Müller, 1900). Although meant to show the neuroglial elements (cf. vol. 3, part I, p. 201), this technique displays here also the neuronal perikarya and perhaps some of the nerve fibers.

Figure 47. Cross-section through spinal cord of Myxine, with types of cell 'which send axons out in the ventral roots' (from Bone, 1963). 1: 'first type of somatic motor cell'; 2: 'second type of somatic motor cell'; 3: 'viscero-motor (?) cell'; x: 'single larger fibre in each root'; m: '*Müller-type axons* in fasciculus longitudinalis medialis, with which dendrites from the first type of somatic-motor cell are synapsing'.

Figure 48. Diagram purporting to illustrate, in ventral view, 'the large-diameter through-conducting systems of the medulla oblongata (from Bone, 1963). Rostral end of system at right, central canal dotted. Arrows point to the position where the axons of these neurons pass downward out of the ventral root. dL: 'large-diameter axon in fasciculus dorsolateralis'; G: 'giant cells'; m: *Müller-type axons*, one terminating at x in 'funiculus' (cf. Fig. 39C). These fibers run, without giving off any detectable collaterals, to the caudalmost portion of the spinal cord and are surrounded by dendrites of motoneurons and perhaps also other nerve cells. Near the caudal end of the spinal cord, the thickness of these fibers decreases and their identification becomes dubious.

Electron microscope pictures recorded by SCHULTZ *et al.* (1956) disclose that the 'giant fibers' are non-medullated. Their finely strippled axoplasm contains neurofilaments and longitudinally oriented mitochondria. Other structures visible in the electron photomicrograph of *Müller's fiber* depicted by the cited authors might be neurotubules and, near, the axon membrane, synaptic vesicles. Spots of increased density along that membrane could represent synaptic contacts with the above-mentioned surrounding neuronal cell processes.

The origin of *Müller's fibers* can be traced with reasonable certainty to large reticular elements *(Müller's cells)* in the basal plate of the brain stem, as already noted by LEYDIG (1847; cf. also above, footnote 68). 'Typical' large *Müller fibers* of the spinal cord have a homolateral origin, with the exception of one or two large lateral respectively dorsolateral fibers which ROVAINEN (1967), because of their contralateral origin, considers to be *Mauthner fibers*,[69] in agreement with previous observations and interpretations by WHITING (1957) and BONE (1963).

ROVAINEN (1967) claims to have identified individual *Müller's fibers*, *Mauthner's fibers* and other large descending fibers of the Petromyzont

[69] AHLBORN (1883), also quoted by WHITING (1957) reported, in addition to several pairs of *Müller cells*, one pair of *Mauthner cells* in young Ammocoetes larvae. *Mauthner fibers* and *Mauthner cells* are dealt with further below in sections 6, 7, 8 of this chapter, as well as in the sections of chapter IX concerning the oblongata of Anamnia. It will here be sufficient to remark that typical *Mauthner fibers* of Ganoids, Teleosts, and Amphibians run through the spinal cord's fasciculus longitudinalis medialis, while those of Cyclostomes are found in lateral or dorsolateral 'funiculus'. Again, in Cyclostomes, their decussation takes place somewhat more caudally than in Gnathostomes, where these fibers commonly decussate at or very near the level of their cells of origin.

a growth cone'; 1: 'first type of somatic-motor neuron'; 2: 'second type of somatic-motor neuron'.

Figure 49. Stereogram, purporting to show some of the types of internuncial neurons in the spinal cord of Myxine (from BONE, 1963). 1: 'large dorsal internuncial neuron'; 2: 'small dorsal internuncial neuron'; 3: 'small dorsal oblique neuron'; 4: 'small ventral commissural neuron'; 5: 'vertical fibre connecting dorsal and ventral funiculi'; 6: 'first type of lateral cell'; 7: 'second type of lateral cell (lateral arcuate neuron)'.

spinal cord, by tracing their origin from specific cells of the brain stem, as indicated in Figures 39D and 190. Figure 39E shows '*giant axons*' in the transilluminated living spinal cord of an Ammocoetes as observed by ROVAINEN (1967).

Although both SAITO (1928) and STEFANELLI (1933) recorded a constant number of *Müller's cells* in Entosphenus (12 pairs, SAITO) respectively in Petromyzon (10 pairs, STEFANELLI), while ROVAINEN (1967) described 8 pairs in Petromyzon,[69a] there remains some doubt whether diverse ambiguous elements should or should not be classified as additional *Müller's cells* (cf. section 3 of the next chapter IX). Much the same doubt obtains about the inclusion of various fairly large fibers into the system of 'true' *Müller's fibers.* Another moot questions concerns the possible decussation, in the oblongata's basal midline, of some fairly large reticulo-spinal fibers which might arise from cells of borderline '*Müllerian type*', and descend, within or outside the fasciculus longitudinalis medialis of the spinal cord, in this latter's ventral or even dorsal funiculus.

WHITING (1948), who also investigated the spinal cord of young larval brook lampreys, stresses the similarities in the rounded, oval outline as well as in the neuronal structure of the spinal cord at that early Ammocoetes stage with the features found at comparable stages of Amblystoma. Moreover, the cited author believes that the probable functional pattern in the young Ammocoetes is likewise rather similar to that obtaining in Amblystoma larvae. Since the divergences from the Vertebrate pattern found in the cord of the adult are not found in the young Ammocoetes, this latter is 'as in so many respects', 'a good prototype of gnathostome Vertebrates' (WHITING, 1948).

The spinal cord of *Myxinoids* (Figs. 40C, 42, 46–51) is, on the whole, similar to that of Petromyzonts, although, within the overall

---

[69a] These 8 pairs do not include one or in some instances 2 pairs of his '*Mauthner cells*'. More recently, ROVAINEN *et al.* (1973) have reported on additional attempts at tracing individual neurites in the spinal cord of Petromyzon marinus. It is claimed that the axon of one 'giant interneuron', located in the dorsal portion of the spinal cord, was followed 'into the lateral medulla of the brain' (medulla oblongata). A formerly unrecognized pair of *Müller fibers* was reported to originate from the reticular cells called I_2 (cf. Fig. 190; the cells I_2 [not here labelled], are the elements located just caudally to I_1). The results of subsequent investigations on the reticulospinal system by means of microelectrodes (ROVAINEN, 1974b) were interpreted as disclosing that dual, namely electrical and chemical synaptic transmission obtains. Most of the recorded unitary EPSPs were subthreshold. The cited author infers that other sources of excitation, 'especially from smaller unidentified neurons, are required for activity'.

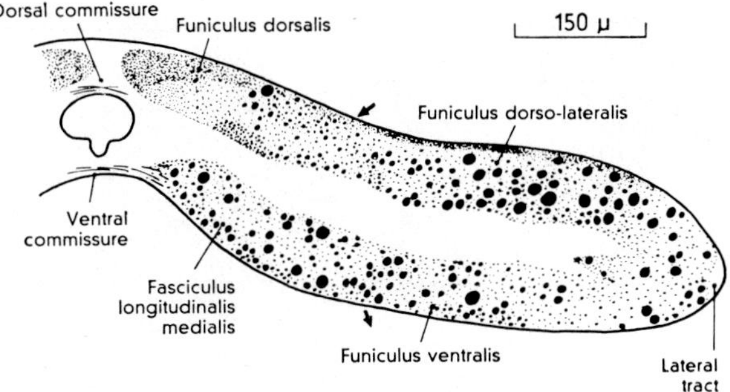

Figure 50. So-called 'fiber spectrum of the white matter' (from BONE, 1963). The larger fibers are said to be accurate in number and position. The arrows indicate the sites where the dorsal and ventral roots are connected with the spinal cord.

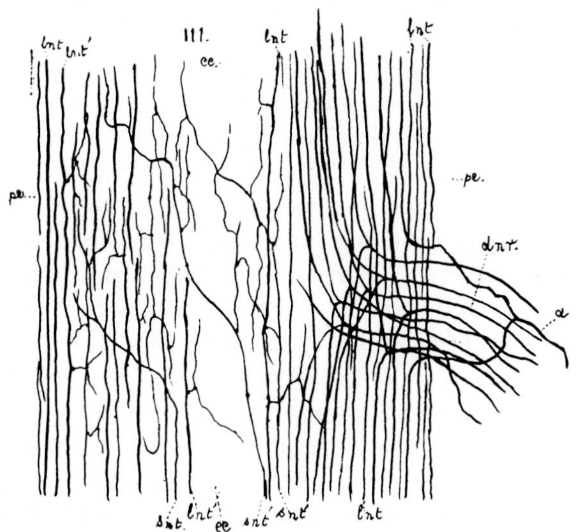

Figure 51. Horizontal longitudinal section through the spinal cord of Myxine in the region of a dorsal root as seen with *Golgi's chrome silver impregnation* (from NANSEN, 1887). This is one of the earliest illustrations showing the dichotomous branching of afferent dorsal root fibers in the posterior funiculus, as characteristic for all Vertebrates, and antedates the findings of CAJAL, who claimed (e.g. RECUERDOS, 1923, Lamina XXV, Fig. 12), that, before his own investigations, '*la bifurcación de las raices posteriores o sensitivas*' was '*desconocida de los sabios*' (cf. also Figs. 86, 108, 130, 131). d, dnr: afferent dorsal root fibers. Other designations not relevant in the aspect here under consideration.

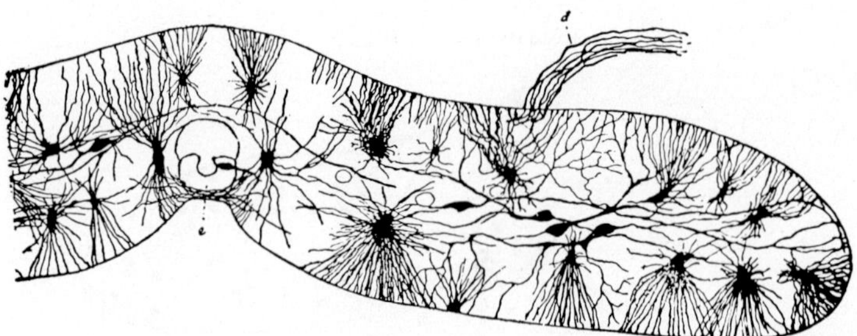

Figure 52. Cross-section through the spinal cord of Myxine, showing, in addition to some neuronal elements and an ependymal cell, various neuroglia cells of astrocytic type (after RETZIUS, from CAJAL, 1909) d: dorsal root; c: central canal.

pattern, a number of significant differences seem to obtain. Thus, KUPFFER (1906) noted that the large dorsal cells characteristic for Petromyzonts, and here already present in early ontogeny, cannot be recognized at corresponding stages of Myxinoid development. He adds that the large dorsal cells 'auch am Rückenmarke der erwachsenen Myxine nicht nachgewiesen sind'. Yet, some cells dorsal to the central canal, and slighthly larger than the neighboring elements, might possibly correspond to the dorsal cells of Petromyzonts (cf. Fig. 49).

BONE (1963), however, in agreement with NANSEN[70] and RETZIUS, observed 'large dorsal internuncial neurons', which he interprets as analogous to the *Rohde cells* of Amphioxus, and claims that their axon joins the medial longitudinal fasciculus (cf. Figs. 42, 48). It will be recalled that BECCARI (whom BONE does not mention), recorded similar internuncial elements in Petromyzonts (cf. above, p. 88). As regards efferent elements, BONE distinguishes, in addition to possibly visceromotor cells, two types of motoneurons, believed to supply 'fast and slow muscle fibers in the myotomes' respectively.

Fibers of Müller are present in adult Myxinoids, but have a much less conspicuous appearance than those of Petromyzonts (cf. Figs. 40C, 42, 47, 48). BONE (1963) prefers to speak of '*Müller-type*' fibers. This author also reports that some of these large descending

[70] It will be recalled (cf. vol. 3, part I, pp. 506, 510) that GOLGI and NANSEN maintained the hypothesis of a particular 'nutritive function' performed by the 'dendrites' of nerve cells. In view of the extensive peripheral dendritic expansions seen in Myxine (and Petromyzonts), BONE (1963) likewise expresses the opinion that, as regards many neuron types with processes passing to the spinal cord's periphery, such dendrites could be 'nutritive in function'.

fibers end in the caudal two thirds of the spinal cord with elongated 'cone-like enlargements', sometimes displaying 'numerous small processes' (cf. Fig. 48), and reminiscent of so-called 'growth cones'.

Internuncial elements and ventral root cells as interpreted by BONE (1963) are shown in Figures 42, 47, and 49. A so-called 'fibre spectrum' in the 'white matter', recorded by that author, is illustrated by Figure 50. Figure 51 shows the bifurcating dorsal root fibers with ascending and descending branches, as demonstrated by NANSEN (1887) and displaying a pattern apparently common to all Craniote Vertebrates. As regards the supporting elements in addition to those of the ependyma proper, cells of astrocytic type[71] were reported and depicted (cf. Fig. 52) in the early studies of RETZIUS (1891).

5. Selachians

In the simplified but in my opinion adequate classification adopted for purposes of comparative neurobiology in volume 1 (pp. 43–71) of this series, Selachians *sensu latiori*, synonymous with Chondrichthyes, Plagiostoma, or Elasmobranchii, are considered to represent a class with three orders, namely Squalidae (Selachians *sensu strictiori*), Rajidae, and Holocephali. From a taxonomic, and perhaps also from a phylogenetic viewpoint, Selachians may be considered the 'lowest' or 'most primitive' group of Gnathostome Vertebrates.

The spinal cord of Selachians displays, accordingly, a number of general features common to all Gnathostomes, and characterized, *inter alia*, by a rounded or oval outline in cross-sections, with a usually rather deep fissura mediana anterior. Also, the occurrence of numerous *medullated nerve fibers*, not present in Agnatha and Acrania, results in a conspicuous distinction between peripheral *white* and central *gray matter*. This latter, moreover, shows a mostly well-defined arrangement into anterior (ventral) and posterior (dorsal) 'horns' respectively columns.

---

[71] Processes of these astrocytic glial elements do not seem to form perivascular pedicles along the capillaries (BONE, 1963), although such structures have been recorded not only in Amphibians but also in Gnathostome fishes (cf. vol. 3, part I, pp. 201–205). Quite evidently, the presence of pedicles in the spinal cord of Petromyzonts would be precluded by the apparent lack of intraspinal blood vessels in these forms.

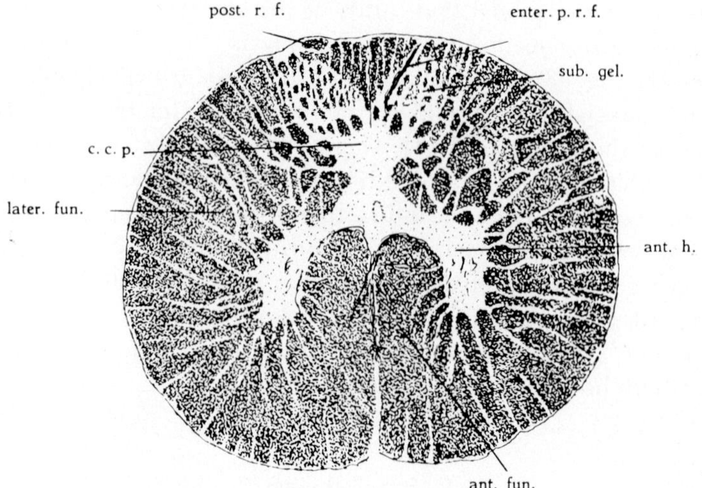

Figure 53. Cross-section through a rostral portion of the spinal cord in Hexanchus griseus (from KEENAN, 1928; myelin-stain, ×17, red. $^4/_5$). ant. fun.: anterior funiculus; ant. h.: anterior horn; c. c. p.: corpus commune posterius; enter. p. r. f.: entering posterior root fibers; later. fun.: lateral funiculus; post. r. f.: posterior root fibers; sub. gel.: *substantia gelatinosa Rolandi.*

Concerning the spinal nerve segments as typically displayed by Selachians and other Gnathostomes, the dorsal roots each include a spinal ganglion and join the ventral roots to form the bilaterally symmetric[72] metameric common spinal nerves from which several rami (e.g. dorsal, ventral, communicating, recurrent) branch off. These rami, whose particular array and development depends on segmental and taxonomic differences, were briefly dealt with in chapter VII, sections 2 and 6 of the preceding volume.

Cross-sections through the spinal cord in representative forms of the various orders of Selachians, illustrating the general configuration of the spinal cord, are shown in Figures 53 to 56. It will be seen that, despite the essentially common features, numerous variations of pattern detail obtain. With regard to the posterior horns, the gray substance dorsal to the central canal forms a common median aggregate

[72] Minor asymmetries, such as slight differences in the level of emergence of dorsal and of ventral roots, reminiscent of conditions obtaining in Cyclostomes, may occur in some Selachian forms.

which bifurcates toward the periphery into the two cornua[73] that become separated by a rather narrow 'posterior funiculus'. KEENAN (1928) designated this unpaired median griseum as the '*corpus commune posterius*'. Thus, the dorsal griseum displays here the outline of a letter Y instead of that corresponding to the upper part of an H. The dorsal cap of the cornua, of translucent or 'gelatinous' appearance in *Weigert preparations*, represents a *substantia gelatinosa* roughly similar to the homologous griseum originally noted by ROLANDO (1773–1831) in Man and other Mammals, but subsequently also described in other forms, including Fishes, by CAJAL (1909). The comparative anatomy of the *substantia gelatinosa Rolandi* and its phylogenetic development as inferred from the morphologic data, was carefully studied by KEENAN (1928, 1929) in an extensive material sampling all major gnathostome Vertebrate groups from Fishes to Mammals.

Concerning the variations in the configuration of the gray substance in the Selachian spinal cord, the differences in width of KEENAN's corpus commune posterius are conspicuous if, e.g., this griseum in Hexanchus (Fig. 53) is compared with its appearance in Acanthias (Figs. 54 A–C), Chimaera (Fig. 56), or Raja (Fig. 55). In this latter form, the 'corpus commune' proper is very narrow, perhaps mainly glial, but dorsal and ventral horns are bilaterally connected by broad grisea of reticular type, while an unpaired, similarly reticulated midline griseum interconnects the two dorsal horns. In addition, it will be seen that conspicuous medullated longitudinal fiber bundles run through the central gray, especially within its dorsal portion.

The *neuronal elements* of the Selachian spinal cord (cf. Figs. 57–59) comprise, within the ventral horn, a commonly ventrolateral group of large *somatic-efferent motoneurons*, a somewhat dorsomedial group of slightly smaller *visceral-efferent*, presumably *preganglionic cells*, and an essentially intermediate group of *commissural internuncials*[74] (cf. e.g. GLEES, 1940; v. LENHOSSÉK, 1892, 1895; KAPPERS *et al.*, 1936). In addition, the gray matter contains *funicular cells*, whose neurites descend, and to some extent perhaps ascend or bifurcate, in the white matter. At rather early ontogenetic stages, transitory primary afferent large

---

[73] '... *dorsalement, cette masse se divise en deux cornes distinctes qui se dirigent vers la surface de la moelle sous l'influence neurobiotactique des racines dorsales*' (KAPPERS, 1947).

[74] No definite data are available concerning the presence of short homolateral internuncials of *Golgi II type*, nor as regards details of the connections made by commissural cells respectively homolateral funicular cells.

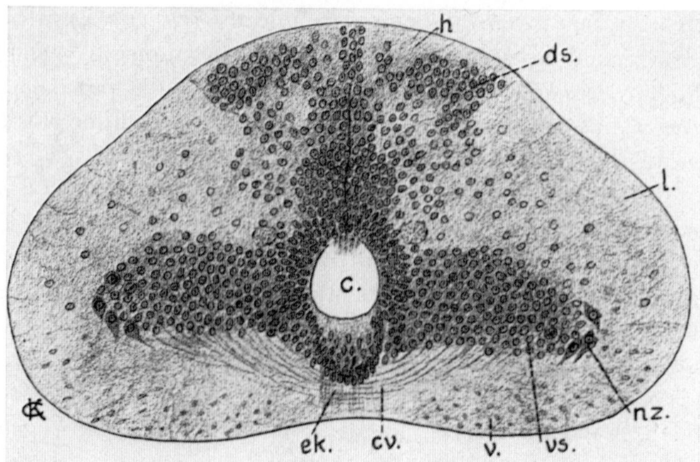

Figure 54A. Cross-section through the spinal cord (pectoral fin region) in a 70 mm embryo of Acanthias (from KUPFFER, 1906; ×150, red. $^4/_5$). c: central canal; cv: ventral commissure; ds: dorsal cell column; ek: ependymal fibers *('ventraler Ependymkeil')*; h: posterior funiculus; l: lateral funiculus; nz: ventral nerve cells (presumably motoneurons); v: anterior funiculus; vs: ventral cell column.

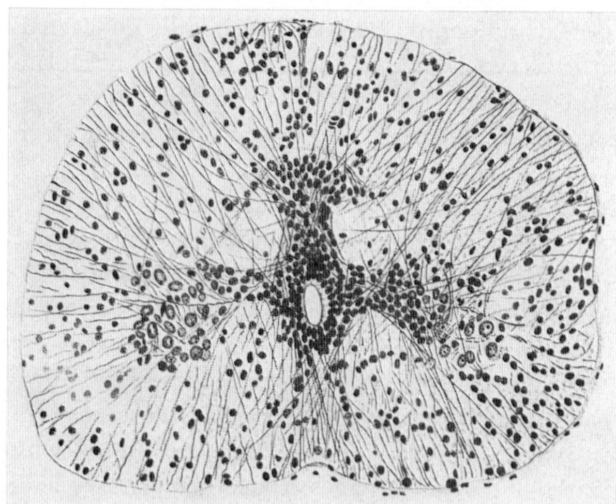

Figure 54B. Cross-section through the spinal cord of Acanthias, as seen with *Weigert's glia stain*, for comparison with Figures 54A and C (after MÜLLER, 1900, cf. Fig. 144 in vol. 3, part I of this series). Although particularly glia fibers are visualized, not only nuclei of glia cells, but also those of neuronal elements are stained, including cytoplasm of some large nerve cells. As regards the bare nuclei in the gray substance, it is not possible to distinguish those of glial and neuronal elements.

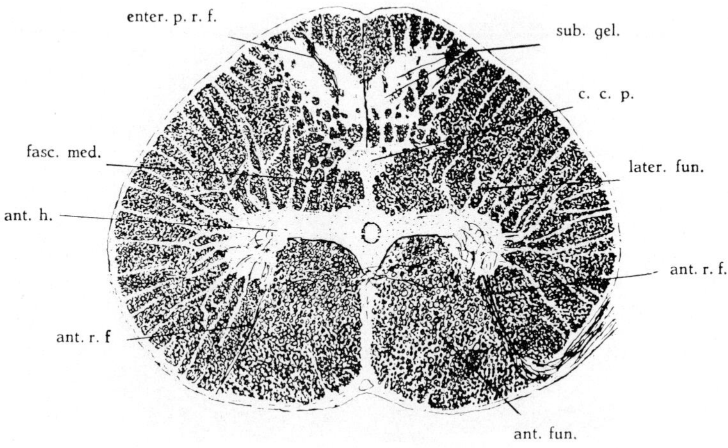

Figure 54C. Cross-section through the rostral portion of the spinal cord in Acanthias vulgaris (from KEENAN, 1928; myelin-stain, ×25, red. $^4/_5$). fasc. med.: fasciculus medianus of STIEDA; other abbreviations as in Figure 53.

dorsal *Rohon-Beard cells* are present (cf. Fig. 57). In older embryos and at the adult stage, all primary afferent neurons seem to be exclusively represented by cells located in the spinal ganglia.

As regards details, the motoneurons and at least some of the internuncial elements, especially in the ventral horns, extend dendrites through the white matter toward the periphery of the spinal cord, forming a *marginal dendritic plexus* (Figs. 58, 59). Other dendrites of motoneurons extend across the midline, particularly in the ventral commissure, and provide a decussation which has been called the '*anterior protoplasmic commissure*'. Still other dendrites run toward the dorsal horns. It can be assumed that a monosynaptic, direct reflex pathway involving dorsal root afferents and motoneurons prevails, supplemented, in a still not properly understood manner, by variously connected internuncials. The neurites of the motoneurons leave the spinal cord through the homolateral ventral roots. Their intraspinal (central) course may run in a slightly caudal direction toward the level of exit. Crossing (contralateral) efferent ventral root fibers do not seem to occur. The neurites of visceral-efferent cells chiefly join the dorsal roots, but some such efferent may also emerge through the ventral roots.

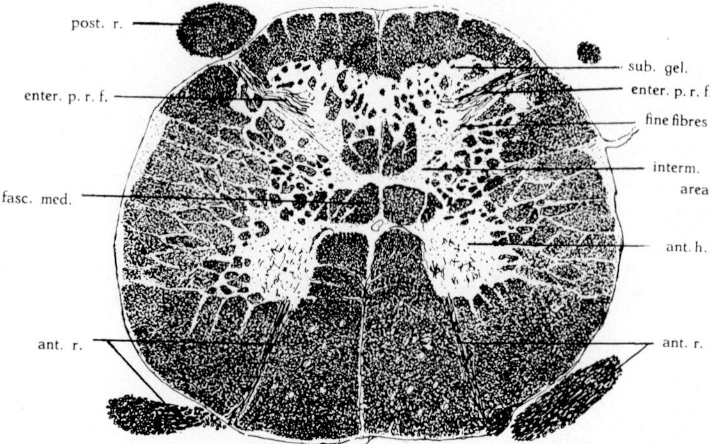

Figure 55. Cross-section through the rostral portion of the spinal cord in Raja clavata (from KEENAN, 1928; myelin-stain, ×18, red. $^4/_5$). ant. r.: extra- and intramedullary portions of anterior root; interm. area: 'intermediate area'; post. r.: extramedullary posterior root; other abbreviations as in Figures 53 and 54C.

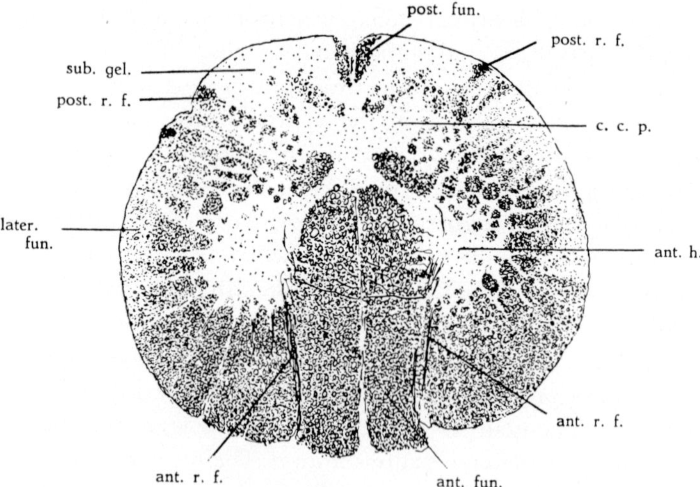

Figure 56 A. Cross-section through the cranial portion of the spinal cord in Chimaera monstrosa (from KEENAN, 1928; myelin-stain; ×25, red. $^4/_5$). post. fun.: posterior funiculus; other abbreviations as in Figures 53, 54C, and 55.

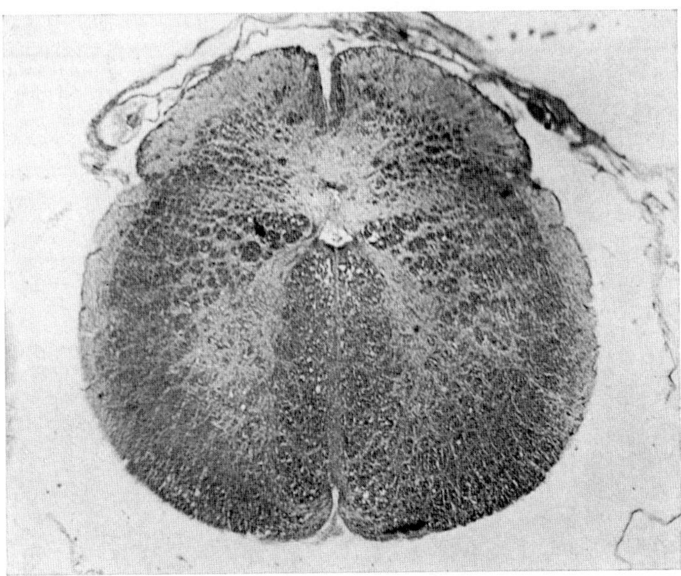

Figure 56B. Cross-section through the cranial portion of the spinal cord in Chimaera colliei for comparison with preceding figure A (hematox.-eosin stain; ×30, red. $^{4}/_{5}$).

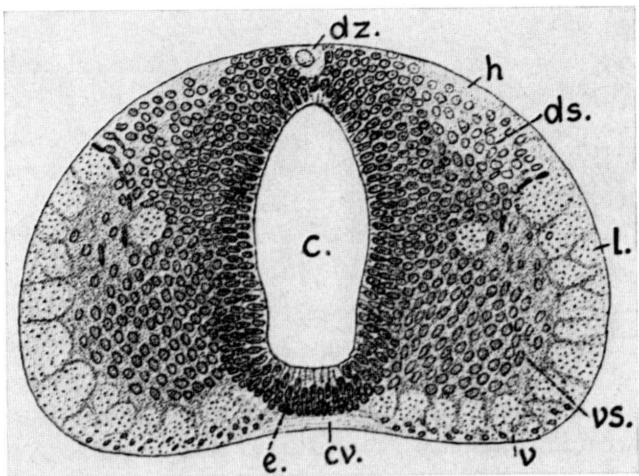

Figure 57. Cross-section through the spinal cord in a fairly early embryo of Scyllium, showing a large dorsal *Rohon-Beard cell* (from Kupffer, 1906; ×150, red. $^{4}/_{5}$). dz: *Rohon-Beard cell ('transitorische grosse Dorsalzelle');* e: ependyma; other abbreviations as in Figure 54A. This early stage of Scyllium, unspecified by Kupffer, substantially precedes that of the Acanthias embryo in Figure 54A.

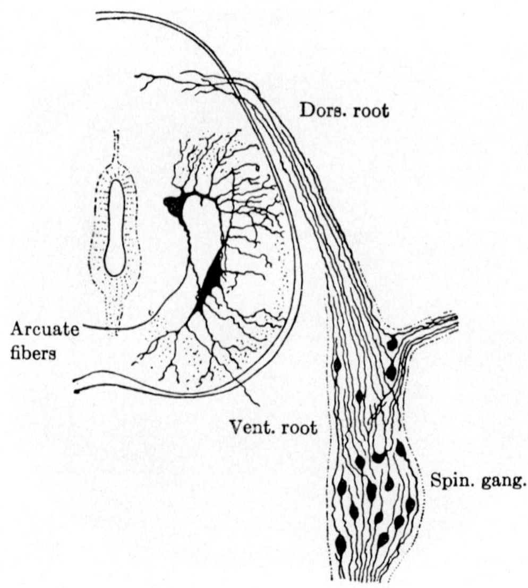

Figure 58. Cross-section through the spinal cord in an embryo of Spinax as seen with the *Golgi method* (after VON LENHOSSÉK, from KAPPERS *et al.*, 1936). The figure shows, besides spinal ganglion and dorsal root fibers, 'cells of origin of arcuate fibers and ventral horn neurons, with peripheral dendritic branches'.

The *supporting* or *accessory cellular elements* of the spinal cord include ependyma, astrocytes, oligodendroglia, and *Hortega cells.*[75] The *Weigert technique* for the staining of neuroglia shows the presence of the characteristic 'glia fibers' displayed by this method in the neuraxis of all Craniote Vertebrates (Fig. 54B)

With respect to the *intramedullary fiber systems*, the dorsal roots join the spinal cord near the apex of the posterior horn and invade the gray substance, spreading out in fan-like fashion (Fig. 59). The dorsal component of the 'fan' displays fine fibers which seem to end 'locally' or segmentally in the '*substantia gelatinosa Rolandi*' and may include 'visceral afferents' (KAPPERS *et al.*, 1936). The perhaps predominantly 'somatic-afferent' ventral component, consisting of coarse fibers, makes contacts with the dorsally directed dendrites of motoneurons, and, moreover, bifurcates into ascending and descending branches. Some such

[75] Although perhaps somewhat less differentiated than in the 'higher' forms, perivascular feet or pedicles are nevertheless present. The comparative histology of glial elements in the Vertebrate series was dealt with in chapter V (vol. 3, part I, pp. 200–211).

root fibers may also take an either exclusively descending or ascending course. The intramedullary longitudinal primary dorsal root fibers run mostly within the gray of the posterior horns, forming 'intracornual' bundles of variable thickness and number whose collaterals or terminal endings have not been observed. It seems, nevertheless, that the ascending course of the root fibers is rather short, so that, at best, few may reach the lower oblongata which lacks clearly recognizable grisea comparable to the dorsal funicular nuclei of higher forms such as Mammals (BROUWER, 1915). WALLENBERG (1907), nevertheless, believes that a rudimentary primordium of these nuclei can be identified in the Selachian oblongata. The narrow posterior funiculi *sensu strictiori*, located on both sides of the midline between the apices of dorsal horns, seem essentially to be formed by descending (and perhaps a few ascending) secondary fibers originating as neurites of funicular cells (v. LENHOSSÉK, 1892). Other dorsal and ventral cells of this type send their neurite into lateral and ventral funiculus. A 'commissura proto-

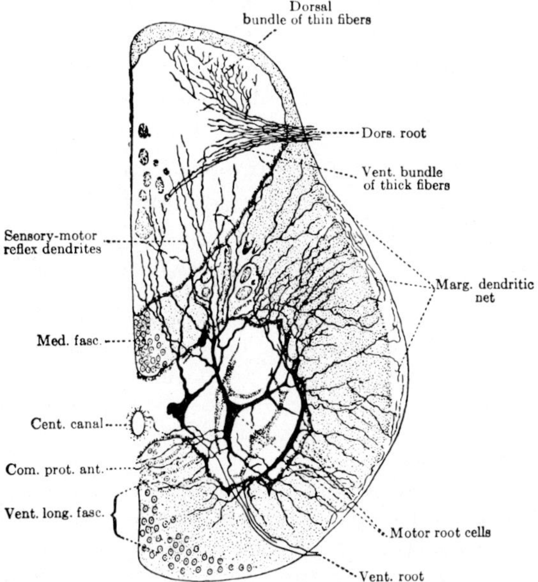

Figure 59. Cross-section through the spinal cord in an adult Ray (Raja, cf. Fig. 55), as seen with the *Golgi technique* (after VON LENHOSSÉK, from KAPPERS *et al.*, 1936). 'Demonstration of the dorsal root, ventral horn cell reflex in a ray. The dendrites of the ventral horn cells branch in the gray substance of the dorsal horns'.

plasmica dorsalis', posterior to the central canal, is said to be formed by dendrites of dorsal funicular cells (KAPPERS *et al.*, 1936).

In contradistinction to the funicular cells *sensu strictiori*, whose neurites remains homolateral, the commissural cells give origin to neurites representing 'arcuate fibers' which cross the midline in the anterior commissure.[76] Most of these fibers which, after bifurcating, descend and ascend in the lateral funiculus, presumably pertain to internuncial connections of the *eigenapparat*. However, some of the ascending components, especially those from the more cranial segments, seem to reach the oblongata and perhaps even the tectum mesencephali (KAPPERS *et al.*, 1936; WALLENBERG, 1907). These fibers, evaluated by KAPPERS as 'the first sensory tract of the cord', and as being analogous to EDINGER's spino-tectal tract in Mammals, would thus be part of the *verbindungsapparat*.[77] A spino-cerebellar component of this extrinsic system comparable to FLECHSIG's dorsal spino-cerebellar tract might be provided by funicular cells restricted to the dorsal grisea of the spinal cord, and reaching the corpus cerebelli (KAPPERS *et al.*, 1936; KAPPERS, 1947). Concerning the *ascending components* of the *verbindungsapparat*, these spino-bulbo-tectal and spino-cerebellar tracts, together with some cranialward directed posterior root fibers reaching the oblongata, as mentioned further above, represent all such components about which a modicum of acceptable evidence has been obtained.

With regard to the *descending channels* of the extrinsic apparatus, one of the most conspicuous components is the *fasciculus longitudinalis medialis* which constitutes the bulk of the anterior funiculus (cf. Figs. 53, 54C, 55, 56). Although lacking 'giant axons' comparable to *Müllerian fibers* of Cyclostomes or to the *Mauthner fibers*[78] of Osteichthyes, Dip-

---

[76] In addition to the main anterior commissure ventral to the central canal, a somewhat more peripheral commissura accessoria or *commissure of Mauthner* is occasionally recognizable (cf. e.g. KAPPERS *et al.*, 1936). The dendritic 'commissura protoplasmica anterior' predominantly within the 'main commissure' may be a component of both. Again, the dendritic 'commissura protoplasmica dorsalis', mentioned above, may or may not include a few crossing neurites.

[77] Although this system seems '*peu développé*' in Selachians, KAPPERS (1947) assumes, on the basis of experimental evidence obtained by TEN CATE, '*qu'il contienne des fibres de tous les niveaux de la moelle*'. On the other hand, assuming that only the more cranial segments of the spinal cord give origin to spino-bulbar and spino-tectal-thalamic fibers, the relevant neural impulses from more caudal segments could also be transmitted to that channel by chains of spinal internuncials.

[78] *Mauthner cells* and fibers may occur as perhaps transitory elements in some Selachian embryos and possibly as occasional rare exceptions or variations in adult forms (cf. also K. and NIIMI, 1969, and chapter IX of this volume).

noans and Amphibians, it contains a number of rather large medullated axons and represents an essentially reticulo-spinal tract. Its fibers are partly (perhaps predominantly) crossed as well as partly uncrossed and originate from the large nerve cells of the reticular formation. This griseum extends, within the deuterencephalic basal plate, from the mesencephalon to the oblongata's caudal boundary neighborhood *(n. reticularis seu motorius tegmenti* of L. EDINGER, 1908, 1912). A relevant investigation of its reticular cells and of the appurtenant fiber tracts in the brain stem of an Acanthias species was undertaken by my associate SAITO (1930) following his study of the *Müllerian cells* in a Cyclostome (1928). Descending vestibular fibers of the second order, originating in the magnocellular lateral vestibular nucleus *(Deiters' nucleus)* and predominantly (or entirely) crossed, are included in the fasciculus longitudinalis medialis and presumably correspond to KAPPERS' tractus octavo-motorius cruciatus, perhaps also to WALLENBERG's tractus octavo-spinalis lateralis cruciatus, located in the ventrolateral funiculus. According to KAPPERS *et al.* (1936), the tractus octavo-motorius cruciatus may be accompanied by fibers from the cerebellum ('tractus cerebello-motorius cruciatus et rectus', whose extension into the spinal cord is, however, rather dubious). It is likewise uncertain whether the crossed and partly uncrossed tectobulbar system extends further caudalward as a tectospinal tract.

Regardless of its special significance as terminal (primary afferent) nucleus of a cranial nerve, and as a derivative of the alar plate, *Deiters' nucleus* is closely associated with the reticular formation by its fiber connections and in respect to some of its functional activities. Other descending components of the *verbindungsapparat* are fairly large uncrossed vestibular fibers included in the fasciculus medianus of STIEDA (1873) located in the latter funiculus on both sides of the narrow 'corpus commune posterius' or median gray stalk of the posterior horn (Figs. 54C, 55). These fibers are believed to originate in the 'ventral vestibular nucleus' of KAPPERS and may also be accompanied by some direct (primary or root) fibers of the vestibular nerve (KAPPERS *et al.*, 1936); moreover, STIEDA's fasciculus presumably contains a modicum of spinal funicular fibers pertaining to the *eigenapparat*.

The observations of DAHLGREN (1914) and of SPEIDEL (1919, 1922) disclosed that, in addition to the ordinary neuronal elements with their interconnections, and to the typical supporting elements such as ependymal and glial cells, a peculiar sort of very large, commonly polynuclear cells of 'glandular' or neurosecretory category occurs in the cau-

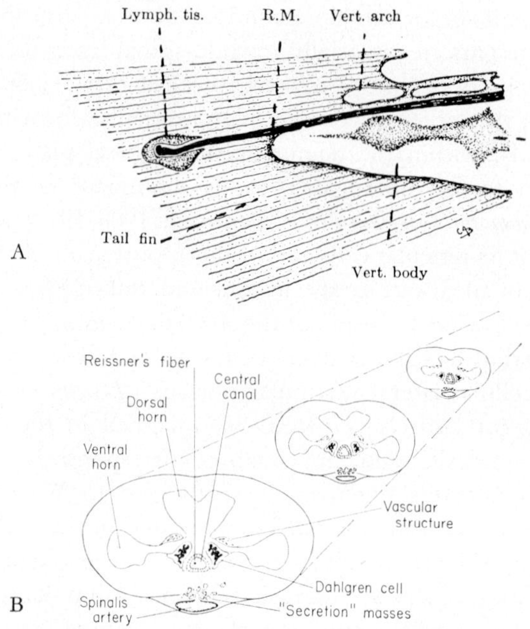

Figure 60A. Posterior end of the spinal cord in the tail fin of a Shark, surrounded by lymphoid tissue (after STERZI, from KAPPERS *et al.*, 1936). R. M.: postchordal spinal cord.

Figure 60B. Diagram depicting posterior part of spinal cord in the Elasmobranch Rhinobatos productus and showing *Dahlgren cells* similar to those described by SPEIDEL in Raia (from BERN and HAGADORN, 1959).

dal part of the spinal cord of Rajidae[79] and Squalidae. In the tail of these latter, moreover, the spinal cord extends beyond the end of the vertebral canal into the tail fin proper as a 'postcaudal cord', whose extremity is surrounded by lymphoid tissue (Fig. 60A).

With regard to Squalidae,[80] 'large irregular glandular cells' were found by SPEIDEL (1922) in Mustelus, Acanthias, and Carcharias. With

[79] DAHLGREN (1914) interpreted these cells as pertaining to the Ray's electric organ, but SPEIDEL (1919) demonstrated their secretory nature and gave thereby the first description of a neurosecretory cell.

[80] No report on presence or absence of these cells in Holocephalians have come to my attention. Although I had intended to undertake a check on my own unprocessed Chimaeroid material, circumstances prevented me from proceeding with that work. Again, it should be mentioned that SPEIDEL (1922) examined, with negative results, the spinal cords of the Cyclostome Petromyzon, and of the urodele Amphibians Diemyctylus and Necturus in search of *Dahlgren cells*.

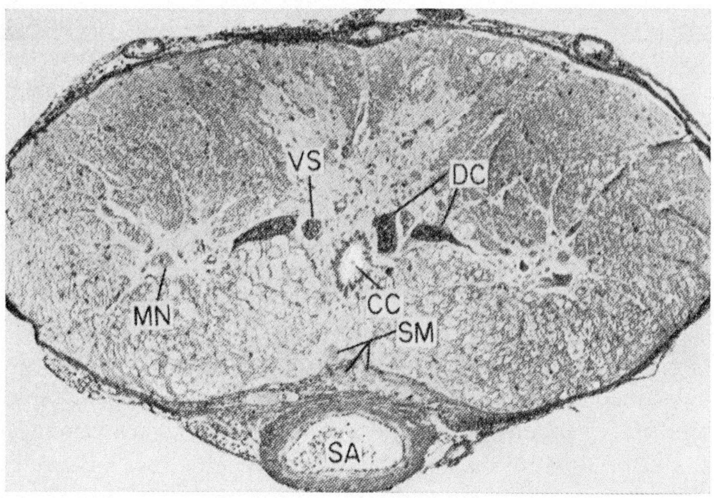

Figure 60C. Cross-section through posterior part of spinal cord in male specimen of Rhinobatos, showing Dahlgren cells (from BERN and HAGADORN, 1959). CC: central canal (containing *Reissner's fiber;* DC: *Dahlgren cell;* MN: motoneuron; SA: anterior spinal artery; SM: 'secretion masses'; VS: vascular structure).

respect to Rajidae, this author, confirming DAHLGREN's findings in the Skate, Raia, also noted such cells in Torpedo.

The largest 'glandular cells' were recorded in Raia, reaching the relatively enormous size of $300 \times 200 \times 176\ \mu$. In Raia, these cells are present in the anterior horn on each side of the central canal, but with rather variable positions, and restricted to a region of spinal cord located in the posterior part of the tail.[81] The *Dahlgren-cells* (Figs. 60B, C) are now interpreted by various investigators as providing a 'major neurosecretory system' situated at the caudal end of the spinal cord in various Plagiostomes and Teleosts (urohypophysis or neurohypophysis spinalis caudalis) and representing a 'counterpart of the neurohypophysis' (ENAMI, 1959). Further remarks on the caudal neurosecretory system of Teleosts are included in section 6 of this chapter. In Plagiostomes, a so-called neurohemal organ, as connected with the urohypophysis of Teleosts, does not seem to be differentiated, and the short processes of the Elasmobranch *Dahlgren cells* appear to terminate diffusely amont the capillaries in the spinal cord's ventral part.

---

[81] In Raia occellata this part extends from the level of the 64th vertebra to that of the 120th vertebra at the tip of the tail.

The innervation of the electric organs located in the tail of Rajidae by spinal cord cells of motoneuron type was discussed in section 5, chapter VII, volume 3, part II of this series. It will also be recalled (cf. footnote 79 of the present section), that the neurosecretory cells discovered by DAHLGREN were originally interpreted by this author as pertaining to the electrical organ system.

6. Ganoids and Teleosts; Crossopterygians: Latimeria

The large and diversified group of Gnathostome Fishes now generally subsumed under the concept *Osteichthyes* presents considerable difficulties for a generally valid and not unduly awkward classification which can be used by the neurobiologist concerned with the relevant aspects of comparative anatomy. Following the taxonomical outline elaborated *ad hoc*[82] in chapter II of volume 1, Osteichthyes are here arbitrarily conceived as a class consisting of the subclass *Ganoids* with three orders, and of the combined subclass and order *Teleosts*, with numerous sub-orders and families. For reasons to be explained further below, a brief discussion of the spinal cord in the 'Crossopterygian' Latimeria will be appended to the present section, preferably to being included in the following section 7, dealing with Dipnoans.

Concerning its overall features, and as repeatedly pointed out by comparative neurologists (e.g. KAPPERS, 1920) the spinal cord of Ganoids is rather similar to that of Selachians. Although, in Teleosts, a

[82] The inclusion of a simplified but reasonably valid, operationally useful outline of taxonomy in volume 1, 'Propaedeutics to Comparative Neurology' was in my opinion appropriate if not outright necessary, because I noticed a widespread and at times disconcerting unfamiliarity with relevant concepts of classification, as well as fundamental zoological data, among advanced neurologists and neurobiologists taking up questions of comparative anatomy or phylogeny. In the course of specialization, many important elementary notions are forgotten or become discarded, and apparently very few scientists will subsequently take the time for a review of the scattered basic data. Sophisticated taxonomists, on the other hand, are as a rule, quite unfamiliar with the meaningful configurational aspects of the neuraxis, which become neglected in the elaborations of their intricate systems. An amusing review on historical developments and on the shortcomings of taxonomy was recently published in 'Science' (RAVEN, P. H. *et al.*: The origins of taxonomy. A review of its historical development shows why taxonomy is unable to do what we expect of it. Science *174:* 1210–1213, 1971). BECCARI (1943, p. 41, footnote) likewise comments on the inherent difficulty in formulating a suitable taxonomic system for the classification of Fishes: '*problema che ciascun zoologo risolve a suo modo*'.

much wider variety of differences in pattern and in certain secondary structural aspects of the spinal cord is manifested, the general similarity of its teleostean organization with that obtaining in Selachians still remains evident.[83]

In Ganoids and 'primitive' Teleosts, the spinal cord extends through the whole length of the vertebral canal. However, with regard to the relationship between length of spinal cord and vertebral canal, briefly discussed in the first section of this chapter, a considerable discrepancy in the ratio of these longitudinal linear dimensions is displayed by some Teleosts such as Lophius piscatorius and Orthagoriscus mola.

In Lophius, the posterior part of the spinal cord is reduced to a fairly thin filum terminale ending as a fine fiber and surrounded, particularly on lateral and ventral side, by densely packed, longitudinally running ventral roots, which form a rather massive cauda equina. Since the union between the dorsal and ventral roots occurs outside the vertebral canal and since the spinal ganglia are also externally located, the elongated portions of the dorsal roots within the canal consist only of the central connections between ganglia and cord (KAPPERS *et al.*, 1936).

In Orthagoriscus mola, the spinal cord appears still more shortened, with a corresponding reduction in the number of spinal nerve segments, of which BURR (1928) recognized 21. The first two, consisting of ventral roots,[84] were evaluated as possibly spino-occipital nerves and three additional spinal rootlets showed connections with other caudally adjacent segments, thus leaving only 16 typical spinal nerves on each side (BURR, 1928; KAPPERS *et al.*, 1936). Figure 61 shows the difference in relative length of the spinal cord obtaining between the Plectognathe Orthagoricus mola and another Teleost, namely the Acanthopterygian Trigla hirundo.

It is, moreover, of interest that the rostralmost portion of the spinal cord in some Ganoids (Polyodon, Lepidosteus, Acipenser and Amia)

---

[83] BECCARI (1943) deals in a short section (4., pp. 40–45) with the spinal cord of Gnathostome Fishes in general and states: '*Assai differente per forma e costituzione appare il medullo spinale nei vasta classe dei Pesci, e notevoli difference esistono tra i Selaci ed i Teleostomi. Tuttavia non riesce difficile enumerare alcuni caratteri generali che si rinvengono in tutte le specie.*' In my elementary introductory '*Vorlesungen*' of 1927, I had followed a similar procedure.

[84] Although, as can be seen from BURR's illustrations and description, these first two pairs consist obviously of ventral roots, they are designated by BURR (1928) and KAPPERS *et al.* (1936) as 'dorsal components of the first spinal nerve'.

is surrounded by a peculiar, darkly pigmented, oblong and spindle-shaped mass of intrameningeal tissue (Fig. 62A). It was described as the 'myelencephalic gland' by CHANDLER (1911) in Lepidosteus, since this formation may extend beyond the spinal cord into the meninges surrounding the caudal oblongata. VAN DER HORST (1925) examined this 'myelencephalic gland' in Polyodon, Acipenser, and Amia. On the basis of its histologic structure (Figs. 62B, C), somewhat similar to that

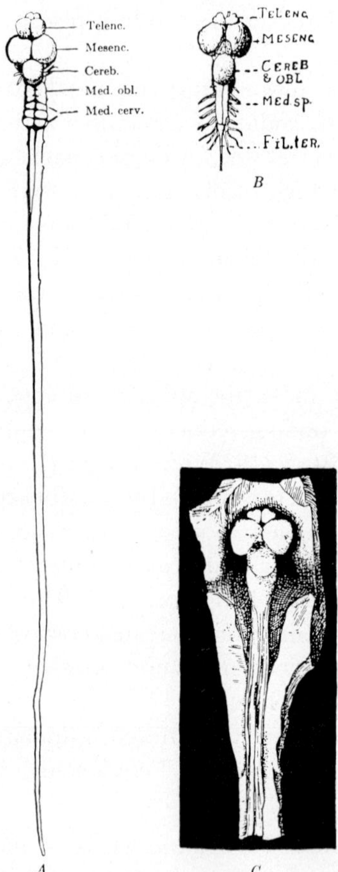

Figure 61. Comparison of the neuraxis in the Teleosts Trigla hirundo and Orthagoriscus mola, illustrating the different relative length of spinal cord (from KAPPERS *et al.*, 1936, A and C after KAPPERS, B after BELA HALLER). A: Trigla hirundo; B: Brain and spinal cord of Orthagoriscus after removal of cauda equina; C: Brain and spinal cord of Orthagoriscus *in situ*, including the cauda equina.

of the spleen, said gland can be considered to consist of myeloid hematopoietic tissue as more recently recorded and studied in Ganoids by SCHARRER (1944); it was dealt with in chapter VI, section 7, p. 674 of volume 3, part II.

Figures 63–67 and 69–71 illustrate the configuration of gray and white substance and the overall outline displayed by cross-sections through the spinal cord in several representative adult Ganoids and Teleosts. For comparison, Figures 62D and 68A–C show a few typical ontogenetic stages.

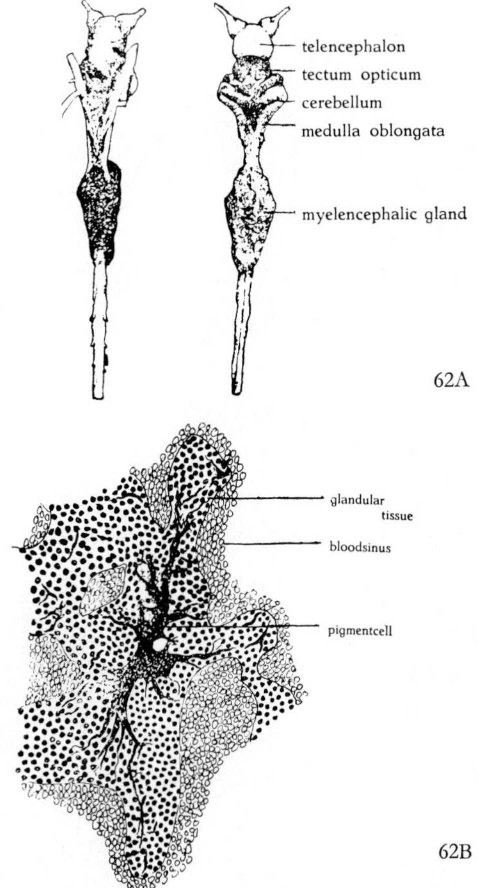

Figure 62A. Brain and rostral part of spinal cord in the Ganoid Polyodon as seen in ventral (left) and dorsal view (from VAN DER HORST, 1925).

Figure 62B. Myeloid tissue, blood sinuses, and pigment cell in the 'myelencephalic gland' of the Ganoid Polyodon (from VAN DER HORST, 1925).

In both Ganoids and Teleosts the corpus commune posterius of KEENAN displays considerable variations in width and dorso-ventral length not only between corresponding regions of different taxonomic forms, but also between levels of one and the same cord, as well as individual variations within a given species. The Ganoid Lepidosteus is particularly mentioned by KEENAN (1928) because of its marked resemblance to the Selachians Spinax niger and Acanthias vulgaris in the arrangement of the corpus commune posterius, whose attachment to the anterior horns is reduced to a narrow strip.

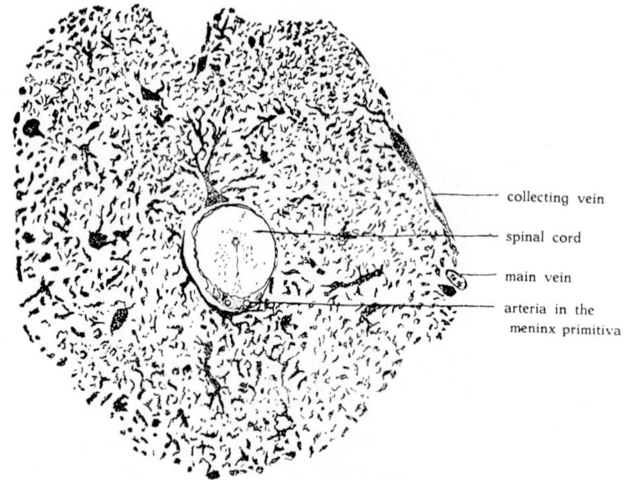

Figure 62C. Cervical spinal cord of Polyodon surrounded by the 'myelencephalic gland'. Only the 'blood cavities' of the 'gland' are indicated (from VAN DER HORST, 1925).

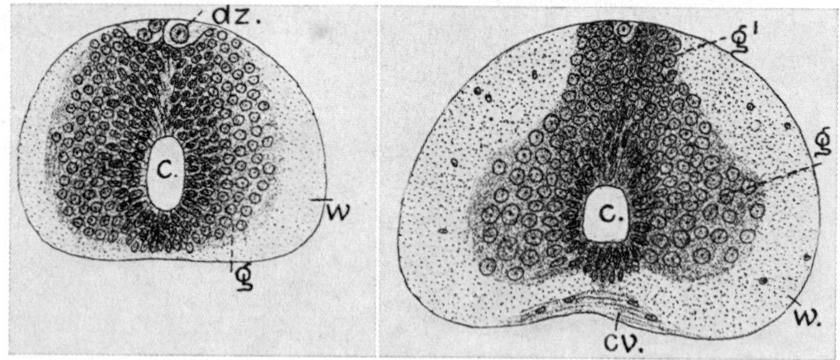

Figure 62D. Cross-sections through the embryonic spinal cord of the Ganoid Acipenser sturio (from KUPFFER, 1906; approx. ×250, red. 4/5). Left: larva of 3 days; right, larva at 9th day; c: central canal; cv: commissura ventralis; dz: dorsal Rohon-Beard cell; g: 'gray substance'; g′: primordium of dorsal horn; w: 'white substance'.

The substantia gelatinosa of Ganoids and Teleosts which is easily distinguished from the adjacent gray matter, and commonly occupies the entire dorsal 'projection' or 'horn' pattern in some Teleosts, but also to some extent in Ganoids, may be characterized by small gray masses which can be partly or completely surrounded by medullated fibers, thus presenting the appearance of an aggregate of several small 'nuclei'. The fibers frequently encircle these latter in a concentric manner. In certain Teleosts, hilus-like (or hilum) arrangements are present, through which fibers can be seen to pass. In other instances the substantia gelatinosa is folded around a central core of fibers. The griseum displays here 'clumps' which, themselves, are not surrounded by medullated fibers. Since, in Gnathostome Fishes, the gelatinous substance forms practically the entire dorsal horns, the body of these latter, as seen in Mammals, is not displayed by said Anamnia. Except perhaps for a small primordium adjacent to the substantia gelatinosa, it appears here as represented only by the corpus commune posterius (KEENAN, 1928). Particularly in some Teleosts, the dorsal portion of the spinal cord becomes substantially enlarged in the region of gradual transition to medulla oblongata and may here display conspicuous swellings. The typical outline of the spinal cord as generally seen in cross-sections can become greatly modified by these lobular protrusions (cf. Fig. 67) whose relationship to dorsal roots, intramedullary or supramedullary dorsal giant cells, radix spinalis trigemini and other neuraxial components shall be pointed out further below.

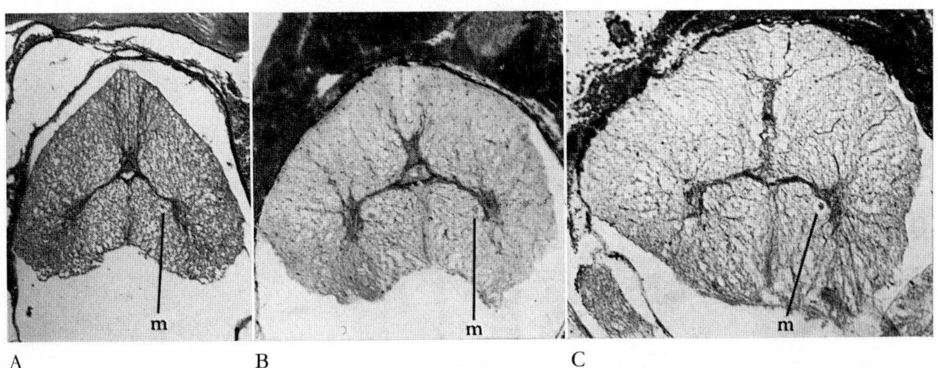

Figure 63. Three cross-sections through rostral and intermediate regions of the spinal cord in the Ganoid Acipenser sturio (hematoxylin-eosin, approx. magn.: A ×20, B ×30, C ×40, all red. $^{3}/_{5}$). m: *Mauthner's fiber.*

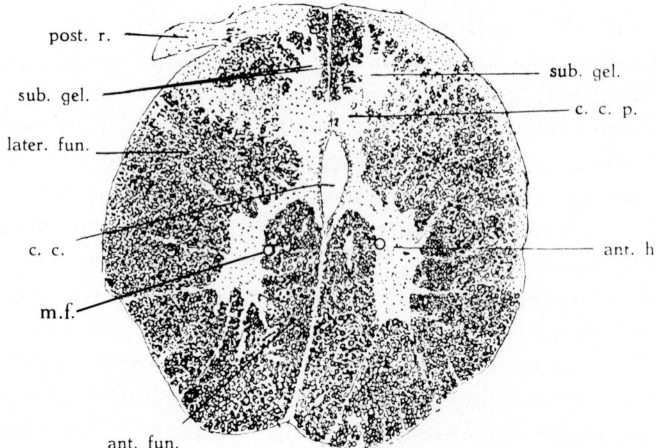

Figure 64. Cross-section through rostral portion of spinal cord in the Ganoid Polyodon folium (from KEENAN, 1928; myelin stain, ×90, red. 4/5). ant. fun.: anterior funiculus; ant. h.: anterior horn; c. c.: central canal; c. c. p.: corpus commune posterius; later. fun.: lateral funiculus; m.f.: *Mauthner's fiber;* post. r.: posterior root; sub. gel.: substantia gelatinosa. *Mauthner's fiber*, not identified by KEENAN, has been added on the basis of my own observations.

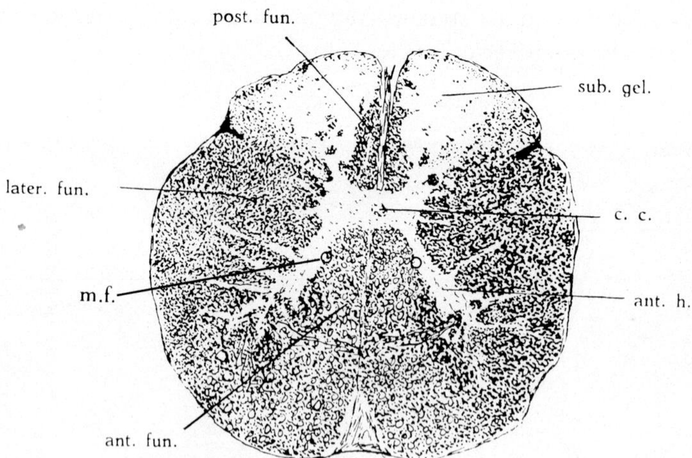

Figure 65. Cross-section through rostral portion of spinal cord in the Ganoid Amia Calva (from KEENAN, 1928; myelin-stain, ×37, red. 4/5). post. fun.: posterior funiculus; other abbreviations as in preceding figure; *Mauthner's fiber*, not identified by KEENAN, has been added on the basis of my own observations.

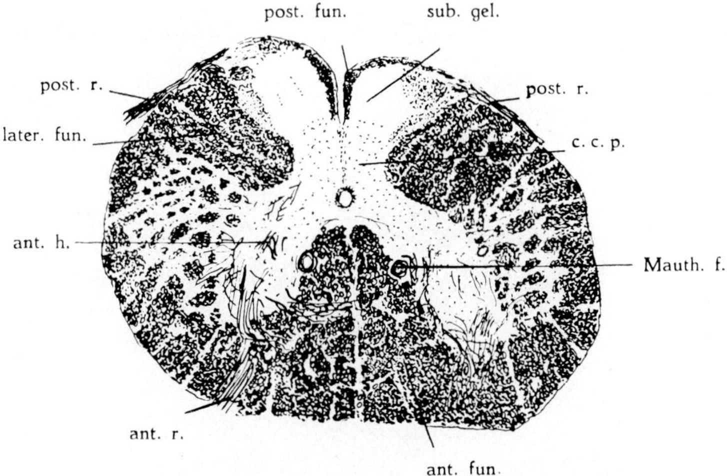

Figure 66. Cross-section through the rostral portion of the spinal cord in the Teleost Esox lucius (from KEENAN, 1928; myelin stain; ×76, red. $^4/_5$). Mauth. f.: *Mauthner's fiber;* other abbreviations as in Figures 64 and 65.

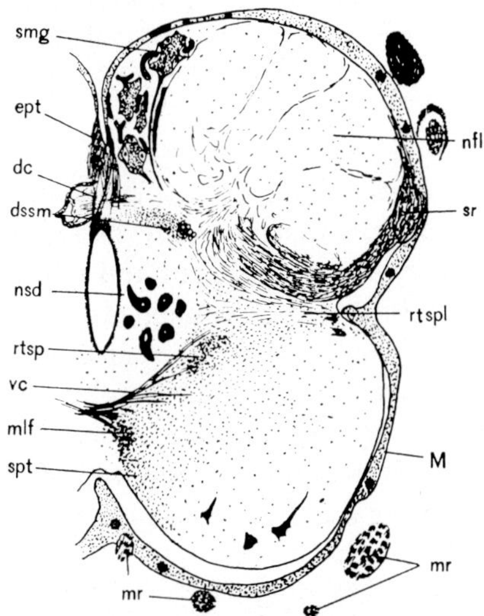

Figure 67A. Cross-section through a rostral portion of the spinal cord in the Teleost Orthagoriscus mola, showing dorsal swelling and large 'supramedullary cells' (from BURR, 1928). dc: dorsal commissure; dssm: descending fibers of supramedullary cells; ept: ependymal thickening; M: meninges; mlf: fasciculus longitudinalis medialis; mr: ventral root; nfl: enlarged dorsal griseum of cord; nsd: 'nucleus spinalis dorsalis'; rtsp: 'tractus reticulo-spinalis'; rtspe: 'tractus reticulo-spinalis lateralis'; smg: 'supramedullary' ganglion cells; spt: 'tractus spino-thalamicus'; sr: dorsal root; vc: ventral commissure.

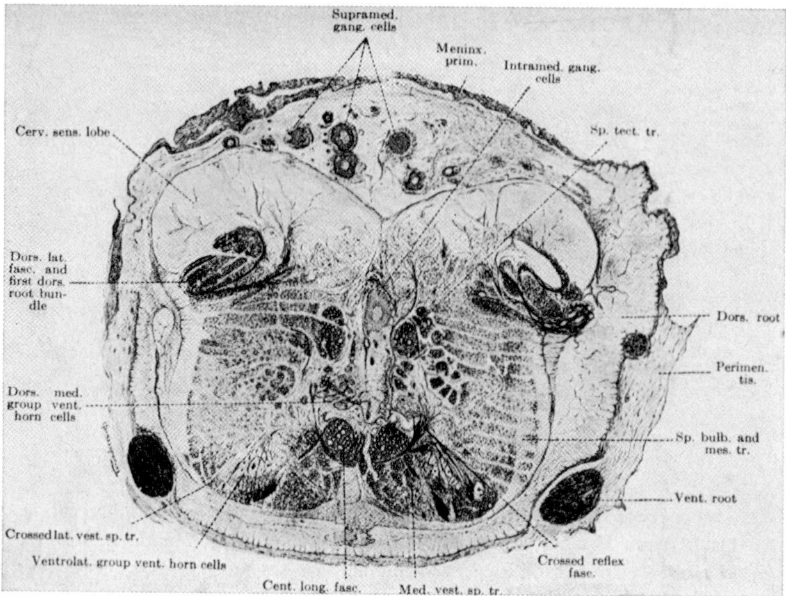

Figure 67B. Cross-section through the rostral portion of the spinal cord in the Teleost Lophius piscatorius with true supramedullary (extramedullary) ganglion cells (from KAPPERS *et al.*, 1936). In the cord's midline, some remnants of the almost obliterated central canal can be recognized.

Concerning the *neuronal elements* of the spinal cord, two groups of large and medium-sized cells are commonly distinguishable in the ventral horn, namely a ventrolateral aggregate and a dorsomedial or dorsal one (cf. Figs. 68C, 69). Both groups contain large motoneurons whose axons emerge through the homolateral ventral roots. According to KAPPERS (1920) the ventrolateral motoneurons, particularly developed in the cervical region, may innervate the musculature of the pectoral fin, while the dorsal group *(gruppo fundamentale delle cellule motrici* of BECCARI) would innervate the trunk musculature.[85] Collaterals arising

[85] BECCARI (1943) justly points out that the available evidence is not sufficiently conclusive, but nevertheless considers KAPPERS' hypothesis as not implausible. KAPPERS' original statement of 1920 is reversed (perhaps inadvertently) in the American edition (KAPPERS *et al.*, 1936, pp. 170–172): 'two groups of efferent cells are clearly distinguishable in the gray matter of the upper cervical cord: a so-called dorsomedial motor and a ventral or ventrolateral group (Figs. 80 and 81). The former group is presumably concerned with the innervation of the cervical fins; the latter group, which extends throughout the body and which frequently consists of somewhat smaller cells, is probably for motor fibers to trunk musculature.' Figure 80 refers here to the spinal cord of Salmo

from motoneurons have been described as turning back into the gray substance.[86]

Some of the medium-sized cells, particularly in the dorsal group, presumably represent visceral efferent (preganglionic) cells. Their axons seem to emerge partly through the dorsal and partly through the ventral roots. In addition, some of the large and medium-sized dorsal group cells appear to be 'associative', namely 'funicular' or 'commissural' elements (cf. Fig. 68C). Although the dendrites of motoneurons and other elements extend with their branches into the white substance, the marginal plexus of Ganoids and Teleosts seems generally less developed than that of Plagiostomes. As in these latter, however, a dendritic commissura ventralis is commonly found in Osteichthyes. Motoneuron dendrites directed toward the entering dorsal root fibers are presumably concerned with monosynaptic sensory-motor reflexes. The neurons of the spinal cord innervating the *electric organs* of Osteichthyes such as Malapterurus and Gymnotus represent specialized ventral horn efferent elements. These electric organs and their innervation were dealt with in chapter VII, section 5 of the preceding volume. Figure 71B illustrates the large 'electromotor' cells located at fairly caudal levels of the spinal cord, and innervating the electric organs in the Mormyrid Gnathonemus. Although presumably representing modified 'somatic efferent' anterior horn motoneurons, their relatively dorsal position is conspicuous. This might be due to a secondary, neurobiotactic shift related to the dorsolateral position of the fiber tract originating in the bulbar '*noyau de commande centrale*' and controlling the electric discharges.

With regard to the internuncial nerve cells in the alar plate derivatives located dorsally to the basal plate component which provides the ventral horns, only scattered and inconclusive data are available. Some of these internuncials can be interpreted as funicular, respectively commissural cells whose neurites join the homolateral or contralateral funiculi. Besides the small and medium-sized cells in the substantia gelatinosa, a few somewhat larger multipolar cells can be seen in the dorsal horn or in corpus commune posterius.

---

fario, while Figure 81, reproduced as Figure 67B in the present treatise, depicts the spinal cord of Lophius.

[86] The conflicting but not particularly enlightening views on this question are discussed by Kappers *et al.* (1936) on pp. 171–172. The topic, nevertheless, is of some interest with respect to feedback activities including the *Renshaw-effect* studied in Mammals and discussed in section 2 (p. 38) of this chapter.

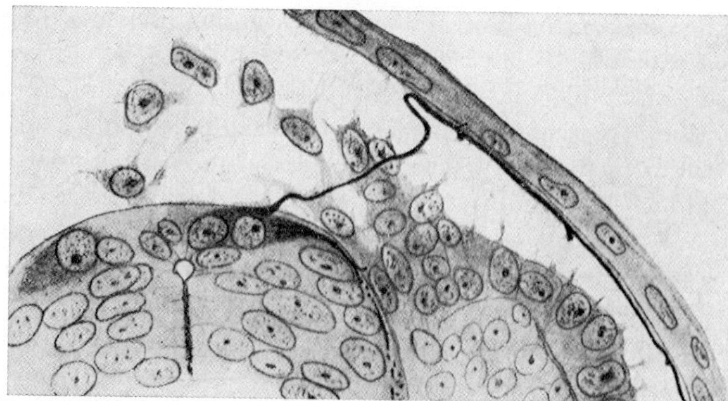

Figure 68 A. Transitory *Rohon-Beard cells* in the spinal cord of a Trout embryo (Teleost Salmo fario) of about 19 days. Drawing combined of several sections, showing a peripheral (afferent) neurite to skin and a central neurite with lateral course in the spinal cord (from HELD, 1909). NEAL and RAND (1936) observed, in a Selachian, an afferent collateral innervating the myotome.

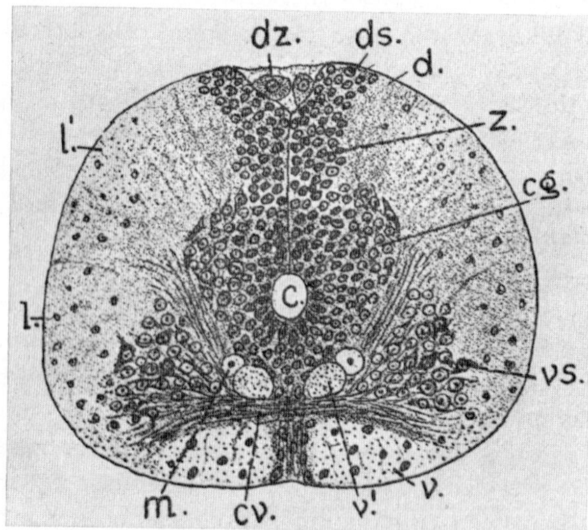

Figure 68B. Cross-section through the spinal cord of a 20 mm Trout larva (Salmo fario) at the level of the pectoral fins, and showing transitory Rohon-Beard cells (from KUPFFER, 1906). cg: 'central gray'; cv: ventral commissure; d: dorsal part of lateral funiculus *('Hinterstrang')*; ds: dorsal horn; dz: large dorsal cells; l: lateral funiculus with thin fibers *('feinfaseriger Teil')*; l': lateral funiculus with coarse fibers *('grobfaseriger Teil')*; m: *Mauthner's fiber;* v, v': dorsal and ventral parts of anterior funiculus; vs: anterior horn; z: *'Zwischenstück der grauen Substanz'* (subsequently designated as 'corpus commune posterius' by KEENAN, 1928).

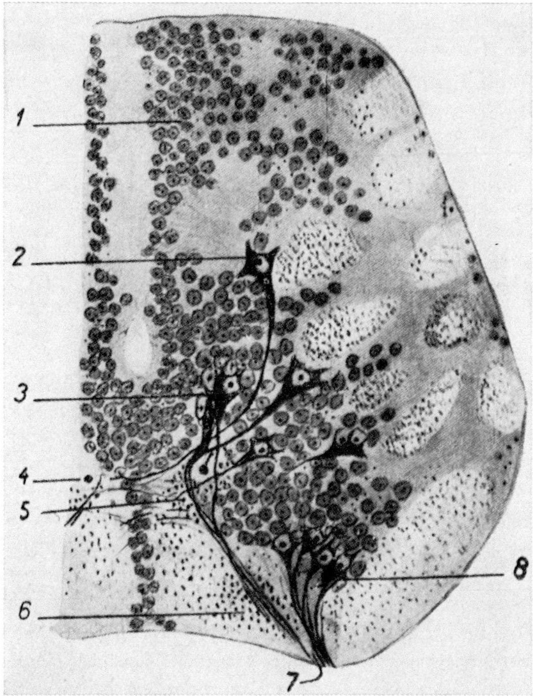

Figure 68C. Cross-section through the rostral portion of the spinal cord in a young Trout (Salmo irideus) of 25 mm length (from Beccari, 1943; *Cajals silver impregnation).* 1: dorsal cell masses *('nuclei funicolari');* 2: dorsal internuncial (commissural?) cell; 3: dorsal group of efferent neurons *('gruppo fondamentale delle cellule motrici');* 4: *Mauthner's fiber;* 5: fasciculus longitudinalis medialis; 6: 'fasciculus octavomotorius cruciatus'; 7: ventral root *('radice motrice');* 8: ventrolateral group of motoneurons *('gruppo ventro-laterale delle cellule motrici').*

Large dorsal transitory afferent *Rohon-Beard cells* are present during the ontogenetic development of Ganoids and Teleosts (cf. Figs. 62, 68A, 68B). Kupffer (1906) and others have shown that these cells disappear rather early in some forms (e.g. Acipenser) and somewhat later in others (e.g. in the Trout, Salmo fario). The peripheral neurite can be traced to the skin (cf. Fig. 68B), but a collateral branch for 'muscle sensitivity' as recorded by Coghill and Herrick in the urodele Amphibian Amblystoma (cf. Fig. 20 in section 2 of this chapter) has not been observed by Held (1909).

In certain adult Osteichthyes, however, as e.g. in the Teleosts Orthagoriscus, Ctenolabrus, and Lophius, peculiar large afferent dorsal nerve cells occur in an intramedullary and supramedullary position. Discovered by FRITSCH (1884, 1886) in Lophius, and therefore also commonly designated as *Fritsch-cells*,[87] neuronal elements of this type were subsequently investigated by numerous authors, including TAGLIANI (1895, 1899), DAHLGREN (1897, 1898), KOLSTER (1898), SARGENT (1898, 1899), JOHNSTON (1900), BURR (1928), KAPPERS *et al.* (1936), GIUSEPPE LEVI (1946, and various earlier publications), and

---

[87] GUSTAV FRITSCH (1838–1927) began his scientific career with an anthropologic expedition to South Africa lasting from 1863 to 1866 and yielding a rich material, upon which his important work '*Die Eingeborenen Südafrikas*' (1873) was based. Soon after his return, he initiated, together with HITZIG, an epochal study on the electrical excitability of the cerebral cortex in the dog, published in 1870. Subsequently, besides discovering the large 'supramedullary' cells of Lophius, FRITSCH completed, *inter alia*, a series of significant studies on the electrical organs of Fishes (Gymnotus, Malapterurus, Gymnarchus niloticus, Mormyrus and Torpedineae) included in several papers and two volumes ('*Die elektrischen Fische*' I, 1887; II, 1890). In connection with these investigations he carried out field trips to Africa. Additional journeys to this and to other Continents were undertaken by FRITSCH in order to gather material for his further anthropologic studies concerning, among other topics, the racial characteristics of the human hair (cf. e.g. '*Das Haupthaar*', 1912, 1915), and of the human eye. He coined the generally accepted term '*fovea centralis*' of the retina, replacing the designation '*macula*'. Partly because of some personal difficulties, he did not become director of an institute, but was nevertheless made *Ordinarius*, with the additional title of '*Geheimer Medizinalrat*' (privy councillor), at the University of Berlin. It is perhaps here apropos to enumerate and to stress some of his considerable scientific accomplishments, covering a wide field, in view of the grossly uninformed and derogatory remarks by KUNTZ in a biographic sketch of HITZIG in '*The Founders of Neurology*' (Thomas, Springfield, 1953, 1970). According to KUNTZ: 'FRITSCH's work with HITZIG was his only important contribution. FRITSCH was a man of wealth, a globe-trotter, who spent about ten years in South Africa. Some time after 1870 he was made *Ausserordentlicher Professor* of physiology at the University of Berlin. He never became *Ordinarius*.' In a compilation '*Some papers on the cerebral cortex*' (Thomas, Springfield, 1960, p. XII), v. BONIN, who should have known better, repeats KUNTZ by stating that FRITSCH 'was a man of independent means, who travelled much and spent, among others, 10 years in South Africa. His work with HITZIG was his only important contribution to medicine'.

In volume 5 of this series, and in dealing there with the cerebral cortex, I intend to come back to the unjustified gratuitous aspersions directed at HITZIG (1838–1907) and to the origin of the totally unfounded as well as rather revolting *canard* concerning both FRITSCH and HITZIG, which, in two different versions displaying uninhibited fanciful imagination, has been elaborated with apparent gusto by HERRICK (An Introduction to Neurology, 1931, p. 336) and by GREY WALTER (*The Living Brain*, 1953, p. 48). One can here merely repeat the old German saw: '*seht, so kommt Geschichte zustande!*'.

KAPPERS (1947). Despite their somewhat specialized features, it appears justified to consider these elements, with KAPPERS and various other investigators, as 'persistent' *Rohon-Beard cells.* In Orthagoriscus (Fig. 67A), although designated as 'supramedullary' cells by BURR (1928), most if not all of them remain within the dorsal portion of the spinal cord. In Ctenolabrus (SARGENT, 1899) such intramedullary or

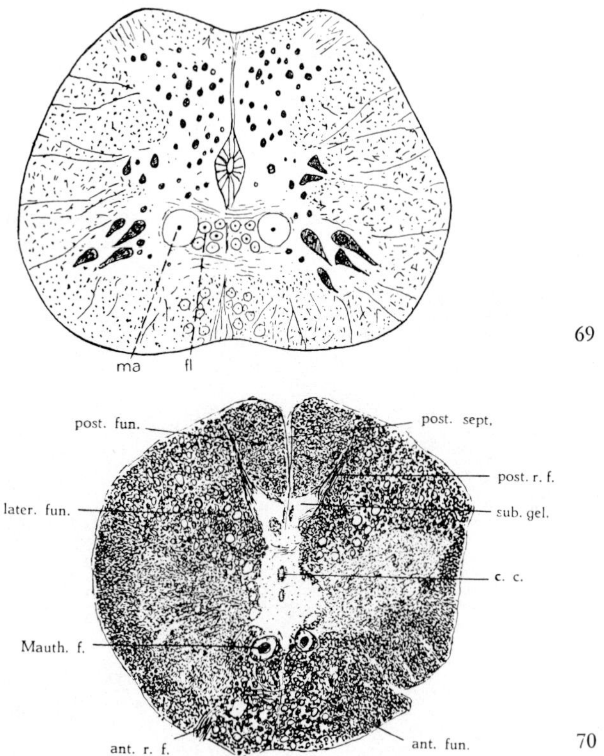

Figure 69. Semidiagrammatic cross-section through the spinal cord of the Teleost Cyprinus auratus (from K., 1927). fl: fasciculus longitudinalis medialis; ma: *Mauthner's fiber;* the two subdivisions of the ventral commissure and the two (ventrolateral and dorsal) groups of anterior horn cells are indicated. The anterior subdivision of the ventral commissure is also known as *commissura accessoria s. Mauthneri.*

Figure 70. Cross-section through the rostral portion of the spinal cord in the Teleost Osmerus eperlanus (from KEENAN, 1928; ×76, red. ¾). Abbreviations as in Figures 64 and 66.

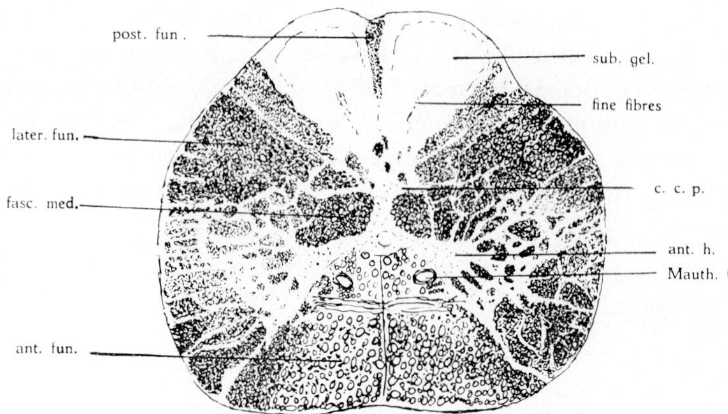

Figure 71 A. Cross-section through the spinal cord of the Teleost albula vulpes (from KEENAN, 1928; ×51, red. $^4/_5$). Abbreviations as in Figures 64 and 66.

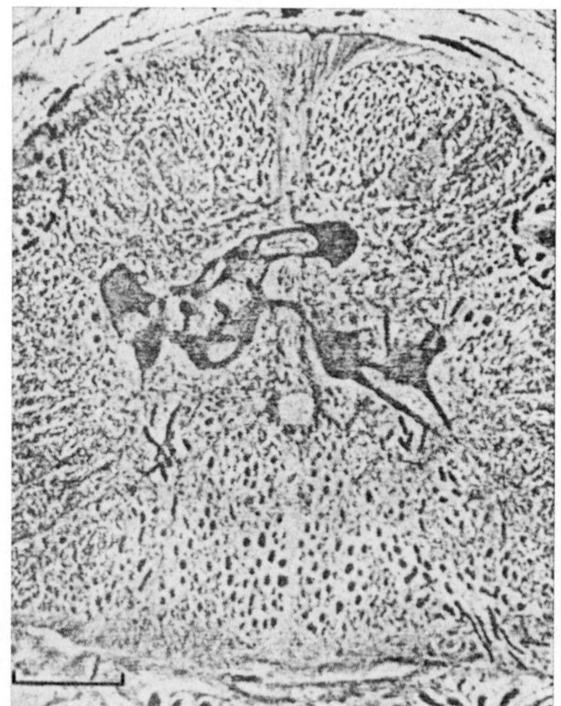

Figure 71 B. 'Electromotor' cells in the caudal medulla spinalis of the Mormyrid Gnathonemus. The dotted lines indicate the location of the descending bulbospinal system controlling the activities of these cells. Scale corresponds to 100 µ (from SZABO, 1961).

supramedullary elements tend to be located near or in the dorsal median sulcus, being either still included within the medulla or, if actually supramedullary, closely adjacent to, respectively contiguous with, the neuraxis. In Lophius (cf. Fig. 67B) external supramedullary cells can be seen scattered within the endomeninx at a non-negligible distance from the cord's dorsal surface.

These large dorsal cells are generally of pseudo-unipolar type with non-medullated neurites. In Lophius and Ctenolabrus, two groups of such elements have been noted, namely a small rostral one, at the transition to the medulla oblongata, and a larger one, whose cells are distributed along the length of the spinal cord.[88]

The centrally directed stem-processes of these cells in the cranial group bifurcate into an ascending and descending branch. This latter is said to ramify within the posterior horn of the cervical spinal cord, while the ascending one joins the radix spinalis trigemini, giving off collaterals emerging with vagus roots or motor components of the trigeminus (KAPPERS, 1947, and others). The analogy of this arrangement with that of the cells in the mesencephalic root of the trigeminus was pointed out by KAPPERS.

The stem-process of the cells in the caudal (or spinal) group likewise dichotomizes into short ascending and long descending components. These latter may alternatingly form an uncrossed and crossed non-medullated bundle dorsolaterally to the central canal, terminating in the dorsal horn, while collaterals seem to emerge with the dorsal roots (KAPPERS, 1947). However, the relevant details concerning course and connections of afferent and efferent axonic branches of these peculiar intra- and supramedullary ganglion cells still remain unclarified.

As regards the intramedullary fiber systems in Osteichthyes, the dorsal roots join the cord laterally to the posterior horns. Many root fibers bifurcate there into ascending and descending branches in the dorsal part of the lateral funiculus, while others may run dorsally toward the posterior funiculus. Others, with a more medial course, enter the substantia gelatinosa. Direct connections between posterior root fibers and efferent elements of the ventral horn seem to be established mainly by the dorsalward extending dendrites of these efferent cells rather than by extensive posterior root collaterals.

---

[88] In his description of the peculiar and very short spinal cord of Orthagoriscus, BURR (1928) does not refer to any particular grouping of the 'supramedullary' cells.

In Teleosts with particularly developed afferent components of the 'cervical' spinal cord segments, the accumulation of root fibers in the dorsolateral funiculus is very pronounced and blends with fibers of the descending trigeminal root. The substantia gelatinosa and related dorsal grisea may form an extensive posterior cap or '*sensory lobe*' which also seems to have some connections with the large supramedullary cells. The presence of a large dorsal 'sensory lobe' greatly modifies the typical cross-sectional outline of the spinal cord (cf. Figs. 67A, B).

In Trigla hirundo, three pairs of swellings can be noted in the 'cervical' portion of the spinal cord (cf. Fig. 61A) and correspond to afferent spinal nerve components innervating three specialized fin rays (KAPPERS *et al.*, 1936). In the related form Prionotus, six pairs of swellings are developed. While KAPPERS *et al.* (1936) suggest that the corresponding fin rays represent 'modified organs of touch', others assume 'special chemical receptors' in the elongated fins (cf. YOUNG, 1955). Should these receptors indeed be 'gustatory', their spinal innervation would significantly differ from the cerebral innervation, by way of the n. facialis, of the cutaneous taste buds in some other Teleosts such as Siluroids (cf. e.g. Fig. 444 in the preceding volume 3, part II).

The secondary fiber systems originating in the spinal cord of Ganoids and Teleosts are neurites of homolateral and contralateral (commissural) funicular cells.[89] Many of these fibers seem to bifurcate, upon reaching the white substance, into ascending and descending branches which run through dorsal and ventral portions of the lateral funiculus, giving off short collaterals. The crossed fibers decussate in the main or in the accessory ventral commissure. While most of these fibers may represent the spinal cord's *eigenapparat*, some of the ascending ones are believed to reach the brain. Thus, KAPPERS and other authors have described a crossed spino-bulbar-mesencephalic tract whose ventrolateral location is indicated in Figure 67B. A 'spino-tectal' tract lateral to the corpus commune posterius, and a dorsal spino-cerebellar tract as well as a more ventral one are also assumed to be present. It remains

[89] Strictly speaking, '*funicular cells*' are neuronal elements pertaining to the *eigenapparat*, whose neurites remain within the spinal cord. '*Tract-cells*', on the other hand, are similar elements, but whose neurites, which also run through the 'funiculi', reach the brain and therefore pertain to the *verbindungsapparat*. Since, in many instances, and particularly in most lower Vertebrate forms, presumed internuncial elements cannot be clearly identified as pertaining to either or even both categories, the term 'funicular cell' *sensu latiori* is here used to subsume either type.

quite uncertain to which extent ascending components of the *verbindungsapparat* are also provided by cranialward directed dorsal root fibers in 'posterior funiculi' and adjacent dorsal neighborhoods of lateral funiculi.

The descending pathways of the *verbindungsapparat* in Ganoids and Teleosts comprise a fasciculus longitudinalis medialis with crossed and uncrossed reticulo-spinal fibers, moreover a crossed lateral and an uncrossed medial vestibulo-spinal tract. Cerebello-spinal, tecto-spinal, and olivo-spinal fibers have been reported, but the evidence for such connections seems rather dubious.

Although the descending systems of the spinal cord in Ganoids and Teleosts are, on the whole, rather similar to those recorded in Plagiostomes, a difference obtains with respect to the fasciculus longitudinalis medialis. In Ganoids and most Teleosts this bundle is characterized by the occurrence, on each side, of a conspicuous single medullated giant fiber, discovered by MAUTHNER (1859), and subsequently found, by GORONOWITSCH (1888), to originate from a paired giant cell of the medulla oblongata, located in the lateral part of the reticular formation at the level of the VIIIth nerve. The neurites of the two *Mauthner cells*, after decussating at the level of their origin, run through the entire length of the spinal cord, giving off short collaterals effecting direct connections with motoneurons.[90] LEGHISSA (1956) described some peculiarities of the myelin sheath of *Mauthner's fiber*, which he found devoid of interruptions comparable to central *Ranvier's nodes* or *Lantermann's clefts*.

There are substantial differences with regard to the size of *Mauthner's fiber* in different taxonomic groups, as can easily be seen by a glance at Figures 63 to 66, and 68C to 71. Moreover, in a number of Teleosts, including Lophius and Orthagoriscus (Fig. 67A, B) *Mauthner's fiber* is missing.[91]

Another significant difference between Plagiostomes and a number of Teleosts concerning the composition of the *verbindungsapparat's* descending components is the presence of descending gustatory tracts

---

[90] *Mauthner fibers* respectively *cells* are not ordinarily found in adult Selachians, although they have been reported in some Squalid embryos. We did not detect these structures in adult Chimaeroids (K. and NIIMI, 1969), but a report is extant which suggests the possibility of their occasional occurrence in these forms.

[91] BECCARI (1943), who dealt with *Mauthner's cell* and *fiber* in several of his investigations, list the absence of this structural arrangement in '*Murenoidi*', '*Anguilloidi*' and a few other groups.

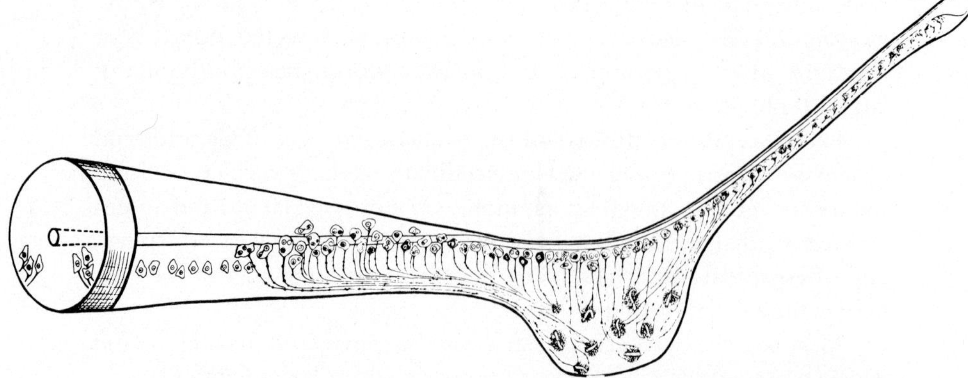

Figure 72. Diagrammatic illustration of the caudal neurosecretory system in the spinal cord of Teleosts, showing tip of spinal cord and the somewhat more rostral vascularized preterminal 'ventral swelling' or 'intumescentia caudalis' (from SANO, 1961).

characteristic for those Teleosts in which the 'gustatory system' is highly developed and may include many taste buds along the entire surface of the body, innervated by the seventh cranial nerve. Here, secondary and tertiary gustatory tracts run back from the facial lobe, or neighboring gustatory centers, into the spinal cord. Rostrally, such fibers are closely joined to the descending trigeminal root and at some levels practically surround it, except laterally. Before the level of the hypertrophied dorsal funicular region is reached, the gustatory path has become a part of the dorsolateral funiculus through which it continues into the spinal cord, effecting connections with its grisea.

The large cells of neurosecretory type, found in the caudal part of the spinal cord of various Selachians, of at least some Ganoids (e.g. Lepidosteus) and of numerous Teleosts, may form, in these latter, a complex caudal neurosecretory system, the *neurohypophysis caudalis* or *urohypophysis*, which has been investigated by FAVARO (1926) and in considerable detail by ENAMI (1958) and SANO (1961).

Previous to the description of its large cells in Plagiostomes by DAHLGREN, and their identification, in Plagiostomes and Osteichthyes, as secretory by SPEIDEL, the nodular swelling or 'warty appendix' of the caudal portion of the Teleostean spinal cord (*'renflement caudal'* of VERNE, 1913–1914) had been noted by numerous authors since more than 160 years ago.[92]

---

[92] Further data of historical interest and comprehensive bibliographies on this topic can be found in the publications by ENAMI (1959) and SANO (1961).

Figures 72 and 73 illustrate the structure of the urohypophysis in the Eel and in a 'generalized Teleost', while Figure 74 brings, in accordance with ENAMI's interpretation, a comparison of 'hypothalamo-hypophysial system' with 'caudal neurosecretory system'. Within this latter, the 'secretion-bearing axons' of the large neuronal elements terminate with bulbous endings in the vicinity of capillary or sinusoid vessels.

Although the functional significance of the Teleostean urohypophysis has not been conclusively ascertained, the investigations of ENAMI and his collaborators seem to indicate a possible participation of

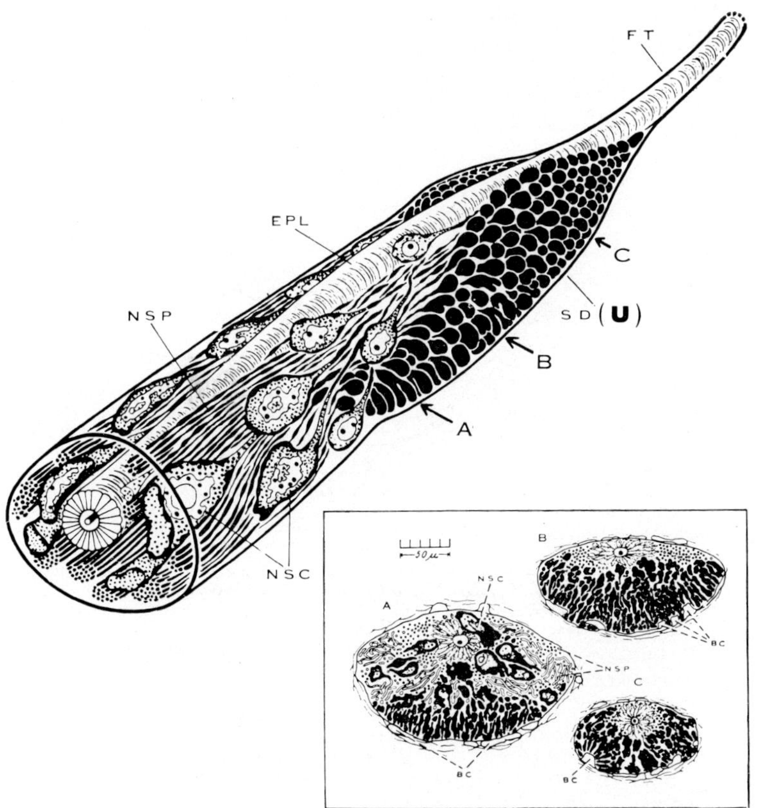

Figure 73. Pattern of arrangement of the caudal neurosecretory system of the Eel (from ENAMI, 1959). The inset shows semidiagrammatic pictures of cross sections at the levels labelled A, B and C. BC: blood capillaries; EPL: ependymal layer; FT: filum terminale; NSC: neurosecretory cells; NSP: neurosecretory pathway; SD (U): storage-depot organ or 'urophysis' sensu strictiori.

this organ in osmoregulatory events, and particularly in the control of sodium exchange. Some effects involving 'gas metabolism' were also suggested by said authors.

The extinct *Crossopterygians*, of which the *Coelacanth Latimeria* is a recently discovered surviving form, have been classified, together with the Dipnoans, as *Sarcopterygii*, representing a 'subclass' of Osteichthyes. They can also be classified as a separate class Choanichthyes, comprising the orders Crossopterygii and Dipnoi. Because of their cerebral morphology, discussed in the previous volume of this series, I prefer to include the Crossopterygians, represented by Latimeria, as a subclass into the class Osteichthyes, and to consider the Dipnoans a separate class, intermediate between Osteichthyes and Amphibia. For this reason, a few comments on the spinal cord of Latimeria, described by MILLOT and ANTHONY (1965) may here be added. Cross-sections (Fig. 75) disclose an overall resemblance with outlines and configura-

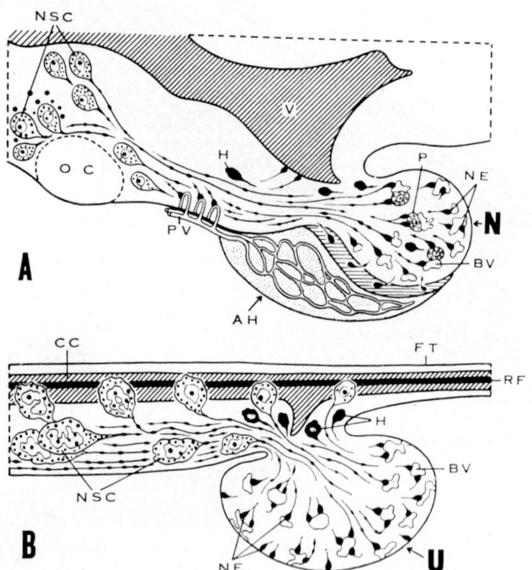

Figure 74. Comparison of the pattern of organization of the Teleostean caudal neurosecretory system B with that of the hypothalamo-hypophysial system A (from ENAMI, 1959). AH: adenohypophysis; BV: bloodvessels; CC: central canal; FT: filum terminale; H: *Herring-body* or similar configurations; N: neurophypophysis; NE: nerve endings; NSC: neurosecretory cells; OP: optic chiasma; P: pituicyte; PV: hypophysio-portal vessel; RF: *Reissner's fiber;* U: urohypophysis; V: third ventricle.

tions found in Plagiostomes and Osteichthyes. The club-shaped ventral horns, containing the large motoneurons, extend farther lateral than the dorsal ones, and are connected by a narrow stalk *('pédicule très mince')* with an unpaired griseal aggregate. The corpus commune posterius is rather short, while the posterior horns are narrow griseal bands in an almost 'horizontal' alignment at some levels. Reticular stripes of grey, extending dorsalward from these horns are regarded by MILLOT and ANTHONY (1965) as a much reduced *'équivalent de la subs-*

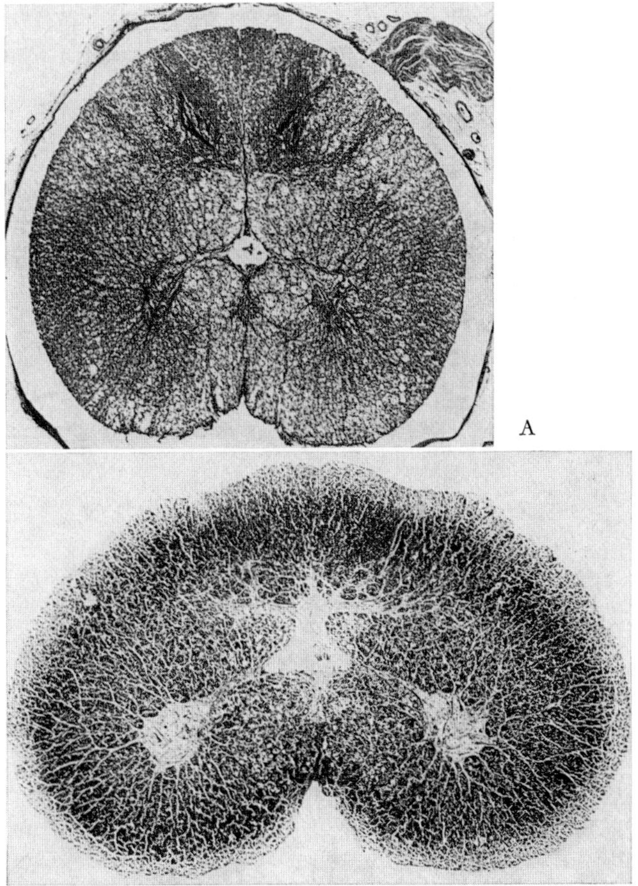

Figure 75. Cross-sections through the spinal cord of the Crossopterygian Latimeria (from MILLOT and ANTHONY, 1965). A Through the 16th segment (protargol impregnation; ×33, red. $^4/_5$). B Through the 3rd segment (osmic acid; ×40, red. $^4/_5$).

tance gélatineuse de Rolando'. On the other hand, the posterior funiculi appear wider than in most Teleosts. However, among these latter, the posterior funiculi are likewise occasionally well developed (cf. Fig. 70, Osmerus). The cited authors report that the spinal cord of Latimeria does not display *Mauthner's fiber* and lacks large dorsal ('supramedullary') cells. A 'caudal hypophysis' was not found to be present. The available data do not include relevant information concerning specific fiber systems or tracts within the spinal cord.

7. Dipnoans

As in the case of Crossopterygians, few observations of significance have been recorded concerning the structure of the Dipnoan spinal cord. In *Ceratodus*, KEENAN (1928) noted a 'more massive' posterior funiculus than in Teleosts, comparable with conditions in Anuran Amphibia. The dorsal horns contain scattered medullated longitudinal fiber bundles and are capped by an irregular mass of substantia gelatinosa (Fig. 76). The ventral horns display a pattern similar to that seen in Latimeria. A thick *fiber of Mauthner* is present.

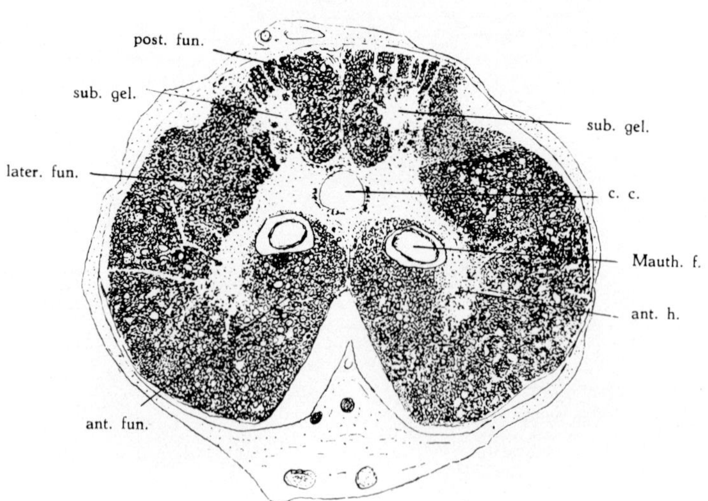

Figure 76. Cross-section through the rostral portion of the spinal cord in the Dipnoan Ceratodus (from KEENAN, 1928; myelin stain; × 33, red. $^4/_5$). Abbreviations as in Figure 66.

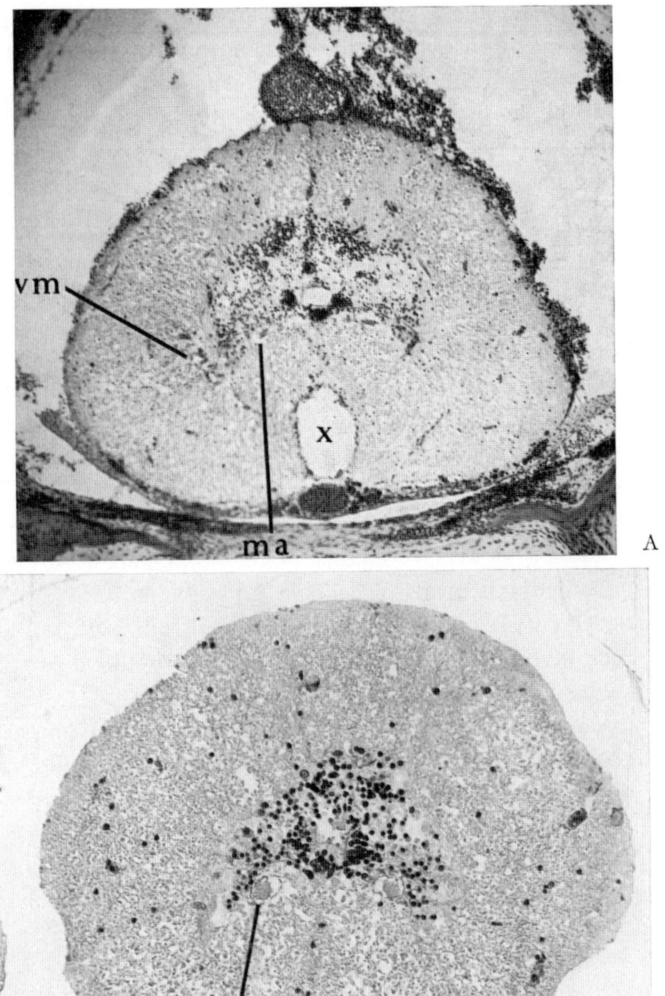

Figure 77. Two cross-sections through fairly rostral regions of the spinal cord in the Dipnoan Protopterus (hematoxylin-eosin stain, A: approx. ×60, B: approx. ×70, red. $^1/_1$). ma: *Mauthner's fiber;* vm: motoneurons at periphery of 'ventral horn'; x: space in ventral septum (fixation-artefact).

In *Protopterus* (Fig. 77) the central gray appears more compact, without formation of distinctive dorsal horns, and the ventral horns are merely faintly indicated by slight protrusions. This configuration of gray substance distinctly resembles conditions commonly found in Amphibians (cf. Figs. 78, 81, and 82).

The multipolar neurons pertaining to the basal plate derivative can be roughly subdivided into a ventromedial and a ventrolateral group (Kölliker, 1896). The largest motoneurons are found in the ventrolateral group located at the margin of the gray substance. According to Burckhardt (1892) the dendrites of these and of other cells of the gray matter are exclusively directed into the white substance. This later encircles the gray core as an all-around thick-walled stratum which, in addition to the longitudinal fiber systems, includes the synaptic neuropil structures. Although 'extragriseal', superficial dendritic plexuses occur in all Gnathostome Fishes, such peripheral neuropil arrangements, essentially remaining in the white substance externally to the periventricular gray which contains the cell bodies, are quite characteristic for the neuraxis of many Urodele Amphibians and were particularly stressed by Herrick. Within the ventral funiculus of Protopterus, a thick *fiber of Mauthner* is included as a component of the fasciculus longitudinalis medialis.

In default of somewhat more detailed observations, comparable to those recorded for other Anamnia, the available data could be interpreted to indicate that the fiber connections pertaining to *eigenapparat* and *verbindungsapparat* in Dipnoans consist of systems and tracts essentially similar to those described for Plagiostomes and Osteichthyes in general.

8. Amphibians

Although some Amphibians are limbless (Gymnophiona), and others possess merely rudimentary extremities (e.g. the Urodeles Proteus and Amphiuma), this class, taken as a whole, is definitely adapted to terrestrial life and can be subsumed under the concept '*Tetrapod Vertebrates*'.[93] The development of fairly large extremities with their special-

---

[93] Among Fishes, partial adaptation to 'terrestrial life' is correlated, in Teleosts, with the development of specialized gill-components, as e.g. in Anabas scandens *(climbing Perch)*, Clarias batrachus *(walking Catfish)* and in Periophthalmus chrysopilos *('Mudskipper')*, who is particularly proficient at climbing trees. In Dipnoans, 'terrestrial adap-

ized musculature in Anurans and most Urodeles is correlated with the presence of an intumescentia cervicalis and an intumescentia lumbalis as a noticeable, but in some forms only faintly indicated spindle-shaped swellings of the spinal cord. This latter, again, extends as a rule through the length of the vertebral canal in Urodeles, but is commonly much shorter in Anurans and can here end into a filum terminale at the level of third or fourth vertebra (cf. above, section 1 of this chapter). The rather long filum terminale, containing a rudimentary central canal and nondescript neuroectodermal cells, mostly of glial type, is here a reduced remnant of the cord's caudal portion which, before metamorphosis, innervated the musculature of the tail at larval stages.

Among the three extant Amphibian orders, the *Urodeles*, particularly the Salamanders, have been frequently regarded as more 'primitive' than the *Anuran* Frogs, but Noble (1931, 1954) justly remarks that this evaluation may be substantiated with respect to only certain features of their anatomy. Moreover, while it is true that the entire neuraxis of numerous Urodeles displays a very simple structural arrangement, the possibility of regressive evolutionary processes cannot be entirely discounted. The *Gymnophiona*,[94] *Apoda* or *Caecilians*, which are considered as more 'primitive' than any other recent Amphibia in many anatomical details, may have directly evolved from the extinct Lepospondyli. If this is true, Gymnophiones followed a quite independent line of evolution from Lower Carboniferous or Devonian times. The many differences between the morphology of Caecilians and that of other recent Amphibia would support such a view (Noble, 1931, 1954). Certain features of the integument, on the other hand, have suggested a possible link of Gymnophiona to the Labyronthodontia of the Carboniferous period.

With regard to the spinal cord of *Urodeles*, v. Kupffer (1906) offers the following comments, which are still essentially valid to-day, after

---

tations' is correlated with a specialization of the air-bladder, which represents the 'precursor' of the Tetrapod lungs. Again, in Teleosts, a peculiar 'adaptation' to rudimentary 'flight' is manifested by forms such as Exocoetus evolans and Dactylopterus volitans.

[94] In 1913, only 19 genera, with 55 species, all included in a single family *(Caeciliidae)*, were known. About 50 years later, Taylor (1968) listed somewhat more than 160 species, distributed upon 34 genera and 3 families *(Ichthyophiidae, Typhlonectidae, Caeciliidae)*. Until recently, no information concerning Apodan fossil representatives was available, but two extinct species have now been reported, namely *Ichthyophis muelleri Brunner* (fragment of mandible, Riss-Würm IG-Period, Pleistocene, Oberfranken, Germany) and *Prohypogeophis tunariensis Marcus* (two fairly large fragments from the Carboniferous, Bolivia).

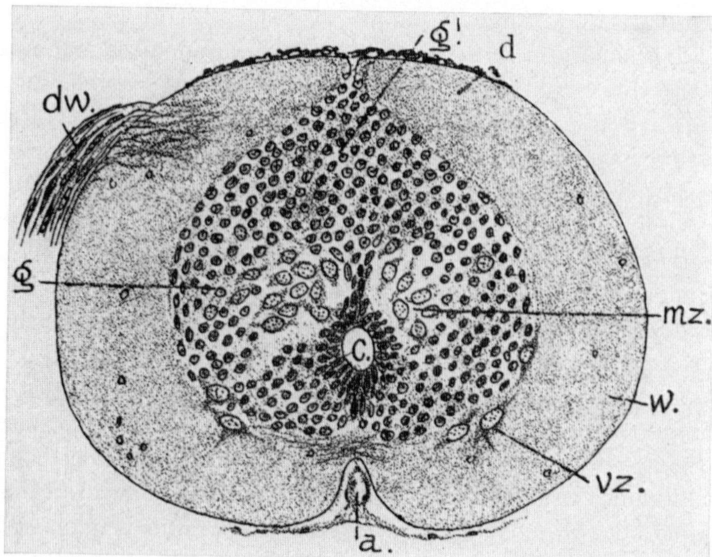

Figure 78 A. Cross-section through the spinal cord in a young specimen of Salamandra atra at the level of anterior extremities (from Kupffer, 1906). a: arteria spinalis anterior; c: central canal; d: dorsal funiculus; dw: dorsal root; g, g′: central and dorsal gray substance; mz: large medial nerve cells; vz: ventral nerve cells (motoneurons); w: ventrolateral funiculus *('Vorderseitenstrang')*. Note the absence of *Mauthner's fiber* (cf. below footnote 96).

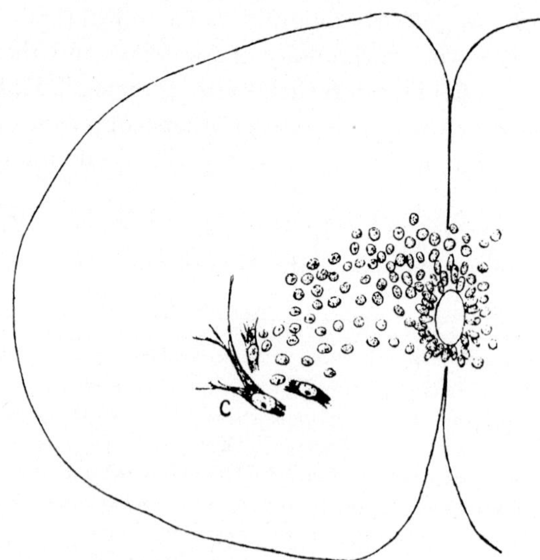

Figure 78 B. Cell-picture shown by a cross-section of the spinal cord in adult Triton cristatus (from Barbieri, 1906). c: motoneurons.

more than 60 years: '*Es ist bekannt, dass bei den Urodelen im erwachsenen Zustande die Form des Rückenmarkes, ganz besonders aber die Gestaltung der grauen Substanz, grossen Wechsel zeigt. Bei Siren lacertina erscheinen die Umrisse der grauen Substanz annähernd so wie bei den Cyclostomen. Die den Centralkanal umschliessende mittlere Portion ist von geringen Dimensionen, die ventralen grauen Säulen aber ragen lateralwärts weit vor und sind im dorso-ventralen Durchmesser schmal* (Kölliker, *1896). Ähnlich verhält es sich bei Proteus, bei Amphiuma und Cryptobranchus. In anderen Formen (Siredon, Triton) übertrifft der quere Durchmesser des Durchschnittes der grauen Substanz nur wenig den dorso-ventralen; ventrale wie dorsale Säulen ragen nicht bedeutend vor. Die Morphogenie des Markes ist bisher noch bei keiner Art zusammenhängend untersucht worden, nur für Triton liegen einige Angaben vor (*R. Burckhardt, *1889).*' A comparison of Figures 78, 79, and 81 will illustrate some aspects of these differences. While, in some forms, dorsal and ventral horns are displayed, the gray substance in others shows a rather compact arrangement which, as e.g. in Figure 78A, Kupffer compared with an inverted 'heart' figure of playing cards (*'kartenherzförmig'*).

Further data on larval and adult spinal cord in Urodeles were provided by the investigations of van Gehuchten (1897, Salamandra), Barbieri (1906, '*Anfibi*'), Herrick and Coghill (1915, Amblystoma), and Herrick (1930, Necturus). The early larval swimming reflex of Amblystoma,[95] studied by the two latter authors, and based on a peculiar transitory arrangement of neuronal circuits, was dealt with

---

[95] *Amblystoma* is now commonly spelled *Ambystoma*, but I do not consider this change justified. Thus, one of the numerous '*wise guys*' of the establishment who palliate their own inferiority by affecting superiority in the pretentious book reviews characteristic for '*Science*', some time ago condescendingly approved 'the correct use of the name Ambystoma rather than the commonly used and erroneous "Amblystoma"; the latter term translates from its Greek roots as "stupid mouth".' To this, one could reply that the Greek ἀμβλὺς and it related form ἀμβλεῖα have the meaning of 'obtuse', 'blunt', 'dull', 'insensitive', 'weak', etc., in several, concrete or figurative senses, as referring e.g. to an obtuse angle, a dull knife, an obtuse person, a stupid fellow, or a dullard. It will here be sufficient to quote Euclid's basic definition in his classical 'Elements': Ἀμβλεῖα γωνία ἐστὶν ἡ μείζων ὀρθῆς. Thus, Amblystoma stands quite well for blunt snout or obtuse mouth as compared to a pointed one (cf. also the mineral *amblygonite* with obtuse angle of cleavage). Likewise, βραδύς, 'slow', etc., may refer to the course of events (e.g. *bradycardia*) or to the sluggish mind of a dolt. Concerning the typical book-reviews in *Science*, one might add, as an aside, the old Greek saw: μωμήσεταί τις μᾶλλον ἢ μιμήσεται.

above in section 2 of the present chapter and need not be reviewed in this context. Neuronal connections at a subsequent stage, namely in the half-grown larva, and illustrating diverse connections of motoneurons with fiber systems of *eigenapparat* and *verbindungsapparat*, are shown in Figure 80 which illustrates the findings and interpretations by HERRICK and COGHILL.

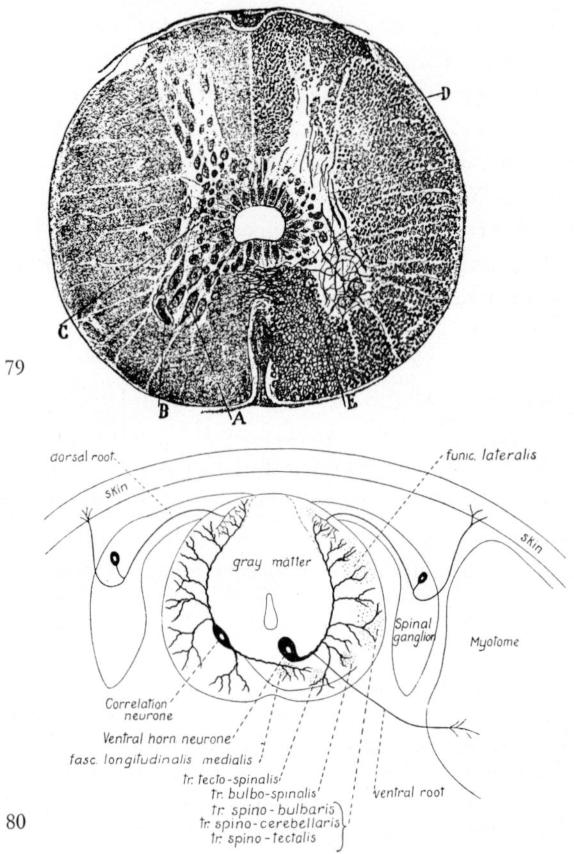

Figure 79. Cross-section through the rostral portion of the spinal cord in the adult newt Pleurodeles Waltii (from CAJAL, 1909, 1952; *Weigert-Pal myelin stain* counterstained with carmine). A: motoneurons; B: capillary in anterior horn; C: neurons in gray substance; D: peripheral zone of dendritic plexures; E: *Mauthner's fiber*. The cross-sectional griseal pattern in this newt, as depicted by CAJAL, differs somewhat from the usual configuration obtaining in Urodeles (cf. the two preceding Figures).

Figure 80. Diagram of the relations of some of the neurones in the spinal cord of larval Amblystoma at a stage following that of the early swimming reflex (from HERRICK and COGHILL, 1915).

A 'substantia gelatinosa' seems, on the whole, poorly recognizable, but some of the dorsal neuronal elements may have connections similar to those of that substance in Fishes. An intermediate group, depicted by KUPFFER (Fig. 78) as '*mediale Nervenzellen*' and by HERRICK (Fig. 81) as 'formatio reticularis grisea', presumed to be continuous with that of oblongata, is located within the boundary neighborhood of alar and basal plate. It may include internuncials as well as 'visceral

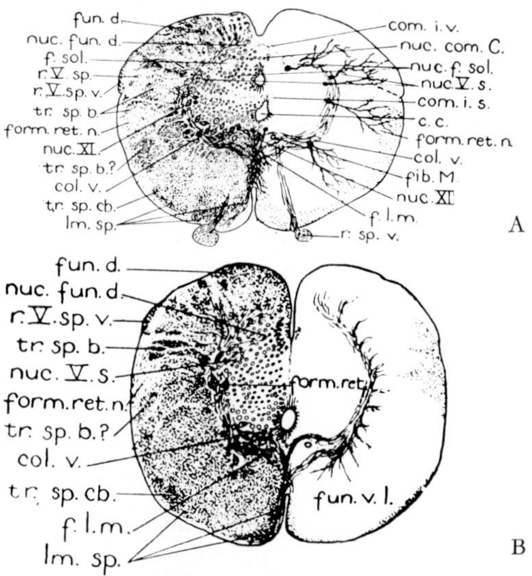

Figure 81. Cross sections through the rostral portion of the spinal cord in adult Necturus (from HERRICK, 1930). A Immediately caudal to calamus scriptorius at exit of rootlet pertaining to first spinal nerve. B Section at a level between first and second spinal nerves. The sections were drawn from a *Weigert series*. On the right side of A, details assembled from corresponding sections with *Golgi impregnation* are added. cc: central canal; col. v.: ventral gray column of spinal cord; com. i.s.: commissura infima Halleri of oblongata, 'somatic component'; com. i. v.: same, 'visceral component'; fib. M.: *Mauthner's fiber;* f. l. m.: fasciculus longitudinalis medialis; form. ret., form. ret. n.: 'formatio reticularis grisea' and its neuropil (n.); f. sol.: fasciculus solitarius of oblongata; fun. d.: funiculus dorsalis; fun. v. l.: funiculus ventrolateralis; lm. sp.: lemniscus spinalis; nuc. com. C.: *nucleus commissuralis of Cajal;* nuc. f. sol.: nucleus of fasciculus solitarius; nuc. fun. d.: nucleus of funiculus dorsalis; nuc. V. s.: nucleus of descending trigeminal root; nuc. XI: 'motor XI nucleus'; nuc. XII: 'nucleus of hypoglossal nerve'; r. sp. v.: ventral root of spinal nerve; r. V. sp.: descending root of trigeminus; r. V. sp. r.: ventral fascicles of descending trigeminal root; tr. sp. b.: tractus spino-bulbaris; tr. sp. cb.: tractus spino-cerebellaris. In B, *Mauthner's fiber*, not labelled, can be seen between the two bundles of 'arcuate fibers' decussating in the ventral commissure.

efferent', 'vegetative' elements. The motoneurons are generally located at the ventral periphery of the gray substance, in a ventrolateral to ventromedial distribution.

Figure 81A shows that in adult Necturus as in larval Amblystoma (cf. Fig. 80), the dendrites of neuronal elements in the central gray branch exclusively within the white substance, where synaptic connections with the fiber systems seem to take place. As regards these fiber systems, nothing definite is known about the details of the circuits provided by the *eigenapparat*, except for the specific larval mechanisms described by COGHILL and HERRICK. Generally speaking, it can be presumed that the overall pattern of connections is comparable to that obtaining in Gnathostome Fishes.

Concerning ascending components of the *verbindungsapparat*, HERRICK (1930) concludes that in Urodeles the dorsal funiculi contain large numbers of ascending branches of (primary) dorsal spinal root fibers 'as in Mammals'. He noted a 'nucleus of the dorsal funiculus in the calamus region' which he compared with nuclei of fasciculus gracilis and cuneatus in Mammals. In Urodeles, however, the secondary

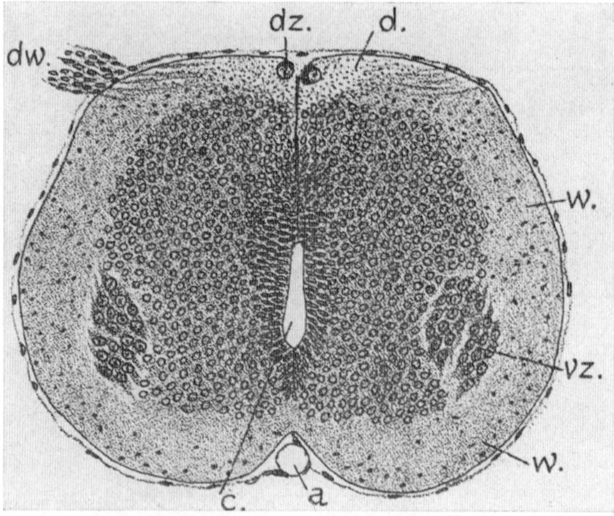

Figure 82. Cross-section through a rostral trunk segment of the spinal cord in Rana fusca at an advanced stage of metamorphosis (from KUPFFER, 1906). a: arteria spinalis anterior; c: central canal; d: dorsal funiculus; dw: dorsal root; dz: degenerating *Rohon-Beard cells;* vz: ventral nerve cells (motoneurons); w: lateral and anterolateral funiculus. *Mauthner's fiber* (perhaps still present) was not recorded by KUPFFER.

fibers arising in the 'nucleus of the dorsal funiculus' take, after their decussation, a more lateral course than in Mammals, to terminate chiefly in the midbrain, a few even ending 'in the dorsal thalamus' (HERRICK, 1930).

Other, secondary ascending systems described by this author include a lateral spino-bulbar tract and a 'spinal lemniscus' in the anterior funiculus. These systems originating from internuncial elements, whose neurites decussate as arcuate fibers in the ventral commissure, are said to be mostly but not exclusively crossed. According to HERRICK, a spino-cerebellar tract may be included in the lateral portion of the 'spinal lemniscus' whose fibers run as far as tectum opticum and perhaps even thalamus. Those of the spino-bulbar tract terminate in the neuropil of the oblongata's reticular formation.

The descending components of the *verbindungsapparat* comprise bulbo-spinal and tecto-spinal tracts consisting of crossed and uncrossed fibers forming the fasciculus longitudinalis medialis, which contains *Mauthner's fiber*.[96] Additional components of this system may be scattered in parts of ventral and ventrolateral funiculus adjacent to the main fasciculus.[97] Figure 81 shows the location of ascending and descending tracts in Necturus according to HERRICK's (1930) interpretation.

The spinal cord of *Anurans* manifests, on the whole, a somewhat higher degree of differentiation than that of Urodeles. Figures 82 and 83 illustrate general aspects of its configuration as seen in cross-sections. The compact gray substance does not display a striking protrusion of ventral and dorsal horns. A substantia gelatinosa is present. In comparing some features of the Anuran spinal cord with those obtaining in Urodeles, BARBIERI (1906) states: '*il midollo adulto degli anuri non corrisponde a tutto quanto quello di urodeli; gli elementi che negli urodeli sono in relazione con la coda e la sua muscolatura (muscolatura laterale) mancano negli anuri, essi sono già scomparsi durante e prima della metamorforsi; gli elementi nervosi in relazione direttamente o indirettamente coi membri, hanno raggiunto negli anuri un più grande sviluppo e differenziamento*'.

---

[96] *Mauthner's fibers*, respectively cells, present in perhaps all larvae, become reduced in those Urodeles (e.g. Geotritons, terrestrial Salamanders) with predominantly terrestrial habitat.

[97] Among the bulbo-spinal system, whose fibers mostly originate in the nucleus reticularis tegmenti of the brain stem, additional vestibulo-spinal fibers, not particularly stressed by HERRICK (1930) are presumably included.

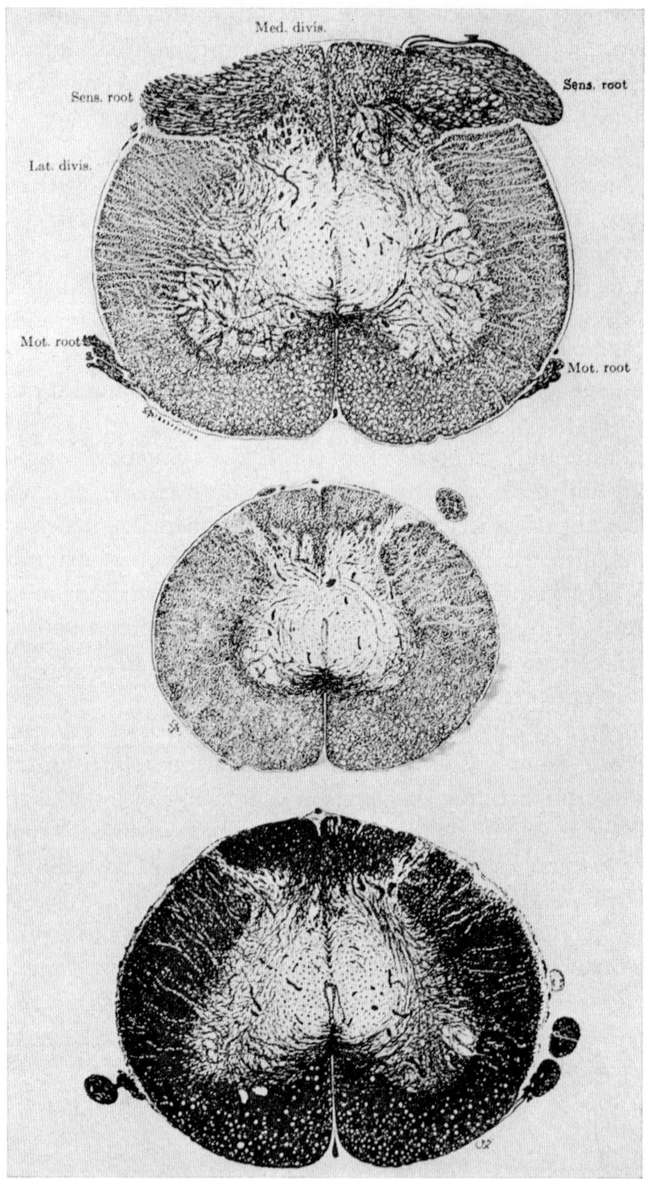

Figure 83. Cross-sections through cervical (top) thoracic, and lumbar segments of the spinal cord in Rana mugiens (from KAPPERS *et al.*, 1936; myelin-stain). Lat. div.: *Lissauer's tract;* Med. divis.: posterior funiculus; Mot. root: ventral root; Sens. root: dorsal root.

Figures 84A–C show distribution and arrangement of various neuronal elements observed by different authors in *Golgi-preparations*. It will be noted that, in contradistinction to conditions obtaining in at least many Urodeles, numerous dendrites ramify within the gray substance proper, although an additional peripheral dendritic plexus is also present. Axonal terminations likewise occur in the gray substance (cf. Fig. 85), which thus contains a neuropil lacking in that of e.g. Necturus (Fig. 81), but rather commonly present in Plagiostomes and Osteichthyes.

Axonal terminations, i.e. junctions between neuronal elements, include typical *synaptic boutons*. Albeit these structures are less easily demonstrable than in Amniota by means of the available techniques, silver impregnations readily show such end feet on dendrites and perikarya of various nerve cells (SILVER, 1942; KENNARD, 1959). There are, in addition, other sorts of contact connections that can be interpreted as functional junctions. Some poorly understood functional significance for processes of impulse transmission is also attributed to the overlap and the frequently close parallel arrangement of dendritic processes displayed by certain neuronal elements.

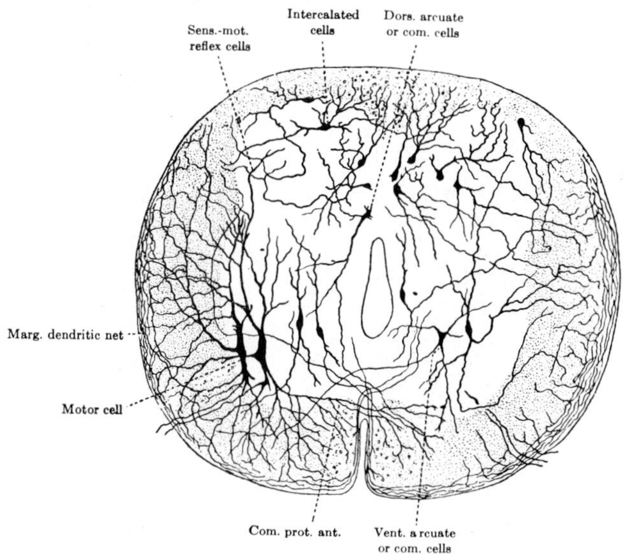

Figure 84A. Neuronal elements of the spinal cord in an advanced larva of the Toad Bufo as seen in *Golgi impregnations* (after SALA, from KAPPERS *et al.*, 1936). Note dorsal and ventral commissures.

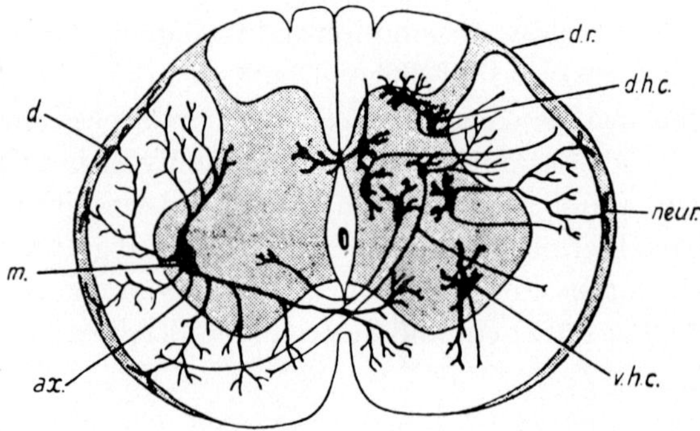

Figure 84B. Cross-section through the spinal cord in a Frog, showing diverse nerve cells and their processes as seen by means of the *Golgi-technique* (after GAUPP, from YOUNG, 1955). ax: neurite of motoneuron; d: dendrite of motoneuron; dhc: small cells of dorsal horn; dr: dorsal root entry; m: cell body of motoneuron; neur: neuropil at periphery of spinal cord (peripheral dendritic plexus); vhc: smaller cell of ventral horn.

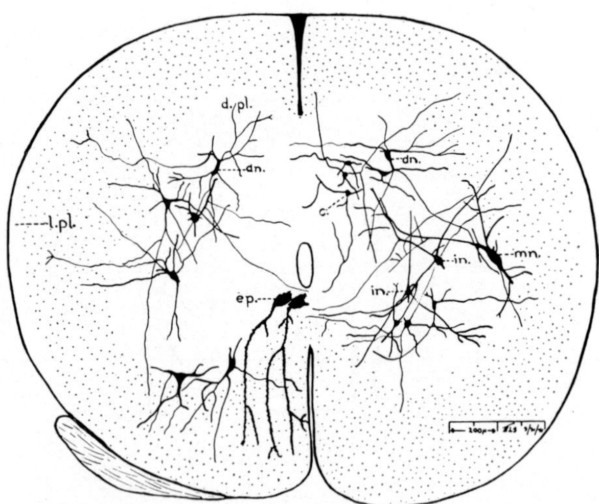

Figure 84C. Drawing of 60 μ thick cross-section *(Golgi-preparation)* through the spinal cord of an adult Frog showing, in a single section, the main cell types (from SILVER, 1942). c: commissural neuron; dn: dorsal horn neuron; d. pl.: dorsal neuropil zone; ep: ependymal cell; in: 'interneuron'; l. pl.: lateral neuropil zone; mn.: motoneuron.

With respect to the terminals of primary dorsal root afferents in the griseum of the Frog's spinal cord it was recently claimed that the convulsant alkaloid bicuculline blocked here presynaptic inhibitions, dorsal root potentials, primary afferent depolarization, and depolarizing effects of gamma-aminobutyric acid (GABA) but did not block effects of other putative amino-acid transmitters. These actions of bicuculline are believed to suggest that GABA may be the transmitter involved in spinal presynaptic inhibition (DAVIDOFF, 1972; cf. also the discussion of transmitter substances in chapter V, pp. 629–636, volume 3, part I of this series).

The large motoneurons of the ventral gray display an arrangement into indistinct clusters or groups. KAPPERS *et al.* (1936) recognize a medioventral one which extends through the whole cord, and a lateral or

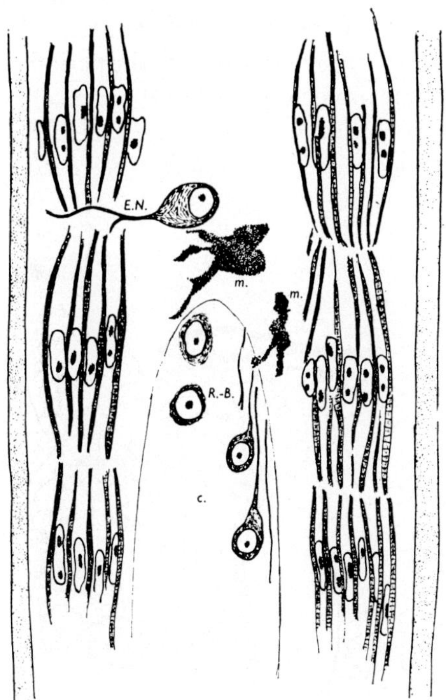

Figure 85 A. Diagrammatic 'near-horizontal' section through the dorsal region of a 58-hours Xenopus larva (from HUGHES, 1957). c: oblique section of the spinal cord's dorsal portion; m: melanoblasts; E.N.: extramedullary (supramedullary) ganglion cell whose peripheral axon ('dendrite') runs in an intermyotomic septum; R.-B.: a group of *Rohon-Beard cells.*

ventrolateral one restricted to the levels of the intumescentiae. This latter group may therefore innervate the extremity musculature, while the former would then innervate the trunk muscles. Results of experiments with electrical stimulation, undertaken by SILVER (1942), support this conclusion based on the anatomical data. Some authors (e.g. SILVER, 1942; NEMEC, 1951) have made attempts at further subdivisions.

Visceral efferent (preganglionic) nerve cells are doubtless present, but these elements have not been identified with sufficient certainty. It seems probable that, at some levels, medium-sized cells in an intermediate region lateral to the central canal represent preganglionic neurons. Their neurites are believed to emerge through both ventral and dorsal roots (cf. section 6, chapter VII in the preceding volume). Funicular, commissural, and short internuncial *(Golgi-type II)* neurons have been variously described or inferred in dorsal and ventral regions of the gray substance. At early larval stages, large dorsal *Rohon-Beard cells* are present, which degenerate during the period of metamorphosis (Fig. 82). Moderately or fairly large dorsal nerve cells occurring within the gray substance of adult Anurans are presumably neuronal elements of a different type, not related to the *Rohon-Beard cells.*

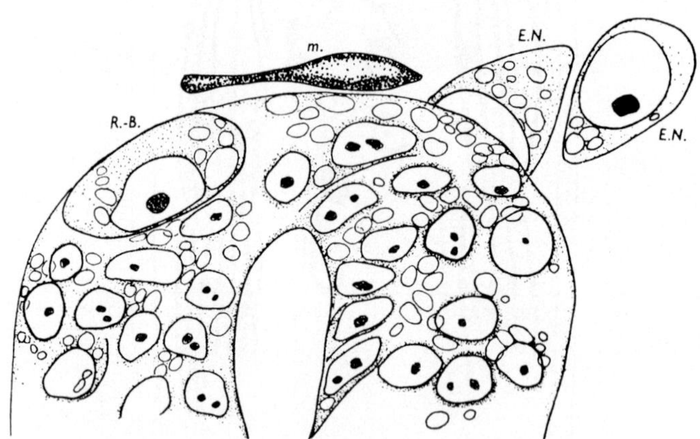

Figure 85B. Semidiagrammatic cross-section through dorsal part of spinal cord in a 58-hours Xenopus larva (from HUGHES, 1957). Abbreviations as in preceding figure; a *Rohon-Beard cell* is shown within the cord, and two cells of extramedullary (supramedullary) type are seen, one of which is emerging from the cord; yolk granules within cytoplasm are still seen.

In the African water Toad *Xenopus laevis*[98] the *Rohon-Beard cells*, which provide the primary sensory system of the trunk at larval stages, are, moreover, supplemented by *extramedullary neurones* (Fig. 85), which appear to migrate from the cord at an early period of neural differentiation (HUGHES, 1957). While the *Rohon-Beard cells* of Xenopus resemble those of other Amphibia so far as they are known, the extramedullary neurones, similar to the supramedullary ganglion cells of the Teleost Lophius, do not appear to have been recognized elsewhere among Tetrapods. This primary sensory system is gradually superseded after the typical spinal ganglia have become functional (HUGHES, 1957).

With respect to the *fiber sytems*, it is noteworthy that, in various Anurans, the entering dorsal root fibers display a distinct subdivision into a heavily medullated *dorsomedial fascicle* and a smaller, thinly medullated and non-medullated *dorsolateral fasciculus* comparable to *Lissauer's tract* or 'marginal zone' in Mammals (CAJAL, 1909; KEENAN, 1929). This dorsolateral bundle, which may also be intermingled with short intersegmental longitudinal internuncial fibers, seems intimately related to the substantia gelatinosa, participating in the formation of the dense 'dorsal neuropil' of SILVER (1942). In KAPPER's opinion (1947), '*les fibres latérales fines qui se terminent dans la corne postérieure, sont les fibres protopatiques ou vitales, tandis que les fibres grosses qui entrent dans les cordons postérieurs, sont des fibres épicritiques ou stéréognostiques. Chez les Amphibiens, la*

---

[98] *Xenopus laevis*, also known as '*südafrikanischer Krallenfrosch*' is classified with the aglossal Pipidae. The genus Xenopus, considered the 'most primitive', is represented by about five species, and has the widest distribution. Some characteristics of skeletal morphology in this peculiar Anuran suggest a relationship to Carboniferous and Permian Branchiosaurs (Phyllospondyli), while Salamanders and Frogs may not have sprung from Branchiosaurs but from Phyllospondyli closely related to the Branchiosauridae (NOBLE, 1931, 1954). Be that as it may, the peculiarities of Xenopidae as well as those of Gymnophiones are suggestive of polyphyletic lines of evolution toward the recent Amphibia since at least the Carboniferous period, such that, e.g. the order Anura might include genera derived from diverse ancestors pertaining to different extinct orders already evolved in the late Silurian. At any rate, the relationship of recent Amphibia to their palaeozoic ancestors remains highly uncertain. Thus, Urodeles and Anurans have also been considered as evolved from aquatic Lepospondyli. With regard to recent Amphibia, the earliest undoubted Anuran is said to be Protobratrachus of the Triassic, while typical Urodeles can be traced back only to the Jurassic (cf. the discussion of Amphibian evolution by YOUNG, 1955).

prédominance appartient à la première catégorie, à l'encontre de ce qui est chez les Mammifères, où elle est aux fibres stéréognostiques des cordons postérieurs'.[99]

The dorsomedial fasciculus joins the well-developed dorsal funiculus, bifurcating in the manner described by NANSEN (1887) in Myxine (cf. Fig. 51) and commonly obtaining in all Vertebrates (Fig. 86). However, besides these ascending and descending primary spinal root fibers with their collaterals, descending *root fibers of cranial nerves* (trigeminus, vestibular, glossopharyngeus and vagus) join the posterior funiculi and substantially contribute to its bulk (WALLENBERG, 1907). Vagus and glossopharyngeal fibers are said to reach the second or third spinal segment, vestibular ones the sixth, and the trigeminal fibers, running within the lateral part of dorsal funiculus, may descend as far as the lumbar enlargement.

According to KAPPERS *et al.* (1936) the dorsal funiculi in 'Fishes' average only about 5 to 6 per cent of the total white matter, while in some Frogs they can make up between 13 and 20 per cent of that substantia alba in the rostral part of the spinal cord. This relatively great size of dorsal funiculi in Anurans is believed essentially to result from the large amount of descending cranial nerve root fibers. If these latter are discounted, only a moderate relative expansion of dorsal funiculi in Anurans might be assumed, and these funiculi, moreover, 'would probably show but small increase in their cephalic as compared with their more caudal portions' (KAPPERS *et al.*, 1936).

There can be little doubt that, also in Anurans, direct (monosynaptic) connections between primary afferent dorsal root fibers and motoneurons obtain. Such connections are effected either by short collaterals reaching the dorsal extension of motoneuron dendrites, or by longer collaterals reaching the motoneuron perikarya in the ventral horn. The presence of such long collaterals has been denied by some authors, but is upheld by CAJAL (1909, 1952) on the basis of his own observations.

---

[99] According to some authors (e.g. HERRICK, 1931), all forms of cutaneous and deep sensibility of 'exteroceptive type' may be classified under two concepts, namely (1) a primitive protective or *protopathic* group, and (2) a more refined 'discriminative' *epicritic* group. This distinction goes back to HENRY HEAD and was discussed on pp. 695–696 in volume 3, part I, chapter V of this series. Although doubtless to some extent valid, its significance, partly based on differences in peripheral end-organs, respectively type of innervation, and partly on the periphery's central connections, still remains controversial (cf. also section 1, chapter VII in the preceding vol. 3, part II).

The central distribution of single brachial or lumbar dorsal roots in some Anurans was investigated by JOSEPH and WHITLOCK (1968) with a modified *Nauta-technique* for terminal degeneration and with other methods (Fig. 87). All dorsal root projections were found to remain homolateral and to be distributed segmentally over approximately five spinal segments. Ascending fibers from caudal segments are located medially to those from more rostral ones (so-called 'somatotopical distribution'). Dorsal root fibers were traced into oblongata and ultimately into the cerebellum. Some evidence for a relay from dorsal column to spinal lemniscus was obtained. With respect to the interesting conclusions and diagrams presented by the cited authors it should be kept in mind that the inferences based on finding with degeneration tech-

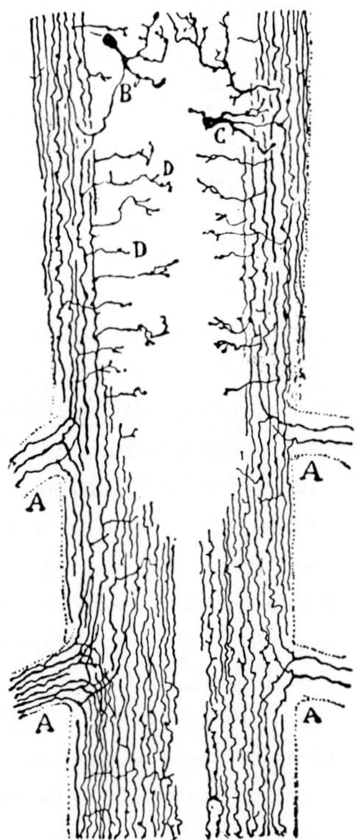

Figure 86. Tangential ('horizontal') section through the dorsal region of the spinal cord in an advanced larva of the Toad Bufo vulgaris (after SALA, from CAJAL 1909; *Golgi impregnation)*. A: dorsal roots with bifurcating fibers; B, C: funicular cells; D: short collaterals *('collatérales rudimentaires')* of posterior root fibers in dorsal funiculus.

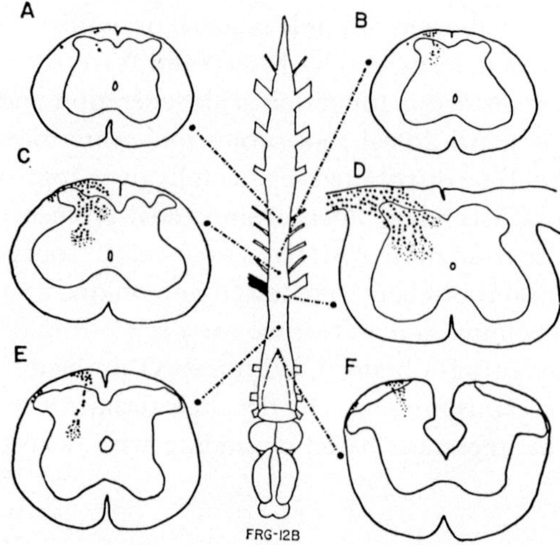

Figure 87. Schematic representation of the segmental distribution of degeneration resulting from transection of the second spinal dorsal root (blackened) in the Frog (from JOSEPH and WHITLOCK, 1968). Levels A–F depict cross-sections from indicated levels of neuraxis. The large and small black dots purport to mark 'the location of degenerating fibers and preterminal-terminal fields, respectively'.

niques are far less certain than commonly made to appear by investigators applying these methods.[100]

The secondary fiber systems originating in the spinal cord of Anurans are neurites of short and long internuncials. Those of *Golgi II type* cells may remain within the central gray, either homolaterally or crossing in dorsal or ventral commissure, both of which could also include 'dendritic' processes. The longer neurites pertain to funicular and commissural cells. The axons of these latter, as dorsal and ventral 'arcuate fibers' (Fig. 84A) decussate in the ventral commissure.

While many of the ascending and descending secondary neurites of funicular and commissural cells are components of the *eigenapparat*, others provide the ascending systems of the *verbindungsapparat*, and the axons of some funicular and commissural cells are doubtless channels giving off connections to both. This, of course, is also the case with regard to some ascending primary dorsal root fibers in dorsal funiculus.

[100] Cf. Chapter V, vol. 3, part I, p. 677.

The main secondary ascending systems to the brain are, as in Urodeles and Fishes, ill-defined spino-bulbar and spino-cerebellar fibers or 'tracts' within antero-lateral and lateral funiculus. The former, providing a spino-bulbar lemniscus, seem mostly crossed and may extend as far as the tectum mesencephali, presumably effecting unspecified connections with the caudal parts of the diencephalon (thalamus, hypothalamus). Such thalamic endings seem to obtain in all Anamnia.

Little is known concerning the descending systems of the spinal cord's *verbindungsapparat* in Anurans. These channels are essentially represented by the fasciculus longitudinalis medialis of the anterior funiculus. Its bulbo-spinal fibers seem to comprise crossed and uncrossed

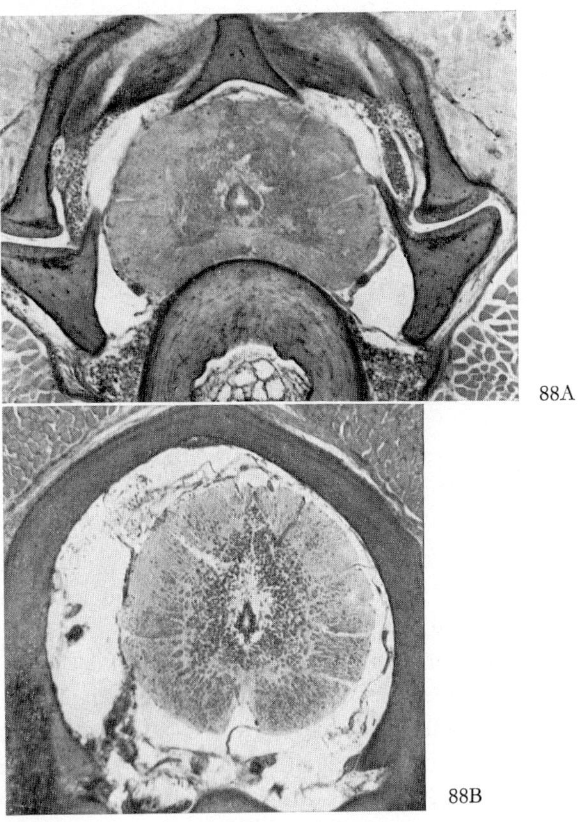

Figure 88A. Cross-section through a fairly rostral region of the spinal cord in the Gymnophione Schistomepum thomense (hematoxylin-cosin stain, approx. ×29; red. $^1/_2$).

Figure 88B. Cross-section through a fairly rostral region of the spinal cord in the Gymnophione Siphonops annulatus (hematoxylin-cosin stain, approx. ×32; red. $^1/_2$).

neurites from the reticular formation of the deuterencephalon as well as vestibulo-spinal fibers. It is uncertain whether direct tecto-spinal or cerebello-spinal connections are present. A *fiber of Mauthner*, occurring in larval Anurans, does not, as far as is known, persist in the examined adult forms.

Detailed data on the spinal cord of *Gymnophiones*, comparable to those available for Urodeles and Anurans, do not seem to be extant. Figure 88, illustrating sections through the spinal cord of Caecilians as seen *in situ* and by a routine cell stain, shows a conspicuous similarity with the overall configuration obtaining in some Urodeles (Fig. 81). In a few other Gymnophiones (Ichthyophis, Hypogeophis), of whose spinal cord various sections were incidentally examined by myself, no significant differences with respect to Schistomepum and Siphonops could be noted.

It is likely that, on the whole, the cellular elements and the fiber systems in the spinal cord of Caecilians conform to the structures described respectively inferred for Urodeles. However, in the adult specimens of Gymnophiones which I could examine, a *fiber of Mauthner* was not seen in the spinal cord, nor could I find a *Mauthner cell* in the oblongata. In larval Gymnophiona, nevertheless, these structures are said to be present (Burckhardt, 1891, also quoted by Tagliani, 1905 and Beccari, 1943).

9. Reptiles

Present-day Reptiles can be classified as comprising four or five orders, namely *Chelonia*, *Squamata*, *Crocodilia*, and *Rhynchocephalia*, respectively *Lacertilia* (Sauria) and *Ophidia*, if these forms are subsumed as orders under a 'subclass' *Squamata*. These groups are, as it were, remnants of a much more diversified population with perhaps 4 subclasses and at least 12 to 17 if not more 'orders' that flourished in the Mesozoic era, and originated in the late Palaeozoic.[101] The Rhynchocephalian Sphenodon of New Zealand (Hatteria, Tuatara) is regarded as a relic surviving with little change from the Triassic beginning of this group, about 200 or more million years ago, and seems to have changed hardly at all since the Jurassic, that is, in the course of perhaps 140 million years (Young, 1955). Sphenodon is a carnivorous, lizard-like reptile up to 2 feet long.

[101] Some early Reptiles may have made their appearance in the late Carboniferous, perhaps well over 200 million years ago (cf. vol. 1 of this series, p. 107).

Generally speaking, and in contradistinction to e.g. Anuran Amphibians, the spinal cord of Reptiles extends through the entire length of the vertebral canal, practically reaching, e.g. in Lacertilians, the tip of the tail. This considerable relative length of spinal cord appears correlated with the retention, in the Reptilian tail, of a 'primitive' muscular metamerism, which becomes superseded or 'fades out' in Mammals. In Reptiles, accordingly, a typically reduced *filum terminale* with conspicuous *cauda equina* does not occur, although, as e.g. in Ophidia (cf. Fig. 89A) a pronounced thinning of the cord's caudalmost portion may be displayed.

Lacertilia and Crocodilia are provided with limb and trunk musculature. Chelonia[102] lack trunk muscles, but have well developed muscles for head, neck, tail and extremities. In said groups cervical and lumbar intumescentiae[103] are present, which are not displayed by Ophidia (cf. Fig. 89). In these latter, only the trunk musculature is present. As can be seen in that figure, the thoracic part of the Chelonian spinal cord, between cervical and lumbar intumescentiae, is quite thin. In Reptiles, as in all Amniota, the spinal nerve roots and the nerve segments formed by the union of dorsal and ventral radices show a symmetrical metameric arrangement.[104]

Figures 90 to 93 illustrate cross-sections through the spinal cord of representative Reptiles. In most forms, the two posterior horns, capped by the *substantia gelatinosa Rolandi*, are well separated from each other, and a funiculus posterior, formed by the heavily medullated dorsomedial fasciculus of posterior root fibers, is rather distinctly developed. In the lower oblongata, at the transition to spinal cord, grisea representing terminal nuclei of the posterior funiculi can be recognized.

---

[102] Perhaps because of the special protection provided by their shells, Chelonia seem to have retained 'primitive' morphologic features. Although Chelonia are much modified in some ways 'they show us several characteristics of the earliest Permian reptiles' (Young, 1955). Hatteria is, qua trunk and limb musculature, here included with the Lacertilia.

[103] In some large fossil Dinosaurs, with extremely developed posterior extremities, the cavity within the vertebral canal occupied by the lumbar intumescentia exceeds the volume of the cranial cavity enclosing the brain.

[104] Kappers *et al.* (1936), following other authors, call attention to the possibility that, as in Amphioxus and presumably also some Selachians, the ventral roots may not be limited to one peripheral segment (myotome) only, but may give some fibers to the adjacent myotomes (respectively myotomic derivatives). Concerning the 'ventral roots' of Amphioxus, cf. above p. 71.

Lissauer's tract or fasciculus dorsolateralis ('lateral root bundle') already pointed out above as evident in some Amphibians, forms the marginal zone of substantia gelatinosa, and frequently shows the tendency to acquire a somewhat lateral position. However, several variations occur. In Crocodilia, KEENAN (1929) described five different layers or zones vaguely displayed by the *Weigert-Pal method* in the apical region of the dorsal horns, beginning at the periphery with the finely medullated or non-medullated fibers of *Lissauer's tract*, and followed by variously patterned strips of gray and white substance, the 'fifth layer' being the 'deeper part of substantia gelatinosa', 'practically devoid

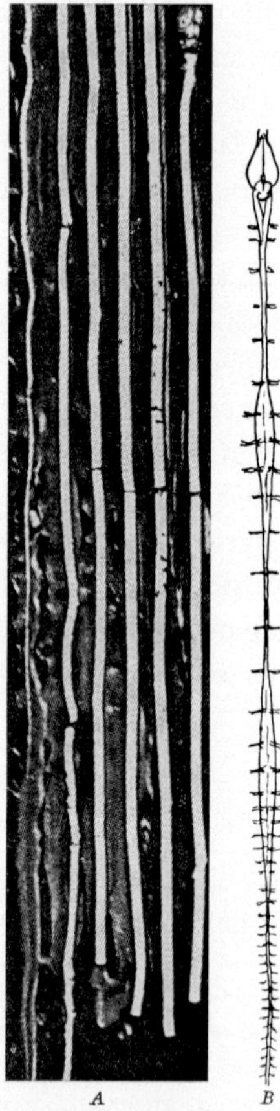

Figure 89. Brain and spinal cord of a 3 m long Python (A) compared with brain and spinal cord of a Turtle (B). The Snake and the Turtle neuraxis are not depicted at an identical scale (from KAPPERS *et al.*, 1936; A after DE LANGE, B after BOJANUS).

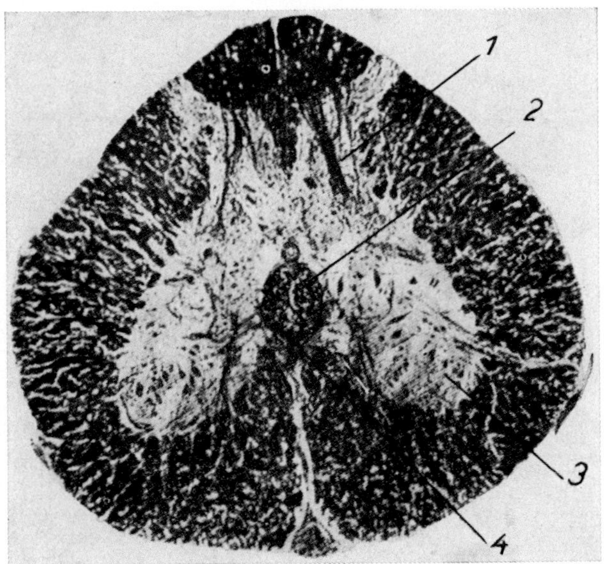

Figure 90 A. Cross-section through the spinal cord of Lacerta muralis (from BECCARI, 1943; myelin stain). 1: dorsal horn; 2: fasciculus longitudinalis medialis; 3: ventral horn; 4: commissura accessoria (the poorly developed commissura ventralis s. anterior proper, not indicated, is faintly recognizable in the narrow layer of central gray ventral to central canal and dorsal to fasciculus longitudinalis medialis).

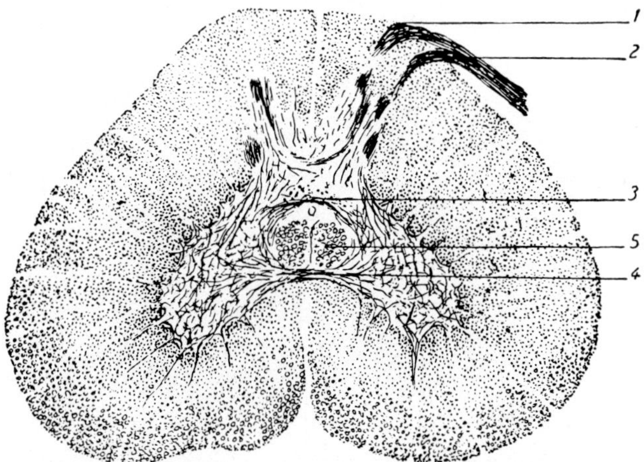

Figure 90 B. Cross-section through the cervical medulla of an adult Gongylus ocellatus (after TERNI, from BECCARI, 1943; myelin stain). 1: dorsomedial (or 'internal') dorsal root bundle; 2: ventrolateral (or 'external' dorsal root bundle (between 1 and 2, not indicated, the poorly medullated '*tract of Lissauer*' related to the dorsolateral posterior root bundle); 3: commissura dorsalis s. posterior; 4: commissura accessoria (cf. preceding figure); 5: fasciculus longitudinalis medialis. Gongylus ocellatus is a skink (Saurian, Lacertilian) commonly found in Sardinia.

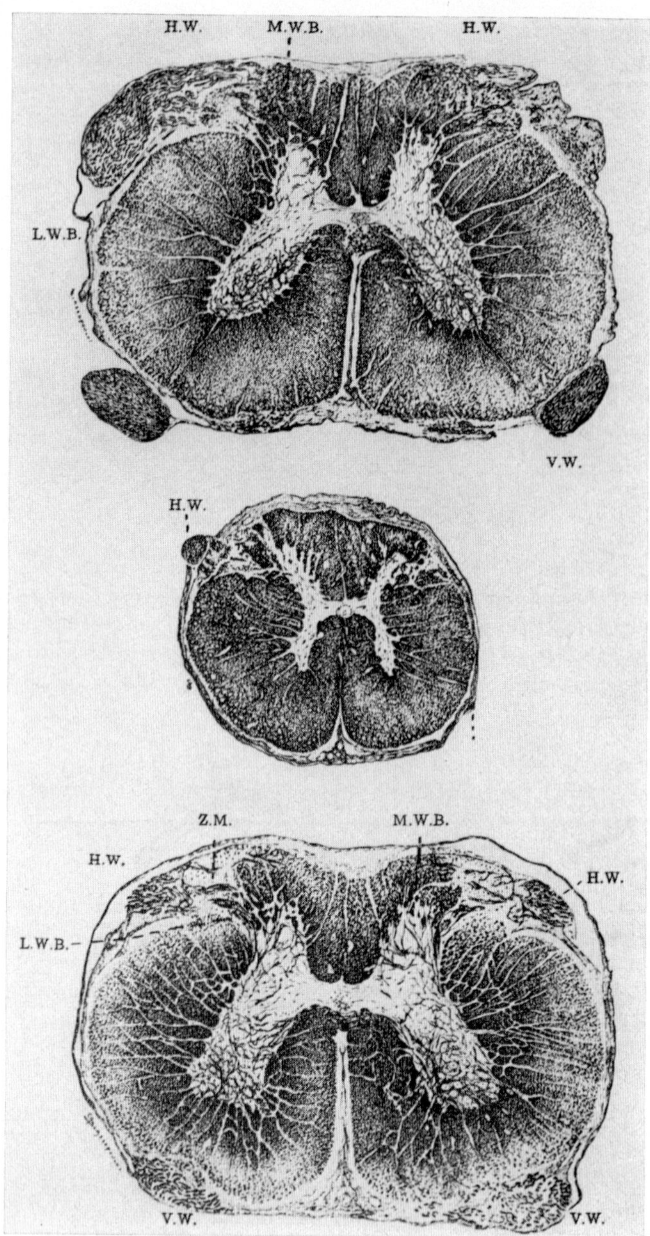

Figure 91. Cross-sections, drawn at the same magnification, through the spinal cord of a 'Turtle' (after DE LANGE, from KAPPERS *et al.*, 1936; myelin stain). Top: cervical cord; middle: thoracic cord; botton: lumbar cord; H.W.: dorsal root; L.W.B.: dorsolateral root bundle; M.W.B.: dorsomedial root bundle; V.W.: ventral root; Z.M.: 'zona marginalis'.

of myelination'. Again, in at least some Saurians, Snakes, and also some Chelonia, a portion of the coarsely medullated dorsal root fibers forms a 'ventrolateral bundle' which turns into the lateral funiculus, ventrolaterally to *Lissauer's tract* and substantia gelatinosa (cf. Figs. 90, 95). With regard to these two latter structures, not only species differences, but also regional differences between cervical, thoracic and lumbar levels of one and the same spinal cord obtain.

Likewise, comparable variations in the arrangement of the ventral horns can be noted. Thus, Figure 91 shows the very narrow ventral horns in the thoracic region of a Turtle. Again, in the Crocodile (Fig. 93), the 'eccentric' ventral position of the central canal is conspicuous. Another peculiarity is the presence, in many Reptilian forms, of marginal grisea or '*nuclei marginales*', presumably derived from a dorsal neighborhood of basal plate (ventral horn). Apparently first described by GASKELL (1885) in the Alligator, they were subsequently shown by v. KÖLLIKER, KAPPERS, STERZI, LACHI, TERNI and others in diverse Reptiles as well as in Birds, and appear thus to be a characteristic and commonly occurring feature of the Sauropsidan spinal cord.[105] Nevertheless, among Mammals, similar peripheral grisea were also noted in Chiroptera by DRÄSEKE (1903) and POLYAK (1924), in the Cat by ANDERSON *et al.* (1964), and in the Cynomolgus monkey by DUNCAN (1953).

With respect to the *neuronal elements* in the spinal cord of Reptiles (Figs. 94, 95, 96), numerous cells, both in ventral and dorsal horns, extend their dendrites into the white substance, a marginal 'dendritic plexus' being quite commonly present at the ventral periphery of the lateral funiculus. The just mentioned marginal nuclei are usually located in the region where the marginal 'dendritic plexuses' are particularly well developed. The motoneurons are generally arranged in two groups, a medial group presumably concerned with innervation of trunk musculature, and a lateral group which seems to innervate the limb musculature, being accordingly developed in cervical and lumbar intumescentiae. The motoneurons in the cord of Ophidia would thus correspond to an expanded medial group.

---

[105] In Birds, v. KÖLLIKER designated these grisea as '*Hofmannsche Kerne*' (according to KAPPERS, 1920, in honor of the technician who directed KÖLLIKER's attention to these configurations). Yet, with regard to their discovery in the Alligator, KÖLLIKER explicitly recognized GASKELL's priority. Additional comments on these grisea, with relevant bibliographies, are given by KAPPERS *et al.*, 1936, and SCHARF, 1958.

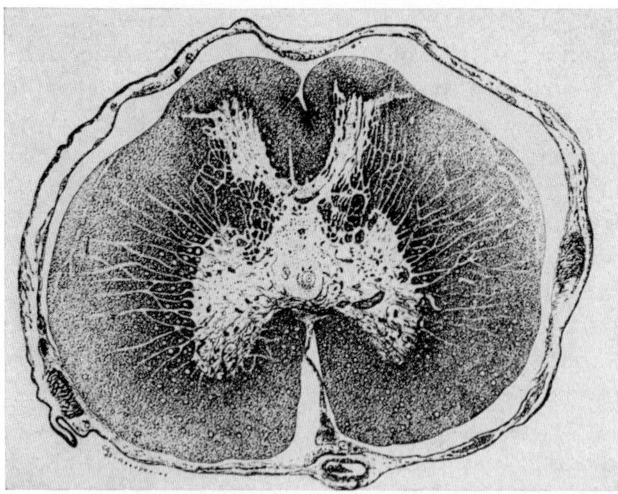

Figure 92. Cross-section through the cervical cord of the Snake Python reticulatus (from Kappers *et al.*, 1936; myelin stain).

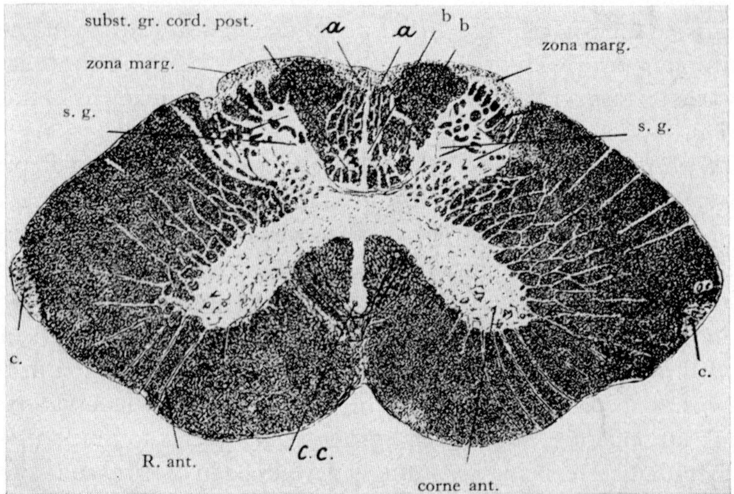

Figure 93. Cross-section through the cervical cord of Crocodilus porosus (from Kappers, 1947; myelin stain). a, b: gray substance in funiculus posterior; c: nucleus marginalis *(Gaskell-Hofmann-Kölliker)*; c. c.: central canal (note its excentric ventral position; corne ant.: ventral horn; Rad. ant.: ventral root; s. g.: substantia gelatinosa; subst. gr. cord. post.: griseum of posterior horn; zona marg.: 'zona marginalis' (part of Lissauer's tract). Note the not labelled dorsal commissure and the anterior commissure, which, although directly ventral to central canal, nevertheless separates part of fasciculus longitudinalis medialis from remainder of anterior funiculus (compare with Figs. 89, 90, 91, 92).

The slightly smaller preganglionic elements are apparently located somewhat more dorsally in the ventral horn. In Turtles, who generally lack a myotomic musculature in the thoracic region, the efferent elements in the narrow ventral horns of the corresponding cord segments are regarded as exclusively preganglionics.

In Reptiles, the preganglionic neurites of the 'visceral-efferent' cells seem to emerge predominantly through the ventral roots, although some such fibers in dorsal roots cannot entirely be excluded. On the other hand, the rather coarse efferent dorsal root fibers originating from multipolar elements in the ventral horn of the cervical spinal cord of Lacerta (BECCARI, 1913) and Chelonia (BANCHI, 1903) have been interpreted as pertaining to the n. accessorius system, that is, as 'branchiomotor' or 'special visceral efferent' rather than preganglionic (BECCARI, 1943). Similar efferent dorsal root fibers were demonstrated by v. LENHOSSÉK (1890) in Birds. However, the functional significance of these fibers, depicted in Figure 96, cannot be regarded as entirely elucidated, especially if, as seems possible, efferent elements of that type may not be restricted to the cervical spinal cord but could occur in lumbar segments (cf. KAPPERS *et al.*, 1936).

The nerve cells in the above-mentioned nuclei marginales *(gruppi cellulari periferici* of TERNI*)* seem to send their neurites into the homolateral, and, by way of the ventral commissure, into the contralateral funiculus lateralis (TERNI, 1926). These neurons are therefore interpreted as funicular cells, respectively perhaps as tract cells of an ascending secondary system which might reach the brain.

In both ventral and dorsal horn, various sorts of internuncial cells, comprising funicular, tract, commissural and shorter interneurons, have been described (cf. Figs. 94, 95, 96). Fairly large elements of this sort are not restricted to the ventral griseum but are also located in dorsal horn and dorsal midline. Little is known about the occurrence of transitory large *dorsal root cells of Rohon-Beard type*, but such elements have been recorded in embryos of Tropidonotus (VAN GEHUCHTEN, 1897).

With respect to the fiber systems, descending and ascending branches of the dorsal root's heavily medullated dorsomedial division form the posterior funiculi such that the ascending fibers pertain to both *eigenapparat*[106] and *verbindungsapparat*. The components involved

[106] Some root fiber collaterals of the *eigenapparat* are said to cross the midline, ending in the contralateral ventral horn (GOLDBY and ROBINSON, 1962).

in this latter terminate in the nuclei of the posterior funiculi within the zone of transition between spinal cord and oblongata. These grisea, whose presence is already suggested in some Amphibians, are more clearly developed in Reptiles, and can be interpreted as relay nuclei for 'muscle sensibility' and 'finer tactile discrimination'. The crossed ascending pathway originating in these nuclei of the posterior funiculi is presumed to form a 'medial lemniscus' reaching the tectum opticum and possibly (as in Mammals) the thalamus.

KAPPERS *et al.* (1936) call attention to the fact that, where the above-mentioned ventrolateral bundle of heavily medullated root fibers is present, a comparison between the size of dorsal funiculi in these Reptiles and in Frogs must take into account (1) the dorsal root fibers in lateral funiculus of such Reptiles, and (2) the large amount of descending cranial nerve root fibers in the Anuran posterior funiculus. In Reptiles the cranial nerve root fibers (essentially of radix descendens trigemini) do not seem to reach caudalward in an appreciable number beyond the uppermost cervical segments.

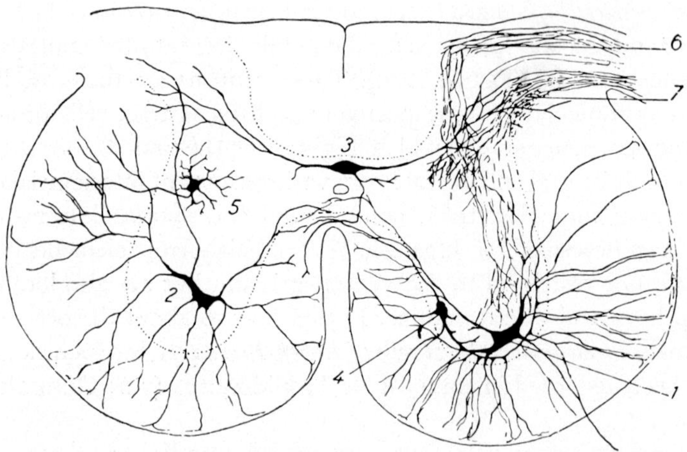

Figure 94. Cross-section showing some neuronal elements and root fibers in the spinal cord of the Chelonian Emys europea as seen by the *Golgi impregnation* (modified after BANCHI, from BECCARI, 1943). 1: motoneuron; 2: dorsal root efferent *(Lenhossék cell and fiber);* 3: median dorsal cell; 4: commissural cell; 5: funicular or tract cell; 6: dorsomedial ('internal') bundle of posterior root; 7: ventrolateral ('external') bundle of posterior root. Between 6 and 7 two fibers of dorsolateral bundle related to *Lissauer's tract*.

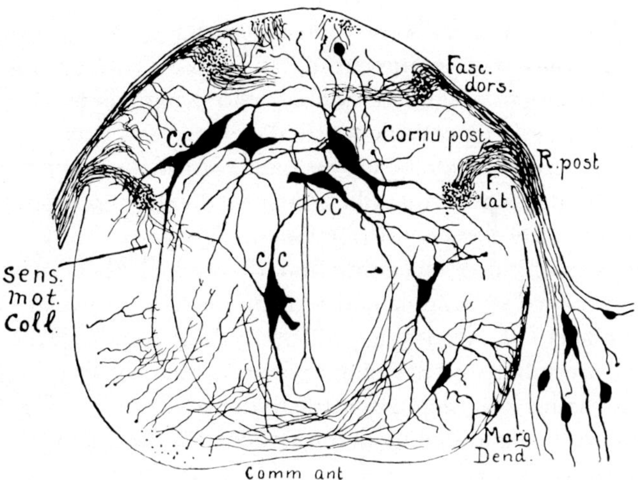

Figure 95. Cross-section showing neuronal elements, including dorsal root components in the late embryonic spinal cord of the Ophidian Tropidonotus natrix as seen by *Golgi impregnation* (after Retzius, from Kappers *et al.*, 1936). C. C.: commissural cells; Comm. ant.: anterior or ventral commissure; Fasc. dors.: dorsomedial posterior root fasciculus; F. lat.: ventrolateral posterior root fasciculus (between the two fasciculi, at C. C. on the left, some fibers of the dorsolateral fasciculus can be recognized); Marg. dend.: marginal dendritic plexus; Sens. mot. Coll.: collaterals of ventrolateral dorsal root fasciculus presumed to effect direct 'sensori-motor' synaptic connections.

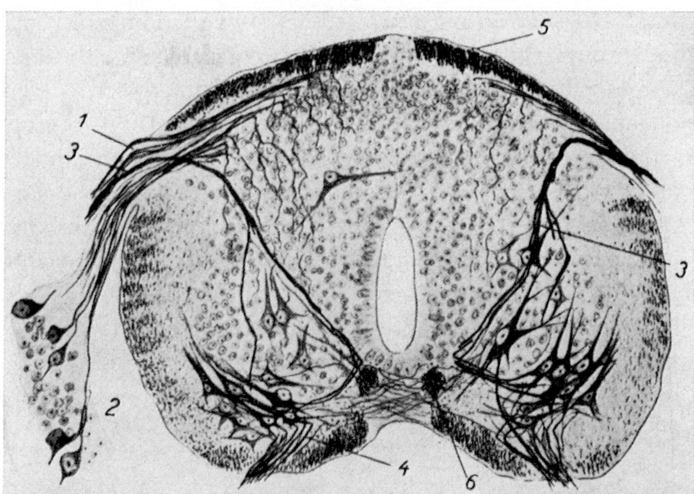

Figure 96. Cross-section through the spinal cord in an advanced embryo of Lacerta muralis at the level of 4th cervical nerve segment (from Beccari, 1943; *Cajal's silver impregnation*). 1 and 2: fibers of posterior root and spinal ganglion cells; 3: dorsal efferent *fibers of Lenhossék;* 4: efferent ventral root fibers; 5: posterior funiculus; 6: fasciculus longitudinalis medialis.

The dorsal root fibers of the ventrolateral bundle are believed to effect connections with motoneurons and large ventral commissural cells. In addition to a single or double ventral commissure,[107] a dorsal commissure is usually conspicuously developed and may also consist of two distinctly separated parts in some forms. It contains not only collaterals of dorsal funicular fibers but also dendrites or even perikarya of commissural elements as pointed out above.

Details concerning ascending and descending fibers of the *eigenapparat* provided by homolateral and commissural funicular cells are poorly understood. According to TERNI (1922) homolateral and contralateral, ascending and descending fibers of this system join the portion of fasciculus longitudinalis medialis dorsal to the accessory ventral commissure, being distributed throughout the length of the spinal cord.

The reasonably well established ascending secondary fiber components of the *verbindungsapparat* seem to comprise a spino-cerebellar and a spino-bulbo-tectal (-thalamic) tract. The former consists of uncrossed as well as crossed spino-cerebellar fibers in the dorsal part of the lateral funiculus. In some forms, a dorsal spino-cerebellar tract is said to be distinguishable from a ventral one, which may partly overlap with the spino-bulbo-mesencephalic system. This latter, also known as the *secondary sensory tract of Edinger*, seems to be predominantly crossed and runs through the ventral portion of lateral funiculus.

The generally acknowledged, crossed and uncrossed descending components of the *verbindungsapparat* are provided by the mesencephalo-bulbo-spinal (reticulo-spinal) fibers of fasciculus longitudinalis medialis and by the secondary vestibulo-spinal fibers. The vestibular components seem to join in part the fasciculus longitudinalis medialis which appears to represent the main 'motor tract' or to form an indistinctly circumscribed separate vestibulo-spinal tract within adjacent neighborhoods of anterior and lateral ('anterolateral) funiculus, here perhaps overlapping with *Edinger's ascending tract*. The presence of a direct (crossed and uncrossed) tectospinal tract and that of a crossed rub-

---

[107] The commissura ventralis is located within the central gray dorsal to the fasciculus longitudinalis medialis. The commissura (ventralis) accessoria passes through fasciculus longitudinalis medialis or, according to some authors, separates the 'true' fasciculus longitudinalis medialis from the remainder of anterior funiculus. If the commissura accessoria is well developed, the commissura ventralis may be much reduced, and vice-versa (cf. Figs. 90–93).

rospinal tract can be assumed. Likewise, cerebellospinal and olivospinal channels may be present, but the available evidence concerning these four systems appears inconclusive.

10. Birds

Recent Birds, which display many morphologic characteristics similar to those of Reptiles, seem to have evolved from these latter as originally toothed forms, perhaps approximately 165 million years ago, about the transition from Triassic to Jurassic Periods. The Triassic Archosaurian Reptiles may have provided the origin of two 'independent' groups which took to the air, the extinct Pterodactyls and the still surviving Birds.[108] At present, more than 20,000 different Avian species seem to be extant, and may be classified into 20 to 30 or more orders, of which most pertain to the superorder *Neognathae* with keeled sternum *(Carinata)*. The superorder *Impennae* includes the flightless Penguins, with a developed crista sterni. The superorder *Palaeognathae* is, *inter alia*, characterized by the lack of a sternal crest, and subsumes flightless *Ratite birds* with several orders, represented by Ostriches (Africa, Southwest Asia), Rhea (South America), and Australasian forms such as Dromaeus (Emu) and Casuarius (Cassowary).

The Avian spinal cord, extending, like the Reptilian one, throughout the length of the vertebral canal, is thereby devoid of 'cauda equina' and typical filum terminale. The cord is, nevertheless, commonly absent in a small part of the tail region, where remnant of atrophied vertebrae are found. Generally speaking, the ventral roots of the Avian spinal cord appear, *prima facie*, to be somewhat larger than the radices dorsales.[109] In contradistinction to most Reptiles, a greater number of

---

[108] Concerning some further details of classification and purported ancestry, cf. the concise discussions in YOUNG, *The Life of Vertebrates* (1955).

[109] Cf. KAPPERS *et al.* (1936, p. 284). Yet, the only more or less exact numerical statement which I could find (GRAF, 1956, also quoted by BLINKOV and GLEZER, 1964) refers to 904 fibers in the Pigeon's dorsal root of the 22nd segment, and to 629 fibers in the 23rd segment's ventral root (BLINKOV and GLEZER, 1964, p. 67 quote here 23rd as 22nd segment). Graf's counts would agree with the overall numerical ratios discussed above in section 1, p. 12 f. However, as can be seen from Figure 97, both segments pertain to the lumbosacral plexus in the region of sinus rhomboidalis, where the Avian posterior roots are 'unusually large' (KAPPERS *et al.*, 1936, p. 203).

cervical and lumbosacral segments is present, but a lesser number of thoracic ones (cf. also KAPPERS *et al.*, 1936). Thus, in the Ostrich (STREETER, 1904), there are 51 spinal nerve segments, including 15 cervical, 8 thoracic, 19 lumbosacral, and 9 coccygeal ones. In the Pigeon, whose neuraxis is illustrated by Figure 97, the vertebral column comprises 14 cervical vertebrae, if two that carry 'cervical ribs' not articulating with the sternum are included. Four or five thoracic vertebrae are present, all except the last united in a single mass. The last thoracic (or rib-bearing) vertebra is united with about five designated as 'lumbar', two 'sacrals' and five caudals, forming a so-called '*synsacrum*' which is fused with the ilium. The short tail contains about six free caudal vertebrae, ending with four that are fused together to form the upturned pygostyle, supporting the tail feathers (cf. YOUNG, 1955).

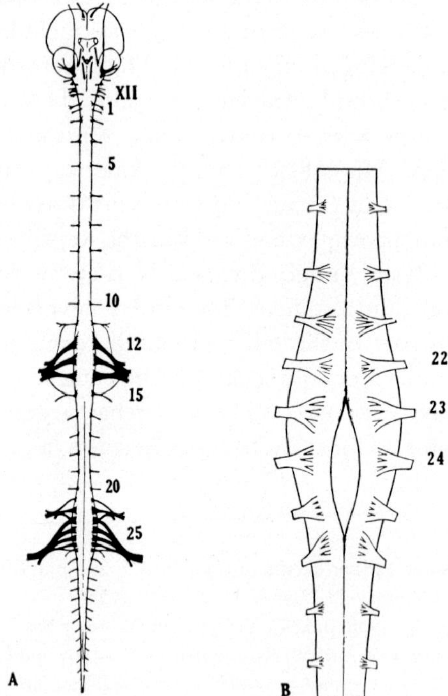

Figure 97. Macroscopic aspects of the spinal cord in the Pigeon (after HUBER, from KAPPERS *et al.*, 1936). A: ventral aspect of neuraxis, showing number of nerve roots and formation of plexuses; B: dorsal view of intumescentia lumbo-sacralis with sinus rhomboidalis.

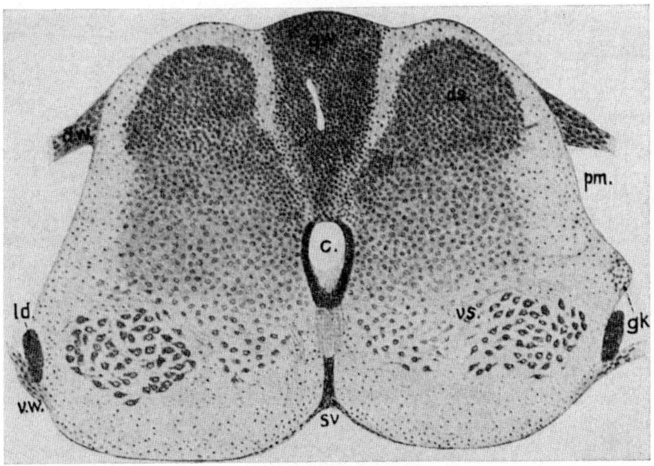

Figure 98. Cross-section through the spinal cord in a 12 to 13 days old Chick embryo at the level of intumescentia lumbalis (from KUPFFER, 1906). c: central canal; ds: dorsal gray column; dw: dorsal root; gk: peripheral nerve cell group *(Hofmann-Kölliker nucleus)*; gw: 'gelatinous' neuroectodermal tissue *('Gallertgewebe')* of sinus rhomboidalis; ld: ligamentum denticulatum; pm: fine lining of endomeninx ('Pia mater'); sv: fissura mediana anterior ('Fissura ventralis'); vs: ventral gray column; vw: ventral root.

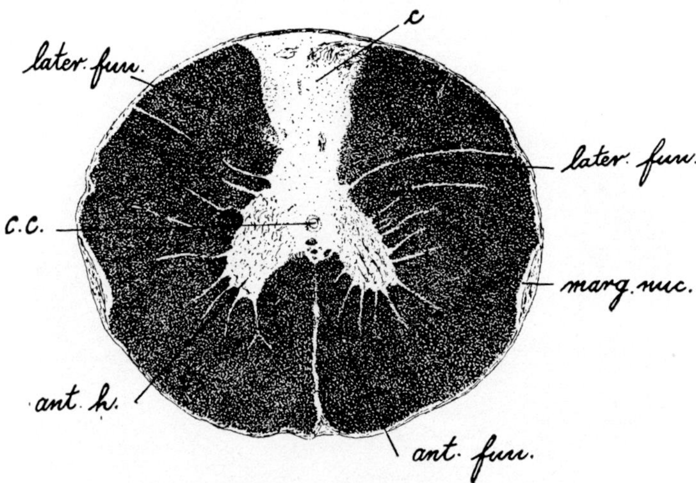

Figure 99. Cross-section through the upper cervical spinal cord in the Gull Larus argentatus (from KEENAN, 1929; myelin stain). c: Common posterior horns, fused in midline; c.c.: central canal; marg. nuc.: nucleus marginalis. Note the apparent nearly complete lack of posterior funiculi. These may already have partly ended in c, and be also partly included in lateral funiculi. The occurrence of individual variations is, moreover, not unlikely.

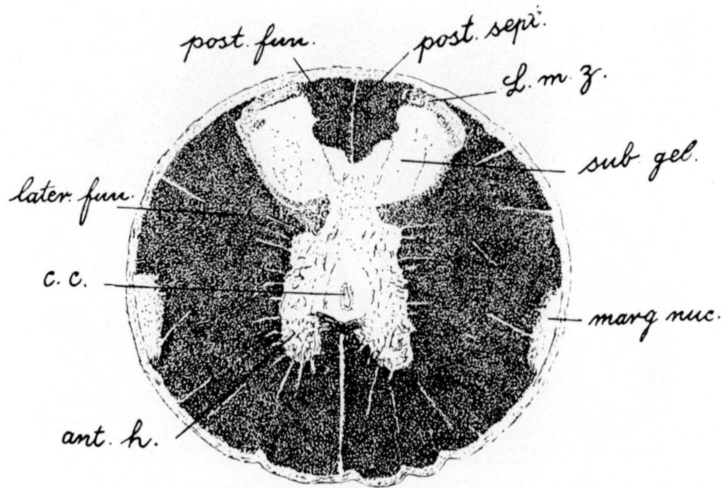

Figure 100. Cross-section through the upper cervical spinal cord of the Chick, Gallus domesticus (from KEENAN, 1929; myelin stain). c.c.: central canal; L.m.z: *Lissauer's marginal zone;* post. sept.: septum posterius. Note posterior funiculi.

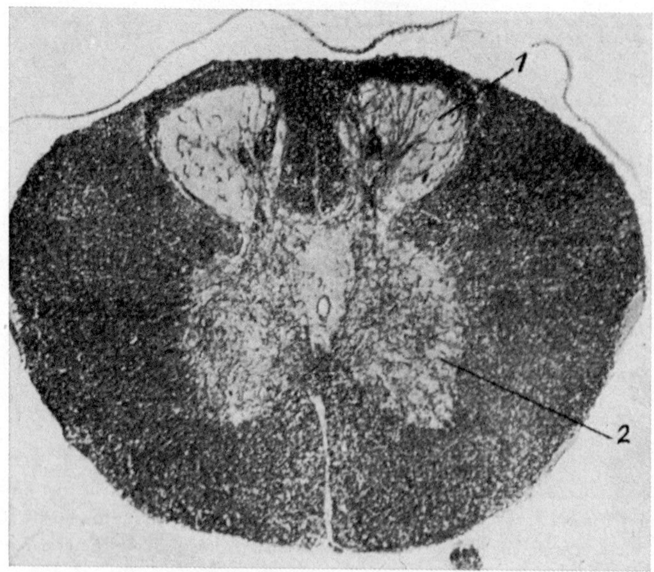

Figure 101. Cross-section through the cervical spinal cord of the Passeriform Blackbird Merula (from BECCARI, 1943; myelin stain). 1: dorsal horn; 2: ventral horn.

Two distinct intumescentiae (cervicalis and lumbalis, respectively lumbosacralis) seem to be displayed by all Birds. In flying forms, the cervical enlargement is usually the larger one, but in flightless forms, particularly in the Ostrich, with considerable development of posterior extremity musculature, the size of the lumbosacral enlargement conspicuously surpasses that of intumescentia cervicalis.

A peculiarity of the intumescentia lumbosacralis in apparently all Birds is the so-called *fossa or sinus rhomboidalis* (Figs. 97, 98, 102). It is formed by a gap between the laterally diverging posterior funiculi inclusive of dorsal horns, perhaps, as Kappers suggested, being a correlated consequence of the increased number of dorsal root fibers at these levels. The resulting dehiscence, also called *sinus lumbosacralis*, develops dorsally to the closed and persistent central canal but remains intraspinal and is filled out by a gelatinous *('gallertartiges')* cellular material which, on the basis of substantial evidence, can be regarded as derived from spongioblastic, glial elements (including the glial septum posterius and ependymal cells migrating from the closed central canal).[110] The sinus with its gelatinous tissue was already noted more than 150 years ago by Emmert (1811) and subsequently investigated by numerous authors (cf. e.g. Terni, 1924). Histochemical techniques disclosed that glycogen is stored in its cellular elements, whose ultrastructure was recently studied by Welsch and Wächtler (1969) in the Pigeon's spinal cord. These authors found the cytoplasm of the sinus-cells crowded with glycogen particles, such that the cellular organelles become confined to narrow spaces at the periphery. Numerous bundles of small filaments extend into most of the abundant cell processes. The electron microscope also disclosed relatively numerous finely medullated nerve fibers forming a loose plexus between the glycogen cells. In addition, Welsch and Wächtler observed non-medullated fibers and synaptic junctions on the surfaces of said cells. The cited authors interpret their findings as suggesting an innervation[111] of the '*glycogen body*'. This latter term is now generally used by histologists and cytologists as the preferable synonym for 'sinus rhomboidalis'.

---

[110] Kappers (1924; cf. also Kappers *et al.*, 1936) as well as Hansen-Pruss (1923, cf. chapter VI of the preceding volume) originally interpreted the gelatinous tissue of the sinus rhomboidalis as a leptomeningeal derivative. However, in his posthumous '*Anatomie du Système Nerveux*' (1947, p. 288), Kappers seems to agree with the opinion that the cells of said tissue '*sont d'origine glieuse*'.

[111] Such innervation would presumably be intraspinal, i.e. by neuronal elements located within the neuraxis.

WELSCH and WÄCHTLER (1969), who point out that '*der Glykogenkörper der Vögel ist bei den Wirbeltieren das Gewebe mit dem höchsten Gehalt an gespeicherten Kohlehydraten*', also noted that the glycogen cells reach, on one hand, the dorsal surface of the spinal cord, and, on the other hand,

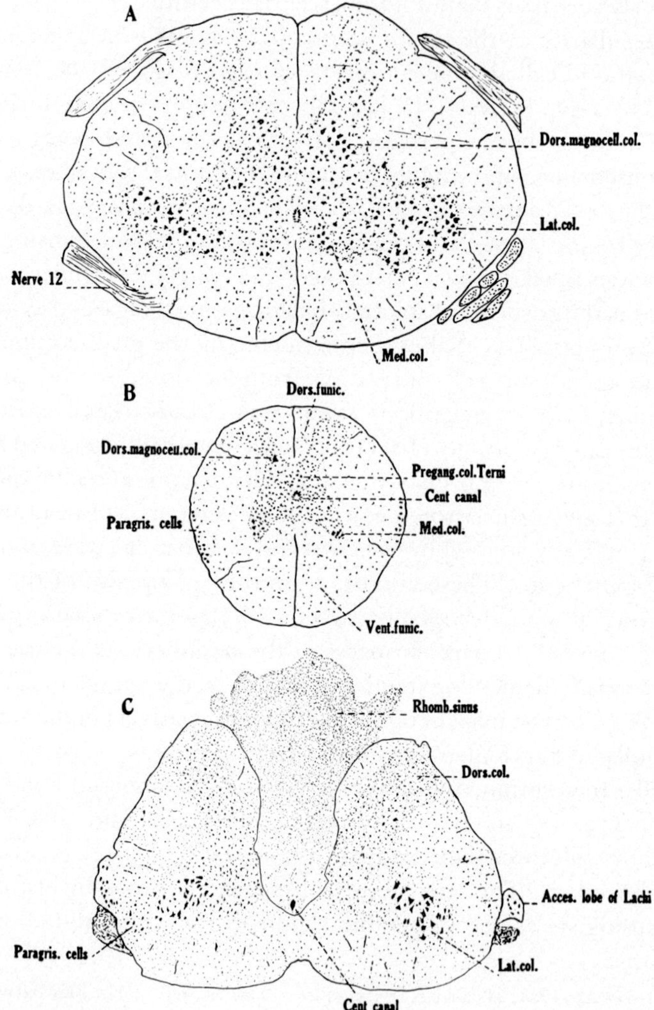

Figure 102. Cross-sections through the Pigeon's spinal cord at representative levels, as drawn from cell-stain preparations (after HUBER, from KAPPERS *et al.*, 1936). A: at level of 12th spinal nerve; B: thoracic cord at level between 16th and 17th nerves; C: at level of sinus rhomboidalis, between emergence of 24th and 25th nerves. Compare with Figures 97 and 98.

may pass through the ependymal lining and invade the lumen of the central canal. No convincing hypothesis concerning function of the richly vascularized glycogen body has been propounded. While some authors regard it as functionally irrelevant, others suspect its significance for not yet elucidated metabolic processes within the nervous system (cf. WELSCH and WÄCHTLER, 1969).

Details of the Avian spinal cord's ontogenetic development are contained in ROMANOFF's (1960) treatise on the Avian embryo and some additional observations have been recorded by STEDING (1962). Again, with regard to recent investigations using the technique of autoradiography, the studies on 'neurogenesis' in the Avian spinal cord undertaken by KANEMITSU (1971, 1972a, 1972b) should be mentioned. This topic, however, pertains to problems of histogenesis as dealt with in volume 3, part I, chapter V, section 1 of this series.

Figures 98 to 103 depict some representative cross-sections illustrating the configuration of grisea and white matter in the Avian spinal cord, including (Fig. 98) an advanced ontogenetic developmental stage of sinus rhomboidalis. In relation to the total volume of white matter, the posterior funiculi, comprising approximately 7 to 8 per cent of the total funicular area seen in the cervical region, are considered small.[112] *Lissauer's zone* caps the substantia gelatinosa at the surface of the cord as a narrow band with variable outlines. In the lower cervical and in the lumbosacral regions the anterior horns greatly predominate in size over the posterior ones. The marginal nuclei (of GASKELL, KÖLLIKER, HOFMANN) are generally conspicuous and form, at least in some instances, a continuous or nearly continuous 'column'. TERNI (1926) recognized about 10 or more 'marginal groups'. Particularly in the lumbar region, these nuclei may protrude from the cord (cf. Fig. 102) as so-called *accessory lobes* of LACHI (1889) near the attachment of the denticulate ligament (as depicted in Figure 98, the marginal nucleus lies here closely dorsal to that ligament). With regard to the overall configuration of gray and white matter, KEENAN (1929) recorded several 'irregularities', respectively 'heterotopies', whose significance as individual or as species variations remains unclarified and which suggest caution in formulating generalized overall descriptions.

Concerning the neuronal elements of the spinal cord, a marginal dendritic plexus has been observed at early embryonic stages but is not

[112] In comparable regions of the Reptilian spinal cord, the posterior funiculi are said to comprise 13 per cent of white matter (BROUWER, 1915; KEENAN, 1929).

apparently well displayed in the adult cord, although dendrites still extend into the white matter. The rather large multipolar motoneurons form, as in Reptiles, a medial (ventromedial) group presumably innervating the trunk musculature, and a large lateral group characteristic for cervical and lumbar intumescentiae. This cell-group, which has been somewhat arbitrarily subdivided into ventrolateral, dorsolateral and lateral subgroups seems thus to innervate the limb musculature. The term '*paragriseal cells*', as used by some authors (cf. Fig. 102) subsumes both the elements of the typical marginal nuclei and the scattered nerve cells within the lateral funiculus. The irregularly distribut-

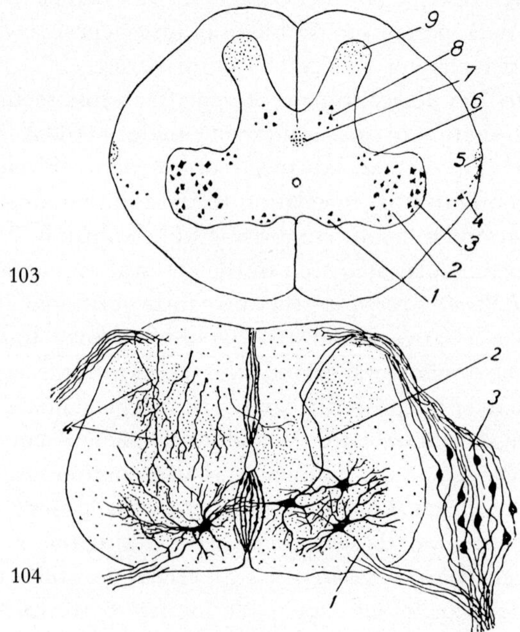

Figure 103. Diagram depicting the overall arrangement of neuronal elements in the avian spinal cord (pigeon) as interpreted by Huber (after Huber, from Beccari, 1943). 1: medioventral column of anterior horn neurons; 2: 'mediolateral' column; 3: 'laterolateral' column; 4: 'paragriseal' cells; 5: marginal nucleus *(Gaskell-Hofmann-Kölliker)*; 6: column of '*Lenhossék cells*'; 7: preganglionic column *(of Terni)*; 8: dorsal 'magnocellular' column; 9: *substantia gelatinosa Rolandi.*

Figure 104. Neuroectodermal elements demonstrated by the *Golgi impregnation* in the cervical spinal cord of a fairly advanced Chick embryo (after van Gehuchten, from Beccari, 1943). 1: efferent ventral root fibers; 2. efferent dorsal root fibers *(of. v. Lenhossék)*; 3: spinal ganglion and afferent dorsal root fibers; 4: collaterals of afferent dorsal root fibers.

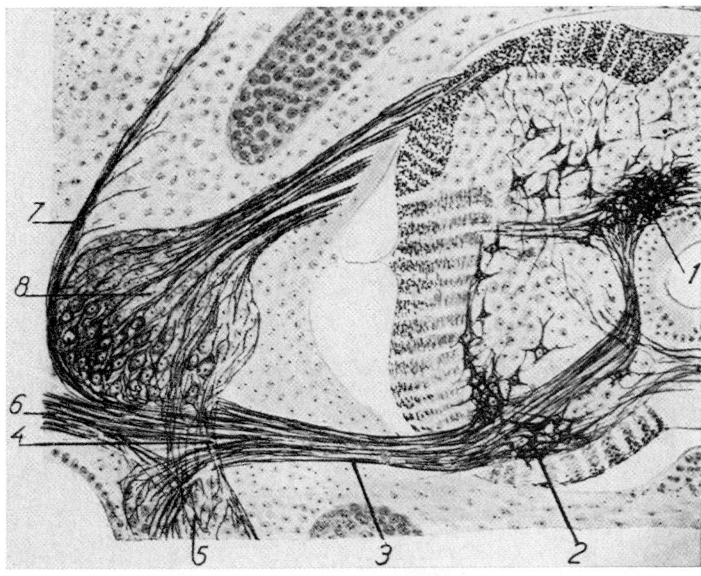

Figure 105. Cross-section, at level of 5th thoracic nerve, through the spinal cord in a Chick embryo of 9–10 days as seen by means of *Cajal's silver impregnation* (after TERNI, from BECCARI, 1943). 1: thoraco-lumbar (preganglionic column; 2: somato-motor column; 3: ventral root; 4: ramus communicans; 5: sympathetic ganglion; 6: ramus ventralis; 7: ramus dorsalis; 8: spinal ganglion with radix dorsalis s. posterior.

ed ventrolateral multipolar paragriseal cells not pertaining to the marginal nuclei probably represent motoneurons located within the white matter. The dorsolateral ones shall be discussed further below.

According to TERNI (1923) and BECCARI (1943) the visceral-efferent (preganglionic) elements are located in a medial or even median group dorsal[113] to the central canal (cf. Figs. 103, 105). A thoraco-lumbar (sympathetic) column of these cells is said to extend from about the last cervical to approximately the third lumbar segment. An additional lumbo-sacral, perhaps parasympathetic column has been noted in

[113] Since, because of the formation of a septum posterius, only the ventral part of the neural tube's original lumen remains as the central canal, this dorsal position of the visceral-efferent cell group is fully compatible with its origin from the dorsal portion of basal plate *('intermedioventral longitudinal zone')*. Further details concerning the ontogenetic development of the Avian spinal cord are contained in ROMANOFF's valuable treatise on 'The Avian Embryo' (1960) which includes a discussion, with exhaustive bibliography, of the relevant investigations by HAMBURGER, LEVI-MONTALCINI, WATTERSON, WENGER, WINDLE and ORR, and others.

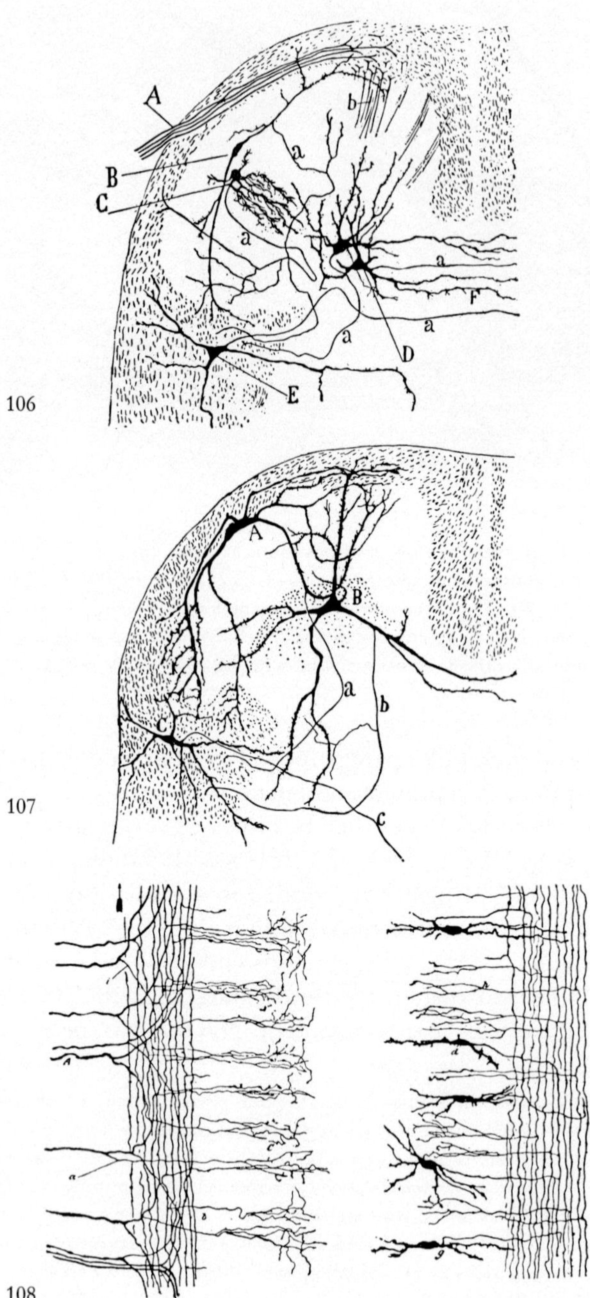

106

107

108

about four or five further, caudally adjacent lumbo-sacral segments. The majority of the viscero-efferent fibers seems to leave the spinal cord through the ventral roots, but, according to KAPPERS *et al.* (1936) some preganglionic fibers may also run dorsad and emerge with the posterior roots. The coarser *fibers of Lenhossék*, which are doubtless efferent, but originate from rather large multipolar elements in the ventral horn, and pass through the dorsal roots (cf. Fig. 104), are not commonly interpreted as pertaining to the vegetative nervous system, but their significance remains uncertain. BECCARI's view, assuming these fibers to represent a system homologous or analogous with the Mammalian spinal accessory complex, was already mentioned above in dealing with the Reptilian spinal cord. In Birds, the *cells of Lenhossék* are predominantly found in upper cervical segments (KAPPERS, 1947), but may occur as far caudad as the lumbosacral enlargement (ROMANOFF, 1960).

The fairly large neuronal elements of the marginal nuclei have been identified as commissural cells by several authors. Their neurites pass through the anterior commissure entering, apparently as (mainly or exclusively ?) ascending components the contralateral anterior funiculus. Some fibers, nevertheless, remain homolateral, ascending at least for some distance. The termination of the neurites originating in the marginal cells has not been ascertained. TERNI (1926), moreover, claims

---

Figure 106. Some neuronal elements in the dorsal horn of a Chick embryo at 15 days of incubation as seen with the *Golgi method* (from CAJAL, 1909). A: posterior root; B: transverse cell near margin of *substantia gelatinosa Rolandi;* C: cell with dense dendritic tuft inside substantia gelatinosa; D: cell of posterior horn with neurite passing through posterior commissure; E: cell of '*noyau interstitiel*' with neurite passing through posterior commissure; F: '*commissure protoplasmique postérieure;* a: neurites; b: posterior root collaterals.

Figure 107. Some neuronal elements in the dorsal horn of a Chick embryo at 19 days of incubation as seen with the *Golgi method* (from CAJAL 1909). A: marginal element of substantia gelatinosa *('grosse cellule marginale');* B: large nerve cell of posterior horn proper; C: '*cellule du noyau interstitiel*'*;* a, b, c: various neurites *('cylindres-axes').*

Figure 108. Longitudinal section (in dorso-ventral sagittal plane) through the spinal cord of a fairly advanced Chick embryo, showing posterior funiculus (left) and lateral funiculus (right) as displayed by *Golgi impregnation* (from CAJAL, 1909). '*On y voit la substance blanche des cordons postérieur et latéral, avec leur collatérales ramifiées dans la substance grise*'. With respect to lateral funiculus '*on voit, également, en d, f, g, que les fibres de ces cordons sont des cylindres-axes directement issus des cellules nerveuses*'.

that no terminal arborizations of extrinsic nerve fibers can be found in the marginal nuclei, thereby postulating a 'spontaneous origin of impulses in these grisea'. KAPPERS *et al.* (1936), however, remain sceptical about the alleged lack of neuronal input to the marginal nuclei. In addition, these grisea are known, since the investigations of TERNI (1926), to be rich in glycogen, which is present both in the neuronal and the glial elements.

Commissural, funicular, and tract cells are found in both ventral and dorsal horns. The presence of short *Golgi II-type* neurons seems probable. Some of the internuncial elements form a so-called magnocellular group within the base of the posterior horn, while the nerve cells of substantia gelatinosa are rather small, except for a few elements along its external periphery. There are, moreover, in the dorsal part of lateral funiculus, between posterior and anterior horn, and particularly at cervical levels, some scattered fairly large multipolar nerve cells, called '*cellules du noyau interstitiel*' by CAJAL (1909, 1952) which appear as an extension of the oblongata's reticular formation, although they may be diverse derivatives of either alar or basal plate. It should also be noted that large transitory cells of *Rohon-Beard type* have not been observed or identified in Avian embryologic material. Some of the sizeable commissural cells related to anterior commissure ('arcuate cells') demonstrable by means of the *Golgi technique* in the dorsal part of the alar plate at early stages (e.g. up to the fifth or sixth day of incubation in the Chick) seem thereafter to migrate farther ventrad within the dorsal horn.[113a]

Figure 103 depicts the cytoarchitectural aspect of a diagrammatic generalized Avian spinal cord section as interpreted by HUBER (1936), while Figures 104–108 ilustrate various further details recorded on the basis of silver impregnations. Concerning the substantia gelatinosa, it can be seen that its small cells possess a dense dendritic arborization directed ventromediad, as well as a few additional dendrites directed at

[113a] As an aside, it may be mentioned that the early differentiation of the Chick's spinal cord has been studied in *organotypic tissue cultures* of explants at the second day of incubation, when few or no cells have formed axons. These cultures were maintained for five to seven days and subsequently studied with light and electron microscopy (LYSER, 1971). Although somewhat ambiguous and inconclusive, the obtained results nevertheless showed 'that initial phases of nerve cell differentiation can proceed in organ cultures' and that there is, at least in this respect 'no specific requirement for interaction with adjacent mesodermal tissues'.

different angles. The 'gelatinous' aspect of the substance may be due to the numerous dense dendritic arborizations with their surrounding terminal neuropil. The neurites of these elements seem, at least with a collateral, to join *Lissauer's tract*. The dorsolateral marginal cells of the substantia gelatinosa, which can be somewhat larger, tend to be spindle-shaped, with long axis running approximately medio-laterad in a transverse plane, and form a peripheral 'arc' along the external curved boundary of said griseum. Their neurites, giving off intragriseal collaterals, seem to join the lateral funiculus.

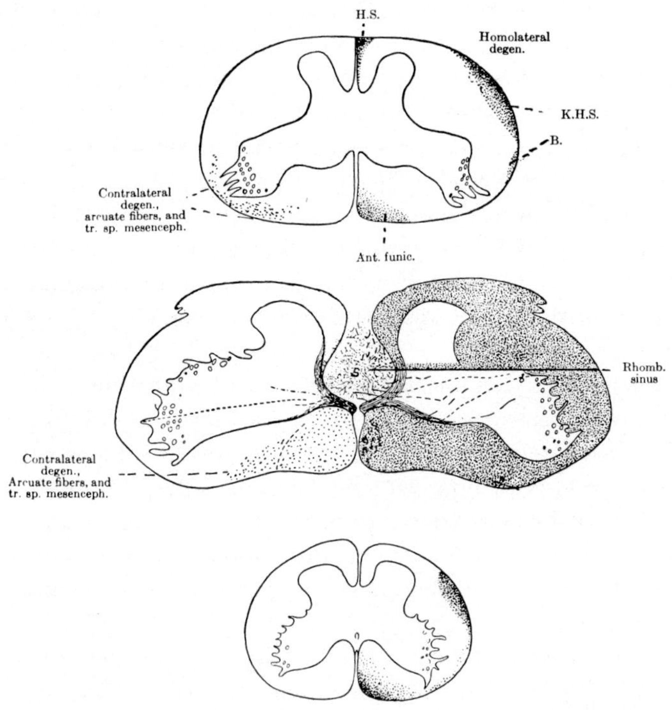

Figure 109. Degeneration of ascending and descending fibers in the spinal cord of the Dove after hemitransection at a lumbar level (after FRIEDLÄNDER, from KAPPERS *et al.*, 1936). Top: cervical region; middle: level of hemitransection; bottom: sacral level; H.S.: degenerated fibers in posterior funiculus *('Hinterstrang')*, showing that the ascending caudal fibers are displaced medialward by the (non-degenerated) more cranial ones; KHS: spino-cerebellar path *('Kleinhirnseitenstrangbahn')*.

The above-mentioned 'interstitial cells', which could also be called 'dorsal or dorsolateral paragriseal cells', appear related to either posterior commissure, anterior commissure, lateral funiculus or to undetermined regions of gray substance (cf. Figs. 106, 107).

With regard to the *fiber systems* of the Avian spinal cord, the dorsal roots display, as in some Amphibians and some Reptiles, the subdivision into a thicker dorsomedial and a thinly medullated or non-medullated dorsolateral fasciculus. The dorsomedial bundle, whose fibers bifurcate into ascending and descending branches, joins the posterior funiculus (Figs. 104, 108), while collaterals provide direct (monosynaptic) and indirect connections with motoneurons, as well as other sorts of connections with internuncials. Although some of the ascending fibers may end within a few segments above their entrance, others doubtless reach the terminal grisea of the posterior funiculi through which their impulses are transmitted to higher brain centers (cf. Fig. 109). It can be seen that the ascending fibers from caudal levels become displaced medialward, toward the median septum, by those from successively more cranial levels, as is the case in Mammals and presumably also in other Vertebrates *('somatotopic distribution')*. Figure 108 shows, in addition to the posterior funiculus, the longitudinal, bifurcating, respectively in part perhaps only ascending or only descending fibers of the lateral funiculus, with their origin from funicular or tract cells, as the case may be, and with collaterals entering the gray substance.

A secondary ascending system, possibly originating in part from the large dorsal cells, and related to the dorsomedial subdivision of the posterior root, is represented by the well developed and apparently mostly or exclusively homolateral (GROEBBELS, 1927) spino-cerebellar fibers. This pathway is known to begin at caudal lumbar levels and is mainly comparable to the Mammalian dorsal spino-cerebellar tract, but may also include an adjacent or overlapping smaller ventral spinocerebellar component.

The dorsolateral bundle becomes related to substantia gelatinosa and *Lissauer's tract*. Some of the commissural cells in the dorsal horn, connected with dorsomedial or dorsolateral root bundles or both, give origin to long ascending contralateral pathways providing a spino-bulbo-tectothalamic tract (spinal lemniscus) which EDINGER, KAPPERS and others regard as the 'phylogenetically oldest secondary sensory tract of the cord's *verbindungsapparat*. Both a ventral and a dorsal commissure are present with variations related to the different levels and to the diverse taxonomic forms. Neurites and also dendrites of posterior horn cells, as well as collaterals of posterior root fibers pass through the dorsal commissure (Figs. 104, 106, 107), whose fibers also traverse the 'gelatinous' tissue of sinus rhomboidalis. Secondary 'arcuate fibers' of spino-bulbar tract, neurites from the marginal nuclei and neurites from other, unidentified sorts of internuncial elements are the significant components of anterior commissures.

The descending components of the *verbindungsapparat* comprise homolateral and contralateral fibers of the fasciculus longitudinalis medialis which seems to contain tegmento-spinal[114] and vestibulo-spinal fibers. A somewhat separate vestibulo-spinal bundle may descend through the zone of transition between anterior and lateral funiculus. GROEBBELS (1927) found evidence that descending primary (root-) fibers of the vestibular nerve reach the spinal cord. In addition, descending cerebello-spinal fibers seem to be included in the ascending spino-cerebellar tract. Moreover, tecto-spinal fibers, crossed and uncrossed, appear to reach various levels of the spinal cord, being presumably included in the anterolateral funicular region, and have been recorded by a number of authors (cf. KAPPERS *et al.*, 1936). A rubro-spinal tract, although not definitely demonstrated, may be present. It is dubious whether, as in the case in Mammals, fibers originating from the telencephalon[115] reach the spinal cord, although this has been claimed by some older and recent authors (e.g. ZECHA, 1962).

Figure 109 shows the pattern of ascending and descending degeneration in the Avian spinal cord (Dove) following a right hemitransection at the lumbar level, and as reported by FRIEDLÄNDER (1898). The ventromedian anterior funiculus seems to include, in addition to the systems mentioned above, 'a bundle of ascending and descending homolateral fibers' whose significance and possible Mammalian homologue (if such exists) remain unknown' (KAPPERS *et al.*, 1936).

---

[114] Some of these tegmento-spinal fibers originate from the deuterencephalon's formatio reticularis, while others are provided by special tegmental grisea such as n. interstitialis and perhaps related pretectal grisea in the diencephalo-mesencephalic boundary zone.

[115] In Mammals, such fibers are represented by the cortico-spinal or pyramidal tract originating in neocortex. Since a comparably developed neocortex is not developed in Birds, whose basal ganglia are particularly differentiated, a descending tract from telencephalon to spinal cord, if at all present in Birds, would presumably originate in the basal grisea of the hemispheres.

Concerning the ascending components, experiments with spinal hemitransections and evaluated by means of the *Nauta technique* were undertaken by KARTEN (1963). Fibers could be followed to terminations (all homolateral 'with few exceptions') in oblongata, cerebellar cortex, tegmentum and tectum mesencephali, and dorsal grisea of the thalamus dorsalis. Thus, as probably 'already' in lower Vertebrates, the spino-bulbo-mesencephalic lemniscus seems here to include endings extending into the diencephalon.

With respect to a general functional evaluation of the Avian spinal cord, YOUNG (1955) considers the relatively small size of the dorsal funiculi and of their grisea at the spino-deuterencephalic transition to be a 'most characteristic feature'. This author assumes that the sense of touch[116] is less developed in the body than it is in Mammals, perhaps even less than in Reptiles. The loose covering of the body by the feathers is believed to have prevented an 'elaborate organization of the sense of touch', the 'finer senses' of Birds being restricted to the eyes, ears, and the bill. On the other hand, there are very large spino-cerebellar tracts, 'presumably proprioceptive and concerned with the delicate adjustments necessary for flight'. The spinal cord is controlled by large descending tracts of the *verbindungsapparat* such as cerebello-spinal, vestibulo-spinal, and tecto-spinal channels. In YOUNG's opinion, there is no direct tract from the telencephalon to the spinal cord, but the influence of the very large basal grisea 'is probably exercised through fibres running to the red nucleus and tegmentum of the midbrain, from which others pass to the cord.' One can agree with the gist of YOUNG's overall appraisal.

In his recent comprehensive treatise on the Avian brain, R. PEARSON (1972), sums up his opinion concerning the spinal cord's *verbindungsapparat* by stating that its overall pattern 'would reflect the type of coordination which one would expect in such bipedal cursorial and flying forms in which vision is a predominant modality'. He then likewise adds the comment: 'It is of great importance to emphasize the absence of any system which would correspond to the crossed pyramidal tracts of mammals, and also to note the apparently very large number of endogenous fibres many of which have a relatively short intersegmental distribution'.

---

[116] It should, however, be kept in mind that although 'touch-fibers' seem to be included in the Mammalian posterior funiculi, these latter are, to a substantial extent, proprioceptive. Yet, this proprioceptive channel seems mainly related, by way of its relays, to the neocortex, which is not comparably differentiated in Birds.

11. Mammals (including Man)

In contradistinction to the conditions displayed by Sauropsidans, and as already pointed out in the introductory remarks on general pattern (section 1 of this chapter), the spinal cord of Mammals does not commonly reach the caudal end of the vertebral canal, regardless whether a given form possesses a large tail, a small tail, or no tail at all. In accordance with that discrepancy in length between spinal cord and vertebral canal, the level of a spinal nerve segment, as designated by its egress through the corresponding bilateral foramina intervertebralia, does not topographically coincide with the level of its actual spinal cord 'segment'. Thus, the spinal nerve roots, running toward their exit, as common spinal nerves, from the vertebral canal, assume a slant course, whose slope gradually increases from cervical to caudal levels The highly oblique lumbosacral roots, surrounding the conus medullaris sive terminalis and its continuation as *filum terminale*, generally form a conspicuous *cauda equina*. Figure 110 illustrates these relationships as obtaining for the adult human spinal cord, while Figure 111 depicts the comparable gross configuration of the spinal cord in a 'lower' Mammal (Rodent, Rat).

The lumbar enlargement is generally not recognizable in forms with lacking or poorly developed posterior extremities (e.g. Dugong and common Whale); nevertheless, it is said to be suggested in Phocaena and Dolphins, being rather far caudal and perhaps related to 'the important part which the tails of these animals play in their activity' (KAPPERS *et al.*, 1936). In the Kangaroo, on the other hand, the lumbar enlargement surpasses the cervical in size, while this latter is commonly the larger in most Mammals.[117] It will be noted that in Figure 111, the Rat's lumbar enlargement is hardly recognizable.

The caudal reduction of Mammalian spinal cord to a filum terminale appears correlated with the vanishing of certain caudal myotomes producing tail musculature in submammalian forms. As shown by GEGENBAUR (1898–1901) and others, the Mammalian tail musculature, irrespective of that organ's size, does not display the primary metameric character seen in 'lower' Vertebrates but results from the differentiation and transformation of its most proximal muscular components,

---

[117] According to VERHAART (1970), the spinal cord of Echidna (now also called Tachyglossus) does not display a 'very marked' intumescentia cervicalis, the lumbar intumescence being 'somewhat more marked'.

the caudal ones thus becoming lost. Nevertheless, some interdependence between tail development and relative length of spinal cord seems to obtain in various instances. As pointed out by GEGENBAUR (1898–1901) and KAPPERS (1920), the spinal cord of the Monotreme Echidna ends about midway within the vertebral canal in contrast to its extension in the Monotreme Ornithorhynchus, whose spinal cord

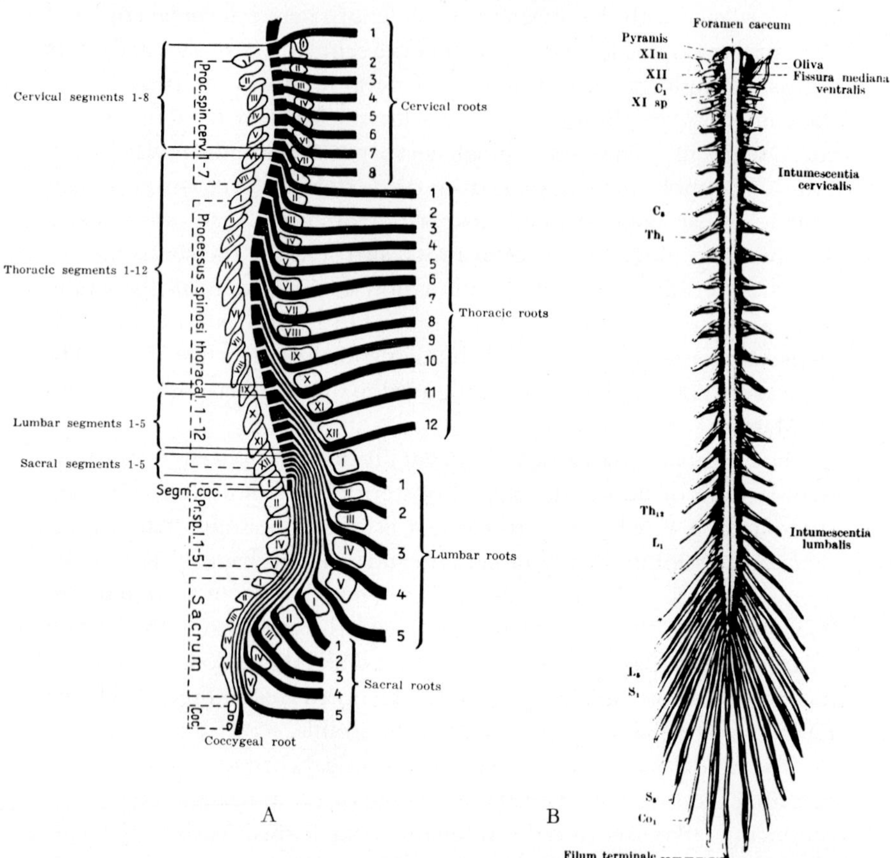

Figure 110A. Schematic representation of topographic relationships between spinal cord segments and vertebral canal in Man (from HAYMAKER-BING, 1956). It will be seen that the human spinal nerves comprise (subject to minor variations) 31 segments, of which 8 are cervical, 12 thoracic, 5 lumbar, 5 sacral, and 1 is coccygeal (the first cervical spinal nerve emerging between os occipitale and first cervical vertebra, and the eighth between seventh cervical and first thoracic ones).

Figure 110B. Ventral aspect of human spinal cord, removed from vertebral canal with intact nerve segments and cauda equina (after NEUMANN from CLARA, 1959).

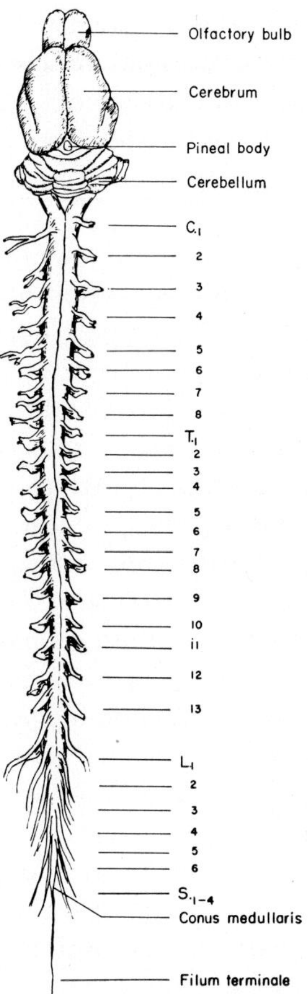

Figure 111. Dorsal aspect of the Rat's neuraxis, showing brain and spinal cord with its nerve roots, as seen in a dissected specimen (from ZEMAN and INNES, 1963).

reaches the level of the sacral vertebral region. Both Prototheria have a tail of about one-quarter of the total body length, but in Ornithorhynchus it is a strong muscular structure used in swimming, while it is much thinner and less functionally significant in Echidna.[118] Yet, in almost tailless Mammals, such as the Lagomorph (formerly 'Rodent') Lepus, in Chiroptera, and in the Insectivore Erinaceus, the spinal cord extends into the sacral region of the first form, but is quite short in the

[118] Concerning Echidna's lumbar intumescence, cf. again the preceding footnote 117.

two others. This, in KAPPERS' opinion, may be attributed to the greater development of posterior extremities in Leporidae, and of the anterior extremities in Chiroptera and Insectivora. Again, in Carnivores, the conus terminalis commonly reaches the caudal lumbar vertebral planes, and in Ungulates it can be found as far caudad as at sacral vertebral levels.

At early Mammalian ontogenetic stages of neuraxis and vertebral column, however, the spinal cord extends through the entire vertebral canal.[119] The subsequent discrepancy in lengths appears related to at least three interconnected factors, namely (1) a difference in further longitudinal growth of axial skeleton and of neuraxis, (2) the above-mentioned phylogenetic changes concerning the tail musculature, and (3) a 'secondary dedifferentiation', for which presumptive evidence has been recorded in some forms. Thus, some seemingly 'regressive' neuroectodermal elements, perhaps even of 'neuronal' type, can be noted in the human film terminale (HARMEIER, 1933, as quoted by KAPPERS *et al.*, 1936). At the caudal end of filum terminale, a small vesicle may persist until rather late embryonic periods and seems, from the morphologic viewpoint, to be *formally* homologous with the terminal vesicle, respectively the more differentiated terminal region of the spinal cord in various Fishes, discussed in preceding sections of this chapter.

Again, rostrally to the filum terminale, a dilatation of the central canal can occur in the conus medullaris. An occasional caudo-dorsal opening of such dilated 'ventricular space' in medullary conus or in filum terminale might represent, according to KAPPERS, a secondary feature rather than a persistent posterior neuropore.

Although, quite evidently, the transverse diameter of vertebral canal must be greater than the width of the enclosed spinal cord, there are, in addition to the discrepancy between length of that canal and longitudinal extent of medulla spinalis, some instances of very consid-

---

[119] In the human newborn, the spinal cord may extend as far caudad as the 3rd or 4th lumbar vertebra, receding to the level of the 2nd by the third year, usually ending at a level between 1st and 2nd lumbar vertebra in the adult. The clinical implications of these topographic relationship for lumbar ('spinal') puncture, etc., are self-evident and well known. The filum terminale usually ends, together with arachnoid and dural sack, at about the 2nd sacral vertebra. The dura then continues as filum terminale durae matris (filum terminale externum) which seems to fuse with the ventral coccygeal periost. This 'filum externum' may include some abortive coccygeal spinal nerve rootlets, but, in contradistinction to the true filum terminale ('internum'), does not appear to be a genuine spinal cord remnant.

erable difference with regard to the transverse diameters of both configurations. Thus, in the Sirenian Dugong and in Cetacea, the canal's diameter may be twelve times that of the cord's. Generally speaking, a substantial difference in relative width is connected with the presence of extensive intradural ('epidural') venous plexuses between periost (external dura) and dura proper. Moreover, in special instances (e.g. in the Dolphin as pointed out in section 7, chapter VI of volume 3, part II) a massive cervico-thoracic *rete mirabile complex* may be present. Moreover, the eccentric position of the cord within the canal has been pointed out by KAPPERS *et al.* (1936) and other authors. In numerous lower and higher Vertebrates, including Man, the spinal cord with its (internal) dural sack is commonly closer to the ventral than to the dorsal wall of the vertebral canal, but in Cetacea and in the Edentate sloth Choloepus this *eccentricity* is particularly conspicuous. A 'dislocation to the side', i.e. an additional *lateral eccentricity* is here 'due to the presence of a large vein' located in one lateral half of the 'epidural' space, depicted by DE BURLET (1911) and KAPPERS (1920).

Figures 112 to 118 illustrate the general configuration of the spinal cord in Man and a few other Mammals, as seen in cross-sections. Generally speaking, the overall arrangement of white and gray substance, and the distribution of nerve cells in this latter, are essentially similar in all Mammals, although a number of secondary differences between the various taxonomic forms[120] obtain. Depending on the arbitrarily assumed viewpoint, such differences may be stressed or downgraded. Quite evidently, in correlation with body-build and behavioral activi-

---

[120] BECCARI (1943), a comparative anatomist with considerable morphologic acumen, justly comments: '*È difficile di stabilire, da un punto di vista sistematico, quali sono gli ordini dei Mammiferi que possiamo ritenere più bassi e quali quelli più elevati, eccettuati i Monotremi con i Marsupiali ed i Primati, collocati agli estremi oppositi.*' He then adds: '*Ecco alcuni esempi di classificazione*', and enumerates differing taxonomic systems proposed by BEDDARD, by CLAUS and GROBBEN, as well as by KINGSLEY. VERHAART (1970), who deals with spinal cord and brain stem of Mammals, has adopted the sophisticated, but by no means convincing, and somewhat pretentious classification by G.G. SIMPSON, based on an interpretation of paleontologic findings. A reasonably useful classification, likewise considering paleontologic data and the presumed 'origin of Mammals', has been propounded, in chapters XIX, XX *et passim*, by YOUNG (1955). For the neurologist or neurobiologist without specialized zoological training, I have adopted a simplified taxonomic system, which I believe to represent a reasonably valid first approximation, suitable for a preliminary approach to these highly complex problems, in vol. 1, pp. 64–71 of the present series.

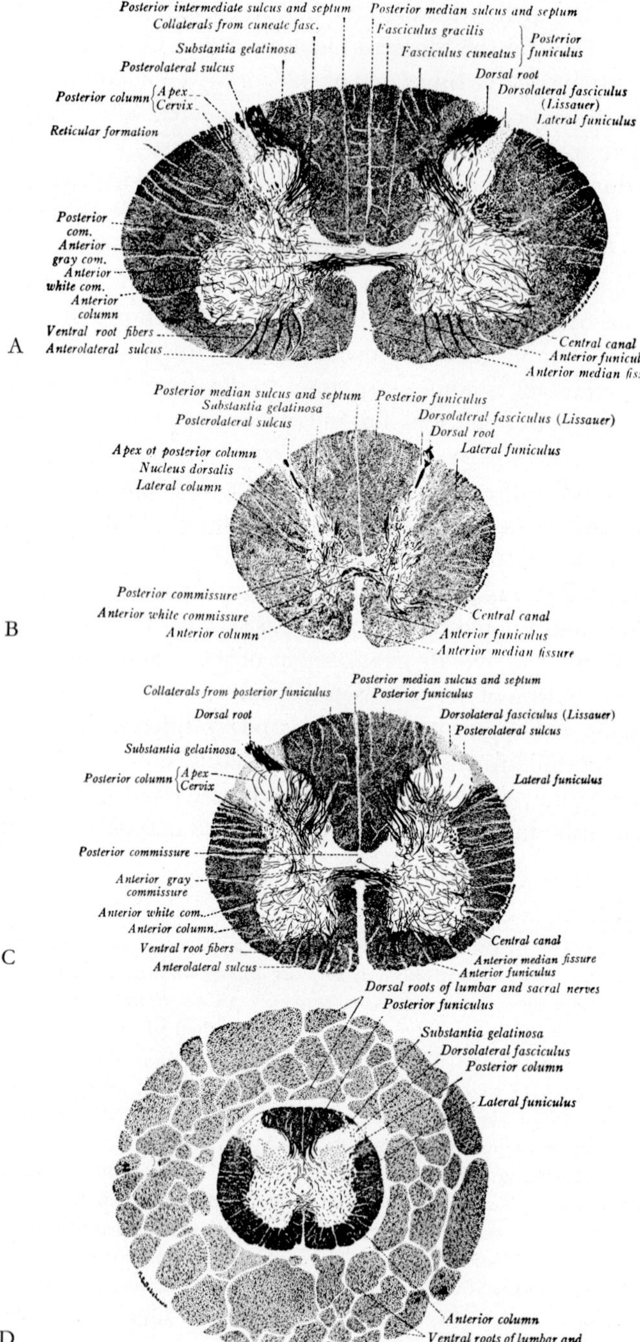
Posterior intermediate sulcus and septum
Collaterals from cuneate fasc.
Substantia gelatinosa
Posterolateral sulcus
Posterior column {Apex Cervix
Reticular formation
Posterior com.
Anterior gray com.
Anterior white com.
Anterior column
Ventral root fibers
Anterolateral sulcus
Posterior median sulcus and septum
Fasciculus gracilis
Fasciculus cuneatus
Posterior funiculus
Dorsal root
Dorsolateral fasciculus (Lissauer)
Lateral funiculus
Central canal
Anterior funiculus
Anterior median fissure
A
Posterior median sulcus and septum
Substantia gelatinosa
Posterolateral sulcus
Apex of posterior column
Nucleus dorsalis
Lateral column
Posterior commissure
Anterior white commissure
Anterior column
Posterior funiculus
Dorsolateral fasciculus (Lissauer)
Dorsal root
Lateral funiculus
Central canal
Anterior funiculus
Anterior median fissure
B
Collaterals from posterior funiculus
Dorsal root
Substantia gelatinosa
Posterior column {Apex Cervix
Posterior commissure
Anterior gray commissure
Anterior white com.
Anterior column
Ventral root fibers
Anterolateral sulcus
Posterior median sulcus and septum
Posterior funiculus
Dorsolateral fasciculus (Lissauer)
Posterolateral sulcus
Lateral funiculus
Central canal
Anterior median fissure
Anterior funiculus
C
Dorsal roots of lumbar and sacral nerves
Posterior funiculus
Substantia gelatinosa
Dorsolateral fasciculus
Posterior column
Lateral funiculus
Anterior column
Ventral roots of lumbar and sacral nerves
D

ties, very substantial differences in the distribution and minute patterns of *eigenapparat* internuncial fibers and their synaptic connections, as well as similar differences concerning the *verbindungsapparat* must obtain, presumably combined with significant biochemical differences. Yet, these highly intricate aspects cannot be properly gaged with the present-day methods and may or may not permanently escape detection as well as satisfactory formulation.

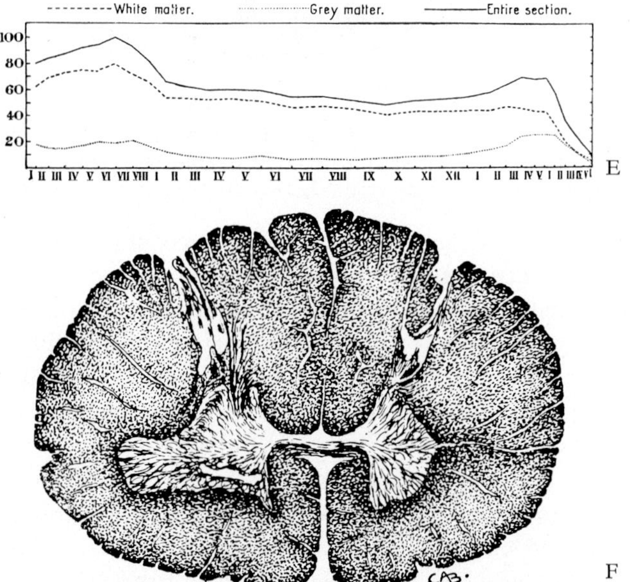

Figure 112E. Graph indicating the variations in cross-sectional area of entire human spinal cord as well as of its gray and white matter at the different levels. The numbers on the ordinate indicate percentage of greatest area (100) displayed between C VI and VII (after Donaldson and Davis, 1903, from Ranson, 1943).

Figure 112F. Cross-section through intumescentia cervicalis at level C VIII of a human spinal cord in a man with congenital lack *(amelia)* of left forearm. Right side of figure corresponds to actual left side of cord (from Kappers *et al.*, 1936, as redrawn after the photomicrograph Figure 11 of Elders, 1910). By an oversight, Kappers *et al.*, in their caption, erroneously refer to 'a man without a right arm'.

Figure 112A–D. Cross-sections through the human spinal cord (child of unspecified age, but showing a definitive stage of overall development) as seen by means of the *Pal-Weigert method* (from Ranson, 1943). A Seventh cervical segment. B Seventh thoracic segment. C Fifth lumbar segment. D Third sacral segment, surrounded by cauda equina.

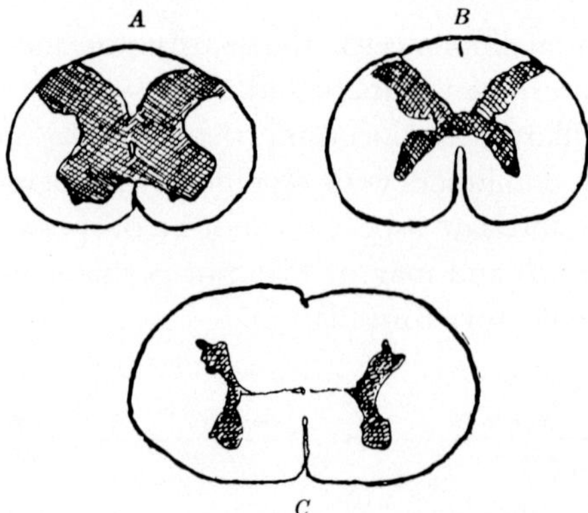

Figure 113A. Rough sketch illustrating configuration of gray core and relationship of white to gray substance as seen in cross-sections through the spinal cord of some smaller and larger Mammals (after DE VRIES, from KAPPERS *et al.*, 1936). A: Mouse (small Rodent); B: Agouti (Dasyprocta, relatively large Rodent); C: Elephant (Proboscidean).

Among the differences that are readily recognized, the following can be mentioned. As e.g. shown in Figure 113, the spinal cord of small Mammals displays a relatively predominant ratio of gray matter, while in large Mammals that of white matter considerably increases. Since this trend is correlated with both numerical increment of required ascending and descending fibers as well as progressive differentiation and complexity of *eigenapparat* and *verbindungsapparat*, not only the animal's size, but also its 'lower' or 'higher' taxonomic status represent here significant parameters. This latter aspect seems particularly relevant with respect to the *verbindungsapparat*'s cortico-spinal ('pyramidal') fiber systems, which shall be discussed further below. Again, the nuclei marginales of the spinal cord in Chiroptera may be mentioned. Scattered nerve cells within the white matter are occasionally noted in a wide variety of Mammals, including Man, in addition to the 'reticular formation'with an appendant 'nucleus cervicalis lateralis' found at cervical levels, particularly near the transition to oblongata, in most forms. However, in the Insectivore Erinaceus, the Ungulate Goat, and in Cetaceans, ganglion cells included in the white matter seem more common. VERHAART (1970) has described and depicted a number of additional, more or less distinctive secondary features characterizing diverse Mammalian taxonomic groups. A short monograph dealing with the spinal cord of Primates was recently prepared by NOBACK and HARTING (1971).

As regards the overall aspect of the Mammalian spinal cord in cross-sections, it can be seen that the anterior median fissure is generally very deep, and commonly reaches the anterior commissure near the boundary of the gray substance. The fissure contains the branches of the anterior spinal vessels discussed in section 7, chapter VI, of the preceding volume. In addition to a usually well recognizable shallow posterior median sulcus, from which a glial septum extends to the gray substance, an accessory posterior intermediate sulcus is frequently present at cervical and upper thoracic levels. It corresponds to a subdivision of posterior funiculi into medial fasciculus gracilis and lateral fasciculus cuneatus. Both may be separated from each other by more or less developed, often branching and irregular glial septa with or without accompanying blood vessels. In the region of dorsal root entrances a sulcus lateralis posterior (BNA, PNA) may correspond to an actual surface depression of the cord. A corresponding sulcus lateralis anterior (BNA) in the region of radix anterior, commonly consisting of more or less discrete rootlets, seems to represent a once standardized figment of the imagination and this term was dropped by the PNA.

Figures 112A–D illustrate the differences in the contour and relative size of the human spinal cord's gray matter at representative levels. The diagram of Figure 112E discloses the relative area of spinal cord, white substance, and gray substance at all levels, the greatest expansion, at about the VIth and VIIth cervical segment, being expressed as 100. Discounting some accumulation of white matter, perhaps related to complexities of the *eigenapparat*, at the level of intumescentia cervicalis, it seems rather evident that the descending fiber masses of *verbindungsapparat* will decrease caudalward as they reach their respective levels, and that the volume of ascending fibers to the brain will increase cranialward by successive additions, both tendencies contributing to the slant of the white matter's curve shown in the diagram.

Again, except for the reticular substance or formation (including a so-called 'lateral cervical nucleus') at some cervical levels, white and gray substances are relatively well demarcated from each other by the apparent outline of a 'boundary'. The difference in width of anterior horns at thoracic levels and at those corresponding to the intumescentiae is conspicuous. The well recognizable *substantia gelatinosa Rolandi* seems, as a rule, particularly developed at lumbosacral levels. In the cranial portion of the spinal cord, the substantia gelatinosa appears relatively extensive at the levels from the first thoracic to the sixth cervical segments, becoming then again enlarged at the first cervical segment, where this griseum blends with the nucleus radicis descendentis s. spinalis trigemini.

The tract of LISSAUER (1885), which accompanies the substantia gelatinosa, commonly includes, in addition to its major dorsolateral portion, a small area located medially to the heavily medullated dorsal root bundle entering the spinal cord. At uppermost cervical levels, *Lissauer's tract* merges with the descending root fibers of the trigeminus.

With respect to some of the other Mammals, the substantia gelatinosa is especially wide in various Artiodactyle Ungulates, e.g. in the Gazelle (BIACH, 1907), where it displays sinuosities or convolutions resembling 'cortical' structures at lumbosacral levels (Fig. 115D). In Cetaceans, on the other hand, the development of substantia gelatinosa is described as 'minimal' (KAPPERS, 1947), the posterior horn being very short, and its 'gelatinous zone' consisting only of a few interrupted areas along the horn's dorsal margin (VERHAART, 1970). It will be recalled that KAPPERS *et al.* (1936) refer to a 'poor development of skin sensibility, including pain' as obtaining in Cetacea (cf. above, section 1, p. 13.

The BNA recognized a columna grisea posterior (dorsal horn) with cervix, apex, and *substantia gelatinosa Rolandi*,[121] moreover a columna lateralis, and a columna anterior (ventral horn). The PNA duplicate columnae griseae (anterior, lateralis, posterior) by adding substantia grisea with cornu anterius, laterale and posterius, and, moreover, substantia intermedia lateralis, 'intervening laterally between the anterior and posterior cornua', the cornu laterale being 'an outgrowth from its lateral aspect'. This lateral horn (Fig. 112B) is part of the intermedioventral zone in morphologic terminology, and extends from lowermost cervical to uppermost lumbar levels. The '*nucleus dorsalis*' *of Stilling and Clarke* (cf. Fig. 112B), located at the medial basis of the posterior horn, is likewise found at the just mentioned levels.[122]

The rather narrow strip of gray interconnecting the 'substantiae intermediae laterales' contains the central canal[123] and was termed 'sub-

---

[121] As can be seen in Figures 112A–C, 'apex' and 'substantia gelatinosa' are generally considered to be synonyms.

[122] EDINGER (1912) and CLARA (1959) use the term '*Stilling-Clarkesche Säule*', thus crediting STILLING with the first mention of this nucleus, (STILLING and WALLACH, 1842, 1843; STILLING, 1859), which was subsequently described as 'dorsal vesicular column' by CLARKE (1851). Other authors, e.g. CROSBY *et al.* (1962) are of the opinion that *Stilling's nucleus* should only refer to' the cervical extension (and sometimes the lumbar)' of *Clarke's column*.

[123] The central canal, occasionally surrounded by a distinctive, essentially glial substantia gelatinosa centralis, becomes not infrequently obliterated in Man and also in

stantia intermedia centralis' by the PNA. The posterior and anterior commissures run within the gray matter dorsally respectively ventrally to the central canal and contain both medullated and non-medullated fibers. In addition, commissural or decussating medullated fibers form a commissural bundle in the white substance ventrally to the central gray matter. Crossing myelinated fibers dorsally to the central gray, i.e. in the bordering white matter of dorsal funiculi are more uncommon, but may be seen at some levels. The terms commissura anterior respectively posterior alba are ambiguous, since merely referring to myelination, but not to location within either white or gray matter.

The neuronal elements in the Mammalian and Human spinal cord have been classified, according to the principles discussed in section 2 of this chapter, as root cells *(Wurzelzellen)* whose neurites pass mainly through ventral or, in far lesser number, through dorsal roots, and as 'immanent cells' *(Binnenzellen)* whose processes do not leave the neuraxis. These latter cells, again, include (a) short internuncials *(Schaltzellen)*, whose processes are believed to remain within the gray substance at a given level and which correspond more or less to *Golgi II-type cells.* Intragriseal elements whose branching ascending and descending neurites may extend through several segments are designated as (b) association cells *(Assoziationszellen)* by CLARA (1959); (c) the commissural cells *(Kommissurenzellen)* extend their processes across the midline in posterior or anterior commissures. Elements whose neurites provide ascending or descending fibers or both within the white matter are (d) funicular cells *(Strangzellen)* which, again, can be subdivided into intrinsic elements of fasciculi proprii, and neurons originating extrinsic connections to the brain (tract cells, *Bahnzellen*). This classification, although quite useful for a first approach, is by no means rigorous, since

---

various other Vertebrates. Although such obliteration may indeed at times be an artefact due to poor fixation or damage upon removal and processing of the cord, there can be little doubt that, at least in Man, a true obliteration can often be seen, being the result of a commonly occurring intravital process, which is *per se* devoid of any pathologic significance. This is evident from preparations obtained by means of careful technical procedures, which, moreover, show displacements and scatterings of ependymal and other spongioblastic elements, pervaded by capillaries, and surrounding the original canal, reduced to a tubule of not more than 10–15 ependymal cells with negligible (about 5 to 10 μ wide) lumen. Pictures of this sort can hardly be considered artifacts (cf. e.g. Fig. 102, p. 157 in vol. 3, part I of this series). Again, with respect to the adult Cetacean spinal cord, VERHAART (1970) remarks: 'the central canal seems to have disappeared entirely and cannot be localized'.

numerous elements may pertain to more than one of these categories, such, e.g. that at least some 'funicular cells', through their axonic branchings might combine, in one neuronal unit, the diverse functions as homolateral 'association cells', as 'commissural cells', as 'intrinsic funicular cells', and as 'tract cells'.

In the ventral horn, the most conspicuous nerve cells are the large multipolar motoneurons, which, as in the other Amniota, can be roughly subdivided into a medial group, innervating the trunk musculature, and into a lateral group related to the musculature of the limbs with shoulder and pelvic girdle. At most thoracic levels, only the medial group is present. Figure 112F (to be compared with 112A) illustrates the significance of the lateral group, which is here shown to be deficient on the left side of the cervical intumescence in a man with congenital lack of left forearm.

More detailed subdivisions have been proposed by various authors and, particularly in Man, attempts have been made to identify delimitable groups of motoneurons concerned with the innervation of specific muscles or muscular groups (e.g. ELLIOT, 1942, 1945). Generally speaking, the medial group can be subdivided into a ventromedial and dorsomedial 'nucleus', perhaps related to ventral and dorsal trunk musculature, respectively. There is, nevertheless, some controversy as to what constitutes the 'medial group'. Thus, the innervation of ventral trunk muscles, supplied by the ramus anterior of the spinal nerve segment, has been attributed to the medial portion of the lateral group, accordingly said to be present at thoracic levels, while the medial group *sensu strictiori* is presumed to innervate only dorsal musculature supplied by the ramus dorsalis. The lateral group may be described as consisting of ventrolateral subdivision for the extremity girdle musculature, of dorsolateral subdivision for upper and lower limbs, and of retrodorsal subdivision for small muscles of hand and foot. A 'central subdivision' includes the so-called 'phrenic nucleus' from about C_4 to C_6, and a somewhat similarly located 'lumbodorsal nucleus' of unknown significance between about L_2 and S_2. Another cell group, occasionally classified as pertaining to the 'central subdivision', is found at cervical levels from the transition toward oblongata to perhaps C_6 (CROSBY *et al.*, 1962). It seems to provide fibers for the spinal accessory nerve and frequently extends laterad with a prominence similar to the 'lateral horn' of thoracic segments. It might be considered 'branchiomotor' or 'special visceral efferent', pertaining to the 'intermedioventral zone' of formanalytic terminology.

Root fibers of the spinal accessory, at least in Man and some other Mammals, take their exit neither through dorsal nor through ventral roots, but in an intermediate position between both. This peculiar course can be interpreted as related to the phylogenetically speaking 'hybrid' nature of the muscles innervated by this nerve, namely of a particular sort of musculature partly derived from caudal branchial arches and partly from musculature of 'myotomic' type.

Generally speaking, the motoneuron clusters seen in the two-dimensional aspect of cross-sections can be traced, with a modicum of accuracy, through a series of successive planes, and thus seem to present a three-dimensional arrangement into more or less continuous longitudinal columns (cf. Fig. 121C). Nevertheless, *qua* detailed subdivisions into 'nuclei' of the anterior horn, there obtains some justifiable doubt concerning the presence of clearly defined continuous columns, formed by these individual so-called nuclei, within the anterior horn

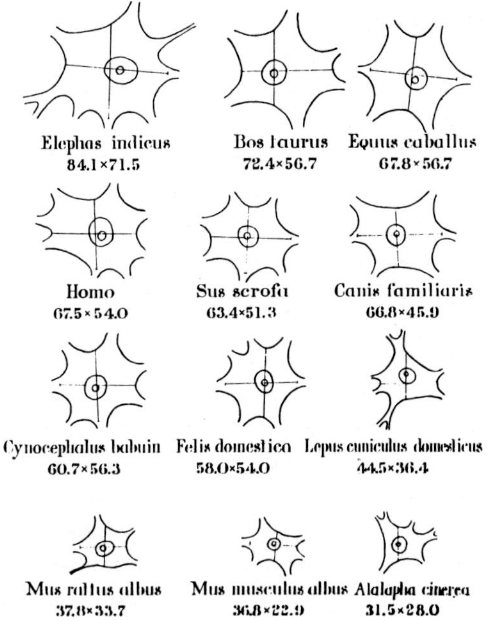

Figure 113B. Cell-body outlines of largest motoneurons found in the intumescentiae cervicales of different Mammals, arranged in the order of the animals average body-weight, respectively body-size, and recording the actual diameters of the cell body in micra, averaged from 10 largest neurons. The lines indicate the measured diameters. *Alalapha cinerea* is the name of a small Bat (from HARDESTY, 1902).

griseum. Hence, there is much disagreement concerning arrangement and significance of such motoneuron 'nuclei' or 'columns' (cf. CROSBY *et al.*, 1962; ELLIOTT, 1942, 1944; KESWANI and HOLLINSHEAD, 1956; NOBACK and HARTING, 1971; ROMANES, 1951, 1961; SPRAGUE, 1948, 1951). Thus, ELLIOTT (1942) stressed the 'basic and irreconcilable differences' among the views of investigators concerned with cytoarchitecture and functional significance of motoneuron groups in the Human and Mammalian spinal cord. It will be seen (Fig. 127) that NOBACK and SIMENAUER indicate a ventrolateral position for the 'phrenic nucleus' in the Chimpanzee, while KESWANI and HOLLINSHEAD (1956), who studied the human spinal cord in cases with unilateral phrenicectomy, located the 'phrenic nucleus' in the most medial portion of the ventromedial cell group at the 3rd, 4th and 5th cervical levels. The relevant cell clusters were found to constitute 'a straight column almost parallel to the ventral longitudinal fissure'. CROSBY *et al.* (1962) appropriately suggest the probability of individual variations. To this, of course, must be added variations among the diverse taxonomic groups, as discussed by ELLIOTT (1944).

Reverting to the distinction between phasic and tonic activities mentioned above in section 2 (p. 30) with respect to reflex action, it might be pointed out that GRANIT (1972) has recently elaborated on the question whether two categories of Mammalian alpha motoneurons, namely elements of phasic and of tonic types, should be distinguished.

In his investigation of the Elephant's spinal cord, HARDESTY (1902) included an interesting comparative study on the size of large motoneurons in diverse Mammals (cf. Fig. 113B). He summarized his results as follows: (1) The size of large cell bodies in the columnae anteriores (as well as in the spinal ganglia) varies appreciably in adult mammals of different sizes. (2) In general the larger mammals have the larger cell bodies, but this variation does not occur in the same ratio as the variations in the size of the animal, the cell bodies varying in much smaller ratios than the body weights. (3) The variations in the volumes of the cell bodies do not occur in higher ratios than the variations in the areas of the transverse sections of the medulla spinalis, and in no higher ratio than the variations in the areas of the substantia grisea contained in the sections. (4) In the larger mammals especially, only a small fraction of the entire neurone is represented in the cell body and its ordinarily visible immediately neighboring processes. (5) The variations in volume of the entire neuron presumably occur in higher ratios and may therefore be more nearly in proportion to the variations in body weight than either the volume of the cell bodies, the areas of the transverse sections of the spinal cord, or the areas of the substantia grisea in this latter.

With respect to the interesting findings and conclusions reported by HARDESTY, it should be recalled that these data concern large motoneurons and spinal ganglion cells within one and the same Vertebrate class (Mammalia). As regards cells of the same type within different Vertebrate classes or, for that matter, certain cell types within one and the same taxonomic group, various complicating factors obtain, such that, as an overall valid or rigorous rule, no definite relation between the size of the cellular elements and the size of the organism could be established in a convincing manner (cf. chapter V, section 2, p. 91f. of volume 3, part I). Even in the series studied by HARDESTY, the dog had frequently motoneurons as large as those of man, and a small bat possessed appreciably larger cell bodies than the mouse.

Within the anterior horns, two other sorts of nerve cells can be considered as closely related to the large notoneurons, namely the *gamma motoneurons* innervating the muscle spindles and the '*Renshaw-cells*' believed to be of *Golgi II-type*.[124] Both are presumed to be smaller than typical motoneurons. These elements, whose significance was discussed above in section 2, have been inferred on the basis of physiologic experiments, but could not be convincingly identified in histologic preparations (cf. e.g. TESTA, 1964). Nevertheless, various nerve cells of relatively small size, scattered within the ventral griseum, may be regarded as comprising gamma motoneurons and possibly '*Renshaw-cells*'.

Besides large motoneurons and gamma motoneurons, the anterior horn *sensu latiori*, i.e. the spinal griseum derived from the basal plate,

---

[124] Muscle spindles seem to be present only in 'Tetrapods', i.e. Amphibians and Amniota. Since the significance of gamma motoneurons and of the '*Renshaw-phenomenon*' has been particularly investigated in Mammals, and nothing certain about the relevant histologic elements in submammalian forms seems to have been ascertained, reference to this question was omitted in the preceding sections dealing with other Vertebrate classes. Thus, it is generally assumed by investigators of that topic (e.g. GRANIT, 1970), that, despite the occurrence of muscle spindles in Amphibia, gamma-cells are not present in this class of Vertebrates (cf. also chapt. VII, section 1, p. 808 in volume 3, part II of this series). In the frog, the same neurons which innervate the ordinary striated muscle fibers are believed to give off collateral branches for the motor innervation of the specialized 'intrafusal' muscle fibers within the muscle spindles.

contains a third sort of root cells, namely preganglionic 'general visceral' efferent elements. These cells are noticeably smaller than typical large 'somatic' or 'special visceral efferent' motoneurons and are located in what has been called the '*intermediate horn*' or 'intermediate column gray' (CROSBY *et al.*, 1962), which, in turn, corresponds to the 'visceral motor' functional division of JOHNSTON (1906) depicted above in Figure 3, and to the intermedioventral longitudinal zone of formanalytic terminology.

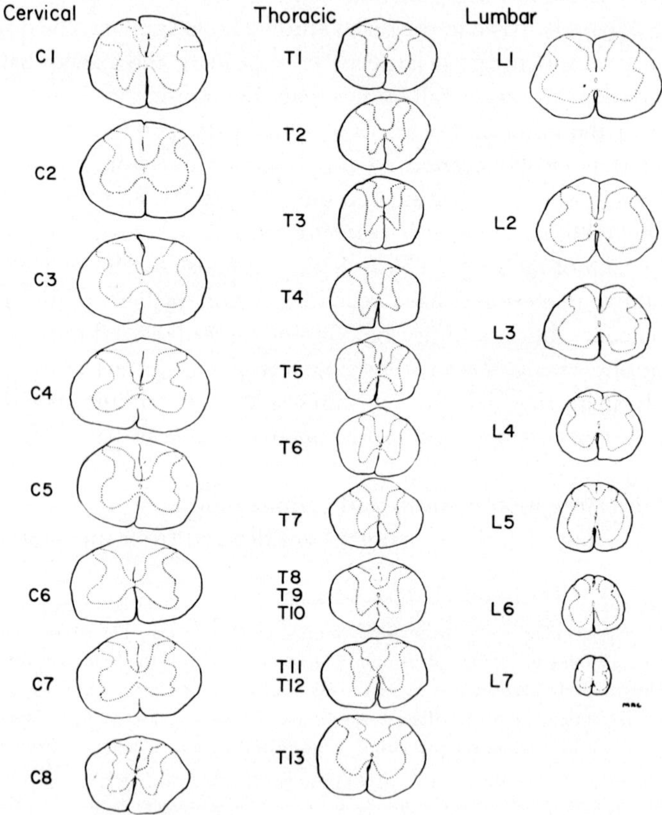

Figure 114. Outlines of cross-sections through the Rat's spinal cord, drawn to scale from histologic sections, and showing configuration as well as area of gray substance at segmental levels (from ZEMAN and INNES, 1963; about × 7, red. $^4/_5$; compare with Fig. 111).

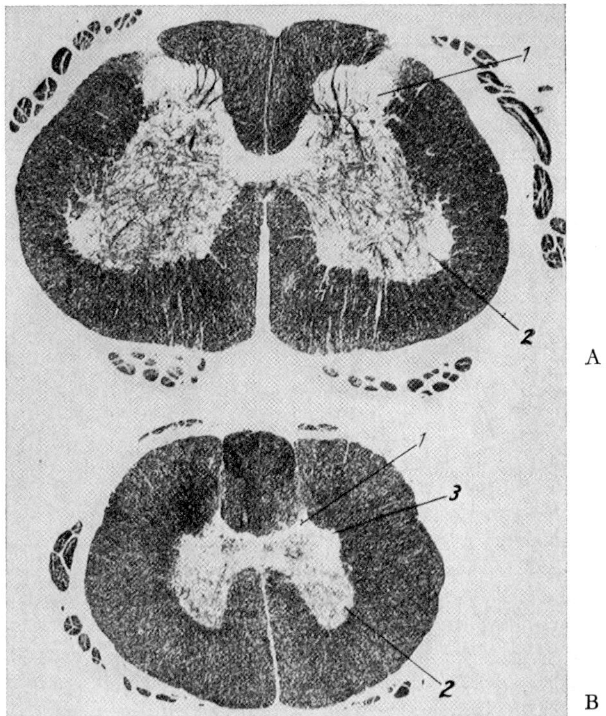

Figure 115A, B. Cross-sections through the Dog's spinal cord as seen by means of a myelin stain (from BECCARI, 1943). A Intumescentia cervicalis. B Thoracic region; 1: posterior horn; 2: anterior horn; 3: lateral horn. In this particular specimen, the thoracic posterior horn, with a small tip of substantia gelatinosa, seems unusually short and presumably represents an individual variation.

In the Mammalian spinal cord, this zone is said to include an intermediolateral nucleus and an intermediomedial one,[125] both of which, however, ontogenetically derive from a more or less homogeneous cell mass in the dorsal portion of the basal plate. It seems fairly certain that the thoracolumbar (sympathetic) preganglionic neurons are located in the intermediolateral cell column, which may slightly protrude as a '*lateral horn*' at some levels, and extends from about C_8 to L_2. POLYAK (1924) and CROSBY *et al.* (1962) assume that the so-called intermedio-

[125] CROSBY *et al.*, 1962; NOBACK and HARTING, 1971. According to the former, the subdivision into intermediomedial and intermediolateral 'columns' obtains from about T_1 to L_3. According to NOBACK and HARTING, the 'intermediomedial nucleus' is 'present at all segmental levels' in Primates.

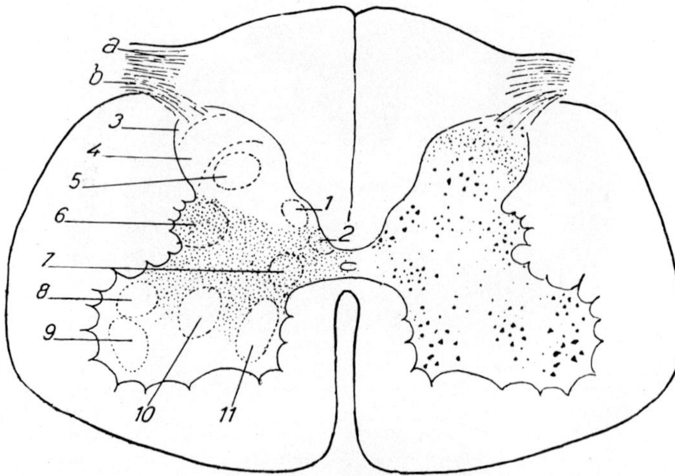

Figure 115C. Semischematic sketch of cross-section through the Dog's spinal cord at the level of first thoracic segment, indicating Beccari's concept of cell groups (from Beccari, 1943). a: dorsomedial posterior root fiber bundle; b: ventrolateral posterior root fiber bundle; 1: nucleus commissuralis posterior; 2: *column* (nucleus) *of Stilling-Clarke;* 3: posteromarginal cells *(cellule marginali dorsali);* 4: *substantia gelatinosa Rolandi;* 5: nucleus proprius of dorsal horn; 6: 'intermediolateral cell column'; 7: 'intermedio-medial' or 'paraependymal' cell column; 8: dorsolateral column; 9: ventrolateral column; 10: 'central' column; 11: ventromedial column; the dotted area on left is said to indicate the so-called 'pars intermedia'.

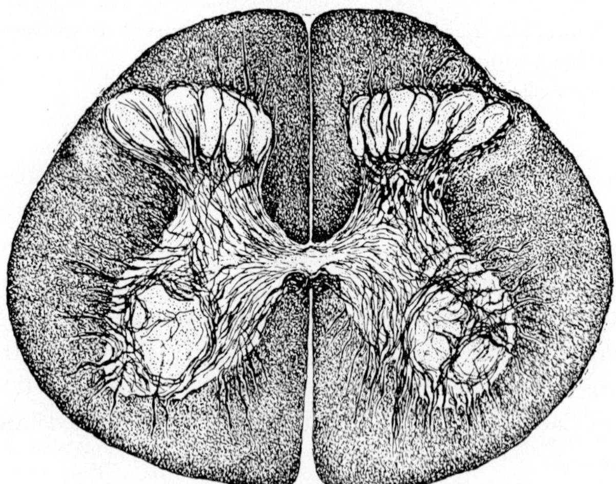

Figure 115D. Cross-section through a sacral level of the spinal cord in the Artiodactyle Ungulate Gazella dorcas, showing convolutions of substantia gelatinosa displayed by a myelin-stain (after Biach, 1907, from Kappers, 1920).

medial column likewise contains sympathetic preganglionics. Yet, this griseum may also include other types of neurons, and its significance remains incompletely understood. The sacral (parasympathetic) preganglionic cells, whose location at the levels from S_2 to S_4 is reasonably well established, appear distributed in a somewhat more scattered fashion within the 'intermediate gray'. Preganglionics related to pupillary dilatation are located in the cervical and upper thoracic segments of the thoracolumbar (sympathetic) intermedioventral zone. Together with their neuropil, respectively possibly present local internuncials, these preganglionics represent the so-called *centrum ciliospinale*.

In Mammals, the preganglionic fibers seem to leave the spinal cord predominantly through the ventral roots, but the exit of some such fibers through posterior roots cannot be excluded and is assumed by diverse authors. CROSBY *et al.* (1962) refer in this connection to the

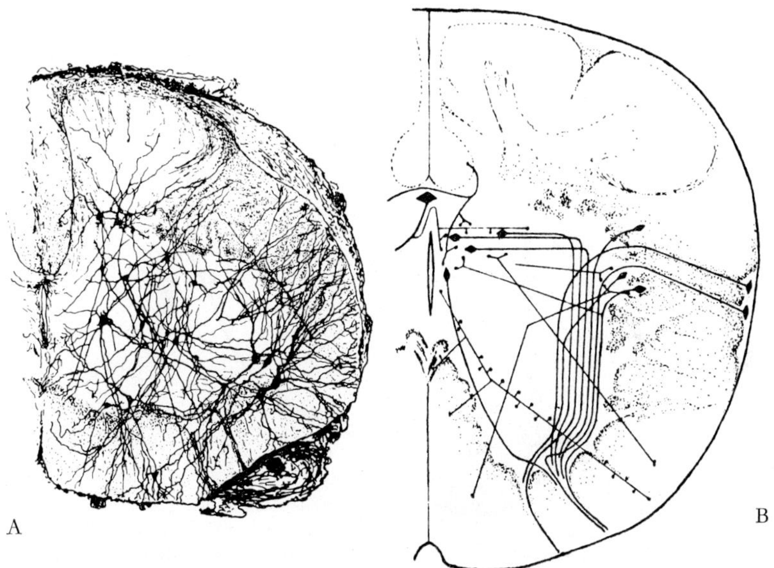

Figure 116A, B. Cross-sections through spinal cord halves in the Bat (Chiropteran) Pterygistes noctula (from POLJAK, 1924). A Combined *Golgi-picture* of cervical intumescentia (the marginal nucleus is recognizable at lower right). The extension of dendrites into white matter should be noted (about × 55, red. $^3/_4$). B POLJAK's interpretation of 'sympathetic centers' in the thoracic spinal cord of Pterygistes. The diagram indicates three efferent cell groups of the intermedioventral zone and includes connecting ('zuführende') collaterals coming from anterior, lateral, and posterior funiculi. A complex 'teledendron' is shown within the anterior horn.

hypotheses formulated by KURÉ, and briefly discussed in section 1, chapter VII of the preceding volume. Although said hypotheses have not been generally accepted and can be evaluated as rather dubious, the inclusion of some 'general visceral' efferent fibers in the Mammalian spinal dorsal roots seems not improbable.

An additional cell group, the so-called 'anterior commissural nucleus', in a medial location, and somewhat ventral to the 'intermedio-medial nucleus' has been described in the Mammalian spinal cord (cf. Fig. 127). It may or may not be identical with the nucleus cornu-commissuralis anterior described by MASSAZZA (1922–24) as bordering on the anterior funiculus.

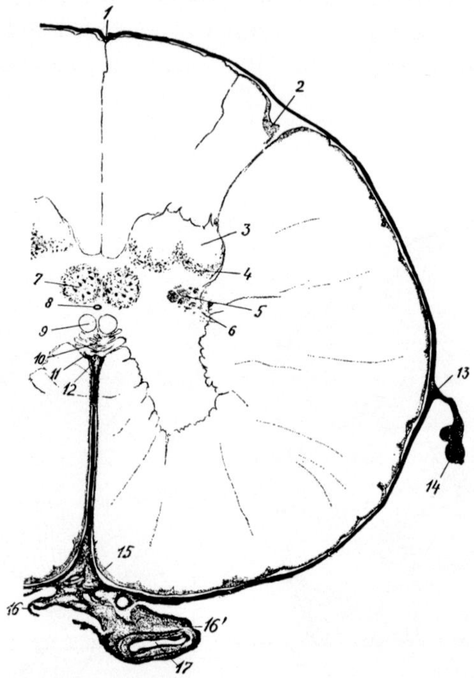

Figure 117. Cross-section through the spinal cord of the Indian elephant at the 6th thoracic level (from DEXLER, 1907; approx. × 10, red. $^{7}/_{10}$). 1: sulcus medianus posterior; 2: surface thickening *('apikale Verdickung des Randschleiers');* 3: *substantia gelatinosa Rolandi;* 4: posterior horn; 5: reticular formation ('processus reticularis'); 6: nerve cells in 'lateral horn'; 7: *column of Stilling-Clarke;* 8: central canal; 9: 'intracommissural ventral bundle' ('intercommissural crossed pyramidal tract', cf. Fig. 118); 10: commissura alba ventralis; 11: branch of 'arteria fissurae ventralis'; 12: end of (leptomeningeal) septum medianum ventrale; 13: insertion of ligamentum denticulatum; 14: edge of ligamentum denticulatum; 15: marginal glia *('marginale Verdickung des Randschleiers');* 16, 16': edges of 'ligamentum ventrale piae matris'; 17: 'arteria ventralis' (a. spinalis anterior).

Marginal nuclei, comparable to those of Sauropsidans, are not seen in the human spinal cord, nor were they commonly recorded in that of other Mammals, with the exception of Chiroptera (DRÄSEKE, 1903; POLYAK, 1924), of the cat (ANDERSON *et al.*, 1964), and of the Cynomolgus monkey (DUNCAN, 1953). In Chiroptera (Fig. 116) these grisea, located at the periphery of the lateral funiculus, are rather well developed, although less so than in some Sauropsidans, and seem to contain a dense dendritic plexus. POLYAK (1924) traced some neurites from the marginal nuclei of Chiroptera into the ventral roots (cf. Fig. 116B) and interpreted their cells of origin as sympathetic efferents. Because of the difficulties inherent in the evaluation of complex *Golgi impregnation* details, this interpretation may be regarded as somewhat uncertain. Again, it is quite possible that, in Chiroptera as in Birds, the rather well developed marginal nuclei are functionally related to flight activity, but the presence of such marginal grisea in the flightless Carnivores, Primates and Reptiles seems to indicate that such relationship, if at all obtaining, would be merely coincidental.

In the Cat, ANDERSON *et al.* (1964) described two types of cells, namely multipolar, flattened elements, and bipolar cells arrayed along the longitudinal fibers of the funiculus. The marginal cells located near the attachment of the denticulate ligament formed a 'beaded column with the enlargements corresponding to each segment'. No marginal cells were found in the sacral segments, although an increased amount of scattered funicular cells were present. Synapses were observed to occur in connection with these elements, but no valid conclusions concerning the significance of these Mammalian 'marginal nuclei' can be drawn. It is not unlikely that a close scrutiny of the spinal cord in other Mammals including man might show a more widespread occurrence of scattered funicular '*paragriseal*' nerve cells than hitherto generally known.

With regard to the posterior horn, CROSBY *et al.* (1962) describe a so-called 'secondary visceral gray', comparable to JOHNSTON's 'visceral sensory zone' (cf. Fig. 3) or to the intermediodorsal longitudinal zone of formanalytic terminology, and located at the posterior horn's base, which borders on the anterior horn's 'intermediate gray'. It is supposed to receive terminal axons of visceral afferent neurons and to originate as well as receive fibers from 'the secondary ascending visceral tracts'. There is, however, no definitive or conclusive evidence for such specifically visceral afferent zone within the Vertebrate spinal cord, particularly since the thinly medullated and non-medullated pos-

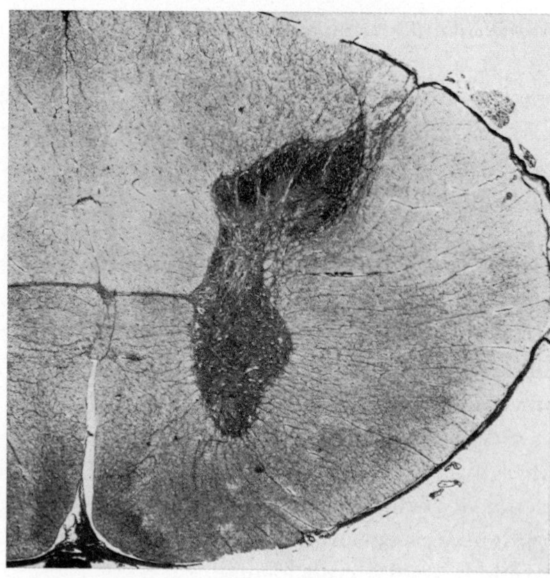

Figure 118A. High cervical segment of the elephant (from VERHAART, 1970; Hägg-quist stain; ×10.5, red. $^{3}/_{5}$).

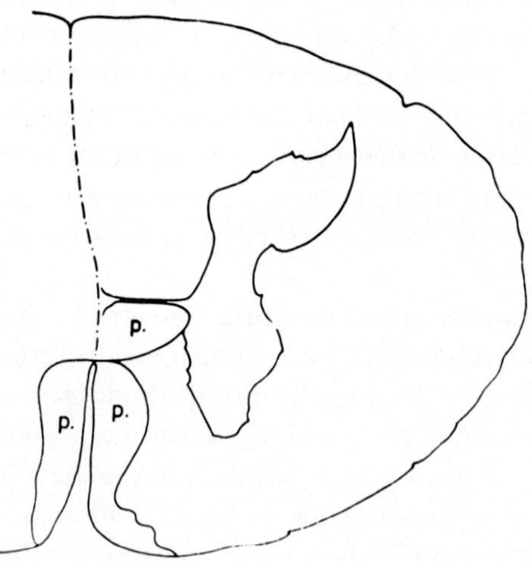

Figure 118B. Outline sketch of Figure A, indicating dorsal, crossed 'intercommissural pyramidal tract' (p) and ventral 'uncrossed pyramidal tract' (p) along surface of fissura mediana anterior (from VERHAART, 1970).

terior root fibers, which are especially related to the more dorsal substantia gelatinosa, may most likely include visceral afferent connections. In contradistinction to the anterior intermediate gray of the intermedioventral zone, I would prefer to designate CROSBY's 'secondary visceral gray' as the 'ventral border zone of the dorsal gray'. Its medial portion seems to include the cell group described by some authors as so-called posterior commissural nucleus (cf. Fig. 127).

The somewhat diffuse griseum of the posterior horn dorsal to the 'secondary visceral gray' of CROSBY *et al.* is called 'dorsal funicular gray' or 'nucleus proprius'. It contains scattered multipolar neurons and smaller stellate cells. Some of these two sorts of elements send dendrites into the substantia gelatinosa (PEARSON, 1952). Within the area of 'nucleus proprius', and at some levels, a medially located 'nucleus cornu-commissuralis', bordering on the posterior funiculus, has been described by MASSAZZA (1922–24).

The *nucleus dorsalis of Stilling-Clarke*, which extends in Man approximately from the 8th cervical to the 3rd lumbar level, seems to be a ventromedial differentiation of the 'nucleus proprius'.[126] It consists of large multipolar cells[127] with profusely branching dendrites and some intermingled smaller elements. The neurites of the large cells provide the essentially homolateral *dorsal spinal-cerebellar tract of Flechsig*. Root fibers from the posterior funiculi, entering this nucleus, end here with dense arborizations (cf. Fig. 119).

The substantia gelatinosa (Fig. 120) contains small nerve cells with extensive dendritic branchings *(Gierke-Virchow cells)*, which, in turn, are surrounded by dense terminal arborizations of posterior root collaterals, particularly of those pertaining to *Lissauer's tract*. As already pointed out above in discussing this griseum of Birds, the gelatinous aspect is doubtless due to the rich dendritic ('protoplasmic') and axonal arborizations together with their accessory glial elements. The external margin of substantia gelatinosa, bordering on *Lissauer's tract*, includes a row of larger neuronal elements, comparable to those de-

---

[126] HOGG (1944) investigated the ontogenetic development of the *nucleus dorsalis of Stilling-Clarke* in Man. It originates fairly dorsally in the alar plate, and as this latter fuses in the midline with its antimere, assumes a ventromedial location in the posterior horn, dorsolaterally to the central canal.

[127] In *Nissl-preparations* of the normal human spinal cord, the large cells of this nucleus commonly display a picture of 'moderate tigrolysis' (cf. p. 658 and Fig. 367 in chapter V, section 8, vol. 3, part I of this series).

scribed above in Birds (cf. Figs. 106, 107), and designated as cellulae posteromarginales or pericornual cells, which form the narrow so-called 'zona spongiosa'.

In addition to the scattered paragriseal cells forming the reticular 'formation', 'process', or 'nucleus' in the cervical spinal cord, a 'lateral cervical nucleus', adjacent to the dorsal portion of posterior horn, has been described by REXED and BRODAL (1957) in the cat, and was interpreted as a 'spino-cerebellar relay nucleus'. NOBACK and HARTING (1971) indicate its position in the Chimpanzee (cf. Fig. 127), and VERHAART (1970) describes that griseum in various other Mammals. It might be considered a dorsal condensation of the 'reticular nucleus'.

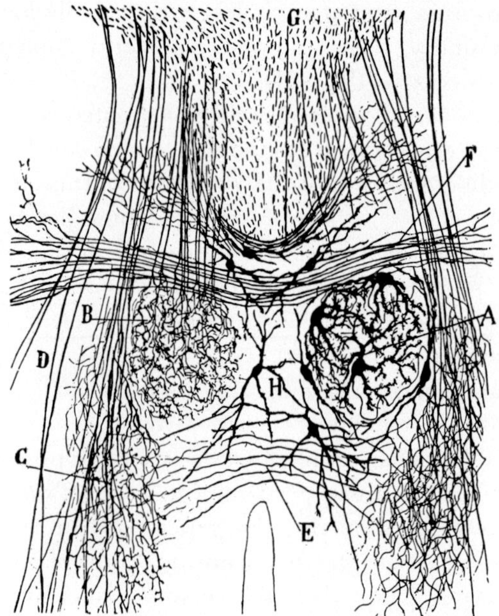

Figure 119. Cross-section through the spinal cord in a new-born dog at a thoracic level, showing details of nucleus dorsalis *(Stilling-Clarke)* as displayed by the *Golgi impregnation* (from CAJAL, 1909). A: nucleus dorsalis, displaying configuration of its cells; B: endings of posterior root collaterals in nucleus dorsalis; C: dorsal root collaterals ending in intermediodorsal and intermedioventral zone; D: dorsal root collaterals reaching anterior horn proper ('motor reflex collaterals'); E: ventral commissure; F: 'middle commissure' (commissura posterior grisea); G: 'posterior commissure' (perhaps commissura posterior alba); H: 'cells of posterior commissure'. The end of fissura mediana anterior can be seen at bottom center. One may wonder what happened to the central canal, which one would expect in the neighborhood of H, and which was perhaps inadvertently omitted or overlooked by CAJAL in preparing this interesting figure.

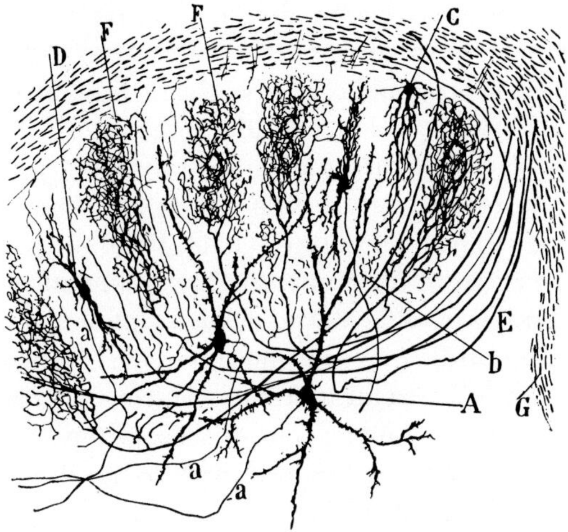

Figure 120. Cross-section through the *substantia gelatinosa Rolandi* in the cervical spinal cord of a newborn Cat, as seen by means of the *Golgi-impregnation* (from CAJAL, 1909). A: cell of nucleus proprius cornus posterioris *('cellules de la tête de la corne postérieure);* C, D: cells of substantia gelatinosa; E: posterior root collaterals entering substantia gelatinosa *('collatérales grosses ou profondes de cette substance');* F: terminal arborizations of fibers E; G: funiculus posterior; a: neurites; b: longitudinal arborizations at tip of 'nucleus proprius' *('arborisations longitudinales du sommet de la corne postérieure').* The pericornual (or posteromarginal cells, cf. Figs. 106, 107 of the Avian spinal cord) were apparently not displayed in the *Golgi-preparations* upon which CAJAL based this drawing.

Apparently transitory intramedullary *neurons of Rohon-Beard type* were independently and simultaneously reported by HUMPHREY (1944) and by YOUNGSTROM (1944). HUMPHREY suggests that, although such neurons have not been identified in Birds (cf. above, p. 172), it seems not impossible that some of the nerve cells depicted by CAJAL (1909) and BOK (1928) in the Avian spinal cord, but otherwise interpreted by these authors, might be comparable elements.

The presence of transitory *Rohon-Beard cells* in the Human (respectively the Mammalian[128] or Sauropsidan) spinal cord does not neces-

---

[128] Previously to the investigations by HUMPHREY (1944) and YOUNGSTROM (1944) ROHON-BEARD type neurons had not been identified in Mammals. Although reports of their presence in additional Mammalian forms have not come to my attention, the occurrence of such cells in at least some other Mammals seems reasonably probable.

sarily imply the presence of an embryonic 'primitive nervous mechanism' comparable to that of some Amphibia. The functional significance of these elements could either be greatly modified, or such cells could merely represent, phylogenetically speaking, an abortive and irrelevant recapitulation of discarded ancestral states. Both these interpretations would be compatible with the erratic behavior of the elements in question. Thus, HUMPHREY (1944, 1950) found the primitive sensory ganglion cells of the human embryo scattered anywhere from the entering dorsal root to the dorsal median septum as well as the roof plate, and even proturding into the interior of the central canal.

Related or similar aberrant morphogenetic events may occur with respect to the heterotopic nerve cells of sensory type displayed along the intramedullary as well as extramedullary course of the ventral root fibers of the adult Human and Mammalian spinal cord. Such findings were reviewed, with bibliographic references, by CROSBY *et al.* (1962), and were also briefly mentioned in sections 2 (footnote 37) and 3, chapter VII of the preceding volume.

With regard to the neuronal elements and their 'nuclear' aggregations in the Mammalian, or for that matter, in the Vertebrate spinal cord, four additional questions are of interest, namely (1) the occurrence of *Golgi II type cells*, (2) the definition of 'nuclei', (3) certain cytoarchitectural aspects, and (4) certain biochemical features.

(1) GOLGI (1894), after whom the short interneurons of 'type II' were named, had recorded numerous such cells in both posterior and anterior horn of various Vertebrates. Findings by KÖLLIKER, LENHOSSÉK and VAN GEHUCHTEN confirmed these observations, and the presence of this sort of elements in the spinal cord of most Vertebrates became generally accepted. CAJAL (1909, 1952), however, expressed a sceptical opinion. He stated (l.c. p. 418): '*Nous avions été surpris, dès nos premières recherches sur la structure de la moelle, de la rareté extrême de ces neurones. Cette surprise, nos observations, à mesure qu'elles se multipliaient, l'ont transformée en quasi-certitude. Aujourd'hui donc, nous avons la conviction qu'il n'existe des cellules à cylindre-axe court ni dans la corne antérieure, ni dans la base et le centre de la corne postérieure.*' CAJAL assumed that the description of such cells is based on incomplete silver preparations. Nevertheless, this author admits the presence of a few *Golgi II type cells* within the substantia gelatinosa: '*A force de la fouiller, nous avons fini par y découvrir à peine quelques neurones à cylindre-axe court.*' Again, CAJAL does not entirely deny the possible occurrence of *Golgi II type commissural cells*, described by GOLGI and LENHOSSÉK, but adds that he has never seen them and

considers '*leur inexistence très probable*'. Because of the uncertainties in the interpretation of an observed impregnation of nerve cell processes as 'complete' or 'incomplete' and despite CAJAL's undisputable technical mastership, I believe that the question concerning occurrence and distribution of *Golgi II type cells* in the Vertebrate spinal cord remains open. I am inclined to accept at least some of the conclusions reached by GOLGI, KÖLLIKER, and LENHOSSÉK.[129]

(2) In standard preparations of the Mammalian spinal cord, particularly in those with myelin stain, the gray core displays a rather compact and fairly homogeneous configuration characterized by posterior and anterior horns, in which, with few exceptions, notably *substantia gelatinosa Rolandi* and *nucleus dorsalis of Stilling-Clarke*, typical 'nuclear' aggregates, such as found in many regions of the brain, are not particularly evident. Yet, in *Nissl*- an other cell-stain preparations, especially in fairly thick ones, more or less circumscribed clusters of nerve cells in anterior horn and other regions of the spinal cord's gray can be noted and, depending on arbitrary criteria, could be designated as '*nuclei*'. Few authors have attempted to define the (neuroanatomical) term 'nucleus'. According to *Stedman's Medical Dictionary*, it is 'a mass of gray matter, composed of nerve cells, in any part of the brain or spinal cord', a '*ganglion*' being 'an aggregation of nerve cells'. *Dorland's Medical Dictionary* defines it as ' a group of nerve cells in the central nervous system', while a 'ganglion' is said to be 'any collection or mass of nerve cells that serves as a center of nervous influence'. Substituting 'within' for 'in', HERRICK's Introduction to Neurology (1931) literally gives the same explanation. This author, while defining 'ganglion' as 'a collection of nerve cells', adds that, in Vertebrates 'the term should be applied only to peripheral cell masses, though sometimes nuclei within the brain are so designated'. In the '*Glossary*' appended to the treatise by BULLOCK and HORRIDGE (1965) a nucleus is said to be 'any of numerous small, anatomically demarcated masses of gray matter in the vertebrate central nervous system'. Although this is somewhat more

---

[129] In view of a certain antagonism between CAJAL and GOLGI, as clearly manifested in their official discourses upon occasion of the *Nobel Prize* ceremony in Stockholm, 1906, one could perhaps suspect that CAJAL was, in this matter, influenced by a subconscious bias. Moreover, the postulated *Renshaw-cells*, mentioned above, might again be pointed out in support of the assumed presence of Golgi-II cells.

explicit, one might here ask 'what is anatomically demarcated'. Evidently, in some instances, gray aggregates are demarcated by more or less distinct surrounding white fiber masses. In other instances, however, the criterion for a demarcation is merely a subjective visual pattern impression of the observer, based on a variety of factors, such as detectable changes in population density, as well as in shapes, sizes, and variety of the group's component elements.

Concerning '*ganglion*', BULLOCK and HORRIDGE (1965) state that it is 'a discrete collection of nerve cells': 'The term is usually applied to a nodular mass defined by connective tissue, in the periphery or in a chain separated by connectives. It is arbitrarily applied to some discrete nuclear masses within the brain, as the basal ganglia of higher Vertebrates'. With regard to Invertebrates, the term is used for some major subdivisions of the central nervous system, as e.g. the 'cerebral ganglion' or some of its main subdivisions, those of the 'optic ganglion' or 'optic lobe' being particular 'neuropilemes'.

Finally, ELLIOTT (1963) speaks of 'ganglia' as large or small separate groups of neurons, situated in different parts of the body, and connected by groups of fibers, a 'ganglionic nervous system' being characteristic for many Invertebrates. He then adds: 'Vertebrates have ganglia too, small groups of nerve cells detached from the central mass and connected to it by nerves; but this detachment is embryologically secondary and minor.' With respect to the term 'nucleus', the cited author states that 'the gray matter presents a mingling of different nerve cells, some of which form fairly compact groups called nuclei (not to be confused with the nuclei of cells), while others seem to be scattered at random'.

These various formulations clearly indicate the semantic difficulties encountered in attempts at reasonably rigorous definitions required for neurobiologic terminology, and the resulting arbitrariness in the application of even fairly standardized common terms. None of these definitions include some significant references to functionally relevant factors, namely to the presence or absence of synaptic connections (neuropil) and to the origin or termination of particular fiber tracts. Thus, although many *ganglia* (e.g. of vegetative plexuses) contain synapses, these latter may perhaps be entirely absent (or rare, respectively doubtful) in most or all Vertebrate spinal ganglia. A 'true' Vertebrate griseal *nucleus*, however, is generally characterized by its synaptic neuropil. Yet, the 'nucleus of the radix mesencephalica trigemini', somewhat comparable to a spinal or cranial nerve ganglion, may perhaps be devoid of synapses, which seem to be located at a distance from the 'nucleus'.[130]

With regard to fiber systems, there are *nuclei of origin* and *terminal nuclei;* yet, many such grisea pertain, depending on the fiber tracts concerned, to both categories. Nevertheless, '*sensory nuclei*', receiving afferents of the first order, and '*motor nuclei*' respectively '*efferent nuclei*' giving origin to either direct muscular efferents, or to preganglionic fibers, can be distinguished. Other 'nuclei' may be designated in accordance with their relevant connections (e.g. 'vestibular nuclei') or their predominant functional significance.[131] Diffusely or 'randomly' arranged nerve cell populations, as justly mentioned by ELLIOTT, have been termed '*formatio*', as in the case of the deuterencephalic formatio reticularis, but even here, the designation 'nucleus' is frequently used. There is, moreover, the question of parcellation: to which extent should certain cell clusters occurring within a gray mass be considered 'subnuclei' of a nucleus, or nuclei within a larger griseal aggregate.

With regard to such aggregates, a relevant classification of nuclei, based upon the arrangement of neuronal cell processes, was suggested by MANNEN (1960). Nuclei, whose nerve cells dendrites extend into, or effect reciprocal exchanges of this sort with, neighboring nuclei, are designated as '*noyaux ouverts*', while nuclei without such interrelationships are '*noyaux fermés*'. This classification is, to some extent, similar to the distinction between open and closed neighborhoods (or sets) in topology. As far as the spinal cord is concerned, cell groups in both 'nuclei proprii' and ventral horn (motoneuron clusters) appear to represent grisea of 'open' type. It seems, moreover, obvious that, if the configuration of dendrites is taken to be a criterion for classification, both MANNEN's '*open*' and '*closed*' nuclei could be further subdivided into a number of additional categories, depending on the amount of dendritic overlap, on the arrangement of 'dendritic fields', or on the distribution of collateral as well as terminal axonic arborizations within the limit of any given, 'open' or 'closed' nucleus. Concomitantly

---

[130] The nucleus radicis mesencephalicae trigemini and its peculiarities shall be discussed further below in chapter IX and XI (Oblongata and Mesencephalon). A brief discussion of its morphologic significance was included in chapter VII of the preceding volume (pp. 827, 850).

[131] In his Spanish contributions, CAJAL frequently employs the term '*foco*' (focus), i.e. active center, or center of origin respectively destination, as synonymous with '*nucleo*'; '*foco*' being here, as in English usage, a center of activity or events *(centro de ciertas actividades o cosas)*.

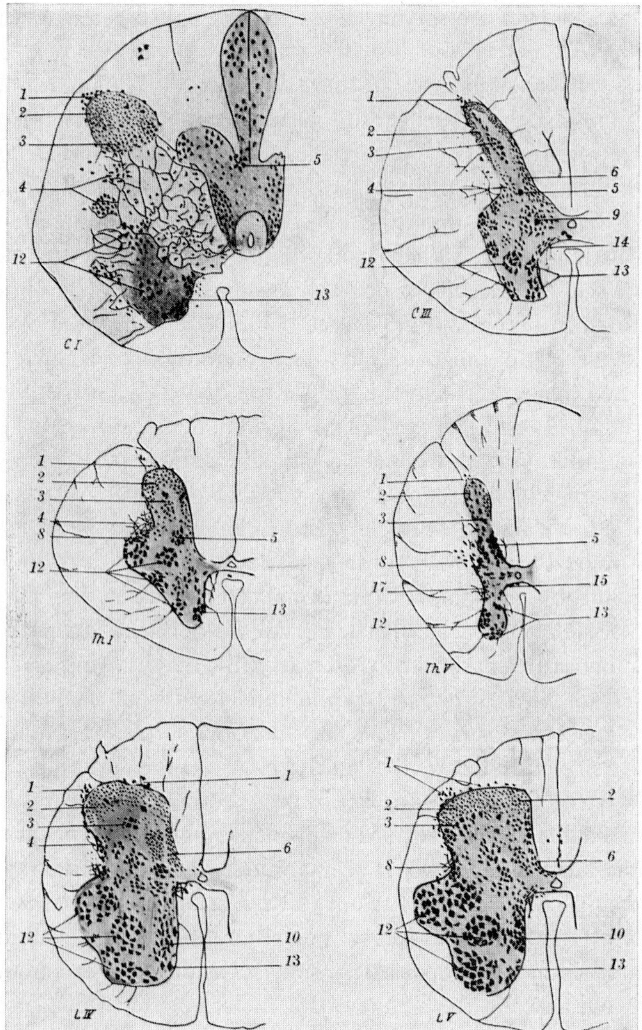

121 A

Figure 121 A–C. MASSAZZA's concept of cell groups in the human spinal cord (after MASSAZZA, 1922–1924, from BOK, 1928). A and B: levels C I to S IV in cross sections; C: graphic reconstruction of longitudinal columnar arrangement; 1: cellulae posteromarginales; 2: *substantia gelatinosa Rolandi;* 3: nucleus proprius cornus posterioris; 4: nucleus reticularis spinalis; 5: *column of Stilling-Clarke;* 6: nucleus cornu-commissuralis posterior; 7: cellulae disseminatae posteriores; 8: nucleus intermedio-lateralis; 9: nucleus intermedio-medialis; 10: nucleus medialis myoleioticus; 11: cellulae disseminatae intermediae; 12: nucleus myorhabdoticus lateralis; 13: nucleus myorhabdoticus medialis; 14: nucleus cornu-commissuralis anterior; 15: nucleus proprius cornus anterioris; 16: nucleus paracentralis anterior; 17: cellulae disseminatae anteriores; 18: cellulae aberrantes.

with refinements of electrophysiologic gadgetry and micro-electrode techniques, a complex terminology might be elaborated. Thus, it has been claimed that certain neurons can influence neighboring neurons in a manner suggestive of 'electrical interaction', such that electrical activity in one nerve cell causes, without involving chemical transmitter agents, a shift in the electrical potential of another nerve cell. With reference to phenomena of this sort, MATTHEWS *et al.* (1971) have described closely aligned bundles of motoneuron dendrites, providing

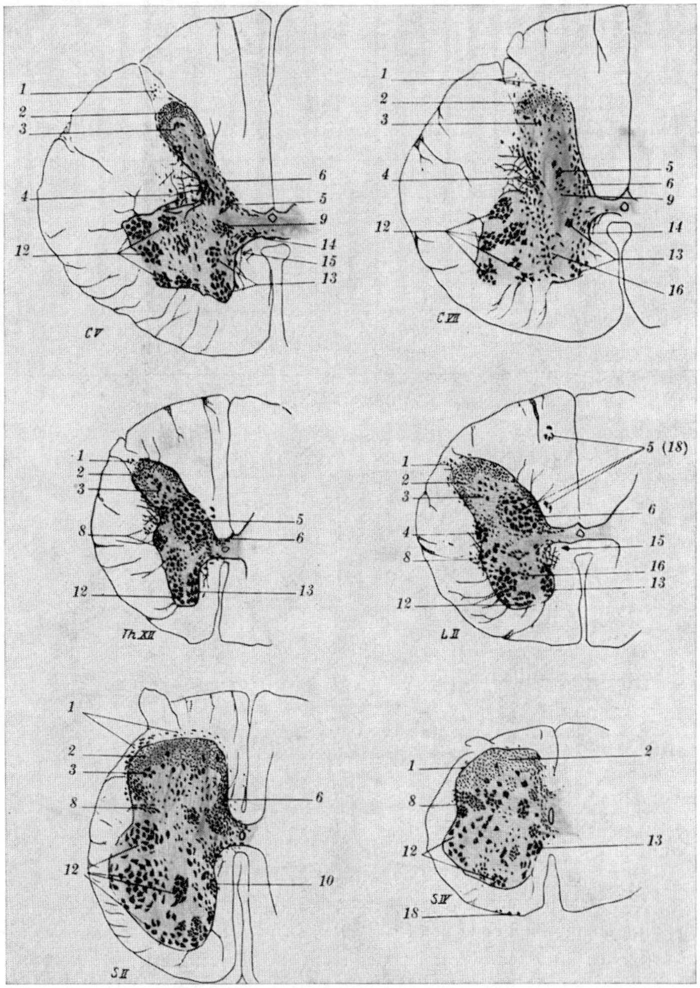

Figure 121 B

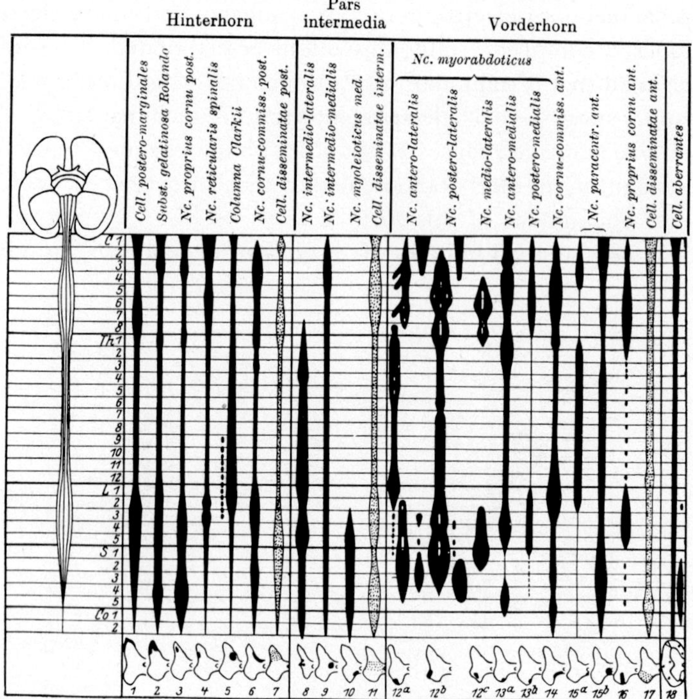

Figure 121 C

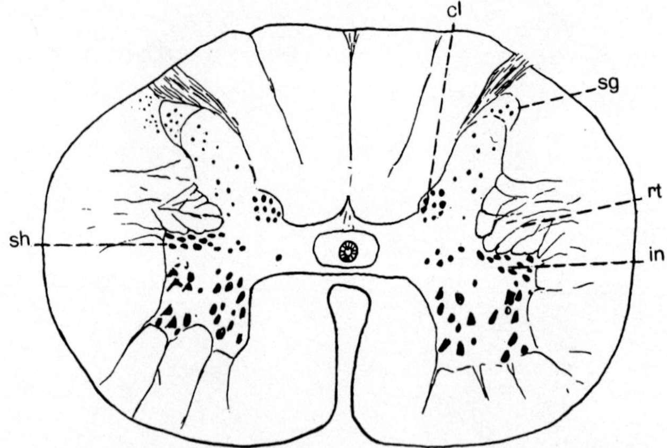

Figure 122. Semidiagrammatic cross-section through the low cervical spinal cord in Man, showing a simplified concept of cell groups (from K., 1927). cl: *nucleus dorsalis of Stilling-Clarke;* in: intermedioventral zone; rt: formatio reticularis; sg: *substantia gelatinosa Rolandi;* sh: 'lateral horn'. The pericornual cells of substantia gelatinosa have been omitted.

an interrelation between a number of motoneurons in the anterior horn griseum of the Cat's spinal cord. The cited authors suggest that the spatial arrangement and appositional relations of these dendrites represent a structural 'substrate for the weak electrical facilitation known to occur between motoneurons in the cat spinal cord'. Again, the recurrent inhibition known as the '*Renshaw phenomenon*' has been regarded as mediated within motoneuron dendrite bundles rather than by the postulated '*Renshaw cells*' (SCHEIBEL and SCHEIBEL, 1971), while others (e.g. WILLIS, 1971) uphold the existence and significance of '*Renshaw interneurons*'. Further clarifications of these poorly understood topics must be awaited. Until now, the various pertinent hypotheses, based on ambiguous findings, and supported by extensive but hardly conclusive argumentations, cannot be considered sufficiently well-founded.

In contradistinction to grisea displaying nuclear configuration, gray substance spread out along the surfaces of telencephalon, mesencephalon, and cerebellum represents 'cortical' formations (e.g. cortex cerebri *sive* telencephali, cortex cerebelli, 'cortex tecti mesencephali').[132] Grisea of this type are commonly characterized by cytoarchitectural 'stratification' or 'lamination' displayed by more or less distinctive 'layers'. Although the gray core of the Vertebrate spinal cord is, as it were, the opposite of a cortex and manifests what could be termed a 'nuclear' arrangement, this is combined with a dorso-ventral succession of cytoarchitectural patterns which form band-like units at approximately right angle to the sagittal plane, and may, with some justification be described as 'laminae'. This concept of griseal laminae in the spinal cord will be discussed further below.

(3) The configurational respectively the cytoarchitectural aspects of the Mammalian spinal cord's gray substance have been studied in particular by JACOBSOHN (1908), MARINESCO (1904), MASSAZZA (1922–1924), SERGI (1926–27), and WALDEYER (1888). These authors were concerned with 'nuclear' griseal aggregations rather than with the 'laminar' sequence which they did not explicitly formulate. The 'nuclear' groups outlined by MASSAZZA, who somewhat overstressed their distinctive outlines in his semidiagrammatic drawings, are shown in Figure 121. Figure 115C illustrates BECCARI's concept of the cellular arrangements in the Dog, and Figure 122 depicts a simplified interpre-

[132] Cf. chapter VI, section 6, footnote 195 of vol. 3, part II.

tation, *qua* cell groups, of a lower cervical segment in Man, as given in my '*Vorlesungen*' of 1927. With regard to such parcellations and classifications as 'nuclei', their commonly ill defined outlines must be taken into consideration. In addition to the above mentioned semantic difficulties, this haziness introduces additional difficulties and uncertainties into the actual analysis of the spinal cord preparations obtained by means of various techniques. These limitations have been repeatedly pointed out by most authors dealing with the topic under consideration. The interpretation of experimental studies, either by tigrolysis or other methods of degeneration, or by electrophysiologic techniques, is subject to comparable ambiguities and unreliabilities.

Concerning the motoneuron cell clusters in the anterior horn, several opinions have been expressed on the basis of the rather ambiguous evidence: (a) each cell group may innervate one muscle; (b) each group subserves a particular type of function, such as flexion, extension, rotation, and adduction-abduction; (c) each group sends its fibers into a particular peripheral nerve.

ROMANES (1964), in a discussion of the 'motor pools' of the spinal cord, points out that, *qua* point (a), there are fewer nuclear cell groups than muscles, and that *qua* point (b) it is difficult to explain how a muscle carrying out several functions would have to be innervated. The cited author, nevertheless, believes that 'each of these interpretations contains some part of the truth, but none of them is capable of general application to the relation between the motor cell groups and the muscles which they innervate'. According to ROMANES, the various columnar cell clusters seem each to innervate a particular muscular group, such, e.g. the hamstrings, while at least some nerves, e.g. obturator and common peroneal originate from a single column. Again, the more lateral subgroups are said to innervate the extensors, and the more medial ones the flexors. In the cited author's view, the obtaining arrangement is best explained by assuming that the nuclear 'columns' are each concerned with the muscles which move the same joint, 'over and above the basic division into those cells which innervate flexor and extensor muscles'. Be that as it may, the foregoing explanation suggested by ROMANES appears to bring a possible and not implausible formulation interpreting, in a very general way, the rather ambiguous available data.

Reverting now to the lamination concept as briefly pointed out above in the introductory section 1 of this chapter, Figure 123 depicts an early attempt to define layers in dorsal spinal grisea (OBERSTEINER

1912, after SANO), and Figures 124–126 illustrate essential features of REXED's (1952, 1964) more recent interpretation, which concerns the entire dorso-ventral extent of the cord's gray matter. Figure 127 shows some nuclear groups and their relations to the laminae of REXED in a cervical segment of the Chimpanzee's cord, as interpreted by NOBACK *et al.* (1971). REXED's lamina I evidently corresponds to the pericornual cells (n. posteromarginalis) of the substantia gelatinosa, which, except for that margin, represents lamina II. Laminae III and IV correspond to the posterior horn's nucleus proprius, although, in some of REXED's outlines, lamina III may arbitrarily include a (perhaps somewhat more 'condensed') strip of substantia gelatinosa. In the ventral portion of the dorsal horn, REXED somewhat unconvincingly distinguishes laminae V and VI. The latter one is said 'to exist in its typical form only in the intumescences'. Lamina VII seems to represent the intermediate gray of the anterior horn, more or less identical with the 'visceral motor subdivision' of JOHNSTON (1906), shown in Figure 3, i.e. with the 'intermedioventral zone' of morphologic formanalysis. 'Intermediomedial nucleus' and 'intermediolateral nucleus' are doubtless part of the so-called lamina VII, which is also supposed to include the *nucleus of Stilling-Clarke*, 'posterior' and 'anterior commissural nucleus', and

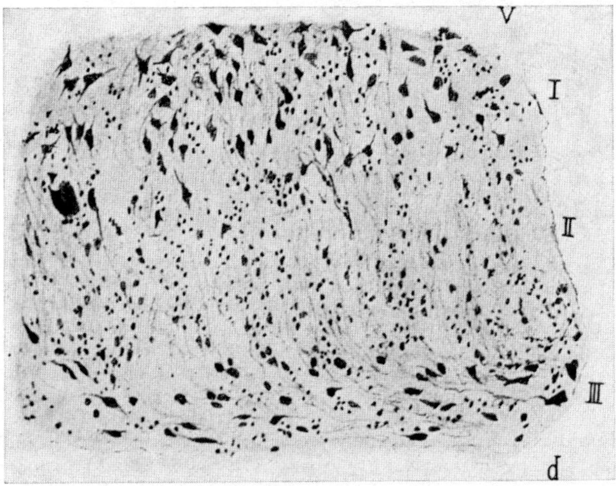

Figure 123. An early concept of lamination *('Schichtung')* at the tip of the Dog's posterior horn (after SANO, from OBERSTEINER, 1912). d: dorsal side; V: ventral side; I: paragelatinous layer of posterior horn; II: substantia gelatinosa propria; III: pericornual or posteromarginal layer.

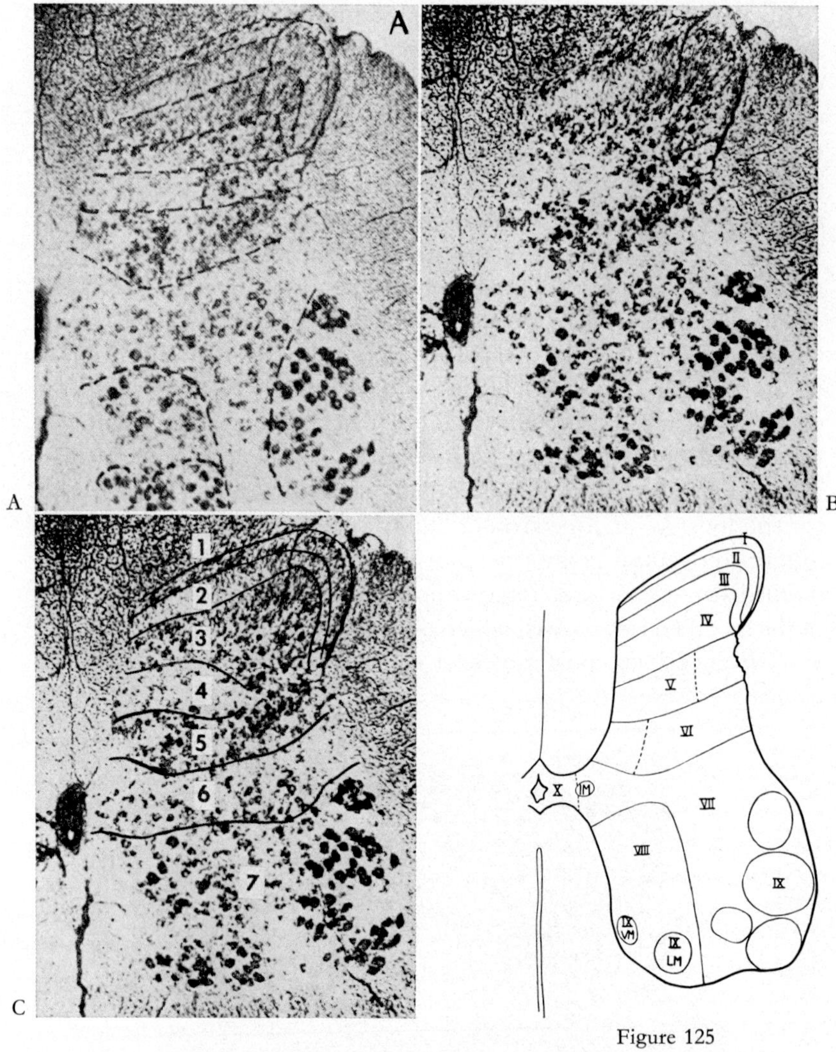

Figure 125

Figure 124 A–C. Cross-section through the 5th cervical segment of the newborn Cat's spinal cord as seen in a thick (100 μ) section stained by a modified *Nissl-method* with toluidine-blue (from Rexed, 1952, 1964). A Rexed's original Figure 16 of 1952, with lines suggesting his lamination concept. B The same section from Rexed, 1964, without lines. C Same section with added lines suggesting a different interpretation as favored by Fukuyama (1954) and myself. All 3 figures reproduced at about × 28.

Figure 125. Schematic drawing at the adjacent 6th cervical segment, for comparison with the preceding figures, and illustrating Rexed's laminae (from Rexed, 1964).

'central cervical nucleus'. Moreover, in the cervical and lumbar enlargements, lamina VII is said to extend ventrally, thereby subdividing lamina IX into a medial and a lateral portion. Lamina VIII is described as located in the anterior horn, surrounding cell groups of Lamina IX, but limited to a medial part of anterior horn in both intumescences. Lamina IX comprises the diverse groups of large motoneurons in cornu anterius.[133] The gray matter surrounding the central canal is alleged to represent an additional lamina X.

Figure 128 shows the concept of stratification elaborated by HIRASAWA (1955). It can be seen that in his scheme layer 1 includes *Lissauer's tract* and nucleus posteromarginalis, layer 2 is substantia gelatinosa proper, and layer 3 comprises the remainder of posterior horn, including the *nucleus of Stilling-Clarke*. Layer 4, with an uncertain subdivision into medial and lateral components, represents the intermedioventral zone. The entire anterior horn proper represents layer 5, which, however, by an intercalated 'radial' strip (VI) can be subdivided into medial and lateral components of V.

Professor FUKUYAMA (1955), while staying as a visiting scientist in the Department of Anatomy of the Woman's Medical College of Pennsylvania, investigated, at my suggestion, the griseal lamination of the Rat's spinal cord. In agreement with the previous authors, he could identify layers 1 and 2 represented by posteromarginal zone and substantia gelatinosa, respectively. He was in doubt whether the 'nucleus proprius' of the posterior horn displayed a single or a double layer (3+4, or 3 *and* 4). His layer 5 corresponded to the intermediodorsal zone. Another doubtful question was the rapport of *Stilling-Clarke's nucleus* dorsalis to either layer 4 or 5. The intermedioventral zone could be identified as represented by layer 6. The anterior horn proper could, at best, be conceived as a single layer (7) characterized by cell clusters representing the cell groups which the classical authors attempted to distinguish. Thus, if layers 3 and 4 were distinguished as separate laminae, a scheme of seven layers resulted, which, as far as layer 5 and with the notable exception of nucleus dorsalis, roughly corresponded to the region included in REXED's laminae I–VI. Taking layers 3 and 4 as a single one (3+4) a scheme of only six layers was obtained. FUKUYAMA,

[133] Discounting the dubious 'central cervical nucleus' mentioned above as supposedly pertaining to lamina VII.

with whose findings I agree, reached his conclusions quite independently of my own views on the topic of 'stratification'.

Concerning REXED's scheme of lamination, I cannot accept the inclusion of *Stilling-Clarke's nucleus dorsalis* into the so-called lamina VII. Said nucleus seems doubtles sto be a ventromedially displaced derivative of the posterior horn's 'nucleus proprius' neighborhood, and should *not* be incorporated into the so-called lamina VII, which can be considered a basal plate component. Even NOBACK *et al.* (1971), who have adopted REXED's scheme, locate the nucleus dorsalis in a neigh-

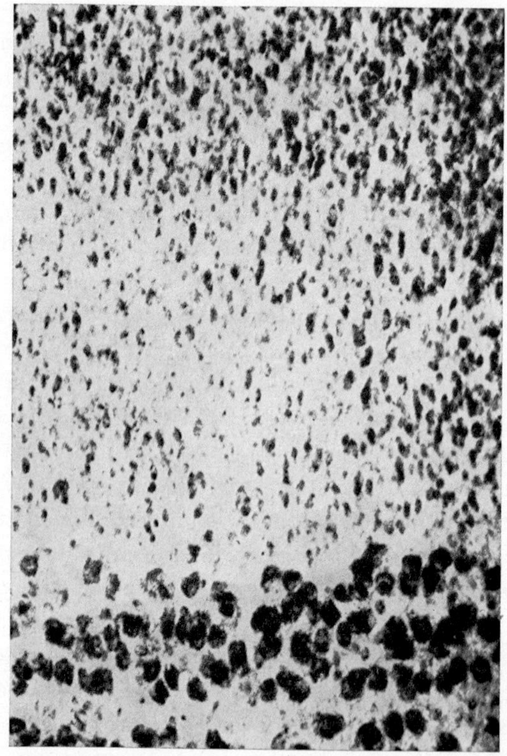

Figure 126. Slightly oblique sagittal (longitudinal) section through the lumbar intumescentia of a 14 days old Cat, displaying REXED's lamina VII (HIRASAWA's wedge-like or radial layer VI) between laminae VI and IX (from REXED, 1964). Although this section displays indeed a very striking 'laminar effect', it is quite obvious, from a consideration of its actual three-dimensional features which can easily be deduced from the cross-sectional aspects (cf. Figs. 4, 124, 125, 127, 128), that, in comparison with a true 'cortical' stratification, REXED's layers represent a 'pseudo-lamination'.

borhood pertaining to the posterior horn, namely that of lamina VI (cf. Fig. 127). As regards other objections to REXED's formulation, I do not believe that the configuration of cell clusters within the anterior horn proper should be subsumed under a concept of distinctive laminae such as VII, VIII, IX, but could, at best, be considered differentiations within a single zone (FUKUYAMA's layer 7). Moreover, while REXED's lamina VII (FUKUYAMA's layer 6) may indeed comprise a par-

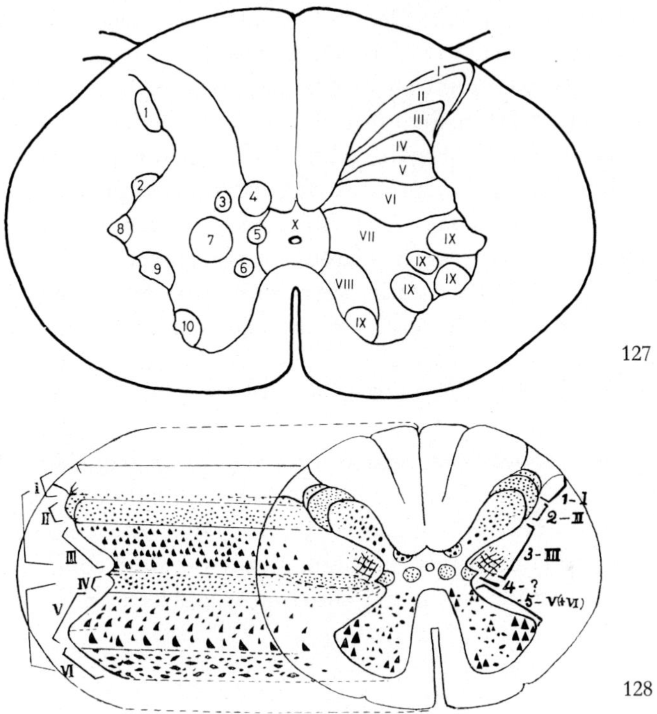

Figure 127. Schematic cross-section through the 8th cervical segment of the spinal cord in a Chimpanzee as interpreted by NOBACK and SIMENAUER. On the left side, several 'nuclei' delimited by the cited authors are indicated, while the right side shows the *laminae of Rexed* (from NOBACK and HARTING, 1971). 1: lateral cervical nucleus; 2: reticular nucleus; 3: posterior commissural nucleus; 4: 'thoracic nucleus' *(column of Stilling-Clarke)*; 5: intermediomedial nucleus; 6: anterior commissural nucleus; 7: central cervical nucleus; 8: intermediolateral nucleus; 9: 'nucleus of the spinal accessory nucleus' *(sic)*; 10: phrenic nucleus.

Figure 128. Diagram of the Mammalian spinal cord showing HIRASAWA's lamination concept (from HIRASAWA, 1955; the original figure is in color). Explanation of notation in text.

ticular 'zone', namely the intermedioventral one, I question the inclusion of particular ventral zone components (HIRASAWA's radial strip VI) into REXED's lamina VII. Also, I cannot agree with the classification of the central gray, surrounding the ependymal canal, as a lamina X comparable with the other laminae.

While REXED's scheme provides indeed some sort of topographic orientation, its parcellation[134] *qua* laminae seems excessive, besides subsuming, in a rather *Procrustean* manner, at least three different sorts of cytoarchitectural configurations under a 'laminar concept' supposed to imply their essentially identical morphologic value. It can hardly be claimed that laminae VII, VIII, IX, and particularly lamina X are form elements *(Formbestandteile)* of the same type as the four or five layers (FUKUYAMA's 1–5) in the posterior horn, or the layer 6 (FUKUYAMA) representing the intermedioventral zone. Finally, it seems quite obvious that the dorso-ventral cytoarchitectural sequences in the gray core of the Mammalian or Vertebrate spinal cord, however much some of their aspects may present a laminar arrangement (cf. Fig. 126), differ significantly from the stratification into layers of true cortical grisea such as cortex telencephali, cortex cerebelli and tectum mesencephali.[135]

(4) Recent developments in neurochemistry and neuropharmacology have led to the elaboration of histochemical techniques providing information about functional aspects of various neural structures, including those of the spinal cord. The large amount of published new data requires further elucidations and has not significantly affected the fundamental concepts of morphologically oriented comparative neurology. Therefore, in the present context and as a preliminary intro-

---

[134] A critique of what I believe to be excessive cytoarchitectural parcellation was given on p. 60 f. of my monograph '*The Human Diencephalon*' (Karger, Basel 1954), and relevant comments can also be found in CHRIST's chapter on 'Derivation and Boundaries of the Hypothalamus' *(The Hypothalamus*, HAYMAKER *et al.* [eds.], Thomas, Springfield 1969, pp. 13–60). A particularly deceiving procedure followed by proponents of cytoarchitectural parcellation is the drawing of lines, purporting to indicate boundaries, upon the cytoarchitectural photomicrographs. It is evident that such lines have, for uncritical readers of their publications, a powerfully suggestive effect.

[135] The cortical griseum of tectum mesencephali with its complex lamination is particularly conspicuous in most submammalian Vertebrates. In Mammals, although a characteristic stratification is still recognizable in the mesencephalic superior colliculi, the 'cortical aspect' of that griseum becomes much less obvious.

duction for readers extraneous to that particular field of investigation,[136] a short mention of two relevant methods of approach may be considered sufficient.

CARLSSON *et al.* (1964) studied the 'cellular localization' of monoamines in the spinal cord of the Mouse and Rat by means of a fluorescent demonstration of certain catecholamines and tryptamines in combination with a pharmacological approach and with transection experiments. Evidence was obtained for the view that noradrenaline (NA) and 5-hydroxytryptamine (5-HT) are localized in special systems of fiber tracts which pertain to monoaminergic neurons, and descend from the brain in the lateral and anterior funiculi to terminate in the gray matter. The amines displayed a massive accumulation in the terminal regions of these fiber tracts, and it can thus be inferred that both amines serve as synaptic transmitters in the spinal cord. Many fiber terminals containing NA or 5-HT were noted in the anterior horns, being more scarce in the posterior ones. Terminals of both kinds were highly concentrated in the intermedioventral zone ('sympathetic lateral column'), where most if not all of the nerve cells seemed surrounded by them. These terminal arborizations apparently originated mainly from a tract descending in the dorsolateral funiculi. According to the cited authors this localization of the descending fibers 'suggests that they are identical with the inhibitory fibres going to the sympathetic column'.

Another approach is aimed at the histochemical demonstration of relevant *enzymes* in developing as well as in adult neural structures. Thus, e.g., DUCKETT and PEARSE (1969) reported on the enzymatic activity in the fetal human spinal cord. Said activity was found to be displayed by neurons of the anterior horn a few weeks before it appears in those of the posterior horn. Again, this activity was noted to occur first at lumbar levels, ascending, in the course of time, gradually ce-

---

136 A textbook on neurohistochemistry, edited by ADAMS (1965), and containing contributions by ten additional authors working in this field, describes the rapidly developing histochemical and cytochemical techniques. This treatise, moreover, dealing with the histochemistry of the normal nervous system, also contains an extensive part concerned with the application of histochemistry to problems of neuropathology. An elementary discussion of, and introduction to, the topic of neurotransmitters, with particular emphasis on cholinergic and adrenergic action, was included in chapter V, section 8, pp. 629 to 636 of volume 3/I. It should here be added that 5-hydroxytryptamine (5HT, also known as *serotonin*), a substance derived from the amino-acid tryptophan, is likewise suspected to be a neurotransmitter, presumably produced and liberated by '*tryptaminergic*' neuronal elements.

phalad. The presence of acetylcholinesterase activity in the 'neuronal cytoplasm' at the lumbar levels of the spinal cord is said to coincide 'with the earliest detectable movement in the lower limbs'.

With respect to the Mammalian spinal cord's *fiber systems*, a short comment on the functional present-day root fiber classification is perhaps appropriate. The efferent fibers of the ventral roots can be subsumed under three categories, namely (1) heavily medullated alpha-fibers of relatively large caliber and rapid conduction, terminating as motor end-plates on 'extrafusal' striated musculature, (2) somewhat thinner medullated gamma-fibers, of generally somewhat less rapid conduction, with 'intrafusal' motor endings for the muscle spindles, and (3) B-fibers, represented by the thinly medullated preganglionic fibers with relatively slow conduction-velocity.[137]

With regard to the axons of alpha motoneurons, another, and somewhat dubious question concerns the presence of *recurrent collaterals*, which may or may not end on 'interneurons', other motoneurons, or even on their cells of origin. Such recurrents are even suspected to reach the contralateral griseum by way of the ventral commissure (cf. e.g. SCHEIBEL and SCHEIBEL, 1970). It must, however, be kept in mind that thin 'nerve fibers' displayed by the *Golgi method* very often cannot be unambiguously identified as 'axons' or as 'dendrites'. In this respect, I harbor the strong suspicion that, *quoad hoc*, even such an undisputed master of that technique as *Don* SANTIAGO not infrequently gave free rein to his imagination. It seems rather certain that the fibers passing through the ventral commissure include dendrites as well as neurites. The presence of at least ipsilateral recurrent collaterals appears, nevertheless, well supported by the concept of *Renshaw cells* (cf. Figs. 16, 18, 22).

The medullated afferent fibers of the dorsal roots[138] can be classified, according to LLOYD (1943) into the groups I, II, III, with an additional group IV subsuming non-medullated afferents pertaining to the

---

[137] Topics concerning fiber caliber and conduction velocity as well as the fiber classification of ERLANGER and GASSER were dealt with in chapter V (vol. 3, part I, p. 579 f. of this series).

[138] Although NOBACK and HARTING (1971) affirm that 'all sensory input to the spinal cord enters via the nerve fibers of the dorsal roots', this perhaps approximately valid statement should be taken with 'a grain of salt' (cf. the discussion of the so-called '*Bell-Magendie law*' in chapter VII, section 2, of the preceding volume 3, part II). The poorly elucidated question concerning the presence of some *afferent* fibers in the spinal ventral roots of Mammals or of Vertebrates in general cannot be entirely discounted.

C-class of ERLANGER's and GASSER's terminology. Group I fibers, heavily medullated and with rapid conduction, include, as subgroup (a) the 'annulospiral' afferents from muscle spindles, and, as subgroup (b), those from '*tendon organs of Golgi*'. Group II fibers, of somewhat smaller caliber and less rapid conduction, comprise 'flower-spray' afferents from muscle spindles, from 'encapsulated receptors' and related specialized input-structures. Groups III and IV gather in the dorsolateral bundle, which joins, or becomes part of, *Lissauer's tract*.

The posterior roots which are generally more compact than the ventral ones, whose fibers may emerge in an aggregation of several rootlets, rather conspicuously display, at their entrance into the spinal cord, the region of transition pointed out, in Man, by REDLICH (1897) and OBERSTEINER (1912). On the basis of myelin stains, it seemed evident that this *zone of Redlich-Obersteiner* (Fig. 129) corresponded to a junction of the peripheral lemnoblastic portion of the nerve fiber with the intraspinal portion imbedded in central neuroglia.[138a] As the root passes through the pia and enters the spinal cord, a constriction of the root, combined with a zone apparently lacking myelin, may be seen, which could be likened to a 'stretched' *node of Ranvier* between peripheral and central myelin sheath. From a viewpoint of neuropathology, some authors interpreted this zone as a *locus minoris resistentiae*.[139] Recently, STEER (1971) investigated the *zone of Redlich-Obersteiner* in the Rat by means of electron-microscopy. According to STEER, a glial zone projects a short distance from the spinal cord along the root. Distad to this zone, the root manifests features of a peripheral nerve characterized by its mesodermal connective tissue components. The central (glial) portion forms a 'dome-shaped junction' with the peripheral ('non-glial', i.e. lemnoblastic) portion of the root. The junction is irregular and glial-invested axons interdigitate with *Schwann cell*-invested nerve fibers. Like the spinal cord with which it is continuous, the proximal part of the root contains marginal glial cells, and this central nervous tissue is provided by a limiting membrane, separating it from the mesodermal connective tissue in the root's distal part.

[138a] This zone of transition is also designated as the '*arch of Fromman*' since, on the basis of silver impregnations, it was already pointed out by said author in 1864 (FROMMANN, C.: Zur Silberfärbung der Axencylinder. Virchows Arch. path. Ann. *31:* 151, 1864, quoted after KAPPERS *et al.*, 1936).

[139] It is, e.g., well known that in order to block a peripheral nerve with a given local anaesthetic, the required dose is larger than if the anaesthetic is administered to the roots within the subarachnoid space.

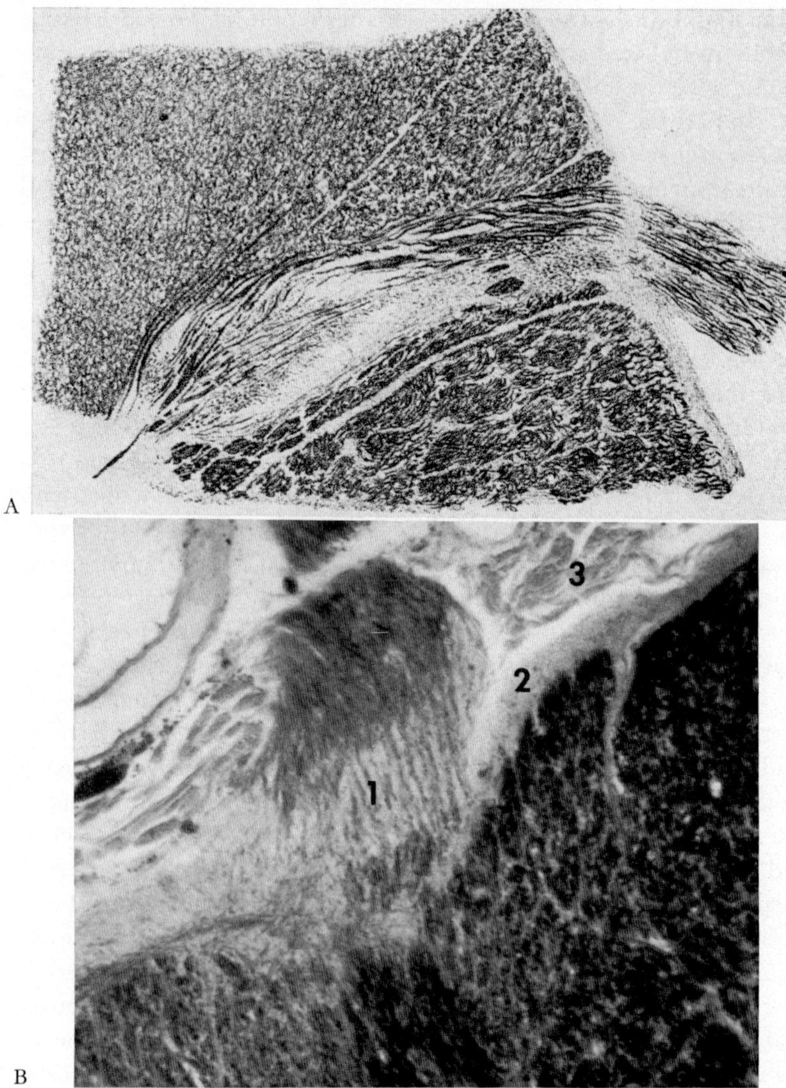

Figure 129A. Entrance of a human cervical posterior root into spinal cord, showing *Redlich-Obersteiner zone* as seen in a *Weigert myelin stain: 'an der Durchtrittstelle durch die gliöse Rindenschichte ist die Wurzel ungefärbt'* (from OBERSTEINER, 1912).

Figure 129B. Entrance of a human posterior root *('Wurzeleintrittszone')* in a cervical spinal cord segment, as seen by means of the *Weil myelin stain* in a specimen from the author's material (×190, red. $^3/_5$). 1: *Redlich-Obersteiner zone;* 2: marginal glia of spinal cord; 3: leptomeninx.

Concerning the mesodermal sheath structures of peripheral nerves, endoneurium, perineurium and epineurium can be distinguished, as elaborated in chapter V, volume 3, part I of this series. The endoneurium seems directly continuous with the leptomeningeal tissue investing the roots along their course through the subarachnoid space. The perineurium may be considered continuous with the dura. An external squamous (mesothelial) lining of the perineurium has been described (cf. Fig. 183B, p. 263 of vol. 3, part I) but remains a controversial feature. SHANTAVEERAPPA and BROWNE (1962) recognized the presence of an allegedly 'ectodermal' 'perineural epithelium' and believe that 'the fasciculi of peripheral nerves are invested by a metabolically active protoplasmic layer composed in mammals of five layers', the 'perineural epithelium' being nothing but the extension of the supposedly 'ectodermal' leptomeninges. 'Thus, the nerve fibres appear to be isolated from body fluids by the membranous epithelial sheath, from their point of origin at the spinal cord to their termination at the motor and sensory end organs', the 'perineural epithelium' being 'equipped to act as a metabolic diffusion barrier' (SHANTAVEERAPPA and BROWNE, 1962). Although I interpret the perineural 'endothelium', if actually present, as a mesodermal lining continuous with the pachymeninx, and consider some of the cited authors' interpretations unconvincing, their overall conclusions might have a modicum of validity with regard to some aspects of peripheral nerve structure.

More recently, AKER (1972) has investigated, by means of fluorescence microscopy, the so-called hematic barriers in peripheral nerves of the rabbit. His findings seem to support the concept of previously suspected 'barriers' between blood and peripheral nerves somewhat comparable to the hemato-encephalic and similar barriers obtaining for the central neuraxis. The peripheral barriers observed by the cited author appear to be provided by the following structures and their particular functional states: (1) plasma membranes of the cells forming inner lamellae of 'perineurial epithelium', (2) the luminal face of the plasma membrane of endothelial cells forming the wall of endoneural capillaries, and (3) in a 'perineurial' sheath surrounding endoneural precapillaries.

Upon entering the spinal cord, the root fibers of the dorsomedial subdivision bifurcate, respectively trifurcate (Figs. 130, 131), thus providing, in addition to the ventralward directed (in Man: 'horizontal') collaterals, ascending and descending branches, in the manner already described in this chapter for the other Vertebrate groups. All

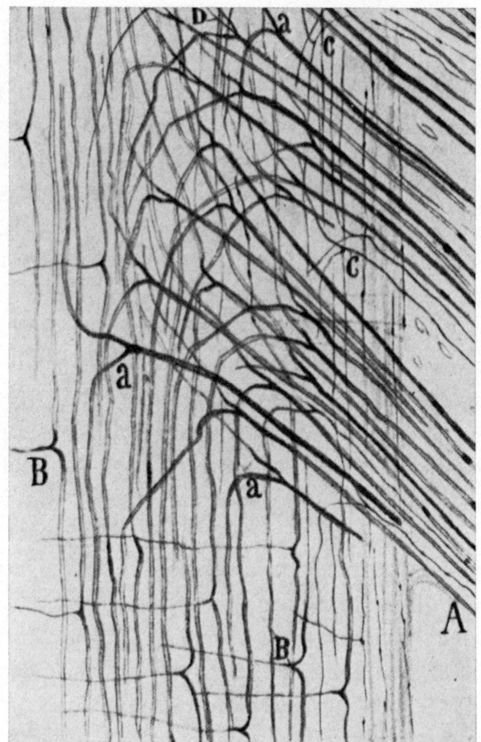

Figure 130. Longitudinal section, at a slight angle to the sagittal plane, through the posterior funiculus and a root entrance in a 15 days old Cat, as displayed by *Ehrlich's methylene blue method* (from CAJAL, 1909). A: dorsal root; B: posterior funiculus with collaterals; a, b: bifurcations and trifurcations of root fibers; c: thin root fibers, which bifurcate in *Lissauer's tract*. The incomplete letter at top is b.

these branches form part of the *eigenapparat*, while only the long ascending branches, ending in the nuclei of posterior funiculi, pertain to both *eigenapparat* and *verbindungsapparat*.

The ventralward directed branches, as shown in Figures 131 and 132A, terminate in both posterior and anterior horn, effecting connections with internuncials, and, in anterior horn, with efferents (motoneurons, presumably also gamma-cells and perhaps also preganglionics[139a] of the intermedioventral zone). A particular sort of collaterals

[139a] Although the *Golgi-pictures* obtained by CAJAL and others strongly suggest terminal endings of dorsal root afferents upon preganglionics, NOBACK and HARTING (1971) deny direct connections of this type and assume that the preganglionic elements are indirectly influenced by terminations upon neurons in the so-called laminae V, VI, VII of REXED.

was described by SOSA (1945) as arising from medial fibers of posterior funiculi and running in a sagittal direction along or within the posterior median septum. Such 'septal collaterals' apparently make connections with posterior horn grisea, including the commissural ones.

As regards the termination of dorsal root fibers in general, PETRAS and CUMMINGS have recently reported on experimental findings obtained by means of rhizotomy in the Rhesus monkey. The authors, whose interpretations are illustrated by Figure 132B, claim that no evidence was found for dorsal root terminal connections with the somata or dendrites of sympathetic or parasympathetic preganglionic neurons. Within the spinal cord, visceral reflex arcs are said to be polysynaptic, and mediated by cell groups such as e.g. nucleus proprius or 'nucleus intermediomedialis'. According to the cited authors 'the impression gained in this study is that visceral motor neurons appear in far greater numbers than somatic motor neurons'. Taking into consideration the numerous sources of error inhering in experimental procedures and in the highly 'subjective' interpretations of techniques such as *Nauta* and *Fink-Heimer methods*, the conclusions presented by PETRAS and CUMMINGS, while of course not altogether implausible, cannot be regarded as very convincing.

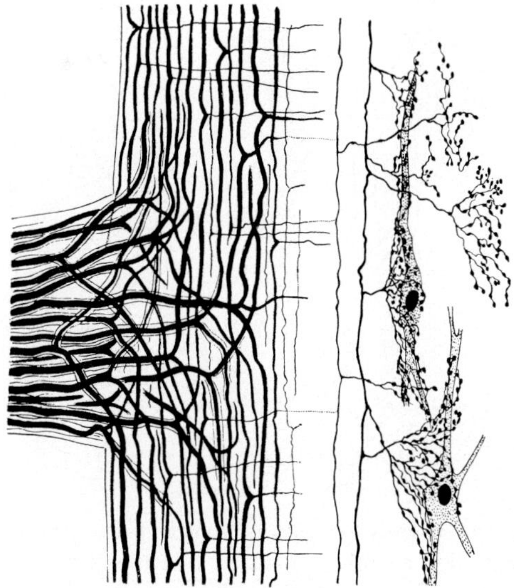

Figure 131. Semidiagrammatic sketch, by EDINGER, based on CAJAL's illustrations, of a dorsal root entrance into the spinal cord and of some of its connections with nerve cells in the gray substance (from EDINGER, 1912).

The long ascending branches assume a '*somatotopic*' arrangement such that those from the more caudal levels are displaced, medialward by, or run medially to, the fibers from more rostral segments (Figs. 133, 134, 143). At the uppermost thoracic and at the cervical levels, the posterior funiculus of Man and many other Mammals becomes thus subdivided into the medial *fasciculus gracilis of Goll* containing root fibers from the extremity, and the lateral fasciculus *cuneatus of Burdach*, carrying fibers from the upper limb. These fasciculi are commonly separated by a rather distinct glial septum, and, on the cord's surface, by the sulcus intermedius posterior (cf. Fig. 112A).

The descending root fiber branches of the dorsomedial subdivision seem to become aggregated in more or less distinctive fasciculi which,

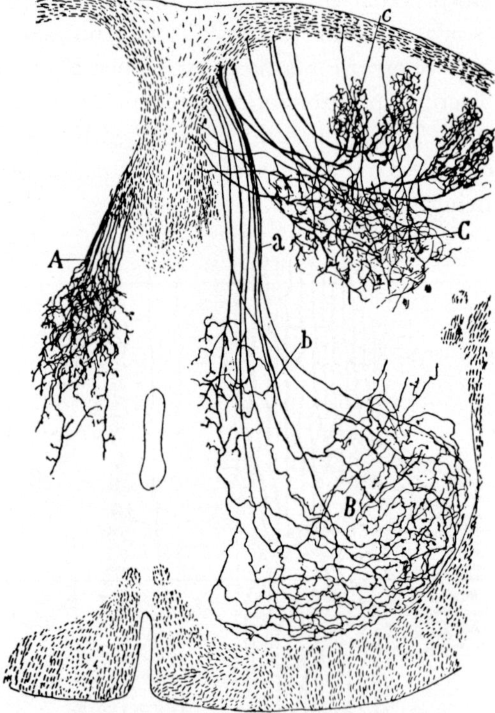

Figure 132A. Various sorts of posterior root collaterals, displayed by the *Golgi method* in the spinal cord of a newborn Rat, and as seen by CAJAL (from CAJAL, 1909). A: collaterals to 'intermediate gray nucleus'; B: terminal arborizations '*embrassant les noyaux moteurs*'; C: extensive arborizations within posterior horns; a: '*faisceau sensitivo-moteur*'; b: collateral reaching the 'intermediate gray nucleus'; c: deep collaterals to substantia gelatinosa.

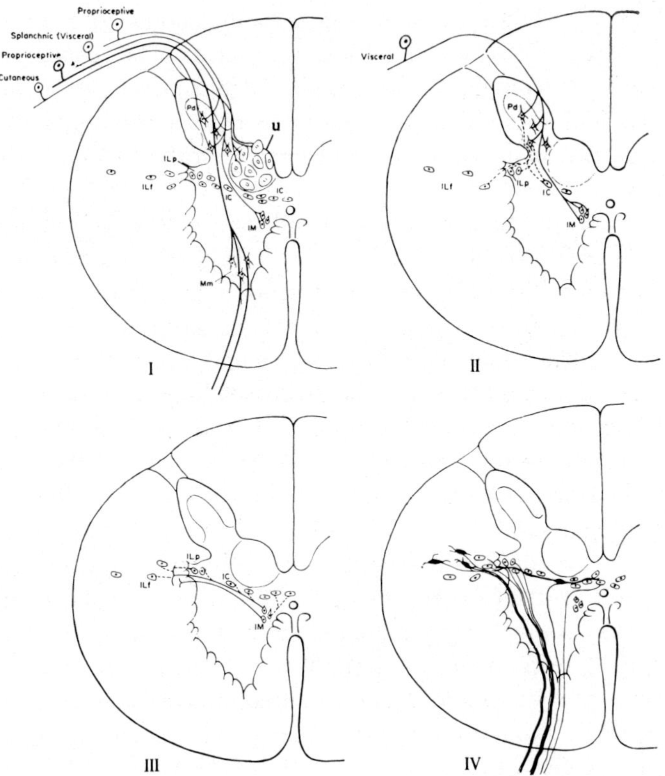

Figure 132B. Schematic drawings purporting to illustrate the termination of dorsal root fibers in the grisea of the spinal cord and the origin of some visceral including preganglionic pathways as inferred to obtain in the Rhesus monkey (from PETRAS and CUMMINGS, 1972). Ic: 'nucleus intercalatus spinalis'; Ilf: 'nucleus intermediolateralis thoracolumbaris pars funicularis'; Ilp: 'nucleus intermediolateralis thoracolumbaris pars principalis'; Im: 'nucleus intermediomedialis'; Mm: somatic motoneurons; Pd: nucleus proprius cornus dorsalis. Added lead u in I indicates *Clarke's column.*

in Man, have been designated as septomarginal, oval, and interfascicular bundles (cf. Fig. 137). The septomarginal fasciculus, also known as the *triangle of Gombault-Philippe,*[140] said to be located in the vicinity of the posterior median sulcus and septum, is presumably formed by descending root fiber components from sacral segments. The *oval field of* FLECHSIG (1876), whose dorsal portion may overlap with the ventral part of septomarginal fasciculus, seems to carry descending root fibers

140 Arch. Med. exp. 6 (1894).

from lumbar and thoracic levels, being located along a dorso-ventrally midway section of septum medianum posterius. The fasciculus interfascicularis or 'comma-shaped bundle of SCHULTZE',[141] located between fasciculus gracilis and cuneatus, is reported to contain the descending root fiber components from cervical and upper thoracic levels.

The thinly myelinated or non-medullated neurites of the dorsolateral (or lateral) root fiber bundle likewise bifurcate upon entering the spinal cord (cf. Fig. 130), but their ascending and descending branches, located in *Lissauer's tract*,[142] are presumably not extending over much more than about one or two segmental levels. These fibers, together with collaterals that may be given off along their ascending and descending course, terminate within the *substantia gelatinosa Rolandi. Lissauer's tract* includes at least two different sorts of fiber systems, namely the aforementioned root fibers of the dorsolateral division, and short 'endogenous' fibers which originate in the gelatinous substance and, either ascending or descending through very few segmental levels, re-enter said substance, thus providing relatively short intersegmental connections for that griseum. According to some authors (e.g. EARLE, 1952, referring to the Cat), the root fibers represent only about 25 per cent of *Lissauer's tract*, and run predominantly within its medial portion, while about 75 per cent of the tract fibers should be considered 'endogenous'.

A definite distinction between the thinly medullated and non-myelinated posterior root fibers *('fibres fines')* on one hand, and the thickly medullated ones *('fibres grosses')* on the other, was already emphasized

---

[141] Arch. Psychiat. Nervenkr. *14* (1883).

[142] The distinction between the extramedullary portion of the dorsolateral root fiber bundle and its intramedullary portion, which becomes a component of *Lissauer's tract*, seems rather self-evident, but can, nevertheless, be further emphasized by the use of RANSON's (1943) terminology. This author restricts the term dorsolateral fasciculus to *Lissauer's tract*, while designating the dorsolateral and dorsomedial extramedullary root fiber bundles as 'lateral' and 'medial' division, respectively, of dorsal root. The terminology concerning this topic is not standardized. Both BNA and PNA ignore *Lissauer's tract* and posterior root subdivisions. It should also be added that these two latter (medial and lateral) subdivisions, although occasionally quite evident, as e.g. shown in the cat by RANSON, are not always, respectively in all forms, sufficiently sharply delimited to allow for an unambiguous clear-cut delimitation. There are probably non-medullated and poorly medullated afferent posterior root fibers, whether running through *Lissauer's tract* or not, respectively pertaining to a definite dorsolateral subdivision or not, which do not end in the substantia gelatinosa, but in the nucleus proprius of the dorsal horn or perhaps even in the ventral horn (cf. above p. 224, Fig. 132A).

by CAJAL (1909). RANSON (1913, 1914, 1943), however, particularly stressed this contrast with regard to the subdivision of dorsal roots and the structure of *Lissauer's tract* (Figs. 135, 136). As the root enters the cord, the unmyelinated and fine medullated fibers turn laterally into said tract. A slight cut in the direction of the arrow (A in Fig. 135) severing the lateral division of the root without injury to the medial one, at once eliminated the pain reflexes obtainable from this root in the cat. The proper extent of such cuts was verified by subsequent microscopic examination. *Per contra*, a long deep cut in the plane indicated by B, severing the medial division as it enters the cord, had little or no effect on the pain reflexes. A series of such experiments (RANSON and BILLINGSLEY, 1916) provided evidence for the assumption that input related to pain sensations is carried by the fibers of the lateral division of the dorsal root. Contributory evidence, suggesting that, in peripheral spinal nerves, such input is mediated by unmyelinated as well as fine medullated fibers, was presented by GASSER (1943).

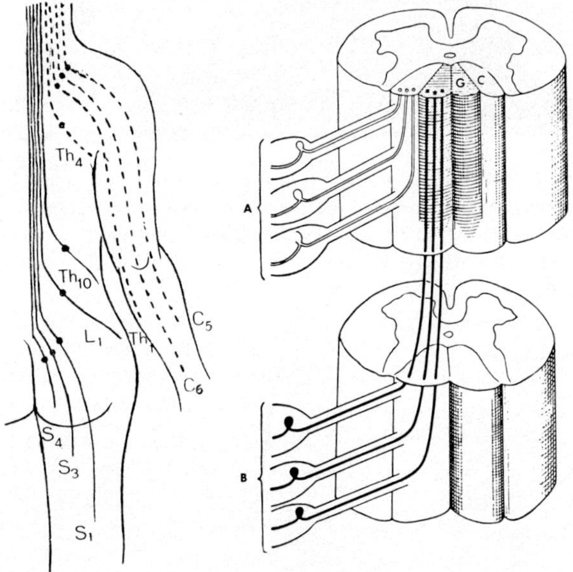

Figure 133. Sketch illustrating segmental arrangement of ascending root fibers in fasciculus gracilis and cuneatus of Man. The black dots represent spinal ganglia (slightly modified after EDINGER, from K., 1927).

Figure 134. Sketch showing further details of fiber arrangement in the human posterior funiculus (from HAYMAKER-BING, 1969). In this and in the preceding semischematic figure, the descending fiber branches (cf. Figs. 130, 131) are purposely ignored.

Since the thinly medullated and non-medullated root fibers of the dorso-lateral subdivision, related to *Lissauer's tract* and to the substantia gelatinosa, are believed to carry input related to pain (and presumably also temperature) sensations, a relevant understanding of this neuronal system appears particularly desirable. Yet, despite numerous studies (e.g. CAJAL, 1909; RANSON, 1913, 1914; EARLE, 1952; PEAR-

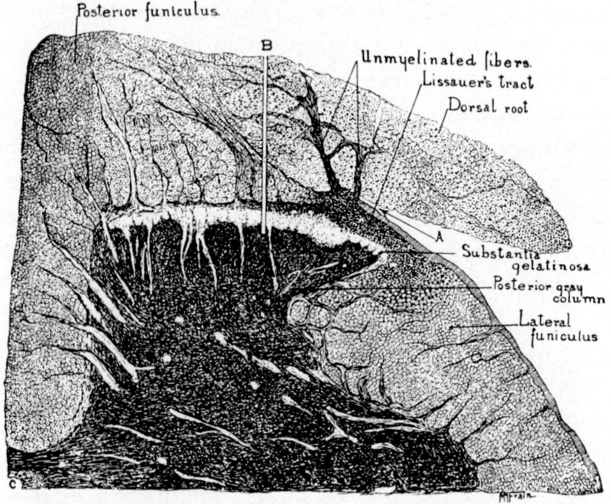

Figure 135. Dorsal portion of seventh lumbar segment of the spinal cord in a Cat, showing the unmyelinated fibers of the posterior root's lateral division entering the *tract of Lissauer*, as displayed by the pyridine-silver technique (from RANSON, 1943). Arrow (A) and line B explained in text.

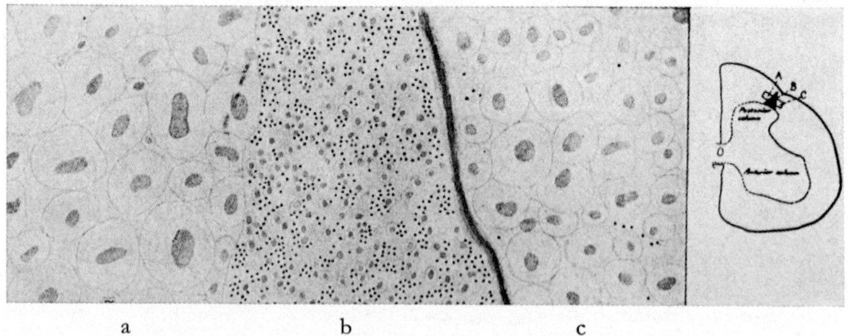

Figure 136. Area from a cross-section through the spinal cord of the Cat as seen by means of the pyridine-silver technique. The location of the area is shown in sketch at right (from RANSON, 1943). A, a: fasciculus cuneatus; B, b: *fasciculus dorsolateralis of Lissauer;* C, c: dorsal spino cerebellar tract.

SON, 1952; SZENTÁGOTHAI, 1964), the pertinent details concerning connections and significance of said system remain rather poorly elucidated.

According to CAJAL's description,[143] the substantia gelatinosa receives input from finely medullated (and non-myelinated) root fibers as well as from the heavily medullated ones which pertain to *Ranson's medial root division* (cf. Figs. 120, 132). As regards output, the neurites of the small cells are said to enter (a) *Lissauer's tract*, (b) the posterior funiculus, and (c) a part of the lateral funiculus called by CAJAL '*faisceau de la corne postérieure*'. My own observations seem to indicate that this latter '*faisceau*' includes numerous scattered longitudinal (i.e. caudocranial) bundles running through the griseum of the posterior horn in the border zone of substantia gelatinosa and 'nucleus proprius' (cf. e.g. Figs. 112A–C and 137). The neurites from the larger pericornual cells of zona spongiosa or marginalis were seen, by CAJAL, to join his '*faisceau de la corne postérieure*'. CAJAL, moreover, pointed out that the dendrites of large cells in the posterior horn's nucleus proprius penetrate the substantia gelatinosa and extend throughout the depth of that griseum (cf. Figs. 107, 120). According to SZENTÁGOTHAI (1964), 'rich connections between the two contralateral gelatinous substances are established over the dorsalmost bundle of the posterior gray commissure'. The cited author interprets the substantia gelatinosa as 'a neuron system closed in itself' because all its axons are believed to 'turn back' and 'terminate in the same structure'. He recognizes, nevertheless, the inclusion of the above mentioned dendrites of nucleus proprius elements within that substance, which, in agreement with similar views expressed by other authors, he regards as 'a system for the modulation of impulse transmission through the larger neurons of dorsal horn'. A further elaboration on such interpretations[144] is the so-called 'gate con-

[143] In this discussion of the *substantia gelatinosa Rolandi*, CAJAL (1909, p. 328 f. and p. 408 f.) supplements his findings in Mammals with observations in Birds, thus assuming that the details recorded in these latter imply identical neuronal relationships in the former. While, at least to some extent, this could perhaps be the case, some justifiable doubts nevertheless remain.

[144] The '*filtering*' and '*modulating*' activities of nervous grisea, suggested by various authors since about 50 years, have been discussed in my monograph on 'The Human Diencephalon' (1954). LHERMITTE, about 1924, stressed the concept of '*analyseurs*', provided by '*filtres selectifs*'. Similar views were subsequently expressed by v. BONIN (1950). The so-called '*gate control theory*' of MELZACK and WALL substitutes the synonymous term 'gate control' for the modulation of a 'filter's' selectivity, and thus merely represents an embellished paraphrase describing a mechanism already suggested, as well as illustrated with a figure, by v. BONIN (1950, pp. 83–86).

trol theory of pain' propounded by MELZACK and WALL, which includes various highly hypothetical and rather unconvincing assumptions. It was briefly discussed in chapter VII, section 1 of the preceding volume.

At uppermost cervical segments, *Lissauer's tract* overlaps with the rhombencephalic radix descendens s. spinalis nervi trigemini. Likewise, the substantia gelatinosa of the spinal cord merges with components of the caudal portions of nucleus radicis descendentis trigemini. These structures, however, shall be dealt with in the following chapter IX.

In addition to *Lissauer's tract*, and to those components of afferent root fibers in the posterior funiculi not concerned with impulse transmission to the brain, the fiber system of the *eigenapparat* includes the so-called *fasciculi proprii* consisting of homolateral and contralateral (commissural) association fibers remaining within the spinal cord (Fig. 137). Generally speaking, the fasciculi proprii (*'Grundbündel'*) are believed to run closely adjacent to the gray substance, except for the descending posterior root fiber bundles enumerated above, and, in the anterior funiculus for some fibers running within the fasciculus longi-

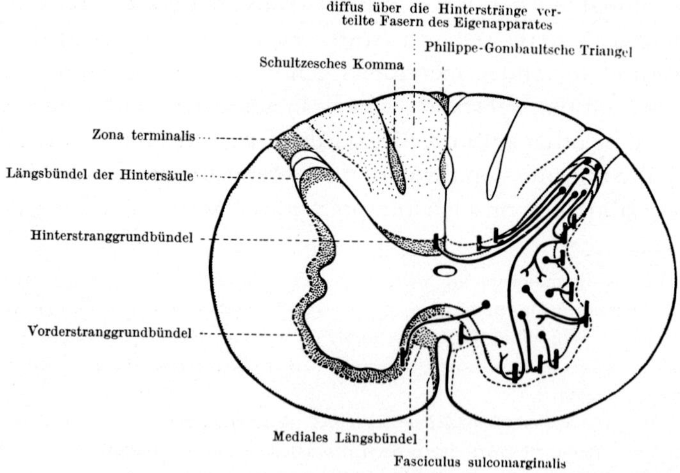

Figure 137. Diagrammatic sketch of the components of *eigenapparat* in the human spinal cord as interpreted by CLARA (from CLARA, 1959). At left, distribution of 'fields' or 'bundles'. At right some details of fiber origin and course. CLARA's *'Längsbündel der Hintersäule'* includes at least part of CAJAL's *'faisceau de la corne postérieure'*. The 'Zona terminalis' is *Lissauer's tract*.

tudinalis medialis and in the so-called 'fasciculus sulcomarginalis which overlaps, at least in Man, with the tractus cortico-spinalis anterior (compare Figs. 137 and 138C with Fig. 141A).

The *ascending components* of the *verbindungsapparat* ar provided by fiber systems originating in the spinal cord's grisea *('endogenous fibers')* and terminating at various levels of the brain up to the diencephalon, and possibly, in some instances, even directly reaching the telencephallon itself. A strictly speaking *'exogenous'* ascending component of the *verbindungsapparat*, taking origin from 'extramedullary' nerve cells in the spinal ganglia, is represented by those fibers of the dorsal roots which end in the nuclei of the posterior funiculi pertaining to the boundary neighborhoods of medulla spinalis and oblongata. These nuclei and their ascending fiber systems shall be dealt with in chapter IX. As regards fasciculus gracilis and fasciculus cuneatus of the posterior funiculus, it is generally assumed that, *qua* channels of the *verbindungsapparat*, these ascending homolateral neurites convey signals related to exteroceptive as well as proprioceptive 'signals' namely to tactile, two-point discriminatory, vibratory, and kinesthetic input.

The *endogenous systems*[145] include spinocerebellar, spino-bulbar, spino-tectal, spino-thalamic, and perhaps a few spino-telencephalic fibers (spino-cortical components reported to be present in the pyramidal tracts). The *dorsal spino-cerebellar tract*, originating from neurons in the *nucleus of Stilling-Clarke*, is also known as *Flechsig's tract*. It was the first long fiber system of the human neuraxis, whose origin, course, and termination could be established, with reasonable certainty, and by means of FLECHSIG's (1876) myelogenetic method. Input from sacral and lower lumbar segments, below the level of nucleus dorsalis, seem to reach this latter by first ascending through the posterior funiculus. The

---

[145] MÜNZER and WIENER (1910) who have provided substantial experimental contributions to the study of the spinal cord's 'endogenous' fiber systems, comment on this topic as follows: '*Die Lehre vom Aufbau des Rückenmarks ist seit Jahren in ihren groben Zügen abgeschlossen. Geht man jedoch zu einem feineren Studium des Rückenmarkaufbaues über, so zeigt sich sofort, dass tatsächlich nur die groben Umrisse feststehen und insbesondere nur jene Züge, welche Rückenmark und Gehirn miteinander verbinden, genau studiert sind.*'

'*Das Studium jener Fasern, welche aus dem Rückenmarke selbst entspringen und teils in diesem selbst endigen, teils zu höheren Zentren aufsteigen, teils selbständige Bahnen einschlagen, teils dem Areale bekannter Rückenmarkssysteme sich anschliessen, ist stark vernachlässigt.*'

Despite multitudinous subsequent studies and publications concerning these questions, it can still be said, after somewhat more than 60 years, that '*tatsächlich nur die groben Umrisse feststehen*'.

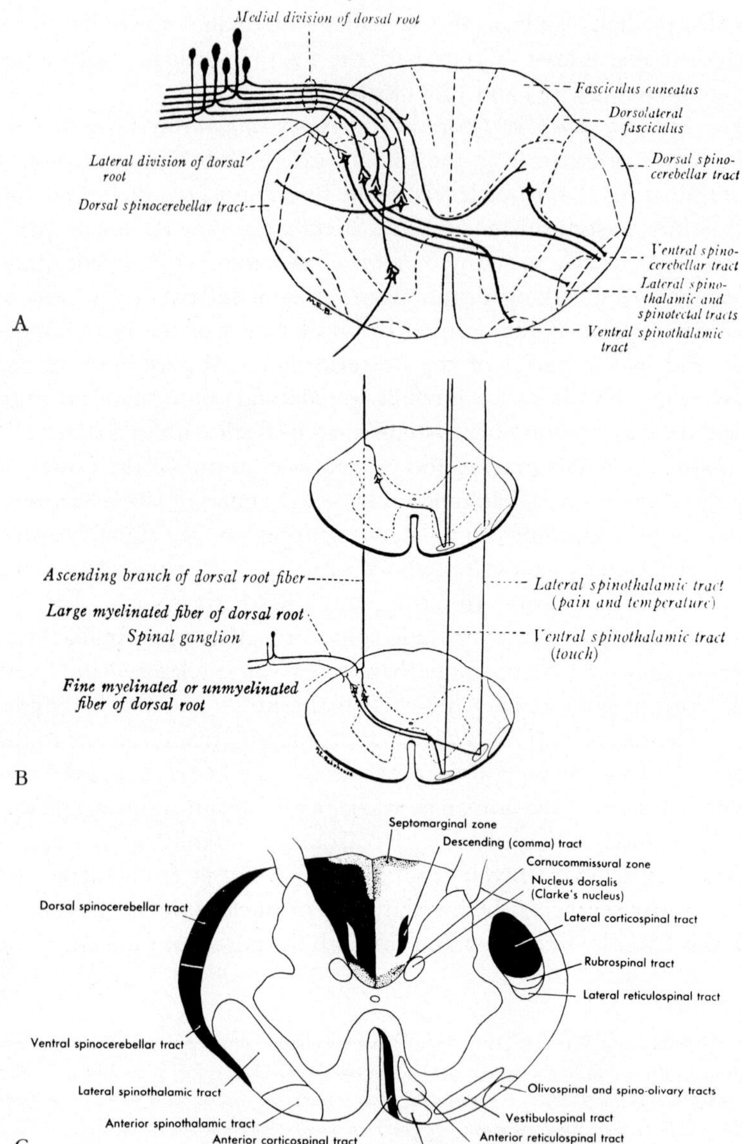

Figure 138A. Diagram of human spinal cord with dorsal root at an undefined lower cervical level, indicating RANSON's concept of origin and location of various ascending tracts (from RANSON, 1943).

Figure 138B. Diagram indicating some special features of ventral and lateral spinothalamic tracts in the human spinal cord according to widely accepted views (from RANSON, 1943).

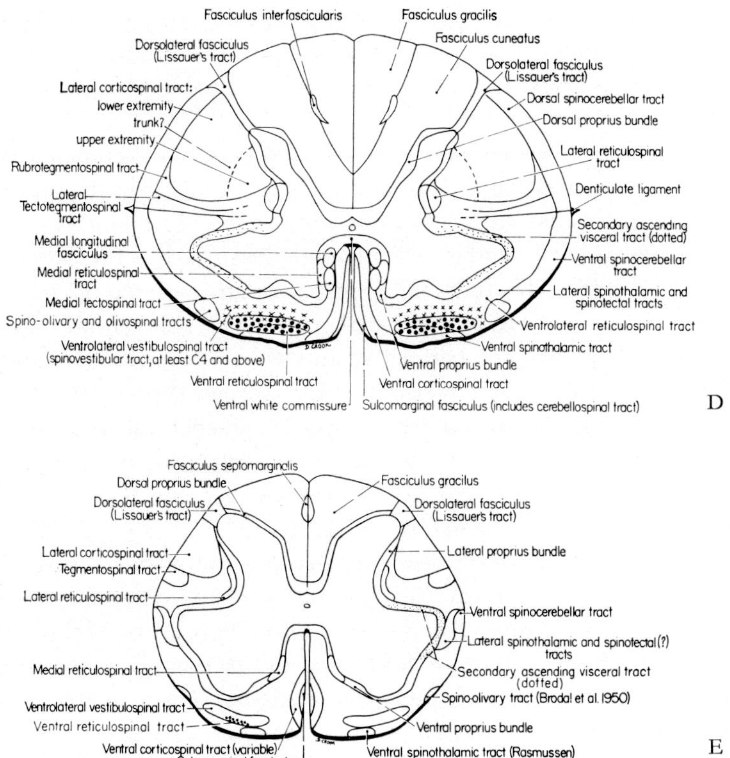

Figure 138D. Diagram indicating Crosby's view of the arrangement of major fiber tracts in the human spinal cord at upper cervical levels (from Crosby *et al.*, 1962).

Figure 138E. Diagram indicating Crosby's concept of tract distribution in the region of the human spinal cord's lumbosacral enlargement (from Crosby *et al.*, 1962). It is evident that all diagrams of this sort must be taken '*cum grano salis*'.

Figure 138C. Diagram showing a widely accepted view concerning the distribution of main 'tracts' in the human spinal cord's white substance. Ascending tracts on left side, descending ones on right. The level of the section is 'generalized' but may correspond to a lower cervical region. 'Compact' tracts indicated in black, and the more loosely arranged, poorly demarcated ones are left white. The dotted areas are supposed to represent intersegmental tracts. It is evident that the sharp boundary lines are purely fictional (from Haymaker-Bing, 1969).

tract of Flechsig, which appears to contain essentially but perhaps not exclusively homolateral fibers, runs along the dorsal surface of the lateral funiculus (Figs. 138, 140A), and begins at the level of nucleus dorsalis (in Man at about L3).

The *ventral spino-cerebellar tract*, which is believed to contain a larger amount of decussating (i.e., of contralateral) fibers than the preceding one, seems to arise from posterior horn cells, beginning perhaps at lower lumbar levels, and is located along the surface of the lateral funiculus (Fig. 138). This tract apparently overlaps with the dorsal spino-cerebellar and the lateral spino-thalamic tracts. The ventral region of overlap, including spino-cerebellar, and lateral spino-thalamic fibers, is also occasionally termed fasciculus anterolateralis superficialis or '*Gowers' tract*'.[146] The dorsal and ventral spino-cerebellar tracts can be classified as 'proprioceptive'. Their connections with the cerebellum, and additional ascending neuronal channels between spinal cord and cerebellum shall be taken up in chapter X.

The *spino-bulbar fiber systems*, originating from insufficiently identified nerve cells, which may or may not be restricted to the posterior horns, are presumed to ascend in the ventral part of the lateral funiculus as well as in the anterior funiculus, and to comprise spino-reticular, spino-vestibular, and spino-olivary fibers. The available data on these channels *qua exact location, termination, uncrossed and crossed components, etc., remain rather sketchy (cf. e.g. the summaries by* CROSBY *et al.*, 1962, and by NOBACK and HARTING, 1971).

The *spino-tectal fibers* may be considered a 'phylogenetically old' system, present as perhaps the main ascending channel of the cord's *verbindungsapparat* in Selachians (cf. above, p. 104). This pathway, from which the spino-thalamic tracts perhaps differentiated as the separate rostral terminations, was possibly 'originally' a more or less unified fiber system with multiple collateral connections within the brain stem. Spino-tectal fibers could, even in higher Mammals, and at least to some extent, be provided by tectal endings of spino-bulbar fibers, as well as by collaterals of spino-thalamic ones, thus remaining, in this respect, parts of said other systems.

The essentially crossed *spino-thalamic pathways* in Man and other Mammals presumably originate from nerve cells in the nucleus proprius

---

[146] Cf. GOWERS, W. R.: The diagnosis of diseases of the spinal cord (Churchill, London, 1879) and: A manual of diseases of the nervous system. 2 vols. (Churchill, London, 1886, 1888).

of the posterior horn. This has been questioned, on not very convincing grounds, by Szentágothai (1964), who suggested an origin from cells in Rexed's so-called laminae VI, VII, and possibly VIII, of which the two last ones doubtless are parts of the anterior horn.[147] Be that as it may, the decussating course of the crossed spino-thalamic fibers in the cord's ventral commissure is generally recognized, although there is some evidence for the presence of uncrossed (homolateral) fibers in the two components of this pathway. Attempts to distinguish a *palaeo-spinothalamic* system from a *neospinothalamic* one, based on vague phylogenetic speculations, do not appear convincing, but may, nevertheless, be construed by postulating evolutionary differentiations of the abovementioned 'primitive ascending channel'.

The *lateral spino-thalamic tract*, located in the ventral part of the lateral funiculus (Fig. 138), and overlapping with the anterior spino-cerebellar bundle, seems to convey impulses related to pain and temperature sensations, but, a least in certain Mammalian forms (Rhesus monkey, as quoted by Noback and Harting, 1971) the presence of some fibers concerned with 'touch' cannot be excluded.

The *anterior or ventral spino-thalamic tract* (Fig. 138) is described as an ascending bundle within the anterior funiculus, consisting mainly of crossed fibers of the second order. It is supposed to 'mediate tactile sensibility' and may originate, like the lateral tract, from cells in the posterior horn's nucleus proprius or from ventrally adjacent neurons of posterior gray columns. As in the case of lateral spino-thalamic tract, collaterals of this pathway terminating in the grisea of spinal

---

[147] Apart from reference to experimental results whose accuracy remains questionable, it is claimed that no cells of the posterior horn's nucleus proprius send their neurites as crossing fibers through the ventral (or anterior commissure). Yet, Cajal, one of the most expert investigators using the *Golgi method*, unequivocally depicted such crossing neurites of nucleus proprius cells in his figure 141, p. 390 (1909) referring to the newborn mouse. Again, Noback and Harting (1971) who accept Szentágothai's interpretation, add the following remark: 'as early as 1952, Pearson demonstrated that the neurons in the proper sensory nucleus do not have axons which cross in the anterior commissure'. Yet, Pearson actually (1952, Fig. 7, p. 525) depicted these axons as decussating in the anterior commissure (perhaps in agreement with Cajal's illustrations), and merely stated that, in his own *Golgi-material*, the neurites of nucleus proprius cells were difficult to trace, becoming lost after a short distance. Pearson then referred, without further comments, to three, not mutually exclusive views concerning the course of these neurites, namely decussation in either anterior or posterior white commissures, and, in some instances, entrance into the anterolateral funiculus of the same side.

cord and brain stem have been assumed. Such connections are supposed to include an accessory collateral system provided by a relay of shorter neurons likewise ending in the thalamus.

The decussating fibers representing neurites of the presumptive 'secondary neurons' of both spino-thalamic tracts seem to cross at a level only slightly cranial to that of their origin, if not actually at that latter segmental level. Clinical observations in man, especially concerning cases of syringomyelia, suggest, however, that the segmental primary neuron input for the lateral spino-thalamic tract is transmitted to the decussating secondary neuron at one level only, while the segmental primary input for ventral spino-thalamic tract may be transmitted both at the level of root entrance and then, again, through ascending fibers of the posterior funiculi, at higher levels (cf. Fig. 138B).

In a functional classification, lateral and anterior spino-thalamic systems are considered essentially '*exteroceptive*'. With respect to 'exteroceptive' input signals, it has been frequently assumed that, in peripheral nerves, separate fibers related to each of the four modalities touch, warmth, cold, and pain are present. Injury to such nerves, carrying said intermingled fibers, affects therefore commonly all four modalities. In the spinal cord, however, a segregation of the different 'sensory' pathways appears to take place. Yet, the details of the manner in which this 'segregation' is effected remain insufficiently clarified. Thus, it is e.g. uncertain whether, in the lateral spino-thalamic tract, separate 'warmth', 'cold', and 'pain' fibers are provided.

The course of '*interoceptive*' or 'visceral afferent' impulses reaching brain stem and prosencephalon likewise remains very poorly understood, and various opinions have been expressed. Thus, homolateral fibers in posterior funiculus, bilateral ones associated with the spino-thalamic tracts, and bilateral relays by chains of short neurons whose axons pass through the fasciculi proprii have been suggested.

Ascending fibers of the Mammalian spinal cord directly reaching the telencephalic cerebral cortex were described and discussed by several authors (Brodal and Walberg, 1952; Brodal and Kaada, 1953; Nathan and Smith, 1955). These fibers may accompany, and be closely related to, the descending cortico-spinal system, perhaps also including spinal-pontine connections (Walberg and Brodal, 1953).

The *descending pathways* of the Mammalian spinal cord's *verbindungsapparat* can roughly be enumerated as comprising the compound fasciculus longitudinalis medialis, and more or less distinctive vestibulo-spinal, reticulo-spinal, solitario-spinal, olivo-spinal, tecto-spinal, rubro-

-spinal, as well as cortico-spinal fiber systems.[148] Generally speaking, one may also, in accordance with CLARA (1959), but in a slightly different sequential order, classify these descending pathways as pertaining to three main categories: (1) systems originating in the grisea of the brain stem (deuterencephalon) and carrying impulses directly or indirectly reaching the 'somatic-efferent' neurons of the spinal cord; (2) systems originating in grisea of the brain stem and carrying (as above) impulses to the cord's 'visceral-efferent' (preganglionic) neurons; (3) a cortico-spinal system, originating in the telencephalic cerebral cortex, directly reaching the cord's grisea, and effecting there direct or indirect (internuncial) connections with 'somatic-efferent' neurons (cf. also Fig. 139).

Seen from a phylogenetic viewpoint, the *fasciculus longitudinalis medialis*, located in the dorsal portion of anterior funiculus in fairly close vicinity of the cord's midline grisea ('substantia intermedia centralis'), could be evaluated as a transformed remnant of the perhaps main primitive 'motor' tract, which was dealt with above in discussing the relevant fiber pathways of the diverse 'lower' Vertebrate forms. In Mammals, including Man, the fasciculus longitudinalis medialis, whose rostral beginning seems to be provided by CAJAL's nucleus interstitialis at the mesencephalo-diencephalic boundary region, represents a very complex fiber system comprising numerous different descending and ascending constituents, details of which shall be taken up in the chapter dealing with the brain stem (IX, XI).

As far as the extension of this bundle into the spinal cord is concerned, its main components are presumably interstitio-spinal, reticulo-spinal, vestibulo-spinal, and tecto-spinal fibers, which may be crossed or uncrossed, or both. A few ascending pathways within the course of fasciculus longitudinalis through the spinal cord were noted

---

[148] Some authors also assume the presence of direct cerebello-spinal fibers originating in cerebellar roof nuclei, but the evidence for such connections is rather inconclusive. According to CROSBY *et al.* (1962) these cerebellospinal fibers 'are thought not to extend beyond the cervical cord'. It will be recalled that the presence of cerebello-spinal fibers in Selachians, Ganoids, Teleosts, and Birds is assumed by some authors. The evidence for such fibers in the Avian neuraxis seems fairly valid. All these Vertebrate forms possess a substantially developed cerebellum.

Again, with regard to additional descending (as well as ascending, i.e. 'reciprocal') tracts, the *paraventricular fasciculus longitudinalis dorsalis of Schütz* (cf. the following chapter IX), taking its course through the tegmental central gray of the brain stem, and probably related to the autonomic systems, is not known to extend, as such, into the spinal cord.

by PROBST (1902), but the presence of such fibers remains a moot question. In the Mammalian brain stem, on the other hand, and as just mentioned above, this bundle contains numerous ascending respectively bifurcating (ascending and descending) components.

Within the cervical portion of the spinal cord, said bundle's descending vestibular fibers, together with reticulo-spinal ones, seem to be part of a neuronal mechanism coordinating eye-movements with motions of the head and neck. Other reticulo-spinal fibers of fasciculus longitudinalis are believed to be involved in the mechanisms of respiration and vegetative functions. Tecto-spinal components may provide one of the various pathway for optic reflexes. Nothing is known concerning the significance of the possibly present interstitio-spinal fibers. Considerable divergence of opinion obtains between the interpretations and the terminologies[149] of authors dealing with these fiber systems, which latter, moreover, may substantially differ not only *qua* diverse Mammalian forms but also *qua* individual variations. It is furthermore quite uncertain how far caudalward the fasciculus longitudinalis medialis should, generally speaking, be assumed to descend within the Mammalian spinal cord.

The lateral *vestibulo-spinal tract* or *bundle of Held*, respectively the vestibulo-spinal tract proper, can be evaluated as a 'phylogenetically old' system about which relatively well substantiated data are available (HELD, 1891, 1893). It is an essentially uncrossed pathway, originating in the large-celled *lateral vestibular nucleus* of DEITERS (1865), descending through the anterior part of the lateral funiculus and extending as far as the caudalmost levels of the spinal cord. The presence of thicker and thinner fibers in this tract may be related to their origin from larger and smaller neuronal elements within the generally large-celled *griseum of Deiters*. A few reciprocal, i.e. ascending connections within the vestibulo-spinal tract were reported by PROBST (1902). Functionally, this tract, perhaps particularly concerned with maintenance of equilibrium,[150] can, still more generally, be considered a component of the 'extrapyramidal motor system'.

---

[149] Thus, the vestibulo-spinal components of fasciculus longitudinalis medialis correspond to the so-called medial vestibulo-spinal tract, the tegmental and tectal components being the medial reticulo-spinal and tecto-spinal tracts, respectively, of miscellaneous authors.

[150] The vestibulo-spinal fibers in fasciculus longitudinalis medialis can, of course, also be interpreted as concerned with the 'maintenance of equilibrium'.

The *lateral reticulo-spinal tract* is known to arise from larger and smaller nerve cells is the brain stem's formatio reticularis, whose crossed and uncrossed fibers descend through the spinal cord's lateral funiculus and may reach its caudal levels. PAPEZ (1926), who reviewed the previously available data on this system, investigated the reticulo-spinal pathways in the Cat by means of the *Marchi method.* PAPEZ and some authors (e.g. CROSBY *et al.*, 1962), distinguish, within the lateral funiculus, a more dorsal lateral from a more ventral reticulo-spinal tract, in addition to the medial reticulo-spinal fibers included, as mentioned above, within the fasciculus longitudinalis medialis. With respect to functional significance, the reticulo-spinal pathway can be regarded as a channel with multipotent or 'multiplex' activities, related to the 'extrapyramidal motor system', to respiration, to coughing, to the vomiting reflexes, and to manifold 'visceral' or 'vegetative' output mediated by sympathetic and sacral parasympathetic nervous system.

With respect to vegetative functions, CARLSSON *et al.* (1964) found descending monoaminergic fibers localized in the dorsal portion of the lateral funiculus, in the 'dorsal superficial zone of the posterior horn', and along the entire surface of lateral and anterior funiculi of small rodents (Mouse and Rat). To a certain extent, and taking into consideration substantial differences between taxonomic forms, some such fibers in lateral or anterior funiculus could be evaluated as pertaining to the lateral and ventral reticulo-spinal tracts of PAPEZ.

A *fasciculus solitario-spinalis*, originating from the nucleus of the tractus solitarius (i.e. from the terminal griseum of afferent glossopharyngeus, vagus, and facialis fibers, was suggested and diagrammatically outlined by CAJAL (1909). It is presumed to play a role in vomiting and coughing reflexes as well as in respiration. More recently, TORVIK (1957) recorded evidence for the existence of this tract, whose fibers may run with the medial and ventral reticulo-spinal ones.

A *tractus olivo-spinalis* is frequently described as originating from the inferior olivary nucleus of the oblongata with crossed and uncrossed fibers, terminating mainly in the upper part of the spinal cord (CROSBY *et al.* 1962). Traditionally, this tract is supposed to pass through a triangular field *('Helwegs Dreikantenbahn')* within the anterior part of the lateral funiculus,[151] together with spino-olivary fibers

[151] This triangular field, named after HELWEG (1888) is occasionally conspicuous by its pale coloration in myelin-stained preparations.

(cf. Figs. 138 C, D, E). Recent authors, e.g. JANSEN and BRODAL (1958), deny the existence of an olivo-spinal pathway.[152]

Tecto-spinal fibers are known to originate in the anterior quadrigeminal bodies (tectum opticum) of the midbrain (cf. further below in chapter XI). Crossed tecto-bulbar and tecto-spinal fibers form, within the rhombencephalic tegmentum, the paramedian fasciculus praedorsalis, whose tecto-spinal fibers then join the spinal cord's fasciculus longitudinalis medialis. It is, moreover, likely that some predominantly uncrossed tecto-spinal fibers within the lateral tecto-bulbar tract of the oblongata reach the spinal cord as a 'lateral tecto-spinal tract' within the lateral funiculus. Tecto-spinal fibers are presumed to terminate at cervical levels, but nothing certain is known about the actual extent of their course. Concerning their functional significance, it seems probable that they mediate responses to optic as well as to therewith associated additional sensory input.

The *rubro-spinal* tract or *bundle of Monakow* takes its origin from the large-celled components in the mesencephalic nucleus ruber tegmenti (v. MONAKOW, 1883, 1910). Its fibers cross in the ventral tegmental decussation, reaching the lateral funiculus of the spinal cord. In Primates, the tract is located ventrally and closely adjacent to, perhaps also overlapping with, the lateral cortico-spinal tract (cf. Fig. 138). The rubrospinal pathway can be evaluated as a significant component of the so-called *extrapyramidal motor system*, and may play a particular role in the control of muscle tonus. It is believed to be better developed respectively more important in 'lower' than in 'higher' Mammalian forms, and may comprise relatively few fiber bundles in Man.[153] The caudal extent of the rubro-spinal tract within the Mammalian spinal cord has not been clearly established. Moreover, as a so-called tract, this pathway seems to be intermingled with reticulo-spinal fibers.[154]

---

[152] CROSBY *et al.* (1962), who bring an elaborate diagram of olivo-spinal and spino-olivary connections (Fig. 194, p. 258 l.c.) interpret it as 'part of a system for coordinating upper extremity and head movements' (p. 104 l.c.). Their added statement that, according to BRODAL, 'some olivospinal fibers reach lower levels of the spinal cord', is presumably a mistake in view of that author's explicit denial. I failed to find BRODAL's alleged statement in his publication quoted by CROSBY *et al.* (1962).

[153] The divergent opinions expressed on this question by different authors are reviewed by NOBACK and HARTING (1971).

[154] It will be recalled that a rubro-spinal tract may, according to some authors, be present in Birds. Its presence in Reptiles is dubious but possible. Whether a rubro-spinal tract can be assumed as occurring in Reptiles as well as in diverse Anamnia depends (a)

The *cortico-spinal pathway*, whose origin from a region of telencephalic neocortex shall be discussed in volume 5 of this series, must be regarded as a unique characteristic of the Mammalian neuraxis, since, only in this latter, a structurally distinctive neocortex is present. From a formanalytic viewpoint, on the other hand, neocortical homologues can be identified in the telencephalon of all other Vertebrates, including the 'lower' Anamnia, as elaborated in the preceding volume 3, part II. Moreover, discounting the possibility of some direct telencephalo-spinal fiber connections in Birds and perhaps even in Reptiles, the Mammalian cortico-spinal system is the only direct pathway from cerebral hemispheres to spinal cord whose existence and general course has been established with reasonable certainty. The term '*pyramidal tract*', commonly applied to said system, was coined in reference to its passage through the gross configurations designated as 'pyramids' of the medulla oblongata *(pyramis medullae oblongatae*, BNA, PNA) in human anatomical nomenclature.

In Man and other Primates, most cortico-spinal fibers running through the 'pyramids' decussate in the region of transition between oblongata and spinal cord. The non-decussating fibers, and perhaps a few of the crossing ones take their course as *anterior* or *ventral cortico-spinal tract* through the cord's anterior funiculus.

The decussating fibers, again apparently together with a few homolateral ones, reach the lateral funiculus, forming the essentially crossed *lateral cortico-spinal tract* (cf. Figs. 138C–E, 140, 141A). As regards the crossing pattern and the course of 'aberrant' cortico-spinal fibers, numerous individual variations obtain in Man[155] (COLLIER and BUZZARD,

---

on a semantic question and (b) on a factual one: (a) should certain vaguely outlined reticular cell groups in the mesencephalic tegmentum of 'lower Vertebrates' be designated as 'nucleus ruber', and if so, in which particular forms; (b) granted that a 'nucleus ruber' in 'lower Vertebrates' can be 'identified', does or does not such 'nucleus' give origin to fibers reaching the spinal cord.

[155] Thus, with respect to the clinical procedure of stereotaxic 'high cervical cordotomy' aimed at destruction of the lateral spino-thalamic tract, and involving the danger of damage to the lateral cortico-spinal tract, a neurosurgeon (TAREN, 1971) recently remarked that said division of the pyramidal pathway 'is more medial and more anterior than we had suspected at this level.' (2nd cervical segment). The cited author also states that 'surprisingly little functional loss may result' if the unintentional lesion of the pyramidal tract remains unilateral. He adds furthermore the following comment: 'not only may the location of the pyramidal tract and number of fibers which it contains vary but also the relative number of decussating fibers. Aberrant corticospinal tracts have been described as well as the case in which the corticospinal tract did not decussate at all'.

1901; RUSSELL, 1898; SPILLER, 1899; ZEMMER, 1898) and apparently also in other Mammals. The reported presence of some reciprocal, ascending spino-cortical fibers within the pyramidal pathway was mentioned above in the enumeration of the *verbindungsapparat*'s ascending components.

In the human spinal cord, the ventral cortico-spinal tract presumably runs caudad at least as far as lumbar levels, and the lateral cortico-spinal tract has been traced to the sacral levels. Much the same may obtain in 'higher' Primates. In Prosimians, a ventral cortico-spinal tract could not be identified (NOBACK and HARTING, 1971).

In the tree shrew Tupaia, variously classified as an Insectivore or as a lemuroid respectively lemuriform Primate according to the interpretations of diverse authors, the 'lateral' or 'crossed' cortico-spinal tract is located in the dorsal funiculus (JANE *et al.*, 1965). This is neither what might be called an Insectivore nor a Primate 'characteristic'. A ventral cortico-spinal tract seems not to be present.

In other Mammalian forms, the caudal extent of the cortico-spinal tracts, about which diverse and in part conflicting data have been presented by numerous authors, appears to vary in accordance with the taxonomic or phylogenetic 'status'. Among reports concerning the comparative anatomy of the pyramidal systems, the older ones by VON LENHOSSÉK (1889) and LINOWIECKY (1914) should be mentioned. VERHAART (1971) has dealt with this topic in a study of Mammalian brain stem and cord, while NOBACK and HARTING (1971) have reviewed such data with respect to the Primate spinal cord. A monograph on the pyramidal tract, with relevant bibliographic references, and summarizing the results of his own extensive investigations on the Mammalian and particularly the human cortico-spinal system, was published by LASSEK (1954).

As regards diverse representative non-Primate Mammalian groups, about whose pyramidal pathway some data are available, the following remarks may be sufficient for purposes of a general orientation.

In the *monotreme Echidna*, the cortico-spinal tract seems to be chiefly a crossed one running through the lateral funiculus, dorsally to the rubro-spinal tract, and apparently extending rather far caudad. In *Marsupials*, the fibers of the pyramidal tract mainly decussate into, and run through, the dorsal funiculus, although in some of these animals a spread course through neck of dorsal horn and adjacent parts of lateral funiculus has been recorded. Nothing certain is here known *qua* presence of uncrossed fibers. The caudal extent of the Marsupial cortico-spinal tract may be relatively short and not reach beyond some thoracic segments.

The pyramidal tract of *Insectivora* seems likewise to be short, being apparently restricted to a ventral cortico-spinal pathway. In *Chiroptera*, a lateral cortico-spinal tract, ventrally to the rubro-spinal tract, has been recorded. In at least some species, it may reach the lumbar intumescence with a few fibers (DRÄSEKE, 1903; VERHAART, 1971).

In *Rodentia*, including Rat, Mouse, Squirrel, and Guinea pig, most of the pyramidal tract appears to cross into, and to run through, the posterior funiculus. Thus, in the Rat, the cortico-spinal fibers occupy a substantial ventral portion of that funiculus, and may extend caudalward at least as far as lumbar levels. In addition, a few anterior cortico-spinal fibers seem to be present, but a distinctive ventral pyramidal tract is believed to be missing in small Rodents. In the *Lagomorph* Rabbit, the pyramidal tract decussates into a dorsal portion of the lateral funiculus. Some fibers may run along the medial side of the dorsal horn and others within the anterior funiculus. A few uncrossed fibers are possibly present within these various components. The caudal extent of the Rabbit's cortico-spinal pathway is believed to be short.

In *Carnivores* a lateral cortico-spinal tract, with mainly crossed fibers, reaches the lateral funiculus, and takes its course in a location approximately comparable to that obtaining in Primates. A ventral cortico-spinal bundle may be present, but seems much less distinct than that of Primates, and could even be lacking in various forms, respectively individual instances.

As regards *Ungulates*, an incomplete decussation, with mainly crossed fibers in lateral funiculus and mainly uncrossed ones in anterior funiculus has been reported for Sheep and Goat. Some fibers may also be present in the dorsal funiculus. The caudal extent of the pyramidal pathway seems to be quite short (ZIEHEN, 1900; DEXLER and MARGULIES, 1906; KAPPERS *et al.*, 1936).

The *Proboscidean Elephant*'s cortico-spinal tract (Figs. 117, 118) partly crosses to the posterior portion of the anterior funiculus, dorsal to the commissura alba ventralis, thus assuming an 'intracommissural' location, and partly descends uncrossed along the surface of fissura mediana anterior (DEXLER, 1907; VERHAART, 1971), but diverse variations seem to obtain. In *Cetaceans*, according to VERHAART, 'a pyramidal tract cannot be seen in the spinal cord, and only in the Odontoceti do some bundles lie along the fissure at the transition to the medulla'

(Odontoceti, i.e. toothed Whales, subsume Dolphin, Porpoise, Narwhal, Sperm Whale).

The pyramidal (cortico-spinal and 'cortico-bulbar') pathway of Mammals brings the primary motor centers of the spinal cord (and of the deuterencephalon) under the direct influence of the neocortex. It can indeed be considered a phylogenetically 'recent' system. The di-

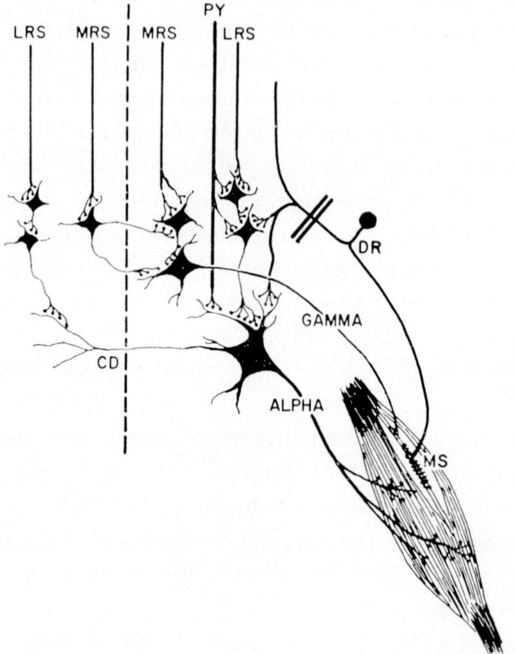

Figure 139. Diagram purporting to illustrate the relationship of the pyramidal and other descending systems to the spinal cord's primary motor neurons (alpha and gamma cells), and including the feedback connections from muscle spindles (after DENNY BROWN, 1966, from BRAIN and WALTON, 1969). CD: dendritic commissure; DR: posterior root with spinal ganglion cell; LRS: lateral reticulo-spinal tract; MRS: medial reticulo-spinal tract; MS: muscle spindle; Py: pyramidal tract; the broken line indicates the cord's midline. A gamma neuron providing innervation of the muscle spindle and thereby 'increasing its sensitivity' is believed to facilitate afferent impulses from the annulo-spiral endings, which, in turn, provide excitation for the alpha motoneurons of the same muscle. The medial reticulo-spinal tracts are supposed to 'drive' the gamma neurons. The lateral reticulo-spinal and the pyramidal tract are assumed to be connected with the 'alpha system'. Section of dorsal root at location indicated by the bars blocks 'activation via the gamma neurone'. This figure may also be compared with the still more generalized and oversimplified Figure 3B on p. 16 of volume 1, purporting to illustrate synaptic connections of a motoneuron (alpha cell).

verse arrangements of the pyramidal tract displayed in taxonomically different Mammalian groups are evidently suggestive of phylogenetic implications. Yet, because of substantial variations within closely related groups, combined with individual variations manifested within a given species, phylogenetic speculations based on the pyramidal system's ambiguous behavior remain rather inconclusive and unconvincing.

Functionally, the cortico-spinal pathway appears to mediate those movements which, in human consciousness, are experienced as 'voluntary', and involve, by way of the spinal nerves, intended motility of the human neck, extremities, and trunk. Nevertheless, the relative size of the pyramidal tract is not believed to be correlated with speed and agility of a Mammal's movements, as e.g. demonstrated by the Squirrel, whose cortico-spinal system seems comparatively exiguous, nor does it seem to depend on the animal's size. Some authors, however, have claimed that the total number of fibers in this pathway depends on body weight, except for very small animals (TADAHICO, 1961, as quoted by BLINKOV and GLEZER, 1968).

Again, as regards the caudal extent of the cortico-spinal system, LASSEK (1954) reaches the following conclusions: animals with short pyramidal tracts are Opossum, Rabbit, Guinea pig, Pangolin (a Pholidote ant-eater formerly included among the Edentata), Bat (cf. however the contradictory statements quoted above), Mole, Hedgehog, and Ungulates. Mammals with cortico-spinal fibers extending the length of the spinal cord are enumerated by the cited author as the Rat, the Mouse, Carnivores, and Primates.

Figure 139 illustrates present-day concepts concerning the endings of a pyramidal tract fiber and of reticulo-spinal ones in the anterior horn of the Human and Mammalian spinal cord. It will be seen that both 'direct' endings on motoneurons, and 'indirect' connections, ending on short internuncials, are postulated.[156] Only one fiber of the pyramidal pathway is here shown, thereby neglecting, in the aspect under consideration, not only their very great number, but also the possible effect of pyramidal fibers terminating in the contralateral gray. With regard to the endings of the pyramidal tracts it should also be mentioned

[156] Some authors assume that direct terminal arborizations of pyramidal tract fibers on motoneurons of the spinal cord's anterior horn occur only in Primates (cf. e.g. BLINKOV and GLETZER, 1968), while, in other Mammalian forms, all cortico-spinal control of said motoneurons is supposed to be mediated by way of intercalated short internuncials.

that the essentially homolateral ventral cortico-spinal tract of Man and perhaps also some other Mammals is believed to end mostly in the contralateral gray of the spinal cord, by decussating, at the levels of termination, through the anterior white commissure.

Many authors have investigated the *number* and the *caliber* of pyramidal tract fibers. Although differing in various details, the results of these studies are in essential agreement with each other and have been concisely tabulated in BLINKOV's and GLEZER's (1968) compilation.

Thus, in Man, the total number of medullated and non-medullated fibers in *one* pyramidal tract immediately rostral to its decussation can be roughly estimated at slightly over one million,[157] of which perhaps more than 20 per cent but presumably less than 30 per cent are non-medullated. The diameter of the medullated fibers is said to vary between about 1 and more than 10, up to about 20 μ while that of the unmyelinated ones varies between 0.4 to somewhat more than 4 μ. Such variations are, moreover, reported to depend on the spinal cord's segmental levels. The number of very coarse medullated fibers (about 30,000 to 40,000 in Man) approximately corresponds to the (unilateral number) of large pyramidal or *Betz cells* in the fifth layer of the neocortical area gigantopyramidalis (Area 4) located in the precentral gyrus, and suggests an origin from these neuronal elements. Some of the fiber diameters differ at various ontogenetic stages, being at first uniformly small and gradually assuming the so-called *fiber-size spectrum* characteristic for the adult condition.

As regard the (unilateral) total number of fibers, LASSEK (1954) gives the following estimate:

Chimpanzee	807,000
Macaque	554,000
'Ungulates'	358,300
Dog	285,300
Cat	186,000
Rabbit	101,700
Opossum	75,700
Mouse	32,300
Bat	8,000

[157] LASSEK (1954 and other publications) reported 1,200,000 fibers, while lower figures down to about 700,000 fibers were recorded by some of the various other authors (quoted by BLINKOV and GLEZER, 1968).

From another computation by TADAHICO (1961), quoted by BLINKOV and GLEZER (1968), a few supplementary and in part somewhat different figures may here be added:

Man	858,000 (78%)
Sea Lion	756,000 (71%)
Bear	580,000 (73%)
Dog	327,000 (56%)
Roe Deer	289,000 (54%)
'Monkey'	282,000 (72%)
Cat	161,000 (55%)
Rabbit	62,000 (55%)

The figures included in parenthesis are supposed to indicate the percentage of medullated fibers within the pyramidal tract at the caudal end of medulla oblongata.[158]

Reverting to the numerical data on cortico-spinal fibers recorded for the human spinal cord, and comparing these figures with the estimated number of motoneurons (cf. section 1, p. 12), one could, very roughly, assess the ratio of pyramidal tract fibers to (alpha) motoneurons as 10 to 1 (i.e. approximately 10^6 to 10^5). Including the terminal arborizations of extrapyramidal and of *eigenapparat* fibers on these motoneurons, there obtains thus a considerable many-one relationship with regard to a motoneuron's input, combined with a substantial one-many relationship of that element's output to its so-called 'motor unit', which may, in some instances, comprise more than 100 striated muscle fibers.[158a]

Again, with respect to ontogenetic development, and particularly in Man, the pyramidal tract, originating in one of FLECHSIG's projection centers of the cerebral hemisphere, is thereby also one of the fiber systems displaying myelogenesis at an early time, usually before birth, namely in the course of the ninth prenatal month, but somewhat later

---

[158] According to BLINKOV and GLEZER, TADAHICO used *Kultschitzky's variation* of the *Weigert stain* for the count of medullated fibers and *Glees' variation* of silver impregnation for the count of axons, including the non-medullated ones. In his studies on location and extent of the Mammalian pyramidal tract, VERHAART (1971) used particularly the *Häggqvist stain* based on a modification of the *Alzheimer-Mann technique* (HÄGGQVIST, 1936, 1937).

[158a] As quoted by NOBACK and HARTING (1971) the motor units of Man are said to include, depending on the type of muscle, between 25 and 2,000 muscle fibers.

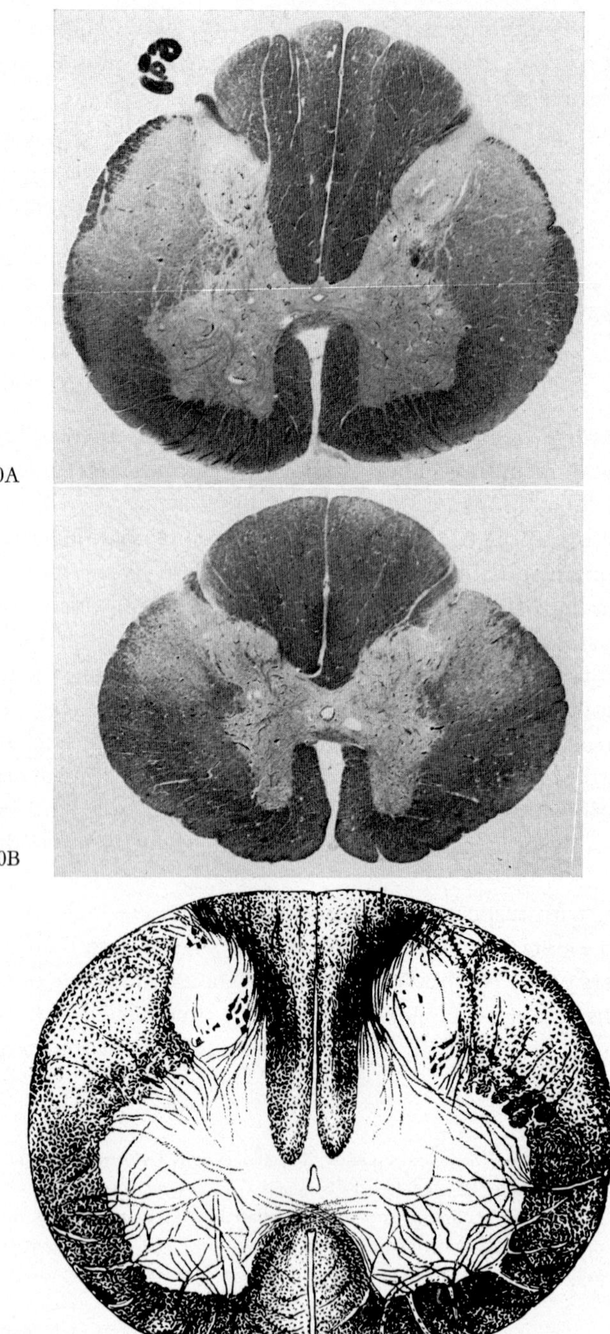

140A

140B

140C

than the main afferent (sensory) projection pathways of the cerebral cortex (FLECHSIG, 1896, 1927). This myelinization proceeds, rather slowly, in a peripheral (caudal) direction. Thus, at birth, the pyramidal system is still essentially non-medullated in the spinal cord, where the lateral cortico-spinal tract becomes clearly outlined by its lack or scarcity of myelin (Fig. 140). At about 2 years of age, the pyramidal pathway appears to have acquired most of its medullation, but this process is believed to continue until at least the fourth year.

According to CLARA (1959), myelinization of the pyramidal pathway seems to proceed relatively more rapidly in Man than in some other Mammalian forms. Generally speaking, however, Mammals which display a substantial degree of effective motility at birth, such as foals and calves, are said to display a well-medullated pyramidal system, while in those forms whose newborns cannot properly move about, the maturation of that fiber system sets in after birth, e.g. in the second week for Dogs and Cats, and in the third week for the Mouse (CLARA, 1959).

In addition to FLECHSIG's myelogenetic method, the study of *Wallerian degeneration*, discussed in volume 3, part I (chapter V, section 8), has provided much information concerning the ascending and descending tracts of the spinal cord. Typical examples of degeneration displaying the location of these fiber systems in Man are illustrated by Figures 141 and 142.

The so-called *somatotopic lamination* within some of the major human spinal pathways, namely the arrangement of their fibers in accordance with peripheral segmental relationships, is indicated in Figure 143.

As regards the spinal sensory innervation of the human body's surface, mediated by the cord's peripheral nerves, it must be kept in mind that the distribution of these latter, except for the sequence of poster-

---

Figure 140. Cross-sections through the spinal cord in the Human newborn and late fetus, showing incompletely medullated respectively still non-medullated cortico-spinal tracts. A Level of intumescentia cervicalis, newborn. Note the asymmetry of spino-cerebellar tracts. B Upper thoracic level below intumescentia (A and B, in which mainly the lateral cortico-spinal tract is indicated by its incomplete myelinization, are from the author's Breslau material. *Weigert-Pal stain,* × ca. 17; red. $^3/_5$). C Unidentified, presumably low cervical level in a 7 months old fetus. Lateral and ventral cortico-spinal tracts as well as fasciculus gracilis and dorsal zone of fasciculus cuneatus are incompletely medullated (from BECHTEREW, 1899, *Weigert-stain*).

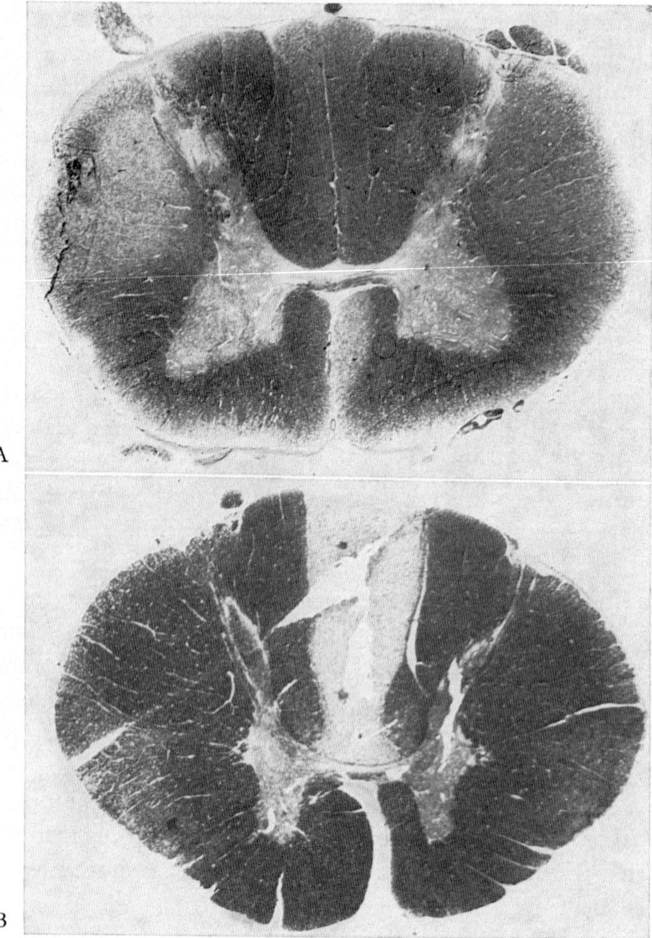

Figure 141. Cross-section illustrating instances of final stages in *Wallerian degeneration.* A Degeneration of uncrossed ventral and of crossed lateral cortico-spinal tracts at a cervical level in an old case of cerebral hemorrhage damaging internal capsule. B Upper thoracic level, several years after a fracture of lower thoracic vertebrae with damage to spinal cord. Only fasciculus gracilis displays prominent degeneration, while some demyelinization is suggested for spino-cerebellar and spino-thalamic tracts on left side (A and B from the author's collection; *Weigert stain;* approx. ×10; red. $^{3}/_{5}$).

ior rami and of thoracic anterior ones (nn. intercostales), does not correspond to the sequence of spinal cord and spinal nerve segments whose peripheral input zones are represented by the overlapping *dermatomes*.

In clinical neurology, particularly for purposes of local or 'topic' diagnosis, the cutaneous metameric segmentation displayed by these dermatomes, as well as the relationships of these latter to the course of peripheral spinal nerves, become of substantial practical importance and are here illustrated by Figures 144A–C. Of similar importance is the relationship between the segmental innervation of trunk and extremity musculature, as indicated by the tabulations of Figure 145, and the innervation of these muscles through their respective peripheral nerves. From an overall morphologic viewpoint, the peripheral spinal nerves were dealt with in chapter VII, section 2, of the preceding volume.

With reference to the discussion of the reflex concept in section 2 of the present chapter VIII, it seems perhaps appropriate to add here a

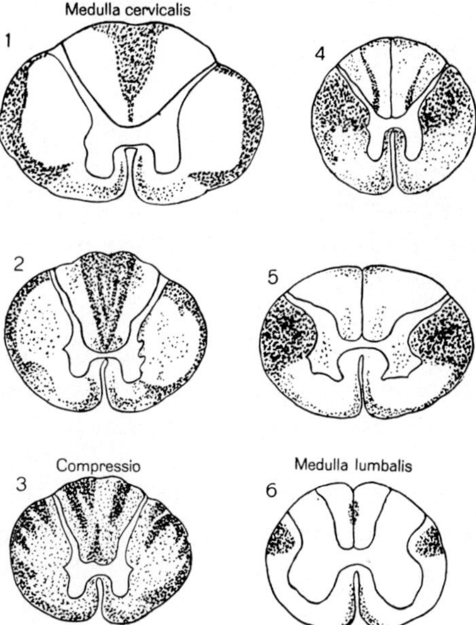

Figure 142. Relatively recent ascending and descending degeneration as displayed by the *Marchi-method* in a human spinal cord, following compression at the 7th thoracic level (from K., 1927, as redrawn by EDINGER, 1912, after HOCHE). For the location of the tracts, Figures 140 to 142 should be compared with Figures 138A, C, D, E, and 143.

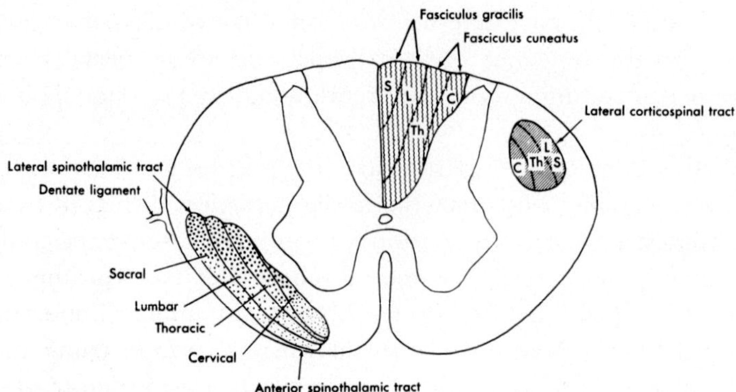

Figure 143. 'Somatotopic segmental lamination' of fiber tracts in the human spinal cord, as presumed to obtain at a generalized cervical level. In the spinothalamic tracts the coarse dots purport to indicate fibers conveying heat and cold impulses, the medium-sized dots those carrying pain impulses, and the fine dots those carrying tactile impulses. The lateral spinothalamic tract is supposed to reach dorsalward as far as the denticulate ('dentate') ligament. The space between spinothalamic tract and cord surface is assumed to contain the ventral spinocerebellar tract. C, Th, L, S refers to cervical, thoracic, lumbar, and sacral segments. It is evident that the lines, drawn as indicating 'boundaries', should not be taken too seriously (from HAYMAKER-BING, 1969, as adapted and combined after various authors).

few comments on some *clinically* significant reflexes manifested by the human spinal cord, and commonly classified as (a) *tendon*, respectively *muscular*, and (b) *cutaneous reflexes*.

Tendon reflexes, characterized by 'a sharp muscular contraction' (BRAIN, 1969), are generally considered to be '*myotatic*', that is, muscular contractions triggered in response to stretch, and mediated by a monosynaptic reflex arc consisting of two neurons. In clinical examination, the stretch is brought about by tapping the tendon, or by suddenly displacing the segment of a limb into which the muscle is inserted. The response is most evident in, but may not be confined to, the muscle directly affected. Tendon reflexes become abolished, diminished, or enhanced ('exaggerated') by lesions involving the afferent, the efferent,[159] or the central neural pathways, which latter 'modulate'

[159] The neuromuscular junction, which may be evaluated as an additional synapse, can be included as part of the 'efferent arc'. Quite evidently, one could insist, with full logical justification, that a 'monosynaptic' reflex is actually 'trisynaptic', if the peripheral neural transducer structure at the receptor level is likewise considered to represent a 'synapse'.

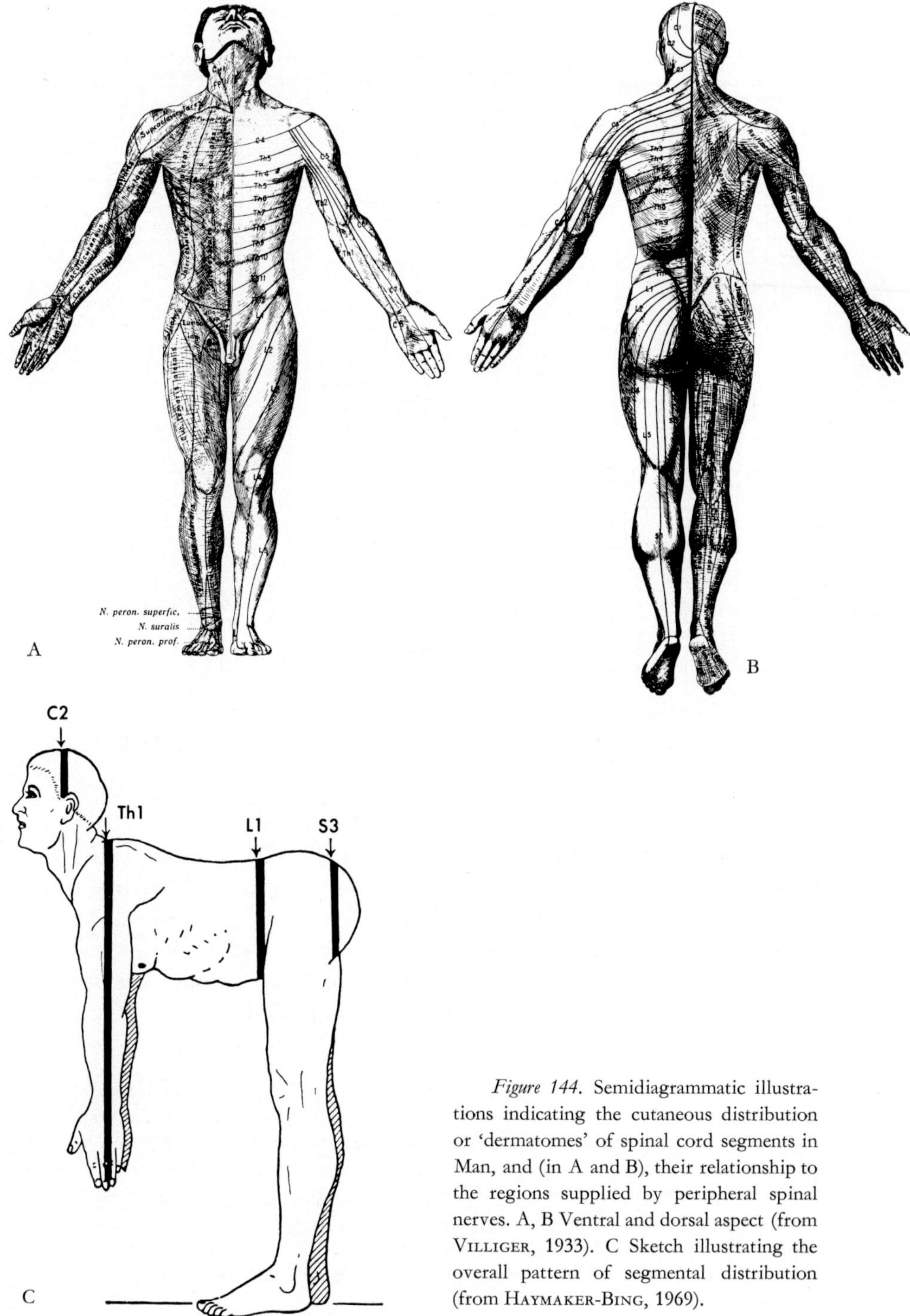

Figure 144. Semidiagrammatic illustrations indicating the cutaneous distribution or 'dermatomes' of spinal cord segments in Man, and (in A and B), their relationship to the regions supplied by peripheral spinal nerves. A, B Ventral and dorsal aspect (from VILLIGER, 1933). C Sketch illustrating the overall pattern of segmental distribution (from HAYMAKER-BING, 1969).

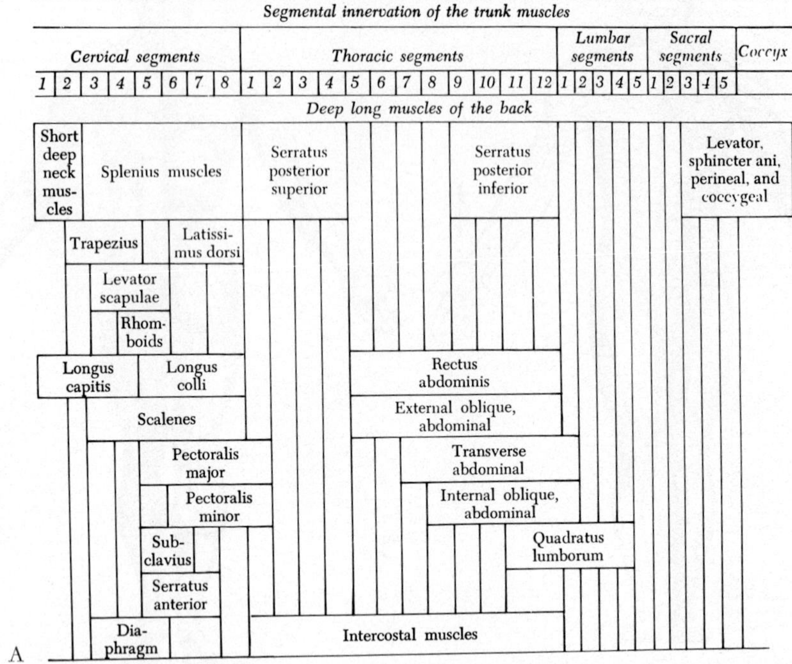

Figure 145. Three tabulations summarizing the segmental innervation of human trunk and extremity musculature (from HAYMAKER-BING, 1969). A Trunk. B Upper extremity. C Lower extremity.

the primary efferent neuron.[160] In addition, a disorder bearing upon the effector, i.e. the muscle, will likewise modify and may completely abolish, its response to a neural impulse.

Early loss of some such reflexes[161] may be due to 'desynchronization' of impulses reaching the motoneurons, i.e. before sensory loss is detectable in cases of polyneuropathy. As regards the influence of the

---

[160] Thus, immediately after complete spinal cord transection, the intact 'monosynaptic reflex arcs' below the lesion's level become, for an indefinite but in Man rather long period, incapable of mediating responses. This sort of 'suppressing' or 'abolishing' effect has been subsumed by v. MONAKOW under his concept of so-called '*diaschisis*'.

[161] Generally speaking, tendon reflexes moreover subsume the responses obtained by stimuli to periosteum, bones, joints, fasciae, muscular and aponeurotic structures, being also called 'deep reflexes' and classified as 'proprioceptive'. Their receptor structures are especially muscle spindles but also presumably tendon spindles and the other sorts of receptor endings in the non-muscular structures.

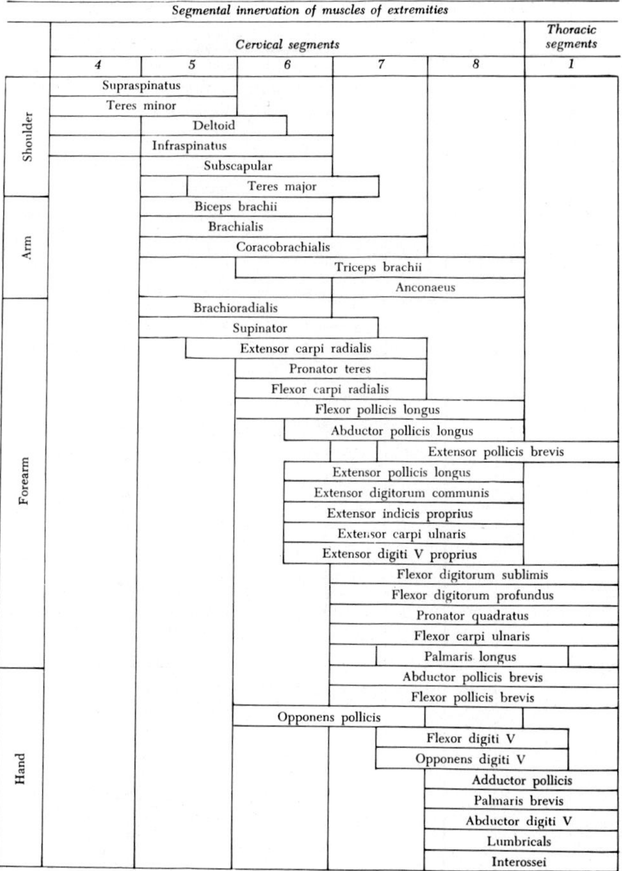

Figure 145B

central 'modulating' pathways upon the 'excitability' of tendon reflexes, lesions of the cortico-spinal system, and functional 'anxiety states' can enhance the responses. These latter are generally diminished by neural shock or by increased intracranial pressure, and may even be congenitally absent (Brain, 1969). 'Anticipation' or 'mental tension' can also, in some instances, by unconscious or subconscious tonic innervation, diminish, respectively suppress, the response instead of enhancing it.[162] A rhythmical series of contractions, related to the

[162] Thus, in eliciting the patellar reflex, clinicians frequently use *Jendrassik's trick or maneuver:* the patient is required to hook his hands together and to pull them apart as hard as he can; while his attention becomes thereby diverted, the patellar jerk is tested.

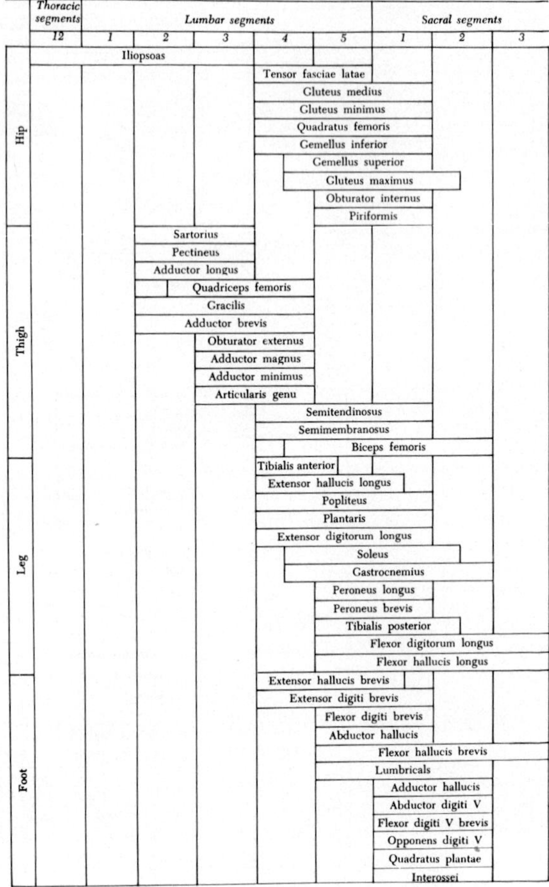

Figure 145C

maintenance of tension in a muscle, is frequently elicited under diverse circumstances, particularly when the tendon reflexes are exaggerated after a corticospinal lesion. *Clonus* of the quadriceps *(patellar clonus)* is then obtained by a sudden sharp downward displacement of the patella, and clonus of triceps surae *(ankle clonus)* by sharp extension ('dorsiflexion') of the ankle joint. Again 'in states of muscular hypertonia a reflex response may spread beyond the muscle stretched, as when a tap on the styloid process of the radius elicits a contraction not only of the brachioradialis, but also of the long flexors of the fingers' (BRAIN,

Reflex	*Mode of elicitation*	*Response*	*Spinal segment*	*Peripheral nerve*
Biceps-jerk	A blow upon the biceps tendon	Flexion of the elbow	Cervical 5–6	Musculo-cutaneous
Triceps-jerk	A blow upon the triceps tendon	Extension of the elbow	Cervical 6–7	Radial
Supinator-jerk or radial reflex	A blow upon the styloid process of the radius	Flexion of the elbow	Cervical 5–6	Radial
Flexor finger-jerk	A blow upon the palmar surface of the semi-flexed fingers	Flexion of the fingers and thumb	Cervical 7–8	Median and ulnar
Knee-jerk	A blow upon the quadriceps tendon	Extension of the knee	Lumbar 2–4	Femoral
Ankle-jerk	A blow upon the tendo calcaneus	Plantar flexion of the ankle	Sacral 1–2	Sciatic

Figure 146. Tabulation listing the principal tendon reflexes relevant for clinical neurology (from Brain and Walton, 1969).

1969). Figure 146 tabulates the principal clinical tendon reflexes, their mode of elicitation, and their innervation as concisely compiled by the just cited author.

Among the *cutaneous reflexes*, the following ones may be mentioned in the present context. The so-called *abdominal reflexes* are manifested as brisk muscular contractions of a part of the ventral abdominal wall in response to a stimulus, e.g. a touch or a light scratch with a pin. In clinical examination, they are usually tested at three levels an each side: just below the costal margin, at the umbilical level, and at the level of the iliac fossa. They are presumably plurisegmental and believed to be 'polysynaptic'. Their apparent dependence upon the integrity of the cortico-spinal system is not understood. Nevertheless, a *lesion of the pyramidal pathway* is indeed usually associated with diminution or loss of the superficial abdominal reflexes. In case of a unilateral impairment of the pyramidal system, this is manifested on the side corresponding to the lateral cortico-spinal fiber bundle. If the pyramidal system's defect is slight, the reflexes may be reduced in vigor, but not completely abolished, the reflexes of the lowest segments being most impaired. Again, lost reflexes can return later, despite persisting pyramidal tract lesion. Nor is the loss of these reflexes always proportional to the severity of said lesion. Thus, in multiple sclerosis, the reflexes may be lost early, at a stage of the disease when other signs of corticospinal lesions are slight. In congenital diplegia, and in some motoneuron diseases (at least during certain 'stages'), the abdominal reflexes 'are usually brisk' (Brain, 1969). The cited author also states that 'little importance can be attached to diminution of the superficial abdominal reflexes in stout people, after repeated pregnancies, and after middle life'. In the course of my own clinical activities, I have occasionally noted the absence of these perhaps partly muscular reflexes in young 'normal' adults.

Since the abdominal reflexes are interpreted as a 'local withdrawal from the stimulus', they are also evaluated as '*nociceptive*' (BRAIN, 1969). Their reflex arcs and synaptic grisea appear to be localized in the 7th to 12th thoracic segments of the spinal cord. Lesions involving the arcs themselves can evidently produce diminution or loss of these reflexes, as e.g. in (so-called 'lower') motoneuron damage caused by poliomyelitis. The *gluteal reflex*, in which a scratch on the buttock evokes contraction of the gluteal musculature, is considered 'physiologically akin to the abdominal reflexes' and involves the 4th and 5th lumbar segments. This, with regard to the aspect here under consideration perhaps unduly lengthy elaboration on a group of relatively 'unimportant' reflexes was purposely presented in order to acquaint the interested non-medical or non-clinical neurobiologist with a typical example of the vagueness and of the numerous uncertainties or ambiguities obtaining in most domains of clinical medicine.[163]

The *cremasteric reflex* is a cutaneous reflex elicited by a light scratch along the medial surface of the upper thigh and resulting in a contraction of the cremaster muscle, with elevation of the testicle. The reflex arc is assumed to run through the first lumbar segment. Its functional interpretation and clinical significance are closely similar to those of the abdominal reflexes. BRAIN (1969) remarks that it is usually extremely brisk in children, in whom it may sometimes be elicited by a stimulus applied to any part of the lower limb, and that it is usually

[163] This 'flimsiness', concomitant with an ever increasing amount of unassimilated detailed data or 'information' is, of course, not restricted to clinical medicine, but also characterizes the status of 'neurosciences' as well as of 'science' in general. Being entirely impervious to brickbats, which, where appropriate, I enjoy heaving myself, the total condemnations decreed upon previous volumes of this series by diverse *petits maîtres*, e.g. in *Science* (158, p. 1443, 1967) and *Brain* (94, p. 193, 1971) merely amused me. Thus, the gentleman sounding off in *Brain* complained that my expressed opinions were 'so hedged about with qualifications that the final effect is often to confuse rather than to inform'. Quite evidently, sceptical qualifications are very much needed for a proper emphasis on the many uncertainties of so-called neurosciences, whose confused status the reviewer is apparently unwilling to admit because of his naïveness and most likely also because he lacks candor. One could here quote an old *Sanskrit* saying: '*Upadeśo hi mûrkhâṇâṃ prakopâya na śântaye*'. It suitably can be translated to mean: *Instruction indeed confuses but does not chasten nincompoops.*

diminished or absent on the affected side in a patient with varicocele. On the basis of my own clinical observations, I suspect that the so-called cremaster reflex involves not only the striated m. cremaster (a derivative of m. obliquus internus abdominis), but also the smooth musculature of the scrotum's tunica dartos (panniculus carnosus) and of the tunica vaginalis propria (which is a derivative of the abdominal fascia transversalis). This reflex may thus not be restricted to somatic motoneurons, but could include the participation of sympathetic (visceral efferent) preganglionics in the cord's lumbar grisea, thereby representing a far more complex pattern of neural responses than commonly assumed.[164]

The *bulbocavernous reflex*, whose spinal cord segments are presumably S2, 3 and 4, consists of contraction of the bulbocavernous muscle, which can be detected by palpation, in response to squeezing the glans penis. This reflex, as noted by BRAIN (1969), is frequently abolished in tabes and in lesions of cauda equina. Sacral segments 4 and 5 are presumed to be involved in the superficial *anal reflex*, manifested by contraction of the sphincter ani externus in response to a scratch upon the skin in the perianal region.[164a]

---

[164] Strong contractions of the scrotum may, for instance, be experienced by some individuals subject to vertigo (dizziness) when confronted with perpendicular respectively steep plunging views of great depth from tall buildings or in mountain climbing. This phenomenon, related to unusual visual perceptions at great height, evidently involves 'conscious' cortical mechanisms with final discharges, perhaps indirectly by way of brain stem centers, into autonomic and other grisea of the spinal cord. In one of the cases which came to my attention, a seasoned mountaineer described that effect with the pithy comment: '*Solch ein Tiefblick zieht mir immer die Eier zusammen*'. Optic responses or reflexes, including those in which the spinal cord participates, shall be dealt with in chapter XI of this volume.

[164a] The *external* or cutaneous anal reflex must be distinguished from the *internal anal reflex*. This latter, manifested by a contraction of the smooth m. sphincter ani internus, is elicited upon introduction of the gloved finger into the anus, and is considered an autonomic, sympathetic reflex, mediated by the hypogastric plexus (presacral nerves). The sympathetic preganglionic neurons of the efferent reflex arc are located in the cord's upper lumbar segments, and the postganglionics seem to be scattered in the so-called inferior mesenteric plexus, which is continuous with 'plexus hypogastricus' and 'presacral nerves'.

It remains a moot question to which extent this reflex is mediated through receptor endings in muscles or mucosa by way of the spinal cord. Direct stimulation of the muscle itself or, again, a local reflex arc through the peripheral plexuses cannot be entirely excluded. If the typical reflex is lost, the anus does not immediately contract upon stimulation.

The *plantar reflex* is or particular importance in clinical neurology. It involves the lower lumbar and upper sacral segments, being evoked by a scratch upon the sole of the foot, particularly along its outer border. A flexor and an extensor response are here distinguished. The former, considered akin to the abdominal reflexes, is normal in adults as well as in children some time after completion of the first year of life. It is characterized by plantar flexion (anatomically true flexion) of the toes, generally associated with 'dorsiflexion' (extension) of the foot at the ankle, and occasionally other variable muscular contractions, e.g. of m. tensor fasciae latae.

The *extensor plantar reflex*, pointed out by BABINSKI (1896) and generally designated as *Babinski's sign* (or as '*positive Babinski*'), consists of a rather slow, tonic hyperextension ('dorsiflexion') of the first toe, frequently accompanied by a sluggish spreading apart ('fanning') of the other four toes and by either their plantar flexion or hyperextension ('dorsiflexion'). *Babinski's sign*[165] is considered pathognomonic for lesions of the corticospinal system. It is, on the other hand, normal in infants, whose pyramidal tract still remains incompletely developed. Nevertheless, a transitory '*positive Babinski*', not related to 'organic' corticospinal damage, can be observed in adults during sleep, in some stages of hypnosis, in deep coma from any cause, and for a short time after epileptic convulsions.

Additional spinal reflexes of some clinical significance are *tonic neck reflexes*, as mentioned above in section 2 (p. 31), and so-called 'associated reactions' involving 'automatic' associated movements of various muscles evoked by motions of others. The '*grasp reflexes*' of hand and foot, normally displayed by infants, subsequently vanish, but may then become abnormal neural manifestations, usually indicating brain damage involving the frontal lobe, with or without direct cerebral lesions of the cortico-spinal system.

As regards the *motor output* of the spinal cord (and deuterencephalon), it is convenient to distinguish, for clinical purposes, an '*upper*' and a '*lower*' *motor neuron*. This simplified and practical concept, which more or less disregards the 'extrapyramidal system' and the presumed end-

---

[165] Additional reflexes which, if positive, display *Babinski's toe sign*, and have the same diagnostic significance, are: *Oppenheim's reflex*, obtained by firm moving pressure on the skin of the tibia, *Chaddock's reflex* obtained by stroking the region of the lateral malleolus, and *Gordon's reflex* upon squeezing the muscles of the leg.

ings of many cortico-spinal fibers on internuncials seems doubtless justified in the aspect under consideration. The '*upper motor neuron*' is here a pyramidal cell of the cerebral cortex, whose neurite discharges (directly or indirectly) upon the '*lower motor neuron*' represented by an efferent root cell. This latter is a motoneuron whose axon innervates striated musculature.

Upper motor neuron lesions generally involve *spastic paralysis* with rigidity, exaggerated tendon reflexes, and *Babinski sign*. The peripheral nerves and the muscles show normal electrical irritability. The factors responsible for the spasticity or 'muscular hypertonia' manifested by established upper motor neuron lesions are poorly understood despite numerous elaborations and speculations on that topic (cf. e.g. BRAIN, 1969). Immediately after such lesions, the paralyzed limbs are usually completely flaccid, and the 'hypertonia' may gradually develop in the course of a few weeks. This spasticity was formerly interpreted as a 'release phenomenon' caused by the loss of the cortico-spinal system's inhibiting mechanisms. At present, it may be regarded as a 'multifactorial effect' including various 'extrapyramidal' components resulting in 'direct facilitation of alpha neurones' and perhaps 'release of gamma innervation'. Commonly occurring upper motor neuron lesions are the various *hemiplegias* caused by diverse sorts of cerebral lesions involving the course of the cortico-spinal tract above the level of the pyramidal decussation. If both antimeric, i.e. right and left cortico-spinal respectively upper neuron pathways are involved, *paraplegia* results, which affects both sides of the body at the corresponding levels. As regards lesions of the spinal cord, a complete transection evidently causes paraplegia, in addition to loss of sensation, below the level of the lesion.

Fully developed *lower motor neuron lesions* are characterized by *flaccid paralysis* and *abolition of all reflexes*. The *Faradic* and *Galvanic excitability* of the peripheral nerve is lost, and the corresponding muscle gradually becomes atrophic, losing its *Faradic irritability* but then displaying, at least for a considerable time, increased *Galvanic irritability*, with slow or sluggish contractions (complete *reaction of degeneration*, RD, cf. vol. 3, part I, p. 575f. and 684f. of this series). In both 'upper' and 'lower motor neuron' damage, the degree of muscular impairment depends, of course, upon the 'severity' of the lesion. Accordingly, loss or substantial impairment of motor function is commonly designated as *paralysis* in contradistinction to *paresis*, which denotes an 'incomplete paralysis'. Again, the term 'paresis' is frequently applied to the result of upper motor neuron lesions.

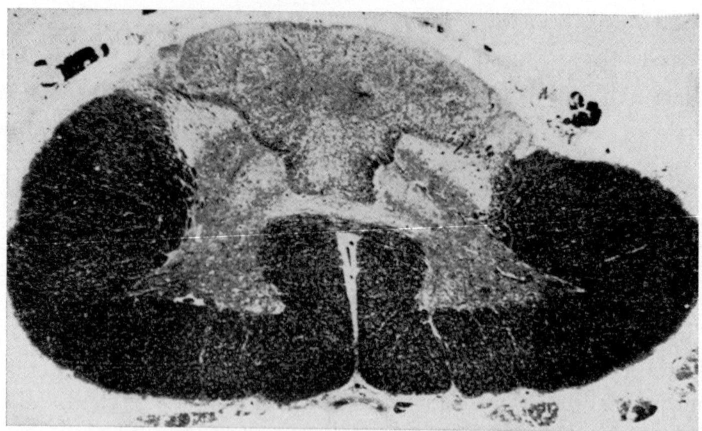

Figure 147A. Fairly typical aspect of the spinal cord in tabes dorsalis at a cervical level. In cross-section, an almost complete demyelinization of the posterior funiculi is displayed. Note an apparent sparing along the border of gray substance, partly corresponding to CLARA's '*Hinterstranggrundbündel*' as shown above in Figure 137 (from WEIL, 1945).

Since, in addition to their clinical importance, various pathologic conditions of the spinal cord may be considered as being of general interest for the neurobiologist, a few comments on some such disorders seem here appropriate.

Tabes dorsalis, together with *general paresis*, was classified as a parenchymatous manifestation of tertiary lues, occurring in a small proportion of persons infected with the treponema pallidum. In tabes, also designated as '*locomotor ataxia*', the essential spinal cord lesion[166] is a degeneration of posterior root fibers,[167] involving demyelinization of

[166] The concomitant involvement of the optic system, with loss of the pupillary reflex to light (*Argyll-Robertson phenomen*) shall be dealt with in chapter XI. With regard to the spinal cord and its nerves it should nevertheless here be recalled (cf. chapter VII, section 6) that the sympathetic preganglionic outflow for the musculus dilator pupillae originates at lower cervical (C VIII) and upper thoracic (Th1, 2) spinal cord levels (*'centrum cilio-spinale.*) Thus, in some lesions, such e.g. as in lower plexus paralysis, *Horner's syndrome* may result on the affected side, characterized by ptosis and perhaps enophthalmos, due to the paralysis of the smooth orbital (palpebral) musculature, miosis (pupillary constriction due to paralysis of musculus dilatator pupillae) and anhidrosis of the face, occasionally combined with vasodilation. The somatic musculature affected in lower plexus paralysis commonly comprises the interossei, the thenar and hypothenar muscles, occasionally also some forearm muscles.

[167] A particular vulnerability of these roots in the *Redlich-Obersteiner zone* (cf. above p. 219) has been suspected.

the posterior funiculi (Fig. 147A). Initial abnormal sensations ('paresthesias') and shooting ('lancinating') pains, particularly in lower extremities, are followed by a variable degree of loss in sensory input, affecting mainly its proprioceptive components. Thus, among others, the patellar reflex becomes abolished *(Westphal's sign)* because of damage to the reflex arc's afferent pathway. Moreover, incoordination of the legs' movements in walking *(ataxia)*, and related symptoms develop. These latter include *Romberg's sign*, which is elicited by requesting the patient to stand erect with feet drawn together and eyes closed: unable to maintain equilibrium without optic input, he then begins to sway. Arthropathies *(Charcot's joints)* and perforating ulcers of the planta pedis are not uncommon, being presumably due to loss of 'nociceptive' dorsal root input. Some of the symptoms characteristic for tabes also occur e.g. in hereditary *Friedreich's ataxia*, whose degenerative lesions generally involve, in addition to the posterior funiculi, also the lateral spino-cerebellar, and the lateral cortico-spinal tracts.

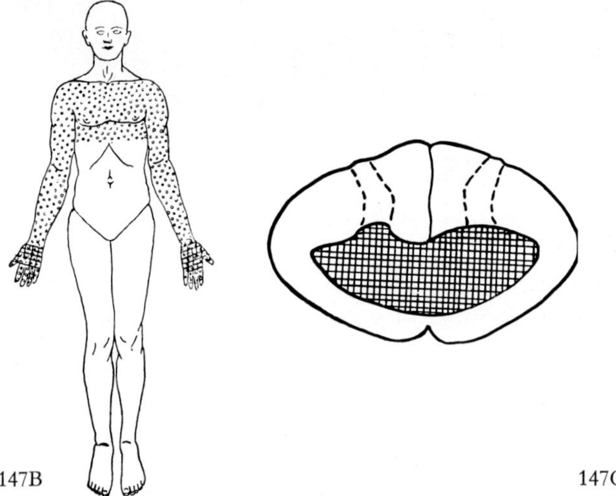

Figure 147B–E. Diagrams and photomicrographs illustrating aspects of syringomyelia (B, C from Ranson, 1943, D from Lichtenstein, 1949, E from Biggart, 1949). B Extent of sensory and motor involvement. Dots represent region of anesthesia for pain and temperature sensations; cross-hatching indicates paralysis and atrophy of small hand muscles. C Rough diagram of lesion affecting lower cervical and upper thoracic segments in B. D Lesion characterized by maximal involvement of lower cervical region. The hollowed spinal cord has collapsed upon histologic processing. E Lesion at cervicothoracic boundary region. The central canal is included in the cavity. In various cases of syringomyelia, the central canal may remain unaffected.

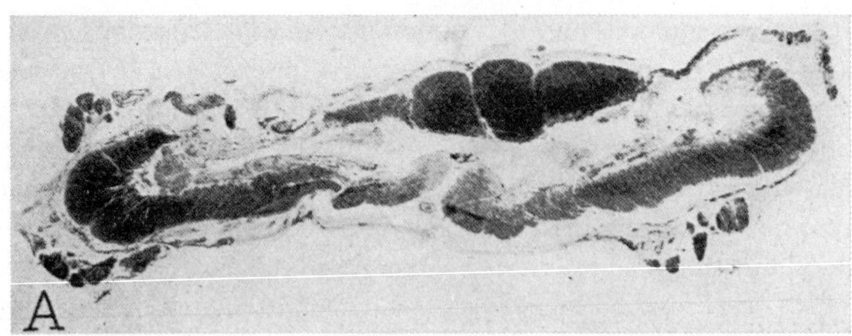

147D

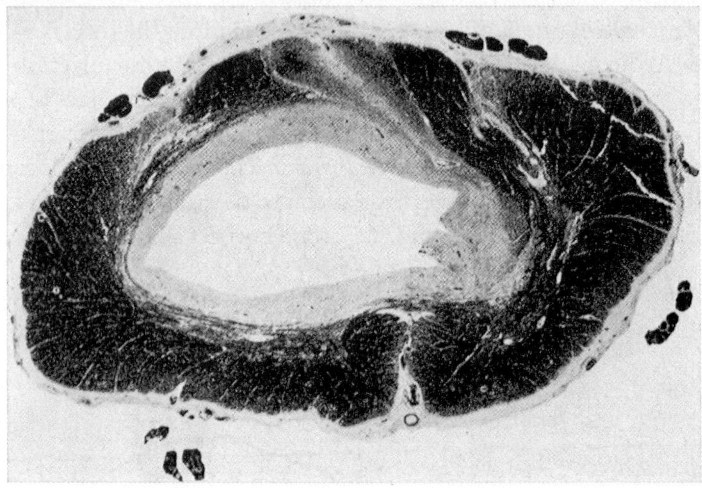

147E

It should be added that present-day clinicians, while subsuming vascular cerebrospinal lues and gummata of the neuraxis under the manifestation of tertiary lues, prefer to consider tabes and general paresis as a further quaternary progression of the infection. This stage *(metalues)* is believed to result from the occurrence of an 'anergy', whereby the 'mesodermal barrier' has become ineffective, thus allowing for an invasion of the neuroectodermal parenchyma. Following the introduction of antibiotic therapy, the incidence of lues showed a conspicuous decline, and cases of metalues are now rare. However, a considerable and widespread increase in the number of luetic infections has *de novo* occurred in recent years, perhaps combined with a higher resistance of the treponema. Accordingly, tabes and general paresis might again become less uncommon.

Sensory dissociation is a peculiar symptom characterized by loss of pain and temperature sensibility without corresponding loss of touch or proprioceptive sensations. It may be caused by lesions affecting either the *central neuraxis* (in this context the spinal cord) or the *peripheral nerves. Central* lesions of this type occur in *syringomyelia* (Fig. 147), a condition in which elongated cavities, commonly accompanied by necrosis, are formed in the neighborhood of the central canal.[168] Decussating fiber systems, particularly in the ventral commissure, are thereby disrupted, and this, as can be seen in Figure 138B, especially affects the pathway for pain and temperature in the lateral spino-thalamic tract, while the pathway for touch still retains alternative channels. Such patients, insensitive to heat and pain in the body regions corresponding to the central lesion, are prone to severe burns and other sorts of injuries which they do not notice at the time. *Peripheral* sensory dissociation can also be caused by certain forms of neuritis or 'neuropathies' e.g. in *leprosy* and in *Morvan's disease*, in which the unmyelinated or thinly medullated 'pain' and 'temperature' fibers, or their respective spinal ganglion cells, are presumably more severely damaged than other afferents.[169]

The peripheral nerve endings presumably related to the transmission of '*pain*' signals were discussed in section 1, chapter VII of volume 3, part II. Concerning the 'control of pain', the following comments on the recently much publicized procedure of *acupuncture* might per-

---

[168] *Syringomyelia* can be regarded as a still poorly understood group of lesions, some of which are doubtless of borderline 'dysplastic' type, while others seem to be 'neoplastic', with central necrosis of a gliomatous or spongioblastic growth. Although it is claimed that syringomyelia does not originate in the central canal, I have myself seen a case which rather obviously represented a neoplastic transformation of the central canal's ependyma. Lesions of syringomyelia are most frequently located in upper thoracic and in cervical spinal cord, occasionally extending into medulla oblongata *(syringobulbia)*. Again, the lesions can involve parts of the anterior horn, causing some 'lower motor neuron damage'. Despite the loss of peripheral pain and temperature sensibility, spontaneous shooting pains may initially occur. Finally, loss of pain sensation can also be a prominent symptom in hysteria. Thus, in another, and quite clear-cut case of syringomyelia of my own observation, the at first consulted '*croaker*' stubbornly insisted on the diagnosis 'hysteria' despite strong evidence to the contrary, which was subsequently fully substantiated by typical wasting of the mm. interossei with resulting '*claw hand*' (concomitant 'lower motor neuron' lesion). We have here a 'parallel' to NISSL's obstinate diagnostic preconception in a case mentioned in footnote 225, p. 522 of chapter V, volume 3, part I. There is also a pertinent old military quip in which a '*medic*' reports to his superior: '*Melde gehorsamst, Herr Stabsarzt: der Simulant in Bett Nummer 5 ist tot.*'

[169] The thereby strongly suggested presence of peripheral 'pain fibers' is difficult to reconcile with some of the ambiguous aspects ol MELZACK's and WALL's so-called '*gate control theory of pain*' (cf. above, p. 229).

haps be appropriate. About 50 years ago, during my first sojourn in Japan (1924–1927), I became acquainted with the two traditional methods of ancient Chinese medicine formerly adopted in Japan, namely *moxa* (moxibustion) and *acupuncture*, which were still widely used by so-called 'native style practitioners'.

Moxa (mogusa, derived from *moe-kusa*, 'burning herb') is a cauterization cure. Small cones prepared from dried leaves of wormwood (Artemisia), applied to the skin, are burned and produce blisters.

Acupuncture is a treatment involving the insertion of fine needles of steel, silver or gold, to a depth varying between 1 to 8 cm at certain spots, usually along diverse systems of lines drawn upon charts of the human body. These lines supposedly indicate the flow of 'vital energy' related to the ancient Chinese concepts of '*Yin*' and '*Yang*' (female and male principles). There is no intelligible relationship between these lines respectively points and any actual anatomical channels or structures respectively functional aspects ascertainable by the data derived from Western scientific investigations.

Those of my Japanese colleagues interested in the history of Sino-Japanese medicine considered said therapeutic methods as devoid of any 'rational basis' and based on mere superstition, but conceded that, as in the case of other '*cults*' (homoiopathy, diverse forms of 'naturopathy', psychoanalysis, etc.) 'subjective relief' might occasionally be experienced by patients so treated. Moreover, in very many conditions recovery or improvement occurs quite naturally regardless of or *despite* medical treatment. At the time, and notwithstanding the wide use of acupuncture, no specific '*anesthetic effect*' was known or claimed to be produced by this then strictly '*therapeutic*' method.

The use of acupuncture for surgical anesthesia in major operations apparently originated during the nineteen-sixties in the 'People's Republic of China' and became widely heralded as well as debated in the American Press (including 'letters' in '*Science*'). The available reports do not seem entirely satisfactory nor fully convincing. In various instances, not only 'twirling' of the needle but also the concomitant transmission of a low voltage electric current were reported. In other instances, the additional administration of analgesics, such as meperidine or morphine was admitted. Moreover, acupuncture anesthesia does not seem to be effective in all cases.

Assuming that some of the reports are nevertheless essentially correct, several and not mutually exclusive tentative explanations are possible, since the actual experience of '*pain*' depends on multifactorial neural events. A 'hypnotic effect', particularly enhanced by 'mass-conditioning' and thorough 'indoctrination' could play the predominant role. Experiments in animals, which may be 'hypnotized' or immobilized by postural manipulations, cannot be adduced against this view. On the other hand, peculiar and not yet understood paradoxical effects on peripheral sensory 'pain' nerve endings do not seem completely impossible. Finally, some amount of fraud related to 'showmanship' and 'propaganda' cannot be entirely excluded (e.g. undisclosed administration of spinal or local anesthesia). Thus, the topic of acupuncture remains a moot question which cannot be dogmatically dismissed but should be approached with a proper degree of scepticism. Evidently, the exotic and cabalistic, or the political aspects of that procedure strongly appeal to the multitudinous '*lunatic fringe*' in medicine and at large. A modernized Chinese text on acupuncture was recently translated and published by KAO (1973), but failed to impress me as being convincing or as elucidating the claimed results.

Sensory and motor symptoms are manifested by the *Brown-Séquard syndrome* occurring in lesions involving more or less approximately one entire half of the spinal cord at a given level (e.g. by stab wounds or partial fractures of the vertebral canal). In its typical form this syndrome may include noticeable *lower motor neuron paralysis* and complete peripheral anesthesia within a *homolateral segmental strip* corresponding to the actual destruction *at the lesion's level*. The here relevant symptoms with respect to the disrupted *long tracts* are *contralateral loss of sensitivity to pain and temperature*, and *homolateral 'upper motor neuron paralysis'*, both below the segment of the lesion (cf. Fig. 148A, B). Together with spastic paralysis and its concomitant signs (*Babinski*, exaggerated tendon reflexes, etc.) there is usually a noticeable homolateral loss of 'deep' or 'proprioceptive' sensibility. Because such severe lesions of the spinal cord seldom involve a true hemisection, but may affect less than one side or encroach upon the opposite one, incomplete respectively modified forms of the *Brown-Séquard syndrome* are more frequent than its typical manifestation. Similarly, diverse symptoms related to the interruption of ascending and descending tracts as well as, to a variable extent, of grisea, occur in *disseminated (multiple) sclerosis* and in the *subacute combined degeneration* concomitant with pernicious anemia.

Degenerative lesions essentially restricted to the spinal cord's *motor system* are those in *amyotrophic lateral sclerosis*, *progressive spinal muscular atrophy*, and *poliomyelitis*. The first two conditions are now frequently considered variations of one, still obscure 'nosologic entity'. In some

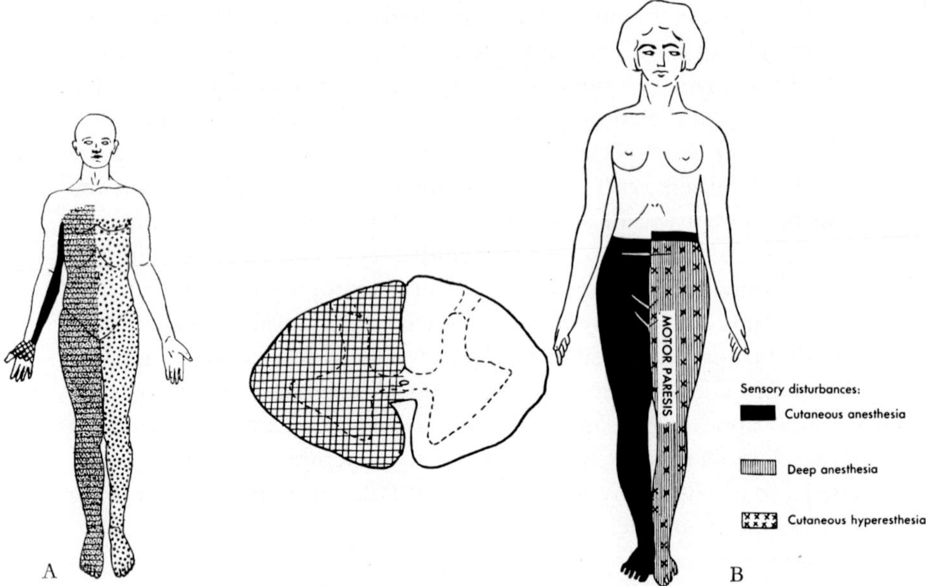

Figure 148. Diagrams indicating features of the *Brown-Séquard syndrome* resulting from right hemisection of the spinal cord (A) at the levels C VIII-Th I and (B) from left one at about Th I (A from Ranson, 1943, B from Haymaker-Bing, 1969). In A, the black strip denotes loss of sensibility along ulnar side of right arm; x-notation: 'lower motor paralysis' of small hand muscles; horizontal lines with small dots: 'upper motor neuron involvement' and some loss of 'deep sensibility'; larger dots: loss of sensibility to pain and temperature. In B, black denotes a strip of complete cutaneous anesthesia on left (homolateral) side of body, and loss of pain and temperature sensations on the body's right side; vertical hatching: deep anesthesia; crosses: cutaneous hyperesthesia. 'Upper motor neuron' involvement as indicated by 'motor paresis' on left side of body.

forms of amyotrophic lateral sclerosis, the lesion is mainly confined to the lateral cortico-spinal tracts. In others, a concomitant degeneration ('neuropathy') of motoneurons occurs. Thus, various combinations of 'upper' and or 'lower' motoneuron involvement are manifested. Typical progressive spinal muscular atrophy is characterized only by 'lower' motor neuron lesions. The virus responsible for *acute poliomyelitis* (infantile paralysis) displays a predilection for the motoneurons in the anterior horns of spinal cord and of those in some of the brain stem nuclei. While a number of such neuronal elements may recover from the stage of initial acute inflammatory processes, others become irreversibly damaged and disappear completely. The end-result is then here a typical 'lower motor neuron paralysis'.

Because motion of the extremities, in particular locomotion or gait, is to a significant extent a function of the spinal cord, a few comments on clinical '*gaits*' may be appended. Conveniently, *atactic*, *spastic*, and *steppage gait* can be distinguished.

The *atactic (or ataxic) gait* is describable as uncertain and reeling. The patient walks on a wide base, slapping the feet and often watching his legs, the ataxia becoming more pronounced with closed eyes. In a simplified classification of further subsets, one could enumerate tabetic-atactic, cerebellar, and drunken gaits. The latter refers to the staggering in acute intoxications, as well as in cases of brain tumor, general paresis, and multiple sclerosis.

In the *spastic (hemiplegic, or spastic-paretic) gait*, the affected leg is rigid, being swung from the hip in a semicircle. The patient commonly leans to the affected side. In incomplete spastic paraplegia (paraparesis) the legs are usually adducted, crossing alternatingly, with knees scraping *(scissor's gait)*.

In *steppage (or foot-drop) gait*, related to 'lower motoneuron damage', there is frequently high knee action, with flopping of the foot, and the toes tend to drag the floor. It occurs, e.g. in paralysis of the n. peronaeus profundus (anterior tibial nerve), in peroneal nerve injuries, neuritis, poliomyelitis, and progressive spinal muscular atrophy.

Reverting, at the conclusion of this chapter on the Vertebrate spinal cord, to the comparative anatomical aspects and their ambiguous *phylogenetic implications*, the following general comments seem appropriate. First, it appears quite evident, as stressed by many authors since L. EDINGER, and also pointed out in my old '*Vorlesungen*' of 1927, that, in the course of evolution, the elementary spinal mechanisms, representing the relatively 'independent' *eigenapparat* of 'primitive forms', has been brought under an increasingly complex control of the brain by means of the *verbindungsapparat*.[170] Although fairly complex and poorly understood neuronal connections obtain in various submam-

[170] '*Die Vergleichung der einzelnen Tiergruppen zeigt als allgemeines Ergebnis, dass bei den höheren Tieren der Verbindungsapparat des Rückenmarks auf einer hohen Differenzierungsstufe steht, während gleichzeitig der Eigenapparat unselbständig geworden ist, dass bei den niederen Tieren dagegen der Verbindungsapparat weniger ausgeprägt ist, der Eigenapparat aber dafür weit selbständiger bleibt.*'

'*Unselbständig und selbständig sind durchaus nicht gleichbedeutend mit weniger differenziert und differenziert. Der unselbständige Eigenapparat eines Säugetierrückenmarkes ist komplizierter als der selbständigere des Rückenmarkes eines Frosches*' (K., 1927, p. 121).

malian Vertebrates, the neural networks within the Amniote and particularly the Mammalian spinal cord grisea are presumably more intricate than those in Anamnia. Nevertheless, in the whole Vertebrate series, from Cyclostomes to Man, the ontogenetically and phylogenetically primitive tubular configuration with a central cell aggregate and peripherally arranged longitudinal fiber tracts has been maintained. This does not exclude the presence of some scattered neuronal elements within the 'white substance' and even directly beneath the cord's surface, as evidenced by the 'marginal nuclei' in Sauropsidans and some Mammalia, referred to above in the relevant descriptions of grisea.

Extension of dendrites into the white matter can occur in lower Vertebrates as well as in Mammals. Thus, a 'shortening of the dendrites' cannot, contrary to some claims, be considered a 'phylogenetic' or 'general evolutionary' trend, but merely seems to represent an 'incidental feature' or '*independent variable*' manifested in 'lower' as well as in 'higher' forms.

Likewise, and with respect to synaptic connections, structurally differentiated 'synaptic knobs' are known to be present in Anamnia such as fishes as well as in the 'higher' Vertebrate forms. It cannot be assumed that an evolution from so-called 'unspecialized contacts' to highly differentiated synapses represents a distinctive trend in overall 'Vertebrate phylogeny'.

On the other hand, a *progressive differentiation* of the relatively simple ascending and descending, crossed and uncrossed components already present in the *verbindungsapparat* of Anamnia seems indeed to have taken place. Thus, a *direct* descending connection of cerebral cortex with spinal cord appears to be present only in Mammals, although an *analogous*, but quite different and rather rudimentary or minor telencephalospinal system might perhaps occur in Birds and possibly even in some Reptiles.

12. References to Chapter VIII

Adams, C.W.M. (ed.): Neurohistochemistry (Elsevier, Amsterdam 1965).

Agduhr, E.: Über ein zentrales Sinnesorgan (?) bei den Vertebraten. Z. Anat. EntwGesch. *66:* 223–360 (1922).

Agduhr, E.: Vergleich der Neuritenzahl in den Wurzeln der Spinalnerven bei Kröte, Maus, Hund und Mensch. Z. Anat. *102:* 194–210 (1934).

Ahlborn, F.: Untersuchungen über das Gehirn der Petromyzonten. Z. wiss. Zool. *39:* 191–294 (1883).

AKER, F.D.: A study of hematic barriers in peripheral nerves of albino rabbits. Anat. Rec. *174:* 21–38 (1972).

ALLEN, WM.F.: Studies on the spinal cord and medulla of cyclostomes with special reference to the formation and expansion of the roof plate and the flattening of the spinal cord. J. comp. Neurol. *26:* 9–77 (1916).

ANDERSON, F.D.; MEADOWS, I., and CHAMBERS, M.M.: The nucleus marginalis of the mammalian spinal cord. Observations on the spinal cord of cat and man. J. comp. Neurol. *123:* 97–100 (1964).

ANGULO (Y GONZALEZ), A.W.: The motor nuclei in the cervical cord of the albino rat at birth. J. comp. Neurol. *43:* 115–142 (1927).

ANGULO (Y GONZALEZ), A.W.: Is myelinogeny an absolute index of behavioral capacity? J. comp. Neurol. *48:* 459–464 (1929).

ANGULO (Y GONZALEZ), A.W.: The prenatal development of behavior in the albino rat. J. comp. Neurol. *55:* 395–442 (1932).

ARMSTRONG, PH.B. and HIGGINS D.C.: Behavioral encephalization in the Bullhead embryo and its neuroanatomical correlates. J. comp. Neurol. *143:* 371–384 (1971).

AUSTIN, G. (ed): The spinal cord: basic aspects and surgical considerations, 2nd ed. (Thomas, Springfield 1972).

AYERS, H.: Ventral spinal nerves in Amphioxus. J. comp. Neurol. *33:* 155–162 (1921).

BABINSKI, J.: Sur le réflexe cutané plantaire dans certaines affections organiques du système nerveux central. C.R. Soc. Biol. *3:* 207–208 (1896).

BAKER, B.C.: The early development of the ventral part of the neural plate of Amblystoma. J. comp. Neurol. *44:* 1–27 (1927/28).

BANCHI, A.: La minuta struttura della midolla spinale dei Cheloni (Emys europaea). Arch. ital. Anat. Embriol. *2:* 291–307 (1903).

BARBIERI, C.: Richerche sullo sviluppo del midollo spinale negli Anfibi. Arch. zoolog. ital. *2:* 79–105 (1906).

BARNES, J.F. and DAVENPORT, H.A.: Cells and fibers in spinal nerves. III. Is a 1:1 ratio in the dorsal roots the rule? J. comp. Neurol. *66:* 459–469 (1937).

BARRON, D.H.: The early development of the sensory and internuncial cells in the spinal cord of the sheep. J. comp. Neurol. *81:* 193–225 (1944).

BAYLISS, W.M.: Principles of general physiology (Longmans, London 1924).

BEARD, J.: On the early development of Lepidosteus osseus. Proc. roy. Soc. London *46:* 108–118 (1889).

BEARD, J.: The transient ganglion cells and their nerves in Raja batis. Anat. Anz. *7:* 191–206 (1892).

BEARD, J.: The history of a transient nervous apparatus in certain Ichthyopsida. An account of the development and degeneration of ganglion cells and nerve fibres. Part I. Raja batis. Zool. Jahrb., Abt. f. Anat. *9:* 319–426 (1896).

BECCARI, N.: Ricerche sulle cellule e fibre del *Mauthner* e sulle loro connessioni in pesci ed anfibi. Arch. ital. Anat. Embriol. *6:* 660–705 (1907).

BECCARI, N.: Neurologia Comparata (Sansoni, Firenze 1943).

BECHTEREW, W.v.: Die Leitungsbahnen im Gehirn und Rückenmark. Deutsch von *R. Weinberg* (Georgi, Leipzig 1899).

BERN, H. and HAGADORN, I.R.: A comment on the Elsamobranch caudal neurosecretory system; in GORBMAN Comparative endocrinology, pp. 725–727 (Wiley, New York 1959).

Best, C.H. and Taylor, N.B.: The physiologic basis of medical practice. A text in applied physiology; 5th ed. (Williams & Wilkins, Baltimore 1950).

Biach, P.: Das Rückenmark der Ungulaten. Arb. neurol. Inst. Univ. Wien *16:* 485–522 (1907).

Biggart, J.H.: Pathology of the nervous system. A student's introduction, 2nd ed. (Livingstone, Edinburgh 1949).

Bing, R.: Kompendium der topischen Gehirn- und Rückenmarksdiagnostik, 5. Aufl. (Urban & Schwarzenberg, Berlin 1922).

Blasius, W.: Rückenmark; in Landois-Rosemann Lehrbuch der Physiologie des Menschen, 28. Aufl., Bd. 2, pp. 663–694 (Urban & Schwarzenberg, München 1962).

Blinkov, S.M. and Glezer, I.I.: The human brain in figures and tables. A quantitative handbook (Plenum Press, New York 1968).

Bodian, D.: Neuron junctions: A revolutionary decade. Anat. Rec. *174:* 73–82 (1972).

Boeke, J.: Das Infundibularorgan im Gehirne des Amphioxus. Anat. Anz. *32:* 473–488 (1908).

Boeke, J.: Neue Beobachtungen über das Infundibularorgan im Gehirne des Amphioxus und das homologe Organ des Craniotengehirns. Anat. Anz. *44:* 460–477 (1913).

Bok, S.T.: Das Rückenmark; in v. Möllendorff Handbuch der mikroskopischen Anatomie des Menschen, vol. IV/1, pp. 478–578 (Springer, Berlin 1928).

Bolk, L.: Über die Neuromerie des embryonalen menschlichen Rückenmarks. Anat. Anz. *28:* 204–206 (1906).

Bone, Qu.: The central nervous system in Amphioxus. J. comp. Neurol. *115:* 27–64 (1960).

Bone, Qu.: The central nervous system; in Brodal and Fänge The biology of Myxine, pp. 50–51 (Universitetsforlaget, Oslo 1963).

Bonin, G. v.: Essay on the cerebral cortex (Thomas, Springfield 1950).

Bossz, J.G. and Ferratier, R.: Studies of the spinal cord of Galago senegalensis, compared with that in man. J. comp. Neurol. *132:* 485–498 (1968).

Brain, W.R.: Diseases of the nervous system, 7th ed., revised by the late Lord Brain and Walton, J.N. (Oxford University Press, London 1969).

Brodal, A. and Kaada, B.R.: Cutaneous and proprioceptive impulses in the pyramidal tract in cats. Acta physiol. scand. *29:* 131–132 (1953).

Brodal, A. and Kaada, B.R.: Exteroceptive and proprioceptive ascending impulses in pyramidal tract of cat. J. Neurophysiol. *16:* 567–586 (1953).

Brodal, A. and Walberg, F.: Ascending fibers in the pyramidal tract of the cat. Arch. Neurol. Psychiat. *68:* 755–775 (1952).

Brouwer, B.: Die biologische Bedeutung der Dermatomerie. Beitrag zur Kenntnis der Segmentalanatomie und der Sensibilitätsleitung im Rückenmark und in der Medulla oblongata. Fol. neurobiol. *9:* 225–236 (1915).

Bullock, T.H. and Horridge, G.A.: Structure and function in the nervous system of invertebrates. 2 vols. (Freeman, San Francisco 1965).

Burckhardt, R.: Histologische Untersuchungen am Rückenmark der Tritonen. Arch. mikr. Anat. *34:* 131–156 (1889).

Burckhardt, R.: Untersuchungen am Hirn und Geruchsorgan von Triton und Ichthyophis. Z. wiss. Zool. *52:* 369–403 (1891).

Burckhardt, R.: Das Centralnervensystem von Protopterus annectens (Friedländer, Berlin 1892).

Burlet, H.M. de: Het aderlijke stelsel der Bradypodidae. Handl. XIII de Ned. en Gen.

Congres, Groningen, Tweede Sectie, p. 268 (1911). Quoted after KAPPERS (1920, 1936, 1947).

BURR, H. S.: The central nervous system of Orthagoriscus mola. J. comp. Neurol. *45:* 33–128 (1928).

CAJAL, S. R.: Histologie du système nerveux de l'homme et des vertébrés. 2 vols. (Maloine, Paris 1909, 1911; Instituto Ramon y Cajal, Madrid 1952, 1955).

CAJAL, S. R.: Recuerdos de mi vida, 3rd ed. (Pueyo, Madrid 1923).

CARLSSON, A.; FALCK, B.; FUXE, K., and HILLARP, N. A.: Cellular localization of monoamines in the spinal cord. Acta physiol. scand. *60:* 112–119 (1963).

CHANDLER, A. C.: On a lymphoid structure lying over the myelencephalon of Lepidosteus. Univers. Calif. Publ. Zool. vol. 9: 85–104 (1911).

CLARA, M.: Das Nervensystem des Menschen, 3. Aufl. (Barth, Leipzig 1959).

CLARKE, J. L.: Researches into the structure of the spinal cord. Philos. Trans. London *141:* 607–621 (1851).

CLARKE, J. L.: Further researches on the grey substance of the spinal cord. Philos. Trans. London *149:* 437–467 (1859).

COGHILL, G. E.: The primary ventral roots and somatic motor column of Amblystoma. J. comp. Neurol. *23:* 121–143 (1913).

COGHILL, G. E.: Correlated anatomical and physiological studies on the growth of the nervous system of Amphibia. I. The afferent system of the Trunk of Amblystoma. J. comp. Neurol. *24:* 161–233 (1914).

COGHILL, G. E.: Anatomy and the problem of behavior (Cambridge University Press, Cambridge 1929).

COGHILL, G. E.: Somatic myogenic action in embryos of Fundulus heteroclitus. Proc. Soc. exp. Biol. Med. *31:* 62–64 (1933).

COGHILL, G. E.: Early embryonic somatic movements in birds and in mammals other than man. Monogr. Soc. Res. Child Devel. *5:* No. 2 (1940).

COLLIER, J. and BUZZARD, F.: Descending mesencephalic tracts in cat, monkey, and man; Monakow's bundle; the dorsal longitudinal bundle; the ventral longitudinal bundle; the ponto-spinal tracts, lateral and ventral; the vestibulo-spinal tract; the central tegmental tract (centrale Haubenbahn); descending fibers of the fillet. Brain *24:* 177–221 (1901).

CRAIGIE, E. H.: An introduction to the finer anatomy of the central nervous system based upon that of the albino Rat (Blakiston, Philadelphia 1925).

CROSBY, E. C.; HUMPHREY, T., and LAUER, E. W.: Correlative anatomy of the nervous system (MacMillan, New York 1962).

CUNNINGHAM, D. J.: The spinal nervous system of the porpoise and dolphin. J. Anat. *11:* 209–228 (1877).

DAHLGREN, U.: The giant ganglion cells in the spinal cord of the order Heterosomata Cope (Acacanthini Pleuronectoidi Guenther). Anat. Anz. *13:* 281–293 (1897).

DAHLGREN, U.: The giant ganglion cell apparatus. J. comp. Neurol. *8:* 177–179 (1898).

DAHLGREN, U.: On the electric motor nerve centers in the skates (Rajidae). Science *40:* 862–863 (1914).

DAVIDOFF, R. A.: Gamma-aminobutyric acid antagonism and presynaptic inhibition in the Frog spinal cord. Science *175:* 331–333 (1972).

DEITERS, O. F. C.: Untersuchungen über Hirn und Rückenmark der Säugethiere. M. SCHULTZE, ed. (Vieweg, Braunschweig 1865).

Dendy, A.: On the parietal sense organ and associated structures in the New Zealand Lamprey (Geotria australis). Quart. J. micr. Sci. *51:* 1–29 (1907).

Dexler, H.: Zur Anatomie des Zentralnervensystems von Elephas indicus. Arb. neurol. Inst. Univ. Wien *15:* 137–181 (1907).

Dexler, H. and Margulies, A.: Über die Pyramidenbahn des Schafes und der Ziege. Morph. Jb. *35:* 413–449 (1906).

Donald, D.: On the incidence and locations of nerve cells in the spinal white matter of two species of primates, man and the cynomologous monkey. J. comp. Neurol. *99:* 103–115 (1953).

Donaldson, H.H. and Davis, D.J.: Description of charts showing areas of the cross section of the human spinal cord at the level of each nerve. J. comp. Neurol. *13:* 19–40 (1903).

Dräseke, J.: Über einen bisher nicht beobachteten Nervenkern (Hofmann-Kölliker) im Rückenmark bei Chiropteren. Anat. Anz. *23:* 571–576 (1903).

Dräseke, J.: Zur mikroskopischen Kenntnis der Pyramidenkreuzung der Chiropteren. Anat. Anz. *23:* 449–456 (1906).

Duckett, S. and Pearse, A.G.E.: Histoenzymology of the developing human spinal cord. Anat. Rec. *163:* 59–66 (1969).

Duncan, D. and Keyser, L.L.: Some determination of the ratio of nerve fibers to nerve cells in the thoracic dorsal roots and ganglia of the cat. J. comp. Neurol. *64:* 303–311 (1936).

Duncan, D. and Keyser, L.L.: Further determinations of the number of fibers and cells in the dorsal roots and ganglia of the cat. J. comp. Neurol. *68:* 479–490 (1938).

Earle, K.M.: The tract of Lissauer and its possible relation to the pain pathway. J. comp. Neurol. *96:* 93–111 (1952).

Easton, Th.A.: On the normal use of reflexes. Amer. Scientist *60:* 591–599 (1972).

Eberl-Rothe, G.: Über den Reissnerschen Faden der Wirbeltiere. Z. mikr.-anat. Forsch. *57:* 137–180 (1951).

Eccles, J.C.: The inhibitory pathways of the central nervous system (Thomas, Springfield 1969).

Eccles, J. and Schadé, J. P., eds.: Organization of the spinal cord. Progr. in Brain Res., vol. 11 (Elsevier, Amsterdam 1964).

Edinger, L.: Einiges vom 'Gehirn' des Amphioxus. Anat. Anz. *28:* 417–428 (1906).

Edinger, L.: Vorlesungen über den Bau der nervösen Zentralorgane des Menschen und der Tiere. I. Das Zentralnervensystem des Menschen und der Säugetiere. II. Vergleichende Anatomie des Gehirns (Vogel, Leipzig 1911, 1908).

Edinger, L.: Einführung in die Lehre vom Bau und den Verrichtungen des Nervensystems (Vogel, Leipzig 1912).

Elders, C.: Die motorischen Centren und die Form des Vorderhorns in den fünf letzten Segmenten des Cervikalmarkes und dem ersten Dorsalsegmente eines Mannes, der ohne linken Vorderarm geboren ist. Mschr. Psychiat. Neurol. *28:* 491–509 (1910).

Elliott, H.C.: Studies on the motor cells of the spinal cord. I. Distribution in the normal human cord. Amer. J. Anat. *70:* 95–117 (1942).

Elliott, H.C.: Studies on the motor cells of the spinal cord. IV. Distribution in experimental animals. J. comp. Neurol. *81:* 97–103 (1944).

Elliott, H.C.: Studies on the motor cells of the spinal cord. V. Position and extent of lesions in the nuclear pattern of chronic and convalescent poliomyelitis patients. Amer. J. Path. *21:* 87–97 (1945).

ELLIOTT, H.C.: Textbook of neuroanatomy (Lippincott, Philadelphia 1963).

EMMERT, A.G.F.: Beobachtungen über einige anatomische Eigenheiten der Vögel. Arch. Physiol. *10:* 377–392 (1811). Quoted after WELSCH and WÄCHTLER (1969).

ENAMI, M.: The morphology and functional significance of the caudal neurosecretory system of fishes; in GORBMAN Comparative endocrinology, pp. 697–724 (Wiley, New York 1959).

ENAMI, M. and IMAI, K.: Caudal neurosecretory system in several freshwater teleosts. Endocrinol. japon. *2:* 107–116 (1955).

ESCOLAR, J.: The afferent connections of the 1st, 2nd, and 3rd cervical nerves in the cat. An analysis by Marchi and Rasdolsky methods. J. comp. Neurol. *89:* 79–92 (1948).

FAVARO, G.: Contributi allo studio morfologico dell'ipofisi caudale (rigonfiamento della medulla spinale) dei teleostei. Atti r. Acad. Lincei, Mem. cl. sc. Ser. 6, vol. 1: 29–72 (1925).

FEARING, F.: Reflex action. A study in the history of physiological psychology (Williams & Wilkins, Baltimore 1930).

FINK, R.P. and HEIMER, L.: Two methods for selective silver impregnation of degenerating axons and their synaptic endings in the central nervous system. Brain Res. *4:* 369–374 (1967).

FLECHSIG, P.: Die Leitungsbahnen im Gehirn und Rückenmark des Menschen auf Grund entwicklungsgeschichtlicher Untersuchungen (Engelmann, Leipzig 1876).

FLECHSIG, P.: Gehirn und Seele, 2. Aufl. (Veit, Leipzig 1896).

FLECHSIG, P.: Anatomie des menschlichen Gehirns und Rückenmarks auf myelo-genetischer Grundlage. (Thieme, Leipzig 1920).

FLECHSIG, P.: Meine myelogenetische Hirnlehre (Springer, Berlin 1927).

FLOOD, P.R.: A peculiar mode of muscular innervation in Amphioxus. Light and electron microscopic studies of the so-called ventral roots. J. comp. Neurol. *126:* 181–218 (1966).

FOX, M.W.; INMAN, O.R., and HIMWICH, W.A.: The postnatal development of the spinal cord of the dog. J. comp. Neurol. *130:* 233–240 (1967).

FRANZ, V.: Haut, Sinnesorgane und Nervensystem der Akranier. (Fauna et anatomia ceylanica nr. 13). Jen Z. Naturw. *59:* 401–526 (1923).

FREUD, S.: Über den Ursprung der hinteren Nervenwurzeln im Rückenmark von Ammocoetes (Petromyzon Planeri). S. Ber. Kaiserl. Akad. Wiss., Wien, math.-nat. Cl. *75:* 15–27 (1877).

FREUD, S.: Über Spinalganglien und Rückenmark des Petromyzon. S. Ber. Kaiserl. Akad. Wiss., Wien, math.-nat. Cl. *78:* 81–167 (1879).

FRIEDLÄNDER, A.: Untersuchungen über das Rückenmark und das Kleinhirn der Vögel. Neurol. Centralbl. *17:* 351–359; 397–409 (1898).

FRITSCH, G.: Über den Angelapparat des Lophius piscatorius. S. Ber. k. preuss. Akad. Wiss. *2:* 1145–1149 (1884).

FRITSCH, G.: Über einige bemerkenswerte Elemente des Centralnervensystems von Lophius. Arch. mikr. Anat. *27:* 13–31 (1886).

FUKUYAMA, U.: On cytoarchitectural lamination of the spinal cord in the albino rat (Abstract). Anat. Rec. *121:* 396 (1955).

GASSER, H.S.: Pain-producing impulses in peripheral nerves. Assn. Res. nerv. ment. Dis. Proc. *23:* 44–62 (1943).

GAULE, J.: Zahl und Vertheilung der markhaltigen Fasern im Froschrückenmark. Abh. mathem.-phys. Cl. d. Kgl. sächs. Ges. d. Wiss. *15:* 737–780 (1889).

Gaule, J. und Lewin, T.: Über die Zahlen der Nervenfasern und Ganglienzellen in den Spinalganglien des Kaninchens. Zbl. Physiol. *10:* 437–440, 465–471 (1897).

Gaupp, E.: *Eckers* und *Wiedersheims* Anatomie des Frosches. 2. Abt. (Vieweg, Braunschweig 1899).

Gegenbaur, C.: Vergleichende Anatomie der Wirbelthiere mit Berücksichtigung der Wirbellosen. 2 vols. (Engelmann, Leipzig 1898–1901).

Gehuchten, A. van: Les cellules de Rohon dans la moelle épinière et la moelle allongée de la truite (Trutta fario). Bull. Acad. roy. Sci. Belg. *30:* 495–519 (1895).

Gehuchten, A. van: La moelle épinière de la truite (Trutta fario). Cellule *11:* 111–174 (1895).

Gehuchten, A. van: Contribution à l'étude des cellules dorsales (Hinterzellen) de la moelle épinière des vertébrés inférieurs. Bull. Acad. roy. Sci. Belg. *34:* 24–38 (1897).

Gehuchten, A. van: Contribution à l'étude de la moelle épinière chez les vertébrés (Tropidonotrix natrix). Cellule *12:* 115–165 (1897).

Gehuchten, A. van: La moelle épinière des larves des Batraciens (Salamandra maculosa). Arch. Biol. *15:* 599–619 (1898).

Gilbert, G. J.: The subcommissural organ. Anat. Rec. *126:* 253–264 (1956).

Glees, P.: Der periphere und zentrale Anteil des sympathischen Nervensystems der Selachier. Acta neerl. morphol. *3:* 209–248 (1940).

Goldby, F. and Robinson, L. R.: The central connexions of dorsal spinal nerve roots and the ascending tracts in the spinal cord of Lacerta viridis. J. Anat. *96:* 153–170 (1962).

Golgi, C.: Untersuchungen über den feineren Bau des zentralen und peripherischen Nervensystems (Fischer, Jena 1894).

Goronowitsch, N.: Das Gehirn und die Cranialnerven von Acipenser ruthenus. Morph. Jb. *13:* 427–574 (1888).

Gottlieb, G. (ed.): Behavioral embryology. Vol. 1. Studies on the development of behavior and the nervous system (Academic Press, New York 1973).

Graf, W.: Caliber spectra of nerve fibers in the pigeon (Columba domestica). J. comp. Neurol. *105:* 335–364 (1956).

Granit, R.: Mechanisms regulating the discharge of motoneurons (Thomas, Springfield 1972).

Groebbels, F.: Die Lage- und Bewegungsreflexe der Vögel. VII. Die Lage- und Bewegungsreflexe der Haustaube nach Läsionen des Rückenmarks und der Oblongata. Arch. ges. Physiol. *218:* 198–208 (1927).

Häggqvist, G.: Analyse der Faserverteilung in einem Rückenmarkquerschnitt (Th 3). Z. mikr. anat. Forsch. *39:* 1–34 (1936).

Häggqvist, G.: Faseranalytische Studien über die Pyramidenbahn. Acta psychiat. neurol. *12:* 457–466 (1937).

Hall, M.: On the reflex function of the medulla oblongata and medulla spinalis. Philos. Trans. *123:* 635–665 (1833).

Hamburger, V.: Anatomical and physiological basis of embryonic motility in birds and mammals; in Gottlieb Behavioral embryology, pp. 52–76 (Academic Press, New York 1973).

Hamburger, V.; Wenger, E., and Oppenheim, R.: Motility in the chick embryo in the absence of sensory input. J. exp. Zool. *162:* 133–160 (1966).

Hardesty, I.: Observations on the medulla spinalis of the elephant with some comparative studies on the intumescentia cervicalis and the neurones of the columna anterior. J. comp. Neurol. *12:* 125–182 (1902).

HARRISON, R.G.: Über die Histogenese des peripheren Nervensystems bei Salmo salar. Arch. mikr. Anat. *57:* 354–444 (1901).

HATSCHEK, R.: Über das Rückenmark des Delphins (Delphinus delphis). Arb. neurol. Inst. Univ. Wien *4:* 286–312 (1896).

HAYMAKER, W.: *Bing's* local diagnosis in neurological diseases. 14th and 15th ed. (Mosby, Saint Louis 1956, 1969).

HELD, H.: Die centralen Bahnen des Nervus acusticus bei der Katze. Arch. Anat. Physiol. (anat. Abt.) *1891:* 271–288 (1891).

HELD, H.: Die centrale Gehörleitung. Arch. Anat. Physiol. (anat. Abt.) *1893:* 201–248 (1893).

HELD, H.: Die Entwicklung des Nervengewebes bei den Wirbeltieren (Barth, Leipzig 1909).

HELWEG, H.K.S.: Studien über den centralen Verlauf der vasomotorischen Nervenbahnen. Arch. Psychiat. Nervenkr. *19:* 104–183 (1888).

HERRICK, C.J.: An introduction to neurology, 5th ed. (Saunders, Philadelphia 1931).

HERRICK, C.J.: The medulla oblongata of Necturus. J. comp. Neurol. *50:* 1–96 (1930).

HERRICK, C.J. and COGHILL, G.E.: The development of reflex mechanisms in Amblystoma. J. comp. Neurol. *25:* 65–85 (1915).

HIRASAWA, K.: Chûsû shinkeikei, toku ni sekizuikaihakushitsu no sôteki kôzô ni tsuite ('concerning laminar structural arrangements in the central nervous system, particularly in the gray substance of the spinal cord' – only Japanese Text). Kyôto Igakkai Zasshi *3:* 428–438 (1955).

HIRT, A.: Über den Aufbau des Spinalganglions und seine Beziehungen zum Sympathicus. Z. ges. Anat., Abt.1 *87:* 275–318 (1928).

HIS, W.: Zur Geschichte des menschlichen Rückenmarks und der Nervenwurzeln. Abh. math.-phys. Cl. kgl. sächs. Ges. Wiss. *13:* 476–544 (1886).

HIS, W.: Die Entwicklung des menschlichen Gehirns während der ersten drei Monate (Hirzel, Leipzig 1904).

HOFFMANN, P.: Untersuchungen über die Eigenreflexe (Sehnenreflexe) menschlicher Muskeln (Springer, Berlin 1922).

HOFFMANN, P.: Die physiologischen Eigenschaften der Eigenreflexe. Ergebn. Physiol. *36:* 15–108 (1934).

HOGG, I.D.: The development of the nucleus dorsalis (Clarke's column). J. comp. Neurol. *81:* 69–95 (1944).

HOOKER, D.: The prenatal origin of behavior (University of Kansas Press, Lawrence 1952).

HOOKER, D.: Early human fetal behavior, with a preliminary note on double simultaneous fetal stimulation. Proc. Assn. Res. nerv. ment. Dis. *33:* 98–113 (1954).

HORST, C.J. VAN DER: The myelencephalic gland of Polyodon, Acipenser and Amia. Proceed. kon. Akad. van Wetensch. Amsterdam *28:* 432–442 (1925).

HUBER, J.F.: Nerve roots and nuclear groups in the spinal cord of the pigeon. J. comp. Neurol. *65:* 43–90 (1936).

HUGHES, A.: The development of the primary sensory system in Xenopus laevis *(Daudin)*. J. Anat. *91:* 323–338 (1957).

HUMPHREY, T.: Primitive neurons in the embryonic human central nervous system. J. comp. Neurol. *81:* 1–45 (1944).

JACOBSOHN, L.: Über die Kerne des Rückenmarkes. Neurol. Centralbl. *27:* 617–626 (1908). Abh. preuss. Akad. Wiss., Berlin, pp. 1–76 (1908).

JACOBSOHN, L.: Über die Kerne des menschlichen Rückenmarks. Anhang z. d. Abh. Kgl. preuss. Akad. Wiss. phys.-math. Kl. *1908:* 1–72 (1909).

JANE, J.A.; CAMPBELL, C.B.G., and YASHON, D.: Pyramidal tract: a comparison of two prosimian primates. Science *147:* 153–155 (1965).

JANSEN, J. und BRODAL, A.: Das Kleinhirn; in BARGMANN Handbuch der mikroskopischen Anatomie des Menschen, Erg. zu Bd. IV/1, Nervensystem (Springer, Berlin 1958).

JOHNELS, A.G.: On the peripheral autonomic nervous system of the trunk region of Lampetra Planeri. Acta zool. *37:* 251–286 (1956).

JOHNELS, A.G.: On the dorsal ganglion cells of the spinal cord in Lampreys. Acta zool. *39:* 201–226 (1958).

JOHNSON, J.I. jr.; WELKAR, W.I., and PUBOLS, B.H. jr.: Somatotopic organization of Raccoon dorsal column nuclei. J. comp. Neurol. *132:* 1–43 (1968).

JOHNSTON, J.B.: The giant cells of Catostomus and Coregonus. J. comp. Neurol. *10:* 375–381 (1900).

JOHNSTON, J.B.: The cranial and spinal ganglia, and the viscero-motor roots in Amphioxus. Biol. Bull. *9:* 112–127 (1905).

JOHNSTON, J.B.: The nervous system of vertebrates (Blakiston, Philadelphia 1906).

JORDAN, H.: Concerning Reissner's fiber in Teleosts. J. comp. Neurol. *30:* 217–227 (1918/19).

JOSEPH, B.S. and WHITLOCK, D.G.: Central projections of selected dorsal roots in Anuran Amphibians. Anat. Rec. *160:* 279–288 (1968).

JOSEPH, B.S. and WHITLOCK, D.G.: Central projections of brachial and lumbar dorsal roots in Reptiles. J. comp. Neurol. *132:* 469–484 (1968).

JOSEPH, H.: Über eigentümliche Zellstrukturen im Zentralnervensystem von Amphioxus. Verh. Anat. Ges., Eng. H. Anat. Anz. *25:* 16–26 (1904).

KANEMITSU, A.: Relation entre la taille des neurones et leur époque d'apparition dans la moelle épinière chez le poulet. Etude autoradiographique et caryométrique. Proc. Japan Acad. *47:* 422–437 (1971).

KANEMITSU, A.: Histogenesis of the chick spinal cord *(Japanese)*. Shinkei Kenkyû no shinpo (Progress in neurologic Research) *16:* 379–388 (1972a).

KANEMITSU, A.: Etude quantitative de la neurogénèse dans la moelle épinière chez le poulet par l'autoradiographie. Proc. Japan Acad. *48:* 758–763 (1972b).

KAO, F.F.: Acupuncture therapeutics. An introductory text (Eastern Press, New Haven 1973).

KAPPERS, C.U.ARIËNS: Die vergleichende Anatomie des Nervensystems der Wirbeltiere und des Menschen, 2 vols. (Bohn, Haarlem 1920–1921).

KAPPERS, C.U.ARIËNS: The lumbo-sacral sinus in the spinal cord of birds and its histological constituents. Psych. en neurol. Bl. *28:* 405–415 (1924).

KAPPERS, C.U.ARIËNS: Anatomie comparée du système nerveux particulièrement de celui des mammifères et de l'homme. Avec la collaboration de E.H. STRASBURGER (Masson, Paris 1947).

KAPPERS, C.U.ARIËNS und HAMMER, E.: Das Zentralnervensystem des Ochsenfrosches (Rana catesbyana). Psych. en neurol. Bladen, Amsterdam, *Feestbundel Winkler:* 368–415 (1918).

KAPPERS, C.U.ARIËNS; Huber, G.C., and CROSBY, E.C.: The comparative anatomy of the nervous system in vertebrates, including man, 2 vols. (Macmillan, New York 1936).

KARTEN, H.J.: Ascending pathways from the spinal cord in the pigeon (Columba livia). Abstr. Proc. 16. int. Congr. Zool. *2:* 23 (1963).

KEENAN, E.: The phylogenetic development of the Substantia gelatinosa Rolandi. Part I. Fishes. Part II. Amphibians, Reptiles, and Birds. Part III. Mammals, General résumé, and conclusions. Proc. kon. Akad. Wetensch. Amsterdam *31:* 837–854; *32:* 300–311, 466–475 (1929).

KENNARD, D.W.: The anatomical organization of neurons in the lumbar region of the spinal cord of the Frog (Rana temporaria). J. comp. Neurol. *111:* 447–467 (1959).

KENNARD, M.A.: The course of ascending fibres in the spinal cord of the cat essential to the recognition of painful stimuli. J. comp. Neurol. *100:* 511–524 (1954).

KESWANI, N.H. and HOLLINSHEAD, W.H.: Localization of the phrenic nucleus in the spinal cord of man. Anat. Rec. *125:* 683–699 (1956).

KIRSCHE, W.: Zur funktionellen Morphologie und Architektonik besonderer synaptischer Apparate in Gehirn von Knochenfischen. J. Hirnforsch. *9:* 3–61 (1967).

KÖLLIKER, A.: Handbuch der Gewebelehre des Menschen. Bd. II/1, 2 (Engelmann, Leipzig 1893, 1896).

KOLMER, W. und MARBURG, O.: Studien über die erste Entwicklung des Zentralkanals des Menschen. Z. Anat. EntwGesch. *89:* 54–70 (1929).

KOLSTER, R.: Über bemerkenswerte Ganglienzellen im Rückenmark von Perca fluviatilis. Anat. Anz. *14:* 250–253 (1898).

KRAUSE, R.: Mikroskopische Anatomie der Wirbeltiere in Einzeldarstellungen. I. Säugetiere. II. Vögel und Reptilien. II. Amphibien. IV. Teleostier, Plagiostomen, Zyklostomen und Leptokardier (De Gruyter, Berlin 1921, 1922, 1923).

KUHLENBECK, H.: Vorlesungen über das Zentralnervensystem der Wirbeltiere (Fischer, Jena 1927).

KUHLENBECK, H.: The human diencephalon. A summary of development, structure, function, and pathology (Karger, Basel 1954).

KUHLENBECK, H.: Brain and consciousness. Some prolegomena to an approach of the problem (Karger, Basel 1957).

KUHLENBECK, H.: Weitere Bemerkungen zur Maschinentheorie des Gehirns. Confin. neurol. *27:* 295–328 (1966).

KUHLENBECK, H.: Gehirn und Bewusstsein. Translated by Prof. J. GERLACH and Dr. U. PROTZER. Erfahrung und Denken: Schriften zur Förderung der Beziehungen zwischen Philosophie und Einzelwissenschaften, Bd. 39 (Duncker & Humblot, Berlin 1973).

KUHLENBECK, H. and NIIMI, K.: Further observations on the morphology of the brain in the Holocephalian Elasmobranchs Chimaera and Callorhynchus. J. Hirnforsch. *11:* 267–314 (1969).

KUO, Z.Y.: (1932–1938), Ontogeny of embryonic behavior in birds. I, J. exp. Zool. *61:* 395–430 (1932a); II, J. exp. Zool. *62:* 453–487 (1932b); III, J. comp. Psychol. *13:* 245–271 (1932c); IV, J. comp. Psychol. *14:* 109–122 (1932d); V, Psychol. Rev. *39:* 499–515 (1932e); VI, J. comp. Psychol. *16:* 379–384 (1933); X, (with T.C. SHEN) J. comp. Psychol. *21:* 87–93 (1936); XI, (with T.C. SHEN) J. comp. Psychol. *24:* 49–58 (1937); XII, Amer. J. Psychol. *51:* 361–379 (1938).

KUPFFER, K. VON: Die Morphogenie des Centralnervensystems; in HERTWIG Handbuch der vergleichenden und experimentellen Entwicklungslehre der Wirbeltiere, 2. Bd., 3. Teil, pp. 1–272 (Fischer, Jena 1906).

LACHI, P.: Alcune particolarita anatomiche del rigonfiamento sacrale nel midollo degli Ucelli. Lobi accessori. Atti Soc. tosc. Sc. nat. *10:* 268–295 (1889).

LANGWORTHY, O.R.: Carnegie Contrib. Embryol. 24, No. 228: 145–157 (1933).

LASSEK, A.M.: The pyramidal tract. Its status in medicine (Thomas, Springfield 1954).

LEGHISSA, S.: Contribution ultérieure à une meilleure connaissance de l'appareil de Mauthner chez les Poissons et observations sur la morphologie de la fibre. Progr. Neurobiol., pp. 45–62 (Elsevier, Amsterdam 1956).

LENHOSSÉK, M. VON: Über die Pyramidenbahnen im Rückenmarke einiger Säugetiere. Anat. Anz. *4:* 208–219 (1889).

LENHOSSÉK, M. VON: Über Nervenfasern in den hinteren Wurzeln, welche aus dem Vorderhorn entspringen. Anat. Anz. *5:* 360–362 (1890).

LENHOSSÉK, M. VON: Beobachtungen an den Spinalganglien und dem Rückenmark von Pristiurusembryonen. Anat. Anz. *7:* 519–539 (1892).

LENHOSSÉK, M. VON: Beiträge zur Histologie des Nervensystems und der Sinnesorgane (Bergmann, Wiesbaden 1894).

LENHOSSÉK, M. VON: Der feinere Bau des Nervensystems im Lichte neuester Forschungen, 2. Aufl. (Fischers med. Buchhandlg., Berlin 1895).

LEVI, G.: Trattato di istologia, 3rd. ed. (Unione Tipografica, Torino 1946).

LHERMITTE, J.: Les fondements biologiques de la psychologie (Gauthier-Villars, Paris n.d., ca. 1924).

LICHTENSTEIN, B.W.: A textbook of neuropathology (Saunders, Philadelphia 1949).

LINOWIECKI, A. jr.: The comparative anatomy of the pyramidal tract. J. comp. Neurol. *24:* 509–530 (1914).

LISSAUER, H.: Beitrag zur Pathologie der Tabes dorsalis und zum Faserverlauf im menschlichen Rückenmark. Neurol. Centralbl. *4:* 245–246 (1885).

LLOYD, D.P.C.: A direct central inhibitory action of dromically conducted impulses. J. Neurophysiol. *4:* 184–190 (1941).

LLOYD, D.P.C.: Neuron patterns controlling transmission of ipsilateral hind-limb reflexes in cats. J. Neurophysiol. *6:* 293–315 (1943).

LLOYD, D.P.C.: Facilitation and inhibition of spinal motoneurons. J. Neurophysiol. *9:* 421–438 (1946).

LLOYD, D.P.C.: On the monosynaptic reflex interconnections of hind-limb muscles; in TOWER and SCHADÉ Structure and function of the cerebral cortex, pp. 289–297 (Elsevier, Amsterdam 1960).

LUCAS, K.: The conduction of the nervous impulse (Longmans, London 1917).

LYSER, K.M.: Early differentiation of the Chick embryo spinal cord in organ culture: light and electron microscopy. Anat. Rec. *169:* 45–64 (1971).

MANNEN, H.: Noyau fermé et noyau ouvert. Arch. ital. Biol. *98:* 333–350 (1960).

MARINESCO, G.: Les localisations médullaires chez le chien et chez l'homme. Sem. méd. *29:* 225–231 (1904).

MASSAZZA, A.: La citoarchitettonica del midollo spinale humane. Arch. Anat. Histol. Embryol. *1:* 323–410 (1922); *2:* 1–56 (1923); *3:* 115–189 (1924).

MATTHEWS, M.A.; Willis, W.D., and WILLIAMS, V.: Dendrite bundles in lamina IX of cat spinal cord: a possible source for electrical interaction between motoneurons? Anat. Rec. *171:* 313–328 (1971).

MAUTHNER, L.: Untersuchungen über den Bau des Rückenmarkes der Fische. S. Ber. Kaiserl. Akad. Wiss. Wien, math.-nat. Cl. *34:* 31–36 (1859).

METTLER, C.C. and METTLER, F.A.: History of medicine (Blakiston, Philadelphia 1947).

Millot, J. et Anthony, J.: Anatomie de Latimeria chalumnae. Tome II. Système nerveux et organes des sens (Editions du Centre National de la Recherche Scientifique, Paris 1965).

Minkowski, M.: Neurobiologische Studien am menschlichen Fötus, in Abderhalden Handbuch der biologischen Arbeitsmethodik, Bd. V, Teil 5B, Heft 5, pp. 511–618 (1928).

Monakow, C. von: Experimenteller Beitrag zur Kenntnis des Corpus restiforme, des äusseren Akustikuskerns und deren Beziehungen zum Rückenmark. Arch. Psychiat. *14:* 1–21 (1883).

Monakow, C. von: Der rote Kern, die Haube und die Regio subthalamica bei einigen Säugetieren und beim Menschen. Arb. hirnanat. Inst. Zürich *4:* 103–225 (1910).

Morin, F. and Catalano, I.V.: Central connections of a cervical nucleus (nucleus cervicalis lateralis of the cat). J. comp. Neurol. *103:* 17–32 (1955).

Müller, J.: Handbuch der Physiologie des Menschen für Vorlesungen, 2 vols. (Hölscher, Coblentz 1834, 1840).

Münzer, E. und Wiener, H.: Experimentelle Beiträge zur Lehre von den endogenen Fasersystemen des Rückenmarkes. Mschr. Psychiat. Neurol. *29:* 1–25 (1910).

Nathan, P.W. and Smith, M.C.: Spinocortical fibers in man. J. Neurol. Neurosurg. Psychiat. *18:* 181–190 (1955).

Nemec, H.: Über die Ausbildung der grauen Substanz im Frosch-Rückenmark. Acta Anat. *13:* 101–118 (1951).

Nicholls, G.E.: The structure and development of Reissner's fiber and the subcommissural organ. Part I. Quart. J. micr. Sci. *58:* 1–116 (1912).

Nicholls, G.E.: Some experiments on the nature and functions of Reissner's fiber. J. comp. Neurol. *27:* 117–200 (1917).

Noback, C.R. and Harting, J.K.: Spinal cord (spinal medulla); in Hofer *et al.* Primatologia, vol. II, Teil 2, Lieferung 2 (Karger, Basel 1971).

Noble, G.K.: The biology of the amphibia (McGraw-Hill, New York 1931; Dover, New York 1954).

Obersteiner, H.: Anleitung beim Studium der nervösen Zentralorgane im gesunden und kranken Zustande (Deuticke, Leipzig 1912).

Olsson, R.: Structure and development of Reissner's fibre in the caudal end of Amphioxus and some lower Vertebrates. Acta zool. *36:* 167–198 (1955).

Olsson, R.: The development of Reissner's fibre in the brain of the Salmon. Acta zool. *37:* 236–250 (1956).

Olsson, R.: Studies on the subcommissural organ. Acta zool. *39:* 71–102 (1958).

Olsson, R. and Wingstrand, K.G.: Reissner's fibre and the infundibular organ in Amphioxus – results obtained with Gomori's chrome alum haematoxylin. Univ. Bergen Årbok 1954, naturvit. rekke Nr. 14, Publ. from the biological Station (Griegs, Bergen 1955).

Orr, D.W. and Windle, W.F.: The development of behavior in chick embryos: the appearance of somatic movements. J. comp. Neurol. *60:* 271–285 (1934).

Owsyannikow, P.: Das Rückenmark und das verlängerte Mark des Neunauges. Mem. Acad. Sci. St. Petersbourg 14, no. 4: 1–32 (1903).

Papez, J.W.: Reticulospinal tracts in the cat. Marchi method. J. comp. Neurol. *41:* 365–399 (1926).

Paton, S.: The reactions of the vertebrate embryo to stimulation and the associated changes in the nervous system. Mitt. a. d. zool. Stat. zu Neapel *18:* 535–581 (1905).

PATON, S.: The reactions of the vertebrate embryo and associated changes in the nervous system. Second paper. J. comp. Neurol. *21:* 345–373 (1911).

PEARSON, A. A.: Role of gelatinous substance of spinal cord in conduction of pain. Arch. Neurol. Psychiat. *68:* 515–525 (1952).

PEARSON, R.: The Avian brain (Academic Press, London–New York 1972).

PETERS, A.: The structure of the dorsal root nerves of Amphioxus. An electron microscope study. J. comp. Neurol. *121:* 287–304 (1963).

PETRAS, J. M. and CUMMINGS, J. F.: Autonomic neurons in the spinal cord of the Rhesus monkey: A correlation of the findings of cytoarchitectonics and sympathectomy with fiber degeneration following dorsal rhizotomy. J. comp. Neurol. *146:* 189–218 (1972).

PODHRADSKY, L. VON: Über die Zahl der Spinalganglienzellen und der Hinterwurzelfasern. Z. Anat. *100:* 281–294 (1933).

POLJAK, S. (later spelled POLYAK): Die Struktureigentümlichkeiten des Rückenmarkes bei den Chiropteren. Z. Anat. EntwGesch. *74:* 509–576 (1924).

PREYER, W.: Specielle Physiologie des Embryo (Grieben, Leipzig 1885).

PROBST, M.: Zur Kenntnis der Schleifenschicht und über centripetale Rückenmarksfasern zum Deitersschen Kern, zum Sehhügel und zur Substantia reticularis. Mschr. Psychiat. Neurol. *11:* 3–12 (1902).

RALSTON, H. J. III.: The fine structure of neurons in the dorsal horn of the cat spinal cord. J. comp. Neurol. *132:* 275–301 (1968a).

RALSTON, H. J. III.: Dorsal root projections to dorsal horn neurons in the cat spinal cord. J. comp. Neurol. *132:* 303–329 (1968b).

RANSON, S. W.: The course within the spinal cord of the non-medullated fibers of the dorsal root. A study of Lissauer's tract in the cat. J. comp. Neurol. *23:* 259–281 (1913).

RANSON, S. W.: The tract of Lissauer and the substantia gelatinosa Rolandi. Amer. J. Anat. *16:* 97–126 (1914).

RANSON, S. W.: The anatomy of the nervous system from the standpoint of development and function, 7th ed. (Saunders, Philadelphia 1943).

RANSON, S. W. and BILLINGSLEY, P. R.: The conduction of painful afferent impulses in the spinal nerves. Amer. J. Physiol. *40:* 571–587 (1916).

RASSMUSSEN, A. T.: Some trends in neuroanatomy (Brown, Dubuque 1947).

REDLICH, E.: Die Pathologie der tabischen Hinterstrangserkrankung. Ein Beitrag zur Anatomie und Pathologie der Rückenmarkshinterstränge (Fischer, Jena 1897).

REED, A. F.: The nuclear masses in the cervical spinal cord of *Macaca mulatta*. J. comp. Neurol. *72:* 187–206 (1940).

REISSNER, E.: Beiträge zur Kenntnis vom Bau des Rückenmarkes von Petromyzon fluviatilis. Arch. Anat. Physiol. wiss. Med. *1860:* 545–588 (1860).

RENSHAW, B.: Influence of discharge of motoneurones upon excitation of neighboring motoneurones. J. Neurophysiol. *4:* 167–183 (1941).

RENSHAW, B.: Central effects of centripetal impulses in axons of spinal ventral roots. J. Neurophysiol. *9:* 191–204 (1946).

RETZIUS, G.: Zur Kenntnis des centralen Nervensystems von Amphioxus lanceolatus. Biol. Unters. N.F. *2:* 29–46 (Fischer, Jena 1891a).

RETZIUS, G.: Zur Kenntnis des centralen Nervensystems von Myxine glutinosa. Biol. Unters. N.F. *2:* 47–53 (Fischer, Jena 1891b).

REXED, B.: The cytoarchitectonic organization of the spinal cord in the cat. J. comp. Neurol. *96:* 415–495 (1952).

Rexed, B.: A cytoarchitectural atlas of the spinal cord in the cat. J. comp. Neurol. *100:* 297–379 (1954).

Rexed, B.: Some aspects of the cytoarchitectonics and synaptology of the Spinal cord. Progr. Brain Res., vol. 11, pp. 58–92 (Elsevier, Amsterdam 1964).

Rexed, R. and Brodal, A.: The nucleus cervicalis lateralis. A spino-cerebellar relay nucleus. J. Neurophysiol. *14:* 399–407 (1951).

Rohde, E.: Histologische Untersuchungen über das Nervensystem von Amphioxus lanceolatus. Schneiders Zool. Beitr. *2:* 169–211 (1888).

Rohde, E.: Muskel und Nerv. II. Mermis und Amphioxus. Zool. Beitr. *3:* 161–178 (1892).

Rohon, J.V.: Zur Histogenese des Rückenmarkes der Forelle. S. Ber. math.-phys. Kl. k. bayr. Akad. Wiss. *14:* 39–56 (1884).

Romanes, G.J.: The motor cell columns of the lumbosacral spinal cord of the cat. J. comp. Neurol. *94:* 313–364 (1951).

Romanes, G.J.: The motor pools of the spinal cord; in Eccles and Schadé Organization of the spinal cord, Progr. Brain Res., vol. 11, pp. 93–119 (1964).

Romanoff, A.L.: The avian embryo. Structural and functional development. (Macmillan, New York 1960).

Rovainen, C.M.: Physiological and anatomical studies on large neurons of central nervous system of the sea lamprey (Petromyzon marinus). I. Müller and Mauthner cells. II. Dorsal cells and giant interneurons. J. Neurophysiol. *30:* 1000–1023; 1024–1042 (1967).

Rovainen, C. M.: Synaptic interactions of identified nerve cells in the spinal cord of the Sea Lamprey. J. comp. Neurol. *154:* 189–206 (1974a).

Rovainen, C. M.: Synaptic interactions of reticulospinal neurons and nerve cells in the spinal cord of the Sea Lamprey. J. comp. Neurol. *154:* 207–224 (1974b).

Rovainen, C.U.; Johnson, P.A.; Roach, E.A., and Mankovsky, J.A.: Projections of individual axons in lamprey spinal cord determined by tracings through serial sections. J. comp. Neurol. *149:* 193–201 (1973).

Russell, J.S.R.: A contribution to the study of some of the afferent and efferent fibers of the spinal cord. Brain *21:* 145–179 (1898).

Sala y Pons: Estructura de la medula espinal de los batracios (quoted after Cajal, 1909, 1952).

Sano, Y.: Das kaudale neurosekretorische System bei Fischen. Ergebn. Biol. *24:* 191–212 (1961).

Sano, Y.; Kawamoto, M., and Hamana, K.: Entwicklungsgeschichtliche Untersuchungen am kaudalen neurosekretorischen System von Salmo irideus (Japanese with German summary). Acta anat. nippon. *37:* 117–125 (1962).

Sargent, P.E.: The giant ganglion cells in the spinal cord of Ctenolabrus adspersus. J. comp. Neurol. *8:* 183–194 (1898).

Sargent, P.E.: The giant cells in the spinal cord of Ctenolabrus coeruleus. Anat. Anz. *15:* 212–225 (1899).

Scharf, J.H.: Sensible Ganglien; in Möllendorff v. und Bargmann Handbuch der mikroskopischen Anatomie des Menschen, vol. IV/3 (Springer, Berlin 1958).

Scharrer, E.: The histology of the meningeal myeloid tissue in the Ganoids Amia and Lepisosteus. Anat. Rec. *88:* 291–310 (1944).

Scheibel, M.E. and Scheibel, A.B.: On the recurrent collaterals of spinal motoneurons (Abstract). Anat. Rec. *166:* 372 (1970).

SCHEIBEL, M.E. and SCHEIBEL, A.B.: Inhibition and the Renshaw cell: a structural critique. Brain, Behav. Evol. *4:* 52–93 (1971).

SCHNEIDER, A.: Beiträge zur vergleichenden Anatomie und Entwicklungsgeschichte der Wirbelthiere (Reimer, Berlin 1879).

SCHULTZ, R.; BERKOWITZ, E.C., and PEASE, D.C.: The electron microscopy of the Lamprey spinal cord. J. Morphol. *98:* 251–273 (1956).

SERGI, S.: Studi sul midollo dello cimpanzè. Riv. Antrop. *24:* 301–390 (1920–21).

SERGI, S.: Studi sul midollo spinale dello cimpanzè. IV. I gruppi cellulari miorabdotici. Riv. Anthrop. *27:* 181–281 (1926–27).

SHANTAVEERAPPA, T.R. and BOURNE, G.H.: The 'perineural epithelium', a metabolically active, continuous protoplasmic barrier surrounding peripheral nerve fasciculi. J. Anat. *96:* 527–537 (1962).

SHERRINGTON, CH.: The integrative action of the nervous system (Yale University Press, New Haven 1906, 1947).

SILVER, M.L.: The motoneurons in the spinal cord of the frog. J. comp. Neurol. *77:* 1–39 (1942).

SIRKEN, C.R. and KUHLENBECK, H.: Preliminary computations of the number of motoneurons in the human spinal cord (Abstract). Anat. Rec. *154:* 489 (1966).

SOSA, J.M.: Collateral nerve fibers within septum dorsale of the spinal cord and medulla oblongata and their connections. J. comp. Neurol. *83:* 157–171 (1945).

SPEIDEL, C.C.: Gland cells of internal secretion in the spinal cord of the skates. Publ. 281 Carnegie Institute of Washington (1919).

SPEIDEL, C.C.: Further comparative studies in other fishes of cells that are homologous to the large irregular glandular cells in the spinal cord of the skates. J. comp. Neurol. *34:* 303–317 (1922).

SPILLER, W.G.: A contribution to the study of the pyramidal tract in the central nervous system of man. Brain *22:* 563–574 (1899).

SPRAGUE, J.M.: A study of motor cell localization in the spinal cord of the rhesus monkey. Amer. J. Anat. *82:* 1–26 (1948).

SPRAGUE, J.M.: Motor and propriospinal cells in the thoracic and lumbar ventral spinal cord of the rhesus monkey. J. comp. Neurol. *95:* 103–123 (1951).

STEDING, G.: Experimente zur Morphogenese des Rückenmarks. Acta anat. *49:* 199–231 (1962).

STEER, J.M.: Some observations on the fine structure of rat dorsal spinal nerve roots. J. Anat. *109:* 467–485 (1971).

STEINER, J.: Die Functionen des Centralnervensystems und ihre Phylogenese. Abt. I–IV (Vieweg, Braunschweig 1885, 1888, 1898, 1900).

STIEDA, L.: Über den Bau des Rückenmarkes der Rochen und Haie. Z. wiss. Zool. *23:* 435–442 (1873).

STILLING, B.: Neue Untersuchungen über den Bau des Rückenmarks (Hotop, Cassel 1859).

STILLING, B. und WALLACH, J.: Untersuchungen über die Textur des Rückenmarks. (Wigand, Leipzig 1842).

STILLING, B. und WALLACH, J.: Untersuchungen über den Bau des Nervensystems. Über die Medulla oblongata (Enke, Erlangen 1843).

STREETER, G.L.: The structure of the spinal cord of the Ostrich. Amer. J. Anat. *3:* 1–27 (1904).

STUDNIČKA, F.K.: Ein Beitrag zur vergleichenden Histologie und Histogenese des

Rückenmarks (Über die sog. Hinterzellen des Rückenmarkes) S. Ber. k. böhm. Ges. d. Wiss. math-nat. Kl. Jahrg. 1895, No. 51: 1–32 (1896).

SZABO, TH.: Anatomo-physiologie des centres nerveux spécifiques de quelques organes électriques; in CHAGAS and PAEZ DE CARVALHO Bioelectrogenesis, pp. 185–201 (Elsevier, Amsterdam 1961).

SZENTÁGOTHAI, J.: Neuronal and synaptic arrangement in the substantia gelatinosa Rolandi. J. comp. Neurol. *122:* 219–239 (1964).

TAGLIANI, G.: Ricerche anatomiche intorno alla midulla spinale dell'Orthagoriscus mola. Monit. zool. ital. *5:* 248–258 (1894).

TAGLIANI, G.: Intorno di cosidetti lobi accessori e alle cellule giganti della midolla spinale di alcuni Teleosti. Boll. Soc. nat. (Napoli) Ser. I, *9:* 59–69 (1895).

TAGLIANI, G.: Über die Riesennervenzellen im Rückenmarke von Solea impar. Anat. Anz. *15:* 234–237 (1899).

TAGLIANI, G.: Le fibre del Mauthner nel midollo spinale dei vertebrati inferiori. Arch. zool. ital. *2:* 386–437 (1906).

TAREN, J.A.: Physiologic corroboration in stereotaxic high cervical cordotomy. Confin. neurol. *33:* 285–290 (1971).

TAYLOR, E.H.: The Caecilians of the world. A taxonomic review (University of Kansas Press, Lawrence 1968).

TEN CATE, J.: Zur Physiologie des Zentralnervensystem des Amphioxus (Branchiostoma lanceolatus). I. Die reflektorische Tätigkeit des Amphioxus. Contribution à la physiologie du système nerveux du Amphioxus (Branchiostoma lanceolatus). II. Les mouvements ondulatoires et leurs innervation. Arch. néerl. Physiol. *23:* 409–415; 416–423 (1938).

TERNI, T.: Ricerche istologiche sul medullo spinale dei Rettili, con particolare riguardo ai componenti spinali del fascio longitudinale mediale (Osservacioni in Gongylus ocellatus Wagl). Arch ital. Anat. Embriol. *18,* Suppl.: 183–243 (1922).

TERNI, T.: Ricerche anatomiche sul sistema nervoso autonomo degli Ucelli. I. Il sistema preganglionare spinale. Arch. ital. Anat. Embriol. *20:* 433–510 (1923).

TERNI, T.: Ricerche sulla cosidetta sostanza gelatinosa (corpo glicogenico) del midollo lombosacrale degli ucelli. Arch. ital. Anat. Embriol. *21:* 55–86 (1924).

TERNI, T.: Sui nuclei marginali del midollo spinale dei Sauropsidi. Arch. ital. Anat. Embriol. *23:* 610–628 (1926).

TESTA, C.: Functional implication of the morphology of spinal ventral horn neurons in the cat. J. comp. Neurol. *123:* 425–444 (1964).

TORVIK, A.: The spinal projection from the nucleus of solitary tract. An experimental study in the cat. J. Anat. *91:* 314–322 (1957).

TRACY, H.C.: Stages in the development of the anatomy of motility of the toadfish (Opsanus tau). J. comp. Neurol. *111:* 27–81 (1959).

TRETJAKOFF, D.: Das Nervensystem von Ammocoetes. I. Das Rückenmark. Arch. mikr. Anat. *73:* 607–680 (1909).

TRETJAKOFF, D.: Das Nervensystem von Ammocoetes. II. Gehirn. Arch. mikr. Anat. *74:* 636–779 (1909).

TUGE, H. (TSUGE): The development of behavior in avian embryos. J. comp. Neurol. *66:* 157–179 (1937).

UNZER, J.A.: Erste Gründe einer Physiologie (Weidmann, Leipzig 1771).

VERHAART, W.J.C.: Comparative anatomical aspects of the mammalian brain stem and the cord, 2 vols. (van Gorcum, Assen 1970).

VERNE, J.: Contribution à l'étude des cellules nevrogliques, spécialement au point de vue de leur activité formatrice. Arch. Anat. micr. Morph. exp. *16:* 149–191 (1913–1914).

VERWORN, M.: Allgemeine Physiologie, 7. Aufl. (Fischer, Jena 1922).

VILLIGER, E.: Die periphere Innervation, 3. Aufl., 6. Aufl. (Engelmann, Leipzig 1919, 1933).

VOLKMANN, A. W.: Über Reflexbewegungen. Arch. Anat., Physiol. wiss. Medicin, Jahrg. 1838: 15–43 (1838).

WALBERG, F. and BRODAL, A.: Spino-pontine fibers in the cat. An experimental study. J. comp. Neurol. *99:* 251–288 (1953).

WALDEYER, H.: Das Gorilla-Rückenmark. Anhang. z. d. Abh. Kgl. preuss. Akad. Wiss. phys.-math. Kl. *1888:* 1–147 (1889).

WALLENBERG, A.: Beiträge zur Kenntnis des Gehirns der Teleostier und Selachier. Anat. Anz. *22:* 289–292 (1907).

WALLENBERG, A.: Die kaudale Endigung der bulbospinalen Wurzeln des Trigeminus, Vestibularis und Vagus beim Frosche. Anat. Anz. *30:* 564–568 (1907).

WARNER, F. J.: The myelinization of the central nervous system of the American water snake (Natrix sipedon). Trans. zool. Soc. London *27:* 307–348 (1952).

WEIL, A.: A textbook of neuropathology, 2nd ed. (Grune & Stratton, New York 1945).

WEISS, P.: An introduction to genetic neurology; in WEISS Genetic neurology, pp. 1–39 (University of Chicago Press, Chicago 1950).

WEISS, P.: Nervous System (Neurogenesis), pp. 346–401, Sect. VII, Special vertebrate organogenesis; in WILLIER, WEISS and HAMBURGER Analysis of Development (Saunders, Philadelphia 1955).

WELSCH, U.: Die Feinstruktur der Josephschen Zellen im Gehirn von Amphioxus. Z. Zellforsch. *86:* 252–261 (1968).

WELSCH, U. and WÄCHTLER, K.: Zum Feinbau des Glykogenkörpers im Rückenmark der Taube. Z. Zellforsch. *97:* 160–168 (1969).

WHITING, H. P.: Nervous structure of the spinal cord of the young larval Brook-Lamprey. Quart. J. micr. Sci. *89:* 359–383 (1948).

WHITING, H. P.: Mauthner neurones in young larval lampreys (Lampetra ssp.). Quart. J micr. Sci. *98:* 163–178 (1957).

WILLIS, W. D.: The case for the Renshaw cell. Brain, Behav. Evol. *4:* 5–52 (1971).

WINDLE, W. F.: Correlation between the development of local reflexes and reflex arcs in the spinal cord of cat embryos. J. comp. Neurol. *59:* 487–507 (1934).

WINDLE, W. F.: Physiology of the fetus. Origin and extent of function in prenatal life (Saunders, Philadelphia 1940).

WINDLE, W. F.: Reflexes of mammalian embryos and fetuses; in WEISS Genetic neurology, pp. 214–222 (University of Chicago Press, Chicago 1950).

WINDLE, W. F. and ORR, D. W.: The development of behavior in chick embryos: spinal cord structure correlated with early somatic motility. J. comp. Neurol. *60:* 287–307 (1934).

WISLOCKI, G. B. and LEDUC, E. H.: The cytology and histochemistry of the subcommissural organ and Reissner's fiber in rodents. J. comp. Neurol. *97:* 515–544 (1952).

WISLOCKI, G. B. and LEDUC, E. H.: The cytology of the subcommissural organ, Reissner's fiber, periventricular glia cells, and posterior collicular recess of the rat's brain. J. comp. Neurol. *101:* 283–310 (1954).

WOLFF, M.: Bemerkungen zur Morphologie und Genese des Amphioxus-Rückenmarkes. Biol. Zbl. *27:* 186–192 (1907).

Yoss, R.E.: Studies on the spinal cord. Part II. Topographic localization within the ventral spino-cerebellar tract in the Macaque. J. comp. Neurol. *99:* 613–637 (1953).

Young, J.Z.: The life of vertebrates (Clarendon Press, Oxford 1950, 1955).

Youngstrom, K.A.: A primary and secondary somatic motor innervation in Amblystoma. J. comp. Neurol. *73:* 139–151 (1940).

Youngstrom, K.A.: Intramedullary sensory type ganglion cells in the spinal cord of human embryos. J. comp. Neurol. *81:* 47–53 (1944).

Zecha, A.: The 'pyramidal tract' and other telencephalic efferents in birds. Acta Morph. neerl.-scand. *5:* 194–195 (1962).

Zeman, W. and Innes, J.R.M.: *Craigie's* neuroanatomy of the rat. Revised and expanded (Academic Press, New York 1963).

Zemmer, P.: Ein Fall von Hirngeschwulst in der linken motorischen Sphäre, linksseitige Lähmung. Abwesenheit der Pyramidenkreuzung. Neurol. Cbl. *17:* 202–203 (1898).

Ziehen, Th.: Morphogenie des Centralnervensystems der Säugetiere; in Hertwig Handbuch der vergleichenden und experimentellen Entwicklungslehre der Wirbeltiere, 2.Bd., 3.Teil, pp.273–394 (Fischer, Jena 1906).

IX. Medulla Oblongata (and Pons)

1. General Pattern and Basic Mechanisms; The Meaning of the Terms 'Deuterencephalon' and 'Brain Stem'

The Vertebrate medulla oblongata *sensu latiori*, as dealt with in the present chapter, includes the entire extent of the rhombencephalon with the exception of the cerebellum. The somewhat different subdivision of rhombencephalon into a rostral metencephalon and a caudal myelencephalon is, nevertheless, fully justified in human or mammalian anatomy.[1] The metencephalon consists here of dorsal cerebellum and ventral pons. In all submammalian forms, however, a morphologically well differentiated pons or basal metencephalic region cannot be recognized as a significant separate major portion of the rhombencephalon. From an overall viewpoint of Vertebrate comparative anatomy, it seems therefore preferable to use a simplified terminology which merely distinguishes the cerebellum from the medulla oblongata in the wider sense.[2]

The *medulla oblongata* is continuous, through two zones of transition or overlap, with the caudally and rostrally adjacent subdivisions of the neuraxis, namely with spinal cord and with mesencephalon. The rostral zone of transition, also designated as '*isthmus rhombencephali*' (BNA), contains components which can be assigned to oblongata, cerebellum, and mesencephalon.[3] A third zone of transition connects a rostral part of oblongata with the dorsally located cerebellum. These three boundary regions represent open neighborhoods in terms of topology.

---

[1] Cf. vol. 3, part II, section 1B, p. 185.

[2] The term bulb (bulbus medullae; French: *bulbe rachidien)* is likewise used as synonym for medulla oblongata, especially in clinical phraseology. The adjective *bulbar* may here refer to the medulla oblongata in the narrower as well as in the wider sense.

[3] With respect to human anatomy, the BNA list an isthmus rhombencephali as a special subdivision with several components. The PNA, however, merely mention that term parenthetically in enumerating 'Sectiones cerebelli'.

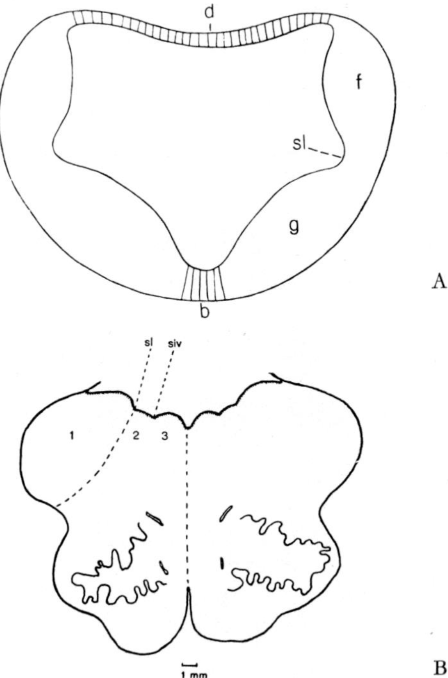

Figure 149. Two semidiagrammatic cross-sections showing the eversion displayed by the Vertebrate medulla oblongata. A Generalized embryonic stage. B Final stage displayed by adult human medulla oblongata (A from K., 1927, B from K., 1929).b: floor plate; d: roof plate; f: alar plate; g: basal plate; sl: sulcus limitans; slv: sulcus intermedius ventralis; 1: alar plate; 2: intermedioventral zone; 3: ventral zone (2 and 3: basal plate).

The peculiar characteristics of the medulla oblongata are substantially conditioned by its connections with the branchial nerves. These latter, which were discussed in section 3 of chapter VII (vol. 3, part II), can be interpreted as being serially homologous with the dorsal roots of spinal nerves.[4]

It is most likely that the predominant morphologic role played by the specialized rhombencephalic dorsal roots represents an important factor in the eversion of the bilateral alar plates.[5] Except for the caudal portion of the oblongata and for its rostral region covered by the cer-

[4] Serial, i.e. metameric homology pertains, of course, to the domain of promorphologic homology.

[5] Cf. e.g. Kappers (1920) and Kuhlenbeck (1927).

ebellum, the alar plates do not 'fuse' in the dorsal midline, as they do in the spinal cord, but diverge and thereby enlarge the central canal which becomes the relatively wide or 'broad' *fourth ventricle* (Fig. 149). The ventricular roof (Fig. 150) is provided by the not only persisting but expanded thin lamina epithelialis of the embryonic roof plate. Nu-

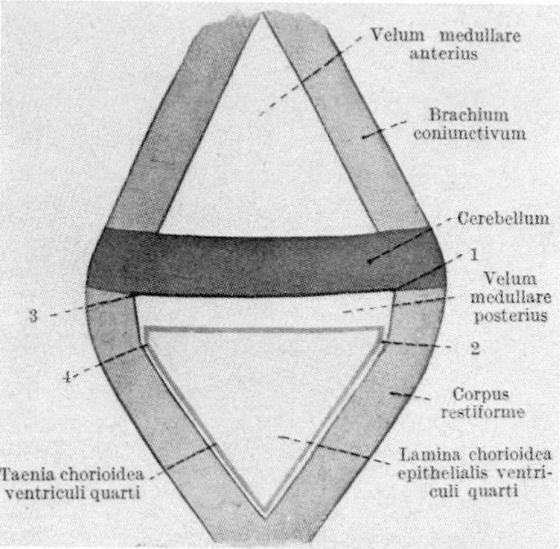

Figure 150. Diagrammatic dorsal view of embryonic human rhombencephalon (after TANDLER, 1929). 1–4: corners of so-called velum medullare posterius, connecting epithelial roof of fourth ventricle to 'cerebellar plate'.

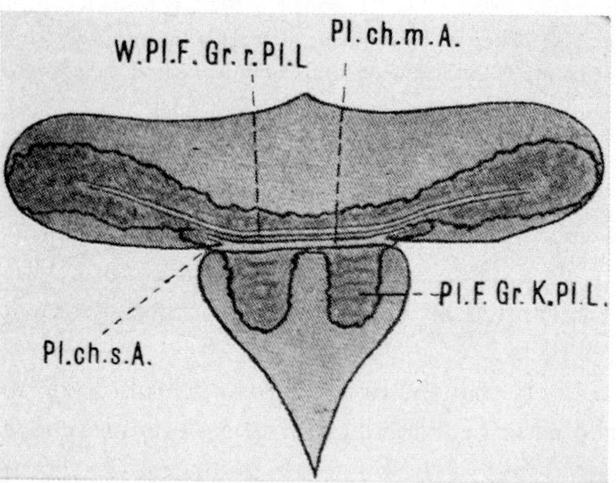

Figure 151. Diagram indicating anlage of plexus chorioidens within epithelial roof of fourth ventricle at a relatively early stage (about 15 mm) of human embryonic development (from HOCHSTETTER, 1929). Pl. ch. m. A.: median part of chorioid fold; Pl. ch. s. A.: lateral part of fold; Pl. F. Gr. K. Pl. L.: caudal group of plexus folds; W. Pl. F. Gr. r. Pl. L.: rostral group of plexus folds.

merous folds of this latter, in combination with the vascularized leptomeninx, become evaginated along certain lines and form the choroid plexus of the fourth ventricle (Figs. 151–152).

The epithelial roof plate covering the ventricular space is attached to a boundary-strip or fringe of roof plate. In dorsal view, this border

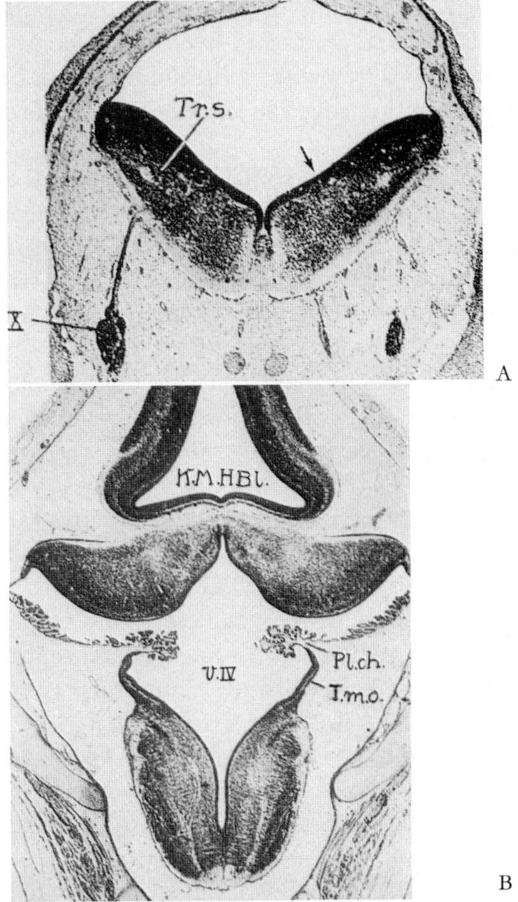

Figure 152. Transverse (A) and 'horizontal' (B) sections through oblongata respectively rhombencephalon in two human embryos between the 15 and 16 mm stages (6th week), showing in A the thin epithelial roof and in B anlage of choroid plexus (from HOCHSTETTER, 1929). KMHBl: mesencephalic ventricle (*'Kaudaler Mittelhirnblindsack'*); Pl. ch.: choroid plexus; Tmo: 'taenia medullae oblongatae'; Tr.s.: tractus solitarius; v.IV: fourth ventricle; X: ganglion and root of n. vagus. The paired cerebellar plate (not labelled) can be recognized below the mesencephalic ventricle. Added arrow in A indicates sulcus limitans.

displays an either well-defined or somewhat modified ('truncated') rhombic or 'lozenge-shaped' outline, which has led to the designations rhombencephalon for the caudal major subdivision of the neuraxis, and *rhomboid fossa*[6] as a synonym for fourth ventricle (Fig. 153). Near the transition of rhombencephalon to mesencephalon, but still caudally to the isthmus region, the dorsal midline fusion of both alar plates results in the production of the *corpus cerebelli*, which, depending on the particular Vertebrate forms, may merely consist of a narrow transverse ridge, frequently with median indentation, or may develop into a large configuration covering an extensive portion of fourth ventricle.

In contradistinction to the spinal cord, in which the *sulcus limitans* becomes effaced in the course of ontogenesis, this groove, approximately delimiting alar from basal plate, generally remains clearly recognizable at various levels of the definitive or 'adult' rhomboid fossa in numerous if not most Vertebrates including man. Additional sulci, namely sulcus intermedius dorsalis and intermedius ventralis are frequently present. The alar and basal plates become thereby roughly subdivided into the dorsal, intermediodorsal, intermedioventral and ventral longitudinal zones dealt with in chapter VI, section 4 of volume 3, part II.

Diverse aspects of these grooves, as seen by dorsal inspection of the rhomboid fossa, and particularly also in cross-sections, are illustrated by Figures 153B, 154, and 155. Regardless of distinctness, indistinctness or even absence of the ventricular sulci, the corresponding longitudinal zones may be conceived as deuterencephalic 'columnar' configurations within the wall of the neural tube and can be traced,

[6] Some authors, e.g. RANSON (1943), use this term as especially denoting the *floor* of the fourth ventricle.

Figure 153 A, B. Dorsal aspect of rhomboid fossa in two anamniote Vertebrates with poorly developed cerebellum. A Petromyzon marinus (from KAPPERS *et al.*, 1936). Cal. scr.: 'calamus scriptorius'; NLA: nerve of anterior lateral line; x: median indentation of cerebellar ridge (rudimentary corpus cerebelli). B Rana mugiens (after a wax model by HAMMER, from KAPPERS *et al.*, 1936). Aur. cereb.: cerebellar auricle; S. centr.: sulcus medianus internus of rhomboid fossa; Sulc. acust.: sulcus intermediodorsalis; Sulc. interm.: an external dorsal sulcus in region of transition to spinal cord; Sulc. parac.: sulcus limitans; Sp. II: spino-occipital nerve roots, of which one (XII) is here interpreted as representing the n. hypoglossus. The dotted line indicates the attachment of removed lamina epithelialis which includes the choroid plexus. Other designations self-explanatory.

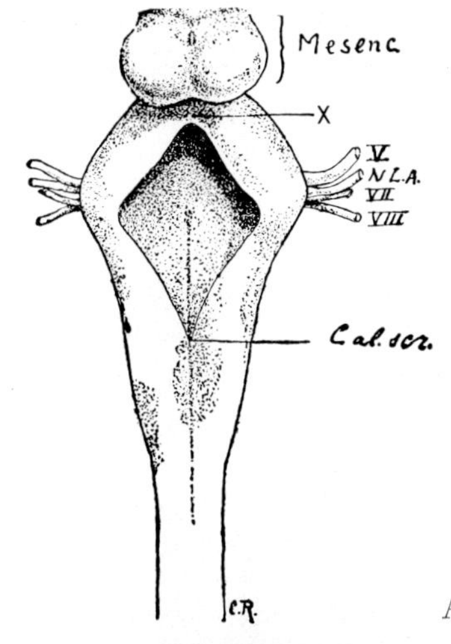
Mesenc
X
V
N.L.A.
VII
VIII
Cal. scr.
C.R.

A

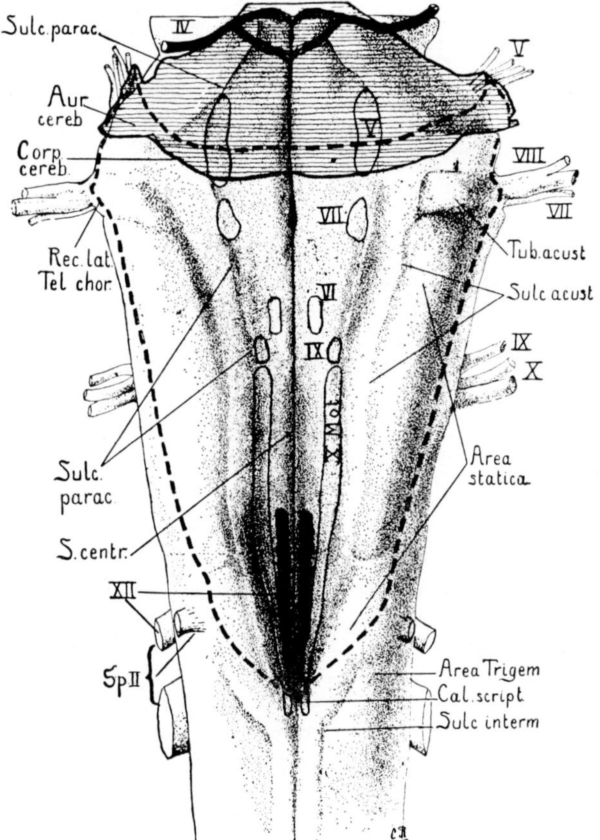
Sulc. parac.
IV
V
Aur cereb
V
Corp. cereb.
VIII
VII
VII
Rec. lat. Tel chor.
Tub. acust
VI
Sulc acust
IX
IX
X
Mot. X
Sulc. parac
Area statica
S. centr.
XII
Sp II
Area Trigem
Cal. script
Sulc interm
C.R.

B

more or less easily, throughout the length of the rhombencephalon into the mesencephalon. In various Vertebrates, certain regions of the rhombencephalon display a considerable degree of eversion, whereby the morphologically dorsal zones assume a topographically lateral position in relation to the now medially located ventral ones.

In discussing structure, fiber connections and mechanisms of the spinal cord, L. EDINGER's distinction of *eigenapparat* and *verbindungsapparat* seems of definite value for an appropriate functional understanding. With regard to the oblongata, it is likewise possible to apply this

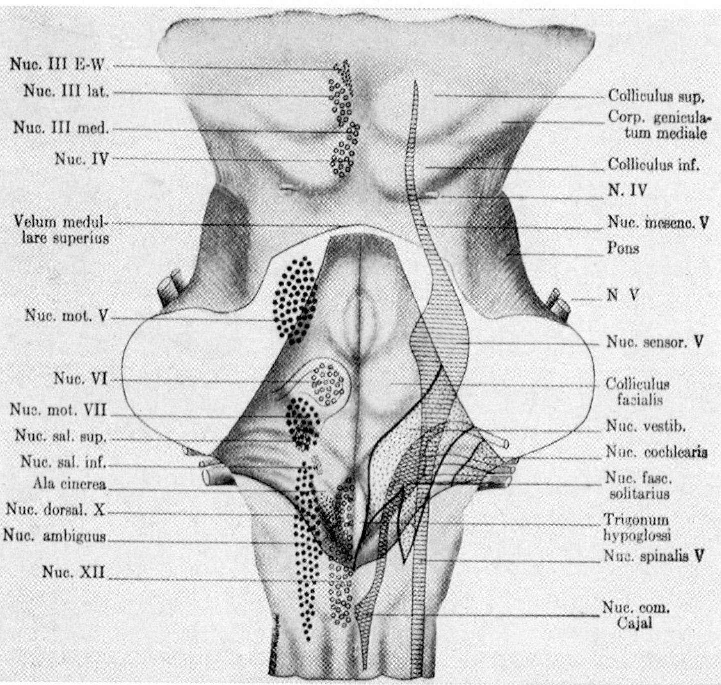

Figure 153 C. Dorsal view of the human brain stem showing the rhomboid fossa. The cerebellum has been removed by transection of its three stalks and the remaining membranous roof of the fourth ventricle has been torn off at its taenial attachment (from HERRICK, 1931). The motor nuclei are indicated on the left and the sensory nuclei on the right. circles: somatic motor nuclei; small dots: general visceral efferent nuclei; large dots: special visceral efferent nuclei; horizontal lines: general somatic sensory area; stipples: special somatic sensory area; double cross-hatching: visceral afferent area; nuc. com. Cajal: nucleus of commissura infima; nuc. III. E-W: preganglionic oculomotor *nucleus of Edinger-Westphal;* nuc. saliv. inf., sup.: inferior salivatory respectively superior salvatory and lacrimal nuclei.

terminology with some qualifications related to the obtaining differences in structure and pattern. Even in the spinal cord, *eigenapparat* and *verbindungsapparat* are closely interrelated and overlapping. This concatenation becomes still more pronounced in the medulla oblongata, where EDINGER's useful distinction is therefore of lesser significance and should not be unduly stressed.

Nevertheless, it could be said that the oblongata's *verbindungsapparat* comprises a caudalward and a rostralward directed component. The former, with efferent as well as afferent[7] fiber systems, connects with the spinal cord and represents a hierarchically superimposed *('übergeordneter')* mechanism. The latter, likewise with afferent and efferent channels, connects the medulla oblongata with the cerebellum and with rostral brain subdivisions. Neural events in cerebellum, mesencephalon, diencephalon and telencephalon partly control activities in the oblongata, but may also, in turn, be subject to bulbar control.

The oblongata's *eigenapparat* is predominantly an intersegmental correlation mechanism. In contradistinction to the segmental pattern of spinal nerves the segmental character of rhombencephalic branchial nerves and of ventral roots, although evident at early development stages, does not retain a substantial significance, and the longitudinally arranged functional systems predominate.

With respect to the oblongata's *external input*, the afferent fibers of the branchial nerves,[8] upon joining the neuraxis, become arranged in separate bundles with ascending and descending components. Most of these components seem to be branches arising at right angles from the dichotomy of the entering root fibers, in a manner similar to that obtaining for the spinal dorsal root fibers in the posterior funiculus. In some instances, however, some afferent bulbar root fibers might not display dichotomy, merely assuming a descending or ascending course, but not both.

The distribution of these afferent branchial root fibers is such, that those of the vestibulo-lateralis system assume the most dorsal location,

---

[7] The terms efferent and afferent refer here to central, and not to peripheral connections, or, in other words, to fiber systems remaining entirely *within* the neuraxis and representing *input or output channels of bulbar grisea*. It should also be kept in mind that the caudal component of the oblongata's *verbindungsapparat* can also be classified as pertaining to the *verbindungsapparat* of the spinal cord.

[8] Including the VIIIth nerve, which, although strictly speaking not a branchial nerve, can nevertheless be classified with these latter (cf. chapters VI, section 4, and VII, section 4 of vol. 3, part II).

while most exteroceptive fibers run in a more ventrally located bundle; both systems, however, take their course through the alar plate's dorsal zone[8a]. The interoceptive fibers from the mucosae or viscera, and the gustatory fibers (even including those from the cutaneous 'taste buds' of certain Fishes) run as a separate bundle through the alar plate's intermediodorsal zone. Thus, with regard to branchial nerves comprising several sorts of afferent components, a *segregation* of these fibers, and their *distribution* upon the corresponding longitudinal tracts take place (Fig. 156). The root fiber bundles, in turn, are accompanied by longitudinally arranged grisea, which represent terminal nuclei containing afferent neurons of the second order. The neurites of these latter not only provide, partly through collaterals, further connections within the rhombencephalon, but also various ascending and descending secondary tracts. As regards such ascending systems, neurites from the terminal nuclei commonly decussate as ventral arcuate fibers, and then turn rostrad to reach, as the case may be, the tectum mesencephali or the thalamus *(fibrae bulbo-tectales* and *bulbo-thalamicae)*. The loops formed by the arcuate fibers and their ascending channels, respectively these latter themselves, are often referred to as *lemnisci*[9] *(lemniscus*, loop, German: *Schleife)*.

The *descending root fibers of the vestibular nerve* (with their concomitant grisea), because of their direction toward the spinal cord, are also known as the so-called *radix spinalis vestibularis* with its 'nucleus'. The descending exteroceptive root fibers, mainly but not exclusively pertaining to the trigeminus, are similarly designated as *radix descendens sive spinalis trigemini*. At the transition to spinal cord this radix may become

---

[8a] It may be recalled that, for purposes of a more detailed form-analysis in chapter VI, section 4, p. 372 of volume 3, part II, the alar plate's dorsal zone was subdivided into (A) Pars superior, lateralis et medialis, and (B) Pars inferior et lateralis. In a simplified manner, (A) can also be described as the dorsal zone's dorsal subzone or part, and (B) as the dorsal zone's ventral and ventrolateral part. Likewise, the basal plate's intermedioventral zone was there subdivided into Pars interna and Pars externa. These two subzones, which could also be called Pars dorsomedialis and ventrolateralis, respectively, are related to a segregation of preganglionics in the former, and of 'branchiomotor' neurons in the latter. A tendency toward this segregation does not seem generally recognizable in Anamnia but is indicated in submammalian Amniota. In Mammals, it appears to have resulted in a definitive and well established segregation.

[9] Ascending spino-tectal and spino-thalamic fibers, discussed in the preceding chapter VIII, and taking their course through the oblongata, are likewise said to represent a spinal lemniscus. The bulbo-tectal and bulbo-thalamic fibers generally join the corresponding ascending spinal fibers. Further details on these systems shall be taken up in the subsequent sections dealing with the diverse Vertebrate groups.

continuous, or overlap, with *Lissauer's tract*, while the nucleus of the descending trigeminal root 'blends' with the spinal cord's substantia gelatinosa. The descending and ascending interoceptive (visceral afferent) and gustatory root fibers, particularly of nerves IX, X, VII, and to a lesser extent perhaps also of V, represent the *tractus solitarius* (or *fasciculus communis*) with its secondary griseum.

Another important secondary ascending system is present in the oblongata of those Vertebrates whose posterior funiculi of the spinal

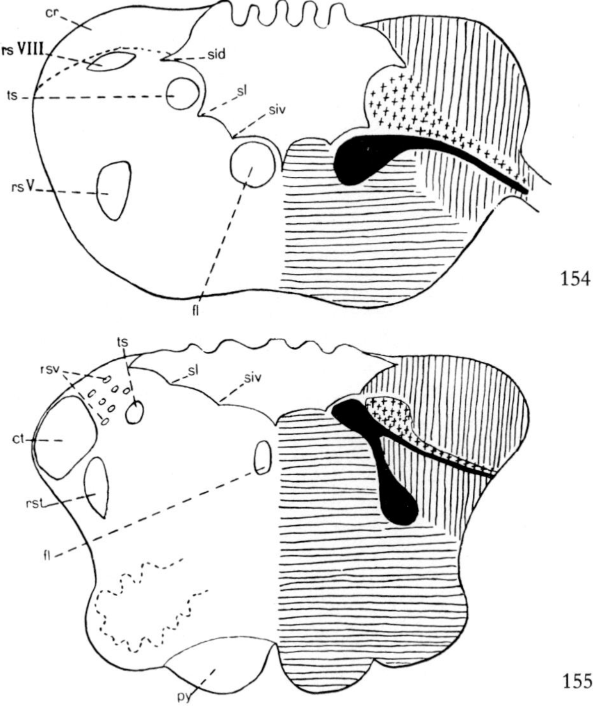

Figure 154. Diagrammatic cross-section illustrating longitudinal zones and representative fiber bundles in the adult Selachian oblongata (from K., 1927). cr: crista cerebellaris; fl: fasciculus longitudinalis medialis; rs: radix descendens seu 'spinalis' trigemini (V) et vestibularis (VIII); sid: sulcus intermedius dorsalis; siv: sulcus intermedius ventralis; sl: sulcus limitans; ts: tractus solitarius; horizontal hatching: ventral zone; vertical hatching: dorsal zone; black: intermedioventral zone; crosses: intermediodorsal zone.

Figure 155. Diagrammatic cross-section illustrating longitudinal zones and representative fiber bundles in the adult human oblongata (from K., 1927). ct: corpus restiforme; py: pyramidal tract; rst: radix descendens trigemini; rsv: radix descendens vestibularis; other designations as in preceding figure.

cord end, at the transition to medulla oblongata, in the grisea designated as *terminal nuclei of the posterior funiculi (nn. of fasciculus gracilis and cuneatus* of Man and Mammals), and briefly mentioned in the preceding chapter VIII. In the present context, dealing with the oblongata's general pattern, it will be sufficient to point out that neurons of the second order, located in these nuclei, send their decussating axons as *fibrae arcuatae internae* to the contralateral side, forming, in the vicinity of the

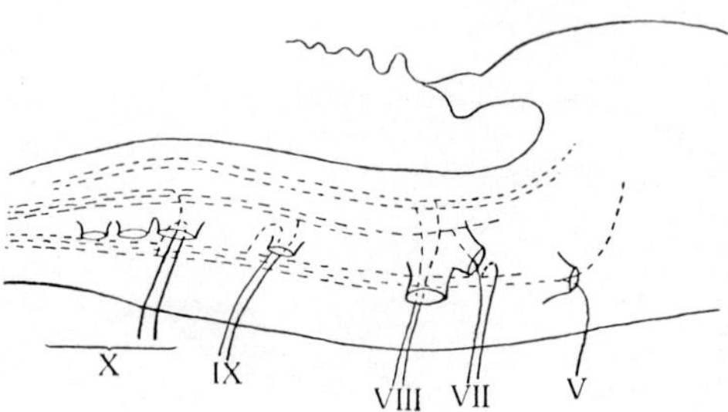

Figure 156. Semidiagrammatic sketch illustrating the longitudinal root fiber bundles and their distribution in the Vertebrate oblongata as displayed in urodele Amphibians (partly based on concepts by HERRICK, and modified from K., 1927). The cranial nerves are designated by their corresponding Roman numerals. Top bundle vestibulo-lateral system; middle bundle 'general and special visceral afferents' *(tractus solitarius)*; lower bundle 'general somatic afferents' (so-called *radix spinalis seu descendens trigemini*. It will be noted that the bifurcating root fibers are here presumed to have ascending as well as descending branches).

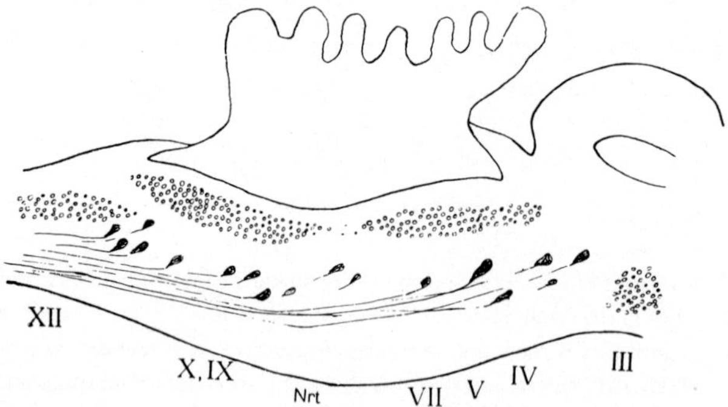

Figure 157. Sketch showing longitudinal arrangement of efferent (motor) cell columns in the Cyclostome (Petromyzont) oblongata (slightly modified from K., 1927). Roman numerals indicate cerebral nerve (XII being, of course, not the hypoglossal, but the spino-occipital column); Nrt: nucleus reticularis tegmenti with true *Müller cells* and elements of *Müllerian* type.

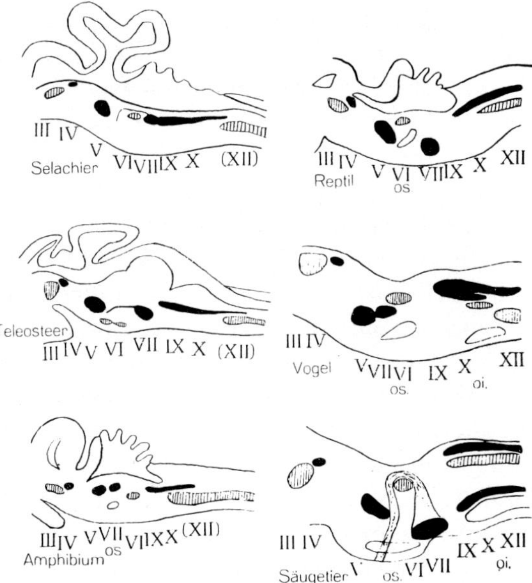

Figure 158. Simplified diagram showing relative position of motor cranial nerve nuclei in the vertebrate series (modified after KAPPERS, from KUHLENBECK, 1927). oi: inferior olivary complex; os: superior olivary complex; Roman numerals as in preceding figure; (XII): spino-occipital column; XII: nucleus of 'true' hypoglossus.

midline, and ventrally to fasciculus longitudinalis medialis and related systems, the so-called *lemniscus medialis*, which particularly, but not exclusively in Mammals, reaches thalamic grisea.[10]

Concerning those functions of the oblongata which may be considered '*motor*' in either a narrower or a wider sense, three systems[11] can be distinguished in all craniote Vertebrates:

(1) The efferent nuclei of the *branchial nerves*.

(2) The efferent or motor nuclei of *bulbar ventral roots*.

(3) The *nucleus motorius tegmenti* and various 'reticular' cell groups derived from or secondarily related to this primordial griseum.

---

[10] The so-called (external) dorsal arcuate fibers and other details concerning the grisea of the posterior funiculi shall be dealt with further below, particularly in discussing the Mammalian oblongata (section 10).

[11] The first two systems can be characterized as concerned with the oblongata's *external* (peripheral) output, while the third one provides an important component of *internal* bulbar output. It should, moreover, be added that the grisea of these three systems are not restricted to the rhombencephalon alone, but, being intrinsic components of the deuterencephalon, also extend into the midbrain.

(1) The nerve cells which give origin to the efferent fibers joining the branchial nerves, i.e. the dorsal roots, are located within the intermedioventral zone. This longitudinal 'nuclear column' comprises the efferent nuclei of trigeminus, facialis, glossopharyngeus and vagus (Fig. 157).

(2) The nuclei from which the ventral root fibers innervating somatic musculature originate, are located within the ventral zone. Only two such nuclei are present in the medulla oblongata, namely those of abducens and hypoglossus.[12] These grisea are 'in line' with the nuclei of trochlearis[13] and oculomotorius pertaining to the mesencephalon. Thus, compared with the intermedioventral zone, the ventral zone of the oblongata contains fewer[14] and less extensive efferent nuclei, which are separated by relatively wide intervals. This circumstance perhaps favored the origin and subsequent phylogenetic differentiation of the nucleus motorius tegmenti[15] (EDINGER, 1908, 1911, 1912) which, because of its particular significance, will be separately discussed further below in section 2 of this chapter.

The overall spatial arrangement *(Lageplan)* of the efferent nuclei in oblongata respectively deuterencephalon is essentially identical for the entire Vertebrate phylum, although, particularly in the more differentiated forms, substantial secondary displacements can be noted. These displacements, which appear correlated with the predominating influence exerted upon said grisea by certain fiber tracts, have been particularly investigated by KAPPERS (1920, 1921) and his students (especially ADDENS, 1928–1933, and BLACK, 1917–1922). The concept of *neurobiotaxis*, formulated and elaborated by the cited author, attempts to provide an explanation for this well-established configurational orderliness, and was discussed in chapter V, section 7, volume 3, part I of this series. Disregarding, in the aspect here under consideration, the

---

[12] 'Lower' Vertebrates do not display a typical nervus hypoglossus, but its 'forerunner' is generally represented by spino-occipital ventral roots whose nucleus can be regarded as homologous to the nucleus nervi hypoglossi of higher forms (cf. sections 3 and 4, chapter VII of vol. 3, part II).

[13] The still poorly clarified problems concerning the morphologic status of nervus trochlearis are briefly dealt with in section 3 of chapter VII.

[14] Two (VI and XII) versus four or five (V, VII, IX, X–XI).

[15] Tegmentum (German: *Haube)* originally referred to the dorsal portion of the mesencephalic so-called pedunculus cerebri (BNA, PNA). Subsequently, the term tegmentum became applied to the region below (i.e. ventrally to) the ventricular floor throughout the deuterencephalon, and containing grisea continuous with those of mesencephalic tegmentum.

unclarified causal implications of that theory, the actually recorded displacements may be summarized as follows (Figs. 158, 159).

In most Vertebrates, the nucleus of the *abducens*[16] maintains a dorsal position, close to the fasciculus longitudinalis medialis. In various Teleosts, however, this nucleus assumes a pronounced basal location, adjoining the ventral tecto-bulbar tract from which it receives substantial input.

In Amniote Vertebrates, the *branchiomotor nucleus of the facial nerve* displays the propensity toward an increasingly ventral shifting (Fig. 160). The intracerebral motor root of the facial nerve tends to form a loop, which may arch around the nucleus of the abducens. This so-called (internal) knee of the facial nerve is present in various Anamnia (Selachians, Teleosts) as well as in Amniota. Again, a caudal position of the branchiomotor facial nucleus obtains in some forms (e.g. Selachians), and a rostral one in others (e.g. Birds), as indicated by Figure 158. Moreover, in those Teleosts with a particularly well-developed gustatory system, a part of the facial as well as of the trigeminal branchiomotor nucleus displays a lateral or ventiolateral shift toward fiber bundles such as 'secondary gustatory tract' or the 'fasciculus lobo-bulbaris' related to posterior hypothalamus (*'gruppo ventrale del nucleo motore'* of V and VII, BECCARI 1943).

In the oblongata of most Anamnia, the intermedioventral longitudinal griseal column comprising the efferent neurons of glossopharyngeus and vagus contains, seemingly in a more or less random fashion, larger 'branchiomotor' and smaller preganglionic elements. In the oblongata of Amniota, however, the branchiomotor cells tend to be displaced in a ventrolateral direction, providing thus a separate griseum

---

[16] The relationships of the nervus abducens and its nucleus in the Cyclostome Petromyzon are incompletely clarified. The reduction of the eyes in Myxinoids seems associated with a corresponding vanishing of eye-muscle nerves and their nuclei. The rudimentary Myxinoid eyes, located on both sides of the unpaired but bilaterally symmetric olfactory organ, lack an identifiable eye musculature, and are separated from the head's surface by body musculature. A thin optic nerve is nevertheless still present. Although essentially represented by a strand of connective tissue, it seems to have retained a few nerve fibers. Various doubts obtain concerning the presence of a nervus abducens and its nucleus in the Gymnophione Amphibians whose eyes are substantially reduced. In the blind Urodele Proteus anguineus, who still retains a rudimentary eye with optic nerve, all eye-muscle nerves and their nuclei are reported missing (BENEDETTI, 1933). In diverse Reptiles, Birds, and Mammals, perhaps also in the Anuran Amphibian Rana catesbyana, a laterally located accessory abducens nucleus has been identified (cf. BECCARI, 1943).

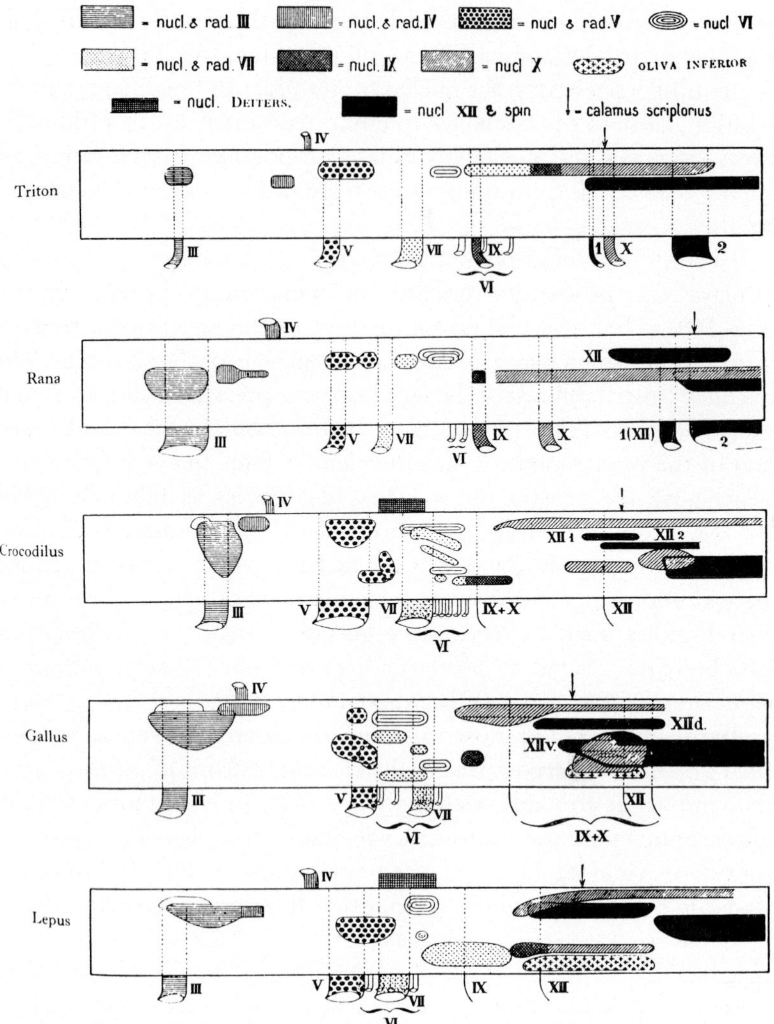

Figure 159. More detailed diagram indicating relationships of motor cranial nerve nuclei in the urodele amphibian Triton (A), the anuran amphibian Rana (B), the reptilian Crocodilus (C), the avian Gallus (D), and the Mammalian Lepus, based on combined studies by KAPPERS, VAN DER HORST, and ADDENS (from KAPPERS, 1947).

which, in the human and Mammalian brain, is known as the *nucleus ambiguus.*[17]

Another significant 'phylogenetic' displacement is manifested by the rostral migration of the motor hypoglossal nucleus of Amniota, respectively its 'spino-occipital' homologon in Anamnia.[18] While in these latter, the 'hypoglossal' nucleus generally reaches the caudal end or a caudal region of the caudal portion of the glossopharyngeal and vagal cell column, it becomes essentially coextensive with that griseal zone in most Mammalia (cf. Figs. 158, 159). Again, diverse secondary subdivisions of the hypoglossal nucleus can become manifested.

With respect to the Vertebrate brain's *basic mechanisms* as e.g. displayed by the neuraxis of a Selachian, and with particular emphasis on the type of peripheral input, HERRICK (1924, 1926, 1931) made use of the functional designations '*nose brain*', '*eye brain*', '*ear brain*', '*skin brain*', and '*visceral brain*'.[19] Roughly speaking, and disregarding the numerous additional complicating aspects, this vividly graphic terminology (Fig. 161), despite its evident oversimplification of highly complex interrelationships, may be accepted as a very useful first approximation.

Concerning the mechanisms listed above (p. 299) under (1) and (2), dealing with external input and output, the medulla oblongata is char-

---

[17] The differentiation of a nucleus ambiguus, in turn, is related to the still incompletely understood evolution of the nervus accessorius, whose significance was pointed out in section 3, chapter VII, of the preceding volume 3, part II. Relevant aspects of that question ('*Accessoriusfrage*', '*problema del nervo accessorio*') have been reviewed and critically discussed by BECCARI (1943).

[18] It could be said that the presence of a 'true' hypoglossal nerve is restricted to Amniota. In Anamnia, which do not possess this nerve, this latter is, nevertheless, represented in diversified ways by the so-called spino-occipital nerves particularly investigated by FÜRBRINGER, BECCARI and many others (cf. section 3, chapter VII of vol. 3, part II, and BECCARI, 1943). It should, moreover, be recalled that a well-defined plane of separation between oblongata and spinal cord cannot be drawn. This boundary zone is by no means intrinsically related to the plane separating cranial cavity from vertebral canal (cf. above, chapter VIII, section 1, p. 10, and K., 1932). '*Die Trennung des Zentralnervensystems der Wirbeltiere am Hinterhauptsloch geschieht durch das Beil des Scharfrichters aber nicht durch die Logik des vergleichenden Biologen*' (HESSE, 1932).

[19] HERRICK (1924) gives the following definitions: *earbrain:* 'the part of the brain concerned with auditory, vestibular, and lateral line reflexes'; *eyebrain:* 'the part of the brain concerned with visual reflexes'; *nose brain:* 'the part of the brain concerned with olfactory reflexes'; *skin brain:* 'the part of the brain concerned with cutaneous sensibility'; *visceral brain:* the lower part of medulla oblongata 'concerned chiefly with visceral reflexes'. Cf. also the comments on this subdivision on p. 641, chapter VI, section 6 of the preceding volume 3, part II.

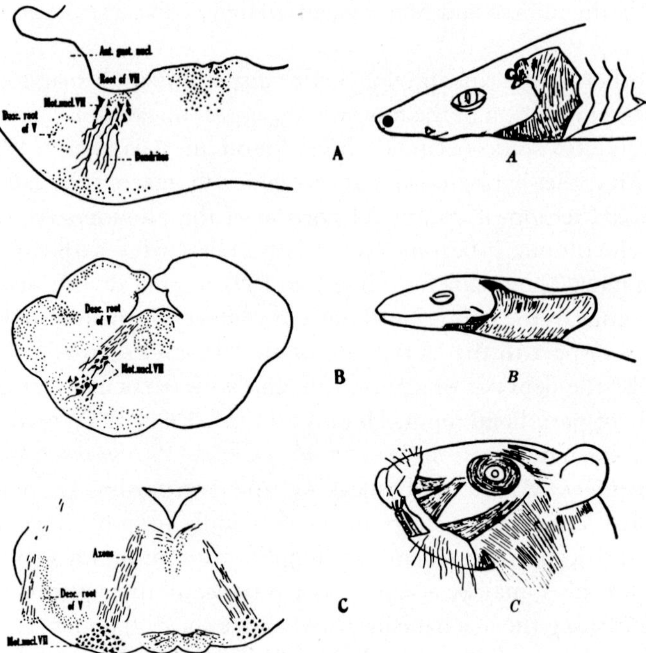

Figure 160. Relationships between branchiomotor facial nucleus and evolutionary differentiation of facial (2nd branchial arch musculature [from KAPPERS *et al.*, 1936]). A: facial nucleus in the selachian Scyllium canicula, and facial musculature of a shark according to RUGE; B: facial nucleus of the reptile Varanus salvator and facial musculature of Varanus according to RUGE; C: facial nucleus of the mammal Mus musculus and facial musculature of Mus according to PARSONS.

acterized by central connections of the cranial nerves V to XII. This includes direct input from one of the so-called higher sense organs, namely the *ear (stato-acustico-lateralis receptor system*, n. VIII), whose primary centers can be regarded as essential components of the 'ear brain', mainly located in the rostral part of the oblongata. A similar location obtains for grisea dealing with 'facial movements' mediated by the Vth and VIIth cranial nerve, and subsumed under the term 'skin brain'. HERRICK (1924) also classified both ear and skin brain as components of the '*face brain*'.

Nevertheless, ear brain and skin brain extend caudalward throughout the medulla oblongata and frequently into the rostral end of the spinal cord. The caudal part of the oblongata is substantially concerned with 'visceral reflexes' and may therefore be termed the visceral brain, although 'somatic reflex centers' are represented here also. Figure 162 shows HERRICK's (1926) concept of an interconnection between 'gustatory' and 'tactile centers' in the medulla oblongata of a Teleost.

In Fishes, the caudal part of the oblongata, related to the roots of nerves IX and X, deals chiefly with the 'visceral functions' of respiration, nutrition, circulation of blood, etc., and the same is also the case in Mammals, although with many secondary changes. 'In air breathing vertebrates the respiratory mechanism is totally different from the gills of fishes and most of the muscles concerned are innervated from the spinal cord, yet even in man the center responsible for maintaining the respiratory rhythm is in its primitive position in the medulla oblongata. Most of the gill muscles of fishes have disappeared in the human body; but some survive in the pharynx and larynx with greatly

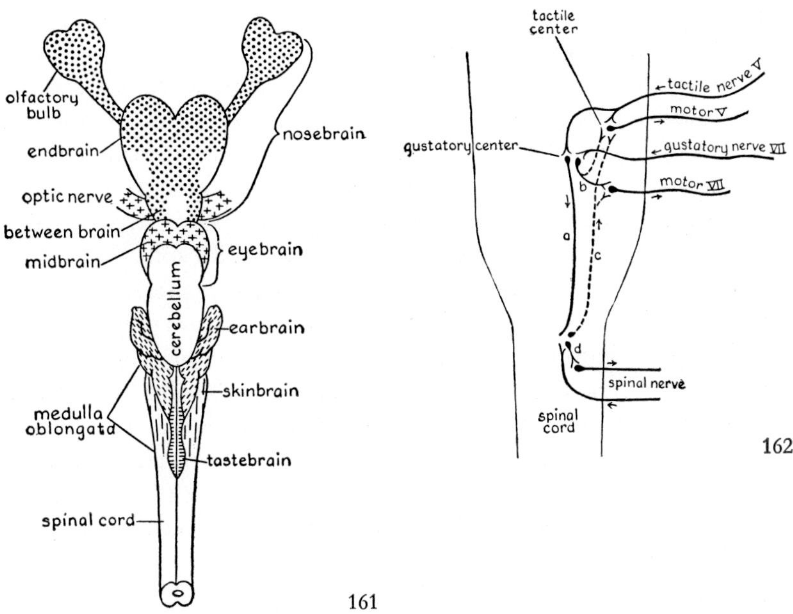

Figure 161. Dorsal view of brain in the selachian Squalus acanthias, shown in only slightly reduced natural size, and illustrating HERRICK's simplified functional subdivision into nose-, eye-, ear-, skin-, and taste-brain (from HERRICK, 1926).

Figure 162. Diagram of some of the neuronal connections between touch and taste input in the oblongata of the teleost catfish Ameiurus, according to HERRICK, and 'as seen in longitudinal section' (from HERRICK, 1926).

changed functions, and these muscles in man are innervated from the visceral motor column as in fishes, even though their functions have been secondarily transformed to the somatic type' (HERRICK, 1926).

Before discussing a few aspects of the significant basic mechanisms related to the oblongata's direct input and output, some comments on the *otic apparatus* and its peripheral nerve endings seem appropriate and are meant to supplement section 1, on peripheral nerve endings, of chapter VII (vol. 3, part II), in which, however, various sorts of receptors, preferably dealt with in connection with particular brain regions, were not considered.[20]

The *otic apparatus* of Vertebrates, which, in Ichthyopsidan forms, displays close relations to the *lateral line system*,[21] is characterized by an arrangement of '*mechanoreceptors*' responding to the pull of gravitation, to acceleration and deceleration of the body or some of its parts, and to sound waves. Vertebrates, or for that matter, a large variety of invertebrate animals are capable of orienting their bodies in space relatively to the gravitational field. Generally speaking, most, or at least many animal forms have a dorsal and a ventral side. This latter does not necessarily lie nearer the earth because it is heavier, since a fish dying in the water turns ventral side up. Again, the long axis of the Vertebrate body is usually kept horizontal, but occasionally, as in man, vertical. Orientation in relation to gravity represents a reflex activity, which involves nervous and muscular mechanisms and special sense organs. Such organs are known as '*static organs*' (NEAL and RAND, 1936).

In free-swimming Urochordates, static organs are present. Thus, in larvae of Ascidians such as Phallusia, a static organ with ciliated sensory epithelium, statolith, and nervous connections, projects into the brain cavity.[22] The Cephalochordate Amphioxus, on the other hand, does not display a recognizable static organ and does not seem to maintain a well-defined balance in swimming.[23] Nevertheless, its swimming attitude with the rostral end downward has been interpreted as suggesting the presence of a gravitational receptor mechanism. The

[20] Cf. vol. 3, part II, p. 783.

[21] The lateral line organs were discussed in chapter VII, section 1 of the preceding volume.

[22] Cf. Fig. 206, p. 295 in vol. 2 of this series.

[23] The 'spiraling' swimming movements of Amphioxus, believed to be mediated by the peculiar arrangement of the 'spinal eyes', was briefly discussed above on p. 69 and 75 of chapter VIII.

larvae of Lampreys are said to assume a similar swimming attitude (Young, 1955).

All Vertebrates have a static, respectively stato-acustic organ, considered by competent morphologists such as, e.g. Neal and Rand (1936), to be a 'novelty in this group and not inherited from invertebrate forbears'. The cited authors share the opinion that the Vertebrate ear represents a modified lateral line organ or a group of such organs. This is suggested by the development of the otic apparatus from a thickened placode of ectoderm on the side of the head. As in the case of the lateral line organs, the ectoderm sinks below the surface, and patches of sensory cells are differentiated. Moreover, the eighth nerve develops as a branch of the seventh, which participates in the innervation of the lateral line organs. Also, in Elasmobranchs, the external apertures of the invagination canal of the statocyst (labyrinth) lies near the openings of the occipital row of lateral line organs.[24]

Among Vertebrates, the Cyclostome *Myxine* displays the simplest (perhaps 'reduced') static organ, roughly resembling an inflated inner tube of a motor car tire (Fig. 163A). Its nerve supply suggests that it corresponds to utriculus and to the two vertical semicircular canals in other Vertebrates.[25] Two ampullae are present, each containing a *crista* with a cluster of hair cells. In addition, a sensory *macula* is located in the basal part of the organ. The tube is filled with fluid *(endolymph)* and a small endolymphatic duct extends dorsally towards the skin.

In *Petromyzon* (Figs. 163B, 164A), a ventral *sacculus* is partly separated from the *utriculus* by a constriction. Two semicircular canals, whose planes are oriented at a right angle to each other, display ampullae with cristae. Additional sensory spots are provided by a papilla in the sacculus, another in the utriculus, and one in the medial part of each ampulla, moreover by a papilla in the so-called dorsal duct (or 'invagination canal'). The sensory hairs in the otic organ of Cyclostomes and other Vertebrates are covered by a gelatinous or mucous matrix which, in the maculae, but not in the cristae, contains calcareous concretions *('statoliths', 'otholiths')*.

---

[24] A comparable separation within the system of lateral line organs is known to occur with respect to the *ampullae of Lorenzini* and the *vesicles of Savi* in Elasmobranchs.

[25] Other authors regard this tube as homologous to a single semicircular canal, namely to the posterior vertical one, joined by a basal chamber corresponding to an 'utriculus-sacculus complex' (Ross, 1963). Yet, in other Vertebrates, each semicircular canal possesses only a single crista and ampulla.

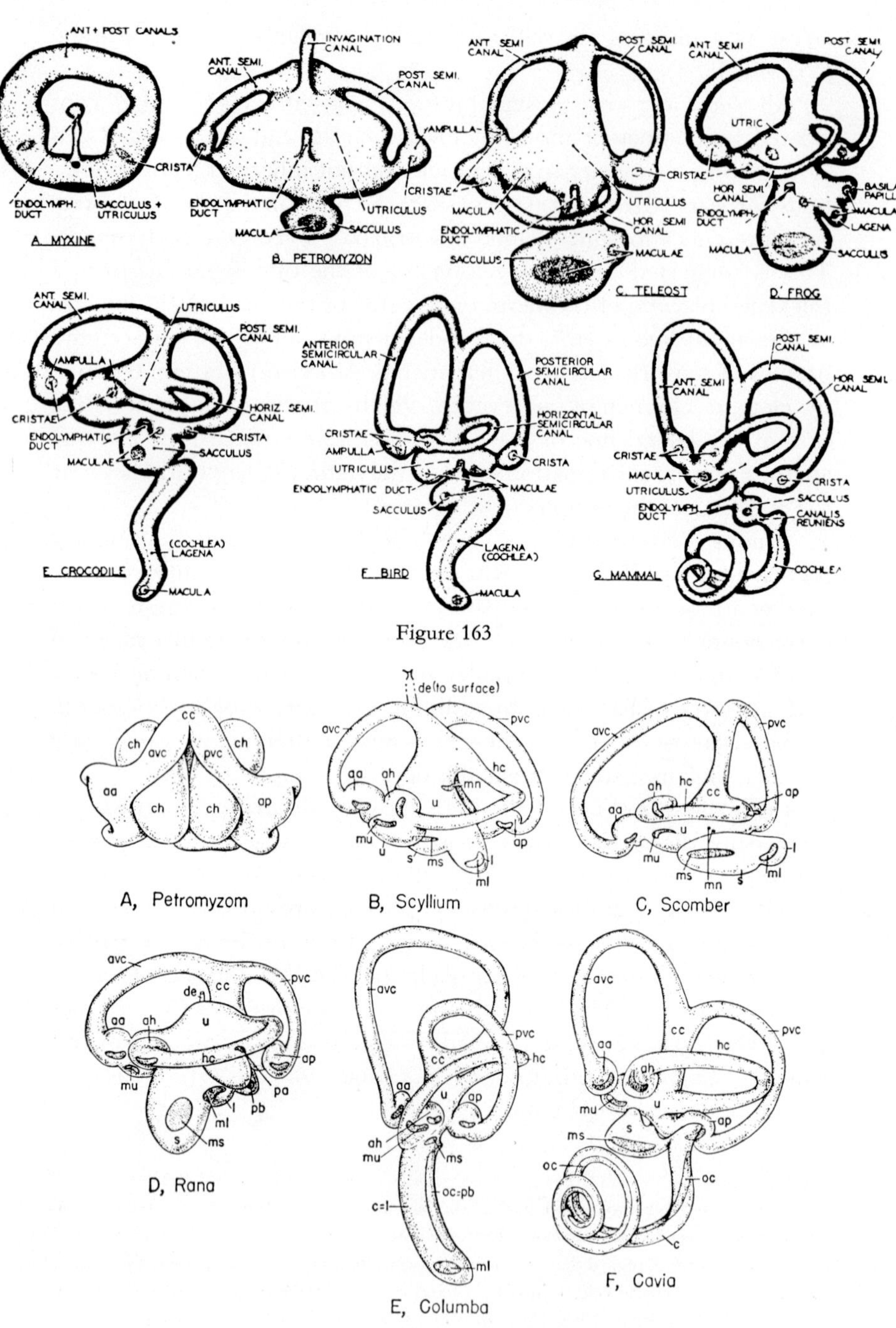

Figure 163

Figure 164

In *Selachians* (Fig. 164B), the cavity of the otic apparatus retains its primary connection with the ectodermal surface through the persistent invagination canal,[26] and is thus filled with sea water instead of endolymph. In some Elasmobranchs, grains of sand replace the calcareous statoliths. As in all Gnasthostome Vertebrates, three semicircular canals, in planes at right angles to each other and with ampullae as well as cristae, are present. The sacculus becomes further separated from the utriculus, the two remaining connected by a ductus or canalis utriculo-saccularis. A so-called *lagena* is formed as an outgrowth of the sacculus. Again, there are diverse sensory maculae. This entire system of canals and sac-like structures, known as the membranous labyrinth, and enclosed in a cartilaginous capsule, which fuses with the cranium, is more or less separated from its surrounding cartilaginous wall by a *perilymphatic* space containing tissue fluid. A comparable perilymphatic space is not clearly recognizable in Cyclostomes. The spiracular slit in many Selachians gives off a dorsal diverticulum which becomes applied to the ventro-lateral wall of the auditory capsule below the prominence formed by the horizontal semicircular canal. According to GOODRICH (1930, 1958), this 'auditory diverticulum' possibly conveys vibrations to the 'auditory labyrinth'. Near the internal opening of the spiracular cleft there is another diverticulum from its anterior wall, containing a neuromast sense organ supplied by the otic branch of the facial. Similar structures were noted in Ganoids and in Dipnoans. The significance of these various spiracular configurations or 'spiracular sense organs' (GOODRICH, 1958) remains obscure.

[26] This canal is commonly designated as 'endolymphatic duct', but NEAL and RAND (1936) consider this term erroneous since the 'true' endolymphatic duct of Teleosts and higher Vertebrates seems to represent a different and secondary outgrowth from the sacculus.

Figure 163. Configurational aspects of the (left) membranous labyrinth in the vertebrate series, as seen in lateral view (after HESSE, from NEAL and RAND, 1936).

Figure 164. Additional configurational aspects of the (left) membranous labyrinth in representative Vertebrates, as seen in lateral view, and based on the studies of RETZIUS. Sensory areas (except in A) are shown as if the labyrinth were transparent (from ROMER, 1950). aa: ampulla of anterior canal; ah: ampulla of lateral canal; ap: ampulla of posterior canal; avc: anterior vertical canal; c: cochlear duct; cc: crus commune; ch: chambers lined with ciliated epithelium in lamprey; de: 'endolymphatic duct'; hc: horizontal canal; l: lagena; ml: macula lagenae; mn: macula neglecta; ms: macula sacculi; mu: macula utriculi; oc: *organ of Corti;* pa: 'papilla amphibiorum'; pb: papilla basilaris; pvc: posterior vertical canal; s: sacculus; u: utriculus.

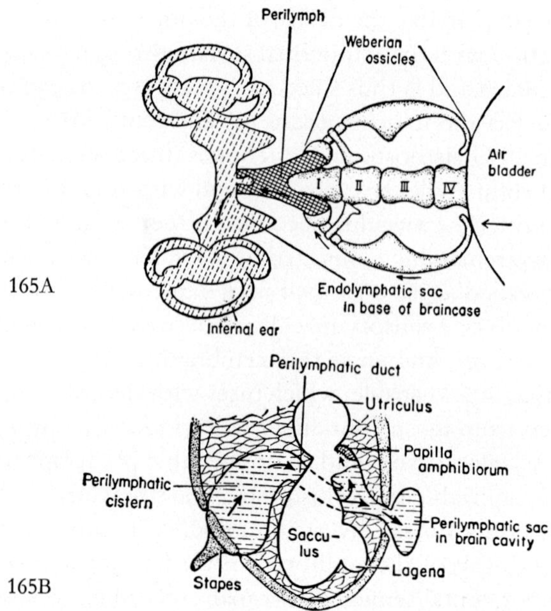

Figure 165 A. Diagrammatic drawing showing the arrangement of the *Weberian ossicles* present in some Teleosts. Roman numerals indicate the vertebrae from which these ossicles are derived. Arrows show direction in which the vibrations are transmitted (after CHVANILOV, from ROMER, 1950).

Figure 165 B. Schematic section of the internal ear in a Salamander. The basilar papilla is absent here, but is found in addition to the papilla amphibiorum in Anurans (after DE BURLET, from ROMER, 1950).

In *Teleosts* (Fig. 163C, 164C), the invagination canal degenerates, being replaced by a 'true' endolymphatic duct[27] which grows out from the sacculus. The cavity of the bony otic capsule with its perilymphatic space opens into the cranial cavity. It seems now well established that at least a number of Teleosts efficiently register sound waves, that is, possess definite sense of '*hearing*' in its behavioristic meaning. The most acceptable evidence points to the *sacculus*, and particularly to the so-called *macula lagenae* (Fig. 164C) as being here the relevant receptor

[27] The endolymphatic duct may terminate with an enlargement, known as the *endolymphatic sac*, and located within the cranial cavity. An endolymphatic sac is absent in some Teleosts, but in others, as well as in various Amphibians and in some Reptiles, the endolymphatic sacs are of substantial size. Right and left ones may connect either dorsally or basally of the brain, and, in Anuran Amphibia, may extend into the vertebral canal.

structure. Accessory arrangements for the transmission of sound waves to the sacculus are provided by various connections with the air bladder (Germ.: *Schwimmblase*). Thus, in Clupeids (Herrings), the air bladder displays a rostral tubular extension reaching the membranous labyrinth. In Ostarophysi, which include Cyprinoids and Siluroids, the air bladder is linked with the sacculus through the *organ of Weber* (Fig. 165). Processes of the rostral vertebrae are here differentiated on each side as the three or four *Weberian ossicles*, which form an articulating chain extending from the air bladder to the labyrinth. These ossicles can be regarded as analogous (but not homologous) to the auditory ossicles of Mammals and other Tetrapods.

In *Amphibians*, the cavity of the otic capsule becomes closed, being thus, except for an extension of the perilymphatic duct, separated from the cranial cavity. The sacculus displays a lagena and a *basilar papilla* (Figs. 163D, 164D). However, several variations in the development of the Amphibian saccular sensory structures obtain, and can be regarded as 'specializations' whose phylogenetic interpretation remains obscure.

The loss of gills in terrestrial Amphibia is concomitant with significant modifications of the visceral arches. The hyomandibular cartilage, losing its function as a suspensory structure of the jaw, is displaced into the spiracular passage as the *columella* or *stapes*. Thus a morphologic pattern corresponding to that of the middle ear of higher Vertebrates becomes manifested.[27a] The columella or stapes plays, within the '*fenestra ovalis*' or '*fenestra vestibuli*' (which is an opening of the otic capsule), against the wall of a perilymphatic space or 'cistern'. The Tetrapod stapes can thereby directly transmit vibrations from the tympanic membrane to the labyrinth's perilymphatic space (Fig. 165B).

In *Sauropsidans*, the lagena and the basilar papilla become part of an elongated tubular structure which, in Crocodiles, displays an outline suggestive of the spirally wound Mammalian cochlea. By the attachment of the Sauropsidan membranous cochlear duct to the wall of its bony case, the perilympathic space becomes here subdivided into a *scala vestibuli* and *scala tympani* comparable to those accompanying the cochlear duct of Mammals (Figs. 165C, 171).

[27a] It is believed that a tympanum and a tympanic cavity are particularly relevant for sound transmission through air, but of lesser import for sound transmission through water or ground. Thus, a middle ear cavity is missing or greatly reduced in diverse aquatic and burrowing Amphibians, including also some Anurans such as burrowing toads (e.g. *Pelobates*).

The scala tympani is related to an aperture of the otic capsule designated as *fenestra cochleae*, through which the perilymphatic space abuts against the cavus tympani. In Mammals, that membranous abutment, separating the perilymphatic scala tympani from the air-filled *tympanic or middle ear cavity*, is also known as the *fenestra rotunda*.

Accessory small sensory patches of the membranous labyrinth, noted by RETZIUS (1880) since about 1872, and subsequently by other authors, in Teleosts, Amphibians, Sauropsidans and Mammals, have been variously described as *macula neglecta* and *papilla amphibiorum* (cf. further below, footnote 129, p. 453, of section 7). Present-day gadgeteers have introduced the term '*sensura neglecta*' for one of these still poorly understood small 'maculae' or 'sensory regions' (cf. e.g. CORREIA *et al.*, 1974).

The overall configuration of the *Mammalian* membranous labyrinth is illustrated by Figures 163G, 164F and 167. The length of the cochlea, whose receptor structures are included in *Corti's organ*, varies from a half turn in the Prototherian Monotreme Echidna to three and a half turns in the Eutherian Ungulate Cervus. Instead of a single auditory bone (stapes), a chain of three ossicles, *malleus*, *incus*, and *stapes* connect the tympanic membrane with the fenestra vestibuli. Two small muscles, *m. stapedius* and *m. tensor tympani* control the linkage effect of

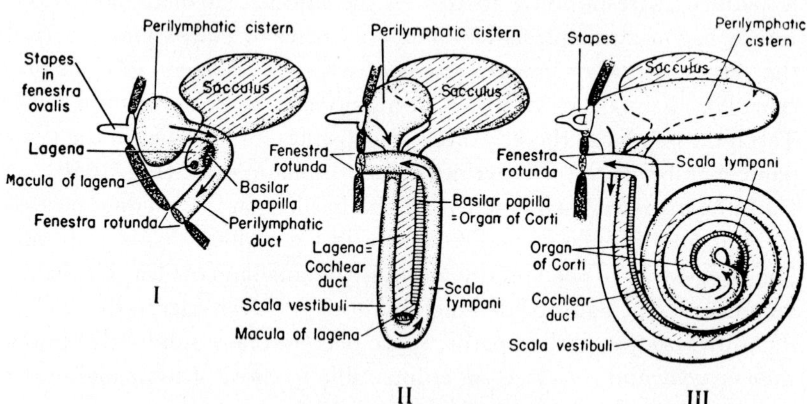

Figure 165 C. Diagrammatic sections through the saccular region to show the evolution of the cochlea in Amniota (from ROMER, 1950). I Primitive Reptile with a small basilar papilla adjacent to the perilymphatic duct. II The Crocodile or Bird type: the lagena has elongated to form a cochlear duct, the basilar sense organ with it, and a loop of the perilymphatic duct follows the cochlear duct in its elongation. III The Mammalian type: the cochlea is further elongated and coiled.

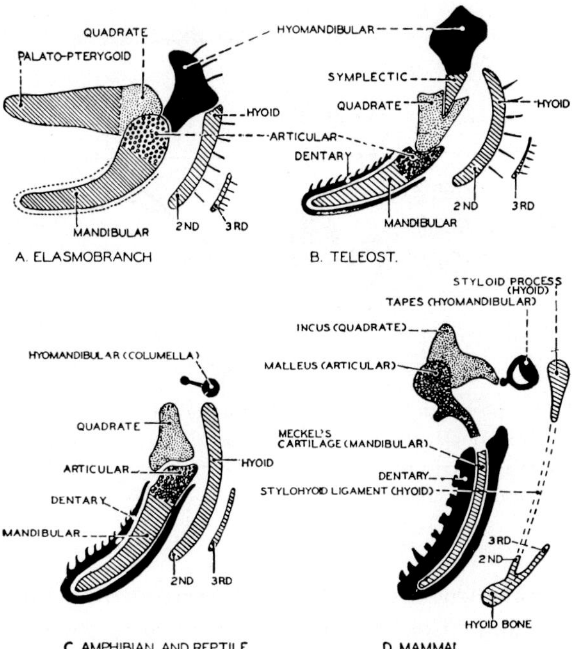

Figure 166. Diagrams of the first and second branchial ('visceral') arches in representative Vertebrates, illustrating the transformation of the mandibular articulation of submammalian forms into the incudo-malleolar articulation of Mammals, as well as the derivation of stapes from the hyomandibular (after GEGENBAUR and STEMPELL from NEAL and RAND, 1936). This figure may be compared with Figure 424, p. 839 in the preceding volume 3, part II.

this chain.[28] The presumed transformation of the mandibular articulation in the jaw of submammalian Vertebrates into the incudo-malleolar articulation of Mammals is illustrated by Figure 166.

In Man and other Mammals, the internal ear can be described as including two rather distinct groups of sense organs, consisting of the so-called *vestibular* structures, which may conveniently be classified as 'proprioceptive', being essentially concerned with body equilibrium, and of the *cochlea*, containing the exteroceptive *organ of Corti*, mediating the sense of hearing.

[28] The *m. stapedius*, pertaining to the hyoid arch, is innervated by a branch of n. facialis. The innervation of *m. tensor tympani*, which pertains to the maxillo-mandibular arch, is provided by the trigeminus through a small branch of the mandibular nerve (V_3).

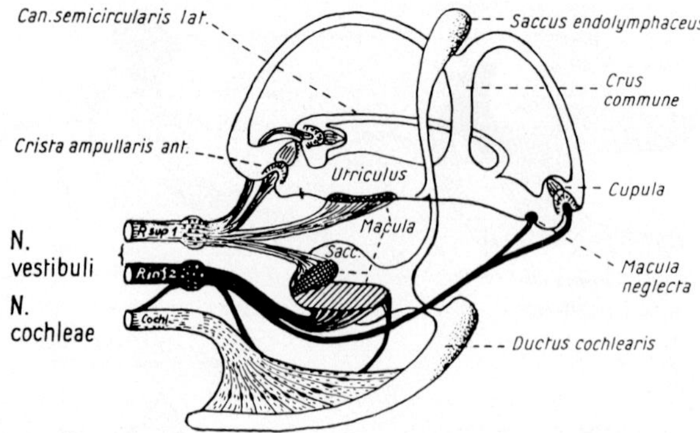

Figure 167. Diagrammatic illustration of the human membranous laybrinth indicating the distribution of vestibular and cochlear components of n. VIII (after DE BURLET, from MIES, 1962). Note again crus commune of anterior and posterior semicircular canals, as also shown in Figures 163 and 164.

The *vestibular* structures consist of the three *semicircular canals* with their ampullae and cristae, and of *utriculus* and *sacculus* with their sensory *maculae*. These structures are innervated by the vestibular division[29] of the eighth cranial nerve *(n. vestibularis)*, while the *organ of Corti* is supplied by the cochlear division *(n. cochlearis)*. The bipolar, primary afferent nerve cells of the vestibular nerve are located within the internal acoustic meatus, forming the *ganglion vestibulare*, commonly represented by closely adjacent swellings of the two main vestibular rami. The likewise bipolar nerve cells of the *cochlear ganglion (ganglion spirale Corti)* lie within spiral canal of the bony *modiolus* (Figs. 167, 171).

A subdivision of the eighth nerve into ramus anterior and ramus posterior obtains in *Amphibians*. The anterior branch generally supplies the lateral and anterior semicircular canals and the utriculus, while the posterior branch joins sacculus and posterior semicircular canal. Similar conditions are also found in *Sauropsidans* where the posterior branch also innervates the *basilar papilla*. In some forms, however, an independent branch for the *macula sacculi* has been described. Minor variations of this innervation pattern can be noted both in

[29] In Man, the vestibular nerve commonly displays two main subdivisions, ramus superior and inferior, whose distribution is shown in figure 167. There are, however, a number of variations, and a superior, inferior and posterior branch may arise distally from a common vestibular ganglion of a single main vestibular nerve.

Amphibians and Sauropsidans.[30] The *cochlear nerve of Mammals* can be interpreted as a derivative of the submammalian ramus posterior nervi octavi, whose vestibular portion is the Mammalian ramus vestibularis inferior. The submammalian ramus anterior nervi octavi corresponds to the ramus vestibularis superior of Mammals.

The planes, in which the three semicircular canals are oriented at a right angle to each other in accordance with the three dimensions of space are illustrated by Figure 168. Anterior (or superior) and posterior semicircular canals have a common limb *(crus commune)*. The ampullae of all three canals open into the utriculus. The structure of a *crista ampullaris* is shown in Figure 169. Peripheral (afferent) neurites of bipolar vestibular ganglion cells end on the sensory receptor elements provided with hair tufts.[31] These latter are imbedded in the gelatinous *cupula ampullaris*, which seems to consist of mucopolysaccharides. In the *maculae* (Fig. 170), the hair cells likewise protrude into a gelatinous layer which, however, contains the *otoliths* or *otoconia*, consisting of a combination of calcium carbonate and protein.

It seems reasonably well established that the adequate stimulus for the *cristae* is an angular acceleration resulting from motions of head respectively body. The inertia of the endolymph leads to initial and terminal displacements of the cupula, causing concomitant deformations of the hair tufts.[32] The stimulus registered by the *maculae* is presumably the gravitational pull on the otoliths, which, in turn, by a shearing or perhaps also by pressure respectively pull, affect the sensory elements.

The cochlear *organ of Corti* (Figs. 171, 172) consists of a highly differentiated sensory epithelium with hair cells and their supporting elements. The relevant cellular structures of the organ rest on the *basilar membrane* separating the endolympathic ductus cochlearis from the perilymphatic scala tympani. A thin vestibular membrane *(Reissner's mem-*

---

[30] Cf. Kappers *et al.*, 1936, who discuss numerous details of this topic with reference to the investigations of Retzius, de Burlet and others, listed in the bibliography of the cited treatise.

[31] Cristae and maculae consist of *supporting cells* and *sensory hair cells*. Recent studies with electron microscopy have disclosed that two sorts of sensory cells with slightly different ultrastructural details and synaptic endings (chalice-type and bouton type) are found in both cristae and maculae. These details, described in contemporary textbooks of histology (e.g. Bloom and Fawcett, 1968) are nor relevant to the aspect here under consideration.

[32] Displacements of the cupulae are also caused by thermal convection currents of the endolymph, as, e.g. elicited by the standardized 'caloric tests'.

brane) separates the perilymphatic scala vestibuli from the cochlear duct. Both perilymphatic scalae communicate through the *helicotrema* at the apex of the cochlea. In Man, the basilar membrane contains about 24,000 fibers *('auditory strings')* which are radially directed from central modiolus to peripheral spiral ligament, but nevertheless display a *nearly* parallel arrangement. Although the canal of the osseous cochlea is wider at the base than at near the apex, the basilar membrane is wider at the apex (about 0.5 mm) than at the basis (about 0.4 mm). Thus, the 'auditory strings' have been compared to piano wires arrayed in order of resonance frequency.

The *hair cells* are located in one inner row of about 3,500 'internal' elements, and an outer alignment of three to five rows comprising about 20,000 'external' hair cells. The hairs (or microvilli) of the sensory cells are imbedded in, or are at least firmly attached to, the tectorial membrane overlapping the *organ of Corti*. Nerve endings of large knob-type are related to the base of the sensory elements. These terminals are provided by the peripheral (afferent) neurites of the bipolar

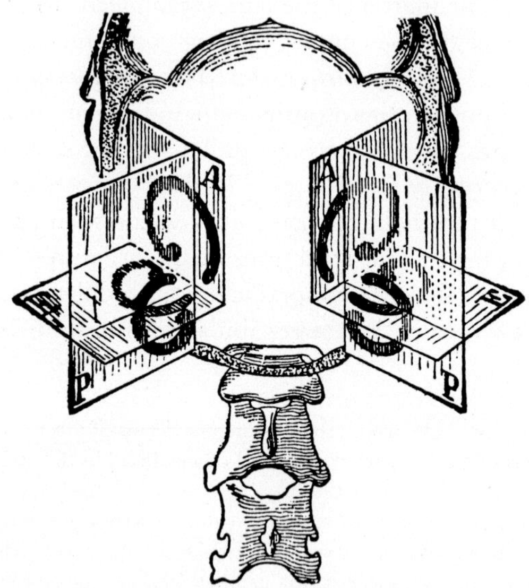

Figure 168. Diagram indicating spatial relationships of the semicircular canals, whose planes lie at right angles to one another, in the human head. Both lateral canals (E) lie in the same plane, while anterior canal (A) of one side lies in a plane parallel to that of the other side's posterior (P) canal (after EWALD, from HERRICK, 1931).

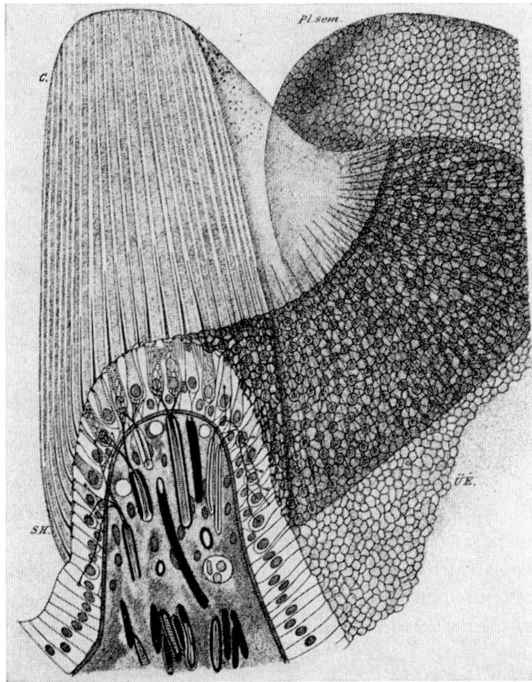

Figure 169. Diagram illustrating three-dimensional aspect of one half of a mammalian ampullar crista as seen in a longitudinal section of a semicircular canal, passing across the crista (after KOLMER, from MAXIMOW, 1930). C: gelatinous mass of the cupula; Pl: sem.: 'planum semilunatum'; SH: hair tufts; UE: transitional epithelium.

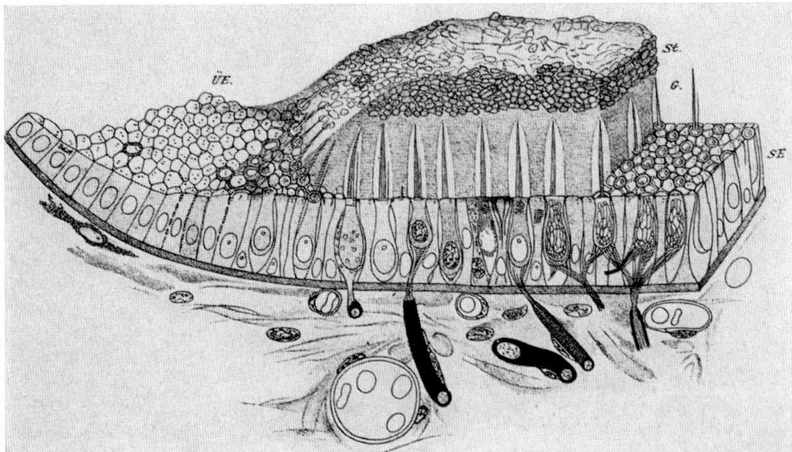

Figure 170. Diagram illustrating three-dimensional aspect of a mammalian macula (after KOLMER, from MAXIMOW, 1930). G: gelatinous layer; SE: sustentacular elements; St: otoconia; UE: transitional epithelium.

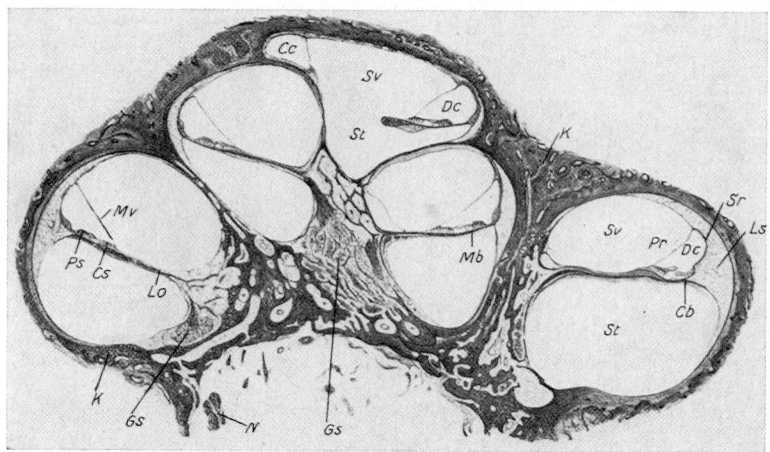

Figure 171 A. Axial section through the human cochlea (after SCHAFFER, from MAXIMOW, 1930). Approx. X 16; red. $^2/_3$. Cb: crista basilaris; Cc: cecum cupulare; Cs: crista spinalis; Dc: ductus cochlearis; Gs: ganglion spirale; K: bony wall of the cochlea; Lo: lamina spiralis ossea; Ls: ligamentum spirale; Mb: membrana basilaris; Mv: membrana vestibularis; N: cochlear nerve; Pr: prominentia spiralis (above lead Cb); Ps: *organ of Corti;* Sr: stria vascularis; St: scala tympani; Sv: scala vestibuli.

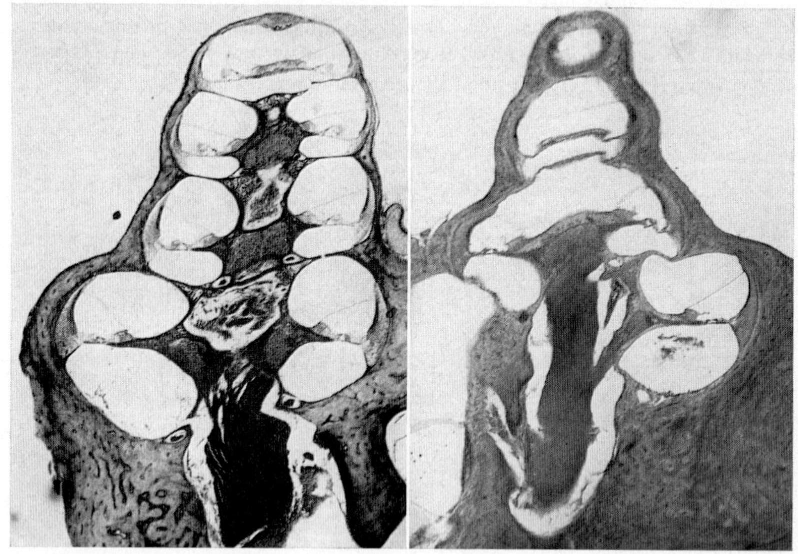

Figures 171 B, C. Paraxial sections through the cochlea of the Guinea pig. B Osmic acid technique. C Iron-hematoxylin stain; ×27; red. $^1/_2$. The unlabelled structures can be easily identified by comparison with Figure 170A.

ganglion cells in the ganglion spirale. In addition, synaptic endings believed to be of efferent type, that is, conveying impulses to the sensory cells, have been described in the *organ of Corti* as well as in the vestibular cristae and maculae. *Centrifugal fibers* reaching the otic apparatus were already recognized by HELD (1893) and, much later, confirmed by RASMUSSEN (1946, 1950). Although it may be surmised that such fibers, perhaps originating in the oblongata at the level of the superior olivary complex, exert some sort of controlling or adjusting effect upon the receptors, nothing certain is known about their function.

With respect to *numerical relationships*, estimates of the number of ganglion cells in the human ganglion spirale range from about 23,000 to 39,000, while the average number of nerve fibers in the trunk of the cochlear nerve has been estimated at 31,000 (BLINKOV and GLEZER, 1968). For the number of (medullated) fibers in the human vestibular nerve, estimates ranging between 14,000 to 24,000 are given. With regard to the *organ of Corti*, each fiber of the cochlear nerve is believed to be connected with several hair cells, while each cell, in turn, has connections with several nerve fibers. According to other estimates, about 90 per cent of all cochlear fibers have connections with the inner hair

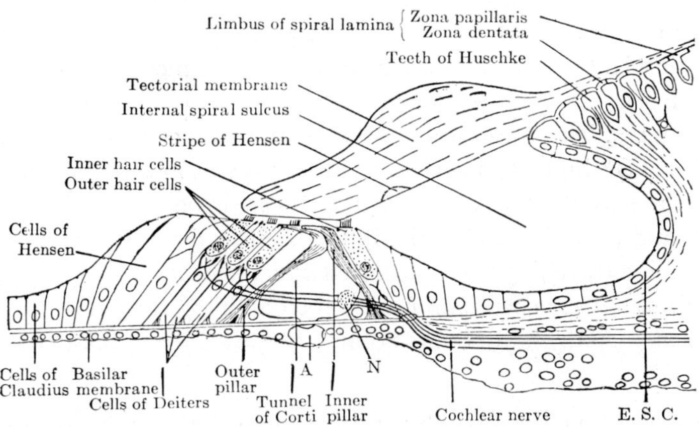

Figure 172. Diagrammatic 'cross-section' (radial section) through the *organ of Corti* (based on findings of HELD and VAN DER STRICHT, from HERRICK, 1931). A: spiral vessel; E.S.C.: epithelium of spiral sulcus; N: 'longitudinal nerve of *tunnel of Corti*'. The right side is oriented toward central axis of cochlea. Although electron microscopic studies have provided additional details, the present diagram, incorporating the findings of the older classical authors, is fully adequate for an illustration of the organ's structure in the aspect here under consideration.

cells, yet 'the number of fibers innervating the outer hair cells appears to be equal to the number of inner hair cells' (ZWISLOCKI and SOKOLICH, 1973). This is supposed to obtain in a Rodent (the Gerbil, *Meriones)*, where there may be about three times as many outer than inner hair cells.

Discounting the mechanisms of tympanic membrane and auditory ossicles as not relevant to the topic here under consideration, a few comments on the transducer function of *Corti's organ* will be sufficient in the present context. It is generally assumed that the sound waves reaching the perilymph of the cochlea[33] affect the basilar membrane, to which the structures including the sensory elements are attached. The *resonance theory* of HELMHOLTZ (1885), which was, despite several shortcomings, widely accepted as the most plausible explanation of sound analysis[34] by *Corti's organ*, is based upon the fact that the basal auditory strings of the basilar membrane are short at the cochlea's base and become longer toward the apex. This, again, agrees with the findings indicating that sound waves of high frequency are registered by the basal portion of *Corti's organ*, and those of low frequency by the apical part. Recent theories, however, based on the investigations of VON BÉKÉSY (1951) have shown that, at very low frequencies, below 50 cps, the basilar membrane vibrates as a whole. When, at higher frequencies, the basilar membrane is driven into vibrations by sine-wave excitation, the vibration is strongest at a particular locus of the membrane. It falls off away from this locus in either direction but more steeply toward the helicotrema. The greatest amplitude is near the oval window (i.e. at the base) for high frequencies, and the locus of maximum amplitude moves toward the apex as the frequencies are lowered. As each peak of a sine waves pushes the oval window against the perilymph of scala vestibuli, a travelling wave is started which reaches its maximum amplitude at a locus corresponding to the obtaining frequency and then falls away rapidly beyond this point. Thus, the thereby affected basilar

---

[33] The motion of the stapes footplate in the fenestra ovalis is presumed to act as a piston, setting up a wave travelling from cochlear base to apex through the scala vestibuli and there returning, through the helicotrema, by way of the scala tympani, to the fenestra rotunda with its 'secondary tympanic membrane'.

[34] The normal range of human sound perception extends from about 16 or 20 to approximately 20,000 cps *(hertz)*. Numerous animals, including dogs, register so-called ultrasonic sounds of much higher frequencies; Bats (Chiroptera) in particular, who use sound waves for echolocation in accordance with the technological 'sonar-principle', are believed to register 'sounds' up to about 100,000 cps (cf. vol. 2, p. 185 f. of this series).

membrane tends to separate spatially the various frequency components of a stimulating wave (cf. VAN BERGEIJK *et al.*, 1960).

When the basilar membrane is deformed by vibration, the sensory hair cells, imbedded in *Corti's organ*, slide with respect to the tectorial membrane, to which the hairs are attached. These latter are affected by the resulting mechanical interactions, which include, *inter alia*, shearing forces. The stimulated hair cells transmit their excitation to the input loci (synapses) of the afferent cochlear nerve fibers. The complex events related to this transduction process remain still poorly elucidated with respect to numerous relevant details. According to ZWISLOCKI and SOKOLICH (1973) most fibers of the auditory nerve respond to both displacement and velocity of the basilar membrane. Except at very high stimulus levels, motion and displacement toward scala tympani produce excitation, while the opposite displacement causes inhibition. The displacement and velocity responses interact. When both are excitatory or inhibitory, they are said to reinforce each other, when of opposite nature, 'a partial cancellation occurs'. The presence of both displacement and velocity responses in single fibers is believed to suggest that outer and inner hair cells of the cochlea interact.

Although differing from the hypothesis propounded by HELMHOLTZ, VON BÉKÉSY's concept of cochlear mechanims also leads to what has been designated a '*place theory*' *of hearing*. This concept implies that the mental quality (or '*qualité pure*') known as '*pitch*' depends upon a place-coordinate in the cochlea, which then becomes encoded by the neural mechanisms (N-events). It seems, however, that both the cochlear and the neural events leading to a perception and discrimination of pitch involve additional poorly understood factors not included in the elementary mechanisms formulated by the 'place theories'. These latter, nevertheless, appear to provide a valid overall approach, while the theory of sound patterns ('*Schallbildertheorie*'), propounded by EWALD (1903, 1922 and other publications), has not found general acceptance. According to this theory, standing waves are generated in the basilar membrane as a whole. To each sound corresponds a specific pattern involving the simultaneous stimulation of numerous and widely scattered hair cells. Not the excitation of a single localized group of nerve fibers, but the mutual spatial relationship of a significant number of separate fiber groups was accordingly assumed to be relevant to the perception of pitch.

Further remarks on the central pathways of audition, their relaying grisea, and additional apposite topics will be included further below in the sections and chapters dealing with the various levels of the brain stem, and, with reference to the diencephalo-cortical connections, shall again be considered in volume 5. The piezo-electric effect displayed by the cochlear 'microphonic potentials and the synchronized potentials (up to about 3,000 cps in the cochlear nerve) displayed by the neural auditory pathway were also briefly discussed in the monograph *Brain and Consciousness* (K., 1957, pp. 226–227).

It should moreover be mentioned that in those Vertebrates which do not possess a differentiated cochlea, but only a basilar papilla (e.g. the frog), the equivalent of a tectorial membrane seems to provide a structure allowing for sound waves to travel on it and to reach maxima of amplitude at different places.

The significance of the *lateralis system* of Anamnia and its specialization in Fishes with *electroreception* was pointed out in sections 1 and 5 of chapter VII of the preceding volume 3, part II. It seems likely that lateral line receptors had a 'natural propensity' to develop responses to electric (respectively electromagnetic) fields either concomitantly with, or independently of, the evolution of discharging electrical organs. Thus, there are indications that some Selachians and Teleosts without electrical organs use '*passive electroreception*' in e.g. prey hunting or navigation. It does not seem improbable that the puzzling long-distance oceanic migrations of Eels may be guided by electric field registration (Rommel and McCleave, 1972). Some data concerning electroreception of Fishes can be found in a recent summary by Bullock (1973). There are, moreover, various conspicuous long-distance migrations of Amniota such as certain South-American Turtles and various Birds, which lack a lateralis system. Although in Birds, the registration of stellar respectively celestial rotation has been suspected, magnetic field detection likewise seems to play a role (cf. e.g. Wiltschko and Wiltschko, 1972). The particular nature of the receptors remains unknown, but a similar type navigation might obtain in the aforementioned Turtles.

The phylogenetic evolution of electrical reception and discharge systems poses complex questions with respect to *Darwinian selection*. Assuming that such systems evolved gradually, it seems difficult to assume a primary 'selective' effect of such variations, which could be considered initially 'neutral', but subsequently may become more or less 'adaptive'.

With respect to the controversy between '*neutralists*' and '*selectionists*' some comments on 'neutral mutations' were given on pp. 27–29, section 1A, chapter VI of volume 3, part. II. It might here be added that the argument citing the so called '*fallacy of omniscience*' cannot logically be invoked against the concept of 'neutralism'. Said fallacy is supposed to assume that, if we cannot immediately see the functions of an organic system, it is therefore functionless. The 'neutralist' viewpoint does by no means claim that a dubious system, structure or feature is functionless, but merely questions, in full agreement with DARWIN's views, its *substantial* 'selective value'. The argument, moreover, is double-edged, since the 'fallacy of omniscience' could just as well delude the 'selectionists'. It seems evident that, in a large number of instances, the presumptive selective or adaptive value of a variation or mutation can only be established *ex post facto* on the basis of available data, and involves here numerous, in part rather uncertain variables respectively parameters.

With respect to *auditory theory* it must also be kept in mind that, because of the constant increase of new data requiring adapted 'modifying hypotheses', 'auditory science' still remains in what has been called a quite 'fluid' state, which makes even a partial synthesis extremely difficult.

Still another moot question is the possible additional or auxiliary 'auditory' function, particularly, but not necessarily exclusively in Anamnia, of the vestibular maculae primarily concerned with the registration of gravitational pull.

Reverting now to several significant aspects of the oblongata's reflex mechanisms, an additional sort of poorly understood receptors deserves brief mention, namely *intracerebral receptors* (cf. vol. 3, part II, chapter VII, p. 783, footnote 6). As far as the medulla oblongata is concerned, *chemoreceptors* of unknown type have been suspected in the *area postrema* (cf. vol. 3, part I, chapter V, p. 365), while the presence of 'medullary chemosensitive receptors' within the *respiratory centers* was inferred by v. EULER and SÖDERBERG (1952) on the basis of experimental evidence in the cat, including the recording of action potentials. It remains, of course, possible that these receptors are merely more or less typical nerve cells with special functional properties, and with or without certain additional structural peculiarities detectable by light respectively electron microscopy.

Generally speaking, the medulla oblongata is of particular importance for the regulation and maintenance of life in the Vertebrate organism. Thus, the grisea of the oblongata include 'centers' concerned with the processes indispensable for normal metabolism, namely res-

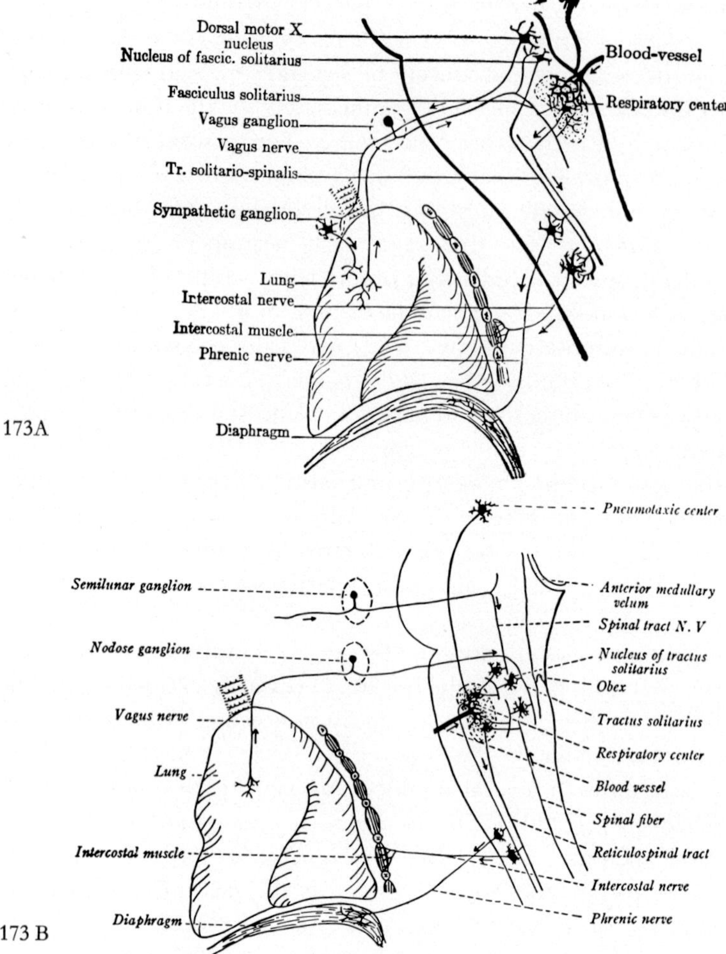

Figure 173 A. HERRICK's adaptation of CAJAL's diagram illustrating the reflectory mechanism of respiration in mammals and man (modified after CAJAL, from HERRICK, 1931).

Figure 173 B. RANSON's adaptation of CAJAL's diagram depicting the reflex mechanism of respiration. In contradistinction to HERRICK, RANSON, perhaps because he included the so-called 'pneumotaxic center', omits a reference to CAJAL's original drawing upon which this modification is based (from RANSON, 1943).

piration, blood circulation, and motility as well as secretory activity of the gastrointestinal tract, etc. Destruction of the oblongata irrevocably involves the death of the organism,[35] while destruction of other parts of the brain is not necessarily incompatible with the continuance of life.

The so-called *respiratory center* of Mammals[36] is contained within the griseum of '*nucleus*' *reticularis tegmenti* and seems to extend approximately from pontine levels to the caudal level of the rhomboid fossa in the region of the calamus scriptorius. Transection of the oblongata caudally to calamus respectively obex stops respiration, but this is not the case in transections which leave even a relatively small part of the calamus region in connection with the spinal cord. The significance of the bulbar brain stem for respiration was already recognized by GALEN (about 130–200 A.D.) on the basis of his vivisection experiments. FLOURENS (1794–1867) designated a rather circumscribed region below the end of the rhomboid fossa at the calamus at the '*nœud vital*', representing the 'respiratory center'. Subsequent experimental studies with improved technical methods suggested the presence of a more extensive griseum in the myelencephalon, including a dorsal subdivision predominantly concerned with expiration, and a ventral, mainly inspiratory subdivision.[37] However, a substantial degree of overlap seems to obtain, combined with a rather diffuse distribution, within the reticular formation, of the relevant multipolar elements taking part in respiratory activities, and the findings as well as the conclusions reached by the diverse contemporary investigators dealing with this topic have remained controversial. The *myelencephalic* respiratory center, in turn, is believed to be influenced by the *pontine* center, which was also designated as '*pneumotaxic center*'.

It seems evident that the bulbar grisea providing the mechanism of the so-called 'respiratory center' are not restricted to the reticular formation but also include at least part of the tractus solitarius with its nu-

---

[35] Unless, of course, life is artifically sustained by laboratory or clinical procedures.

[36] The respiratory mechanisms in the brain stem of submammalian Vertebrates have been less intensively investigated than those of Mammals. In contradistinction to air-breathing Vertebrates, the respiratory activity of aquatic ones, such as Fishes, is commonly correlated with gill movements controlled by branchial nerves and thereto related grisea of the medulla oblongata.

[37] The dorsal subdivision is said to extend into the ventral neighborhood of tractus solitarius and its nucleus, while the ventral subdivision may extend lateralward into the surroundings of nucleus ambiguus.

cleus, moreover the nucleus dorsalis vagi, and presumably additional cell groups with their neuropil. CAJAL (1909, Fig. 313, p. 752 l.c.) elaborated an instructive diagram of respiratory mechanisms, subsequently used by HERRICK (1931) and RANSON (1943) with slight modifications, as shown here in Figure 173.

The rhythmic activity of the 'respiratory center' doubtless depends on a number of different physiologic events. These latter include the *Hering-Breuer reflex*, a well-known feedback effect demonstrated more than hundred years ago, and designated by BREUER (1868) as '*Selbststeuerung der Athmung durch den Nervus vagus*'. Inspiratory inflation of the lung inhibits further inspiration through stimulation of afferent vagus fibers and contributes to initiate expiration. Expiratory collapse of the lung has the opposite effect, presumably through another set of recep-

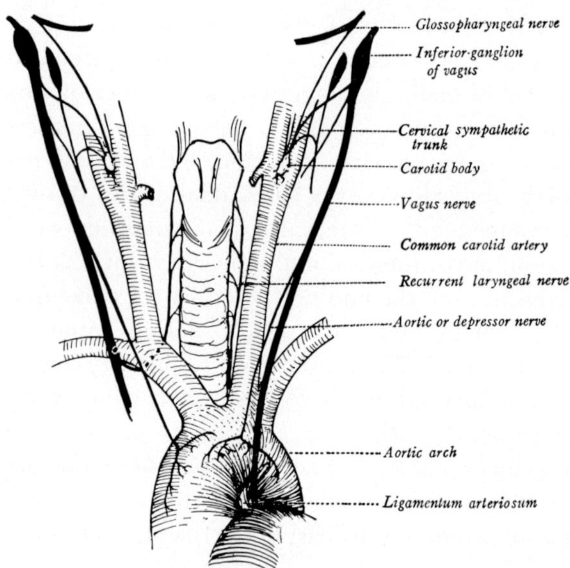

Figure 174. Diagrammatic sketch illustrating location of carotid body, some gross aspects of its innervation, as well as innervation of aortic arch (from RANSON-CLARK, 1959). The aortic body (or bodies), not shown, is (or are) located within the concavity of the aortic arch. The carotid sinus, not labelled, is represented by the initial segment of internal carotid, here laterally to external carotid. Actually, the internal carotid is here located dorsally to the a. carotis externa. Carotid sinus and carotid body are innervated by branches from both glossopharyngeal and vagus nerves, and moreover, by thin sympathetic twigs. Numerous variations obtain with respect to the detailed features of this general configurational arrangement.

tor nerve endings and afferent vagus fibers. This double system of negative and positive feedback operates through the respiratory centers, which, in turn, control the motoneurons of essential and auxiliary respiratory muscles. The 'channel' closing the feedback loop is provided by extraneural physical events, namely by inflation respectively deflation of the lung. Fiber tracts carrying impulses from the oblongata to the spinal motoneurons of respiratory muscles (diaphragm, intercostal and abdominal muscles, mm. scaleni etc.) do not seem to comprise a single '*respiratory bundle*', but rather to include reticulo-spinal and solitario-spinal fibers as well as components of the fasciculus longitudinalis medialis. In addition, parasympathetic and sympathetic outflow[38] to the smooth musculature and to the glands of the respiratory tract may be influenced by the activities of the 'respiratory center'.

Among afferent impulses affecting the 'respiratory center' those originating from the *vascular chemoreceptors* and *pressoreceptors* discussed in section 1 of the chapter VII, volume 3, part II should again be mentioned in this context. The action of the pressoreceptors was described, more than hundred years ago, about 1866, by Cyon and Ludwig, who identified the nervus depressor (an afferent vagus branch) in the rabbit. Although stimulation of the pressoreceptors mainly affects the circulatory system by reducing heart rate and blood pressure, respiration may also concomitantly be slightly inhibited. As regards the vascular chemoreceptors, their stimulation increases rate and depth of respiration, and causes a rise in blood pressure. The location of *carotid body*, its innervation, and that of the pressoreceptors in *carotid sinus* and *aortic arch* is indicated by the diagram of Figure 174, which refers to the arrangement obtaining for human anatomy. Figure 175 shows the general histologic features of the mammalian *glomus caroticum*, which are also characteristic for *glomus aorticum*.

The activity of the respiratory center appears, moreover, to be affected by the *intracerebral chemoreceptors* mentioned above on p. 323. It is believed that these elements respond to H-ion increase in the blood, possibly also directly to fluctuations of the blood's CO_2 or O_2 concentration. Said intracerebral receptors are commonly presumed to represent the primary mechanism responding to anoxemia, while carotid and aortic bodies might represent auxiliary mechanisms providing an

[38] It is generally assumed that sympathetic outflow dilates, and parasympathetic (vagus) outflow constricts the bronchi.

additional drive when the respiratory 'center' is impaired and does not sufficiently respond to altered conditions of the blood flowing through it.

In addition, *intrinsic rhythmic activities* of nerve cell groups have been assumed, which may act in combination with self-reexciting and self-inhibiting events as well as with the various sorts of peripheral stimulations with their reciprocal feedback innervation discussed in the foregoing comments.

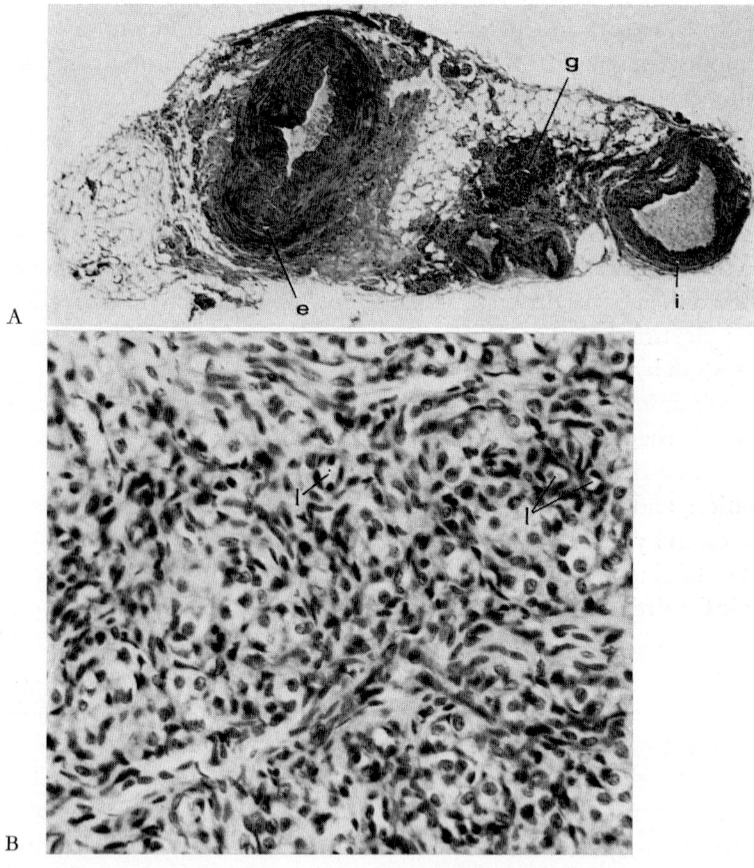

Figure 175. Histologic features of glomus caroticum. A Low power view showing location of glomus in the guinea pig between internal and external carotid (hematoxylin-eosin; ×70; red. $^2/_3$). B Details, at somewhat higher power, of glomus caroticum in the cat (hematoxylin-eosin; ca. 300; red. $^2/_3$). e: external carotid; g: glomus caroticum; i: internal carotid; l: lumen of sinusoid in glomus.

Section of both nn. vagi is followed by deep and slow respiration, presumably due to the elimination of the *Hering-Breuer effect*. A similar type of respiration, however, results from a transection of the brain stem at pontine levels, thus eliminating the influence of the pontine 'respiratory center' upon the myelencephalic one, or the interaction between both. If, in addition, both vagi are transected, breathing is arrested in a position of deep inspiration. This has been explained by assuming that a pontine or perhaps even mesencephalic 'pneumotaxic center' might act like the vagus nerves by limiting inspiration and allowing expiration to begin. If either one of these mechanisms remains, breathing continues, although being deep and slow; if both mechanisms are eliminated, the inspiratory state becomes 'jammed' and alternating respiratory movements are thereby prevented (STELLA, 1938; PITTS, 1940; RANSON, 1943).

Heart action and blood pressure are likewise controlled by grisea located in the oblongata and frequently designated as *heart and vasomotor 'centers'*. These latter comprise, in addition to reticular cell groups more or less coextensive or perhaps even in part identical with the so-called 'respiratory' centers, the nucleus dorsalis vagi, the tractus solitarius and its griseum as well as an undefined number of other neural structures (cf. also BAUMGARTEN, 1962). MAREY's well-known *law of the heart* is an expression of the activities displayed by these centers and by the above-mentioned vascular pressoreceptors. It states that, generally speaking, pulse rate is inversely related to arterial pressure, or, in other words, that rise or fall in blood pressure cause decrease or increase of heart rate, respectively. The *Bainbridge reflex*, on the other hand, is characterized by an increase in heart (pulse) rate upon rise of venous blood pressure in the right atrium. Within the left heart ventricle, not only pressoreceptors but also chemoreceptors seem to be present, both of which cause bradycardia and vasodilation upon stimulation.[39]

Reflex mechanisms, concisely suggested by CAJAL (1909) with respect to the human and mammalian oblongata, are those concerned with *coughing* and *vomiting*.[40] These behavioral activities, including their

---

[39] This may even lead to vasovagal syncope. Further details, with bibliographic references, have been summarized by GAUER (1960).

[40] Of these two 'reflectory' activities, coughing can easily be performed by 'voluntary' (cortical) action, while vomiting is not, as a rule, readily brought about by 'will'. It can, nevertheless, to some extent be 'voluntarily' triggered. Again, both activities may also, to a variable degree, be inhibited by the 'will'.

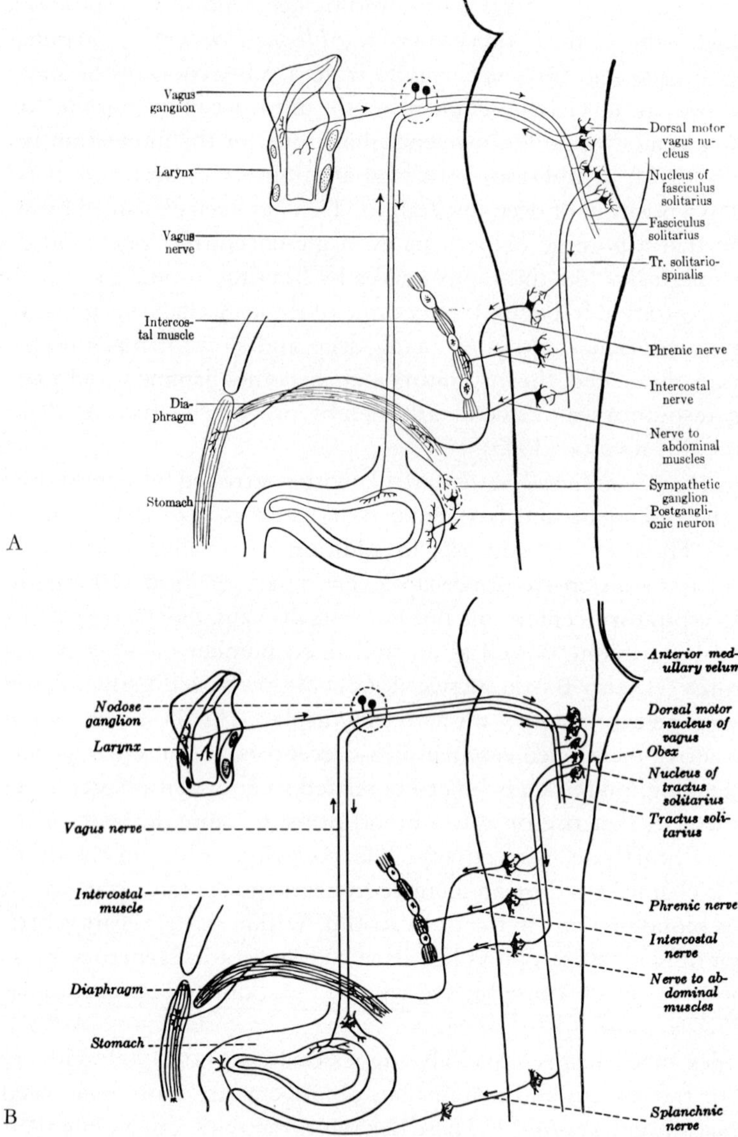

Figure 176. Diagrams showing two modified versions of CAJAL's original concept of the neural mechanisms mediating coughing and vomiting. A HERRICK's version (from HERRICK, 1931, after CAJAL). B RANSON's version (from RANSON, 1943; this author, and, in the later editions, S. L. CLARK, do not refer to CAJAL's original scheme).

relevant neuronal networks involve, in a partially overlapping fashion, the respiratory and the gastro-intestinal tract.

Figure 176 illustrates two adaptations of CAJAL's original diagram (1909, Fig. 312, p. 751) by HERRICK (1931) and RANSON (1943). In ordinary reflectory *cough*, an irritation of the mucous membrane of the larynx is transmitted to the nucleus of tractus solitarius, from which further impulses are directly or indirectly transmitted to the grisea of the spinal cord concerned with innervation of diaphragm, intercostal, and abdominal muscles. In addition to a 'tractus solitario-spinalis', undefined 'reticulo-spinal' connections are most likely also involved. The diagrams, moreover, omit the necessary participation of nucleus ambiguus, which mediates closure of the glottis, while expiratory pressure is built up. Sudden opening of the glottis then results in an explosive discharge of air.

In ordinary *vomiting*, and discounting the particular 'trigger-zones' in the pharyngeal mucosa, an 'irritation of the stomach' is carried by afferent vagus fibers to the griseum of tractus solitarius, from which, as in the cough reflex, direct and indirect descending pathways reach the spinal grisea. In this case, moreover, there is an excitation of the nucleus dorsalis vagi, from which parasympathetic[41] preganglionic fibers go into the vagus nerve, ending in the terminal gastric plexuses, 'from which, in turn, postganglionic fibers pass to the muscles of the stomach which participate in the ejection of its contents' (HERRICK, 1931). A comparable effect, likewise through the tractus solitarius, is mediated by the nervus glossopharyngeus innervating the above-mentioned 'trigger-zones'.

Besides '*reflectory vomiting*' caused by stimulation of visceral afferent fibers (mechanical stimuli, chemical stimuli by emetics or toxic substances), '*central vomiting*' may result from activation of the so-called 'vomiting center'. Substances such as apomorphine, emetine and others, if administered intravenously, directly act upon the relevant bulbar grisea. Related conditions are the '*pernicious vomiting*' in pregnancy, the '*projectile vomiting*' in elevated intracranial pressure (e.g. in case of brain tumors), etc., moreover the vomiting in *motion sickness*, triggered by the *vestibular system*. Finally, vomiting of cortical ('*psychic*') origin might be mentioned, occurring e.g. upon some strong emotions. Particular optic, olfactory or gustatory percepts can likewise induce vomiting,

[41] It will be recalled (cf. vol. 3, part II, chapter VII, p. 925), that, in a generalized formulation, the parasympathetic system is presumed to have an 'emptying' function.

also involving to a greater or lesser extent the participation of the cerebral cortex. In the 'parallel events' of consciousness, vomiting is commonly (but not necessarily always) accompanied by the experience of *nausea*, which is a *qualité pure*. The behavioristic aspect of nausea is characterized by a more or less definable pattern of vegetative and somatic neuronal circuit processes.

As regards additional central connections of cranial nerves provided by the oblongata, CAJAL's diagram of the mammalian tongue innervation by n. trigeminus, n. glossopharyngeus, and n. hypoglossus, shown below in Figure 268C, likewise represents an appropriate illustration for the understanding of basic neuronal mechanisms pertaining to the deuterencephalon. It should be recalled that n. glossopharyngeus, n. vagus, and n. facialis[42] mediate here input originating in the chemoreceptors (taste buds), while the n. trigeminus is concerned with the other sorts of sensory input.

Still another significant component of the oblongata's neuronal mechanisms is made up by the grisea of the VIIIth nerve with their connections and anatomically distinctive fiber tracts. The relevant details concerning these vestibular and cochlear systems shall be discussed further below in dealing with the various Vertebrate forms, respectively with the apposite subdivisions of the deuterencephalon.

Because the term '*brain stem*' (*truncus cerebri*) has not been included in the standardized nomenclature, but is nevertheless frequently used with different denotations by various authors, some comments on its meaning seem appropriate in concluding the introductory section of this chapter.

According to VILLIGER (1920) the brain stem comprises the base of the telencephalon ('stem of the telencephalon' with '*Stammganglien*'), the diencephalon, the mesencephalon, the pons, and the medulla oblongata. Thus, only the 'pallium' telencephali and the cerebellum are excluded. HERRICK (1931) defines as brain stem 'all of the brain except the cerebellum and the cerebral cortex and their dependencies, *i.e.* the segmental apparatus'. This latter concept, however, requires some additional comments, which will be given further below.

GLOBUS (1937) applies the term brain stem 'to the axial divisions of the brain, which include the medulla oblongata, pons, midbrain (with

---

[42] For purposes of simplification, the participation of n. facialis and n. vagus, and the cranial parasympathetic output to the glands of the lingual mucosa were omitted in CAJAL's diagram shown on p. 533.

overlying quadrigeminal plate), and the interbrain (with the basal ganglia) as far forward as the anterior commissure. Laterally the stem extends as far as the most medial portion of the *island of Reil*'. According to ELLIOTT (1954, 1963),[43] 'the tubular portion of the brain containing the third and fourth ventricles and aqueduct is called the brain stem because it acts as a stem from which other, suprasegmental structures sprout'.

Both HERRICK (1931) and ELLIOTT (1954, 1963) stress the distinction between segmental and suprasegmental structures, apparently first suggested by A. MEYER about 1898. It is evident that, in respect to its peripheral nerves, most of the longitudinal extent of the tubular neuraxis displays a *segmental* arrangement. Short interconnections, confined to the open neighborhood of a segment, are *intrasegmental;* longer connections are *intersegmental.* Very large 'coordination and correlation mechanisms', which are considered 'of later evolutionary origin' (HERRICK,1931) and 'not included in this simple plan', namely 'the cerebellum and the cerebral hemispheres' (ELLIOTT, 1954), respectively 'the cerebellar cortex and the cerebral cortex' (HERRICK, 1931), are *suprasegmental structures.*[44] It seems, however, evident that the tectum mesencephali, particularly in lower vertebrates, likewise represents a very significant suprasegmental structure (K., 1927, p. 195) which commonly displays what could reasonably be called a 'cortical' arrangement. One could, moreover, contend that the highly important *reticular formation* (or n. motorius tegmenti), briefly discussed in the next section of this chapter, should also be considered a suprasegmental griseum.

In dealing with the concept brain stem I originally favored a definition[45] similar to that adopted by ELLIOTT (1954, 1963), namely as in-

---

[43] ELLIOTT (1954, 1963) comments on both the variety and the lack of sufficiently specific definitions of the term '*brain stem*' in the literature. He follows the usage which extends said term 'to include the median portion of the forebrain enclosing the third ventricle'. According to the cited author 'this has proved convenient in many circumstances and does not seem to result in any confusion or ambiguity' (ELLIOTT, 1954). Yet, despite my esteem for this very competent author, I would maintain that 'the median portion of the forebrain enclosing the third ventricle' represents a rather vague concept. While perhaps convenient, the quoted definition does not, in my opinion, eliminate the prevailing ambiguity and confusion.

[44] It is interesting to note that one author refers to cerebellar and cerebral *cortex*, and the other to cerebellum and cerebral *hemispheres*.

[45] '*Dem Grosshirn stellen wir gegenüber den Hirnstamm (Truncus cerebri), Diencephalon, Mesencephalon und Rhombencephalon (ohne Cerebellum) umfassend. Das Einbeziehen der basalen Ganglien des Grosshirns in die Bezeichnung Hirnstamm, wie es so oft üblich ist, halten wir aus entwicklungsgeschichtlichen Gründen für bedenklich.*' (K., 1927).

cluding diencephalon, mesencephalon and rhombencephalon without cerebellum (K., 1927, p. 68). Subsequently, however, I became convinced that it seems preferable to designate as *'brain stem' only the deuterencephalon (mesencephalon and rhombencephalon) exclusive of cerebellum* (HAYMAKER and K., 1955, 1962, 1971). There are sound anatomic reasons for the employment of this limited meaning, particularly with respect to human gross anatomy respectively brain-dissection. Thus, the mesencephalon can easily be separated, in an approximately accurate manner, from the forebrain (diencephalon) by a curved incision through the diencephalo-mesencephalic boundary zone. Through other appropriate incisions, the three cerebellar peduncles can be cut, whereby the cerebellum is removed. The resulting preparation, as e.g. shown further below in Figures 272A, 273A and 274, represents what could roughly be called a significant anatomical 'unit'. The term 'brain stem' is thus restricted to the deuterencephalon or, in other words, to a subdivision of the brain from which all cranial nerves (III to XII) comparable to spinal nerve roots emerge. Thus, the *deuterencephalon includes all of the brain stem and, in addition, the cerebellum.* Topographically, the entire human brain stem in this definition is *infratentorial.* Although, in view of the limitations intrinsic to all semantic models, said definition appears to me the least objectionable, I do not claim that confusion, ambiguity, and inconsistency are thereby completely eliminated.[46]

The ambiguity of many generalized neurobiologic formulations may be illustrated by the following comments made by an otherwise quite competent physiologist (BAUMGARTEN, 1962). This author states: *'Phylogenetisch ist das Rautenhirn ein alter Hirnteil'.* Unless elaborately specified, this remark is evidently quite meaningless. With reference to the entire Vertebrate phylum, the rhombencephalon is in no way 'older' than telencephalon or diencephalon.[47] On the other hand, taking a

---

[46] *Confusion* will doubtless remain because of the various definitions based on different viewpoints. *Ambiguity* becomes evident in attempting to separate cerebellum (e.g. 'auricles' and 'crista cerebellaris') from the oblongata *sensu latiori* of lower vertebrate forms, in some of which, moreover, (e.g. Cyclostomes and Amphibians) the corpus cerebelli may be rather insignificant. *Inconsistency* is evidenced by excluding the suprasegmental cerebellum while including the likewise suprasegmental tectum opticum or tectum mesencephali. Obviously, in e.g. human or mammalian gross dissection, it is more convenient and less mutilating to detach the cerebellum than to slice off the tectum mesencephali.

[47] Clearly, it becomes here necessary to ask: which particular configurations *within* telencephalon, diencephalon, and deuterencephalon can be considered old, respectively new, and why so? Or: in which respect is, e.g. a Mammalian telencephalon or rhombencephalon new, respectively old?

stage comparable to that of Amphioxus as phylogenetically primitive, then, of course, the deuterencephalon is phylogenetically older than the secondary differentiations of archencephalon, namely telencephalon and diencephalon. Likewise, as a secondary differentiation of the deuterencephalon, the cerebellum is a 'phylogenetically new' configuration of the *Vertebrate* series.

BAUMGARTEN (1962) furthermore remarks: '*Die Brücke und die Oliven sind neencephale Strukturen, die mit der fortschreitenden Entwicklung des Neocerebellums Schritt halten. Die Pyramiden an der Basis des Rautenhirns sind ebenfalls später, im Zusammenhang mit dem Grosshirn entstanden*'. Now, if the cited author means by '*die Oliven*' the inferior olivary complex, this latter cannot possibly be evaluated as 'neencephalic' since it is doubtless present in Selachians. Pons and pyramids, on the other hand, can be justifiably classified as pertaining to EDINGER's 'neencephalon'. Yet, this merely implies that the Amniote or Mammalian rhombencephalon, in the same manner as telencephalon and diencephalon, displays complexly interwoven patterns of 'old' and 'new' neuronal configurations. Although phylogenetic interpretations including designations such as 'palaeencephalic' and 'neencephalic' may be considered fully justified in an appropriate context, such evaluations, in order to avoid meaningless as well as misleading '*clichés*', require far more careful logical and semantic procedures than most neurobiologic authors seem to realize.

2. Some Remarks on the so-called Formatio Reticularis Tegmenti

In the wider sense, the terms *substantia reticularis* or *formatio reticularis* refer to a sort of nervous parenchyma characterized by a more or less dense network of medullated fibers, within whose meshes neuronal pericarya and their synaptic neuropil are located. In the narrower sense, usually qualified as *formatio reticularis tegmenti*,[48] this term de-

---

[48] '*Tegmenti*' as emphasized by BECHTEREW (1898), with reference to the fact that the rostral portion of this formation is found in the mesencephalic tegmentum.

Recent authors have the tendency to include in their 'reticular formation' diencephalic grisea which either display a reticular aspect (e.g. the nucleus reticularis thalami ventralis) or are supposed to be functionally correlated with the deuterencephalic reticular formation, respectively with the vaguely defined 'centrencephalic' or 'activating system'. From a rigorous morphologic viewpoint, however, there is no justification to consider the diencephalic grisea in question as serially homologous with the formatio reticularis

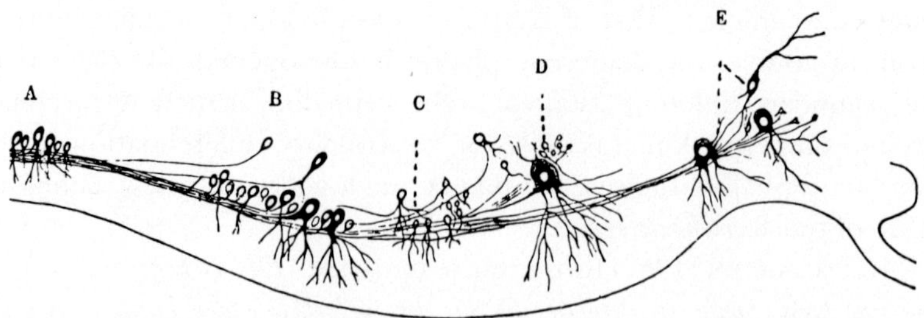

Figure 177. The reticular elements, *Müller's fibers*, and the arrangement of the fasciculus longitudinalis medialis in an Ammocoetes larva of Petromyzon (after Tretjakoff, from Kappers *et al.*, 1936. A: nucleus reticularis inferior; B: nucleus reticularis medius; C: trigeminal reticular group; D: isthmus group (C, D: nucleus reticularis superior); E: nucleus reticularis mesencephali.

notes a diffuse and elongated deuterencephalic neuronal aggregate, which includes, among its elements, large multipolar cells of 'motor' type, and is intermingled with a great number of longitudinal and transverse medullated fiber bundles or tracts.[49] Said griseum, which Kölliker (1896) designated *in toto* as *nucleus magnocellularis diffusus*, extends throughout the basal plate of the brain stem in all Vertebrates. This was particularly pointed out by Edinger (1908, 1911, 1912), who coined the term *nucleus motorius tegmenti*. Among other investigators of that griseum's comparative anatomy, Beccari (1921, 1943), Castaldi (1923, 1928), van Hoevell (1911), and Tretjakoff (1909) deserve special mention. Figures 177 and 178A show the overall arrangement of the *nucleus reticularis tegmenti* in the neuraxis of a Cyclostome and in the human brain.

In lower vertebrate forms, e.g. in Petromyzon, this nucleus is represented by a column of loosely spaced sizeable neuronal elements located within the basal plate and (in Petromyzonts) including the large *Müller cells*. Neurites originating in this aggregate run predominantly caudalward and transmit impulses to primary efferent elements of

---

tegmenti. Still more incongruous, with respect to morphologic concepts, is the designation '*formatio reticularis telencephalica*' recently applied to basal telencephalic grisea (including e.g. nucleus accumbens and region of diagonal band). At most, these designations represent a neurologic pun appealing to neurologists disregarding well established morphologic pattern concepts.

[49] Although this cell aggregate is present in Cyclostomes, the corresponding longitudinal and transverse fiber systems are here non-medullated (cf. footnote 3, p. 2 of chapter VIII).

brain stem and spinal cord. The nucleus motorius tegmenti can here be evaluated as an essentially secondary 'motor' center, whose elements receive their input from various functional systems.

Generally speaking, the diverse axonal end-arborizations connecting with *individual nerve cells* of nucleus motorius tegmenti in Anamnia may be provided by:

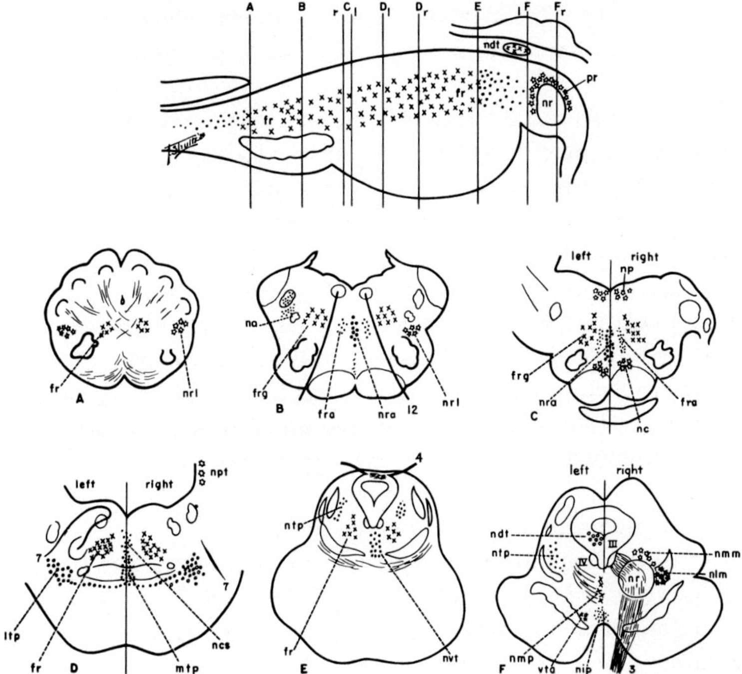

Figure 178 A. Semidiagrammatic sketch illustrating the extent of formatio reticularis tegmenti in the human brain (designed on the basis of original material in cooperation with Dr. Webb Haymaker at the Armed Forces Institute of Pathology while preparing a chapter on disorders of the brain stem for Baker's Clinical Neurology; use was made, for comparison, of Jacobsohn's 1909 survey). fr: formatio reticularis; fra: f.r.alba; frg: f.r. grisea; ltp: lateral tegmental process of pontile nuclei; mtp: medial t.p. of pont.n.; na: nucleus ambiguus; nc: nucleus conterminalis; ncs: nucleus centralis superior; ndt: nucleus dorsalis tegmenti; nip: nucleus interpeduncularis; nlm: nucleus reticularis lateralis mesencephali; nmn: nucleus reticularis medialis mesencephali; nmp: nucleus mesencephali tegmento-peduncularis; np: nucleus praepositus hypoglossi; npt: nucleus pigmentosus tegmento-cerebellaris; nr: nucleus ruber tegmenti; nra: nucleus of the raphe; nrl: nucleus reticularis lateralis; nvt: nucleus ventralis tegmenti; pr: prerubral tegmentum; vta: ventral tegmental area; Arabic numbers refer to cranial nerves; r and l in upper sketch indicate levels of right and left sides in the depicted sections; III, IV: nuclei of oculomotor and trochlear nerves.

1. Fibers from the forebrain (telencephalon and the subdivisions of diencephalon, including hypothalamus);
2. Fibers from tectum mesencephali;
3. Fibers from the trigeminus system;
4. Fibers from the stato-acustico-lateralis system;
5. Fibers from the cerebellar apparatus;
6. Fibers from the glossopharyngeus-vagus (and facialis) system;
7. Fibers from the ascending spinal systems, which include direct connections with nucleus motorius tegmenti as well as indirect ones by way of tectum mesencephali or cerebellum.

Thus, by this sort of arrangement, olfactory, optic, cutaneous, vestibulo-static or proprioceptive, acoustic, and visceral input, in various combinations or permutations, as the case may be, can converge upon a *single 'motor' element ('polyvalent transmitting neuron'* of KAPPERS), which, in turn, discharges into 'primary' efferent cell groups, in accordance with SHERRINGTON's principle of '*common final path*'. The nucleus reticularis tegmenti, moreover, gives origin to *ascending systems*, which, by way of several still insufficiently identified channels, convey impulses to tectum mesencephali, diencephalic (and perhaps also mesencephalic) cell groups of pretectal region as well as to various additional diencephalic grisea. These *ascending 'reticular channels'* seem to be of particular significance in the brain of 'higher' Vertebrates, particularly of Mammals as briefly pointed out further below.

In Fishes, Amphibians, and perhaps to some extent also in Reptilians, *Edinger's nucleus motorius tegmenti* seems to represent the most important 'motor center' of the brain, presumably corresponding to what J. STEINER (1885, 1888, 1900), on the basis of extensive experimental studies, but without relevant knowledge of, or reference to, structural details,[50] designated as '*allgemeines Bewegungszentrum*' respectively '*Hirnzentrum*'. With respect to the above-mentioned Vertebrates, STEINER localized this 'center' in rostral parts of the medulla oblongata at the transition to the base of mesencephalon, and defined it functionally '*als das einzige Lokomotionszentrum des Körpers, welches alle komplizierten Bewegungen desselben nach Massgabe der ihm aus mehreren Quellen zufliessenden Erregungen ausführt*'. Among such 'sources', STEINER mentions the telencephalon, the diencephalon ('*Sehhügel*', thalamus), the tectum mes-

[50] Despite various unconvincing or dubious experimental findings and interpretations as well as some obvious errors (such e.g. as the denial of the presence of a diencephalon in Teleosts) STEINER's treatises are of considerable interest, containing numerous valid original observations and concepts.

encephali, and the cerebellum. In numerous experiments, the cited author particularly investigated thereby obtained *'forced movements' (Zwangsbewegungen)* which he interpreted as resulting from the disturbed activities of his 'general motor center'.

The available data suggest that, from this secondary motor *center* or cell pool provided by the nucleus motorius tegmenti of lower Vertebrates there arises a compound common final path which reaches the primary efferent ('motor') elements, and thus transmits the converging impulses. These latter, originating from all functional systems, may act jointly or singly, simultaneously or consecutively, with mutual 'reinforcements' ('summations', 'facilitations') or 'inhibitions', in the most diversified combinations and permutations.

In gnathostome Anamnia (Chondrichthyes and Osteiichthyes and Amphibia) such common final path is doubtless represented by the *fasciculus longitudinalis medialis* which presumably constitutes the main, if not the only central motor tract,[51] and which, moreover, from a functional viewpoint, could in some respects be compared with the pyramidal pathway of higher Mammalia. In Cyclostomes, a 'fasciculus longitudinalis medialis' is roughly suggested, but the descending fibers are, on the whole, more diffusely scattered in the spinal cord's non-medullated 'white substance' which does not clearly display the typical 'funiculi' of higher vertebrates. Thus, descending fibers of the 'reticular pathway' may run through lateral and even posterior 'funiculus' (cf. p. 89, section 4, chapter VIII). Although the fasciculus longitudinalis medialis is doubtless a two-way channel, its spinal cord extension can be regarded as an essentially if not exclusively descending pathway.

Concomitantly with the differentiation of additional central motor tracts in the phylogenetic or taxonomic vertebrate series, the fasciculus longitudinalis medialis undergoes functional as well as structural changes, thereby assuming a different significance in 'higher' forms, becoming, in particular, more closely related to the vestibularis system and to the eye-muscle nerve nuclei.

Moreover, already in Anamnia, first order afferent[52] grisea of the

---

[51] Auxiliary 'motor' tracts in at least some Anamnia are provided by vestibulo-spinal fibers not included in the fasciculus longitudinalis medialis and perhaps also by dubious tecto-spinal or still more dubious cerebello-spinal ones.

[52] Afferent centers of the *first order* are those receiving synaptic terminal arborizations of primary afferent neurons. If such grisea are also termed *'primary afferent'* it becomes necessary clearly to distinguish *'primary afferent neurons or fibers'* from *'primary afferent centers'*. The axons originating in these latter are *'secondary'*. Evidently some such fibers, as in the case of descending vestibular ones, are both *'secondary sensory'* and *'secondary motor'*.

vestibularis system, especially the ventral or lateral vestibular nucleus, give origin to descending fibers, which can be classified as secondary 'motor' and reach the spinal cord either through the fasciculus longitudinalis medialis or as independent vestibulo-spinal fasciculi.

Again, the single (bilaterally paired) large *Mauthner cell*, whose crossed fiber, discussed in the preceding chapter, joins the fasciculus longitudinalis medialis, could perhaps be evaluated as a specialized differentiation of the lateral vestibular nucleus, i.e. *as an alar plate derivative*. Yet, for practical purposes, and from a functional viewpoint, it may be considered a component of the 'reticular' nucleus motorius tegmenti.[53] Thus, although the origin of primary efferent neurons seems to be rather definitely restricted to the basal plate, and the 'nucleus motorius tegmenti' *sensu strictiori* is likewise a basal plate derivative, the well substantiated 'motor' functional significance of the basal plate and the conversely primary 'sensory' import of the alar plate should not be interpreted as 'obeying' a rigid 'law' but rather as expressions of a generally prevailing overall tendency, which does not preclude a 'secondary motor' function of alar plate derivatives, nor the well-established endings of primary afferent channels within basal plate grisea.

In addition, with respect to a morphologic evaluation of the basal plate, which represents a topologic grundbestandteil of the developing embryonic as well as of the adult Vertebrate brain, an ontogenetic primary and a definitive secondary stage must be distinguished, this latter being characterized (at least in some Amniota) by the inclusion of cell aggregates derived from the alar plate. Substantial '*group migrations*'[54] and displacements of cell masses from alar into basal plate in the course of ontogeny, as e.g. disclosed by ESSICK (1907), were pointed out in

---

[53] Cf. the comments by BECCARI (1943) on pp. 187–188 of his '*Neurologia comparata*'. Whether the *Mauthner cell* is, ontogenetically respectively phylogenetically, a derivative of alar or of basal plate remains, in my opinion, a still wide open question.

[54] In addition to such '*group migrations*' of neuroblastic or neuronal elements from alar plate into basal plate, migrations and displacements remaining within either basal or alar plate likewise occur and were discussed in section 4, p. 386, chapter VI of the preceding volume. In accordance with HAMBURGER and LEVI-MONTALCINI (1950), three main types of ontogenetic neuronal '*migrations*' may be distinguished: (1) centripetal migrations of individual undifferentiated elements toward the inner matrix, preparatory to mitosis; (2) centrifugal migration of elements toward the mantle; (3) 'group migrations' of griseal aggregates.

section 4 of chapter VI (vol. 3, part II). The so-called *rhombic lip*[55] of the alar plate seems to provide an important source of such migrating elements. More recently, investigations by HARKMARK (1954) in chick embryos, and autoradiographic studies by TABER PIERCE (1966) in mouse embryos, have attempted to provide further data on this topic. However, because the avaliable methods of observation involve definite limitations and ambiguities, the hitherto obtained results cannot be evaluated as conclusive with regard to specific details. There is, nevertheless, little doubt that the *nuclei pontis* and at least most of the *nuclei arcuati* of Mammals derive from such migration, whose pathway is indicated by the corpus pontobulbare of *Essick*, representing an embryonic remnant. Migrated alar plate elements likewise participate in the formation of *reticular raphé nuclei*, of the *inferior olivary complex*, and of *Bechterew's nucleus reticularis tegmenti pontis*. It seems possible that those grisea which are secondarily located in a tegmental neighborhood but originate in the alar plate predominantly constitute neuronal aggregates with substantial cerebellar connections. In addition, a relevant participation of migrating alar plate elements in the formation of the *superior olivary complex*, which pertains to the cochlearis system, can be suspected. Finally, a conspicuous cell group of 'reticular' type in the dorsolateral tegmentum of Man and Mammals has been designated as *nucleus reticularis lateralis*. It is located between the griseum of the descending trigeminal root dorsally, and the inferior olivary complex ventrally. On account of the specifically cerebellar connections of the lateral reticular nucleus, which are briefly considered in the following chapter X, and because of its dorsolateral location, this griseum may be classified as not pertaining to 'nucleus motorius tegmenti' respectively 'formatio reticularis tegmenti' *sensu strictiori*. There remains also some doubt whether a few and more compact grisea of uncertain significance (such e.g. as *Roller's nucleus*, cf. below p. 546) which are topographically related to the diffuse 'reticular formation' should be 'included' in this latter.

Experimental studies undertaken with a variety of electrical recording and other techniques on the mammalian brain stem in the course of the last fifty years have greatly elaborated on the functional significance of the nucleus reticularis tegmenti which is now also frequently designated as the '*reticular system*'. Particularly, in addition to the re-

---

[55] The somewhat ambiguous term '*rhombic lip*' was dealt with in section 1B, p. 184 and footnote 58, chapter VI of the preceding volume.

spiratory, cardiovascular, and 'vegetative' activities of that system, and to its effect on 'phasic' and 'tonic' muscular performances, the importance of that system for the maintenance of a state of 'wakefulness' and 'attention', as well as for 'sensory perception' has been emphasized.

The close relationships between vestibular and reticular systems were especially investigated by SPIEGEL and his collaborators (SPIEGEL, 1934; SPIEGEL and INABA, 1927; SPIEGEL and SOMMER, 1944; BERNIS and SPIEGEL, 1925). These studies also dealt with the influence of the reticular system on the state of wakefulness *(Wachzustand)*. An ascending reticular pathway leaving the rostral end of the tegmental reticular formation and joining the lemniscus medialis at its entrance into the thalamus was identified by FLECHSIG (1881, 1896) in the human brain and designated as '*Haubenstrahlung*'.[56] This tegmental radiation presumably represents the terminal portion of what is now generally but loosely called the ascending reticular system, which is said to provide or to include an 'extralemniscal sensory system' within the brain stem (FRENCH *et al.*, 1953), but whose details remain poorly elucidated. It can merely by inferred that fibers of the various ascending sensory systems either end in, or give off collaterals to, the reticular formation. Their impulses may then be transmitted, by way of short relays as well as by direct longer fibers, through FLECHSIG's *Haubenstrahlung* which reaches the diencephalic grisea and thereby can exert an influence upon the cerebral cortex.[57] The presumed relationships of the 'ascending reticular' pathway, which may also include descending feedback connections, to a system of diencephalic and rhombencephalic so-called '*centrencephalic grisea*' shall be dealt with in chapters XII and XV of volume 5.

As regards the 'reticular system's' influence upon the cerebral cortex, the following remarks will be sufficient in the present context. The normal *alpha* or *Berger rhythm* of the EEG is generally considered to be

---

[56] FLECHSIG (1898, p. 63) stated: '*Da, wo die Schleifenschicht in den Thalamus eintritt (hinter dem centre médian, an welches sie Fasern abgiebt), schliessen sich ihr überdies Fasern an, welche aus dem oberen Ende der Formatio reticularis austreten, so dass auch die hierin etwa enthaltenen centripetalen Leitungen (es gelangen zahlreiche Fasern aus sensiblen Nervenkernen in dieselbe hinein) sich den Schleifenfasern auf ihrem Verlauf zur Hirnrinde zugesellen würden. Ich habe diesen ganzen Complex sensibler Leitungen mit dem Namen "Haubenstrahlung" belegt. (Archiv für Anatomie und Physiologie 1881. Anatom. Abth. S. 49ff.)*'.

[57] According to FRENCH *et al.* (1953), this ascending system 'does not function specifically'. These authors apparently mean to say that, in their opinion, this system does not transmit signals encoding characteristics pertaining to specific sensory modalities.

the manifestation of a cortical 'synchronization' concomitant with a 'resting' state in 'relaxed' wakefulness. The effect of 'activating' ascending reticular impulses on the EEG have been interpreted as a '*desynchronization*' characterizing a cortical state of '*alertness*'. Thus, on the basis of animal experiments, the terms '*arousal reaction*' or '*alerting reaction*' were coined (cf. e.g. MORUZZI and MAGOUN, 1949). Yet, numerous uncertainties in the interpretations of either EEG or otherwise recorded potentials remain.

Altogether, from the multitudinous and in part ambiguous or contradictory observations, a complex facilitating and inhibiting effect, or, in other words, a 'composite inhibitor-excitor influence' (ELLIOTT, 1963) of the tegmental reticular formation upon other systems can be inferred.

Thus, reticular terminals in sensory grisea may either block or facilitate synaptic transmission. In addition, the centrifugal nerve fibers ending in sense organs such as cochlea, labyrinth,[58] and retina, and believed to 'control' or 'adjust' the receptor structures, may originate either in the reticular formation itself or in neuronal groups under that formation's direct control. Again, in contradistinction to the 'alerting reaction', an opposite effect, namely 'diversion of attention' or 'sensory habituation' has also been attributed, either wholly or in part, to 'reticular activities'. An example of such 'habituation' is, e.g., manifested if the perception of a continuous noise becomes suppressed.

Normal *sleep*[59] may be considered a cyclic 'regulation process' controlled by diverse subcortical grisea, interacting in various inhibiting as well as facilitating patterns, suppressing the cortical events concomitant with wakefulness, and presumably interfering with, or 'blocking' the input through the main sensory channels. There are valid reasons to assume that the reticular formation plays a significant role in the mechanism of the sleep-wakefulness rhythm. A fairly low arousal threshold distinguishes 'natural' or 'normal' dreamless sleep from closely related states such as 'paradoxical sleep', trance, coma or anesthesia, in which said threshold may be considerably higher.

---

[58] Cf. above in section 1, p. 319 of this chapter. GRANIT (1955), in discussing 'centrifugal control of sensory messages', concludes that 'spontaneously active' sense organs are 'controlled from nervous centers in the brain', and that 'spontaneous activity' in the intact organism is not wholly at the mercy of 'biological noise', the organism itself being able to 'adjust the level of permanent firing to its needs'.

[59] Cf. JUNG (1965), OSWALD (1962, 1966), K. (1972).

With respect to the action of anesthetics, narcotics or CNS depressants (e.g. ether and barbiturates), a primary effect on the cerebral cortex was formerly believed to obtain. At present, in accordance with the data and hypotheses concerning the 'reticular formation', it is assumed that most of these substances initially affect the 'ascending reticular system', causing 'desynchronization' and thereby involving the cortical activities. Yet, numerous complicating variables seem to obtain, and it cannot be claimed that the recent theories attempting to explain the relevant action mechanisms of neuropharmaca are sufficiently adequate.

Among 'motor' performances of the reticular formation, the *respiratory activities* and the participation in the *maintenance of muscular tonus* should be emphasized. According to SPIEGEL (1929), it can be assumed '*dass die Aufrechterhaltung der statischen Innervation ausser durch spinale propriozeptive Reflexbögen und durch labyrinthäre Reflexe durch Zellen der Subst. reticularis ermöglicht wird, die afferente Impulse aus der Muskulatur durch den Vorderseitenstrang empfangen und ihrerseits wieder Reticulospinalbahnen zu den Vorderhornzellen entsenden*'. The experiments of SPIEGEL and his collaborators had shown that complete loss of tonus occurred only if, together with lesions of the vestibular grisea the adjacent lateral portions of the reticular formation were destroyed, while after destruction of vestibular grisea loss of tonus remained incomplete.

Concerning the definition of tonus, SPIEGEL (1929) states: '*Wir bezeichnen als Tonus den unwillkürlich aufrechterhaltenen Spannungswiderstand der Muskulatur, der so lange währt, als die Körperteile, zwischen welchen die betreffenden Muskeln gespannt sind, nicht durch willkürliche oder Reflexreize in Bewegung versetzt werden*'. Instead of '*unwillkürlich*', one could perhaps here substitute '*spontaneous*' or '*intrinsic*' *('spontan', 'idiopathisch')* and for '*willkürlich*', '*cortically determined*' *('zusätzlich kortikal bedingt')*.

Reverting to the morphological respectively structural aspects of the mammalian and human *formatio reticularis tegmenti*, a clear semantic distinction between the functional (physiologic) concepts '*reticular system*' or '*reticular centers*', and the anatomically definable neuronal aggregates should be stressed. Quite evidently the so-called reticular system involves particular simultaneous activities of nerve cells in a number of different, topographically diversified, and disjoint cell groups. Again, some authors include the respiratory and cardiovascular centers in their 'reticular system' or 'reticular formation' (e.g. BRODAL, 1957), while others do not (e.g. ROSSI and ZANCHETTI, 1957; O'LEARY *et al.*, 1958).

An anatomically expedient subdivision of the formatio reticularis tegmenti remains in many respects unsatisfactory because of the 'uniform' and to some extent random distribution of its nerve cells, and this arrangement was strikingly indicated by KÖLLIKER's (1896) original designation '*nucleus magnocellularis diffusus*'. Although a number of more or less describable cell groups can be distinguished,[60] their boundaries are undefinable, and lines encircling such 'nuclei' in some of the atlases (e.g. OLZEWSKI and BAXTER, 1954) can be evaluated at best as overemphasizing some topographic neighborhoods but otherwise as potentially misleading figments of the imagination. Using highly questionable criteria of excessive parcellation, perhaps up to 100 different reticular 'nuclei' might be 'delimited' in the mammalian brain stem.

With regard to the multitudinous functional activities of the extensive tegmental reticular aggregate, an early comment by CAJAL (1909) and a more recent one by BAUMGARTEN (1962) appear pertinent. CAJAL (1909), in discussing the structure of the '*substance réticulée*', deals with the problem of an ascending reticular sensory pathway, which FLECHSIG (1881, 1896) had already previously included in his *Haubenstrahlung*. CAJAL, who does not refer to FLECHSIG's observations on this topic, summarizes his opinion as follows: '*On ne peut émettre que des hypotheses sur l'extension longitudinale et sur les connexions des voies sensitives de troisième ordre qui entrent dans la constitution de la substance reticulée. Les fibres qui les forment se raréfient à mesure qu'on s'approche des tubercules quadrijumeaux, d'où l'on conclut avec une certaine vraisemblance que la plupart de leurs fibres ascendantes se terminent dans les noyaux moteurs du pont ou dans les tubercules quadrijumeaux eux-mêmes.*

Quant à la terminaison des fibres du reste, des cordons antérieur et latéral de la moelle, on en est réduit aussi à des hypothèses. BECHTEREW *admet qu'elle a lieu dans la protubérance, au niveau du noyau réticulé de la calotte, et même plus haut, dans le noyau central supérieur, du moins, pour une partie de ces fibres.* KÖLLIKER *accepte cette opinion. Il ajoute que ces mêmes noyaux recoivent aussi les arborisations ultimes des tubes sensitifs de troisième ordre issus des cellules de la substance réticulée. S'il en était réellement ainsi, il faudrait de toute nécessité accepter l'existence de voies sensitives de quatrième ordre, constituées par les axones des cellules du noyau réticulé de la calotte et d'autres foyers. Une telle complication paraîtra excessive à plus d'un car elle implique une organisation bul-*

[60] These 'nuclei' or grisea shall be dealt with further below in the subsequent sections of the present chapter.

baire si inextricable que l'on hésite à l'admettre; d'autant plus que les voies sensitives de second et de troisième ordre et leurs connexions avec les noyaux moteurs semblent suffire, a priori, à toutes les exigences des réflexes et des combinaisons de mouvements.'

It is here of particular interest that CAJAL, in contradistinction to FLECHSIG, who propounded the existence of an ascending reticular sensory pathway, 'hesitates' to admit such system, because it would imply 'an excessively complicated and inextricable bulbar organization'. Unfortunately, however, such type of organization doubtless does actually obtain and, whatever the unknown or dubious details, FLECHSIG's ascending reticular pathway into the diencephalon has been substantially corroborated by the present-day gadgeteers.

Concerning the diversified activities of the 'reticular system' *in toto*, BAUMGARTEN (1962) remarks: '*Es leuchtet ein, dass eine so komplexe Leistung nicht allein von ein oder zwei wohlumschriebenen Nervenkernen mit Zellen gleicher oder ähnlicher Funktion ausgeübt werden kann oder gar von einer "unspezifisch" arbeitenden Retikularformation. Es muss sich vielmehr um einen höchst komplizierten, aber sinnvoll geschalteten und weitausgebreiteten nervösen Apparat in der SR handeln, in welchem jedes Neuron seine besonderen anatomischen Verbindungen und seine besondere funktionelle Aufgabe hat. Da es nahezu hoffnungslos erscheint, diesen Apparat im einzelnen zu entwirren, wird sich die Aufgabe des Neurophysiologen darauf beschränken müssen, die gröbsten Bahnverbindungen und Erregungsbeziehungen zu klären.*' Although I concur with the gist of the cited author's statement, the formulation: '*in welchem jedes Neuron seine besonderen anatomischen Verbindungen und seine besondere funktionelle Aufgabe hat*', can be regarded as rather ambiguous. I am inclined to believe that, depending on the pattern of neural activities, one and the same neuron may have a variety of different 'functions' (i.e. may be '*polyvalent*'). With regard to such a neuron's 'particular anatomical connections' it could be assumed that, quite in general, they may be (1) not rigorously but rather probabilistically (statistically) determined, (2) redundant (i.e. including more connections than are commonly 'in use', thus providing the possibility of new, 'additional' synaptic activities, and (3) not restricted to a final or definitive configurational pattern, but perhaps allowing at least under some conditions, and within certain limits, for additional 'new' connections by actual further outgrowths.

BRODAL (1965) evaluates the reticular formation not as a 'single entity', but as consisting of numerous minor cell groups or '*units*' which again differ in their connections and are interconnected more or less directly with many other parts of the brain. According to the cited author, these connections 'create abundant possibilities for an extensive interplay and the anatomical features must be assumed to reflect functional aspects'.

Contemporary investigations concerned with neurochemistry and pharmacology of the neuraxis have emphasized the occurrence and distribution of *catecholamines* in the nervous system as well as its physiological role (CARLSSON, 1959; DAHLSTRÖM and FUXE, 1964). Thus, M. VOGT (1954, 1959) demonstrated that, in the diencephalon and brain stem of the Dog, the maximum quantity of noradrenaline is located in the hypothalamus, with somewhat less in mesencephalon and reticular formation (cf. Fig. 178B). In the telencephalon, the hippocampal formation likewise seems to contain a substantial amount of noradrenaline, much less being present in the neocortex. It is believed that, in all regions related to the sympathetic subdivision of the autonomic system, the concentration of noradrenaline is high. VALDMAN (1964) states that 'it is completely unknown whether there is noradrenaline inside the nerve cells, the non-myelinated fibers or the glial elements'. DAHLSTRÖM and FUXE (1964), however, demonstrated, by means of a fluorescence technique, the location of three monoamines, including noradrenaline, *in the cytoplasm of nerve cells*, *in nerve fibers*, and *in synaptic terminals*. It is, of course, also possible that, in lesser concentration, some of these substances may be present in glial elements or intercellular 'ground substance'.

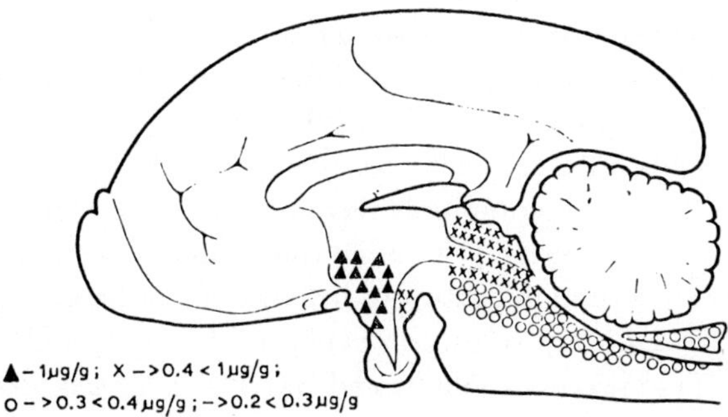

Figure 178 B. Diagrammatic sketch of a sagittal section through the Dog's brain, indicating the distribution of noradrenaline in reticular formation, other brain stem areas, and in diencephalon (after M. VOGT, 1954).

With respect to *dopamine*, regarded as one of the mediators of a group of catecholamines, CARLSSON (1959) has studied its quantitative localization and showed that there is no agreement between the distribution of noradrenaline and dopamine. This topic shall again briefly be considered in section 10 of the present chapter in dealing with the nucleus loci caerulei, as well as in section 9 of chapter XI in a discussion concerning some aspects of the mesencephalic substantia nigra. It is here of interest that *dopamine*, which appears to play an important role in the so-called extrapyramidal motor system of Mammals, does not seem to be present in the neuraxis of Amphibians, where *adrenaline* and *noradrenaline* were the relevant catecholamine recorded in studies of this sort. Such findings seem to suggest that, despite a very similar basic arrangement of communication channels, and despite the morphological homologies of grisea, considerable biochemical as well as functional differences may obtain in the diverse taxonomic forms.

A report on the reticular formation's pharmacology and physiology, summarizing the still unconclusive data concerning neurochemistry and similar topics resulting from the investigations in this rapidly expanding field, has been published by VALDMAN (1967).

Reverting to the above-mentioned recent findings concerning the neurochemistry of the Mammalian nucleus loci caerulei, it might be added that this griseum, although perhaps functionally related to the reticular formation, can be considered as representing a separate cytoarchitectural unit *sui generis*.

3. Cyclostomes

The medulla oblongata of *Petromyzonts* displays, in a rather simple manner, the basic topological configuration of the longitudinal zones dealt with in section 4, chapter VI of the preceding volume. The *sulcus limitans* of the ventricular wall is particularly conspicuous and the *sulcus intermedioventralis* is commonly recognizable. A *sulcus ventralis accessorius*, pointed out by SAITO (1930) can frequently be seen. The *sulcus intermediodorsalis*, although in various instances identifiable, tends to flatten out. The region of transition between oblongata and spinal cord represents TRETJAKOFF's (1909) and SAITO's *'Übergangsgebiet'*.[61] This

[61] SAITO (l.c.) subdivides the Petromyzont oblongata into a (1) rostral part, roughly corresponding to the extent of the anterior 'visceral column', (2) a caudal part approximately coextensive with the rostral two thirds of posterior 'visceral column', and (3) a region of transition *('Übergangsgebiet')* into spinal cord.

transitional zone begins rostrally at the approximate level of 'obex' or calamus scriptorius, whose ventricular edge provides the caudal attachment of lamina epithelialis respectively choroid plexus. Said *Übergangsgebiet* corresponds thus to the rostral end of the spinal cord's central canal and to its 'opening' into the fourth ventricle. The *commissura infima Halleri* passes through the obex region just dorsocaudally to that opening. Figure 179 illustrates the general arrangement and the trough-like configuration of the oblongata in a fairly advanced Ammocoetes larva, and Figures 180–184 depict representative levels of the Petromyzont oblongata, whose rostral limit is roughly indicated by the decussation of the trochlear nerve in the isthmus region.

Despite the apparent simplicity of the Petromyzont deuterencephalon, the detailed interpretation of many structural relationships remains uncertain and has resulted in various contradictory opinions expressed in the literature. Particularly dubious questions involve the nucleus and the fibers of nervus abducens, moreover, with respect to caudal mesencephalon and to cerebellum, the nucleus of the trochlear nerve and the presence or absence of *Purkinje cells*. Summarized data on the deuterencephalon of Petromyzonts can be found in the texts on comparative neurology by EDINGER (1908), BECCARI (1943), JOHNSTON (1906), KAPPERS (1947), KAPPERS *et al.* (1936), PAPEZ (1929) and myself (K., 1927). Among relevant investigation reports are those by ADDENS (1928, 1933), AHLBORN (1883), HEIER (1948), JOHNSTON (1902), OWSJANNIKOW (1903), SAITO (1928, 1930), SCHILLING (1907), STEFANELLI (1933, 1934), and TRETJAKOFF (1909).[62]

With regard to the oblongata's overall pattern, the roots of vestibular nerve, of anterior and posterior lateral line nerves, and of the branchial nerves are connected with the *alar plate*, which contains the terminal nuclei of the afferent fibers. In addition, the alar plate of the *Übergangsgebiet* displays a rostral group of the '*large dorsal cells*' characteristic for the spinal cord as mentioned in section 4 of the preceding chapter VIII.

The main components of the *basal plate* are (1) the 'viscero-motor cell column giving origin to the efferent branchial nerve fibers, (2) the

[62] A more recent paper by NIEUWENHUYS (J. comp. Neurol. *145*, p.165 f., 1972) confirms the generally recognized configurational features of the Petromyzont deuterencephalon and purports to introduce topologic viewpoints. The cited author also attempts to clarify some of the various unsettled or controversial questions, but in this respect his interpretations can still be assessed as inconclusive.

'somatic motor' cell column, essentially represented by the motoneurons of the spino-occipital nerve roots located in the Übergangsgebiet, and (3) the 'nucleus reticularis tegmenti', containing the large *cells of Müller* in addition to smaller and less conspicuous neuronal elements.

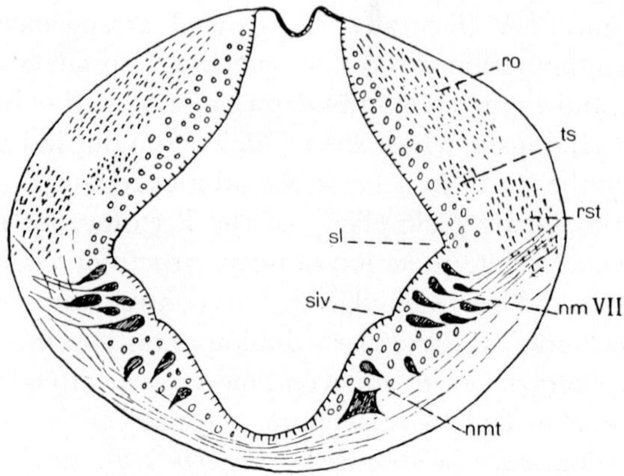

Figure 179. Simplified semidiagrammatic cross-section through the oblongata of an Ammocoetes at the level of the motor facial nerve nucleus, displaying the fundamental configurational components (from K., 1927). nmVII: nucleus motorius nervi facialis; nmt: nucleus motorius tegmenti with one large cell of *Müllerian type;* ro: radix nervi octavo-lateralis; rst: radix descendens nervi trigemini; siv: sulcus intermedius ventralis; sl: sulcus limitans; ts: tractus solitarius.

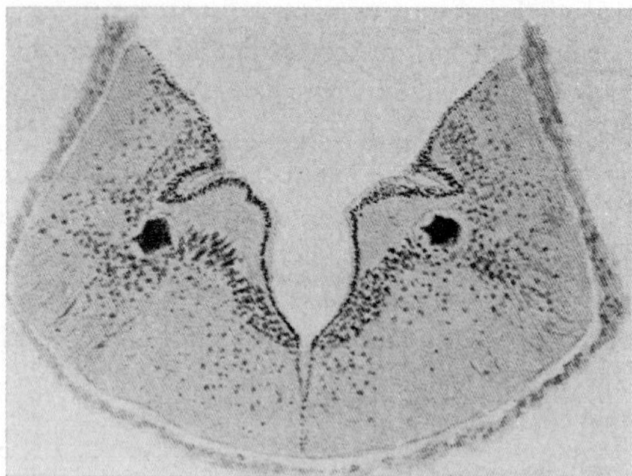

Figure 180. Cross-section through the oblongata of Entosphenus japonicus at a level just caudal to the isthmus region, showing SAITO's fourth pair of *Müller cells* lateral to the 'motor trigeminal column' (from SAITO, 1926).

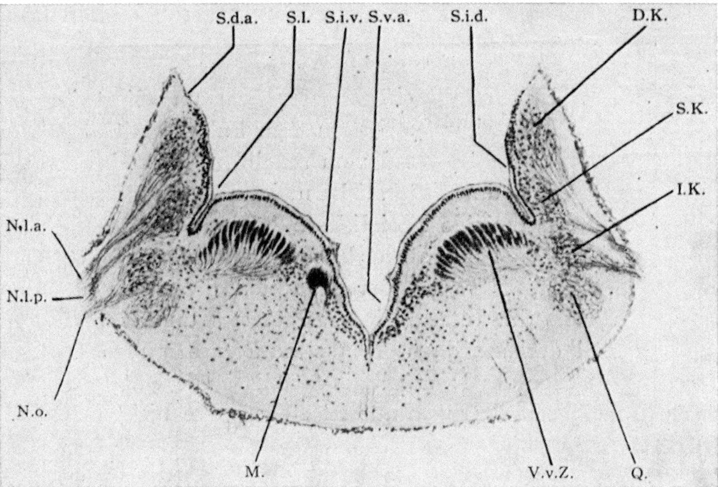

Figure 181. Cross-section through the oblongata of Entosphenus japonicus at the level of octavo-lateralis root entrance, showing at left a *Müller cell* of SAITO's seventh pair (from SAITO, 1930). DK: dorsolateral lateralis nucleus; IK: ventral octavus nucleus; M: seventh left *Müller cell;* N.l.a.: nervus lateralis anterior; N.l.p.: nervus lateralis posterior; N.o.: nervus octavus; Q: descending root of trigeminus with 'nucleus'; S.d.a.: 'sulcus accessorius dorsalis'; S.i.d.: sulcus intermedius dorsalis; S.i.v.: sulcus intermedius ventralis; S.K.: tractus and 'nucleus' lateralis respectively octavolateralis intermedius; S.l.: sulcus limitans; S.v.a.: sulcus ventralis accessorius; V.v.z.: anterior 'visceromotor column', at level of transition between V and VII.

Beginning with the configurations pertaining to the *alar plate*, and with regard to some relevant aspects of the different neuronal systems, various features of the vestibulo- or octavo-lateralis complex may be considered first. Two lateral nerves, n. lateralis anterior and n. lateralis posterior, are generally distinguished. The former, also known as n. lateralis facialis, joins the oblongata at the level of facial and vestibular nerves, slightly dorsal and rostrad to the vestibular root. It innervates the lateral line sense organs of the head.[63] The posterior lateral nerve enters the oblongata somewhat dorsally and rostrally, at the level of the glossopharyngeal nerve.

---

[63] The lateral line sense organs of Cyclostomes are less well developed than those of Gnathostome fishes and are, in this respect, comparable to those in Amphibians (cf. KAPPERS *et al.*, 1936, p. 439, and section 1, chapter VII, p. 794 of volume 3, part II in this series). Petromyzonts display pit organs, while thin canals have been described in Myxinoids.

The *vestibular root*, a little more ventral than the lateralis bundles, displays an anterior and a slightly more dorsal posterior subdivision. The intracerebral root fibers of the octavo-lateralis complex generally bifurcate into ascending and descending rami, although non-bifurcating, either ascending or descending branches may also be present. Some vestibular root fibers seem to decussate in the ventral midline. Again, various differences in fiber caliber between as well as within

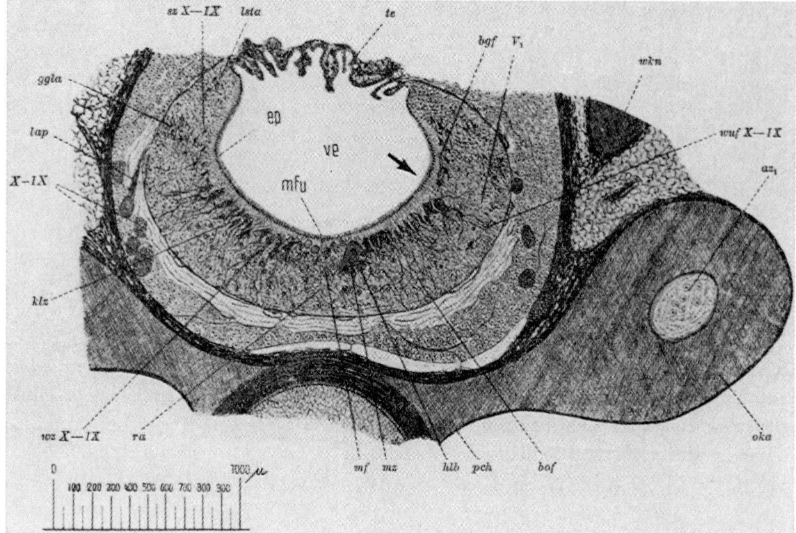

Figure 182A. Cross-section through the oblongata of Petromyzon fluviatilis at the rostral level of the glossopharyngeus-vagus column (probably *Bielschowsky impregnation*, from Krause, 1923). azl: leptomeningeal cells within otic capsule; bgf: blood vessel; bof: arcuate fibers; ep: ependyma (approximately at sulcus limitans); ggla: octavus root fiber tract and 'nucleus'; hlb: 'fasciculus longitudinalis medialis'; klz: group of smaller cells in caudal 'visceromotor column'; lap: caudal group of posterior lateralis fibers; lsta: dorsal lateralis 'area' respectively 'nucleus'; mf: *Müllerian fiber;* mfu: sulcus medianus internus of fourth ventricle; mz: fairly large 'reticular cell' identified by Krause as '*Müller cell*' but apparently not corresponding to one of Saito's nomenclature; oka: otic capsule; pch: parachordal cartilage; ra: raphé; te: tela respectively plexus chorioideus; szX-IX: tractus solitarius and afferent glossopharyngeus-vagus 'nucleus'; ve: fourth ventricle; wkn: rostral cartilaginous neural arch; wufX-IX: root fibers of glossopharyngeus-vagus group; wzX-IX: probably part of nucleus reticularis tegmenti (Krause identified this group as '*motorische Wurzelzellen der Nn.IX-X*, but I believe that these latter are represented by the element at klz and slightly dorsal to it); V_1 radix descendens trigemini; X-IX: bundle of glossopharyngeus-vagus root fibers. Added arrow: sulcus limitans.

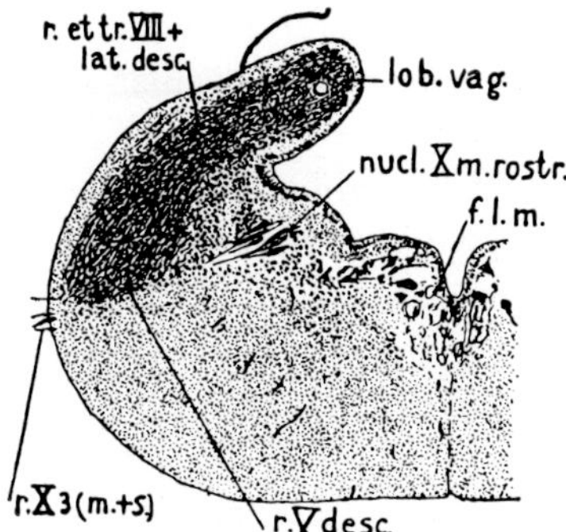

Figure 182 B. Cross-section through the so-called lobus vagi of Petromyzon fluviatilis, presumably somewhat caudal to the level of the preceding figure. Sulcus limitans and sulcus intermedioventralis (not labelled) are in this instance very easily recognizable (after ADDENS, from KAPPERS *et al.*, 1936). f.l.m.: fasciculus longitudinalis medialis; lob.vag.: 'lobus vagi'; nucl. X. m. rostr.: nucleus motorius rostralis nervi vagi; r. V. desc.: descending root of V; r. et. tr. VIII + lat. desc.: descending root fibers and 'tracts' of octavo-lateralis; r. X 3 (m. + s.): bundles of 'motor' and 'sensory' vagus root fibers.

the various lateralis and vestibularis root subdivision can be noted and are presumably of functional significance.

The entire dorsal portion of the alar plate, constituting the bulk of this configuration, particularly in the rostral two thirds of the oblongata, represents the '*area statica*' (also called '*tuberculum acusticum*', JOHNSTON, 1906, KAPPERS *et al.*, 1936). Roughly three fiber fascicles, with adjacent correlated cell groups, are distinguishable, namely a dorsal, an intermediate, and a ventral bundle respectively 'nucleus' (cf. Fig. 181). This latter bundle consists essentially of the vestibular fibers. Nevertheless, as KAPPERS *et al.* (1936) justly remark, there is, within the 'area statica', an intimate relation between terminals of primary afferent vestibular fibers and those of the (particularly anterior) lateral line nerves.

The external stratum of the area statica, especially in its dorsal portion, is constituted by a layer of dense neuropil which forms a 'molecular' layer and represents the so-called *crista cerebelli*. It contains neuronal processes of dorsal and intermediate lateralis 'nuclei' together with terminals from lateralis and perhaps also a few vestibularis fibers. The crista cerebelli[64] is thus mainly related to the lateral line tracts. It ends

[64] KAPPERS (1947) points out that '*cette crête n'existe que chez les animaux qui possèdent des organes sensoriels des lignes latérales, c'est-à-dire chez les Poissons et les Batraciens urodèles. Avec la vie terrestre, les organes latéraux et la crête cérébelleuse de la moelle allongée disparaissent*'.

caudally at about the entrance level of posterior lateral nerve root and is rostrally continuous with the molecular layer of the cerebellum (LARSELL, 1967).

An anterior, poorly delimitable cell group, close to cerebellum and isthmus, and perhaps provided by elements pertaining to all three above-mentioned 'nuclei', has been designated as *nucleus octavo-motorius anterior* (e.g. KAPPERS, 1947, and others). The more caudal portions of

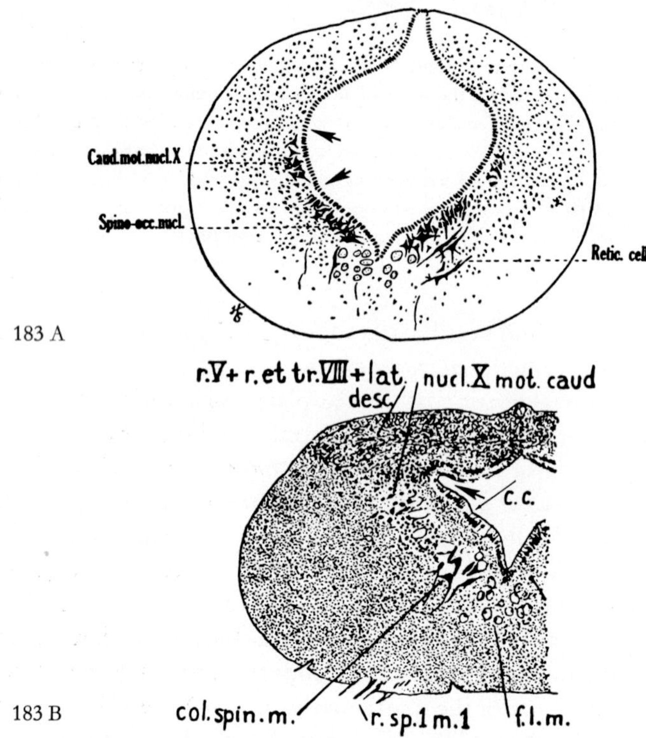

Figure 183 A. Cross-section through the 'calamus' region of Petromyzon marinus, showing caudal part of efferent vagus nuclear column, spino-occipital nuclear column, and reticular cells (after HUBER and CROSBY, from KAPPERS *et al.*, 1936).

Figure 183 B. Cross-section through 'calamus' region of Petromyzon fluviatilis near rostral end of *commissura infima Halleri* in Petromyzon fluviatilis (after ADDENS, from KAPPERS *et al.*, 1936). c.c.: central canal; col.spin.m.: spino-occipital motor column; f.l.m.: fasciculus longitudinalis medialis; nucl. X mot. caud.: caudal part of efferent vagus nucleus; r.sp.lm.1: ventral root of first spinal nerve (spino-occipital nerve); r V+r.et tr. VIII+lat.desc.: fiber bundles of descending trigeminal and octavo-lateral roots. In the roof of the ventricle some fibers (not labelled) of commissura infima can be recognized.

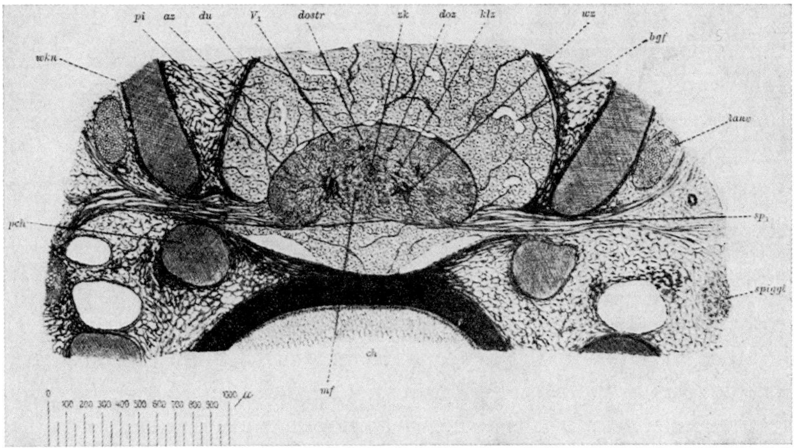

Figure 184. Cross-section through the *'Übergangsgebiet'* of the oblongata in Petromyzon fluviatilis (from KRAUSE, 1923). az: leptomeningeal cells; bgf: meningeal bloodvessel; ch: notochord; dostr: posterior funiculus; doz: large dorsal cells; du: 'dura mater'; klz: 'small dark cells' (probably caudal end of viscero-motor vagus column); lanv: nervus lateralis; mf: *Müller fibers;* pch: parachordal cartilage; pi: 'pia mater'; sp_1: ventral root of first spinal nerve (spino-occipital nerve); spiggl: 'first spinal ganglion'; wkn: first cartilaginous neural arch; zw: spino-occipital root cells (somatic-motor column); zk: central canal; V_1: radix descendens trigemini.

the ventral (predominantly or exclusively vestibular) 'nucleus' represents a *nucleus vestibularis posterior*, of which a medial portion *(noyau vestibulaire postérieur medial*, KAPPERS) and a lateral subdivision *(noyau vestibulaire postérieur latéral*, KAPPERS) might be distinguished, in addition to a still more lateral interstitial cell group *(noyau vestibulaire latéral tangentiel*, KAPPERS) within the entering root fiber bundle.

In addition to longitudinal ascending or descending primary root fibers, and to secondary ones arising in *area statica*,[65] cells in the 'nuclei' of this latter give origin to transverse decussating 'arcuate fibers' with manifold but insufficiently clarified connections (cf. Fig. 185). Some of these arciform fibers reach the mesencephalon (torus semicircularis and tectum), forming a contralateral ascending 'tract' designated as *lemniscus lateralis*. Fibers of this type, believed to end in the reticular formation of the mesencephalic tegmentum, provide what has been called the *'tractus octavo-motorius anterior cruciatus.*

[65] According to KAPPERS (1947), the tangential nucleus of Petromyzonts *'ne donne que des fibres réflexes ascendantes qui sont accompagnées de quelques fibres radiculaires ascendantes.*

Other arciform secondary vestibular fibers join a poorly delimited *fasciculus longitudinalis medialis* and reach the spinal cord. Moreover, the diverse arciform fibers, together with the uncrossed ascending and descending secondary and primary ones, seem to give off collaterals to most cell groups and in particular to nucleus reticularis tegmenti. In addition to the crossed vestibulo-spinal fibers within the fasciculus longitudinalis, somewhat more lateral uncrossed secondary vestibular fibers likewise seem to provide a vestibulo-spinal 'system'.

The *posterior lateral vestibular nucleus*, in which crossed and uncrossed vestibulo-spinal are presumed to originate, contains a number of rather large multipolar neuronal elements and might be considered homologous to the *nucleus vestibularis of Deiters* in higher vertebrates. Yet, a definite boundary between posterior lateral and posterior medial vestibular 'nucleus' cannot be recognized, and both seem to form a more or less common aggregate including larger and smaller nerve cells. It is likely, as KAPPERS *et al.* (1936) point out, that among the elements of this nuclear group may lie the 'forerunners' of the large *Mauthner cells* characteristic for the oblongata of various gnathostome Anamnia, particularly of many Osteichthyes and of larval or aquatic Amphibia. Although perhaps representing a derivative of the alar

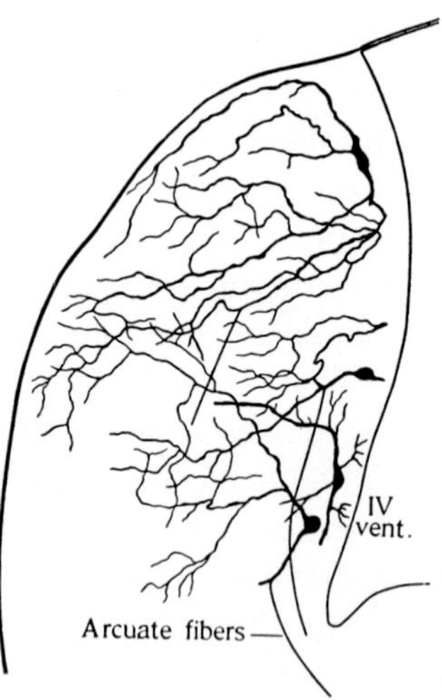

Figure 185. Cross-section through the 'tuberculum acusticum' of Lampetra wilderi as seen in a *Golgi impregnation*, and showing the origin of two arcuate fibers (from JOHNSTON, 1906).

plate's vestibular neighborhood, the *Mauthner cell* shows a more or less medialward directed displacement and may give the impression as being a lateral component of nucleus reticularis tegmenti.[66] Typical *Mauthner cells* can be easily recognized by their conspicuous size, their occurrence as a single pair, and the distinctive large caliber of their crossed descending fiber in the fasciculus longitudinalis medialis of the spinal cord. Yet, it seems not possible to define a fully acceptable criterion allowing for a clear-cut distinction between the 'true' *Mauthner cells* of Anamnia commonly considered to possess these elements, and similar large cells in the vestibular griseum of forms not generally regarded as displaying the paired *Mauthner cells*.

Thus, Whiting (1957) identified a pair of *Mauthner neurones* in young larval lampreys, and Rovainen (1967) described a double pair, consisting bilaterally of a larger and a slightly more caudal smaller one, in the oblongata of Petromyzon marinus. Rovainen (1967) defines *Mauthner cells* 'on the basis of (a) their location near the eighth nerve, and (b) their decussated axons extending into the spinal cord'. The cited author, nevertheless, enumerates certain differences related to behavior under experimental conditions, between his so-called *Mauthner cells* of Petromyzon and the typical *Mauthner cells* of Teleosts.[67] In addition, he points out that the axon of the lamprey cell passes into the cord's lateral funiculus rather than remaining in the medial longitudinal fasciculus. Neither the functional differences nor the dissimilarity in the course of the fiber would, *per se*, be valid argu-

---

[66] In my '*Vorlesungen*' of 1927, I still considered *Mauthner's cell* to be a component of 'nucleus motorius tegmenti' and my collaborator Gerlach (1933) likewise interpreted said cell, together with n. reticularis tegmenti, as a derivative of the basal plate. Since that time, further studies and observations have led me to the opinion that the origin of the *Mauthner cell* from either basal or alar plate still remains an open question. The close association of the *Mauthner cell* with the aggregate of vestibulo-lateralis 'nuclei' and fibers is evident. On the other hand, although the lateral vestibular *nucleus of Deiters* may display a 'reticular' appearance and is doubtless functionally closely related to the nucleus motorius tegmenti, its derivation from the alar plate seems rather well established. Thus, in the definitive ontogenetic ('adult') configuration, a 'blurring' of interpenetrating alar and basal plate neighborhoods may perhaps have resulted.

[67] Among the several functional differences cited by Rovainen (1967) are e.g. the following: the lamprey's *Mauthner cell* does not display 'laterality' with respect to the vestibular nerve. Stimulation of that nerve on either side is said to elicit EPSPs and IPSPs, while the *Mauthner cell* of Teleosts is primarily excited by homolateral nerve stimulation and inhibited by the contralateral one. Again, in the lamprey, mutual inhibition does not occur between the antimeric cells, but such a mechanism is reported as well developed in Teleosts.

ments against the rather unusual morphological homologization of single cellular elements. It remains thus a question of arbitrary choice whether one should or should not consider *Mauthner cells* to be present in Petromyzonts.[68]

As regards the *afferent components* of the branchial nerves (V., VII., IX., X.), those of the trigeminus form a conspicuous, essentially descending bundle medially accompanied by a columnar cell aggregate, and located in the ventrolateral region of the alar plate (cf. Figs. 179, 180, 181). As a topologic neighborhood, said region pertains, nevertheless, to the alar plate's *dorsalgebiet*, of which the 'sensory' trigeminal 'nuclear complex' can be considered a derivative. The descending root fibers seem to reach the rostral end of the spinal cord. Although a few ascending fibers may be present (either as non-bifurcating ascending root components or as rostral branches of bifurcating ones), a *radix mesencephalica trigemini* with its large intracerebral primary afferent nerve cells seems to be missing.[69] The *radix descendens trigemini* is joined by the 'somatic afferent' (cutaneous) root fibers of facial, glossopharyngeus and vagus nerve.

Arcuate fibers originating in the 'nucleus' radicis descendentis trigemini give off collaterals in a manner similar to those of the vestibulo-lateral *fibrae arciformes* mentioned above, and form an indistinct ascending *bulbar lemniscus* joining the spino-bulbar and spino-tectal fibers (*'spinal lemniscus'*), and apparently running rostrad as far as the tectum mesencephali and presumably the diencephalon.

The *'visceral afferent' root fibers* ('gustatory' as well as 'general') of facial, glossopharyngeus and vagus nerves form a rostrally rather thin and poorly recognizable but caudalward increasingly wider *'tractus solitarius'* with its concomitant second order neurons, from which, *inter alia*, arcuate fibers take their origin. The tractus solitarius with its 'nucleus' pertain to the alar plate's intermedioventral zone, which, being rostrally much less developed than the vestibulo-lateral and the trigem-

---

[68] Thus, KAPPERS *et al.* (1936) as quoted above, speak of *Mauthner cell* forerunners. Since recent Cyclostomes can hardly be the ancestors of Gnathostome Anamnia, it is perhaps better to say in partial agreement with ROVAINEN (1967), that the so-called *Mauthner cell* of lampreys might closely resemble the postulated phylogenetic precursor of that cell as subsequently highly differentiated in some Gnathostome Anamnia.

[69] '*Il est important de noter que la racine mésencephalique du trijumeau, racine proprioceptive par excellence, présente chez tous les autres Vertébrés, manque chez les Cyclostomes. Ce fait s'explique probablement par la fonction tout-à-fait différente de la musculature trigéminale de ces animaux*' (KAPPERS, 1947).

inal neighborhoods in the dorsal zone, becomes crowded into a narrow space medially to the basal components of '*area statica*' and dorsomedially to the fibers and cells of *radix descendens trigemini*, in the vicinity of the sulcus limitans. At the levels of the *efferent* 'posterior visceral column' however, the corresponding '*visceral afferent*' cell aggregate and its concomitant root fiber bundles expand in size and extend gradually further dorsad (cf. Figs. 182A, B), and may even, in some instances, protrude mediad as a so-called *lobus vagi*.

The frequent complete disappearance of the at most faintly indicated sulcus intermediodorsalis is presumably related to the distorting interplay of formative processes leading to *opposite* pattern deformations. Rostrally, there obtains a ventrad 'cramping' of the 'visceral afferent zone' by the predominating neighboring 'somatic afferent zone' with its more bulky cell aggregations and fiber tracts. Caudally, on the other hand, a more substantial 'visceral afferent zone' expands dorsalward and tends toward formation of a 'lobus vagi'.

The caudal extremities of the antimeric (bilateral) visceral afferent cell columns become joined respectively fused in the dorsal midline at the calamus scriptorius, above the central canal of the oblongata's *Übergangsgebiet*. The merging cell aggregates form here an either paired or unpaired (median), more or less interstitial nucleus, related to fibers decussating within the obex, and known as the *commissura infima Halleri* or '*commissure calamique*' (KAPPERS, 1947). This commissure and its nucleus are present in all Vertebrates from Cyclostomes to Mammals, but are variously differentiated as regards size and details of structural components. In Petromyzonts, the commissura infima doubtless contains crossing primary fibers of tractus solitarius and perhaps also of radix descendens trigemini. The nucleus commissurae infimae is considered by KAPPERS (1920) to represent a griseum correlating the vagus-system with more caudal (spinal) neural mechanisms.

The alar plate of the oblongata's 'closed portion', i.e. within TRETJAKOFF's and SAITO's 'Übergangsgebiet' may or may not include, in its dorsal portion, a group of cells comparable to the nuclei of the posterior funiculi, from which arcuate fibers of an ascending system (*e.g.* a bulbo-tectal lemniscus) could arise. The alar plate of this region, however, contains a number of large rounded *dorsal cells*, which are similar to, or identical with, the large dorsal elements of the spinal cord dealt with in section 4 of the preceding chapter VIII. SAITO (1930) found that these elements, also known as '*sensible Dorsalzellen*' display, like the *Müller cells* dealt with further below, an instance of *constant cell number*. With respect to Entosphenus japonicus, and in the rostral part of the *Übergangsgebiet*, 16 cells were recorded, namely 8 on each side, forming 8 symmetrically arranged antimeric pairs. In the caudal part of that transitional zone, which might or might not be considered rostral spinal cord rather than caudal oblongata, SAITO found 14 cells constituting 7 antimeric, but much less symmetrically arranged pairs. SAITO attempted, in agreement with my own views, to determine the topographic relationship of the large dorsal cells to the alar plate's dorsal and intermediodorsal zones, but I do not any longer believe that the conclusions which we (jointly but rather independently) reached can be considered sufficiently convincing.[70] Concerning these zones, not only as regards Petromyzonts, but with respect to the entire Vertebrate series, it should be stressed that they represent, topologically, not closed but rather *open neighborhoods* without definite boundary 'lines', or, in other words, not including any 'points' of their boundary. Thus, in elementary topology, an *open set* represents *its own interior*. Accordingly, although dorsal and intermediodorsal zones are subdivisions of the alar plate, and intermedioventral and ventral zones subdivisions of the basal plate, neither set of subdivisions necessarily preempt the entire alar or basal plate, respectively.[71] Thus, some components of alar or basal plate need not, or cannot, be explicitly classified as component of one of its two 'specialized zones'.[72] Again, while SAITO (1930), on the basis of the accessory sulci which he observed, and with my explicit agreement,[73] attempted further to subdivide the alar plate's dorsal zone

---

[70] It should be noted that SAITO (1930), here again in full agreement with our joint observations, remarked that '*im Übergangsgebiet ist, wegen der sehr schwachen Entwicklung des Sulcus intermedius dorsalis, die Unterscheidung des Dorsalgebietes und des Intermediodorsalgebietes nicht so deutlich zu erkennen*'.

[71] In logical notation: $dz \cdot idz \supset al$, $ivz \cdot vz \supset b$, but not $dz \cdot idz \equiv al$ nor $ivz \cdot vz \equiv b$. In some of our early diagrams depicting the *bauplan* of the vertebrate neural tube (e.g. fig. 158 of vol. 3/II) we have indeed drawn linear boundaries giving the impression that $x \cdot y \equiv z$ rather than $x \cdot y \supset z$ might obtain. These diagrams, however, represent useful fictions in the manner of a first approximation and are not meant to be topologically rigorous, using lines merely to depict open boundary zones. Still more complex linear grids such as e.g. displayed by figs. 42A–D of the present volume, are not topologic, but *metrically topographic*, in accordance with D'ARCY THOMPSON's *coordinate transformations* (cf. chapter VI, section 1A, p. 66 of vol. 3, part II).

[72] Cf. chapter VI, section 4, p. 373 of the preceding volume 3, part II.

[73] The late Dr. TAMESUKE SAITO was one of my most brilliant and promising graduate students *(Kenkyûsei)* at Keiô University (Tôkyô) since about 1926. To my deep regret, he prematurely succumbed to tuberculosis shortly before my second sojourn in Japan 1933–1934.

and the basal plate's ventral zone each into a *pars propria* and *pars accessoria*, I was unable, so far, to formulate, in a satisfactory and convincing manner, any relevant morphologic or 'functional' orderliness characterizing such parts or 'subzones' within the framework of the morphologic brain pattern.

Turning now to the configurational aspects of the *basal plate*, the intermedioventral zone, rather well delimited from the alar plate by the sulcus limitans, contains rostrally the efferent nerve cells of the trigeminus group, and caudally those of the vagus group, both groups, also known as anterior and posterior visceral efferent cell columns, being generally separated from each other by an interval lacking the sizeable typical efferent elements. Within the anterior column, a definable boundary between the longer rostral trigeminal 'nucleus' and the shorter caudal facial 'nucleus' is not usually recognizable. KAPPERS (1947) states that the former '*n'est séparé du noyau moteur du facial que par une cellule géante de Müller*'. According to our observations in Entosphenus, this 'demarcating' cell might perhaps be that of SAITO's eighth pair. There is likewise no clear-cut boundary between the shorter rostral glossopharyngeal cell aggregate and the longer caudal vagal one in the posterior visceral efferent column.[74] The relation of the branchial nerve roots to the external surface of the oblongata (in Entosphenus japonicus) is illustrated by Figure 192.

Both visceral efferent columns innervate striated branchial musculature. In addition, their facial, glossopharyngeal and vagal components presumably contain preganglionic neurons of the cranial parasympathetic system, perhaps provided by the slightly smaller nerve cells within the columns. Concerning the vagal portion of the column, KAPPERS *et al.* (1936), refer to the opinion of ADDENS, according to which the rostral vagal subdivision should be considered branchiomotor, and the caudal one preganglionic.

As regards the *basal plate's ventral zone* characterized by efferent nerve cells providing ventral root fibers, only the column pertaining to the *spino-occipital nerves*, and located within the oblongata's caudal transitional region, can be unambiguously recognized. Several rootlets emerge at the ventrolateral surface of the neuraxis and form one or two ventral spino-occipital roots. It is difficult to draw a distinction between these roots and those of the caudally following true spinal

---

[74] A subdivision between 'frontal' and 'caudal' vagus 'nucleus' as referred to by KAPPERS *et al.* (1936, p. 523) could not be detected in our material.

nerves. Thus, ADDENS designated the first spino-occipital root as the first spinal one (cf. Fig. 183B). Although we found no dorsal roots within the transitional region, some such reduced or rudimentary roots seem occasionally to occur in that region as depicted by KAPPERS *et al.* (1936) in accordance with ADDENS. The spino-occipital roots of Petromyzonts and their nucleus of origin, like those of other Anamnia, can be evaluated as kathomologous with the hypoglossal nerve and nucleus of Amniota, despite the fact that Cyclostomes do not possess a tongue comparable to that of higher Vertebrates.[75]

In contradistinction to Gnathostomes, whose oblongata generally displays the clearly recognizable nucleus of a distinctive nervus abducens, the identification of that nucleus in Petromyzonts has remained highly uncertain. It is well established that the nervus abducens of lampreys emerges from the oblongata jointly with the trigeminus, from which it finally separates to innervate the m. rectus inferior (comparable to the m. rectus lateralis of Gnathostomes, cf. NISHI, 1922). As KAPPERS (1920) [76] suggested, this close relationship between trigeminal and abducens nerves would not seem altogether incompatible with (a perhaps secondary) 'inclusion' or 'absorption' of the abducens nucleus into the caudal portion of the trigeminal cell column,[77] regardless of the fact that this latter is 'branchiomotor', while the abducens nucleus can be evaluated as 'somatic-motor'. On the other hand, JOHNSTON (1902) cautiously suggested that the abducens fibers joining the trigeminus root arise from the 'ventral motor column', i.e. the somatic motor ventral zone of basal plate, at the levels of VII and VIII. He added that 'this isolated portion of the ventral motor column occupies the same position as the nucleus of VI in Acipenser'. This was later essentially confirmed by STEFANELLI (1935), who likewise located the nucleus of the abducens '*nella colonna somatomotoria*', although slightly more

---

[75] According to KAPPERS (1947), '*il n'y a pas de nerf hypoglosse chez les Pétromyzontes*'. KAPPERS is evidently right with respect to orthohomology. He emphasizes that '*la Lamproie ne possède pas de langue analogue*' (he might also have added: '*homologue*') '*à celle des Vertèbres supérieurs. Sa "soi-disant langue" est innervée par le trijumeau. La colonne motrice cervicale antérieure ne contient que les cellules radiculaires des racines ventrales spino-occipitales*'. Cf. also chapter VII, section 3, p. 859 of the preceding volume 3, part II.

[76] '*Es ist daher wahrscheinlich, dass die entsprechenden Wurzelzellen dem Trigeminuskern einverleibt oder wenigstens angrenzend sind*' (KAPPERS, 1920).

[77] Further problems related to the eye-muscle nerve nuclei of Petromyzonts, including the claim of ADDENS (1933) that some oculomotor fibers originate from the rostral end of the trigeminal cell column, shall be discussed further below in chapter XI, dealing with the mesencephalon.

caudad than Johnston, namely '*all' altezza del nucleo motore del VII e del IX, più medialmente sottostante alle cellule reticolari del nucleo reticolare medio mediale*'.

Neither Saito (1930) nor myself were able to recognize the nucleus of the abducens with a sufficient degree of certainty in our material. I remained unconvinced by the various specific elaborations on this topic found in the literature, but believe that, while the above-mentioned interpretation by Kappers cannot be entirely discounted, those by Johnston and Stefanelli seem perhaps more acceptable approximations concerning the location of the elusive nucleus abducentis of Petromyzonts.

The *nucleus reticularis tegmenti* of the Petromyzont deuterencephalon is characterized by the presence of the large *Müller cells* whose neurites *('giant fibers')* reach the spinal cord and were dealt with in section 4 of the preceding chapter VIII. Ahlborn (1883), Johnston (1902), Tretjakoff (1909) and other authors had already noted the rather regular distribution and the relatively small number of cells amounting to more or less two dozen. Considering only a well identifiable set of the largest elements to be true *Müller cells*, distinguishable from other reticular cells by their order of magnitude, Saito (1928) recorded their remarkably constant number, position, and shape. He identified altogether 24 cells respectively 12 pairs in Entosphenus japonicus. Three of these pairs (grouped as a double and a single pair) were found in the mesencephalic tegmentum, two pairs just caudal to the isthmus level, and seven pairs distributed along various levels of the rostral 'visceromotor column' (cf. Figs. 186, 187). Their position within the basal plate is either along the intermedioventral or the ventral longitudinal zone. However, if the set of *Müller cells* is considered to represent a morphologic unit *sui generis*, then it could be classified as a relevant component of the basal plate not necessarily being an intrinsic part of either intermedioventral or ventral zone or both. In other words, these two zones may be conceived as open topologic neighborhoods penetrated by, but not intrinsically including the nucleus reticularis *qua* component of the larger set represented by the basal plate (cf. also the footnotes 71 and 72 above).

Subsequently to Saito's studies (1928, 1930), Stefanelli (1933) investigated the arrangement of *Müller cells* in Petromyzon planeri and its Ammocoetes larva, in Petromyzon fluviatilis, and in Petromyzon marinus. Stefanelli describes 10 symmetric cell pairs, as here shown in Figures 188 and 189. He also suggested that two of his 'giant cells'

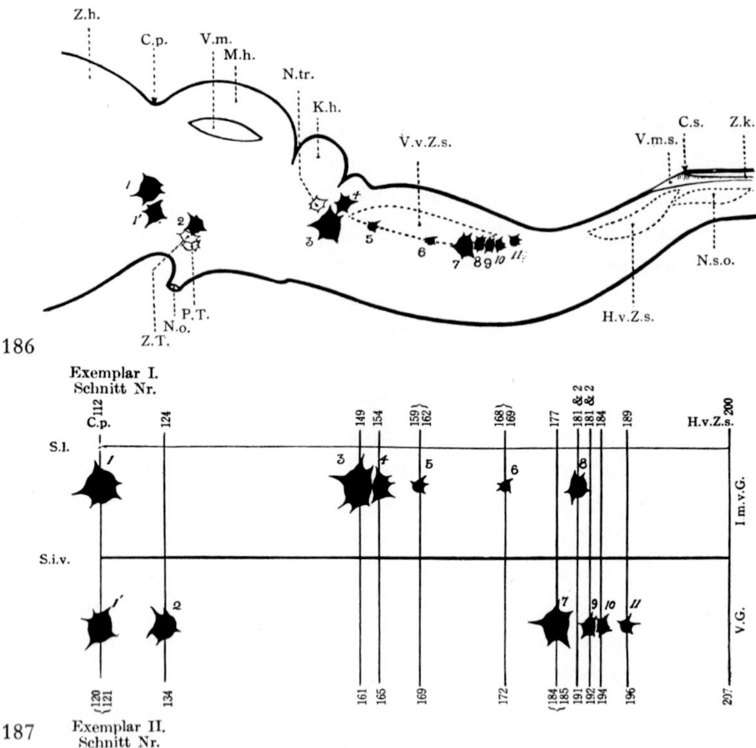

Figure 186. Semidiagrammatic sketch showing location and relative size of the Müller cells in the deuterencephalon of Entosphenus japonicus (from Saito, 1928). C.p.: commissura posterior; C.s.: calamus scriptorius; H.v.Z.s.: caudal viscero-motor column; K.h.: cerebellum; M.h.: mesencephalon; N.o.: nervus oculomotorius; N.tr.: nucleus nervi trochlearis; N.s.o.: Nucleus spino-occipitalis; P.T.: peripheral division of oculomotor nucleus; V.m.: ventriculus mesencephali; V.m.s.: 'Ventriculus medullospinalis' (transition of fourth ventricle to central canal); V.v.Z.s.: rostral viscero-motor cell column; Z.H.: diencephalon; Z.K.: central canal; Z.T.: central division of oculomotor nucleus. The numerals indicate the *Müller cell* pairs, 1 and 1' being the first '*Doppelpaar*' of Saito.

Figure 187. Diagrammatic sketch of localization and relative size of *Müller cells* in Entosphenus japonicus, indicating the distances between these cells with reference to the numbered sections of two different series (from Saito, 1928). C.p.: commissura posterior; H.v.Z.s.: level of rostral end of caudal viscero-motor column; I.m.v.G.: intermedio-ventral zone; S.i.v.: sulcus intermedius ventralis; S.l.: sulcus limitans; V.G.: ventral zone of basal plate.

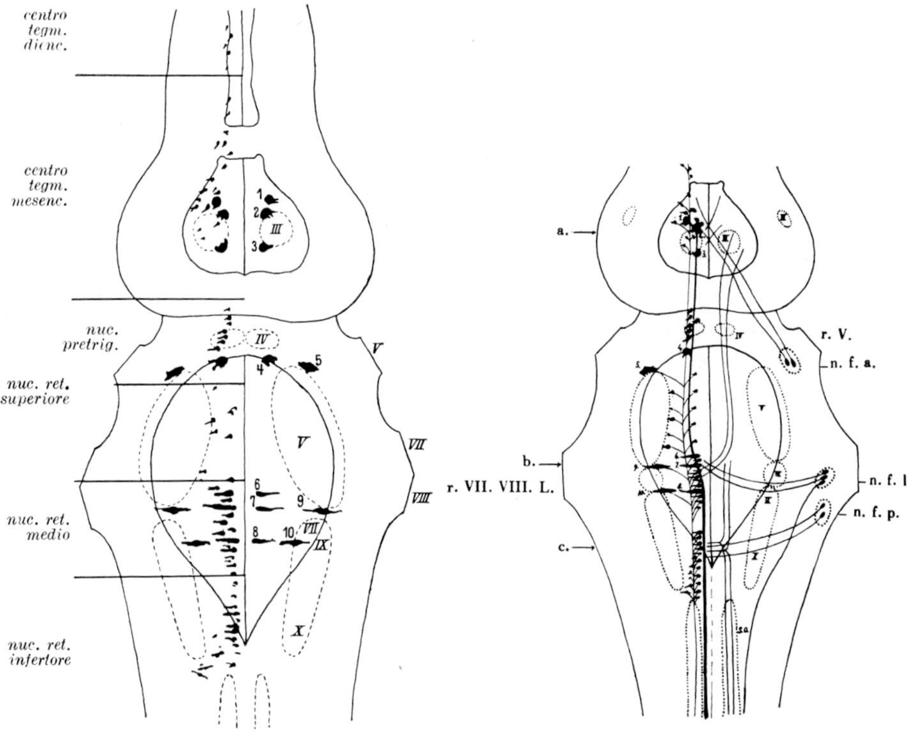

Figure 188

Figure 189

Figure 188. Semidiagrammatic drawing illustrating the tegmental reticular system in Petromyzon marinus. All tegmental neuronal elements are shown in horizontal projection on the left, and only the 'giant cells' on the right (from STEFANELLI, 1933a). 1, 2, 3: giant cells of the mesencephalic tegmental center; 4: giant cell of '*centro pretrigeminale mediale*'; 5: giant cell of '*nucleo pretrigeminale laterale*'; 6, 7, 8: giant cells of '*nucleo reticolare medio-mediale*'; 9, 10: giant cells of '*nucleo reticolare medio-laterale*'. The efferent nuclei of III, IV, V, VII, IX, X, and the spino-occipital column (at bottom, not labelled) are indicated by dash-outlines.

Figure 189. Semidiagrammatic sketch corresponding to preceding figure, but indicating some fiber connections in STEFANELLI's interpretation (from STEFANELLI, 1933b). On the right side, three groups of fusiform cells pertaining to 'area statica' are outlined; their axons are shown to decussate at levels a, b, c; n.f.a.: '*nucleo anteriore delle cellule a fuso*'; n.f.l.: '*nucleo laterale delle cellule a fuso*'; n.f.p. '*nucleo posteriore delle cellule a fuso*'; r.V: trigeminal root; r.VII.VIII.L.: root of facial, acoustic, and anterior lateral nerves.

except for the homolateral course of their descending axons, might correspond to the *Mauthner cells* found in various other Anamnia.

STEFANELLI moreover reported that the number and bilateral-symmetric distribution of the *Müller cells*, but not their size, are identical in all four Petromyzont forms which he examined. He had made extensive and detailed measurements, recording, e.g. the major and minor diameters of these elements, and found that they were largest in Petromyzon marinus, smallest in Petromyzon planeri, and intermediate in Petromyzon fluviatilis. To a certain extent, but not in direct proportion, this corresponds roughly to the size differences obtaining between the three species.[78]

The divergence between STEFANELLI's (1933) and SAITO's (1928, 1930) numerical findings may either be related to species differences or to the fact that a rigorous distinction between the smaller *Müller cells* and some of the other reticular elements is exceedingly difficult. Supposedly precise measurements, as e.g. expressed in μ, may, because of numerous interfering variable, be highly inaccurate. A grading of these cells by inspection based on a comparison of their outlines obtained with calibrated microscopic projection on a screen or by photomicrographs seems, in accordance with my experiences in using the relevant procedures, more accurate than the deceivingly precise numerical 'measurements'.

More recently, ROVAINEN (1967) described in Petromyzon marinus 16 *Müller cells* (8 pairs), namely 3 mesencephalic, 1 isthmic, and 4 bulbar pairs, and, in addition 1 or 2 *Mauthner cell* pairs (cf. Fig. 190). The *Müller cells* are here defined 'as the cells of origin of the eight pairs of uncrossed large axons *(Müller fibers)* in the medial longitudinal fasciculus of the spinal cord'. His claim to have exactly traced and identified the 8 fibers within the spinal cord (cf. Figs. 190) does not seem in all respects convincing, particularly in view of individual variations and irregularities in the arrangement of these fibers, but can nevertheless be accepted, with due reservations, as at least partially valid. Rather non-

---

[78] '*La grandezza delle cellule non è tuttavia proporzionale alle grandezze somatiche poichè le differenze tra le grandezze cellulari di cellule omologhe delle tre forme sono assai minori delle rispettive grandezze somatiche*' (STEFANELLI, 1933). In view of this author's detailed *quantitative* measurements, the statement in a recent paper on the lamprey's brain stem in J. comp. Neurol. *145*, 1972, namely that STEFANELLI made 'a detailed, although *qualitative* study' of these cells (l.c. p. 173) seems to include an oversight. As regards comparable relationships between body size and that of certain neuronal perikarya with regard to some other animals and nerve cell types, cf. above chapter VIII, section 11, p. 190.

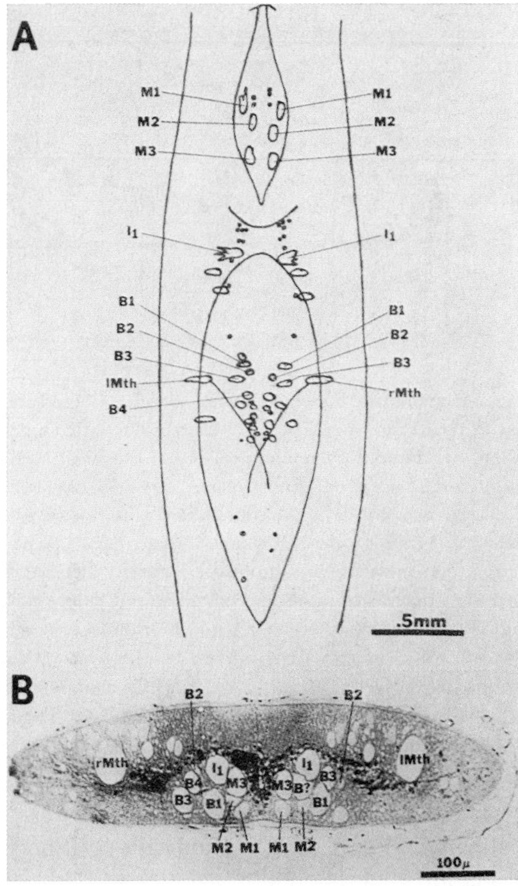

Figure 190. Müller cells in an Ammocoetes of Petromyzon marinus and their 'cell body-axon relations determined by histologic tracings' (from ROVAINEN, 1967). A Brain reconstruction from transverse serial sections. B Photomicrograph of spinal cord 11 mm beyond end of fourth ventricle. B 1–4: bulbar *Müller cells*; I 1: isthmic *Müller cell;* M 1–3: mesencephalic *Müller cell;* lMth and rMth: left and right *Mauthner cell.* In B, this notation indicates the respective axons. One *Müller fiber* (B ?), apparently from the bulbar group, 'could not be traced to its cell body'. 'The dark objects in the section are cells or parts of cells. The round body above the central canal is a dorsal cell, and the more irregular body to the left is probably a motoneuron. It is only by chance that both cells happen to be on the left side of this section'.

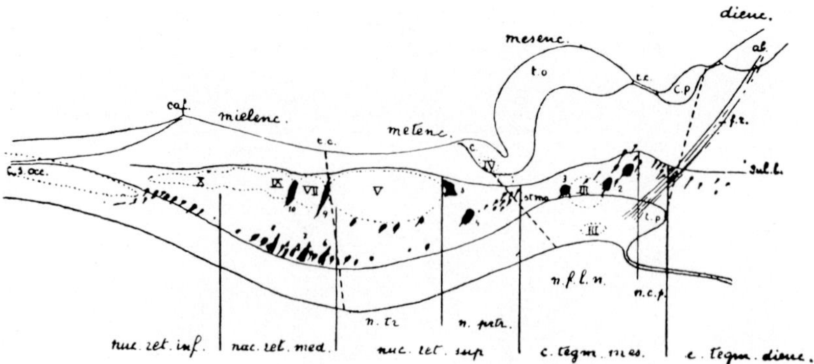

Figure 191. Semidiagrammatic drawing showing a unilateral projection of the cells of nucleus reticularis tegmenti and of cranial nerve efferent nuclear columns upon the midsagittal plane in the Petromyzont deuterencephalon and caudal diencephalon according to STEFANELLI's interpretation (from STEFANELLI, 1934). ab: habenular ganglion; c: cerebellum; cal: calamus scriptorius; c.p.: commissura posterior; c.tegm.dienc.: '*centro tegmentale diencefalico*'; c.tegm.mes.: '*centro tegmentale mesencefalico*'; dienc.: diencephalon; fr.: fasciculus retroflexus; mesenc.: mesencephalon; meten: metencephalon; mielenc.: myelencephalon; n.c.p.: nucleus of posterior commissure; n.f.l.n.: nucleus interstitialis of fasciculus longitudinalis medialis; n.prtr.: '*nucleo pretrigeminale;* n.tr.: '*nucleo reticolare trigeminale;* nuc.ret.inf.: 'nucleus reticularis inferior'; nuc.ret.med.: 'nucleus reticularis medius'; nuc.ret.sup.: 'nucleus reticularis superior'; t.c.: tela chorioidea; t.p.: tuberculum posterius; s.occ.: spino-occipital column; sul.l.: sulcus limitans. The Roman numbers indicate efferent cranial nerve nuclei.

descript reactions which the cited author calls 'distinctive' were obtained by stimulation experiments.[79]

In contradistinction to the possible '*Mauthner cells*' with homolateral axons suggested by STEFANELLI (1933), ROVAINEN's (1967) single or double pair of '*Mauthner cells*' gives origin to decussating axons and accordingly conform to his definition of '*Mauthner cells*'. The question must be left open whether STEFANELLI's cells 9 and 10 (cf. Figs. 188 and 189) respectively SAITO's cell 8 (cf. Figs. 186 and 187) might actually correspond to ROVAINEN's two '*Mauthner cells*' Mth and Mth' (cf. Fig. 190, in which only Mth is indicated).

---

[79] 'Stimulation of individual *Müller* and *Mauthner axons* in situ through intracellular electrodes led to distinctive movements in both larvae and adults. Such movements included flexion of the body and tail, movements of the fins, rotations of the body, and propagated undulations' (ROVAINEN, 1967).

The *nucleus reticularis tegmenti* of Petromyzonts which extends caudalward into the *übergangsgebiet*, contains, moreover, an undefined number of medium sized or smaller elements, including, among others, the cell aggregate designated by ROVAINEN (1967) as that of the vagal region. All of the decussating fibers within the oblongata's basal plate do not seem to be arcuate fibers arising in the primary afferent cell groups but appear also to include fibers from reticular elements, providing crossed ascending and descending pathways. Particularly from the caudal portion of the oblongata, such crossed descending fibers may run through the various 'funiculi' of the spinal cord.

Following his investigation on *Müller cells*, STEFANELLI (1934) summarized his results and his views on the 'reticular' 'tegmental centers' of Petromyzonts with particular emphasis on their general significance and with reference to the various opinions expressed in the literature. He distinguishes a diencephalic, a mesencephalic, and a rhombencephalic center, again subdividing this latter in general agreement with VAN HOEVELL and KAPPERS, into '*nucleo reticolare superiore*', '*nucleo reticolare medio*', and '*nucleo reticolare inferiore*' (cf. Fig. 191).

As regards STEFANELLI's '*centro diencefalico*',[80] I cannot agree with the interpretation of that author, who, moreover, assumes that the rostral end of sulcus limitans extends through the diencephalon. This, of course, would justify the inclusion of some larger elements, located ventrally to that extension, into the tegmentum. Since the sulcus limitans may display apparent junctions with either sulcus diencephalicus medius or diencephalicus ventralis (cf. Fig. 63, I in vol. 3, part II), it is possible that STEFANELLI refers to a few scattered larger elements in thalamus ventralis or hypothalamus, which may send neurites into rostral neighborhoods of the mesencephalic tegmentum, but which cannot be classified, on convincing grounds, as either '*tegmental*' or '*reticular*'. The mesencephalic *basal plate*, i.e. the '*true tegmentum*', moreover, extends at the *tuberculum posterius*, as the *tegmental cell cord* or *tegmental cell plate* into the posterior hypothalamic (so-called 'infundibular' or 'mammillary') recess[81] of all Vertebrates (cf. Fig. 185 in vol. 3, part II). Derivatives of this tegmental cell cord are, e.g. the *nucleus interstitialis of Cajal* and the interstitial tegmental nucleus of the posterior commissure. Although 'forerunners' of these grisea might be postulated in the tegmental cell cord of Petromyzonts, I do not believe that such

[80] The cited author adds: '*da me osservata per primo nei Petromyzonti*'.

[81] Cf. Chapter VI, section 5, footnotes 160 and 161, p. 399 of volume 3, part II.

specific nuclei can be recognized in Petromyzonts. Again, at certain stages of ontogenetic development, the most rostral *Müller cell* can be seen in the tegmental cell cord at the tuberculum posterius (cf. e.g. Fig. 186i in vol. 3, part II). It is thus also quite possible that STEFANELLI's '*centro diencefalico*' merely corresponds to the rostral end of his mesencephalic center as represented by the tegmental cell plate at the mesencephalo-diencephalic boundary.

The oblongata of Petromyzonts (Fig. 192), regardless of its phylogenetic relationship to that of 'higher' forms, displays, despite some complicating factors, a rather primitive and relatively simply overall

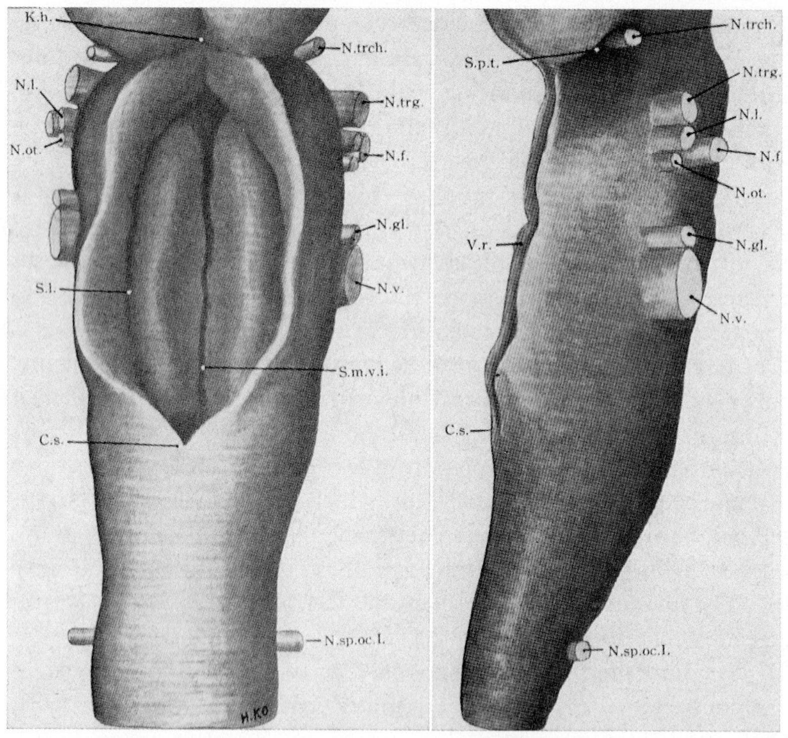

Figure 192. Wax-model of rhombencephalon of Entosphenus japonicus reconstructed from serial sections and showing location of cranial nerve roots (from SAITO, 1930). A Dorsal view. B Lateral view. C.s.: calamus scriptorius; K.h.: cerebellum; N.f.: nervus facialis; N.gl.: nervus glossopharyngeus; N.l.: nervus lateralis; n.ot.: nervus octavus; N.sp.oc. 1: first spino-occipital nerve; n.trch.: nervus trochlearis; N.trg.: nervus trigeminus; N.v.: nervus vagus; S.l.: sulcus limitans; S.m.v.i.: 'sulcus medianus ventralis internus'; S.p.t.: sulcus post-tectalis; V.r.: ventriculus rhombencephali.

arrangement, which is of considerable interest for purposes of comparative neurology. It illustrates both some of the similarities and some of the differences between deuterencephalon and spinal cord. The *similarities* include: (1) the *longitudinal afferent root fiber tracts*, whose input, through connections with primary afferent centers, is transferred to ascending secondary afferent pathways; (2) the location, in the basal plate, of *efferent nuclei*, which receive terminals of direct or indirect re-

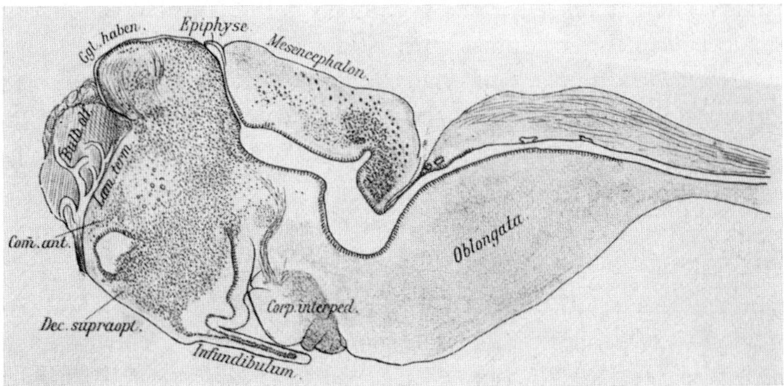

Figure 193A. Midsagittal section through the brain of Myxine glutinosa showing the reduced ventricular spaces (from EDINGER, 1906). The designation 'Epiphyse' merely indicates the topologic neighborhood of this non-developed structure.

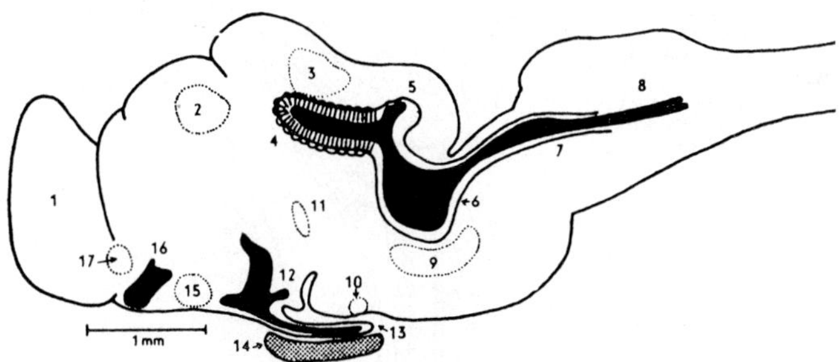

Figure 193B. Semidiagrammatic midsagittal section through the brain of Myxine glutinosa, showing the remnants of the ventricular system in black (from ADAM, 1963). 1: telencephalon; 2: epithalamus; 3: 'posterior commissure' (and mesencephalic roof); 4: 'subcommissural organ'; 5: posterior mesencephalic recess; 6: 'main part of mesencephalic ventricle'; 7: rhombencephalic ventricle; 8: central canal; 9: 'commissural system of the trigeminal nuclei'; 10: decussation of fasciculus retroflexux; 11: 'commissure of the posterior tubercle'; 12: posterior hypothalamic ventricular space; 13: neurohypophysis; 14: adenohypophysis; 15: postoptic commissure; 16: ventricular space of preoptic recess; 17: anterior commissure.

flex arcs, and, in turn, give origin to the efferent root fibers of basically 'segmental' peripheral nerves.

As regards relevant *dissimilarities*, one could mention the following: (1) the fact that some of the descending afferent root fiber components in the oblongata (e.g. V and VIII) form more massive bundles than the ascending ones, while, in the spinal cord, the ascending components are, generally speaking, more conspicuous and longer than the descending branches. (2) Both the terminal (afferent) nuclei and the primary efferent nuclei of the peripheral nerves display peculiarities not obtaining for the spinal cord, and related to the subdivision of these rhombencephalic neuronal configurations into branchial nerve aggregate, octavus system, eye muscle nerve aggregate, and spino-oc-

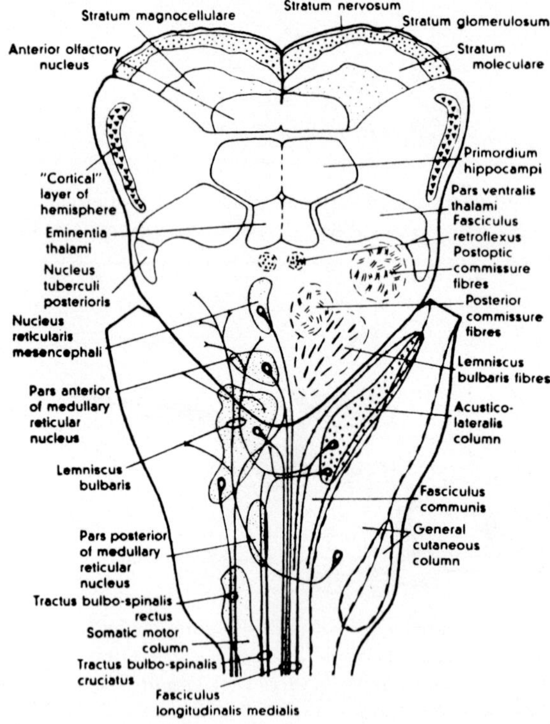

Figure 194. Schematic horizontal section of adult Myxinoid brain, showing telencephalic, diencephalic and mesencephalic neighborhoods in addition to configurational features of the oblongata, whose rostral 'horns' (not labelled) are seen on both sides of the caudal mesencephalic region (combined after figures of JANSEN, from BONE, 1963).

cipital system.[82] (3) The basal plate, in addition, is characterized by the 'nucleus motorius s. reticularis tegmenti' which, although perhaps not intrinsically differing from somewhat similar but much simpler intraspinal neuronal mechanisms, is, nevertheless, in any significantly comparable manner, not displayed by the vertebrate spinal cord.

Turning now to the oblongata of *Myxinoids*, it should be recalled that spinal cord and brain in this order of Cyclostomes display many peculiarities,[83] some of which are rather difficult to interpret. Several relevant problems concerning ontogenetic features of Myxinoid brain morphology were pointed out chapter VI, sections 1B (p. 192), and 6 (p. 479) of volume 3, part II.

From a functional viewpoint, some of the peculiarities manifested by the brain of Myxinoids are doubtless related, as various authors have pointed out, to a considerable reduction of the octavo-lateralis and optic systems (including atrophy of the eyes), combined with a hypertrophy of the olfactory and cephalic cutaneous systems.

The morphogenesis of the Myxinoid brain involves '*gestaltungsvorgänge*' producing a considerable pattern deformation or distortion,[84] whereby telencephalon, diencephalon, mesencephalon, and rhombencephalon become 'telescoped' into each other, such that telencephalon is shoved into diencephalon, and this latter toward mesencephalon. Rostral rhombencephalon and mesencephalon are also 'squeezed' to-

---

[82] Cf. chapter VI, section 4 and chapter VII, sections 2, 3 and 4 of the preceding volume 3, part II.

[83] '*I Mixinoidi, costituenti l'altro Ordine dei Ciclostomi, presentano un encefalo con molti centri ridotti o mancanti a causa di notevole atrofie dipendenti della vita parassitaria che questi animali conducono*' (STEFANELLI, 1934).

[84] BONE (1963) refers to the opinion expressed by CONEL (1929, 1931), namely that, because the Myxinoid embryo develops under a tough inelastic shell, and lies upon a dense mass of yolk, it becomes gradually more and more flattened dorsoventrally. As the embryo grows rostrad toward the yolk, it is said also to be compressed rostro-caudally. The combination of a dorsoventrally flattened ontogenetic stage, already recognized by DEAN (1899), with rostro-caudal compression and subsequent 're-orientating' morphogenetic changes, are interpreted as causative factors resulting in the pattern distortion. Such simplified mechanical explanations remain rather unconvincing, since the primary '*field-effects*' resulting in these configurational peculiarities could just as well be independent variables intrinsic to neuraxial growth, of which yolk and shell behavior might then be the dependent ones. A considerable pattern distortion involving both 'stretching' and 'telescoping', and resulting in an aberrant configuration, considerably differing, however, from that displayed by Myxinoids, is likewise characteristic for the brain of the Holocephalian *Chimaera* (K. and NIIMI, 1969).

gether or 'interlocked', such that rhombencephalic 'horns' protrude rostralward on both sides of the mesencephalon (cf. Figs. 193, 194).

Among studies dealing with the adult brain are those by AYERS and WORTHINGTON (1907, 1908, 1911), EDINGER (1906), HOLM (1901), HOLMGREN (1919), JANSEN (1930), RETZIUS (1893), RÖTHIG and KAPPERS (1914), and WORTHINGTON (1905). A monograph on the biology of Myxine, edited by BRODAL and FÄNGE (1963), contains a chapter, by several contributors, on nervous system and sense organs.

The brain of adult Myxinoids is characterized by a pronounced *reduction of the ventricular system*, although a more or less typical and communicating lumen of the neural tube's different subdivisions is present during ontogenetic development. At the final or adult stage, however, the rather narrow ventricular cavities of the mesencephalon and rhombencephalon still communicate with each other and with the spinal cord's central canal, but those of the telencephalon have disappeared while the diencephalic ones are strikingly reduced and commonly entirely isolated (cf. Fig. 193). In addition, this reduction displays various individual variations, manifested by recesses, by narrow ependymal 'tubes' or by 'channels' connected with the generally prevailing remnants of the ventricular configuration whose regionally differing ependymal elements have been described by ADAM (1956, 1963) and others.

It is generally assumed that all Cyclostomes, including Myxinoids, lack medullated nerve fibers. EDINGER (1906), however, believed that some of the thicker fibers central of cranial nerves and neuraxis were provided with a myelin sheath. He stated: '*Nur in ihnen kommen auch, was den bisherigen Untersuchern entgangen ist, echte markhaltige Nervenfasern vor. Sie liegen in den zumeist marklosen Nervenwurzeln und ausserdem in den Kreuzungsfasern, welche, aus den Nervenkernen hervorgehend, überall von dem caudalen Mittelhirnende bis ins Rückenmark hinein nachweisbar sind*'. Although both in Petromyzonts and in Myxinoids certain nerve fibers, as seen with the light microscope, give indeed the impression of being medullated, and may even assume a bluish hematoxylin tinge in *Weigert stains* (cf. the figures in EDINGER's colored plates of 1906), I believe that, as can be inferred from the available electron photomicrographs, true myelin sheaths are probably entirely missing in these forms (cf. also p. 2 and footnote 3 in chapter VIII of this volume).

The medulla oblongata constitutes the largest subdivision of the Myxinoid brain (Fig. 194), and, in dorsal view, has been likened by BONE (1963) to 'an arrowhead with a truncated point which passes into the spinal medulla and the barbs (formed by the two large anterior horns resembling the auricles of lower gnathostomes) lying around the mesencephalon.' Into these 'horns' pass the trigeminal nerves representing one of the main peripheral afferent pathways to the Myxinoid oblongata. The much reduced fourth ventricle within the thick-walled oblongata communicates caudally with the central canal of the spinal cord[85] and rostrally with the mesencephalic ventricle. This latter communication is, as a more or less general rule, complicated by the presence of a thin so-called 'isthmic canal' dorsally to the main passage. In the region of this 'isthmic canal' a rather short, somewhat thin stretch of roof plate has been interpreted as a rudimentary choroid plexus by EDINGER (1906) and JANSEN (1930), but can hardly be evaluated as a structure *functionally* comparable to that plexus in other Vertebrates.[86]

Figures 195 to 198 illustrate representative levels of the Myxinoid oblongata; a comparison with Figures 180 to 184 will clearly show the striking differences between the two orders of Cyclostomes as regards the oblongata's general configuration.

In the region of the *alar plate*, the predominating afferent branchial nerve system is provided by the *cutaneous trigeminal root components* which, inter alia, innervate the characteristic rostral tentacles (6 in Myxine and 4 in Bdellostoma). Whether the more caudal roots designated as nervus buccalis pertain to trigeminus or facial nerve remains a moot question. It is likewise uncertain whether the facial nerve proper includes cutaneous fibers which, moreover, might entirely be lacking

---

[85] Although the Myxinoid central canal may be represented by a simple very narrow ependymal channel, its outline can, not infrequently, display some variations, as, e.g. described in the following manner by JANSEN (1930): 'In the spinal cord the ventricle consists of two intimately related parallel canals, which in cross-sections give the lumen a characteristic 8-shape. The two canals stay in open connection along the entire length. At the transition to the medulla this double canal is converted into a single one, as the dorsal canal ends in a rather large diverticle which lies dorsal to the main ventricle'. It is, in this as well as in other respects, of interest to compare the figures of EDINGER's (1906) investigation with those of JANSEN's (1930) and ADAM's (1963) contributions. Cf. also Figs. 40, 41, 46, 47, 49, 52 in the preceding chapter VIII of this volume.

[86] EDINGER (1906) justly states: '*Dann sind die Plexus chorioidei und überhaupt die häutigen Anhangsgebilde des Gehirns so auffallend gering entwickelt, wie bei keinem einzigen im Wasser lebenden Wirbelthiere. Bei allen diesen finden wir ja über dem Vorderhirne und Zwischenhirne, bei vielen auch im Mittelhirnbereich und wieder bei allen im Nachhirne mächtige epithelbedeckte Gefässschlingen, Apparate, die nach mancherlei histologischen Befunden höchst wahrscheinlich der Absonderung einer intracerebralen Flüssigkeit dienen. Myxine hat, abgesehen von einer ganz kurzen Platte am Anfang des Oblongatadaches, gar nichts von all diesem!*'

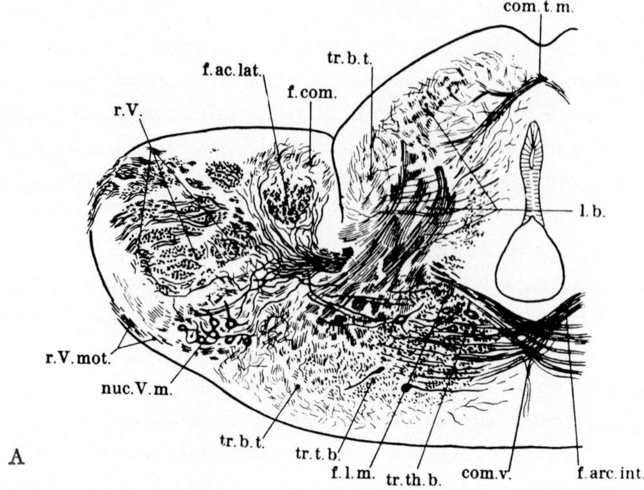

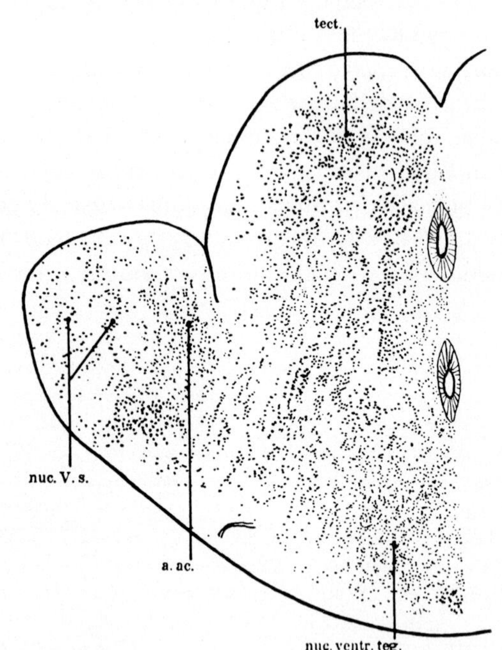

Figure 195. Cross-sections through rostral levels of the oblongata in Myxine glutinosa, showing, on the left, the rhombencephalic horn protruding on the lateral side of the mesencephalon (from JANSEN, 1930). A Silver impregnation. B Cytoarchitectural picture. a.ac.: area octavo-lateralis; com.t.m.: commissura tecti mesencephali; com.v.: commissura ventralis; f.ac.lat.: fasciculus octavo-lateralis; f.arc.int.: fibrae arcuatae internae;

in the glossopharyngeus-vagus group. The second order nerve cells accompanying the intracerebral afferent root fibers and their arborizations form a long 'cutaneous nucleus' which extends through the length of the oblongata and corresponds to the nucleus of the descending trigeminal root in Petromyzonts. From the small cells of this aggregate, whose dendrites have extensive connections with other alar and basal plate neuronal groups, numerous arcuate fibers arise, while other axonal branches join the descending root fiber bundles.

The *octavo-lateralis area* (cf. Fig. 194) is much more reduced than in Petromyzonts.[87] It is located dorsomedially between trigeminus system and tractus solitarius as described by JANSEN (1930) and others (Figs. 195, 196). The lateralis fibers are mainly included, as so-called nervus lateralis anterior and posterior, within nervus trigeminus and nervus buccalis. A lateralis component of the facial nerve is doubtful. The vestibular nerve (n. acusticus) includes an anterior subdivision from the utriculus, and a posterior one from the sacculus.[88] A portion of the utricular subdivision, designated as 'general cutaneous root of ganglion utriculare' (JANSEN, 1930), joins the ventral part of the descending trigeminal root (Figs. 196, 197A).

The *'visceral afferent area'*, provided by the tractus solitarius, which is also frequently called 'fasciculus communis system', receives its root fibers in part from the facial nerve, but mainly from the glossopharyngeus-vagus group.[89] These fibers seem to sweep around the lateral surface of the oblongata until reaching a dorsomedial position, medially to the octavo-lateral are (Figs. 195, 197). The secondary neurons of this system may be scattered about the fiber bundles, and like those of the trigeminal and octavo-lateral area, give origin to arcuate fibers.

[87] The lateral line system is exceedingly reduced in Myxinoids, and little information is here available on its inconspicuous structures (cf. ROSS, 1963).

[88] The reduced inner ear of Myxine, consisting of two fused semicircular canals joined to an utriculo-saccular complex, was depicted above in Figure 163 of section 1.

[89] There is some doubt as to the presence of a distinct glossopharyngeal nerve, which may have been substantially 'absorbed' by the vagus, but the complex of branchial nerve roots caudal to n. facialis can be safely termed glossopharyngeus-vagus group.

f. com.: tractus solitarius ('fasciculus communis'); f. l. m.: fasciculus longitudinalis medialis; l. b.: lemniscus bulbaris; nuc. ventr. teg.: nucleus ventralis tegmenti; nuc. Vm.: efferent trigeminal nucleus; nuc. Vsens.: afferent trigeminal nucleus; rV: radix nervi trigemini; tect.: tectum mesencephali; tr. b. t.: tractus bulbo-tectalis; tr. th. b.: 'tractus thalamo-bulbaris'.

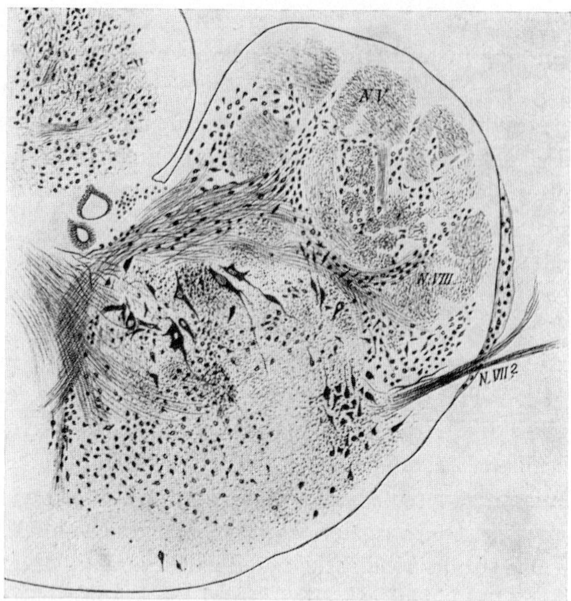

Figure 196. Cross-section through the oblongata of Myxine glutinosa at the level of root fiber exit presumably representing efferent facialis components (from EDINGER, 1906). Some giant cells can be recognized in the basal plate region. Dorsally, a caudal portion of tectum mesencephali can be seen. Note also the ventricular remnants.

Again, the fiber systems of all three areae merge caudally with ascending root fibers from the spinal cord (Figs. 197B, 198). EDINGER (1906) suggested here, as indicated on his figure, the presence of a '*nucleus*' *funiculi posterioris*.

The *arcuate fibers* arising from the various afferent areas form a very conspicuous *commissura ventralis*[90] which gives off collaterals to the motor nuclei and the reticular formation. While some collateral fibers may descend toward the spinal cord, their bulk provides an ascending lemniscus bulbaris, presumably accompanied by lemniscus fibers of spinal origin, and mainly reaching the tectum mesencephali.

Within the *basal plate*, the *motor nuclei of the branchial nerves* are found in a lateral location. They form a rostral column pertaining to trigemi-

[90] EDINGER (1906), who was impressed by the size of this decussation, states: '*Diese letzteren ganz enormen Kreuzungsfasern bilden etwas ganz Charakteristisches in der Oblongata von Myxine. Kreuzungen liegen ja hier bei allen Vertebraten, aber so mächtige Züge wie bei Myxine hat kein einziges*'.

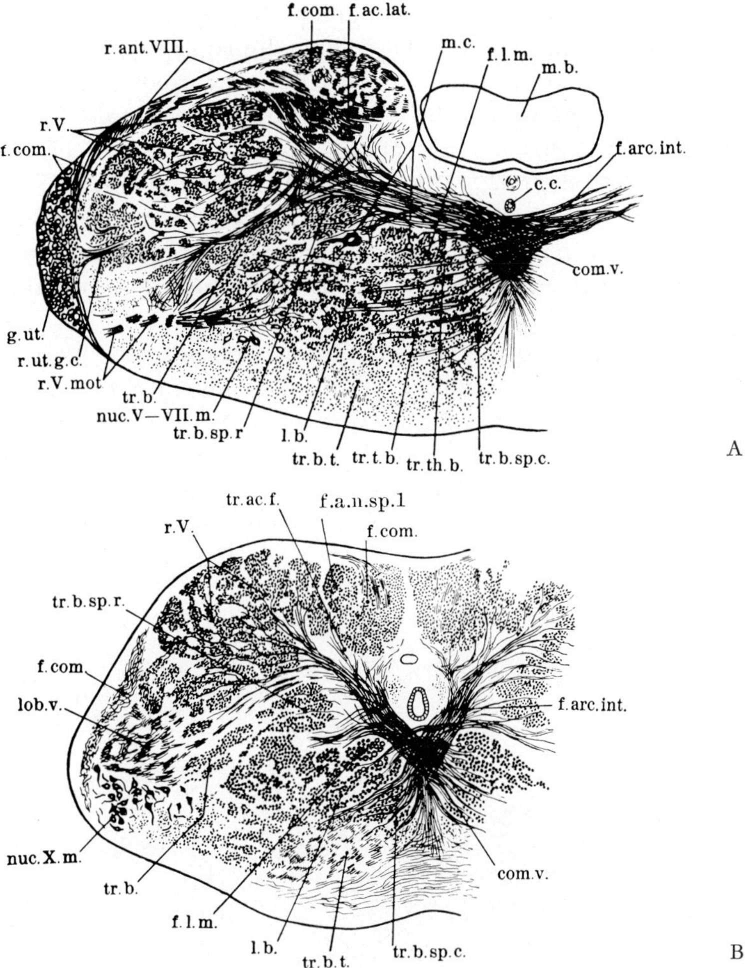

Figure 197. Cross-sections through intermediate levels of the oblongata in Myxine glutinosa as seen in silver impregnations (from Jansen, 1930). A Rostral level of efferent V–VII nucleus. B Level of efferent X nucleus. g.ut.: ganglion utriculare nervi octavi; f.a.n.sp.1: ascending (afferent) fibers of first spinal nerves; lob.v.: lobus vagi; m.b.: caudal end of midbrain roof; m.c.: '*Müller cell*'; nuc.V/VII mot.: efferent cell column of fifth and seventh cranial nerve; nuc.X mot.: efferent vagus nucleus; r.ant. VIII: radix anterior nervi octavi; r.utr.g.c.: general cutaneous root from the ganglion utriculare; r.V mot.: efferent trigeminal root; tr.b.: 'tractus brevis'; tr.b.sp.c.: tractus bulbo-spinalis cruciatus; tr.b.sp.r.: tractus bulbo-spinalis rectus. Other abbreviations as in Figure 195.

nus and facialis, and a caudal column pertaining to the glossopharyngeus-vagus group. The anterior column displays a rostral magnocellular, exclusively trigeminal portion, and a more caudal trigeminal and facial one, with smaller cells. The caudally adjacent motor nucleus of the glossopharyngeus-vagus group forms an elongated cell column along the lateral surface of the oblongata. According to JANSEN (1930), some of the intracerebral efferent vagus root fibers run at first dorsomediad and then turn lateralward, describing a loop around bundles of the general cutaneous afferent system before emerging from the neuraxis.

With respect to the ventral ('somatic motor') zone of the basal plate, a nucleus of the abducens is, of course, absent, since eye muscles and their nerves are missing in Myxinoids. Nevertheless, the ventral motor zone is represented in the caudal part of the oblongata, where a rostral extension of the spinal cord's somatic motoneuron column can be seen medially to the efferent vagus nucleus (Fig. 198). This cell

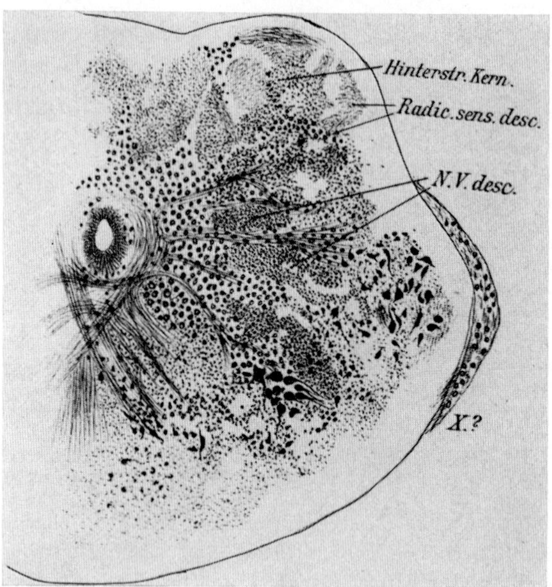

Figure 198. Cross-section through a caudal level of the oblongata in Myxine glutinosa (from EDINGER, 1906). The figure shows what EDINGER interpreted as nucleus of the posterior funiculus. The large lateral cell group is presumably the efferent vagus nucleus, medially to which the spino-occipital column, representing the rostral end of spinal cord ventral horn, can be seen. Still more medially, a group of smaller 'reticular' elements.

group of the ventral zone pertains to the first spinal nerve root which may be evaluated as a modified *spino-occipital nerve*.

The '*nucleus motorius s. reticularis tegmenti*', located medially to the 'viscero-motor columns', displays a rather diffuse arrangement, but has been subdivided into pars anterior and posterior (Fig. 194). Both parts contain large multipolar nerve cells, some of which are the largest ones of the entire brain, and could be interpreted as corresponding to the *Müller cells* of Petromyzonts, discounting the rather intricate semantic details of their definition, discussed above in connection with these latter elements and with the 'giant axons' within brain stem as well as spinal cord. Relatively thick '*Müllerian-type fibers*' respectively '*Müller fibers*', although somewhat less conspicuous than in Petromyzonts, are present in brain stem and spinal cord of Myxinoids. Some such fibers originate from large reticular cells, while others, despite their thickness, may be axons of smaller neuronal elements. Again, while some of these fibers remain uncrossed, others decussate before joining the 'fasciculus longitudinalis medialis'. Depending on differing viewpoints, some of these crossed fibers and their cells of origin could be likened to '*Mauthner fibers*' and '*Mauthner cells*', while, on the other hand, crossed bulbo-spinal fibers arising in the 'acustico-lateralis column' might be considered the equivalents of the '*Mauthner cell mechanism*'.

As regards *ascending fiber systems*, arising in, passing through, or ending in the oblongata, there are presumably some spino-bulbar root fibers ending in EDINGER's '*Hinterstrang-Kern*' mentioned above, as well as secondary spino-bulbo-tectal fibers joining the 'bulbar lemniscus'. This latter communication channel may include some fibers reaching the thalamus. A particular ventrolateral and ventral component of the bulbar lemniscus system was described by JANSEN (1930) as 'tractus bulbo-tectalis' (Figs. 195A, 197A, B).

The descending systems include fibers arising in more rostral brain subdivisions and either ending in the oblongata or running through it to reach the spinal cord. Among the former, ill-defined thalamo-bulbar and hypothalamo-bulbar fibers seem to be present. A perhaps predominantly crossed descending mesencephalic tectal system may include tecto-spinal in addition to the tecto-bulbar fibers. The likewise somewhat ill-defined fasciculus longitudinalis medialis arises in the mesencephalic tegmentum and extends through the length of the spinal cord. It receives crossed and uncrossed additions from the bulbar tegmentum, which consist of partly ascending and partly descending fibers. Frequently, these axons bifurcate into rostral and caudal branches as they enter the fasciculus, which thus does not represent an exclusively descending tract but rather, *at least within the brain stem*, a pathway for reciprocal bulbo-mesencephalic connections.

An uncrossed descending system arising in the bulbar nucleus motorius tegmenti, and running laterally to the fasciculus longitudinalis has been described by JANSEN (1930) as tractus bulbo-spinalis rectus. Still another descending system arising in the oblongata is provided by the ventral tractus bulbo-spinalis cruciatus arising, perhaps as descending collateral branches of arcuate fibers pertaining to the bulbar lemniscus, from the octavo-lateral and other afferent 'nuclei'. A 'tractus acustico-funicularis' formed by descending secondary octavo-lateral fibers is described by JANSEN (1930) as running along the descending octavo-lateral root fiber fascicle, and ending at the caudal levels of the viscero-motor vagus column. The cited author also noted a bundle which he called 'tractus brevis', believed to arise from the octavo-lateral area and also from 'the general cutaneous nucleus'. It is said to descend homolaterally, and dorsally to the 'motor nuclei', decreasing in size caudalward, and apparently not reaching the spinal cord, but giving off some collaterals to the ventral commissure.

Although rather diffuse, non-descript and very incompletely elucidated *qua* exact details, the fiber pathways and cellular arrangements in the Myxinoid oblongata display a very definite orderly overall configuration easily comparable with that obtaining in Petromyzonts. Again, the oblongata of these latter embodies, *in nuce*, most of the relevant features characteristic for that subdivision of the brain in all Vertebrates. Thus, the oblongata of Cyclostomes provides a useful paradigm for a general description of bulbar configurations and mechanisms.

Evaluated from a 'functional viewpoint', which emphasizes the significance of fiber systems[91] *qua* basic communication channels between 'grisea' concerned with particular aspects of input, output, or signal processing, and discounting substantial differences related to a blurred, deformed morphologic pattern (*'Bauplanverzerrung'*), the Myxinoid oblongata is indeed of considerable interest as pertaining to a much reduced and very rudimentary, yet 'generalized vertebrate brain'.

---

[91] Hence, JANSEN (1930) reaches the conclusion that 'in spite of striking morphologic differences, an analysis of the functional relations of the nuclear masses in the brain of Myxine reveals great concordance with the conditions in generalized vertebrate brains, like those of Petromyzonts and Amphibia'.

Ludwig Edinger, in his pioneering search for a relatively simple Vertebrate central nervous system displaying, *in nuce*, all relevant anatomical as well as functional features further developed in higher forms, and whose understanding would provide a suitable introduction to the problems of neurobiology, pointed out the particular significance of the urodele Amphibian neuraxis. In his monograph on the brain of *Myxine* (1906) he added the following comments:

'*Vor Jahren schon habe ich das Gehirn der Salamandrinen als das in mikroskopisch-anatomischer Beziehung am einfachsten gebaute bezeichnen müssen, auch heute, nachdem ich Myxine und Petromyzon nun genauer kenne, ist dies aufrecht zu halten. Bei Myxine fehlen die unnöthigen Theile oder sie sind stark atrophisch, es sind aber die Theile, welche das Thier brauchen kann, nicht schlechter ausgebildet wie bei anderen Vertebraten. Die Oblongata, das Rückenmark lassen deshalb auch heute noch keine völlige Entwirrung zu.*

Dass wir hier ein niederes Gehirn von besonderer Vollkommenheit in einigen Richtungen, mit regressiven Processen an anderen Hirntheilen vor uns haben, das scheint keinem Zweifel zu unterliegen. Für das Regressive kann auch geltend gemacht werden, dass keine zwei Exemplare mit völlig gleichem Gehirne existiren und dass die Unterschiede immer darin bestehen, dass der eine oder andere Hirntheil schlechter ausgebildet ist als bei anderen Exemplaren; wie denn die Vorderhirnantheile überhaupt und immer schlechter entwickelt sind als bei anderen Vertebraten'.

4. Selachians

The medulla oblongata of Selachians[92] may be roughly delimited as extending from the trochlear decussation, that is, from the isthmus rhombencephali, to the origin of the first 'true' spinal roots, thus excluding these latter but including the origin of the spino-occipital nerve roots. Again, this rhombencephalic bulb can be subdivided into an upper or rostral portion, related to the extracerebral roots of the trigemino-facial branchial nerve group, and into a lower or caudal portion, related to the extracerebral roots of the vagus group

---

[92] Various summaries on the Selachian, and, for that matter, on the Vertebrate medulla oblongata, elaborated on the basis of original observations and in accordance with a diversity of overall viewpoints, can be found, together with the relevant bibliographic references, in the texts on comparative neurology cited above in section 3, p. 349 of this chapter. Investigations by my collaborators and myself, containing data on basic aspects of the Selachian oblongata, are those by Saito (1930), Gerlach (1947), and K. and Niimi (1969).

(Figs. 199–204). Although nervus octavus and lateralis anterior pertain, together with the trigemino-facial group, to the upper oblongata, the posterior lateralis system as well as the area statica extend caudalward into the bulb's lower part.

The transitional levels between upper oblongata and mesencephalon are represented by the isthmus rhombencephali whose dorsal wall, forming the *velum medullare anterius s. superius* contains decussation and exit of the trochlear nerve. Caudally to this velum, the roof of the upper oblongata is constituted by the cerebellum, and still more caudally by lamina epithelialis with its choroid plexus. The transverse caudal

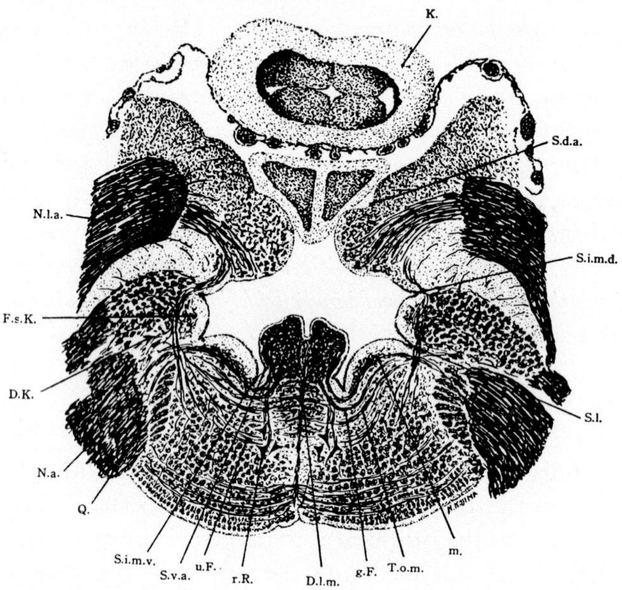

Figure 199. Cross-section through the oblongata of Acanthias at the level of *Deiters' nucleus* (from SAITO, 1930). D.K.: *Deiters' nucleus;* D.l.m.: decussation of arciform fibers; F.s.K.: afferent nucleus of facial nerve; g.F.: fibers joining fasciculus longitudinalis medialis; K.: cerebellum; m: efferent fibers of facial nerve; N.a.: nervus vestibularis; N.l.a.: nervus lateralis anterior; Q.: radix descendens trigemini; r.R.: rostral rhombencephalic subgroup of reticular cells; S.d.a.: sulcus dorsalis accessorius; S.i.m.d.: sulcus intermedius dorsalis; S.i.m.v.: sulcus intermedius ventralis; S.l.: sulcus limitans; S.v.a.: sulcus ventralis accessorius; T.o.m.: tractus octavo-motorius; u.F.: fibers of 'tractus reticularis'. The fasciculus longitudinalis medialis (not labelled) can easily be identified and protrudes into the ventricle, displaying an additional accessory groove. The indentation between the two antimeric longitudinal fasciculi is the sulcus medianus basalis ventriculi quarti.

edge of the cerbellum providing the attachment of lamina terminalis is the *velum medullare posterius s. inferius*. Since the lamina epithelialis is easily removed, the part of fourth ventricle covered by this thin structure can be loosely called the open part of the rhomboid fossa. However, there does not seem to occur in Selachians, nor for that matter generally in Anamnia, a normally present actual opening comparable to the Mammalian *foramina of Luschka or Magendie*, through which a free communication between ventricular fluid and leptomeningeal spaces is provided.

The roof of the oblongata's lower portion is formed by nervous tissue, namely by the two alar plates fused together in the midline, including here an obliterated remnant of the roof plate, and containing,

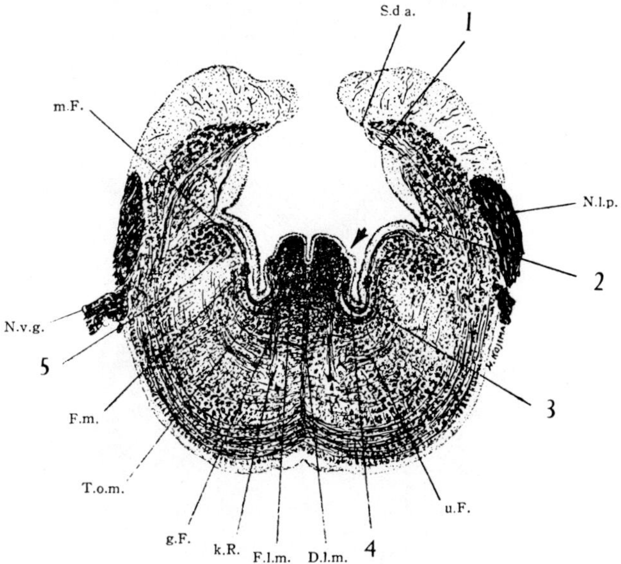

Figure 200. Cross-section through the oblongata of Acanthias at the level of the vagus group (from Saito, 1930, with some changes in labelling). D. l. m.: decussation of arciform fibers; F. l. m.: fasciculus longitudinalis medialis; F. m.: fasciculus vestibulo-spinalis dorsalis ?; g. F.: fibers of reticular system; k. R.: caudal rhombencephalic reticular subgroup; m. F.: arcuate fibers from tractus solitarius system ?; N. l. p.: nervus lateralis posterior; N. v. g. rootlet of glossopharyngeus-vagus group; S. d. a.: sulcus dorsalis accessorius; T. o. m.: tractus octavo-motorius; u. F.: arcuate fibers; 1: sulcus dorsalis accessorius inferior; 2: sulcus intermediodorsalis; 3: sulcus limitans; 4: sulcus intermedioventralis fused with sulcus ventralis accessorius; 5: tractus solitarius. Arrow points to sulcus accessorius fasciculi longitudinalis medialis. Radix descendens V, unlabelled, ventrolateral to 5.

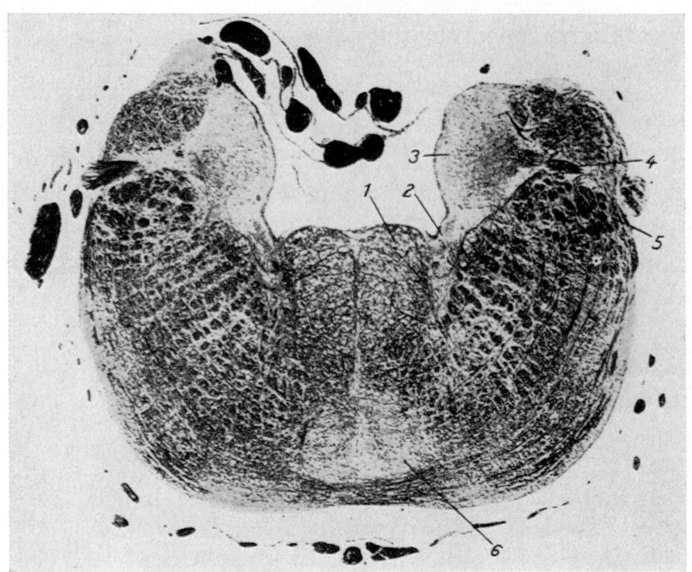

201

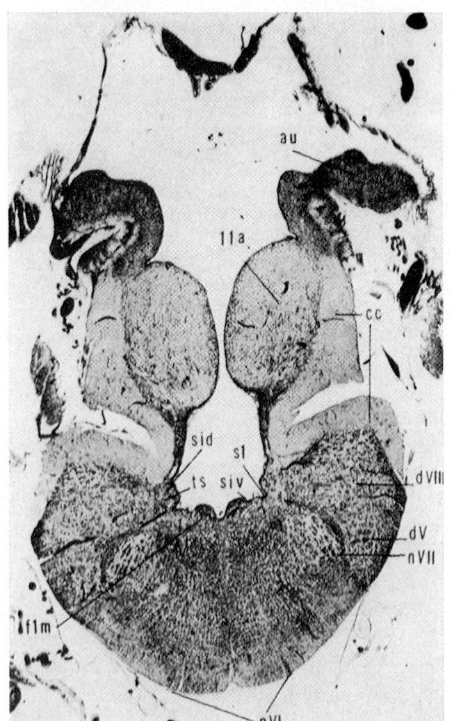

202

at the levels of calamus scriptorius, a more or less well developed *commissura infima Halleri.*

The lumen of this closed part of the fourth ventricle, continuous with the central canal of the spinal cord, becomes gradually narrower in a caudal direction. Generally speaking, the grisea of the Selachian medulla oblongata do not display distinct boundaries suitable for a clear-cut delimitation of circumscribed nuclei, but, as in Cyclostomes, represent longitudinal columns or arrays of cell populations whose elements are scattered between fiber bundles. Nevertheless, in contradistinction to Cyclostomes, a few somewhat more distinctive clustered cell groups or 'nuclei' can be recognized.

Along the ventricular surface of the oblongata, not only the main longitudinal sulci, but also a number of accessory ones are commonly well developed (Figs. 154, 199, 200, 201–205). The *sulcus limitans* can usually be traced throughout the oblongata from the isthmus region to the most caudal levels continuous with the spinal cord. Dorsally to sulcus limitans, the *sulcus intermedius dorsalis* is recognizable at levels caudal to the auricles. The intermediodorsal zone, representing the morphologic equivalent of the functionally ill-defined visceral afferent longitudinal system, is roughly bounded by these two sulci. In the region of the auricles and of the octavo-lateral alar plate components, several variable *accessory sulci* can be seen, including SAITO's (1930) *sulcus dorsalis accessorius.*[93] The alar plate regions dorsal to intermediodorsal sulcus

[93] Cf. e.g. Figs. 200, 202. In Figure 200 an unlabelled accessory sulcus can be seen dorsally (in the inferior leaf of auricle) as well as ventrally to SAITO's *sulcus accessorius dorsalis*. If this latter should be called s. a. d. *principalis*, the two others might be designated as s. a. d. *superior* and *inferior*, respectively. Caudalward, these sulci join, in a variable manner, the sulcus *principalis* which may run as far as the calamus region.

Figure 201. Cross-section through the oblongata in Scyllium canicula at the level of the inferior olivary nucleus, as seen in a *Weigert-Loyez stain* (from BECCARI, 1943). 1: fasciculus longitudinalis medialis; 2: viscero-efferent column; 3: viscero-afferent column (tractus solitarius); 4: afferent root fibers of vagus; 5: radix descendens trigemini; 6: inferior olivary nucleus *('oliva bulbare')*.

Figure 202. Cross-section through the oblongata of Chimaera Colliei at the level of nervus abducens exit. Although a hematoxylin-eosin stain, the medullated fiber tracts were conspicuously displayed by the hematoxylin (from K. and NIIMI, 1969). au: auricle of cerebellum; cc: crista cerebellaris; dV: radix descendens trigemini; dVIII: radix s. tractus descendens octavo-lateralis; flm: fasciculus longitudinalis medialis; lla: lobus of nervus lineae lateralis anterior; nVI: root fibers of nervus abducens; n. VII: root fibers of nervus facialis; sid: sulcus intermedius dorsalis; siv: sulcus intermedius ventralis; sl: sulcus limitans; ts: tractus solitarius.

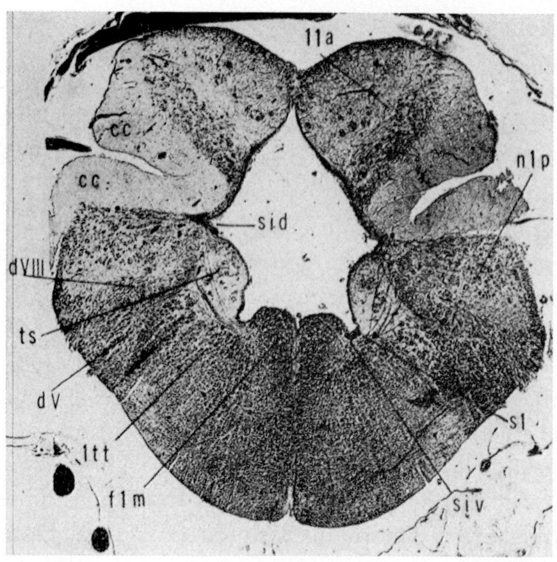

Figure 203. Cross-section through the oblongata of Chimaera Colliei at level of nervus lineae lateralis posterior and glossopharyngeus roots (from K. and NIIMI, 1969). ltt: lateral tegmental tract(s); nlp: ascending root of nervus lineae lateralis posterior. Other abbreviations as in Figure 202.

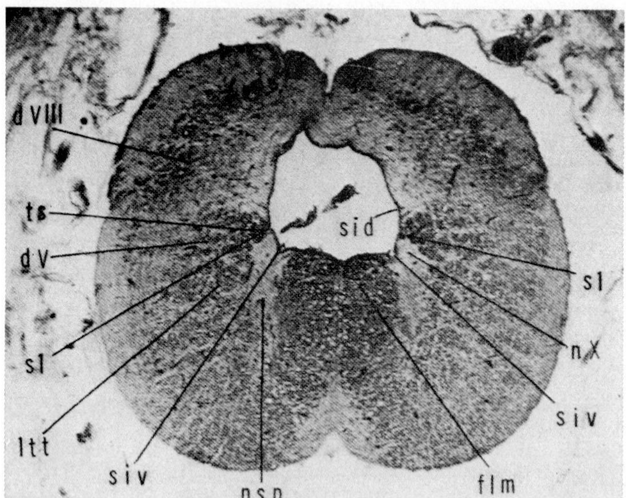

Figure 204. Cross-section through the oblongata of Chimaera Colliei at level of vagus and spino-occipital nerve nuclei (from K. and NIIMI, 1969). nsp: nucleus of spino-occipital ventral nerve roots; nX: efferent nucleus of vagus (and accessorius). Other abbreviations as in Figure 202. The inferior olive, not labelled, can easily be identified (cf. Fig. 201).

represent the dorsal zone which contains the so-called somatic afferent longitudinal systems.

Ventrally to sulcus limitans, the *sulcus intermedius ventralis* extends from caudal levels of the oblongata to the auricular region, where it becomes indistinct, being rostrally replaced by a very vaguely outlined 'system' of accessory grooves. The ventricular wall strip between sulcus limitans and sulcus intermedius ventralis pertains to the basal plate's intermedioventral longitudinal zone, and that between midline *(sulcus medianus basalis ventriculi quarti)* and sulcus intermedius ventralis pertains to the ventral longitudinal zone. This latter corresponds approximately to the somatic-efferent zone, and the intermedioventral zone corresponds roughly to the so-called viscero-efferent zone in terms of the doctrine of functional nerve components. A *sulcus ventralis accessorius* (SAITO, 1930) may be found within the ventral zone (cf. Fig. 200) and is perhaps related to a protrusion, into the ventricular lumen, of a conspicuously bulging fasciculus longitudinalis medialis in diverse but not all Selachians. Variable slanting connections between sulcus limitans and sulcus intermedioventralis as well as between this

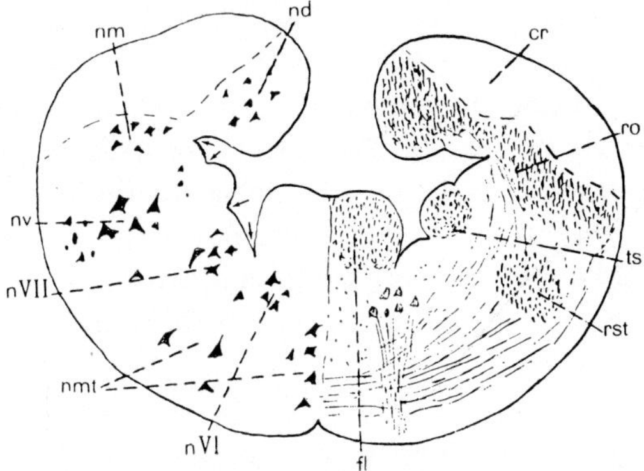

Figure 205. Simplified sketch, illustrating a cross-section through the oblongata in Acanthias at the level of efferent facial and of abducens nucleus, based on data by KAPPERS and on personal observations (from KUHLENBECK, 1927). cr: crista cerebellaris; fl: fasciculus longitudinalis medialis; nd: nucleus nervi lateralis; nm: nucleus octavo-lateralis; nmt: nucleus motorius tegmenti; nv: 'nucleus vestibularis ventralis' *(Deiters' nucleus;* the more laterally located smaller cells represent the nucleus tangentialis); n. VI: nucleus nervi abducentis; n VII: nucleus motorius nervi facialis; ro: radix et tractus nervi octavo-lateralis; rst: radix descendens nervi trigemini (the nucleus of this tract is merely suggested by a single cell above the lead to n VII on left); ts: tractus solitarius (the sulcus limitans is represented by the ventricular groove ventral to this bulging bundle).

latter and sulcus ventralis accessorius system obtain, which may display, at least in some Selachian forms, a rather confusing pattern. Postmortal or fixation artefacts can doubtless introduce further distortions, but the occurrence of the enumerated sulci in well preserved and carefully processed material seems reasonably well established.

The comments on the overall pattern of the Cyclostome medulla oblongata made on p. 348 in the introductory part of section 3, also essentially apply, *mutatis mutandis*, to the Selachian oblongata. In the alar plate of this latter's *übergangsgebiet*, the 'large dorsal cells' of Cyclostomes are missing. Other differences involve the basal plate, whose 'nucleus reticularis tegmenti' does not display the typical *Müllerian cells* of Cyclostomes. On the other hand, as specified further below, the caudal oblongata of Selachians includes a rudimentary *inferior olivary griseum* not unequivocally demonstrable in Cyclostomes.

With respect to details of the *alar plate's neuronal systems*, the octavolateralis complex, located in the dorsal longitudinal zone, displays a conspicuous degree of differentiation. The crista cerebelli is well developed, being continuous with the cerebellar auricle. Dorsal fiber bundles of the nervus lateralis anterior are related to a griseum with ovoid outline in cross-sections and forming a distinctive 'lobus' or 'nucleus' lineae lateralis anterioris (cf. e.g. Fig. 202) which apparently corresponds to the nucleus dorsalis of LARSELL (1967). More ventral fiber bundles join an intermediate area, designated as nucleus ventralis lineae lateralis anterioris, presumably the nucleus medialis of LARSELL (1967), lateral to the tractus solitarius and dorsal to the vestibular 'nuclei' (Fig. 206). This ventral 'nucleus' presumably also receives vestibular fibers. The griseum of the nervus lateralis posterior[94] is caudally and laterally adjacent to the ventral nucleus of the anterior lateral line (cf. Fig. 203).

---

[94] The anterior lateral nerve supplies the sensory structures of the head's three lateral lines, and the associated *ampullae of Lorenzini* respectively *vesicles of Savi* (cf. chapter VI, section 1, p. 794 of vol. 3, part II). It enters the oblongata, approximately between trigeminal and facial roots, becoming subdivided into a dorsal and a ventral bundle. The nervus lateralis posterior supplies the lateral line structures of the trunk and enters the oblongata approximately at, and slightly dorsally to, the glossopharyngeal root level. Peripherally, it runs, for some distances, in partially close association with the vagus. In discussing the Selachian lateralis system, KAPPERS (1947) refers to observations by HOAGLAND (1933) and RUBIN (1935) indicating that, in addition to the generally accepted mechanoreceptor function of that system registering water flow and vibrations, a thermoreceptor function might obtain (cf. also the comments, including references to electroreceptor functions, in the above-mentioned section 1 of chapter VI).

The nervus vestibularis s. octavus supplies, with an anterior branch, the ampullae of anterior and lateral semicircular canals, and the macula utriculi. Its posterior branch innervates ampulla of posterior semicircular canal, macula sacculi, macula neglecta, and macula lagenae (cf. Fig. 164B). The rather diffuse vestibular grisea of the oblongata include a large-called *nucleus vestibularis ventralis* (Figs. 199, 205, 206), whose homology with the *nucleus vestibularis lateralis of Deiters* in higher Vertebrates is generally recognized. The lateral and ventralateral portion of nucleus vestibularis ventralis contains likewise some rather large cells, and may be distinguished as *nucleus tangentialis*. It extends somewhat farther caudad than the main ventral vestibular nucleus. Again, a dorsal portion of this latter, extending as far as the lower oblongata, and accompanying descending root fiber bundles, could be designated as *nucleus of the descending vestibular root* (Figs. 202–204, 212B).

The nucleus vestibularis ventralis receives many end-arborizations of fibers from the anterior vestibular branch. The posterior vestibular

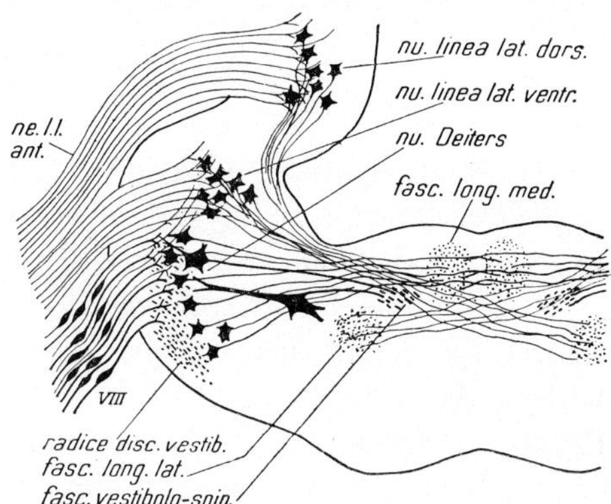

Figure 206. Diagrammatic sketch of a cross-section illustrating terminal octavo-lateralis nuclei and origin of secondary arciform octavo-lateralis fibers in Selachians, based on sections through the oblongata of advanced Scyllium canicula embryos processed by *Cajal's silver impregnation* (from BECCARI, 1943). ne. l. l. ant.: nervus lineae lateralis anterior; fasc. long. lat.: fasciculus longitudinalis lateralis (tractus octavo-motorius et vestibulo-spinalis cruciatus); radice disc. vest.: descending vestibular root. Other abbrev. self-evident. The three nuclei correspond to nd, nm, and nv in Figure 205 which uses a slightly different terminology. If the elongated large cell in *Deiters' nucleus* were sending its axon into contralateral fasc. long. med. instead of into homolateral vestibulo-spinal tract, as indicated by BECCARI, it could be considered a '*Mauthner cell*' in accordance with that element's current definition.

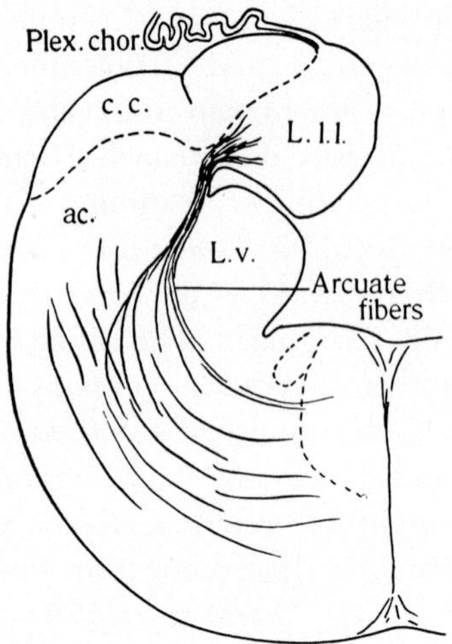

Figure 207. Simplified cross-section through the oblongata in Scyllium at the level of octavo-lateral nuclei, showing arcuate (or arciform) fibers (from JOHNSTON, 1906). ac: 'tuberculum acusticum' (corresponding to nm and nv in Fig. 205); cc: cerebellar crest; L.l.l.: lobus lineae lateralis (of anterior lateral nerve); L.v.: 'lobus vagi' (tractus solitarius, probably more at VII than at X level); Plex. chor.: lamina epithelialis and choroid plexus of fourth ventricle.

branch contributes substantially to the descending vestibular root, but both roots, upon their entrance, seem to give off ascending as well as descending fibers. Some root fibers from the radix anterior are also said to cross the ventral midline, but, despite various and in part vague or conflicting reports, the details of these fiber distributions and arrangements remain insufficiently clarified.

As regards the rostral boundaries of the octavo-lateral grisea and their fiber bundles, the lateralis grisea are continuous with the auricles, while those of the vestibularis extend toward the corpus cerebelli. Among the secondary fibers arising in these grisea, arcuate fibers (Fig. 207) are conspicuous and provide the essentially crossed *lemniscus lateralis* or *fasciculus longitudinalis lateralis autorum*, which may, however, also contain some uncrossed ascending secondary or even primary (root) fibers. KAPPERS (1947) believes that the lemniscus lateralis contains mostly lateralis components, the vestibular grisea being predominantly involved with connections not mediated by that lemniscus, which, nevertheless, in my opinion doubtless also includes vestibularis components.

Other octavo-lateralis connections are fiber bundles joining that system's grisea with the cerebellum, moreover a *tractus vestibulo-spinalis rectus* from *Deiters' nucleus*, a *tractus vestibulo-spinalis cruciatus lateralis*, a *fasciculus vestibulo-spinalis dorsalis*, and (perhaps crossed as well as uncrossed) vestibulo-spinal fibers (tractus vestibulo-spinalis medialis) within the *fasciculus longitudinalis medialis*, moreover shorter octavo-lateralis links with nucleus reticularis tegmenti and other grisea, such as efferent cranial nerve nuclei.

It should be added that identification, exact location, and tracing of these various fiber systems remain rather uncertain with respect to many details, although the available data can be assessed as providing, in the manner of a useful first approximation, a reasonably valid overall account of the general connection-pattern. The complex and poorly understood relationships of *Deiters' nucleus* to the nucleus motorius tegmenti and its fiber systems shall be pointed out further below in dealing with the reticular formation.

In addition to the octavo-lateralis grisea, the afferent nuclei within the oblongata's alar plate comprise those of the *branchial nerves*, namely of trigeminus, facialis, and of the glossopharyngeus-vagus group. The *main sensory trigeminal nucleus* is located at the entrance level of the trigeminal root, its (secondary) afferent cells being scattered among the fiber bundles. The caudal extension of this griseum is the *nucleus of the descending trigeminal root*, whose cells are likewise distributed along or within that fiber tract which runs basally to the descending vestibular root as far as the transition to spinal cord (Figs. 199–205). Afferent, presumably proprioceptive trigeminus fibers, originating from rather large intracerebral vesicular cells, roughly comparable to the *Rohon-Beard cells* discussed in the preceding chapter, form the *mesencephalic root of the trigeminus*, which shall be dealt with further below in chapter XI (mesencephalon). All afferent grisea of the trigeminus are derivatives of the alar plate's dorsal zone, in which the main sensory nucleus of that nerve and the nucleus of its descending root assume a ventrolateral position.

The *sensory (viscero-afferent) nucleus of the facial nerve* is located within the intermediodorsal zone of the alar plate. Its cells are scattered between the ascending and descending root fiber bundles forming the rostral part of the *tractus solitarius*. In various Selachian species, this bundle, together with its griseum, may conspicuously protrude into the ventricle between sulcus limitans and sulcus intermedius dorsalis (cf. e.g. Fig. 205, and right side of Figs. 210A and B). The facial sensory nuclear column and its longitudinal fiber system are directly continuous with the more caudal tractus solitarius and griseal components pertaining to *glossopharyngeal and vagus nerves* (Figs. 202–204). The viscero-afferent fibers of VIIth, IXth and Xth nerves comprise 'gustatory' fibers, which are believed to be more numerous than those of Cyclostomes, and 'unspecialized', so-called general viscero-afferent ones. Cutaneous (somatic-afferent) fibers are said to be scarce in facial and glossopharyngeal nerves, but more numerous in the vagus. These fibers are presumed to join the descending root of the trigeminus and its 'nucleus'.

Second order fibers originating in the afferent nuclei of the branchial nerves cross the ventral midline as *arcuate fibers*, giving off collaterals, and form an essentially crossed *tractus bulbo-tectalis* which, together with similar secondary fibers from the spinal cord, represents a bulbar and spinal '*lemniscus medialis*', some of whose fibers may also reach the thalamus. A conspicuous separate so-called tractus gustatorius ascendens secundarius, which is characteristic for many Osteichthyes, and includes a special relay-griseum at the isthmic levels, has been mentioned by KAPPERS (1947) and others, but, if indeed at all present in Selachians, remains here very dubiously recognizable. It is possible that, at caudal levels of the oblongata (Fig. 204) a nondescript cell group in the dorsal portion of the alar plate could represent a *nucleus of the posterior funiculus*, contributing ascending secondary spinal fibers to the bulbar and spinal lemniscus.

Ascending homo- or contralateral fibers from the trigeminal grisea may join the *spino-cerebellar tract*, while some descending fibers from sensory branchial nerve grisea are believed to reach rostral portions of the spinal cord either through fasciculus longitudinalis medialis or bulbo-spinal tract. Descending fibers of trigeminal and of tractus solitarius systems decussate in the *commissura infima Halleri* mentioned further above.

In the *basal plate*, the *motor or efferent cranial nerve nuclei* are those of the branchial nerves V, VII, IX, X, of the nervus abducens, and of the spino-occipital roots. The *motor nucleus of the trigeminus* consists of loosely grouped cells within the intermedioventral region of the oblongata's rostral tegmentum at the level of the emerging extracerebral trigeminus roots. The *motor nucleus of the facial nerve*, likewise pertaining to the intermedioventral zone, is represented by a cell group in the lateral tegmentum at caudal levels of the upper oblongata, in which the *nucleus of the abducens* is also situated. This nucleus, however, has its location within the *ventral* longitudinal zone of the dorsal tegmentum,

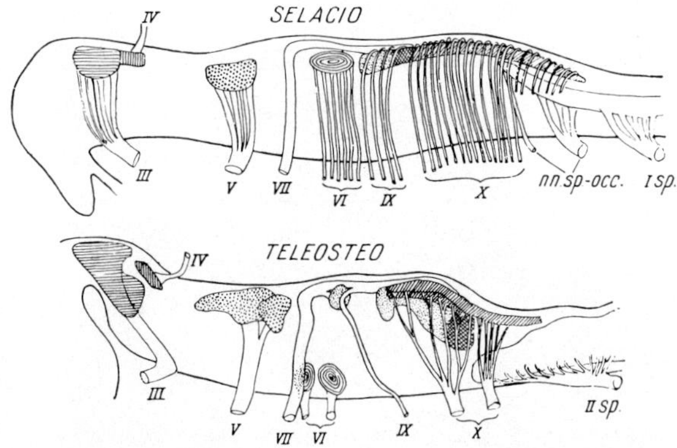

Figure 208. Diagrammatic sketch showing efferent (motor) nuclei and roots of deuterencephalic cranial nerves in a Selachian (Acanthias vulgaris) and in a Teleost (Tetrodon), projected upon a sagittal plane (simplified after VAN DER HORST, from BECCARI, 1943). Fine dots: facial nucleus; cross-hatching; glossopharyngeus nucleus; oblique hatching: vagus nucleus. Other designations self-evident. BECCARI, in agreement with a comment by KAPPERS, points out that considering the sequences obtaining in lower Vertebrates, the facial nerve could be designated as n. VI, and the abducens as n. VII, since the conventional sequence abducens VI and facial VII essentially holds only for Mammals.

close to the fasciculus longitudinalis medialis (Fig. 205). As apparently in most or all Selachians, proximal root fibers of the facial nerve form a loop or knee crossing parts of the abducens nucleus; the distal part of the loop then runs ventrolaterad to emerge from the brain. Figure 202 shows root fibers of both the facial and abducens nerves.

The *motor nuclear column of glossopharyngeus and vagus*, pertaining to the intermedioventral zone, is diffusely located in dorsal and dorsolateral regions of the tegmentum and includes the *accessory component* of vagus nerve.[95] The medial portion of the efferent IXth and Xth nerve nucleus is situated near the ventricle, between sulcus limitans and sulcus intermedius ventralis (Fig. 204). The entire cell population of this griseum forms the caudal part of the so-called viscero-motor cell column whose elements comprise branchiomotor and preganglionic neurons. However, as pointed out by KAPPERS (1947) for Plagiostomes, it

---

[95] '*Le nerf accessoire (XIème) des Plagiostomes, qui innerve un petit muscle trapézoide, prend son origine dans la partie caudale de la colonne motrice du pneumogastrique. Les cellules d'origine de ce nerf ne se distinguent de celles du pneumogastrique sous aucun rapport*' (KAPPERS, 1947).

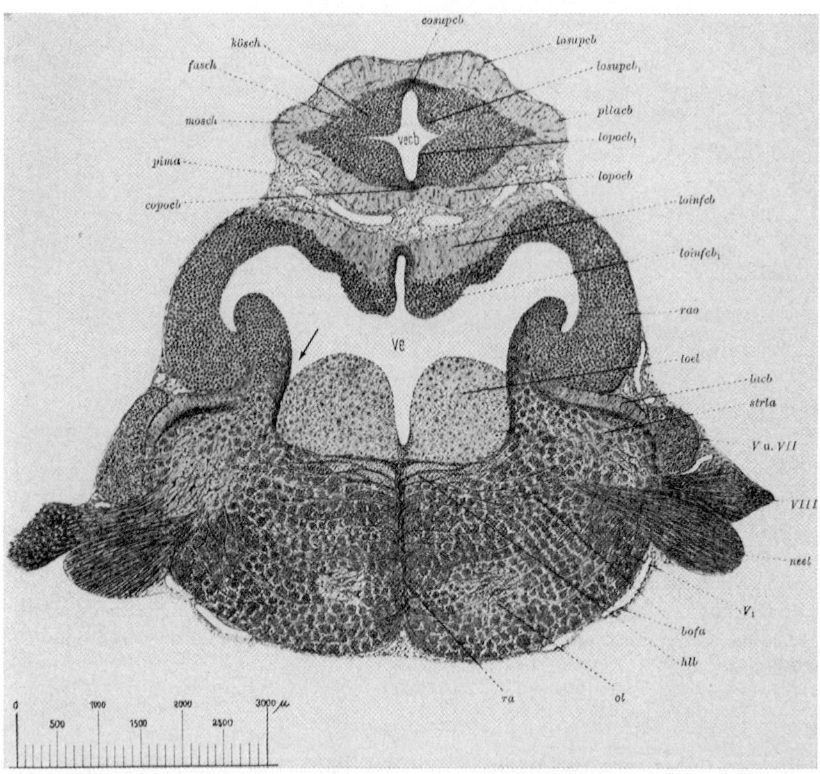

Figure 209. Cross-section through the rostral oblongata in Torpedo occelata (probably *Bielschowsky silver impregnation*, from KRAUSE, 1923). bofa: arcuate fibers; copacb, cosupcb: cerebellar commissures; fasch: fiber layer of cerebellum; hlb: fasciculus longitudinalis medialis; kösch: granular layer (eminences) of cerebellum; lacb: crista cerebellaris; loel: lobus electricus; loinfcb, losupcb: 'inferior' and 'superior' lobes of cerebellum; mosch: molecular layer of cerebellum; neel: 'nervus electricus'; ol: tegmental reticular cell group (interpreted as '*obere Olive*' by KRAUSE); pima: leptomeninx; pllacb: 'plica lateralis cerebelli'; ra: raphe; rao: auricle; strla: octavo-lateralis nuclei (called 'stratum laterale' by KRAUSE); ve: fourth ventricle; vecb: cerebellar ventricle; V u. VII: part of extracerebral roots of trigeminus and facial; VIII: part of octavo-lateralis roots. Added arrow indicates sulcus limitans.

is not possible to distinguish with reasonable certainty these two sorts of cells from each other. It seems likely that they are intermingled within their nuclear column, perhaps somewhat differing in size, such that the branchiomotor cells are represented by the larger elements.[96]

---

[96] Much the same could perhaps be said not only with respect to Plagiostomes, but also with regard to Anamnia in general.

The *nuclear column of the 'somatic motor' spino-occipital ventral nerve roots* pertains to the basal plate's ventral zone and is located basally to the caudo-median portion of motor vagus nucleus, laterally to fasciculus longitudinalis medialis (Fig. 204). A diagram elaborated by BECCARI (1943) on the basis of his own studies in combination with the data recorded by VAN DER HORST (1918), and depicting the Selachian deuterencephalic motor nuclei in sagittal view for comparison with those of Teleosts, to be dealt with in the next section, is shown in Figure 208.

By and large, the medulla oblongata of Plagiostomes displays a rather uniform configuration with respect to the various orders and genera of this Vertebrate class, the obtaining differences being, as a rule, of a more or less trivial nature. There obtains, nevertheless, at least one conspicuous exception, which concerns the *lobus electricus of Torpedinidae*, discussed and depicted in section 5, chapter VII of the preceding volume 3, part II. The lobus electricus represents here a specialized development and expansion of the branchiomotor nuclear column, resulting in a considerable pattern distortion which, nevertheless, preserves the fundamental topologic connectedness of its component neighborhoods. These latter can be easily recognized by inspection (Figs. 209, 210).

The *tegmentum* of both upper and lower medulla oblongata contains the rhombencephalic portion of *nucleus reticularis s. motorius tegmenti*, which is characterized by irregular groups of multipolar and in part rather large nerve cells scattered between the numerous, mainly longitudinal and mostly medullated fiber bundles of the basal plate. The resulting reticular formation can be conceived as a single griseum extending from the tegmental cell plate in the region of tuberculum posterius at the diencephalo-mesencephalic boundary to the caudal end of the oblongata. BECCARI (1930), SAITO (1930) and GERLACH (1947) have reported on various details of this griseum and distinguished various subdivisions or nuclei. Thus, BECCARI described a mesodiencephalic center,[97] a metencephalic center with two subgroups, and a myel-

---

[97] The diencephalic portion of BECCARI's mesodiencephalic center apparently corresponds to the *tegmental cell plate* at the mesencephalo-diencephalic boundary. Whether, in this transitional neighborhood *(Übergangsgebiet)*, the tegmental cell plate, located within the concavity of the rostral end of sulcus limitans, should, or should not be considered part of the diencephalon, is evidently an arbitrary semantic choice. The intrinsic diencephalic bauplan with its four longitudinal zones does not include a tegmental 'reticular' component, or, in this respect, as justly stated by GERLACH (1947): *'jedoch ist von einem diencephalen reticulären Kern nicht die Rede'*.

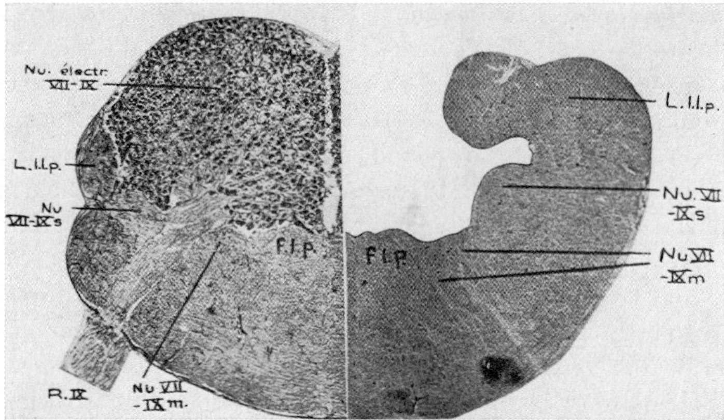

Figure 210 A. Cross-section through left half of oblongata in Torpedo marmorata, at the glossopharyngeal level, compared with corresponding right half of oblongata in Scyllium canicula (from KAPPERS, 1947). f.l.p.: fasciculus longitudinalis medialis; L.l.l.p.: lobus lineae lateralis posterioris; Nu. électr. VII–IX: lobus electricus, containing efferent cells of VII–X innervating electric organ, and derived from the branchiomotor (visceromotor) intermedioventral zone of basal plate; Nu. VII–IX s.: afferent facial and glossopharyngeus-vagus nucleus (tractus solitarius, intermediodorsal zone of alar plate); Nu. VII–IX: branchiomotor (special and general visceral efferent) nucleus of nerve and glossopharyngeus-vagus group; R IX: ramus electricus of n. glossopharyngeus.

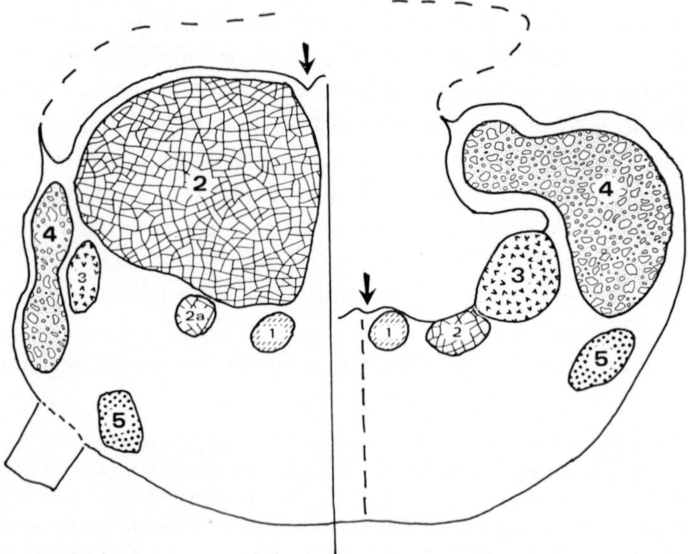

Figure 210 B. Outline sketch based on the preceding figure and illustrating displacements and pattern distortion caused by the expansion of the intermedioventral zone related to the differentiation of lobus electricus. 1: fasciculus longitudinalis medialis;

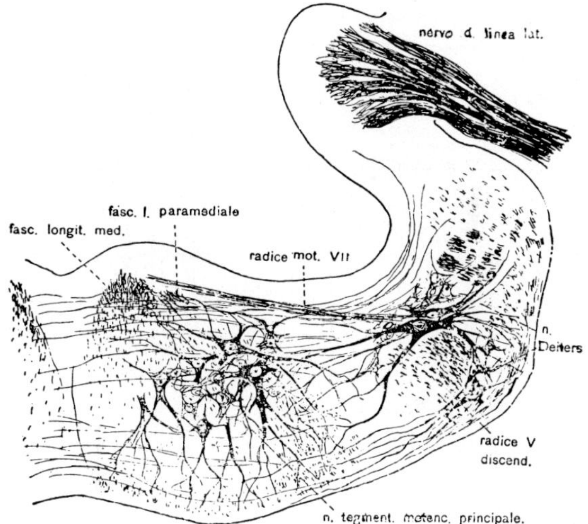

Figure 211 A. Cross-section through the oblongata in an advanced embryo of Scyllium canicula at the level of *Deiters'nucleus* as displayed by *Cajal's silver impregnation* (from Beccari, 1930). The thick dorsal nerve root is probably that of the anterior nerve of the lateral line. Beccari's 'fasciculo longitudinale paramediale' represents a subdivision of fasciculus longitudinale mediale.

encephalic one. Saito distinguished a mesencephalic, an isthmic, and a rhombencephalic reticular group, the latter displaying a rostral and a caudal subgroup. Gerlach recorded a mesencephalic and a rhombencephalic '*Hauptgruppe*', of which the rhombencephalic one was further subdivided into an isthmic, a praetrigeminal, and a caudal group.[98] Figures 211–213 illustrate some aspects of the Selachian reticular nuclei.

[98] Each of the slightly different classifications used by the three cited and by other authors appears defensible. The differences of interpretation are easily explainable by (1) the rather diffuse, nondescript arrangement of the 'reticular' cells, (2) frequent 'individual variations', and (3) some degree of diversity *qua* particular species. Thus, Beccari particularly studied Scyllium canicula, Saito the East-Asiatic Acanthias mitsukurii, and Gerlach predominantly Galeus canis.

2: intermedioventral zone; 2a: part of intermedioventral zone in Torpedo not included in lobus electricus; 3: tractus solitarius and its griseum; 4: octavo-lateral grisea; 5: radix descendens trigemini and its griseum. Arrow indicates midline. The lamina epithelialis (choroid plexus) and its attachment have been added in order to show the high degree of eversion and ventricular deformation obtaining in Torpedo.

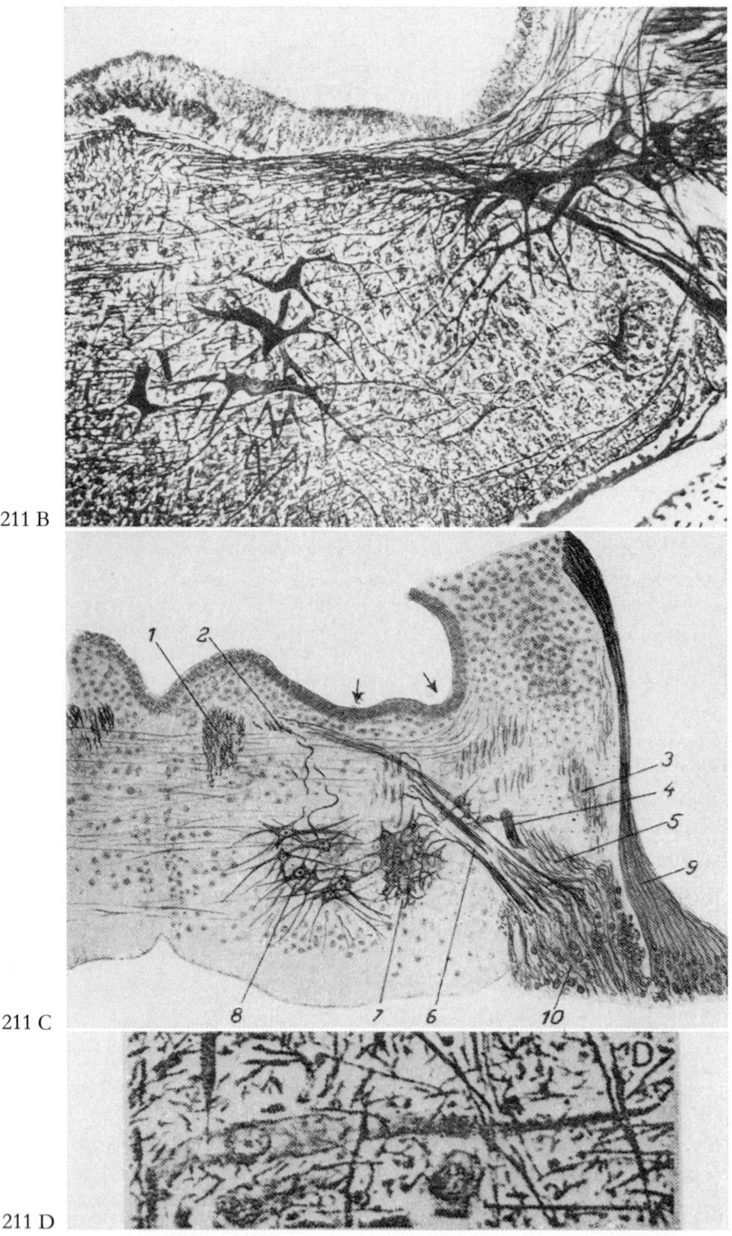

Figure 211 B. Cross-section through the oblongata in an advanced embryo of Scyllium canicula, showing *Deiters' nucleus*, reticular cells, and intracerebral efferent root of VII, as displayed by *Cajal's silver impregnation*. Unlabelled photomicrograph (from BECCARI, 1930).

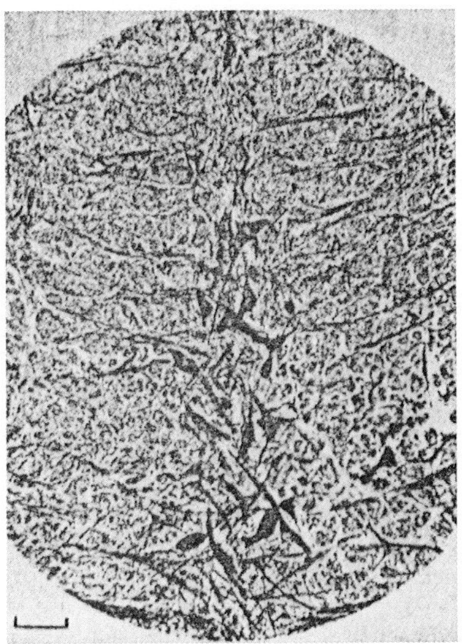

Figure 211E. Raphé nucleus in the rhombencephalon of Raja clavata, presumed to represent the '*noyau de commande*'. The scale indicates 100 μ (from SZABO, 1961).

Within the overall rostral subdivision of the rhombencephalic main group, SZABO (1961) has identified grisea assumed to control the discharges of the electric organs in Torpedo and in Raja *(noyau de commande centrale*, cf. section 5 in chapter VII of volume 3, part II). In Torpedo, '*deux amas symétriques de cellules allongées*' are described within the reticular formation (Fig. 211D), while in Raja clavata, the relevant griseum appears as a raphé nucleus (Fig. 211E).

The neurites of the reticular elements display a variety of not yet sufficiently well understood connections by descending and ascending,

---

Figure 211C. Cross-section through the oblongata in an advanced embryo of Scyllium canicula at the level of the trigeminal nucleus, as displayed by *Cajal's silver impregnation* (from BECCARI, 1943). 1: fasciculus longitudinalis medialis; 2: 'fasciculus longitudinalis paramedialis'; 3: ascending vestibular root fibers; 4: radix mesencephalica trigemini; 5: afferent trigeminal root; 6: efferent trigeminal root; 7: efferent trigeminal nucleus; 8: upper rhombencephalic reticular group; 9: root and ganglion of anterior nerve of lateral line; 10: trigeminal ganglion. Added arrows: sulcus limitans (right), sulcus intermedius ventralis.

Figure 211D. Cell and neuropil of '*noyau de commande*' in Torpedo marmorata (silver impregnation, sagittal section). At upper left, the tip of an inserted microelectrode is schematically indicated (from SZABO, 1961).

crossed and uncrossed pathways. The reticulo-spinal components provide a substantial portion of the fasciculus longitudinalis medialis and also reach the spinal cord through fibers not included in that bundle and apparently running through the anterolateral funiculus.

Generally speaking, the rhombencephalic reticular nucleus receives input from the cerebellum through a *tractus cerebello-tegmentalis* or '*cerebellomotorius*' (e.g. KAPPERS, 1947), through the arcuate fibers mentioned above, through mesencephalo-bulbar[99] (particularly tecto-bulbar) systems, and spino-bulbar connections.

SAITO (1930) very appropriately deals with the close relationship between *Deiters' vestibular nucleus* and the bulbar reticular formation. *Deiters' nucleus* is doubtless a vestibular terminal (i.e. afferent) griseum, whose elements derive from the dorsal zone of *alar plate*. Said nucleus remains topologically dorsal to sulcus limitans, being lateral, i.e. again topologically dorsal to intermediodorsal zone. Despite these functional and morphological characteristics, *Deiters' nucleus* contributes 'motor', descending crossed and uncrossed fibers to the bulbo-spinal systems. Some such fibers join the fasciculus longitudinalis medialis, others form a more lateral tractus vestibulo-spinalis s. octavo-motorius rectus et cruciatus. Still other descending vestibular fibers run in a medial direction and gather within a longitudinal tract of the dorsal tegmentum lateral to fasciculus longitudinalis medialis. This *lateral tegmental tract* (Figs. 203, 204, 206) can be followed into the spinal cord where its fibers are located near the midline between ventral and dorsal gray horns. Said tract corresponds to the system known as *fasciculus (vestibularis) medianus* of STIEDA or *fasciculus vestibulo-spinalis dorsalis* of KAPPERS (1947).

Notwithstanding its 'sensory' function and its derivation from the alar plate it seems therefore permissible to subsume *Deiters' nucleus*, in accordance with views of EDINGER (1912), under the concept of a nucleus motorius tegmenti *sensu latiori* (SAITO, 1930). It appears perhaps possible that in the course of the ontogenetic formative processes related to the differentiation of vestibular and tegmental reticular nuclei, diverse functional parameters combine or overlap with configurational ones, such that a 'blurring' of the 'boundary' between alar and basal

---

[99] Mesencephalo-bulbar pathways presumably relay telencephalic and diencephalic (including hypothalamic) impulses. It is doubtful whether (except for some fibers from the tegmental cell plate) direct descending diencephalic fibers reach the oblongata; still more doubtful are direct telencephalo-bulbar connections.

plate results. Again, one might also call this a 'blurring' of various characteristic features distinguishing alar from basal plate derivatives. Figure 213 illustrates various relationships of the Selachian reticular formation as recorded by SAITO.

Although, as a rule, typical *Mauthner cells* are not seen in adult Plagiostomes, it remains uncertain whether some large elements of reticular type, either derived from the tegmentum or from *Deiters' nucleus* (cf. e.g. Fig. 206) and giving off a contralateral fiber descending in or

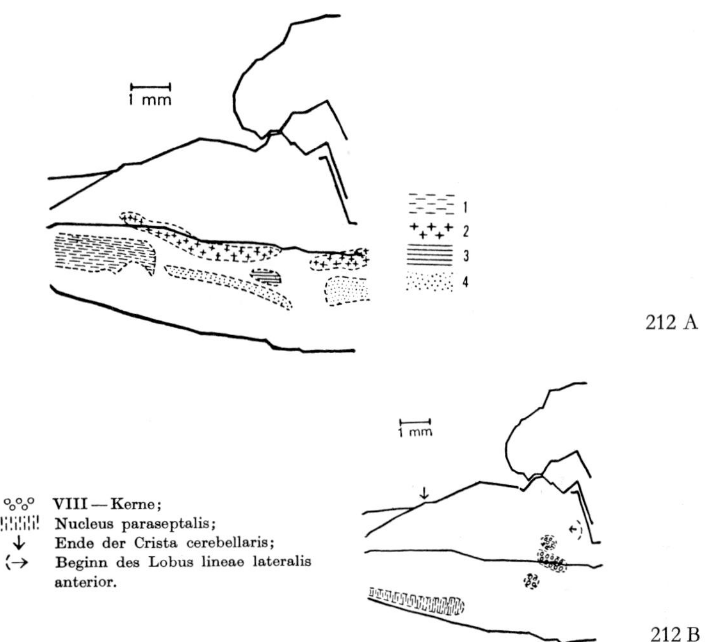

Figure 212A. Scalar *(massstabgetreue)* projection of some grisea in the oblongata of Galeus canis upon the sagittal midline plane (from GERLACH, 1947). 1: spino-occipital cell column; 2: rostral and caudal 'visceromotor column'; 3: nucleus of descending trigeminal root (in order to avoid excessive overlap, only a portion of this nucleus, showing that griseum's general position, has been indicated); 4: reticular nuclei of oblongata.

Figure 212B. Scalar projection of some grisea in the oblongata of Galeus upon the sagittal midline plane (from GERLACH, 1947). Three octavo-lateral nuclei are indicated. These are, in dorso-ventral sequence, n. of nervus anterior lineae lateralis, rostral part of n. radicis vestibularis (or octavo-lateralis) descendentis, *Deiters' nucleus.* The n. paraseptalis is generally interpreted as representing the inferior olivary griseum of Anamnia.

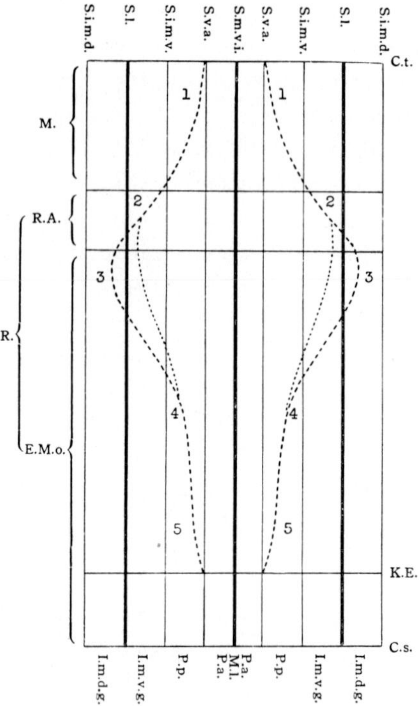

Figure 213. Diagram indicating relationships of deuterencephalic nucleus reticularis tegmenti in the brain of Acanthias mitsukurii in horizontal projection (from SAITO, 1930). The outline in dashes delimits the extent of nucleus motorius tegmenti *sensu latiori* of EDINGER, including *Deiters' nucleus*. The dotted outline separates *Deiters' nucleus* from nucleus reticularis tegmenti *sensu strictiori*. The numerals indicate the levels of particular cell groups; 1: mesencephalic reticular group; 2: isthmus group; 3: *Deiters' nucleus;* 4: rostral rhombencephalic group; 5: caudal rhombencephalic group. C.t.: commissura transversa; C.s. calamus scriptorius; E.m.o.: medulla oblongata; I.m.d.g.: intermedio-dorsal zone; I.m.v.g.: intermedioventral zone; K.E.: approximate caudal end of nucleus motorius tegmenti; M.: mesencephalon; M.l.: ventral midline (raphe); P.a.: 'pars accessoria of ventral zone'; P.p.: 'pars propria of ventral zone'; R.: rhombencephalon; R.A.: isthmus region (which SAITO includes in the rhombencephalon); S.i.m.d.: sulcus intermedius dorsalis; S.i.m.v.: sulcus intermedius ventralis; S.l.: sulcus limitans; S.m.v.i.: sulcus medianus ventralis internus (sulcus medianus basalis ventriculi quarti); S.v.a.: sulcus ventralis accessorius.

near the fasciculus longitudinalis medialis, should, or should not be regarded as *Mauthner cells*.[100]

At the levels roughly corresponding to the caudal half of the oblongata, a conspicuous elongated griseum of moderate size is found in the medial and median neighborhood of the basal plate, and is generally interpreted as the primordial *inferior olivary nucleus* (KOOY, 1917; BECCARI, 1943). Because of its median position, it was also named *nucleus paraseptalis* (EDINGER, 1908; GERLACH, 1947). According to KAPPERS (1947) this griseum (Figs. 201, 212B, also, but not labelled, in Fig. 204) may be homologous with the medial accessory inferior olivary of higher Vertebrates.[101] The Selachian and, for that matter, the Osteichthyan inferior olive should not be confused with paramedian or raphé-groups of reticular cells, from which it differs by a more basal location near the oblongata's ventral surface and by a slightly gelatinous appearance related to its neuropil. Nevertheless, a clear-cut demarcation between the two griseal aggregates is occasionally difficult to detect, and BECCARI (1943) justly remarks in his description of the primordial inferior olives that '*non è improbabile que da qualche Autore esse siano state confuse, nei bassi Vertebrati, con il nucleo reticolare inferiore*'. In addition to evident, predominantly crossing arcuate fibers connecting this inferior olive with corpus cerebelli, poorly elucidated olivo-spinal respectively spino-olivary pathways are believed to be present.

With regard to the various fiber systems of the Selachian oblongata mentioned in the preceding discussion, it should be added that it is most difficult to disentangle these tracts from those more or less related ascending and descending pathways which, passing through the oblongata, connect spinal cord with other brain subvidivions such as mesencephalon and cerebellum. Thus, e.g., fibers of bulbar origin provide conspicuous additions to the spino-cerebellar system and correspond to the olivo-cerebellar and reticulo-cerebellar fibers described by LARSELL (1967) in Plagiostomes.

---

[100] Thus, LARSELL (1967) found *Mauthner cells* in some 40 mm embryos of Acanthias but not in adults. Cf. also the comments in K. and NIIMI (1969). The still open question whether *Mauthner cells* represent a vestibular or tegmental derivative was pointed out above in footnote 66, p. 357.

[101] Although JOHNSTON (1902) interpreted a somewhat similarly located tegmental (perhaps 'reticular') cell group in Petromyzon as inferior olivary nucleus, its presence as a definable 'inferior olive' in Cyclostomes with very rudimentary (Petromyzonts), respectively for practical purposes absent (Myxinoids) cerebellum remains highly doubtful.

5. Ganoids and Teleosts; Latimeria

In conformity with the simplified classification adopted for purposes of comparative neurology in the introductory volume 1 of this series, *Ganoids* are here conceived to represent a subclass of the class *Osteichthyes*, and to subsume the three orders *Palaeoniscidae* (Polypterus), *Chondrostei* (Acipenser, Polyodon), and *Holostei* (Amia, Lepidosteus). Basic investigation of the Ganoid brain are those by GORONOWITSCH (1888) and JOHNSTON (1901) on Acipenser, KAPPERS (1906) on Amia and Lepidosteus (now generally spelled Lepisosteus) and HOOGENBOOM (1929) on Polyodon. Additional first-hand data are contained in the treatises on comparative neurology listed above in section 3, p. 349 of this chapter, which also include the relevant findings in Teleosts.[102] Data on the oblongata of the *Crossopterygian* Latimeria are presented in the monograph by MILLOT and ANTHONY (1965) dealing with the nervous system of that rather peculiar fish.

Generally speaking, the oblongata of *Ganoids* resembles that of Selachians, by displaying the same essential features, and being, on the whole, less modified by secondary pattern distortions than the oblongata of many Teleosts. Figures 214 to 220 illustrate representative cross-sectional levels in Acipenser and Polyodon comparable with Figures 199 to 205 of the preceding section on the Plagiostome oblongata. Upon inspection of these figures, it will be seen that, in the ventricular wall, sulcus limitans, sulcus intermedius dorsalis, and sulcus intermedius ventralis can be identified, although said grooves may become blurred in some regions. Moreover, as in Selachians, accessory sulci are likewise seen to be present.

In the aspect here under consideration, a shortened and more concise description of the neuronal structures in the Ganoid oblongata will be sufficient, since, *mutatis mutandis*, these grisea and overall pathways correspond rather closely to those obtaining in Plagiostomes.

Within the *alar plate*, the *octavo-lateralis system* includes the grisea related to the distribution of an anterior and a posterior nervus lateralis. A *crista cerebelli* and a *dorsal lobus nervi lineae lateralis anterioris*, continuous with the cerebellar auricle, are well developed. Ventrally to the

[102] As regards older authors who investigated some of the here significant features of the Teleostean brain, the following might be mentioned in this context: CATOIS (1901), C.J. HERRICK (1905, 1907), VAN DER HORST (1918), KAPPERS (1906), MAYSER (1881), and TELLO (1909).

cerebellar crest, the area statica displays an *intermediate* octavo-lateralis 'nucleus', the 'tuberculum acusticum' of JOHNSTON (1906), related to anterior and posterior lateral line nerves, and in some degree also to vestibular nerve. It is not impossible that in some forms, such as Amia, in which a dorsal nucleus nervi lineae lateralis is not clearly evident, this griseum may be included in the medial part of the 'intermediate nucleus'. The close relationship between 'tuberculum acusticum' and crista cerebellaris is indicated in Figure 221. Still more *ventrally*, an essentially vestibular griseum is located, which can be further subdivided

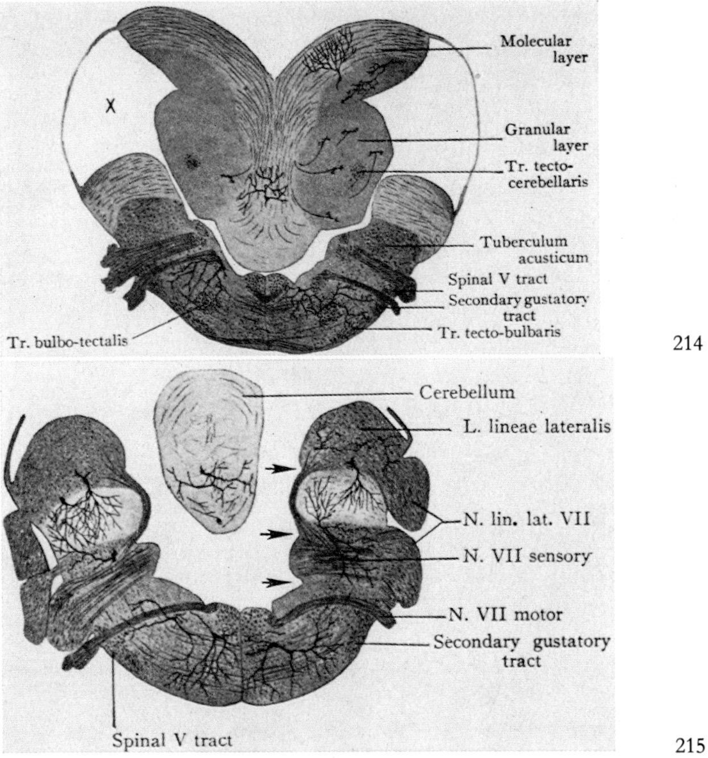

Figure 214. Cross-section through the oblongata of the Sturgeon (Acipenser) at the level of the trigeminal root (from JOHNSTON, 1906). The added x indicates the ventricular space of the auricle. The structure above the 'tuberculum acusticum' is the oblongata's cerebellar crest. Figures 214 to 216 are drawings based on Golgi preparations.

Figure 215. Cross-section through the oblongata of Acipenser at the level of the facial nerve root (from JOHNSTON, 1906). The added arrows indicate from below upward, sulcus limitans, sulcus intermediodorsalis, and sulcus dorsalis accessorius.

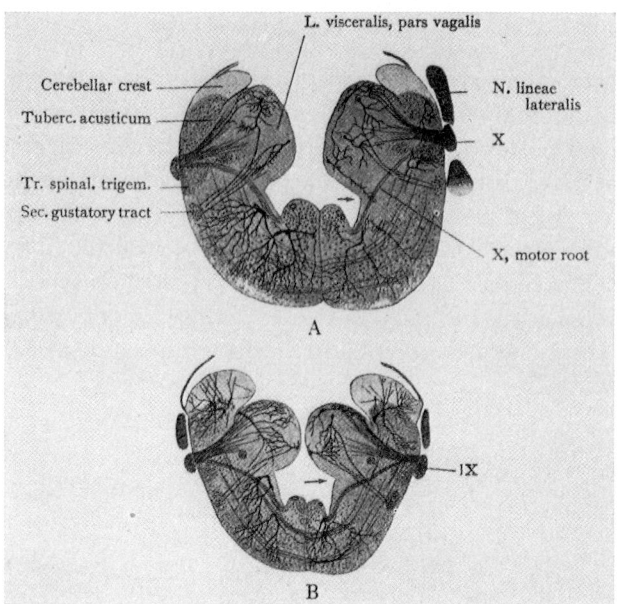

Figure 216. Cross-sections through the oblongata of Acipenser at the level of the vagus (A) and at that of glossopharyngeus (B). Section A is drawn at a higher magnification than B (from JOHNSTON, 1906). The added arrow indicates sulcus limitans. Below this, sulcus intermedius ventralis can easily be identified. Above lobus visceralis, sulcus intermedius dorsalis. The sulcus accessorius dorsalis has disappeared (by junction with s. int. dors.).

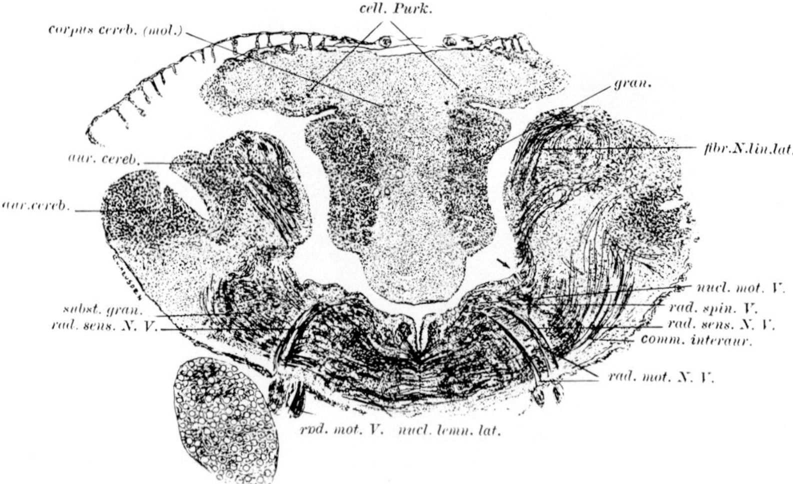

Figure 217. Cross-section through the oblongata of Polyodon at the level of the trigeminus (from HOOGENBOOM, 1929). cell. Purk.: *Purkinje cells* of cerebellum; subst. gran.: 'migrated granular cells' from auricle. Other abbr. self-explanatory. The added arrow indicates sulcus limitans. Figures 217 to 220 are drawn from *Weigert-Pal preparations.*

in the same manner as that of Selachians. It includes the cell aggregate generally interpreted to represent *Deiters' nucleus* (cf. above, p. 391). Some details of 'tuberculum acusticum' and of *Deiters' nucleus*, with its associated 'reticular' *Mauthner cell*, are shown in Figures 221 and 222 A.

At the level of the *trigeminus root entrance* (Figs. 214, 217), the alar plate contains an ill-defined main terminal griseum, ventrolaterally continuous with that of the descending root, as well as a fiber bundle pertaining to the mesencephalic root.

At levels caudally to the entrance of afferent facialis fibers, the intermediodorsal zone forms the '*visceral lobe*' appurtenant to VIIth, IXth, and Xth nerves. It contains the corresponding terminal griseum and a more or less rudimentary tractus solitarius. Figures 215 and 218 show that the facial part of the 'visceral lobe' is more developed in Acipenser than in Polyodon, while in both forms the vagal part of that lobe, bulging into the ventricle, is particularly prominent, being, again, more developed in Acipenser than in Polyodon (Figs. 216, 219). As in Selachians, and, for that matter, quite generally in Vertebrates, the cutaneous fibers in the facial, glossopharyngeal and vagus nerve presumably join the radix descendens trigemini and its griseum. The cutaneous components seem to be more numerous in the vagus group than in the facial nerve.[103] Again, as in Selachians, the pathways originating in the alar plate include *arciform fibers* pertaining to the different system such as lemniscus lateralis, tractus octavo-motorius cruciatus et rectus, and bulbar lemniscus. In contradistinction to Selachians, a homolateral *ascending secondary gustatory tract*, running ventrally to the descending root of the trigeminus, is very clearly recognizable (Figs. 214, 215, 216). It reaches a griseum in the isthmus region.

The *basal plate* contains, within the *intermedioventral zone*, the efferent nuclear columns of the branchiomotor nerves V, VII, IX and X in an arrangement essentially similar to that obtaining for Plagiostomes and shown in Figures 158 and 208. As in Selachians the efferent facial nerve nucleus is in continuity with that of the vagus group, but separated by an interval from the motor trigeminal nucleus, and the intracerebral efferent facial root fibers describe a rostrally directed loop before emerging.

---

[103] According to Hoogenboom (1929), a cutaneous component may be missing in the facial and in the glossopharyngeus nerves of Polyodon.

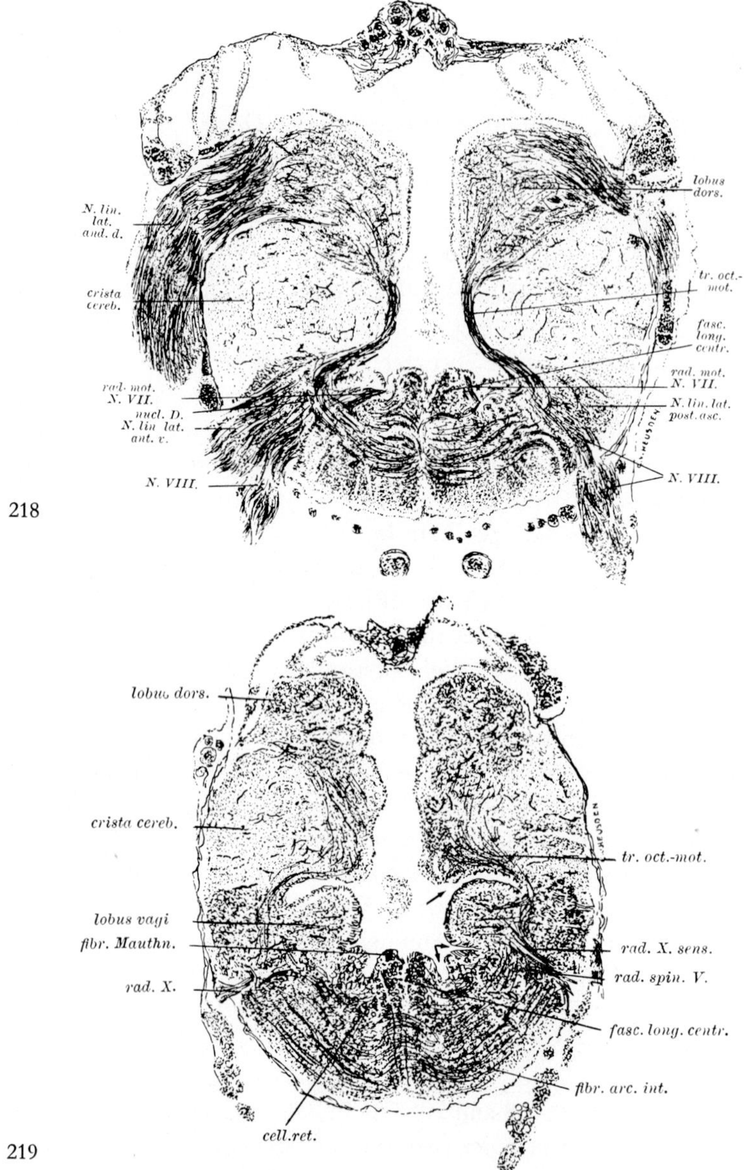

Figure 218. Cross-section through the oblongata of Polyodon at the level of anterior lateral line and vestibular nerves (from HOOGENBOOM, 1929). lobus dors.: nucleus nervi lineae lateralis anterioris; nucl. D.: *Deiters' nucleus;* fasc. long. centr.: fasciculus longitudinalis medialis.

Figure 219. Cross-section through the oblongata of Polyodon at the level of nervus vagus (from HOOGENBOOM, 1929). Arrows to sulcus intermediodorsalis and sulcus limitans have been added. Basally to sulcus limitans, to sulcus intermedioventralis.

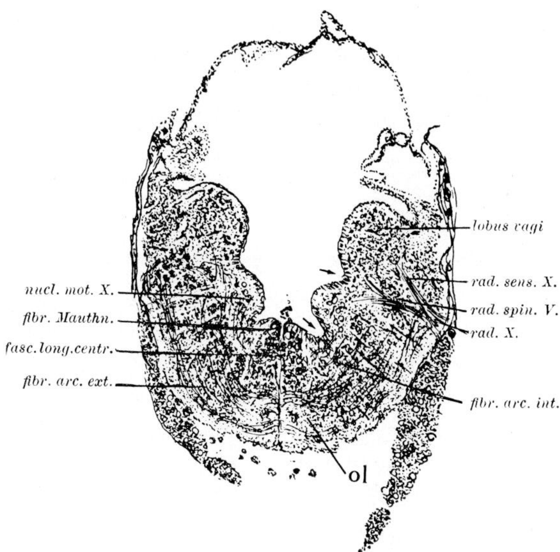

Figure 220 A. Cross-section through the oblongata of Polyodon at the level of nervus vagus caudally to end of crista cerebellaris and nucleus of posterior lateral line (from HOOGENBOOM, 1929). A lead to inferior olive (ol) has been added. Arrows: sulcus limitans, sulcus intermedioventralis.

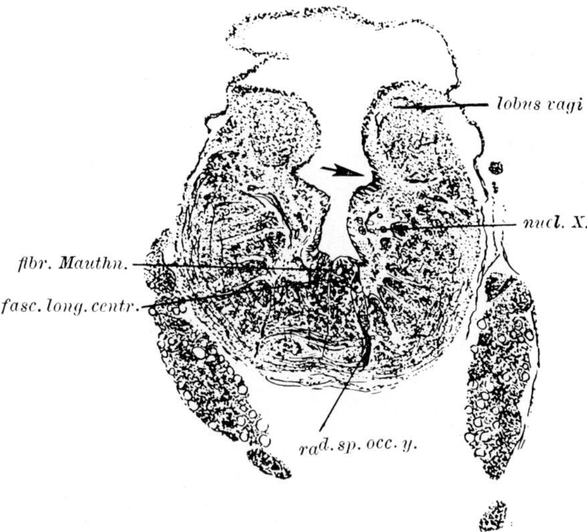

Figure 220 B. Cross-section through the oblongata of Polyodon at the level of a spino-occipital nerve root, just rostral to calamus scriptorius (from HOOGENBOOM, 1929). Added arrow points to sulcus limitans.

Within the basal plate's *ventral zone*, the nucleus nervi abducentis consists of a scattered array of motoneurons, intermingled with some reticular cells, and in fairly close proximity to the fasciculus longitudinalis. The nucleus of the abducens can easily be identified by tracing its intracerebral rootlets to their origin. These rootlets emerge caudally to the level of the emerging efferent facial roots, but the nucleus itself, although partly coextensive with the anterior part of motor facial nucleus, extends also in part rostrally to this latter.

In the caudal part of the oblongata, basally to the visceral-efferent column of the vagus, a somatic efferent spino-occipital nuclear column is present (cf. Fig. 220B). HOOGENBOOM (1929) recorded two groups of spino-occipital rootlets (y and z) in Polyodon. The cited author justly emphasizes the difficulties, already pointed out by FÜRBRINGER, in attempting a generally valid classification of spino-occipital nerves. It might be added that similar difficulties obtain with respect to a rigorous distinction between spino-occipital and 'true' spinal nerves within their level of transition.

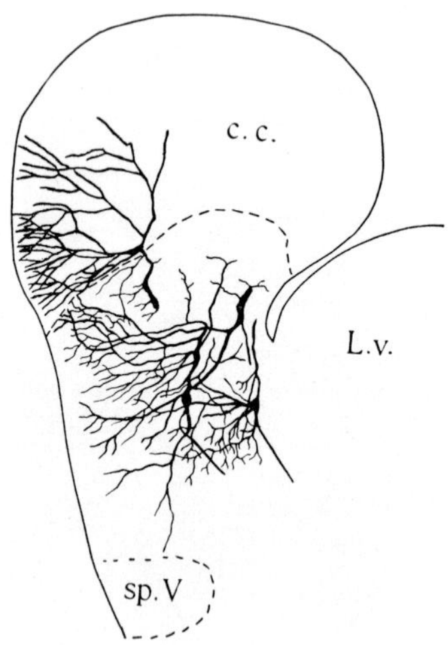

Figure 221. Cross-section through the 'tuberculum acusticum' of Acipenser at a level approximately corresponding to that of Figure 216B, and as seen by means of the Golgi method (from JOHNSTON, 1906). c.c.: cerebellar crest; L.v.: lobus glossopharyngei et vagi; sp.V: radix descendens trigemini. At top, a large cell, interpreted by JOHNSTON as a '*Purkinje cell*', sends dendritic processes (and its neurite ?) into the crista cerebellaris. At right bottom, origin of an arcuate fiber from the 'tuberculum acusticum'.

The *nucleus motorius tegmenti* of the oblongata's basal plate consists of intermingled large and small multipolar 'reticular' nerve cells, whose rather diffuse arrangement nevertheless suggests a caudal group roughly coextensive with the vagus levels, a middle group at the vestibular level, and an anterior group at some distance rostrally to the *Mauthner cells*, and approximately extending to the rostral level of the trigeminus roots. In accordance with VAN HOEVELL, HOOGENBOOM designates these subdivisions as nucleus reticularis inferior, n. reticularis medius, and n. reticularis superior. The paired *Mauthner cell* is located within the caudal portion of nucleus reticularis medius. At caudal levels of the oblongata, an *inferior olivary nucleus* corresponding to the olivary griseum described above in Selachians, can be identified (Fig. 220A).

With respect to ostensible but morphologically trivial variations displayed by the different taxonomic forms of Ganoids (as well as of Selachians) the following could be enumerated: (1) variations involving depth, distinctiveness, and topographic (but not topologic) position of the different ventricular sulci; (2) degree of prominence, 'bulging', or protrusion of lateral line lobe, visceral lobe, fasciculus longitu-

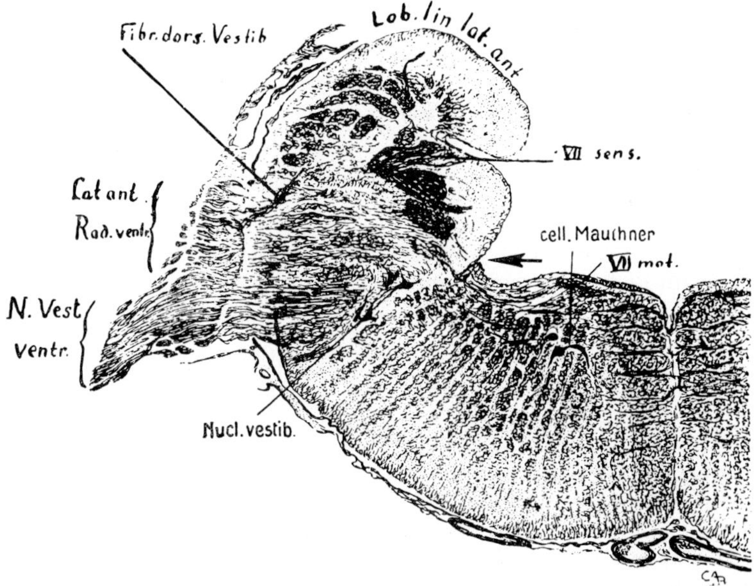

Figure 222A. Cross-section through the oblongata of Amia calva at the level of the vestibular root's entrance, showing Deiters' nucleus and the Mauthner cell (after SCHEPMAN, from KAPPERS *et al.*, 1936).

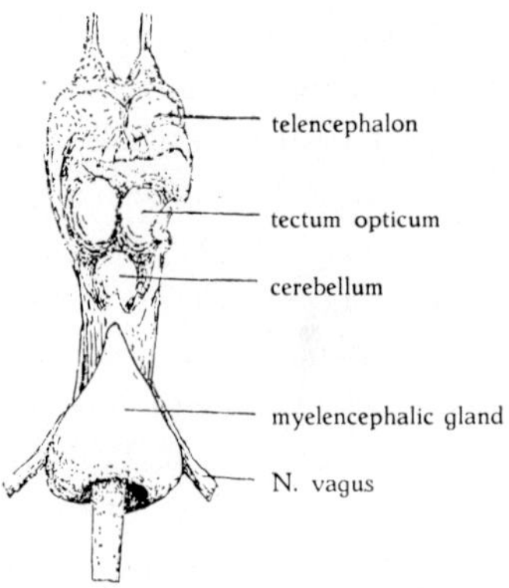

Figure 222 B. Sketch showing the so-called 'myelencephalic gland' of the Ganoid Amia (from VAN DER HORST, 1925).

dinalis medialis, etc.; (3) extension and folding of crista cerebellaris as well as its dorsal 'covering' by auricles[104] and lobus lineae lateralis.

An additional feature recorded in various Ganoids is the presence of hematopoietic myeloid tissue in the rostral spinal and the caudal 'myelencephalic' meninges (CHANDLER, 1911; VAN DER HORST, 1925; SCHARRER, 1944), as dealt with in chapter VI, section 7, p. 674 of volume 3, part II, and in chapter VIII, section 6, p. 110 of the present volume. The gross aspect of this intrameningeal so-called 'myelencephalic gland' of CHANDLER and VAN DER HORST in Amia is depicted by Figure 222B.

Turning now to the oblongata of *Teleosts*, the bewildering polymorphism of this extremely multiform group, subsuming more than 20,000 different species, must be taken into consideration.[105] Teleosts may be conceived either as an order, or perhaps better, like the Gan-

---

[104] HOOGENBOOM (1929) described in Polyodon an apparent basalward migration of granular cells from the auricles, somewhat analogous to that displayed e.g. in Mammals by the *corpus pontobulbare of Essick*. The location of such 'displaced' granular cells is shown in Figure 217. The 'nucleus lemnisci lateralis' indicated by HOOGENBOOM in that figure is presumably part of the reticular formation.

[105] Because of this variety, the interrelationships and the presumptive evolution of Teleosts is only understood in a very rudimentary manner. Their classification has not significantly improved since the period before World War I. An interesting survey and attempt at classification was recently published by GOSLINE (1971).

oids, as a subclass of Osteichthyes, and are commonly believed to have evolved from a Ganoid Holostean stock.

With respect to the configuration of the oblongata, some Teleosts, such e.g. as the Esocid[106] pike (Esox lucius), display an arrangement whose details differ very little from that obtaining in most Ganoids, and which is thus also easily comparable with the oblongata configuration in Selachians. Many Teleosts, on the other hand, exhibit highly specialized as well as diversified features related to the predominance of certain tracts and grisea. The particular development of these neuronal structures leads to considerable pattern distortions which can easily be recognized by comparing a rostro-caudal sequence of cross-sections through the oblongata of, in this respect representative, Teleosts (Figs. 223A to 232) with roughly corresponding oblongata levels of Ganoids (Figs. 214 to 221) and of Selachians (Figs. 199 to 206).

As regards the relevant details, it will here be sufficient to bring merely a short account of the Teleostean oblongata referring to the main components discussed above in connection with their arrangement in Selachians and Ganoids.

Inspection of the above-mentioned figures will disclose that, in contradistinction to the oblongata of most Plagiostomes and Ganoids, the ventricular grooves have been 'levelled' or 'planished' (German: '*sind verstrichen*') in correlation with the formative processes leading to an hypertrophy or expansion of various grisea. Except for the sulcus limitans, which remains identifiable at the adult stage in at least numerous instances, most other grooves have here disappeared, although an occasional secondary or accessory ventral or dorsal groove may be present (cf. Figs. 224B, 226A, 227).

With respect to the *alar plate*, the connections of the vestibular nerve, display, in correlation with the peculiarities of the Teleostean labyrinth (cf. above in section 1), greater complexities[107] than in Plagiostomes and Ganoids, while the arrangements of the lateralis system essentially conforms to that obtaining in the other Gnathostome fishes. As in these latter, the entire *area octavo-lateralis* (a derivative of dorsal zone) includes a dorsal griseum, the nucleus nervi lineae lateralis ante-

---

[106] Esocidae, a family belonging to the suborder (or subclass) Physostomi, have been also classified as included in the suborder Mesichthyes (cf. e.g. YOUNG, 1955).

[107] Cf. the more detailed account given by KAPPERS (1947).

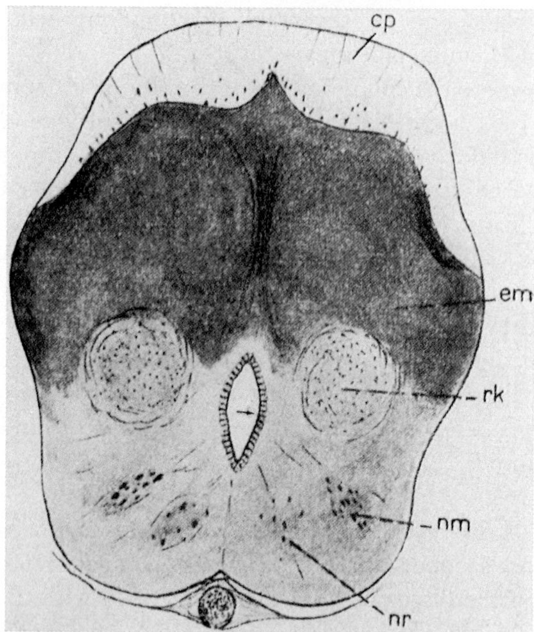

Figure 223 A. Simplified drawing of a cross-section through the oblongata of Cyprinus auratus at the level of the motor trigeminal nucleus (from K., 1927). cp: corpus cerebelli; em: eminentiae granulares of cerebellum, fused in the midline; nm: nucleus motorius trigemini; nr: nucleus reticularis tegmenti; rk: nucleus of ascending secondary gustatory tract ('superior secondary gustatory nucleus'; '*Rindenknoten*' of MAYSER); arrow to sulcus limitans has been added to the 1927 figure.

Figure 223 B. Cross-section through the oblongata of the Siluroid Teleost Arius at the level of the motor trigeminal nucleus nucleus (from KAPPERS *et al.*, 1936). The added arrow indicates sulcus limitans, above which the caudal end of superior secondary gustatory nucleus (not labelled) can be identified.

Figure 223 C. Comparison of two cross-sectional levels through the rostral oblongata of Arius. At the right, level of 'nucleus motorius trigemini anterior'; at left, level of 'nucleus motorius trigemini posterior', illustrating some further details of the level depicted in the preceding Figure 223 B (after VAN DER HORST, from KAPPERS *et al.*, 1936).

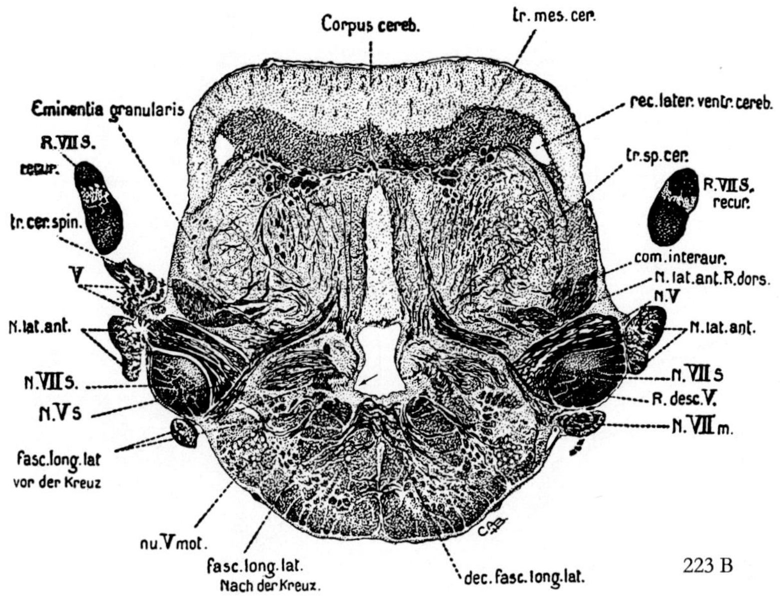
Corpus cereb.
tr. mes. cer.
Eminentia granularis
rec. later. ventr. cereb.
R. VII S. recur.
tr. sp. cer.
R. VII S. recur.
tr. cer. spin.
V
com. interaur.
N. lat. ant. R. dors.
N. V
N. lat. ant.
N. lat. ant.
N. VII S.
N. VII S
R. desc. V.
N. V S
N. VII m.
fasc. long. lat vor der Kreuz
nu. V mot.
fasc. long. lat. Nach der Kreuz.
dec. fasc. long. lat.

223 B

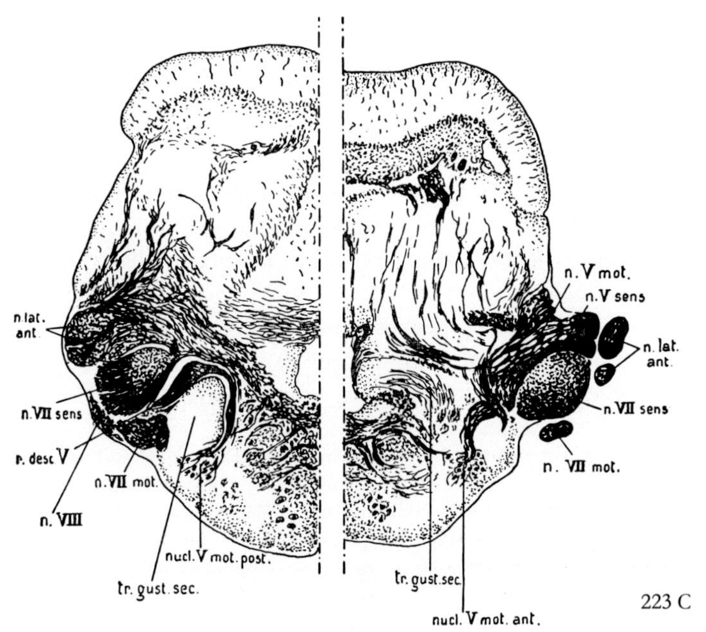
n. V mot.
n. V sens
n. lat. ant.
n. lat. ant.
n. VII sens
n. VII sens
r. desc V
n. VII mot.
n. VII mot.
n. VIII
nucl. V mot. post.
tr. gust. sec.
tr. gust. sec.
nucl. V mot. ant.

223 C

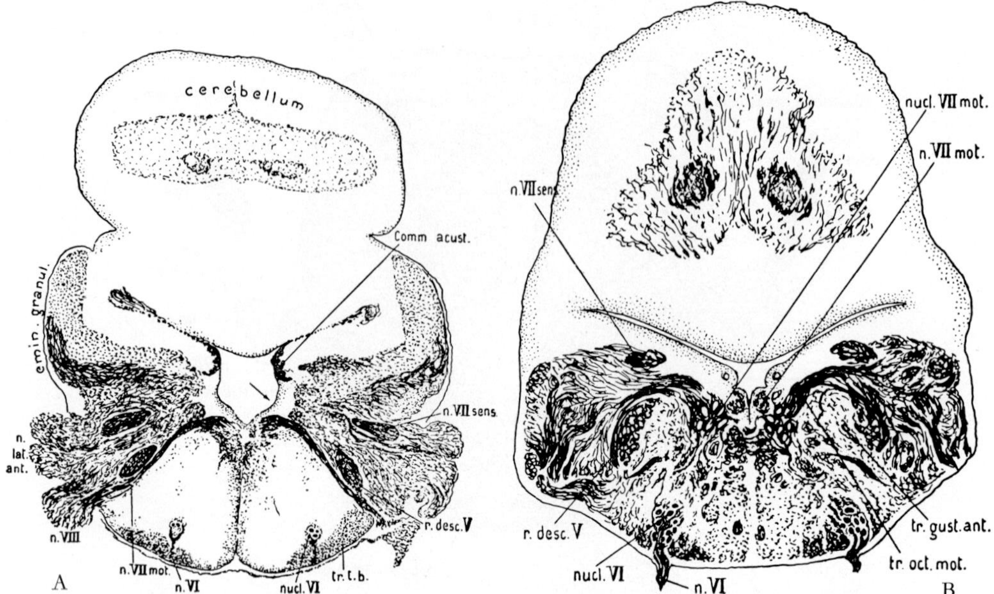

Figure 224 A. Cross-section through the oblongata of the Teleost Mugil chelo at the level of nucleus abducentis (after VAN DER HORST, from KAPPERS *et al.*, 1936). comm. acust.: 'commissura acustica' related to 'area octavo-lateralis'; tr. t. b.: tractus tecto-bulbaris. Other abbr. self-explanatory. Added arrow indicates faintly suggested sulcus limitans on right side.

Figure 224 B. Cross-section through the oblongata of the Teleost 'Tetrodon speciosus' at the level of the abducens nucleus (after VAN DER HORST, from KAPPERS *et al.*, 1936). It will be noted that the sulcus limitans has disappeared. The conspicuous groove on both sides of the midline is an accessory sulcus.

rioris[108] (Fig. 232B), an intermediate acustico-lateral griseum (the tuberculum acusticum of JOHNSTON), which may be subdivided into medial and lateral parts, and is related to the crista cerebellaris, moreover a ventral vestibular griseum including *Deiters' nucleus*. A substantial lemniscus lateralis, reaching mesencephalic structures (nucleus isthmi, torus semicircularis, tectum mesencephali) originates in these grisea, which, moreover, possess cerebellar connections.

---

[108] Because of the expansive development of the Teleostean cerebellar eminentiae granulares, in combination with that of a lobus facialis, a prominent isolated lobus nervi lineae lateralis anterioris (or lobus dorsalis) as e.g. conspicuous in some Plagiostomes (Figs. 202, 203) and Ganoids (Figs. 218, 219) is not commonly displayed, since that griseum becomes encompassed by neighboring configurations (cf. Fig. 223B). As in some Ganoids this 'nucleus dorsalis' of the area octavo-lateralis might become fused with or included in the medial part of the 'intermediate' griseum (cf. above p. 407).

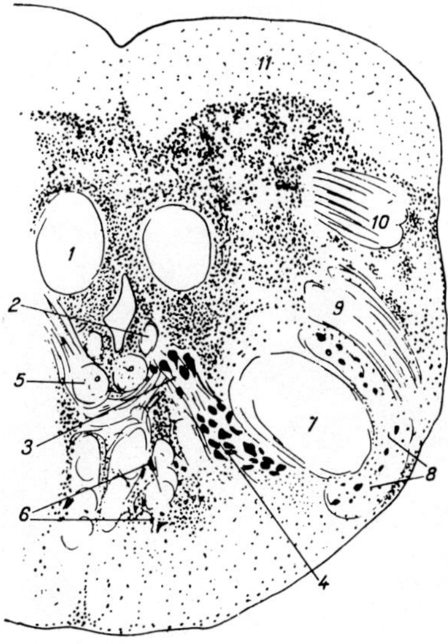

Figure 224 C. Cross-section through the oblongata of the Teleost Cyprinus carpio at the level of the motor facial nucleus as seen in a *Nissl stain* (from BECCARI, 1943). 1: afferent root of facial; 2: motor root of facial; 3: dorsal cell group of motor facial nerve; 4: ventral cell group of motor facial nerve; 5: fasciculus longitudinalis medialis; 6: reticular cells of nucleus motorius tegmenti; 7: tractus gustatorius secundarius anterior; 8: radix descendens trigemini; 9: radix vestibularis; 10: radix nervi anterioris lineae lateralis; 11: crista cerebellaris.

The bulbar afferent terminal grisea of the *trigeminal nerve* are likewise derivatives of the dorsal zone and comprise, as in other fishes, the main sensory nucleus of the trigeminus and the nucleus of the descending root. Cutaneous fibers of the glossopharyngeus and vagus join the radix descendens trigemini. Such somatic afferent respectively exteroceptive fibers seem to be lacking in the facial nerve of most Teleosts, but may be present in relatively few instances. In Teleosts with a large head, such e.g. as in Lophius piscatorius, the afferent root fibers of the ophthalmic division are more heavily medullated than those of the maxillo-mandibular portion. The descending ophthalmic root fiber bundle, located ventrally to the maxillo-mandibular components of the radix spinalis trigemini can therefore be easily distinguished and traced caudalward, where it ends in a large dorsal funicular nucleus of the rostral spinal cord (cf. Fig. 67B). From this griseum originate essentially crossed spino-bulbo-mesencephalic and spino-bulbo-cerebellar pathways. The descending ophthalmic root fibers become separated, at lower levels of the oblongata, by transverse entering vagus root fibers,

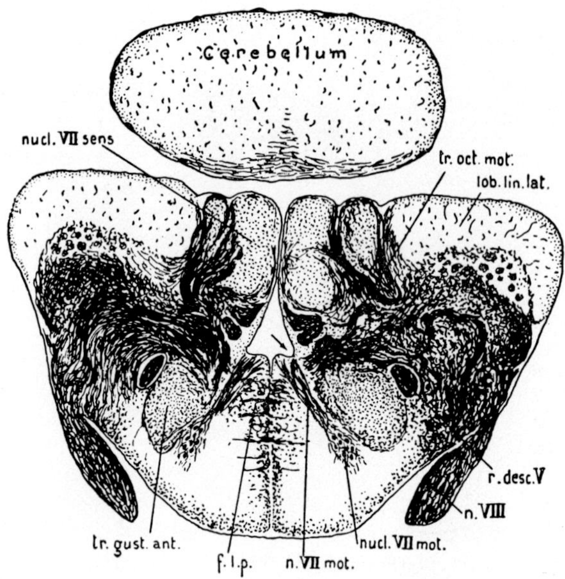

Figure 225 A. Cross-section through the oblongata of Arius at the level of the afferent (gustatory) facial nucleus (after VAN DER HORST, from KAPPERS *et al.*, 1936). f. l. p.: fasciculus longitudinalis medialis; lob. lin. lat.: crista cerebelli.

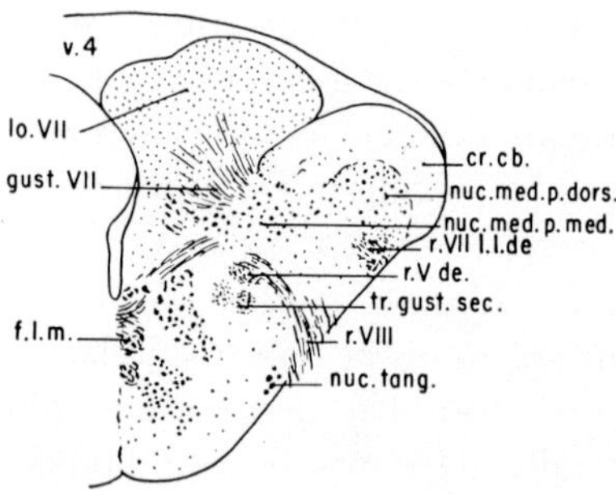

Figure 225 B. Cross-section through the oblongata of the Siluroid Teleost Ameiurus at the level of the facial nerve root (slightly modified after PEARSON, 1936, from LARSELL, 1967). cr. cb.: crista cerebellaris; lo. VII: nucleus facialis gustatorius; nuc. med. p. dors., nuc. med. p. med.: lateral and medial portions of intermediate acustico-lateralis griseum; nuc. tang.: nucleus tangentialis VIII (?); r. VII. l. l. de.: dorsal (descending ?) lateral line root of VII.

from the maxillo-mandibular portion, most of whose caudal components reach the nucleus commissurae infimae and may also decussate in that commissure. Nevertheless, some ophthalmic root fibers may also be related to the system of commissura infima. Again, in contradistinction to other Teleosts, the caudal part of radix descendens trigemini is here accompanied by intra- and supramedullary large afferent ganglion cells of a type comparable to the cells of radix mesencephalica trigemini and to *Rohon-Beard cells* (cf. also chapter VIII, section 6).

In many Teleosts, especially in Cyprinoids and Siluroids, the afferent roots of the *branchial nerves VII, IX and X* contain numerous '*gustatory*' components, particularly related to the development of *cutaneous taste buds* innervated by a branch of the facial, while the gustatory fibers of glossopharyngeus and vagus are seemingly restricted to the innervation of branchial, pharyngeal and oral taste buds (cf. also vol. 3, part II, chapter VII, sections 1 and 4). The greatly expanded bulbar respectively deuterencephalic gustatory centers and pathways of many Teleosts have been especially investigated by HERRICK (1905, 1906, 1907, 1908). The separate antimeric afferent grisea of the nervus facialis (Fig. 227) may commonly fuse in the midline to form a conspicuous tuberculum impar which, although topologically rostral to the glossopharyngeus-vagus lobe, becomes partially enclosed or surrounded by the likewise greatly expanded vagal lobes (Figs. 228–230).

The secondary fibers arising in these grisea[109] which are derivatives of the intermediodorsal zone, form two main bundles, *tractus gustatorius secundarius ascendens and descendens* (Figs. 228–230, 233A, B). The former ascends, predominantly on the homolateral side, to a large caudal (isthmic) midbrain griseum, the *nucleus gustatorius secundarius superior*, or '*Rindenknoten*' of MAYSER (1881) closely contiguous to the nucleus isthmi (cf. chapter XI, section 4). A tertiary tract originating in the upper nucleus gustatorius reaches the inferior lobes of the hypothalamus. The descending secondary gustatory tracts ends in the funicular grisea[110] of the *übergangsgebiet* and to some extent in the grisea of *commissura infima Halleri*. Commissural pathways also interconnect the two

[109] There seem to be some differences in the relative amount of fibers provided by facial and by vagal lobes to either main bundle. This may depend on the Teleostean form involved or merely represent differences in the interpretation of the various authors concerned with these pathways.

[110] Thus, the lateral funicular nucleus of some Teleosts was designated as nucleus gustatorius secundarius inferior by HERRICK.

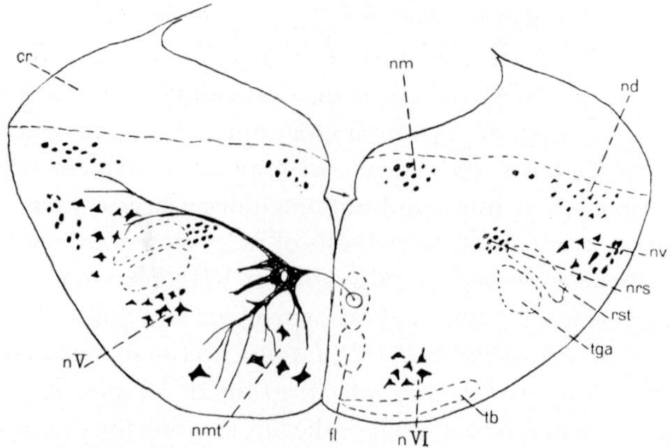

Figure 226A. Simplified drawing of a cross-section through the oblongata of Ameiurus, depicting the left *Mauthner cell* (in part after BARTELMEZ, 1915, from K., 1927). cr: crista cerebellaris; fl: fasciculus longitudinalis medialis; nV: nucleus motorius trigemini; n. VI: nucleus nervi abducentis; nd: 'nucleus dorsalis nervi octavi' (lateral part of intermediate acustico-lateralis griseum); nm: 'nucleus medialis nervi octavi' (medial part of intermediate acustico-lateralis griseum); nmt: nucleus motorius tegmenti; nrs: nucleus radicis descendentis trigemini; nv: 'nucleus ventralis nervi octavi' *(Deiters' nucleus,* the lateral portion of this griseum being the nucleus tangentialis); rst: radix descendens (s. spinalis) trigemini; tb: tractus tecto-bulbaris; tga: tractus gustatorius secundarius ascendens.

sides of lobus facialis and the two vagal lobes, and may include some crossed fibers for the gustatory pathways. In addition, some crossed secondary gustatory fibers seem to join the general spino-bulbo-tectal lemniscus. There are, moreover, doubtless various sorts of connections between this 'gustatory system' and the nucleus motorius tegmenti. A typical tractus solitarius, formed by visceral afferent root fibers of the branchial nerves VII to X gathered in a distinctive bundle, is, at best, barely recognizable in most Teleosts.[111]

As regards the structure of the gustatory lobes, that of the lobus facialis is on the whole much less complex than that of the lobus vagi. This latter displays a conspicuous sequence of layers, already pointed

[111] Thus, BECCARI (1943) is doubtful about the presence of a 'true' fasciculus solitarius in Fishes, and KAPPERS (1947) stresses the development of that fasciculus in '*Vertébrés à respiration pulmonaire*'. Be that as it may, one is, nevertheless, justified to maintain that a fasciculus or tractus solitarius, containing descending and perhaps also some ascending visceral-afferent VII, IX and X root fibers, is present in all Vertebrate classes from Cyclostomes to Mammals. Even in these latter, the tractus solitarius is by no means exclusively related to functions of pulmonary respiration.

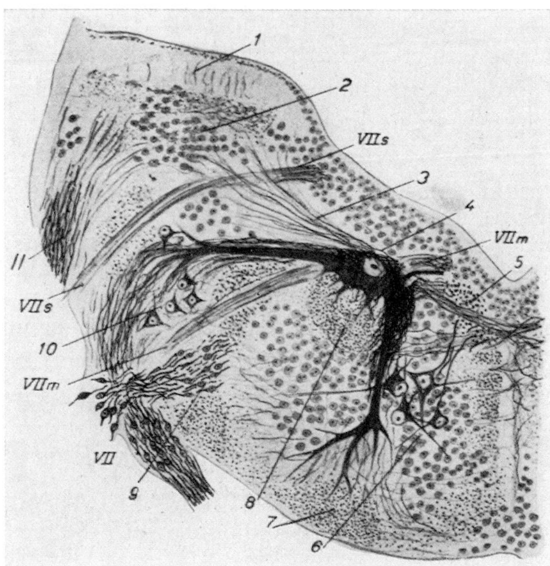

Figure 226 B. Cross-section showing the *Mauthner cell* and its connections in a young specimen *('avanotto, forma giovanile')* of Salmo fario as displayed by *Cajal's silver impregnation* and reconstructed from three consecutive serial sections (from BECCARI, 1943). 1: cerebellar crest; 2: '*nucleo della linea laterale*' (intermediate octavo-lateralis griseum); 3: 'dorsal' arcuate fibers; 4: *Mauthner's cell;* 5: fasciculus longitudinalis medialis; 6: nucleus motorius tegmenti *('nucleo reticolare medio');* 7: tractus tecto-bulbaris; 8: fasciculus longitudinalis lateralis (octavo-lateral lemniscus); 9: nucleus tangentialis (in comparison with Figure 226A it will be seen that this cell aggregate is here located ventrally to *Deiters' nucleus);* 10: *Deiters' nucleus;* 11: root of nervus anterior lineae lateralis; VII m, s: efferent respectively afferent root of facial; VIII: root of octavus.

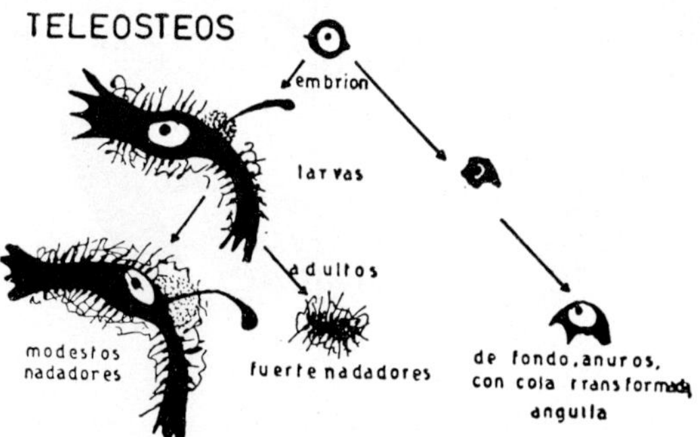

Figure 226 C. Diagram indicating the evolution of the Teleostean *Mauthner cell* in relation with swimming activities involving trunk respectively tail (from STEFANELLI, 1962). The term '*anuros*' at right refers to 'tailless' Teleostean forms.

out by MAYSER (1881), and here shown in Figures 230A and B. It will be seen that HIRASAWA distinguished nine layers, while BECCARI enumerates four. A simplified version is indicated by Figure 229. In addition to the intermediodorsal alar plate components, the lobus vagi contains an intermedioventral basal plate component provided by a dorsal expansion of the efferent glossopharyngeus-vagus nuclear column. This cell group, also designated as *nucleus dorsalis vagi*, apparently innervates muscles of palato-pharyngeal structures representing a '*Gaumenorgan*' particularly supplied with taste buds.

The caudal part or *übergangsgebiet* of the oblongata contains dorsal components which cannot easily be classified as either pertaining to dorsal zone or to intermediodorsal zone but can be regarded as general alar plate derivatives, and comprise the grisea of *commissura infima Halleri* and the funicular nuclei (Figs. 231, 232A). A lateral and a medial funicular nucleus can be distinguished in many Teleosts. Although these grisea may be considered homologous with the nuclei of funic-

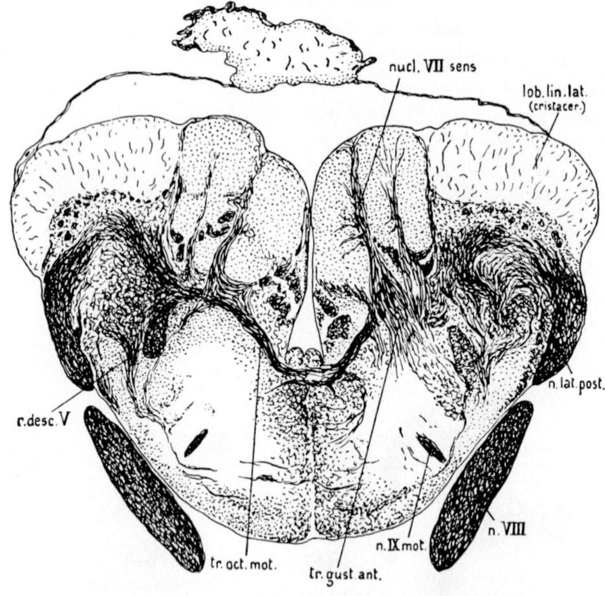

Figure 227. Cross-section through the oblongata of Arius at the entrance of nervus lateralis posterior into its griseum (after VAN DER HORST, from KAPPERS *et al.*, 1936). The griseum covered by the cerebellar crest is the intermediate octavo-lateral griseum. It will be noted that the sulcus limitans has disappeared, the grooves on both sides of the midline being secondary or 'accessory' sulci.

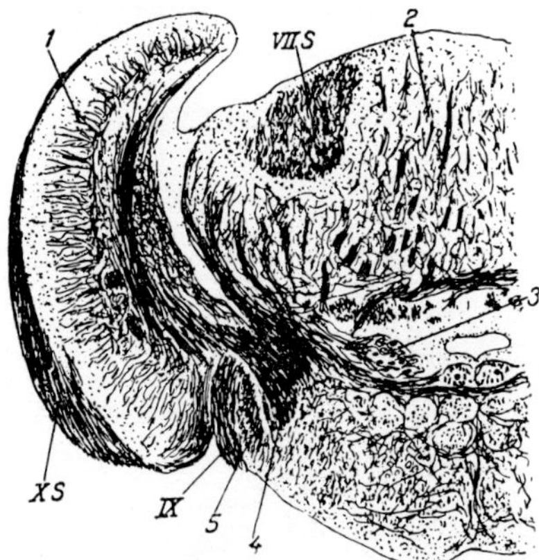

Figure 228. Cross-section through the oblongata of the carp, Cyprinus carpio, at the level of the (median) gustatory facial and the (lateral) gustatory vagus lobes as seen in a modified *Cajal silver impregnation* which displays here mainly the nerve fibers (from BECCARI, 1943). 1: lobe of afferent vagus (*'lobo sensitivo del vago'*, gustatory vagus lobe); 2: lobe of afferent facial (gustatory lobe of VII); 3: 'viscero-motor column' of VII, IX and X; 4: tractus gustatorius secundarius ascendens; 5: radix descendens V; VIIs: afferent facial root fibers; IX: glossopharyngeus root fibers; Xs: layer of afferent vagus root fibers. The locus of 'smoothened out' *(des 'verstrichenen')* sulcus limitans can be recognized.

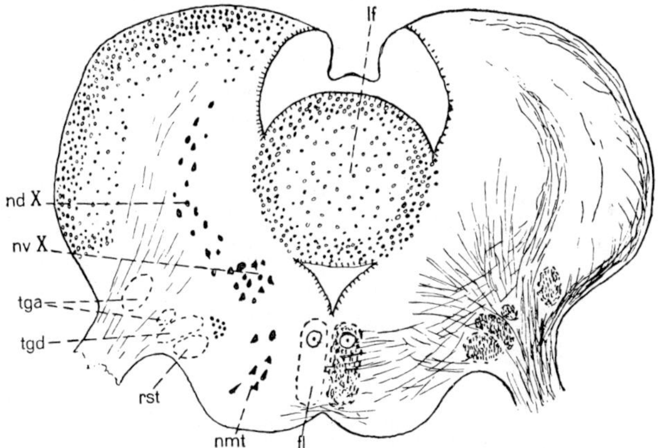

Figure 229. Simplified drawing of cross-section through lobus vagi and caudal part of lobus facialis in Cyprinus auratus (from K., 1927). fl: fasciculus longitudinalis medialis; lf: lobus facialis; ndX: nucleus dorsalis nervi vagi; nmt: nucleus motorius tegmenti; nv: nucleus ventralis nervi vagi; rst: radix descendens trigemini (slightly dorsomedially to the tract, its nucleus is indicated; tga: tractus gustatorius secundarius ascendens; tgd: tractus gustatorius secundarius descendens. The bilaterally symmetric groove in the ventricle's inferior part is the remnant of sulcus limitans. The *Mauthner fiber* (not labelled) can be easily identified. The slight differences in the position of the tracts as compared with Figures 228 and 230B are presumably related to the differences in levels or to species differences (Cyprinus auratus *vs.* Cyprinus Carpio) or both.

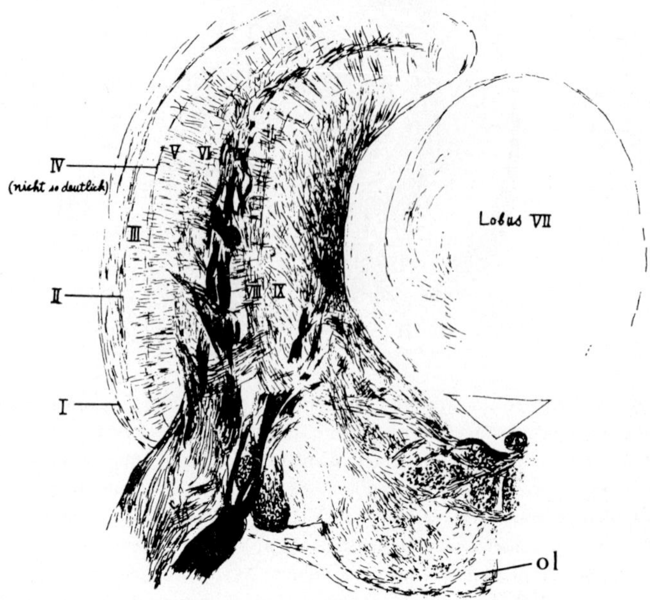

Figure 230 A. Cross-section through the oblongata of the goldfish Carassius auratus at a level roughly corresponding to that of Figure 229 and showing fibers respectively sequence of layers of lobus vagi in HIRASAWA's notation (from HIRASAWA, 1955) The lead to inferior olivary nucleus (ol) has been added on the basis of my own observations.

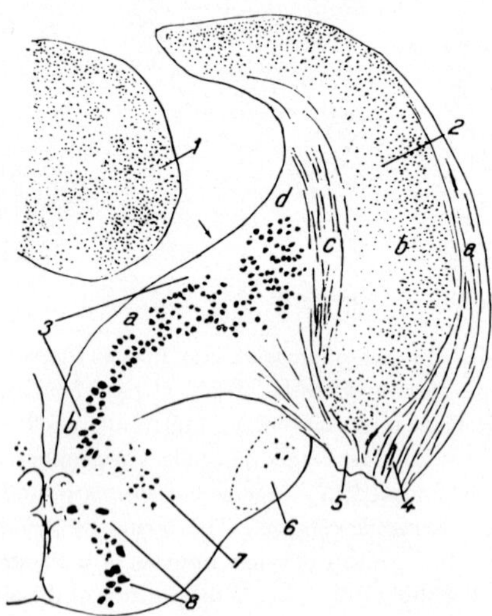

Figure 230 B

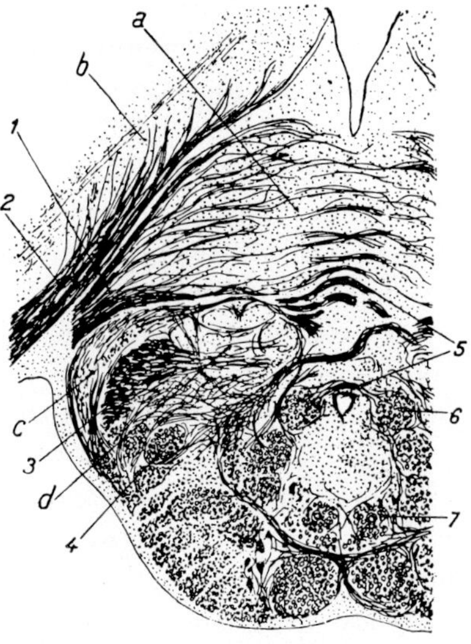

Figure 231. Cross-section through the lower oblongata of Cyprinus carpio near caudal end of lobus vagi and through *commissura infima Halleri* (based on a figure by HERRICK, from BECCARI, 1943). 1: radix descendens vagi; 2: afferent root of vagus; 3: tractus gustatorius descendens lobi facialis; 4: radix descendens trigemini; 5: 'somatic portion' of commissura infima; 6: 'fasciculus funiculo-ambiguus' (pertaining to 'visceral portion' of commissura infima); 7: fasciculus longitudinalis medialis; a: nucleus commissuralis visceralis of commissura infima; b: nucleus funicularis lateralis; d: nucleus funicularis medialis. It will be seen that the 'visceral component' of commissura infima consists of a dorsal portion related to lobus vagi, and of a ventral portion related to tractus solitarius, both portions being separated by the 'somatic portion' (upper part of lead 5).

Figure 230 B. Cross-section through the oblongata of Cyprinus carpio at caudal end of lobus facialis and approximately midway through lobus vagi, based on a *Nissl-preparation* (from BECCARI, 1943). 1: caudal end of lobus facialis; 2: lobus vagi; 3: 'nucleus motorius vagi'; 4, 5: afferent respectively efferent root of vagus; 6: radix descendens trigemini; 7: nucleus motorius tegmenti; 8: spino-occipital motor column (perhaps only at lower lead, upper lead indicates possibly reticular elements). In 2, a: external fiber stratum; b: internal cell stratum; c: internal fiber stratum; d: internal cell stratum; in 3, a: parvocellular group; b: magnocellular group. Added arrow indicates possible locus of eliminated sulcus limitans; if d should pertain to 3 rather than to 2, this locus would approximately correspond to level of letter d.

ulus gracilis and cuneatus of higher forms, such as Mammals, their still poorly understood connections are doubtless quite different and, despite homology, functional analogy does not obtain. Descending trigeminal and gustatory, respectively viscero-sensory fibers as well as short ascending spinal pathways end in these grisea, which have also relations to commissura infima. No significant data concerning the output channels of the Teleostean funicular nuclei are available, but some of these may join the bulbar lemniscus.

An overall survey of the central relations provided by the afferent components of the cranial nerves in the Teleosts Silurus glanis and Mormyrus caschive is included in a study by BERKELBACH VAN DER SPRENKEL (1915).

As regards the configuration of the *basal plate*, the efferent nuclei of the branchial nerves, pertaining to the *intermedioventral zone*, display a few peculiarities. Thus, in some Teleosts, the motor nucleus of the trigeminus is displaced toward the tractus gustatorius secundarius ascendens. The nucleus may also be split into an anterior and posterior cell

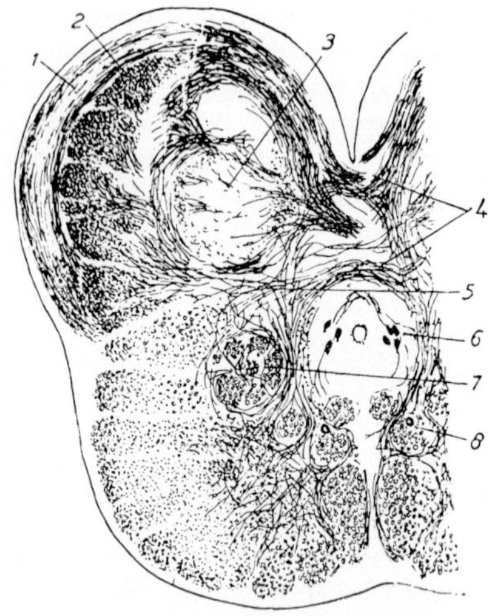

Figure 232A. Cross-section through the bulbo-spinal Übergangsgebiet in Ameiurus nebulosus, showing the substantially developed 'funicular nuclei' (based on data by HERRICK, from BECCARI, 1943). 1: nucleus funicularis lateralis within 'dorsolateral funiculus'; 2: radix descendens trigemini; 3: nucleus funicularis medialis; 4: *commissura infina Halleri;* 5: 'reticular' griseum (perhaps rostral end of spinal cord lateral 'reticular process'); 6: nucleus motorius vagi; 7: fasciculus spino- and bulbo-tectalis; 8: fasciculus longitudinalis medialis with Mauthner fiber.

group (Figs. 223B, C). In Lophius, on the other hand, a part of the motor trigeminal nucleus has shifted ventrally toward the tractus tecto-bulbaris.

The motor nucleus of the facial nerve can likewise undergo displacements, and become, in part, shifted toward the ascending secondary gustatory tract (Fig. 224B), or, in Lophius, together with a portion of the motor glossopharyngeus nucleus, toward the tecto-bulbar tract. The peculiar differentiation of the efferent glossopharyngeus-vagus column into a nucleus ventralis remaining at the usual location within intermedioventral zone, and into a nucleus dorsalis extending into lobus vagi (Fig. 229) was already mentioned above in connection with the specialized 'gustatory' centers. Generally speaking however, the visceral efferent nuclear column of the Teleostean glossopharyngeus and vagus has the tendency to retain its 'primitive' position within the paraventricular intermedioventral zone as in Selachians and Ganoids. A vagus branch comparable to the accessory or XIth nerve does not seem to be present in Teleosts.

Within the *ventral zone* of the *basal plate* the Teleostean nucleus nervi abducentis generally displays a marked basal shift toward the ventrally located crossed and uncrossed tecto-bulbar tracts (Figs. 224A, 224B), in contradistinction to Selachians, where the abducens nucleus remains in rather close vicinity to the particularly massive dorsally located fasciculus longitudinalis medialis. Moreover, the Teleostean abducens nucleus tends to be subdivided into two groups, of which the caudal one apparently receives more descending ventral vestibular fibers than the rostral group. The displacements of abducens together with the previously mentioned ones of motor trigeminal, facial and glossopharyngeus nuclei manifest an evident correlation with the position of functionally significant fiber tracts and were repeatedly pointed out by KAPPERS as manifestations of *neurobiotaxis*. In the caudal *übergangsgebiet* of the Teleostean oblongata, a rostral extension of the spinal cord's 'somatic' ventral horn can be found, from which spino-occipital ventral roots emerge.[112] Figure 208 compares, in a simplified version, the Selachian and the Teleostean pattern of deuterencephalic motor nerve roots and nuclei.

---

[112] The number of 'assimilated' spinal nerve roots taken up into the caudal oblongata varies in the different groups of fishes, and the detailed systematizations as well as the terminology concerning spino-occipital nerves remain inconclusive as well as unsatisfactory. In Teleosts, according to KAPPERS (1947), '*le premier nerf cervical est spino-occipital au sens de* FÜRBRINGER'.

Within the overall basal plate, the Teleostean *nucleus reticularis seu motorius tegmenti* displays an arrangement essentially similar to that obtaining for Selachians and Ganoids. However, since the nuclear pattern of the Teleostean deuterencephalon is generally somewhat more salient than that of Plagiostomes, rather distinct, but nevertheless variable rhombencephalic reticular cell groups can be recognized, which may roughly be subdivided into a rostral, intermediate, and caudal group. Again, a tendency toward segregation into a lateral and into a medial or paramedian column is evident.

The rostral or 'metencephalic' subgroup, with further secondary subdivisions, including an 'isthmic' one, is approximately coextensive with the motor nuclei of the trigeminus, its caudal subdivision, at the level of *Deiters' nucleus*, incorporating the two *Mauthner cells*.[113] The intermediate subgroup is found at the levels of motor and sensory facial nuclei, and the caudal ('posterior' or 'inferior') subgroups corresponds to the levels of glossopharyngeus and vagus nuclei.

The *Mauthner cell* of Teleosts is, in general, but nevertheless with some exceptions,[114] particularly well differentiated (Figs. 226A–C). Not only its often strikingly large size, which has attracted the attention of numerous authors, but also its numerical constancy and its rather fixed position are of particular interest. It represents one of the relatively few instances in which the concept of morphologic homology can be applied to a single cellular element.

The dendrites of the *Mauthner cell* extend far toward the periphery of the oblongata. A large lateral dendrite has connections with primary and secondary afferent vestibular fibers. Moreover, many other homolateral and contralateral octavo-lateral connections are provided by numerous dendritic ramifications. Dorsal, ventral, and medial dendrites are related to such systems as trigeminus and its descending tract with primary and secondary fibers, tractus tecto-bulbaris, tractus cerebello-motorius, fasciculus longitudinalis medialis, and presumably also to the 'gustatory system'. The peculiar types of synapses mediating the *Mauthner cell's* input were pointed out and depicted in volume 3, part I, chapter V, section 4, pp. 286, 290, 291. The neurite of this cell origi-

---

[113] The unsettled questions whether the *Mauthner cell* should be regarded as a derivative of the alar plate vestibular grisea or of the basal plate nucleus reticularis tegmenti, and in which respect the *Deiters' nucleus* should or should not be considered as pertaining to the 'reticular system', were pointed out above in section 3, p. 357, footnote 66, and section 4, p. 402.

[114] Cf. chapter VIII, section 6, p. 125, and footnote 91 of that chapter.

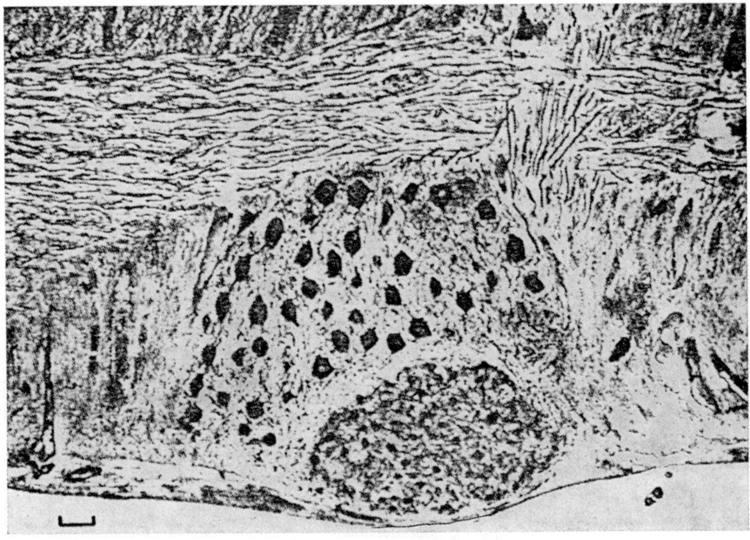

Figure 232 B. Approximately midsagittal section (silver impregnation) through dorsal and ventral 'noyau de commande' in the rostral portion of the oblongata of Gymnarchus niloticus. The scale indicate 100 μ (from SZABO, 1961).

nates on the medial side ('*angulo mediale della cellula*', BECCARI, 1943), here commonly surrounded by an especial synaptic 'axon cap', and forthwith decussates to join the contralateral fasciculus longitudinalis medialis. According to BECCARI, the *Mauthner fiber* emits '*a brevi intervalli*' short collaterals connecting with bulbar reticular elements and with spinal motoneurons.[115] It appears traceable to the caudal end of the spinal cord. It is possible that, within the oblongata, some of the collaterals seen by BECCARI are provided not only with output loci, but also with input loci receiving additional impulses from reticular elements. The *Mauthner cell* with its fiber thus represents a typical example of '*common final path*' in SHERRINGTON's sense and is generally regarded, as emphasized by KAPPERS *et al.* (1936) to be particularly significant for the control of tail movements in swimming, being besides of general importance 'in the preservation of equilibrium'.

---

[115] There are, nevertheless, some doubts about the histologic details pertaining to the *Mauthner fiber's* output structures, which, moreover, might vary in accordance with the different taxonomic forms.

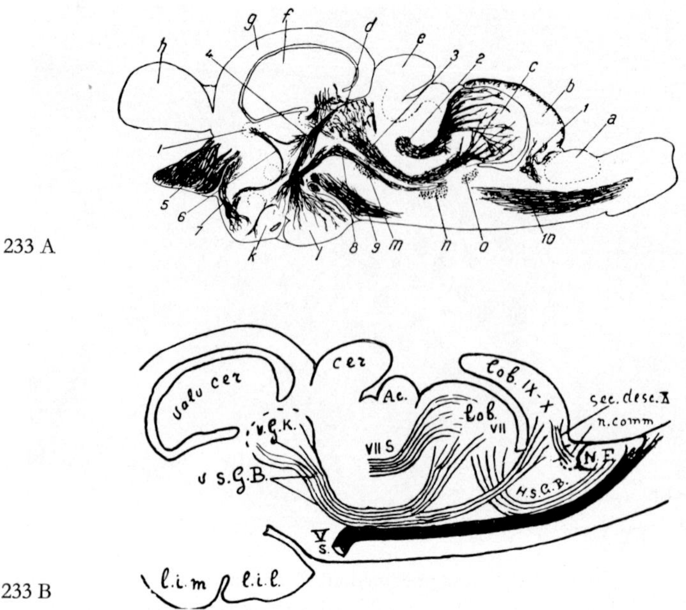

Figure 233A. Drawing of parasagittal section through the brain of the Teleost Minyotrema melanops (a Remora), showing gustatory tracts and grisea. Based on reconstructions by HERRICK from sagittal serial sections and additional data by JOHNSTON (from BECCARI, 1943). 1: tractus gustatorius secundarius descendens; 2: afferent (gustatory) root of facial; 3: tractus gustatorius secundarius ascendens; 4: tertiary gustatory fasciculus combined with 'cerebello-lobar fibers' ('lobar' refers here to the hypothalamic 'lobi'); 5: nervus respectively tractus opticus; 6: *commissura horizontalis of Fritsch;* 7: *tractus tecto-lobaris of Johnston,* related to commissura transversa; 8: tractus lobo-bulbaris; 9: tractus tecto-bulbaris; 10: fasciculus longitudinalis medialis; a: nucleus funicularis (including griseum of nucleus gustatorius inferior; b lobus (sensitivus) vagi; c: lobus (sensitivus) facialis; d: nucleus gustativus secundarius (superior); e: cerebellum; f: valvula cerebelli; g: tectum mesencephali ('lobus opticus'); h: telencephalon; i: 'nucleus praetectalis' *('n. corticalis' of Fritsch?);* k, l: medial respectively lateral lobuli of lobus inferior (lateralis) hypothalami; m: nucleus motorius trigemini; n: nucleus motorius facialis; o: nucleus motorius glossopharyngei (?).

Figure 233B. Simplified drawing, by KAPPERS (1920), of a parasagittal section through the brain of a Cyprinoid, based on data by HERRICK (1905) and illustrating some of the tracts and structures depicted in the preceding Figure 233A (from KAPPERS *et al.*, 1936). Ac: area octavo-lateralis s. statica; cer.: cerebellum; H.S.G.B.: tractus gustatorius secundarius descendens *('hintere sekundäre Geschmacksbahn');* l.i.l.: lobulus lateralis lobi inferioris hypothalami; l.: m.lobulus medialis lobi inferioris; lob.VII: lobus facialis; lob.IX–X: lobus vagi; n.com.: nucleus commissurae infimae; N.F.: nucleus funicularis; sec.desc.X: descending secondary gustatory fibers of X; valv.cer.: valvula cerebelli; v.g.k.: nucleus gustatorius secundarius superior *('vorderer Geschmackskern');* v.s.G.B.: tractus gustatorius secundarius ascendens; V.s.: radix descendens trigemini.

Stressing the importance of the *Mauthner neuron*[116] for tail action in swimming activities, STEFANELLI (1962) interprets the '*aparato mauthneriano*' as '*tipicamente larval*' and said to be undergoing atrophy in those Teleosts without tail (Orthagoriscus), or with rigid bodies ('Cofferfishes', Ostracion), or with a tail not significantly effective for swimming movements (Hippocampus). On the other hand, '*Mauthner's apparatus*' seems to be retained in those forms in which metamorphosis does not involve relevant changes *qua* swimming activities.

The rostral part of the reticular formation in Teleosts with electric organs includes, as in Selachians (cf. above p. 401) a '*noyau de commande*' presumed to control the electric discharges. SZABO (1961) has depicted this griseum in Gymnarchus niloticus, where it assumes a basal location in or about the midline, and consists of a more dorsal magnocellular nucleus associated with a ventral nucleus containing smaller cells (Fig. 232B). In addition to these unpaired nuclei, a lateral group of reticular cells seems to represent a paired lateral 'command nucleus'.

Within the caudal portion of the Teleost oblongata, the basal plate[117] includes a '*nucleus olivaris inferior*' essentially similar to that found in the Plagiostome oblongata. It is described as a rather medial or median group of relatively small spindle-shaped cells which forms a griseum of variable compactness and occasionally of gelatinous appearance (Fig. 230A). In addition to its poorly understood cerebellar and possibly spinal connections, this primordial inferior olive, which displays variable relationships to the actual midline, may receive ('external') arcuate fibers from the funicular nuclei and tecto-bulbar fibers (cf. KAPPERS *et al.*, 1936).

As regards the oblongata of the *Crossopterygian* Latimeria,[118] some general features have been recorded in the monograph by MILLOT and ANTHONY (1965). The available data seem to indicate a fairly simple overall pattern, similar to that obtaining in Plagiostomes and Ganoids rather than to that in Teleosts (Figs. 234A–D). It can easily be seen

---

116 '*Este aparato binaural representado por las células de Mauthner es el más precoz en diferenciarse en el embrión y representa la primera via que, extendiendóse a lo largo de la médula espinal, entra en juego de la coordinación del movimiento de la cola en la natación, en conexión con los centros vestibulares y, sobre todo, con los de la linea lateral.*' (STEFANELLI, 1962).

117 The location of inferior olivary griseum in the adult basal plate does not preclude a possible origin of an undefined proportion of its elements by early 'migration' from the alar plate (cf. above, section 2, p. 340/341).

118 Cf. the comments on the taxonomic interpretations of Latimeria in section 6, p. 128 of the preceding chapter VIII.

that the three main longitudinal sulci, sulcus limitans, sulcus intermediodorsalis and sulcus intermedioventralis are conspicuously developed.

The *alar plate* includes in its *dorsal zone* an octavo-lateral area with crista cerebellaris, a rostral nervus lobus lineae lateralis anterioris, and a transition into the cerebellar auricle. A rostral main trigeminal nucleus is present and, ventrolaterally at more caudal levels, a radix descendens trigemini can be identified in the photomicrographs published by the cited authors.

The alar plate's *intermediodorsal zone* contains the viscero-afferent nuclei of facialis, glossopharyngeus and vagus, including some fibers that could be interpreted as a primordial tractus solitarius.

The *basal plate* displays, in its *intermedioventral zone*, the efferent nuclei of the Vth, VIIth, IXth, and Xth nerves, but precise details as to their specific 'separation' or boundaries are not available. A typical ra-

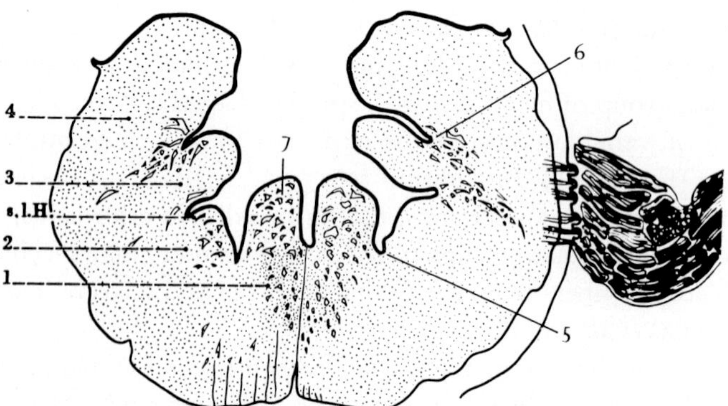

Figure 234A. Semidiagrammatic cross-section through the oblongata of Latimeria at the level of an undefined branchial nerve root entrance (from ANTHONY and MILLOT, 1965, with some added leads). 1: ventral zone *('zone somato-motrice')*; 2: intermedioventral zone *('zone viscéro-motrice')*; 3: intermediodorsal zone *('zone viscéro-sensitive')*; 4: dorsal zone *('zone somato-sensitive')*; 5: sulcus intermedioventralis; 6 sulcus intermediodorsalis; 7: bulging fasciculus longitudinalis medialis. Leads 5 to 7 have been added to the original figure. s.l.H.: sulcus limitans.

Figures 234B–D. Cross-sections through the oblongata of Latimeria (from MILLOT and ANTHONY, 1965). The unlabelled (photographic) plate figures (apparently silver proteinate impregnation, ×14, here reproduced $^2/_3$) have been labelled in accordance with my interpretation. B Rostral level, showing cerebellum and part of auricles. The extracerebral nerve roots are presumably trigeminal and anterior lateralis branches. C Intermediate level, perhaps at posterior portion of facial root. D Caudal level, presumably

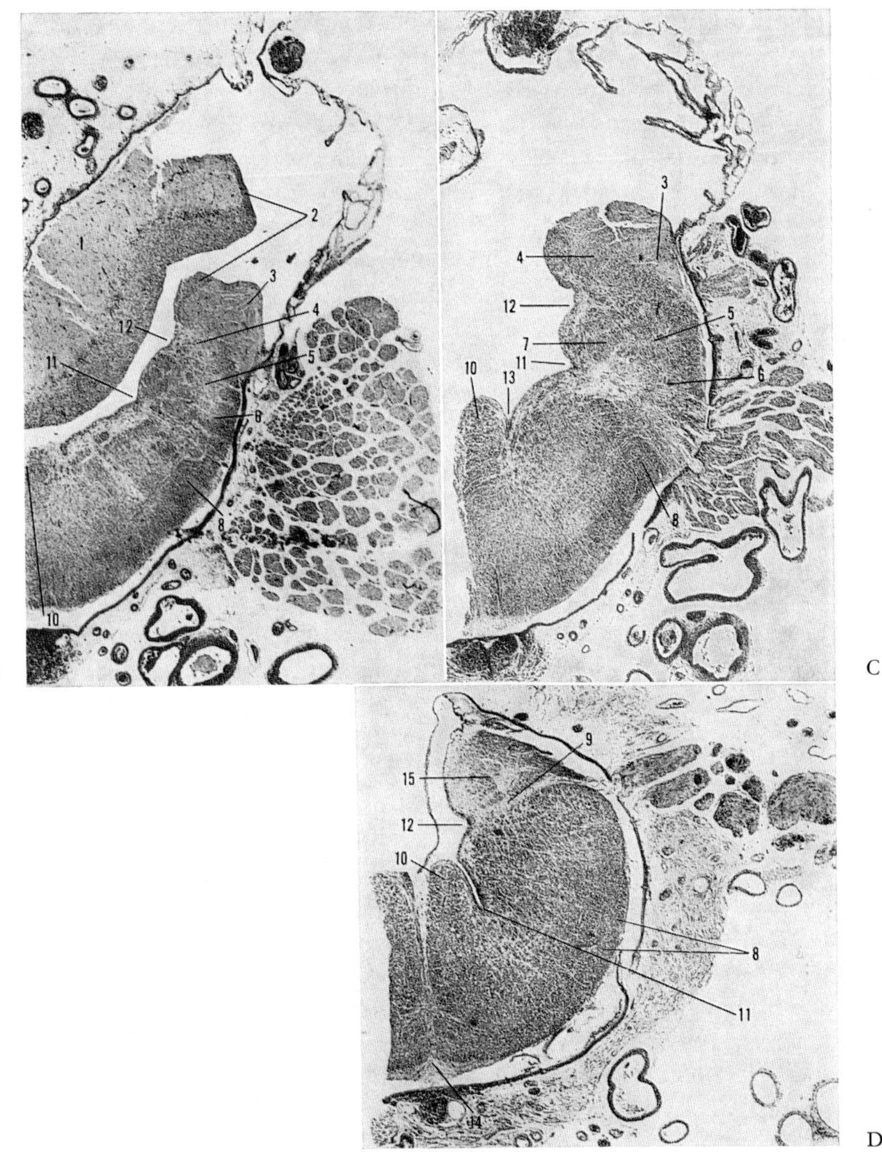

corresponding to caudal vagus rootlets. 1: cerebellum; 2: superior and inferior auricle leafs; 3: presumably cerebellar crest; 4: probably lobus nervi lineae lateralis anterioris; 5: intermediate octavo-lateralis area; 6: perhaps ventral octavus area; 7: afferent facialis griseum; 8: probably radix descendens trigemini; 9: afferent vagus griseum; 10: fasciculus longitudinalis medialis; 11: sulcus limitans; 12: sulcus intermedius dorsalis; 13: sulcus intermedius ventralis; 14: perhaps inferior olivary nucleus.

mus accessorius vagi (XIth nerve) does not seem to be present. Within the *ventral zone*, a 'nucleus' of the VIth nerve was identified; its rootlets emerge as ventral roots caudal to the level of the glossopharyngeus root. Still more caudally, the ventral zone contains the 'nucleus' of the spino-occipital nerves, of which three pairs were recorded, the first pair emerging as ventral roots approximately at the level of the last vagus rootlet.

Other *basal plate* components include a *reticular*[119] and an *inferior olivary nucleus*. Concerning these grisea, MILLOT and ANTHONY (1965) state that in the posterior region of the rhombencephalon '*un large noyau réticulaire longe l'axe de symétrie. En dehors de lui et assez ventralement une courte et étroite parolive bulbaire, à bords irréguliers, marque un modeste relais de certaines fibres ascendante en direction du cervelet (fibres spino-olivaires, directes; fibres olivo-cérébelleuses, croisées).*'

In addition to the just mentioned fiber systems, the cited authors noted the general spino-bulbo-tectal lemniscus *(faisceau spino-mésencéphalique)* and the fasciculus longitudinalis medialis, considered to be the main motor tract. In a manner similar to that displayed by many Plagiostomes, this bundle, particularly in the caudal two thirds of the oblongata, protrudes into the ventricle on both sides of the midline.[120] The authors also briefly mention tecto-bulbar and tecto-spinal pathways.

6. Dipnoans

Relevant data on the oblongata of Dipnoans are contained in the publications by HOLMGREN and VAN DER HORST (1925) and by GERLACH (1933). The latter author investigated the brain of *Protopterus*, while HOLMGREN and VAN DER HORST dealt with the brain of *Ceratodus*. Both papers contain the pertinent bibliographic references to previous studies on the Dipnoan brain, among which those by BURCKHARDT (1892) on Protopterus, and by BING and BURCKHARDT (1904) on Ceratodus are of particular significance.[121]

---

[119] '*La formation réticulaire, qui infiltre tout le trone cérébral*'. (MILLOT and ANTHONY, 1965).

[120] '*Le faisceau longitudinal posterieur, reconnaissable sur les coupes transversales à la forte saillie qu'il détermine sur le plancher du 4e ventricule*'. (MILLOT and ANTHONY, 1965).

[121] As far as I could ascertain, specific data on the oblongata of Lepidosiren are not available. Except for the telencephalon, investigations on the brain of that Dipnoan have not come to my attention.

Generally speaking, the brain of *Protopterus* appears to be somewhat less complex than that of *Ceratodus*, and, on the whole, more similar to the Amphibian brain. As regards the oblongata, both Dipnoan forms display a configuration which can also easily be compared with that obtaining for Plagiostomes and Ganoids.

Figures 235 to 238 depict cross-sections through representative levels of the oblongata in Protopterus, and Figures 239 and 242 illustrate such sections through the oblongata of Ceratodus.[121a] The choroidal folds of the lamina epithelialis covering the fourth ventricle are evident in Figures 236, 237 (Protopterus) and 239 (Ceratodus). The sulcus limitans can easily be recognized in both forms; sulcus intermedius dorsalis (called sulcus lateralis by HOLMGREN and VAN DER HORST), and sulcus intermedius ventralis are likewise present.[122] Some accessory dorsal sulci within the alar plate's dorsal zone of Ceratodus might perhaps be fixation artefacts.

In the *dorsal zone* of the *alar plate*, the *octavolateralis area* displays an arrangement comparable to that in Plagiostomes and Ganoids, including a transition to the cerebellar auricles and the presence of a crista cerebellaris which seems much better developed in Ceratodus than in the available material of Protopterus. A ventrolateral portion of the dorsal zone includes rostrally the main afferent nucleus of the trigeminus and, caudally, the radix descendens trigemini with its griseum. The *intermediodorsal zone* contains rostrally the afferent nucleus of nervus facialis, and more caudally the corresponding nuclei of glossopharyngeus and vagus. A tractus solitarius is more conspicuous than in other fishes and has some similarities with that of Amphibians.[123] In the region of calamus scriptorius, the two alar plates, fusing in the dorsal

---

[121a] Figures 235–237 represent sections stained by an iron-hematoxylin and picric acid technique; Figures 239 and 240 are silver impregnation pictures, and Figures 241 and 242 are pictures obtained with a myelin stain.

[122] Because of a locally unsatisfactory state of conservation, the identification of these sulci in the available Protopterus material involved some difficulties. Again, the ventricular groove in the oblongata of Ceratodus which I am inclined to regard as sulcus intermedius ventralis was interpreted by GERLACH (1933) as corresponding to SAITO's sulcus ventralis accessorius. It is, however, not unlikely that these two sulci display mutual interconnections and combinations.

[123] KAPPERS (1920) suggests that the phylogenetic development of the postvagal portion of tractus solitarius, which may contain primary as well as secondary fibers, is correlated with the development of pulmonary respiration involving the inclusion of afferent fibers from the respiratory tract into fasciculus solitarius.

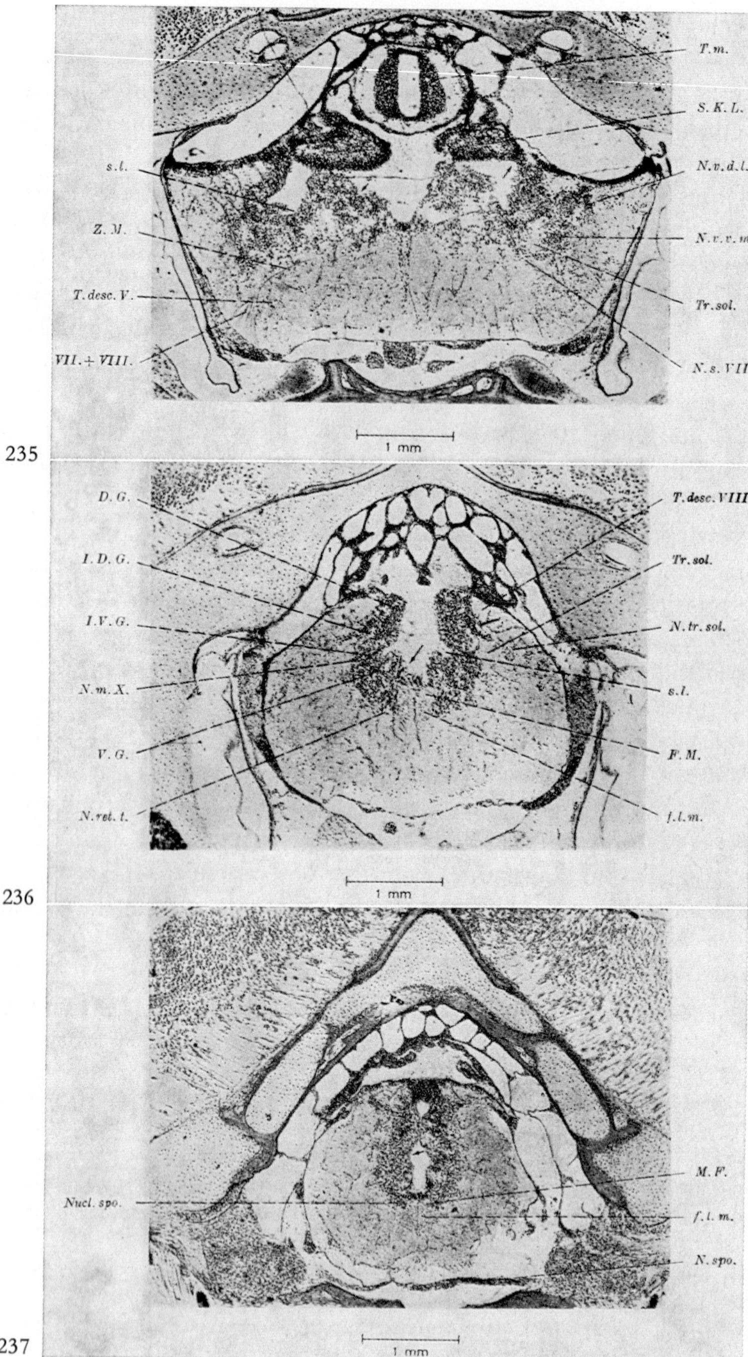
T.m.
S.K.L.
s.l.
N.v.d.l.
Z.M.
N.v.v.m.
T.desc.V.
Tr.sol.
VII.+VIII.
N.s.VII.
1 mm
235
D.G.
T.desc.VIII
I.D.G.
Tr.sol.
I.V.G.
N.tr.sol.
N.m.X.
s.l.
V.G.
F.M.
N.ret.t.
f.l.m.
1 mm
236
M.F.
Nucl.spo.
f.l.m.
N.spo.
1 mm
237

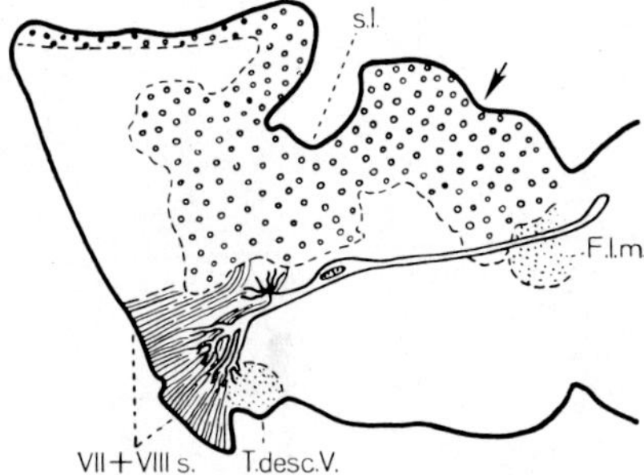

Figure 238. Semidiagrammatic sketch depicting the *Mauthner cell* and its location in Protopterus (from GERLACH, 1933). Abbreviations as in the three preceding figures. Added arrow indicates sulcus intermedius ventralis. The sulcus to the left of sulcus limitans may bé sulcus intermedius dorsalis or an accessory sulcus.

Figure 235. Cross-section through the oblongata in a young specimen of Protopterus annectens at the level of a *Mauthner cell* (from GERLACH, 1933). N.v.d.l.: nucleus octavolateralis intermedius; N.v.v.m.: nucleus vestibularis; N.s.VII.: sensory nucleus of nervus facialis; S.K.L.: upper auricular lip of cerebellum; S.l.: sulcus limitans; T.desc.V.: radix descendens trigemini; T.m.: Tectum mesencephali; tr.sol.: tractus solitarius; Z.M.: *Mauthner cell;* VII. VIII.: intracerebral facial and vestibularis roots. Added arrows point to sulcus intermedius dorsalis and to sulcus intermedius ventralis.

Figure 236. Cross-section through the oblongata of Protopterus at the level of caudal vagus roots (from GERLACH, 1933). D.G.: dorsal zone of alar plate; F.M.: *Mauthner fiber;* I.D.G.: intermediodorsal zone; I.V.G.: intermedioventral zone; N.m.X.: nucleus motorius vagi; N.ret.t.: nucleus reticularis tegmenti; N.tr.sol.: nucleus of tractus solitarius; s.l.: sulcus limitans; T.desc.VIII: descending vestibular tract; Tr.sol.: tractus s.fasciculus solitarius; V.G.: ventral zone of basal plate. Added arrow indicates sulcus intermedius ventralis. The numerous spaces above the choroid plexus pertain to saccus endolymphaticus, which is very extensive in this Dipnoan (cf. p. 310).

Figure 237. Cross-section through the oblongata of Protopterus at level of calamus scriptorius (from GERLACH, 1933). f.l.m.: fasciculus longitudinalis medialis; M.F.: *Mauthner fiber;* N.spo.: part of spino-occipital nerve rootlet; Nucl.spo.: nucleus of spino-occipital nerves. Note: dorsocaudal recess of fourth ventricle formed by lamina epithelialis and choroid plexus. The row of spaces above this recess pertains to saccus endolymphaticus. Arrow: sulcus limitans.

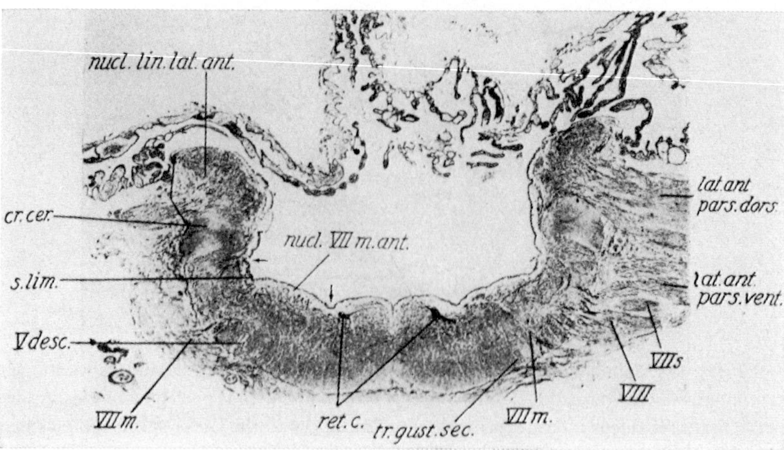

Figure 239. Cross-section through the oblongata of Ceratodus at the level of facialis root and anterior motor facial nerve nucleus (from HOLMGREN and VAN DER HORST, 1925). cr. cer.: crista cerebellaris; VII m.: efferent facialis root fibers; VII s.: afferent facialis root fibers; ret. c.: cells of nucleus reticularis tegmenti. Other abbreviations self-explanatory. Added arrows indicate (left) sulcus intermedius dorsalis and (right) sulcus intermedius ventralis. Along the alar plate's ventricular surface, some nondescript accessory sulci can be noted.

midline, provide a well developed commissura infima.[124] In Ceratodus, moreover, HOLMGREN and VAN DER HORST (1925) believe to have identified a fiber system comparable to the ascending and descending secondary gustatory tracts of Teleosts.

Within the *basal plate*, the *intermedioventral zone* displays the 'visceromotor column' of the efferent trigeminus, facialis, vagus and glossopharyngeal nuclei. An indistinctly delimited interval seems to separate a rostral subdivision of that column, pertaining to V and VII, from a caudal glossopharyngeal-vagal subdivision.

The motor nuclei located in the *ventral zone* are those of the abducens and of the spino-occipital rootlets. The root fiber bundles of the abducens emerge as ventral roots, caudally to the level of facialis root and rostrally to that of the glossopharyngeus root. The nucleus of the abducens was tentatively identified in Ceratodus by HOLMGREN and VAN DER HORST as located close to the ventricular floor, just laterally to

---

[124] The ependymal lining of this commissure is conspicuously columnar and, according to GERLACH (1933) suggests similarities with the ependyma of the subcommissural organ.

fasciculus longitudinalis medialis. Its cells seemed 'very much scattered' within the just mentioned level between VII and IX, and could not be unequivocally distinguished from the reticular cells of that region. GERLACH was unable to identify either rootlets or nucleus of the abducens in his material of Protopterus, presumably because of the difficulties mentioned above in footnote 122.

The nuclear column of the spino-occipital nerves is easily recognizable laterally to fasciculus longitudinalis, in both Protopterus (Fig. 238) and Ceratodus (Fig. 242). In the former, one to three ventral rootlets seem to occur, emerging rostrally to the level of calamus scriptorius. In Ceratodus, likewise rostrally to that level, about four to six rootlets may be present.

The *nucleus reticularis s. motorius tegmenti* of Dipnoans seems to extend rather diffusely throughout the oblongata, without a very clearcut subdivision into specific groups, although some lateral and some medial cell populations could perhaps be distinguished, the medial cell groups being located laterally and ventrally to the fasciculus longitudinalis medialis.

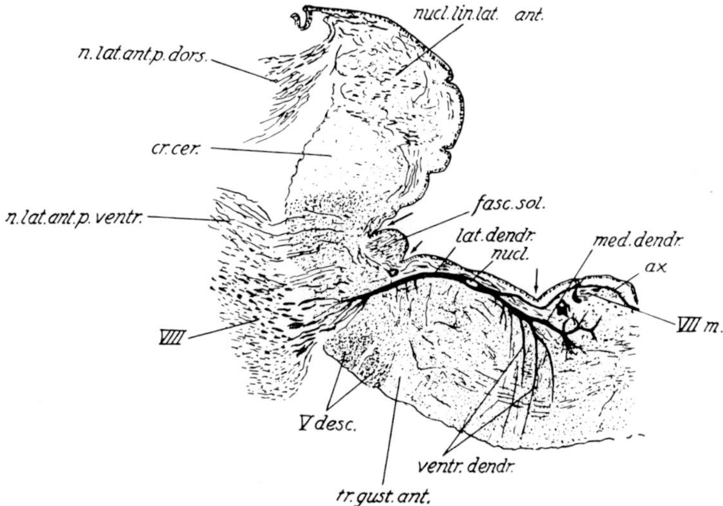

Figure 240. Cross section through the oblongata of Ceratodus at the level of the *Mauthner cell.* Drawing based on several consecutive sections (from HOLMGREN and VAN DER HORST, 1925). ax: axon of *Mauthner cell;* fasc. sol.: fasciculus solitarius; nucl.: nucleus of *Mauthner cell* body; tr. gust. ant.: tractus gustatorius anterior (ascendens); VII m.: intracerebral root fibers pertaining to 'loop' of efferent facial nerve component. Other abbreviations self-explanatory. Added arrows indicate, from left to right, sulcus intermedius dorsalis, sulcus limitans, and sulcus intermedius ventralis.

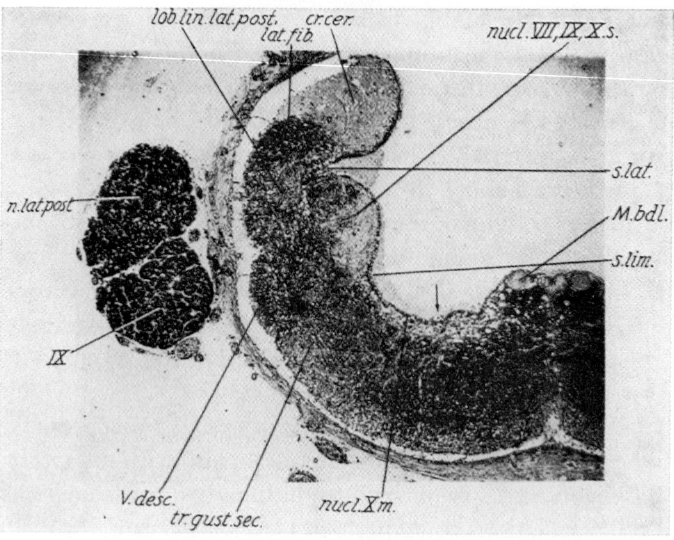

Figure 241. Cross-section through the oblongata of Ceratodus at the level of the first motor vagus rootlets (from HOLMGREN and VAN DER HORST, 1925). lat.fib.: fibers of nervus lateralis; M.bdl.: *Mauthner fiber* (so-called '*Mauthner bundle*'); nucl. VII, IX, X s: afferent (terminal branchial nerve nuclei; nucl. X m.: motor (efferent) vagus nucleus; s.lat.: sulcus intermedius ventralis of my terminology. Added arrow: sulcus intermedius ventralis, to the right of this sulcus, a sulcus accessorius ventralis can be recognized.

At the level of the octavus root entrance, the nucleus reticularis tegmenti includes the *Mauthner cell* (Figs. 235, 238, 240). In Protopterus, its medial dendrite[125] may reach as far as the contralateral side just across the midline. The cell's axon runs dorsomedialward and decussates at or just behind the cell's level, running caudad at the dorsal edge of fasciculus longitudinalis medialis.

In Ceratodus, HOLMGREN and VAN DER HORST (1925) describe a peculiar relationship of the *Mauthner fiber* to some axons of reticular cells. These neurites are said to accompany, and, toward the caudal end of the oblongata, to join the *Mauthner fiber* within a common sheath such that the *Mauthner fiber* actually becomes '*Mauthner's bundle*' in the terminology of the cited authors. Although this curious and unusual ar-

[125] GERLACH in Protopterus, as well as HOLMGREN and VAN DER HORST in Ceratodus stress the difficulty in making a distinction between cell body and main dendrites of the Dipnoan *Mauthner cell*. The cell body is here formed only by a slight swelling around the nucleus and is not much thicker than the main dendrites, of which the lateral one is said to be especially thick.

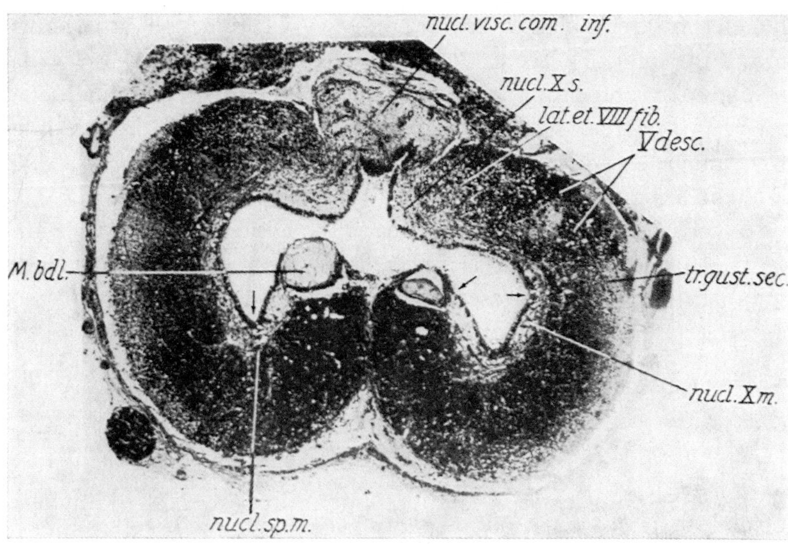

Figure 242. Cross-section through the oblongata of Ceratodus at the level of commissura infima (from Holmgren and van der Horst, 1925). Nucl. sp. m.: spino-occipital nuclear column; nucl. visc. com. inf.: (interstitial) 'visceral' nucleus of commissura infima. Other abbr. as in preceding figs. Added arrows (left) sulcus intermedius ventralis, probably fused with sulcus accessorius ventralis system and (right) an accessory sulcus related to *Mauthner fiber* (also clearly recognizable on left side).

rangement according to the interpretation of Holmgren and van der Horst cannot be summarily dismissed as unlikely,[126] it is not impossible, in view of the many deceiving artifacts, distortions, coagulates and precipitates intrinsic to silver impregnation pictures, that the cited authors may have been led astray into a misinterpretation.[127]

A well-distinguishable inferior olivary nucleus, comparable with that in Plagiostomes, Ganoids, and Teleosts, does not seem to be present in Dipnoans, but some scattered nerve cells of small reticular type can be detected in the ventral portion of the basal plate at caudal oblongata levels, that is in a topologic neighborhood corresponding to

---

[126] Kappers *et al.* (1936) justly compare this arrangement with the somewhat similar structural relationships obtaining for the so-called 'compound', composite or multicellular giant fibers of certain Polychaete and Oligochaete Annelids (cf. vol. 2, pp. 80–87 of this series).

[127] Be that as it may, this open question should easily be decidable through further investigations by means of electron microscopy, which, despite some shortcomings, is particularly applicable to problems of this sort.

the inferior olive's locus. Lack or at best very rudimentary differentiation of the inferior olive can be interpreted as related to the fact that the Dipnoan cerebellum, which displays a higher degree of differentiation than the Cyclostome and the Amphibian one, is, on the other hand, much less developed than the cerebellum of Selachians, Ganoids, and Teleosts.

As regards the main fiber systems of the Dipnoan oblongata, sufficiently detailed information is not available, but the recorded data indicate that these pathways essentially correspond to the basic pattern of fiber tracts obtaining for other Fishes.

7. Amphibians

Since Amphibians represent, taxonomically as well as in phylogenetic speculations, the most primitive group of those Vertebrates which are characterized by typically *tetrapod* and *terrestrial* forms, their neuraxis is of evident theoretical interest. *Urodeles*, in particular, display a very simple, clear-cut, and rather unspecialized neural configuration, whose significance for the understanding of more complex central nervous systems was justly pointed out by LUDWIG EDINGER (cf. above section 3, p. 383). Relevant early studies on the Amphibian brain are those by OSBORN (1888) and by KINGSBURY (1895). In accordance with EDINGER's viewpoint, the Urodele central nervous system was intensively studied for many years by HERRICK, who summarized his findings and interpretations in a monograph on the brain of the Salamander *Amblystoma tigrinum* (1948).

In addition to these reports and to the descriptions contained in the previously cited texts on comparative neurology, detailed data on the oblongata of Urodeles can be found in the numerous publications by HERRICK (e.g. 1914, 1930 and many others). As regards *Anurans*, the accounts by GAUPP in ECKER's, WIEDERSHEIM's und GAUPP's *Anatomie des Frosches* (1899), by KAPPERS and HAMMER (1918), and by LARSELL (1934) should be mentioned. Data on the oblongata of *Gymnophiona* are scarce and sketchy; a few incidental observations were reported by myself (K., 1922, 1969).

Figures 243 to 245 illustrate representative cross-sectional levels of the Urodele oblongata, and Figures 246A–C depict the course of some main fiber tracts in these forms. The Anuran oblongata is shown in Figures 247 and 248, while some features of the Gymnophione oblongata are indicated in Figures 249A–C. It can be seen that, generally speaking, the sulcus limitans is well recognizable. In many instances sulcus intermedius dorsalis and sulcus intermedius ventralis are likewise unambiguously displayed.

The *dorsal zone* of the *alar plate*, as is the case in Fishes, contains the grisea of the octavus and lateralis systems. However, the diversity among adult Amphibians between an entirely aquatic and a partially or wholly terrestrial habitat is correlated with various degrees of retention respectively reduction or even complete disappearance of the *lateralis system*, which is present in larval stages of Urodeles, Anurans, and Gymnophiona.

According to NOBLE (1931, 1954) the lateral line organs and their neuronal apparatus can be found in all thoroughly aquatic Urodeles and their larvae. They are present in some mountain-brook species such as the larger forms of Desmognathus, but are inconspicuous or lacking in the more terrestrial forms of the same genus. They are lacking in terrestrial Plethodontids. Although present in larvae of all Amblystoma species, they show various stages of degeneration in the adults. Among Anurans, the lateralis system, present in the aquatic larvae, is generally absent in the adults, but nevertheless found in adults of Pipidae (aglossal toads such as Xenopus), of Discoglossidae (Bombina), and of some true toads (Bufonidae) such as Ceratophrys. However, the lateral line system is not present in some aquatic toads, nor in the thoroughly aquatic Gymnophione Typhlonectes which is believed to have had 'fossorial ancestors'.

An interesting paper by ESCHER (1925) deals with the various phylogenetic problems concerning disappearance of the Vertebrate lateral line organs in connection with the transition from aquatic to terrestrial life. These problems involve among other questions, a possible transformation of the Anamniote lateral line organs into other cutaneous structures characteristic for Amniota.[128]

As in Fishes, the Amphibian *lateral line nerves* include a *rostral* ('preauditory') component related to the seventh cranial nerve respec-

---

[128] Thus, my teacher MAURER (1895) elaborated the hypothesis that the Mammalian hairs represent phylogenetic derivatives of transformed lateral line organs. While there is little doubt, as ESCHER admits, concerning a general homology of hair-anlagen and lateral line sensory buds, such general homology (discussed on pp. 200–203 of volume 1) does not necessarily imply a direct phylogenetic derivation. As regards MAURER's hypothesis, I would agree with ESCHER's comment: '*Doch hat man sie so wenig widerlegen können, wie es gelungen ist, sie streng zu beweisen*'.

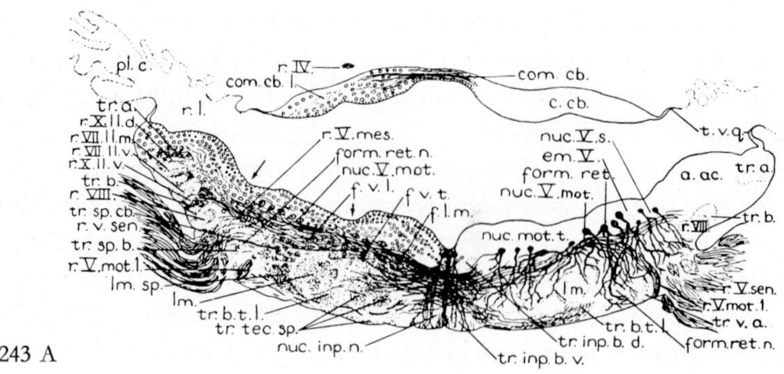

243 A

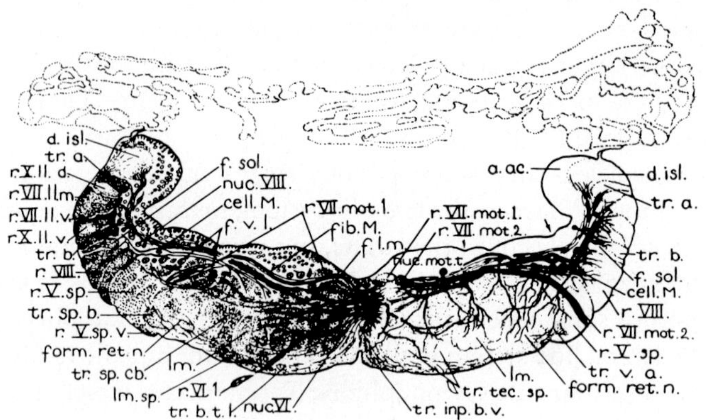

243 B

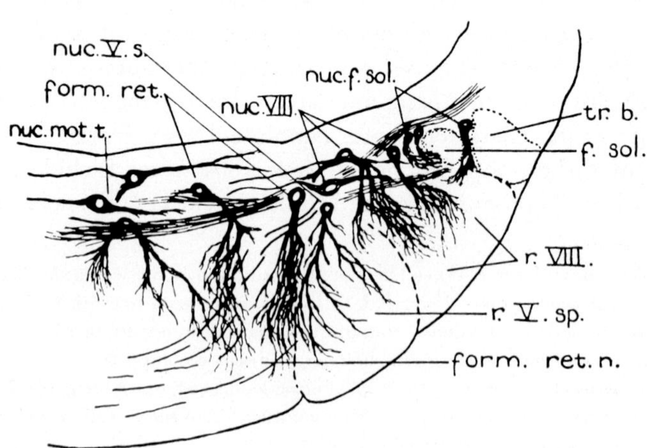

243 C

tively acustico-facialis root complex, and a *posterior* ('postauditory') component related to the glossopharyngeus-vagus group. Further subdivisions of the Urodele lateral line nerve roots at their entrance into the oblongata were described in various ways by several authors and are discussed in LARSELL's monograph on the cerebellum (1967). The ascending and descending fibers of the dorsal anterior lateral line nerve are shorter than those of the other roots. They are connected with a zone of neuropil which KINGSBURY (1895), in Necturus, designated as '*dorsal island of alba*'. In addition to lateralis and other afferent root fibers, the dorsal zone of the alar plate displays two longitudinal tracts, described by KINGSBURY (1895) as tracts a and b, of which the latter is located more ventrally than the former. Both run as far as the rostral spinal cord and are interpreted as 'correlation fascicles', essentially consisting of secondary octavo-lateralis fibers.

Figure 243 A. Cross-section through the oblongata of Necturus at the level of the trigeminal roots (from HERRICK, 1930). Left side of figure based on a *Weigert section*, on the right sides findings from *Golgi and Cajal impregnations* have been included. The two arrows, which I have added, point to sulcus limitans and sulcus intermedius ventralis. a.ac.: area acustica-lateralis; c.cb.: corpus cerebelli; com.cb.: commissura cerebelli; e.V: eminentia trigemini; f.l.m.: fasciculus longitudinalis medialis; f.v.l.: 'ventrolateral fasciculi'; f.v.t.: ventral tegmental fascicles; form.ret.: 'formatio reticularis grisea'; form. ret.n.: neuropil of formatio reticularis; lm.: 'general bulbar lemniscus; lm.sp.: spinal lemniscus; nuc.inp.n.: neuropil of nucleus interpeduncularis (extending caudad from mesencephalon); nuc.mot.t.: nucleus motorius tegmenti; pl.c.: plexus chorioideus; r.l.: recessus lateralis of 4th ventricle; r (with Roman numeral): radix of designated cranial nerve; r.V mes.: radix mesencephalica trigemini; r.ll.d.: dorsal lateralis root; r.ll.m.: middle lateralis root; r.ll.v.: ventral lateralis root; tr.a.: tractus a (dorsal longitudinal tract of area acustico-lateralis); tr.b.: tractus b (ventral longitudinal tract of area acustico-lateralis); tr.b.t.l.: tractus bulbo-tectalis lateralis; tr.inp.b.d.: tractus interpedunculo-bulbaris dorsalis; tr.inp.b.v.: tractus interpedunculo-bulbaris dorsalis; tr.sp. b.: tractus spino-bulbaris; tr.sp.cb.: tractus spino-cerebellaris; tr.tec.sp.: tractus tecto-spinalis; tr.v.a.: tractus visceralis (gustatorius) ascendens; t.v.q.: taenia ventriculi quarti.

Figure 243 B. Cross-section through the oblongata of Necturus at the level of the two *Mauthner cells* (from HERRICK, 1930). Added arrows (right) indicate sulcus intermedius dorsalis and sulcus limitans. d.isl.: *dorsal neuropil island of Kingsbury;* cell. M.: *Mauthner cell;* f.i.b.M.: *Mauthner fiber;* f.sol.: fasciculus (s.tractus) solitarius. Other abbreviations as in Figure 243A.

Figure 243 C. 'Reticular formation' with 'neurons of the motor tegmentum' (nucleus motorius tegmenti) and part of the 'sensory field' in a cross-sectional plane slightly caudad to that of Figure 243B. Composite drawing in which the cells have been assembled from several sections processed by *Cajal's silver impregnation* (from HERRICK, 1930). Abbreviations as in Figures 243A and B.

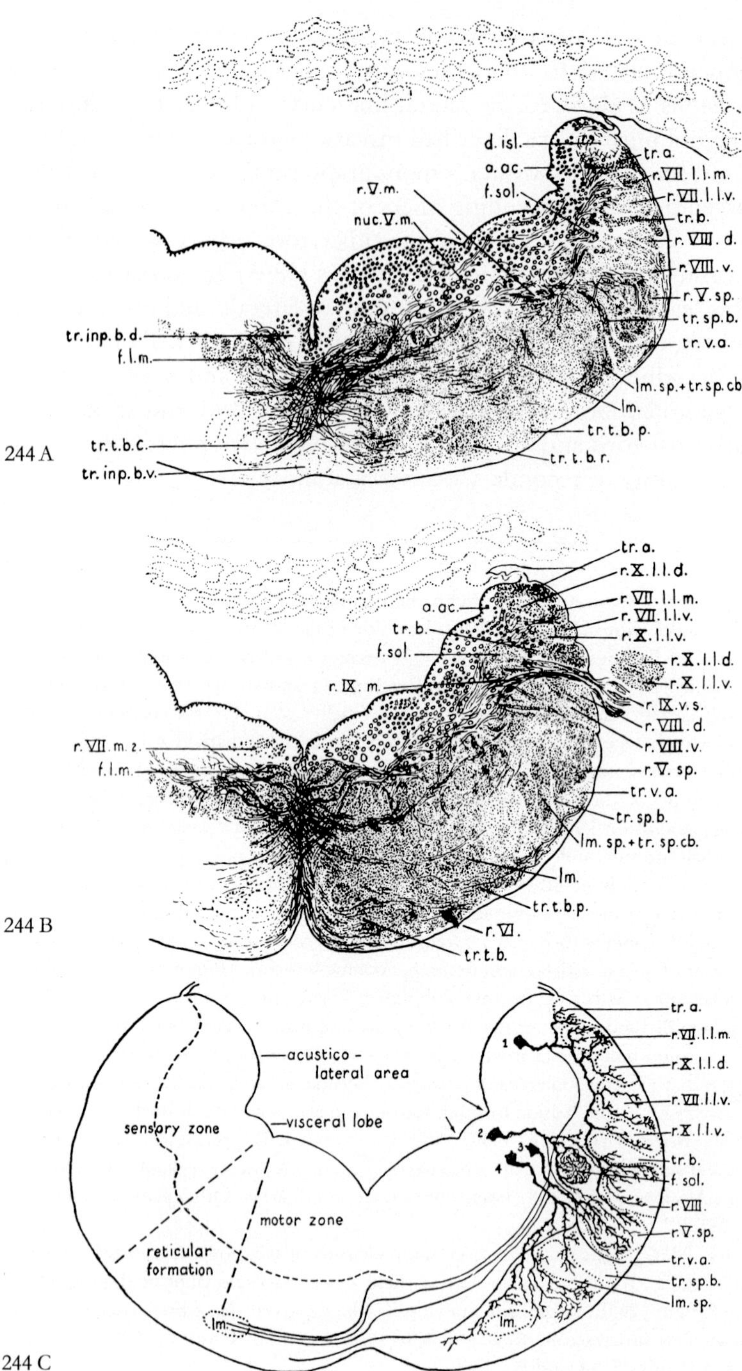

d. isl.
a. ac.
f. sol.
r. V. m.
nuc. V. m.
tr. a.
r. VII. l. l. m.
r. VII. l. l. v.
tr. b.
r. VIII. d.
r. VIII. v.
r. V. sp.
tr. sp. b.
tr. v. a.
lm. sp. + tr. sp. cb.
lm.
tr. t. b. p.
tr. t. b. r.
tr. inp. b. d.
f. l. m.
tr. t. b. c.
tr. inp. b. v.
244 A
tr. a.
r. X. l. l. d.
r. VII. l. l. m.
r. VII. l. l. v.
r. X. l. l. v.
a. ac.
tr. b.
f. sol.
r. IX. m.
r. X. l. l. d.
r. X. l. l. v.
r. IX. v. s.
r. VIII. d.
r. VIII. v.
r. V. sp.
tr. v. a.
tr. sp. b.
lm. sp. + tr. sp. cb.
lm.
tr. t. b. p.
r. VI.
tr. t. b.
r. VII. m. 2.
f. l. m.
244 B
acustico -
lateral area
visceral lobe
sensory zone
motor zone
reticular
formation
lm.
lm.
1
2
3
4
tr. a.
r. VII. l. l. m.
r. X. l. l. d.
r. VII. l. l. v.
r. X. l. l. v.
tr. b.
f. sol.
r. VIII.
r. V. sp.
tr. v. a.
tr. sp. b.
lm. sp.
244 C

Generally speaking, the lateral line system of *Anuran* larvae is essentially similar to that of Urodeles. The disappearance of this system in adult forms becomes correlated with the disappearance of the crista cerebellaris which is characteristic for the oblongata of Fishes and aquatic Urodele Amphibians.

The nervus octavus *seu* vestibularis (VIII) of *Urodeles* corresponds, on the whole, rather closely to that of Fishes, but displays modifications in peripheral distribution related to the particular differentiations

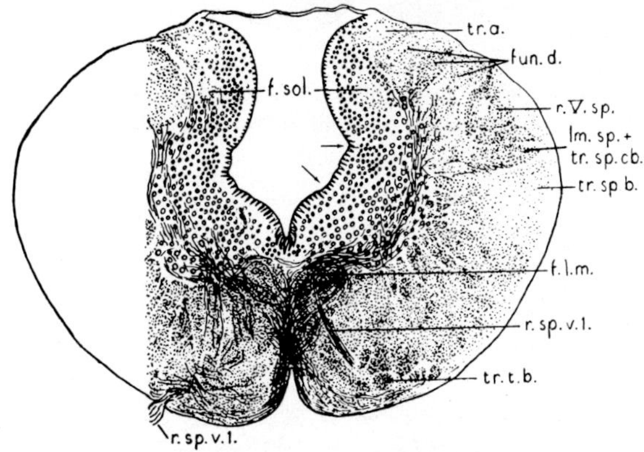

Figure 244D. Cross-section through the oblongata of Amblystoma tigrinum at a lower vagus level corresponding to exit of hypoglossal root (from HERRICK, 1948). fun. d.: rostral end of dorsal funiculus of spinal cord; r. sp. v. 1: rootlet of nervus hypoglossus.

Figure 244A. Cross-section through the oblongata of Amblystoma tigrinum at the level of the motor trigeminal nucleus (from HERRICK, 1948). Added arrows indicate sulcus limitans and intermedioventralis. tr.t.b.c.: tractus tecto-bulbaris cruciatus; tr.t.b.p.: tractus tecto-bulbaris posterior; tr.t.b.r.: tractus tecto-bulbaris rectus. Other abbreviations as in Figures 243A–C.

Figure 244B. Cross-section through the oblongata of Amblystoma tigrinum at the level of the glossopharyngeus root. An emerging rootlet of the abducens can be seen (from HERRICK, 1948).

Figure 244C. Diagrammatic drawing of a cross-section through the oblongata of an advanced larval Amblystoma tigrinum near the level of the glossopharyngeus roots, showing four types of neurons in the 'sensory zone' (from HERRICK, 1948). Added arrows indicate sulcus intermedius dorsalis and sulcus limitans. Neuron 1 in synaptic connection with all afferent cranial nerve components; neuron 2 mainly connected with visceral-afferent tractus solitarius, and less extensively with VIII and V fibers; neuron 3 related to V fibers; neuron 4 related to V fibers and formatio reticularis. Elements of type 1–4 remain present in adult Urodeles.

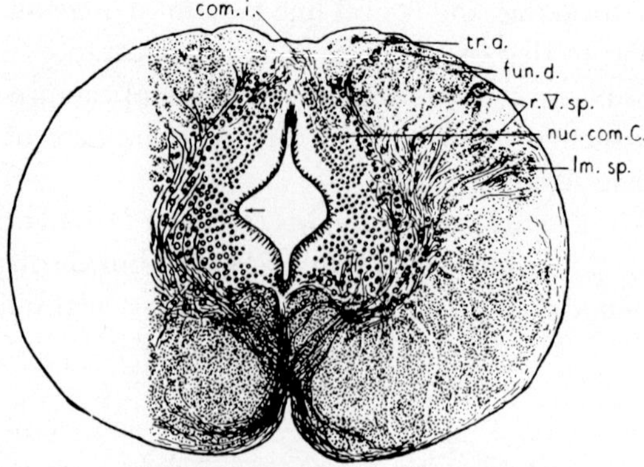

Figure 244 E. Cross-section through the oblongata of Amblystoma tigrinum immediately caudally to calamus scriptorius, showing *commissura infima Halleri* (Herrick, 1948). nuc. com. C.: *nucleus commissuralis of Cajal.*

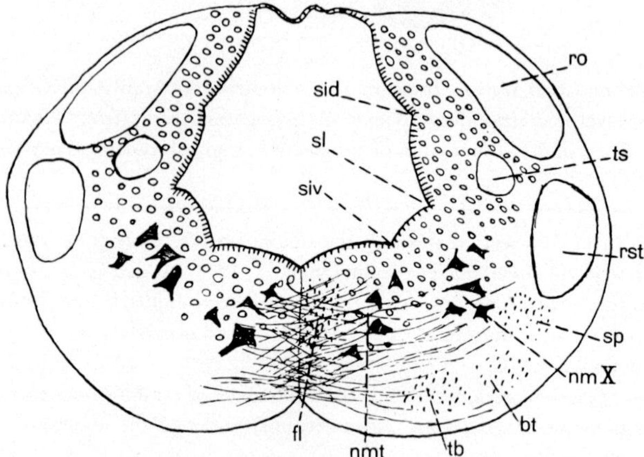

Figure 245 A. Simplified drawing of cross-section through the oblongata of Triton at the level of efferent vagus nucleus (from K., 1927). bt: tractus bulbo-tectalis (bulbar lemniscus); fl: fasciculus longitudinalis medialis; nmX: nucleus motorius vagi; nmt: nucleus motorius tegmenti; ro: radix descendens nervi octavi (et lateralis?); rst: radix descendens trigemini; sid: sulcus intermedius dorsalis; siv: sulcus intermedius ventralis; sl: sulcus limitans; sp: tractus spino-tectalis? (spinal lemniscus); tb: tractus tecto-bulbaris; ts: tractus solitarius.

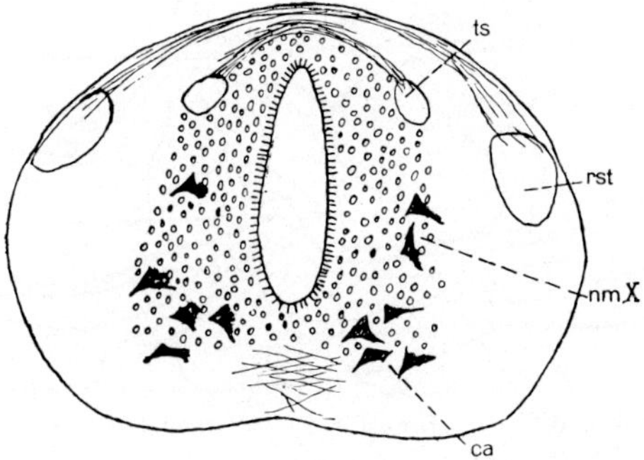

Figure 245 B. Simplified drawing of cross-section through the oblongata of Triton at the level of *commissura infima Halleri* (from K., 1927). ca: cornu anterius medullae spinalis (nucleus nervi hypoglossi). Other abbreviations as in Figure 245A.

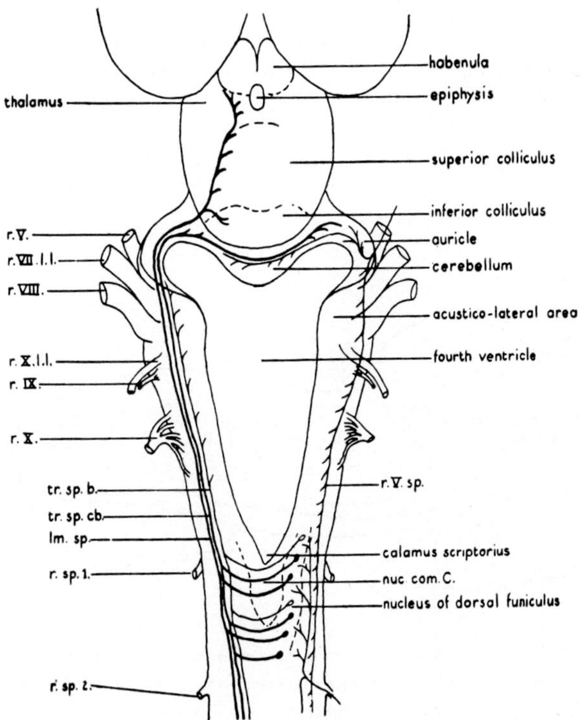

Figure 246 A. Outline drawing of brain stem of Amblystoma tigrinum in dorsal view, indicating the course of spinal and bulbar lemnisci, including spino-cerebellar tract (from Herrick, 1948). Abbreviation as in Figures 243 and 244.

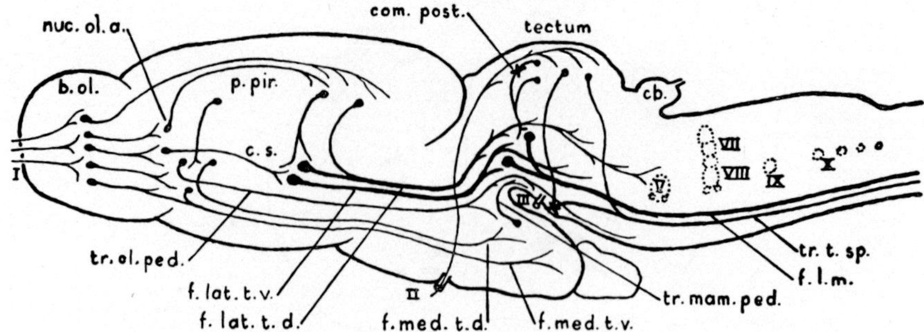

Figure 246 B. Outline drawing of brain of Amblystoma tigrinum in lateral view, indicating course of various main descending pathways relevant to oblongata and spinal cord (from HERRICK, 1948). b.ol.: bulbus olfactorius; cb: cerebellum; com.post.: commissura posterior; c.s.: so-called 'corpus striatum' (nuclei basimediales); f.lat.t.d.: lateral forebrain bundle, dorsal fascicles; f.lat.t.v.: ventral fascicles of lateral forebrain bundle; f.l.m.: fasciculus longitudinalis medialis; f.med.t.d.: medial forebrain bundle, dorsal fascicles; f.med.t.v.: medial forebrain bundle, ventral fascicles; nuc.ol.a.: 'nucleus olfactorius anterior'; p.pir.: so-called 'primordium piriforme' (pallial zone D_1); tr.mam. ped.: tractus mammillo-peduncularis; tr.ol.ped.: tractus olfacto-peduncularis; tr.t.sp.: tractus tecto-spinalis.

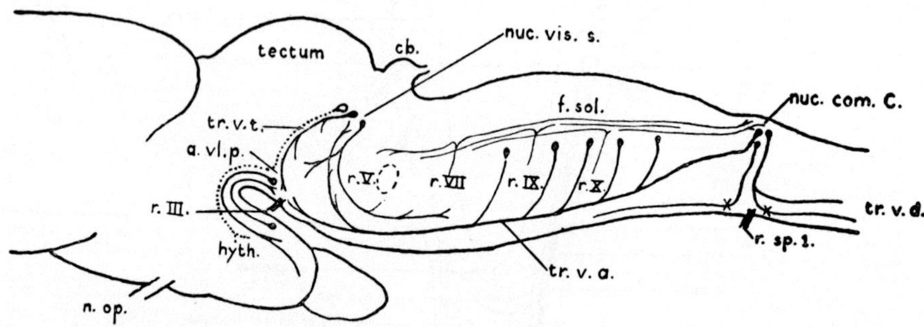

Figure 246 C. Outline drawing of brain stem of Amblystoma tigrinum in lateral view, indicating connections of 'visceral-gustatory system' (from HERRICK, 1948). a.vl.p.: 'area ventrolateralis pedunculi' (region of mesencephalic basal plate); cb: cerebellum; f.sol.: fasciculus s. tractus solitarius; hyth.: hypothalamus; n.op.: nervus opticus; nuc.com.c.: *nucleus commissurae infimae Halleri of Cajal;* nuc.vis.s.: nucleus visceralis secundarius; r: radix; r.sp.l.: radix nervi hypoglossi; tr.v.a.: tractus visceralis secundarius ascendens; tr.v.d.: tractus visceralis descendens; tr.v.t.: tertiary visceral tract.

of the Amphibian labyrinth briefly pointed out above in section 1, p. 311. The anterior octavus branch is said to supply the anterior and lateral ampullae and the macula utriculi. The posterior branch supplies the posterior ampulla, the macula sacculi (as in Selachians, but with some species variations), the papilla lagenae, the papilla basilaris, and the so-called papilla amphibiorum.[129] The posterior branch of the octavus or vestibular can thus be regarded as including the primordium of the nervus cochlearis in higher Vertebrate forms.

With regard to the Amphibian octavus system, the extensive development of the *saccus endolymphaticus* is of particular interest. This structure arises from the sacculus of the membranous labyrinth and grows out into the cranial cavity where its location can best be described as intrameningeal. The sacs may be relatively short in some Urodeles, greatly expand in others, and even become joined by midline fusion in Amblystoma. In some Anurans, the fused endolymphatic sacs form a ring around the brain, continuing caudalward into the vertebral canal at least as far as the seventh vertebra. The sac commonly displays numerous diverticula, and seems to contain a calcareous fluid. Nothing certain is known about the functional significance of this structure, whose comparative morphology together with other pertinent problems has been discussed in a detailed investigation by Dempster (1930). In Selachians, an endolymphatic duct opens directly on the body surface (cf. above, section 1, p. 309). In some Ganoids, the duct may barely reach the cranial cavity, but does not extend as far as this latter in Teleosts. In Dipnoans, on the other hand, such intracranial extension occurs, particularly in Protopterus, where numerous diverticula are present (cf. Figs. 236, 237), which may contain otolith crystals. The similarity of the endolymphatic sac in Protopterus and in some Amphibians has been interpreted as indicative of a phylogenetic relationship (Whiteside, 1922).

---

[129] The Amphibian utriculus is equipped with a macula utriculi and a smaller 'macula neglecta'. De Burlet (1928, 1929) demonstrated the 'auditory character' of this macula, covered with a tectorial membrane, and now generally called '*papilla amphibiorum*', differing from an otolith carrying '*macula s. papilla neglecta*'. A 'true' papilla neglecta seems to be lacking in Urodele and Anuran Amphibians but is reported to be present in at least some Gymnophiona (Ichthyopis), where both papilla neglecta and papilla amphibiorum are said to be present. The papilla basilaris, which can be regarded as the primordium of the cochlea, is absent in various Urodeles (Proteus, Menobranchus, Amphiuma, Necturus) and, moreover, rather small in most Amphibians. As regards Gymnophiona, cf. Sarasin and Sarasin (1892).

In *Anurans*, following metamorphosis, and disappearance of the lateralis system,[130] the alar plate region corresponding to the acustico-lateral area receives only root fibers of the rostral or ventral, and of caudal or dorsal branches of the VIIIth nerve. The nuclei in this area become transformed into a predominantly small celled dorsal subdivision and into a more ventral magnocellular subdivision of the dorsal acoustic nucleus. The former subdivision seems to be a transformation of the nucleus dorsalis nervi lineae lateralis anterioris and the ventral subdivision appears to represent a transformation of the intermediate octavo-lateralis nucleus. The ventral octavus area, as in Urodeles and Fishes, includes *Deiters' nucleus* (cf. Figs. 247B, 248A, B) with perhaps additional ill-defined differentiations. It should be added that *Deiters' nucleus* of BECCARI, respectively nucleus ventralis nervi octavi of my '*Vorlesungen*' (K., 1927) correspond to the griseum called *noyau ventral magnocellulaire* by KAPPERS (1947). Again, a clearly recognizable nucleus tangentialis does not seem to be present in Amphibians, although some lateral cell groups of nucleus ventralis nervi octavi might correspond to a nondescript 'tangential' component.

Dorsal (internal) and ventral (external) arcuate fibers originating from the dorsal acoustic nucleus join the lemniscus lateralis and also become related to a cell group located in the tegmental basal plate representing a primordial *superior olivary nucleus* generally recognized as present in most or all Anurans. It also has been identified in *Gymnophiona* (K., 1922) as shown in Figure 249A. The superior olivary nucleus, considered to be a new feature in the Anuran oblongata, seems to develop in connection with the transformation of part of the central lateral line apparatus into an acoustic system (LARSELL, 1934, 1967).

The ventral and ventrolateral part of the alar plate's *dorsal zone* contains the afferent grisea of the fifth nerve, namely a rostral main sensory or superior trigeminal nucleus (Figs. 243A, 247A) and more caudalward, extending as far as the *übergangsgebiet*, the radix descendens trigemini with its related cell groups and fiber systems (Figs. 244C, 245A, B, 248B, 249A, B). As in other Vertebrates, the radix descen-

---

[130] Although the transformation of the aquatic acustico-lateralis system into a terrestrial cochlearis system seems rather convincingly established, some doubts remain as expressed by BECCARI (1943): '*Ma la cosa non risulta ancora definitivamente chiarita e resta quindi ancora da stabilire se, como hanno supposto il* KAPPERS *ed il* RÖTHIG, *le vie centrali dell'area acustico-laterale degli Ittiopsidi diventino vie cocleari negli Anfibi terrestri e negli Amnioti, o se piuttosto si tratti di due sistemi differenti*'. Cf. also footnote 135a, p. 464.

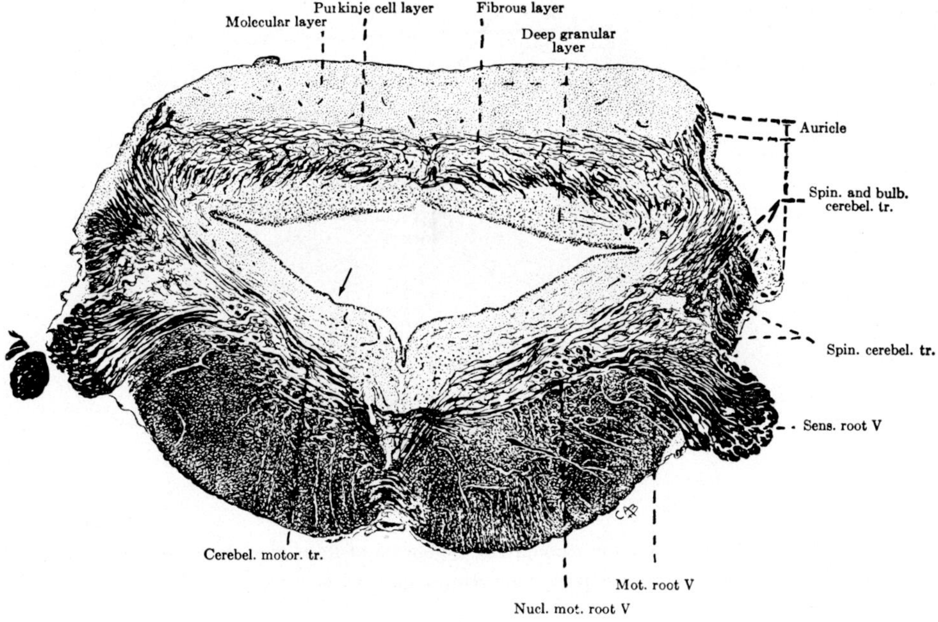

Figure 247A. Cross-section through the oblongata of Rana mugiens at the level of the motor trigeminal nucleus and corpus cerebelli (from KAPPERS *et al.*, 1936). Added arrow points to approximate locus of poorly manifested sulcus limitans.

dens trigemini is joined by cutaneous (somatic-afferent) root fibers pertaining to other branchial nerves, especially to the vagus.

The neighborhood occupied by the main sensory nucleus may somewhat protrude into the ventricle, forming an eminentia trigemini. The mesencephalic nucleus of the trigeminus, to be dealt with in chapter XI, is present in all three orders of Amphibia.[131]

As regards the caudal end of the oblongata and the region of transition to the spinal cord, it does not seem possible to recognize definable cell groups of the dorsal zone which could represent *funicular nuclei.* These latter, according to KAPPERS and HAMMER (1918), might, however, be suggested in Anurans by medio-dorsal expansions (*'unregelmässige Fortsätze'*) of the spinal cord's posterior gray horns in the above-mentioned region of transition.

[131] The cells of the *radix mesencephalica trigemini* in Gymnophiona escaped my attention as a novice more than fifty years ago, and were not mentioned in my paper of 1922. However, I believe to have identified these elements in the mesencephalic tectum of my present, still incomplete Gymnophione material.

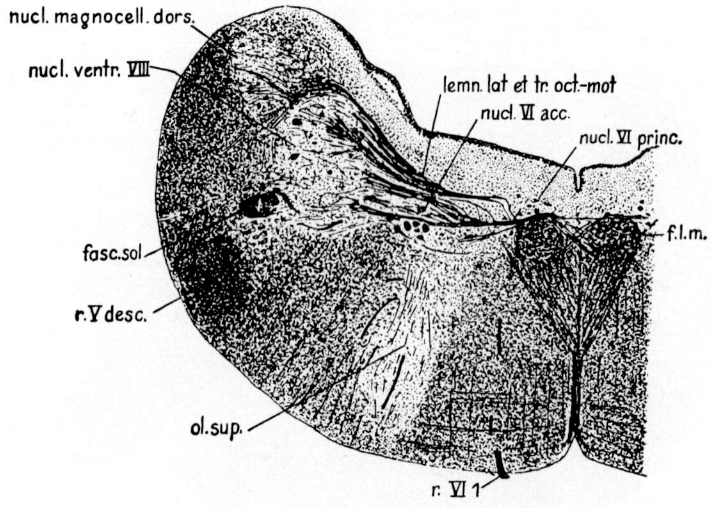

Figure 247 B. Cross-section through the oblongata of Rana mugiens at the level of the superior olivary nucleus as seen in a myelin stain (from KAPPERS *et al.*, 1936).

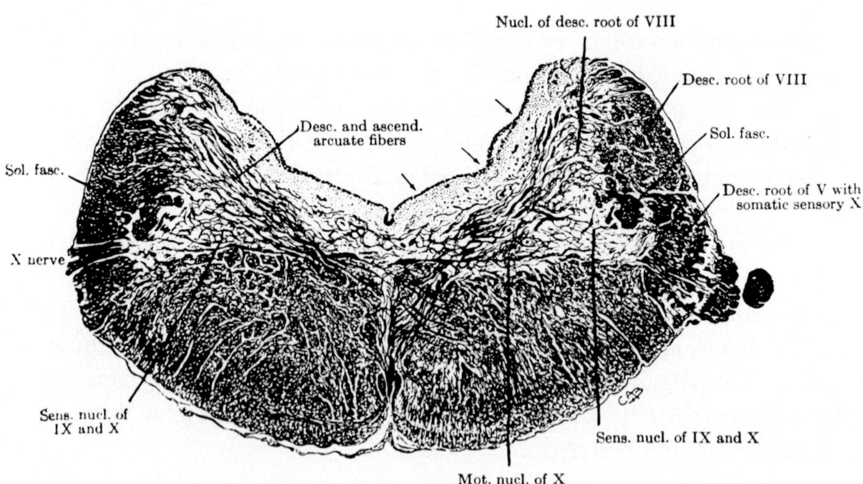

Figure 247 C. Cross-section through the oblongata of Rana mugiens at a rostral vagus level (from KAPPERS *et al.*, 1936). Added arrows indicate sulcus intermedius dorsalis, sulcus limitans, and locus of poorly manifested sulcus intermedius ventralis.

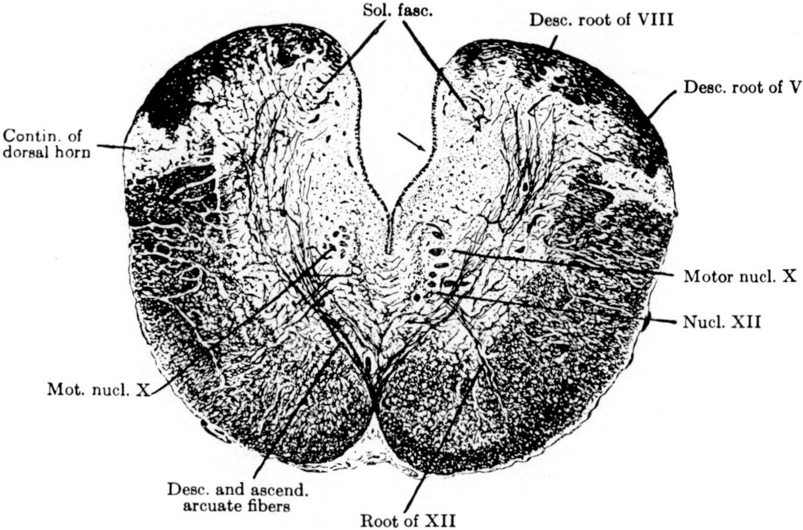

Figure 247D. Cross-section through the oblongata of Rana mugiens at a lower vagus level including the hypoglossal nucleus (from KAPPERS *et al.*, 1936). Contin. of dorsal horn: funicular nuclei. Added arrow indicates sulcus limitans.

The *intermediodorsal zone* of the alar plate contains, in conformity with that of other Anamnia respectively Vertebrates in general, the grisea related to the 'visceral afferent' root fibers of the branchial nerves VII, IX, and X. In Urodeles, according to HERRICK (1944), the facialis fibers are chiefly gustatory, but those of glossopharyngeus and vagus seem to be essentially 'general visceral afferent'. Bifurcating ascending and descending root fibers of these nerves form a rather well-defined *tractus solitarius* sive *fasciculus communis*,[132] which presumably also includes secondary fibers. In the region of the calamus scriptorius, this system is related to the commissura infima (Figs. 244E, 245B) and to a dorsal cell group representing the *commissural nucleus of Cajal*. Both the commissure and the correlated dorsal griseum seem to include a 'somatic afferent' (trigeminal) and a 'visceral afferent' *(communis*, VII, IX, X) component.

In the *basal plate* of the Amphibian oblongata, the efferent nuclei of the branchial nerves, located in the intermedioventral zone, display

[132] At least with respect to Urodeles, the majority of X root fibers, however, are said to descend without bifurcating.

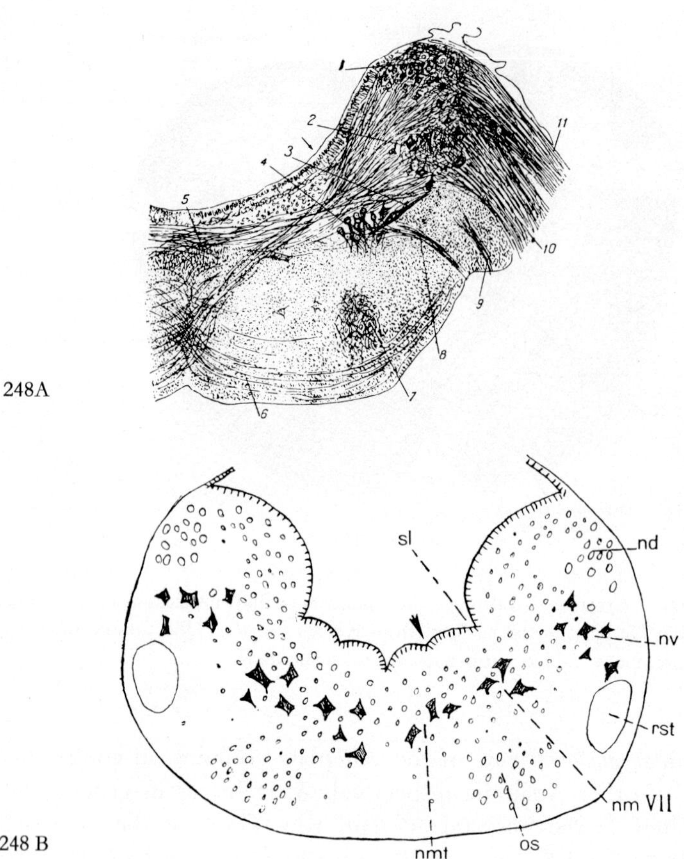

Figure 248 A. Cross-section through the oblongata of Rana esculenta at the level of upper olive and motor facial nucleus, as seen by means of a modified *Cajal silver impregnation* (from BECCARI, 1943). 1: dorsal (magnocellular and 'parvocellular') octavus nucleus; 2: *Deiters' nucleus;* 3: large nerve cell considered by BECCARI to be a questionable '*Mauthner cell*'; 4: efferent facial nucleus; 5: fasciculus longitudinalis medialis; 6: fibers of 'primordial corpus trapezoideum'; 7: superior olivary nucleus; 8: efferent root of facial; 9: afferent root of facial; 10: anterior trunk of nervus octavus; 11: posterior trunk of octavus. Added arrow indicates approximate locus of sulcus limitans.

Figure 248 B. Simplified drawing of cross-section through the oblongata of a Frog (Ranid in general) at the level of superior olive (from K., 1927). nd: nucleus dorsalis nervi octavi (the lead points to the 'magnocellular subdivision, the 'parvocellular' one being shown dorsal to that cell group); nm VII: nucleus motorius nervi facialis; nmt: nucleus motorius tegmenti; nv: nucleus ventralis nervi octavi *('Deiters' nucleus');* os: oliva superior; rst: radix descendens nervi trigemini; sl: sulcus limitans. Added arrow indicates sulcus intermedius ventralis.

some difference with respect to their arrangement in Urodeles and Anurans. In the former, as in Plagiostomes, the nucleus of the facial nerve, separated from the motor nucleus of trigeminus, is continuous with the nuclear column of glossopharyngeus and vagus. In Anurans, the facial nucleus is separated from that of the IXth and situated more rostrally, rather close behind the motor trigeminal nucleus (cf. Fig. 159).

In addition to fairly large motoneurons innervating striated musculature, the efferent grisea of VII, IX, and X contain smaller nerve cells, presumed to be preganglionics, but these elements are not distinctly aggregated in definable autonomic nuclei separated from the column of the branchiomotor nuclear grisea. The musculus trapezius of Anurans is supplied by caudal efferent fibers of the vagus gathering in a branch which can be considered to be the *accessorius vagi* or XIth nerve.

Within the ventral zone, the nucleus of the abducens is located rather close to the fasciculus longitudinalis medialis, lying rostrally to nucleus facialis in Urodeles, but caudally to that nucleus in Anurans. In both instances, however, the root fibers of the abducens emerge caudally to those of the facial. Several minor variations occur in the position and differentiation of the abducens nucleus,[133] which may be situated rather close to the midline and is even believed to be, at least partially, a floor plate derivative in some Urodeles (cf. the discussion by Herrick, 1948).

In the caudal part of the oblongata, a rostral extension of the spinal cord's 'somatic-motor'ventral horn is located ventrally to the end of the vagus 'visceromotor' column and gives origin to ventral roots innervating the tongue musculature. These roots can be evaluated as representing a 'true' *hypoglossal* or XIIth nerve. It will be recalled that the tongue of Cyclostomes is a muscular piston innervated by the trigeminus and quite unrelated to the hypobranchial musculature, being

---

[133] I was, so far, unable to locate the abducens nucleus in my Gymnophione material, nor could I identify a nervus abducens in the 1922 paper on the Caecilian brain, although this nerve may be present in at least some Gymnophiones (cf. K. 1927, p. 146, footnote). Because of the difficulties in obtaining properly preserved Gymnophione material, especially with regard to the brain stem, details concerning the oblongata in this Amphibian order are much less well known than those obtaining in Anurans and particularly in Urodeles. Again, my own studies on the Gymnophione brain concerned mainly the prosencephalon. Additional incidental findings related to the brain stem were merely included because of the scarcity of reports on that peculiar and highly interesting 'aberrant' Amphibian brain.

in no way homologous with the tongue of Gnathostome Vertebrates (cf. above, section 3, p. 362, footnote 75). Gnathostome fishes possess an immovable tongue, forming a swelling in the floor of the mouth, and supported by the basihyal. Despite its lack of musculature, this fish tongue is a configuration kathomologous with the tetrapod tongue and corresponds to the root of this latter. *Amphibia* possess a *motile tongue*, whose body originated from the mandibular arch and whose musculature is of hypobranchial derivation. The tongue of Amphibia exhibits numerous variations as regards its freedom of movement, being capable of a mushroom-like extensive forward projection *('boletoid tongue')* in some Urodeles. In certain Anurans, it is more freely

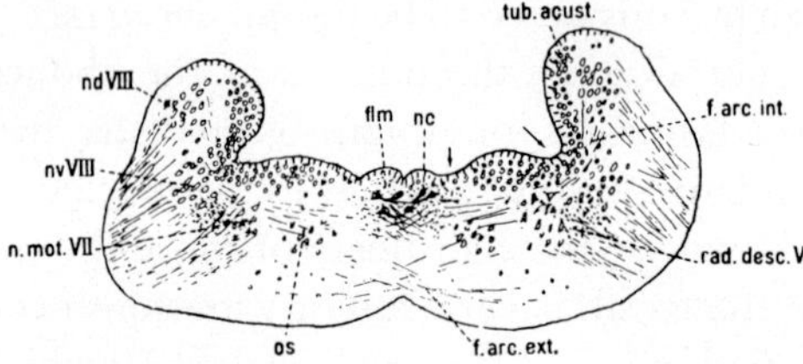

Figure 249A. Simplified drawing of cross-section through the oblongata of the Gymnophione Siphonops annulatus (from K., 1922). f. arc. ext., int.: external and internal arcuate fibers; flm: fasciculus longitudinalis medialis; nc: nucleus centralis; nd VIII: nucleus dorsalis octavi; nv: nucleus ventralis octavi *('Deiters' nucleus')*; os: oliva superior; tub. acust.: 'tuberculum acusticum (part of n. dorsalis octavi). Added arrows indicate sulcus limitans and sulcus intermedius ventralis.

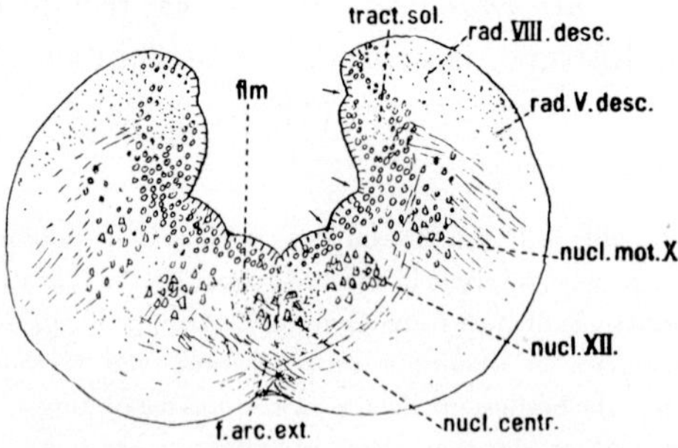

Figure 249B. Simplified drawing of cross-section through the oblongata of Siphonops at the level of nucleus hypoglossi and efferent vagus nucleus (from K., 1922). tract. sol.: fasciculus s. tractus solitarius. Added arrows indicate sulcus intermedius dorsalis, sulcus limitans, and sulcus intermedius ventralis.

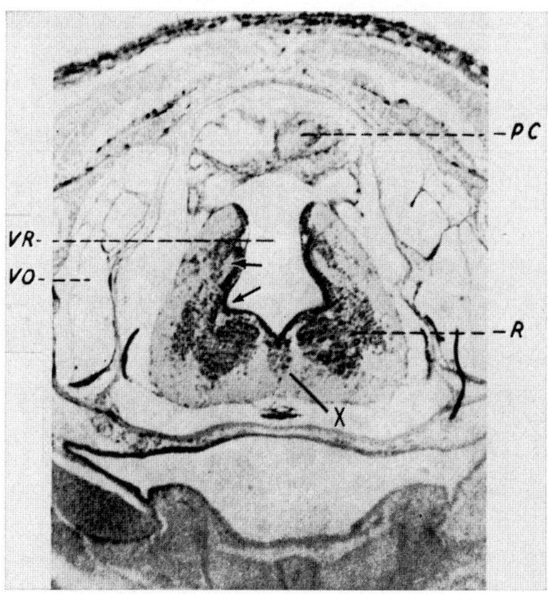

Figure 249 C. Cross-section through the oblongata of an advanced larva (55 mm) of the Gymnophione Ichthyophis (from KRABBE, 1962). PC: 'choroid plexus'; R: basal plate grisea; VO: 'otic vesicle'; VR: fourth ventricle. X: nucleus centralis (not identified by the cited author); upper arrow: tractus solitarius below sulcus intermediodorsalis; lower arrow: sulcus limitans (X and arrows are added designations). Lead PC actually ends in saccus endolymphaticus.

movable than in others. The hypoglossal nerve of Urodeles corresponds presumably to the first occipito-spinal nerve rootlets of Fishes.[134] The Anuran hypoglossus apparently represents a second spino-occipital root (*'nervus secundus'*), since a more rostral root, displayed by tadpoles, disappears in the course of metamorphosis. The caudal part of the oblongata, from which the XIIth nerve arises, is, as a rule, not included within the cranial cavity.

The *overall basal plate* of the Amphibian oblongata contains a reticular *nucleus motorius tegmenti* comparable to that found in Fishes, but much less clearly subdivided into distinctive subgroups or 'reticular nuclei'. Thus, HERRICK (1930) stated that, in Urodeles such as Necturus (Fig. 243C), the neurons of the reticular formation do not form a

---

[134] It should be kept in mind that, as BECCARI (1943) justly emphasizes '*il numero dei nervi spinali assimilati della testa è molto variabile nelle differenti specie di Pesci; e varia anche notevolmente, a sviluppo ultimato, la loro repartizione in nervi occipitali ed occipito-spinali*'.

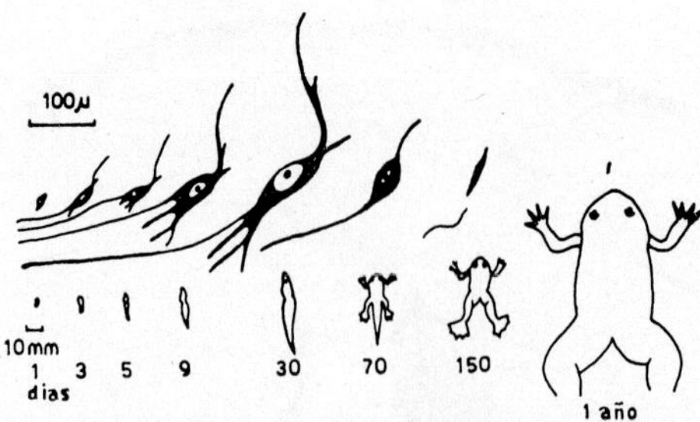

Figure 250 A. Drawing indicating 'evolution and involution' of the *Mauthner cell* in an Anuran Amphibian (Xenopus). The maximal development of the cell is shown to coincide with the maximal development of the tail (from STEFANELLI, 1962).

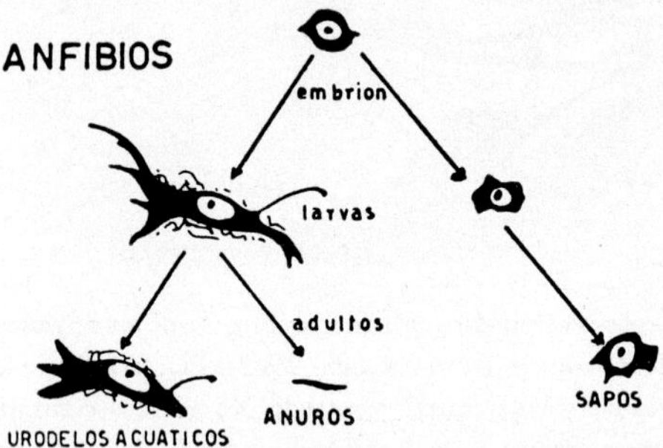

Figure 250 B. Diagram indicating evolution of the *Mauthner cell* in Amphibia (from STEFANELLI, 1962). The designation '*sapos*' means toads.

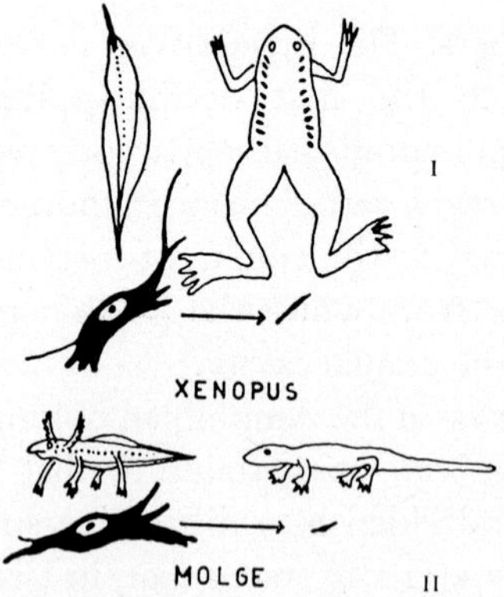

Figure 250C. Comparison of the *Mauthner cell's* evolution in the aquatic Anuran Xenopus (I) and in the terrestrial Urodele Geotriton Molge (II). Despite persistence of lateral line, the *Mauthner cell* atrophies in the aquatic Xenopus concomitantly with disappearance of the tail. In the terrestrial Molge, the *Mauthner cell* atrophies concomitantly with disappearance of lateral line despite persistence of the Tail (from STEFANELLI, 1962).

definite nucleus. They may send their dendrites laterally into the 'sensory field' or medially into the 'motor field'. In the oblongata of Gymnophiones, however, a well defined 'reticular' *nucleus centralis*, consisting of rather large multipolar cells related to the fasciculus longitudinalis medialis, can be found on both sides of the median raphé (Figs. 249A, B, C, 330G). This nucleus, identified by the author in 1922, extends along said fasciculus throughout the entire length of the *rhombencephalon* and seems to be a characteristic feature of the Gymnophione brain (K., 1922, 1969). It represents here the most conspicuous griseum of the entire formatio reticularis tegmenti, which, moreover, is rather poorly developed in the Caecilian *mesencephalon* (cf. chapter XI, section 6).

At the level of nucleus vestibularis, a *Mauthner cell* is present in adult aquatic Urodeles[135] as well as in Anuran larvae (Figs. 250A, B, C). In the aquatic Anuran Xenopus, the lateral line system persists, but the *Mauthner cell* becomes atrophic concomitantly with the loss of the larval tail. The location, within the basal plate, of a *superior olivary nucleus* pertaining to the cochlear system of Anurans and Gymnophiona was mentioned above. The decussating arciform fibers related to this griseum and to the lateral lemniscus in these Amphibians can be evaluated as a primordial *trapezoid body* comparable to that of Mammals (Beccari, 1943). On the other hand, an *inferior olivary nucleus*, present in the oblongata of Fishes, cannot be convincingly identified in Amphibians. The lack of a definable griseum representing that nucleus may be related to the low degree of cerebellar development characteristic for Amphibia.

As regards the main distinctively organized fiber systems arising from, ending in, or passing through the Amphibian oblongata, these pathways essentially correspond, *mutatis mutandis*, to those discussed above in dealing with the oblongata of Fishes.

Among publications dealing with the hodology of the Amphibian oblongata in addition to Herrick's contributions, the investigations by Benedetti (1933), Kreht (1930, 1931) and by Röthig (1927) de-

---

[135] It has disappeared in adult terrestrial Salamanders and Geotritons. As regards Gymnophiona, a *Mauthner cell* seems to be present at larval stages in at least some, but perhaps also all forms (Tagliani, 1906). In the adult specimens which I examined, a *Mauthner cell pair* could not be identified. This, of course, would agree with Stefanelli's (1962) interpretation relating this cell to swimming activities involving substantial participation of tail motions.

serve special mention. Figures 246A and B illustrate the arrangement of certain main fiber tracts in Urodeles as mapped by HERRICK (1948), and inspection of Figures 242A to 244E will disclose the meticulous identification, by the cited author, of numerous pathways at different levels of the Urodele oblongata. Although comparable details are not available for Anurans, a rather similar distribution of the relevant fiber systems can be assumed. With respect to Gymnophiona, the location and presumptive course of only a few main neural channels appears reasonably certain.

In concluding the discussion of the Amphibian[135a] oblongata, it is perhaps pertinent to emphasize the following particularly 'primitive' and 'generalized' histologic features of the Urodele brain respectively oblongata as repeatedly pointed out by HERRICK, subsequently summarized by LARSELL (1967), and well illustrated in Figure 244C. The inner gray layer is continuous, rostro-caudally, from the isthmus to the spinal cord, and, transversely, from the attachment of lamina epithelialis on one side of the fourth ventricle to the other. The outer white layer is likewise uninterrupted, but displays, in the transitional zone between alar and basal plate, a lighter zone in myelin-stained sections, due to the presence of fewer and smaller medullated fibers than in the more dorsal or ventromedial neighborhoods.

The cell bodies of nearly all neurons are located rather close to the ependymal lining in a deep layer of periventricular gray. Their dendrites project into a superficial layer of fibers constituting both a 'white matter' and synaptic fields of neuropil. The axons of the nerve cells commonly arise from a large 'stem dendrite' or occasionally from the cell body. In the synaptic fields, which display an arrangement in more

---

[135a] In contradistinction to the central neuraxis, however, ontogenetic remnants of the lateralis system seem to be manifested in Amniota by components of the transitory dorsolateral placodes appearing during early ontogenetic stages in all Vertebrates (cf. chapter VI, section 3, p. 844 of vol. 3, part II). Both epibranchial and dorsolateral placodes are somehow related to the morphogenesis of cranial nerves. At least some of these evanescent structures represented by the dorsolateral placodes, as described in the Reptilian Gecko by EVANS (1935) as well as in the Mammalian guinea pig by DA COSTA (1923) have been specifically interpreted as being related to the cranial lateralis system of Anamnia (cf. EVANS, l. c. p. 389). Other Vertebrate placodes are olfactory and otic placodes. Because of the 'blurred' aspect of many early ontogenetic processes, there is still considerable divergence of opinion with regard to the origin of the extracerebral octavus ganglia from either placodes (including wall of otic vesicle) or neural crest, or both (cf. KAPPERS, 1947, pp. 63 and 434).

or less distinctive longitudinal zones, terminals of long and short fibers interweave with dendritic branches.

The long axons, both medullated and non-medullated, collect in fascicles or tracts connecting cells in various regions with neuropil fields of more or less distant parts of the neuraxis. Some of these homolateral or contralateral tracts can easily be compared with the main neuronal pathways generally present in the Vertebrate central nervous system.

In *Urodeles*, fiber connections of various neuropil fields correspond to those of particular nuclei in more differentiated Vertebrate brains, but these fields, as a rule, do not contain the nerve cell bodies. Thus, an arrangement quite similar to that obtaining in the central nervous system of many Invertebrates is here displayed (cf. vol. 2, p. 250–251).

Well-defined nuclei are therefore few in the grisea of Urodeles, but precursors of many nuclei of 'higher' Vertebrates are represented by 'locally specialized' elements or by connections of fiber fascicles recognizable as corresponding to specific fiber tracts in the more differentiated Vertebrate brains. In *Anurans*, however, the localized synaptic 'neuropil fields' show all stage in the differentiation of true '*nuclei*' by migration of cells from the periventricular gray into the surrounding 'white substance' (HERRICK 1948, LARSELL, 1967).

As can be seen in Figure 244C, HERRICK (1948) distinguished three bilateral longitudinal zones of the oblongata, a dorsal 'sensory zone', a ventral 'motor zone', and an intermediate correlative and integrative or 'reticular' zone. All these include both gray and white substance. Quite obviously, HERRICK's 'sensory zone' corresponds to the alar plate, and his 'motor zone' to the basal plate. His 'reticular formation' or intermediate zone, however, is evidently a rather arbitrary postulate, since the 'correlative' nucleus motorius tegmenti extends throughout the width of the basal plate, as also clearly displayed by his own findings (cf. Figs. 243B and C).

HERRICK (1935), and under his direction ROOFE (1935) have also undertaken a very extensive and minute investigation of the endocranial blood vessels, the meninges, and the membranous parts of the brain in Amblystoma. Numerous details of their description of cerebral blood vessels are beyond the scope of the generalized presentation of this topic given in chapter VI, section 7 of the preceding volume.

As regards the urodele Amphibion oblongata, however, some of the findings by the cited authors seem pertinent in the aspect here under consideration. Figure 250D shows the extensive rhombencephalic

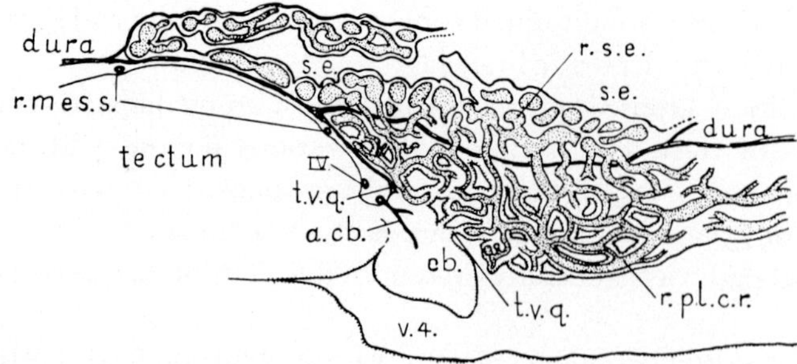

Figure 250D. Approximately sagittal section through the 'endolymphatic organ' and rhombencephalic chorioid plexus in Amblystoma tigrinum, illustrating the vascular connections between them. Combined from 3 *Golgi sections* inclined about 45° to the horizontal plane (about ×20, red. $^3/_4$). The outer wall of the endolymphatic sac is torn away caudally (from HERRICK, 1935). a.cb.: arteria cerebelli; cb.: cerebellum; r.mes.s.: (arterial) ramus mesencephali superior; r.p.l.c.r.: rete of rhombencephalic choroid plexus; r.s.e.: rete of endolymphatic sac; s.e.: saccus endolymphaticus; t.v.q.: taenia ventriculi quarti; v.r.: fourth ventricle; IV: decussation of nervus trochlearis. It should be added that what HERRICK designates here as '*dura*', represents the '*limitans arachnoideae*' in the terminology which I favor (cf. vol. 3, part II, chapter VI, section 7, Figs. 362, 363).

plexus overlaid by the endolymphatic sac (cf. section 1, p. 310, footnote 27; section 7, p. 453, also Figs. 236, 237 of the Dipnoan Protopterus). Both structures contain a complex rete of large venous sinusoids which also receives nearly all of the venous blood from the prosencephalon, including the cerebral hemispheres. All blood derived from the arterial capillary bed of the prosencephalic choroid plexuses and most of that from the telencephalic walls reach an intricate sinusoidal venous rete (termed '*nodus vasculosus*' by GAUPP, 1899) arising and located within opposite external surfaces of the velum transversum, namely caudal surface of paraphysis and rostral surface of (diencephalic, parencephalic) dorsal sac. This rete is the 'first unit' of the sinusoidal system. Its blood then passes caudalward through larger venous channels into the above-mentioned rhombencephalic rete, representing the 'second unit', which discharges into the jugular sinuses and jugular veins. This arrangement is, in effect, a double endocranial portal system of two units connected in series, which, as HERRICK (1935) remarks, 'presents some interesting physiological problems'.

8. Reptiles

Although the griseal configuration of the Reptilian oblongata and the arrangements of its fiber systems display numerous secondary differences, if representative Lacertilia, Ophidia, Chelonia, and Crocodilia are compared with each other, the dorsal zone components of the alar plate in all these forms can be regarded as generally more developed and structurally differentiated than those of the intermedioventral zone. Again, in the basal plate, the nucleus motorius tegmenti contains, on the whole, more large reticular cells than in lower forms, and its cell populations are more distinctly grouped than in Amphibians. The three morphologically significant chief ventricular sulci, namely, sulcus limitans, sulcus intermedius dorsalis, and sulcus intermedius ventralis, can be identified in the wall of the fourth ventricle of most Reptilian forms, particularly at some ontogenetic stages or at least in some regions of the adult oblongata. Figures 251A to 254B illustrate representative bulbar levels as displayed in four Reptilian orders.

Investigations dealing with bulbar neural structures of the Reptilian brain are, in particular, those by BECCARI (1912, 1914), BLACK (1920), DE LANGE (1916–1917), FREDERIKSE (1931), GILLASPY (1954), VAN HOEVELL (1911), TUGE (1932) and WESTON (1936). Again, various summaries including further bibliographic references are found in the previously cited relatively few extant texts on comparative neurology.

The *dorsal zone* of the *alar plate* in the Reptilian oblongata, if compared with the corresponding neighborhood in the Amphibian bulb, is characterized by the complete disappearance of the lateralis system, which is here not even recognizably indicated at any ontogenetic stage, and by a further differentiation of both cochlear and vestibular grisea.

The ramus posterior nervi octavi becomes related to three terminal nuclei (Fig. 251A), and includes the fibers supplying the lagena with its macula and papilla basilaris, which can be regarded as a primordial cochlea comparable to that of Mammals (cf. section 1, Figs. 163, 165C). The dorsomedially located *nucleus (cochlearis) dorsalis magnocellularis* corresponds to the cochlear griseum of Amphibians. The *nucleus angularis* is found in a dorsolateral position, near the angle formed by the attachment of lamina epithelialis to alar plate, or, in other words, near the so-called taenia. Ventrally to these two cochlear grisea, there is, moreover, a transversely arranged, plate-like aggregate of slightly smaller neuronal elements, the *nucleus laminaris*, which is particularly

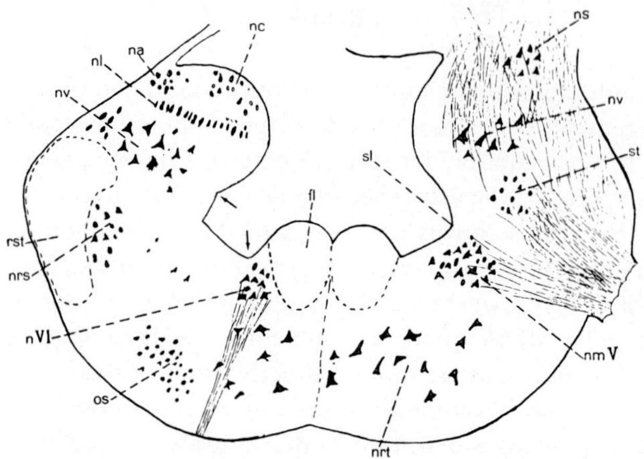

Figure 251 A. Simplified drawing representing a cross-section through the oblongata of a Lizard, showing, at right, the level of motor trigeminal nucleus, and, at left the level of cochlear grisea and of abducens nucleus (from K., 1927). fl: fasciculus longitudinalis medialis; n. VI: nucleus nervi abducentis; na: nucleus angularis; nc: nucleus magnocellularis dorsalis (cochlearis); nl: nucleus laminaris; nm V: nucleus motorius trigemini; nrs: nucleus radicis descendentis trigemini; nrt: nucleus reticularis tegmenti; na: nucleus vestibularis superior seu anterior; nv: nucleus vestibularis ventralis *(Deiters' nucleus,* the group of laterally adjacent cells on the figure's left side represents the nucleus tangentialis); os: oliva superior; sl: sulcus limitans; st: main afferent trigeminal nucleus. The added arrows indicate sulcus intermedius dorsalis at left, and sulcus intermedius ventralis at right.

conspicuous in Crocodilia, but can also be found in representatives of other orders, although, because of its lesser development, some authors did not identify that nucleus in all investigated Reptilian forms. In Lacerta, according to BECCARI (1943), '*e probabilmente nella maggior parte degli altri sauri, tutto il sistema cocleare centrale si presenta più semplice*'.

Secondary pathways originate as essentially decussating dorsal and ventral arcuate fibers (Figs. 251B, 253A, C) which join the lemniscus lateralis and also become related to the intercalated griseum of the *superior olivary complex* located in the basal plate, whose components are discussed further below.

In addition to the problems of identification caused by the variable degree of differentiation and distinct separation of the three cochlear cell aggregates in the diverse Reptilia, another still poorly elucidated problem concerns the undetermined amount of vestibular root fiber components which may end in the cochlear grisea. As regards the

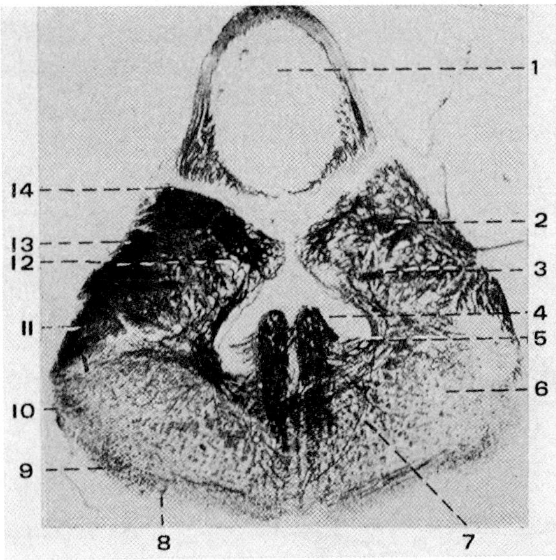

Figure 251 B. Cross-section through the rostral oblongata of Lacerta vivipara stained by the *Weigert-Pal method* (from FREDERIKSE, 1931). 1: cerebellum; 2: descending vestibular root; 3: afferent facial root (?); 4: fasciculus longitudinalis medialis; 5: efferent facial root (?); 6: superior olivary griseum; 7: arcuate fibers; 8: abducens root; 9: tractus vestibulo-spinalis lateralis; 10: tractus spino-tectalis et cerebellaris; 11: afferent trigeminus root; 12: cochlear grisea; 13: radix vestibularis; 14: radix cochlearis.

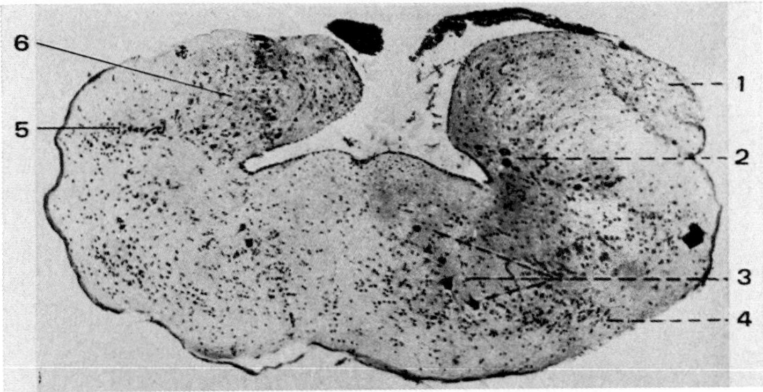

Figure 251 C. Cross-section through the oblongata of Lacerta at the level of *Deiters' nucleus, Nissl stain* (from FREDERIKSE, 1931). 1: radix vestibularis; 2: *Deiters' nucleus;* 3: nucleus reticularis tegmenti; 4: n. olivaris superior; 5: n. laminaris; 6: dorsal cochlear grisea. Leads 5 and 6 have been added. The small cell group above the 6 lead may be the nucleus angularis.

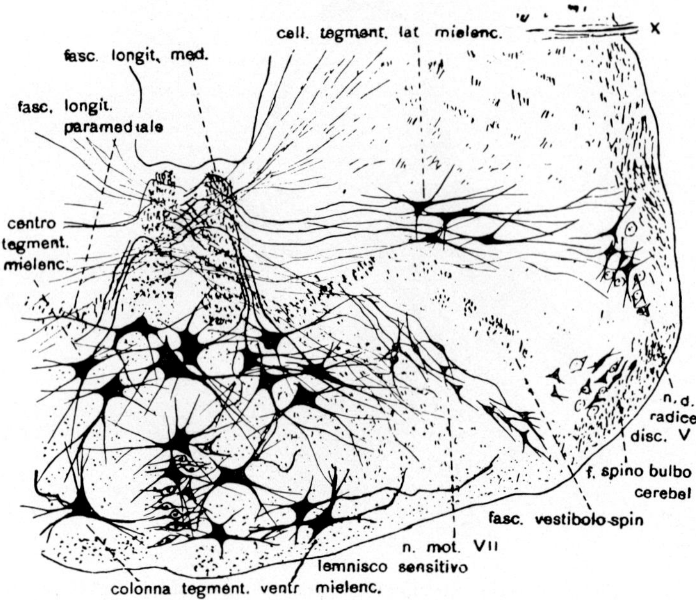

Figure 251 D. Cross-section through the oblongata in an advanced embryo of Lacerta muralis at the level of the motor facial nerve nucleus, and as seen by means of *Cajal's silver impregnation.* This section shows the large multipolar elements of the nucleus reticularis tegmenti (from BECCARI, 1943).

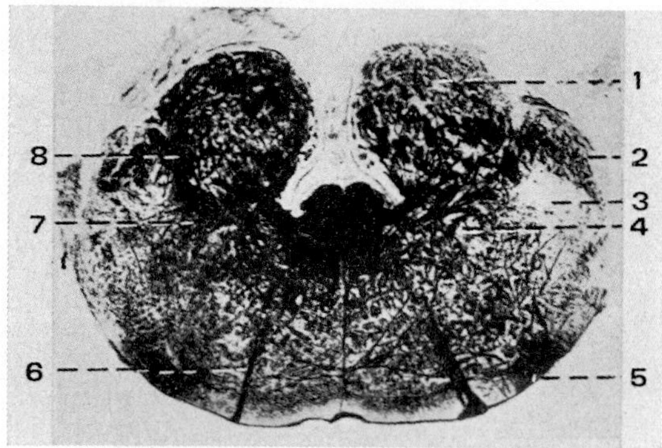

Figure 251 E. Cross-section *(Weigert-Pal stain)* through the oblongata of Lacerta vivipara at the level of the rostral extremity of hypoglossal nucleus (from FREDERIKSE, 1931). 1: radix descendens vestibularis and nucleus; 2, 3: radix descendens trigemini and its griseum (3); 4: efferent nucleus of vagus; 5: tractus spino-tectalis et cerebellaris, perhaps including lateral vestibulo-spinal tract; 6: radix hypoglossi; 7: 'tractus vestibularis descendens medialis'; 8: tractus solitarius (lead is too short, cf. Fig. 252).

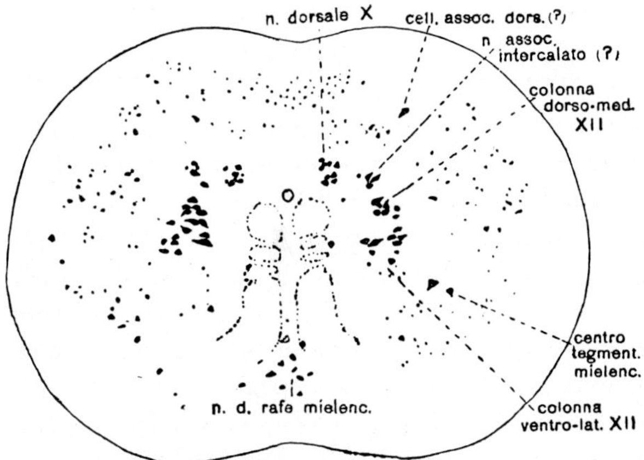

Figure 251F. Cross-section *(Nissl-stain)* through the caudal (closed) portion of the oblongata in Lacerta muralis (from BECCARI, 1943). n. assoc. intercalato (?): perhaps primordium of *nucleus intercalatus Staderini.*

question of homologies, the configurational arrangement of the three Reptilian cochlear nuclei represents a typical Sauropsidan feature which is displayed, in essentially identical manner, by the Avian oblongata, discussed in the next section of the present chapter. With respect to Mammalian homologies, it is generally assumed that the Reptilian nucleus angularis corresponds to the tuberculum acusticum (nucleus cochlearis dorsalis) and the nucleus dorsalis magnocellularis to the Mammalian nucleus cochlearis ventralis[136] (cf. also KAPPERS, 1947). A griseum homologous to the Sauropsidan nucleus laminaris, which may even be lacking or very poorly developed in some Reptiles, cannot be identified in the Mammalian oblongata and is thus perhaps entirely lacking in this Vertebrate class (cf., however, pp. 486, 549).

The *vestibular* component of the octavus consists of an anterior root bundle related to utriculus, anterior and lateral ampulla, and of a ventral portion of the posterior root bundle, related to posterior ampulla, macula neglecta utriculi and sacculus except for the 'cochlear' apparatus, which is supplied by the dorsal (cochlear) portion of the posterior root bundle.

---

[136] This would imply a ventrolateral shifting of nucleus dorsalis magnocellularis. Assuming that the pattern of fiber connections obtaining in Reptiles is somewhat comparable to the Mammalian one, such displacement does not seem improbable and would imply neurobiotactic parameters.

The terminal nuclei of the vestibularis system,[137] located ventrally, rostrally and caudally to the cochlear grisea, are likewise well developed (cf. Fig. 251A). They include, as in Fishes, a nucleus tangentialis, a nucleus vestibularis superior seu anterior, a nucleus vestibularis ventralis *(Deiters' nucleus)*, and a caudalward extending nucleus of the descending vestibular root. These cell groups are, depending on the taxonomic forms, more or less distinctly demarcated, and may also display vaguely outlined further subdivisions. Thus, WESTON (1936) separates in Chelonia a nucleus vestibularis ventromedialis from a nucleus vestibularis ventrolateralis representing *Deiters' nucleus*, which is again subdivided into lateral, dorsal, and ventral parts (Fig. 252A). The cited author, moreover, describes a nucleus vestibularis dorsolateralis as rostral continuation of *Deiters' nucleus*, and a more rostrodorsal nucleus vestibularis superior. Together, these latter two grisea apparently correspond to BECCARI's nucleus vestibularis superior.

The vestibular grisea receive, in addition to their vestibular input, descending fibers from the cerebellum as well as fibers from other afferent bulbar nuclei. The efferent pathways of the vestibular grisea include crossed and perhaps also homolateral descending and ascending contributions to the fasciculus longitudinalis medialis, homolateral vestibulo-spinal fibers mainly originating in *Deiters' nucleus*, essentially homolateral vestibulo-cerebellar connections, and essentially decussating contributions to the lateral lemniscus. Numerous poorly understood, but doubtless functionally very relevant reciprocal connections between vestibular grisea and the closely related nucleus motorius tegmenti are evident. Moreover, there obtains direct or indirect vestibular output into efferent cranial nerve nuclei.

---

[137] A small amount of vestibular root fibers may decussate and reach the contralateral vestibular nuclei (BECCARI, 1943). In addition, some vestibular root fibers ascend directly to the cerebellum (cf. chapter X).

Figure 252A. Cross-section *(Nissl-stain)* through the oblongata of the Chelonian Chrysemys marginata at the level of motor trigeminus nucleus (from WESTON, 1936). a.coch. area cochlearis; cell. Purk. *Purkinje cells* of cerebellum; f.l.m.: fasciculus longitudinalis medialis; gr.per.ventr.: 'periventricular griseum'; l.aur.: 'lobus auricularis of cerebellum'; n.mv.N.V. motor nucleus of trigeminus; n.sens.prim. N.V.: nucleus sensibilis principalis of trigeminus; n.vest.dors.lat. nucleus vestibularis dorsolateralis; n.vest.vent.lat.p.dors.; nucleus vestibularis ventrolateralis, pars dorsalis; n.vest.vent.

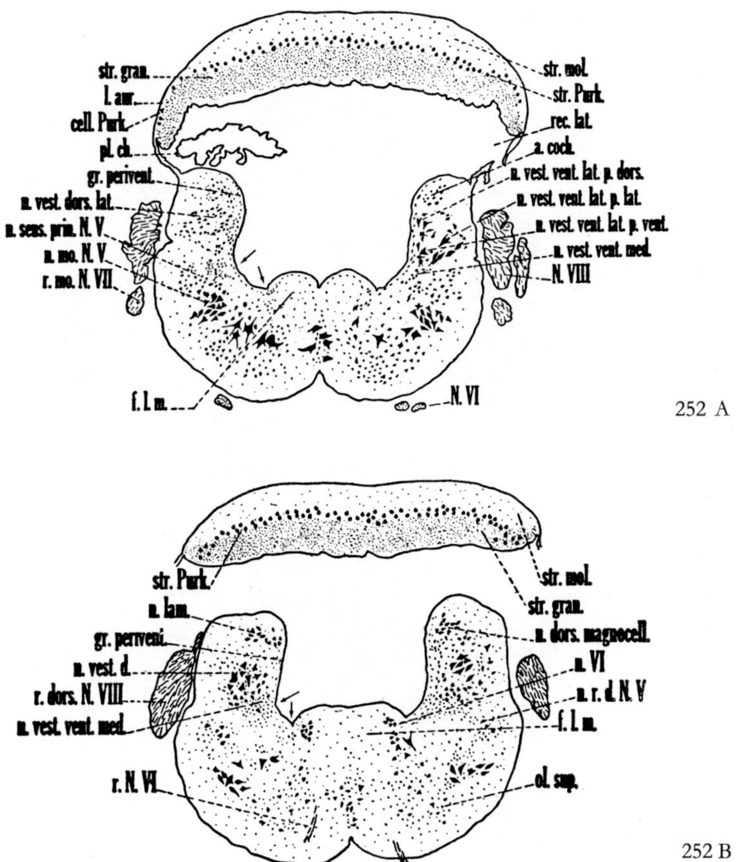

252 A

252 B

lat. p. lat.: nucleus vestibularis ventrolateralis pars lateralis; n. vest. vent. lat. p. vent.: nucleus vestibularis ventrolateralis pars ventralis; n. vest. vent. med.: nucleus vestibularis ventromedialis; pl. ch.: choroid plexus; rec. lat.: lateral recess of fourth ventricle; r. mo. N. VII: motor root of facial nerve; str. gran.: granular layer of cerebellum; str. mo.: moleculary layer of cerebellum; str. Purk.: Purkinje cell layer of cerebellum. Added arrows for sulcus limitans and intermedius ventralis.

Figure 252 B. Cross-section *(Nissl-stain)* through the oblongata of Chrysemys at a level just caudal to the entrance of nervus octavus (from WESTON, 1936). n. dors. magnocell.: nucleus dorsalis magnocellularis (cochlearis); n. lam.: nucleus laminaris; n. r. d. N. V: nucleus radicis descendentis trigemini; n. vest. desc.: nucleus radicis vestibularis descendentis; n. VI: nucleus abducentis; ol. sup.: nucleus olivaris superior; r. dors. N. VII: dorsal portion of octavus root; r. N. VI: radix nervi abducentis. Other abbreviations as in preceding figure.

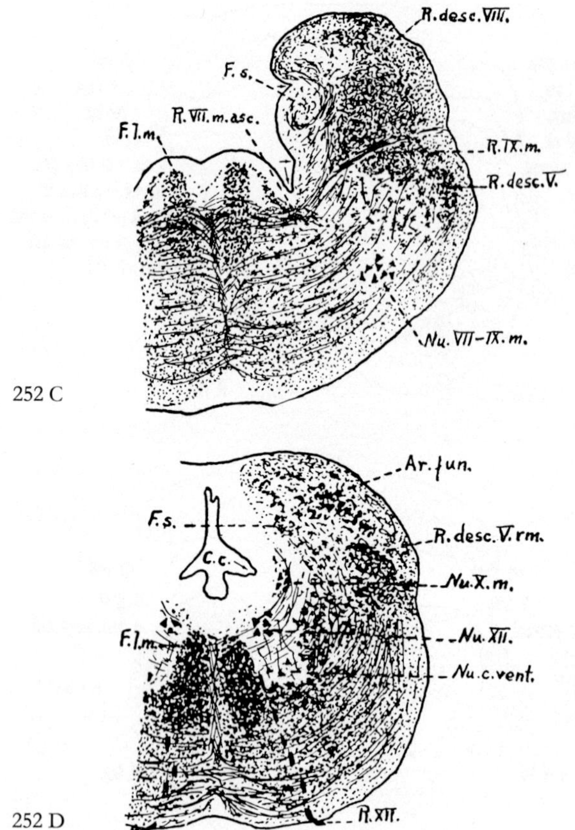

Figure 252C. Cross-section through the oblongata of the Chelonian Damonia subtrijuga (myelin stain) at the level of the glossopharyngeus root (from BLACK, 1920). F.l.m.: fasciculus longitudinalis medialis; F.s.: tractus solitarius; Nu.VII–IX M.: 'frontal end of the combined facial and glosspharyngeal motor nucleus'; R.desc.V: radix descendens trigemini; R.VII m.asc.: part of intracerebral loop of efferent facial root fibers; R.desc.VIII: radix descendens vestibularis; R.IX m.: efferent root fiber bundle of glossopharyngeus. Sulcus intermedius dorsalis and sulcus limitand (arrows) can easily be identified above and below ventricular wall bulge related to tractus solitarius; lowest arrow: sulcus intermedius ventralis.

Figure 252D. Cross-section through the caudal (closed) portion of the oblongata in Damonia (from BLACK, 1920). Ar.fun.: region of funicular nuclei; cc.: central canal; Nu.Xm.: efferent nucleus of vagus; Nu.XII: nucleus of hypoglossus; Nu.c.vent.: 'nucleus cornu ventralis' (probably ventral subdivision of hypoglossal nucleus; R.XII: root bundle of hypoglossus. Other abbreviations as in preceding figure. Added arrow indicates sulcus limitans.

Within the *ventrolateral portion of the alar plate's dorsal zone*, the rostral main sensory trigeminal nucleus and the nucleus of the descending root, which extends far caudalward, are generally well developed in Reptiles and particularly voluminous in Crocodilia (Fig.s. 253A–D) and some Ophidia.[138] A rostral bundle of secondary trigeminal fibers in these forms may constitute a conspicuous, predominantly homolateral quinto-cerebellar tract. Other mainly contralateral secondary pathways are provided by arcuate fibers joining the general spino-bulbo-tectal lemniscus.

The alar plate's intermediodorsal zone, which contains the terminal grisea of the 'visceral-afferent' cranial nerve components, is much less bulky than the dorsal zone, but nevertheless displays a rather substantial tractus solitarius (Figs. 251E, 252C, D) despite the scarcity of gustatory fibers in Reptiles. Only the facial and the glossopharyngeal nerves are said to include a small number of such fibers. On the other hand, as stressed by KAPPERS, the progressive 'general visceral-afferent' innervation of pharynx, larynx, and lungs is correlated with a corresponding development of the tractus solitarius, particularly in its caudal portion. Some of its fibers decussate in the commissura infima, and homolateral as well as contralateral components of that system are believed to descend as far as the spinal cord's second cervical segment. The griseal column of the intermediodorsal zone is formed by the terminal nuclei of the cranial nerves VII, IX, and X. It accompanies the tractus solitarius and gives origin to diverse secondary fibers with various connections, such e.g. as arcuate fibers presumably joining the general bulbar lemniscus. Specific secondary ascending and descending gustatory tracts, comparable to those of Fishes and at least some Amphibia, have not been identified with any degree of certainty. This would agree with the generally reduced or minor development of the 'gustatory system' in Reptiles.

Some exteroceptive (cutaneous, somatic) fibers may be present in the branchial nerves VII, IX, X, and have been reported, at least in the vagus. As seems to be generally the case in all Vertebrates, these fibers

---

[138] '*Chez les Crocodiles, cette hypertrophie se manifeste surtout dans la racine sensorielle qui innerve la surface énorme de la gueule. Chez les Serpents, cette hypertrophie se montre aussi dans la racine motrice et dans la racine mésencéphalique, ce qui est à rapporter au développement considérable des muscles des mâchoires et de leurs fibres proprioceptives qui proviennent en partie du noyau mésencéphalique de ce nerf*' (KAPPERS, 1947). The mesencephalic root and its cells, here mentioned by KAPPERS, shall be dealt with in chapter XI.

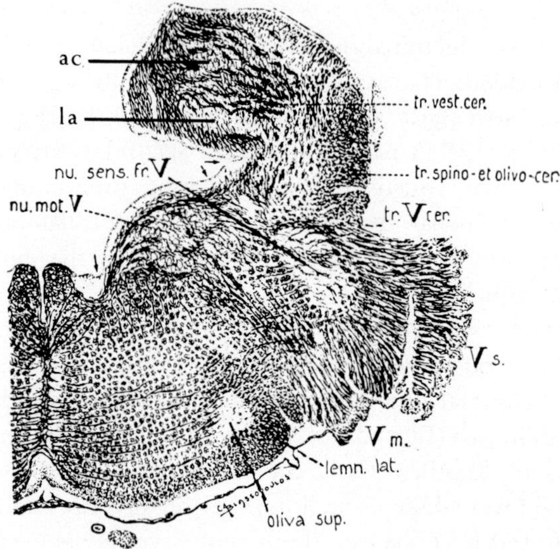

Figure 253 A. Cross-section (myelin stain) through the oblongata of a Crocodile at the level of motor and main sensory trigeminal nuclei (from KAPPERS, 1947). nu. sens. fr. V: 'noyau sensoriel frontal du trijumeau' (main afferent trigeminal nucleus); tr. V. cer.: tractus quinto-cerebellaris. The leads ac (area cochlearis) and la (nucleus laminaris) have been added; arrows indicate sulcus intermediodorsalis, sulcus limitans and sulcus intermedioventralis.

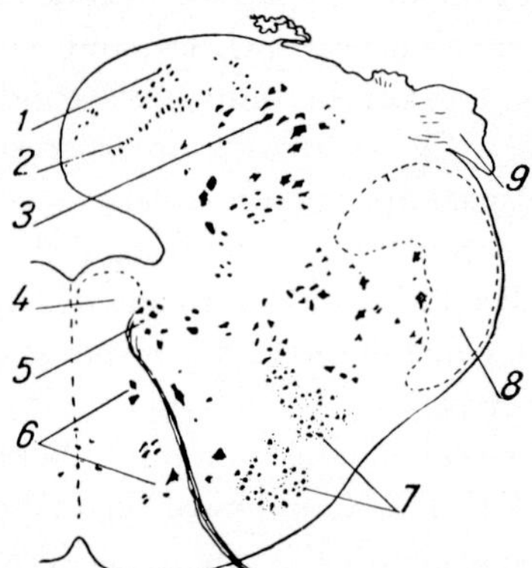

Figure 253 B. Cross-section *(Nissl stain)* through the oblongata of Alligator sclerops at the entrance of the posterior acusticus root (after VAN HOEVELL, from BECCARI, 1943). 1: nucleus cochlearis dorsalis magnocellularis; 2: nucleus laminaris; 3: n. vestibularis ventrolateralis (the cell group below the lead is presumably likewise a component of the vestibular grisea); 4: fasciculus longitudinalis medialis; 5: nucleus of abducens; 6: nucleus reticularis tegmenti; 7: superior olivary complex (the upper lead points to what has also been termed 'nucleus of lemniscus lateralis'); 8: radix descendens trigemini; 9: radix of posterior (ventral) trunk of nervus octavus.

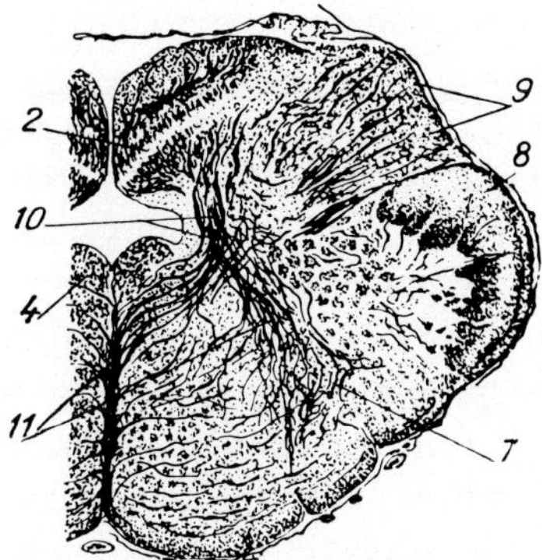

Figure 253 C. Cross-section (myelin stain) through the oblongata of an Alligator, approximately corresponding to the level of B, but not including nucleus abducentis and its intracerebral root (after DE LANGE, from BECCARI, 1943). 10: fibrae arcuatae cochleares dorsales; 11: fibrae arcuatae cochleares ventrales. Other designations as in B.

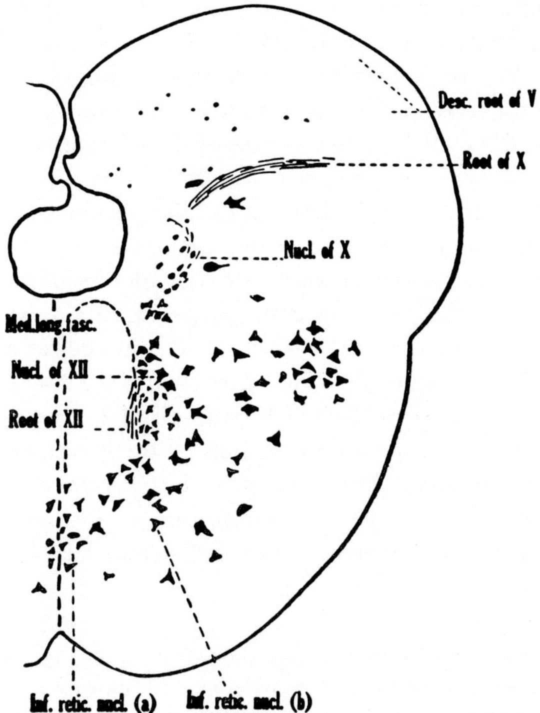

Figure 253 D. Cross-section *(Nissl stain)* through the caudal oblongata of a Caiman, at the level of the hypoglossal nucleus (after VAN HOEVELL, from KAPPERS *et al.*, 1936). The cell group between nucleus dorsalis vagi and nucleus hypoglossi is perhaps the nucleus intercalatus.

presumably do not join the fasciculus solitarius, but rather the radix descendens trigemini and its griseum.

In addition to the terminal nuclei of cranial nerves, the alar plate of the caudal oblongata, at the transition to the spinal cord, includes a '*dorsal funicular area*' (Fig. 252D) continuous with the dorsal funiculus of the spinal cord and apparently containing cell groups representing poorly or very moderately developed funicular nuclei which may contribute to the general spino-bulbo-tectal lemniscus.

With respect to the neuronal configurations of *the basal plate's intermedioventral zone*, the motor nucleus of the trigeminus, located near the ventricular wall in some forms (Fig. 253A), tends to be displaced ventrolateralward in others (Fig. 252A), particularly at its more caudal level. This displacement may result in a more or less distinct separation of two cell aggregates, variously described as dorsal and ventral groups by different authors (cf. e.g. Beccari, 1943; Kappers, 1947).

A ventrolateral displacement of the efferent facial nucleus is still more pronounced (Figs. 251D, 254A). A dorsal and a ventral subgroup can likewise be present. In at least some Reptiles, e.g. in Varanus, the motor facial nucleus is definitely caudal to the nucleus of the abducens, and the intracerebral course of the efferent facial root fibers displays a loop very similar to that seen in Mammals (cf. Fig. 269). In other Reptiles, motor facial nucleus and abducens nucleus occur approximately at the same transverse levels. Again, the motor facial nucleus may be quite close to, or overlap with, the efferent glossopharyngeal nucleus (cf. Fig. 252C).[139] This latter, however, or at least part of it, is said to be represented by the rostral end of the nuclear column from which the efferent vagus fibers originate (Kappers, 1947).

The efferent vagus nucleus of Reptiles displays, depending on the taxonomic forms and the interpretations by the investigators, various degrees of ventrolateral displacement, such that a dorsal and a ventrolateral cell group, respectively nucleus, can be distinguished. As regards this ventrolateral expansion of the efferent VII, IX, and X nu-

---

[139] The various diagrams by Kappers, Addens, Black, and others, depicting the longitudinal arrangement of the efferent nuclei in the brain stem (cf. Figs. 158, 159), although quite acceptable and instructive, do not agree with each others in various details. This is due to the difficulties inherent in any attempt at an accurate mapping of grisea with poorly defined outlines. Some of the discrepancies and uncertainties concerning the nuclear 'columns' of the intermedioventral zone in Reptiles are discussed by Beccari (1943, pp. 151–153).

clei, KAPPERS (1947) expresses the opinion that the smaller neuronal elements which retain a more dorsal position, might represent the autonomic (preganglionic) component of these branchial nerves.

The Reptilian vagus nerve includes a cranial accessory branch or XIth nerve, whose fibers originate from the caudal portion of the vagus nuclear column. In addition, spinal accessory roots have been de-

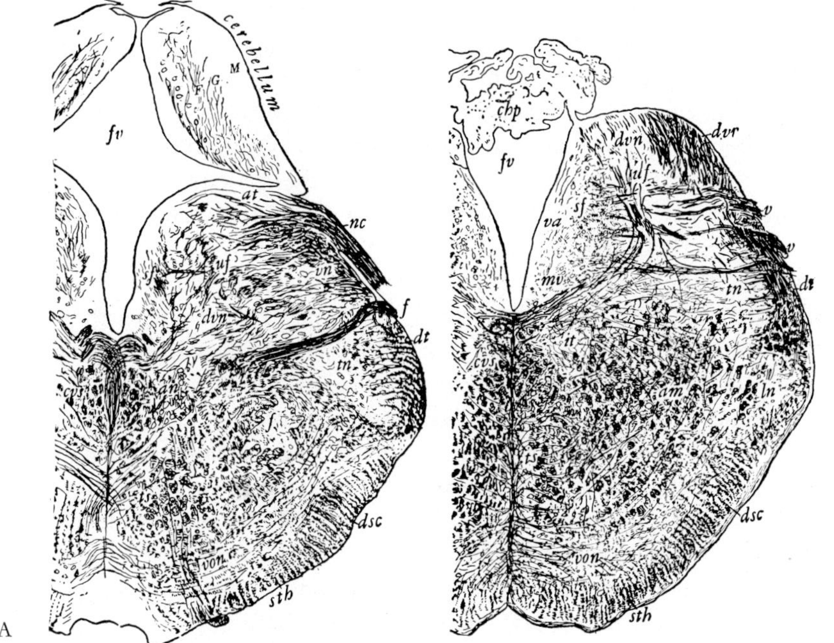

Figure 254 A. Cross-section (myelin stain) through the oblongata of the water snake Natrix sipedon at the level of the efferent facial nerve nucleus (from PAPEZ, 1929). am: 'nucleus ambiguus'; an: nucleus of abducens; at: 'acoustic tubercle'; chp: choroid plexus; cvs: fasciculus longitudinalis medialis; dsc: spino-cerebellar tracts; dt: descending root of trigeminus; dvn: nucleus of descending vestibular root; dvr: descending vestibular root; FGM: fiber, granular, and molecular layers of cerebellum; f: nucleus and root of facial nerve; fv: fourth ventricle; it: 'interstitial nucleus' (perhaps part of reticular formation); mv: motor nucleus of vagus; nc: nervus cochlearis; sf: tractus solitarius; sth: 'trigemino-thalamic tract' (probably general spino-bulbar lemniscus system); tn: nucleus of descending trigeminal root; ts: tecto-spinal tract; uf: arcuate fibers (described as 'cerebellar bundle'); v: vagal roots; va: afferent nucleus of vagus; vn: vestibular grisea; von: 'inferior olive' (in Figure 254A probably a reticular cell group).

Figure 254 B. Cross-section through the oblongata of Natrix sipedon at the level of the 'vagal region' (from PAPEZ, 1929). Abbreviations in legend of preceding figure A.

scribed (in Lacerta) by BECCARI (1943). These fibers seem to originate from a cell group dorsally adjacent to the hypoglossus nucleus, and continuous with the dorsal part of the spinal cord's ventral horn.

Medially to the dorsal efferent vagus nucleus, KAPPERS (1947) mentions a griseum recorded by his collaborator MOFFIE in all Reptilian orders, and located, near the ventricle, between the dorsal efferent vagus nucleus and the hypoglossal nucleus. KAPPERS suggests that said griseum might be a predecessor of the Mammalian *nucleus intercalatus Staderini*. It was likewise depicted by BECCARI (1943) in Lacerta (Fig. 251F). PAPEZ (1929, p. 396) had also noted a comparable 'small cell nucleus' in the lower oblongata of a Water Snake.

Within the *ventral zone* of the oblongata's *basal plate*, the nucleus of the abducens displays diverse secondary variations with respect to the different orders. In Chelonia and Crocodilia, this griseum is diffuse and elongated, extending between the levels of glossopharyngeal and facial nucleus.[140] In Lacertilia and Ophidia, the abducens nucleus is more compact and, *in toto*, has reached a rostral location such that its root fibers emerge rostrally to those of the facial (in Varanus), or partly rostrally and at the same level, but not caudally to the facial as in Anamnia. Most Reptiles seem to possess an accessory abducens nucleus which innervates the musculus retractor bulbi and the musculus bursalis (m. of nictitating membrane), and is located medially to the nucleus radicis descendentis trigemini. Its fibers ascend to join the main root fiber bundle of the abducens. KAPPERS (1947) cites the shifting of this accessory nucleus toward the descending trigeminal root as a manifestation of neurobiotaxis. BECCARI (1943) points out that in Reptilian species lacking a m. retractor bulbi and a nictitating membrane (e. g. Chamaeleon vulgaris), the accessory nucleus of the abducens is not present.

In the caudal portion of the oblongata, the nucleus nervi hypoglossi represents a rostral extension of the spinal cord's ventral horn. In many, if not most Reptiles, the hypoglossal nucleus displays a dorsal and a ventral cell group or subdivision[141] (Figs. 251F, 252D).

Pertaining to the basal plate in general, the *nucleus reticularis seu motorius tegmenti* whose characteristic multipolar elements appear to be somewhat more numerous than in Amphibians, also displays a more

---

[140] The abducens root fibers emerge here caudally to those of the facial and in part only at the same level, without extending rostralward of this latter.

[141] BECCARI (1943) designates these subdivisions as '*colonna dorsomediale*' and '*colonna ventrolaterale*' (Fig. 251F).

distinct cell-grouping, but is, nevertheless, not significantly different from the Amphibian reticular formation as regards its overall configuration.

The superior olivary nucleus, likewise located in the general basal plate, manifests a higher degree of differentiation than in Amphibians, in accordance with a progressive development of the cochlearis system (Figs. 251A, 251C, 252B, 249A). Dorsal and ventral arcuate fibers originating in the three arcuate nuclei form a transverse basal system designated as '*trapezoid body*' and running through or near the bilateral superior olivary nuclei.[142] The dorsal arcuate fibers take a course near the ventricular floor and then bend basalward on both sides of the midline, while the ventral arcuate fibers sweep ventrad toward the homolateral superior olivary nucleus. More laterally, these fibers gather to form the ascending lateral lemniscus, which reaches mesencephalic grisea, and is believed to contain, in addition to the decussating fibers, some homolateral ones. The detailed synaptic relationship of 'trapezoid fibers' to the superior olive, and the further connection of this latter are still insufficiently elucidated. In connection with the cochlearis system, an efferent cochlear bundle of uncertain origin, joining the VIIIth nerve, and transmitting impulses to the peripheral receptor structures, should briefly be mentioned. It was identified by BOORD (1961) in Crocodilians, and shall again be considered in the next section 9, dealing with Birds. This bundle represents the centrifugal fiber system recognized eighty years ago in Mammals by HELD (1893) and mentioned above in section 1 (p. 319).

In addition to the superior olive, a more lateral cell group has been described as nucleus of the lateral lemniscus (cf. e.g. WESTON, 1936; BECCARI, 1943). It is uncertain whether this griseum is here a derivative of alar or of basal plate, and whether it should, or should not be considered an intrinsic component of the Reptilian superior olivary complex.[142a]

Within the overall basal plate of a more caudal portion of the oblongata, an inferior olivary griseum has been identified by some au-

---

[142] As far as Sauropsida are concerned, this arrangement, characteristic for some Amphibia, and all Amniota, is particularly conspicuous in Birds (cf. Figs. 255E and 256A).

[142a] The question of decussating root fibers remains disputed. In the opinion of ADDENS (1934) those vestibular fibers which CAJAL interpreted as decussating root fibers actually represent secondary fibers. ADDENS states that 'there is no primary vestibular root crossing beneath the ventricle', but this author admits a partial decussation of primary vestibulo-cerebellar root fibers dorsally to the ventricle.

thors (e.g. PAPEZ, 1929) but there remains some doubt whether this configuration does not merely represent a reticular cell group. Be that as it may, the 'inferior olive' of Reptiles, like that of Amphibians, is not particularly well developed. This would agree with the relatively minor development of the cerebellum in these forms, as compared with the cerebellum of Selachians, Teleosts, Birds, and Mammals. BECCARI (1943) and others point out that the primordial inferior olivary nucleus of Fishes, Amphibians, and Reptiles corresponds to the medial accessory olivary nucleus *('paraoliva mediale')* in Mammals.

Concerning the fiber systems originating, ending, or passing through the Reptilian oblongata, a comparison of bulbar cross-sectional levels with corresponding ones in Amphibia discloses an obvious greater complexity. Nevertheless, these pathways can be said to represent, in the aspect here under consideration, the same main channels as obtaining in Anamnia and dealt with in the preceding sections. It may, however, here be added that, since the Reptilian mesencephalic tegmentum contains a griseum comparable to the nucleus ruber tegmenti, the mesencephalo-bulbar fiber bundles contain, in the lateral tegmental field, a poorly delimited crossed rubro-bulbar tract, which, according to BECCARI (1943), apparently does not extend into the spinal cord as a rubro-spinal tract. Yet, as e.g. KAPPERS *et al.* (1936) suspect, the presence of a perhaps inconspicuous rubrospinal channel cannot be entirely excluded.

9. Birds

The Avian oblongata displays, on the whole, various features similar to those obtaining in the taxonomically and phylogenetically related Sauropsidan Reptiles, but manifests, at least in certain respects, a higher degree of differentiation. Among studies particularly dealing with bulbar configurations in Birds are those by BARTELS (1925), BLACK (1922), BRANDIS (1893–95), BRODAL *et al.* (1950), CAJAL (1908), CRAIGIE (1928, 1930), GROEBBELS (1922), KARTEN and HODOS (1967), KOOY (1917), SANDERS (1929), SINN (1913), STINGELIN (1965), and WALLENBERG (1898–99, 1900, 1902, 1904).

Since Birds represent the most diversified Amniote vertebrate group and include far more species and orders than the class Reptilia, reasonably detailed data on the Avian brain are available for only a relatively very small proportion of these multitudinous forms. In addition to the summaries given by the texts on comparative neurology, a

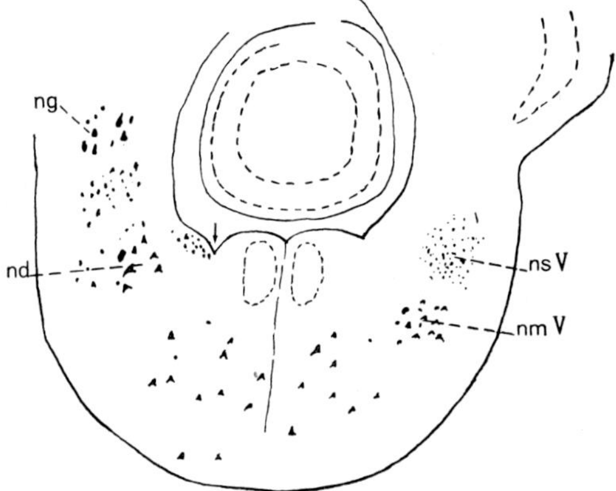

Figure 255A. Simplified drawing of a cross-section through the oblongata in a Pigeon (Columba) at the level of the motor trigeminus nucleus, showing the arrangement of cell groups (from K., 1927). nd: *Deiters' nucleus* (with n. tangentialis); ng: 'vestibulo-cerebellar' cell group (which includes the cell cluster ventral to the indicated one); nmV: motor nucleus of trigeminus; nsV: main sensory nucleus of trigeminus. The scattered cells in the basal plate pertain to nucleus reticularis tegmenti, the cell group adjacent to sulcus limitans (arrow) are presumably a medial vestibular group (n. triangularis).

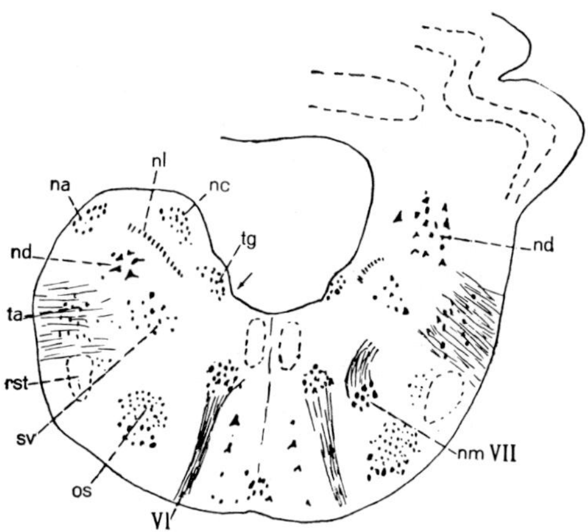

Figure 255B. Simplified drawing of a cross-section through the oblongata in a cross-section through the oblongata in a Pigeon at the level of cochlear and abducens nuclei. Left half somewhat more caudal than right (from K., 1927). na: nucleus angularis; nc: nucleus cochlearis dorsalis; nd (left): *Deiters' nucleus;* nd (right): vestibulo-cerebellar cell group; nmVII: nucleus motorius facialis; os: superior olivary griseum; rst: radix descendens trigemini and its adjacent griseum; sv: nucleus radicis descendentis vestibularis; ta: nucleus tangentialis; tg: nucleus vestibularis medialis s. triangularis.

relevant digest on the oblongata, with numerous references, can be found in R. PEARSON's recent comprehensive treatise on the Avian brain (1972). Moreover, ROMANOFF (1960), in his encyclopedic text on the Avian embryo, deals with the ontogenetic development of cranial nerves and their bulbar relationships.

Figures 255A to 259B illustrate cross sections through representative bulbar levels in the brains of diverse Avian forms. It can be seen that the three main longitudinal ventricular sulci are more or less smoothened out *('verstrichen')*, and, except for the sulcus limitans, cannot be easily identified. Even this latter may become hardly recognizable (cf. Figs. 259A–B).

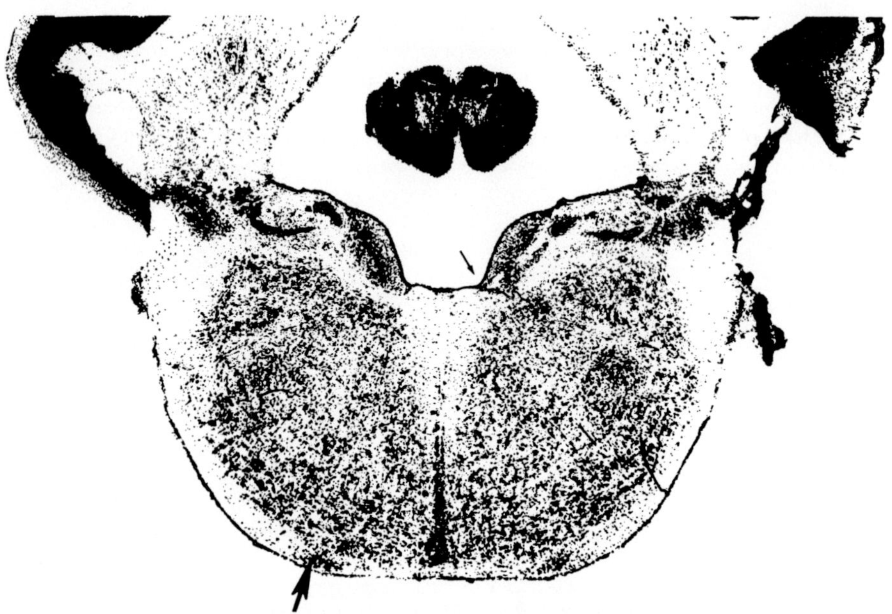

Figure 255 C. Cross-section through the oblongata of a Pigeon (Columba livia) at the level of cochlear and of some vestibular nuclei, as seen in a *Nissl preparation* (from KARTEN and HODOS, 1967). In comparing this section with the simplified figure 255B it should be taken into consideration that, on the left side of this latter, the cochlear nuclei, as in a plane with caudodorsal-rostroventral inclination, are shown together with superior olive and abducens nucleus. With regard to the plane of KARTEN's and HODOS' stereotaxic section, the abducens and facial nuclei as well as the superior olive are located in planes rostrally preceding the here depicted one. The two added arrows indicate sulcus limitans (dorsally) and a 'pontine' nucleus (ventrally).

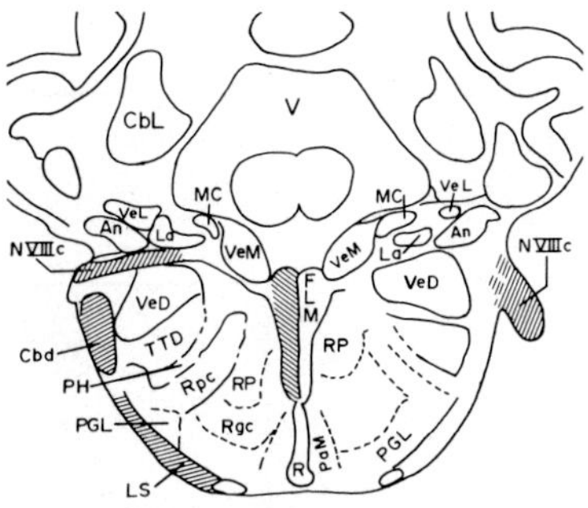

Figure 255 D. Outline drawing, indicating KARTEN's and HODOS' interpretation of the structures shown in the preceding figure (from KARTEN and HODOS, 1967). An: nucleus angularis; Cbd: tractus spinocerebellaris dorsalis; Cbl: 'nucleus cerebellaris lateralis'; FLM: fasciculus longitudinalis medialis; La: nucleus laminaris; LS: lemniscus spinalis; MC: nucleus (cochlearis) magnocellularis; NVIIc: nervus octavus, pars cochlearis; PaM: 'nucleus paragigantocellularis lateralis'; PH: 'plexus of Horsley' (formed by collaterals which branch off from the spino-cerebellar tracts and reach diverse bulbar grisea; also called 'plexus collateralis'); R: nuclei of raphé; RP: 'nucleus reticularis pontis caudalis'; Rpc: 'nucleus reticularis parvocellularis'; TID: nucleus of radix descendens trigemini; V: fourth ventricle (including an obliquely cut 'lingular' portion of cerebellum; VeD: nucleus radicis descendentis vestibularis; Vel: portion of nucleus vestibularis lateralis; VeM: nucleus vestibularis medialis.

As in Reptiles, the *dorsal zone* of the Avian oblongata's *alar plate* is dominated by the substantial development of the primary terminal grisea pertaining to the *nervus octavus*. It seems most likely that this aspect of differentiation is related to both the peculiar acoustic behavior (song as well as sound imitation), and to the flight of birds. Some primary root fibers of both cochlear and vestibular octavus subdivisions may *perhaps* cross the midline and reach contralateral bulbar octavus grisea.

The *cochlear nuclei* are represented by the three cell groups already dealt with in the preceding section 8, namely nucleus magnocellularis, nucleus angularis, and nucleus laminaris. Cochlear and lagenar fibers[143] of the ramus posterior nervi octavi are distributed in a somewhat dif-

[143] R. PEARSON (1972) quotes the number of about 7,500 fibers in ramus cochlearis and lagenaris for an 'average bird' weighing around 100 g. Somewhat more than 10% of these fibers may be related to the lagena. The cited author also quotes additional figures concerning cell numbers in the three cochlear nuclei and the superior olivary griseum, based on quantitative investigations by WINTER (1963) and WINTER and SCHWARZKOPFF (1961).

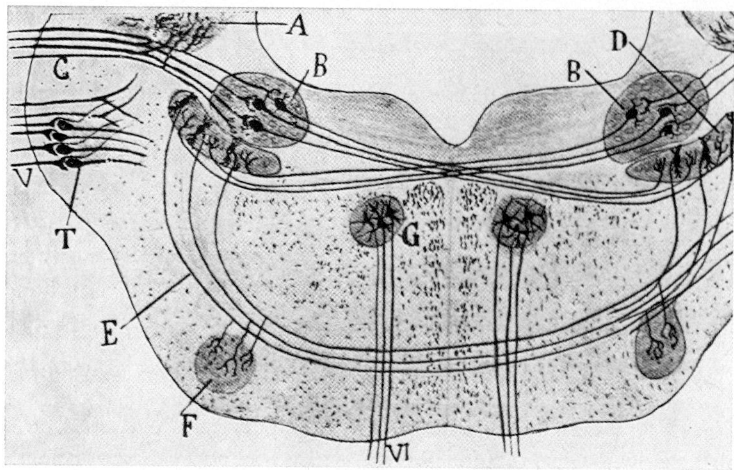

Figure 255 E. Semidiagrammatic sketch of the bulbar cochlear pathways in the Avian brain (from CAJAL, 1923). A: nucleus angularis; B: nucleus cochlearis dorsalis (magnocellularis); C: radix cochlearis; D: nucleus laminaris; E: fibrae arcuatae internae ventrales becoming 'trapezoid body' (*'cuerpo trapezoido o via acústica secondaria'*); F: superior olivary griseum; G: nucleus abducentis; T: nucleus tangentialis; V: radix vestibularis.

ferent manner in so far as lagenar fibers seem to reach mainly the vestibular and reticular grisea, although a not negligible lagenar component appears to end in both the cochlear angular and magnocellular nucleus. The true cochlear fibers, at a short distance from their entrance into the medulla, bifurcate into lateral and medial branches, the former ending in n. angularis and the latter in n. magnocellularis. The angular nucleus, presumably corresponding to the tuberculum acusticum of Mammals, and the magnocellular nucleus, comparable to the Mammalian ventral cochlear nucleus,[144] are doubtless the main prim-

[144] Since the Sauropsidan magnocellular cochlear nucleus is generally considered homologous to the magnocellular nucleus present in some Amphibia, it is commonly evaluated as the 'phylogenetically older' cochlear griseum. It should, however, be kept in mind that the cochlear griseum of Anurans can be subdivided into a predominantly magnocellular ventral portion and into a more small-celled dorsal one, which latter might perhaps be comparable with the Sauropsidan nucleus angularis. As regards the nucleus laminaris, CAJAL (1909) states that its nerve cells *'ressemblent fort à celles de l'olive supérieure accessoire des mammifères'* and suggests that this latter griseum represents the Mammalian homologue of the Avian (respectively Sauropsidan) laminar nucleus (*'noyau à petites cellules'*). Yet, as mentioned above in connection with the Reptilian oblongata (p. 471), a griseum homologous to nucleus laminaris cannot be reliably identified in Mammals.

ary cochlear centers, while the laminar nucleus is generally regarded as an essentially secondary center, although this griseum may also include endings of primary cochlear fibers. Figures 255E and 256A depict CAJAL's (1909, 1923) and BOK's (1915) interpretations of some basic bulbar cochlear connections. It can be seen that, in both rather similar concepts, the laminar nucleus is regarded as a secondary, and the superior olive as a tertiary center. This latter, which will again be referred to in dealing with the basal plate, is also said to receive substantial projections from the angular nucleus (SCHWARTZKOPFF, 1968). Generally speaking, the observations of BOORD and RASMUSSEN (1963) seem to indicate an orderly distribution of cochlear and lagenar fibers upon the bulbar primary acoustic centers, such that a so-called '*point-to-point*' connection between apical, middle, and basal thirds of cochlea, and specific regions of the cochlea nuclei obtains. It may be inferred that this type of distribution is maintained by the secondary and tertiary pathways, thereby implying a 'central tonotopic projection from the cochlea'.[145] As regards nerve cell numbers in the cochlear nuclei, the following approximate figures may be quoted: magnocellular nucleus from 3,400 to 1,500; angular nucleus from 2,700 to 10,000; laminar nucleus from 2,500 to 10,000 (WINTER and SCHWARTZKOPFF, 1961; cf. also R. PEARSON, 1972). These numbers seem to vary with respect to the diverse Avian species. In the Kiwi (Apteryx australis), CRAIGIE (1930) has described a peculiar, medially displaced subdivision of the angular nucleus.

As in other Vertebrates, the *vestibular nerve*, upon entering the oblongata, divides, perhaps mainly through bifurcation of its fibers, into an ascending bundle, taking its course toward the vestibulo-cerebellar grisea and the cerebellum, and into a descending bundle, extending caudalward as radix descendens vestibularis (cf. also Fig. 156).

The vestibular grisea are profusely differentiated, and some authors (BARTELS, 1925; SANDERS, 1929) have described up to 11 separate cell

---

[145] A more detailed evaluation of these relationships can be found in R. PEARSON's (1972) treatise. The question of 'point-to-point' projection will again be taken up with regard to the Mammalian cochlear system in the following section 10. It should also be recalled that, in respect to afferent centers, the terms 'primary', 'secondary', 'tertiary' as here used refer to grisea in which fibers of said categories end. Since 'primary' centers contain the perikarya of 'secondary' neurons, they can of course also be called 'secondary' from a different viewpoint. Thus, *mutatis mutandis*, some terminologic confusion *qua* classification of 'centers' may result, if the semantic procedures and definitions adopted by the various authors are not clearly understood.

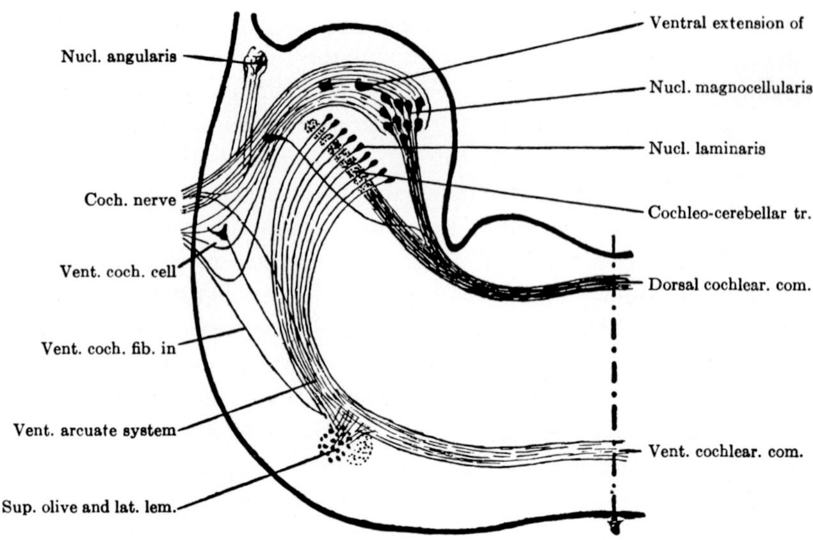

Figure 256A. Diagram indicating the course of primary and secondary cochlear fibers in the fowl, Gallus domesticus, according to the interpretation by Bok (from Kappers *et al.*, 1936).

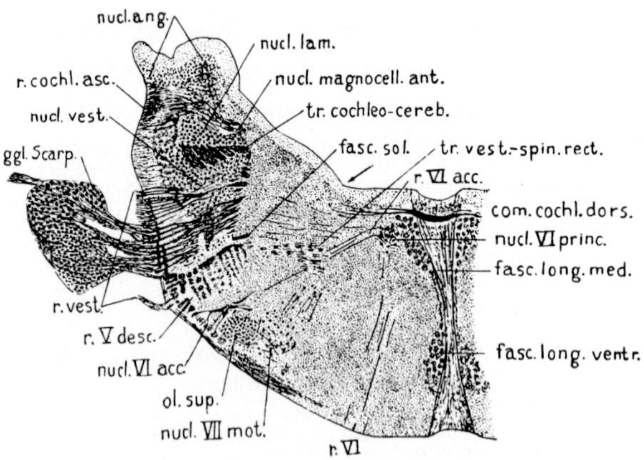

Figure 256B. Cross-section through the oblongata in a chick embryo of seven days at the level of abducens and motor facial nuclei, as seen by means of *Cajal's silver impregnation.* The general configuration at this stage is essentially identical with the adult arrangement (after Addens, 1933, from Kappers, 1947). Arrow: sulcus limitans.

groups or nuclei. Again, numerous species differences obtain, some of which were pointed out by STINGELIN (1965). For a convenient first approximation it is here perhaps sufficient to distinguish the following six grisea: nucleus tangentialis, nucleus ventrolateralis (*Deiters' nucleus*), nucleus radicis descendentis vestibularis, nucleus vestibularis medialis s. triangularis, nucleus rostralis s. oralis, and nucleus superior. The two last named might also be designated as the *vestibulo-cerebellar cell groups* (cf. e.g. Figs. 255A, B).

The *tangential nucleus* was originally described by CAJAL (1908) in Teleosts and in Birds as a '*noyau spécial du nerf vestibulaire des poissons et des oiseaux*'. This cell group is located at the entrance of a bundle of large vestibularis root fibers, which, upon passing through the tangen-

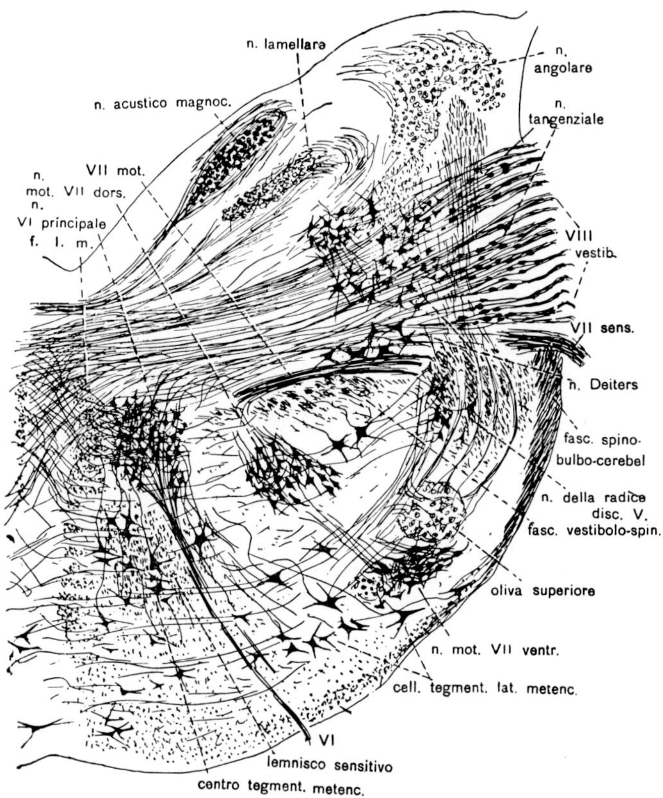

Figure 257A. Cross-section through the oblongata in a sparrow (Passer Italiae) embryo just before hatching, at the level of abducens and motor facial nuclei, as seen by means of *Cajal's silver impregnation* (from BECCARI, 1943).

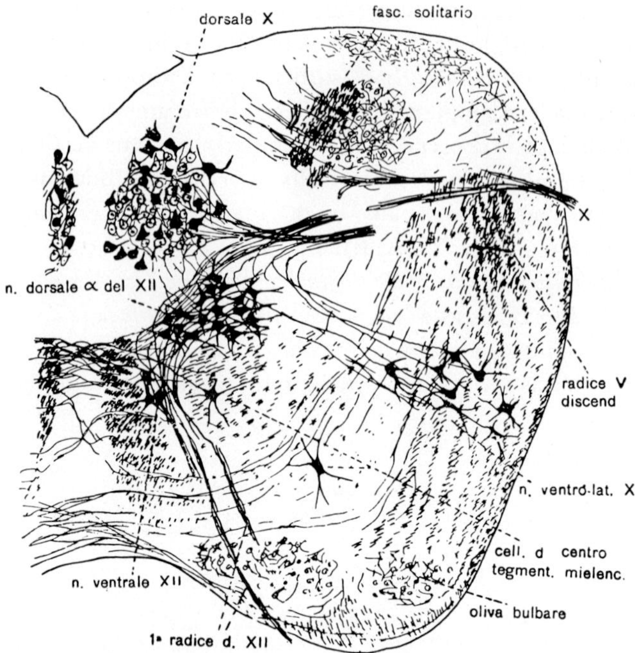

Figure 257B. Cross-section through the oblongata in an advanced embryo of Passer at the level of the hypoglossus root, as seen by means of *Cajal's silver impregnation* (from BECCARI, 1943).

tial nucleus, give off short collateral synaptic endings with a characteristic calyx-shape, depicted in Figure 194C, p. 286, chapter V, volume 3, part I of this series, and essentially representing '*boutons de passage*'. The neurites of nerve cells in the tangential nucleus are given off on the perikaryon's medial side, run parallel to the primary vestibularis fibers, and, according to CAJAL, seem to lose themselves in the neuropil of *Deiters' nucleus* and in medially adjacent bilateral reticular midline regions.

The large vestibular griseum corresponding to *Deiters' nucleus sensu latiori* is represented by a neighborhood containing a rather diffusely distributed nerve cell population, which may, however, display a conspicuous fragmentation in some forms.[146] Rostralward and dorsal-

[146] BARTELS (1925) emphasized that the fragmentation or parcellation of *Deiters' vestibular griseum* is particularly pronounced in song-birds (e.g. Alanda and Fringilla), with about three 'nuclei' each in both its dorsal and its ventral subdivision.

ward, this nondescript griseum seems to merge with, or perhaps rather to include, the vestibulo-cerebellar nuclear group. One can, therefore, in accordance with the appropriate title of BARTELS' paper (1925) designate this entire ill-defined nuclear complex as '*die Gegend des Deiters- und Bechterewkernes der Vögel*'.

The parcellations of this extensive overall neighborhood undertaken by CAJAL (1908), BARTELS (1925), SANDERS (1929) and others have resulted in mutually somewhat differing schemes encumbered by

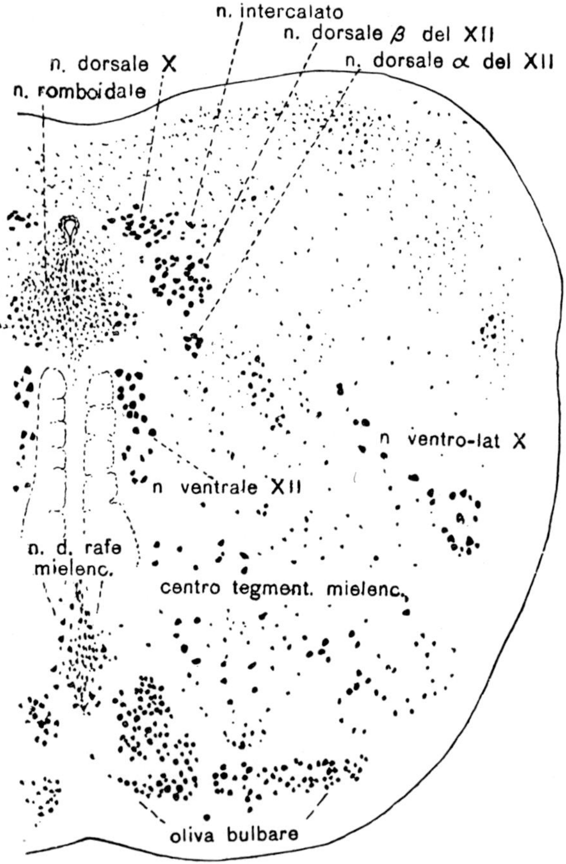

Figure 257C. Cross-section through the oblongata in an adult Passer *(Nissl stain)* at the level of vagus and hypoglossal nuclei and central canal (from BECCARI, 1943). The grisea of the posterior funiculus (nuclei funiculares) are not labelled but identifiable.

a diversified and thereby complex terminology. The variously recorded details are difficult to interpret or to reconcile with each other.[147]

There is, nevertheless, little doubt that a substantial portion of said griseum may be designated as nucleus vestibularis ventrolateralis and corresponds to *Deiter's nucleus sensu strictiori*, in which numerous vestibular fibers terminate. Neurites originating in that nucleus provide the homolateral vestibulo-spinal tract. Other axons (descending, ascending or bifurcating) join the contralateral and homolateral fasciculus longitudinalis medialis. Reciprocal cerebellar connections are likewise assumed to obtain.[148] In addition, contributions to the lateral lemniscus systems and unspecified synaptic links with reticular formation and other bulbar grisea can be inferred. Direct descending and perhaps ascending connections between telencephalon and vestibular grisea have been suspected (cf. e.g. KAPPERS, 1947) but the evidence for such linkage still remains inconclusive.

The nucleus of the descending vestibular root appears to be a direct caudal extension of *Deiters' nucleus*, but consisting of smaller neuronal elements, and reaches the levels of the bulbar calamus. The nucleus vestibularis medialis *seu* triangularis is adjacent to the ventricle (cf. e.g. Figs. 255B, C, D). It extends not only rather far caudalward, but also in a rostral direction, where it merges with the vestibulo-cerebellar group. Relevant details concerning the neuronal linkage of nucleus radicis descendentis vestibularis and of nucleus vestibularis medialis are not available.

As regards the vestibulo-cerebellar grisea, I have been unable, in the diversified material which I could examine, to identify with sufficient certainty the subdivisions delimited by CAJAL, BARTELS, and SANDERS. I merely noted a diffuse dorsal and ventral group (nucleus vestibularis rostralis). Both groups, which appear to have mainly cerebellar relations, might each perhaps again be subdivided into medial and lateral subgroups.[149]

---

[147] A partial attempt at comparing and interpreting the terminologies respectively the nuclear parcellations proposed by the above-mentioned authors has been made by STINGELIN (1965) and by LARSELL (1967).

[148] It should also here be mentioned that a perhaps substantial amount of vestibular *root fibers* ascends into the cerebellum (cf. chapter X).

[149] The cell group which I would designate as nucleus vestibularis superior corresponds approximately to the dorsal part of nucleus vestibularis superior in the atlas of KARTEN and HODOS (1967) at level AP 0.00 on their figures p. 138 and 139, while my 'nucleus rostralis' here seemingly corresponds to the ventral part of their nucleus vestibu-

Concerning the ontogenetic development of the cochlear and vestibular nuclei, LEVI-MONTALCINI (1949) showed that, following removal of the otocyst in chick embryos at the 40 hour stage, the early differentiation of cochlear and vestibular grisea in the oblongata was, on the whole, not prevented, despite lack of inner ear and cochlear respectively vestibular ganglia. Only in later developmental stages various degrees of hypoplasia were displayed by the cochlear nuclei, particularly by n. angularis, and least by n. laminaris. Among the vestibular nuclei, only the nucleus tangentialis (which is intimately related to the entering vestibular root) was appreciably affected. Extirpation of various brain regions did not affect the development of either cochlear or vestibular grisea.

The *afferent bulbar trigeminal nuclei*, located within the *ventrolateral subdivision of the alar plate's dorsal zone*, comprise the main sensory nucleus and the nucleus of the descending root. The *main sensory nucleus* is found close to the region of trigeminal root entrance into the brain. The afferent portion of that root, whose central neurites originate in the ophthalmic and maxillo-mandibular trigeminal ganglionic complex, includes also the peripheral afferent neurites of the mesencephalic root, originating from the large vesicular cells located in the mesencephalon's alar plate, and dealt with in chapter XI.

Depending on the different taxonomic forms, the size of the main sensory nucleus *(n. principalis)* varies greatly, and, according to STINGELIN (1965), seems to be correlated with the size of the beak. This latter structure is evidently an important area of sensory trigeminal innervation. In some forms, a dorsal and a ventral subdivision of the nucleus principalis can be recognized.

Secondary fibers arising in nucleus principalis join the contralateral and perhaps homolateral bulbar lemniscus, some collaterals providing connections with other bulbar grisea. In addition, secondary ascending trigeminal fiber bundles reaching the cerebellum have been recorded.

---

laris superior together with part of their n. vestibularis lateralis. The nucleus vestibularis dorsolateralis at KARTEN's and HODOS' levels PO 0.25 (together with 'processus cerebellovestibularis'), and 0.50 (p. 142 to 145 l. c.) might be the caudal end of the entire vestibulocerebellar complex. Again, STINGELIN's (1965) nucleus vestibularis dorsolateralis, pars lateralis, and *nucleus Deiters dorsomedialis*, shown in Figure 259 of the present chapter, presumably correspond to the lateral respectively medial part of the griseum which I have tentatively designated as nucleus (vestibularis) rostralis *seu* oralis. This latter may not exactly correspond to the cell group called nucleus oralis by BARTELS (1925).

Still another ascending system is said to be represented by the partially crossed *quinto-frontal tract*. This bundle, which appears to be a perhaps direct secondary trigeminal projection upon basal ganglia of the telencephalon, was first described by WALLENBERG (1903, 1904).

The *radix descendens trigemini and its nucleus* extend caudalward where the nucleus merges with the spinal cord's substantia gelatinosa. Details of the fiber connections effected by nucleus of the descending root and by the primary root fibers themselves are poorly elucidated, but doubtless include contributions to the bulbar lemniscus and connections with various bulbar grisea. It is generally assumed that the mesencephalic root has a proprioceptive function, while the nucleus principalis is a station of the exteroceptive discriminatory pathway (*'sensibilité cutanée gnostique*, KAPPERS, 1947), and that the descending root with its nucleus, presumably in addition to other functional components, includes predominantly fibers related to nociceptive and temperature sense conduction (pain, '*fibres protopathiques*', KAPPERS, 1947). Generally speaking, the trigeminal nerve is said to be less developed in most Birds than in Reptiles.[150]

At the transition to the spinal cord, the *overall dorsal region of the alar plate* contains the *nuclei of the posterior funiculus* (cf. Fig. 257C). These grisea, in which ascending spinal dorsal root fibers end, are moderately developed and presumably contribute secondary fibers to the bulbar lemniscus, possibly also to the cerebellum. In addition to nucleus gracilis and cuneatus, a nucleus cuneatus externus can be identified in at least some Avian representatives.

The *intermediodorsal zone of the alar plate* comprises the cell column formed by the *viscero-afferent terminal nuclei of the facial, glossopharyngeal and vagus nerves*. The facial nerve of Birds is commonly assumed to contain general visceral afferent fibers, but no cutaneous exteroceptive nor gustatory ('special visceral afferent') ones. The *sensory facial nucleus* is rather small and accompanied by the thin rostral extremity of the tractus solitarius[151] (cf. Fig. 256B).

---

[150] '*Le nerf trijumeau est moins développé chez la plupart des Oiseaux que chez les Reptiles*' (KAPPERS, 1947). 'The trigeminal (cranial V) nerves are rather small in birds' (ROMANOFF, 1960). Cf. also the comments by J. Z. YOUNG on cutaneous sensibility in Birds, quoted at the end of section 10 in the preceding chapter VII. While doubtless essentially correct, this evaluation must be qualified by taking into account the well developed sensitivity of the Avian beak region.

[151] '*Faisceau présolitaire*' of KAPPERS (1947). '*Les centres gustatifs de la moelle allongée sont atrophiés, tandis que le fasc. solitaire et la substance grise parasolitaire des réflexes gastriques et respiratoires sont plus développés*' (KAPPERS, l. c.).

The *glossopharyngeal nerve* includes most, if not all, Avian taste fibers, in addition to some general visceral afferent components. Its terminal nucleus, caudally adjacent to that of the facial, is accompanied by the enlarged *tractus solitarius*, whose bulk increases caudalward and displays a conspicuous size at vagus levels (Fig. 257B).

The *vagus nerve* contains few, if any, gustatory components. General visceral afferent fibers predominate, although some exteroceptive, cutaneous fibers are also present. These latter, however, join the *radix descendens trigemini and its nucleus.* Many of the general visceral afferent components of the vagus are related to the control of respiration.[152] As in other Vertebrates, a bundle of tractus solitarius fibers decussates by way of the *commissura infima.* Little is known in detail and with reasonable certainly about the secondary fibers originating in the viscero-afferent terminal grisea. Some such fibers may join the bulbar lemniscus, others could descend toward the spinal cord, while a substantial amount of connections with reticular and other bulbar centers doubtless obtains.

Within the *basal plate of the oblongata*, the most rostral efferent griseum of the intermedioventral zone is the *motor trigeminal complex.* In the Ratite Casuarius, KAPPERS (1920) described a main motor nucleus in a dorsal position similar to that obtaining in Chelonian and Lacertilian Reptiles. He considered that location as 'primary'. Nevertheless, even in Casuarius, the cited author recorded an additional, but much smaller ventrolaterally displaced separate subdivision. In most Birds, however, the motor trigeminal griseum is split into at least three distinctive nuclei[153] (cf. e.g. Fig. 259A).

With respect to the *motor facial nucleus*, and particularly in connection with his theory of neurobiotaxis, KAPPERS (1947 *et passim*) stresses the rostral shift of that griseum. While, in Reptiles, the motor facial nucleus is located caudally to the exit of its root fibers, this griseum has assumed a position rostrally to that level of exit, and a subdivision of the facial nucleus is in close proximity to the efferent trigeminal grisea, forming a *trigemino-facial complex.* In Birds, two or three efferent facial nuclear groups are present, while in Reptiles, a single or a double effer-

---

[152] A concise summary of the available data on vagal control of Avian respiration is included in the treatise by R. PEARSON (1972).

[153] The inferred specific relationships of these motor trigeminal nuclei to particular muscles or muscle groups, as based on experimental evidence obtained by KÔSAKA and HIRAIWA (1905) and subsequent authors, are discussed by STINGELIN (1965) and R. PEARSON (1972).

ent facial nucleus can be found, as compared with *always* at least two and frequently three nuclei in Birds. The nucleus innervating the posterior portion of the digastric muscle (n. facialis dorsalis) is generally in continuity with the intermediate trigeminal nucleus (n. motorius V, pars ventromedialis), which innervates the anterior part of the digastric (cf. Fig. 259A). The facial nerve also includes general visceral efferent

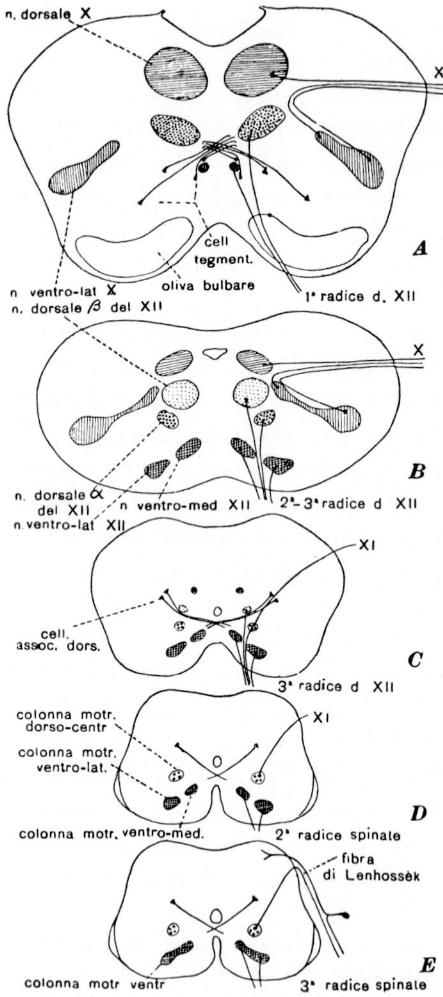

Figure 257D. Diagrammatic cross-sections through the oblongata in an advanced embryo of Passer, indicating the arrangement of vagus, accessory, and hypoglossus nuclei (from Beccari, 1943).

(preganglionic, cranial autonomic) fibers. The location of their cells of origin has not been ascertained, but may be within or adjacent to the dorsal portions of the motor facial grisea.

Both the *nervus glossopharyngeus and vagus* display a dorsal and a ventral (ventrolateral) efferent nucleus (Figs. 257A, B, D), but the dorsal glossopharyngeus nucleus seems to be only a rather short portion of the

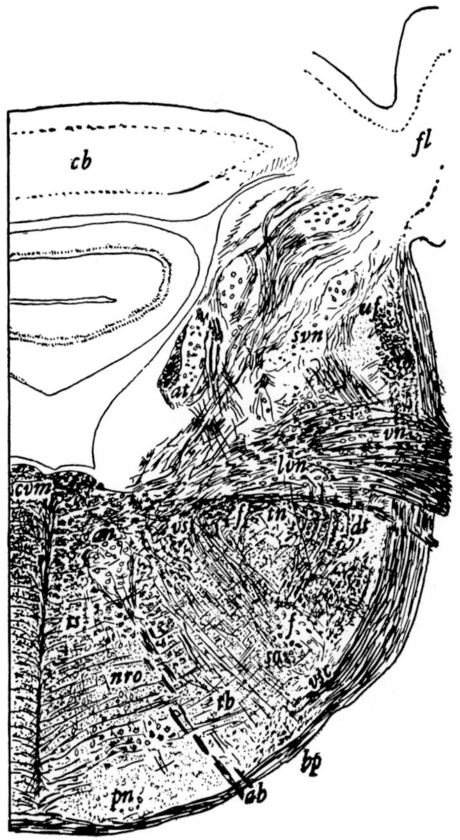

Figure 258. Cross-section (myelin stain) through the oblongata in the robin (Merula migratoria) at the abducens and facial level (from Papez, 1929). ab: abducens nerve; an: abducens nucleus; at: n. laminaris; bp: 'brachium pontis' ('pontine fibers'); cb: cerebellum; cvm: fasciculus longitudinalis medialis; dt: radix descendens trigemini; f: facial nerve and motor nucleus; fl: cerebellar flocculus; lvn: lateral vestibular nucleus; nro: formatio reticularis tegmenti; pn: 'pontine nuclei'; so: superior olive; svm: superior vestibular nucleus; tb: 'trapezoid body'; tn: caudal end of main sensory trigeminal nucleus; ts: tecto-spinal tract; uf: 'cerebellar fiber bundle'; vn: vestibular nerve; vsc: lemniscus system (?); vs: vestibular bulbo-spinal tract (?).

common dorsal nuclear column pertaining to both IXth and Xth. It is likely that the general visceral efferent (preganglionic) cells of these nerves are located in the dorsal nucleus.

The *ventral or ventrolateral nuclear cell column of IXth and Xth nerves* is also designated as *nucleus ambiguus* in conformity with human respectively mammalian anatomical nomenclature. Near the transition to spinal cord, the caudal part of that cell column represents the nucleus of the *cranial accessory nerve (accessorius vagi, n. XI cranialis)*. A *spinal acces-*

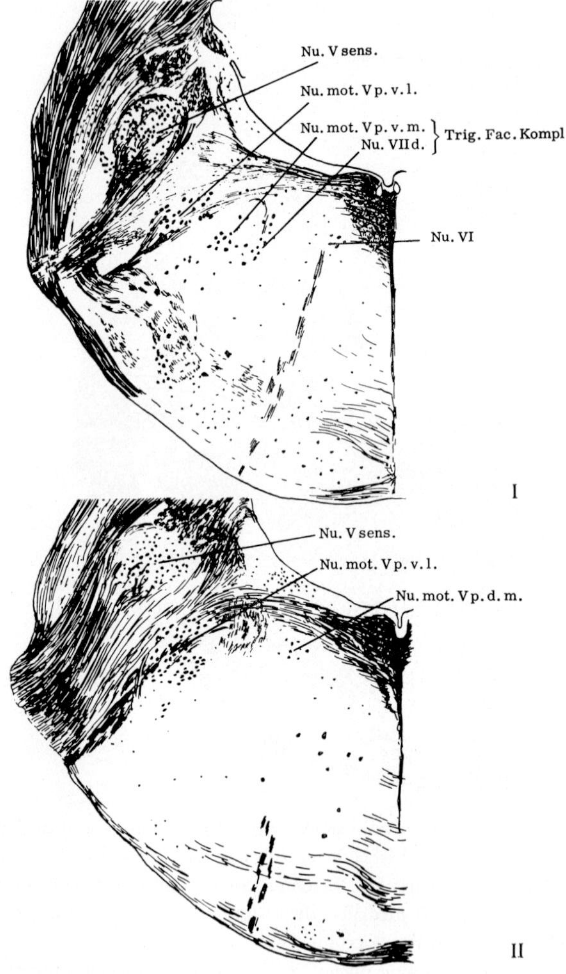

Figure 259 A. Cross-section through the oblongata in the Strigiform owl Strix aluco at levels of trigeminal, abducens, and facial nuclei (from STINGELIN, 1965). Level I is more caudal than level II.

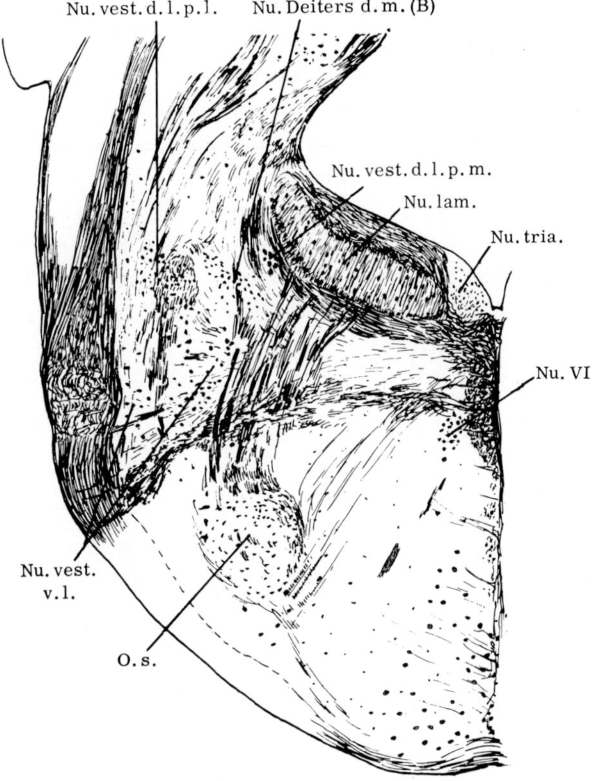

Figure 259 B. Cross-section through the oblongata in the crow Corvus corone at the vestibular level (from Stingelin, 1965).

sory component[154] is likewise present, and arises as modified dorsal roots (including axons of the type of *Lenhossek's fibers*, cf. chapter VIII section 9, p. 157, section 10, p. 171) from a neighborhood of the spinal cord's intermedioventral zone at some of the uppermost cervical segments. This neighborhood can be regarded as a caudal (spinal) extension of the nucleus ambiguus column, and the spinal rootlets originating therefrom join the cranial accessory as in Mammals.[155] The ontoge-

[154] Cf. chapter VII, section 3, p. 857 of volume 3, part II.

[155] The main difference between the Avian and the Mammalian spinal accessory seems to be the greater caudal extension of that nerve in Mammals. Further comments on the Avian spinal accessory can be found in the treatises of Romanoff (1960) and R. Pearson (1972). Detailed discussions of the 'accessorius-problem', in addition to those listed in chapter VII, volume 3, part II, are the papers by Lubosch (1899) and Beccari (1942).

netic development of the Avian spinal accessory has been described by CHIARUGI (1889), WINDLE and ORR (1934), and ROGERS (1965). The joint spinal and cranial accessory nerve seems to supply pharynx and some shoulder muscles including a presumable kathomologon of the Mammalian trapezius. The spinal accessory may essentially contribute to the innervation of these latter only.

Of the cranial nerve nuclei pertaining to the ventral or 'somatic-efferent' zone of the basal plate, the grisea giving origin to the *nervus abducens* are located at a level corresponding to that of the motor facial nuclei (Figs. 255B, 257A). The abducens root fibers, as is also the case

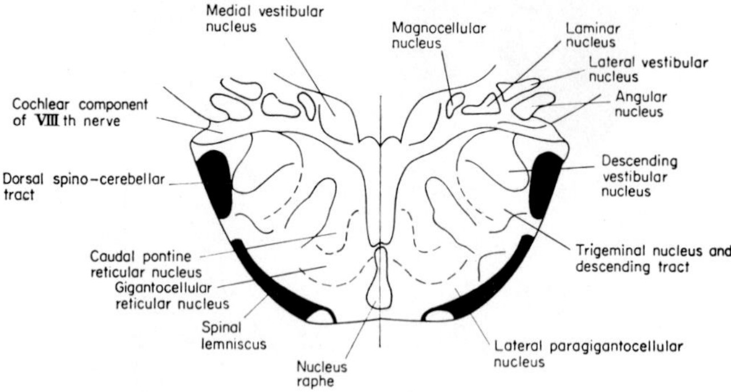

Figure 260 A. Diagram showing the general cross-sectional aspect of the Avian oblongata at the cochlear and vestibular level (from R. PEARSON, 1972).

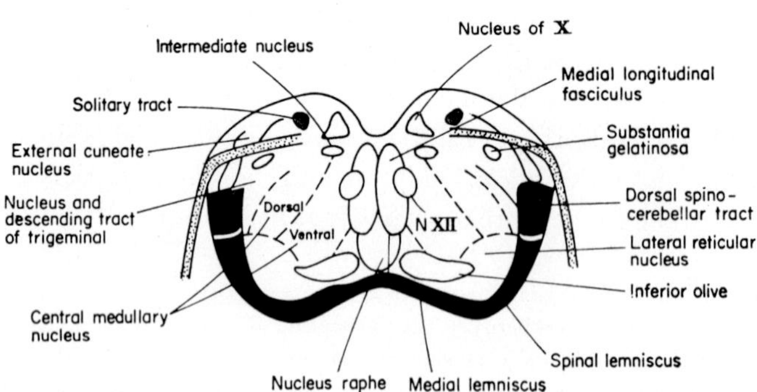

Figure 260 B. Diagram showing the general cross-sectional aspect of the Avian oblongata at the hypoglossal level (from R. PEARSON, 1972).

in some Reptiles (cf. p. 480), emerge rostrally to those of the facial.[156] The nucleus principalis abducentis assumes a medial and rather dorsal position, close to the fasciculus longitudinalis medialis. In numerous forms, a smaller, ventrolaterally displaced accessory nucleus has been described (ECKHARDT and ELLIOT, 1935, LEVI-MONTALCINI, 1942, and others), which lies at a considerable distance from the main nucleus, near the surface of the medulla, in close vicinity to the descending trigeminal root and to the superior olive (Fig. 256B). It is believed that this accessory nucleus innervates the nictitating membrane of the eye, and KAPPERS (1947) considers the position of the accessory nucleus adjoining the descending fibers of the trigeminal ophthalmic branch to be '*un example classique de l'influence neurobiotactique*'. ADDENS (1933) suggests the possible presence of additional Avian accessory abducens nuclei.[157]

The Avian *hypoglossal nerve* originates from at least two nuclear grisea, of which the ventral one lies close to the fasciculus longitudinalis medialis, while the dorsal nucleus appears to have been displaced toward the vagus region (Figs. 257B, C). Additional fragmentations of both dorsal and ventral hypoglossal nuclei, as described by BECCARI (1943) and others, can be noted. The complexity of the Avian hypoglossal griseum, whose details still remain poorly clarified, might be related to the fact that a branch of nervus hypoglossus innervates the *syrinx*. This latter, whose muscles are believed to derive from the sternohyoid musculature, represents the Avian vocal organ, and occurs exclusively in that Vertebrates class. The syrinx, concerned with voice production is thus functionally analogous to the larynx, but not homologous, being instead a modified tracheobronchial junction. A morphologically homologous larynx, characteristic for tetrapod Vertebrates, is also present in Birds, but lacks here the vocal folds and a thyroid cartilage.

In concluding the enumeration of efferent cranial nerve components, the *centrifugal cochlear bundle*, whose griseum of origin has not been exactly ascertained, but which may arise in the vicinity of the superior olive, should also be mentioned. In the Pigeon, this bundle is

---

[156] In the Ratite *Casuarius*, however, the abducens roots emerge caudally to the facial nerve (cf. e.g. STINGELIN, 1965).

[157] According to ADDENS (1933) an accessory abducens nucleus is also said to occur in some Amphibians, but the evidence does not seem conclusive. However, a nictitating membrane does make its appearance in that Vertebrate class.

said to decussate rostrally to the abducens nucleus and to follow a ventral course from the raphé, intersecting the facial genu. The efferent cochlear fibers accompany the facial ones for a short distance and join the vestibular nerve as small fascicles, finally passing into the cochlear nerve (BOORD, 1941). Efferent systems of this type, ending in receptor structures such as e.g. labyrinth (HELD, 1893; cf. also section 1, p. 319), or retina (CAJAL, 1911) seem to provide excitatory or inhibitory feedback loops or both, 'calibrating', as it were, the degree of receptor activities. In this respect, their function might be compared with that of the gamma fibers innervating muscles spindles (cf. vol. 3, part II, chapter VII, section 1, p. 809), although these latter fibers are of true 'motor' type.

According to BOORD (1961) the efferent cochlear bundle potentiates the cochlear microphonic potential which is possibly generated at the apex of the hair cells. It simultaneously reduces the voltage of the auditory nerve's response to sound. This produces a marked reduction in the acoustic input to the ascending auditory pathway as indicated by direct recordings of the evoked potentials following stimuli (cf. also the discussion of that system by R. PEARSON (1972).

Within the overall basal plate of the Avian oblongata, the *nucleus reticularis tegmenti* is conspicuously displayed by diffuse cell groups of large and smaller elements and their neuropil, imbedded in a 'grid' of medullated fiber tracts. As in Reptiles and other Vertebrates, it is possible to distinguish with BECCARI (1943) a superior reticular 'nucleus' at the trigeminal level, an intermediate 'nucleus' at the octavus level, and an inferior 'nucleus' at the vagus level. At caudal levels of the oblongata, BECCARI recorded in the ventral central gray a median small-celled *nucleus rhomboidalis* (Fig. 257C) which may or may not be related to the reticular formation. A 'forerunner' of *nucleus intercalatus*, to be dealt with in the discussion of the Mammalian bulb, was also noted at these levels by BECCARI.

KARTEN and HODOS (1967), on the other hand, have adopted a somewhat different scheme of parcellation in their atlas of the Pigeon's brain. Six of the more than 10 reticular nuclei delimited by the cited authors (n. reticularis gigantocellularis, n. reticularis parvocellularis, etc.) are shown in Figures 255C, D and 260A. Farther caudally, KARTEN and HODOS describe a *nucleus centralis oblongatae* with further subdivisions, and a *nucleus reticularis lateralis* (Fig. 260B). At rostral levels, toward and within the isthmus region, they recorded a *nucleus reticularis pontis oralis*, a *nucleus tegmenti ventralis*, a '*nucleus of the locus ceruleus*' comparable to the homonymous Mammalian griseum to be dealt with in section 10, and a '*nucleus subceruleus*' (*dorsalis et ventralis*). In addition, they noted a *nucleus tegmenti dorsalis* within the isthmic ventral central gray caudal to the trochlear nucleus.

The overall connections of the Avian reticular formation are, *mutatis mutandis*, essentially similar to those pointed out in dealing with Anamnia and Reptilia. These channels form a grid of interwoven transverse and longitudinal bundles which include collaterals of arcuate fibers, reticulo-spinal bundles, reciprocal reticulo-cerebellar pathways, contributions to fasciculus longitudinalis medialis,[158] descending and ascending systems linking the bulbar reticular formation with more rostral grisea of the neuraxis, and numerous undefined shorter intrabulbar links.

The ventral surface region of the basal plate in the rostral portion of the oblongata contains cell groups which PAPEZ (1929) identified as *pontine nuclei* (Figs. 255C, D, 258). CRAIGIE (1930) and KAPPERS (1936, 1947) likewise recognized these grisea as primordial pontine structures, and further evidence substantiating this interpretation was obtained by BRODAL *et al.* (1950). Generally speaking, a medial and a lateral pontine nucleus may be distinguished. As in Mammals, the location of these grisea within the basal plate is secondary, being the end-result of a cell migration from their ontogenic site of origin within the rhombic lip of the alar plate (HARKMARK, 1954). Again, as in Mammals, the pontine nuclei project upon regions of cerebellar cortex (cf. chapter X). These fiber tracts represent a bilateral '*brachium pontis*'. Although it seems likely that the Avian pontine nuclei receive descending fibers from the telencephalon, perhaps in addition to other fibers of uncertain origin, it appears improbable that such descending telencephalic projection is of cortical origin. In this respect, the Avian pontine grisea[159] presumably differ from the Mammalian ones. Again, if

[158] According to BECCARI (1943) uncrossed neurites, originating from his nucleus reticularis intermedius in Selachians, Reptilians, and Birds, provide a distinctive descending bundle along the lateral side of the fasciculus longitudinalis medialis. It was designated by this author as the *fasciculus longitudinalis paramedialis*, and described as reaching the spinal cord.

[159] In view of the much lesser cerebellar development in Reptiles, it seems less likely, although not entirely impossible, that some 'pontine' cell groups might be present in this Sauropsidan group. Instead of migrated pontine cells derived from the rhombic lip, some local reticular elements might have comparable cerebellar connections, but the available data are still inconclusive.

the term 'pons' is taken in the sense of macroscopic anatomy, then a 'true' pons as displayed by Mammals is not present in Birds, whose *nuclei pontis*, regardless of some differences in their input channels, are nevertheless *orthohomologous* with the Mammalian ones. Whether, *in toto*, the Avian 'pons' should or should then be evaluated as orthohomologous or kathomologous with respect to the Mammalian pons remains thus an arbitrarily decidable question (cf. also vol. 3, part II, chapter VI, section 1B, p. 185 and section 4, p. 375).

Another griseum located within the bulbar general basal plate is the *superior olivary nucleus*, intercalated in that section of the cochlear pathway designated as '*trapezoid body*' (Figs. 255B, E, 256A, B, 257A, 258, 259B). KAPPERS (1920) and others distinguished a dorsal and a ventral portion of the Avian superior olive. Grisea of the trapezoid body, corresponding to the trapezoid nuclei of Mammals, have not been identified, but KAPPERS *et al.* (1936) suggest that they may be represented by scattered reticular elements. A somewhat more dorsolateral cellular group, adjacent to or within the gathering fibers of the lateral lemniscus can be designated as *nucleus lemnisci lateralis* and might be regarded as pertaining to the superior olivary complex (KAPPERS *et al.*, 1936). It is, however, doubtful whether this scattered cell population, which has also been recorded in Reptiles (FREDERIKSE, 1931) can be convincingly delimited, and whether at least part of it is not an interstitial nucleus of the lateral lemniscus located in, as well as derived from, the alar plate. Even the superior olivary complex itself, although located in, and being a component of, the adult basal plate, may be partly or wholly an alar plate derivative.[160] Depending on the taxonomically different Avian forms, the superior olivary complex proper is said to contain between about 2,000 and 12,000 neuronal elements (WINTER and SCHWARTZKOPFF, 1961).

---

[160] The evidence for a ventralward directed cell migrations from the alar plate (rhombic lip and its neighborhood), particularly studied by ESSICK (1907, 1912) and HARKMARK (1954) seems convincing as well as conclusive, but its extent is only incompletely ascertained. According to HARKMARK, it involves the nuclei pontis, the inferior olivary complex, the 'nucleus raphe' and 'some other structures'. It seems possible but by no means certain that these latter include parts or even the whole of the superior olivary griseum. HARKMARK (l.c.) stresses the difficulties of accurately identifying cell migrations in static histologic preparations. Experimental methods introduce various intrinsic sources of error or ambiguity. No recent investigation using contemporary 'tagging' methods, e.g. by means of radioautography, have come to my attention, but even these improved techniques are not entirely free of certain limitations.

Except for the fiber pathways originally observed by CAJAL and depicted in Figures 255E and 256A, nothing certain is known about the fiber connections of the Avian superior olivary complex.[161] It is not believed that its neuronal elements contribute to the lemniscus lateralis, which arises from nucleus angularis, nucleus magnocellularis, and nucleus laminaris. Extrapolating from observations made by YOSHIDA (1925) in Mammals, KAPPERS *et al.* (1936) believe that the superior olive contributes fibers to fasciculus longitudinalis medialis and adjacent reticular bundles ('fasciculus praedorsalis'), some of which may reach the spinal cord, and, in addition, distributes fibers to reticular and motor bulbar grisea.

With regard to the cochlear pathway the manifold but still incompletely analyzed combinations of homolateral and of contralateral (crossing, decussating) primary[162] as well as secondary fibers should be emphasized. Reciprocal interaction (Germ. *Wechselwirkung*) between sound reception of both ears can thus be established at diverse unspecified levels. Such interaction is a prerequisite for the at least partially possible registration of sound localization and distance by the neural mechanisms (cf. also KAPAL, 1962).

The *inferior olivary complex* of Birds (Figs. 257B, D, 260C), located near the basal bulbar surface at caudal levels of the oblongata, was already noted by the early investigators of the Avian brain. It is conspicuously developed in correlation with the expanded cerebellum characterizing this vertebrate class. Generally speaking, a dorsal and a ventral lamella can be dinstinguished, each of which again displays additional subdivisions. According to KOOY (1917), who undertook a comprehensive comparative and phylogenetic survey of the inferior olivary griseum in the entire Vertebrate series, lateral parts of the dorsal lamella correspond to the dorsal accessory olive in Mammals, while medial portions represent the Mammalian medial accessory olivary nucleus, which presumably is the 'phylogenetically oldest' olivary gri-

---

[161] SCHWARTZKOPFF (1968), in contradistinction to CAJAL, assumes that the principal projections to the superior olive originate from the nucleus angularis, while many of the fibers from the laminar nucleus merely pass through the superior olivary connections. R. PEARSON (1972) refers to the controversial aspect of this question.

[162] The question concerning the crossing of (afferent or efferent) root fibers in the Vertebrate neuraxis is still incompletely elucidated and remains controversial. Some aspects of this problem have been discussed by ADDENS (1934). Cf. also above footnote 142a, p. 481.

seum, already found in Selachians. Most of the ventral lamella, on the other hand, seems to be homologous with the main Mammalian olivary complex which, to some extent, can be evaluated as a 'neencephalic structure'. In this respect, it is of interest that 'neencephalic' pontine grisea likewise make their first clear-cut appearance in Birds. Subsequently to KOOY's study, VOGT-NILSEN (1954) presented a detailed investigation of the Avian inferior olivary complex in more than a dozen

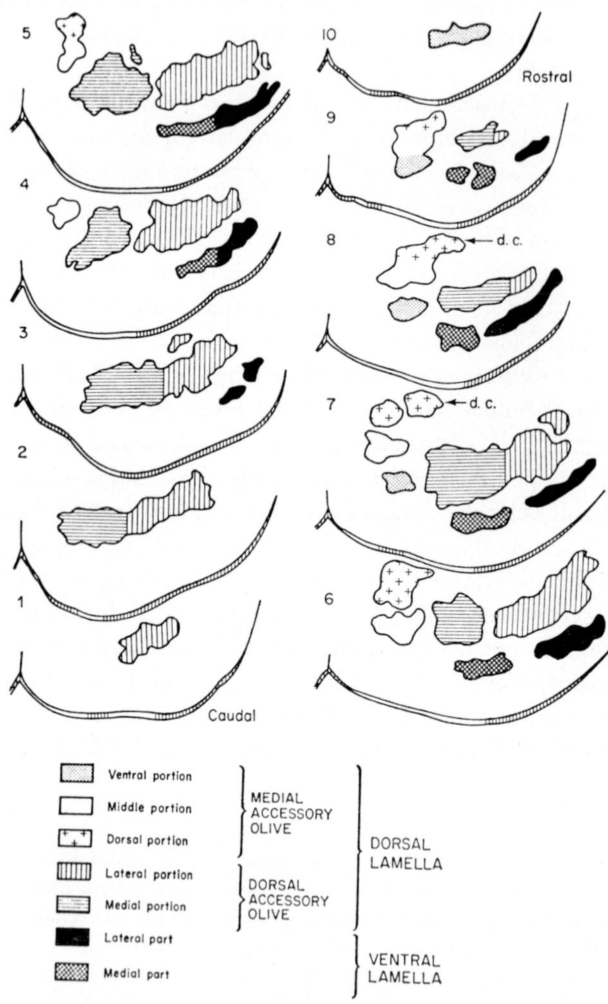

Figure 260 C. Semidiagrammatic sketches illustrating configuration and components of the Avian inferior olivary complex (after VOGT-NILSEN, 1954, from R. PEARSON, 1972).

species. The gist of his results, which are fully compatible with KOOY's overall conclusions, is indicated by the diagrams of Figure 260C. The alar plate origin of all or much of the inferior olivary grisea was mentioned above in footnote 160 and also referred to in section 4, p. 386 of chapter VI, volume 3, part II. The connections of the olivary complex with rostral regions and with the spinal cord are still poorly understood. The somewhat better elucidated, essentially crossed olivo-cerebellar connections shall briefly be dealt with in chapter X.

As regards important *descending channels* providing input to efferent nuclei, components of fasciculus longitudinalis, other descending fibers within reticular formation, moreover the crossed tractus rubrobulbaris (with its presumptive rubrospinal extension), and the tectobulbar system should again be mentioned. With respect to *ascending channels* providing bulbar input, it will here be sufficient to mention the assumed bulbar connections of the spinal lemniscus system. Ascending channels carrying bulbar output are represented by bulbotectal and bulbothalamic fibers joining the two main lemniscus systems, as well as by intrinsic fiber components of the reticular formation (including fasciculus longitudinalis medialis) transmitting impulses to the mesencephalic tegmentum and presumably also to diencephalic, respectively telencephalic, grisea.

10. Mammals (Including Man)

The oblongata *sensu latiori* of Mammals, comprising the subdivisions of the rhombencephalon designated as *pons* (ventral part of metencephalon) and as *myelencephalon* (medulla oblongata *sensu strictiori*) has been dealt with by many investigators whose studies provided a profusion of details. In the aspect here under consideration, namely within the framework required for a 'general survey' of the Vertebrate central nervous system's comparative anatomy including an introduction to 'pertinent biologic and logical concepts' it is obviously impossible to give a reasonably complete account of the manifold information available in the multitudinous literature. As regards this latter, it must here suffice to point out a few publications specifically containing, together with numerous bibliographic references, data relevant for the comparative anatomy of the Mammalian oblongata.

VERHAART's (1970) comprehensive treatise concerns the topography of main fiber systems and major grisea in about 18 different Mammalian taxonomic groups. Data on the oblongata of *Monotremes* are contained in the papers by ABBIE (1934) and HINES (1929). SCHNEIDER (1957), in his investigation of the brain of *Chiroptera*, likewise brings some observations on the oblongata. A description of the oblongata in *Rodents* is given in CRAIGIE's (1929) Neuroanatomy of the Rat, revised and expanded by ZEMAN and INNES (1963).

Much information about the configuration of the Mammalian oblongata can also be found in the treatise by FLATAU and JACOBSOHN (1899) and in the contribution by HALLER V. HALLERSTEIN (1934) to vol. 2 of the *Handbuch der vergleichenden Anatomie der Wirbeltiere* (BOLK *et al.*).

A very careful, detailed, and critical description of the *human oblongata* has been presented by ZIEHEN (1903, 1913, 1920, 1926) in his unfortunately not completed contribution to BARDELEBEN's *Handbuch der Anatomie des Menschen*. Other particular and pertinent aspects of the human oblongata not dealt with by ZIEHEN have been considered in the papers by GAGEL and BODECHTEL (1930) and by my associate HIKIJI (1933). Among atlases specifically dealing with the Mammalian brain stem are those by MEESEN and OLSZEWSKI (1949) for the *Rabbit*, and, for *Man*, by JACOBSOHN (1909), by OLSZEWSKI and BAXTER (1954) and by TILNEY and RILEY (1943). A wealth of fundamental data concerning generally Vertebrate and also specifically Mammalian bulbar structural aspects has been elucidated by the fundamental studies of CAJAL (1909 and many others), which are summarized in that author's classic Treatise.[163]

Figures 261A to 266E illustrate representative cross-sectional bulbar levels in the brain of a few 'lower' and 'intermediate' Mammalian forms.[164] Comparing e.g. Figure 261A *(Monotreme Echidna)* with Fig-

[163] In addition to the here given short selection and to the previously cited extant texts on comparative neurology, the numerous investigations of the *Obersteiner School*, contained in '*Arbeiten aus dem Neurologischen Institut an der Universität Wien*', and those by G. FUSE and his collaborators, contained in '*Arbeiten aus dem Anatomischen Institut der Kaiserlich-Japanischen Universität zu Sendai*' should be mentioned as especially dealing with the Mammalian brain and including many relevant contributions to the comparative anatomy of brainstem respectively oblongata in this Vertebrate class. With respect to the oblongata of Monotremes and Marsupials, the significant contributions by ZIEHEN (1897, 1908) in *Semon's zoologische Forschungsreisen)* should also be particularly mentioned. Thus, various figures in ABBIE's (1934) paper are based on the *Semon-Ziehen series* of the Echidna brain.

[164] The reader interested in further details and illustrations concerning the oblongata in different Mammalian forms is referred to the cited publications by VERHAART (1970), HALLER V. HALLERSTEIN (1934), and FLATAU and JACOBSOHN (1899).

ure 253A *(Reptilian Crocodile)*, the overall similarity of griseal and fiber arrangements in Mammalian and Sauropsidan oblongata is clearly recognizable, but it can be seen that the presence of a well developed pons and of a pyramidal tract represent here two significant characteristics of 'progressive' differentiation. Concomitantly with the development of a *pons* as a *macroscopically* significant bulbar subdivision, the alar plate neighborhood connecting cerebellum with brain stem becomes, on each side, more or less segregated into three distinctive stalk-like configurations, of which two essentially pertain to the rhombencephalic bulb, namely the *brachium pontis* or middle cerebellar peduncle, and the inferior cerebellar peduncle or *corpus restiforme*. A particular medial subdivision of this latter, containing nucleo-cerebellar fibers from the

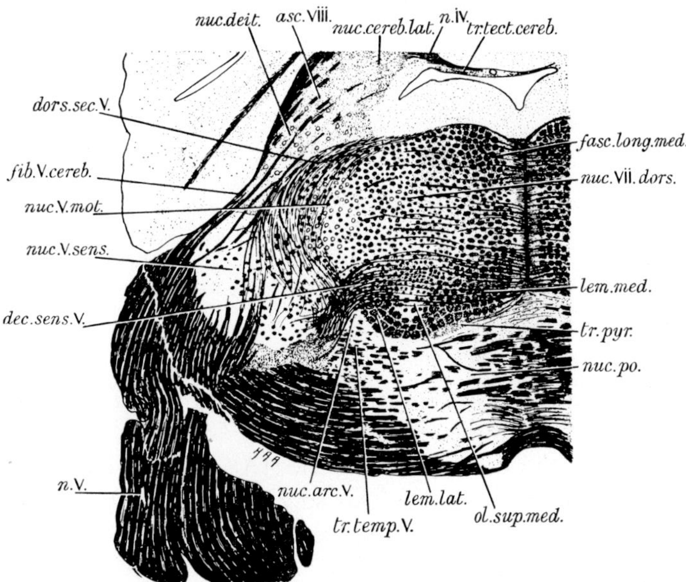

Figure 261 A. Cross-section through the oblongata of the Monotreme Echidna at the level of entrance of the fifth nerve as seen in a myelin stain (from Abbie, 1934). asc. VIII: vestibulo-cerebellar fibers; dec. sens. V: 'trigeminal sensory decussation'; dors. sec. V: 'dorsal secondary trigeminal fibers'; fib. V. cereb.: 'trigeminal fibers passing to cerebellum'; lem. med.: lemniscus medialis; n. IV: trochlear nerve; nuc. arc. V: 'nucleus arcuatus trigemini'; nuc. cereb. lat.: nucleus lateralis cerebelli; nuc. deit.: *Deiters' nucleus;* nuc. po.: nuclei pontis; ol. sup. med.: medial nucleus of superior olivary complex; tr. pyr.: pyramidal tract; tr. tect. cereb.: tractus tecto-cerebellaris; tr. temp. V: 'tractus temporo-trigeminalis. Other abbreviations self-explanatory.

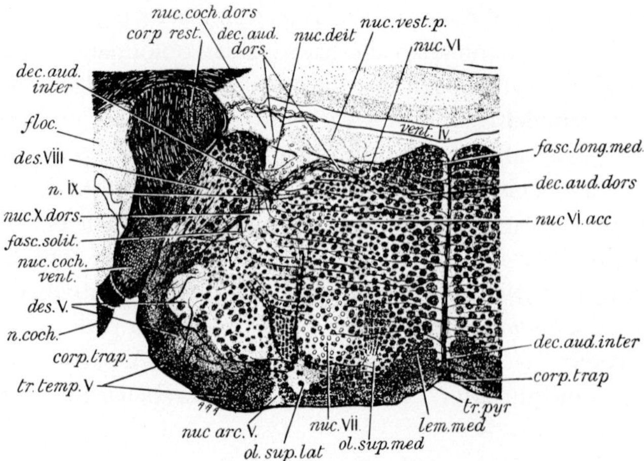

Figure 261B. Cross-section through the oblongata of Echidna at the level of abducens and motor facial nerve nucleus (after ABBIE, 1934). corp. rest.: corpus restiforme (inferior cerebellar peduncle); corp. trap.: corpus trapezoideum; dec. aud. dors.: 'dorsal auditory decussation *(of von Monakow)*'; dec. aud. inter.: 'intermediate auditory decussation *(of Held)*'; desc. V: descending trigeminal root and its nucleus; desc. VIII: descending vestibular root and its nucleus; floc.: cerebellar flocculus; n (IX, resp. coch.): nerve root; nuc. (coch., deit. etc.): nucleus; nuc. VI acc.: accessory abducens nucleus; ol. sup. lat., med.: lateral and medial superior olivary grisea. Other abbreviations as in preceding figure or self-explanatory.

vestibular grisea, as well as some other ascending and descending cerebellar fiber systems, is the so-called *juxtarestiform body* or internal segment of the inferior cerebellar peduncle (IAK: *innere Abteilung des unteren Kleinhirnstieles autorum)*. The superior cerebellar peduncle or *brachium conjunctivum*, connecting cerebellum with midbrain, runs through the isthmus rhombencephali.

The *ventricular sulci* related to the longitudinal zonal system display a variable degree of distinctness in the different adult forms, the *sulcus limitans* usually remaining recognizable as regards most of the zonal system's length, while sulcus intermedius ventralis and sulcus intermedius dorsalis become smoothened out and are frequently even barely or not at all indicated at diverse ontogenetic stages.[165]

---

[165] In the human oblongata, however, to be dealt with further below, both dorsal and ventral intermediate sulci can be clearly displayed at caudal levels in the adult stage.

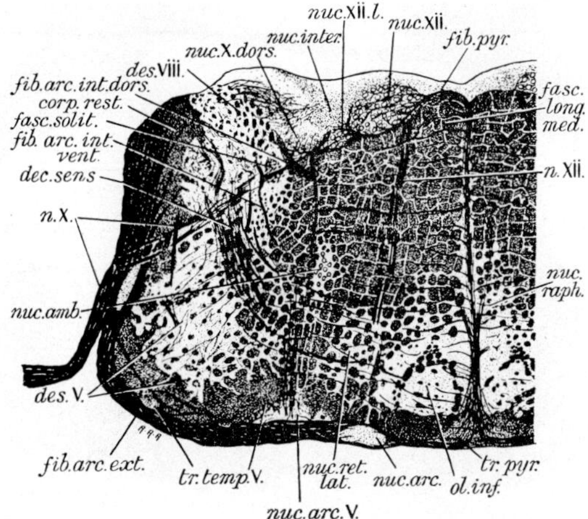

Figure 261 C. Cross-section through the oblongata of Echidna at the level of the hypoglossus nucleus (from ABBIE, 1934). dec. sens.: 'decussating sensory fibers' (fibrae arcuatae sive arciformes internae); des. V, VIII: radix descendens trigemini respectively vestibularis and their nuclei; fib. arc. ext.: fibrae arcuatae externae; fib. arc. int. dors. vent.: fibrae arcuatae internae dorsales respectively ventrales; fib. pyr.: pyramidal tract fibers; nuc. amb.: nucleus ambiguus; nuc. arc.: nucleus arcuatus; nuc. arc. V: 'nucleus arcuatus trigemini'; nuc. inter.: *nucleus intercalatus Staderini;* nuc. raph.: (reticular) nucleus of the raphé; nuc. ret. lat.: lateral reticular nucleus; ol. inf.: nucleus olivaris inferior; tr. pyr.: pyramidal tract; tr. temp. V.: 'tractus temporo-trigeminalis'. Other abbreviations as in preceding figures or self-explanatory.

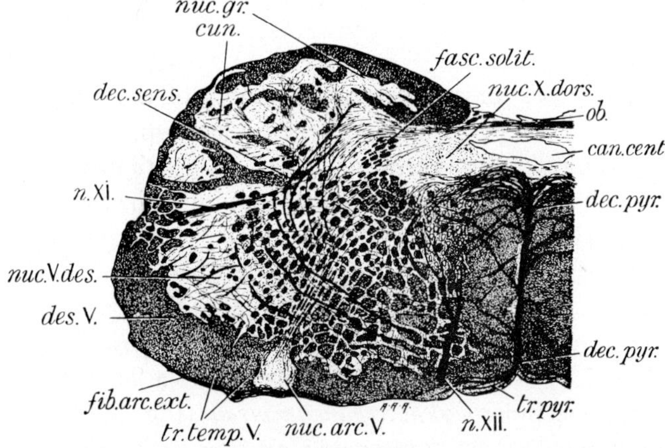

Figure 261 D. Cross-section through the closed part of the oblongata of Echidna (from ABBIE, 1934). dec. pyr.: pyramidal decussation; nuc. cun. grac.: nucleus cuneatus respectively gracilis; n. XI: nervus accessorius (cranialis); n. XII: nervus hypoglossus (its nucleus of origin, not labelled, can easily be recognized; ob: obex (with commissura infima). Other abbreviations as in preceding figures or self-explanatory.

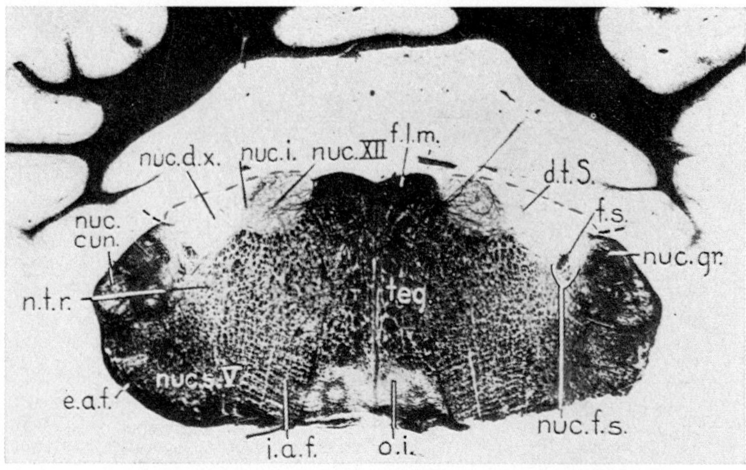

Figure 261E. Cross-section through the oblongata of the Monotreme Ornithorhynchus at the level of hypoglossal nucleus and inferior olive as seen by means of the *Weigert stain* (from HINES, 1929). d.t.S.: dorsal tegmental bundle (fasciculus longitudinalis dorsalis) *of Schütz;* e.a.f.: external arcuate fibers; f.l.m.: fasciculus longitudinalis medialis; f.s.: fasciculus s. tractus solitarius; i.a.f.: internal arcuate fibers; nuc.XII: nucleus hypoglossi; nuc.cun.: nucleus cuneatus; nuc.d.X: nucleus dorsalis vagi; nuc.grac.: nucleus gracilis; n.f.s.: nucleus fasciculi solitarii; nuc.i.: *nucleus intercalatus Staderini;* nuc.s.V: nucleus radicis descendentis trigemini; n.t.r.: 'nucleus triangularis reticularis'; o.i.: inferior olivary nucleus; teg.: tegmentum.

Relevant gross and microscopic features of the *human oblongata*, which, again in the aspect here under consideration, is essentially similar to that of most other Primates, shall be dealt with further below following a general discussion of bulbar structures in the 'subprimate' forms.

The *dorsal zone* of the Mammalian *bulbar alar plate* contains, as in Sauropsidans, the *cochlear* and *vestibular grisea* of the octavus. The cochlear nerve reaches two griseal aggregates, the *nucleus cochlearis dorsalis (tuberculum acusticum)*, which commonly shows a 'laminar' structure, and the magnocellular *nucleus cochlearis ventralis* (Fig. 267A). Their Sauropsidan homologa are presumably nucleus angularis and nucleus magnocellularis, respectively (cf. above p. 485). In Marsupials, the ventral cochlear nucleus has retained a rather dorsomedial position, more or less comparable to the location of the magnocellular nucleus in the Sauropsidan oblongata.

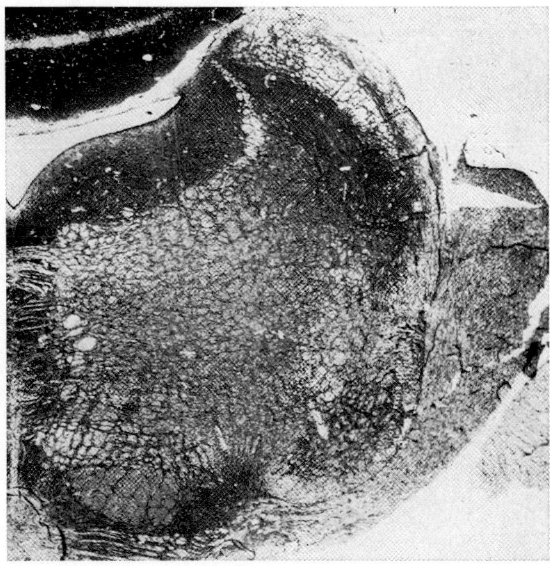

Figure 262 A. Cross-section through the oblongata of the marsupial Opossum at a middle pontine level. *Häggqvist stain;* ×27; red. $^2/_3$ (from VERHAART, 1970).

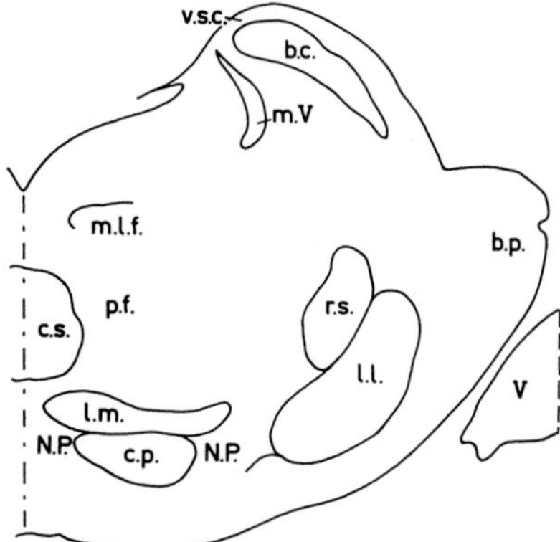

Figure 262 B. Explanatory outline sketch of Figure 262 A (from VERHAART, 1970). b.c.: brachium conjunctivum; b.p.: brachium pontis; c.p.: pedunculus cerebri; c.s.: (reticular) 'central superior nucleus'; l.l.: lateral lemniscus; l.m.: medial lemniscus; m.l.f.: fasciculus longitudinalis medialis; m.V.: radix mesencephalica trigemini; N.P.: basal pontine nuclei; p.f.: 'predorsal fascicle; r.s.: rubrospinal tract; v.s.c.: ventral spino-cerebral tract; V: extracerebral root of trigeminal nerve; VII, VIII: facial respectively cochlear nerve roots.

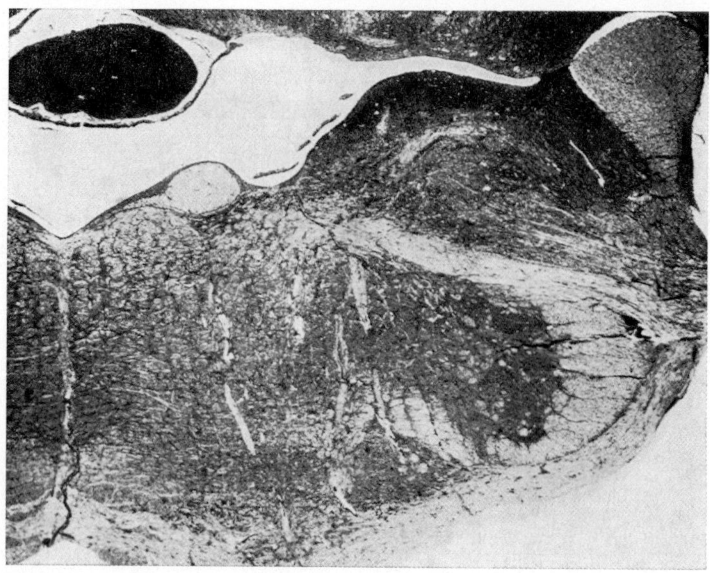

Figure 262C. Cross-section through the oblongata of the Opossum at the level of trapezoid body and superior olivary complex. *Häggqvist stain;* ×21; red. $^2/_3$ (from Verhaart, 1970).

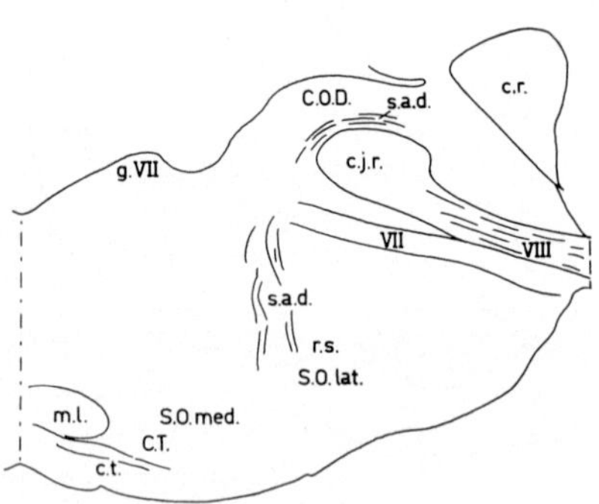

Figure 262D. Explanatory outline sketch of Figure 262C (from Verhaart, 1970). c.j.r.: corpus juxtarestiforme; C.O.D.: dorsal cochlear nucleus; c.r.: corpus restiforme; C.T.: nucleus of the corpus trapezoideum; c.t.: corpus trapezoideum; g.VII: (internal) genu of facial nerve; m.l.: medial lemniscus; r.s.: rubrospinal tract; s.a.d.: dorsal acoustic stria; S.O.lat.: 'lateral part of the superior olive'; S.O.med.: 'medial part of the superior olive'.

Recent interpretations concerning the cochlear nuclei in the Cat, based on patterns of degeneration observed by means of light- and particularly electron-microscopy, can be found, together with the pertinent bibliographic references, in a paper by KANE (1974). The cited author distinguishes, in the caudal portion of the nucleus cochlearis ventralis, a so-called '*octopus cell area*' into which descending fibers of the cochlear nerve, proceeding to the dorsal cochlear nucleus, seem to give off collaterals. In accordance with the present-day *Newspeak*, the shortened designation OCA is introduced for said neologism.

The entering cochlear root fibers separate or bifurcate into an ascending and descending branch. This pattern of bifurcation, which includes a distinction between dorsal and ventral branches, is rather complex, and, although illustrated by CAJAL (1909) on the basis of *Golgi preparations*, and subsequently studied by numerous others, cannot be considered sufficiently elucidated. Be that as it may, the primary cochlear root fibers reach the ventral and the dorsal cochlear nucleus. In the latter, and in parts of the ventral cochlear nucleus, these fibers seem to end with diffuse, neuropil-like plexuses, while in at least a large part of the ventral nucleus the synapses are of a peculiar end-bulb or basket type (Fig. 267B) described by HELD (1891, 1893). This latter arrangement can be interpreted as providing one to one or many-one connections essentially compatible with a so-called '*point to point projection*' from the cochlea. Such topologic distribution was shown to obtain by several authors. Thus, LEWY and KOBRAK (1936) found that, in the Rabbit, the fibers from the basal part of the cochlea, carrying impulses presumably corresponding to high frequency tones, ended dorsally in both ventral cochlear nucleus and tuberculum acusticum, while the fibers from the apical part, believed to be activated by low frequency tone, reached the ventral parts of both grisea, especially the medioventral portion of the ventral nucleus. However, the projection of the most basal cochlear neighborhood, particularly in the vicinity of fenestra rotunda and assumed to transmit the tones of highest pitch, differed from that of the rest of the cochlea. The total number of degenerated fibers obtained by lesions of this basal extremity of *Corti's organ* was comparatively small; their endings split up in the deepest layers of tuberculum acusticum, and, rather diffusely, in the ventral cochlear nuclous.[166] It seems, moreover, that the topical representation of the

[166] This might, e.g., perhaps to some extent agree with CAJAL's (1909, 1952) Figure 330 on p. 782, based on an interpretation of his *Golgi preparations* of nucleus cochlearis ventralis in the newborn dog.

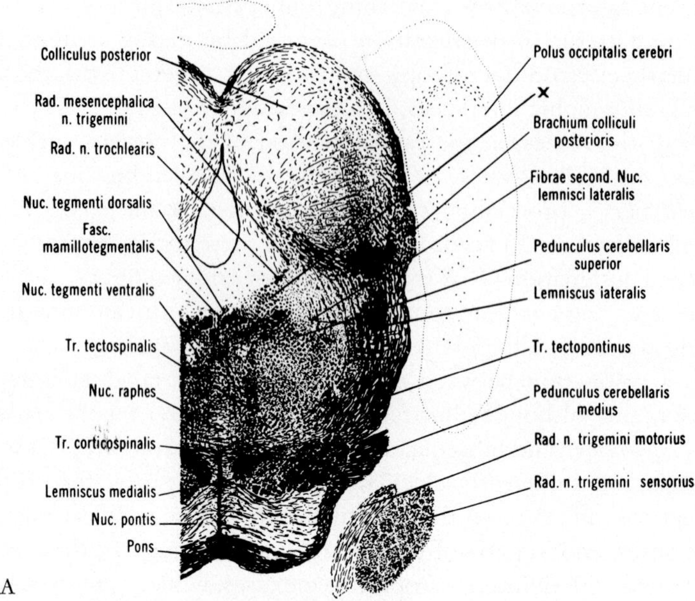
Colliculus posterior
Rad. mesencephalica n. trigemini
Rad. n. trochlearis
Nuc. tegmenti dorsalis
Fasc. mamillotegmentalis
Nuc. tegmenti ventralis
Tr. tectospinalis
Nuc. raphes
Tr. corticospinalis
Lemniscus medialis
Nuc. pontis
Pons
Polus occipitalis cerebri
x
Brachium colliculi posterioris
Fibrae second. Nuc. lemnisci lateralis
Pedunculus cerebellaris superior
Lemniscus iateralis
Tr. tectopontinus
Pedunculus cerebellaris medius
Rad. n. trigemini motorius
Rad. n. trigemini sensorius

263 A

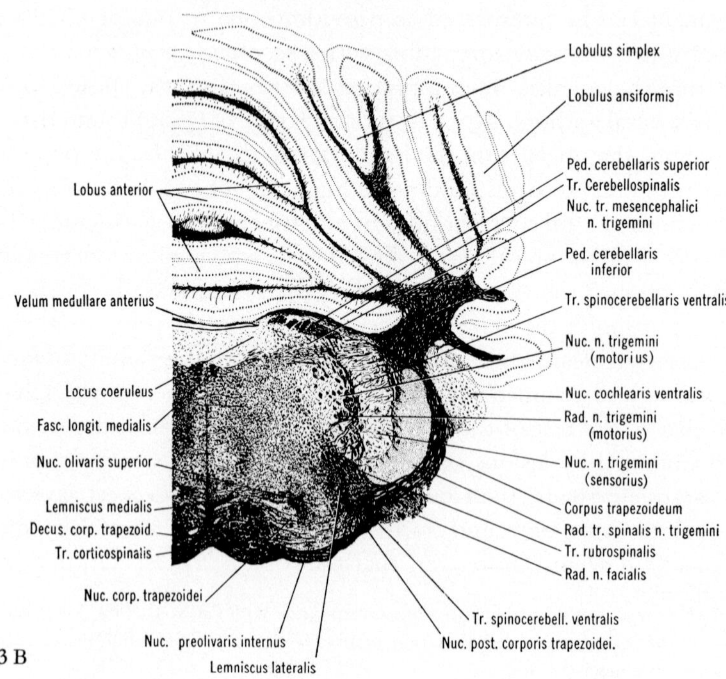
Lobulus simplex
Lobulus ansiformis
Lobus anterior
Ped. cerebellaris superior
Tr. Cerebellospinalis
Nuc. tr. mesencephalici n. trigemini
Ped. cerebellaris inferior
Velum medullare anterius
Tr. spinocerebellaris ventralis
Nuc. n. trigemini (motorius)
Locus coeruleus
Nuc. cochlearis ventralis
Fasc. longit. medialis
Rad. n. trigemini (motorius)
Nuc. olivaris superior
Nuc. n. trigemini (sensorius)
Corpus trapezoideum
Lemniscus medialis
Rad. tr. spinalis n. trigemini
Decus. corp. trapezoid.
Tr. rubrospinalis
Tr. corticospinalis
Rad. n. facialis
Nuc. corp. trapezoidei
Tr. spinocerebell. ventralis
Nuc. preolivaris internus
Nuc. post. corporis trapezoidei.
Lemniscus lateralis

263 B

cochlea in the primary acoustic grisea might perhaps somewhat differ in the various Mammalian taxonomic groups. Divergent or contradictory statements, together with further bibliographic references, can be found on pp. 146–148 of CROSBY's *et al.* (1962) 'Correlative Anatomy of the Nervous System'. Although the relevant problems still remain poorly elucidated, topical cochlear representation as such can, nevertheless, be considered reasonably well established.

Crossed and presumably also some homolateral secondary fibers arising in the cochlear nuclei ascend to higher brain levels and gather in

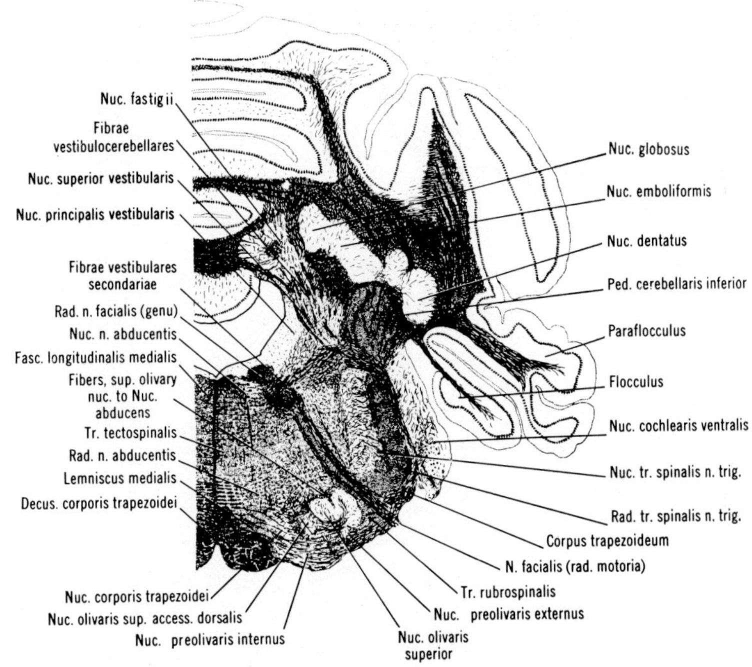

Figure 263 C. Cross-section through the oblongata of the Rat at the level of abducens nucleus and motor facial root exit (from CRAIGIE, ZEMAN, and INNES, 1963).

Figure 263 A. Cross-section through the oblongata of the Rat (Eutheria, Rodentia) at the pontine level (from CRAIGIE, ZEMAN, and INNES, 1963). Figures 263A–F are taken from *Weigert series*, ×15; red. $^{2}/_{3}$. The plane of the section A passes through pons ventrally, isthmus below the ventricle (aqueduct), and mesencephalic posterior coliculus above ventricle. Added lead x indicates approximate position of nucleus loci coerulei.

Figure 263 B. Cross-section through the oblongata of the Rat at the level of main trigeminal nuclei (from CRAIGIE, ZEMAN, and INNES, 1963).

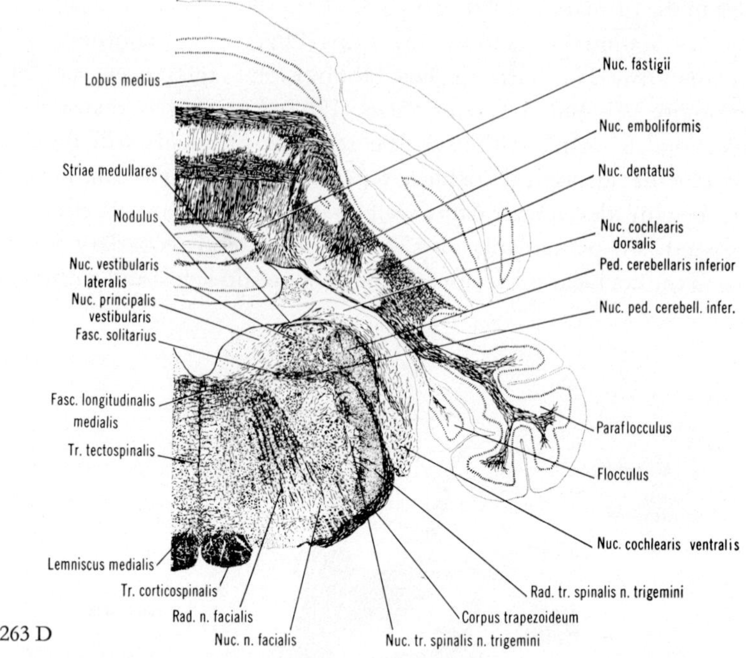
Lobus medius
Nuc. fastigii
Nuc. emboliformis
Striae medullares
Nuc. dentatus
Nodulus
Nuc. cochlearis dorsalis
Nuc. vestibularis lateralis
Ped. cerebellaris inferior
Nuc. principalis vestibularis
Nuc. ped. cerebell. infer.
Fasc. solitarius
Fasc. longitudinalis medialis
Paraflocculus
Tr. tectospinalis
Flocculus
Nuc. cochlearis ventralis
Lemniscus medialis
Tr. corticospinalis
Rad. tr. spinalis n. trigemini
Rad. n. facialis
Corpus trapezoideum
Nuc. n. facialis
Nuc. tr. spinalis n. trigemini

263 D

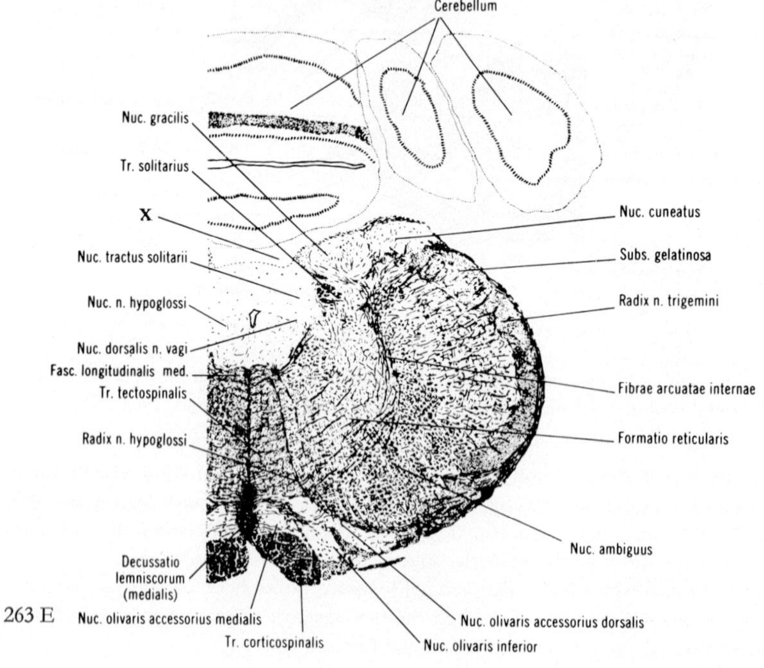
Cerebellum
Nuc. gracilis
Tr. solitarius
X
Nuc. cuneatus
Nuc. tractus solitarii
Subs. gelatinosa
Nuc. n. hypoglossi
Radix n. trigemini
Nuc. dorsalis n. vagi
Fasc. longitudinalis med.
Fibrae arcuatae internae
Tr. tectospinalis
Radix n. hypoglossi
Formatio reticularis
Nuc. ambiguus
Decussatio lemniscorum (medialis)
Nuc. olivaris accessorius medialis
Nuc. olivaris accessorius dorsalis
Tr. corticospinalis
Nuc. olivaris inferior

263 E

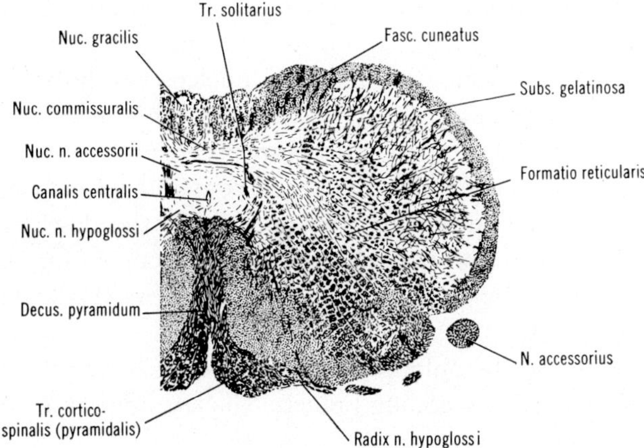

Figure 263F. Cross-section through the oblongata of the Rat at the level of pyramidal tract decussation (from CRAIGIE, ZEMAN, and INNES, 1963). The commissura infima (not labelled) can be seen below its interstitial nucleus commissuralis.

the lemniscus lateralis, besides effecting connections with superior olivary complex and other bulbar grisea. The existence of a secondary, perhaps reciprocal pathway from dorsal cochlear nucleus to cerebellum is also not improbable.[167] Figure 267C illustrates an early concept of cochlear pathways suggested by CAJAL (1909) and still remaining essentially valid as a useful first approximation. Figure 267D shows a simplified but more recent interpretation, by LEWY and KOBRAK (1936), of three respectively four (dorsal, two intermediate, and ventral) systems of crossing secondary cochlear fibers in the Rabbit's bulb. This topic will again be taken up further below in dealing with the superior olivary complex.

[167] Comments on additional details concerning the acoustic pathway will be given further below in dealing with the superior olivary complex, and with the human oblongata, moreover in chapter XI (mesencephalon) and in volume 5 (prosencephalic levels of the neuraxis). A summary and interpretation of the then available data on neural mechanisms has been presented by GALAMBOS (1954).

Figure 263D. Cross-section through the oblongata of the Rat at the level of motor facial nerve nucleus (from CRAIGIE, ZEMAN, and INNES, 1963).

Figure 263E. Cross-section through the closed portion of the Rat's oblongata at the level of nucleus nervi hypoglossi (from CRAIGIE, ZEMAN, and INNES, 1963). The lead with the changed designation x indicates the dorsal surface of the oblongata in the obex region.

The bulbar terminal grisea of the *vestibular nerve* are represented by four nuclei, namely *Deiters' nucleus* (n. vestibularis lateralis s. magnocellularis), nucleus vestibularis medialis, nucleus vestibularis superior, and nucleus radicis descendentis vestibularis. This latter griseum can be considered to be a caudal extension of *Deiters' nucleus*, differing, however, *qua* details of connections, and containing mostly somewhat smaller elements. Caudalward, the nucleus of the descending root seems to merge with the external nucleus of funiculus cuneatus, which has been regarded, by some authors, as an accessory vestibular nucleus (cf. BECCARI, 1943).

Upon entering the oblongata, the vestibular nerve bifurcates into an ascending and a descending branch. The ascending bundle runs to

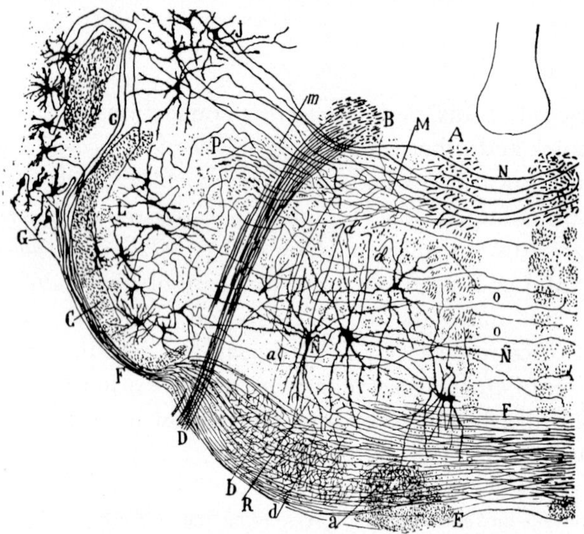

Figure 264A. Cross-section through the oblongata of the Mouse at the level of abducens nucleus and motor facial root as seen in a *Golgi impregnation* (from CAJAL, 1909). A: fasciculus longitudinalis medialis; B: motor root of facial nerve at internal genu; C: radix descendens trigemini; D: exit of facial nerve; E: pyramidal tract; F: fibers of trapezoid body; G: nucleus cochlearis ventralis; H: tuberculum acusticum; J: *Deiters' nucleus;* L: nucleus radicis descendentis trigemini; M: abducens nucleus; N: dorsal internal arcuate fibers from *Deiters' nucleus* joining contralateral fasciculus longitudinalis medialis; P: ascending secondary trigeminal fibers; a, b, d: apparently collateral trapezoid fiber terminals in trapezoid and superior olivary grisea; c: trapezoid fibers originating in tuberculum acusticum. Other designations refer to unclarified details of reticular formation.

the superior vestibular nucleus which may predominantly receive collaterals, and proceeds into the cerebellum. The descending branch reaches *Deiters' nucleus*, a part of which is located at the levels of vestibular root entrance. The descending branch, moreover, also connects with the medial vestibular nucleus. The caudal extension of this branch forms the descending vestibular root, accompanied by its griseum.

Secondary fibers arising in superior, medial, descending ('spinal') vestibular nuclei, and probably also in *Deiters' nucleus*, contribute to the fasciculus longitudinalis medialis. The findings of different authors concerning contralateral, homolateral, ascending and descending fibers, as well as details of their provenance from the different vestibular nuclei, although agreeing *qua* overall features, are mutually at variance

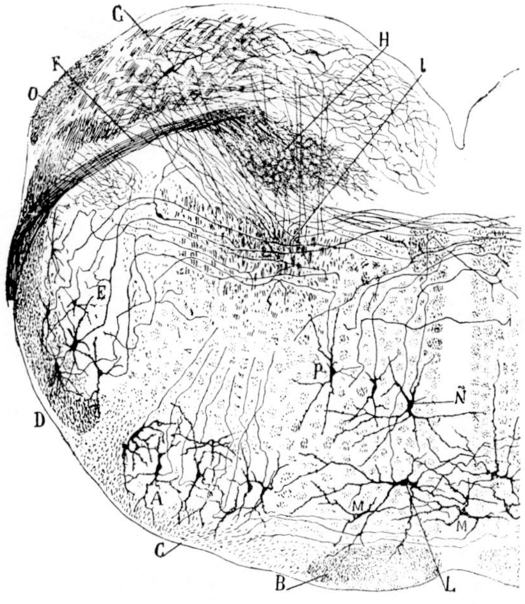

Figure 264 B. Cross-section *(Golgi impregnation)* through the oblongata of the Mouse (newborn) at the level of motor facial nerve nucleus (from CAJAL, 1909). A: motor facial nerve nucleus; B: pyramidal tract; C: *'reste du cordon lateral'* (probably tr. bulbocerebellaris fibr. arc.); D, E: radix descendens trigemini and its nucleus; F: root fibers of nervus glossopharyngeus (and vagus); G: radix descendens vestibularis and its nucleus; H: tractus solitarius and its nucleus; I: *'voie centrale du nerf vestibulaire'*; O: corpus restiforme. Other designations refer to unclarified details of reticular formation.

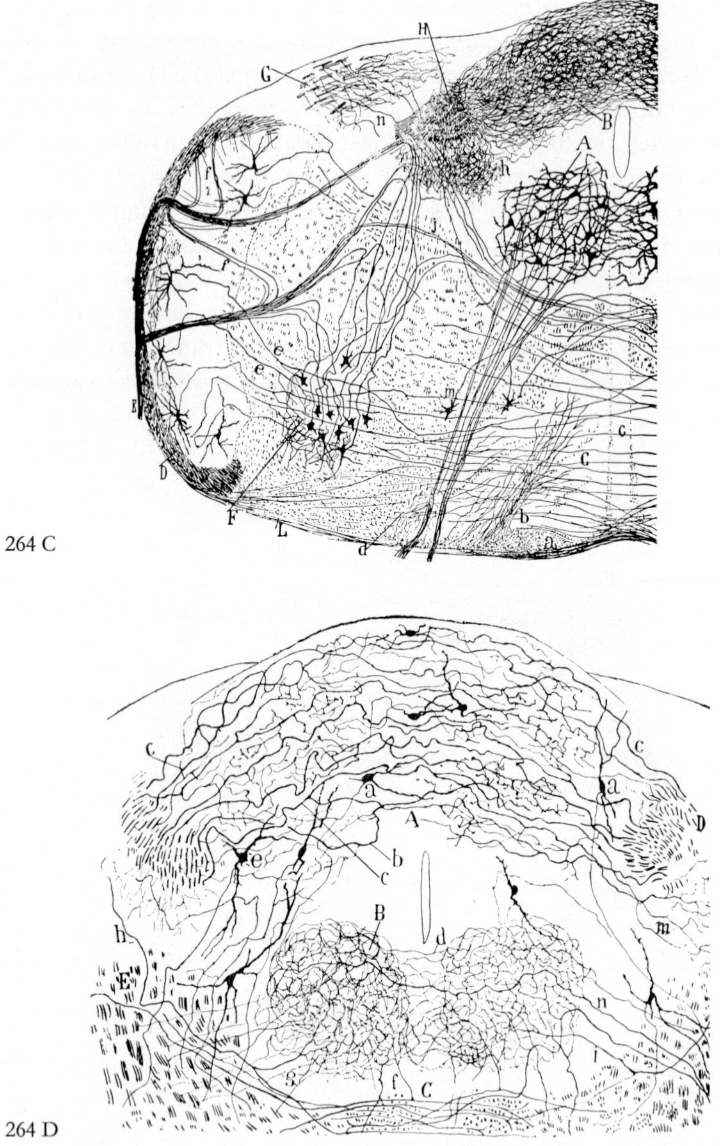

Figure 264 C. Cross-section *(Golgi impregnation)* through the oblongata of a four days old Mouse at the level of hypoglossal nucleus and commissura infima (from CAJAL, 1909). A: nucleus nervi hypoglossi; B: interstitial nucleus of commissura infima; C: inferior olivary complex; D: radix descendens trigemini; E: root fibers of vagus (and glosso-pharyngeus); F: nucleus ambiguus; G: caudal end of radix descendens vestibularis and its nucleus; H: tractus solitarius; L: ventral external arcuate fibers related to inferior

with respect to many particulars. This fasciculus,[168] either directly, or indirectly by way of the formatio reticularis, mediates connections with eye muscle nerve and other cranial nerve nuclei, and with an undefined number of additional brain stem grisea besides extending into the spinal cord. Moreover, a rather well corroborated, essentially homolateral *vestibulo-spinal tract* originates in *Deiters' nucleus* and runs caudad among the longitudinal fasciculi of the reticular formation. Spino-vestibular connections seem likewise to be present.

The *tractus vestibulo-cerebellaris* or'nucleo-cerebellar tract' is provided by a presumably reciprocal fiber system whose ascending components seem mainly to originate in nucleus vestibularis superior and *Deiters' nucleus*, but the two other vestibular nuclei might perhaps also be linked to that system. The nucleus vestibularis medialis, moreover, apparently contributes to and receives fibers from, the *fasciculus longitudinalis dorsalis* of SCHÜTZ, a thinly medullated and partly non-medullated fiber system located within the periventricular gray of the basal plate, and briefly dealt with further below in the description of that plate's anatomical components. A pathway connecting the vestibular grisea with the cerebral cortex has not been clearly identified as regards

[168] It can be assumed that the vestibular and related components of the fasciculus longitudinalis medialis are particularly concerned with the reflectory coordination of eye and head movements. The composition and function of fasciculus longitudinalis medialis will again be dealt with in chapter XI (Mesencephalon).

---

olivary complex; a: pyramidal tract; b: collaterals of pyramidal tract fibers and of other systems; c: fibrae arcuatae internae; d: '*collatérales du reste du cordon latéral*'; e: secondary trigeminal collaterals reaching nucleus ambiguus; f: '*fibres récurrentes de la racine motrice allant à la racine du trijumeau*' (in contradiction to CAJAL's interpretation, it seems most likely that these are somatic afferent, cutaneous fibers of IX and X joining the descending trigeminal root. CAJAL designated E above as '*racines motrices*', but it seems likely that they represent efferent and afferent root fibers); j: '*radiculaires motrices croisées des nerfs vague et glosso-pharyngiens*' (this interpretation, although quite possible, has been contested); h: fibers ending (or originating?) in tractus solitarius system *('collatérales de la racine sensitive de ces nerfs destinées au noyau attenant au faisceau solitaire')*.

Figure 264D. Details of commissura infima in a cross-section *(Golgi impregnation)* through the lower oblongata of the Mouse (from CAJAL, 1909). A: interstitial commissural nucleus; B: hypoglossal nucleus; C: decussation of medial lemniscus *(ruban de Reil)*; D: fasciculus solitarius; a: cells of commissural nucleus; b, c: terminals of afferent glosso-pharyngeal and vagus fibers; d: commissure of collaterals from hypoglossal nucleus; f, g: second order afferent collaterals entering hypoglossal nucleus; h, i, m, n: unclarified fibers, of which at least i and n connect with hypoglossal nucleus.

its anatomical features, which might be provided by secondary or higher order vestibular fibers joining the lemniscus lateralis and by ascending fiber systems of the reticular formation. The vestibulo-cortical linkage, however, can be considered to be well substantiated on the basis of functional and clinical data; it will again be considered with respect to the relevant topics dealt with in volume 5. A detailed review of the Mammalian vestibular nuclei and their relations to spinal cord and cerebellum has been presented by BRODAL (1964) and stresses a number of still unresolved questions.

As regards the homologies of Mammalian vestibular grisea in comparison with those of Sauropsidans, the nucleus vestibularis superior of Reptiles and the vestibulo-cerebellar group of Birds doubtless correspond to the superior vestibular griseum of Mammals. There is likewise little doubt about the homology of the nucleus radicis descendentis and that of *Deiters' nucleus* in all three classes, although, in Birds, this latter nucleus is represented by a more complex griseal aggregate. In both Mammals and Birds, a nucleus vestibularis medialis s. triangularis is present, but this griseum, presumably differentiated from a dor-

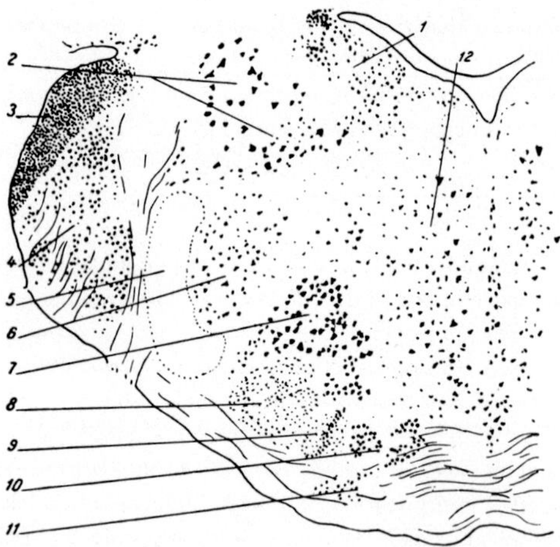

Figure 265. Cross-section *(Nissl stain)* through the oblongata of the Guinea pig (Cavia cobaya, Rodentia) at the level of the motor facial nerve nucleus (from BECCARI, 1943). 1: nucleus vestibularis medialis; 2: *nucleus vestibularis (Deiters);* 3: nucleus cochlearis dorsalis (tuberculum acusticum); 4: nucleus cochlearis ventralis; 5, 6: radix descendens trigemini and its nucleus; 7: nucleus motorius facialis; 8: superior olivary nucleus; 9: accessory superior olive; 10: nucleus of trapezoid body; 11: *nucleus praeolivaris medialis (s. internus) of Cajal;* 12: 'nucleus reticularis inferior' *(centro tegmentale mielencefalico).*

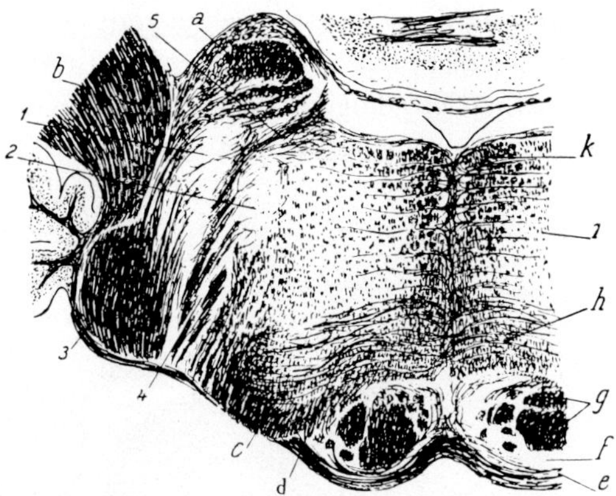

Figure 266A. Cross-section *(Weigert stain)* through the oblongata of the Dog (Canis fam., Carnivora) at the level of the two main trigeminal nuclei (from BECCARI, 1943). 1: main afferent trigeminal nucleus; 2: motor trigeminal nucleus; 3: afferent trigeminal root; 4: motor trigeminal root; 5: mesencephalic trigeminal root; a: brachium conjunctivum; b: brachium pontis; c: lemniscus lateralis; d: trapezoid body; e: pontine fibers; f: pontine nuclei; g: pyramidal tract bundles; h: lemniscus medialis; i: reticular formation; k: fasciculus longitudinalis medialis.

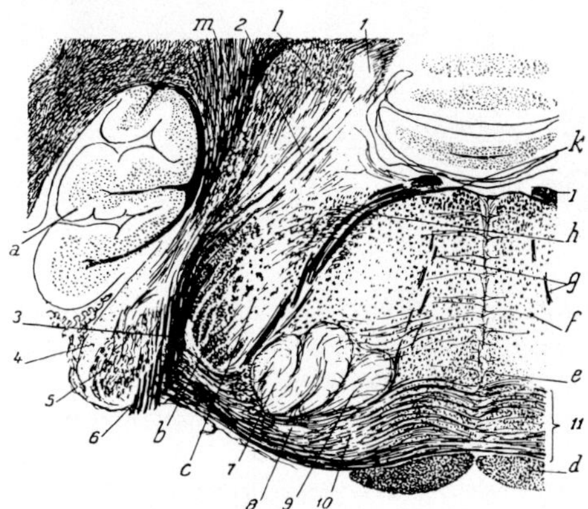

Figure 266B. Cross-section *(Weigert stain)* through the oblongata of the Dog at the level of principal octavus nuclei (from BECCARI, 1943). 1: nucleus vestibularis superior *(Bechterew)*; 2: nucleus vestibularis lateralis *(Deiters)*; 3: radix nervi vestibularis; 4: nucleus cochlearis dorsalis (tuberculum acusticum); 5: nucleus cochlearis ventralis; 6: radix nervi cochlearis; 7: nucleus olivaris superior (principalis); 8: nucleus praeolivaris lateralis; 9: nucleus olivaris superior accessorius; 10: nucleus of trapezoid body; 11: trapezoid body; a: flocculus of cerebellum; b, c: radix descendens trigemini and its nucleus; d: pyramidal tract; e, f: reticular formation; g: intracerebral rootlets of nervus abducens; h, i: intracerebral root fibers and internal knee of facial nerve; k: perhaps crossed facial root fibers (according to BECCARI: *'fascetto crociato del VII, salivatorio')*; l: vestibulo-cerebellar bundle (juxtarestiform body or 'IAK'); m: corpus restiforme.

somedial neighborhood of *Deiters' nucleus*, cannot be recognized with sufficient certainty as a distinctive unit in Reptiles. In Mammals, on the other hand, a typical nucleus tangentialis is not differentiated. At most, it may be represented by some nondescript lateral cell groups pertaining to *Deiters' nucleus*, and scattered among fiber bundles of the vestibular root at their entrance into the bulb.

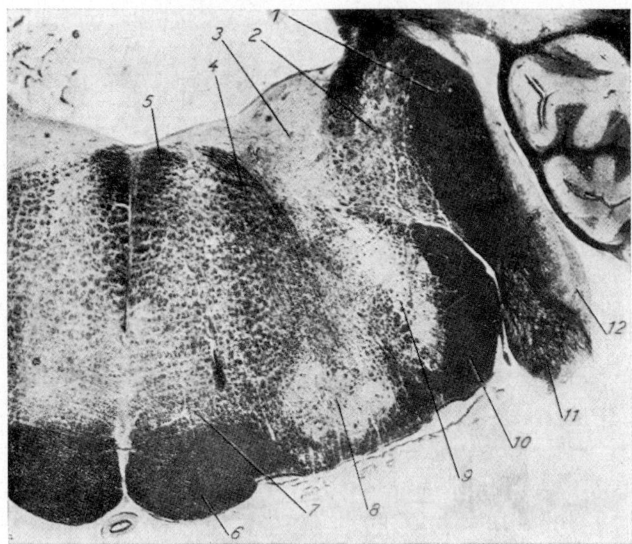

Figure 266 C. Cross-section *(Weigert stain)* through the oblongata of the Dog at the level of motor facial nerve nucleus, caudally to trapezoid body (from BECCARI, 1943). 1: corpus restiforme; 2: nucleus vestibularis lateralis *(Deiters)*; 3: nucleus vestibularis medialis; 4: motor facial nerve root (medial or proximal limb of root fiber loop); 5: fasciculus longitudinalis medialis; 6: pyramidal tract; 7: lemniscus medialis; 8: motor facial nerve nucleus; 9, 10: nucleus radicis descendentis trigemini and radix; 11: nucleus cochlearis ventralis; 12: nucleus cochlearis dorsalis.

Figure 266 E. Cross-section *(Weigert stain)* through the oblongata of the Dog at the level of decussatio lemnisci bulbaris and caudal extremity of inferior olivary complex (from BECCARI, 1943). 1: medial internal arcuate fibers; 2: nucleus gracilis; 3: nucleus cuneatus; 4: nucleus cuneatus externus s. accessorius; 5: fibrae arcuatae internae; 6: nucleus radicis descendentis trigemini; 7: nucleus ambiguus; 8: nucleus (reticularis) lateralis; 9: root fibers of hypoglossus; 10: nucleus olivaris inferior accessorius medialis; 11: fibrae arcuatae externae ventrales; 12: pyramidal tract; 13: fibers of lemniscus medialis intermingled with those of the pyramidal tract.

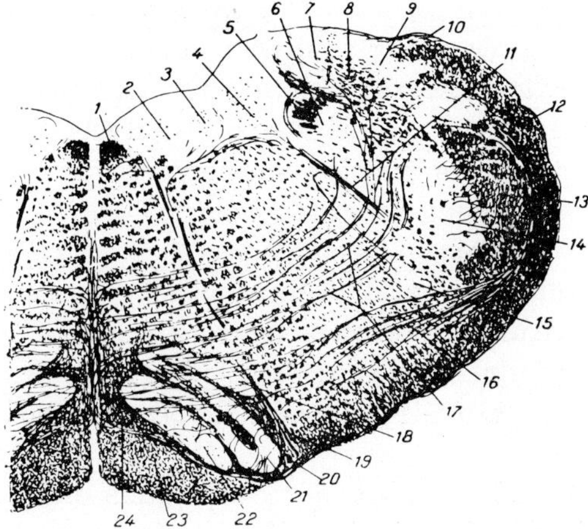

Figure 266D. Cross-section *(Weigert stain)* through the oblongata of the Dog at the level of vagus nuclei and inferior olivary complex (from BECCARI, 1943). 1: fasciculus longitudinalis medialis; 2: nucleus nervi hypoglossi; 3: *nucleus intercalatus Staderini;* 4: nucleus dorsalis vagi; 5, 6: tractus solitarius and its nucleus; 7: nucleus vestibularis medialis; 8: radix descendens vestibularis and its nucleus; 9: nucleus cuneatus; 10: fibrae arcuatae externae dorsales; 11: root fibers of nervus vagus; 12: tractus spino-cerebellaris dorsalis; 13, 14: radix descendens trigemini and its nucleus; 15: nucleus ambiguus; 16, 17: internal arcuate fibers; 18: dorsal accessory inferior olivary nucleus; 19: root fibers of hypoglossal nerve; 20: fibrae arcuatae externae ventrales; 21: nucleus olivaris inferior principalis; 22: medial accessory inferior olivary nucleus; 23: pyramidal tract; 24: lemniscus medialis.

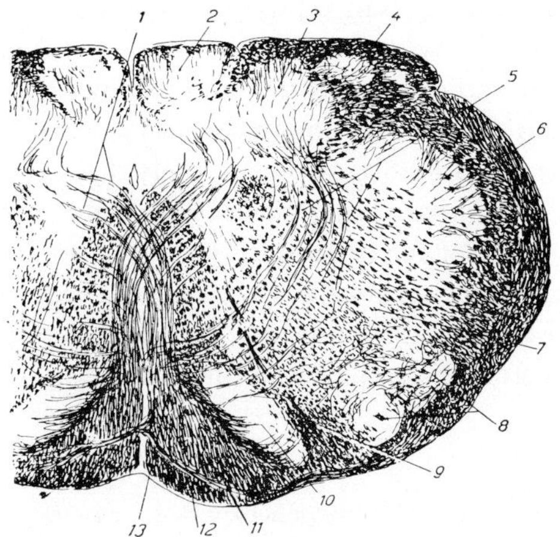

Figure 266 E

Concerning the *peripheral portion of the nervus octavus*, KAPPERS (1947) reviews in detail its relation to vestibular and spiral ganglia as well as to the different parts of the labyrinth, and discusses various still at present controversial topics, including the question whether the vestibular nerve does or does not contain some fibers related to the basal part of the cochlea and forming a so-called *bundle of Oort*.

The *ventrolateral subdivision of the bulbar alar plate's dorsal zone* contains two of the afferent trigeminal nuclei, namely the main sensory nucleus and the nucleus of the descending trigeminal root. A diagram of the entire primary trigeminal system, including some of its secondary connections, and as particularly elucidated by CAJAL (1909) is shown in Figure 268A.

Although most afferent fibers of the trigeminus originate as central neurites from the pseudo-unipolar cells of the 'semilunar' *Gasserian trigeminal ganglion*, those pertaining to the mesencephalic root are peripheral neurites of the large vesicular cells located in the isthmic and mesencephalic alar plate, and forming the diffusely arranged 'nucleus' of the mesencephalic root.[168a] Except for this component, most afferent trigeminal root fibers bifurcate within the bulb, and near their entrance, into an ascending and a descending branch. The former reaches the main sensory nucleus, and the latter represents the radix descendens trigemini, which, accompanied by its griseum, runs caudad as far as the rostral portion of the spinal cord's posterior horn. Thinly medullated and non-medullated root fibers may bend caudalward into the descending root without such bifurcation (WINDLE, 1926).

Within the descending bundle, the root fibers of the ophthalmic division (V_1) seem to be ventrally located, those of the mandibular division (V_3) dorsally and dorsomedially, and those of the maxillary division (V_2) between the two. Although this has been reasonably well ascertained for the human bulb, it appears likely that, with perhaps some minor variations, a similar arrangement obtains in the other Mammals. The descending root of the trigeminus is, moreover, joined by the exteroceptive fibers of facial, glossopharyngeal and vagus nerves.

The fibers of the *mesencephalic root* are doubtless proprioceptive, being mainly related to the muscles of mastication and possibly to al-

---

[168a] A few of the large vesicular cells of nucleus radicis mesencephalicae trigemini (to be dealt with in chapter XI) are usually found along the mesencephalic portion of that root caudally to the isthmus.

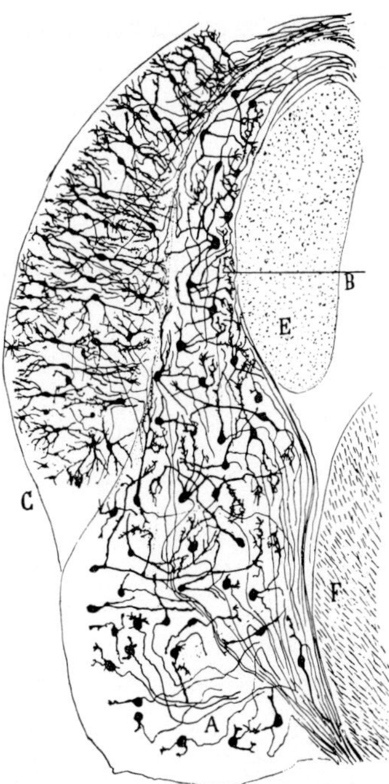

Figure 267 A. Cross-section *(Golgi impregnation)* through the terminal cochlear nuclei in a four days old Rabbit (from CAJAL, 1909). A: nucleus cochlearis ventralis; B: dorsal 'tail' of ventral cochlear nucleus; C: tuberculum acusticum (dorsal cochlear nucleus); E: corpus restiforme; F: radix descendens trigemini.

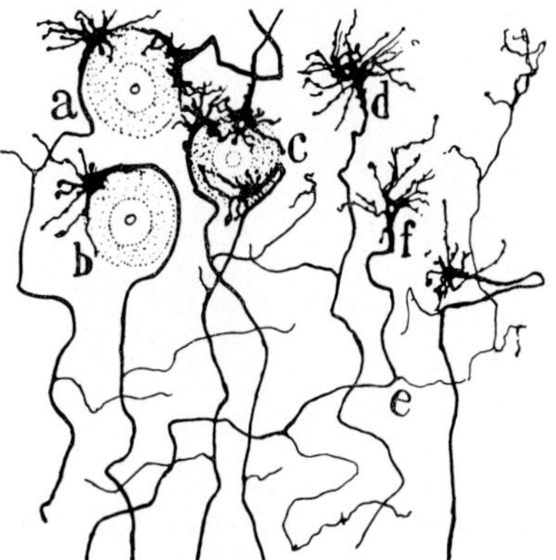

Figure 267 B. *Golgi impregnation* details of primary cochlear (root) fibers ending with *bulbs or calices of Held* in the ventral cochlear nucleus of a Rabbit a few days after birth (from CAJAL, 1909). a: cell connected with two bulbs; b: cell connected with a single bulb; c: cell receiving three bulbs; e: ultimate collateral of an ascending cochlear root fiber; d, f: 'asteriform' end bulbs with numerous appendages.

veolar pressoreceptors of the jaws. Some of the fibers pertaining to cells in the *Gasserian ganglion* are likewise suspected to be proprioceptive, but most of the *Gasserian root fibers* seem to carry tactile, pain, and temperature impulses.

The nucleus *sensibilis principalis trigemini* represents a rather large griseum, whose medium sized respectively small cells tend to be grouped in irregular clusters, and is located at pontine levels, laterally

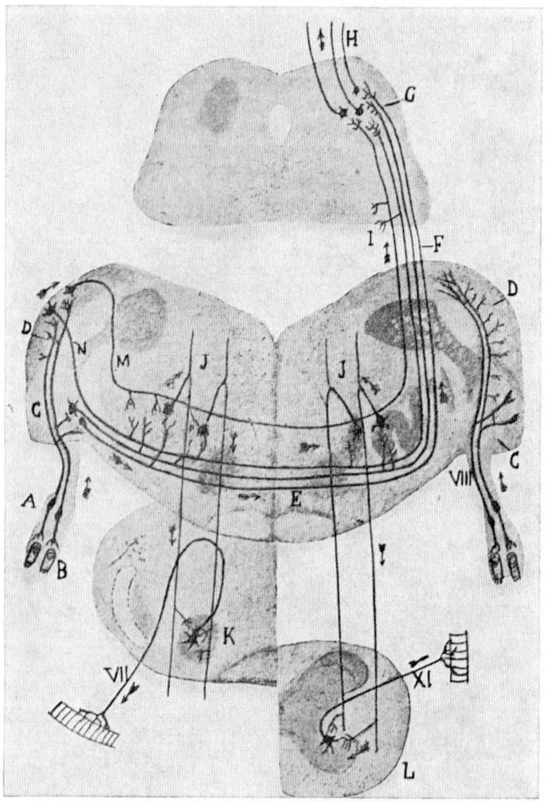

Figure 267 C. CAJAL's original concept of ascending and descending cochlear pathways (from CAJAL, 1909). A: ganglion spirale of cochlea; B: sensory cochlear hair cells; C: nucleus cochlearis ventralis; D: nucleus cochlearis dorsalis (tuberculum acusticum); E: trapezoid body; F: lateral lemniscus (ascending cochlear pathway); G: griseum of colliculus inferior tecti mesencephali; H: central (cortical) ascending pathway; I: collateral to superior nucleus of lateral lemniscus; J: short cochlear pathways arising from superior olivary complex; K: motor nucleus of facial nerve; L: cervical spinal cord with nucleus of spinal accessory nerve. The arrows indicate direction of impulse transmission.

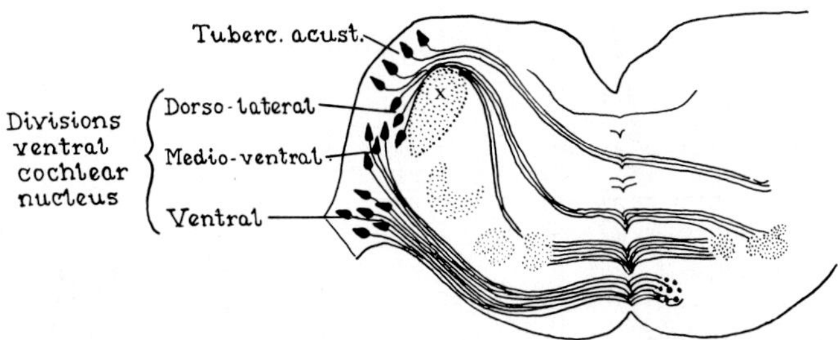

Figure 267D. Diagram showing LEWY's and KOBRAK's concept of primary cochlear nuclei and their initial secondary systems as obtaining in the Rabbit's oblongata (from LEWY and KOBRAK, 1936). 'Dorsally are shown *Monakow's striae acusticae* to the nuclei of the midbrain; ventrally follows *Held's intermediate system* for the superior olives and the interolivary field, and most ventrally the trapezoid body to its crossed nucleus.' The added designation x indicates the restiform body.

respectively dorsolaterally to the motor trigeminal nucleus. The available data indicate that the main sensory trigeminal nucleus chiefly mediates tactile sensibility but might possibly also be concerned with some aspects of proprioceptive functions, supplementing, in this respect, the mesencephalic root and its nucleus.

The *nucleus of the descending trigeminal root* begins rostrally without a well-definable demarcation from the main sensory nucleus and, receiving collateral endings (cf. Fig. 268A), accompanies the entire length of the root fiber bundle on this latter's medial aspect. In rostro-caudal sequences, three portions of said griseum can be roughly distinguished (cf. e.g. OLSZEWSKI and BAXTER, 1943), namely, *pars rostralis*, *pars intermedia* ('subnucleus interpolaris'), and *pars caudalis*. In the rostral and intermediate portions, the neuronal elements of medium and small size are rather diffusely scattered. Both parts do not strikingly differ from each other nor from the main sensory nucleus.[169]

The pars caudalis, *per contra*, displays a definite lamination, essentially similar to that obtaining for the spinal cord's *substantia gelatinosa Rolandi*. Three more or less distinctive layers can here be recognized (Fig. 268B). An outer layer of fusiform or irregular cells is located among the root fibers. The long axes of these cells are commonly par-

[169] Pars intermedia contains a few larger cells. The medium sized cells of the nucleus principalis are, on the whole, slightly smaller than those of pars rostralis and intermedia nuclei radicis descendentis.

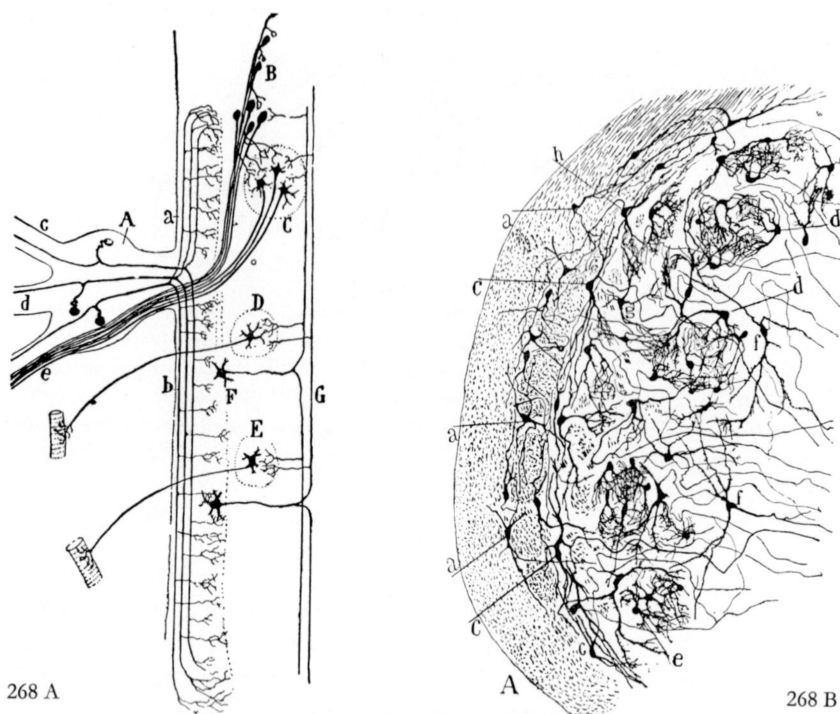

Figure 268 A. Diagram depicting the overall arrangement of the Mammalian trigeminal nuclei, including a few secondary fiber pathways (from CAJAL, 1909). A: trigeminal ganglion *(ganglion semilunare sive Gasseri)*; B: nucleus of the mesencephalic trigeminal root; C: motor trigeminal nucleus; D: motor facial nerve nucleus; E: hypoglossal nucleus; F: nucleus of descending ('spinal') trigeminal root; G: ascending and descending secondary trigeminal fibers *('voie centrale du trijumeau)*; a: ascending, primary afferent root fibers of the trigeminus, ending in main sensory nucleus (which is not labelled); b: radix descendens (s. 'spinalis') trigemini; c: nervus ophthalmicus (V_1); d: nervus maxillaris (V_2); e: nervus mandibularis (V_3). The long peripheral (afferent) neurites originating in nucleus radicis mesencephalicae join the mandibular nerve; their short, central cellulifugal (efferent) neurites effect connections with motoneurons of the motor trigeminal nucleus, thereby providing a monosynaptic proprioreceptive reflex arc, which is presumably modulated, as suggested in CAJAL's scheme, by ascending secondary fibers. Whether the peripheral proprioceptive fibers of the mesencephalic root are restricted to the nervus mandibularis, or may also be distributed to the other trigeminal or even additional cranial nerve branches, still remains an open question.

Figure 268 B. Cross-section *(Golgi impregnation)* through pars caudalis (s. gelatinosa) of the nucleus radicis descendentis trigemini in the newborn Rabbit (from CAJAL, 1909). A: radix descendens trigemini (the label A corresponds to the rostral, ophthalmic or V_1 component of the root); a: 'interstitial cells'; c: marginal cells; d: cellular islets of gelatinous substance; e: small cells within the islets; f: large cells not included in the islets; g:

allel to the descending tracts' curved dorso-ventral cross-sectional border. This layer corresponds to the 'pericornual cells' or cellulae posteromarginales of the spinal cord's dorsal horn. Internally or medially adjacent there extends a rather wide 'gelatinous' layer, whose small neuronal elements are provided with extensive dendritic plexuses. Still more internally or medially, and, as it were, in the 'hilus' of the gelatinous substance, an array of larger elements is arranged as a magnocellular layer. In comparison with the spinal cord, the 'gelatinous layer' corresponds to the *substantia gelatinosa Rolandi*, and the 'magnocellular layer' to the nucleus proprius of the posterior horn. The pars caudalis nuclei radicis descendentis trigemini merges caudalward with the corresponding structures of the spinal cord. Figure 268B should be compared with Figures 106, 107, 120, and 123 of chapter VIII. The nucleus of the descending trigeminal root is believed to be concerned with pain and temperature sensibility and to some extent with aspects of touch sensibility. It is likely that this latter is mediated by the rostral and intermediate portions, while the caudal portion seems to be concerned

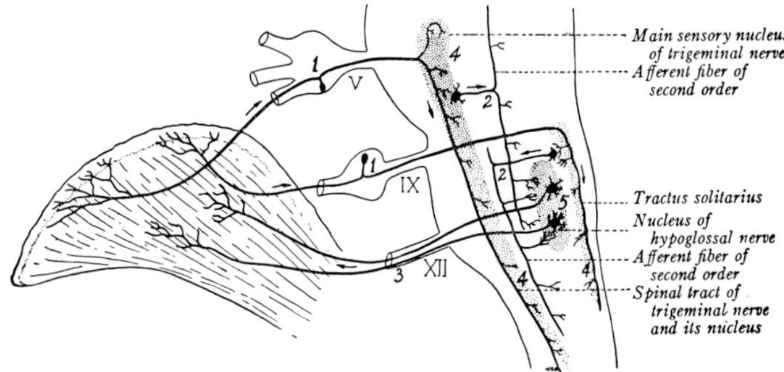

Figure 268 C. Diagram showing CAJAL's concept of relationships between trigeminal, glossopharyngeal, hypoglossal bulbar grisea in the innervation of the tongue (after CAJAL, 1909, from RANSON, 1943). 1: first order afferent neurons of V and IX; 2: second order afferent neurons; 3: motor fibers of XII. For greater simplicity, the afferent (gustatory) components of facial and vagus nerves have been omitted.

---

isolated cells between islets; h: marginal cell whose axon seems to run 'into white substance or rostral portion of root'. a, c and h can be evaluated as corresponding to the posteromarginal cells of the spinal cord's dorsal horn; d, e, g are comparable to the elements of the cord's gelatinous substance, and those designated as f to the cell's of the 'nucleus proprius' cornus posterioris medullae spinalis. Figure 268 B should be compared with Figures 106, 107, 120 and 123 of the preceding chapter VIII.

with the transmission of pain impulses and with at least some components of temperature sensibility.[170]

Secondary pathways arising in the afferent trigeminal grisea include arcuate fibers from the main sensory nucleus and from the nucleus of the descending root joining the contralateral lemniscus medialis. Another system, arising from the caudal part of the descending nucleus seems to reach the medial part of the lateral spino-thalamic tract, which, at caudal levels of the bulb (medulla oblongata *sensu strictiori*) runs in a rather superficial position, before joining the medially located bulbo-thalamic tract (lemniscus medialis *sensu strictiori*) at pontine levels. A third secondary tract is said to arise in the main sensory nucleus and to ascend in a position below and not far from the periventricular gray, forming the dorsal secondary ascending tract (cf. e.g. PAPEZ and RUNDLES, 1937). These secondary tracts are sometimes collectively designated as the trigeminal lemniscus, which reaches thalamic and mesencephalic grisea. There is still considerable uncertainty concerning details of course and significance of the fibers in the trigeminal lemniscus system, which may also include some homolateral pathways. A few secondary trigeminal fibers are shown in Figure 268A depicting CAJAL's diagram.

Homolateral secondary trigeminal fibers from the main sensory nucleus and from the griseum of descending root are said to enter the cerebellum among the rostral components of the (vestibular) nucleo-cerebellar tract. The possibility that direct (primary) trigeminal root fibers join this system cannot be excluded. Collaterals from the mesencephalic root are also believed to be included among the trigemino-cerebellar fibers.

The *intermediodorsal zone of the bulbar alar plate* contains the *tractus s. fasciculus solitarius* formed by the descending visceral afferent root fibers of the facial, glossopharyngeal and vagus nerves. The inclusion of some afferent trigeminal root fibers from the mucosae of oral cavity with tongue, and from the nasal cavity (somehow related to respiration) seems probable. A small celled terminal griseum accompanies and surrounds this distinctive rounded fiber bundle and represents the joined afferent nuclei of the just mentioned nerves. With respect to various minor variations, some of that griseum's cell groups may be

---

[170] With respect to the spinal cord, the substantia gelatinosa and its probable relationships to pain mechanisms were dealt with in chapter VIII, p. 199 and 229 with footnote 144.

located within the tractus itself, while others can surround the fiber bundle in an annular fashion. Again, a portion of the nucleus tractus solitarii may extend so as to become laterally or dorsolaterally directly adjacent to the efferent (preganglionic) nucleus dorsalis vagi. Still another cell group displaying a slight ventrolateral displacement can be designated as *nucleus parasolitarius*.

The rostral part of the tractus solitarius with its nucleus becomes indistinct and tapers out at diverse transverse levels of the motor facial nerve nucleus. In this region, the tractus and its nucleus shift to a greater distance from the ventricular floor and may be located dorsomedially and closely adjacent to the nucleus radicis descendentis trigemini. The afferent facial root fiber bundles commonly descend in the tractus solitarius without giving off conspicuous ascending branches.[171] Nevertheless, in some instances, the rostral end of fasciculus solitarius can be traced as far rostrad as the level of motor trigeminal nucleus.

The caudal end of tractus solitarius and its griseum corresponds, with some variations, approximately to the level of the caudal extremity of the nucleus dorsalis vagi in the closed portion of the oblongata, but may slightly extend spinalward beyond that region. At these levels, primary and perhaps also secondary fibers of tractus solitarius complex, commonly accompanied by interstitial cells, cross the dorsal midline through the obex and caudally to that structure, forming the *commissura infima* with its nucleus (cf. e.g. Figs. 261D, 263F, 264D).

Concerning the entire length of fasciculus s. tractus solitarius, it is commonly assumed that the taste fibers of facial, glossopharyngeus and vagus are restricted to the bundle's rostral third or half, while the 'general visceral afferent fibers', particularly those of the vagus, associated with respiratory, cardiac, vascular, and gastrointestinal reflexes, constitute the caudal part of the tract, and have their relevant synaptic links with the corresponding caudal regions of the accompanying griseum.

The secondary fibers arising in the grisea of tractus solitarius are incompletely elucidated with respect to their details. An ascending gustatory respectively visceral pathway (*'lemniscus visceralis'*) may join the lemniscus medialis and thereby reach mesencephalon as well as dience-

---

[171] In BECCARI's (1943) opinion, the tractus solitarius of Mammals '*parebbe esclusivamente discendente postvagale. Il fascicolo, discendendo, lascia molte fibre nel proprio nucleo; ma in parte arriva fino al nucleo della commessura infima, situato caudalmente all'obex*'.

phalon, besides giving off collaterals within the reticular formation. Visceral, including gustatory projections to the cerebellum doubtless obtain, and are presumed to be mediated by channels derived from, or related to, this 'visceral lemniscus' (cf. e.g. KAPPERS, 1947). A descending *tractus solitario-spinalis* with perhaps predominantly respiratory function has been inferred by various investigators (e.g. KOSAKA and YAGITA, 1905). Components of this system seem to arise in the nucleus commissuralis related to the commissura infima. The significance of the tractus solitarius and its associated grisea for cardiovascular, respiratory and gastrointestinal activities was discussed in section 1 of this chapter, dealing with particular mechanism, and illustrated by the diagrams of Figures 173 and 176.

In the *region of transition* to the spinal cord, and extending from there to variable caudal levels of the oblongata, the *alar plate* contains the *funicular nuclei*, which are essentially derivatives of the plate's *dorsal zone* and constitute the terminal grisea of ascending spinal posterior root fibers dealt with in chapter VIII (cf. e.g. Figs. 133, 134, and 143). These cell groups include, on both sides of the midline, the medially located *nucleus gracilis of Goll*, and the laterally adjacent *nucleus cuneatus of Burdach*, both within the cranial end of the respective fasciculi of the spinal cord's funiculus posterior.

A second order pathway originating in the funicular nuclei is provided by the conspicuous system of decussating *fibrae arcuatae internae* which form, on both sides of the midline, the *medial lemniscus sensu strictiori*, and represent a *tractus bulbo-thalamicus* mediating proprioceptive as well as tactile input. Connections with tectum mesencephali, either through collaterals or separate bulbo-tectal fibers *('tractus bulbo-tectalis')* obtain, and collaterals to other grisea, such e.g. as reticular formation, seem probable.[172]

In various taxonomic forms, a *nucleus cuneatus accessorius s. externus*, also known as *nucleus of Blumenau or Monakow*, becomes separated from the main griseum (cf. Fig. 266E). The external cuneate nucleus is characterized by somewhat larger multipolar cells and gives off *dorsal external arcuate fibers* which reach the cerebellum by way of the inferior cerebellar peduncle or corpus restiforme. This cerebellar stalk also gives passage to the dorsal spino-cerebellar tract and to the olivo-cerebellar fiber system.

---

[172] KAPPERS (1947), however, maintains that '*à l'encontre du f. secondaire protopatique, f. spino-thalamique ou f. antéro-lateral, les fibres lemniscales n'émettent pas de collatérales pendant leur passage dans la moelle allongée*'.

Both the main cuneate nucleus and the nucleus gracilis also give off a variable system of *ventral external arcuate fibers* running along the bulb's surface and, at least in greater part, representing cerebellar connections.

As regards the configuration and the relative size of the funicular nuclei, numerous variations related to the diverse taxonomic forms have been described. An unpaired *median funicular nucleus (nucleus of Bischoff)*, intercalated between the two nuclei graciles, is found in the Insectivore Erinaceus and some other lower Mammals.

The *calamus scriptorius region* of Mammals is characterized by the presence of the *area postrema*, a vascular cushion representing a '*paraependymal organ*' (cf. Figs. 279A, B). Together with other structures of this type, the area postrema was discussed *qua* structure and significance in chapter V, section 5, p. 362f. of volume 3, part I.

Turning now to the *basal plate configurations* of the Mammalian oblongata, the efferent cranial nerve nuclei of the *intermedioventral zone* shall be considered first. The most rostral griseum of this series is the *nucleus motorius trigemini* or n. masticatorius, located at pontine levels, and ventromedially adjacent to the main sensory trigeminal nucleus, being, however, usually separated from this latter by a fiber bundle which includes the radix mesencephalica and other root fibers of the fifth (cf. e.g. Figs. 261A, 263B, and 266A). The branchiomotor ('special visceral efferent') trigeminal nucleus innervates the muscles of the maxillo-mandibular arch, mainly represented by the muscles of mastication.[173] A 'somatotopic' localization of cell groups innervating the different muscles seems to obtain within the nucleus, as, e.g., described for the Cat by SZENTÁGOTHAI (1949). In addition to fibers from various sources (cortical, tectal, tegmental, etc.) controlling its activities, the nucleus masticatorius receives collaterals from the mesencephalic root, which apparently represent the central neurites, and presumably mediate proprioceptive reflexes of 'monosynaptic' type (cf. Fig. 268A). The nucleus motorius trigemini, pertaining to the intermedioventral zone's ventrolateral subzone, does not include preganglionic elements.

[173] Further details are considered below in dealing with the human oblongata. It will here be sufficient to enumerate musculi masseter, temporalis, pterygoideus internus and externus as muscles of mastication *sensu strictiori*, and m. tensor veli palatini, m. mylohyoideus, m. tensor tympani, moreover anterior belly of m. digastricus as additional muscles innervated by the mandibular branch of nervus trigeminus. To some extent, the m. buccinator, innervated by n. facialis, could also be considered a muscle of mastication *sensu latiori*.

The preganglionic fibers carried by some peripheral branches of the trigeminal are provided through extracerebral anastomoses with other branchial nerves.

The *nucleus motorius nervi facialis* is likewise a branchiomotor griseum consisting of large motoneurons, and innervating striated musculature of head and neck derived from the hyoid arch. The nucleus does not contain preganglionic elements and is rather ventrally located in the tegmentum at levels of transition between pons and oblongata *sensu strictiori*, being more or less caudodorsally adjacent to the superior olivary complex, and adjoining the ventromedial border of nucleus radicis descendentis trigemini (cf. Figs. 261B, 263D, 265, 266C, 269). Numerous variations *qua* details of position and of secondary subdivisions obtain in the Mammalian series. The efferent root fibers originating in the motor facial nucleus stream dorsomediad through the teg-

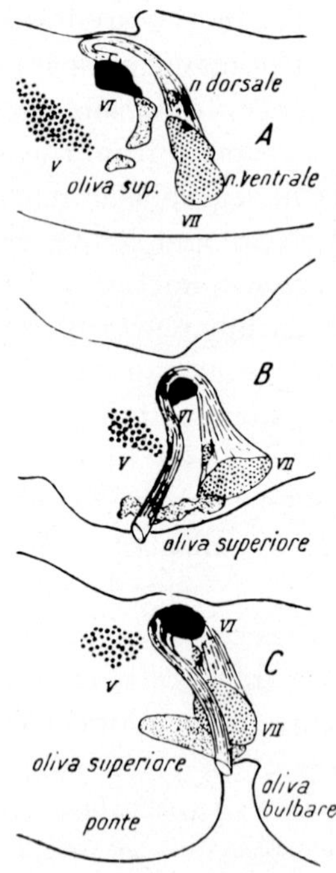

Figure 269. Diagram illustrating configurational relationships of internal facial knee and relevant bulbar grisea in a Reptile, a lower Mammal, and in Man (modified after KAPPERS, from BECCARI, 1943). I Varanus (Lacertilian Reptile). II Erinaceus (Insectivore Eutherian Mammal). III Homo (Primate Mammal).

mentum, gather dorsolaterally to fasciculus longitudinalis medialis, and form a loop over or near the abducens nucleus. The root fibers then pass laterad toward their exit (Fig. 263C). This bend about or above the abducens nucleus is also designated as the *inner knee (genu internum)* of the facial nerve (Fig. 269). KAPPERS (1947 *et passim*) has analyzed and interpreted in terms of neurobiotaxis the diverse manifestations of the internal facial knee and their relationship to the various displacements or shifts displayed by the motor facial nucleus. Again, there is some evidence for a 'somatotopic' representation within the motor facial nucleus, of the muscles innervated from this griseum (SZENTÁGOTHAI, 1948), but the conclusions reached by different authors remain conflicting.

Besides the platysma, important facial muscles innervated by the facial nerve are m. orbicularis oculi, mm. quadrati labii superioris et inferioris, m. orbicularis oris, and m. buccinator. Other relevant muscles innervated by this nerve are m. stapedius, m. stylohyoideus, and posterior belly of m. digastricus.

The branchiomotor *nuclei of the glossopharyngeal, vagus, and cranial accessory nerves* are provided by a joint, but variably interrupted and not very wide cell column of motoneurons, located in the ventrolateral tegmentum, and forming the *nucleus ambiguus* (Figs. 261C, 263E, 266D, 266E). The efferent fibers originating in this griseum innervate the striated musculature of larynx and pharynx. Attempts at discovering a 'somatotopic' representation within the nucleus ambiguus were undertaken by a few investigators (cf. SZENTÁGOTHAI, 1943). The root fibers of this nucleus run first dorsad and then bend laterad to join the other root fibers of vagus and glossopharyngeus which emerge on the oblongata's lateral surface. The efferent fibers from the caudalmost levels of nucleus ambiguus, in the closed portion of the oblongata, form the *cranial accessory nerve*.

The *spinal accessory nerve* originates from the dorsolateral part of the ventral horn at several upper cervical spinal cord segments. Its roots emerge laterally, in an intermediate linear array between dorsal and ventral roots. The most rostral cell groups of its nucleus may reach caudal levels of the oblongata, and appear as a transition to the caudal end of the nucleus ambiguus.[174] The still not entirely clarified mor-

[174] This seems to be the 'nucleus retroambiguus' depicted by OLSZEWSKI and BAXTER (1954) in Man. The so-called '*nucleus supraspinalis*' represents presumably the rostral end of a ventral cell group of the spinal cord's anterior horn, from which some ventral root fibers of the first spinal nerve take their origin.

phologic and phylogenetic problems concerning the XIth nerve with its cranial and spinal subdivisions[175] are briefly discussed in chapter VII, section 3, p. 857, volume 3, part II.

While the cranial accessory nerve, joining the vagus, innervates laryngeal musculature, the spinal accessory innervates m. sternocleidomastoideus and m. trapezius. Both these muscles, moreover, also receive motor branches from the plexus brachialis.

The *dorsomedial subzone of the basal plate's intermedioventral zone* contains the *preganglionic* (cranial parasympathetic) *nuclei of facial, glossopharyngeus, and vagus nerves.* These grisea, whose nerve cells are generally smaller than those of nucleus ambiguus, and mostly of 'fusiform' shape, form an uninterrupted column, which is most conspicuous at vagus levels, where it lies close to the periventricular gray and is designated as *nucleus ('motorius') dorsalis vagi* (Figs. 261C, D, 263E, 266D). Rostralward, beginning at glossopharyngeal levels, this column generally tends to shift ventrolateralward away from the periventricular gray and gradually tapers out near the caudal level of nucleus motorius facialis, or at a short distance dorsomedially to the caudal extremity of that nucleus. It is likely that the rostral portion of nucleus dorsalis vagi et glossopharyngei includes the *nucleus salivatorius inferior* of the IXth, as well as the *nucleus salivatorius superior et lacrimalis* of the VIIth. These nuclei, whose existence is reasonably well established by numerous physiologic data, have not been accurately identified and were vaguely localized within the reticular formation on the basis of questionable findings based on experiments with chromatolysis. The morphologic evidence for inclusion of these nuclei into the griseal column of nucleus dorsalis vagi, glossopharyngei et facialis might appear far more convincing. The origin of cardiac, bronchopulmonal, and gastrointestinal preganglionic fibers from the caudal region represented by nucleus dorsalis vagi *sensu strictiori* is generally accepted. An attempt at identifying a 'somatotopic localization' in the nucleus dorsalis vagi of the Rabbit was made by GETZ and SIRNES (1949). Regardless of still unclarified details, it can thus be presumed that the preganglionic fibers originating in the entire 'general visceral efferent' column of vagus, glossopharyngeus and facialis provide the cranial parasympathetic (or 'autonomic') root fibers in these three cranial nerves.

---

[175] Both subdivisions, which are essentially efferent, may include a small, variable and undefined amount of afferent fibers. The transitory or rudimentary ganglion of the spinal accessory was pointed out in the here cited section 2 of chapter VII.

The *ventral zone* of the Mammalian oblongata's *basal plate* includes, as in Sauropsidans, two somatic efferent cranial nerve nuclei, namely those of nervus abducens and nervus hypoglossus. The *nucleus nervi abducentis* (Figs. 261B, 263C, 269) is dorsally located, close to the fasciculus longitudinalis medialis, from which it receives terminals, and adjacent to, or, as the case may be, within, the internal knee of the efferent facial root. As in Reptiles and Birds, a ventrolaterally displaced accessory nucleus of the abducens can be found in numerous Mammals (cf. above, p. 480 and 501). In Primates including Man, which do not possess a nictitating membrane, this accessory nucleus is not present.[176] The nervus abducens innervates the musculus rectus lateralis *sive* externus of the eye bulb.

The *nucleus nervi hypoglossi* has a medial location near the floor of the fourth ventricle at levels of the calamus scriptorius, and, within the closed portion of the oblongata commonly extends as far as the transition to spinal cord (Figs. 261C, D, 263E, F, 264C, D, 266D, E). The hypoglossal root fibers, which innervate the intrinsic and extrinsic musculature of the tongue, run toward the ventral surface of the oblongata, emerging laterad to the pyramidal tract. The morphologic and phylogenetic significance of the XIIth nerve, and the transitory ontogenetic appearance of a dorsal hypoglossal root with *Froriep's ganglion* at caudalmost levels of the Mammalian oblongata were dealt with in section 3 of chapter VII (vol. 3, part II).

The intrinsic tongue muscles innervated by the n. hypoglossus are mm. longitudinalis superior, inferior, verticalis, and transversalis linguae. Extrinsic tongue muscles innervated by this nerve are mm. styloglossus, hyoglossus, and genioglossus.

A simplified concept of relationships between the motor innervation of the tongue through the nervus hypoglossus, and the afferent tongue nerves is shown in Figure 268C. Although this diagram, elaborated more than sixty years ago by CAJAL, does not take into account various details concerning the tongue's innervation and the presumed subdivisions of tractus solitarius, etc., it still represents a very useful and instructive first approximation.

---

[176] This accessory nucleus is not to be confused with a 'parabducens nucleus' of small cells located laterally to the abducens nucleus and presumably pertaining to the reticular gray, but perhaps relaying impulses to the abducens nucleus (cf. e.g. CROSBY *et al.*, 1962). Again, an 'accessory abducens nucleus' occasionally referred to as present in Man is presumably an accessory facial nucleus (cf. CROSBY *et al.*, 1962).

In addition to the enumerated efferent components of branchial nerves and to the bulbar cranial nerves of ventral root type (VI and XII) the *efferent component of the cochlear nerve* ('efferent cochlear bundle') already dealt with on p. 319 of section 1 and also referred to in connection with the Avian oblongata (section 9, p. 502) remains to be mentioned as pertaining to the cranial nerve fibers arising in the basal plate. Nothing certain, however, has been elucidated about the location of their cells of origin. It can merely be surmised that these efferent neurons might lie at cochlear bulbar levels somewhere in the tegmentum, perhaps near the superior olivary complex.

Among other insufficiently clarified and controversial problems the two following ones[177] should be mentioned: (1) the question whether a few efferent root fibers of branchial nerves, of abducens and of hypoglossus are crossed, i.e. emerge on the contralateral side;[178] (2) the question to which extent proprioceptive fibers[179] are included in the just mentioned nerves, particularly also with respect to some afferent cells suspected to be included in either nuclei or roots of abducens and hypoglossus.

Still another topic about which the available data are rather incomplete, concerns the so-called '*supranuclear innervation*' of the diverse efferent cranial nerve nuclei by fibers originating in numerous other grisea. Besides many other connections, important fiber tracts are, in this respect, the various bulbar components of pyramidal, rubro-spinal, and tecto-spinal tracts. The question whether such 'supranuclear innervation' by the pyramidal tract is crossed, uncrossed, or both, plays, e.g., a significant role in clinical neurology and shall again be considered further below in dealing with the human bulbar configurations.

Within *the overall basal plate*,[180] the Mammalian reticular formation is well developed and has been parcellated, by diverse investigators, into

---

[177] Comments on these questions, with further bibliographic references, can be found in the treatises by Beccari (1943), Crosby *et al.* (1962), Kappers (1947) and Kappers *et al.* (1936).

[178] In this respect only the presence of a crossed component in the mesencephalic oculomotor nerve III, and the complete (or at least essentially complete) decussation of the likewise mesencephalic trochlear (IV) nerve can be considered certain.

[179] As regards the trigeminus, the proprioceptive functional significance of the mesencephalic root can be considered as reasonably certain.

[180] 'Forerunners' of distinctively Mammalian basal plate nuclei, particularly of n. intercalatus (cf. Fig. 257C) and perhaps also of n. caeruleus, seem, nevertheless, to be present in Sauropsidans, especially in Birds.

numerous but poorly delimitable cell groups, for which no generally accepted standardized nomenclature obtains. Roughly speaking, a medially located '*substantia reticularis alba*', in which the medullated fiber systems predominate, can be distinguished from a lateral region, the '*substantia reticularis grisea*', in which the cell groups and their neuropil are somewhat more conspicuous between the grid of myelinated fiber bundles. A caudal and rostral myelencephalic subdivision, and a pontine subdivision may likewise be recognized. Some of the descriptive terms refer to cell size, e.g. to magnocellular groups, and others to location, e.g. to paramedial or raphe nuclei, and, in the pons, to a nucleus ventralis tegmenti *(ganglion tegmenti profundum of Gudden)*, which lies actually in the dorsal tegmentum but basally to the central gray, as well as ventrally and ventrolaterally to fasciculus longitudinalis medialis. A pontine group is also the nucleus centralis superior, located on both sides of the raphe and extending in a lateral direction. Figure 270A illustrates the structural aspect of neuropil and synapses in the Mammalian reticular formation as seen by means of *Cajal's silver impregnation*. It seems possible, moreover, that, as has been inferred by investigators using the *Golgi method*, some neuronal elements of the reticular

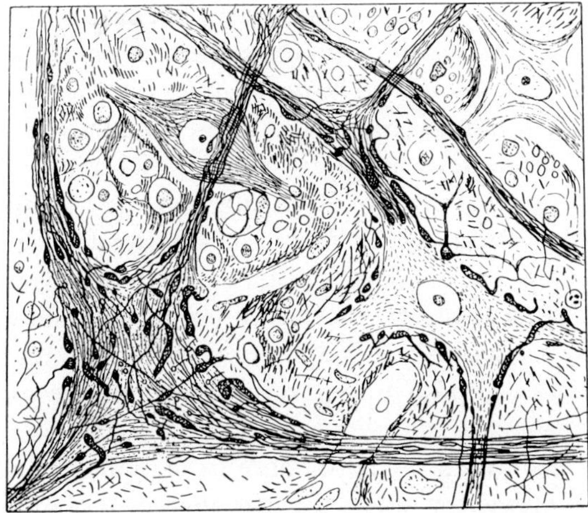

Figure 270 A. Structural details of Mammalian formatio reticularis grisea as displayed by *Cajal's silver impregnation* in the oblongata of an adult Rabbit (after CAJAL, 1909, from EDINGER, 1912). In addition to neuropil, two large multipolar neurons and two somewhat smaller ones, various sorts of axosomatic and axodendritic synapses can be seen.

formation possess axons which, running through the entire length of this aggregate give off collateral endings to practically all grisea of the brain stem (cf. e.g. in JASPERS *et al.*, 1958). At any rate, many of the diffusely scattered medullated fiber bundles running through the length of the reticular formation presumably represent intrinsic channels pertaining to that griseum.

An irregular group of large, medium, and small sized reticular cells is located in the ventral part of the myelencephalic trigeminal root region and dorsally to inferior olivary complex, representing the *nucleus reticularis lateralis*. This griseum is not distinctly recognizable in Sauropsida, where it might or might not be included in the inferior nucleus of the reticular formation. In Mammals, however, it constitutes a distinctive special cerebellar relay nucleus, which apparently does not pertain to the reticular formation *sensu strictiori*. It can be parcellated into two or three subdivisons and receives fibers of ascending spinal systems such as tractus spino-cerebellaris in accordance with a somatotopic pattern. The nucleus reticularis lateralis, in turn, projects propricceptive and tactile impulses essentially by way of the homolateral corpus restiforme to the cerebellum which also sends descending fibers into this griseum (cf. CROSBY *et al.*, 1962, JANSEN and BRODAL, 1958).

There are, moreover, within the basal plate, adjacent to, or in the vicinity of the efferent cranial nerve nuclei, various grisea less clearly outlined in Sauropsidans, but representing rather distinctive nuclei in Mammals, namely *nucleus of locus caeruleus*, *nucleus praepositus hypoglossi*, *nucleus intercalatus Staderini*, and *nucleus of Roller* (cf. fn. 180).

The *nucleus loci caerulei* (coerulei) is located ventrally to sulcus limitans, at rostral levels of the main trigeminal nuclei (Fig. 263B), close to the central gray, and rostrally extending into the isthmus region. It also lies closely adjacent to the more dorsally situated radix mesencephalica trigemini. The nerve cells of the 'nucleus caeruleus' which has an unusually dense capillary vascularization (FINLEY and COBB, 1940), are commonly medium sized and multipolar. In Man (cf. further below), to a lesser extent also in some other Mammals, these cells contain a blackish-brown melanotic pigment, and the designation of the nucleus is derived from its gross appearance in the human brain. Caudalmost elements of nucleus radicis mesencephalicae trigemini and nucleus caeruleus may intermingle in a zone of overlap, but can rather easily be distinguished from each other. Connections and functional significance of nucleus caeruleus have not been elucidated by the traditional methods of neuroanatomy, but its involvement in respiratory control was suggested (JOHNSON and RUSSELL, 1952). However, recent investigations by means of techniques based on fluorescence histochemistry of monoamine-containing neuronal elements (cf. vol. 3, part I, p. 630, and chapter VIII, section 10, p. 217 of the present volume) have not only shown that the nucleus loci caerulei is composed of catecholamine containing cells, in contradistinction to the neighboring nucleus radicis mesencephalicae trigemini, but have led to surprising interpretations and conclusions, suggesting a rather unique significance of that griseum (cf. HUBBARD and DICARLO, 1973).

It is believed, on the basis of results with these techniques, that a single neuron of the nucleus loci caerulei may be capable of transmitting impulses to a large number of grisea throughout the brain, including efferent and reticular brain stem nuclei, the cerebellar cortex, the mesencephalic colliculi, the geniculate bodies, parts of the thalamus, the hypothalamus, the cerebral cortex (e.g. the hippocampus). The diffuse monosynaptic noradrenergic system 'having its cell bodies (or most of its cell bodies) in the locus caeruleus' is said to be important for cortical arousal, as well as for 'paradoxic sleep' (cf. CHU and BLOOM, 1973; FUXE and HANSSEN, 1967; HUBBARD and DICARLO, 1973; JOUVET, 1967; OLSON and FUXE, 1971; LOIZU, 1969; UNGERSTEDT, 1971). As regards the cerebellum, stimulation of the nucleus loci caerulei is supposed to result in prolonged inhibition of the *Purkinje cells* after a long latent period (SIGGINS *et al.*, 1971).

Although the demonstration of catecholamine containing grisea by means of fluorescence can be regarded as rather conclusive, the inferences *qua* detailed specific direct fiber respectively synaptic connections seem to be based on still not entirely convincing interpretations. Again, effects of nucleus loci caerulei stimulations upon the cerebellar cortex may well be related to a variety of diverse factors related to interference with other neuronal structures closely adjacent to the relatively small griseum of the locus caeruleus. It should finally be added that cells of both nucleus loci caerulei and of nucleus radicis mesencephalicae trigemini may become dorsally displaced along the ventricular wall of brachium conjunctivum into the deep (periventricular) medullary center of the cerebellum.

The *nucleus praepositus hypoglossi* extends from the rostral pole of hypoglossal nucleus to the neighborhood of the abducens nucleus. Connections and significance of nucleus praepositus have not been ascertained. Morphologically, this griseum appears to be a rostral expansion and extension of the *nucleus intercalatus Staderini*. This latter griseum (Figs. 261C, 266D), consisting of medium sized and small multipolar cells, is intercalated between hypoglossal nucleus and nucleus dorsalis vagi. It seems to be connected with both neighboring nuclei as well as with the medial vestibular nucleus and the *dorsal longitudinal bundle of Schütz* in the periventricular central gray. STADERINI, who described this nucleus about 1894, subsequently again discussed the various still dubious questions concerning its significance (1938).

At rostral levels, where the preganglionic column of nucleus dorsalis vagi et glossopharyngei and the grisea of tractus solitarius tend to be displaced away from the periventricular gray in a ventral or ventrolateral direction, the nucleus praepositus hypoglossi directly borders on the nucleus vestibularis medialis s. triangularis.

The *nucleus of Roller* is a group of medium sized to small cells located ventrally to the hypoglossal nucleus, but apparently not giving off hypoglossal root fibers.[181] It seems to be a correlating griseum (*'nucleo associativo'*, cf. e.g. BECCARI, 1943) and might perhaps be considered part of the reticular formation.

In the oblongata of Mammals, the periventricular central gray is commonly more clearly delimited than in submammalian Vertebrates from the various afferent and efferent nuclei of cranial nerves and from the reticular formation. In addition to the subependymal cell plate, which represents an embryonic remnant, it contains a layer with small to medium sized nerve cells. In the bulbar basal plate, an ill-defined *nucleus paramedianus dorsalis* extends from the obex levels rostralward as far as the pons, merging, at isthmus levels, with the *nucleus dorsalis tegmenti* (fig. 263A). Within this layer runs the thinly medullated and partly non-medullated *fasciculus longitudinalis dorsalis s. posterior of Schütz* (1891), which represents a complex ascending and descending periventricular fiber system extending from the diencephalon as far caudad as at least the transition of oblongata to spinal cord. The fasciculus longitudinalis dorsalis of the central gray appears essentially related to the central component of the autonomic (vegetative) nervous system and presumably effects connections with numerous brain stem grisea. Nucleus paramedianus dorsalis, nucleus dorsalis tegmenti and similar periventricular grisea may be considered interstitial respectively accompanying nuclei of said fasciculus.

---

[181] It was originally described in 1881 by ROLLER (Arch. mikr. Anat. *19*, p.383) as '*Kleinzelliger Hypoglossuskern*'.

The *subependymal cell plate*, whose significance was discussed in section 1 of chapter V (vol. 3 part I) extends through the entire length of the cerebral neuraxis, but its appearance and thickness in the adult brain is not only very variable with respect to brain regions and taxonomic forms, but also displays considerable individual variations. It was first pointed out by my collaborator HIKIJI (1933) in an investigation concerning the oblongata of the human newborn (cf. Figs. 279B, C, 284F).

The ventral portion of the Mammalian bulbar basal plate includes, moreover, three important griseal aggregates, some of which, although being basal plate components at the definitive ontogenetic stage, derive, either wholly or in part, from the alar plate, namely *nuclei pontis* with *nuclei arcuati*, *superior olivary complex* and *inferior olivary complex*.

As a gross subdivision in macroscopic anatomy, the *pons* can be considered a specifically Mammalian configuration. It consists of the pontine nuclei, which receive essentially homolateral cortico-pontine fibers and give off predominantly crossed fibers reaching the cerebellum by way of the *brachium pontis* or middle cerebellar peduncle. The pontile grisea commonly appear to be randomly scattered between the grid of longitudinal cortico-pontine and transverse ponto-cerebellar fibers. The longitudinal fibers of the *pyramidal tract* likewise pass through the pons (Fig. 266A), or closely adjacent to its dorsal boundary (Figs. 261A, 263A).

Although evidently components of the adult basal plate, these nuclei are derivatives of the alar plate, as shown by ESSICK (1907 etc.) and confirmed by others (cf. also above footnote 160). An embryonic remnant of this ontogenetic migration is, in at least some Mammalian forms, quite conspicuous and represents the *corpus pontobulbare*, to be again pointed out further below in discussing the human oblongata. In the different Mammalian forms, the relative size of the pons seems directly proportional to the development of neopallium and neocerebellum (i.e. roughly speaking, cerebellar hemispheres). In 'lower' Mammals, the pons does not cover the entire trapezoid body, whose transverse fibers can then be seen on the bulbar surface caudally to the boundary of pons (cf. Fig. 263C). In Monotremes, the pons is described as a 'retrotrigeminal' configuration (ABBIE, 1934), since the trigeminal root emerges at its rostral border. The variously developed nuclei arcuati or nuclei arciformes, found on the surface of the oblongata laterally respectively ventrally to, or within the pyramids can be in-

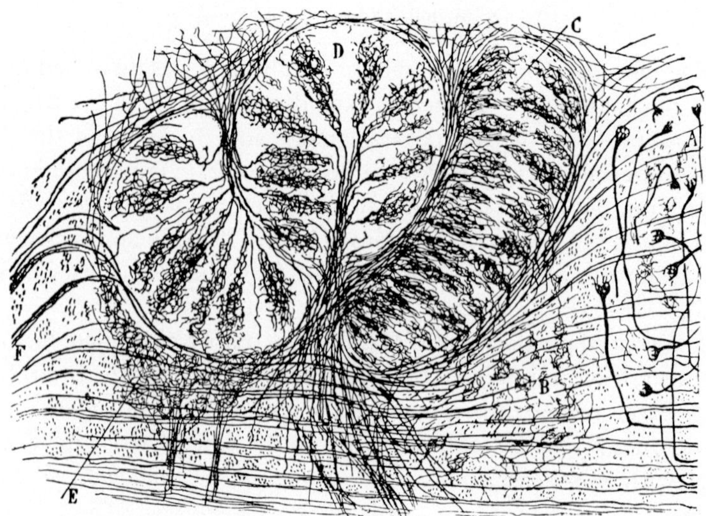

Figure 270B. Structural details of Mammalian superior olivary complex as displayed by the *Golgi impregnation* in a cross-section through the oblongata of a Cat a few days after birth (from CAJAL, 1909). A: nucleus of trapezoid body; B: nucleus praeolivaris internus; C: (medial) accessory superior olivary nucleus; D: (principal) superior olivary nucleus; E: nucleus praeolivaris externus *(foyer semilunaire ou préolivaire externe);* F: fibers of trapezoid body.

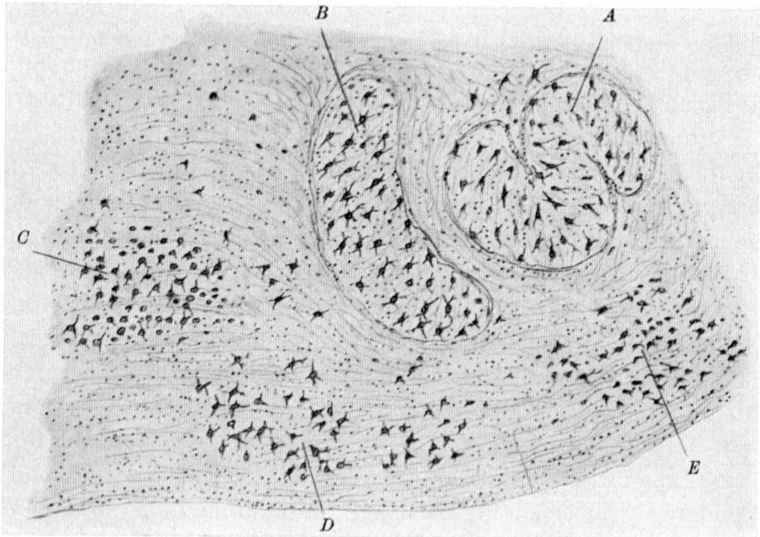

Figure 270C. Superior olivary complex (cross-section, *Nissl stain)* in an adult Rabbit (after CAJAL from MINGAZZINI, 1928). A: main superior olivary nucleus; B: nucleus olivaris superior accessorius; C: nucleus of trapezoid body; D: internal preolivary nucleus; E: external preolivary nucleus. B and C are oriented in mirror symmetry to each other.

terpreted as caudally displaced pontine nuclei,[182] mostly connected with the cerebellum by fibrae arcuatae externae ventrales (cf. Fig. 261C).

The *superior olivary complex* of Mammals is related to the decussating secondary cochlear fibers of the *trapezoid body* which gather as *lemniscus lateralis* on the lateral side of that complex (cf. e.g. Figs. 261B, 263B, 266A). The Mammalian superior olive generally consists of a lateral convoluted nucleus (n. olivaris superior lateralis s. principalis), and of a medial cell plate (n. olivaris superior medialis s. accessorius).[183] Both nuclei contain a dense neuropil. Farther mediad a more diffuse cell group is designated as the nucleus of the trapezoid body. Its structure resembles that of the ventral cochlear nucleus. Another, more indistinct cell group is located laterally to n. olivaris superior principalis, and could be designated as n. corpus trapezoidis lateralis, in contradistinction to the above-mentioned medial one. It might, however, also be regarded as one of the interstitial grisea of the lemniscus lateralis. Two additional cell groups, n. praeolivaris medialis and lateralis, are located on the ventral side of the superior olivary complex, whose details are illustrated by Figures 270B and C.

The superior olivary complex is considered to be a reflex and possibly also a relay center intercalated in the auditory pathway (YOSHIDA, 1925). A bundle described as superior olivary peduncle is believed to reach the abducens nucleus and the fasciculus longitudinalis medialis. According to RASMUSSEN (1953), however, this peduncle is essentially formed by a bundle arising in a 'retro-olivary' reticular cell group and containing the efferent cochlear pathway referred to further above. The contribution of superior olivary complex to the lateral lemniscus, observed by some authors, is denied by others (cf. the comments by

---

[182] ABBIE (1934) describes in Echidna, besides typical arcuate nuclei, a group of arcuate grisea which he called '*nucleus arcuatus trigemini*'. His interpretation, however, does not appear convincing, and these structures might likewise represent aberrant pontine grisea (cf. Figs. 261A to D) such as perhaps nucleus conterminalis (Fig. 178A).

[183] With respect to this cell plate, which commonly appears 'semilunar' in cross-sections, CAJAL (1909) believes that it might be homologous with the nucleus laminaris of Birds (cf. above footnote 144). Although this latter is evidently an alar plate derivative, and the Mammalian accessory olive is clearly located within the basal plate, a secondary ventral displacement of nucleus laminaris does not seem impossible. The homology suggested by CAJAL still remains, however, a moot question. As regards the term 'semilunar', it should be added that, in CAJAL's terminology, the nucleus praeolivaris externus is also called '*noyau (or "foyer") semilunaire*'.

BECCARI, 1943, CROSBY *et al.*, 1962, and KAPPERS, 1947). Recurrent fibers from the superior olivary complex *('fibers of Held')* seem to terminate in the primary cochlear nuclei. Be that as it may, there is little doubt that many collateral connections are effected by the superior olivary griseum, directly or indirectly reaching various motor cranial nerve nuclei and other cell groups. Elements of the reticular formation can presumably act as relevant intermediate links.

Reverting to the secondary (arcuate or arciform) fibers originating in the cochlear nuclei, LEWY and KOBRAK (1936), who have reviewed and discussed, on the basis of their own investigation, the various pathways described by previous authors, indicate their interpretation by the diagram of Figure 267D, as mentioned further above. The dorsalmost bundle (in the Rabbit) forms the *striae acusticae of Monakow*, which originate from the tuberculum acusticum and run dorsally through the tegmentum below the central gray. More ventrally, *Held's intermediate system* consists of two subdivisions, originating from dorsolateral and medioventral divisions of the ventral cochlear nucleus. These fibers run through superior olivary nuclei and interolivary field, representing the dorsal portion of the trapezoid body. Most ventrally, a system which LEWY and KOBRAK call the trapezoid body (*sensu strictiori*, but better perhaps conceived as the ventral portion of that body) originates from medioventral and ventral divisions of the ventral cochlear nucleus. It is described as passing to the contralateral (medial) nucleus of the trapezoid body. These three pathways constitute the essential feeders of the lateral lemniscus which reaches mesencephalic and diencephalic grisea.[183a]

The inferior olivary complex of Mammals, ventrally located in the caudal part of the oblongata's basal plate is, perhaps either wholly or to a substantial degree, a ventrally displaced ontogenetic derivative of the alar plate, and consists of three subdivisions (Fig. 271). The *principal olivary nucleus* is extensively developed in 'higher' Mammala and assumes here the shape of a convoluted, bent plate with medial hilus. The two other subdivisions are *a dorsal and a medial accessory olivary nucleus*, this latter being generally considered the phylogenetically 'oldest' component (*'oliva primordiale'*, BECCARI, 1943). The essentially crossed

[183a] With respect to origins, course, and distribution of the dorsal and intermediate acoustic striae in the Rhesus monkey, a recent, not very convincing and rather inconclusive report, based on results with the *Fink-Heimer* and similar techniques has been published by STROMINGER (1973).

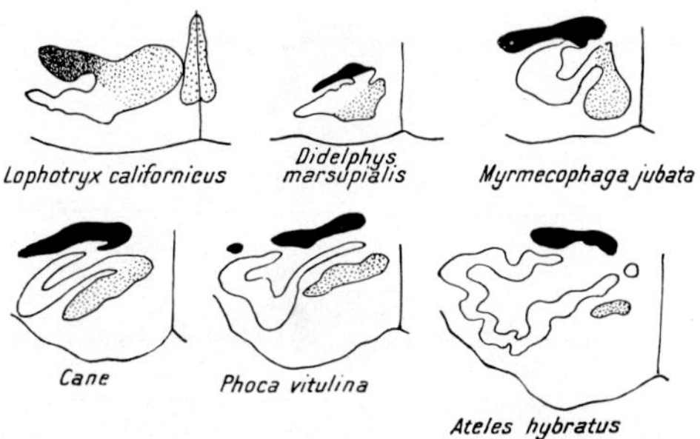

Figure 271. Diagrammatic cross-sections through the inferior olivary complex of a Bird (Lophotryx californicus) and of several Mammals: Marsupial, Edentate, Carnivore Canid Dog, Carnivore Pinniped Phoca, and Primate Ateles (after Kooy, 1917, from Beccari, 1943). Stipples: primordial inferior olive, respectively in Mammals the medial accessory inferior olive; black: dorsal accessory inferior olive, assumed to be derived from the darker stippled dorsolateral extension of the submammalian primordial griseum; blank outline: ventrolateral submammalian extension respectively mammalian principal inferior olivary nucleus.

olivo-cerebellar fibers join as decussating fibrae arcuatae internae the contralateral corpus restiforme. Roughly stated, it can be said that nucleus olivaris inferior principalis projects upon the cerebellar hemispheres ('neocerebellum'), and that the fibers from the accessory olives reach vermis and flocculus. The inferior olivary complex receives, in turn, ascending spino-olivary fibers and descending fibers through the *central tegmental (so-called 'thalamo-olivary') tract.* This pathway seems to include a number of different components, which shall be briefly mentioned in section 9 of chapter XI. Further reference to the olivocerebellar projections (Fig. 351C) will be found in sections 1 and 9 of chapter X.

The localization of olivocerebellar projections, as elaborated by Jansen, Brodal, Walberg and others (cf. Jansen and Brodal, 1958; Larsell, 1970, 1972) shall briefly be pointed out further below in chapter X. Olivospinal fibers, formerly presumed to be present, are generally denied by recent authors. Because of the diverse sources of errors and of the ambiguities inherent in experimental methods, combined with possible differences related to the various taxonomic forms, many particulars remain uncertain and one could still remark with Beccari (1943): '*si conoscono soltanto in parte le connessioni dell'oliva bulbare*'.

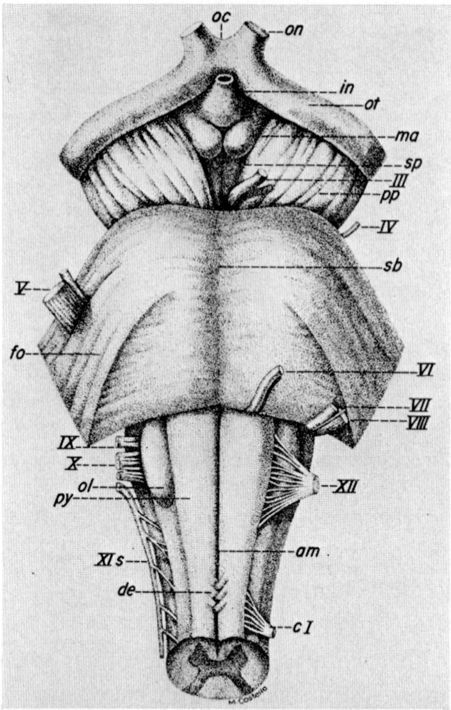

Figure 272A. Basal aspect of human brain stem (from an unpublished study in collaboration with Dr. WEBB HAYMAKER). am: fissura mediana anterior; cI: ventral roots of first cervical nerve; de: decussatio pyramidum; fo: fasciculus obliquus pontis; in: infundibular stem (neurohypophysis); ma: mammillary bodies; pp: pes pedunculi; py: pyramid of medulla oblongata; oc: chiasma opticum; ol: inferior olive; on: nervus opticus; ot: tractus opticus; sb: sulcus basilaris; sp: substantia perforata posterior (within interpeduncular fossa); III to XII roots of cranial nerves; XIs nervus accessorius spinalis.

With regard to the alar plate derivation of pontine, inferior and perhaps also some superior olivary grisea,[184] the ontogenetic migration of their cell masses into the basal plate might be considered a particularly pronounced manifestation of neurobiotaxis in the overall interpretation presented in section 7, chapter V, volume 3, part I.

Caudally to the pons, the *cortico-spinal* or *pyramidal tract* is located on both sides of the midline under the basal surface of the oblongata and decussates at the levels of transition to the spinal cord (Figs. 261D,

[184] Particularly if, in accordance with CAJAL's hypothesis (cf. the preceding footnote 183), the Mammalian medial accessory olivary nucleus should indeed correspond to the Sauropsidan nucleus (cochlearis) laminaris.

263F). Details concerning this decussation and other aspects of the cortico-spinal tract were dealt with in section 11 of the preceding chapter VIII.

The essentially crossed *rubrospinal tract* appears to run, with some variations, through a lateral region of the tegmentum, as e.g. shown in Figures 262B, 262D, 263B, C, 280C, D, 281C, 282C, in part perhaps closely joined to the *lateral tectospinal tract.*

Following the preceding condensed summary concerning overall features of the Mammalian oblongata, some relevant gross and microscopic aspects of that brain subdivision in *Man* shall be considered. In basal view (Fig. 272A) the myelencephalon appears as a cone-shaped enlargement of the spinal cord and, like this latter structure, presents an *anterior median fissure*. In the region of the pyramidal decussation the crossing fasciculi are seen obliterating the fissure which again becomes deeper craniad and ends with a depression *(foramen caecum)* at the caudal border of the pons. On both sides of the rostral anterior median fissure are located the columnar swellings of the *pyramids* which taper toward their caudal ends and contain the cortico-spinal or pyramidal tracts.

Lateral to each pyramid and separated from it by a *sulcus lateralis anterior* there is an oval prominence, the *olive (oliva)*, which corresponds to the principal nucleus of the inferior olivary complex. Through the

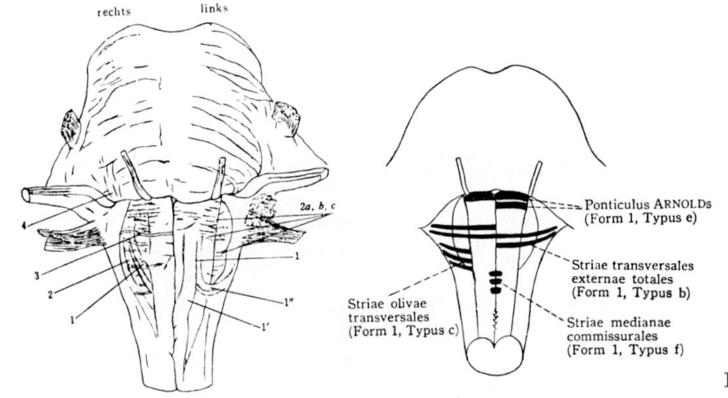

Figure 272B. Externally visible, highly variable bundles of fibrae arcuatae externae ventrales or basal bulbar striae of the human oblongata (from OGAWA, 1933). I Sketch illustrating a specific case (the leads are not relevant in the present context). II Sketch illustrating diverse types of basal bulbar striae in accordance with a classification suggested by OGAWA.

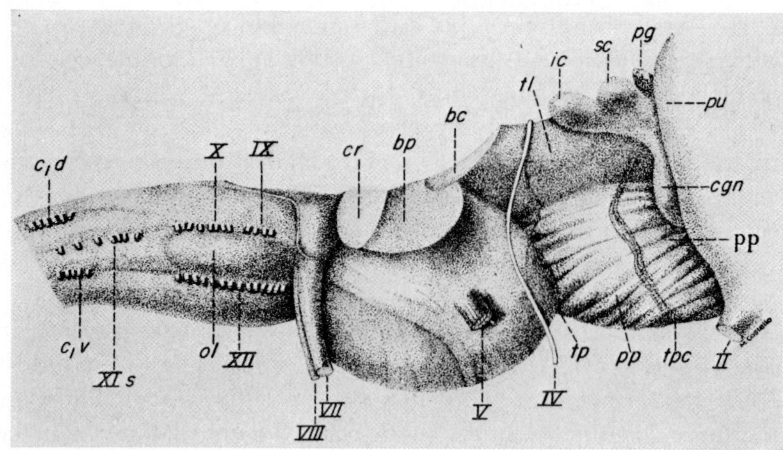

Figure 273A. Lateral aspect of human brain stem (from an unpublished study in collaboration with Dr. WEBB HAYMAKER). bc: brachium conjunctivum; bp: brachium pontis; cgm: corpus geniculatum mediale; cr: corpus restiforme; c_1d: dorsal roots of first cervical nerve; c_1v: ventral roots of first cervical nerve; ic: inferior colliculus; ol: inferior olive; pp: pes pedunculi; pu: pulvinar thalami; sc: superior colliculus; tl: trigonum lemnisci; tp: taenia pontis (fila lateralia pontis); tpc: tractus peduncularis transversus; XIs: nervus accessorius spinalis. It will be noted that in dissections of this sort (cf. also Figs. 272A and 274A) some neighboring diencephalic structures, including the pineal (pg) remain connected with the 'brain stem'.

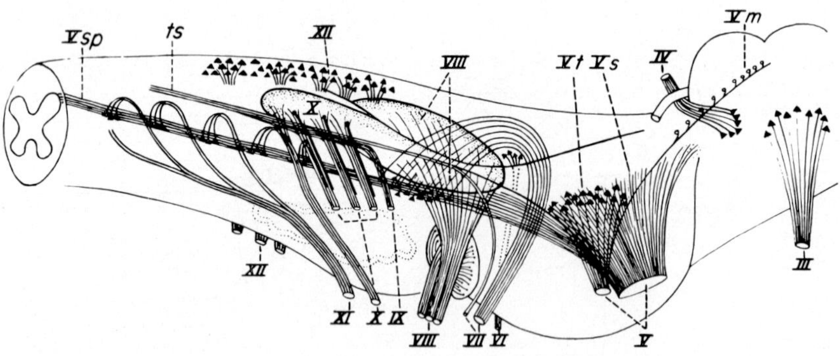

Figure 273B. EDINGER's concept of a simplified transparent diagram illustrating relationships of cranial nerve nuclei and root fibers in lateral view, to be compared with preceding figure (slightly modified after EDINGER, 1912). Vm: mesencephalic trigeminal root; Vs: main sensory nucleus; Vsp: descending root; Vt: motor nucleus; ts: tractus solitarius.

anterolateral sulcus the rootlets of the hypoglossal nerve (XII) emerge, and at the rostral end of this nucleus, at the caudal boundary of the pons, the nervus abducens (VI) leaves the brain. Dorsolaterally to the olive, and approximately in a furrow called *sulcus postolivaris*, the root fibers of nervus glossopharyngeus (IX), vagus (X), and accessorius cranialis (XIc) emerge. The caudal roots of the nervus accessorius (n. accessorius spinalis, XIs) emerge in a lateral alignment intermediate between the dorsal and the ventral spinal nerve roots (Fig. 273A). Crossing the surface of the olives, of the pyramids, and the region caudal to the olive, individually very variable thin fiber bundles, *fibrae arcuatae externae ventrales* or *basal bulbar striae* may be found (OGAWA, 1933, and other authors), as shown in Figure 272B.

The *basal aspect of the pons* is characterized by a system of transverse fiber bundles, separated by the superior pontile sulcus from the pes pedunculi of the midbrain, and by the inferior pontile sulcus from the myelencephalon. In the midline of the pons and in continuation of the fissura mediana anterior, but separated from this latter by the foramen caecum, there is commonly a shallow *median basilar sulcus*, approximately corresponding to the course of the basilar artery. This sulcus, however, is caused by the bilateral bulging of the pyramidal tract which passes, under cover of the pontile fibers, from the pes pedunculi of the mesencephalon to the pyramid of the medulla oblongata.

Several superficial bundles of pons fibers can be recognized and roughly distinguished as the rostral fasciculus superior and the caudal fasciculus inferior pontis. These fibers extend laterad to form the *brachium pontis* or *middle cerebellar peduncle* entering the cerebellar hemispheres. In the ill-defined boundary zone between the brachium pontis and the pons proper the root of the trigeminal nerve emerges, bounded rostrad by the fasciculus superior pontis and caudad by a frequently slightly oblique bundle, designated as fasciculus obliquus pontis, which blends caudad with the fasciculus inferior. The motor component of the trigeminal root may be seen as a distinct bundle *(portio minor)* on the rostromedial side of the main or sensory portion *(portio major)* of the emerging root.

In the angle between brachium pontis, myelencephalon, and cerebellum *(cerebello-pontine angle)*, the nervus acusticus (VIII) emerges with its rostral and medial vestibular, and its caudal and lateral cochlear components. Medially adjacent to the acustic nerve the root of nervus facialis (VII) emerges. The autonomic (vegetative) and the afferent component of the facial nerve may be visible as a separate fila-ment, called *nervus intermedius*, between the main part of the facial nerve root and the acoustic nerve. Both the facial and the acoustic nerve emerge approximately in line with the glossopharyngeal and vagus nerves. At the rostral border of the pons, an individually very variable bundle of fibers *(fila lateralia pontis s. taenia pontis)* may run parallel to the margin of the pons and then either join the brachium pontis or disappear into the cerebellum along the *upper peduncle (brachium conjunctivum)*.

In *dorsal aspect* (Fig. 274A), the closed portion of the medulla oblongata, continuous with the spinal cord, displays both sulcus intermedius posterior and sulcus medianus posterior medullae spinalis. The *median sulcus* deepens to become a median '*fissure*' ending at the caudal tip of the rhomboid fossa. The rostral continuation of the fasciculus gracilis enlarges to a swelling known as the *clava*, which contains the *nucleus of Goll*. The upper end of fasciculus cuneatus, extending somewhat farther rostrad than funiculus gracilis and its clava, likewise enlarges to form the tuberculum cuneatum containing the grisea of funiculus cuneatus. Lateral to tuberculum cuneatum and to the corresponding posterolateral sulcus, another slight eminence may be visible, the tuberculum cinereum, which corresponds to the descending tract of the trigeminus and its nucleus, this latter being continuous with the *substantia gelatinosa Rolandi* of the spinal cord. Further rostrad the tuberculum cinereum is replaced by the *restiform body* (inferior cerebellar peduncle) whose external surface obliterates the posterolateral sulcus. An elevated strip of gray matter, which is frequently grossly recognizable, extends caudad from the pons, basally to the root of the acoustic nerve, and covers the dorsomedial part of the restiform body, running parallel to the *ponticulus of Ziehen*. This strip is a remnant of the embryonic *corpus pontobulbare* (Fig. 278F).

The fourth ventricle extends from the posterior end of the cerebral aqueduct at the level of isthmus rhombencephali to the caudal part of the oblongata, where its lumen narrows to become a central canal continuous with that of the spinal cord. The rostral part of the ventricle's roof is formed by the thin anterior (or superior) medullary velum upon which the lingula cerebelli lies, by the ventricular surface of brachia conjunctiva, and by the paired tectal or fastigial recesses of the corpus cerebelli. Caudally to this it is represented by the velum medullare posterius (or inferius), which is formed by the taenia cerebelli bordering on nodulus and floccular stalk. Attached to the velum medullare posterius is the thin, membranous caudal part of the ventricular roof provided by the lamina epithelialis derived from the roof plate and covered by the leptomeningeal tela chorioidea. This latter structure invaginates the epithelial roof and thereby forms the paired choroid plexus of the fourth ventricle. *The choroid plexus* consists of two longitudinal strips which may fuse in the midline and are continuous with a transverse choroid plexus strip attached to velum medullare posterius and extend-

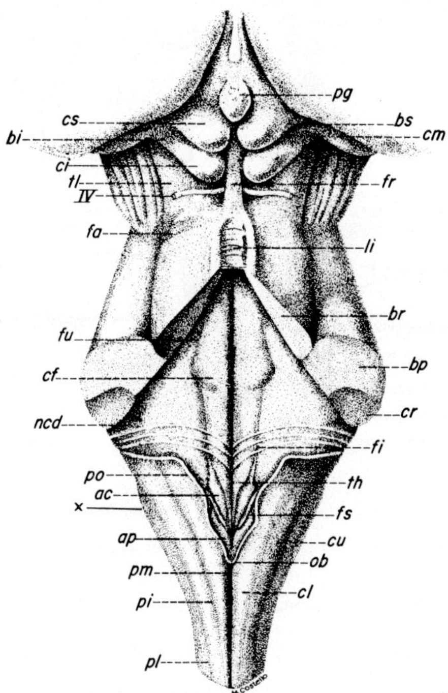

Figure 274A. Dorsal aspect of human brain stem with exposed rhomboid fossa. the cerebellum has been removed by transection of its three peduncles (from an unpublished study in collaboration with Dr. Webb Haymaker). ac: ala cinerea; ap: area postrema; bi: brachium quadrigeminum inferius; bp: brachium pontis; br: brachium conjunctivum; bs: brachium quadrigeminum superius; cf: colliculus facialis; ci: colliculus inferior; cl: clava; cm: corpus geniculatum mediale; cr: corpus restiforme; cs: colliculus superior; cu: tuberculum cuneatum; fa: fila lateralia pontis; fi: fovea inferior; fr: frenulum veli medullaris anterioris; fs: funiculus separans; fu: fovea superior; li: lingula cerebelli; ncd: nucleus cochlearis dorsalis (tuberculum acusticum); ob: obex; pg: pineal body; pi: sulcus intermedius posterior; pl: sulcus lateralis posterior; pm: sulcus posterior medianus; po: ponticulus; th: trigonum hypoglossi; tl: trigonum lemnisci; X: tuberculum cinereum; IV root of trochlear nerve.

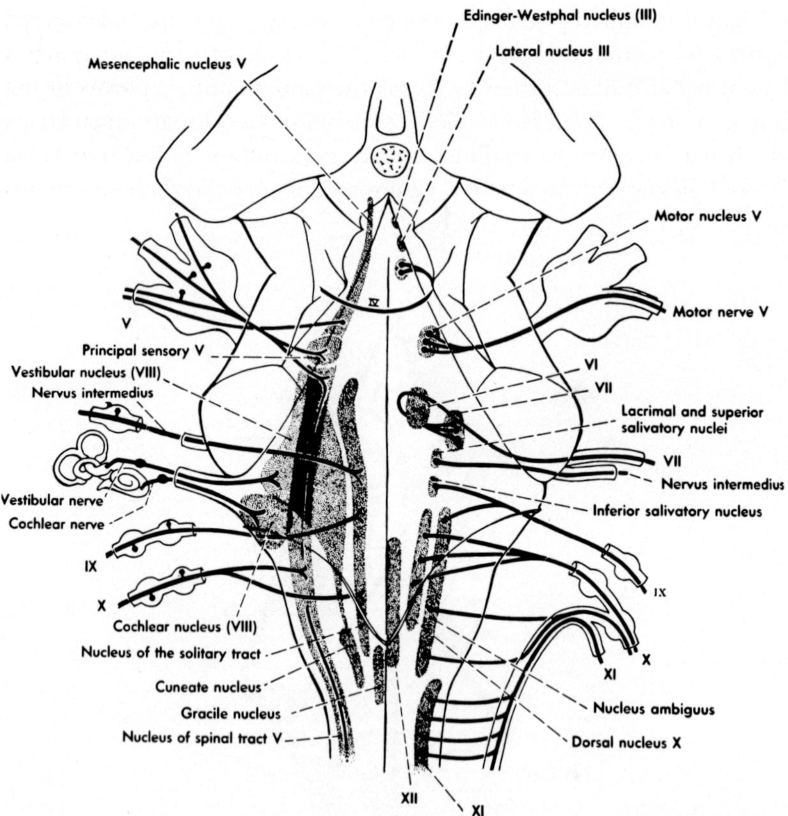

Figure 274 B. Dorsal aspect of the human rhomboid fossa and brain stem showing, in diagrammatic fashion, relationships of the cranial nerves. Afferent nuclei on left, efferent nuclei on right (from HAYMAKER-BING, 1969).

ing laterad to the tip of the fourth ventricle's lateral recesses (cf. also Figs. 150, 151 and 152B). The caudal border of the evaginated lateral recess is formed by a glious membrane lined by ependyma and representing the stretched pars rostralis taeniae medullae oblongatae *(velum medullare inferius of Henle)*. The lamina epithelialis and the choroid plexus are attached to the edge of this taenia which extends caudad as far as the obex. The thickened, essentially glious edge of the alar plate which tapers out into the taenia, is the *ponticulus of Ziehen* and others.[184a]

[184a] The *ponticulus* of ZIEHEN and HENLE, namely the essentially glial shelf or ledge at the attachment of rhombencephalic lamina epithelialis chorioidea (taenia ventriculi quarti) should not be confused with the ponticulus or *propons* of OBERSTEINER, represented by transverse fiber bundles and arcuate grisea externally to the pyramids caudally adjacent to the pons (cf. also footnote 58, p. 184 of vol. 3, part II).

Toward the obex, however, the ponticulus becomes a thin membranous taenia.

At the tip of the lateral recesses, portions of the lamina epithelialis have disappeared through resorption of tissue during ontogenetic development, and tufts of choroid plexus protrude freely into the subarachnoid space through this paired opening, which is known as the *apertura lateralis ventriculi quarti* or *foramen of Luschka*. Another, but unpaired median and dorsocaudal communication between the fourth ventricle and the subarachnoid space of the cisterna cerebello-medullaris (cisterna magna) is commonly found in the epithelial roof of the ventricle in the region of the caudal end of the choroid plexus. This is the *apertura mediana ventriculi quarti* or *foramen Magendii*, which is likewise caused by tissue resorption during embryonic development, starting perhaps about the seventh week of intrauterine life (HOCHSTETTER, 1929). The configuration of the *foramen of Magendie* is individually very variable. It may merely be a rather narrow passage or a cribriform plate with a number of sieve-like openings. A wide foramen as frequently seen in gross dissection may or may not be an artefact or the enlargement of a smaller communication by the tearing of delicate tissue strands. The actual presence of a *foramen Magendii* corresponding to the conventional anatomic descriptions has thus been doubted by some authors,[185] but although it may occasionally be missing,[186] the presence of a caudal opening of the fourth ventricle seems well substantiated by careful investigations.

The lozenge-shaped floor of the fourth ventricle, also designated as the *rhomboid fossa*, is exposed by removing the cerebellum after severing the three cerebellar peduncles and the velum medullare superius, and by tearing off the lamina epithelialis along the taenia ventriculi quarti, which represents the edge of *Henle's velum medullare inferius* and the ventricular border of ponticulus as well as of obex region. The rhomboid fossa is conventionally subdivided into three parts, *pars superior*, *pars intermedia*, and *pars inferior*. The triangular superior part is located rostrad to the lateral recesses and its rostral apex is the transition of fourth ventricle into cerebral aqueduct at the level of isthmus

---

[185] A dissertation by VAN DER MEULEN (Over het foramen van *Magendie*, Amsterdam 1935, with English summary) completed under the sponsorship of C. U. ARIËNS KAPPERS, reviews some of the problems here at issue.

[186] Failure of *Luschka's and Magendie's foramina* to develop, or complete secondary obstruction of these openings occasions internal hydrocephalus (cf. p. 277, chapter VI, vol. 3, part II).

rhombencephali. The intermediate part corresponds to the region of the ventricular floor located between, and extending into, the antimeric lateral recesses. The triangular inferior part is situated caudally to the lateral recesses, its caudal end being continuous with the central canal of the lower oblongata and of the spinal cord. The caudal portion of the pars inferior, somewhat resembling the point of a writing quill, is also known as the *calamus scriptorius*. Its apex is dorsally bridged by a medullary lamina of individually variable thickness, designated as the obex, and containing the commissura infima.

The floor of the rhomboid fossa displays a number of relevant landmarks which, depending on individual variations, may have a more or less conspicuous appearance. Just rostrally to the obex there is a dorsal paired oblong reddish vascular cushion, the *area postrema*, medially bounded by the *funiculus separans* from a likewise oblong, tapering eminence known as the *ala cinerea* and containing the nucleus dorsalis vagi. The medial boundary of the funiculus separans corresponds to the *sulcus limitans*. The funiculus itself appears to be a glial ridge which may also contain some longitudinally running non-medullated nerve fibers. Again, the dorsal respectively dorsolateral boundary of the ala cinerea corresponds to the sulcus limitans which commonly deepens rostrad to form the *fovea inferior*, into which the rostral, tapering tip of ala cinerea extends. This rostral portion of the ala usually contains the caudal part of the dorsal nucleus of the glossopharyngeal nerve, but a clear-cut delimitation between dorsal vagal and dorsal glossopharyngeus nucleus is not recognizable, since both nuclei are parts of a continuous griseal column.

Mediad, the ala cinerea is separated by the *sulcus intermedius ventralis* from a triangular area, the *trigonum hypoglossi*, which corresponds to the rostral portion of the nucleus nervi hypoglossi. Rostrally to this trigone, the sulcus intermedius ventralis disappears, being absorbed by the sulcus limitans, and the ventricular floor mediad to this latter forms the eminentia medialis which extends rostrad to the entrance into the aqueduct. Laterally to ala cinerea, i.e. to sulcus limitans, and reaching caudalward as far as laterally to rostral end of funiculus separans and area postrema, the *area acustica* can be seen as a roughly triangular field extending into the lateral recess and into the caudal region of pars superior fossae rhomboideae (Fig. 275B). The rostral and caudal as well as the medial surface of area acustica corresponds to the location of the *vestibular nuclei*, while its portion extending into the lateral recess corresponds to the *dorsal cochlear nucleus* forming the *tuberculum acusticum*, and to the *ventral cochlear nucleus* in the ventral, everted part of the lateral recess.

The pars intermedia fossae rhomboideae may be crossed by individually very variable transverse or somewhat oblique fibers designated as *striae medullares* (Fig. 274A, but not present in the specimen depicted by Figs. 275A–C). These striae, which sometimes pass over the region of tuberculum acusticum and disappear in the sulcus medianus, represent various cerebellar connections with reticular, arcuate, and other nuclei (Fuse, 1912; Rasmussen and Peyton, 1946; and other authors). Aberrant cochlear fibers are not, or at most very rarely if ever, included in these grossly visible superficial striae. In the older literature (Villiger, 1920, and others) the striae medullares were named *striae acusticae* and an occasionally conspicuous oblique bundle was designated as *conductor sonorus*, but these misleading terms were subsequently dropped. Ziehen (1934) had recommended the term *striae cerebellares.*

In the superior part of the rhomboid fossa the *eminentia medialis* commonly displays a circumscribed elevation known as *colliculus facialis* which is formed by the loop *('internal knee')* of facial root fibers passing over the nucleus of the abducens nerve, whose position is thus

Figure 275A. Dorsal view of a wax-plate reconstruction of the oblongata in a human newborn from a complete set of serial sections (from Hikiji, 1933).

indicated by the facial colliculus. Laterally and rostrally to this latter, the sulcus limitans merges with a depression designated as *fovea superior* which corresponds to the location of the motor and more laterad to that of the main sensory nucleus of the trigeminus. Both nuclei, however, are situated at some distance away from the surface of that depression in the ventricular floor (cf. Fig. 282F). The rostral end of the fovea superior, reaching the isthmus region, often displays a bluish discoloration due to the subependymal course of a vein, and this area is designated as the *locus caeruleus*. The discoloration, which may also assume a blackish or brownish-rusty tinge, is, moreover, related to the presence of a griseum, the *substantia ferruginea* or *nucleus loci caerulei*, whose nerve cells contain pigment granules (cf. above, p. 544).

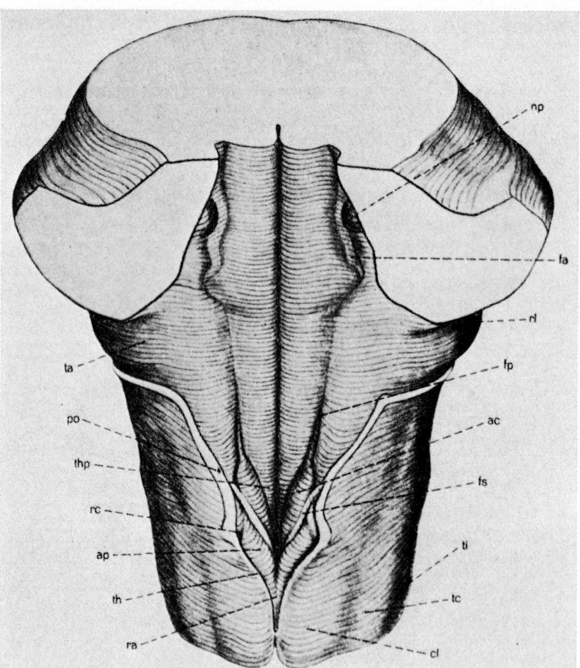

Figure 275 B. Drawing illustrating main features of the reconstruction shown in the preceding figure (from Hikiji, 1933). ac: ala cinerea; ap: area postrema; cl: clava; fa: fovea superior s. anterior; fp: fovea inferior s. posterior; fs: funiculus separans; np: nucleus parietolateralis; po: ponticulus; ra: recessus caudalis ventriculi quarti; rc: recessus lateralis calami scriptorii; rl: recessus lateralis ventriculi quarti; ta: tuberculum acusticum; tc: tuberculum cuneatum; th: thin taenia of fourth ventricle; thp: trigonum hypoglossi; ti: tuberculum cinereum.

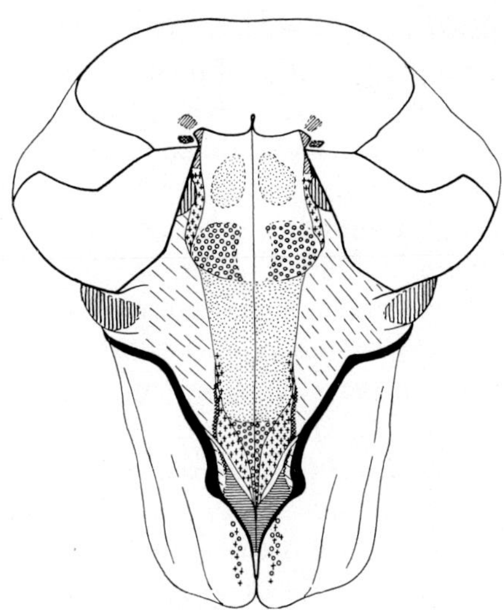

Figure 275 C. Semischematic drawing of the rhomboid fossa in the preceding two figures, with projection of several griseal areas (from Hikiji, 1933). Circles: caudally hypoglossal nucleus, rostrally nucleus of abducens within colliculus facialis; crosses: caudally ala cinerea (nucleus dorsalis vagi et glossopharyngei), rostrally nucleus motorius trigemini; crosses and circles in caudal (closed) oblongata indicate that, in dorsal projection, nucleus hypoglossi and nucleus dorsalis vagi become mapped upon approximately one and the same surface area; dots: caudally nucleus praepositus hypoglossi, rostrally a dorsal component of nucleus reticularis tegmenti; interrupted oblique hatching: vestibular grisea; close oblique hatching: nucleus loci caerulei; crossed hatching: caudally grisea of tractus solitarius, rostrally nucleus radicis mesencephalicae trigemini; vertical hatching: caudally cochlear nuclei, rostrally nucleus parietolateralis.

The ependymal floor of the fourth ventricle frequently displays irregular folds, know as *rugae*, which present individually variable patterns, especially in the fovea superior and in the region of the trigonum hypoglossi, whose lateral part can be demarcated as a distinct longitudinal ridge, named *area plumiformis* by Retzius (1896). Figures 275A–C show relevant configurational features of the human rhomboid recorded in the newborn by my associate Hikiji (1933) on the basis of a careful mapping combined with a reconstruction obtained from a complete and well preserved set of serial sections. Figures 273A and 274A illustrate additional relationships of human deuterencephalon and cranial nerves as corroborated by an investigation undertaken some years ago at the then Army Institute of Pathology by Dr. Haymaker and myself in preparation for our chapter on disorders of the brain stem in *Baker's Clinical Neurology* (1955, 1962, 1971).

With regard to the *cranial nerve roots*, it should here be pointed out that at the transition of intracerebral to extracerebral course of their fibers, an *Obersteiner-Redlich zone*, described for the spinal nerves on p. 219 of the preceding chapter VIII (Figs. 129A, B) is likewise present.

Turning now to the *internal and microscopic aspect of the oblongata*,[186a] some relevant features of transverse planes shall be considered in caudo-rostral sequence. Figures 276A–C depict levels pertaining to the region of transition between oblongata and spinal cord. These levels are characterized by the *decussation of the pyramidal tract*. As regards its pattern, the decussation displays numerous individual variations, concerning both the mutual interlacing of the crossing fiber bundles and the ratio of decussating fibers to those remainly uncrossed in the ventral corticospinal tract. In this respect, considerable asymmetry between right and left sides may occasionally occur, as pointed out and depicted by OBERSTEINER (1912). A so-called *bundle of Pick* (or *Henle-Pick)*, likewise discussed by OBERSTEINER, may occasionally be found on the ventromedial side of the nucleus radicis descendentis trigemini, in a rather dorsal position, and is generally interpreted as an aberrant[187] contralateral pyramidal bundle whose decussation may have occurred at higher bulbar levels (Fig. 276C). Because of the wide range of variations, there is no conclusive evidence about a definite localization of particular (arm, respectively leg) fibers within the bulbar pyramid and its decussation,[188] although such localization is believed to obtain in the spinal cord's lateral corticospinal tract (cf. Fig. 143).

The *rubrospinal tract* has here presumably reached a dorsal neighborhood of the lateral funiculus (Fig. 276B).

As the level of the pyramidal decussation, a group of motoneurons in the intermediate portion of the spinal cord's anterior horn remnant

---

[186a] In addition to the studies by OLZEWSKI and BAXTER (1954) and by others, a further attempt was recently made by BRAAK (1971 and other publications) to classify and parcellate human brain stem nuclei. This author, inter alia, stresses the significance of an aldehyde-fuchsin stain by which 'lipofuscin granules in the nerve cells are stained selectively'.

[187] Similar heterotopias may involve other pathways such, e.g. as the dorsal spino-cerebellar tract. Moreover, some still not entirely resolved doubts about the actual significance of *Pick's bundle* have been expressed (cf. OBERSTEINER, 1912).

[188] The very rare syndrome of *hemiplegia cruciata*, briefly discussed further below in connection with relevant clinical implications, is related to the spatial distribution of fibers for upper and lower extremity within the pyramidal decussation. Figure 280D (at the level of recessus lateralis) indicates a highly conjectural but of course possible fiber localization within the pyramid according to the interpretation of CROSBY *et al.* (1962).

represents the rostral part of the *nucleus of the spinal accessory nerve*, contributing to the innervation of sternocleidomastoid and trapezius muscles. Another group of spinal accessory motoneurons can usually be found dorsally to the pyramidal decussation, likewise in the intermedio-dorsal zone of the anterior horn remnant. The root fibers originating from these groups of motoneurons take a dorsolateral course, emerging ventrally to the tip of the posterior horn (cf. Figs. 276A and D). Cell groups in the ventral and medial part of the anterior horn remnant appear to represent anterior root motoneurons of the first spinal nerve and may be designated as the true '*supraspinal nucleus*'.

The caudal end of the *hypoglossal nucleus*, bilaterally below the central canal, moreover the caudal tip of nucleus dorsalis vagi within the central gray dorsally to hypoglossal nucleus, and the still slightly more dorsal, rather exiguous end of tractus solitarius with its griseum can usually be recognized at levels of pyramidal decussation.

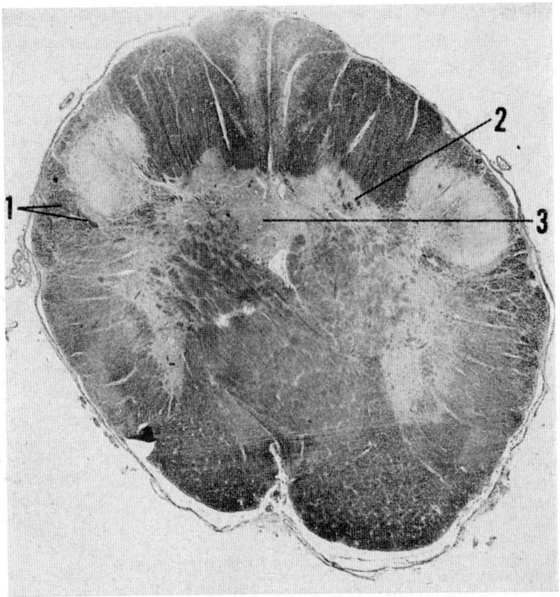

Figure 276A. Slightly oblique cross-section through the human oblongata at a level of the pyramidal decussation's caudal third *(Weigert stain*, about ×10, red. $^{7}/_{10}$). 1: rootlet of spinal accessory nerve; 3: locus of almost obliterated central canal; 2: caudal end of tractus solitarius. For identification of other structures, cf. Figures 276B and C.

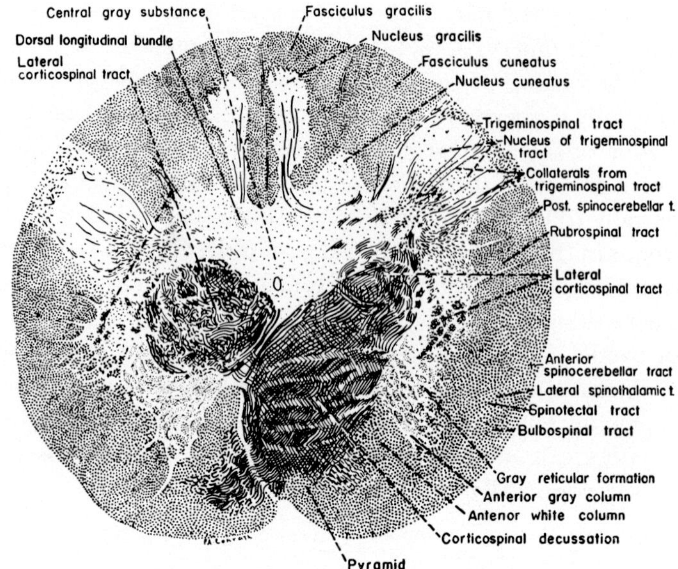

Figure 276 B

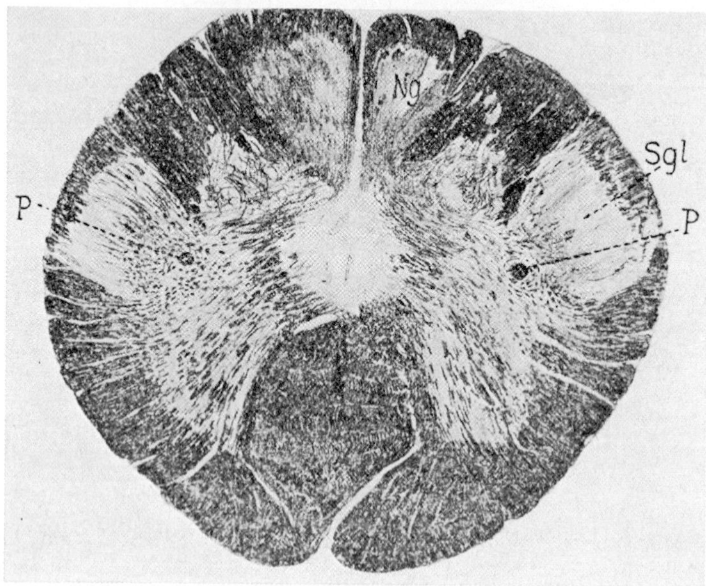

Figure 276 C

The next levels with characteristic configuration and still pertaining to the closed part of the oblongata are those including the *decussatio lemnisci* (Figs. 277A–D). At these levels, the uncrossed corticospinal tracts form two compact and prominent fiber masses on both sides of the basal midline, and represent the bulbar pyramids of gross anatomical nomenclature. The *inferior olivary complex* begins to unfold, the two *funicular nuclei*, including an additional *nucleus cuneatus externus s. lateralis*, are conspicuous, and further grisea, such as *nucleus reticularis lateralis*, *formatio reticularis tegmenti*, *nucleus ambiguus*, and *arcuate nuclei* can be recognized.

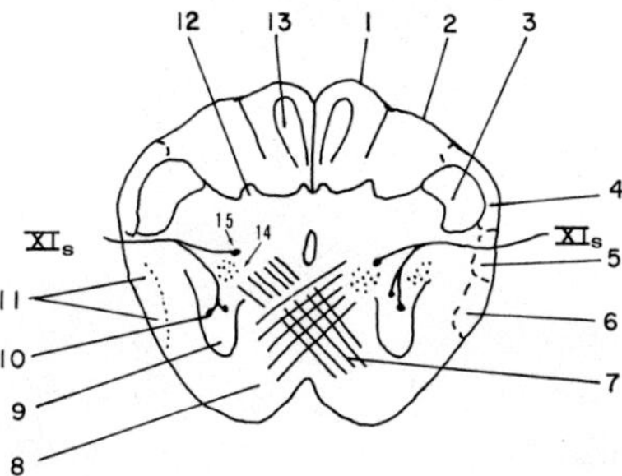

Figure 276D. Outline sketch of cross-section through lowermost human oblongata showing origin of spinal accessory, decussation of pyramidal tract, and approximate location of some tracts (adapted from an unpublished investigation of the human brain stem in collaboration with Dr. Webb Haymaker). 1: fasciculus gracilis; 2: fasciculus cuneatus; 3: nucleus radicis descendentis trigemini near its junction with substantia gelatinosa medullae spinalis; 4: radix descendens trigemini; 5: dorsal spinocerebellar tract; 6: ventral spinocerebellar tract; 7: approximate locus of ventral spinothalamic tract as its longitudinal fibers become intermingled with decussating pyramidal fibers running in a roughly transverse plane; 8: pyramid; 9: locus of 'supraspinal nucleus' in rostral end of cornu anterius medullae spinalis; 10: ventral portion of spinal accessory nucleus; 11: lateral spinothalamic tract (note overlap with 5 and 6); 12: nucleus cuneatus; 13: nucleus gracilis; 14: tractus corticospinalis lateralis (pyramidal tract) distal to decussation; 15: dorsal portion of spinal accessory nucleus.

Figure 276B. Cross-section through human oblongata near caudal border of pyramidal decussation, showing, on right side, an alternating interweaving of decussating fibers (from Kuntz, 1950). It will be noted that Kuntz uses the term 'trigeminospinal tract' (and its nucleus) instead of radix descendens trigemini. The interpretations of this author concerning some fiber systems ('dorsal longitudinal bundle', 'bulbospinal tract') and the location of rubrospinal tract, although permissible, remain quite doubtful.

Figure 276C. Bilateral *bundle of Pick* at level of pyramidal decussation in the human oblongata (after Stern, from Obersteiner, 1912). Ng: *nucleus of Goll;* P: *Pick's bundle;* Sgl: substantia gelatinosa (nucleus radicis descendentis trigemini).

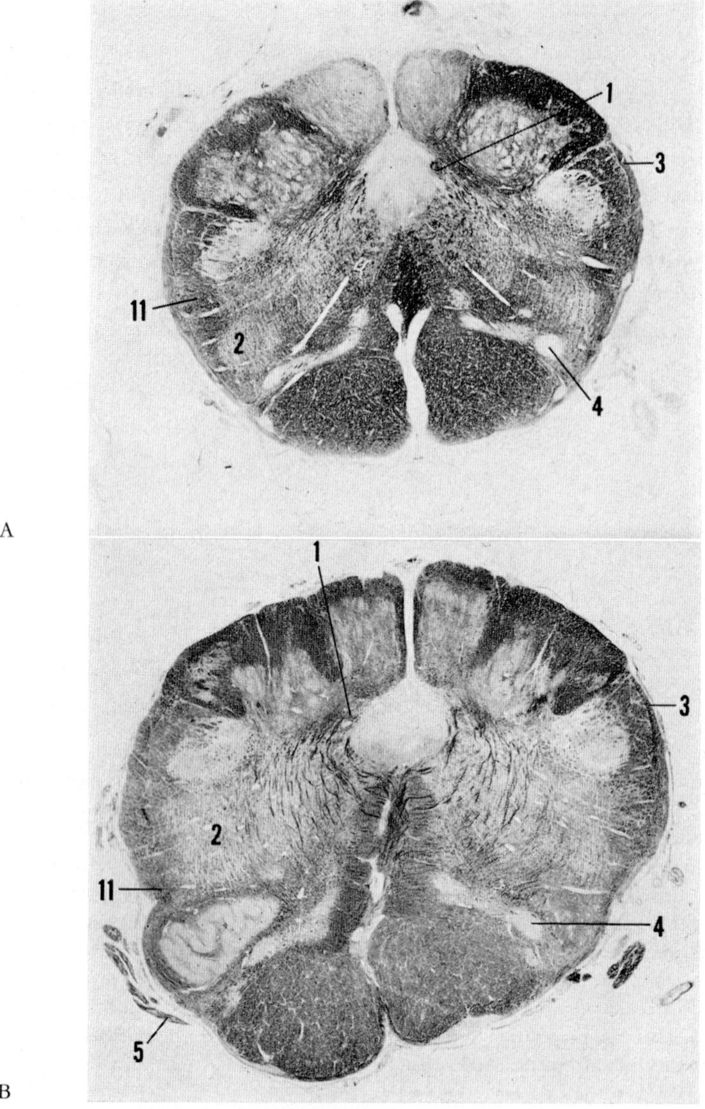

Figure 277A, B. Two cross-sections *(Weigert stain)* through levels of decussatio lemnisci in the human oblongata (approx. ×10; red. $^{7}/_{10}$). 1: caudal portion of tractus solitarius; 2: nucleus reticularis lateralis; 3: dorsal spino-cerebellar tract passing dorsad externally to descending trigeminal root; 4: caudal end of main inferior olivary nucleus; 5: extracerebral rootlet of hypoglossus nerve. Other structures can be easily identified by comparison with Figures 276D, 277C and D. The cross-section of Figure B is slightly rostral to that of Figure A and somewhat obliquely cut to show on left side the rostralward increasing expansion of main inferior olive. Note also external cuneate nucleus and lateral spinothalamic tract (11).

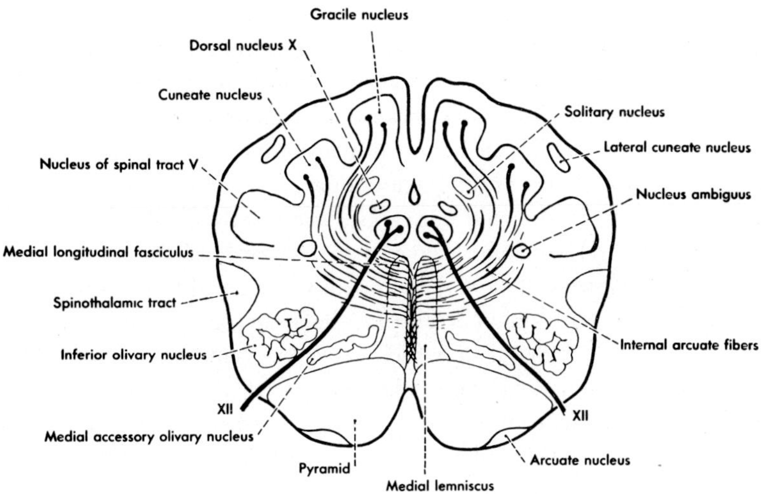

Figure 277C. Semidiagrammatic cross-section through the human oblongata at levels of decussatio lemnisci corresponding to those of Figures 277A and B (from HAYMAKER-BING, 1969).

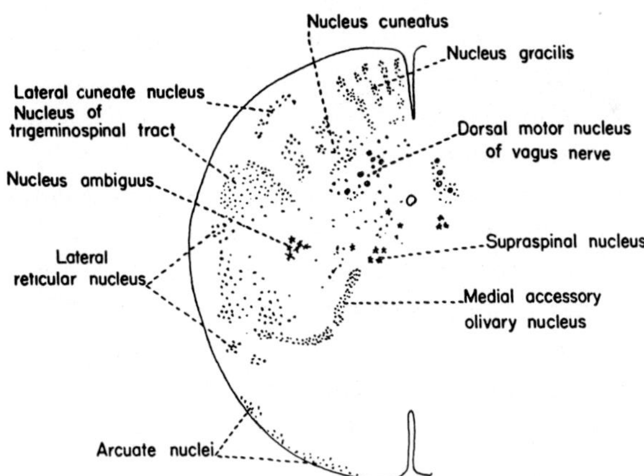

Figure 277D. Semidiagrammatic cross-section at the level of the three preceding figures, showing the cytoarchitectural configuration (slightly modified after JACOBSOHN, 1909, from KUNTZ, 1950). What is here designated as 'supraspinal nucleus' probably pertains to the nucleus hypoglossi. It will also be noted that the radix descendens trigemini is here designated by the term 'trigeminospinal tract' ('spinal tract of V' in the preceding Figure 277C).

At some caudal levels of the bulbar pyramids, the rostral beginning of pyramidal decussation overlaps and intermingles with the caudal end of decussatio lemnisci. This latter is mainly formed by the secondary internal arcuate fibers originating in nucleus gracilis and main cuneate nucleus, to which are added similar fibers from the grisea of tractus solitarius and descending trigeminal root. These *internal arcuate* fibers decussate in the midline raphe, commonly displaying a sharp, ventrally directed bend and then run rostralward. Most of the crossed fibers gather dorsally to the pyramid and form, within the interolivary space, together with the ventral spinothalamic tract, the *lemniscus medialis (medial fillet of Reil)*, of which they represent the bulbothalamic component. Dorsalward, the medial lemniscus blends with the essentially descending tectobulbar and tectospinal tract or *fasciculus praedorsalis*. This latter, in turn, blends dorsally with the *fasciculus longitudinalis medialis*. The lemniscus medialis can be regarded as an ascending pathway mediating proprioceptive, touch, and presumably also taste[189] impulses.

Crossing internal arcuate fibers from the nucleus of the descending trigeminal root, mediating pain and temperature sensations, are presumed to reach the *lateral spinothalamic tract* on the opposite side. This latter bundle remains in a superficial location, overlapping or intermingling with the ventral spinocerebellar tract. The *dorsal spinocerebellar tract* tends to shift dorsalward, and begins to pass externally over the descending trigeminal root (Figs. 277A, B).

The cross-sectional levels at the caudal third of the fourth ventricle (pars inferior fossae rhomboideae) are illustrated by Figures 278A–F. On both sides of the midline, just externally to the subventricular gray, lies the nucleus hypoglossi within the *hypoglossal trigone*. The adjacent *ala cinerea* with nucleus dorsalis vagi is laterally bounded by sulcus limitans, and medially by sulcus intermedius ventralis. The *nucleus of Roller* and the *nucleus intercalatus Staderini* can be recognized. The *nucleus ambiguus*, inconspicuous in myelin-stained sections, is seen in cell-stains as a well-defined cell group approximately situated between reticular for-

---

[189] The secondary taste fibers of the bulbothalamic system arise in the griseum of tractus solitarius. It is assumed that only the more rostral part of this griseum is concerned with gustatory function, the caudal part being restricted to cardiovascular and general visceral input. Thus, gustatory internal arcuate fibers to medial lemniscus would not be present at the here depicted caudal levels (Figs. 277A–D). Nevertheless, since inferences based on the available evidence may be quite deceptive, the presence of caudal gustatory internal arcuate fibers cannot be excluded.

mation and lateral reticular nucleus, dorsally to the dorsal accessory inferior olive. Two *vestibular nuclei*, namley nucleus vestibularis medialis seu triangularis, and nucleus of the descending vestibular root, make their appearance at these levels. The descending vestibular root consists of numerous small compact bundles between which their griseum is located.

The *restiform body* appears as a large dorsolateral fiber mass, whose major components are dorsal spinocerebellar tract and especially the olivocerebellar fibers. These latter decussate as a massive system of internal arcuate fibers. In addition, the restiform body receives *dorsal external arcuate fibers* from the external cuneate nucleus, and presumably also contributions from nucleus gracilis, main cuneate nucleus and other undefined grisea. The cuneate nuclei extend farther rostrad than nucleus gracilis, which disappears at levels such as illustrated by Figures 278A, C, E.

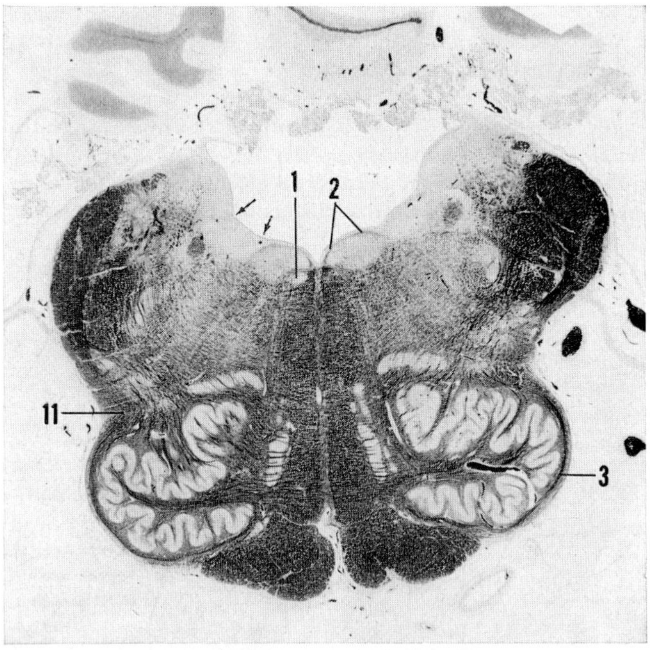

Figure 278A. Cross-section *(Weigert stain)* through the oblongata at a level of pars inferior fossae rhomboideae (approx. ×10; red. $^{7}/_{10}$). 1: *nucleus of Roller;* 2: *fasciculus longitudinalis dorsalis of Schütz* (perhaps in part intermingled with other undefined subventricular fiber systems); 3: fibrae arcuatae externae ventrales (periolivares); 11: lateral spinothalamic tract. Arrows indicate sulcus intermedius ventralis and sulcus limitans. Other structures easily identified by comparison with Figures B to F.

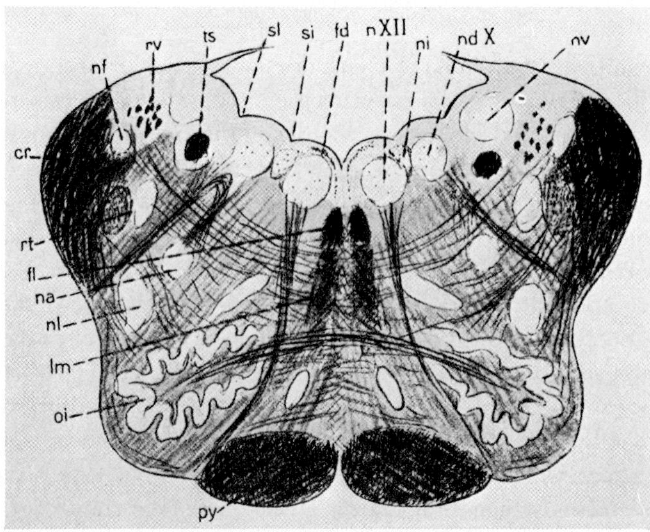

Figure 278 B. Simplified sketch of cross-section corresponding to the level shown in A (from K., 1927). cr: corpus restiforme; fd: *fasciculus longitudinalis dorsalis of Schütz;* fl: fasciculus longitudinalis medialis; lm: dorsal boundary of lemniscus medialis (between end of this lead and fasciculus longit. medialis lies the fasciculus praedorsalis s. tractus tecto-bulbo-spinalis; the ventral part of lemniscus medialis extends to the interolivary dorsal boundary of pyramidal tract); n. XII: nucleus nervi hypoglossi; ndX: nucleus dorsalis vagi; nf: rostral end of cuneate griseum; ni: nucleus intercalatus; nl: nucleus reticularis lateralis; nv: nucleus vestibularis medialis; oi: main inferior olivary nucleus with adjacent and medial accessory inferior olivary nuclei; py: pyramid; rt: radix descendens trigemini with its griseum; rv: radix descendens vestibularis with its griseum; si: sulcus intermedius ventralis; sl: sulcus limitans; ts: tractus solitarius with its griseum.

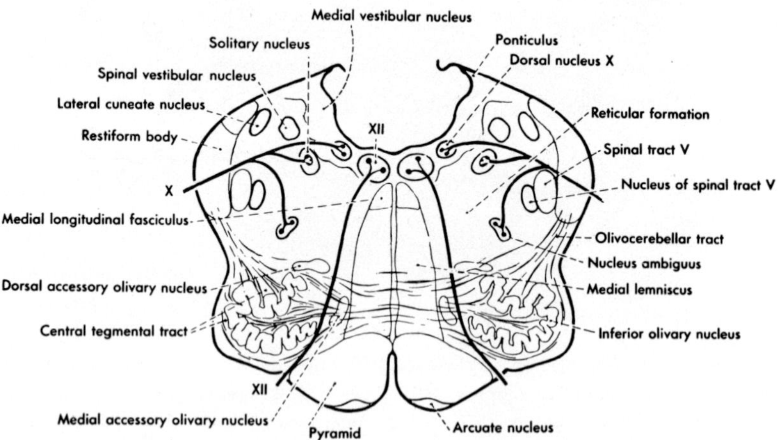

Figure 278 C. Semidiagrammatic cross-section corresponding to levels of the two preceding figures (from HAYMAKER-BING, 1969).

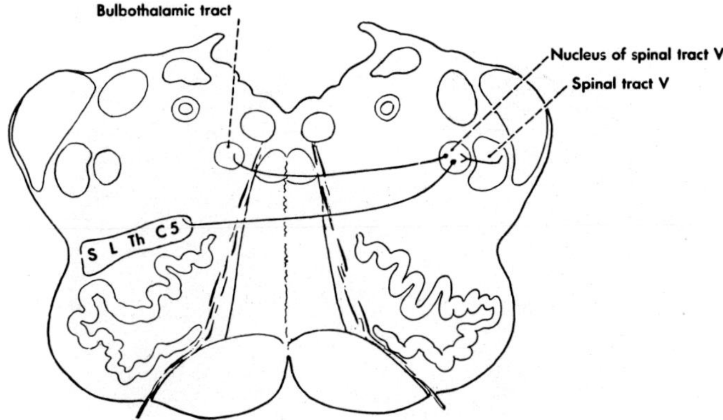

Figure 278 D. Diagrammatic cross-section corresponding to level C and showing assumed course of secondary ascending trigeminal fibers originating in nucleus of descending trigeminal root (adapted after various authors, from HAYMAKER-BING, 1969). S, L, Th, C, 5 designate somatotopic levels in lateral spinothalamic tract, namely sacral, lumbar, thoracic, and cervical, 5 being the trigeminal pain and temperature fibers. The 'bulbothalamic' tract, referred to on p. 587, is rather problematic. It seems perhaps more likely that, instead of, or besides, joining the 'bulbothalamic tract', the pathway in question joins a dorsal portion of the medial lemniscus.

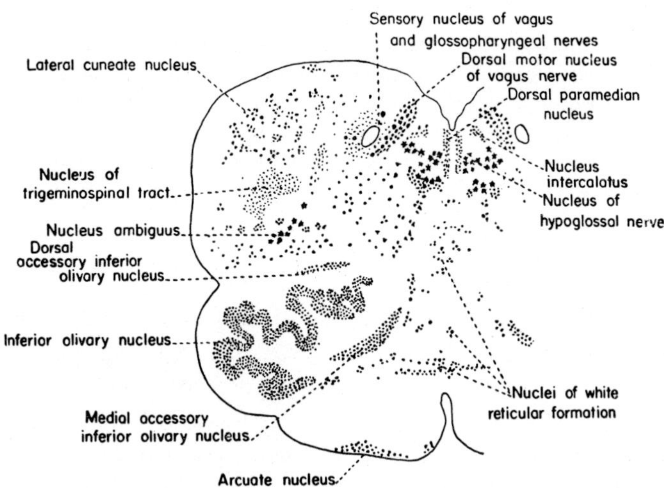

Figure 278 E. Semidiagrammatic cross-section depicting cytoarchitectural arrangement at levels of Figures A, B, C (slightly modified after JACOBSOHN, 1909, from KUNTZ, 1950). Note the minor variations in the non-standardized terminology used for the different Figures A–D.

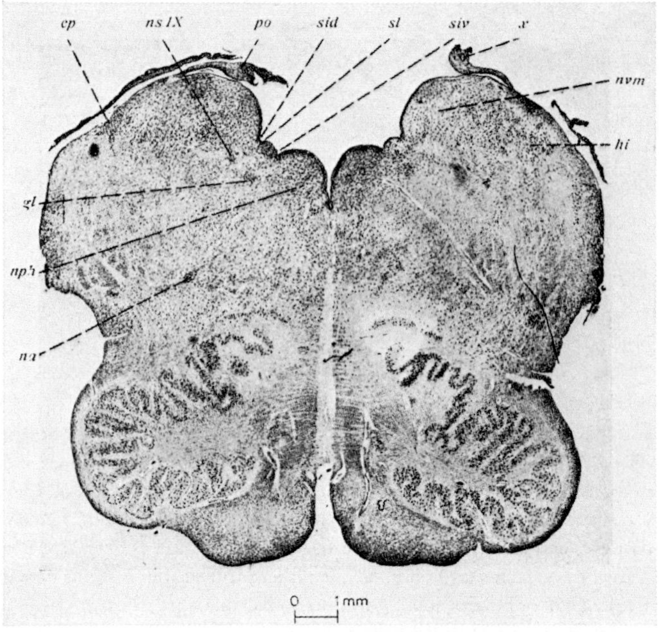

Figure 278 F. Cross-section *(Nissl stain)* through pars inferior fossae rhomboideae of the newborn at a level slightly rostral to the preceding ones. The nucleus hypoglossi is here replaced by nucleus praepositus hypoglossi (from Hikiji, 1933). cp: corpus pontobulbare; gl: part of nucleus dorsalis glossopharyngei displaced ventrally to ala cinerea; hi: rostral remnant of funicular nuclei (n. cuneatus); po: ponticulus; na: nucleus ambiguus; nph: nucleus praepositus hypoglossi; nsIX: griseum of tractus solitarius (glossopharyngeal portion); nvm: nucleus vestibularis medialis; sid: sulcus intermedius dorsalis; siv: sulcus intermedius ventralis; sl: sulcus limitans; x: ependymal proliferation within ponticulus sulci sid and sl tend to fuse.

The highly variable, essentially crossed *ventral arcuate fibers*, presumably originating in the arcuate nuclei, likewise join the restiform body. A similar system, running dorsalward in the raphe, and reaching the ventricular floor, is represented by arcuatocerebellar fibers which may form the striae medullares or so-called striae acusticae at the levels of recessus lateralis. In connection with the level here under consideration, the caudal end of the central tegmental tract on the lateral and dorsolateral border of main inferior olivary nucleus (cf. Fig. 278C), and the *fasciculus longitudinalis dorsalis of Schütz* within the periventricular gray (cf. Figs. 278A, B) should also be mentioned.

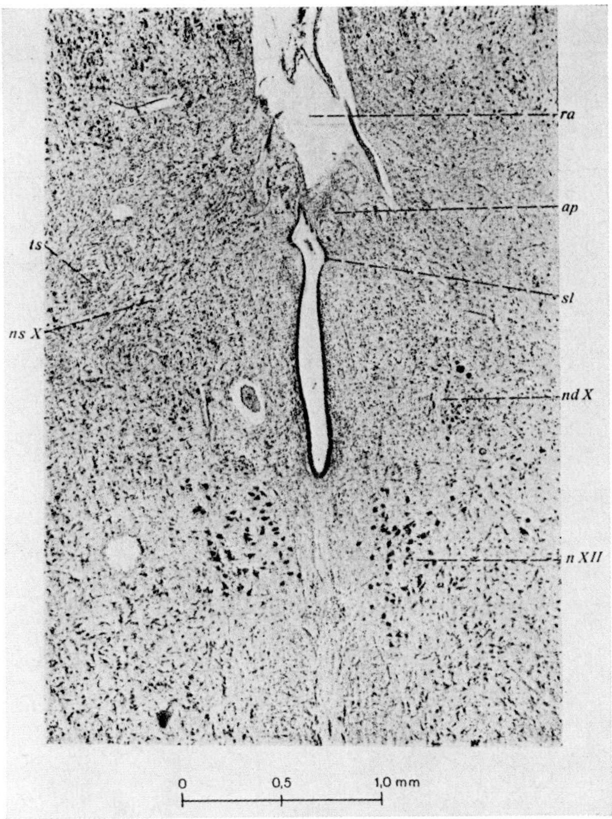

Figure 279 A. Cross-section *(Nissl stain)* showing rostral portion of canalis centralis in the oblongata of the newborn (from Hikiji, 1933). ap: caudal extremity of area postrema; n XII: hypoglossal nucleus; ndX: nucleus dorsalis vagi; nsX: nucleus of tractus solitarius; ra: recessus caudalis (supracommissuralis) ventriculi quarti; sl: sulcus limitans; ts: locus of tractus solitarius.

With regard to some features concerning the opening of the central canal into the fourth ventricle, Figures 279A–C, which require no further comments besides their explanatory legends, illustrate the location of *area postrema*, *funiculus separans*, and of the nuclei in the vicinity of the central gray. Figure 279D depicts structural details of the *main inferior olivary nucleus*,[190] whose morphologic features are shown in Figure 279E.

---

[190] The olivary grisea (inferior and superior) are commonly referred to as 'olives'. Strictly speaking, however, the term '*olive*' *(oliva medullae oblongatae)* originally designated merely the grossly visible oval configuration located on the lateral side of the macroscopic pyramid at the basal surface of the oblongata (cf. Fig. 272A), and containing the main inferior olivary nucleus.

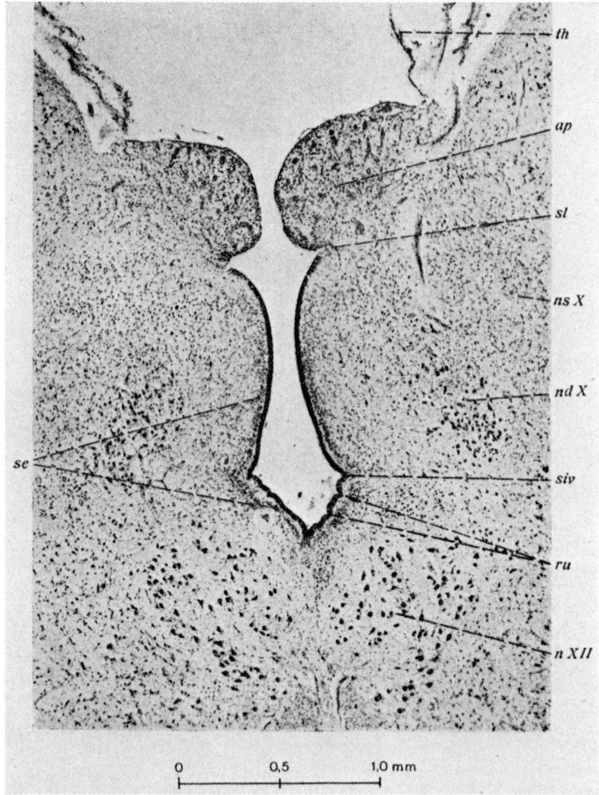

Figure 279 B. Cross-section *(Nissl stain)* rostral to that of preceding figure, showing opening of central canal into fourth ventricle (from HIKIJI, 1933). ru: rugae of ependymal lining; se: subependymal cell plate; siv: sulcus intermedius ventralis; th: thin taenia (lamina epithelialis, roof plate) of fourth ventricle. The area postrema (ap) is here conspicuous. Other abbreviations as in preceding figure.

Figure 279 D. Details of main inferior olivary nucleus, as in a cross-section seen by means of the *Golgi impregnation* (combined from drawings by CAJAL, from EDINGER, 1912). At left, fibers of central tegmental tract and their end arborizations in a convolution of the olivary griseum can be seen. Neurites of olivary cells run toward the right (in the direction of the olive's hilus) as olivocerebellar fibers. In addition, unspecified fibers, coming from the medial side, terminate in the olivary gray.

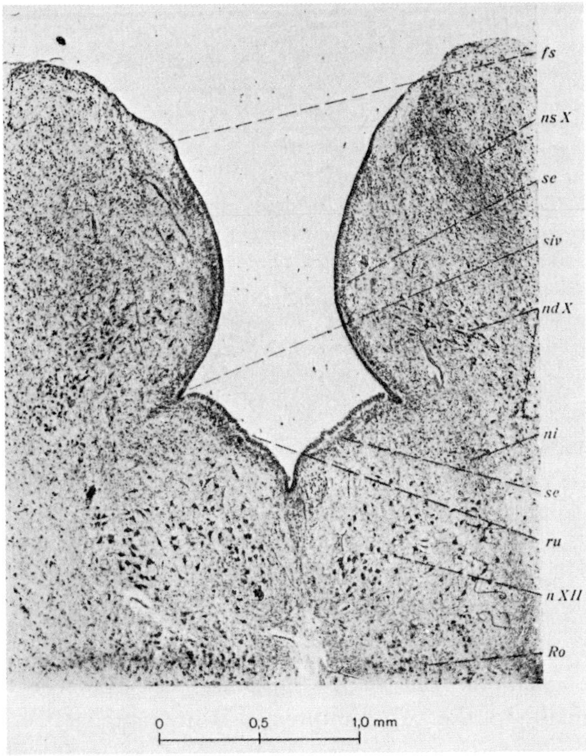

Figure 279 C. Cross-section *(Nissl stain)* rostral to that of preceding figure, showing the gradual widening of fourth ventricle in calamus region and the location of funiculus separans (from HIKIJI, 1933). fs: funiculus separans (the locus of sulcus limitans roughly corresponds to ventromedial border of funiculus); ni: nucleus intercalatus; Ro: *nucleus of Roller*.

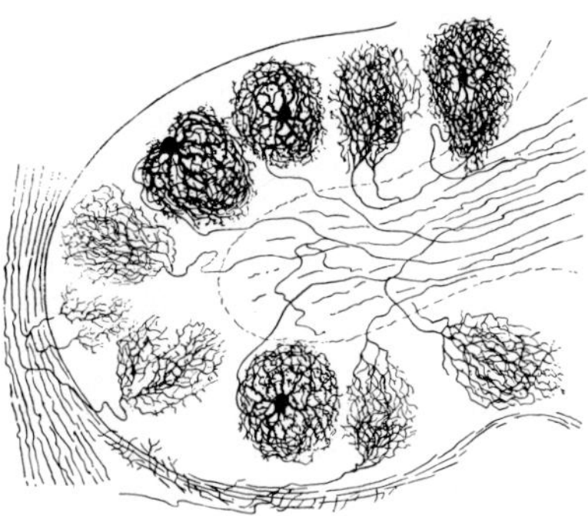

Figure 279 D

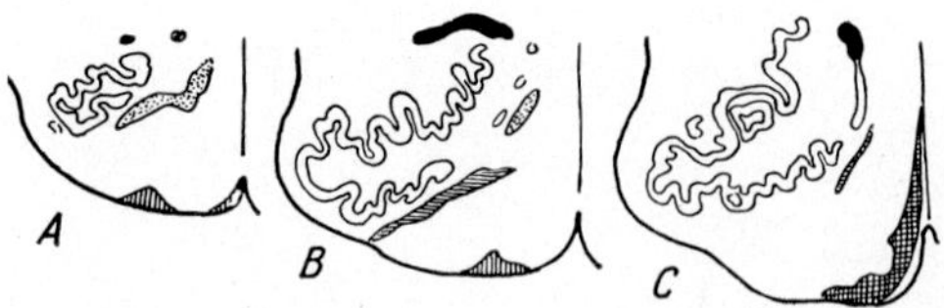

Figure 279E. Semidiagrammatic cross-sections through the inferior olivary complex of adult man (from BECCARI, 1943; C redrawn after KOOY, 1917). A Near caudal extremity; B Through middle third; C Through rostral third; black: dorsal accessory inferior olive; dotted: medial accessory olive; blank: main inferior olive; horizontal and vertical hatching: arcuate nuclei; cross-hatching: 'nucleus precursorius pontis' ('propons', presumably a transitional griseum between true nuclei pontis and arcuate nuclei). Compare with Figure 271.

The cross-sectional levels corresponding to the fourth ventricle's lateral recesses, respectively to the pars intermedia fossae rhomboideae are illustrated by Figures 280A–H. These levels are characterized by the two *cochlear nuclei*[191] which form a griseal cap dorsally to the fiber masses of the restiform body. The dorsomedial part of this cap is the *tuberculum acusticum*, and the ventral portion pertains to *nucleus cochlearis ventralis*, medially to which the *corpus pontobulbare* is located. The *nucleus vestibularis medialis s. triangularis of Schwalbe* lies below the ventricular floor medially to the tuberculum acusticum and is itself medially bounded by the sulcus limitans which, moreover, represents the approximate lateral boundary of nucleus praepositus hypoglossi.

In diverse instances a bundle of *arcuatocerebellar fibers*, ascending through the tegmentum in the midline raphe, emerges on the floor of the rhomboid fossa and runs laterad to reach the cerebellum along the fiber systems of the restiform body. The course of these arcuatocerebellar bundles across the ventricular floor forms the variable striae medullares[191a] mentioned above on p. 561.

---

[191] Readers interested in recent descriptions respectively viewpoints concerning the human cochlear nuclei may be referred to a paper by BACSIK and STROMINGER (1973).

[191a] These *striae medullares* are also occasionally referred to as *striae Piccolomini* (cf. e.g. FUSE, 1912).

Figure 280A, B. Cross-section s *(Weigert stain)* through the oblongata at the level of lateral recesses and cochlear nuclei (A appr. ×10, red. $^{2}/_{3}$; B appr. ×8; red. $^{2}/_{3}$). 1: nervus glossopharyngeus; 2: radix descendens vestibularis and its griseum; 3: tractus solitarius and its griseum; 4: nucleus dorsalis glossopharyngei (nucleus salivatorius inferior); 5: striae medullares; 6: tuberculum acusticum; 7: arcuatocerebellar fibers in raphé; 8: corpus pontobulbare (laterally to this oval griseum lies the nucleus cochlearis

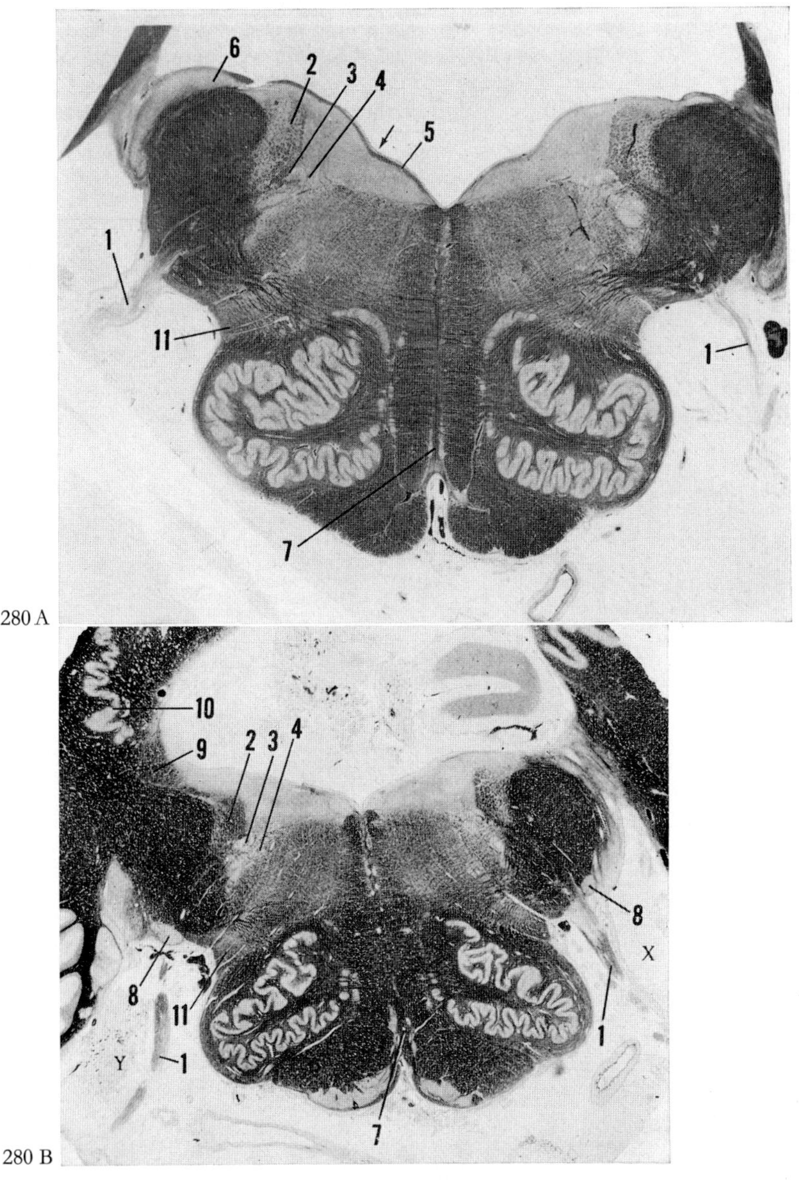

ventralis, to which the taenia of recessus lateralis is attached); 9: nucleo-cerebellar tract (IAK, inner portion of inferior cerebellar stalk or restiform body); 10: nucleus dentatus cerebelli with origin of brachium conjunctivum fibers on its medial side; 11: lateral spinothalamic tract; x: lateral recess in neighborhood of *foramen Luschkae;* y: tuft of choroid plexus protruding through *foramen Luschkae*. Other structures easily identified by comparison with Figures C, D, E. Note also the large arcuate griseum ('propons') in B.

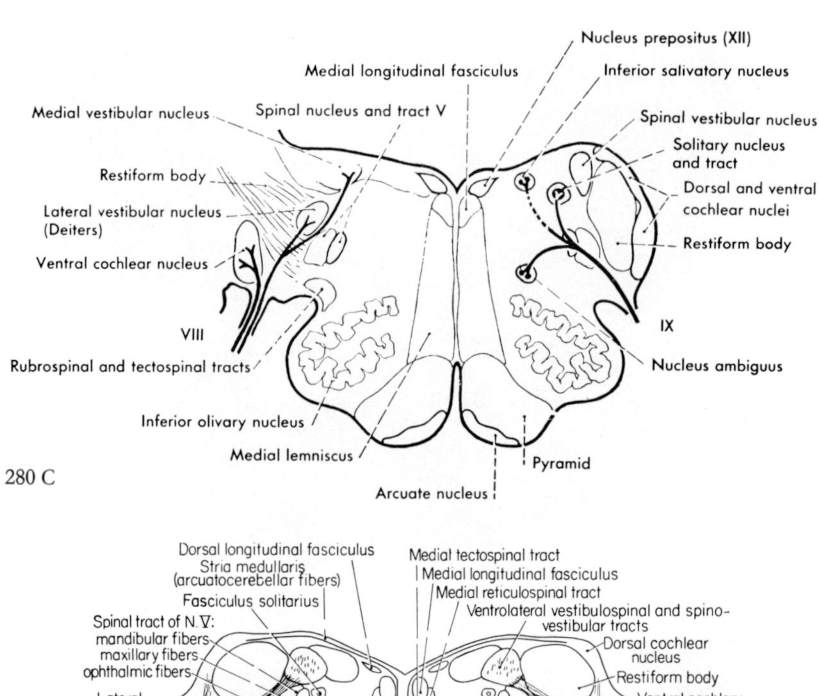

280 C

Dorsal longitudinal fasciculus
Stria medullaris (arcuatocerebellar fibers)
Fasciculus solitarius
Spinal tract of N. V:
mandibular fibers
maxillary fibers
ophthalmic fibers
Lateral reticulospinal tract
Rubrotegmentospinal tract
Lateral tectotegmentospinal tract
Ventral reticulospinal tract
Ventral spinothalamic tract (Dejerine)
Medial lemniscus
Arcuatocerebellar fibers
Ventral spinothalamic tract of comparative neurologists
Medial tectospinal tract
Medial longitudinal fasciculus
Medial reticulospinal tract
Ventrolateral vestibulospinal and spino-vestibular tracts
Dorsal cochlear nucleus
Restiform body
Ventral cochlear nucleus
N. VIII, cochlear
N. IX
N. VIII, vestibular
Ventral spinocerebellar tract
Lateral spinothalamic tract and spinotectal tract
Central tegmental tract
Ventral secondary ascending tract of V
Lateral and ventral corticospinal tracts and corticobulbar tract
Arcuate nucleus
T C G F U L T

280 D

Medial vestibular nucleus
Spinal vestibular nucleus
Dorsal cochlear nucleus
Nucleus ambiguus
Ventral cochlear nucleus
Pontobulbar nuclei
Nucleus of trigeminospinal tract
Retrofacial nucleus
Inferior olivary nucleus
Nucleus of tractus solitarius
Dorsal motor nucleus of vagus
Dorsal paramedian nucleus
Interfascicular hypoglossal nucleus
Magnocellular nucleus of reticular formation
Nucleus of the raphe
Arcuate nuclei

280 E

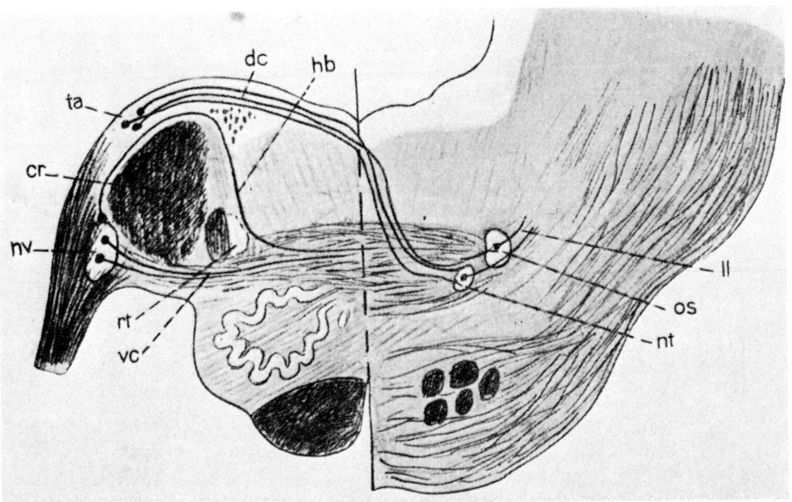

Figure 280 F. Simplified sketch depicting the course of secondary and higher order cochlear fibers gathering in lemniscus lateralis (from K., 1927). Left: level of cochlear nuclei; right: level of pons and superior olivary complex; cr: corpus restiforme; dc: dorsal cochlear pathway (below periventricular gray and not to be confused with the strieae medullares in the ventricular floor shown in Figure 280A); hb: intermediate cochlear pathway *(bundle of Held)*; ll: lemniscus lateralis; nt: medial nucleus of trapezoid body; nv: nucleus cochlearis ventralis; os: superior olivary griseum; rt: radix descendens trigemini and its nucleus; ta: tuberculum acusticum; vc: ventral cochlear pathway. This sketch should be compared with Figure 267D.

Figure 280 C. Semidiagrammatic cross-section corresponding to levels of A and B (from Haymaker-Bing, 1969). The depicted plane is slightly oblique, being somewhat more rostral on the left side (cf. also Fig. B). The nucleus salivatorius inferior is here drawn a little more dorsally than its locus as identified in the actual section shown in Figure B.

Figure 280 D. Semidiagrammatic cross-section corresponding to the preceding levels and indicating the location of relevant fiber tracts and of some other landmarks (from Crosby *et al.*, 1962). The location and boundaries of some of these tracts as here shown in the interpretation by Crosby *et al.* remain more or less conjectural. T, C, and G in the medial lemniscus refer to the presumed gustatory (T), cuneate, and gracile components. L, T, U, F in pyramid refer to the assumed location of cortico-spinal fibers for leg, trunk, and arm (U) fibers, also for cortico-bulbar fibers (F) in the pyramid. What is here labelled 'ventral spinothalamic tract of comparative neurologists' is, in my opinion, merely a component of the Mammalian lateral spinothalamic tract.

Figure 280 E. Semidiagrammatic cross-section depicting cytoarchitectural arrangement at the level of Figures A–D (slightly modified after Jacobsohn, 1909, from Kuntz, 1950). What Kuntz designates here as 'interfascicular hypoglossal nucleus' doubtless represents the nucleus praepositus hypoglossi. At this level, the 'Dorsal motor nucleus of the vagus' pertains to the glossopharyngeal segment, being presumably the nucleus salivatorius inferior. There is some doubt whether the 'Retrofacial nucleus' still represents a separate part of glossopharyngeal 'nucleus ambiguus' or the caudalmost portion of the motor nucleus facialis complex.

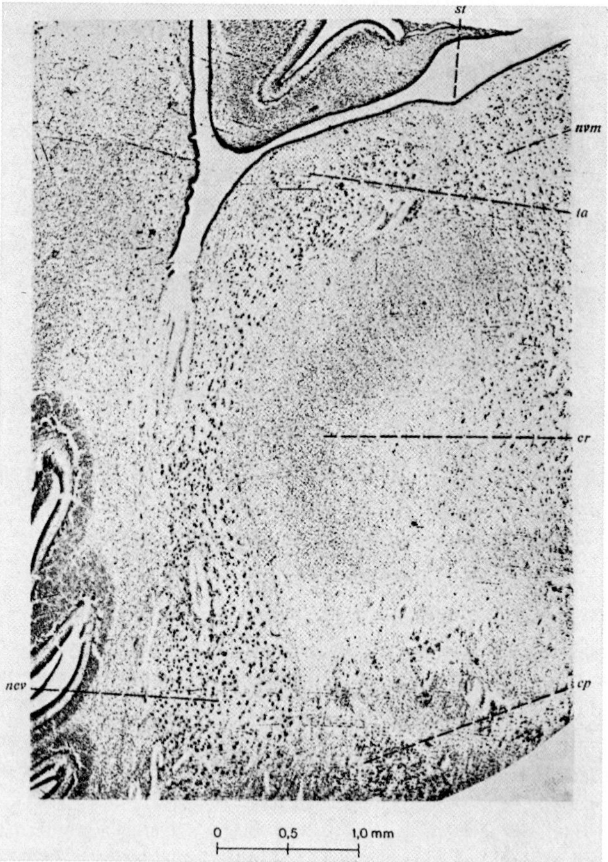

Figure 280 G. Details of dorsal and ventral cochlear nuclei of the newborn *(Nissl stain)* as seen in a cross-section approximately corresponding to the level of A (from HIKIJI, 1933). cp: corpus pontobulbare; cr: corpus restiforme; ncv: nucleus cochlearis ventralis; nvm: nucleus vestibular medialis; st: medial boundary sulcus of tuberculum acusticum; ta: tuberculum acusticum.

The cell column of nucleus dorsalis vagi et glossopharyngei (n. alae cinereae) is displaced away from the ventricular floor, lying medially to tractus solitarius respectively its griseum, and presumably represents the nucleus salivatorius inferior.

The cross-sectional levels rostrally adjoining those through the lateral recesses pertain to pars superior fossae rhomboideae respectively pons, and are illustrated by Figures 281A–G. The *pars basilaris pontis* includes the *pontine nuclei* with their transverse fibers which gather on

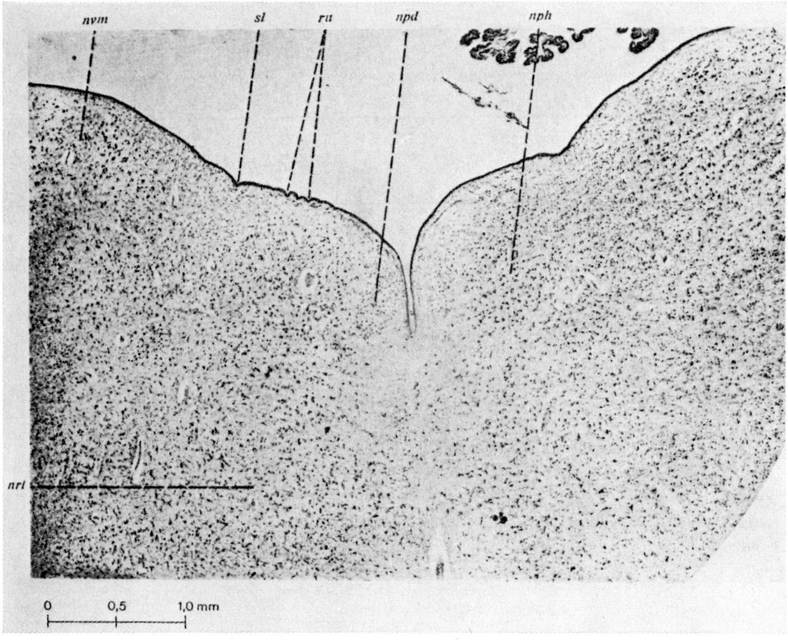

Figure 280H. Details of ventricular floor *(Nissl stain)* in a cross-section at levels of pars intermedia fossae rhomboideae of the newborn (from Hikiji, 1933). npd: 'nucleus paramedianus dorsalis'; nph: nucleus praepositus hypoglossi; nrt: nucleus reticularis tegmenti; nvm: nucleus vestibularis medialis; ru: rugae of ependymal lining; sl: sulcus limitans.

both sides laterally to form the brachium pontis or middle cerebellar peduncle. At a right angle to these fibrae pontis the thick but variably distributed fibers of the corticopontine and corticospinal (together with some corticobulbar) systems run through the pars basilaris. The size of the corticopontine bundles decreases caudalward, since this pathway, except for the fibers reaching propons and arcuate nuclei, ends in the homolateral pontine grisea, from which the essentially crossed pontocerebellar fibers originate. The corticospinal system, presumably still with some corticobulbar fibers, emerges caudally to form the massive pyramids of the oblongata.

The *basal plate region* located dorsally to pars basilaris is the *pars tegmentalis pontis*, which represents the direct rostral continuation of the myelencephalic tegmentum. It contains the reticular formation and other grisea. The flattened and lateralward expanding band of the

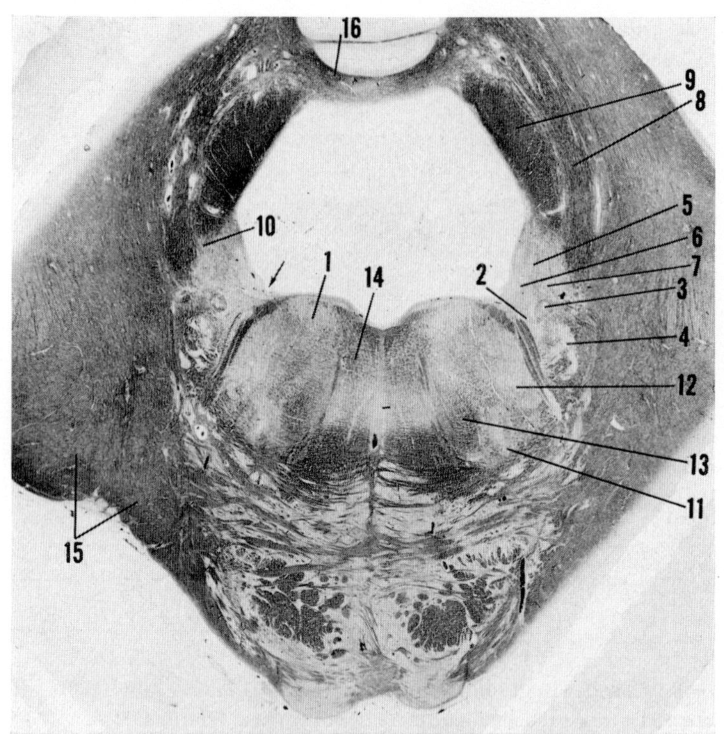

281 A

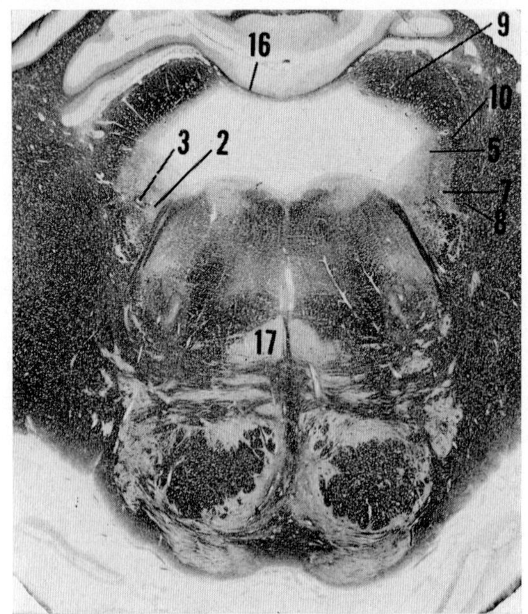

281 B

medial lemniscus, crossed by the transverse fibers system of trapezoid body, approximately provides, with its ventral limit, a boundary line between pars tegmentalis and pars basilaris pontis.

The *tegmentum* displays, laterally to the fasciculus longitudinalis medialis, the *nucleus of the abducens* within the *colliculus facialis* formed by the loop of the facial nerve root fibers. The motor (or special visceral efferent) nucleus of that nerve is located dorsolaterally to the superior olivary complex. *The preganglionic (superior salivatory and lacrimal) nucleus of the facial nerve* is presumably represented by the rostral end of the general visceral efferent column which, farther caudad, includes the nucleus dorsalis glossopharyngei et vagi. At the level of nervus facialis, this griseum is located on the medial side of, and rather close to, tractus solitarius and its nucleus.

The *superior vestibular nucleus* appears at these levels, which also include *the lateral magnocellular nucleus of Deiters* and a rostral part of the *medial vestibular nucleus* (Fig. 281F).

The *superior olivary grisea* are shown in figure 281G. The *lateral lemniscus*, intermingled with fibers of the neighboring *lateral spinothalamic tract*, gathers at the lateral edge of the superior olivary complex. The *medial lemniscus*, at its lateral edge, approaches the lateral lemniscus and tends here to overlap with bundles of that pathway.

Cross-sectional levels through the rostral portion of pars superior fossae rhomboideae including the transition of fourth ventricle to cerebral aqueduct, that is to say the region of isthmus rhombencephali, are illustrated by Figures 282A–F. Caudally, these levels are characterized

---

Figure 281 A, B. Cross-sections *(Weigert stain)* through the pons at levels of abducens nucleus, motor facial nucleus, and superior olivary complex (A ca. ×8, B ca. ×6; red. $^{2}/_{3}$). 1: nucleus abducentis; 2: nucleus salivatorius superior et lacrimalis; 3: tractus solitarius and its griseum; 4: radix descendens trigemini and its griseum; 5: nucleus vestibularis superior; 6: rostral end of nucleus vestibularis medialis; 7: nucleus vestibularis lateralis *(Deiters)*; 8: corpus restiforme; 9: brachium conjunctivum; 10: juxtarestiform body (IAK); 11: superior olivary complex; 12: nucleus motorius facialis; 13: central tegmental tract; 14: tractus tecto-bulbaris et spinalis; 15: brachium pontis; 16: velum medullare superius s. anterius with tractus tecto-cerebellaris (in B the lingula cerebelli can be seen above the velum); 17: so-called medial 'reticulotegmental pontile process' (perhaps extension of pontine gray blending with 'central tegmental nucleus'). Arrow points to sulcus limitans. Other structures or configurations can easily be identified by comparison with Figures 281C–E. The lateral spinothalamic tract (11 in Figs. 277A–280B) has now joined the area of lateral lemniscus shown in C.

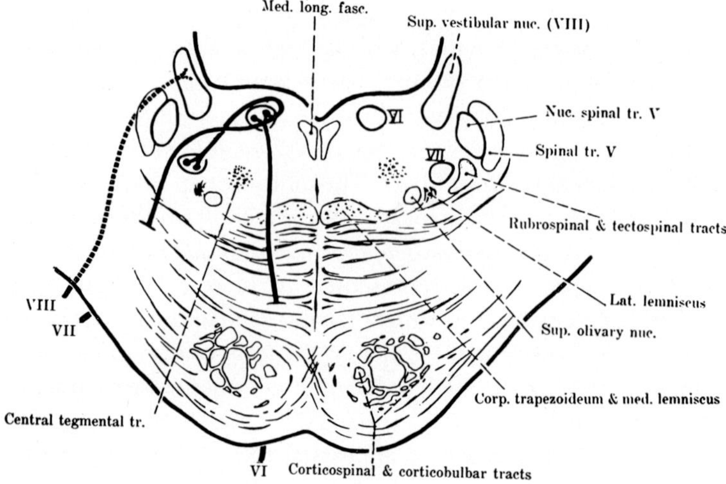

Figure 281 C. Semidiagrammatic cross-section through the pons approximately corresponding to the levels shown in A and B (from HAYMAKER-BING, 1956).

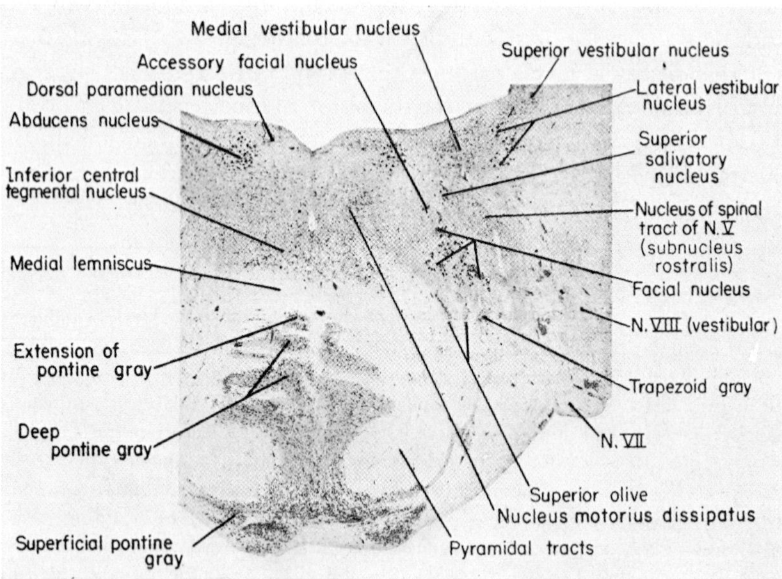

Figure 281 D. Cross-section through the pons *(Nissl stain)* at the level of abducens nucleus, motor facial nucleus, and superior olivary complex, showing the cytoarchitectural configuration (from CROSBY *et al.*, 1962).

by the location of *the two main trigeminal nuclei*, namely *nucleus motorius* and *nucleus sensibilis principalis*.

The essentially crossed secondary afferent fibers of the trigeminus decussate in scattered bundles and join the contralateral medial lemniscus in a manner comparable to the decussation of functionally similar fibers from the nucleus of the descending trigeminal root. There seems to obtain, moreover, an additional ascending secondary tract originating in the main sensory nucleus of the trigeminus, formed by mainly crossed fibers running in a more compact dorsal bundle (cf. Fig. 282B) and gathering in the so-called dorsal secondary ascending trigeminal tract or '*bulbothalamic*' tract believed to end in thalamic grisea (nucleus

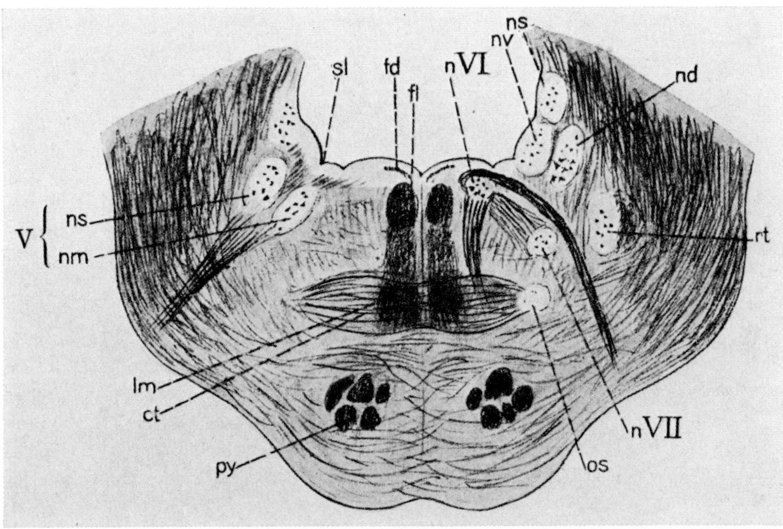

Figure 281 E. Simplified sketch depicting cross-sections through pons, showing on the right the level of internal facial genu, and on the left the level of main trigeminal nuclei (from K., 1927). ct: corpus trapezoideum (s. trapezoides, also c. 'trapezoidum'); fd: *fasciculus longitudinalis dorsalis of Schütz;* fl: fasciculus longitudinalis medialis; n. VI: nucleus nervi abducentis; n. VII: nucleus motorius facialis; nd: *nucleus of Deiters;* nmV: nucleus motorius trigemini; nsV: nucleus sensibilis s. afferens trigemini; ns: nucleus vestibularis superior *(of Bechterew)*, also shown, but not labelled, on left side; nv: nucleus vestibularis medialis s. triangularis *(of Schwalbe);* os: superior olivary complex; py: fasciculi pyramidales (cortico-spinal and in part corticobulbar tract); rt: nucleus radicis descendentis trigemini; sl: sulcus limitans.

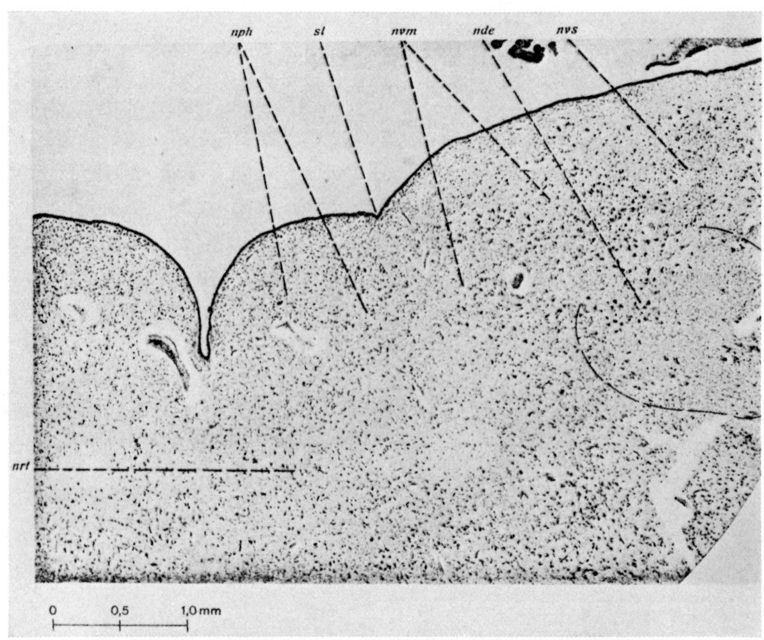

281 F

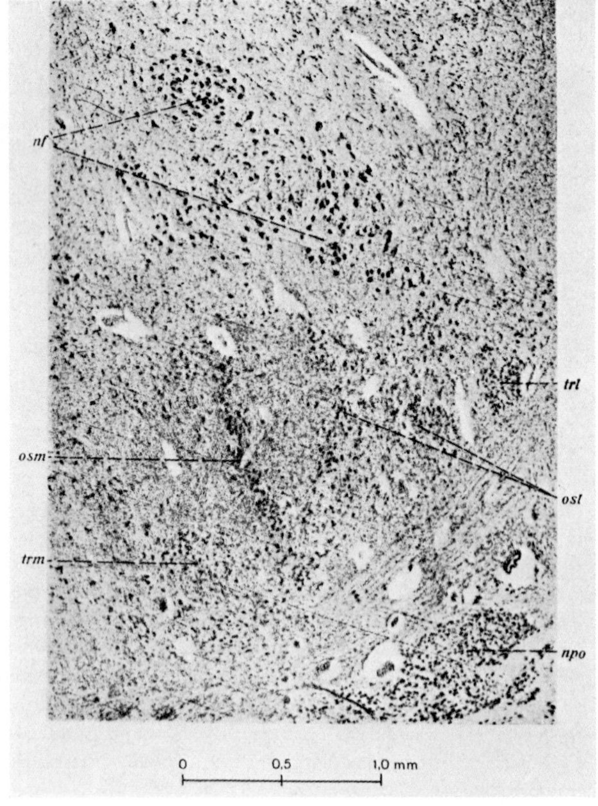

281 G

ventralis posteromedialis) together with the trigeminal component of medial lemniscus.[192] There are, at these levels, besides particular nondescript subdivisions of the reticular formation such as 'superior central tegmental nucleus', and other arbitrarily delimited parcellations with nonstandardized nomenclature, several further grisea whose significance remains obscure. Among these latter, the *nucleus loci caerulei ('nucleus pigmentosus pontis'* of JACOBSOHN, 1909, and KUNTZ, 1950), located in the basal plate and rather closely adjacent to some of the trigeminal grisea, was mentioned above in dealing with the comparative anatomy of the mammalian oblongata and with the gross features of the human rhomboid fossa. The suspected, but quite insufficiently authenticated relationship of that griseum to the 'upper respiratory center' (so-called *'pneumotaxic center')* and the more recent speculations based upon theories of neurochemistry were mentioned above on p.545.

Likewise in the neighborhood of the trigeminal grisea, a *nucleus ovalis* and a *nucleus parietolateralis* can be noted (Fig. 284F). The latter cell group, at least in the newborn, may protrude into the ventricle.[193] Both nuclei, although close to the basal plate, appear still to be *alar*

---

[192] The evidence concerning the dorsal secondary ascending tract (bulbothalamic tract) remains inconclusive (cf. also CROSBY *et al.*, 1962). There is little doubt that rostrally to the facial knee (Fig. 282A) a conspicuous dorsal bundle of secondary afferent trigeminus fibers is present (Fig. 282B). Nevertheless, some or most of its fibers could join the 'ventral secondary ascending trigeminal tract' included in the medial lemniscus system.

[193] Cf. the comments on these nuclei by HIKIJI (1933).

Figure 281F. Details *(Nissl stain)* of ventricular floor in the newborn at a pontine level caudal to colliculus facialis, and including vestibular grisea (from HIKIJI, 1933). nde: *nucleus of Deiters* (n. vestibularis lateralis); nph: nucleus praepositus hypoglossi (nucleus eminentiae teretis of diverse authors); nrt: formatio reticularis tegmenti; nvm: nucleus vestibularis medialis; nvs: nucleus vestibularis superior *(n. of Bechterew)*; sl: sulcus limitans.

Figure 281G. Details *(Nissl stain)* of motor facial nerve nucleus and of superior olivary complex in the newborn (from HIKIJI, 1933). nf: motor facial nerve nucleus; npo: nuclei pontis; osl: nucleus olivaris superior, pars lateralis (main superior olivary nucleus, *rotula of Ziehen)*; osm: nucleus olivaris superior, pars medialis (medial accessory nucleus, *pararotula of Ziehen)*; trl: nucleus trapezoidalis lateralis; trm: nucleus trapezoidalis medialis (corresponding to cell groups called internal and external preolivary nuclei by *Cajal)*. Left side medial, right side lateral. It can be seen, comparing this figure with Figures 270B and C, that the superior olivary complex is relatively smaller in Man than in some other Mammals.

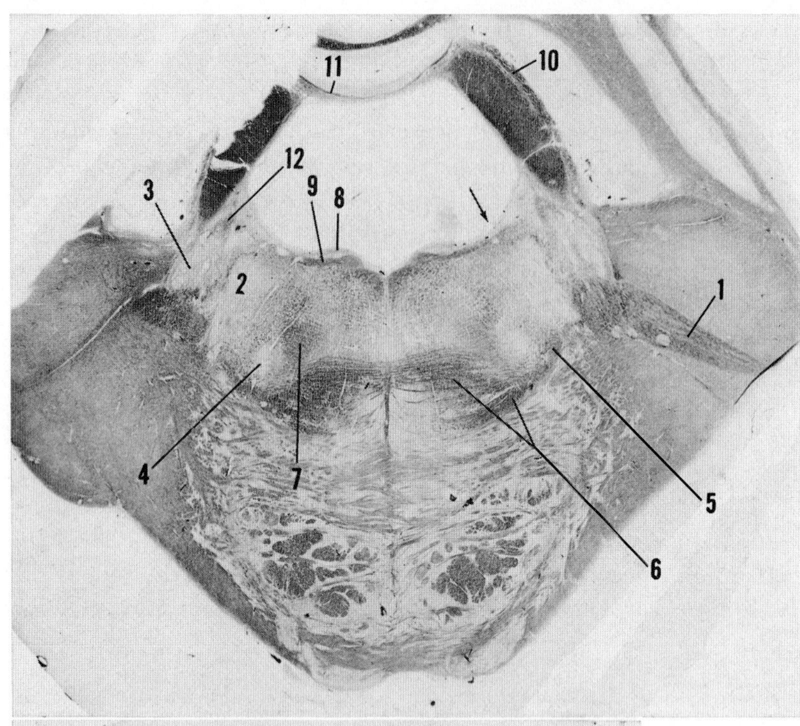

282 A

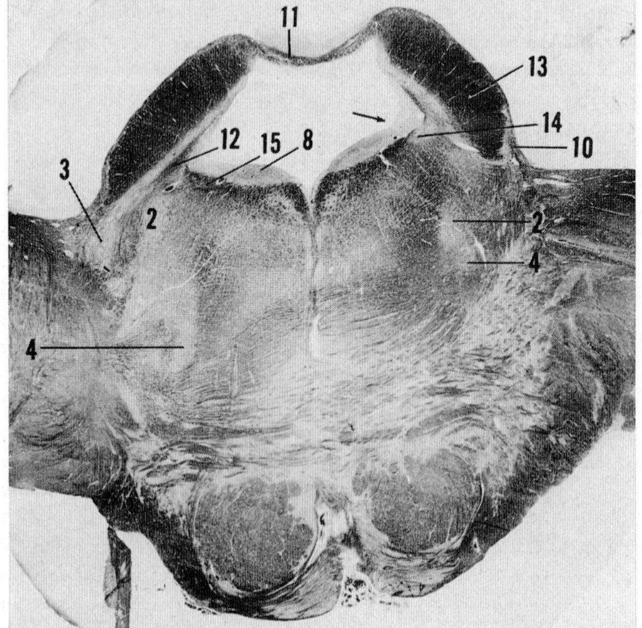

282 B

plate components. Rostrally adjacent to the main sensory trigeminal nucleus, and extending toward the cerebellum, scattered rather large pigmental cells have been described as a '*nucleus pigmentosus tegmentocerebellaris*' (JACOBSOHN, 1909). The cited author also recorded, in the basal plate, and in the lateral region of the tegmentum, a '*nucleus pigmentosus tegmentopontilis*'. In the isthmus region, in part traversed by ventrally directed crossing fibers of the brachium conjunctivum, a cell aggregate represents the 'pedunculopontine tegmental nucleus' of Jacobsohn (cf.

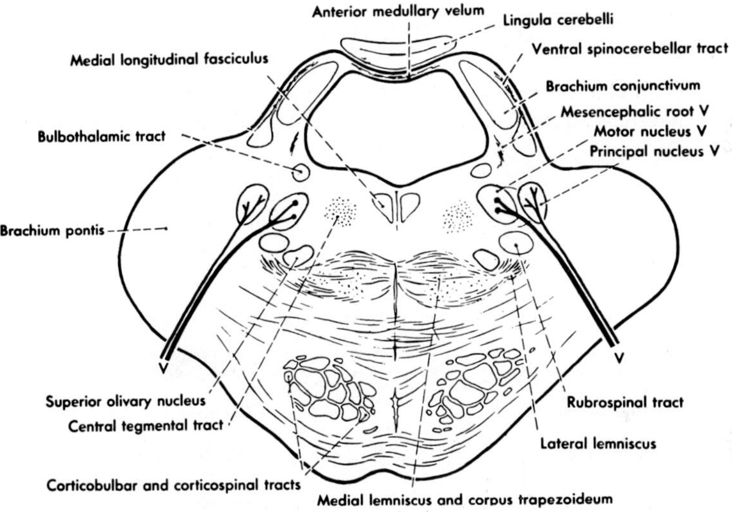

Figure 282C. Semidiagrammatic cross-section through the pons approximately corresponding to level shown in A (from HAYMAKER-BING, 1969).

Figure 282A, B. Cross-sections *(Weigert stain)* through the pons at levels of main trigeminal nuclei. B is somewhat more rostral than A (A ca. ×8, B ca. ×7; red. 2/3). 1: intracerebral trigeminal root passing through brachium pontis; 2: motor trigeminal nucleus; 3: main sensory trigeminal nucleus ('convolutio trigemini'); 4: superior olivary complex; 5: lemniscus lateralis; 6: lemniscus medialis; 7: central tegmental tract; 8: *fasciculus longitudinalis dorsalis of Schütz;* 9: rostral portion of facial loop; 10: dorsal spinocerebellar tract encircling brachium conjunctivum; 11: velum medullare anterius s. superius with tractus tecto-cerebellaris; 12: radix mesencephalica trigemini; 13: brachium conjunctivum; 14: nucleus loci coerulei; 15: dorsal secondary afferent fibers of nervus trigeminus. Arrow points to approximate locus of sulcus limitans.

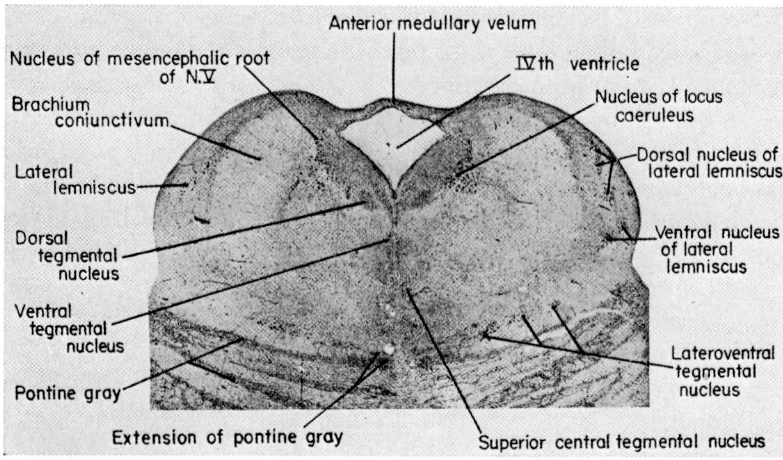

Figure 282D. Cross-section *(Nissl stain)* through pars tegmentalis pontis at a level somewhat rostral to that of B, and near rostral end of fourth ventricle (from Crosby *et al.*, 1962).

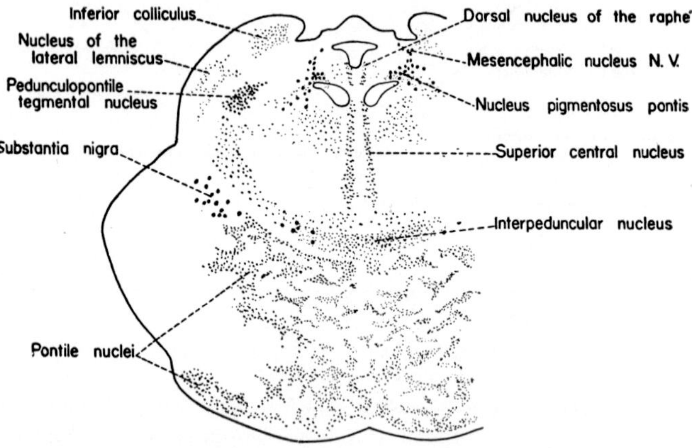

Figure 282E. Semidiagrammatic cross-section depicting cytoarchitectural arrangement of pons in the isthmus region at caudal end of *aquaeductus Sylvii* (slightly modified after Jacobsohn, 1909, from Kuntz, 1950). The 'nucleus pigmentosus pontis' is the nucleus loci caerulei. Several structures (inferior colliculus, substantia nigra, interpeduncular nucleus) already pertain to mesencephalon.

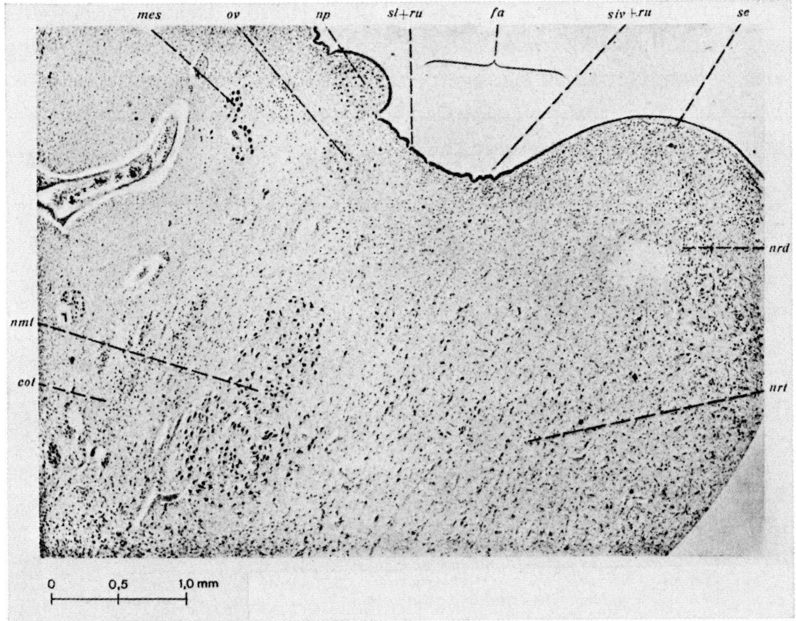

Figure 282 F. Details *(Nissl stain)* of main trigeminal nuclei as seen in a cross-section through the level of the fovea superior in the newborn (from HIKIJI, 1933). cot: convolutio trigemini (main sensory nucleus); fa: fovea anterior s. superior fossae rhomboideae; mes: cells of nucleus radicis mesencephalicae trigemini; nmt: nucleus motorius trigemini; np: 'nucleus parietolateralis'; nrd: dorsal subdivision of nucleus reticularis tegmenti; nrt: nucleus reticularis tegmenti; ov: 'nucleus ovalis; ru: *'rugae G. Retzii';* se: subependymal cell plate; siv: sulcus intermedioventralis; sl: sulcus limitans.

Fig. 283E).[194] The 'nucleus tegmenti dorsalis', already noted by v. GUDDEN more than 90 years ago, is located in the basal central gray of uppermost pons, of isthmus and caudal mesencephalic aqueduct. It is doubtless related to the *fasciculus longitudinalis of Schütz* which shall be dealt with in chapter X.

---

[194] A perusal and comparison of the atlases by JACOBSOHN (1909), OLSZEWSKI and BAXTER (1954), and RILEY (1943) in connection with a detailed actual study of the configurations concerned, will provide the critical observer with striking evidence of the confusion resulting from a complex, nonstandardized nomenclature based upon in part arbitrary and frequently unconvincing parcellation. The difficulties connected with the untangling and operationally valid description of the brain stem's intricate organization are obviously considerable.

Summarizing the reasonably well understood grisea of the Human, and, generally speaking of the Mammalian oblongata, the following list may be given, in an arrangement based upon configurational (morphologic, or bauplan) relationship.[195] It is understood that the oblongata *sensu latiori* includes here the pons.

A. *Alar plate*

I. *Dorsal zone*

1. Primary afferent cochlear nuclei.
   a) Nucleus cochlearis dorsalis (tuberculum acusticum).
   b) Nucleus cochlearis ventralis.
2. Primary afferent vestibular nuclei.
   a) Nucleus vestibularis superior *(Bechterew)*.
   b) Nucleus vestibularis lateralis *(Deiters)*.
   c) Nucleus vestibularis medialis *(Schwalbe)*.
   d) Nucleus radicis vestibularis descendentis.
3. Primary afferent trigeminal nuclei.
   a) Nucleus sensibilis principalis trigemini.
   b) Nucleus radicis descendentis trigemini.
   c) Nucleus radicis mesencephalicae trigemini.[196]
4. Primary afferent funicular ('relay') nuclei.
   a) Nucleus fasciculi gracilis *(Goll)*.
   b) Nucleus fasciculi cuneati *(Burdach)*.
   c) Nucleus fasciculi cuneati externus *(Blumenau, Monakow)*.

---

[195] Cf. K., 1927, pp. 167–168. One could, of course, also apply different rules of classification. Thus, OLSZEWSKI and BAXTER (1954) whose atlas has substantial merits despite numerous extravagances and arbitrary as well as ambiguous parcellation procedures, distinguish: (1) cranial nerve nuclei, (2) relay nuclei, and (3) nuclei of unknown connections. In the entire brain stem, category (1) subsumes 27 nuclei, while 20 and 46 (!) are included in categories (2) respectively (3). This makes a total of 93 different grisea. JACOBSOHN (1909) recorded approximately 83 nuclei, discounting a few diencephalic grisea depicted in his still very useful and valuable brain stem atlas.

[196] While the typical primary afferent cranial nerve nuclei contain neuronal elements of the second order receiving first order terminals, the 'nucleus' radicis mesencephalicae contains the first order neurons and pertains therefore to a distinctive category rather similar to the status of a peripheral spinal or cranial nerve ganglion.

II. *Intermediodorsal zone*

1. Primary afferent facial, glossopharyngeal and vagal cell column s. grisea tractus solitarii.
   a) Nucleus tractus solitarii.
   b) Nucleus parasolitarius.

B. *Basal plate*

I. *Intermedioventral zone*

1. Nucleus motorius trigemini (branchiomotor).
2. Nucleus motorius facialis (branchiomotor).
3. Nucleus ambiguus (IX, X, XI, branchiomotor).
4. Nucleus accessorii spinalis (modified branchiomotor).
5. parasympathetic (preganglionic) cell column.
   a) Nucleus salivatorius superior et lacrimalis (VII).
   b) Nucleus dorsalis glossopharyngei et vagi (includes nucleus salivatorius inferior and nucleus alae cinereae).

II. *Ventral zone*

1. Nucleus nervi abducentis.
2. Nucleus nervi hypoglossi.

III. *Basal plate in general* (including secondarily basal grisea of alar plate origin as pointed out in text).

1. Formatio reticularis tegmenti.
2. Nucleus reticularis lateralis (presumably a cerebellar relay nucleus).
3. Nuclei pontis (cerebellar relay nuclei).
4. Nuclei arcuati (cerebellar relay nuclei).
5. Superior olivary complex (mainly cochlear relay grisea).
6. Inferior olivary complex (cerebellar relay grisea).

In a comparable simplification the reasonably well understood major channels either passing through, or ending, respectively originating in, the Human oblongata can be enumerated as follows. It is understood that at least some of these pathways presumably contain both ascending and descending fibers.

I. *Descending pathways*

1. Pyramidal system.
   a) Tractus corticospinalis.
   b) Tractus corticobulbaris.
2. Corticopontine system.
3. Tractus tectobulbaris and perhaps tectospinalis *(fasciculus praedorsalis)*.
4. Rubrobulbar and rubrospinal system.[197]
5. Central tegmental tract (so-called 'tractus thalamo-olivaris').
6. Fasciculus longitudinalis medialis (descending components).
7. Reticulospinal system.
8. Vestibulospinal tract.
9. *Fasciculus longitudinalis dorsalis of Schütz* (descending components).
10. Solitariospinal fibers.
11. Descending primary afferent root fibers.
    a) Radix descendens trigemini.
    b) Radix descendens vestibularis.
    c) Tractus s. fasciculus solitarius (VII, IX, X).

II. *Ascending pathways*

1. Lemniscus medialis (proprioceptive, touch and presumably taste channel).
   a) Pars spinalis (tractus spinothalamicus anterior s. ventralis).
   b) Pars bulbaris (tractus bulbothalamicus from various grisea).
2. Tractus spinothalamicus et bulbothalamicus lateralis (pain and temperature channel).
3. Bulbotectal pathways (either as collaterals of or independent fibers within pathway 2 and perhaps also pathway 1).
4. Lemniscus lateralis (cochlear channel), also the less well understood ascending vestibular system.
5. Ascending reticular systems.
6. Tractus spinocerebellares.
   a) Tractus spinocerebellaris dorsalis *(Flechsig)*.
   b) Tractus spinocerebellaris ventralis *(Gower)*.

---

[197] Although perhaps approximately correct, the apparently precise locations assigned by most authors to the rubrospinal respectively rubrobulbar tract in oblongata and pons can be considered highly conjectural.

7. Tractus nucleocerebellares (IAK).
8. Fibrae olivocerebellares.
9. Tractus s. fibrae spino-olivares.
10. Fibrae laterocerebellares (from nucleus reticularis lateralis).
11. Fasciculus longitudinalis medialis (ascending components).
12. *Fasciculus longitudinalis dorsalis of Schütz* (assumed ascending components).
13. Ascending primary afferent root fibers.
    a) Fibers of radix mesencephalica trigemini.
    b) Some ascending trigeminal fibers of main sensory trigeminal root.
    c) Various ascending components of the vestibular nerve.
    d) A few ascending components of tractus solitarius.

The anatomical features of the human oblongata are evidently of particular import with respect to clinical neurology and neuropathology. Clinical and pathologic observations, on the other hand, have again significantly contributed to an understanding of complex neuroanatomical and functional relationships. Details of this mutual interdependence have been especially stressed in the treatises on local (topic) diagnosis by BING (1922) and HAYMAKER-BING (1956, 1969). Again, in our chapter on diseases of the brain stem and its cranial nerves for *Baker's handbook of clinical neurology*, we have emphasized the relevant anatomical aspects (HAYMAKER and K., 1950, 1962, 1971). Also, the elementary text on neuroanatomy by RANSON (1943) contains numerous well chosen discussions and illustrations of clinical applications. As in the case of the preceding chapter VIII (spinal cord) it is perhaps appropriate to conclude the present discussion of the human oblongata with some *generalized* and *introductory* clinical respectively neuropathologic topics.[198]

[198] The present series, which intends to counteract the narrow specialization and compartmentalization of neurobiology, is directed at a relatively small group of serious readers, who are either *connaisseurs* with considerable insight, or, if beginners, earnestly, and with the necessary stamina, intend to acquire this wide outlook. In such diversified group, some readers may not be adequately familiar with the manifold overall aspects of clinical, neuropathologic, and other interdisciplinary topics. Thus, an introductory generalized and correlated discussion of these matters, pointing out the many facets and still imperfectly understood problems at issue, obviously becomes a necessary requirement. A more detailed elaboration, as is evident from the extant literature, would fill endless volumes without providing further significant answers. Perhaps understandably irked by well deserved quips which I have directed at the many foibles of the 'sophisticated' but

The nuclei and intracerebral root fiber bundles of cranial nerves have a close spatial relationship with the main longitudinal descending and ascending pathways at various levels of the brain stem. The thus resulting diverse anatomical patterns provide the key to localization in terms of syndromes, whose neuropathologic features, in turn, by a here operationally valid circular reasoning, yield further anatomical and functional evidence.

Hemiplegia alterna (crossed paralysis in the wider sense) is a not altogether uncommon manifestation of brain stem lesions.[199] On the

---

decadent scientific establishment, an irate critic in *Brain* (94, p. 193, 1971) has censured my procedure as '*superficial*'. Now, because, e.g., many if not most clinical neurologists have only a rather hazy knowledge of basic neuroanatomy, neuropathology and neurophysiology, the late RUSSELL BRAIN (*Lord* BRAIN), a very competent clinician of unusually wide outlook, includes numerous much needed but most elementary neuroanatomical, neuropathological, and neurophysiological data in his excellent treatise on diseases of the nervous system. BRAIN's likewise most commendable text '*Clinical Neurology*' (1960), directed at non-neurologists or young students, is restricted to very generalized bare essentials. Yet, since *Lord* BRAIN was a (very uncommon) weighty representative of the establishment, said British critic would certainly never have dared to decry that author's procedure as 'superficial'. Again, other *petits maîtres* chose to censure me by churlish reviews in various journals, *inter alia* also particularly objecting to my inclusion of neuropathologic and various other seemingly extraneous but actually pertinent matters in the present series. Being, by now, a rather seasoned hand at the 'scientific game', whose vainglorious ambitions should not be taken too seriously, I was merely mildly amused. Quite serenely proceeding with my task in accordance with the maxim of a congenial old warrior: '*Je n'ai pas besoin d'espérer pour entreprendre, ni de réussir pour persévérer*', I shall nevertheless quote an appropriate epigram from the *Palatine Anthology:*

Ἡράκλειτος ἐγώ· τί μ' ἄνω κάτω ἕλκετ' ἄμουσοι;
    οὐχ ὑμῖν ἐπόνουν, τοῖς δ' ἔμ' ἐπισταμένοις.
εἷς ἐμοὶ ἄνθρωπος τρισμύριοι, οἱ δ' ἀνάριθμοι
    οὐδείς. ταῦτ' αὐδῶ καὶ παρὰ Περσεφόνῃ.

Palat. Anth. VII, 128.

It can be translated as follows:

'*Heraclitus am I. What if ye drag me up and down, ye nitwits?*
Not for you did I toil, but for those who understand me.
One such man I deem equivalent to thirty thousand,
But the innumerable rabble counts as nobody.
This I shall proclaim, yea in Persephone's domain.'

[199] Generally speaking, lesions of the neuraxis involving 'topical' syndromes can be vascular, neoplastic, and of other miscellaneous sorts such as luetic gumma, tuberculoma, abscess, circumscribed meningitis, etc., respectively caused by external trauma.

side of the damage, there are disturbances of cranial nerve function, combined with contralateral symptoms caused by the involvement of long pathways (pyramidal tract or sensory channels) which, although affected on the lesion's side, have essentially contralateral relationships by way of decussations (cf. Figs. 283–288).

Thus, a lesion as depicted in Figure 283, and damaging the right

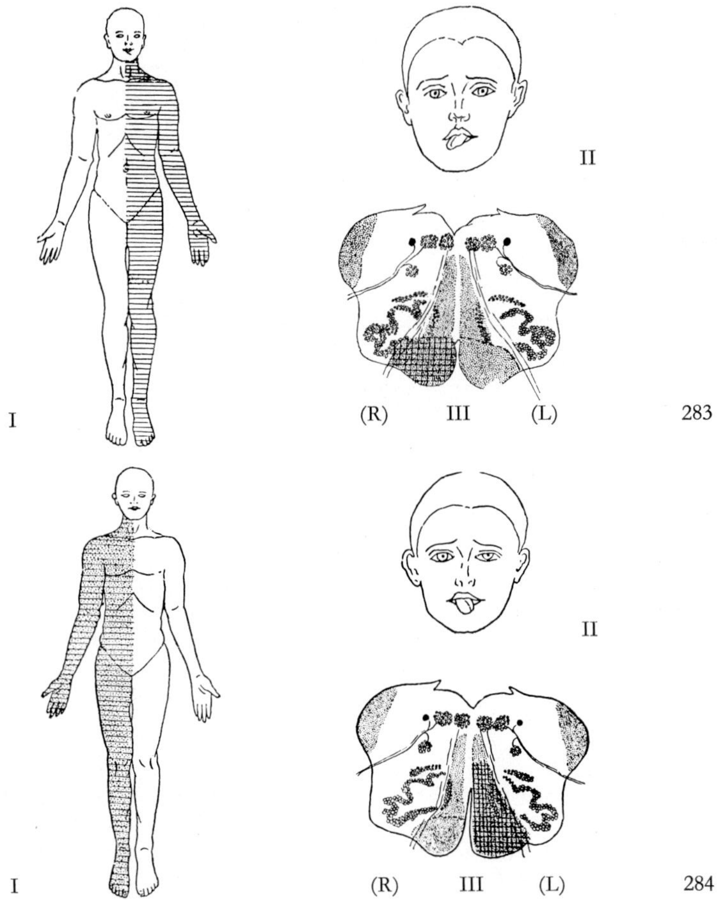

Figure 283. Diagram illustrating a case of hemiplegia alterna involving the nervus hypoglossus (from Ranson, 1943). I Extent of hemiplegia. II Deviation of protruded tongue toward side of hypoglossal involvement. III Site of lesion (cross-hatching).

Figure 284. Diagram illustrating a case of hypoglossal hemiplegia alterna with involvement of medial lemniscus causing some sensory loss as explained in text (from Ranson, 1943). I Extent of hemiplegia (horizontal hatching) and of sensory involvement (fine dots). II and III as in preceding figure.

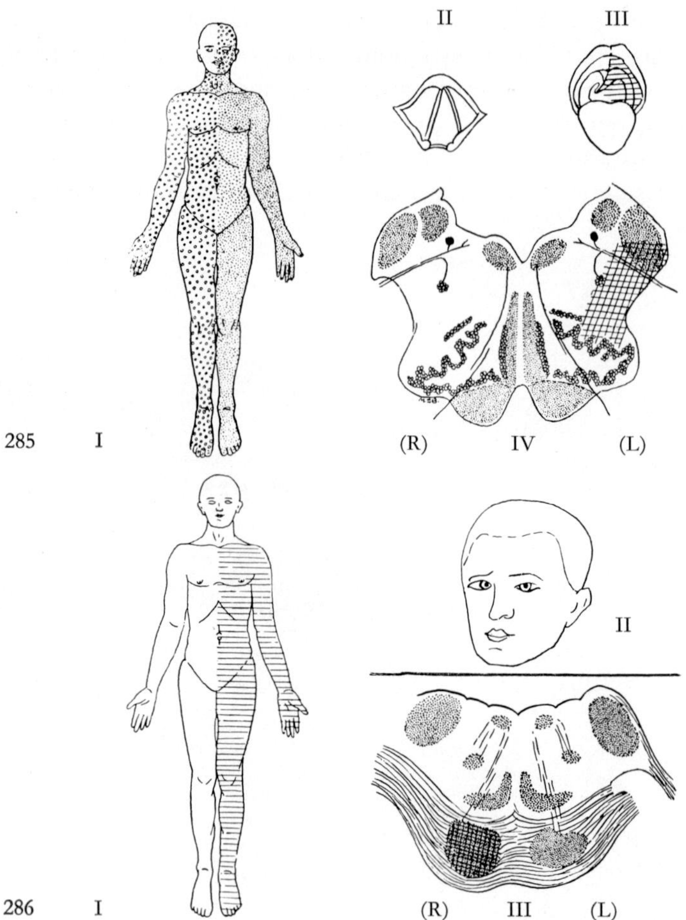

Figure 285. Diagram illustrating a case of 'posterior inferior cerebellar artery syndrome' (thrombosis) involving left lateral region of oblongata (after Ranson, 1943). I Extent of thermal and pain anesthesia (coarse dots or circles) and of slightly impaired muscular coordination (fine dots). Although here also indicated on the trunk, such impairment affects mainly the finer movements of upper and lower extremities. II Laryngoscopic picture (during inspiration) indicating so-called 'cadaveric position' of left vocal cord (it should be noted that the (indirect) laryngoscopic picture is a mirror image, also showing the epiglottis at top and the tubercula corniculata at bottom). III Extent of palatine paralysis (horizontal hatching). IV Site of lesion (cross-hatching).

Figure 286. Diagram illustrating a case of hemiplegia alternans involving the nervus abducens (from Ranson, 1943). I Extent of hemiplegia. II Strabismus internus of right eye. III Seat of lesion.

pyramid together with the neighboring hypoglossal root, will cause contralateral, i.e. left *spastic hemiplegia ('upper motor neuron lesion')* combined with right hypoglossal '*lower motor neuron paralysis*'. On protrusion, the tongue deviates toward the side of the lesion through action of the unopposed contralateral genioglossal muscle, which pulls the root of the tongue toward the spina mentalis, while the paralyzed half of the tongue lags behind. Since the tongue is essentially a complex muscle covered by mucosa, considerable atrophy of the involved half (hemiatrophy) will occur with the passage of time. This syndrome is rather rare because neighboring structures are usually involved. Thus, Figure 284 illustrates a similar condition complicated by contralateral loss of the sense of posture and of passive movements, together with impairment of tactile sensibility. Pain and temperature sensibility remains normal. The loss is restricted to the body region innervated by spinal nerves and does not involve the trigeminal field. In cases of this sort, the lesion extends from the pyramid into the medial lemniscus.

Figure 285 illustrates the results of a lesion in the lateral region of the oblongata, affecting the nucleus ambiguus. Paralysis of the homolateral vocal cord and soft palate is here combined with loss of pain and temperature sensitivity over the contralateral side of the body and the homolateral side of the trigeminal field, caused by damage to lateral spino-thalamic tract (already crossed), and nucleus of the (uncrossed) descending trigeminal root. Since medial lemniscus (with anterior spino-thalamic tract) and main sensory trigeminal nucleus are not affected, no loss of tactile sensibility occurs. Some uncoordination of finer movements by homolateral extremities may become noticeable, and *Romberg's sign* (cf. chapter VIII, p. 263) with a tendency to fall toward the left can be manifested, caused by damage to spinocerebellar tracts and corpus restiforme.[200] Vertigo and conjugate deviation of the eyes

[200] With regard to vascular lesions, such as, e.g. thrombosis, a condition about corresponding to that of Figure 285 is the so-called 'posterior inferior cerebellar syndrome' characterized by (1) ipsilateral paralysis of palate, pharynx, vocal cords, (2) analgesia and thermoanesthesia, ipsilateral for face and contralateral for extremities and trunk, and (3) some homolateral cerebellar deficiency (corpus restiforme); (4) homolateral *Horner's syndrome* (paralysis of the ocular sympathetic) may likewise be included, resulting from interruption of supranuclear sympathetic fibers in the lateral or dorsolateral tegmentum. Such fibers convey impulses to the preganglionics in the intermedioventral zone of spinal segments C_8 to ca. Th_2 *('centrum ciliospinale')*. *Horner's syndrome* involves *ptosis* (smooth musculus palpebralis superior and orbitalis), some degree of *enophthalmos* (sunken eyeballs), and *miosis* (constricted pupil due to paralysis of dilator pupillae). Cf. Figure 291 illustrating the approximate regions supplied by the arteries of the brain stem.

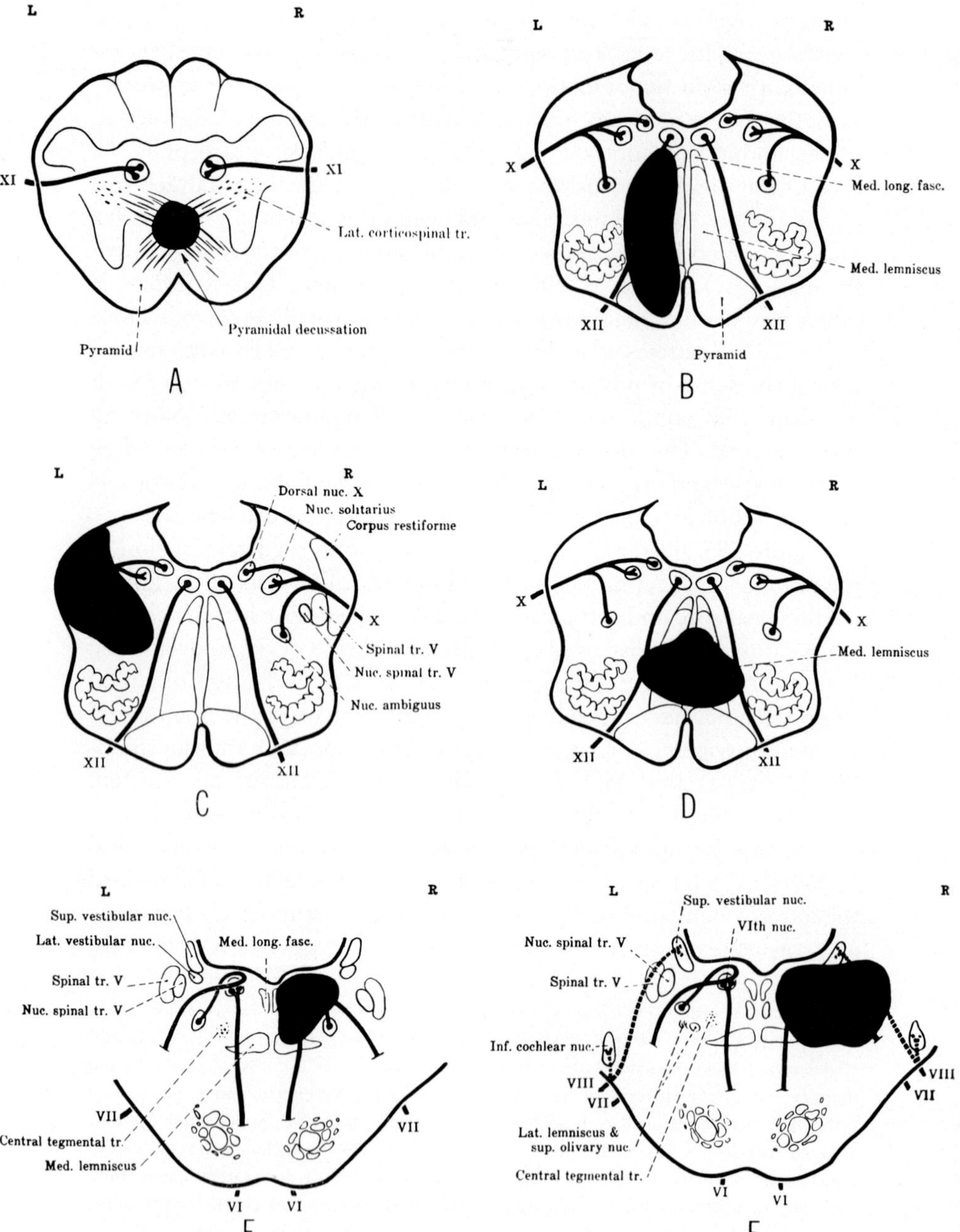

Figure 287. Additional diagrams illustrating lesions of the oblongata involving cranial nerves or main fiber systems or both (from HAYMAKER-BING, 1956).

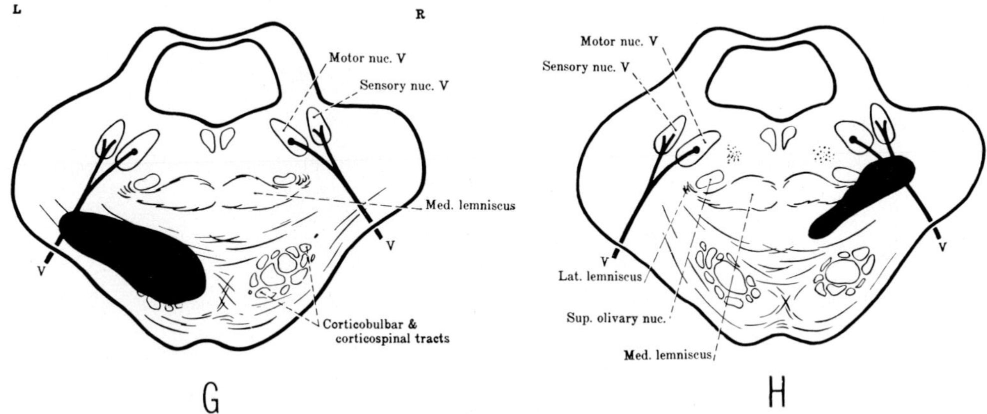

toward the side of the lesion may initially occur in connection with a partial affection of vestibular grisea (n. of descending vestibular root and perhaps n. vestibularis medialis).

A lesion in the caudal pons as indicated by Figure 286 will cause contralateral spastic hemiplegia combined with homolateral internal strabismus, this latter being caused by 'lower motor neuron lesion' of nervus abducens paralyzing the m. rectus lateralis and leading to the predominant pull by m. rectus medialis.

Pontine lesions of similar type can also involve the root fibers of the facial nerve, causing an alternate hemiplegia with homolateral facial paralysis, with or without concomitant involvement of nervus abducens. Damage to afferent facial root fibers can result in loss of taste sensations on the homolateral anterior two thirds of the tongue. If the motor fibers innervating the m. stapedius are affected, hyperacusis (intensification of loud noises) generally results.

Bell's palsy, commonly involving a complete paralysis of the facial muscles innervated by temporo-facial (upper) and cervico-facial (lower) branches of the seventh nerve is a peripheral lesion, usually due to a non-suppurative inflammation of the nerve within the stylomastoid foramen. Bulbar lesions affecting the facial nerve nucleus and its intracerebral root frequently do not include all the components of that nerve. In unilateral supranuclear central lesions interrupting the corticobulbar fibers for the facial nucleus *('upper motor neuron damage')*, the

Signs Resulting from Lesions Illustrated in Figures 287A–H

		Left	Right
A	Head		
	Trunk and Limbs	Spastic paralysis of all four limbs with intact sensibility	
B	Head	Paralysis and atrophy of the tongue	
	Trunk and Limbs		Hemiplegia and reduction of deep sensibility
C	Head	Anesthesia, analgesia and thermo-anesthesia of face	
	Trunk and Limbs	Hemiataxia and hemiasynergia of cerebellar type	
D	Head	Bilateral paralysis and atrophy of the tongue	
	Trunk and Limbs	Bilateral ataxia and loss of deep sensibility	
E	Head		Paralysis of the face and lateral rectus, and of conjugate lateral ocular deviation to right, possibly hyperacusis
	Trunk and Limbs	Reduction of superficial sensibility	
F	Head	probably hypacusis	Anesthesia of trigeminal distribution, paralysis of lateral ocular deviation to right, paralysis of face, probable hyperacusis
	Trunk and Limbs	Reduction of superficial sensibility	
G	Head	Trigeminal anesthesia and weakness and atrophy of the masticatory muscles, possibly hypacusis	Weakness but no atrophy of face and tongue
	Trunk and Limbs		Spastic hemiplegia
H	Head		Trigeminal anesthesia and weakness and atrophy of the masticatory muscles, probably hypacusis
	Trunk and Limbs	Reduction of pain and thermal sense	

Figure 288. Table indicating symptoms resulting from the lesions depicted in preceding figure (slightly modified after HAYMAKER-BING, 1956). Hyperacusis in E (R) and F (R) is related to paralysis of musculus stapedius. Hypacusis in G and in H (R) may be caused by paralysis of musculus tensor tympani. Possible hypacusis in F and H might result from unilateral damage to lateral lemniscus system. Both hyperacusis and hypacusis are, in such lesions, rather variable and nondescript symptoms whose recognition may require refined otologic tests.

upper branch of the facial is known to be spared. Its griseum of origin seems to have a bilateral innervation by crossed and uncrossed corticobulbar fibers, while the lower facial apparently lacks uncrossed supranuclear input and must depend on its crossed corticobulbar connections which become interrupted by a contralateral lesion at relevant rostral levels.

Symptoms caused by damage to the vestibular system include *nystagmus*,[201] other eye muscle disturbances such as conjugate deviation, and, moreover, *vertigo*. Lesions of cochlear nerve and its nuclei may result in unilateral deafness, e.g. in acustic 'neuromas' or 'neurinomas' at the *cerebellopontine* angle *('angle tumors')*. Bulbar involvement of the ascending auditory pathways is not uncommon, but remains frequently unilateral or only partial because of the wide and bilateral distribution of the cochlear channel. *Tinnitus* (humming, whistling or roaring noises) is another symptom of cochlear system disturbances, most frequently due to lesions of the internal ear, or of the cochlear root at the cerebellopontine angle, but occasionally also of bulbar of still more central origin.

Bulbar damage to the trigeminus system may result in paralysis or paresis of muscles of mastication, mylohyoid, anterior belly of the digastric, tensor tympani and tensor veli palatini if the motor nucleus or its root fibers are affected. Unilateral supranuclear interruption of the trigeminal corticobulbar fibers seems not to result in functional disturbances of the muscles innervated by this nerve since a bilateral distribution of these fibers appears to obtain.

---

[201] *Nystagmus* is a conjugate involuntary rhythmic motion of the eyeballs (horizontal, vertical, rotatory or mixed), which can also be restricted to only one eye. It is usually rapid in one direction and slow in the opposite, the rapid phase being arbitrarily taken to describe the nystagmus as, e.g. 'right' or 'left'. Nystagmus may be of retinal, labyrinthine, vestibular, cerebellar, or generally 'bulbar' respectively 'central' origin. It can be produced and observed in '*caloric tests*', douching the external auditory meatus with cold or warm water, thereby inducing convections of the endolymph in the semicircular canals. In '*galvanic tests*', nystagmus is induced by an electric current stimulating the vestibular labyrinth. In addition, '*rotation tests*' by means of a whirling chair can be performed. *Conjugate deviation* is characterized by the 'involuntary' deflection of the eyes in the same direction at the same time. In lesions of the pons, involving the abducens nucleus or its neighborhood, conjugate deviation is directed away from the side of the lesion, while in telencephalic lesions affecting the cortical eye muscle 'centers' deviation toward the lesion's side is generally observed. A rare form of nystagmus is the so-called *retractory nystagmus* in which the eye bulbs (or still more rarely one bulb) move forward and backward. This nystagmus is believed to be caused by disturbances involving the fasciculus longitudinalis medialis mechanism, including perhaps also the pretectal region in some of these cases.

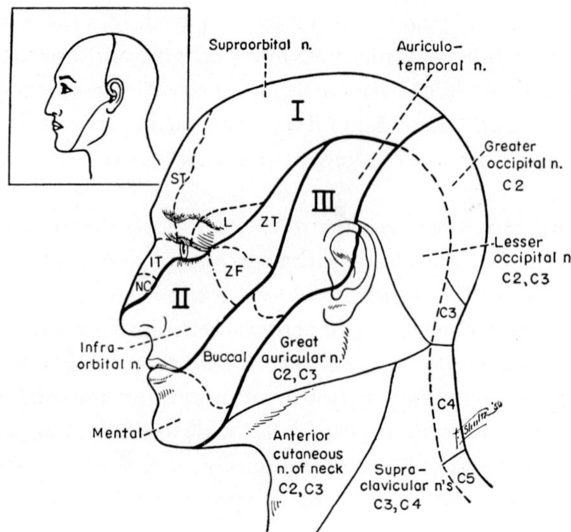

Figure 289. Sensory distribution of ophthalmic (I), maxillary (II), and mandibular (III) subdivisions of trigeminal nerve (after various authors, from HAYMAKER and K., 1955). Ophthalmic division branches: ST, supratrochlear; L: lacrimal; IT: infratrochlear; NC: nasociliary. Maxillary branches: ZT: zygomatico temporal; ZF: zygomaticofacial. Insert: caudal boundary of anesthesia after *retrogasserian* neurotomy. Note that the cervical cutaneous nerves supply back of head and angle of jaw.

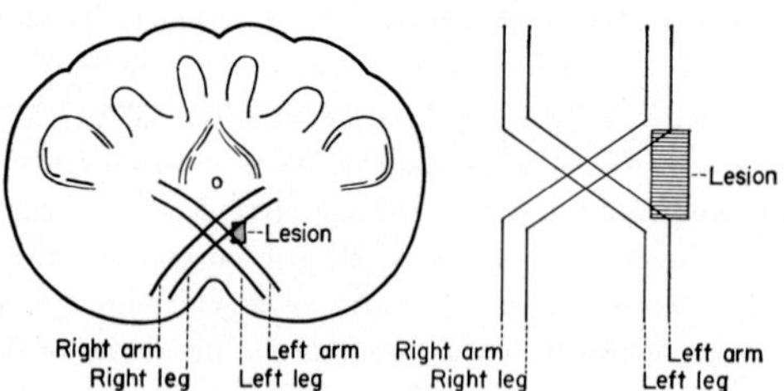

Figure 290. Diagrammatic representation of possible site of a lesion causing hemiplegia cruciata of left upper and right lower extremity (from HAYMAKER and K., 1955). Because of the apparently variable types of pyramidal decussation, different diagrams, presuming other patterns of fiber distribution, can be drawn to illustrate lesions causing crossed hemiplegia of this sort.

Figure 289 illustrates the afferent peripheral distribution of the trigeminus. An important reflex mediated by the ophthalmic division is the *corneal reflex*, resulting in closure of both eyelids (m. orbicularis oculi) through the facial nerve upon a slight stimulus (e.g. by a wisp of cotton wool) to the cornea. The *jaw jerk*, elicited by tapping the chin, the mouth being slightly opened and relaxed, corresponds to the *tendon reflexes* discussed on p. 144, section 11 of chapter VIII. Still another reflex is the *sneeze reflex* produced by tickling the nasal mucosa. These reflexes become relevant for diagnostic purposes in clinical neurology.

Other significant bulbar reflectory activities, such e.g. as respiration, coughing, vomiting, and cardiovascular control were briefly dealt with in section 1 of this chapter. Topics concerning the reticular formation have been included in section 2.

Cavities of *syringobulbia*, surrounded by glial proliferation or reaction, and comparable to those of *syringomyelia* (cf. chapter VIII, section 11, p. 265, and footnote 168) may occur not only in the closed portion of the oblongata characterized by a central canal, but can also extend as far as the trigeminal levels of the pons or even still more rostralward. Such cavity may be unilateral in the neighborhood of nucleus ambiguus and even nucleus radicis descendentis trigemini, thus being quite unrelated to central canal respectively raphe. Although the syndromes of syringobulbia and syringomyelia have been found associated with anomalies in the craniovertebral junction area (including *Arnold-Chiari types of deformation)*, respectively with spina bifida (cf. also GARDNER, 1965; HAYMAKER and K., 1962), this correlation may be incidental to a concomitant multiple genome disturbance. Nevertheless, it has been suggested that the condition is related to an initial hydromyelia with secondary closure or separation of diverticula. It will be recalled (cf. volume 3/II, chapter VI, section 1C, p. 223), that some authors relate syringomyelia to the so-called 'dysraphic disturbances'. There is, however, little doubt that syringomyelia or syringobulbia subsume a variety of processes with similar end effects, including borderline neoplastic conditions, not necessarily related to the 'dysraphism' of spina bifida.

As regards the *pyramidal system*, the analysis of lesions at its diverse brain stem levels has provided some clues from which the pattern of *'supranuclear' ('upper motor neuron') innervation* of various cranial nerve nuclei may be deduced. Quite apart from the uncertainty related to the numerous individual variations, many of the thus reached inferences remain highly conjectural. A fairly large lesion at the level of the pyr-

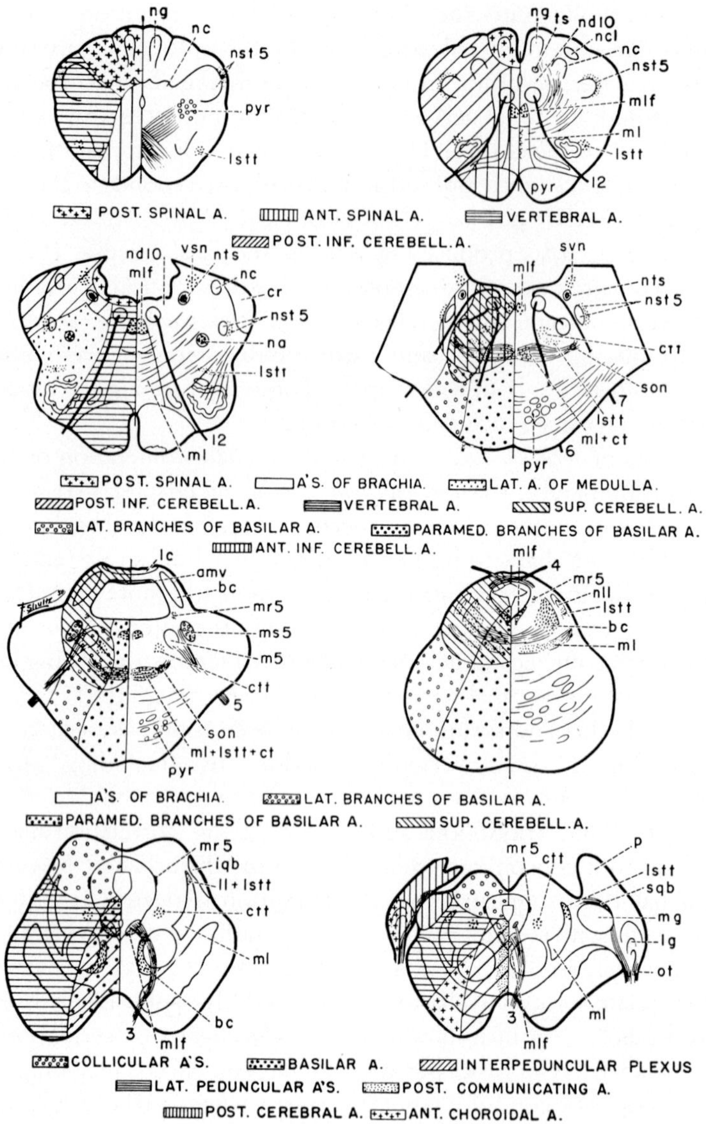

Figure 291. Fields of blood supply through various arteries of the human brain stem (from HAYMAKER and K., 1955, 1971). The 'arteries of brachia cerebelli' are the three cerebellar arteries and lateral branches of the basilar artery. Collicular arteries are superior cerebellar and posterior cerebral. Lateral peduncular arteries are superior cerebellar, posterior cerebral, and anterior choroidal. The interpeduncular plexus is formed by posterior cerebral, posterior communicating, and basilar arteries. Cf. also section 7,

amidal decussation, as indicated in Figure 287A can result in spastic paralysis involving all four extremities *(tetraplegia)* but, fortunately, is a very rare condition. Likewise rare is the *crossed hemiplegia sensu strictiori (hemiplegia cruciata)*, in which the arm of one side, and the leg of the opposite side are affected. Figure 290 illustrates the site of a lesion which can cause this syndrome. This latter, however, depends on the apparently variable grouping of fibers within the decussation, and the diagram indicates only one of several different possibilities compatible with a syndrome of this sort. As a rule, moreover, a lesion damaging the bulbar pyramid will also affect neighboring structures.

As regards clinically relevant reflexes or signs correlated with 'pyramydal disturbances' (cf. chapter VIII, p. 260, footnote 165), a reference to *Rossolimo's* and *Mendel-Bechterew's signs* may here be added. The former is obtained by tapping the pulps of the toes, and, in such lesions, triggers flexion and abduction of the toes. *Mendel-Bechterew's sign* is elicited by tapping the lateral (cuboid) region of the dorsum pedis, thereby causing plantar flexion of the toes in cases of upper motor neuron damage.

Although various apparently clear-cut bulbar syndromes, such as pointed out above, have been described and named in the clinical literature, few of the actual cases encountered in medical practice conform exactly to the standardized textbook descriptions. Because of the close spatial relationships of diverse neural structures such as grisea and pathways within the brain stem, of obtaining individual variations, and the wide diversity in type, spatial extent, and temporal course of

---

p. 711–716 of chapter VI in the preceding volume 3, part II. It must, moreover, be kept in mind that, because of the innumerable variations, illustrations by figures such as here proffered, although quite helpful, should cautiously be taken *cum grano salis*. amv: anterior medullary velum; bc: brachium conjunctivum; cr: corpus restiforme; ct: corpus trapezoideum; ctt: central tegmental tract; iqb: inferior quadrigeminal brachium; lc: lingula cerebelli; lg: lateral geniculate body; ll: lemniscus lateralis; lstt: lateral spinothalamic tract; m5: nucleus motorius trigemini; ms5: nucleus sensibilis principalis trigemini; nst5: nucleus of the descending trigeminal root; mg: medial geniculate body; ml: lemniscus medialis; mfl: fasciculus longitudinalis medialis; na: nucleus ambiguus; nc: nucleus cuneatus; ncl: nucleus cuneatus lateralis (s. externus); nd 10: nucleus dorsalis vagi; ng: nucleus gracilis; nll: nucleus of lateral lemniscus; nts: nucleus tractus solitarii; ot: optic tract; p: pulvinar thalami; pyr: pyramidal pathway; son: superior olivary complex; sbq: superior quadrigeminal brachium; svn: superior vestibular nucleus; ts: tractus solitarius (and its griseum); vsn: nucleus of the descending vestibular root.

the relevant lesions, numerous deviations from and diverse combinations of, textbook syndromes do occur. These latter, and their illustration by diagrams such as Figures 283 to 288, and 290 nevertheless represent valid and instructive first approximations. Again, because vascular lesions are here particular common, an approximate delineation of the regions supplied by the brain stem arteries, as e.g. shown in Figure 291 provides, despite the wide range of variations, a significant help for an appropriate understanding of the occurring syndromes.

11. References to Chapter IX

ABBIE, A. A.: The brain-stem and cerebellum of Echidna aculeata. Phil. trans. roy. Soc. London Ser. B. *224:* 1–74 (1934).

ADAM, H.: Der III. Ventrikel und die mikroskopische Struktur seiner Wände bei Lampetra (Petromyzon) fluviatilis L. und Myxine glutinosa L., nebst einigen Bemerkungen über das Infundibularorgan von Branchiostoma (Amphioxus) lanceolatum Pall; in KAPPERS Progr. in Neurobiol., Proc. first intern. Meetg. of Neurobiologists, pp. 146–158 (Elsevier, Amsterdam 1956).

ADAM, H.: Brain ventricles, ependyma, and related structures; in BRODAL and FÄNGE, The biology of Myxine, pp. 137–143 (Universitetsforlaget, Oslo 1963).

ADDENS, J. L.: The eye-muscle nerves of petromyzonts, especially in their general morphological significance. Proc. kon. Akad. Wetensch. Amsterdam *31:* 733–748 (1928).

ADDENS, J. L.: The motor nuclei and roots of the cranial and first nerves of vertebrates. Z. Anat. EntwGesch. *101:* 307–410 (1933).

ADDENS, J. L.: A critical review of the occurrence of crossing root fibres in the facialis, vestibular, glossopharyngeal and vagus nerves. Psych. en neurol. Bladen *1934:* 274–291 (1934).

AHLBORN, F.: Untersuchungen über das Gehirn der Petromyzonten. Z. wiss. Zool. *39:* 191–294 (1883).

AYERS, H. and WORTHINGTON, J.: The skin end-organs of the trigeminus and lateralis nerves of Bdellostoma dombeyi. Amer. J. Anat. *7:* 327–336 (1907).

AYERS, H. and WORTHINGTON, J.: The finer anatomy of the brain of Bdellostoma dombeyi. 1. The acustico-lateral system. Amer. J. Anat. *8:* 1–17 (1908).

AYERS, H. and WORTHINGTON, J.: The finer anatomy of the brain of Bdellostoma dombeyi. 2. The fasciculus communis system. J. comp. Neurol. *21:* 593–617 (1911).

BACSIK, R. D. and STROMINGER, N. L.: The cytoarchitecture of the human anteroventral cochlear nucleus. J. comp. Neurol. *147:* 281–289 (1973).

BARNARD, J. W.: The hypoglossal complex of Vertebrates. J. comp. Neurol. *72:* 489–524 (1940).

BARTELMEZ, G. W.: Mauthner's cell and the nucleus motorius tegmenti. J. comp. Neurol. *25:* 87–128 (1915).

BARTELS, M.: Über die Gegend des Deiters- und Bechterewkernes bei Vögeln. Z. Anat. EntwGesch. *77:* 726–784 (1925).

BAUMGARTEN, R. VON: Rautenhirn (Rombencephalon); in LANDOIS-ROSEMANN, Lehrbuch der Physiologie des Menschen, 28. Aufl., vol. 2, pp. 695–715 (Urban & Schwarzenberg, München 1962).

Beccari, N.: La costituzione, i nuclei terminali e le vie di connessione del nervo acustico nella Lacerta muralis. Arch. ital. Anat. Embriol. *10:* 646–649 (1912).

Beccari, N.: Il IX, X, XI e XII pajo di nervi cranici negli embrioni di Lacerta muralis (Contribuzioni allo studio del significato morfologico dei nervi della testa). Arch ital. Anat. Embriol. *13:* 1–78 (1914).

Beccari, N.: Studi comparativi sulla struttura del Rombencefalo. I. Nervi spino-occipitale e nervo ipoglosso. II. Centri tegmentali. Arch. ital. Anat. Embriol. *19:* 122–291 (1922).

Beccari, N.: I centri tegmentali nell'asse cerebrale dei Selacei. Arch. zool. ital. *14:* 411–429 (1930).

Beccari, N.: Revisione del problema del nervo accessorio (con nuove osservazioni nel Teleostei). Arch. ital. Anat. Embriol. *47:* 8–71 (1942).

Beccari, N.: Neurologia Comparata (Sansoni, Firenze 1943).

Bechterew, W. von: Die Leitungsbahnen im Gehirn und Rückenmark. Ein Handbuch für das Studium des Aufbaues und der inneren Verbindungen des Nervensystems (Georgi, Leipzig 1898).

Bechterew, W. von: Die Funktionen der Nervencentra. 3 vols. (Fischer, Jena 1908, 1909, 1911).

Békésy, G. von and Rosenblith, W. A.: The mechanical properties of the ear; in Stevens, Handbook of Experimental Psychology, pp. 1075–1115 (Wiley, New York 1951).

Benedetti, E.: Il cervello e i nervi cranici del Proteus anguineus Laur. Memorie Ist. ital. Speleologia. Ser biologica. Mem. 1–79 (1933).

Bergeijk, W. A. van; Pierce, J. R., and David, E. E., Jr.: Waves and the ear (Anchor Books, Doubleday, New York 1960).

Berkelbach van der Sprenkel, H.: The central relations of the cranial nerves in Silurus glanis and Mormyrus caschive. J. comp. Neurol. *25:* 5–63 (1915).

Bernis, W. J. and Spiegel, E. A.: Zentren der statischen Innervation und ihre Beeinflussung durch Gross- und Kleinhirn. Arb. neurol. Inst. Univ. Wien *27:* 197–225 (1925).

Bing, R.: Kompendium der topischen Gehirn- und Rückenmarksdiagnostik, 5. Aufl. (Urban & Schwarzenberg, Berlin-Wien 1922).

Bing, R. und Burckhardt, R.: Das Zentralnervensystem von Ceratodus Fosteri. *Semons* zool. Forschungsreisen, Bd. 1. Jenaische Denkschriften *4:* 513–584 (1905).

Black, D.: The motor nuclei of the cerebral nerves in phylogeny: a study of the phenomena of neurobiotaxis. I. Cyclostomi and Pisces. J. comp. Neurol. *27:* 467–564 (1917). II. Amphibia. J. comp. Neurol. *28:* 379–427 (1917). III. Reptilia. J. comp. Neurol. *32:* 61–98 (1920). IV. Aves. J. comp. Neurol. *34:* 233–275 (1922).

Blinkov, S. M. and Glezer, I. I.: The human brain in figures and tables. A quantitative handbook (Plenum Press, New York 1968).

Bloom, W. and Fawcett, D. W.: A textbook of histology, 9th ed. (Saunders, Philadelphia 1968).

Bok, S. T.: Die Entwicklung der Hirnnerven und ihrer zentralen Bahnen. Die stimulogene Fibrillation. Folia neurobiol. *9:* 475–565 (1915).

Bone, Qu.: The central nervous system; in Brodal and Fänge, The biology of Myxine, pp. 50–91 (Universitetsforlaget, Oslo 1963).

Boord, R. L.: The efferent cochlear bundle in the caiman and pigeon. Exp. Neurol. *3:* 225–237 (1961).

Boord, R.L.: Ascending projections of the primary cochlear nuclei and nucleus laminaris in the pigeon. J. comp. Neurol. *133:* 523–542 (1968).

Boord, R.L. and Rasmussen, G.L.: Projection of the cochlear and lagenar nerves on the cochlear nuclei of the pigeon. J. comp. Neurol. *120:* 463–475 (1963).

Boss, K.: Studien über die Entwicklung des Gehirns bei Fringilla canaria und Chelydra serpentina. Gegenbaurs morphol. Jb. *45:* 337–390 (1913).

Braak, H.: Über die Kerngebiete des menschlichen Hirnstammes. IV. Der Nucleus reticularis lateralis und seine Satelliten. Z. Zellforsch. *122:* 145–159 (1971).

Brain, R.: Clinical neurology (Oxford University Press, London 1960).

Brain, R.: Diseases of the nervous system, 7th ed., revised by the late Lord Brain and J.N. Walton (Oxford University Press, London 1969).

Brandis, F.: Untersuchungen über das Gehirn der Vögel. I. Theil: Übergangsgebiet vom Rückenmark zur Medulla oblongata. Arch. mikr. Anat. *41:* 168–194 (1893).

Brandis, F.: Untersuchungen etc. II. Theil: Ursprung der nerven der Medulla oblongata. Hypoglossus und Vagusgruppe. Arch. mikr. Anat. *41:* 623–649 (1893).

Brandis, F.: Untersuchungen etc. II. Theil: Acusticusgruppe. Arch. mikr. Anat. *43:* 96–116 (1894).

Brandis, F.: Untersuchungen etc. III. Theil: Der Ursprung des N. Trigeminus und der Augenmuskelnerven. Arch. mikr. Anat. *44:* 534–555 (1895).

Breuer, J.: Die Selbststeuerung der Athmung durch den Nervus vagus. S. Ber. Akad. Wiss., Wien. math-nat. Cl. 58, II: 909–937 (1868).

Brodal, A.: The reticular formation of the brain stem (Oliver and Boyd, Edinburgh 1957).

Brodal, A.: Anatomical observations on the vestibular nuclei, with special reference to their relations to the spinal cord and the cerebellum. Acta oto-laryng. Suppl. *192:* 24–51 (1964).

Brodal, A.: Some points of view on the anatomy of the brain stem with particular reference to the reticular formation. Excerpta med. *110:* 449–460 (1965).

Brodal, A. and Fänge, R. (eds.): The biology of Myxine (Universitetsforlaget, Oslo 1963).

Bullock, T.H.: Seeing the world through a new sense: electroreception in Fish. Amer. Scient. *61:* 316–325 (1973).

Burckhardt, R.: Das Centralnervensystem von Protopterus annectens (Friedländer, Berlin 1892).

Burlet, H.M. de: Über die Papilla neglecta. Anat. Anz. *66:* 199–209 (1928).

Burlet, H.M. de: Zur vergleichenden Anatomie der Labyrinthinnervation. J. comp. Neurol. *47:* 155–169 (1929).

Cajal, S.R. y: Sur un noyau spécial du nerf vestibulaire des poissons et des oiseaux. Travaux Lab. Rech. biol. Univ. Madrid *6:* 1–20 (1908).

Cajal, S.R. y: Les ganglions terminaux du nerf acoustique des oiseaux. Travaux Lab. Rech. biol. Univ. Madrid *6:* 195–224 (1908).

Cajal, S.R. y: Histologie du système nerveux de l'homme et des vertébrés. 2 vols. (Maloine, Paris 1909, 1911; Instituto Ramon y Cajal, Madrid 1952, 1955).

Cajal, S.R. y: Recuerdos de mi vida, 3rd. ed. (Pueyo, Madrid 1923).

Carlsson, A.: The occurrence, distribution and physiological role of catecholamine in the nervous system. Pharmacol. Rev. *11:* 490–493 (1959).

Castaldi, L.: Studi sulla struttura e sullo sviluppo del mesencefalo. Ricerche in Cavia cobaya. Arch. ital. Anat. Embriol. *20:* 23–225 (1923); *21:* 172–263 (1924); *23:* 481–609 (1926); *25:* 157–306 (1928).

Catois, E.H.: Recherches sur l'histologie et l'anatomie microscopique de l'encéphale chez les poissons. Bull. Soc. France Belg. *36:* 1–166 (1901).

Chiarugi, G.: Lo sviluppo dei nervi vago, accessorio, ipoglosso e primi cervicali nei Sauropsidi e nei Mammiferi. Atti Soc. Toscan. Sci. nat. *10:* 149–245 (1889).

Chu, N. and Bloom, F.E.: Norepinephrine-containing neurons: change in spontaneous discharge patterns during sleeping and waking. Science *179:* 908–910 (1973).

Clara, M.: Das Nervensystem des Menschen, 3. Aufl. (Barth, Leipzig 1959).

Conel, J.L.: The development of the brain of Bdellostoma stouti. I. External growth changes. J. comp. Neurol. *47:* 343–403 (1929).

Conel, J.L.: The development of the brain in Bdellostoma stouti. II. Internal growth changes. J. comp. Neurol. *52:* 365–499 (1931).

Correia, M.J., Landolt, J. P. and Young, E. R.: The sensura neglecta in the pigeon: a scanning electron and light microscope study. J. comp. Neurol. *154:* 303–315 (1974).

Da Costa, C.: As problemas morfologicas da cabeca dos vertebrados e a formacaõ dos ganglios nos mamiferos. Arq. Anat. Anthrop. *26:* 487–523 (1923).

Craigie, E.H.: An introduction to the finer anatomy of the central nervous system based upon that of the albino Rat (Blakiston, Philadelphia 1925).

Craigie, E.H.: Observations on the brain of the humming bird *(Chrysolampis mosquitus* Linn. and *Chlorostilbon caribeus* Lawr.). J. comp. Neurol. *45:* 377–481 (1928).

Craigie, E.H.: Studies on the brain of the kiwi *(Apteryx australis)*. J. comp. Neurol. *49:* 223–357 (1930).

Crosby, E.C.; Humphrey, T., and Lauer, E.W.: Correlative anatomy of the nervous system (Macmillan, New York 1962).

Dahlström, A. and Fuxe, K.: Evidence for the existence of monoamine-containing neurons in the central nervous system. I. Demonstration of monoamines in the cell bodies of brain stem neurons. Acta physiol. scand. 62, Suppl. *232:* 5–53 (1964).

Dean, B.: On the embryology of Bdellostoma stouti. A general account of Myxinoid development from the egg and its segmentation to hatching. Festschrift *Carl von Kupffer:* 221–277 (Fischer, Jena 1899).

Dejerine, J.: Anatomie des centres nerveux, 2 vols. (Rueff, Paris 1901).

Dempster, W.: The morphology of the amphibian endolymphatic organ. J. Morphol. *50:* 71–126 (1930).

Eckardt, L.B. and Elliott, R.: The development of the motor nuclei of the hindbrain in the chick, Gallus domesticus. J. comp. Neurol. *61:* 83–99 (1935).

Ecker, A.; Wiedersheim, R. und Gaupp, E.: Anatomie des Frosches, 2 vols, 2. Aufl. (Vieweg, Braunschweig 1899).

Edinger, L.: Über das Gehirn von Myxine glutinosa. Phys. Abh. Königl. preuss. Akad. Wiss. 1906.

Edinger, L.: Vorlesungen über den Bau der nervösen Zentral-Organe des Menschen und der Tiere. I. Das Zentralnervensystem des Menschen und der Säugetiere. II. Vergleichende Anatomie des Gehirns (Vogel, Leipzig 1911, 1908).

Edinger, L.: Einführung in die Lehre vom Bau und den Verrichtungen des Nervensystems (Vogel, Leipzig 1912).

Elliott, H.Ch.: Textbook of the nervous system. A foundation for clinical neurology, 2nd ed. (Lippincott, Philadelphia 1954).

Elliott, H.Ch.: Textbook of neuroanatomy (Lippincott, Philadelphia 1963).

Escher, K.: Das Verhalten der Seitenorgane der Wirbeltiere und ihrer Nerven beim Übergang zum Landleben. Acta zool. *6:* 307–413 (1925).

Essick, C.R.: The corpus ponto-bulbare – a hitherto undescribed nuclear mass in the human hindbrain. Amer. J. Anat. *7:* 119–135 (1907).

Essick, C.R.: On the embryology of the corpus ponto-bulbare and its relation to the development of the pons. Anat. Rec. *3:* 254–257 (1912).

Essick, C.R.: The development of the nuclei pontis and the nucleus arcuatus in man. Amer. J. Anat. *13:* 25–54 (1912).

Euler, C. von and Söderberg, U.: Medullary chemosensitive receptors. J. Physiol., Lond. *118:* 545–554 (1952).

Euler, C. von and Söderberg, U.: Slow potentials in the respiratory centers. J. Physiol., Lond. *118:* 555–564 (1952).

Evans, L.Th.: Epibranchial and lateral line placodes of the cranial ganglia in the gecko, Gymnodactylus Kotschyi. J. comp. Neurol. *61:* 371–393 (1935).

Ewald, J.R.: Zur Physiologie des Labyrinths. VII. Mittheilung. Pflügers Arch. ges. Physiol. *93:* 485–500 (1903).

Ewald, J.R.: Schallbildtheorie und Erkenntnistheorie. Z. Sinnesphysiol. *53:* 213–217 (1922).

Finley, K.H. and Cobb, S.: The capillary bed of the locus caeruleus. J. comp. Neurol. *73:* 49–58 (1940).

Flatau, E. und Jacobsohn, L.: Handbuch der Anatomie und vergleichenden Anatomie des Zentralnervensystems der Säugetiere. Makroskopischer Teil (Karger, Berlin 1899).

Flechsig, P.: Zur Anatomie und Entwicklungsgeschichte der Leitungsbahnen im Grosshirn des Menschen. Arch. Anat. Physiol., anat. Abt. (1881).

Flechsig, P.: Gehirn und Seele, 2. Aufl. (Veit, Leipzig 1896).

Flechsig, P.: Anatomie des menschlichen Gehirns und Rückenmarks auf myelogenetischer Grundlage (Thieme, Leipzig 1920).

Frederikse, A.: The Lizard's brain. An investigation of the histological structure of the brain of Lacerta vivipara (Callenbach, Nijkerk 1931).

French, J.D.; Verzeano, M., and Magoun, H.W.: An extralemniscal sensory system in the brain stem. Arch. Neurol. Psychiat. *69:* 505–519 (1953).

Fuse, G.: Die innere Abteilung des Kleinhirnstiels (Meynert, IAK) und der Deiterssche Kern. Arb. hirnanatom. Inst. Zürich *6:* 29–267 (1912a).

Fuse, G.: Über die Striae am Boden des 4. Ventrikels (Bodenstriae: Striae medullares acusticae der älteren Autoren; 'Klangstab' von Bergmann). Neurol. Zbl. *31:* 403–413 (1912b).

Fuxe, K. and Hanssen, L.: Central catecholamine neurons and conditioned avoidance behavior. Psychopharmacologia *11:* 439–447 (1967).

Gagel, O. und Bodechtel, G.: Die Topik und feinere Histologie der Ganglienzellgruppen in der Medulla oblongata und im Ponsgebiet, mit einem kurzen Hinweis auf die Gliaverhältnisse und die Histopathologie. Z. Anat. EntwGesch. *91:* 130–250 (1929–1930).

Galambos, R.: Neural mechanisms of audition. Physiol. Rev. *34:* 497–528 (1954).

Gardner, W.J.: Hydrodynamic mechanisms of syringomyelia; its relation to myelocele. J. Neurol. Psychiat. *28:* 247–259 (1965).

Gauer, O.H.: Kreislauf des Blutes, in Landois-Rosemann, Lehrbuch der Physiologie des Menschen, 28. Aufl., vol. 1, pp. 65–186 (Urban & Schwarzenberg, München 1960).

Gerlach, J.: Über das Gehirn von Protopterus annectens. Ein Beitrag zur Morphologie des Dipnoerhirnes. Anat. Anz. *75:* 310–406 (1933).

Gerlach, J.: Beiträge zur vergleichenden Morphologie des Selachierhirnes. Anat. Anz. *96:* 79–165 (1947).

Getz, B. and Sirnes, T.: The localization within the dorsal motor vagal nucleus. An experimental investigation. J. comp. Neurol. *90:* 95–110 (1949).

Gillaspy, C.C.: Experimental study of the cranial motor nuclei in reptilia. J. comp. Neurol. *100:* 481–510 (1954).

Globus, J.H.: Practical neuroanatomy. A textbook and guide for the study of form and structure of the nervous system. Adapted to the needs of the student and practicing physician (Wood, Baltimore 1937).

Goodrich, E.S.: Studies on the structure and development of vertebrates, 2 vols. (Dover, New York 1958).

Goronowitsch, N.: Das Gehirn und die Cranialnerven von Acipenser ruthenus. Morph. Jb. *13:* 515–574 (1888).

Gosline, W.A.: Functional morphology and classification of Teleostean fishes. (University of Hawaii Press, Honolulu 1971).

Granit, R.: Receptors and sensory perception. The aims, means, and results of electrophysiological research on the process of reception (Yale University Press, New Haven 1956).

Granit, R.: The basis of motor control. Integrating the activity of muscles, alpha and gamma motoneurons and their leading control systems (Academic Press, New York 1970).

Groebbels, F.: Der Hypoglossus der Vögel. Zool. Jb. Anat. *43:* 465–484 (1922).

Groebbels, F.: Die untere Olive der Vögel. Anat. Anz. *56:* 296–301 (1923).

Haller v. Hallerstein, Graf V.: Äussere Gliederung des Zentralnervensystems; in Bolk *et al.*, Handbuch der vergleichenden Anatomie der Wirbeltiere, Bd. II, pp. 1–318 (Urban & Schwarzenberg, Berlin 1934).

Hamburger, V. and Levi-Montalcini, R.: Some aspects of neuroembryology; in Weiss, Genetic neurology, pp. 128–160 (University of Chicago Press, Chicago 1950).

Harkmark, W.: Cell migrations from the rhombic lip to the inferior olive, the nucleus raphe and the pons. A morphologic and experimental investigation on chick embryos. J. comp. Neurol. *100:* 115–209 (1954).

Haymaker, W.: Bing's local diagnosis in neurological diseases, 14th ed., 15th ed. (Mosby, St. Louis 1956, 1969).

Haymaker, W. and Kuhlenbeck, H.: Diseases of the brain stem and its cranial nerves; in Baker's Clinical neurology, Vol. 2, chapter 23, pp. 1260–1334; 2nd ed., Disorders of the brain stem and its cranial nerves, vol. 3, chapter 29, pp. 1456–1526; 3rd ed., chapter 30, vol. 3, pp. 1–82 (Hoeber-Harper, Harper & Row, New York 1955, 1962, 1971).

Heier, P.: Fundamental principles in the structure of the brain. A study of the brain of Petromyzon fluviatilis. Acta anat. suppl. VI (Hakan, Lund 1948).

Held, H.: Die centralen Bahnen des Nervus acusticus bei der Katze. Arch. Anat. Physiol., Anat. Abt.: 207–288 (1891).

Held, H.: Die centrale Gehörleitung. Arch. Anat. Physiol., Anat. Abt.: 201–248 (1893).

Helmholtz, H.L.F.: Sensations of tone (Longmanns, Green, London 1885).

HERRICK, C.J.: The central gustatory paths in the brain of fishes. J. comp. Neurol. *15:* 375–456 (1905).

HERRICK, C.J.: On the centers for taste and touch in the medulla oblongata of fishes. J. comp. Neurol. *16:* 403–439 (1906).

HERRICK, C.J.: A study of the vagal lobes and funicular nuclei in the brain of the codfish. J. comp. Neurol. *17:* 67–87 (1907).

HERRICK, C.J.: On the phylogenetic differentiation of the organs of smell and taste. J. comp. Neurol. *18:* 159–166 (1908).

HERRICK, C.J.: The medulla oblongata of larval Amblystoma. J. comp. Neurol. *24:* 343–427 (1914).

HERRICK, C.J.: Neurological foundations of animal behavior (Holt, New York 1924).

HERRICK, C.J.: Brain of Rats and Men (University of Chicago Press, Chicago 1926).

HERRICK, C.J.: The medulla oblongata of Necturus. J. comp. Neurol. *50:* 1–96 (1930).

HERRICK, C.J.: An introduction to neurology, 5th ed. (Saunders, Philadelphia 1931).

HERRICK, C.J.: The membranous parts of the brain, meninges and their blood vessels in Amblystoma. J. comp. Neurol. *61:* 297–346 (1935).

HERRICK, C.J.: The fasciculus solitarius and its connections in amphibians and fishes. J. comp. Neurol. *81:* 307–331 (1944).

HERRICK, C.J.: The brain of the Tiger Salamander Ambystoma tigrinum (University of Chicago Press, Chicago 1948).

HESSE, R.: Über die Abgrenzung des Gehirns. Sitz. Ber. preuss. Akad. Wiss. phys. math. Kl. 1932, IV: 23–32 (1932).

HIKIJI, K.: Zur Anatomie des Bodens der Rautengrube beim Neugeborenen. Anat. Anz. *75:* 406–442 (1933).

HINES, M.: The brain of Ornithorhynchus anatinus. Phil. Trans. roy. Soc. London Ser. 13, *217:* 155–287 (1929).

HIRASAWA, K.: Laminar structural arrangements in the central nervous system (Japanese). Kyoto Igakkai Zasshi *3:* 428–438 (1955).

HIROSE, K.: Über eine bulbospinale Bahn. Fol. neurobiol. *10:* 371–383 (1916–1917).

HOAGLAND, H.: Electrical responses from the lateral-line nerves of catfish. J. gen. Physiol. *16:* 695–714 (1933).

HOCHSTETTER, F.: Beiträge zur Entwicklungsgeschichte des menschlichen Gehirns. III. Teil (Deuticke, Wien 1929).

HOEVELL, J.J.L.D. VAN: Remarks on the reticular cells of the oblongata in different vertebrates. Proc. Soc. Sc. Kon. Akad. Wetensch. Amsterdam *13:* 1047–1065 (1911).

HOLM, J.F.: The finer anatomy of the nervous system of Myxine glutinosa. Morph. Jb. *29:* 365–401 (1901).

HOLMGREN, N.: Zur Anatomie des Gehirns von Myxine. Kungl. svensk. vet. Akad. Handl. 60, pt. 7: 1–96 (1919).

HOLMGREN, N. and VAN DER HORST, C.J.: Contributions to the morphology of the brain of Ceratodus. Acta zool. *6:* 59–165 (1925).

HOOGENBOOM, K.J.H.: Das Gehirn von Polyodon folium Lacép. Z. mikr.-anat. Forsch. *18:* 311–392 (1929).

HORST, C.J. VAN DER: Die motorischen Kerne und Bahnen in dem Gehirn der Fische, ihr taxonomischer Wert und ihre neurobiotaktische Bedeutung. Tijdschr. d. nederl. Dierk. Vereen *16:* 168–270 (1918).

HUBBARD, J.E. and DI CARLO, V.: Fluorescence histochemistry of monoamine-containing cell bodies in the brain stem of the squirrel monkey *(Saimiri sciureus)*. J. comp. Neurol. *147:* 553–565 (1973).

JACOBSOHN, L.: Über die Kerne des menschlichen Hirnstamms (Medulla Oblongata, Pons, und Pedunculus cerebri). Abh. d. K. preuss. Akad. Wiss. phys.-math. Kl., Anhang (1909).

JANSEN, J.: The brain of Myxine glutinosa. J. comp. Neurol. *49:* 359–507 (1930).

JANSEN, J. und BRODAL, A.: Das Kleinhirn. Erg. zu Bd. IV/1, Nervensystem. Handb. d. mikr. Anat. des Menschen, *Bargmann, W.* ed. (Springer, Berlin 1958).

JASPERS, H.H.; PROCTOR, L.D.; KNIGHTON, R.S.; NOSHAY, W.C., and COSTELLO, R.T.: Reticular formation of the brain (Little, Boston 1958).

JOHNSON, F.H. and RUSSELL, G.V.: The locus caeruleus as a pneumotaxic center (Abstract). Anat. Rec. *112:* 438 (1952).

JOHNSTON, J.B.: The brain of Acipenser. Zool. Jb. *15:* 59–260 (1901).

JOHNSTON, J.B.: The brain of Petromyzon. J. comp. Neurol. *12:* 87–106 (1902).

JOHNSTON, J.B.: The nervous system of vertebrates (Blakiston, Philadelphia 1906).

JOUVET, M.: Neurophysiology of the states of sleep; in QUARTON *et al.*, The neurosciences: a study program, pp. 529–544 (Rockefeller University Press, New York 1967).

JUNG, R.: Physiologie und Pathophysiologie des Schlafes. Referat, Verh. d. deutschen Gesellschaft f. innere Medizin, pp. 788–797 (Bergmann, München 1965).

KANE, E.C.: Patterns of degeneration in the caudal cochlear nucleus of the Cat after cochlear ablation. Anat. Rec. *179:* 67–91 (1974).

KAPAL, E.: Gehörsinn; in LANDOIS-ROSEMANN, Lehrbuch der Physiologie des Menschen, 28. Aufl., vol. 2, pp. 903–938 (Urban & Schwarzenberg, München 1962).

KAPPERS, C.U.A.: The structure of the Teleostean and Selachian brain. J. comp. Neurol. *16:* 1–109 (1906).

KAPPERS, C.U.A.: Untersuchungen über das Gehirn von Amia calva und Lepidosteus osseus. Abh. Senkenberg. naturf. Ges. *30:* 449–500 (1907).

KAPPERS, C.U.A.: Die vergleichende Anatomie des Nervensystems der Wirbeltiere und des Menschen, 2 vols. (Bohn, Haarlem 1920–1921).

KAPPERS, C.U.A.: The evolution of the nervous system in Invertebrates, Vertebrates, and Man (Bohn, Haarlem 1929).

KAPPERS, C.U.A.: Anatomie comparée du système nerveux particulièrement de celui des mammifères et de l'homme. Avec la collaboration de E. STRASBURGER (Masson, Paris 1947).

KAPPERS, C.U.A. und HAMMER, E.: Das Zentralnervensystem des Ochsenfrosches (Rana catesbyana). Psych. en neurol. Bl. (Feestbundel *Winkler*): 368–415 (1918).

KAPPERS, C.U.A.; HUBER, G.C., and CROSBY, E.C.: The comparative anatomy of the nervous system in vertebrates including man, 2 vols. (Macmillan, New York 1936).

KARTEN, J.H. and HODOS, W.: A stereotaxic atlas of the brain of the pigeon *(Columba livia)*. (The Johns Hopkins Press, Baltimore 1967).

KINGSBURY, B.F.: On the brain of Necturus maculatus. J. comp. Neurol. *5:* 139–205 (1895).

KÖLLIKER, A.: Handbuch der Gewebelehre des Menschen, Bd. II, Teil 1 und 2 (Engelmann, Leipzig 1893, 1896).

KOOY, F.H.: The inferior olive in Vertebrates. Fol. neurobiol. *10:* 205–269 (1916–1917).

KOSAKA, K. and HIRAIWA, K.: Über die Facialiskerne beim Huhn. Jahrb. Psychiat. *25:* 57–69 (1905).

Kosaka, K. und Yagita, K.: Experimentelle Untersuchungen über den Ursprung des N. vagus und die centrale Endigung der dem Plexus nodosus entstammenden sensiblen Vagusfasern, sowie dem Verlauf ihre sekundären Bahnen. Shinkeigaku Zasshi, Tokyo: *4:* 29–49 (1905).

Krabbe, K.H.: Studies on the morphogenesis of the brain in some Urodeles and Gymnophions ('Amphibians'). (Munksgaard, Copenhagen 1962).

Kreht, H.: Über die Faserzüge im Zentralnervensystem von *Salamandra maculosa* L. Z. mikr.- anat. Forsch. *23:* 239–320 (1930).

Kreht, H.: Über die Faserzüge im Zentralnervensystem von *Proteus anguineus* Laur. Z. mikr.-anat. Forsch. *25:* 376–427 (1931), *48:* 192–285 (1940).

Kuhlenbeck, H.: Zur Morphologie des Gymnophionengehirns. Jen. Z. f. Naturw. *58:* 453–484 (1922).

Kuhlenbeck, H.: Vorlesungen über das Zentralnervensystem der Wirbeltiere (Fischer, Jena 1927).

Kuhlenbeck, H.: Die Grundbestandteile des Endhirns im Lichte der Bauplanlehre. Anat. Anz. *67:* 1–51 (1929).

Kuhlenbeck, H.: Buchbesprechung zu *Hesse, R.:* Über die Abgrenzung des Gehirns. Anat. Anz. *75:* 124–127 (1932).

Kuhlenbeck, H.: Observations on the morphology of the rhombencephalon in the gymnophione Siphonops annulatus (Abstract). Anat. Rec. *163:* 311 (1969).

Kuhlenbeck, H.: Schopenhauers Satz 'Die Welt ist meine Vorstellung' und das Traumerlebnis. Schopenhauer Jahrb. 53 (Festschrift Hübscher): 376–392 (1972).

Kuhlenbeck, H. and Niimi, K.: Further observations on the morphology of the brain in the Holocephalian Elasmobranchs Chimaera and Callorhynchus. J. Hirnforsch. *11:* 267–314 (1969).

Lange, S.J. de: Das Hinterhirn, das Nachhirn und das Rückenmark der Reptilien. Fol. neurobiol. *10:* 385–422 (1916–1917).

Larsell, O.: The differentiation of the peripheral and central acoustic apparatus in the frog. J. comp. Neurol. *60:* 473–527 (1934).

Larsell, O.: The comparative anatomy and histology of the cerebellum from myxinoids through birds. Edited by Jansen, J. (University of Minnesota Press, Minneapolis 1967).

Larsell, O.: The comparative anatomy and histology of the cerebellum from Monotremes through Apes. Edited by Jansen, J. (University of Minnesota Press, Minneapolis 1970).

Larsell, O. and Jansen, J.: The comparative anatomy and histology of the cerebellum. The human cerebellum, cerebellar connections and cerebellar cortex (University of Minnesota Press, Minneapolis 1972).

Levi-Montalcini, R.: Origine ed evoluzione del nucleo accessorio del nervo abducente nell'embrione di pollo. Act. pontif. Sc. *6:* 335–345 (1942).

Levi-Montalcini, R.: The development of the acustico-vestibular centers in the chick embryo in the absence of the afferent root fibers and descending fiber tracts. J. comp. Neurol. *91:* 209–241 (1949).

Lewy, F.H. and Kobrak, H.: The neural projection of the cochlear spirals on the primary acoustic centers. Arch. Neurol. Psychiat. *35:* 839–852 (1936).

Loizu, L.A.: Projections of the nucleus locus caeruleus in the albino rat. Brain Res. *15:* 563–566 (1969).

LUBOSCH, W.: Vergleichend-anatomische Untersuchungen über den Ursprung und die Phylogenese des N. accessorius *Willisii*. Arch. mikr. Anat. *54:* 514–602 (1899).

MARBURG, O.: Mikroskopisch-topographischer Atlas des menschlichen Zentralnervensystems (Deuticke, Leipzig 1927).

MAURER, F.: Die Epidermis und ihre Abkömmlinge (Reinicke, Leipzig 1895).

MAXIMOW, A. A.: A text-book of histology. W. BLOOM, ed. (Saunders, Philadelphia 1930).

MAYSER, P.: Vergleichend anatomische Studien über das Gehirn der Knochenfische mit besonderer Berücksichtigung der Cyprinoiden. Z. wiss. Zool. *36:* 259–364 (1881).

MEESSEN, H. and OLSZEWSKI, J.: A cytoarchitectural atlas of the rhombencephalon of the rabbit (Karger, Basel 1949).

MEYER, A.: Critical review of the data and general methods and deductiona of modern neurology. J. comp. Neurol. *8:* 113–148; 249–313 (1898).

MIES, H.: Vestibularapparat; in LANDOIS-ROSEMANN, Lehrbuch der Physiologie des Menschen, 28. Aufl., pp. 772–785 (Urban & Schwarzenberg, München 1962).

MILLOT, J. and ANTHONY, J.: Anatomie de Latimeria chalumnae. Tome II. Système nerveux et organes des sens. (Editions du Centre National de la Recherche Scientifique, Paris 1965).

MINGAZZINI, G.: Medulla oblongata und Brücke; in MÖLLENDORFF Handbuch der mikroskopischen Anatomie des Menschen, vol. IV/1, pp. 579–643 (Springer, Berlin 1928).

MORUZZI, G. and MAGOUN, H. W.: Brain stem reticular formation and activation of the EEG. Electroenc. clin. Neurophysiol. *1:* 455–473 (1949).

NEAL, H. V. and RAND, H. W.: Comparative Anatomy (Blakiston, Philadelphia 1936).

NISHI, S.: Beiträge zur vergleichenden Anatomie der Augenmuskulatur. Arb. anat. Inst. kaiserl.-japan. Univ. Sendai *7:* 65–82 (1922).

NOBLE, G. K.: The biology of Amphibia (McGraw-Hill, New York 1931; Dover, New York 1954).

OBERSTEINER, H.: Anleitung beim Studium des Baues der nervösen Zentralorgane im gesunden und kranken Zustande, 5. Aufl. (Deuticke, Leipzig 1912).

OGAWA, T.: Mikroskopische Untersuchungen über den Verlauf der bulbären Basalstriae *(G. Fuse)* oder der bulbären Basalbündel *(K. Schaffer)*. Arb. anat. Inst. d. kaiserl. japan. Univ. Sendai *15:* 213–325 (1933).

O'LEARY, J. and COHEN, L. A.: The reticular core. Physiol. Rev. *38:* 243–276 (1937).

OLSON, L. and FUXE, K.: On projections from the locus caeruleus noradrenaline neurons: the cerebellar innervation. Brain Res. *28:* 165–171 (1971).

OLSZEWSKI, J. and BAXTER, D.: Cytoarchitecture of the human brain stem (Karger, Basel 1954).

OSBORN, H. F.: A contribution to the internal structure of the amphibian brain. J. Morphol. *2:* 51–96 (1888).

OSWALD, J.: Sleeping and waking. Physiology and psychology (Elsevier, Amsterdam 1962).

OSWALD, J.: Sleep (Penguin books, Harmondsworth 1966).

OWSJANNIKOW, P.: Das Rückenmark und das verlängerte Mark des Neunauges. Mém. Acad. imp. St. Pétersburg *14:* 1–32 (1903).

PAPEZ, J. W.: Comparative neurology (Crowell, New York 1929).

PAPEZ, J. W. and RUNDLES, W.: The dorsal trigeminal tract and the centre médian nucleus of Luys. J. nerv. ment. Dis. *85:* 505–519 (1937).

PEARSON, A.A.: The acoustico-lateral nervous system of fishes. J. comp. Neurol. *64:* 201–294 (1936).

PEARSON, R.: The Avian Brain (Academic Press, London–New York 1972).

PITTS, R.F.: The respiratory center and its descending pathways. J. comp. Neurol. *62:* 605–625 (1940).

RANSON, S.W.: The anatomy of the nervous system from the standpoint of development and function, 7th ed.; 10th ed., revised by S.L. CLARK (Saunders, Philadelphia 1943, 1959).

RASSMUSSEN, A.T. and PEYTON W.T.: The location of the lateral spinothalamic tract in the brainstem of Man. Surgery *10:* 689–710 (1941).

RASMUSSEN, A.T. and PEYTON, W.T.: Origin of the ventral external arcuate fibers and their continuity with the striae medullares of the fourth ventricle of Man. J. comp. Neurol. *84:* 325–337 (1946).

RASMUSSEN, G.L.: The olivary peduncle and other fiber projections of the superior olivary complex. J. comp. Neurol. *84:* 141–219 (1946).

RASMUSSEN, G.L.: Further observations on the termination of the so-called olivary peduncle (Abstract). Anat. Rec. *106:* 235 (1950).

RASMUSSEN, G.L.: Further observations on the efferent cochlear bundle. J. comp. Neurol. *99:* 61–74 (1953).

RETZIUS, G.: Zur Kenntnis des inneren Gehörorgans der Wirbeltiere. Arch. Anat. Physiol. anat. Abt. 1880: 235–244 (1880).

RETZIUS, G.: Das Gehirn und das Auge von Myxine. Biol. Unters. N. F. *5:* 55–68 (1893).

RETZIUS, G.: Das Menschenhirn. Studien in der makroskopischen Morphologie (Fischer, Jena 1896).

RILEY, H.A.: An atlas of the basal ganglia, brain stem, and spinal cord (Williams & Wilkins, Baltimore 1943).

ROGERS, K.T.: Development of the XIth or spinal accessory nerve in the chick. With some notes on the hypoglossal and upper cervical nerves. J. comp. Neurol. *125:* 273–286 (1965).

ROMANOFF, A.L.: The Avian Embryo. Structural and Functional Development. (Macmillan, New York 1960).

ROMER, A.S.: The vertebrate body (Saunders, Philadelphia 1950).

ROMMEL, S.A. Jr. and MCCLEAVE, J.D.: Oceanic electric fields: perception by American Eels? Science *176:* 1233–1235 (1972).

ROOFE, P.G.: The endocranial blood vessels of Amblystoma tigrinum. J. comp. Neurol. *61:* 257–293 (1935).

ROSS, D.M.: The sense organs of Myxine glutinosa L.; in BRODAL and FÄNGE, The biology of Myxine, pp. 150–160 (Universitetsforlaget, Oslo 1963).

ROSSI, G.F. and ZANCHETTI, A.: The brain stem reticular formation. Anatomy and physiology. Arch. ital. Biol. *95:* 199–435 (1957).

RÖTHIG, P. and KAPPERS, C.U.A.: Further contribution to our knowledge of the brain of Myxine glutinosa. Proc. Kon. ned. Akad. Wet. *17:* 2–12 (1914).

RÖTHIG, P.: Beiträge zum Studium des Zentralnervensystems der Wirbeltiere. 11. Über die Faserzüge im Mittelhirn, Kleinhirn und der Medulla oblongata der Urodelen und Anuren. Z. mikr.-anat. Forsch. *10:* 381–472 (1927).

ROVAINEN, C.M.: Physiological and anatomical studies on large neurons of central nervous system of the sea lamprey (Petromyzon marinus). I. Müller and Mauthner cells.

II. Dorsal cells and giant interneurons. J. Neurophysiol. *30:* 1000–1023; 1024–1042 (1967).

RUBIN, M.A.: Thermal reception in fishes. J. gen. Physiol. *18:* 643–647 (1933).

SAITO, T.: Über die Müllerschen Zellen im Gehirn des japanischen Flussneunauges (Entosphenus japonicus Martens). Fol. anat. japon. *6:* 458–473 (1928).

SAITO, T.: Über das Gehirn des japanischen Flussneunauges (Entosphenus japonicus Martens). Fol. anat. japon. *8:* 189–263 (1930).

SAITO, T.: Über die retikulären Zellen im Gehirn des japanischen Dornhaies (Acanthias mitsukurii Jordan et Fowler). Fol. anat. japon. *8:* 323–343 (1930).

SANDERS, E.B.: A consideration of certain bulbar, midbrain and cerebellar centers and fiber tracts in birds. J. comp. Neurol. *49:* 155–222 (1929).

SARASIN, P. and SARASIN, F.: Über das Gehörorgan der Caeciliden. Anat. Anz. *7:* 812–815 (1892).

SCHILLING, K.: Über das Gehirn von Petromyzon fluviatilis. Abh. Senckenberg naturf. Ges. *30:* 423–446 (1907).

SCHNEIDER, R.: Morphologische Untersuchungen am Gehirn der Chiropteren (Mammalia). Abh. Senckenberg. naturf. Ges. *495:* 1–92 (1957).

SCHÜTZ, H.: Anatomische Untersuchungen über den Faserverlauf im zentralen Höhlengran und den Nervenfaserschwund in demselben bei der progressiven Paralyse der Irren. Arch. Psychiat. Nervenkr. *22:* 527–587 (1891).

SCHWARTZKOPFF, J.: Structure and function of the ear and of the auditory brain areas in birds. Pp. 41–63 in DE REUK, A.V.S. and KNIGHT, J., Hearing mechanisms in Vertebrates, Ciba Symposium (Little, Brown and Co., Boston 1968).

SIGGINS, G.R.; HOFFER, B.I.; OLIVER, A.P., and BLOOM, F.E.: Activation of a central noradrenergic projection to cerebellum. Nature *233:* 481–483 (1971).

SINN, R.: Beitrag zur Kenntnis der Medulla oblongata der Vögel. Monatschr. Psychiat. Neurol. *33:* 1–39 (1913).

SPIEGEL, E.A.: Zur Physiologie und Pathologie des Skelettmuskeltonus, 2. Aufl., Der Tonus der Skelettmuskulatur (Springer, Berlin 1923, 1927).

SPIEGEL, E.A.: Tonus; in BETHE; BERGMANN; EMBDEN und ELLINGER, Handbuch der normalen und pathologischen Physiologie, Bd. 9: Allgemeine Physiologie der Nerven und des Zentralnervensystems, pp. 711–740 (Springer, Berlin 1929).

SPIEGEL, E.A.: Labyrinth and cortex. The electroencephalogram of the cortex in the stimulation of the labyrinth. Arch. Neurol. Psychiat. *31:* 469–482 (1934).

SPIEGEL, E.A. und INABA, C.: Experimentalstudien am Nervensystem. Zur zentralen Lokalisation von Störungen des Wachzustandes. Z. ges. exp. Med. *55:* 164–184 (1927).

SPIEGEL, E.A. and SOMMER, I.: Neurology of the eye, ear, nose and throat (Grune & Stratton, New York 1944).

STADERINI, R.: Nucleus praepositus nervi hypoglossi e nucleo intercalato. Anat. Anz. *87:* 101–105 (1938).

STEFANELLI, A.: Numero, grandezza e forma di alcuni peculiari elementi nervosi dei Petromyzonti. Z. Zellforsch. *18:* 146–165 (1933a).

STEFANELLI, A.: Le cellule e le fibre di Müller dei Petromyzonti. Arch. ital. Anat. Embriol. *31:* 519–548 (1933b).

STEFANELLI, A.: I centri tegmentali dell'encefalo dei Petromizonti. Arch. zool. ital. *20:* 117–202 (1934).

STEFANELLI, A.: Neurologia ecologica de los centros estáticos de los vertebrados. Biologica (Facultad de Medicina de la Universidad de Chile) *33:* 49–61 (1962).

STEINER, I.: Die Functionen des Centralnervensystems und ihre Phylogenese. I. Untersuchungen über die Physiologie des Froschhirns. II. Die Fische. III. Die wirbellosen Thiere. IV. Reptilien, Rückenmarksreflexe, Vermischtes (Vieweg, Braunschweig 1885, 1888, 1898, 1900).

STELLA, G.: On the mechanisms of production, and the physiological significance of 'apneusis'. J. Physiol. *93:* 10–23 (1938).

STINGELIN, W.: Qualitative und quantitative Untersuchungen an Kerngebieten der Medulla oblongata bei Vögeln. Bibl. anat., Nr. 6 (Karger, Basel 1965).

STROMINGER, N.L.: The origins, course and distribution of the dorsal and intermediate acoustic striae in the Rhesus monkey. J. comp. Neurol. *147:* 209–233 (1973).

SZABO, TH.: Anatomo-physiologie des centres nerveux spécifiques de quelques organes électriques; in CHAGAS and PAEZ DE CARVALHO, Bioelectrogenesis, pp. 185–201 (Elsevier, Amsterdam 1961).

SZENTÁGOTHAI, J.: Die Lokalisation der Kehlkopfmuskulatur in den Vaguskernen. Z. Anat. EntwGesch. *112:* 704–710 (1943).

SZENTÁGOTHAI, J.: The representation of facial and scalp muscles in the facial nucleus. J. comp. Neurol. *88:* 207–220 (1948).

SZENTÁGOTHAI, J.: Functional representation in the motor trigeminal nucleus. J. comp. Neurol. *90:* 111–120 (1949).

TABER PIERCE, E.: Histogenesis of the nucleus griseum pontis, corporis pontobulbaris and reticularis tegmenti pontis *(Bechterew)* in the mouse. An autoradiographic study. J. comp. Neurol. *126:* 219–239 (1966).

TAGLIANI, G.: Le fibre del *Mauthner* nel midollo spinale dei vertebrati inferiori. Arch. zool. ital. *2:* 386–437 (1906).

TANDLER, J.: Lehrbuch der systematischen Anatomie, 4. Bd.: Nervensystem und Sinnesorgane (Vogel, Leipzig 1929).

TELLO, F.: Contribución al conocimiento del encephalo de los teleósteos. I. Los nucleos bulbares. Trabajos Lab. Invest. biol. Univ. Madrid *7:* 1–29 (1909).

TILNEY, F. and RILEY, H.A.: The form and functions of the central nervous system, 3rd ed. (Hoeber, New York 1938).

TRETJAKOFF, D.: Das Nervensystem von Ammocoetes. II. Das Gehirn. Arch. mikr. Anat. *74:* 636–779 (1909).

TUGE (TSUGE), H.: Somatic motor mechanisms in the midbrain and medulla oblongata of Chrysemys elegans (Wied). J. comp. Neurol. *55:* 185–271 (1932).

UNGERSTEDT, U.: Stereotaxic mapping of the monoamine pathways in the rat brain. Acta Physiol. Scand. Suppl. *367:* 1–48 (1971).

VALDMAN, A.V. (ed.): Pharmacology and physiology of the reticular formation. Progr. Brain Res., vol. 20 (Elsevier, Amsterdam 1967).

VERHAART, W.J.C.: Comparative anatomical aspects of the mammalian brain stem and the cord, 2 vols. I text, II illustrations and tables (van Gorcum, Assen 1970).

VILLIGER, E.: Gehirn und Rückenmark. Leitfaden für das Studium der Morphologie und des Faserverlaufs, 7. Aufl. (Engelmann, Leipzig 1920).

VOGT, M.: The concentration of sympathin in different parts of the central nervous system under normal conditions and after the administration of drugs. J. Physiol. *123:* 451–481 (1954).

VOGT, M.: Catecholamines in brain. Pharmacol. Rev. *11:* 483–489 (1959).

VOGT-NILSEN, L.: The inferior olive in birds. J. comp. Neurol. *101:* 447–481 (1954).

WALLENBERG, A.: Die secundäre Acusticusbahn der Taube. Anat. Anz. *14:* 353–369 (1898).

WALLENBERG, A.: Untersuchungen über das Gehirn der Tauben. Anat. Anz. *15:* 245–271 (1899).

WALLENBERG, A.: Über centrale Endstätten des Nervus octavus der Taube. Anat. Anz. *17:* 102–108 (1900).

WALLENBERG, A.: Eine zentrifugal leitende direkte Verbindung der frontalen Vorderhirnbasis mit der Oblongata (Rückenmark?) bei der Ente. Anat. Anz. *22:* 289–292 (1902).

WALLENBERG, A.: Der Ursprung des Tractus isthmo-striaticus oder bulbo-striaticus der Taube. Neurol. Centralbl. *22:* 98–101 (1903–1904).

WALLENBERG, A.: Neue Untersuchungen über den Hirnstamm der Taube. Anat. Anz. *24:* 357–369 (1904).

WHITESIDE, B.: The development of saccus endolymphaticus in *Rana temporaria* L. Amer. J. Anat. *30:* 231–266 (1922).

WHITING, H.P.: Mauthner neurons in young larval lampreys. Quart. J. micr. Sci. *98:* 163–178 (1957).

WILTSCHKO, W. and WILTSCHKO, R.: Magnetic compass of European Robins. Science *176:* 62–64 (1972).

WINDLE, W.F.: Non-bifurcating nerve fibers of the trigeminal nerve. J. comp. Neurol. *40:* 229–240 (1926).

WINDLE, W.F. and ORR, D.W.: The development of behavior in chick embryos: spinal cord structure correlated with early somatic motility. J. comp. Neurol. *60:* 287–307 (1934).

WINTER, P.: Vergleichende qualitative und quantitative Untersuchungen an der Hörbahn von Vögeln. Z. Morphol. Oekol. Tiere *52:* 365–400 (1963).

WINTER, P. und SCHWARTZKOPFF, J.: Form und Zellzahl der akustischen Nervenzentren in der Medulla oblongata von Eulen (Strigen). Experientia *17:* 515–516 (1961).

WORTHINGTON, J.: The descriptive anatomy of the brain and the cranial nerves of Bdellostoma dombeyi. Quart. J. micr. Sci. *49:* 137–182 (1905).

YOSHIDA, I.: Über die funktionelle Bedeutung der oberen Olive nebst ihren Faserbahnen. Fol. anat. japon. *3:* 111–136 (1925).

YOUNG, J.Z.: The life of Vertebrates (Clarendon Press, Oxford 1952, 1955).

ZEMAN, W. and INNES, J.R.M.: *Craigie's* neuroanatomy of the Rat, revised and expanded (Academic Press, New York 1963).

ZIEHEN, TH.: Das Zentralnervensystem der Monotremen und Marsupialier. I. Teil. Semons Zoologische Forschungsreisen in Australien und dem Malayischen Archipel, Bd. 3/I, pp. 1–188 (Fischer, Jena 1897).

ZIEHEN, TH.: Anatomie des Centralnervensystems. 1. Abtg. I, II. 2. Abtg. I–IV. *Bardeleben's* Handbuch der Anatomie des Menschen. Vierter Band (Fischer, Jena 1899, 1903, 1913, 1920, 1926, 1934).

ZIEHEN, TH.: Das Zentralnervensystem der Monotremen und Marsupialier. *Semons* Forschungsreisen in Australien und dem Malayischen Archipel, Bd. 3/II, pp. 789–921 (Fischer, Jena 1908).

ZWISLOCKI, J.J. and SOKOLICH, W.G.: Velocity and displacement responses in auditory nerve fibers. Science *182:* 64–66 (1973).

X. The Cerebellum

1. General Pattern and Basic Mechanisms

The Vertebrate cerebellum develops ontogenetically, and likewise can be regarded as phylogenetically evolved, from the dorsal zone of the rhombencephalic alar plate's rostral portion. These antimeric alar plate components, thus originally representing a bilateral anlage, fuse in the dorsal midline, thereby eliminating or 'absorbing' the epithelial roof plate. At subsequent significant ontogenetic stages in forms with a complex cerebellum, and in the adult state of those with a simple cerebellum, this latter then displays an *unpaired median subdivision*, the primordial *corpus cerebelli* or *transverse cerebellar plate*, which may or may not still retain a bilaterally symmetric pattern, and a *paired lateral auricular subdivision*, corresponding to the lateral recess of the fourth ventricle[1] (cf. Figs. 292–294). The '*auricular lobes*' *(auriculae cerebelli)* and the primordial '*corpus cerebelli*' have been especially pointed out by LARSELL (1967 and numerous other publications) as the basic morphologic components of the Vertebrate cerebellum, whose differentiation and complexity are not directly correlated with the organism's status in the ascending taxonomic and presumptive phylogenetic scale (BECCARI, 1943).

Petromyzont Cyclostomes indeed display a very simple cerebellar plate flanked by the two moderately developed auricles, while in gnathostome fishes such as Selachians and Teleosts, the cerebellar plate becomes greatly thickened and folded.[2] In Ganoids and Dipnoans the expansion of corpus cerebelli is, to various degrees, much less pronounced. Amphibians quite generally possess a very simple cerebellum consisting of auricles and a rather plain cerebellar plate (corpus). The cerebellum also remains poorly or at best very moderately developed

[1] Cf. vol. 3, part II, pp. 183–185, 373–375.

[2] This folding represents inversion, or alternating inversions and eversions in the longitudinal plane. Inversion can be combined with invagination into the mesencephalic ventricle (Teleosts). An everted cerebellar plate is present in Amphibians and some Reptilians.

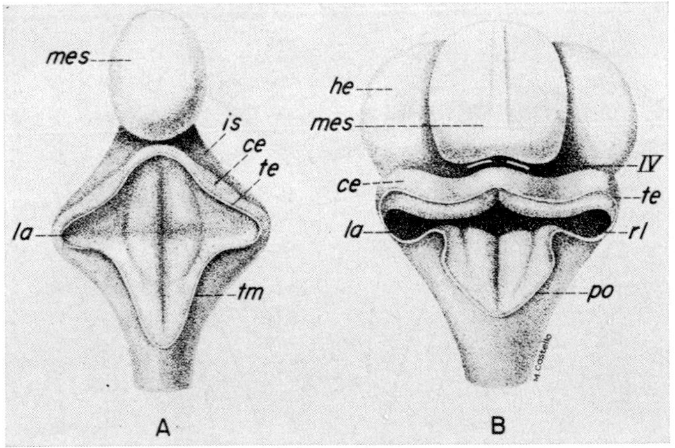

Figure 292. Dorsal view of rhombencephalon and mesencephalon in human embryos of about 14 mm (A) and 27 mm (B) CR length, showing cerebellar anlage. The roof plate of the fourth ventricle has been removed (modified after HOCHSTETTER, 1929). ce: cerebellar plate; he: cerebral hemisphere; is: isthmus rhombencephali; la: lateral recess of fourth ventricle; po: anlage of ponticulus; rl: rhombic lip; te: taenia cerebelli; tm: taenia medullae oblongatae; IV: nervus trochlearis.

in the Amniote Reptilia, becoming again greatly expanded in Birds and Mammals.[3]

From a viewpoint stressing the doctrine of functional nerve components, the cerebellum represents a particular differentiation of the *alar plate's dorsal zone* which includes the *octavolateral* (special somatic afferent) and the *trigeminal* (general somatic afferent) subzones. The former subzone extends mainly into the auricular portion and the latter into the primordial corpus, but this relationship, becoming somewhat blurred, cannot be considered as defined by a clear-cut, linear demarcation.

Elaborating on views emphasizing the significance of the octavolateral component, as e.g. propounded by JOHNSTON (1906), HERRICK

[3] According to BECCARI (1943): '*Particolarmente nei Selaci, pesce dotati di grande mezzi di locomozione, e nei Pesci in genere, il cerveletto raggiunge uno sviluppo di gran lunga superiore a quello che si osserva in forme zoologicamente più elevate, quali gli Anfibi ed i Rettili squamati. Cià si attribuisce alla sua importanza funzionale, che si ritiene proporzionata allo sviluppo delle masse muscolari dalle quali dipendono i movimenti del corpo in costante equilibrio nell'ambiente*'. One might here add that 'functional' aspects rather than those of actual 'muscular mass' presumably represent the relevant parameters.

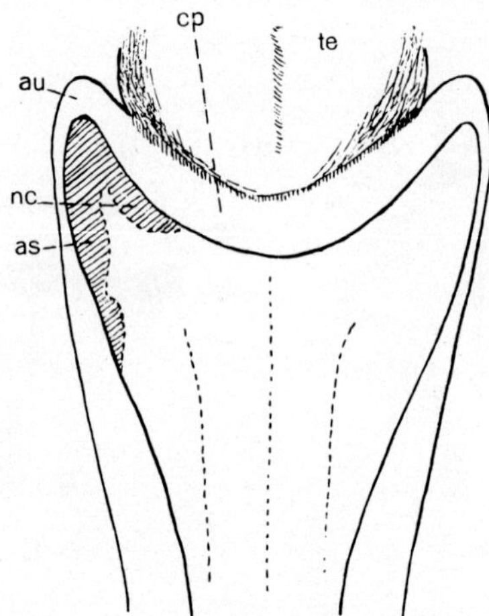

Figure 293. Semidiagrammatic sketch of the cerebellum of the urodele Amphibian Necturus (modified after HERRICK, from K., 1927). as: area statica; au: auricula cerebelli; cp: corpus cerebelli; nc: nucleus cerebelli; te: tectum mesencephali (s. tectum opticum).

(1914, 1931) and others, the comprehensive investigations of LARSELL documented the concepts of a *lobus auricularis* or *archicerebellum*, essentially related to the special somatic afferent (or octavolateralis) system, and of a primordial *corpus cerebelli (palaeocerebellum)* essentially related to the general somatic afferent system. In Mammals, a part of the corpus cerebelli receiving neocortical input by way of the pontine nuclei becomes the *neocerebellum* in EDINGER's sense. Since the most conspicuous input to the cerebellum may be classified as proprioceptive, SHERRINGTON (1906, 1947) concluded that 'the cerebellum is the *head ganglion of the proprioceptive system*.'

The foregoing simplified formulations can be regarded as providing an useful first approximation which requires a number of further qualifications. There is little doubt that, even in many or most Anamnia, and particularly in Amniota, the cerebellum, especially its corpus, receives some input from practically all sensory (or afferent) systems, including optic and presumably visceral ones.[4] Moreover, vestibular (or octavolateral) input is not exclusively restricted to the lateral auricular portion, but may also reach portions of the corpus. In Anam-

[4] Optic input seems to be mediated by tecto-cerebellar connections (cf. also K., 1927, p. 172).

nia, this octavolateral input to the corpus seems to involve a perhaps narrow caudal (posterior) region which, in the course of phylogenetic evolution, appears to become the so-called nodulus, delimited from the remainder of the corpus (or corpus cerebelli *sensu strictiori)* by a morphologically significant transverse groove, the *posterolateral fissure*. The nodulus remains lateralward continuous with the auricular derivative. The thus resulting (vestibular) *flocculo-nodular lobe* of Birds and Mammals represents LARSELL's *archicerebellum*, the flocculus of these Amniota approximately corresponding to the rostral auricular lip of Anamnia. It should be recalled that, in all Amniota, the lateralis system, which is still to some extent present in the tetrapod Amphibia (cf. the preceding chapter IX), has completely disappeared. In addition to the flocculo-nodular lobe, a rostral component *(lingula)* of the

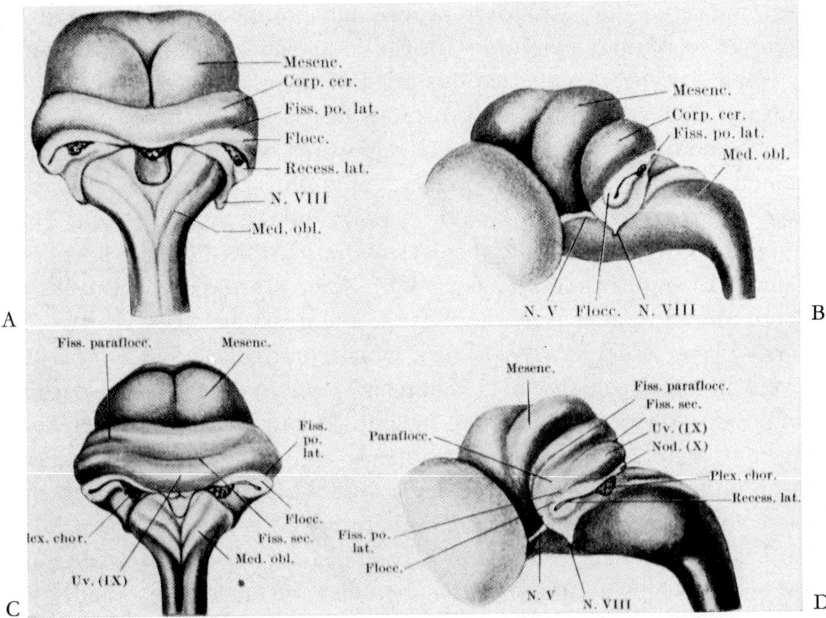

Figure 294. Aspects of the brain in embryos of the Chiropteran Mammal *Corynorhinus sp.*, showing gross features of the developing cerebellum in caudodorsal and lateral view (modified after LARSELL and DOW, from JANSEN and BRODAL, 1958). A, B Embryo of 12 mm length. C, D Embryo of 16 mm length. Fiss. paraflocc.: fissura parafloccularis; Fiss.post.lat.: fissura posterolateralis; Fiss.secunda: fissura secunda; Nod. (X).: nodulus ('folium X'); Paraflocc.: 'paraflocculus'; Uv (IX): uvula ('folium IX'); other abbr. self-explanatory.

mammalian and presumably also avian cerebellar corpus seems to receive a vestibular projection.

Again, if the term '*neocerebellum*' is not restricted to a cerebellar region under specifically neocortical influence, but taken to include an influence exerted by other telencephalic connections and mediated by '*pontine nuclei*', then it becomes evident that a definite 'neocerebellum' is also present in Birds. The BNA and PNA term 'metencephalon', subsuming cerebellum and pons, applies, *sensu strictiori*, only to Mammals and, with some reservations, to Birds. In Reptiles and Anamnia, it can, however, be used as merely another designation for cerebellum. More recently, BRODAL (1972) has subdivided, on the basis of its afferent connections, the Mammalian cerebellum into (1) *vestibulocerebellum*, (2) *spinocerebellum*, and (3) *pontocerebellum*. These three subdivisions do not exactly correspond to flocculonodular lobe, vermis, and hemispheres, but display substantial overlap. With some qualifications, the concepts vestibulocerebellum, spinocerebellum and pontocerebellum can also be used, in BRODAL's opinion, for the cerebellar output. This author, in particular, stresses the various possible channels between cerebellum and cerebral cortex ('cerebro-cerebellar pathways').

The neuronal elements respectively grisea of the Vertebrate cerebellum comprise the *cerebellar cortex*, commonly evaluated as a '*suprasegmental structure*' (cf. e.g. HERRICK, 1931) and the *cerebellar nuclei*. The cerebellar cortex[5] generally displays a stratification into three layers, namely an outer molecular layer with some scattered nerve cells, a dense inner granular layer, and, at the boundary between the two, a narrow layer of large *Purkinje cells*, which, particularly but not exclusively in Sauropsidans and Mammals, tend to form a single row (Fig. 295). In lower Vertebrates, including many Reptiles, the cerebellar input and output fibers run merely through granular and molecular layer, an inner medullary layer being absent. The cerebellar cortex in these forms thus actually represents an ependymal 'praecortex'.[6] In Birds and Mammals, however, the cerebellar cortex is separated from the ependyma by a rather massive medullary layer, which contains, in

[5] The cerebellar cortex with its intracortical connections could be, loosely speaking, considered to represent the *eigenapparat* of the cerebellum, but the distinction of *eigenapparat* and *verbindungsapparat*, significant in the discussion of segmental portions of the neuraxis, does not seem particularly helpful in dealing with the cerebellum (cf. also K., 1927, p. 172).

[6] Cf. vol. section 6, p. 615 of chapter VI in the preceding volume 3, part II. Some Reptiles, however, such as the Alligator, display a medullary layer of the cerebellar cortex.

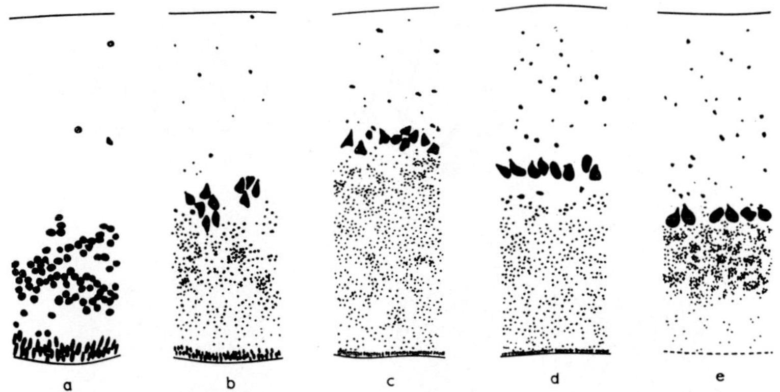

Figure 295A. Diagrams purporting to illustrate layers of cerebellar cortex in representative Vertebrates (from NIEUWENHUYS, 1967). a: lamprey; b: 'lungfish'; c: turtle; d: lizard; e: pigeon.

addition to afferent and efferent fibers as well as to some intrinsic ones, the cerebellar nuclei.

Beginning about 1837 with PURKINJE[7] many authors have investigated the histologic structure of the cerebellar cortex, whose features are perhaps best elucidated for the Mammalian and Avian cerebellum on the basis of the fundamental contributions initiated about 1888 by S. R. Y CAJAL (1911). Subsequent electron-microscopic studies, as e.g. by NIKLOWITZ (1962), FOX *et al.* (1967). MUGNAINI (1969, 1972) and others (cf. Fig. 299), have added numerous interesting ultrastructural details, which, however, can be considered as of secondary significance in the aspect here under consideration, since these observations did not, so far, alter the well-established fundamental concepts elaborated by CAJAL.

Generally speaking, the cerebellar cortex differs from that of the cerebral hemispheres not only by a much less complex structural organization but also by an essentially uniform regional overall cytoarchitectural structure, which, moreover, at least *qua* basic lamination, is manifested throughout the Vertebrate phylum. Yet, particularly with respect to Mammals, slight regional structural differences, which might be functionally significant, have been pointed out by various authors (e.g. JANSEN and BRODAL, 1958; BRODAL, 1967; VOOGD, 1967).

[7] Cf. vol. 3, part I, pp. 502–503, with Figure 300 on p. 504.

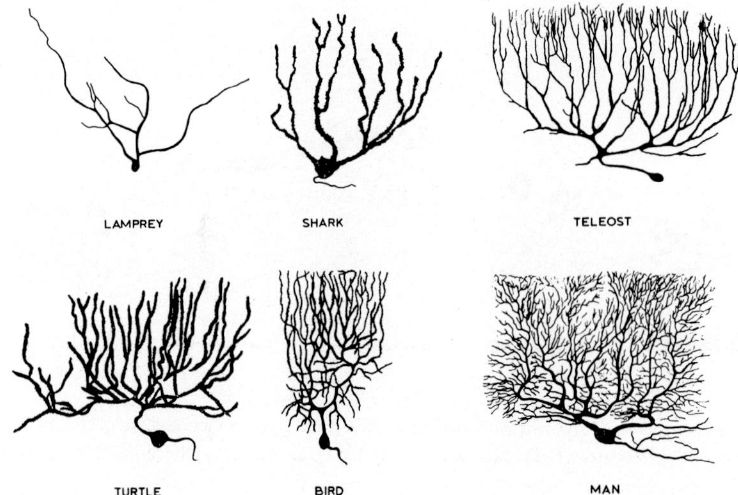

Figure 295B. Purkinje cells of representative Vertebrates, as seen by means of the *Golgi impregnation* (after CAJAL, JOHNSTON, SCHAPER, LARSELL, and KÖLLIKER, from NIEUWENHUYS, 1967). A: JOHNSTON (1902); B: SCHAPER (1898); C: SCHAPER (1893); D: LARSELL (1932); E: CAJAL (1911); F: KÖLLIKER (1896).

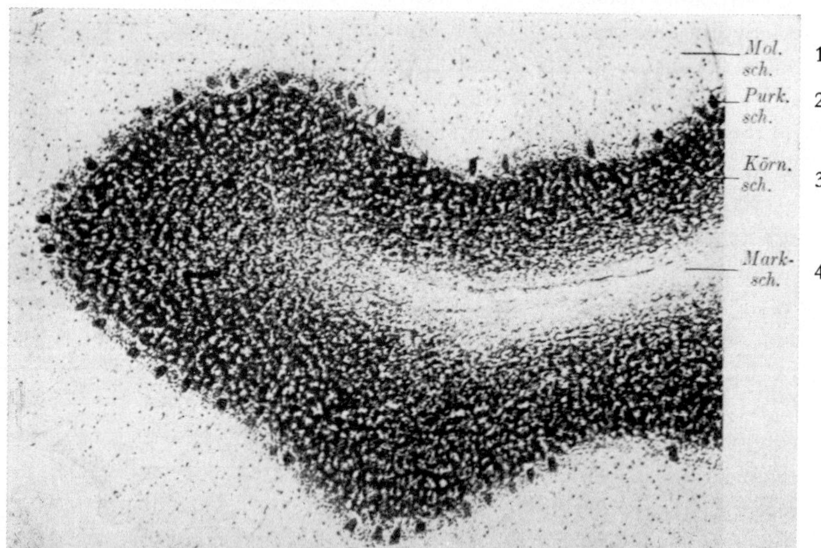

Figure 296A. Layers of the adult human cerebellar cortex as seen in a section at right angle to the long axis of a subfolium, and by means of modified *Nissl stain* (from JAKOB, 1928; ×70; red. $^2/_3$). 1: molecular layer; 2: layer of *Purkinje cells;* 3: granular layer; 4: medullary core. The irregular gaps in the *Purkinje cell* layer, presumably caused by the normal cell loss of aging, will be noted.

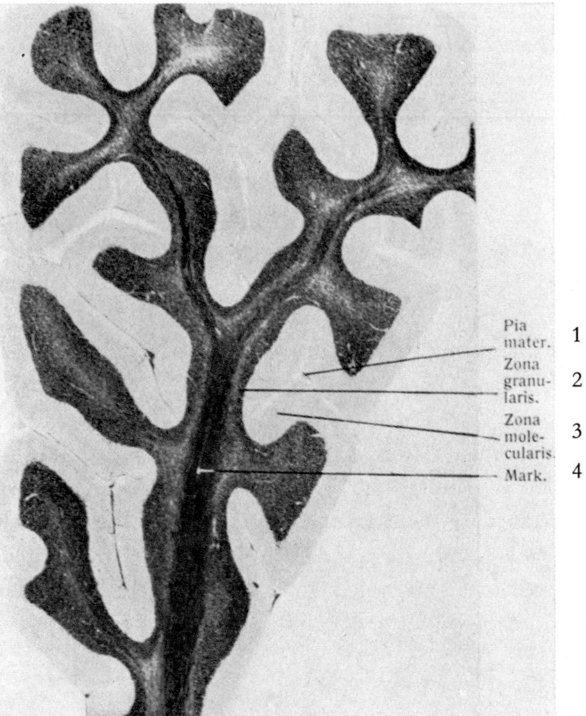

Figure 296B. Section at right angle to the long axis of a human cerebellar folium, showing layers of cerebellar cortex as seen in a *Weigert stain* (from Stöhr-Möllendorff 1933; ×6; red. $^2/_3$). 1: leptomeninx (pia mater); 2: granular layer; 3: molecular layer; 4: medullary core. *The Purkinje cells* cannot be distinguished at this low magnification.

The surface of the Mammalian cerebellum and to some extent also that of Birds, shows numerous transverse sulci which separate the narrow parallel convolutions called *folia cerebelli.*[8] A section in a sagittal plane, taken at a right angle to a transverse folium, displays a central medullary core *(lamina medullaris)* surrounded or rather covered by the cortical griseum (Fig. 296).

The large *Purkinje cells* (Figs. 295B, 297A), whose number in Man is estimated at about 15×10^6 or more, are flask-, beet-, or pear-shaped, being arrayed in a single, moderately spaced row at the bound-

[8] The transverse sulci, respectively fissures, are fairly deep furrows of the cerebellar surface. Together with short side branches, they produce, in sagittal gross sections of the cerebellum, the leaf-like pattern known as *arbor vitae.*

ary of granular and molecular layers. The *Purkinje cell* neurites originate at the base of their perikarya and represent the output channels of the cerebellar cortex. They become medullated close to their origin, giving off recurrent collaterals which reach neighboring *Purkinje cells*, apparently also their own cell of origin, and perhaps further neuronal elements of the granular or molecular layer.[9] At the top (or peripheral extremity) of a *Purkinje cell*, one to three main dendrites extend throughout the width of the molecular layer, branching repeatedly into dense arborizations, studded with spines.[10] This branching takes place in a fan-like pattern, comparable to a *trellis-work*, whose plane is oriented at a right angle to the transverse long axis of the folium (cf. Figs. 297B, 298A–D).

The *granular layer* consists mainly of very densely packed small neuronal elements designated as granule cells, whose perikarya are almost entirely filled by their nuclei. Several (about 4 to 7) short, very thin dendrites with claw-like terminals radiate in various directions. At the peripheral pole, each granule cell gives off a nonmedullated neurite, which ascends into the molecular layer and there bifurcates in T-fashion, both branches taking a course parallel to the long axis of the fol-

---

[9] Various authors (cf. e.g. JAKOB, 1928) describe three plexuses related to the *Purkinje cells*, and formed by their recurrent collaterals, which, at least near their origin, generally begin as medullated fibers. Plexus infraganglionaris runs along the base of the *Purkinje cells*, plexus supraganglionaris is related to the apex along the origin of main dendrites, and plexus superior s. intradendriticus is said to pass along the branching dendrites in the molecular layer. These plexuses, however, especially the two more peripheral ones, doubtless contain fibers of other provenance, to be dealt with further below, namely axons of basket cells, of stellate cells, and perhaps also branches of climbing fibers.

[10] Cf. volume 3, part I, pp. 70, 468–469, 624, with Figure 352. Several aspects of *Purkinje cells* (mitochondria, centrosomes, synaptic connections) are depicted in Figures 78, 87, 190, 191 of the cited volume. Investigators using electron microscopy have recently described a *Purkinje cell* cilium protruding into the neuropil. 'Whether all or only some *Purkinje cells* posses a cilium remains undetermined' (DEL CERRO *et al.*, 1969). Fluorescence microscopy is said to disclose intensively fluorescent granules, perhaps lysosomes, throughout the cytoplasm (SARNAT, 1968, in Mouse and Cat).

With respect to the dendritic 'spines' FOX *et al.* (1957, 1967) claim that the central portions of the *Purkinje cells*, as far as the primary, secondary and even the beginning of some of the tertiary branches, are 'smooth' (i.e. without 'spines'), while the peripheral branches are 'spiny'. The climbing fibers and the basket terminations would thus have contacts with the 'smooth' *Purkinje cell* surfaces, while the parallel fibers (granule cell neurites) would connect with the 'spiny branches'. The cited authors also described internal structural differences between 'spiny branchlets' and 'smooth branches', involving, *inter alia*, the arrangement of mitochondria.

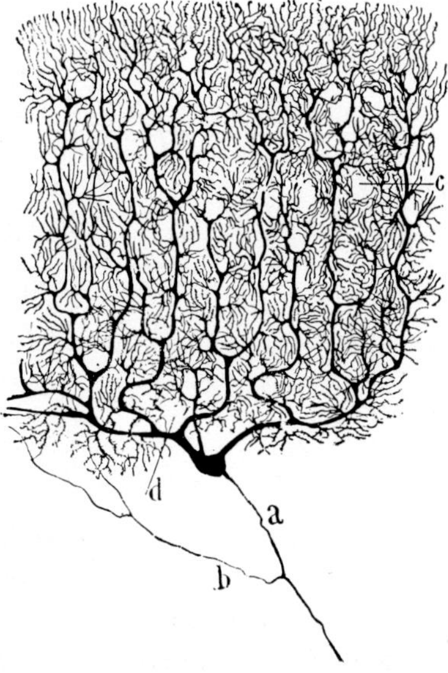

Figure 297A. Golgi impregnation of *Purkinje cell* in the human cerebellum, showing recurrent collateral, branches of which seem to reach dendrites of their cell of origin (from CAJAL, 1933). a: neurite; b: collateral; d: expansion of dendritic tree.

ium and thus at a right angle to the trellis-plane of the *Purkinje cell* dendrites (cf. Figs. 297B, 298). The extremely large number of small granule cells, estimated at between 10 to 100 billion (10^{10} to 10^{11}) in adult Man,[11] is related to a cellular density of the granular layer which perhaps exceeds that of any other neuraxial griseum.[12] It is thus possible that the number of small cerebellar granule cells not only equals but significantly surpasses that of the cortical neurons in the entire human cerebral cortex, variously estimated at between 9 to 18 billion (9×10^9 to 1.8×10^{10}).

[11] Cf. BLINKOV and GLEZER (1968).

[12] The greatest density of cellular packing in the human and mammalian body is perhaps displayed by the blood cells consisting mainly of erythrocytes (somewhat more than 4×10^6 per mm^3 while the primate cerebellar granule cells might average 2.4×10^6 per mm^3). Thus, if the total number of cells in the human body is very conservatively estimated at about 32×10^{12}, then approximately 25×10^{12} would be blood cells and about 7×10^{12} tissue cells, the total number of nerve cells in the brain amounting to perhaps 1.5×10^{10} or more. If the high figures for cerebellar granule cells and cortical nerve cells mentioned in the text are accepted, these estimates might have to be revised upwards. All such estimates are, of course, both inaccurate and inexact, but the cited figures can be regarded as useful first approximations.

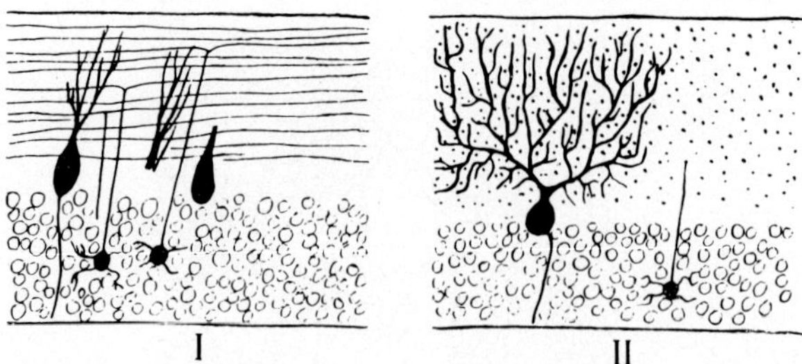

Figure 297B. Diagram illustrating spatial relationships of *Purkinje cells* to granule cell T-axons (from K., 1927). I Plane parallel to long axis of convolution. II Plane at right angle to long axis of convolution.

Mostly in the peripheral stratum of the granular layer, close to or even within the *Purkinje cell* row, scattered rather large *Golgi II* type cells ('large granular cells', 'large stellate cells', '*grandes cellules étoilées*' of CAJAL) are present. Their dendrites extend into the molecular layer, while the axon, originating from the basal part of the cell body, ends with complex delicate ramifications in the adjacent region of the granular layer.

The molecular layer is to a large extent formed by the dendritic branches of the *Purkinje cells* and the neuritic parallel T-fibers of the small granule cells. It contains relatively few perikarya of neuronal elements, represented by *basket cells* and *(small) stellate cells.* The former have numerous dendrites whose branches are distributed in the molecular layer. The axons of these cells run parallel to the surface of the folium and to the plane of the branching *Purkinje cell* dendrites, and thus at a right angle to the course of the T-fiber branches. The basket cell axons give off numerous descending collaterals whose end-arborizations form basket-like structures surrounding the perikarya of *Purkinje cells (Endkörbe of* KÖLLIKER). The ensemble of fibers on a *Purkinje cell* body 'condenses like the point of an artist's brush and skirts the first section of the axone of the cell at the point in which the latter still lacks the myelin sheath' (CAJAL, 1933). The more superficially located small stellate cells display relatively few as well as small dendrites, and fine branching, essentially nonmedullated axons which generally run parallel to the surface and also to the axons of the basket cells, but without descending to the *Purkinje cell* bodies (CAJAL, 1933; LARSELL, 1942).

Although at present (cf. e.g. ECCLES *et al.*, 1967) the Mammalian cerebellar cortex is generally considered to contain the five different neuron types dealt with in the preceding paragraphs, some authors have described additional types such as fusiform, intercalated, and synarmotic cells. The two former may be aberrant or displaced *Purkinje* respectively basket or *Golgi cells*. Even *Bergmann neuroglia cells* could have been mistaken for neuronal elements. The synarmotic cells of LANDAU (1928), in the deeper portion of the granular layer or even in the medullary core are believed to provide an 'association' between the

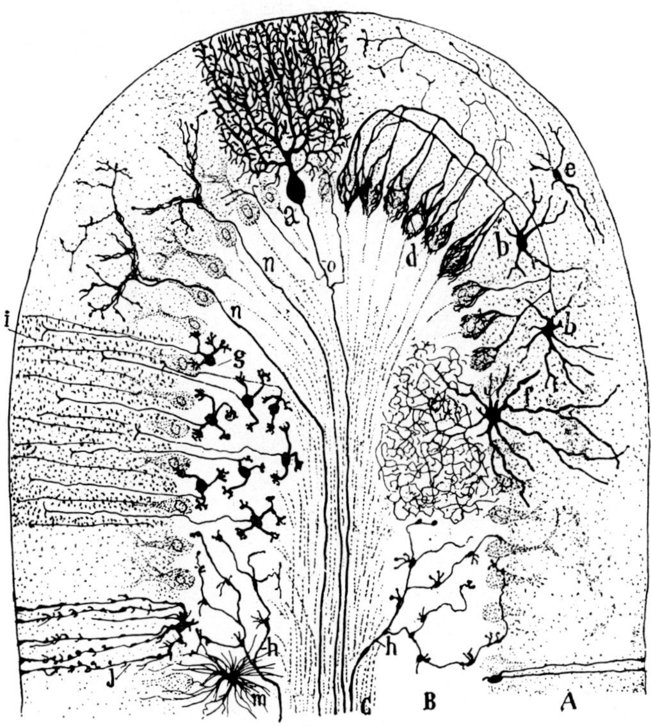

Figure 298 A. Semidiagrammatic transverse section through a Mammalian cerebellar convolution illustrating the structural relationships as displayed by the *Golgi impregnation* (after CAJAL, 1911, 1933). A: molecular layer; B: granular layer; C: medullary core; a: *Purkinje cell;* b: basket cell *('petites cellules étoilées de la couche moléculaire');* d: end-arborizations of basket cell neurites; e: (small) stellate cell *('cellules étoilées superficielles');* f: large granule cell *(Golgi cell, 'grandes cellule étoilées de la couche des grains';* g: (small) granule cells *('grains avec leur cylindre-axe ascendant et bifurqué en i');* h: mossy fibers; i: granule cell axons in molecular layer; j: *Bergmann cell ('cellule épithéliale ou névroglique en panache');* m: astrocyte *('cellule névroglique de la couche des grains');* n: climbing fiber; o: collaterals of *Purkinje cell* axon.

cortex on both sides of a folium, since their processes, which are difficult to interpret *qua* neuritic or dendritic expansions, seem to bridge the medullary core of a folium. The significance of these synarmotic cells, however, remains uncertain (cf. ECCLES *et al.*, 1967 p. 28, and BRODAL and JANSEN, 1958, p. 122).

With respect to the submammalian cerebellar cortex, *Purkinje*, granular, *Golgi II*, and stellate cells can be found in all Vertebrates, while typical basket cells may be absent in various Anamnia, although being present in Teleosts (CAJAL, 1911). Generally speaking, an increasingly complex differentiation of the Purkinje cells, roughly corresponding to the taxonomic and presumptive phylogenetic status, could be said to obtain (cf. Fig. 295B).

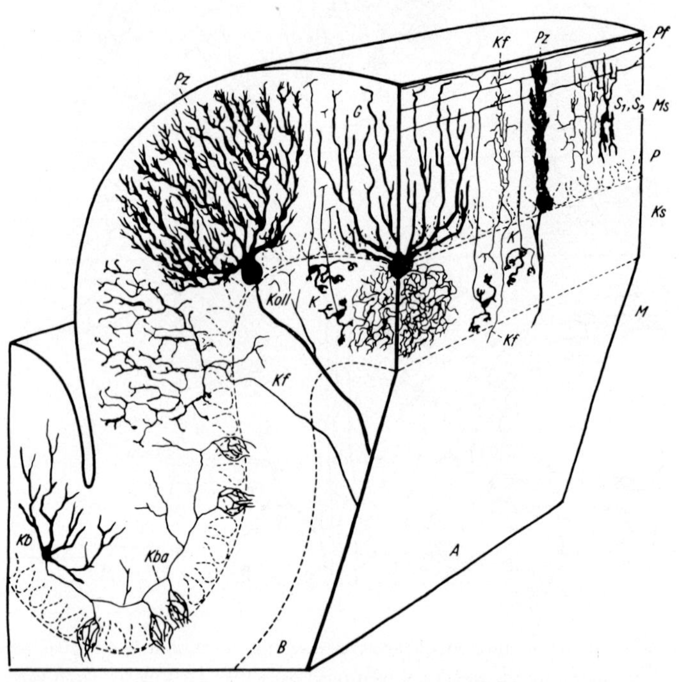

Figure 298B. Semidiagrammatic drawing of subfolium of human cerebellar cortex, illustrating three-dimensional arrangement of relevant neuronal structures in planes orthogonal (B) and parallel (A) to long axis of folial convolution (modified after JAKOB, 1928, from JANSEN and BRODAL, 1958). K: granule cells; Kb: basket cell; Kba: axon of basket cell; Kf: climbing fiber; Koll: collateral of *Purkinje cell* axon; Ks: granular layer; Ms: molecular layer; P: layer of *Purkinje cells;* Pf: parallel fibers (axons of granule cells); Pz: *Purkinje cell;* S_1, S_2: stellate cell whose axon may be either short (S_1) or relatively long (S_2) with ascending and descending collaterals.

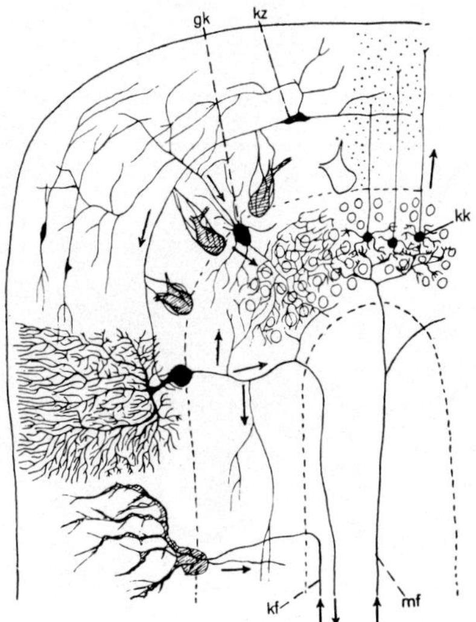

Figure 298C. Semidiagrammatic drawing of an orthogonal section through a subfolium of human cerebellar cortex with emphasis on directions of impulse conduction (from K., 1927). gk: large granule cell *(Golgi cell)*; kf: climbing fiber; kk: small granule cell; kz: basket cell; mf: mossy fiber. Although the sketch unduly stresses relationships of *Purkinje cell* axon collaterals to granule cells, neglecting the collateral plexuses at various *Purkinje cell* levels, this diagram is of particular interest, since it emphasized the *feedback* (or '*reverberating*') circuits discussed in the cited '*Vorlesungen*' of 1927 long before their concepts became generally recognized and fashionable among neurobiologists.

As regards the glial elements of the Mammalian and Avian cerebellar cortex there occur in addition to the three common sorts of such cells, namely astrocytes, oligodendroglia, and mesoglia, the characteristic *Bergmann* and *Fañanas cells*. The *Bergmann cells* are also occasionally referred to as '*Golgi epithelial cells*'. Electron microscopists have reported that some astrocytes, as well as *Bergmann* and *Fañanas cells* may display a cilium (cf. e.g. Mugnaini in Larsell and Jansen, 1972). Electron microscopic observations concerning the occurrence of a *Purkinje-cell* cilium were mentioned above on p. 632 (footnote 10).

Further comments on the neuroglia of the cerebellar cortex will be found in section 3, chapter V of volume 3, part I together with illustrations depicting the various cell types (Figs. 132, p. 190, Fig. 137, p. 196, Fig. 138, p. 197, loc. cit.). Some recent observations and comments on Mammalian cerebellar neuroglia, stressing the still rather inconclusive histochemical aspects, are discussed in a paper by Sotelo (1967).

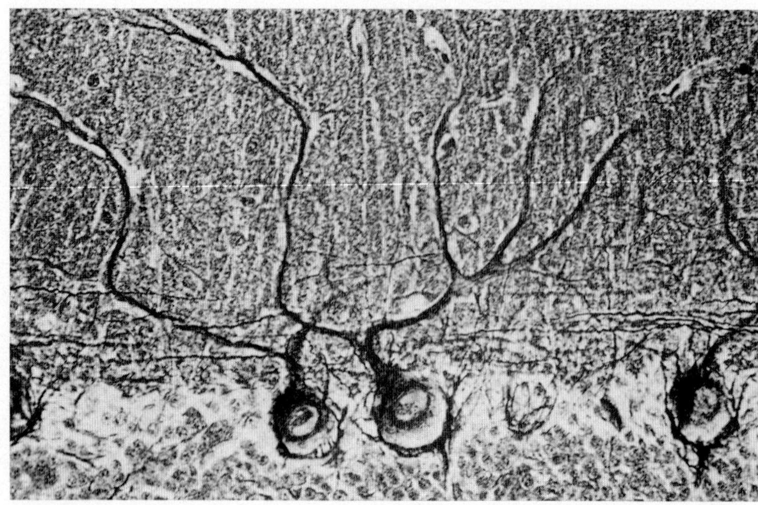

Figure 298 D. Detail of orthogonal section through a Mammalian cerebellar subfolium showing, by means of *Cajal's neurofibrillar impregnation*, *Purkinje cells* with climbing fibers, end-arborizations of basket cell axons, and fiber plexuses formed by basket cell neurites as well as presumably by *Purkinje cell* axon collaterals (original from the author's 1919–1920 Jena material, Carnivore or Man, detailed record lost).

In Reptiles and expecially in Anamnia, the ependymal cells with their processes may significantly contribute to the neuroglial framework of the cerebellar cortex (CAJAL, 1911). However, the above-mentioned three types of neuroglia cells are here likewise present in diverse ratios related to the particular taxonomic forms.

Turning now to the *extrinsic (efferent and afferent) fiber connections* of the cerebellar cortex, and discounting the poorly understood respectively dubious synarmotic cells, the *cortical output channels* seem to be provided exclusively by the *Purkinje cell* neurites. Their target loci will be considered further below, in dealing with the cerebellar nuclei, and in the discussion of basic cerebellar mechanisms concluding the present section.

As regards the *cortical input channels*, two sorts of afferent fibers were demonstrated by the fundamental studies of CAJAL, namely mossy fibers *(fibras musgosas, fibres moussues)* and climbing fibers *(fibras trepadoras, fibres grimpantes)*. The still unsettled questions concerning the differences in the origin of these two fiber types with respect to the various input channels reaching the cerebellar cortex will again be taken up in the concluding discussion of cerebellar mechanisms, together with further comments on their functional significance. Both types of fibers have been recorded in Anamnia as well as in Amniota.

The *mossy fibers* are generally described as coarse and dividing several times in the medullary layer before their rami enter the cortex, where they subdivide again in finer branches. These latter give off short terminal twigs ending as 'mossy' tufts which provide synaptic connections of glomerular type with the claw-like terminals of the granule cell dendrites. Further details concerning the appearance of these synapses can be found in section 4 of chapter V (vol. 3, part I, p. 282 and Fig. 196, p. 289).[13]

The *climbing fibers*, of smaller caliber than the mossy fibers, pass apparently without branching through the granular layer. As a climbing fiber reaches and contacts a *Purkinje cell* body, it finally losses its myelin sheath near the origin of the large dendrites and branches into twigs which run, in the manner of a climbing vine, along the dendritic ramifications, thereby providing extensive axodendritic synaptic connections (cf. Figs. 298A–D, 299).[14] According to CAJAL, the terminal distribution of a climbing fiber is restricted to a single *Purkinje cell*, thus representing a one-to-one type of connection. Recent authors, however, have maintained that climbing fibers may branch in the upper portion of the granular layer and can contact two, three or more *Purkinje cells* separated from each other by some interval. Moreover, it is claimed that retrograde collaterals return to the granular layer, and that other collaterals may reach stellate cells (cf. e.g. FOX *et al.*, 1967, 1969). Because of the limitations inherent in the various relevant technical procedures, these questions concerning the terminal distribution of a climbing fiber cannot be answered with certainty, except for its well documented contact with at least one *Purkinje cell.*

According to the original views of CAJAL (1911), the *mossy fibers* were suspected to be provided by the inferior cerebellar peduncle (corpus restiforme *sensu strictiori*), while the *climbing fibers* were conceived as pertaining (a) to the pontocerebellar system, and (b) to the vestibulocerebellar input (via the nucleocerebellar bundle or I.A.K.). At present (cf. e.g. FOX *et al.*, 1967), the mossy fibers are regarded by various

---

[13] These synapses are also characterized by the presence of many mitochondria cf. Figure 78, p. 116, vol. 3, part I), and correspond to the so-called eosin-bodies *(îlots éosinophiles ou protoplasmiques* of CAJAL, 1911).

[14] Cf. also vol. 3, part I, Figure 191, p. 281.

authors as terminals of spinocerebellar, pontocerebellar, and vestibulocerebellar systems; climbing fibers, on the other hand, are believed to arise from the inferior olivary complex, which is supposed to be 'the major source' of these fibers. According to BRODAL (1972), however, 'the assumption that all olivary axons end as climbing fibers may not be tenable'. This author cites physiologic evidence that even a considerable proportion of climbing fibers could come from the pontine nuclei.

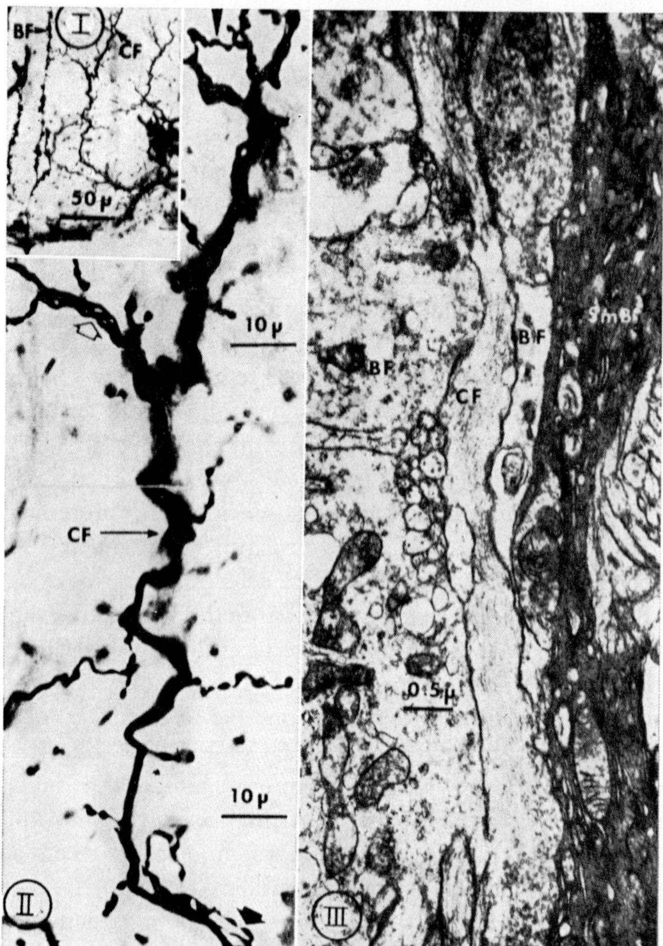

Figure 299. Climbing fibers in the cerebellar cortex of Macaca mulatta as identified in *Golgi-impregnations* and in electron microscopic pictures (from Fox *et al.*, 1967). I Medium power view. II Oil immersion montage. III Electron micrograph of a climbing fiber on a smooth dendritic branch. BF: Bergmann cell fiber; CF: climbing fiber; SmB: 'smooth dendritic branch'.

From another viewpoint, the cerebellar input systems can be said to consist of afferent root fibers (fibers of the first order) and of fibers pertaining to afferent channels of second, third, and higher orders. In addition, the pontocerebellar fibers of Birds and Mammals represent efferent (or motor) channels of at least second order[14a]. CAJAL (1911) already emphasized that cerebellar input fibers *('fibres centripètes')*, such as his *'fibres grimpantes'*, included two sorts of fibers: *l'une*, *motrice venue du bulbe*, *l'autre*, *sensorielle ou vestibulaire*, *émanée du bulbe*'. The olivocerebellar fibers are rather difficult to classify as pertaining either to the 'motor' or to the 'sensory' system.

The above-mentioned *root fibers* (or afferents of the first order) are essentially provided by fibers of the vestibular system, but also include some trigeminal and possibly also some 'visceral afferent' ones. With respect to these various input components, considerable differences in their degreee of development, prominence, or even *qua* presence or absence, are doubtless manifested by the diverse Vertebrate taxonomic forms.

On the basis of recent studies with techniques based on fluorescence histochemistry of monoamine containing neurons, cerebellar input from the nucleus loci caerulei has been inferred, and was briefly discussed in section 10 of the preceding chapter IX. This input is supposed to exert an inhibiting influence on *Purkinje cells* (via mossy or climbing fibers ?).

A peculiar type of medullated fibers was described by CAJAL (about 1895), somewhat later by SMIRNOW (about 1898), and subsequently by a number of additional authors, in the cerebellar cortex of various Mammals, including Man, and of chick embryos (cf. CAJAL, 1911; JANSEN and BRODAL, 1958).

These fibers, usually occurring as scarce scattered single elements, and most rarely, if at all, in bundles, seem to pass from the medullary center into the molecular layer, where they may follow an erratic wavy course or run parallel to the surface at the outermost level of the molecular layer, being apparently restricted to the cortex of 'vermis' and flocculus. Although nothing certain has been elucidated concerning their provenance, the *Cajal-Smirnow fibers* may represent aberrant axons (either of *Purkinje cells* or of afferent systems) and can be regarded

[14a] 'Motor' or 'efferent' in so far as the impulses directed through these channels originate in the telencephalon (cerebral cortex of Mammals, respectively undefined telencephalic grisea of Birds).

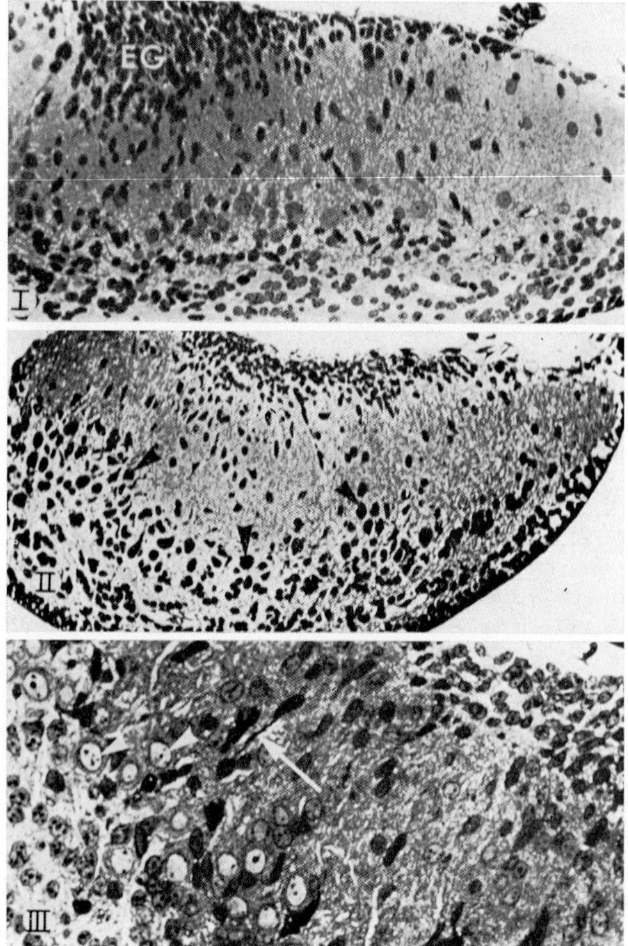

Figure 300. Aspects of the transitory superficial granular layer in the cerebellum of the frog as seen in sagittal sections (from Gona, 1972). I Cerebellum of a bullfrog tadpole (Rana catesbiana) on the day of emergence of the forelegs (EG indicates 'cone' of superficial granular layer 'migrating down toward the ependymal region'). II Cerebellum of Rana pipiens tadpole three days after foreleg emergence (arrowheads are supposed to indicate 'indentations' in the *Purkinje cell* layer). III Cerebellum of Rana pipiens six days after emergence of the forelegs (arrow shows 'long process of a migrating granule cell', and arrowheads indicate 'prominent *Purkinje cells*'). I ×250, II ×200, III ×400; red. $^2/_3$.

as common anomalies related to borderline dysplastic (or dysontogenetic) events.

As regards ontogenetic *morphogenesis sensu strictiori* of the vertebrate cerebellum, thoroughly reviewed by LARSELL (1967–1969) on the basis of original investigations, the detailed contribution by PALMGREN (1921), which also concerns the midbrain, should likewise be pointed out.

The *histogenesis* of the cerebellar cortex in many different Vertebrates is characterized by the presence of a *transitory superficial granular layer* which provides an additional secondary matrix supplementing the primary periventricular or 'ependymal' matrix. This embryonic granular layer *('embryonale Körnerschicht')* was apparently first described by HESS (1858) who already recognized its transitory nature. It was then investigated in a more detailed manner by OBERSTEINER (1883) and by CAJAL (1911), who designated it as the *'couche d'Obersteiner'*. Among the observations on this layer by other authors, those of SCHAPER (1894), who especially investigated the cerebellum of Teleosts, are of particular significance.

The transitory superficial granular layer appears to arise mainly from the caudal edge of the cerebellar plate near the attachment of the epithelial roof plate (velum medullare posterius) and seems to expand along the membrana limitans externa over the cerebellar surface. Following SCHAPER's investigations, it was commonly assumed that a transitory superficial granular layer appears only in Vertebrates possessing a voluminous cerebellum (Teleosts, Birds, and Mammals) but does not occur in those with a simple cerebellum (Cyclostomes, Amphibians, Reptiles), nor in Selachians, although the cerebellum of these latter assumes a considerable development but is formed by a folded plate *('gefaltete Lamelle')* less massive than the cerebellum of Teleosts (JAKOB, 1928).

A recent investigation of cerebellar morphogenesis in the Frog (GONA, 1972) disclosed, however, that a transitory superficial granular layer of the cerebellum is likewise formed at least in this Amphibian, but appears only in the prometamorphic period (accelerated hind leg growth), being previously absent, except for some scattered cells. At early stages, a well defined *Purkinje cell* zone is likewise lacking, but large cells, presumably immature *Purkinje neurons*, are found in the periventricular matrix. With the onset of the prometamorphic period, a fairly thick transitory superficial granular layer becomes established (cf. Fig. 300) apparently by migration from the base (i.e. the rostral region) of the cerebellar plate.[15] In the course of a second migration toward the ventricular layer the transitory superficial granular layer provides granule cells, while the *Purkinje cells*, which derive from the ventricular matrix, concomitantly mature.

CAJAL (1911), who investigated the histogenesis of the cerebellum particularly in Mammals and Birds, had previously recorded various stages in the centralward directed migration of developing granule cells (cf. Fig. 301). According to CAJAL, the superficial granular layer, however, also gives origin to small stellate cells and to basket cells. Recent observations on the evolution of the transitory superficial granular layer in the cerebellum of chick and mouse have been reported by FUJITA (1969) who made use of the autoradiographic tritiated thymidine technique. Subsequently, ALTMAN (1972) described the postnatal development of the transitory superficial granular layer and the maturation of *Purkinje cells* in the Rat on the basis of histological, histochemical, autoradiographic, and electron microscopic techniques, adding sundry details to the previously available data.

It seems thus well established that in Mammals, Birds, Teleosts, and at least in some Amphibians, the definitive cerebellar cortex arises both by peripheral migration from the ventricular matrix and by central migration from the transitory superficial granular layer which gradually becomes exhausted in the course of this process, and can therefore be assessed as an accessory matrix. Yet, despite the numerous studies by a large group of investigators, some questions remain unclarified or controversial. It is, e.g., uncertain whether this layer might not be present at some stages during the ontogenesis of various Vertebrate forms in which it was not previously recorded. The recent findings of GONA (1972) concerning the Frog are here a case in point.

As regards the histogenetic details, SCHAPER (1897), KAPPERS (1921) and others believe that both external and internal matrix may take a roughly equal part in the formation of the definitive cerebellar

---

[15] This, if actually the case, would represent an exception from the apparently more common origin of the transitory superficial granular layer at the caudal (taenial) edge of the cerebellar plate. It should, moreover, be mentioned that, according to HANAWAY (1967), who used the technique of autoradiography in the chick, the elements of the superficial granular layer are said to originate from the periventricular ependymal matrix and to migrate through the previously formed periventricular layer toward the surface. The conclusions of the cited author do not seem entirely convincing. Nevertheless the possibility of such origin of the external superficial granular layer in Birds and perhaps other Vertebrates must be kept in mind as a possibility.

cortex. According to such views, the derivation of at least some *Purkinje cells* from the external matrix appears possible, while contemporary authors, as cited above, restrict the origin of these cells to the internal matrix.

Schaper introduced the concept of indifferent cells, subsequently designated as medulloblasts by Bailey and Cushing, and as bipotential mother cells by Globus and myself (cf. section 1 of chapter V, vol. 3, part I). If, as seems likely, the superficial external matrix consists of bipotential mother cells, then the derivation of some spongioblastic, i.e. glial elements from said matrix could be expected. The observations of Fujita (1969) in the Mouse, and of Hayashi (1966) in the Chick confirm this suspicion, although it was noted that the transitory superficial granular layer produces rather few neuroglial cells, these latter originating predominantly within the internal matrix. My own first-hand acquaintance with the behavior of the transitory superficial granular layer is essentially restricted to its evolution in the human cerebellum, supplemented by merely incidental observations in other Vertebrate forms. Upon careful consideration of all available evidence I am nevertheless inclined to favor the views of Kappers and Schaper, with the qualification that the superficial (external) matrix might well display, with regard to various details, significantly different modes of differentiation related to the diverse taxonomic forms, thus precluding the formulation with respect to Vertebrates in general,

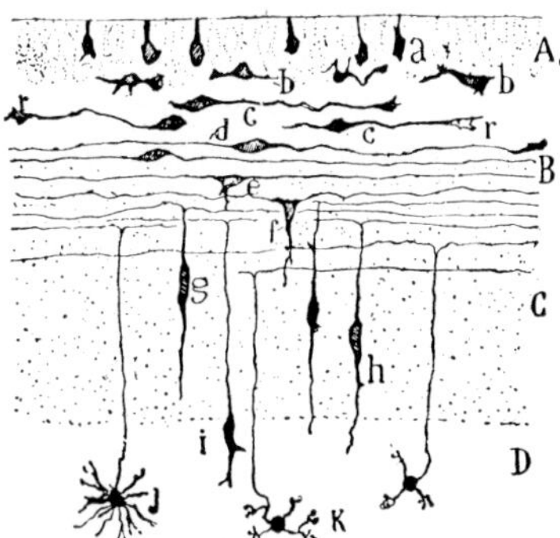

Figure 301. Schematic figure depicting ontogenetic evolution of granule cells derived from the transitory superficial granular layer (from Cajal, 1911, 1929, 1960). A: 'indifferent cells'; B: granule cells at 'horizontal bipolar stage'; C: molecular ('plexiform') layer; D: granular layer; g, h: stage of 'vertical bipolarity'; i, j: 'embryonic granule cells'; k: 'almost fully developed granule cell'.

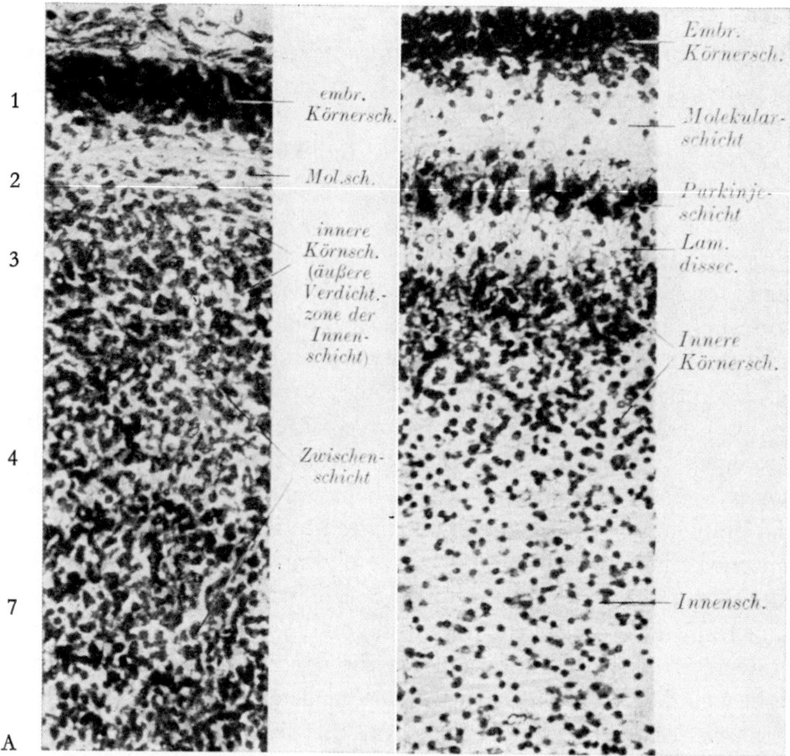

Figures 302A–D. Transitory superficial granular layer in the human cerebellum (from Jakob, 1928). A At 3rd intrauterine month. B At 8th intrauterine month. C At birth. D At third postnatal month (All ×260, red. $^4/_5$). 1: superficial granular layer; 2: molecular layer; 3: granular layer; 4: intermediate layer (*'Zwischenschicht'*, migration from ependymal matrix); 5: *Purkinje cell* layer; 6: 'lamina dissecans'; 7: late intermediate layer (*'Innenschicht')*; x: large neuroblasts with long dendritic expansions (possibly ontogenetic precursors of *Purkinje* and large granule cells). The so-called *'lamina dissecans'* (6) appears during Mammalian (Human) ontogenesis as a sparsely populated narrow layer basally adjacent to the layer of *Purkinje cell* perikarya.

Figure 302E. Normal transitory superficial granular layer in the cerebellar cortex of a newborn infant (I) compared with abnormal persistent superficial granular layer in the cerebellum of a 22-years-old adult (II). The abnormal superficial granular layer was directly continuous with a neoplasm of spongioneuroblastic type (from K., 1950). g: granular layer; m: molecular layer; p: layer of *Purkinje cells* (whose pale pericarya are not clearly visible in the photomicrographs); s: normal superficial granular layer; ps: abnormal persistent superficial granular layer.

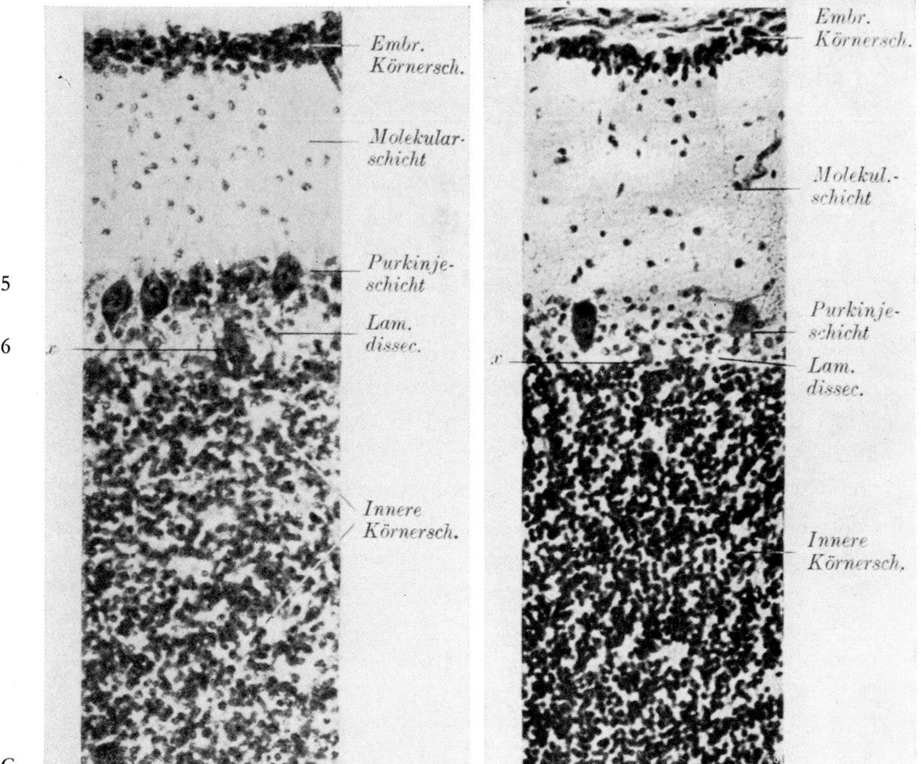
Embr.
Körnersch.
Molekular-
schicht
5
Purkinje-
schicht
6
Lam.
dissec.
x
Innere
Körnersch.
C
Embr.
Körnersch.
Molekul.-
schicht
Purkinje-
schicht
x
Lam.
dissec.
Innere
Körnersch.
D

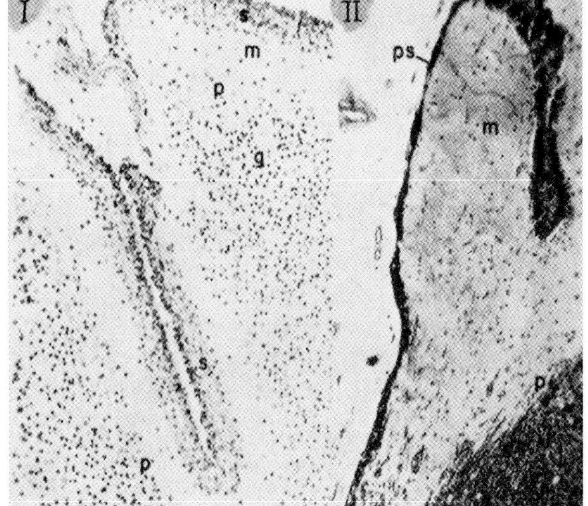
I
s
m
p
g
s
p
II
ps
m
p

Figure 302 E

of valid overall statements *qua* production of neuroectodermal cell types, except perhaps *qua* well documented origin of small granule cells in substantial amount.[16]

In human ontogenesis (Figs. 302A–E) that transitory superficial granular layer is fairly wide at birth, when it may have a thickness of 10 to 20 cells (K., 1950). In the postnatal period, the layer gradually decreases in width and disappears between the 9th and 20th month (cf. e.g. JAKOB, 1928; RAAF and KERNOHAN, 1944). Considerable individual variations obtain, but at the end of the first year the external granular layer, if still present, is commonly very narrow. Nevertheless, in some cases (cf. Fig. 302E), an abnormally persistent respectively dysplastic external granular layer of variable extension or thickness may be retained during adult life (cf. e.g. BIACH, 1910; K. and HAYMAKER, 1946, 1947; K., 1950).

Disturbances of the peculiar histogenesis of the cerebellar cortex result in various forms of dysplasias, hamartomas, hamartias or choristomas, as described by BARTON (1934), BIELSCHOWSKY and SIMONS (1930), FOERSTER and GAGEL (1933), HEINLEIN and FALKENBERG (1939), K. and HAYMAKER (1946), K. (1950), and others. The significance of embryonic remnants as a potential source of neuroectodermal tumors has been repeatedly stressed and can be regarded as well established.[17] In conformity with these views there are many observations indicating that several types of cerebellar neoplasms may arise from remnants of the transitory superficial granular layer. Such tumors are medulloblastoma, spongioneuroblastoma, glioneuroma, and granuloblastoma.[18]

---

[16] In Mammals, the origin of at least some stellate, basket, and large *Golgi II cells* from this layer seems likewise well documented by the observations of CAJAL (1929, 1960).

[17] Cf. the comments on neoplastic growth and classification of neuroectodermal tumors in volume 3, part I, chapter V, section 9, p. 751f.

[18] The term 'granuloblastoma' was specifically suggested by STEVENSON and ECHLIN (1934) as well as SACCONE and EPSTEIN (1948) for primary neuroectodermal tumors originating from the transitory superficial granular layer of the cerebellar cortex. Such neoplasms may consist of both neuroblastic and spongioblastic elements.

Figures 303A–C. Cross-sections through the human rhombencephalon showing configuration and arrangement of the cerebellar nuclei (Myelin stain, approx. ×3; red. $^1/_3$). 1: nucleus dentatus; 2: nucleus emboliformis; 3: nuclei globosi; 4: nucleus tecti seu fastigii; 5: corpus restiforme; 6: corpus juxtarestiforme (IAK); 7: brachium conjunctivum; 8: brachium pontis. Cf. also Figures 280–281.

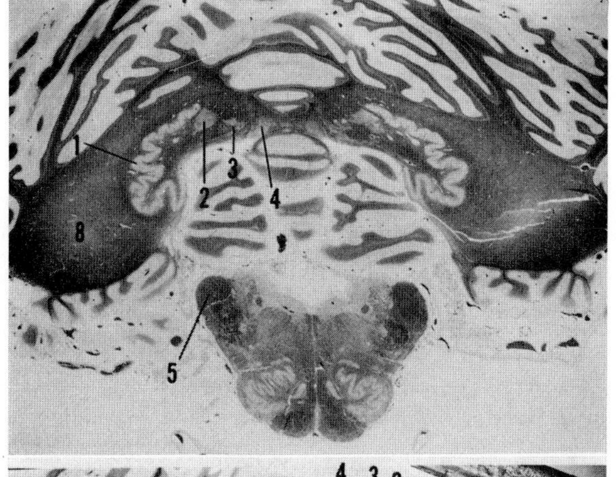

A

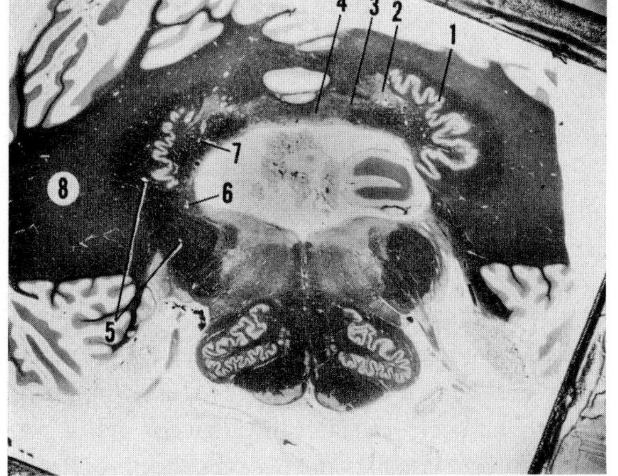

B

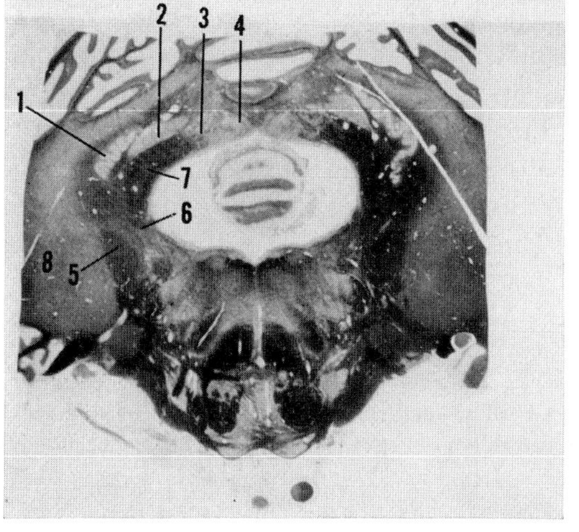

C

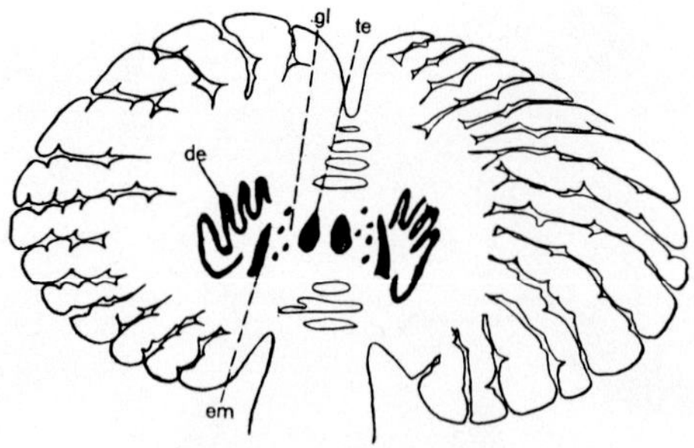

Figure 303D. Simplified diagram of a horizontal section through the human cerebellum showing arrangement of cerebellar nuclei (from K., 1927). de: nucleus dentatus; em: nucleus emboliformis; gl: nuclei globosi; te: nucleus tecti seu fastigii. At top, incisura cerebelli posterior, at bottom velum medullare anterius with part of brachium conjunctivum; in midline above and below nuclei, sections through some sulci of the vermis have been indicated.

Turning now to the second set of cerebellar grisea, namely the *cerebellar nuclei*, it could be stated that, at least in higher Vertebrates, they are essentially concerned with cerebellar output and seem to derive, both ontogenetically and phylogenetically, from rostral vestibular grisea, respectively from the dorsal zone of the alar plate.[19] In Man, whose cerebellar cortex was taken above as a paradigm, four paired gray masses are enclosed in the medullary substance (cf. Figs. 303A–D). The *nucleus tecti seu fastigii* lies close to the midline in the roof of the fourth ventricle, the laterally adjacent *globose griseum (nucleus globosus* or *nuclei globosi)* consists of several rounded clusters, which lie medially to the elongated cell plate of *nucleus emboliformis*. This latter is located in the medial hilus of the laterally adjacent large nucleus dentatus, which represents as folded layer, very similar to the main inferior olivary nucleus. Further details concerning structure and connections of these nuclei shall be taken up further below in the discussion of cerebellar mechanisms as well as in section 9, specifically dealing with the Mammalian and Human cerebellum. It will here be sufficient to men-

---

[19] The ontogenetic relationship of the cerebellar nucleus respectively nuclei to the vestibular grisea has been disputed by Rüdeberg (1961). This author described the formation of cerebellar nuclei by migrations from his 'columna dorsalis' interpreted as a component of the cerebellum itself. Nevertheless, the interpretations and descriptions propounded by the cited author can be considered rather unconvincing and I do not share the opinion expressed by Larsell (1967) that these studies 'appear to have definitely set aside any relationship of these nuclei to the vestibular nuclei'.

tion that the neurites of *Purkinje cells* terminate in these nuclei which, in turn, give origin to all major efferent cerebellar tracts, with the exception of some *Purkinje cell* neurites directly reaching with their terminals the vestibular nuclei, perhaps mainly *Deiters'* and *Bechterew's* nucleus. Such *Purkinje fibers* seem mainly to originate in the flocculonodular lobe.

Whether and to which extent the cerebellar nuclei receive collaterals from the input systems reaching the cerebellar cortex, and whether said nuclei also originate some retrograde fibers or collaterals directed toward the cerebellar cortex, remain poorly elucidated questions.

The cerebellar nuclei of *Anamnia* are less well developed, being represented by a single griseum which can be interpreted as a rostral extension of the vestibular or octavomotor complex. This, often rather diffuse primordial *nucleus lateralis cerebelli* (EDINGER, 1901; VAN HOEVELL, 1916) is commonly located in a basal portion of the cerebellum, frequently in a so-called *eminentia ventralis cerebelli* which may slightly protrude into the fourth ventricle (EDINGER, 1901). CAJAL (1911), who also quotes EDINGER, depicts this nucleus in Teleosts and states that it sends neurites into the cerebellar cortex.[20]

In *Amniota*, the cerebellar nuclei tend to be displaced dorsomedialward, thus becoming included into the medullary core of the cerebellum. In Reptiles, two cerebellar nuclei, a medial and a lateral one, are displayed, three or perhaps more are present in Birds, and two or four cerebellar nuclei can be found in Mammals.[21]

Generally speaking, it can be said that in Anamnia, respectively in those Vertebrates whose cerebellar nuclei are not extensively developed, the cerebellar output is chiefly provided by the neurites of the

[20] CAJAL (l.c.) states: '*Nous avons trouvé chez les poissons, au-dessous du cervelet, un ganglion dont les cellules envoient leurs cylindre-axe jusqu'à l'écorce cerebelleuse. Mais ce noyau correspond-il au ganglion du toit des oiseaux et des mammifères? c'est que nous ne saurions dire. Il pourrait représenter tout aussi bien l'olive bulbaire, seul noyau dont, en toute certitude, les cylindre-axes vont au cervelet.* EDINGER *donne, dans ses monographies, le nom de noyau latéral du cervelet au noyau que nous venons de mentionner chez les poissons*'. Be that as it may, it is evidently impossible, on morphologic grounds, to homologize said nucleus with the inferior olive *(olive bulbaire)*, quite apart from the fact that a reasonably well definable inferior olivary griseum, which seems to send neurites into the cerebellum, can be recognized at least in Selachians and Teleosts.

[21] Variable statements concerning the number of cerebellar nuclei in Birds and Mammals can be attributed to the difficulties inherent in a suitable but arbitrary parcellation of closely adjacent grisea with indistinct, coalescent respectively overlapping boundary neighborhoods.

Purkinje cells which form the main efferent cerebellar pathways, supplemented by the output mediated by the primordial cerebellar nuclei.[22] In higher Vertebrates, these nuclei become the main grisea for cerebellar output as pointed out by HERRICK (1924).

In connection with the overall pattern of cerebellar fiber pathways, the various *cerebellar commissures* should here be mentioned. Some of the decussating fibers pertain to afferent, and others to efferent pathways. A *commissura lateralis* respectively *vestibulo-lateralis*, moreover a *commissura cerebellaris* containing trigeminal and spino-cerebellar components, have been described by HERRICK, LARSELL and others. The commissura vestibulo-lateralis may also contain fibers interconnecting the two auricles. In modified form, these two commissures, as e.g. discussed by BECCARI (1943), have also been recorded in higher Vertebrates including Mammals (the more rostral *decussazione del cerveletto*, and the more caudal *decussazione laterale o flocculare)*.

As regards the functions and basic neuronal mechanisms of the cerebellum it could be stated that this subdivision of the brain represents a suprasegmental 'organ' concerned with *coordination and regulation of motility* as well as with *maintenance of posture and of equilibrium*. The cerebellum, however, does not seem to initiate motions of body or limbs in any significant manner, but rather to control or smoothen movements in progress of execution, providing, *inter alia*, muscular synergy, i.e. coordination of synergists and antagonists. The overall arrangement of input and output channels as illustrated by Figures 304A–E indicates that the cerebellum is connected *in parallel (im Nebenschluss)* with the main input and output pathways of the neuraxis. Although these diagrams refer to the Mammalian cerebellum, the preceding statements can be taken to hold, *mutatis mutandis* (e.g. lack of a pons in Anamnia and presumably also in Reptiles), for the entire Vertebrate phylum.

In Mammals, three *peduncles*, namely, *corpus restiforme sensu latiori*, *brachium pontis*, and *brachium conjunctivum sensu latiori* (with median velum medullare superius) connect the cerebellum with the adjacent brain subdivisions, which, in the same sequence, are medulla oblongata *sensu strictiori*, pons, and mesencephalon. All three peduncles contain cere-

---

[22] The questions to which extent some output from the cerebellar nuclei returns or does not return to the cortex, and to which extent these nuclei also receive extracerebellar input, are still poorly elucidated. It is commonly assumed that afferent pathways to the cerebellar cortex may give off some collaterals to cerebellar nuclei.

bellar *input fibers*, but those taking their course through the superior peduncle (ventral spino-cerebellar tract, and tecto-cerebellar fibers, these latter mostly in velum medullare) are a relatively small component of that stalk. Again, all three peduncles give passage to cerebellar *output fibers*, which reach grisea of the motor system. These output

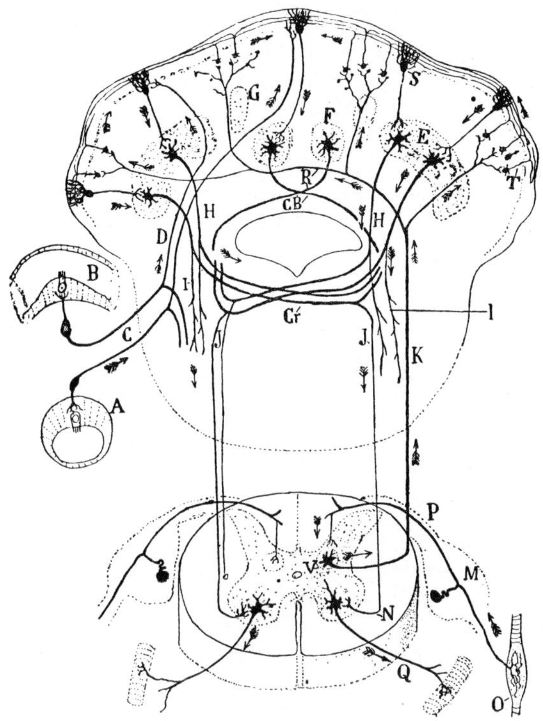

Figure 304A. Diagram of main Mammalian cerebellar input and output channels in the original concept of CAJAL (from CAJAL, 1911). A, B: primary vestibular neurons whose central neurites reach the cerebellar cortex; C: vestibular root fibers; CB: crossed cerebello-bulbar fibers; Cr: brachium conjunctivum system *('entrecroisement du pédoncule cérébelleux supérieur')*; D: ascending vestibular root fiber branch *('articulée probablement avec les cellules de Purkinje')*; E: nucleus dentatus *('olive cérébelleuse')* giving origin to superior cerebellar peduncle; F: nucleus tecti, giving origin to *'voie cérébello-bulbaire croisée'*; G: nucleus emboliformis and nn. globosi; H: superior cerebellar peduncle; I: *'branche descendante directe de ce pedoncule'*; I: *'branche descendante croisée de ce pédoncule;* K: dorsal spinocerebellar tract (from *Clarke's column*, *'articulée peut-être avec les grains par l'intermédiaire des fibres moussues')*; M: spinal ganglion; N: assumed termination of crossed direct brachium conjunctivum fibers upon spinal motoneurons; O: input from muscle spindle; P: posterior root, partly connecting with *Clarke's column;* Q: ventral root; R: spinocerebellar input; S: *Purkinje cell;* T: mossy fiber; V: *Clarke's column ('donnant naissance à la voie sensitive ascendante cérébelleuse)*.

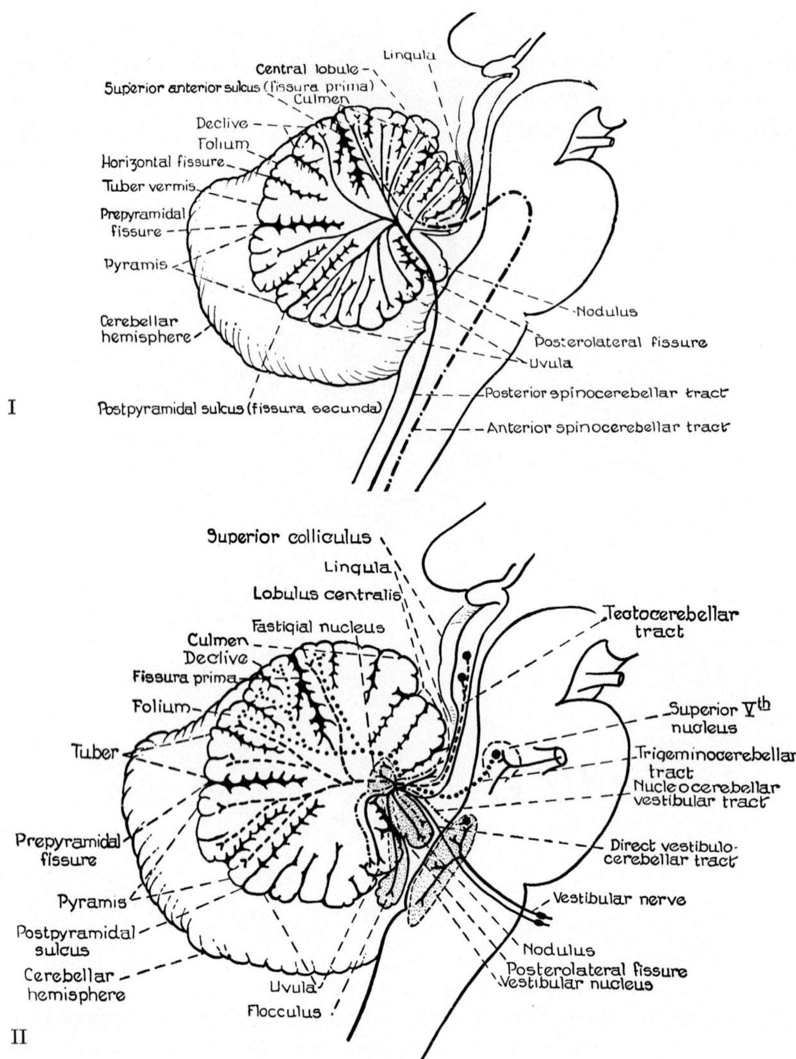

Figure 304B. Diagrams showing some Mammalian cerebellar input channels according to more recent views. The pontocerebellar input, shown in Figures 304C and 304D, and the olivocerebellar system shown in Figure 304C have been omitted (from LARSELL, 1942). I Spinocerebellar connections. II Vestibulocerebellar, trigeminocerebellar, and tectocerebellar connections.

channels are cerebello-tegmental and cerebello-vestibular tracts, moreover, through the brachium conjunctivum, a rather massive tract reaching, after a decussation ventrally to the aqueduct, the nucleus ruber tegmenti and, more rostrally, ventrolateral grisea of the dorsal thalamus.[23] The collateral cerebellar discharge into the motor system, mediated by these and other grisea, assists in 'regulating' or 'modulating' the activities performed by the main motor system of the brain. The cerebellum thus plays a role in 'graduating and harmonizing muscular contractions, both in voluntary movement and in the maintenance of posture' (BRAIN, 1969). This remark should, however, be modified to include not only 'voluntary' but also 'non-voluntary' movements. The activity of the Mammalian and Human cerebellar cortex, in turn, becomes 'modulated' by events, some of which are doubtless 'voluntary', taking place in the telencephalic neocortex.

The significance of the cerebellum for motility, suspected by WILLIS (1622–1675), was confirmed by the studies of ROLANDO (1773–1831), FLOURENS (1794–1867), and VULPIAN (1826–1887). Among subsequent authors STEINER (1885, 1888, 1900) investigated cerebellar function in Selachians, Teleosts, Amphibians and Reptiles. With respect to Mammals and Man, however, the studies by LUCIANI (1891, 1893) are of fundamental importance. Clinical aspects of cerebellar function were further elucidated by HOLMES (1922) and others. A fairly recent treatise on generally accepted views about cerebellar physiology and pathology was compiled by DOW and MORUZZI (1958). It also includes the results of some original research undertaken by these authors.

Removal of the entire cerebellum as performed by STEINER (1885–1900) in Selachians, Teleosts, Amphibians and Reptiles seems to

[23] Further details concerning these various fiber systems and their homolateral or contralateral course shall be included in the subsequent sections, dealing with the various Vertebrate classes. Generally speaking, many of the tracts connecting with the cerebellum are uncrossed. However, even some of the homolateral ones may decussate within the cerebellum. Again, a serial sequence of two alternating decussations can result in an essentially homolateral relationship of certain input and output channels. Thus, while symptoms of prosencephalic and brain stem damage are frequently manifested on the opposite side of the lesion, symptoms of cerebellar affections are generally displayed on the lesion's side or may be only 'obscurely localized' (cf. also HERRICK, 1931). Thus, in unilateral lesions, the tendency to fall toward the side of the lesion, or, *qua* forced movements, the tendency to 'roll' in that direction, are frequently quite conspicuous. Again, the 'forced movements' *(Zwangsbewegungen)* seem to be particularly, but perhaps not exclusively related to asymmetric lesions involving a bilaterally uneven distribution of effects due to either deficiency (negative) or irritation (positive) effects.

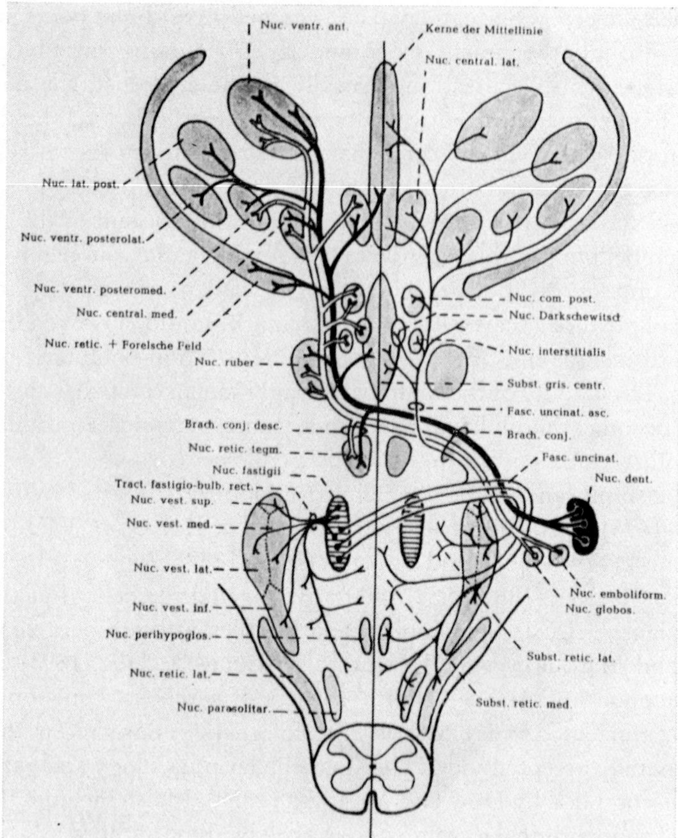

Figure 304C. Diagram showing the complexity of significant cerebellar output systems according to recent views (from JANSEN, 1961).

Figure 304D. Simplified diagrams of Mammalian cerebellar input (I) and output (II). The vestibular system (cf. Fig. 304B, II) has been omitted (from HERRICK, 1931).

Figure 304E. Simplified diagrams of human cerebellar input (I) and output (II) with emphasis on the three cerebellar stalks. The vestibulocerebellar and some other input has been omitted (from K., 1927). bc: brachium conjunctivum; bf: pontocerebellar fibers; bp: brachium pontis; cr: corpus restiforme; fb: fibrae cerebellotegmentales bulbi; nd: nucleus dentatus; sd: tractus spinocerebellaris dorsalis *(Flechsig)*; sv: tractus spinocerebellaris ventralis *(Gowers)*; te: nucleus tecti; tp: tractus cerebello tegmentalis pontis; tu: tractus (s. fasciculus) uncinatus.

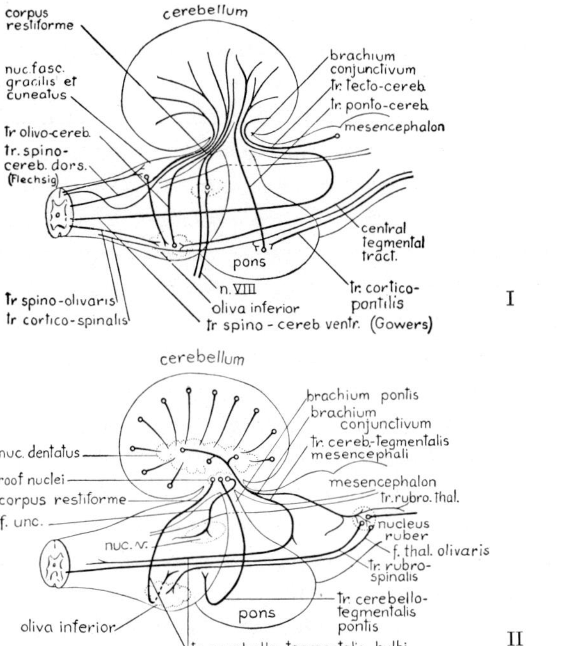
corpus
restiforme
cerebellum
nuc.fasc.
gracilis et
cuneatus
tr olivo-cereb.
tr. spino-
cereb. dors.
(Flechsig)
brachium
conjunctivum
tr. tecto-cereb.
tr. ponto-cereb.
mesencephalon
central
tegmental
tract.
pons
n. VIII
tr. cortico-
pontilis
tr spino-olivaris
tr cortico-spinalis
oliva inferior
tr spino - cereb ventr. (Gowers)
I
cerebellum
brachium pontis
brachium
conjunctivum
nuc. dentatus
tr. cereb.-tegmentalis
mesencephali
roof nuclei
mesencephalon
corpus restiforme
tr. rubro. thal.
f. unc.
nucleus
ruber
nuc. V.
f. thal. olivaris
tr. rubro-
spinalis
tr. cerebello-
tegmentalis
pontis
pons
oliva inferior
tr. cerebello-tegmentalis bulbi
II

304 D

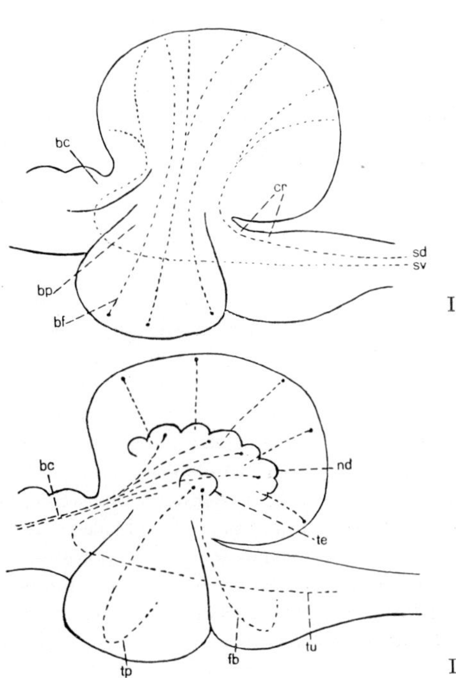
bc
cr
sd
sv
bp
bf
I
bc
nd
te
tp
fb
tu
II

304 E

have rather little effect. These animals, although deprived of that subdivision, are able to swim, respectively jump or run in an apparently normal manner as far as 'overall' or 'gross motility' is concerned.[24] Even the removal of the valvula cerebelli in Teleosts is compatible with subsequent normal swimming activities. STEINER (1885–1900) remarks here: '*Bringt man nämlich so operirte Fische ins Wasser zurück, so pflegen einige derselben ganz normal zu schwimmen. Andere aber liegen auf dem Rücken mit völlig guter Athmung; indess nur für kurze Zeit, denn nach* $^1/_4$ *bis* $^1/_2$ *Stunde erholen auch diese sich und schwimmen ganz äquilibrirt. Wenn man solche Fische etwas näher betrachtet, so findet man, dass sie während der Bewegung, die eine durchaus normale Locomotion ist, ganz leicht seitlich schwanken. Je besser die Operation geglückt ist, um so geringfügiger sind diese Schwankungen; sie stehen auf derselben Stufe mit jenen geringfügigen seitlichen Schwankungen, welche man bei Fischen beobachtet, denen sämtliche Flossen angeleimt worden sind. Stehen diese Fische auf dem Grunde, so ist nichts Abnormes an ihnen zu bemerken. Dass sie blind sein müssen, möge der Vollständigkeit halber noch erwähnt sein; fehlt ihnen doch mit der Decke des Mittelhirns das Sehcentrum.*» TUGE (1934), on the other hand, reported that the decerebellate Teleost 'always swims with inclination to either side and rolls its body constantly when resting, and that this phenomenon does not disappear before killing in most of the cases, although its degree decreases with time'. This author also noted, as immediate reactions to the operation of 'total' cerebellar extirpation, 'rapid spiral, up and down, and disordered movements. With mechanical stimulation the animals reacted violently with these movements.'

---

[24] The interpretation of the reports by STEINER and others on cerebellar ablation is beset with two difficulties. The first concerns lack of precision about the total amount of cerebellum removed (corpus, valvula, auricles respectively flocculus) as well as about damage to adjacent brain parts. The second is of a semantic nature, dealing with a verbally sufficiently clear-cut description of the motor disturbances. Some authors may present reports of nondescript symptoms with a verbal precision entirely inapplicable to the actual findings. Thus, in an instance unrelated to cerebellar function I remember that, at a scientific meeting, a well-known neurobiologist, endowed with the '*gift of gab*', presented a long film on experimental striatal disturbances, accompanied by most detailed classifying statements on types of disturbances which I failed to recognize. Thinking that this was perhaps due to insufficient familiarity with functional aspects, I asked for further explanations from a friend of mine who is one the world's foremost neurophysiologists with wide clinical experience, and who had repeatedly seen that film. This gentleman gave the following laconic answer: 'all what one could here really say is that these monkeys were very sick'.

The cited author ascribes the disturbances produced by cerebellar ablation in Teleosts to 'hyposthenia and ataxia; that is to say, function of the cerebellum is to reinforce and maintain the appropriate muscular tonus in the animal' (TUGE, 1934).

STEINER, who had acquired a very refined technique in many years of experimental work on the neuraxis of Cyclostomes, Selachians, Ganoids, Teleosts, Frogs, and Lizards as well as on the central nervous system of Amphioxus and Invertebrates, pointed out the great difficulties in removing the Teleostean cerebellum together with its valvula, an operation which unavoidably affects the tectum mesencephali. It is not unlikely that the somewhat more pronounced functional disturbances observed by TUGE (1934) were due to damage inflicted upon mesencephalic tectal and mesencephalo-rhombencephalic tegmental grisea as well as fiber systems. Yet, it is rather probable that the removal of the cerebellum should at least, and particularly in the period following the operation, produce some effects. Thus, even the frog, according to the observations of STEINER (1885), is said to display a certain awkwardness in jumping, occasionally overshooting or undershooting what could be regarded as the 'intended mark'.

It might also be added that, despite concomitant extirpation of tectum opticum (to a variable and undefined degree!), the operated Teleosts were presumably not entirely 'blind' as claimed by STEINER in the comment quoted above. The remaining optic input to lateral geniculate complex, pretectal grisea, hypothalamus and mesencephalic tegmentum doubtless still provided some amount of optic 'information'.

Ablation of the cerebellum in Birds likewise appears to cause only transitory and nondescript postural and locomotory disturbances, depending on extension and depth of the lesion (GROEBBELS, 1928, and others). Nevertheless, on the whole, the resulting symptoms of cerebellar deficiency recorded in Birds seem to be somewhat better recognizable than those in Reptiles and Anamnia. Cerebellar ablation studies in Birds undertaken by KARAMYAN (1956) and quoted by PEARSON (1972) disclosed, in addition to the muscular symptoms 'severe alimentary disorders' ascribed to 'autonomic dysfunctions'. Whether these disturbances were actually related to cerebellar lesions, or to damage inadvertently inflicted upon brain stem structures, must remain an open question. Both explanations are, of course, possible.

With regard to Mammals, the careful and well-documented experimental work by LUCIANI (1891) established three different generalized types of disturbances due to cerebellar lesions, namely (1) *tonus anomalies* [24a] (atony, hypertony), and disturbances of posture, (2) *ataxy-asynergy* (disorders of movements), and (3) *asthenia*. Again, in experimental work, a temporal sequence of three stages can be roughly recognized, namely (a) a '*stage of exaltation*' with restlessness and forced movements, (b) a *stage of deficiency*, and (c) a *stage of compensation* with functional restitution.

Although the tonus disturbances are commonly characterized by atony, hypertony may likewise obtain, especially in fits of seizures with extensor rigidity in periods immediately following the ablation. The atactic asynergic disturbances involve jerky movements, and overshooting of mark *(dysmetria)*. *Tremor*, increased during execution or movements (so-called intention tremor) and (cerebellar) nystagmus, more pronounced in Man[25] than in other Mammals, are likewise observed. The asthenia generally becomes much less pronounced in the chronic condition.

Of all the conventional major subdivisions of the brain, the cerebellum is doubtless the least 'relevant' or 'important' and seems to represent a typical example of neurobiologic '*redundancy*'[26]. This evaluation can be supported by the following arguments.

(1) The cerebellum is the only one of these 'major' subdivisions which, for practical purposes, can be found completely missing in some Vertebrate brains (e.g. Myxinoids and Gymnophiona), discounting structurally and functionally negligible topologic neighborhood vestiges.

(2) The degree of cerebellar differentiation is, in part, not definitely correlated with the taxonomic or phylogenetic status of the diverse Vertebrate groups. There is a high degree of cerebellar differentiation in Selachians and Teleosts, a rather low degree in Dipnoans, a still lower degree in Amphibians, and a variable but on the whole low to

---

[24a] Concerning the here relevant problems of muscular tonus, the critical review by Spiegel (1929), based on considerable first-hand experience, deserves particular mention. The cited handbook chapter includes references to the pertinent cerebellar activities and to various differences in the results and evaluations by diverse investigators, some of which even disagree with a few of the observations reported by Luciani.

[25] Further comments on clinical symptoms of cerebellar disturbances will be found below in section 9 dealing with the Mammalian and Human cerebellum.

[26] With respect to the neurobiologic significance of 'redundancy', cf. vol. 1, chapter 1, section 4, p. 20 in this series.

very moderate degree in Reptiles. Birds and Mammals, on the other hand, display a high degree of cerebellar differentiation.

(3) The insignificant or nondescript effects of cerebellar ablation in Selachians, Teleosts, Amphibians, Reptiles, and Birds.

(4) The marked tendency toward compensation and functional restitution following cerebellar ablations in Mammals.

(5) The lack of any conspicuous deficiency symptoms in some of the recorded human cases of near complete cerebellar agenesis (cf. Fig. 100, p. 274 of vol. 3, part II).

On the other hand, if, in Mammals and Man, the redundant cerebellar participation has become a well-established component of the neuraxial motor output pattern, it seems evident that subsequent deprivations or distortions of that component might lead to considerable disturbances with variable degrees of restitution. Again, in some conditions such as tabes dorsalis and *Friedreich's ataxia*, the damage involves fiber systems providing input to cerebellum as well as to extracerebellar grisea, and the resulting symptoms are only in part results of impaired cerebellar function. Generally speaking, it could be stated with RADEMAKER (1924) that destruction of the cerebellum does not result in the loss of any single function but rather in an overall inadequacy of responses (cf. also CROSBY *et al.*, 1962).

The essentially 'stereotyped' and relatively simple neuronal arrangement of the cerebellar cortex, disclosed by the studies of CAJAL and others, is characterized by a rectangular lattice construction of rather striking 'geometrical precision' with regard to the spatial orientation of neuronal elements along the axes of a 'semicartesian' coordinate system. A definite relationship of that peculiar 'construction' was already emphasized long ago by OBERSTEINER (1912) before recent authors stressed this particular aspect in attempting to elucidate the cerebellar neuronal mechanisms.[27] The fiber connections within the cerebellar cortex display most conspicuous instances of *feedback circuits*, which I pointed out in 1927, long before the feedback concept became

[27] OBERSTEINER (1912) stated: '*Zieht man noch die grobanatomische Tatsache heran, dass die Kleinhirnwindungen der Mehrzahl nach im wesentlichen eine zur Körperachse quere Richtung einhalten, dass also auch die Parallelfasern meistens den gleichen Vorlauf nehmen müssen, so scheint diese überwiegend durchgeführte Orientierung der Nervenzellen in der Kleinhirnrinde, wie wir ähnliches in keinem anderen Hirnteil wiederfinden, in irgendeiner Relation zu der Funktion des Kleinhirns zu stehen. Einen Fingerzeig geben uns vielleicht die innigen Beziehungen des Vestibularapparates zu diesem Hirnteile.*'

fashionable. The feedback effect was, in fact, 'downgraded'[28] by CAJAL, who apparently subsumed it under his '*avalanche de conduction*'. It is most likely that said feedback circuits are related to the peculiar high frequency discharges of the cerebellar cortex described by SPIEGEL (1937), and persisting after section of the cerebellar peduncles.

In the model of a cerebellar 'neuronal machine' elaborated by ECCLES *et al.* (1967) one of the most striking features, inferred from microelectrode studies, is the extent to which said 'machine' depends on *inhibition*. Of the five conventional types of cerebellar cortical cells, four (basket cells, stellate cells, large granule cells, and *Purkinje cells*) are supposed to have exclusively inhibitory function, only the small granule cells being exclusively excitatory (essentially for the inhibitory *Purkinje cells*). Basket cells, stellate cells and large granule cells (*Golgi cells*) are presumed to 'control' and 'focus', entirely by inhibitory action, the excitatory effect on *Purkinje cells* exerted by the small granule cells. Since the entire output of the cerebellar cortex, transmitted exclusively by axons of *Purkinje cells*, is supposed to be inhibitory, it follows that all imput into the cerebellar cortex is transformed into inhibition. The axons of the *Purkinje cells* transmit this inhibition to the cerebellar nuclei and to those vestibular (or possibly other neuraxial) grisea in which some *Purkinje cell* neurites terminate.

Recent authors, using electron-microscopy combined with a '*freeze-fracture technique*', which displays certain details related to pre- and postsynaptic membranes, claim that, by means of this technique, excitatory and inhibitory synapses can be distinguished (LANDIS and REESE, 1974). The cited authors have described ultrastructural peculiarities of various cerebellar synapses, such as of parallel fibers to *Purkinje spines*, mossy or climbing fiber to granule cell dendrite, etc., in accordance with their not altogether convincing or conclusive interpretations of the observed ultrastructural data, which also include particulars of synaptic cleft widths, subsynaptic *fuzz*, *puncta adhaerentia*, and related adhering junctions.

Most of the actual mammalian cerebellar output (in contradistinction to cerebellar *cortical* output) is provided by the cerebellar nuclei, whose neuronal elements are excitatory. Since these neurons are supposed to be inhibited by the *Purkinje cells*, the excitatory 'final output' of the cerebellum is presumed to represent, as it were, 'a negative image of the output from the cerebellar cortex'. It is, moreover, assumed

[28] Cf. vol. 1, p. 11f.

that collaterals of climbing and mossy fibers, ending in the cerebellar nuclei, have an excitatory effect on these latter. Despite a few contradictory reports, most contemporary investigators believe that the cerebellar nuclei do not send axons respectively collaterals to the cerebellar cortex (MUGNAINI, 1972).

Although ECCLES *et al.* (1967) have presented their concept of the cerebellum as a 'neuronal machine' with considerable aplomb and assurance, numerous doubts nevertheless remain. Data concerning localization in the cerebellar cortex[29] (cf. MUGNAINI, 1972), and observations indicating that, depending on the stimulus parameters, opposite effects may be obtained on cerebellar stimulation (cf. e.g. SPIEGEL, 1968), justify a sceptical attitude toward the claim by ECCLES *et al.* that all *Purkinje cells* are inhibitory, respectively that these cells are exclusively inhibitory in accordance with '*Dale's principle*'.[30]

FOX *et al.* (1967) who essentially accept the concepts propounded by ECCLES *et al.* (1967) have elaborated, on the basis of their structural investigations by means of the *Golgi impregnation* and of electron microscopy, a simplified summary of the 'circuitry' of the cerebellar cortex, incorporating the older anatomical and the more recent electrophysiological data available in the literature.

According to the cited authors, 'olivocerebellar fibers, conceivably, drive the intracerebellar nuclei and continuing on as climbing fibers, stimulate the *Purkinje cells*. The climbing fiber – *Purkinje cell* ratio may be one to one. *Purkinje cells* actively inhibit the cerebellar nuclei and *Deiters' nucleus*. This inhibition is continuously modified by circuits activated by mossy fibers, arriving from spinal, reticular, vestibular, pontine and tegmental sources'.

'A single mossy fiber has a wide distribution. Some distribute to at least two adjacent folia and their ramifying branches in the granular layer have many nodal points (rosettes) each contacted by a group of granule cells. In turn each granule cell, having 3 to 6 dendrites, is contacted by 3 to 6 different mossy fibers. Therefore, at the input stage in the cerebellar islands, there is much divergence and some convergence'.

'The dense beam of parallel fibers generated by granule cells activates longitudinal series of *Purkinje*, upper stellate, basket and *Golgi*

---

[29] The problems of localization shall be briefly dealt with in connection with the Mammalian cerebellum (section 9 of the present chapter).

[30] Cf. vol. 3, part I, p. 631f.

cells in the molecular layer. The activity induced by the parallel fibers on *Purkinje cell* spiny dendritic branchlets must differ qualitatively[31] from that induced by climbing fibers on *Purkinje cell* smooth dendritic branches: the spiny branchlets are more distant from the *Purkinje cell* axon hillock; furthermore, the numerous and varying calibered parallel fibers allow for a greater temporal dispersion of arriving impulses'.

'The basket and stellate cells, activated by a beam of parallel fibers, inhibit by means of their transversely running axons rows of *Purkinje cells* flanking the excitatory beam of parallel fibers'.

'*Golgi cells* with their nest-like endings in the cerebellar islands have a negative feedback to the granule cells'.

'The recurrent collaterals of *Purkinje cells* inhibit other *Purkinje cells* and inhibit basket and *Golgi cells*'.

'The output stage of the cerebellar cortex feeds back to the input stage of the cerebellar cortex by means of *Purkinje cell* axon collaterals contacting *Golgi cells*, which in turn contact granule cells in the cerebellar islands'.

Be that as it may, and in particular with respect to the postulate that '*Dale's principle*' holds for the various neuronal elements of the cerebellar cortex, a sceptical attitude remains justified. It could be maintained that, despite the relative simplicity and uniformity, not to say dullness manifested by the structure of the cerebellar cortex, many questions of its functional activity remain poorly elucidated, thereby precluding a sufficiently detailed and overall valid theory of its actual performance.

With respect to the various sorts of input processed by the cerebellar cortex and transformed into an output 'modulating' the motor and other efferent activities of the neuraxis, there is little doubt that, in some Mammals (e.g. Cetaceans and Chiroptera) who have developed a *sonar system*,[32] the cerebellum plays an important role for the processing of the relevant signals. Likewise, in those Fishes (e.g. Rays and Mormyridae etc.) who possess specific *electrical guidance systems*,[33] the cerebellum seems to be significantly concerned with registration and further processing of the input data pertaining to these 'radar' systems. In particular, various peculiarities of the hypertrophic mormyrocere-

---

[31] One might here quite generally ask '*why?*', and more specifically what do the cited authors really mean by the postulated 'qualitative' difference? Greater or lesser 'temporal dispersion', as mentioned further below, would imply 'quantitative' rather than 'qualitative' differences.

[32] Cf. vol. 2, p. 186f. of this series.

[33] Cf. vol. 3, part II, chapter VIII, section 5 of this series.

bellum, originally studied by FRANZ (1912, 1914, 1920–21) and other authors, can now be conceived as related to said method of navigation by 'electrical guidance'.

The symposium on cerebellar neurobiology and evolution edited by LLINÁS (1969) contains various sophisticated but not altogether convincing nor particularly lucid papers purporting to elaborate on these complex and poorly understood topics.

Finally, the participation of cerebellar activities in *biologic clock mechanisms*[34] has been suspected by some authors (e.g. BRAITENBERG, 1967; FREEMAN, 1969). Thus, the regular spacing of *Purkinje cells* in the cerebellum of many Vertebrates suggested a definite sequential 'firing' of such cells lying along a beam of parallel fibers, this sequence being comparable to the 'ticking' of a biological clock. Although, during spontaneous 'firing', *Purkinje cells* along the same beam were found to show little tendency to discharge in a correlated sequence, a precise sequence is said to occur in response to a 'synchronous afferent volley' produced by stimulation of afferent pathways, and in response to 'angular acceleration of the animal' (FREEMAN, 1969). The cited author has presented a model purporting to show how the precise sequential firing of the *Purkinje cells* might form the basis of a 'physiological timing mechanism'[35] 'which could be utilized in the control of muscle activity'.

---

[34] Cf. vol. 3, part II, chapter VI, section 1, pp. 12–15 of this series.

[35] Some rather naive discussion remarks to the cited paper by FREEMAN (l.c. p. 417) were made by BULLOCK, who commented on the question 'of what is meant by a timing device' and stated that 'normally, it is a mechanism for converting temporal difference into a spatial difference, as in a watch'. BULLOCK then adds: 'maybe there is another meaning hidden in the proposition of timing devices that has not been articulated clearly or that I have missed, and I have a feeling that the central contribution of significance of the phrase "timing device" has yet to be made explicit'. BULLOCK does not seem to realize that the concept '*timing device*' subsumes two opposite functional aspects, namely (1) the recording and measurement of time by transformation into space (length relationship), and (2) the determination of temporal sequences (sequence programming) by the transformation of spatial into temporal relationships. It is well known and has been repeatedly emphasized (e.g. v. BONIN, 1950; K., 1957, 1961) that the *modus operandi* of neuronal networks is highly suitable for the conversion of temporal patterns into spatial ones and *vice-versa*. In contradistinction to BULLOCK, several participants in the cited discussion seem to have been familiar with the relevant concepts. Some pertinent comments on 'biologic clocks' and 'sequence programming' are included in chapter 6, section 1, pp. 12–15 of the preceding volume 3, part II.

Braitenberg (1967) refers to the following 'possible uses' of a neurobiological 'chronometric device' provided by the cerebellar 'machinery'.

(a) Delays may be interposed between the various components of rapid voluntary movements. Critical timing with a precision approaching milliseconds must here play an important role. The cerebellum, according to the cited author, may supply a temporal definition 'better than milliseconds'.

(b) Delays are essential components of devices which perform auto- and cross-correlation of functions of time, i.e. 'operations of fundamental importance in the detection of signals in noise, in the detection of movement, etc.'.

(c) 'Delays may be used to measure other delays, e.g. between successive spikes in spike sequences emitted by a neuron, for which a principle of pulse interval coding has sometimes been postulated (MacKay, 1962)'.

2. Cyclostomes

The cerebellum of *Petromyzont Cyclostomes* consists of a median transverse corpus which is laterally continuous with the vestibulo-lateral area (cf. Figs. 153A, 192A). The transition between corpus cerebelli and vestibulo-lateral region is not clearly delimited, nor is this latter region configurated as a typical auricula. Nevertheless, the Petromyzont cerebellum can be said to display rather clearly the two basic morphologic components stressed by Larsell (cf. above p. 624). The crista cerebelli within the area statica or vestibulo-lateral region was described in section 3 of the preceding chapter IX.

As regards *neuronal elements*, small granule cells, rather nondescript medium sized cells, and rudimentary *Purkinje cells* are present. A primordial *nucleus cerebelli*, closely related to the rostral vestibulo-lateral area, can be recognized. Figures 305A–C illustrate some of the overall aspects of the Petromyzont cerebellum.

Concerning the *cerebellar input and output channels*, the former are represented by primary and secondary fibers of the octavolateral and trigeminal systems, as well as by some spinocerebellar and tectospinal fibers. In addition, hypothalamo-cerebellar (lobo-cerebellar) fibers arising in the caudal inferior portion (pars inferior s. postoptica hypothalami) seem to gather in a rather vaguely discernible 'tract' which may include all or some of the 'general visceral' input presumably reaching the cerebellum.

Cerebellar output is provided by the neurites of the *Purkinje cells* which form a system of internal arcuate fibers continuous with those of the area statica. These fibers radiate caudoventrally and rostrally from the basal part of the cerebellum. According to LARSELL (1967), the caudal part of that cerebello-tegmental radiation includes crossed and uncrossed fibers descending in oblongata and perhaps even reaching the

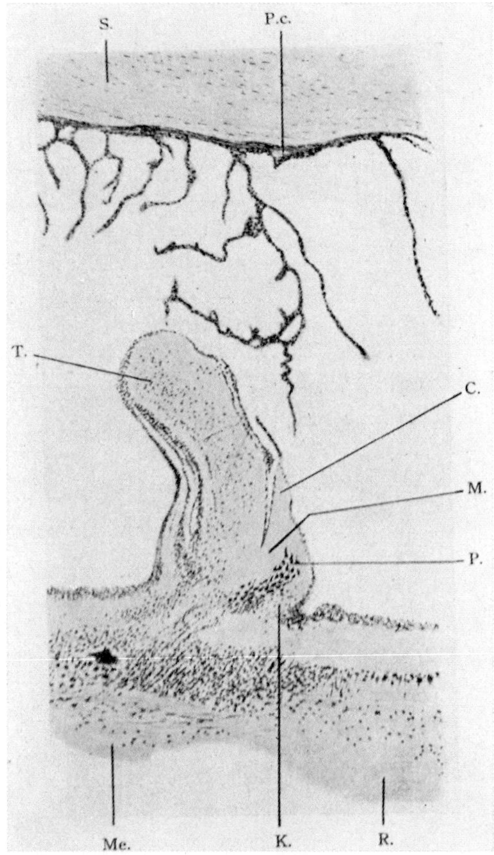

Figure 305 A. Sagittal section through caudal mesencephalon and rostral rhombencephalon of Entosphenus japonicus *(Nissl stain)* showing corpus cerebelli with its layers (from SAITO, 1930). C.: cerebellum; K.: granular layer; M.: molecular layer; Me.: tegmentum mesencephali; P.: *Purkinje cells;* P.c.: choroid plexus; R.: rhombencephalon. S.: cranial cartilage; T.: tectum mesencephali.

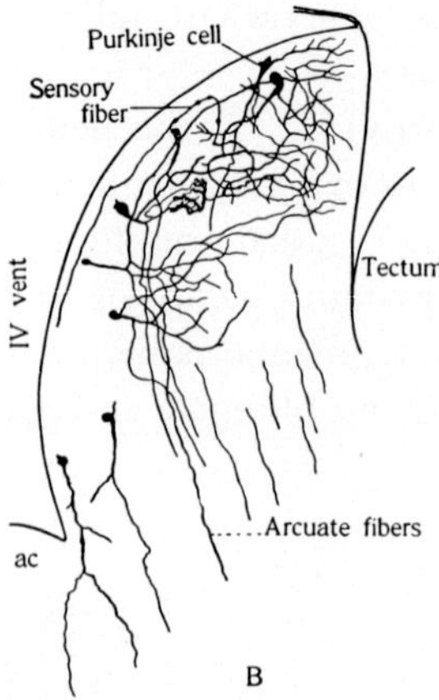

Figure 305B. Sagittal section *(Golgi impregnation)* through the corpus cerebelli of the Petromyzont Lampetra (from JOHNSTON, 1906). ac: area acusticolateralis.

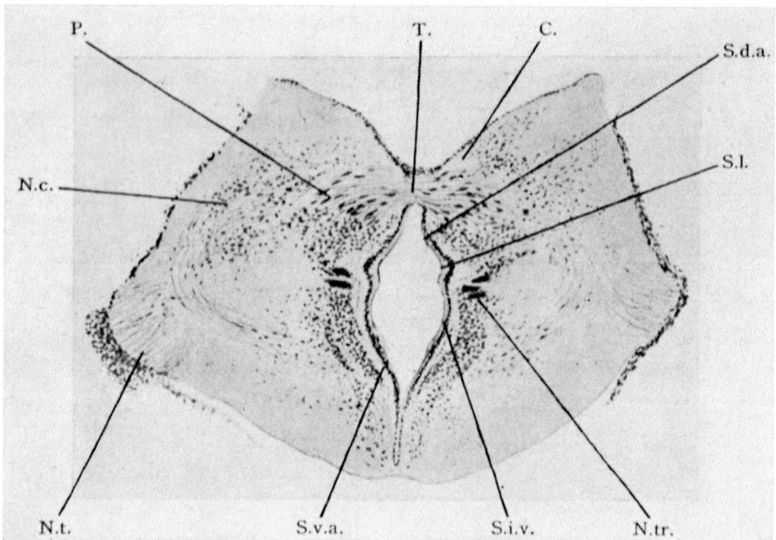

Figure 305C. Cross-section through isthmus region and base of corpus cerebelli of Entosphenus japonicus *(Nissl stain)*. The plane of the section is slightly oblique in a dorsorostral to ventrocaudal inclination (from SAITO, 1930). C.: cerebellar plate (corpus cerebelli); N.c.: nucleus cerebelli; N.t.: nervus trigeminus; N.tr.: nucleus nervi trochlearis; P.: *Purkinje cells;* S.d.a.: sulcus dorsalis intermedius; S.i.v.: sulcus intermedioventralis; S.l.: sulcus limitans; S.v.a.: sulcus ventralis accessorius; T.: decussating trochlear root fibers mingled with cerebellar commissural fibers.

spinal cord (so-called 'tractus cerebello-motorius' including a 'tractus cerebello-spinalis'). Some cerebellar output within the tractus cerebello-motorius appears augmented by fibers arising in the nucleus cerebelli, which receives direct or collateral terminals of *Purkinje cell* neurites. Intermediate and more rostral cerebello-tegmental fibers accompany the anterior octavomotor fascicle and reach the mesencephalic tegmentum, thereby including a 'forerunner' of the brachium conjunctivum. Additional rostral cerebellar output fibers are said to reach the tectum and torus semicircularis mesencephali as well as the diencephalon (hypothalamus and thalamus).

The *commissural system* of the Petromyzont cerebellum has been variously described by different authors who endeavoured to distinguish distinctively definable components (cf. LARSELL, 1967). Discounting these not very convincing attempts it could be said that the well recognizable but nondescript cerebellar commissure doubtless contains vestibulo-lateral, trigeminal, spino-cerebellar and various categories of efferent cerebellar fibers, being, moreover, as again to be mentioned further below, intermingled with decussating root fibers of the trochlear nerve.

Despite numerous studies concerning the Petromyzont brain (e.g. JOHNSTON, 1902; TRETJAKOFF, 1909; SAITO, 1930; HEIER, 1948; cf. also LARSELL, 1967) many relevant topics remain poorly elucidated and controversial. Thus, even the evaluation of the large cerebellar neuronal elements as *Purkinje cells* has been contested, but, as I believe, with not much justification. However, it does not seem that the dendrites of these primordial Purkinje cells are definitely arranged in a single sagittal plane as is generally the case in gnathostome Vertebrates (cf. section 1, p. 632 and section 3, p. 674). Another moot question is the identification of the nucleus nervi trochlearis. According to ADDENS, KAPPERS, LARSELL, TRETJAKOFF and other authors, the trochlear nucleus lies dorsally to the sulcus limitans[36] at the basis of the

[36] Such dorsal location of a motor nucleus, although apparently contradicting a fairly well established 'rule' based on actual observations, should, of course, not dogmatically be considered 'impossible', since, depending on multiple parameters, most or all 'laws' or 'rules' include exceptions. On the other hand, despite meticulous and successful techniques, the identification of the small trochlear nucleus in many lower Vertebrates remains rather difficult and uncertain, particularly in forms with poorly developed eye muscles, and the interpretations of authors claiming an alar plate location of said nucleus (be it 'primary' or 'secondary') have failed to convince me. It should also be mentioned that a caudal and rather dorsal location of the trochlear nucleus has been reported for the Dipnoan Ceratodus and for some urodele Amphibians (cf. the discussion by BECCARI, 1943, pp. 275–278).

cerebellar plate or within the velum medullare anterius, and at a considerable distance caudally to the oculomotorius nuclei. According to SAITO (1930), whose findings agree with my own observations, the nucleus nervi trochlearis has indeed a caudal position, being located within the isthmus region, but lying, however, ventrally to the sulcus limitans in the dorsal portion of the intermedioventral zone and apparently representing the rostral extremity of the trigeminal motor column (cf. Fig. 305C).[37] It is most likely that, at least in some instances (e.g. in Fig. 208, p. 460 of KAPPERS, 1920) *Purkinje cells* have been mistaken for trochlearis motoneurons. Likewise, since the decussating trochlearis fibers are intermingled with those of the cerebellar commissure, the fibers depicted by KAPPERS (l.c.) might include both true trochlearis root fibers[38] and fibers of the cerebellar commissure.

In contradistinction to Petromyzonts, *Myxinoids* do not possess a morphologically and functionally significant cerebellum, although, from the formanalytic viewpoint, the homologa of both cerebellar plate (corpus cerebelli) and vestibulo-lateral region (auricular lobes) are present.[39] *Purkinje cells* as well as other histologic characteristics of cerebellar structure cannot be recognized with any reasonable degree of certainty and thus appear to be, for practical purposes, entirely missing. The corpus cerebelli is a narrow ridge caudal to the mesencephalon (cf. Fig. 193A) which, however, contains what could be called a nondescript 'cerebellar commissure' formed by octavolateral and trigeminal fibers.[40] Again in contradistinction to Petromyzonts, a conspicuous bilateral auricle-like configuration is present, containing trigeminal and octavolateral grisea[41] (cf. Figs. 194, 195B). This 'horn'

---

[37] Further reference to the still poorly elucidated 'trochlearis problem' can be found in chapter VII, Section 3, p. 843 of the preceding volume 3, part II.

[38] Although in most Vertebrates practically all trochlearis root fibers decussate in the velum medullare anterius before their exit, a few homolateral, non-decussating trochlearis fibers are said to occur in Petromyzon and some Teleosts (cf. the discussion by BECCARI, 1943, cited above in footnote 36).

[39] According to BONE (1963) 'the outstanding problem of the brain of Myxine is undoubtedly the vexed question of the presence of a cerebellum'. One could reply to this remark that we have here merely a question of arbitrary semantics, depending on what one chooses to define as a 'cerebellum'. Thus both the authors affirming the presence of a Myxinoid cerebellum and those denying its presence are doubtless 'right'.

[40] Rather irrelevant minor differences between the cerebellum of Myxine and that of Bdellostoma are discussed, with reference to the pertinent literature, by LARSELL (1967).

[41] In contradistinction to the trigeminal system, the octavolateral system of Myxinoids displays a substantial 'reduction'.

or 'auricle', however, does not have a cerebellar structure, and differs from a true auricle by being solid, i.e. by not representing a rostrolateral or lateral extension of the rhombencephalic ventricle.

3. Selachians

The well-developed cerebellum of *Selachians* consists of a relatively large dorsally everted *corpus* and a likewise substantial *auricle*. This latter includes a dorsal lip, continuous with the caudal end of corpus cerebelli, and a ventral (or ventrolateral) lip, continuous, as *crista cerebelli*, with the octavolateral area (Figs. 306A, B, C, D). The auriculae, together with the dorsal lip portion and caudal end of corpus cerebelli presumably correspond to the lobus vestibulo-lateralis or archicerebellum of LARSELL, separated from the palaeocerebellum by the lateral sulci para-auriculares and the transverse '*sulcus postremus*' which, together, may represent the '*fissura posterolateralis*' of higher forms (e.g. Birds, and Mammals). Again, a rather deep furrow may separate the cerebellar crest corresponding to the level of LARSELL's 'nucleus dorsalis' from that related to his 'nucleus medialis'[42] (cf. e.g. Figs. 202 and 203). Rostralward, however, both bands of crista merge (LARSELL, 1967).

Considerable differences obtain, in accordance with the different Chondrichthyan taxonomic groups, as regards relative development and configuration of corpus cerebelli and of auricles. KAPPERS *et al.* (1936) justly point out that both subdivisions exhibit a certain independence of each other with respect to their morphologic relations and in their fiber connections. This independence and the just mentioned differences, in turn, seem mainly to depend upon the degree of development displayed by the various main cerebellar input channels (spinocerebellar, olivo-cerebellar, and octavolateral systems).

In some instances, the dorsally everted portion of the corpus is entirely smooth (e.g. in Scilliorhinus canicula), while in some forms (e.g. Acanthias) a single transverse sulcus is present; other forms (e.g. Mustelus and still more pronouncedly Lamna) display a number of additional transverse sulci of varying depth resulting in the formation of 'lobi' and 'lobuli' (cf. Fig. 307). All these folds, however, essentially affect the entire wall of the cerebellar plate in contradistinction to the

---

[42] Cf. section 4, p. 390 of the preceding chapter IX.

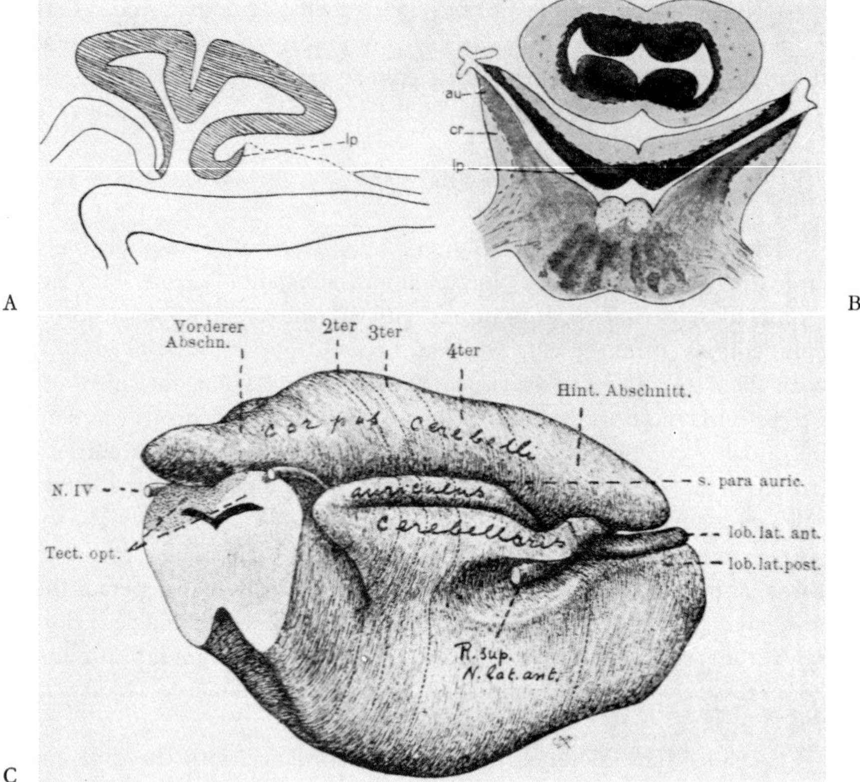

Figure 306 A. Simplified sketch illustrating the configuration of a Selachian cerebellum (Acanthias) as seen in midsagittal section (from K., 1927). lp: posterior cerebellar lip; the transverse groove between this lip and the main corpus is the 'sulcus postremus' perhaps representing the 'fissura posterolateralis' in Birds and Mammals.

Figure 306B. Simplified sketch illustrating the configuration of a Selachian cerebellum as seen in cross-section (from K., 1927). au: auricula; cr: crista cerebellaris; lp: posterior cerebellar lip, laterad continuous with upper leaf of auricle; above the lip, the corpus cerebelli with its granular eminences can easily be identified.

Figure 306C. Wax model of the cerebellum in the shark Spinax niger (after Voorhoeve, from Kappers, 1921). Four rostro-caudal subdivisions of the corpus are here distinguished, although not separated by well definable sulci. Other abbreviations self-explanatory. Read auricula instead of auriculus, and note sulcus para-auricularis separating its upper leaf from corpus. This sulcus crosses the midline as sulcus postremus (cf. Figure 306A).

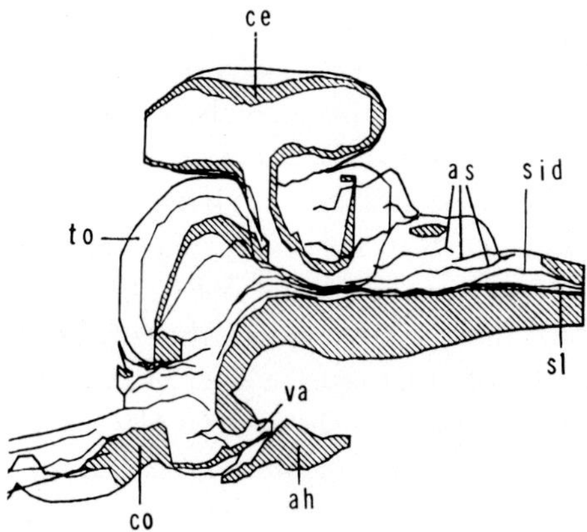

Figure 306D. Part of graphic reconstruction, projected upon midline plane, of the brain in Chimaera Colliei, showing configuration of corpus cerebelli (from K. and NIIMI, 1969). ah: adenohypophysis; as: accessory sulci; ce: cerebellum; co: chiasmatic ridge; sid: sulcus intermedius dorsalis; sl: sulcus limitans; to: tectum mesencephali; va: attachment (rostral end) of saccus vasculosus; a shallow transverse sulcus corresponds to the lead ce.

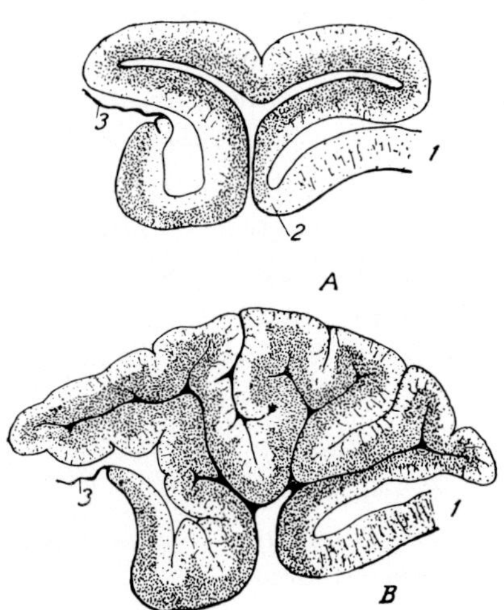

Figure 307. Parasagittal sections through the cerebellum of Acanthias vulgaris (A) and of Mustelus laevis (B) showing species differences with respect to transverse sulci of corpus cerebellum (modified after STERZI, from BECCARI, 1943). 1: tectum mesencephali; 2: velum medullare anterius; 3: attachment of epithelial roof of fourth ventricle.

cerebellar sulci displayed by Birds and Mammals. In these Amniota, the sulci do not involve the ventricular surface.

Some asymmetries of the corpus cerebelli likewise occur in Plagiostome forms where this subdivision shows a substantial expansion (e.g. in Lamna and Myliobates, as illustrated by KAPPERS, 1921). A longitudinal (midsagittal) sulcus[43] is likewise displayed in forms, and may be combined with a transverse one (Figs. 308A, B).

As regards the cellular arrangements, the corpus cerebelli displays a prominent external molecular layer, a layer of *Purkinje cells* whose dendrites are generally arranged in the sagittal plane, a modified granular layer, and an essentially medullated fibrous layer at the level of *Purkinje cell* and granular layer. In the lateral part of the corpus, the granular layer becomes greatly reduced by the prevalence of the fibers, and in some instances seems even almost negligible. The bulk of the granular layer is provided by thick longitudinal bilateral medial folds appearing dorsally and ventrally as granular eminences (Figs. 309A–D).

The *Purkinje cells* become scarce or disappear near the midline and the compactness of the fibrous layer is reduced as its axons spread out within the granular eminences. Auricles and crista cerebelli display, in addition to molecular and *Purkinje cell* layers, a substantial granular layer. With respect to neuronal elements, stellate cells are present in the molecular layer, but typical basket cells have not been demonstrated. The Plagiostome *Purkinje cells,* described in detail by LARSELL (1969) and others, may be somewhat more scarce and primitive in auriculae and crista than in corpus. The granular layer includes *Golgi II* elements ('large granule cells'). Among the afferent terminals climbing fibers and 'primitive' mossy fibers have been identified, but their glomerular synapses with granule cells, as characteristic for higher Vertebrates, have not been demonstrated (LARSELL, 1969). The supporting elements may include primitive *Bergmann cells.*[44]

The *nucleus cerebelli*[45] *(nucleus lateralis cerebelli* of EDINGER, 1901) is usually found in the so-called eminentia ventralis cerebelli (Figs. 310A,

---

[43] This sulcus may, as KAPPERS *et al.* (1936) point out, be related to the bilateral (alar plate) origin of the corpus cerebelli.

[44] Those interested in further details concerning the cerebellar neuroglia of Plagiostomes and lower Vertebrates in general, may be referred to LARSELL's comprehensive monograph (1967) which contains an extensive bibliography.

[45] The interpretation of its origin by RÜDEBERG (1961), purporting to disclaim a relationship of this griseum to the octavolateral area, was cited above in section 1, p. 650, footnote 19 of the present chapter.

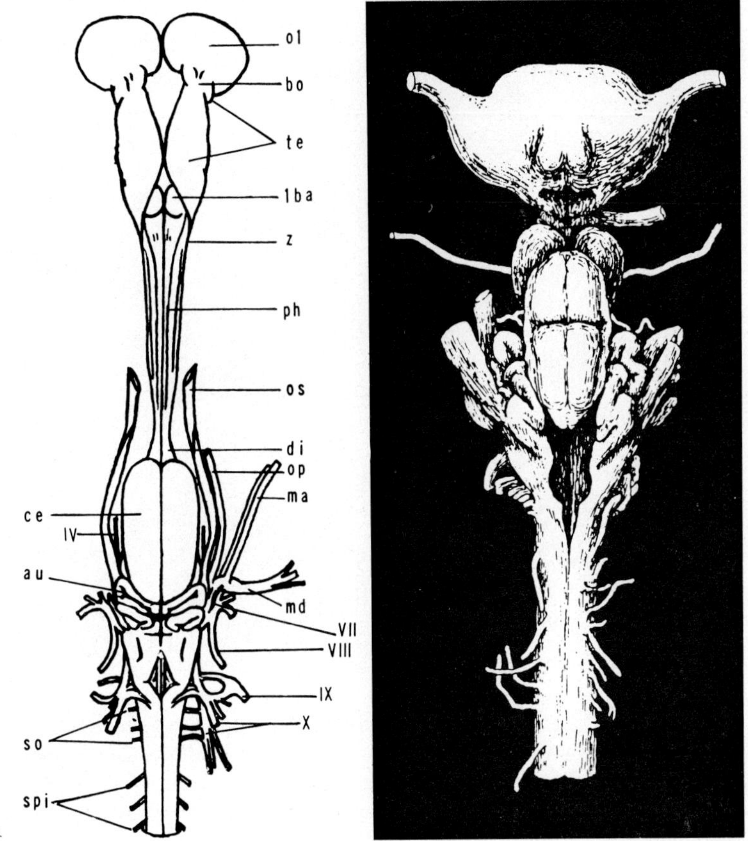

Figure 308A. Semidiagrammatic sketch of the brain of Chimaera Colliei in dorsal view, showing the longitudinal median sulcus of corpus cerebelli (from K. and Niimi, 1969). au: cerebellar auricle; ce: corpus cerebelli; di: diencephalon; lba: bulges of telencephalic basal lobes; ma: maxillary branch of V; md: mandibular branches of V; ol: olfactory organ (nasal cavity); op: nervus ophthalmicus profundus; os: nervus ophthalmicus superficialis; ph: preoptic portion of hypothalamus; so: spino-occipital nerve roots; spi: spinal nerve roots (ventral); te: telencephalon; z: zone of telo-diencephalic boundary. The corpus cerebelli covers the tectum mesencephali, cf. Figure 306D. Compare also with Figures 305A, B, 308B, and 309D. bo: bulbus olfactorius.

Figure 308B. Dorsal view of the brain in the Rajid Plagiostome Raja clavata, showing transverse and median longitudinal sulcus of corpus cerebelli (from Kappers, 1921).

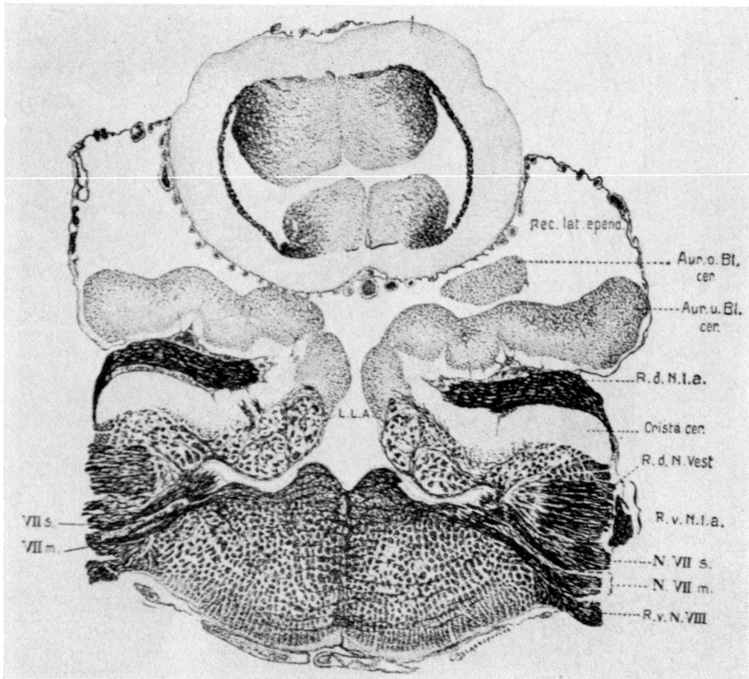

Figure 309A. Cross-section (myelin stain) through the caudal portion of the cerebellum in Scyllium canicula (from Kappers, 1921). Aur. o. Bl., Aur. n. Bl.: upper respectively lower lip of auricula; LLA: lobus lobi lateralis anterioris; R. d. N. l. a.: ramus dorsalis nervi lateralis anterioris; R. d. N. V.: ramus dorsalis nervi octavi s. vestibularis; R. v. N. l. a.: ramus ventralis nervi lateralis anterioris; N. VII s.: afferent facialis root fibers; N. VII m.: efferent facialis root fibers.

Figure 309B. Low power transverse section showing corpus cerebelli in Acanthias (hematoxylin-eosin; ×12; red. $^2/_3$). 1: corpus cerebelli; 2: mesencephalon; 3: n. oculomotorius; 4: saccus vasculosus; 5: posterior hypothalamus; 6: adenohypophysis; arrow: sulcus limitans.

Figure 309C. Details of corpus cerebelli from section of preceding figure (×48; red. $^2/_3$). a: molecular layer; b: *Purkinje cell layer;* c: fibrous layer within reduced granular layer; d: eminentia granularis. Although some of the cells in c are glial elements, others are doubtless granule cells as can be substantiated by means of suitable *Golgi preparations.*

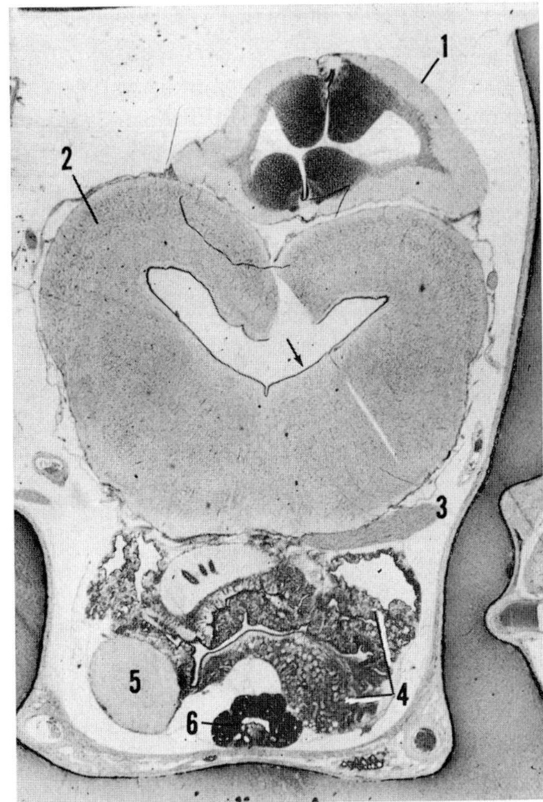

309 B

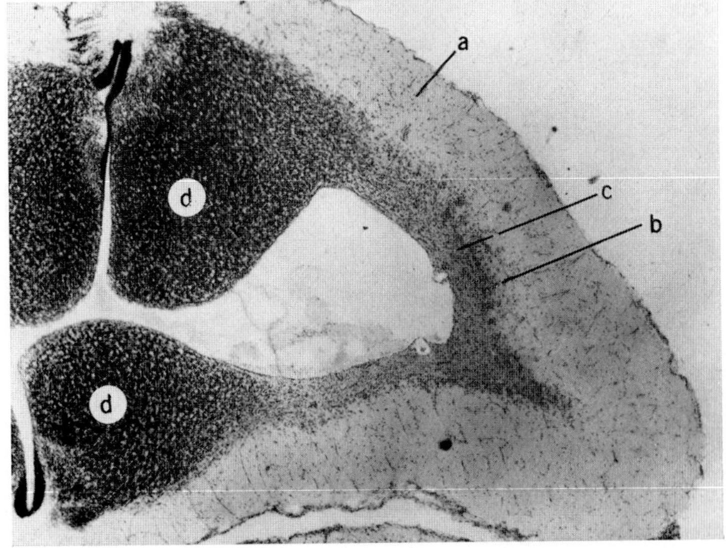

309 C

B). It doubtless receives efferent cortical cerebellar fibers *(Purkinje cell* neurites), presumably from corpus but perhaps also from the auricular portion. However, the details of both its input and output remain unclarified. It can merely be surmised that this nucleus from which fibers of the 'tractus cerebello-motorius' may arise pertains to the overall cerebellar output system, although it is also believed to receive some lateralis input (KAPPERS, 1921).

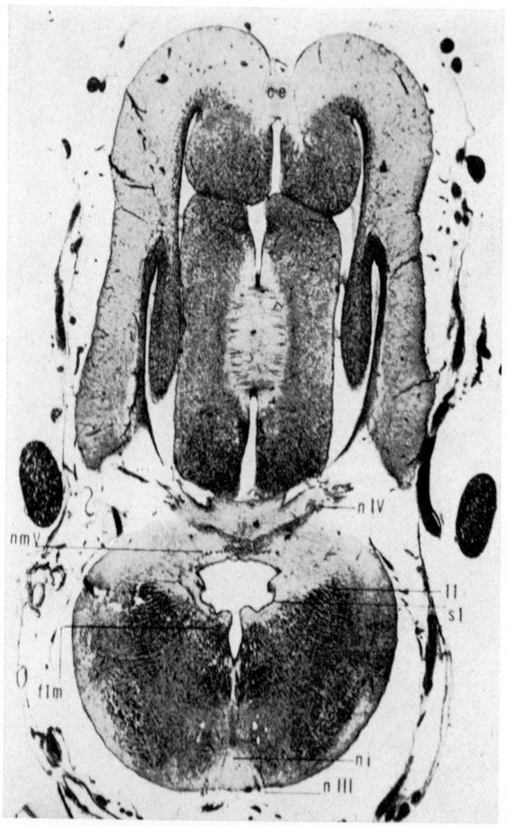

Figure 309D. Cross-section through the isthmus region in the brain of Chimaera Colliei showing corpus cerebelli with granular eminences (from K. and NIIMI, 1969). ce: corpus cerebelli (the label is located at the bottom of the longitudinal sulcus); flm: fasciculus longitudinalis medialis; ll: lateral lemniscus; nmV: nucleus radicis mesencephalicae V; ni: nucleus interpeduncularis; nIII: oculomotor root; nIV: trochlear root; sl: sulcus limitans.

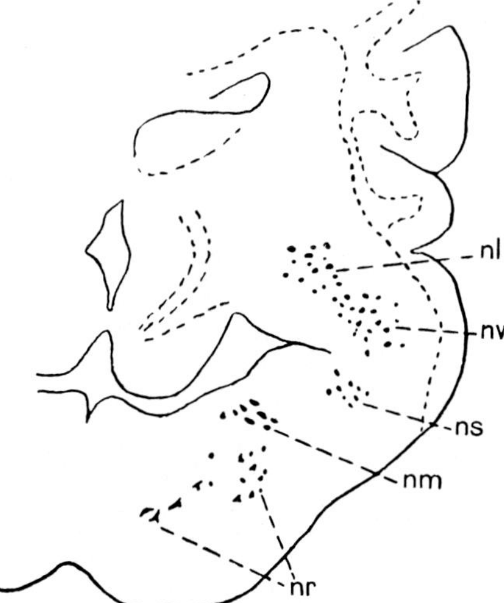

Figure 310A. Sketch showing the cerebellar nucleus in a Selachian (simplified after VAN HOEVELL, 1916, KAPPERS, 1921, and on the basis of personal observations, from K., 1927). nl: nucleus (lateralis) cerebelli; nm: nucleus motorius trigemini; nr: nucleus reticularis tegmenti; ns: nucleus sensibilis trigemini; nv: nucleus vestibularis. The original figure of VAN HOEVELL referred to Selache maxima.

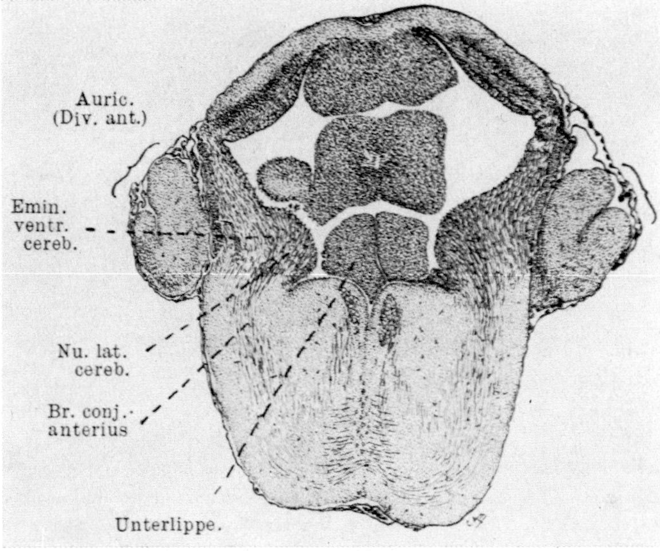

Figure 310B. Cross-section (myelin stain) through the cerebellum of the Plagiostome Spinax niger at the level near rostral end of auriculae (after VOORHOEVE, from KAPPERS. 1921).

The cerebellar input (Figs. 312A, B) comprises a spinocerebellar tract with various nondescript subdivisions, an olivocerebellar system intermingled with some of the spinocerebellar channels, a trigeminocerebellar system, a mesencephalo-cerebellar, and a 'lobocerebellar sys-

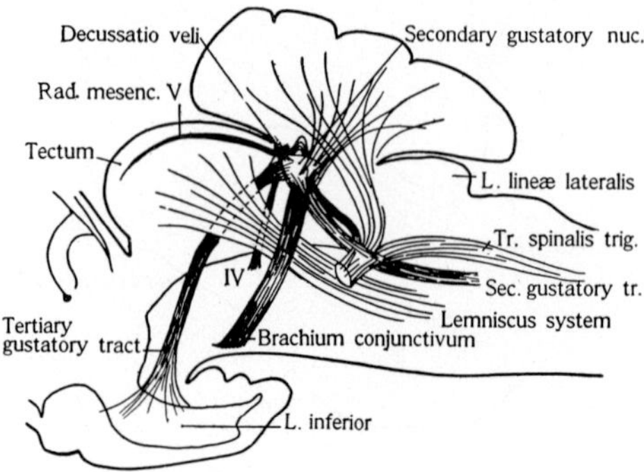

Figure 311A. Relations of cerebellum, 'brachium conjunctivum' and some other fiber tracts in Selachians, particularly based on findings in Scyllium, as interpreted by JOHNSTON and projected upon the midsagittal plane (from JOHNSTON, 1906).

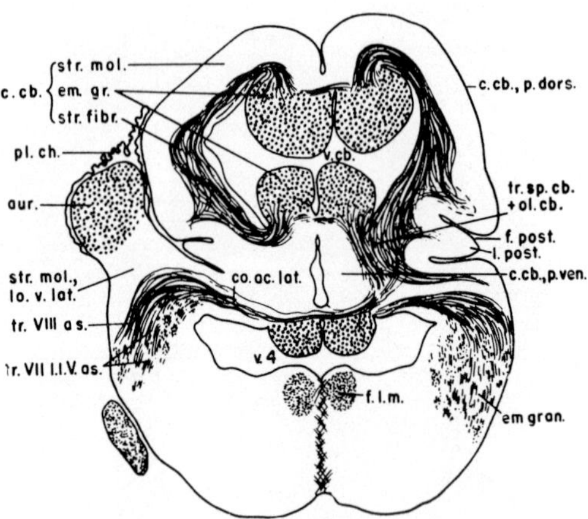

Figure 311B. Cross-section through the cerebellum of Scyllium, showing, on right side, entrance of tractus spinocerebellaris and olivocerebellaris into corpus cerebelli (from LARSELL, 1967). c.cb.: corpus cerebelli; c.cb.p.ven.: ventral part of corpus cerebelli; em.gran.: so-called 'eminentia granularis' of vestibulo-lateral lobe, related to crista cerebelli; f.post.: 'posterior fissure'; l.post.: 'posterior lobe of cerebellum'; tr.sp.cb.+ ol.cb.: tractus spinocerebellaris and olivocerebellaris; tr.VII as.: ascending lateral line tract of VII; tr.VIII as.: ascending vestibular tract. Other abbreviations self-explanatory.

tem' (from the posterior hypothalamus). These channels[46] reach the corpus, which also seems to receive some octavolateral input and perhaps also (primary or secondary) visceral afferent fibers from the branchial nerve system (e.g. from IXth and Xth nerve). The input of the auricular portion appears to consist mainly of primary and secondary lateralis and vestibularis fibers.

Recent data on spinocerebellar systems in Selachians (as well as Teleosts and Amphibians) can be found in a paper by HAYLE (1973).

The *cerebellar output*, apparently mainly mediated by the *Purkinje cell* neurites, with some contribution from the nucleus cerebelli, includes rostral, intermediate, and caudal channels. The rostral component, which becomes essentially crossed and roughly corresponds to the brachium conjunctivum of higher Vertebrates, reaches, *inter alia*, the reticular formation of the mesencephalic tegmentum (including a 'precursor' of the nucleus ruber), and the posterior hypothalamus. Some of its fibers are believed to connect with thalamic grisea, and others are said to join the fasciculus longitudinalis medialis in which they may, in part, then run caudalward. The intermediate crossed and uncrossed cerebellobulbar component reaches the octavolateral grisea (e.g. the so-called tractus cerebellovestibularis) and tegmental bulbar grisea. The caudal component or tractus cerebello-motorius *sensu strictiori* seems to take its course through the homolateral and contralateral tegmentum, near or even in the fasciculus longitudinalis medialis. These fibers are said to end in relation with reticular cells and motoneurons of the oblongata. A caudal extension of this channel reaches the spinal cord (LARSELL, 1967, and others), representing a tractus cerebello-spinalis. Details concerning the question to which extent corpus and auricles participate in the composition of the efferent channels remain uncertain.

4. Ganoids and Teleosts; Latimeria

The cerebellum of *Ganoids* and *Teleosts* differs from that of Plagiostomes by the peculiar inversion of a rostral portion of corpus cerebelli designated as *valvula cerebelli*, which becomes invaginated, to a lesser or

[46] Cf. e.g. EDINGER (1901), KAPPERS (1921), BECCARI (1943), GERLACH (1947), LARSELL (1967), and K. and NIIMI (1969). Again, the question, to which extent this input is homolateral or contralateral, remains incompletely elucidated. Some of the input fibers doubtless decussate in the cerebellar commissures mentioned above in the present section and in section 1 of this chapter.

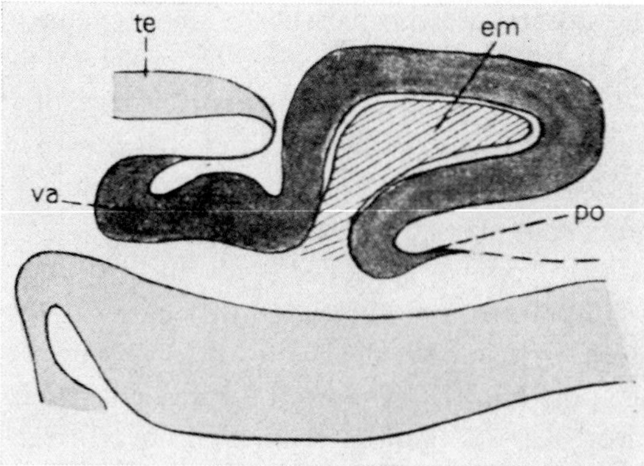

Figure 312A. Semidiagrammatic sagittal section through the cerebellum of a young Teleost shortly after hatching, showing valvula cerebelli and almost completed secondary fusion (concrescence) of the eminentiae granulares (based on figures of SCHAPER and personal observations, from K., 1927). em: secondary fusion *(Verwachsung)* of the eminentiae granulares; po: 'pars postrema' (caudal lip of corpus), separated from corpus proper by the 'sulcus postremus' (not labelled); va: valvula cerebelli.

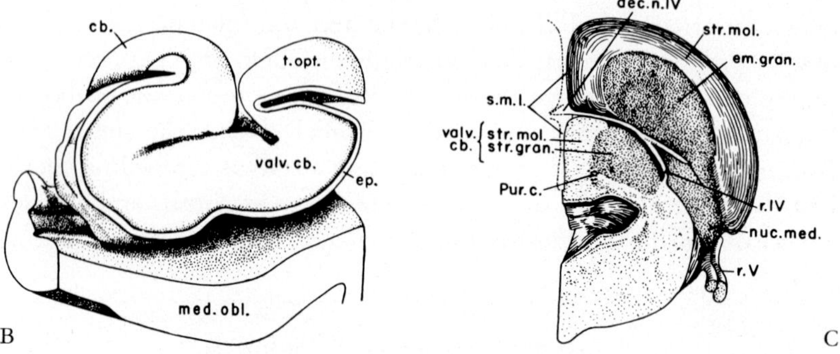

Figure 312B. Midsagittal aspect of cerebellum in the Ganoid Polypterus bichir, as seen in a wax model (after VAN DER HORST, 1919, from LARSELL, 1967). cb.: posterior part of corpus cerebelli; ep.: ependyma; t.opt.: tectum mesencephali; valv.cb.: valvula cerebelli.

Figure 312C. Frontal view of wax model in preceding figure (after VAN DER HORST, 1919, from LARSELL, 1967). dec.n.IV: decussation of trochlear nerve; em.gran.: eminentia granularis; nuc.med.: 'nucleus medialis' (nucleus ventralis lineae lateralis anterioris); Pur.c.: *Purkinje cells;* r.IV, r.V: roots of trochlear and of trigeminus nerves; s.m.l.: sulcus medianus longitudinalis; valv.cb.: valvula cerebelli; str.mol., gran.: molecular respectively granular layer.

greater degree, into the mesencephalic ventricle (Figs. 312, 316A). Early ontogenetic stages, as e.g. depicted in Figures 52, 62I, and 71 of volume 3, part II, indicate that the valvula originates from a transverse cerebellar plate fully comparable to the anlage of corpus cerebelli in other Vertebrates.

In *Ganoids*, the valvula cerebelli is of relatively moderate extension, but *Teleosts* display a substantial and in some instances even an extreme development of this configuration (cf. Figs. 317, 319, 320). Concomitantly with the expansion of the Teleostean valvula cerebelli, the antimeric parts of the tectum mesencephali may undergo a lateralward displacement. The commonly obtaining midline fusion of the alar plates, combined with disappearance of the epithelial roof plate, becomes thereby impossible, and a thin roof plate is thus retained (cf. Fig. 316A). Rostralward, however, the midline fusion of mesencephalic alar plates usually still obtains and even commonly includes a secondary midline fusion of the mesencephalic so-called *torus longitudinalis* (cf. chapter XI, section 4). In addition, the development of valvula cerebelli may encroach upon or invade the velum medullare anterius

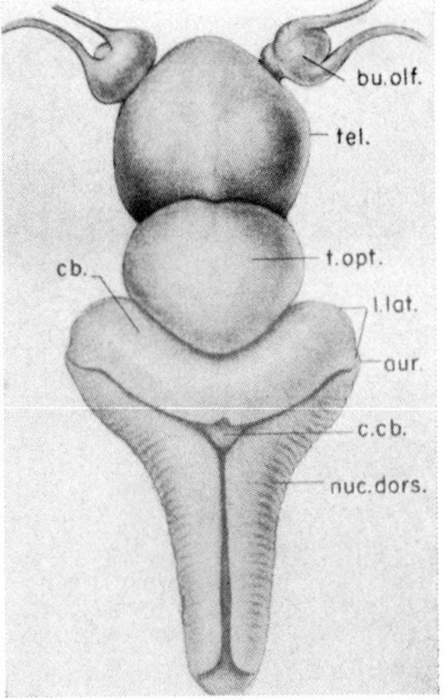

Figure 312D. Dorsal view of brain in the Ganoid Polyodon folium lacep, showing some configurational relationships of cerebellum (after HOCKE HOOGENBOOM, 1929, from LARSELL, 1967). aur.: auricula; bu.olf.: olfactory bulb; cb.: cerebellum (posterior lip); c.c.b.: corpus cerebelli; l.lat.: 'lateral lobule of cerebellum'; nuc.dors.: 'nucleus dorsalis' (n. lineae lateralis anterioris); tel.: telencephalon; t.opt.: tectum mesencephali.

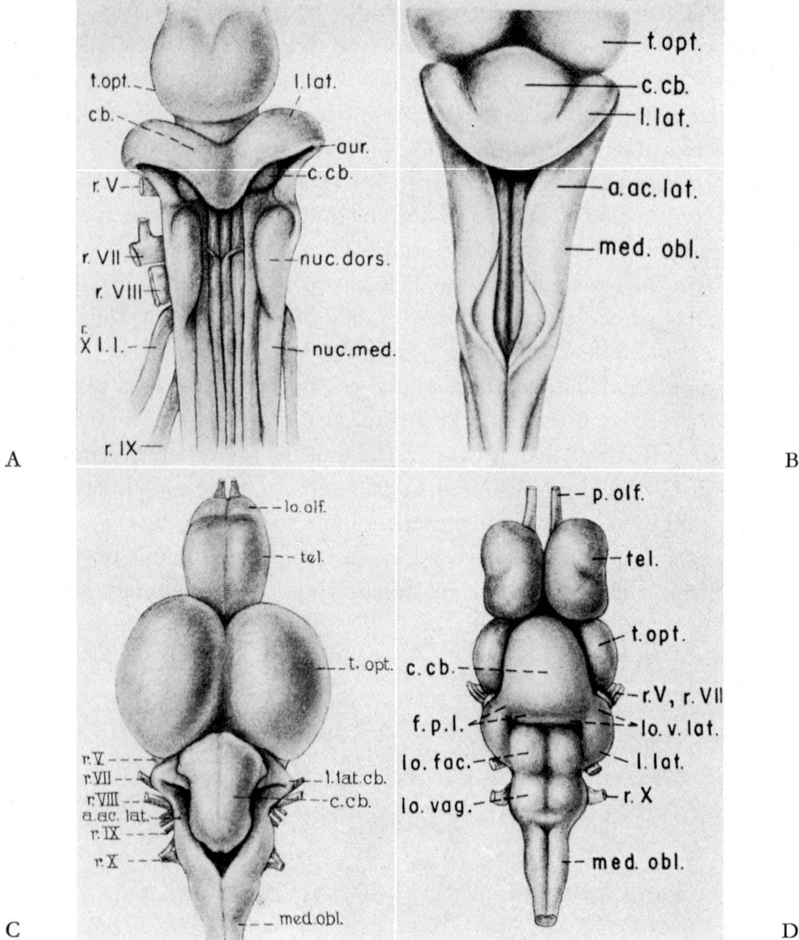

Figure 313 A. Dorsal view of brain stem and cerebellum in the Ganoid Acipenser (after JOHNSTON, 1901, from LARSELL, 1967). r: roots of nerves V, VII, VIII, IX, and of lateral line component of X.

Figure 313 B. Dorsal view of brain stem and cerebellum in the Ganoid Amia (after KINGSBURY, 1897, from LARSELL, 1967). a.ac.lat.: octavolateral area; med.obl.: medulla oblongata *(sensu strictiori)*. Other abbreviations as in preceding figures from LARSELL.

Figure 313 C. Dorsal view of brain in a young (Teleost) Salmon, Onchorhynchus spec. (from LARSELL, 1967). Abbreviations as in preceding figures.

Figure 313 D. Dorsal view of brain in the Teleost Ameiurus (after C.L. HERRICK, 1891, from LARSELL, 1967). f.p.l.: 'posterolateral fissure'; lo.fac.: lobus facialis; lo.vag.: lobus vagi; lo.v.lat.: 'vestibulolateral lobe. Other abbreviations as in preceding figures.

which can be replaced by a caudal extension of the mesencephalic epithelial roof plate. A recess of extracerebral (extraventricular) intracranial space may thereby protrude into the mesencephalic ventricle (cf. e.g. Fig. 316A). While the decussation of the trochlear nerve of at least some Ganoids is located in a still fairly typical velum medullare anterius, this decussation assumes a position within a fairly caudal part of valvula cerebelli in apparently numerous Teleosts (cf. Fig. 378 F). With respect to its main input, the valvula cerebelli is substantially related to the lateralis system, apparently receiving root fibers of the anterior lateral nerve in addition to secondary and perhaps higher order more caudal lateralis input, some of which seems to be mediated by the so-called nucleus lateralis valvulae.[47]

Rather typical *auriculae*[48] are present in Ganoids and some Teleosts (Figs. 313A–C), while in other Teleosts this region becomes massive (Fig. 313D) with comcomitant obliteration of its ventricular recess. According to KAPPERS *et al.* (1936) this change is related to a great increase of the Teleostean eminentiae granulares or granular masses both of corpus cerebelli and auriculae. The eminentiae granulares, as particularly shown by SCHAPER (1894), tend to fuse in the midline of the corpus cerebelli thereby obliterating most of the cerebellar ventricle (Fig. 312A) and extend rostralward toward the valvula. Caudalward the conjoint eminentiae granulares become continuous with auricular region and cerebellar crest, usually forming a lateral protrusion (lateral granular eminence) of the cerebellum's surface (Figs. 223A, B, 316C). In Teleosts, this eminence commonly lacks a molecular layer, which still overlies the granular eminence in at least various Ganoids (Fig. 314B).

Among *Ganoids*, the cerebellum of Polypterus, described by VAN DER HORST (1919), displays a corpus divided into bilateral halves by a

---

[47] The fiber system arising in this griseum is commonly designated as 'tractus mesencephalo-cerebellaris posterior' (cf. KAPPERS *et al.*, 1936).

[48] Several differences in the evaluation of valvula cerebelli, auriculae, granular eminences and fiber tracts have been expressed by the various authors who investigated this topic (ADDISON, 1923; BERKELBACH VAN DER SPRENKEL, 1915; FRANZ, 1911, 1913, 1920; GORONOWITSCH, 1888; HERRICK, 1924; HOOGENBOOM, 1929; VAN DER HORST, 1919, 1926; JOHNSTON 1901, 1906; KAPPERS *et al.*, 1936; LARSELL, 1967; A. A. PEARSON, 1936). These differences, although in part due to the ambiguity of the actual findings provided by the diverse technical methods, are to a substantial degree of a purely semantic nature, related to the differing morphologic views and definitions adopted by the individual authors.

'sulcus medianus longitudinalis' (Figs. 312, B, C). Only the caudal portion of corpus is plainly visible in dorsal view. Somewhat more of the corpus is exposed in Polyodon and in Acipenser; the posterior corpus cerebelli of this latter also shows a shallow midsagittal sulcus (Figs. 313A, 314A). Comparing Figure 313C with 313D it can be seen that in some *Teleosts* (perhaps the majority), the corpus cerebelli extends, or is bent, caudalward, while in others, such as Siluroids, it extends rostralward, covering part of the tectum mesencephali. In certain forms, this rostral protrusion may extend beyond the mesencephalon (cf. ARONSON, 1963). The peculiarities of the extremely large *Mormyrid* cerebellum will be dealt with further below.

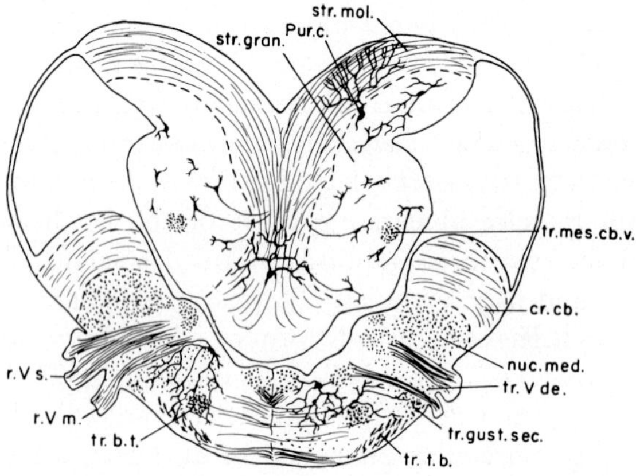

Figure 314A. Cross-section (Golgi impregnation) through the rhombencephalon of the Ganoid Acipenser at the level of the sensory (s) and motor (m) trigeminal roots (after JOHNSTON, 1906, from LARSELL, 1967). cr.cb.: cerebellar crest; tr.b.t.: tractus bulbotectalis; tr.gust.sec.: tractus gustatorius secundarius; tr.mes.cb.v.: tractus mesencephalocerebellaris ventralis; tr.t.b.: tractus tectobulbaris; tr.V de.: descending trigeminal root. Other abbreviations as in preceding figures.

Figure 314B, C. Cross-sections through the rhombencephalon of Acipenser in rostralward sequence from preceding figure (from JOHNSTON, 1906). Added lead x shows 'primitive eminentia granularis' covered by molecular layer.

Figure 314D. Cross-section through brain stem and cerebellum of Acipenser at caudal level of tectum mesencephali, showing fusion of valvula cerebelli with portion of tectum (from JOHNSTON, 1906). Added lead x indicates extracerebral (extraventricular) space.

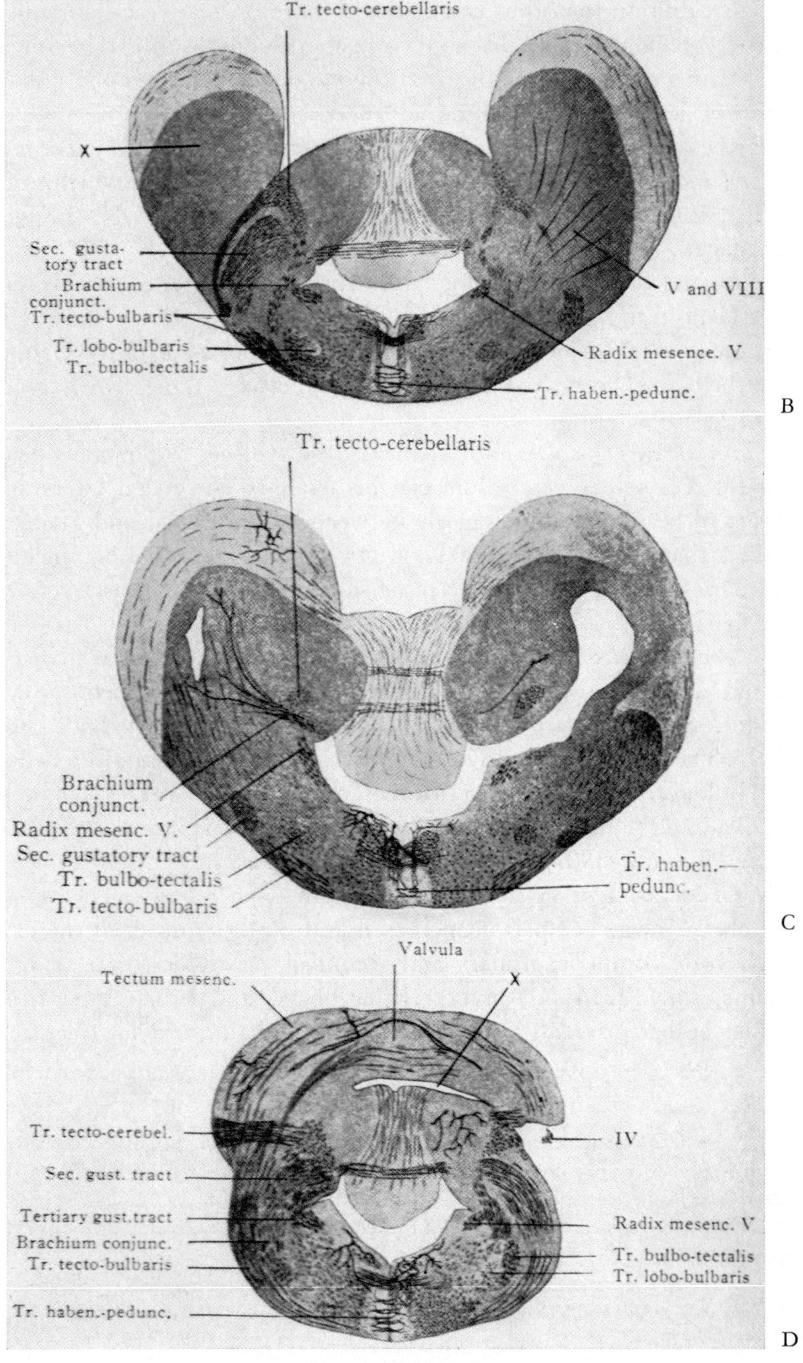
Tr. tecto-cerebellaris
X
Sec. gustatory tract
Brachium conjunct.
Tr. tecto-bulbaris
Tr. lobo-bulbaris
Tr. bulbo-tectalis
V and VIII
Radix mesence. V
Tr. haben.-pedunc.
B
Tr. tecto-cerebellaris
Brachium conjunct.
Radix mesenc. V.
Sec. gustatory tract
Tr. bulbo-tectalis
Tr. tecto-bulbaris
Tr. haben.—pedunc.
C
Valvula
Tectum mesenc.
X
Tr. tecto-cerebel.
IV
Sec. gust. tract
Tertiary gust.tract
Radix mesenc. V
Brachium conjunc.
Tr. bulbo-tectalis
Tr. tecto-bulbaris
Tr. lobo-bulbaris
Tr. haben.-pedunc.
D

In addition to corpus cerebelli and more or less typical auriculae, the cerebellum of Ganoids and Teleosts includes, as in Plagiostomes, but with a number of minor variations *qua* detailed topography and degree of development, a somewhat 'subcerebellar' *nucleus cerebelli* which may be related to an eminentia ventralis cerebelli[49] (Fig. 315). Terminals of *Purkinje cell* axons may end in this griseum through which some still poorly defined efferent cerebellar channels presumably become mediated.

On the whole, the histologic elements and the fiber connections of the Ganoid and Teleostean cerebellum are similar to those obtaining in Plagiostomes. According to KAPPERS *et al.* (1936) some of the Ganoid cerebella may be evaluated as transitional between those of Selachians and Teleosts.

As regards the histologic elements, the *Purkinje cells* in many parts of the Teleostean cerebellum can be arranged in several layers and seem to be placed less regularly between the molecular and granular layers than in Plagiostomes (KAPPERS, 1921). Again, the presence of true basket cells, apparently lacking in Plagiostomes, has been reported in Teleosts by CAJAL (1911).

The *fiber systems*, including a *cerebellar commissure*, essentially correspond to those described for Selachians, but display, nevertheless, some differences particularly related to the development of a valvula cerebelli. The *input channels* can be roughly classified as (1) spinocerebellar, (2) bulbocerebellar, (3) posterior mesencephalocerebellar, (4) tectocerebellar, and (5) lobocerebellar. With regard to the spinocerebellar system, HERRICK (1907), KAPPERS (1921) and others have distinguished a dorsal and a ventral spinocerebellar tract in some Teleosts. The bulbocerebellar channels, partly uncrossed and partly crossed, as the case may be,[50] comprise primary and secondary vestibulolateral components, trigeminal input, olivocerebellar fibers[51] and perhaps input from other bulbar grisea. The posterior mesencephalocerebellar tract, not

[49] Nucleus lateralis cerebelli of EDINGER (cf. section 3). This griseum is indeed lateral with respect to corpus cerebelli, but may be quite medial in its close relation to the rhombencephalic ventricle.

[50] Cf. the comments in section 3 of this chapter and in the relevant sections of the preceding chapter IX.

[51] The apparently entirely crossed olivocerebellar component may be less conspicuous in various Ganoids and Teleosts than in Plagiostomes because of a lesser development of the inferior olive in the former groups (cf. e.g. KAPPERS *et al.*, 1936).

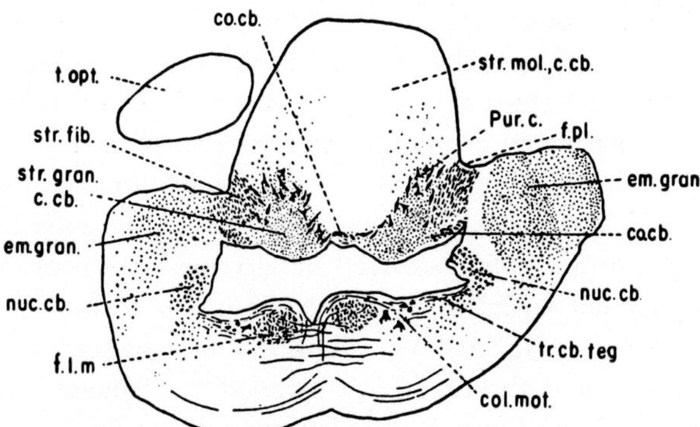

Figure 315. Cross-section through the rhombencephalon of the Ganoid Amia, showing the nucleus cerebelli (from LARSELL, 1967). co.cb.: cerebellar commissure; col.mot.: 'motor cell column' (apparently ventral zone of basal plate); em.gran.: eminentia granularis; f.l.m.: fasciculus longitudinalis medialis; f.pl.: 'posterolateral fissure'; nu.cb.: nucleus cerebelli; str.fib.: stratum fibrosum of cerebellum; tr.cb.teg.: cerebellotegmental tract. Other abbreviations as in previous figures from LARSELL.

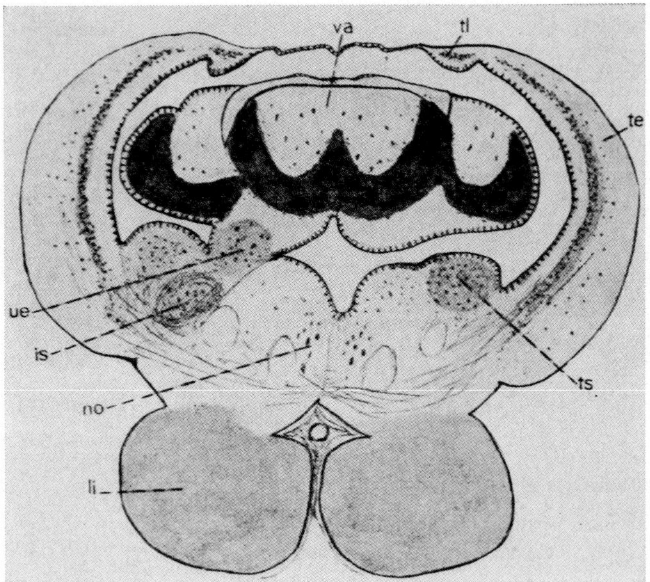

Figure 316 A. Simplified drawing of cross-section through mesencephali ventricle of the Teleost Cyprinus auratus, showing some relationships of valvula cerebelli (from K., 1927). is: nucleus isthmi; li: lobi inferiores hypothalami; no: nucleus nervi oculomotorii; te: tectum mesencephali; tl: torus longitudinalis; ts: torus semicircularis; ue: so-called '*Übergangsganglion*' or nucleus lateralis valvulae; va: valvula cerebelli. The lead va passes, just above the valvula, through an extracerebral (extraventricular) space.

present in Plagiostomes, but displayed by Ganoids, is particularly developed in Teleosts, and related to the evolution of a valvula cerebelli. The tract originates from a griseum located medially to mesencephalic torus semicircularis and essentially receiving secondary (perhaps also some primary) lateralis input which is then transmitted to the valvula. Especially in Teleosts, this griseum, the *nucleus lateralis valvulae*, is located in a neighborhood characterized by a 'concresence' or 'fusion' of valvula and medial aspect of torus semicircularis, forming a so-called '*Übergangsganglion*' (Fig. 316A). Although considered a tegmental component by some authors, I am inclined to evaluate it as an alar plate derivative, representing a specialized subdivision of torus semicircularis (nucleus lateralis mesencephali), located laterally (i.e. 'dorsal-

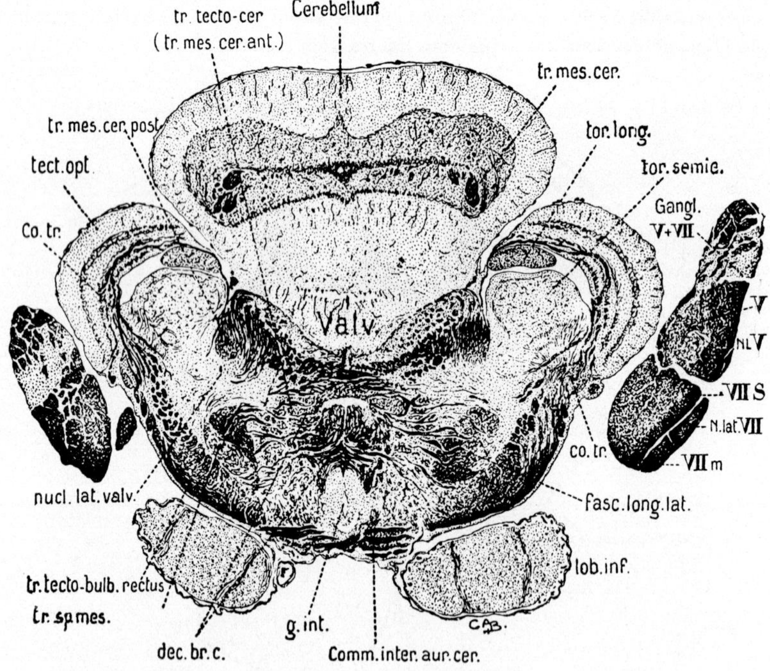

Figure 316B. Cross-section (myelin stain) through the brain of the Siluroid Teleost Arius at the isthmus level, showing cerebellar relationships (from KAPPERS, 1921). comm. inter. aur. cer.: so-called (ventral) interauricular commissure of WALLENBERG (1907); dec. br. c.: decussation of 'brachium conjunctivum'; co. tr.: commissura transversa (diencephalic); g. int.: interpeduncular nucleus (or 'ganglion'); lob. inf.: lobi inferiores hypothalami. Other abbreviations self-explanatory.

ly') to sulcus limitans. It is not impossible that fibers from the somewhat more caudally located nucleus visceralis secundarius or '*Rindenknoten*' join the overall system of posterior mesencephalocerebellar tract.

The *tectocerebellar system* or tractus mesencephalocerebellaris anterior which predominantly but not exclusively provides optic input for the cerebellum originates from various regions of mesencephalic tectum including torus longitudinalis and pretectal grisea at the diencephalo-mesencephalic boundary regions, and perhaps also from torus semicircularis. Fibers of this system may partly run through the tegmentum, entering the cerebellum at the valvula, or through the velum medullare anterius if this is suitably developed. Numerous minor variations and diverse components of that system have been described in different forms by the authors dealing with that topic.

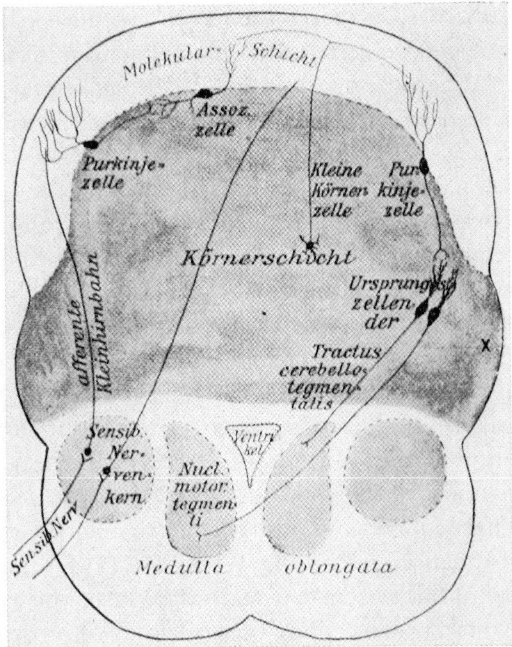

Figure 316C. Diagrammatic cross-section through a generalized Teleostean cerebellum at a level caudal to that of the two preceding figures A and B, showing the general configurational arrangement together with some fiber connections in the interpretations of Franz. Some of these latter are questionable but nevertheless interesting. This figure should be compared with Figure 223A in chapter IX (from Franz, 1911). Added x indicates protrusion of granular eminence.

The *tractus lobocerebellaris* is commonly represented by a fairly compact medullated essentially uncrossed bundle running from the posterior lobes of the hypothalamus through the mesencephalic tegmentum into the corpus cerebelli (cf. e.g. KAPPERS *et al.*, 1936). It may include a diversity of functional components, including olfactory and 'visceral' ones in addition to output from the saccus vasculosus, which is present in most Osteiichthyes.

The cerebellar *output channels* can be roughly classified as (1) cerebello-tectal, (2) anterior cerebello-tegmental, (3) cerebello-vestibulolateral, and (4) cerebello-motor systems which are partly crossed and partly uncrossed, and have been described in somewhat different ways by a number of authors (e.g. ADDISON, 1923; KAPPERS *et al.*, 1936; LARSELL, 1967; PEARSON, 1936; TUGE, 1935). The abovementioned tractus lobocerebellaris presumably also includes ascending, i. e. cerebellar output fibers (cerebellohypothalamic channels).

(1) *Cerebello-tectal fibers* reach the tectum mesencephali, perhaps including the pretectal region and the torus semicircularis, in diverse ways, partly through the tegmentum. (2) *The anterior cerebello-tegmental system*, also designated as *brachium conjunctivum*, seems to be essentially crossed, with a well developed decussation ventral to the fasciculus longitudinalis medialis (Fig. 316B). Besides connecting with the mesencephalic reticular formation which contains the 'forerunner' of nucleus ruber tegmenti, as well as with the oculomotor nucleus, it may run as far rostrad as some diencephalic grisea. (3) The fibers of the *cerebello-vestibulolateral system* are especially conspicuous in the region of cerebellar crest, but some of its components, particularly crossed ones, may also be present in the next enumerated subdivision. (4) *The cerebello-motor system*, also designated as tractus cerebello-motorius seems to take its origin in corpus, including valvula,[52] and consists of internal arcuate fibers, most of which decussate through the tegmentum and presumably effect connections with the motor cranial nerve nuclei and the reticular formation. According to KAPPERS (1947) some of the caudal component of this system may reach the spinal cord as a *tractus cerebellospinalis*, whose presence, however, is doubted or denied by some other authors. Thus TUGE (1935), who subsumes the 'cerebellifugal connections with the underlying motor mechanisms' under the term

[52] To which extent this and the other efferent cerebellar systems are provided by *Purkinje cell* neurites or neurites of the cerebellar nucleus remains an open question. It seems likely, as LARSELL (1967) assumes, that the *Purkinje cell* neurites represent the predominating component.

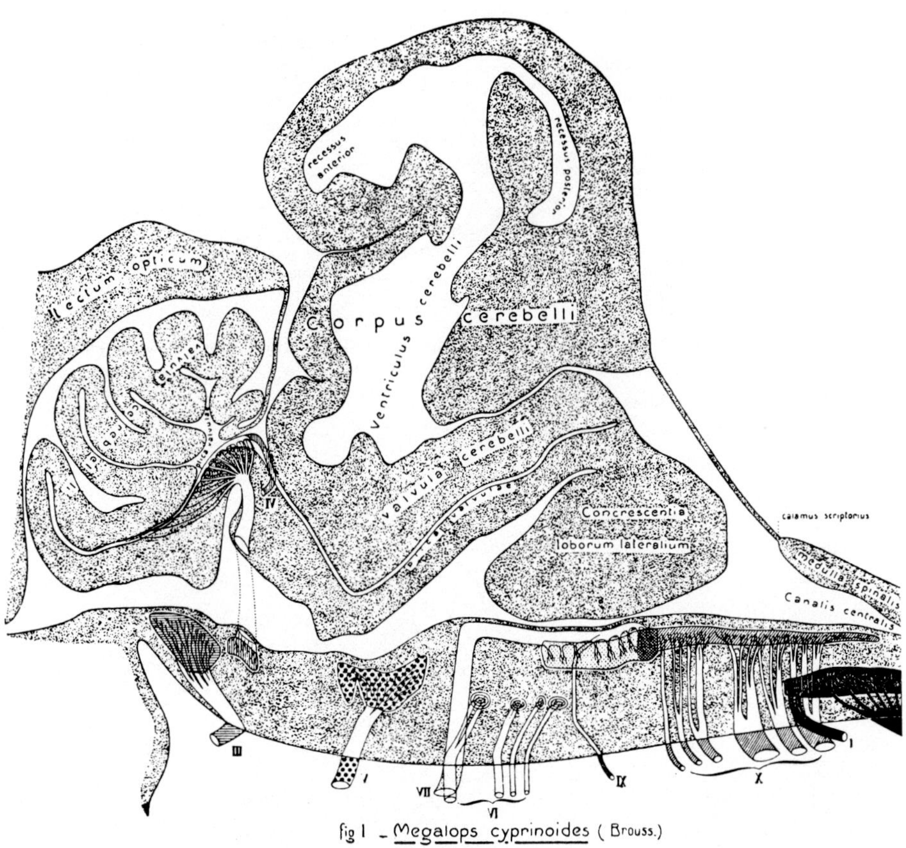

Figure 317. Slightly diagrammatic sagittal section, with projection of structures on the midline, of the deuterencephalon in the Teleost Megalops cyprinoides, showing relationships of corpus and valvula cerebelli (from VAN DER HORST, 1926).

'cerebellar motor radiation', states, on the basis of his observation with the *Marchi method*, that 'no fibers from the cerebellum reach the spinal cord'.

Figures 314 to 317 illustrate various aspects of Ganoid and Teleostean cerebellar morphology respectively structure.[53] Some of the assumed cerebellar connections are shown in Figures 316C, 318A and 318B.

[53] Cf. also Figures 214, 217, 223A–C, and 224A, B.

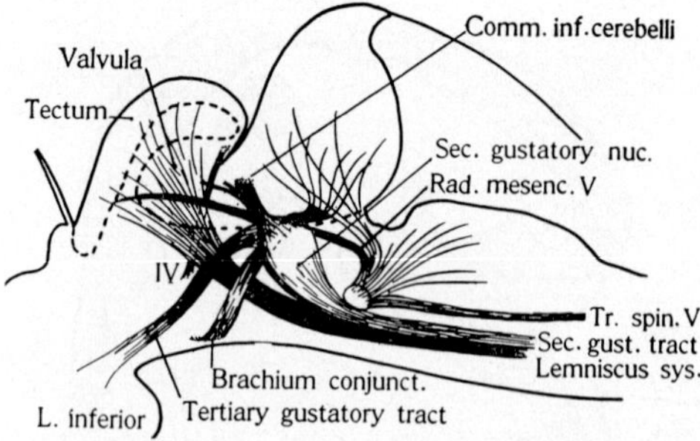

Figure 318A. Semidiagrammatic drawing showing relationships of cerebellum, 'brachium conjunctivum', and gustatory tracts in the Ganoid Acipenser, as interpreted by JOHNSTON and projected upon the midsagittal plane (from JOHNSTON, 1906).

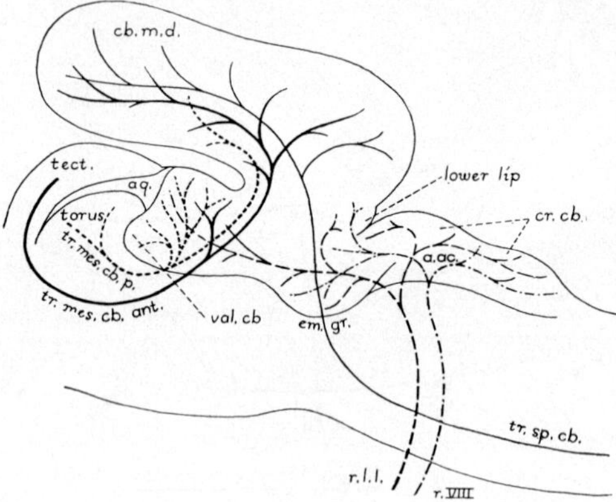

Figure 318B. Diagram of parts of the cerebellum and some of its fiber connections in the Teleost Siluroid Ameiurus, projected upon the midsagittal plane (from HERRICK, 1924). a.ac.: octavolateral area; aq: mesencephalic ventricle; cb.m.d.: dorsal median body of cerebellum; cr.cb.: cerebellar crest; em.gr.: eminentia granularis; r.l.l.: roots of lateral line nerve; torus: torus semicircularis of mesencephalic alar plate; tr.mes.cb.ant.: anterior mesencephalocerebellar tract; tr.mes.cb.post.: posterior mesencephalocerebellar tract; tr.sp.cb.: spinocerebellar tract.

In concluding the discussion of the Ganoid and Teleostean cerebellum, a few remarks on its very peculiar if not outright freakish development in the Teleostian *Mormyrids* are perhaps appropriate.[54] Various species of these fishes possess 'weakly electric organs' providing a 'guidance system' for navigation, as discussed in section 5, chapter VII

[54] The highly unusual features of the Mormyrid brain were pointed out by SANDERS (1882), STENDELL (1914), BERKELBACH VAN DER SPRENKEL (1915), NAWAR (1961), moreover by the contributors to the symposia conducted by FOX and SNIDER (1967) and by

of the preceding volume 3, part II. In contradistinction to other, 'electric' or 'non-electric' fishes, however, at least some of the Mormyrids have an extraordinarily high brain-body ratio,[55] which, in *Mormyrus caschive*, is comparable to that obtaining for the human brain (FRANZ, 1912). This ratio is due to the size of the cerebellum, which covers the entire remainder of the brain, and, in dorsal view, completely hides it, except for olfactory bulbs and tracts (Figs. 319, 320, 322). Discounting the cerebellum, however, the relative size of the Mormyrid brain does not significantly differ from that of other Teleosts. It should, moreover, be mentioned that the strongly electric fish Electrophorus electricus, whose brain was described by DE OLIVEIRA CASTRO (1961) likewise displays a rather large cerebellum, which, however, does not reach the relative dimensions of that brain subdivision in Mormyrids (cf. Fig. 319B). In Electrophorus, moreover, the size of corpus cerebelli greatly exceeds that of the rather moderately developed valvula.

Again, the enormous size of the cerebellum is mainly due to the excessive development of *valvula cerebelli*, which is repeatedly folded, forming numerous convolutions (Figs. 319, 320). FRANZ (1911) called the hypertrophied portion of the valvula '*Mormyrocerebellum*'. For the caudal parts of the cerebellum, including an intermediate somewhat thinner valvular plate, caudally adjacent configurations which may or may not still belong to the valvula, and the corpus cerebelli, which is subdivided into several lobes, together with the modified auriculae he used the term '*Ichthyocerebellum*'.

The characteristic *lamellae* of the Mormyrocerebellum as particularly investigated by FRANZ are represented by ridges, into which the granular layer does or does not significantly extend. This layer, however, forms minor superficial protrusions into the sulci separating the lamellae (Figs. 321A, B). The *Purkinje cells* display a very regular straight

---

LLINÁS (1969). Said features were particularly stressed in the studies of FRANZ (1911, 1914, 1920, who remarked 1914): '*Bekanntlich sind die Mormyriden vor allen anderen Teleostiern, auch ihren nächsten Verwandten, den Cypriniden, durch ein enorm grosses Kleinhirn ausgezeichnet. Es gibt im ganzen Bereich der Wirbeltiere keinen zweiten Fall eines derartig besonders ausgebildeten und von den Verhältnissen bei den nächstverwandten Tieren so weitgehend abweichenden Gehirns*'.

[55] Brain-body weight ratio, expressed in percentage (i.e. $E/P \times 100$) for Mormyrus caschive varies between 1.42 for small specimens and 0.41 for large ones (NAWAR, 1961), in agreement with the general rule that the intraspecific average brain-body ratio decreases gradually as the body weight of the specimen is increased. Comparable figures for Man may vary between roughly 3.7 (body weight of about 30 kg) and 1.59 (body weight of about 80 kg). It can be seen that 1.42 (Mormyrus) comes fairly close to 1.59 (Man). Cf. also the comments on brain weight in section 8, chapter VI of volume 3, part II.

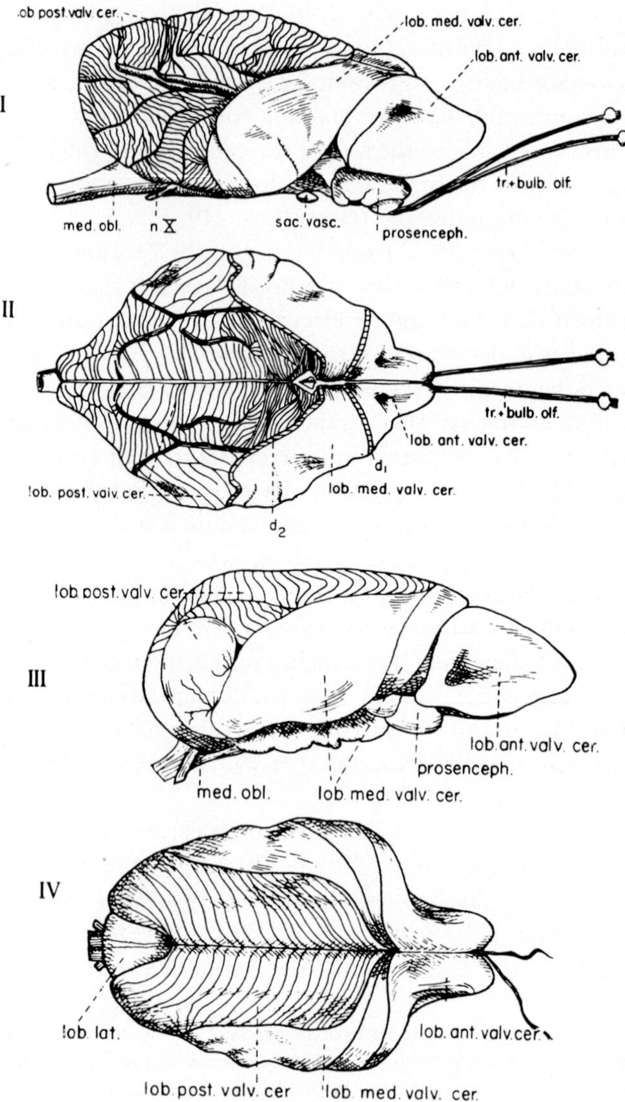

Figure 319 A. Lateral and dorsal views of the brain in two Mormyrids. The cerebellum with its hypertrophic valvula covers the whole brain (from NAWAR, 1961). I, II Mormyrus caschive. III, IV Hyperopisus bebe. d_1: depression between anterior and middle lobe of valvula cerebelli; d_2: depression between its middle and posterior lobe; sac. vasc.: saccus vasculosus. Other abbreviations self-explanatory.

orientation of their dendrites, which results in an orthogonal striation of the molecular layer, particularly in sagittal sections, but also in transverse sections, where the somewhat less conspicuous striation is crossed at a right angle by the course of the granule cell neurites demonstrable by *Golgi impregnations*. Since these neurites all enter from one direction (as it were, from 'below' the lamellae), they run already parallel to the lamellar surface, and do not bifurcate. In addition to the *Purkinje cells* and the glial elements, the neuronal elements are represented by stellate cells, *Golgi cells*, and some other poorly understood cells (e.g. 'basal cells', cf. the descriptions in the symposium of LLINÁS, 1969).

Various 'subcerebellar' grisea, related to valvula and corpus cerebelli, have been described by FRANZ (1914 *et passim*) as Ganglia I, II, III and IV (Figs. 322A–C). Ganglion I is the most caudal one and may represent a rostral portion of eminentia granularis. Ganglion II is presumably a specialized caudal component of ganglion mesencephali laterale (torus semicircularis) although some authors interpret it as part of the mesencephalic tegmentum. Ganglion III, located medially to the preceding, seems likewise to be a specialized portion of the torus semicircularis. Ganglion IV is most probably the nucleus lateralis valvulae or 'Übergangsganglion' (cf. Fig. 316A). It is closely adjacent to the so-called '*Rindenknoten*' or nucleus gustatorius secundarius superior (cf. Fig. 223A) pertaining to the mesencephalon.

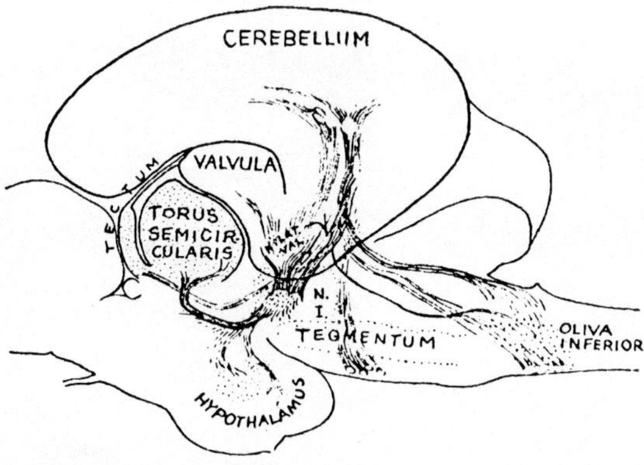

Figure 319B. Sketch illustrating general configuration of cerebellum in Electrophorus electricus as projected on the midsagittal plane, together with some main cerebellar connections (from DE OLIVEIRA CASTRO, 1961). NI: nucleus isthmi; N.LAT.VAL: nucleus lateralis valvulae.

As regards *functional significance* and *fiber connections* of the cerebellum, HERRICK (1905) and others believed that the hypertrophy of the Mormyrid valvula was mainly due to the development of the gustatory system. FRANZ (1911), while agreeing with HERRICK's view, emphasized the significance of the afferent facialis components (cf. e.g. Fig. 322C), but apparently failed to realized that these 'facialis compo-

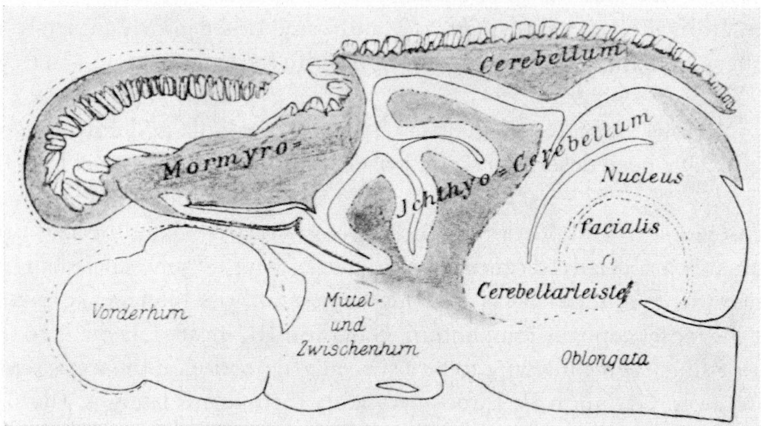

Figure 320A. Paramedian sagittal section through the brain of the Mormyrid Gnathonemus. The granular layer of the cerebellum is indicated by gray shading. It should be recalled that 'facialis' refers here to n. posterior lineae lateralis (from FRANZ, 1911).

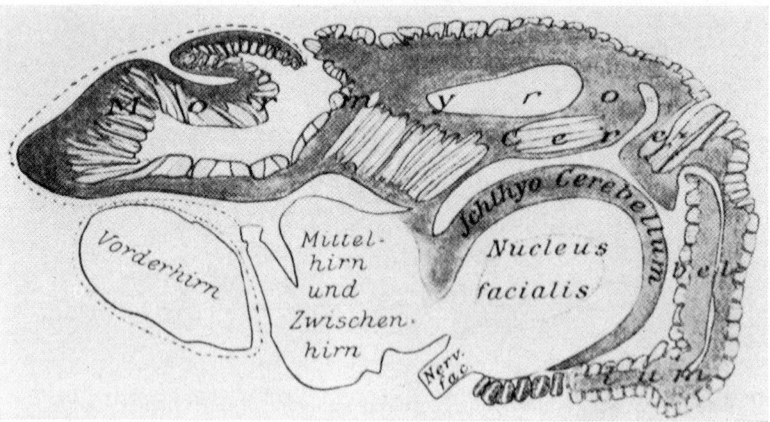

Figure 320B. Paramedian sagittal section, slightly more lateral than the preceding through the brain of Mormyrus caschive (from FRANZ, 1911).

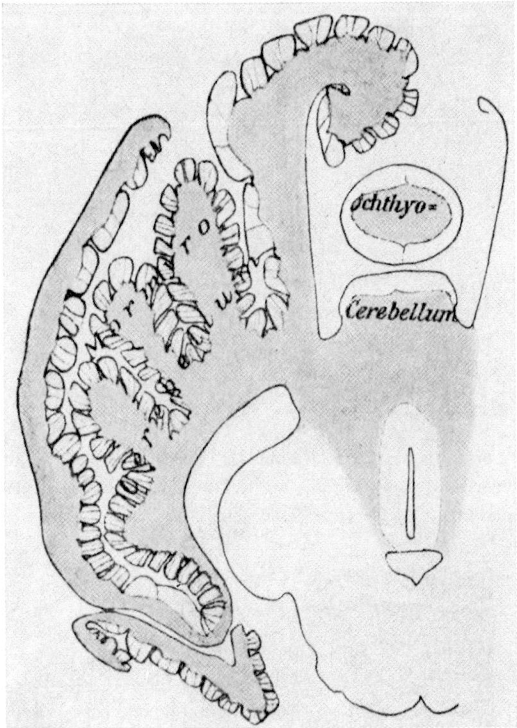

Figure 320C. Cross-section, at level of corpus cerebelli ('Ichthyocerebellum'), through the brain of Mormyrus caschive (from FRANZ, 1911).

nents' represent the input of *nervus lateralis posterior*. The relationships of the octavolateralis system to the Mormyrid cerebellum were particularly pointed out by BERKELBACH VAN DER SPRENKEL (1915). FRANZ, however, had already stressed the role of the cerebellum in Teleosts, and especially in Mormyrids as a mechanism for the integration of input from the different sensory systems and, moreover, had already assumed its connection with the *electric activities* of the Mormyrids.[56] Al-

[56] FRANZ (1911) stated: '*Doch lehrte uns schon die Faseranatomie, dass das Kleinhirn bei Fischen eine universellere Aufgabe hat, es assoziiert Eindrücke aus den verschiedensten Sinnesgebieten zu Impulsen an die motorischen Zentren der Oblongata*'.

'...*die vielfältig assoziierten Facialisreize kombinieren sich noch im Cerebellum mit Reizen anderer Sinnesgebiete und veranlassen dann stete, sehr schwache, aber äusserst fein dosierte effektorische Impulse, die zweierlei bezwecken mögen: die genaue Koordination der Bewegungen und der Austeilung schwacher elektrischer Schläge, welche geeignet sind, herannahende Angreifer schon aus einiger Entfernung zu verjagen*'.

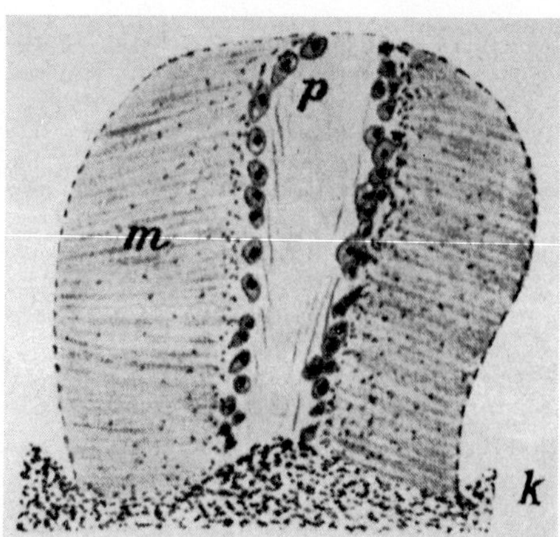

Figure 321 A. Simplified sketch of lamella (*'Oberflächenformation'*) of the Mormyrocerebellum in Petrocephalus, as seen in a sagittal section (from FRANZ, 1911). k: granular layer; m: molecular layer; p: Purkinje cells.

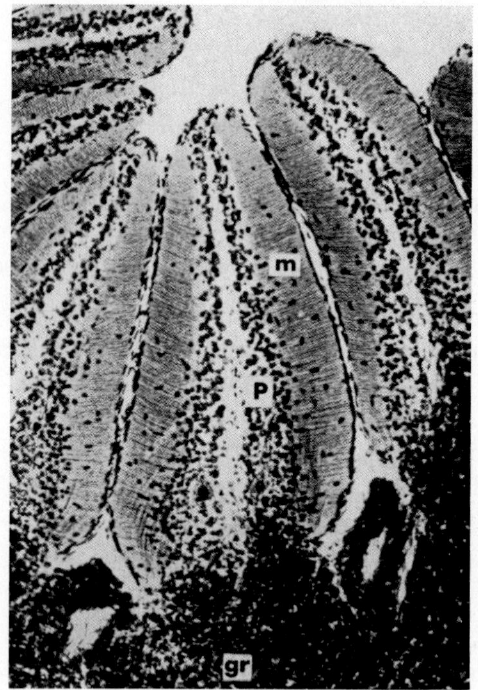

Figure 321 B. Photomicrograph (sagittal section, silver proteinate impregnation) of valvular ridges (lamellae) in Petrocephalus bovei (from NIEUWENHUYS, 1967). gr: granular layer; m, p as in A.

though FRANZ could not, on the basis of the then available data, suspect the significance of said activities as a '*guidance system*', his remarkable hypothesis, that the Mormyrid facialis (posterior lateralis) impulses were related to the electrical discharges, became subsequently confirmed by LISSMANN (1958) by means of experimental observations. Since then, various authors (e.g. BENNETT, 1965; BENNETT and GRUNDFEST, 1961; SZABO, 1961, 1965, 1967; SZABO and FESSARD, 1965) have reported numerous details on electric organs, electroreceptors, and electric activities in Mormyrids as well as in other electric fishes (cf. also section 5, chapter VII of vol. 3, part II).

The electric discharges may be controlled by a 'central command nucleus' presumably located in the mesencephalic tegmentum medially to the ganglion III of FRANZ, and directly or indirectly connected with the valvula *('Mormyrocerebellum' sensu strictiori)*. This command griseum, in turn, may control a medullary relay nucleus included in the reticular formation of the rostral rhombencephalon. Bulbo-spinal fibers from this griseum reach the modified motoneurons (or 'electromotor neurons'), numbering perhaps circa 250, and located at approximately the segmental levels of the electric organ.[57] An 'electrotonic coupling'[58] of neuronal elements in the 'relay' and 'electromotor' grisea has been inferred.

The fiber connections of the cerebellum in Mormyrids display, to some extent, an overall arrangement identical with that obtaining in other Teleosts.[59] Perhaps in some sort of correlation with an habitat in muddy waters, the eyes of Mormyrids are, as a general rule, rather small, and FRANZ (1914) points out that the tractus tectocerebellaris (or mesencephalo-cerebellaris of his terminology), reaching the corpus cerebelli after a partial decussation, is poorly developed *('sehr schwach')*.

---

[57] *Mutatis mutandis*, 'command nuclei' (CN; NCD of FESSARD, 1958) relay nuclei ('medullary relay nuclei', MRN), and actual discharge grisea ('electromotor neurone nucleus', EMN; CMD of FESSARD, 1958) are, or can be presumed to be, present in all electric fishes (cf. section 5, chapter VII of volume 3, part II and, for greater, but in part still unsatisfactorily elucidated specific details, BENNETT, 1968 or SZABO, 1961).

[58] Cf. p. 638, section 8, chapter V of volume 3, part I.

[59] FRANZ (1914) comments: '*Sonstige Faserzüge im Mormyridenhirn weisen verhältnismässig wenig Besonderheiten auf. Wir finden alle uns bekannten afferenten Kleinhirnbahnen der Teleostier*'. Concerning efferent tracts, which he subsumes under a single category, FRANZ adds: '*Das einzige efferente Kleinhirnfasersystem, der Tractus cerebello-tegmentalis, ist im Mormyridengehirn wiederum leicht auffindbar, und zwar ist er stärker entwickelt, als ich früher annahm, doch keineswegs auffallend stark*'.

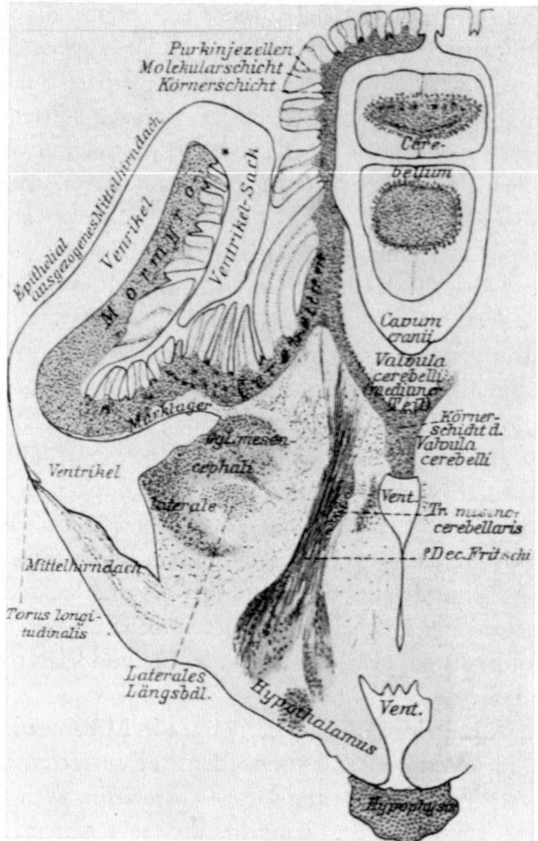

Figure 322A. Cross-section through the brain of the Petrocephalus at a rostral mesencephalic level overlapping with caudal hypothalamus, showing eversion of tectum mesencephali and part of 'lobocerebellar' fiber systems (labelled as *Dec. Fritschi)*, connecting with hypothalamus (from FRANZ, 1911).

Figure 322B. Somewhat more caudal cross-section through the brain of the Mormyrid Marcusenius longianalis showing 'Ganglia II–IV' of Franz and commissural systems (from FRANZ, 1911).

Figure 322C. Cross-section through the brain of Petrocephalus at a still more caudal level, showing eminentia granularis and fibers of n. posterior lineae lateralis, labelled 'Nervus facialis' (from Franz, 1911).

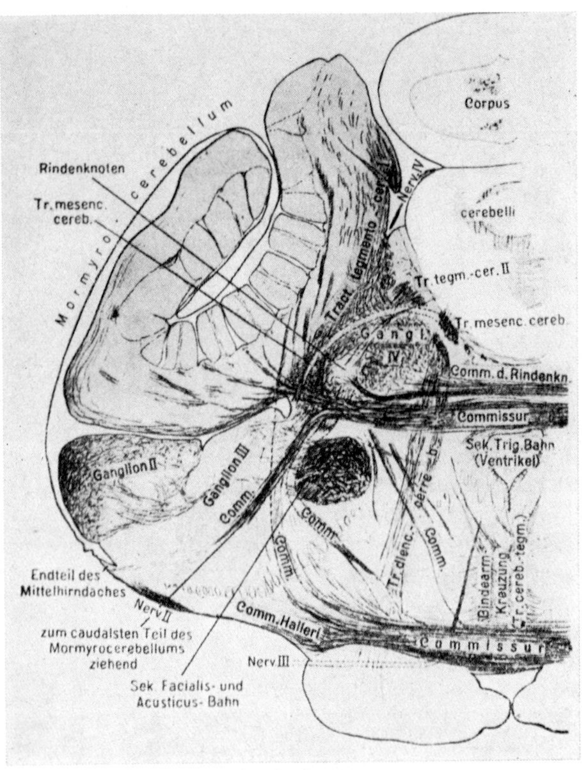
Corpus
cerebelli
Mormyrocerebellum
Rindenknoten
Tr. mesenc. cereb.
Tract. tegmento-cereb.
Nerv. IV
Tr. tegm.-cer. II
Tr. mesenc. cereb.
Gangl. IV
Comm. d. Rindenkn.
Commissur
Sek. Trig. Bahn (Ventrikel)
Ganglion II
Ganglion III
Comm.
Comm.
Comm.
Tr. dienc.-cereb.
Comm.
Bindearm-Kreuzung (Tr. cereb.-tegm.)
Endteil des Mittelhirndaches
Nerv II
zum caudalsten Teil des Mormyrocerebellums ziehend
Comm. Halleri
Commissur
Nerv III
Sek. Facialis- und Acusticus- Bahn

322 B

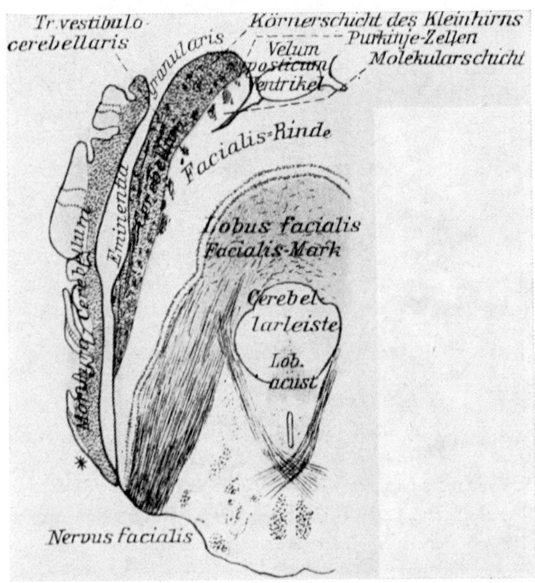
Tr. vestibulo-cerebellaris
Körnerschicht des Kleinhirns
Purkinje-Zellen
Molekularschicht
Velum posticum
Ventrikel
granularis
Eminentia
Facialis-Rinde
Lobus facialis
Facialis-Mark
Cerebellarleiste
Lob. acust.
Mormyro-Cerebellum
Nervus facialis

322 C

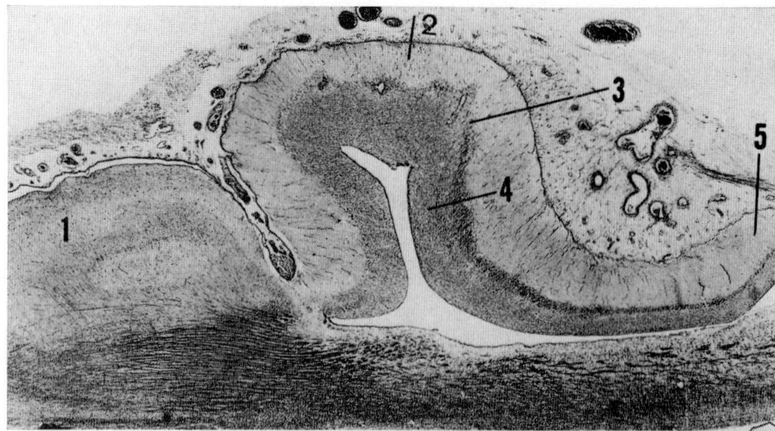

Figure 323 A. Parasagittal section (silver impregnation), close to the midline, through the cerebellum of Latimeria (from Millot and Anthony, 1965). 1: tectum mesencephali; 2: molecular layer; 3: *Purkinje cell* layer; 4: granular layer; 5: posterior lip of cerebellum (the designations have been added to the original figure).

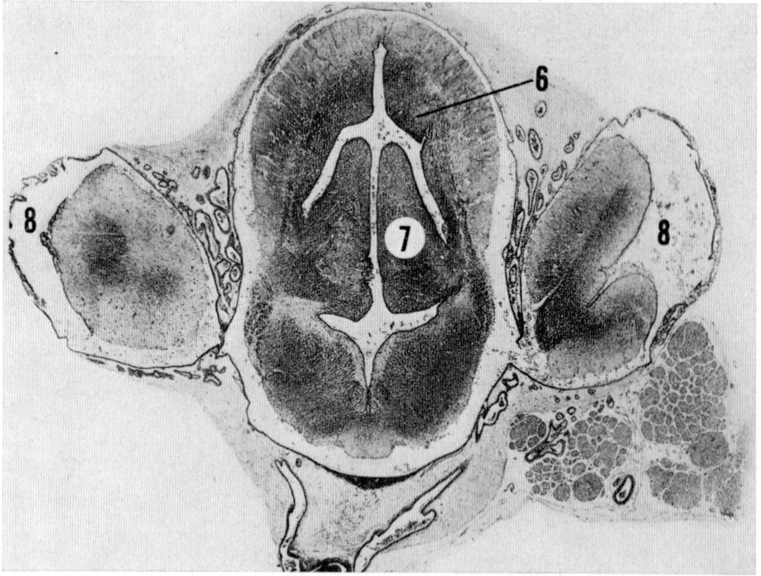

Figure 323 B. Transverse section through the metencephalon of Latimeria, showing corpus cerebelli and rostral end of auriculae (from Millot and Anthony, 1965). 6: superior granular eminence; 7: inferior granular eminence; 8: ventricular recess of auricula (the designations have been added).

On the other hand, concomitantly with the peculiar development of that brain subdivision, a few particular specializations have been recorded. These include a suspected diencephalocerebellar channel connecting with dorsal and ventral thalamic grisea and a separate second tegmentocerebellar system, both related to the corpus cerebelli. As regards the valvula, its particular connections with the ganglia I–IV of FRANZ should be mentioned, in addition to valvular components of the cerebellothalamic channel, and to presumably mainly efferent crossed and uncrossed valvulotegmental ('mormyrocerebello-mesencephalic') fiber systems. FRANZ (1914) also emphasizes the conspicuous pattern of diverse, still at present poorly analyzed *cerebellar commissures* (cf. Fig. 322B).

Concerning the cerebellum of the *Crossopterygian Latimeria*, some general data have been recorded in the publication by MILLOT and ANTHONY (1965). These authors report that '*le métencéphale comporte un cervelet bien développé*'. While this statement is not entirely unjustified, it seems nevertheless quite evident that this cerebellum is less developed than that of Plagiostomes and Teleosts. Discounting the differences in configuration, its status can be evaluated as intermediate between Ganoids and Dipnoans. The cerebellum of these latter, in turn, stands, *qua* development, between the cerebellum of fishes and that of Amphibia.

The morphologic features of the cerebellum in *Latimeria* (Figs. 323A, B) display a dorsally evaginated corpus with a fairly long caudal lip, and a narrow cerebellar ventricle whose lumen shows a paired lateral recess basally to an unpaired dorsal one. Superior and inferior granular eminences, somewhat similar to those in Plagiostomes (cf. Figs. 306B, 309A, B) are present, with a slightly different pattern, including the suggestion of minor accessory sulci. Ventrally to the inferior eminence, a recess of the fourth ventricle extends dorsolateralward. A valvula cerebelli, characteristic for Ganoids and Teleosts, is lacking. The auriculae, however, are rather well developed, and essentially resemble those in Plagiostomes.

As regards the histologic structure, MILLOT and ANTHONY noted a molecular layer with its fiber system, a granular layer, and, between these two, a regularly arranged layer of *Purkinje cells*. In the auriculae the *Purkinje cells* are likewise present but show a less regular arrangement and smaller size than in the corpus.[60]

[60] The cited authors remark: '*En conclusion l'étude de Latimeria vient confirmer la remarquable unité de la structure cérébelleuse dans la série des Vertébrés.*'

A *nucleus cerebelli* was not mentioned by the cited authors, but could presumably be included among the nondescript cell masses ventrally adjacent to the inferior granular eminence (cf. Fig. 322B).

As regards the fiber systems, no specific details have been recorded, but it can be presumed that the main cerebellar input and output channels do not significantly differ from those obtaining in other fishes.

5. Dipnoans

The cerebellum of Dipnoans, as investigated in *Ceratodus (Neoceratodus)* by VAN DER HORST (1925), by HOLMGREN and VAN DER HORST (1925), and in *Protopterus* by GERLACH (1933), is less developed than that of other Gnathostome fishes.[61] Ceratodus displays a more extensive corpus cerebelli than Protopterus, whose auriculae, on the other hand, are relatively larger, while the corpus is entirely covered by the caudal end of tectum mesencephali. Neither Ceratodus nor Protopterus display a valvula cerebelli, although a rostral transition of cerebellar plate to velum medullare anterius has been interpreted as the mere suggestion *('Andeutung')* of a valvula by GERLACH.[62] A crista cerebellaris is present in both forms.

Some features of the cerebellum in *Ceratodus* are illustrated by Figures 324A–C. It can be seen that the corpus forms a caudalward tilted plate, whose posterior lip is continuous with the auriculae. In cross-section (Fig. 324B), the corpus cerebelli projects as a prominent ridge, consisting of a median extension of molecular layer, into the ventricle. This ridge is flanked by granular cell masses representing a *'granular eminence'*. A portion of the lateral corpus surface lacks a molecular layer, and forms an external eminentia granularis (Fig. 324A).

Few specific details concerning the histologic structure are available, except for the arrangement of the three typical layers. It remains uncertain to which extent *Purkinje cells* are present in auriculae and crista cerebellaris. A nondescript cell mass ventrolaterally to the medi-

---

[61] Previous investigators of the Dipnoan brain including its cerebellum are BING and BURCKHARDT (1904) for Ceratodus, BURCKHARDT (1892) and FULLIQUET (1886) for Protopterus. Their publications together with other relevant papers and findings are cited in the papers by HOLMGREN and VAN DER HORST (1925) and by GERLACH (1933), respectively.

[62] Cf. Figure 325B and also the sagittal sections through the brain of Ceratodus depicted in Figures 72 I and II, p. 197, of volume 3, part II.

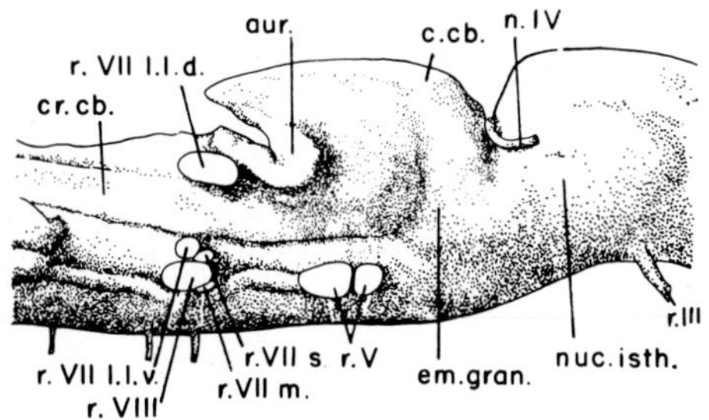

Figure 324A. Lateral view of the cerebellum in Ceratodus (modified after HOLMGREN and VAN DER HORST, 1925, from LARSELL, 1967). aur.: auricula; c.cb.: corpus cerebelli; cr.cb.: cerebellar crest; em.gran.: (external) eminentia granularis; l.l.d.: dorsal lateral line root; l.l.v.: ventral lateral line root; m.: motor; s.: sensory; r.: nerve root; nuc.isth.: nucleus isthmi.

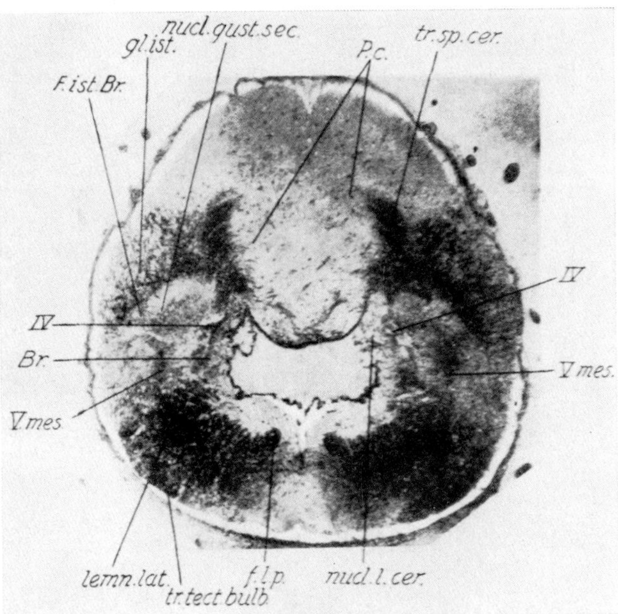

Figure 324B. Cross-section through the brain of Ceratodus at a rostral level of corpus cerebelli (from HOLMGREN and VAN DER HORST, 1925). Br.: brachium conjunctivum; F. ist. Br.: 'fibres from ganglion isthmi to brachium conjunctivum'; f. l. p.: fasciculus longitudinalis medialis; lemn. lat.: lemniscus lateralis; nucl. gust. sec.: nucleus gustatorius secundarius; nucl. l. cer.: nucleus lateralis cerebelli; P. c.: *Purkinje cells;* tr. sp. cer.: tractus spinocerebellaris; tr. tect. bulb.: tractus tectobulbaris; IV: root fibers of n. trochlearis; V mes.: radix mesencephalica trigemini.

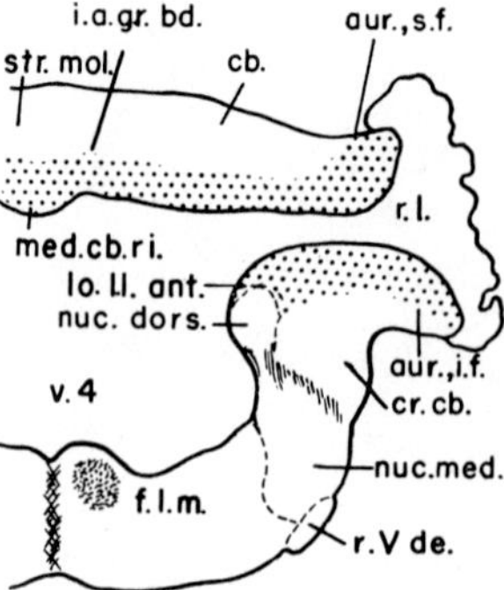

Figure 324C. Cross-section through rhombencephalon of Ceratodus at the level of lateral recess of fourth ventricle (modified after VAN DER HORST, 1919, from LARSELL, 1967). au.s.f., au.i.f.: superior and inferior fold (lip) of auricle; cb.: cerebellum; cr.cb.: cerebellar crest; i.a.gr.bd.: 'interauricular granular band'; f.l.m.: fasciculus longitudinalis medialis; lo.ll.ant.: 'anterior lateral line lobe; med.cb.ri.: 'median cerebellar ridge; nuc. dors.: 'dorsal nucleus of acousticolateral lobe'; nuc.med.: 'medial nucleus of acousticolateral lobe'; r.l.: lateral (auricular) recess of fourth ventricle; r.V de.: radix descendens trigemini; str.mol.: molecular layer; v.4: fourth ventricle.

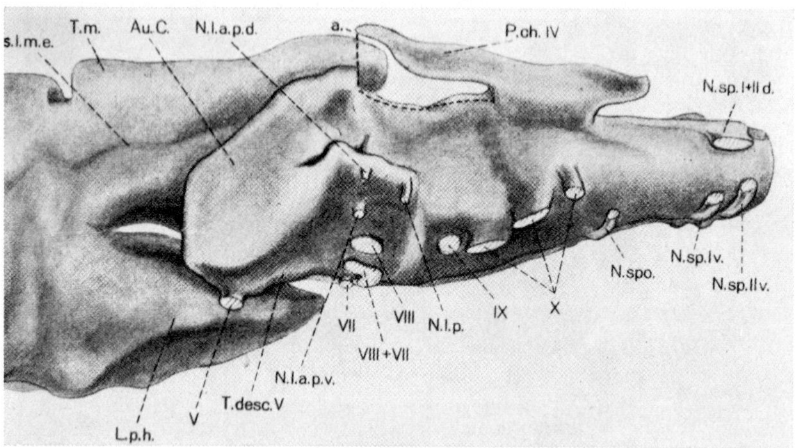

Figure 325A. Lateral view of deuterencephalon in Protopterus as displayed by a wax model (from GERLACH, 1933). a: attachment of choroid plexus; Au.c.: auricula cerebelli; L.p.h.: lobus posterior hypothalami; N.l.a.p.d., p.v.: nervus lateralis anterior, pars dorsalis respectively ventralis; N.l.p.: nervus lateralis posterior; N.spo.: nervus spino-occipitalis; n.sp. I, II., d., v.: nervus spinalis I, II, dorsal respectively ventral root; P.ch.IV: choroid plexus of fourth ventricle; s.l.m.e.: sulcus lateralis mesencephali externus; T.desc.V: radix descendens trigemini; T.m.: tectum mesencephali.

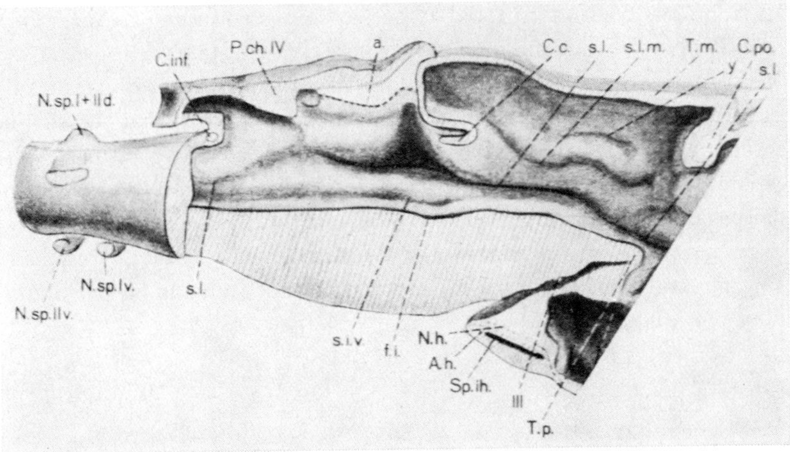

Figure 325B. Medial (ventricular) view of deuterencephalon in Protopterus as shown by the preceding wax model (from GERLACH, 1933). a: attachment of roof plate lamina epithelialis (plexus chorioideus); A.h.: adenohypophysis; Coc.: corpus cerebelli; C.po.: commissura posterior; f.i.: fovea isthmi; N.h.: neurophypophysis; S.i.v.: sulcus intermedioventralis; S.l.: sulcus limitans; S.l.m.: sulcus lateralis mesencephali; Sp.ih.: interhypophyseal space; T.m.: tectum mesencephali; T.p.: tuberculum posterius; y.: accessory (skrinkage ?) sulcus; III: oculomotor nerve. Other abbreviations as in preceding figure.

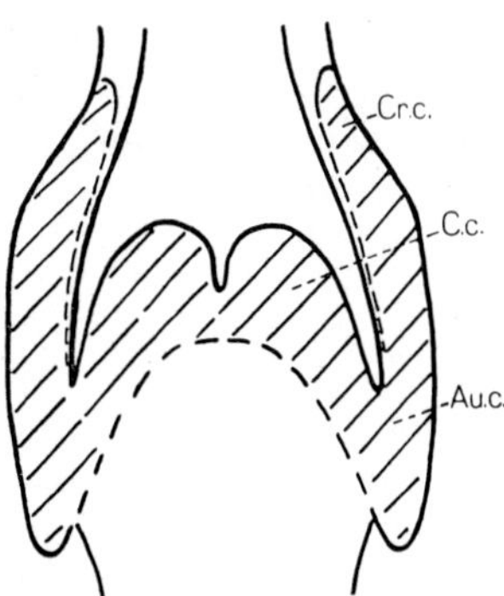

Figure 325C. Simplified diagram illustrating the cerebellar configuration in Protopterus (from GERLACH, 1933). Au.c.: auricula cerebelli; C.c.: corpus cerebelli; Cr.c.: crista cerebellaris.

an ridge has been interpreted as perhaps corresponding to the *nucleus cerebelli (*HOLMGREN and VAN DER HORST, 1925; LARSELL, 1967).

Some features of the cerebellum in *Protopterus* are illustrated by figures 325A–E. In the cross-sections it can be noted that the corpus cerebelli has a bilateral symmetric configuration whose antimeres are separated by a longitudinal median ventricular sulcus. A group of caudal paramedian granular cells was designated by GERLACH (1933) as '*nidus cerebelli*' and may represent a rudimentary '*eminentia granularis*'. The general stratification into molecular and granular layer obtains,

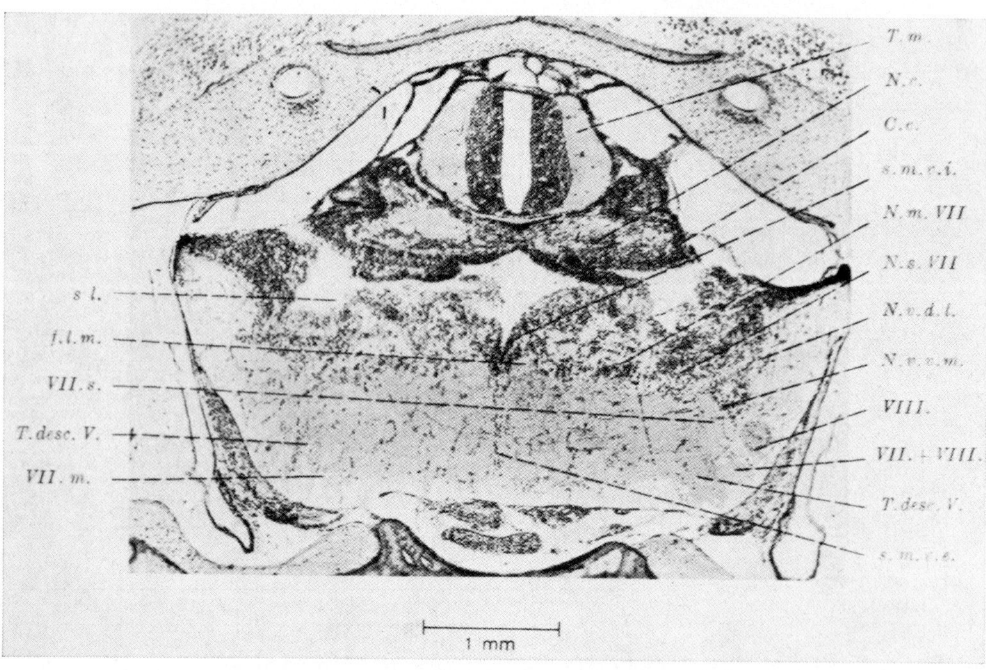

Figure 325D. Cross-section through the deuterencephalon of Protopterus at a level near caudal end of corpus cerebelli, corresponding to entrance of octavolateralis and facialis root fibers (from GERLACH, 1933). f.l.m.: fasciculus longitudinalis medialis; m.: motor respectively efferent (referring to cranial nerve root or nucleus); N.c.: 'nidus cerebelli'; N.v.v.m.: nucleus vestibularis ventromedialis; s.: sensory respectively afferent (cf. above m.); s.l.: sulcus limitans; s.m.v.e.: raphe, ventrally ending in sulcus medianus externus rhombencephali; s.m.v.i.: sulcus medianus internus rhombencephali; T.desc.V.: descending trigeminal root. Other abbreviations as in preceding Protopterus figures.

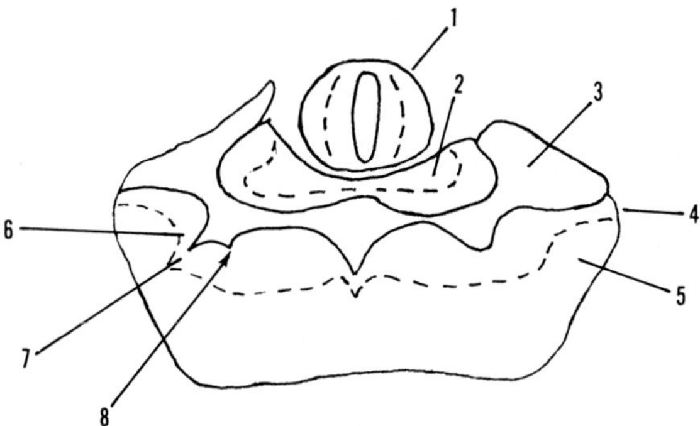

Figure 325E. Simplified outline sketch based on the preceding and on adjacent cross-sections of our Breslau Protopterus material. 1: tectum mesencephali; 2: corpus cerebelli (the lead ends in the molecular layer); 3: recessus auriculae (recessus lateralis ventriculi quarti); 4: auricula (inferior or lower lip); 5: cerebellar crest; 6: lobus lineae lateralis anterioris; 7: sulcus intermediodorsalis; 8: sulcus limitans. The numerous epithelial folds shown in the preceding figure 325D between the roof plate extensions covering the recessus auriculae (3) pertain to the saccus endolymphaticus (cf. chapter IX, section 7, p. 453).

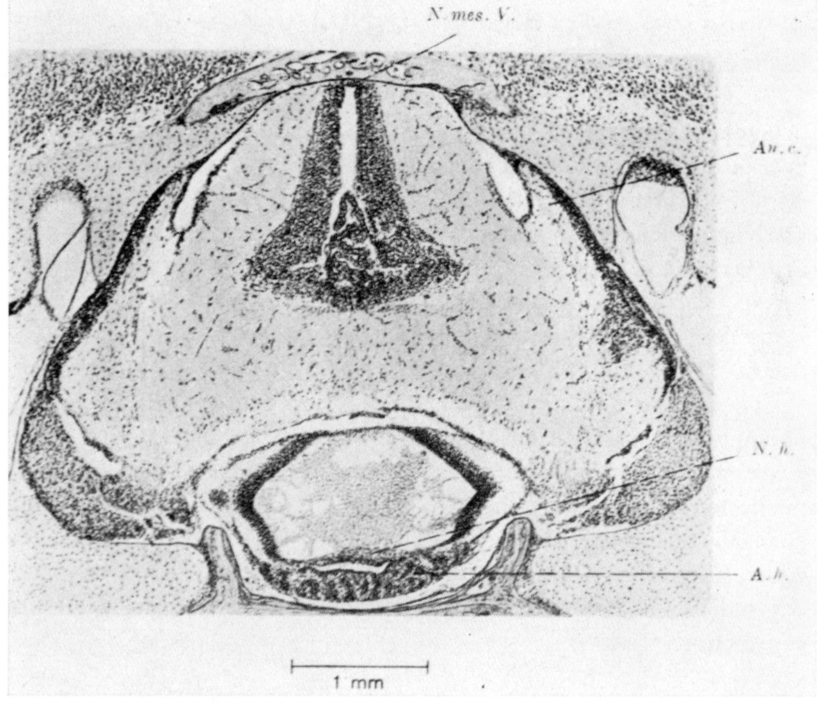

Figure 325F. Cross-section through the isthmus region of Protopterus, showing the rostral extension of the cerebellar auricle with its granular layer (from GERLACH, 1933). Cf. also Fig. 384 D. A.h., N.h.: adeno- and neurohypophysis.

but the *Purkinje cells* show a somewhat irregular distribution, and were not seen to be present in auriculae nor in the crista cerebelli which is ventrolaterally adjacent to the inferior or caudal auricular lip. A *nucleus cerebelli* was not identified, but could perhaps be included in the periventricular cell masses near the rostral end of the auricular recess.

In evaluating the morphological status of the Dipnoan cerebellum within the group of Anamnia, GERLACH (1933) grades the steps of cerebellar differentiation in the following sequence: 1. Petromyzon, 2. Amphibia, 3. Protopterus, 4. Ceratodus, 5. Selachians, 6. Teleosts. The overall similarity of many brain features in Protopterus and Amphibia is evident, and was also pointed out in my old '*Vorlesungen*' of 1927. As regards Protopterus, the corpus cerebelli might be compared with that of Anura rather than of Urodela, while the well developed auriculae suggest a comparison with Urodela.

As regards the *fiber connections* of the Dipnoan cerebellum, only fragmentary data are available.[63] It can, nevertheless, be presumed that all significant channels mentioned in the preceding enumerations of cerebellar fiber systems characteristic for fishes are here present at least in a rudimentary manner.

Among *afferent connections* in Ceratodus as well as in Protopterus, octavolateralis channels reaching crista, auricles, and also corpus, have been noted. Trigeminocerebellar fibers, observed by HOLMGREN and VAN DER HORST (1925) as well as by GERLACH (1933) seem to be particularly conspicuous, and mainly connect with the corpus which is also assumed to receive spinocerebellar input. Rostral afferents include tectocerebellar fibers and probably also rudimentary diencephalocerebellar (particularly hypothalamocerebellar or lobocerebellar) connections.

The efferent systems, whether *Purkinje cell* neurites or fibers originating in the poorly definable 'nucleus cerebelli' appear mainly to form a rostral nondescript 'brachium conjunctivum' reaching the mesencephalic tegmentum and perhaps other rostral grisea. These efferent channels presumably also include ill-defined more caudal components equivalent to the so-called tractus cerebello-motorius. A probably present rudimentary system of cerebellar commissures remains quite inconspicuous.

[63] Nothing certain is here known *qua* crossed or uncrossed distribution of these systems.

6. Amphibians

Generally speaking, the Amphibian cerebellum is poorly developed and assumes, as pointed out by GERLACH (1933), cf. section 5, p. 712, an intermediate position between that of Dipnoans and Petromyzont Cyclostomes.[64] It comprises an everted plate-like corpus[65] which may be quite rudimentary in some forms (e.g. in Gymnophiona or in the blind urodele Proteus), and the laterally adjacent auriculae. Both components display variable relative dimensions. A crista cerebellaris is present, but relatively much less developed than in most fishes (Fig. 327A).

Volume 1 of the comprehensive monograph on the cerebellum by LARSELL (1967) contains detailed references to the authors who reported on the Amphibian metencephalon. Among these investigators, in addition to LARSELL, BENEDETTI (1933), HERRICK (1924, 1948), KAPPERS (1921, 1947 and other publications), KINGSBURY (1895), KREHT (1930, 1931, 1940), and RÖTHIG (1927) should here be mentioned.

Figures 293, 326A and B illustrate configurational aspects of the cerebellum in Urodela. In some of these forms, particularly in those evaluated as 'lower', the corpus cerebelli assumes a bilateral (i.e. faintly paired) pattern (Fig. 326B), while in others (Figs. 293, 326A), an apparently unpaired transverse plate is displayed. Again, in some Urodela, a faint transverse sulcus rostral to the ventricular edge of the cerebellar plate is interpreted by LARSELL (1967) as representing the fissura posterolateralis delimiting the lobus vestibulolateralis including its athwart located component, from the corpus cerebelli *sensu strictiori*. The fissura posterolateralis, as identified in Birds and Mammals, delimits there the flocculonodular lobe from the corpus.

---

[64] According to HERRICK (1924), who also refers to JOHNSTON (1902) and TRETJAKOFF (1909), Petromyzonts 'possess a cerebellum whose organization is in the main similar to that of the lower Amphibia, although perhaps more primitive'. The similarity of the cerebellar development in Urodela and in Dipnoans was pointed out above, particularly as regards Protopterus, but obtains thereby likewise with respect to Ceratodus, whose corpus cerebelli is somewhat more developed.

[65] There is no distinct boundary between the lateral neighborhoods representing the auriculae and the median corpus. Thus, slightly different uses of the term corpus cerebelli as well as auriculae of Urodeles can be noted in the literature and are discussed by LARSELL (1967). This may also be said to be the case with respect to the cerebellum in other Anamnia.

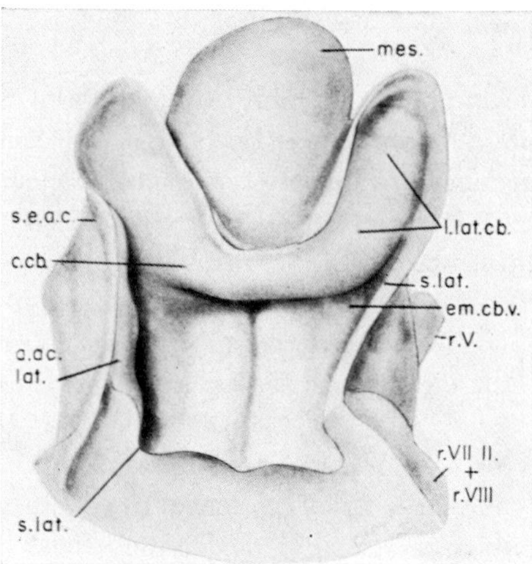

Figure 326A. Model of cerebellum in the Urodele Triturus torosus (from LARSELL, 1967). a.ac.lat.: area vestibulolateralis; c.cb.: corpus cerebelli; em.cb.v.: 'ventral cerebellar eminence'; l.lat.cb.: 'lateral lobule of cerebellum'; mes.: mesencephalon; r.V: trigeminal root; r.VII ll. + r.VIII: lateral nerve root of facial and root of vestibular nerve; s.e.a.c.: external sulcus of vestibulolateral area; s.lat.: 'lateral sulcus' (lower lead: probably sulcus intermediodorsalis; upper lead: probably an accessory sulcus of auricula).

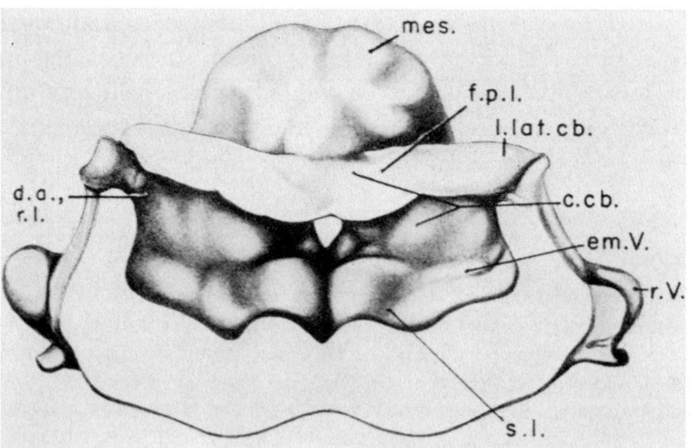

Figure 326B. Model of cerebellum of the Urodele Cryptobranchus (from LARSELL, 1967). d.a., r.l.: 'anterior diverticulum', recessus lateralis of fourth ventricle; em.V.: ventral cerebellar eminence; f.p.l.: 'posterolateral fissure'. Other abbreviations as in preceding figure. s.l.: here probably sulcus limitans.

Figures 243A and 327A–C illustrate aspects of the Urodele cerebellum in cross-sections. A molecular layer and a granular layer adjacent to the ependyma can be distinguished. The fiber systems run essentially through the molecular layer but may also encroach upon the layer of granule cells. In the median portion of the corpus, which affords passage to the cerebellar commissure, the cellular elements may be scarce. The *Purkinje cells*, at the boundary of granular and molecular layers, are

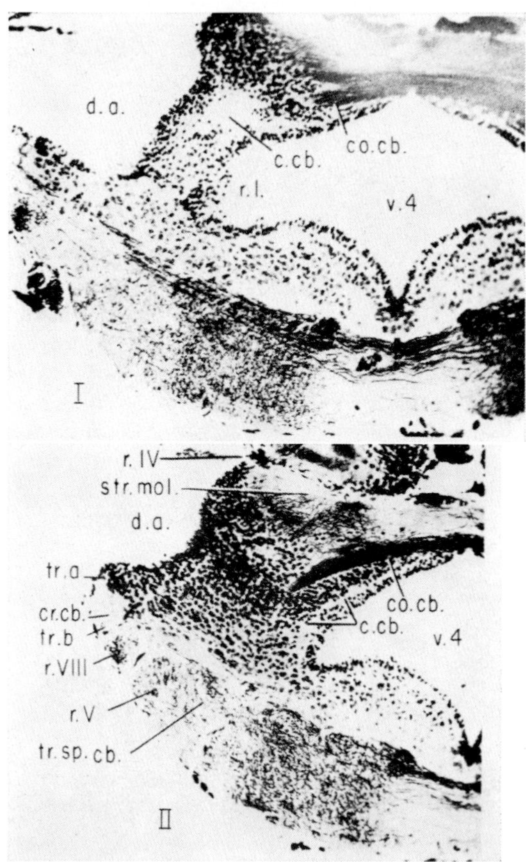

Figure 327A. Cross-sections *(Weigert stain)* through vestibulolateral area and portion of corpus cerebelli in Triturus (from LARSELL, 1967). Section I is slightly caudal to section II. c.cb.: corpus cerebelli; co.cb.: commissura cerebelli; cr.cb.: crista cerebellaris; d.a.: 'anterior diverticulum'; r.IV, V, VIII: roots of cranial nervus; r.l.: lateral recess of fourth ventricle (lateral to this label in I: nucleus cerebelli); str.mol.: stratum moleculare; tr.a., b.: dorsal and ventral 'longitudinal tract of acusticolateral area'; tr.sp.cb.: tractus spinocerebellaris.

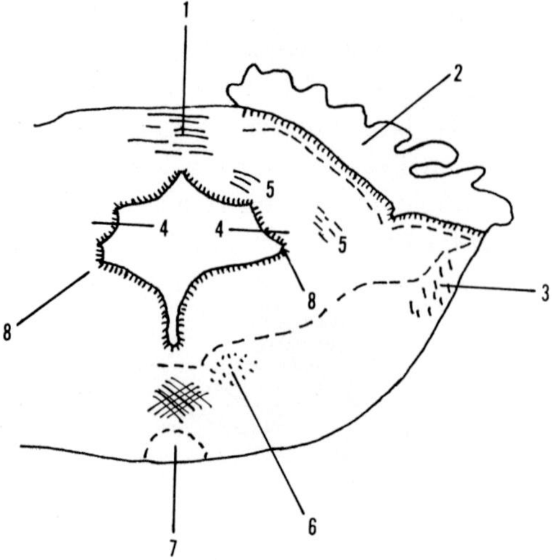

Figure 327B. Simplified sketch of a cross-section through the deuterencephalon of Amblystoma, showing cerebellar commissure in corpus, rostral portion of auricular recessus lateralis, and location of nucleus cerebelli (adapted from HERRICK, 1948, fig. 91). 1: corpus cerebelli with commissure; 2: recessus lateralis; 3: lateralis component of afferent facialis root, adjacent to root fibers of n. VIII; 4: nucleus cerebelli; 5: portions of tractus spinocerebellaris; 6: fasciculus longitudinalis medialis; 7: caudal portion of nucleus interpeduncularis; 8: sulcus limitans.

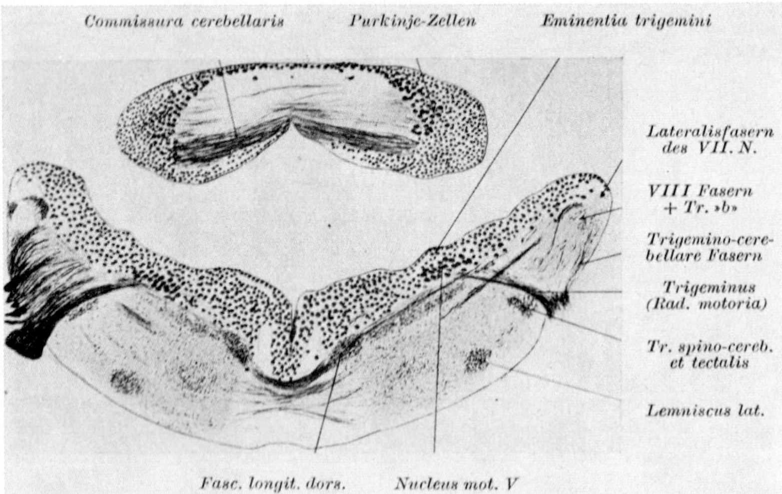

Figure 327C. Cross-section through the rhombencephalon of Salamandra maculosa (myelin stain), showing caudal portion of cerebellar plate (from KREHT, 1930).

rather irregularly arranged and considered to be 'reduced' respectively 'primitive' by KAPPERS *et al.* (1936) and other authors, and the dendritic expansions of these neuronal elements do not display a very definite orientation at a right angle to the transverse plane. LARSELL (1967) identified granule cells and stellate cells, but did not recognize basket nor *Golgi cells*. Yet, it seems at least probable that neuronal elements of *Golgi II* type may be present. A *nucleus cerebelli* comparable to that of other Anamnia is located in the ventral cerebellar eminence (Figs. 327A, B).

The lateral cerebellar subdivision designated as auricle in Urodeles, and, for that matter, in most other Anamnia, includes grisea related to both vestibulolateral and trigeminal input. LARSELL (1967) discusses the difficulties in a definition of 'auricles' based upon functional fiber connections. I prefer to use the designation '*auriculae*' for the more or less ear-like topologic neighborhoods of the 'rhombic lip' forming a transition between dorsolateral (alar plate) wall of oblongata and transverse cerebellar plate.

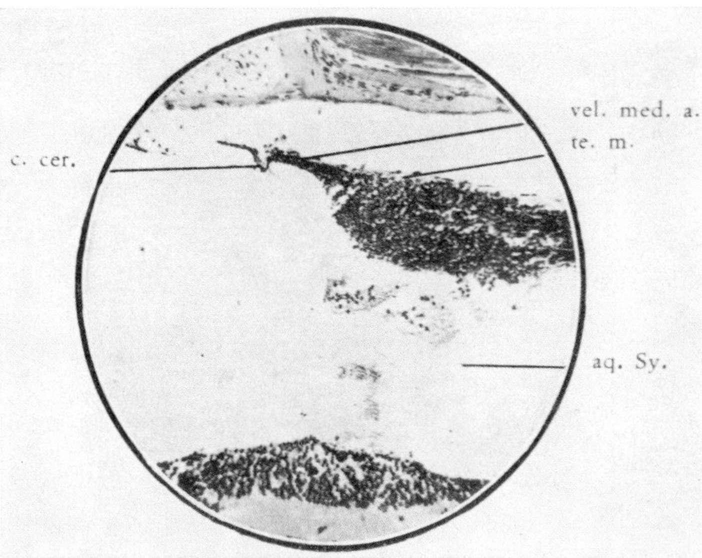

Figure 328. Paramedian sagittal section through caudal part of tectum mesencephali in Proteus anguineus, showing the cerebellar plate as a rudimentary appendage (from BENEDETTI, 1933). aq. Sy.: mesencephalic ventricle; c. cer.: commissura cerebellaris; te. m.: tectum mesencephali; vel. med. a.: velum medullare anterius.

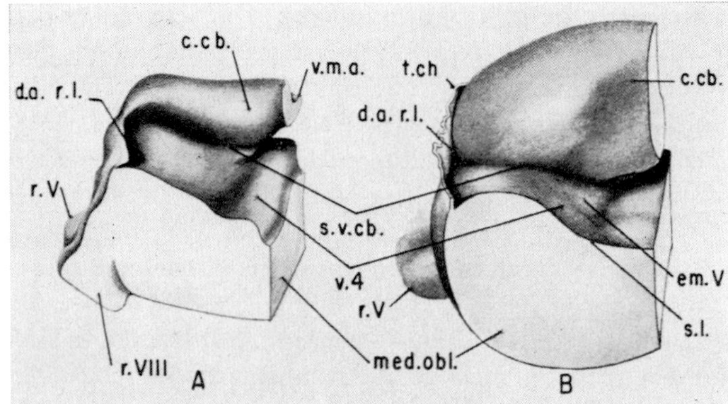

Figure 329A, B. Two models of left halves of the frog's cerebellum viewed from the caudal side (from LARSELL, 1967). A Larval frog of 36 mm length with well developed legs (Hyla regilla). B Adult frog, Rana pipiens. c.cb.: corpus cerebelli; d.a.r.l.: 'anterior diverticulum of lateral recess'; em.V: 'trigeminal eminence'; r.V, VIII: roots of trigeminus respectively octavus; s.l.: sulcus limitans; s.v.cb.: 'sulcus ventralis cerebelli'; v.m.a.: anterior medullary velum; v. 4: fourth ventricle.

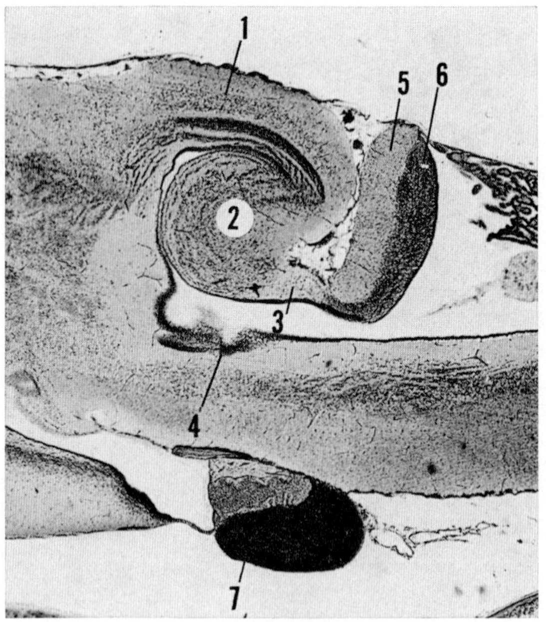

Figure 329C. Paramedian sagittal section through mesencephalon and metencephalon of the frog *(Nissl stain)*, showing cerebellar plate (×9; red. $^2/_3$). 1: tectum mesencephali; 2: torus semicircularis; 3: velum medullare anterius; 4: fovea isthmi; 5: molecular layer of corpus cerebelli; 6: granular layer; 7: hypophyseal complex.

Figure 328 shows, in sagittal section, the extremely reduced corpus cerebelli in the blind urodele Proteus anguineus, whose auriculae are, however, relatively well developed. The transverse plate of the cerebellar corpus contains here essentially only a rather thin commissural bundle (cerebellar commissure). If, in a perhaps unjustified arbitrary concept, overemphasizing the corpus cerebelli, the auricles are not considered to be 'true' cerebellar structures, then one might claim that a cerebellum is not present in Proteus, nor, *mutatis mutandis*, in Myxine (cf. section 2) nor in Gymnophiona (cf. further below).

In order to avoid repetitions, the cerebellar input and output channels of Urodela shall briefly be dealt with at the end of this section, in a short discussion of these systems with regard to Amphibians in general.

Figures 153B, 247A, 300, and 329A–E illustrate some aspects of the *Anuran* cerebellum. It can easily be seen that, at the adult stage, the corpus cerebelli is better developed than that of Urodela, while the auriculae are relatively reduced.

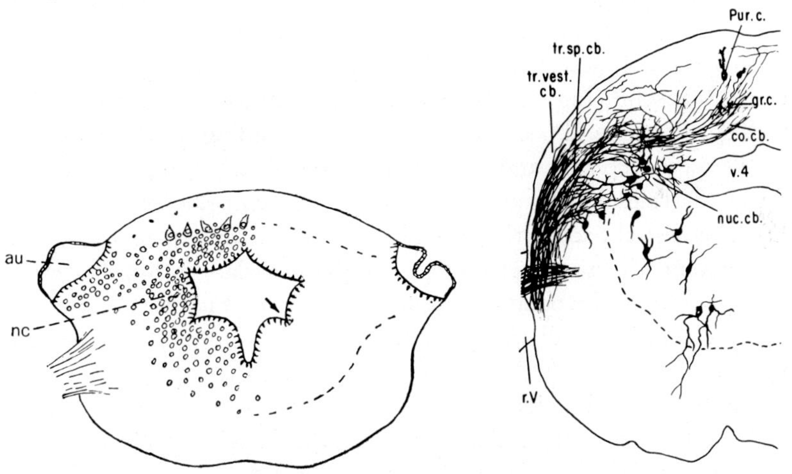

Figure 329D. Simplified drawing illustrating a cross-section through the cerebellum of an advanced frog larva (from K., 1922) au: auricular recess; nc: nucleus cerebelli; arrow indicates sulcus limitans.

Figure 329E. Cross-section through the cerebellum *(Golgi impregnation)* at the level of nucleus cerebelli in an advanced larva (with well developed legs) of the frog Hyla regilla (from Larsell, 1967). co.cb.: cerebellar commissure; gr.c.: granule cells; nuc.cb.: nucleus cerebelli; Pur.c.: *Purkinje cell;* r.V: trigeminal root; tr.sp.cb.: spinocerebellar tract; tr.vest.cb.: vestibulocerebellar tract; v.4: fourth ventricle.

Components of the neuraxis, particularly the vestibulolateral area and the adjacent auriculae, undergo important changes during metamorphosis from the aquatic larval stage to the land-adapted stage of the adults of many species. The lateral line system, which is the most substantial constituent of the auricles, disappears, except in some species such e.g. as the frog Xenopus, in which part of it remains in the adult (cf. section 7 of chapter IX). Details of these transformations have been investigated and discussed by LARSELL (1967).

As regards the histologic elements, the *Purkinje cells* are, in general, more regularly arranged than in Urodeles, and their dendrites spread at a right angle to the transverse fibers. Although stellate cells can be noted, and *Golgi cells* may be present, basket cells seem to be lacking, but climbing fibers have been described (GLEES *et al.*, 1958, LARSELL, 1967).

Few data are available on the cerebellum of *Gymnophiona*, which is illustrated by Figures 330A–G. The fiber systems and histologic elements in this order of Amphibia have not been investigated in detail nor with the relevant techniques.

With respect to the morphologic aspect, it can be said that the corpus cerebelli is reduced to a thin transverse commissure at the caudal edge of tectum mesencephali and, roughly speaking, may thus be considered 'absent' (Fig. 330A). The auriculae, however, are particularly well developed, and because of the peculiar bendings of the basal plate in the mesencephalic and rhombencephalic flexures, reach rostrad as far as the caudal poles of the telencephalic hemispheres (Figs. 330C, D). The rostral part of the auriculae are presumably mainly related to the afferent trigeminal system,[66] while their caudal portion may essentially pertain to the octavus system which, at least in most aquatic larvae, includes a lateralis component apparently not present in the adult forms.

In the material which I examined in 1922, I believed to have identified relatively large neuronal elements in the dorsomedial lip of the auricle as *Purkinje cells*, and adjacent smaller elements as granule cells (Fig. 330B). An interauricular commissure was tentatively depicted as commissura intertrigemina (K., 1922). Gymnophione material subsequently studied a few years ago showed a similar arrangement compat-

---

[66] A cell mass extending from a rostral portion of auricle toward the mesencephalon was tentatively identified as ganglion isthmi (K., 1922, Fig. 25), but might perhaps merely pertain to the auricular complex.

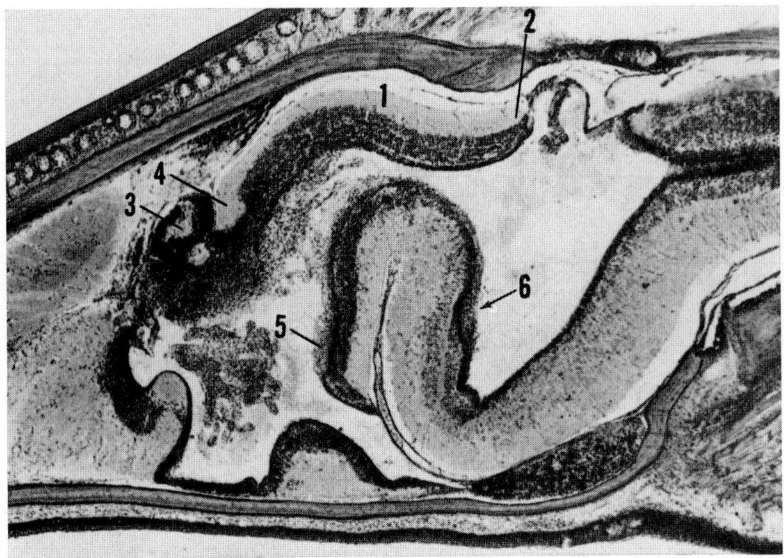

Figure 330A. Midsagittal section through mesencephalo-rhombencephalic region in the brain of the Gymnophione Schistomepum. The corpus cerebelli is here reduced to a thin commissural bundle (cf. Fig. 328). 1: tectum mesencephali; 2: commissura cerebellaris; 3: ganglion habenulae; 4: commissura posterior; 5: rostral end of basal plate (tegmentum); 6: fovea isthmi.

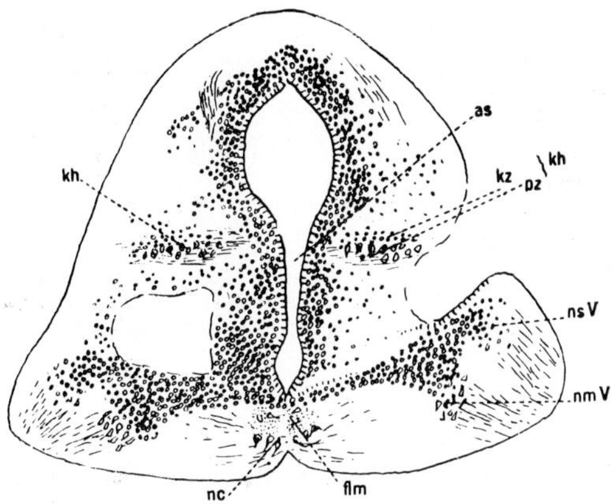

Figure 330B. Slightly oblique cross-section through mesencephalon and rostral rhombencephalon of the Gymnophione Siphonops, showing transition of 'aqueduct' to floor of fourth ventricle and slightly damaged auricular complex, dorsal to which a row of rather large neuronal elements was interpreted as representing *Purkinje cells.* The basal portion of the auricular complex contains trigeminal grisea (from K., 1922). as: '*aqueductus Sylvii*'; flm: fasciculus longitudinalis medialis; kh: cerebellar component (presumably pertaining to auricular complex); kz: granule cells; nc: nucleus centralis; nm V: efferent trigeminal nucleus; ns V: afferent trigeminal nucleus; pz: 'Purkinje cells'.

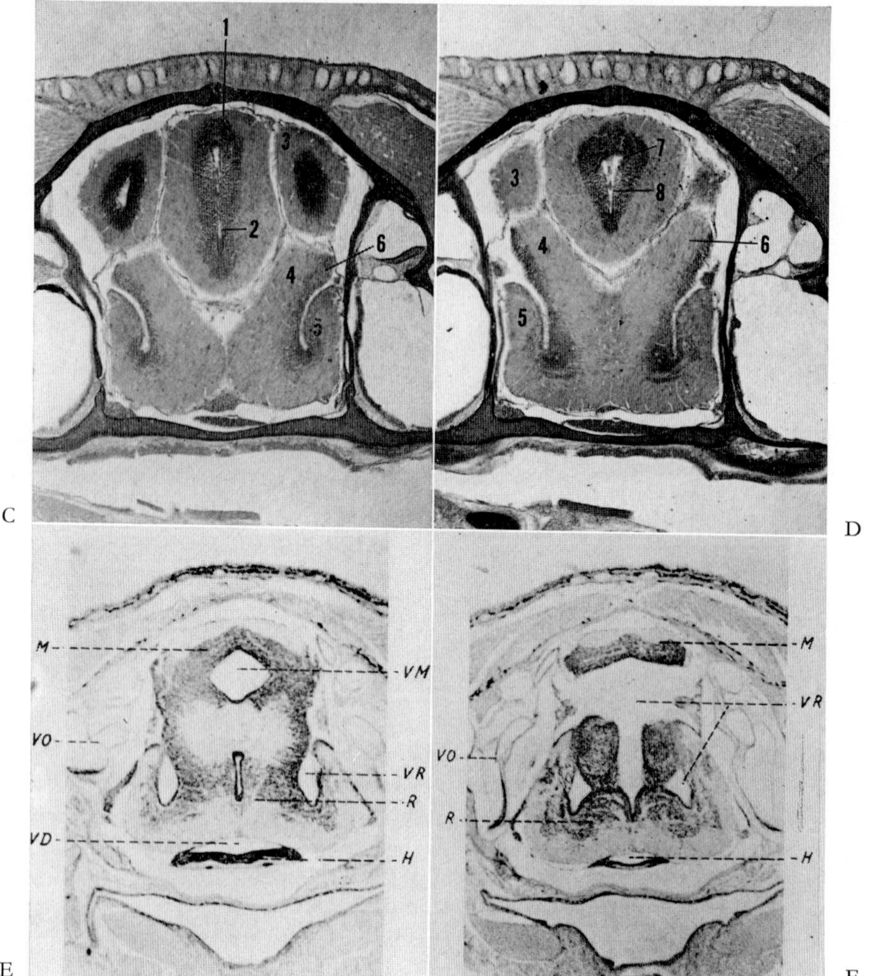

Figure 330C. Cross-section through the brain of Siphonops, showing rostral extent of auricular complex. 1: rostral mesencephalon, near caudal end of 'subcommissural organ'; 2: sulcus limitans; 3: caudal pole of telencephalic hemisphere; 4: upper lip of auricular complex; 5: lower lip of auricular complex; 6: larger cells interpreted as '*Purkinje cells*' in 1922.

Figure 330D. Cross-section through the brain in Siphonops, somewhat caudal to level of preceding figure. 7: sulcus lateralis mesencephali; 8: presumptive locus of sulcus limitans, which is not manifested (transition of alar to basal plate). Other abbreviations as in preceding figure.

Figure 330E. Cross-section through the brain in a fairly advanced larva (55 mm length) of the Gymnophione Ichthyophis, showing connection between mesencephalon

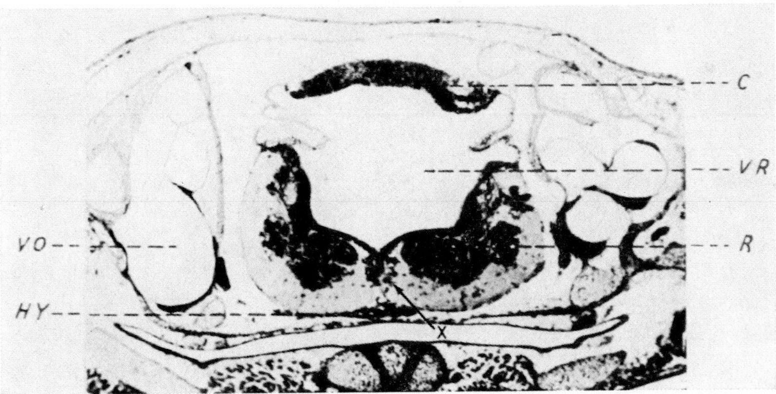

Figure 330G. Cross-section through the brain in a larva of Ichthyophis at a level caudal to that of the preceding figure, but in another specimen, said to be somewhat younger than the 55 mm stage (from KRABBE, 1962). C: 'cerebellum' (this, in my opinion, is not the corpus cerebelli, but the caudal end of mesencephalic tectum); Hy: hypophysial complex. The conspicuous cell group on both sides of the rhombencephalic raphe is the tegmental nucleus centralis which I described in 1922, and which is here indicated by the added lead x. The unlabelled sulcus limitans and sulcus intermedius dorsalis can easily be identified.

ible with that interpretation (Figs. 330C, D), but being by now somewhat more skeptical, I prefer to leave that question open.[67] Figures 330E–G (from suitable late larval material of Ichthyopis depicted in the publication by KRABBE, 1962) show the caudal aspects of the auriculae whose recess opens into the main lumen of the fourth ventricle.

[67] The author's paper of 1922 was mainly concerned with morphologic aspects of the telencephalon, but, because of the scarcity of reports on the Gymnophione neuraxis, included incidental findings on the other brain subdivisions. The material, although adequate for a morphologic survey of telencephalon and diencephalon, was, moreover, unsuited for detailed histologic studies, and, in some instances, poorly preserved or damaged (cf. Fig. 330B). Illustrations from very badly preserved, poorly fixed Gymnophione material are also depicted on plate XXVII of KRABBE's (1962) fairly recent publication.

---

and rhombencephalon (from KRABBE, 1962). H: hypophysial complex; M: mesencephalon; R: rhombencephalon; VD: infundibular recess of third ventricle; VM: ventriculus mesencephali; VO: 'otic vesicle'; VR: 'ventriculus rhombencephali' (auricular recess).

Figure 330F. Cross-section through the brain of a larva of Ichthyophis, slightly caudal to level of preceding figure (from KRABBE, 1962). The lead H points to infundibular recess; the lower lead of VR points to caudal end of auricular recess. Figures 330E and F should be compared with Figures 330A and B with regard to the ventricular configura tion

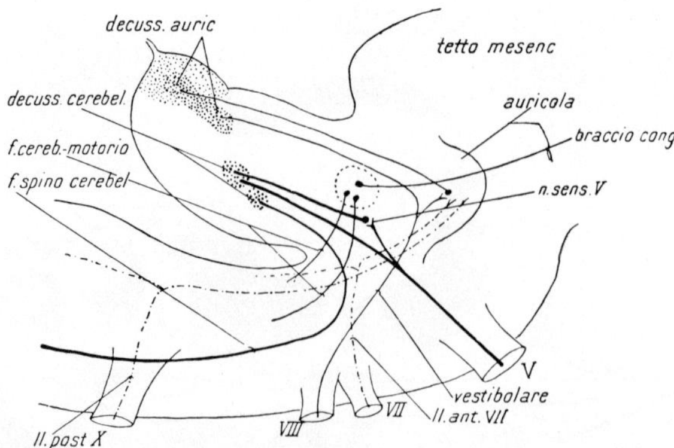

Figure 331. Diagram, based on interpretations by LARSELL and by BECCARI, of main cerebellar connections in an urodele Amphibian (from BECCARI, 1943). braccio cong.: 'brachium conjunctivum'; ll. ant. VII: anterior lateral line nerve component of facialis; ll. post. X: posterior lateral line nerve component of vagus. It will be seen that the cerebellar decussations of auriculae and of other systems are here indicated as quite separate. This, however, does not always seem to be the case.

As regards the *fiber connections* of the Amphibian cerebellum in general, the *input channels* comprise octavolateral and trigeminal as well as other bulbocerebellar fibers to the auricular complex. This input includes root fibers and presumably also secondary fibers in an undetermined proportion. The *interauricular commissure* joins the commissural system of corpus cerebelli. The spinocerebellar, tectocerebellar,[68] and hypothalamocerebellar (lobocerebellar) tracts are considered the main input channels of the *corpus*, which also seems to receive trigeminal, octavolateral and undefined bulbocerebellar input.[69] Although many details of the cerebellar commissure have been described, its components remain poorly elucidated (cf. e.g. LARSELL, 1967).

The cerebellar *output channels* seem to comprise *Purkinje cell* neurites as well as fibers originating in nucleus cerebelli. Some of the output is represented by internal arcuate fibers arising in auricular complex as well as in corpus and either cross the midline or remain homolateral. A

[68] Tectocerebellar *sensu latiori*, that is, including fibers from tectum mesencephali proper and from torus semicircularis.

[69] The presence of olivocerebellar fibers remains questionable, since a clearly identifiable inferior olivary complex does not seem to be present in Amphibians (cf. section 7 of chapter IX).

'primitive brachium conjunctivum', decussating in the isthmic tegmentum, ventrally to fasciculus longitudinalis medialis, but also including homolateral fibers, is directed rostrad. It contains, however, components which turn caudad as a tractus cerebello-motorius. This efferent system may include a cerebellospinal tract. Cerebellotectal fibers are presumably present in the tectocerebellar channel. Figure 331 depicts, in the interpretation of BECCARI (1943), and in accordance with LARSELL's views, the main cerebellar connections characteristic for Amphibia.

Discounting, because relevant observations are not available, the conditions obtaining in Gymnophiona, whose 'cerebellum' is essentially an auricular complex, the differences between the cerebellar connections of Urodeles and Anurans concern essentially the following points: (1) the lateralis input is lacking in practically all adult Anurans, except, of course, at larval stages. (2) The spinocerebellar and tectocerebellar systems are generally much better developed in Anurans. Evidently, an undefined number of additional differences can be assumed, but remain, as KAPPERS *et al.* (1936) stated, rather 'of emphasis than of kind'.

7. Reptiles

Although the Reptilian cerebellum significantly differs in the several orders of this Vertebrate class, it can be generally evaluated as poorly differentiated, both in comparison with taxonomically 'lower' forms

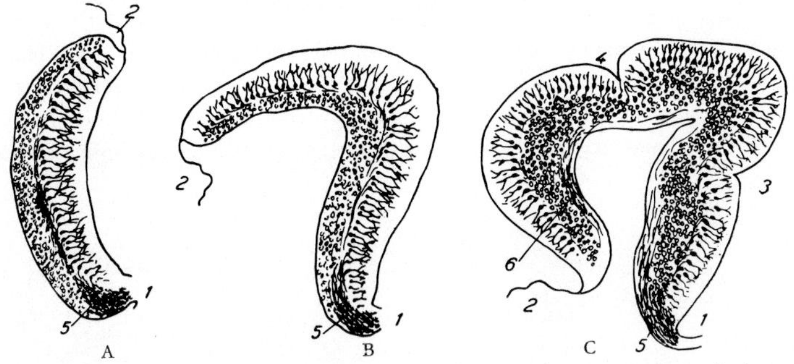

Figure 332A–C. Semidiagrammatic sagittal sections showing three typical configurational aspects of the Reptilian corpus cerebelli (from BECCARI, 1943). A Lacertilian. B Chelonian. C Crocodilian (Alligator). 1: velum medullare anterius; 2: attachment of epithelial roof of fourth ventricle ('velum medullare posterius'); 3: rostral sulcus ('fissura prima'); 4: caudal sulcus ('fissura secunda'); 5: cerebellar commissure; 6: 'floccular commissure'.

such as Selachians or Teleosts, or with the 'higher Amniota' (Birds and Mammals). The Lacertilian cerebellum, although, on the whole, somewhat more developed, is similar to that of anuran Amphibia, while the Crocodilian cerebellum displays the highest degree of Reptilian differentiation including features leading to a direct comparison with certain aspects of that brain subdivision in Birds and Mammals.

With regard to its overall configurational features, the Lacertilian corpus cerebelli is an everted plate (Fig. 332A). This eversion is still more pronounced in certain forms such as the *Gila monster* Heloderma (Fig. 332D) or in the Rhynchocephalian Sphenodon. The corpus of

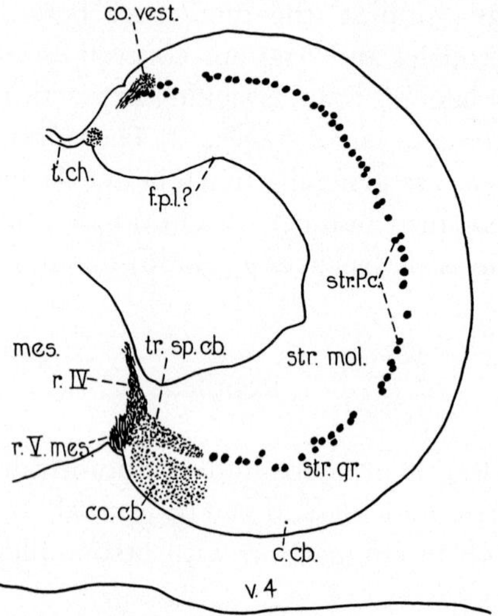

Figure 332D. Paramedian sagittal section through the highly everted cerebellum of the Gila monster, Heloderma (from LARSELL, 1967). c.cb.: corpus cerebelli; co.vest.: commissura vestibularis; f.p.l.?: possible 'fissura posterolateralis'; r.IV: root of trochlear nerve; r.V.mes.: mesencephalic trigeminal root; tr.sp.cb.: tractus spinocerebellaris.

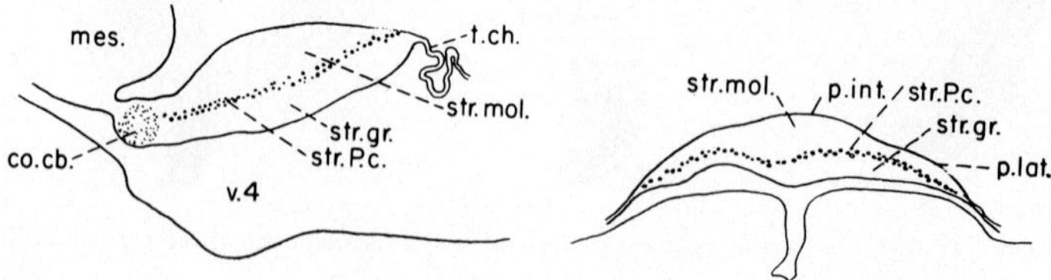

Figure 332E. Midsagittal section through cerebellum of Garter snake, Eutania sirtalis (from LARSELL, 1967). co.cb.: cerebellar commissure; mes.: mesencephalon; str.gran., mol., Pc.: granular, molecular, and *Purkinje cell* layers; t.ch.: attachment of epithelial roof plate; v. 4: fourth ventricle. Other abbreviations as in preceding figure.

Ophidia tends to tilt dorsocaudalward from a broad base (Fig. 332E). In Chelonia the cerebellum can be described as evaginated, forming a vaulted plate (Fig. 332C, cf. also Figs. 252A, B). This is still more the case in Crocodilia, whose corpus shows two or three characteristic sulci (Fig. 332B, 333A, B), and also displays a stratum medullare between ependyma and granular layer[70] (cf. Fig. 335 compared with Fig. 334).

In all Reptilia, the corpus cerebelli is much better developed than the auricles, which represent the *flocculi* of the Amniote cerebellum. This condition is related to the entire absence of the lateral line system and to the presence of a more substantial spinocerebellar system than found in Anamnia (KAPPERS *et al.*, 1936). The flocculus, nevertheless, remains as an essentially vestibular subdivision of the cerebellum.

The nucleus cerebelli of Anamnia has shifted into the Reptilian corpus cerebelli, and is represented by two often indistinctly separated grisea, nucleus medialis and n. lateralis cerebelli (Figs. 334, 335).

The neuronal elements in the Reptilian cerebellar cortex are known to include fairly typical basket cells, and all histologic features characterizing that cortex in Birds and Mammals, as dealt with in section 1, seem to be displayed in more or less typical manner. Relatively minor variations *qua* thickness respectively arrangement of molecular and granular layers occur with respect to the different forms. Again, the *Purkinje cell* layer may be several cells thick in Lizards and Snakes as well as in smaller Turtles, but a single layer obtains in Chelone midas. One to several layers may be found in Crocodilia (LARSELL, 1967).

As regards the Reptilian cerebellar fiber connections, the best known afferent systems are the spinocerebellar tracts, of which most authors distinguish a dorsal and a ventral one. The dorsal spinocerebellar tract may also include undefined bulbocerebellar[71] fibers. An olivocerebellar tract has been recognized by a number of investigators and seems to occur either as a diffuse system or as a more defined bundle joining the dorsal spinocerebellar tract. The octavocerebellar system includes vestibulocerebellar fibers and fibers from the cochlearis

---

[70] This arrangement can thus be evaluated as a significant step from paraependymal 'praecortex' to true 'cortex' (cf. section 1, p. 628).

[71] The bulbocerebellar system may include reticulocerebellar fibers as noted by WESTON (1936). With regard to the olivocerebellar fibers, it should again be pointed out that the inferior olive of Reptiles, if indeed present, is, at best, not particularly well developed. It will be recalled that the presence of this griseum in Amphibians and Dipnoans is rather questionable, although it might perhaps be represented by some cells within the caudal portion of the oblongata's reticular griseum.

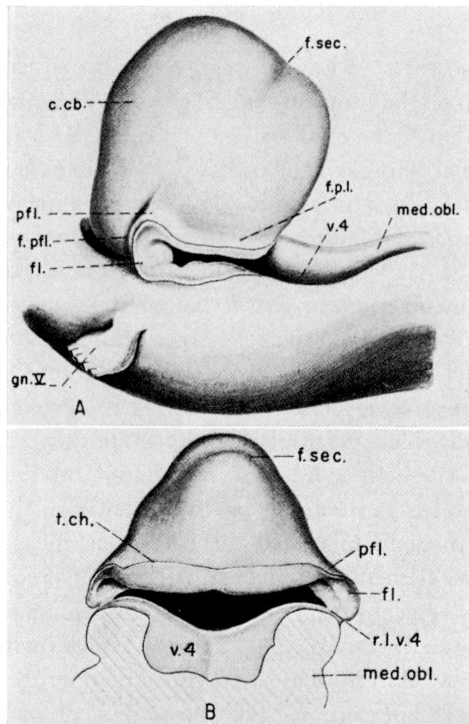

Figure 333. Cerebellum of a 1 m long Alligator mississippiensis (from LARSELL,1967). A Lateral view. B Posterior view, somewhat tilted upward. fl.: flocculus; f. pfl.: 'parafloccular fissure'; f. p. l.: 'posterolateral fissure'; f. sec.: 'fissura secunda'; gn. V: ganglion trigemini; med. obl.: medulla oblongata; p. fl.: paraflocculus; r. l. v. 4: lateral recess of fourth ventricle. Other abbreviations as in preceding figures from LARSELL.

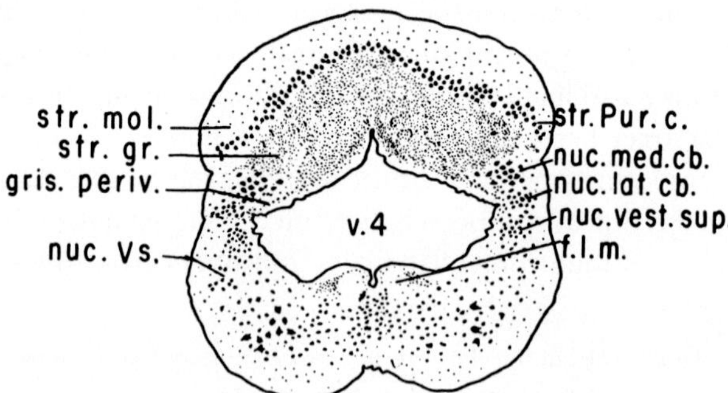

Figure 334. Cross-section through metencephalon of the Chelonian Chrysemys marginata, showing cerebellar nuclei (after WESTON, 1936, from LARSELL, 1967). gris. periv.: periventricular gray; f. l. m.: fasciculus longitudinalis medialis; nuc. lat. cb.: nucleus lateralis cerebelli; nuc. med. cb.: nucleus medialis cerebelli; nuc. vest. sup.: nucleus vestibularis superior; nuc. Vs.: afferent nucleus of trigeminus; str. gr., mol., Pur. c.: granular, molecular, and *Purkinje cell* layers.

component. Again, the octavocerebellar connections are partly root fibers and partly of the second order (cf. e.g. BECCARI, 1943). A comparable trigeminocerebellar system is likewise present. The tectocerebellar channel includes fibers from tectum *sensu strictiori* (tectum opticum) and from torus semicircularis. Additional input from rostral brain region, such as hypothalamus and thalamus can be surmised, but the identification of a lobocerebellar tract, ascribed to LARSELL by KAPPERS (1947), is specifically denied by the former author (LARSELL, 1967). There is no reasonably valid evidence, as far as I could ascertain, that efferent channels from the telencephalon are relayed to the cerebellum by cell groups which might be considered pontine grisea comparable to those present in Birds and particularly developed in Mammals,

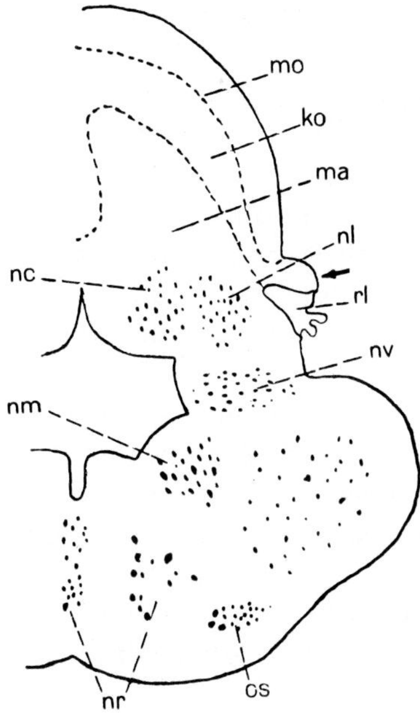

Figure 335. Cross-section through the metencephalon of the Alligator, showing the cerebellar nuclei (simplified after VAN HOEVELL, 1916, from K., 1927). ko: granular layer; ma: medullary layer; mo: molecular layer; nc: nucleus cerebelli medialis; nl: nucleus cerebelli lateralis; nm: efferent nucleus of trigeminus; nr: nucleus reticularis tegmenti; nv: nucleus vestibularis anterior; os: oliva superior; rl: recessus lateralis (auricular recess, above its lumen, the slightly protruding flocculus).

but this possibility cannot be entirely dismissed.[72] The vestibulocerebellar channel reaches mainly the flocculus and the caudal portion of corpus, but some connections with 'nucleus medialis cerebelli' have been claimed. The other systems seem essentially or predominantly related to corpus, although some connections with flocculus and medial cerebellar nucleus have been claimed.

The *efferent channels* can be subsumed, according to Beccari (1943) under the concepts of (1) a cerebellobulbar respectively cerebello-bulbotegmental system which may include direct or indirect cerebellospinal fibers, and (2) a rostral brachium conjunctivum system.

The *cerebellobulbar channels*,[73] which seem to originate mainly but perhaps not exclusively in the medial cerebellar nucleus, include the cerebellovestibular fibers, some of which may also be neurites of *Purkinje cells*. The presence of such neurites in the cerebellotegmental systems appears likewise possible. The cerebellospinal fibers are believed to be included in the fasciculus longitudinalis medialis (Beccari, 1943) or in the vestibulospinal tract. The brachium conjunctivum seems to arise in the nucleus lateralis cerebelli and decussates ventrally to fasciculus longitudinalis medialis. A substantial amount of its fibers reach the nucleus ruber tegmenti, which is undoubtedly present as such in Reptiles. Other fibers reach the more diffuse components of mesencephalic tegmentum, and presumably additional undefined grisea rostral to the cerebellum, as well as the eye muscle nerve nuclei. Some fibers take a caudalward course toward the bulbar tegmentum.

Although the brachium conjunctivum fibers predominantly decussate, the ratio of uncrossed and crossed fibers in the cerebellobulbar channel remains uncertain. Crossing components in this latter decussate as arcuate fibers in the bulbar tegmentum.

The *cerebellar commissures* include essentially the crossed components of the afferent channels, but may also carry some efferent and intrinsic components. A rostral commissura cerebelli, adjacent to velum medullare anterius[74] and to decussating trochlear root, and a caudal 'commissura vestibularis' can be distinguished.

---

[72] Kappers (1947, p. 234) mentions a '*faisceau strio-cerebelleux, décrit par* Huber *et* Crosby ('26)', but I could not find any reference to that tract in the paper quoted by Kappers.

[73] In part comparable to the 'tractus cerebellomotorius' of Plagiostomes and other fishes.

[74] Beccari (1943) comments on the term 'decussatio veli' which may include fibers of nucleus isthmi and other non-cerebellar systems. According to the cited author said decussation '*non è un'entità anatomica definita*'.

Before proceeding with a discussion of the Avian and Mammalian cerebellum, which have much in common,[75] it seems appropriate to reiterate the significance of some morphologic features displayed by the *Crocodilian cerebellum* (Figs. 332C, 333A, B). Larsell (1967, 1970) believes that the folia of the Avian and of the Mammalian cerebellum display an homologous pattern, consisting of components I to X in rostrocaudal sequence. Three significant sulci or fissures, namely fissura prima, fissura secunda, and fissura posterolateralis, separate components V and VI, VIII and IX, and IX and X, respectively (cf. Figs. 337D, 344A). Again, the transverse sulci displayed by the cerebellum of the Alligator are presumed to represent fissura prima, fissura secunda, and fissura posterolateralis of the cerebellum in the two 'higher' classes of Amniota.

In this respect, the *fissura posterolateralis* is of particular importance, since it separates the *flocculonodular lobe* (the nodulus being component X) from the corpus cerebelli *sensu strictiori*. The flocculonodular lobe, in turn, represents the main component of what Larsell termed archicerebellum, the palaeocerebellum being the corpus *sensu strictiori* of Anamnia and Reptiles, while the neocerebellum became that part of the Mammalian (and Avian) corpus receiving telencephalic input mediated by the nuclei pontis.

Elaborating on the concepts of *palaeocerebellum* and *neocerebellum*, propounded by Edinger (1910), Larsell had introduced the term *archicerebellum* for the cerebellar regions predominantly receiving vestibulolateralis or vestibular input, that is, for the auriculae of lower Vertebrates respectively the flocculi of Amniota, together with certain parts of the corpus. The term palaeocerebellum was restricted to that part of the corpus considered 'the chief terminus of the spinocerebellar and other phylogenetically early connections, such as trigeminal and tectal' (exclusive of vestibulolateral or vestibular). Larsell (1951) stated, however, that there are no sharp boundaries between the stages of development, phylogenetic or ontogenetic, although there is ground for a convenient subdivision into archicerebellum, palaeocerebellum, and neocerebellum, in accordance with the various subdivisions. It should be recognized, as he added, that in the phylogenetic development of the cerebellum, all the afferent connections, save those from

[75] The overall morphologic similarity of Avian and Mammalian cerebellum is particularly remarkable in view of the entirely different pattern of telencephalic, diencephalic and mesencephalic differentiation obtaining in these two higher classes of Amniota.

telencephalon by way of the pons, and possibly the olivocerebellar fibers (perhaps lacking in Cyclostomes, Dipnoans, and Amphibians) are represented in the most primitive Vertebrates. It is, moreover, of interest that LARSELL, in the three volumes of his posthumous monograph (1967, 1970, 1972), which reflect his final views on cerebellar morphology, seems to have abandoned the use of the term 'archicerebellum' and of the cognate phylogenetic designations.

8. Birds

The cerebellum of Birds is much more highly developed than that of Reptiles. It displays a massive *corpus* with a conspicuous transverse fissuration (Fig. 336). Distinctive *flocculi*, closely related to the vestibular system, are present. The development of these flocculi varies somewhat in accordance with the diverse taxonomic forms, and is complicated by the close apposition of *paraflocculi* which do not pertain to the flocculonodular lobe, as pointed out further below following a brief discussion of the folia.

Following earlier investigations by other authors, and particularly by INGVAR (1918), whose study suggested a significant first approximation[76] to the proper interpretation of sulci or fissures, respectively cerebellar lobes, LARSELL (1948, 1967) concluded that the fissuration of the Avian cerebellum results in the formation of 10 primary folia, designated, in rostro-caudal sequence, as folia I to X. Again, some of these folia become secondarily subdivided into subfolia by additional sulci. Figures 337A–D and 338A–D illustrate LARSELL's fundamental concepts and terminology. This author, moreover, reached the conclusion that the Avian cerebellar folia respectively lobuli can be directly homologized with those of the Mammalian cerebellum. SAETERSDAL

---

[76] INGVAR (1918), following, adapting, and elaborating an earlier terminology suggested by BROUWER (1913), recognized fissures *x*, *y*, *z*, and *un* as fundamental transverse furrows occurring during Amniote embryonic development of the cerebellum. These four sulci correspond in that order, to primary and secondary fissures, first uvular sulcus, and posterolateral fissure of LARSELL's (1967) terminology. INGVAR, however, evaluated fissure y as sulcus praepyramidalis. INGVAR's three lobes of the entire cerebellum of Birds correspond to the three lobes of the Crocodilian cerebellum (Fig. 332C). The nodulus and the uvula of Birds, however, are deeply separated by the posterolateral fissure, whereas in Crocodilians the nodulus is part of an uvulonodular segment which has not separated into two subdivisions (LARSELL, 1967).

(1959a), however, evaluated the intraculminary fissure (between IV and V), as representing the true homologue of the Mammalian primary fissure, and thereby in contradistinction to LARSELL, interpreted folia III and IV as representing the culmen of the Mammalian cerebellum, II being the central lobe, and I the lingula.

A lateral portion of folium IX (subfolium IXc) extends toward the flocculus (Fig. 338A, B). LARSELL has shown that it corresponds to the Mammalian (accessory) paraflocculus and forms, together with the flocculus, the so-called Avian auricle, of which only the floccular component is directly comparable to the Anamniote auricula. Moreover, LARSELL (1948) described a region of unfoliated or only slightly foliated cortex forming the ventrolateral surface of the cerebellum (cf. Fig. 339) and constituting an incipient cerebellar hemisphere or so-called 'neocerebellum', essentially characterized by input from telencephalon mediated through pontine nuclei.

With respect to an overall gross subdivision of the Avian and Mammalian cerebellum, one could perhaps distinguish four main por-

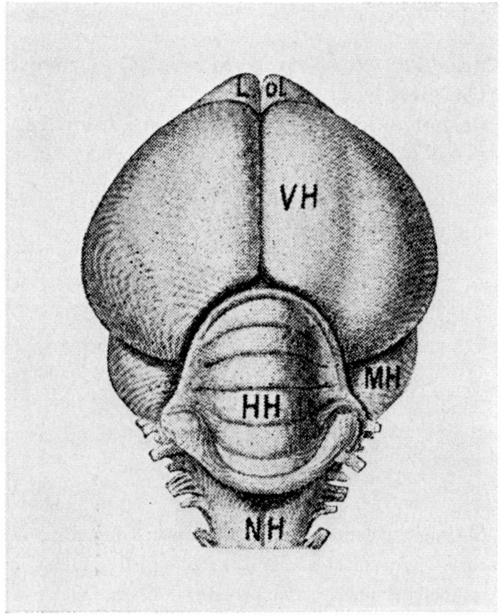

Figure 336. Dorsal view of the Pigeon's brain, showing overall aspect of cerebellum (after WIEDERSHEIM from HERTWIG, 1912). HH: cerebellum; Lol: olfactory bulb; MH: mesencephalon (tectum opticum); NH: medulla oblongata; VH: telencephalon.

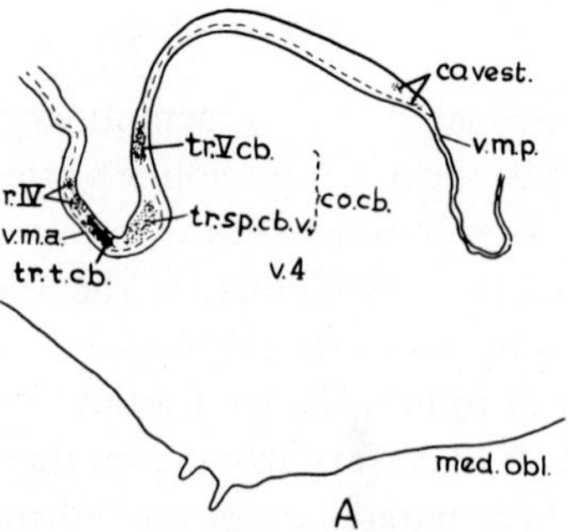

Figure 337A. Midsagittal section through the cerebellum in a Duck embryo of 10 days (from LARSELL, 1967). co.cb.: cerebellar commissure; co.vest.: vestibular commissure; med.obl.: medulla oblongata; r.IV: root of trochlear nerve; tr.sp.cb.: spinocerebellar tract; tr.V cb.: trigeminocerebellar tract; v.m.a.: velum medullare anterius; v.m.p.: velum medullare posterius; v.4: fourth ventricle.

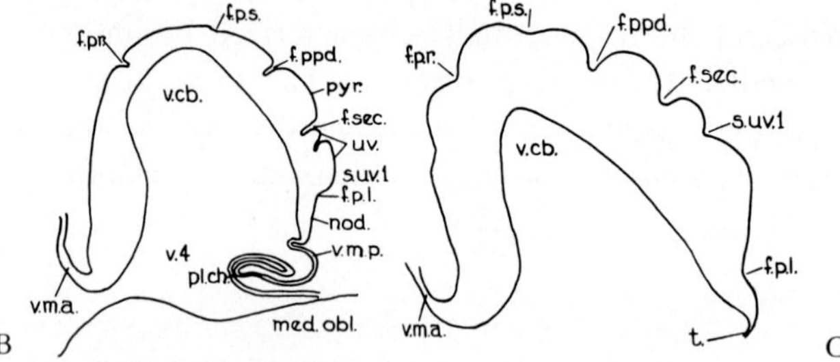

Figure 337B. Midsagittal section of cerebellum in a Duck embryo of 11½ days (from LARSELL, 1967). Abbr. given in D.

Figure 337C. Midsagittal section of cerebellum in a Cormorant embryo of 65 mm length (from LARSELL, 1967).

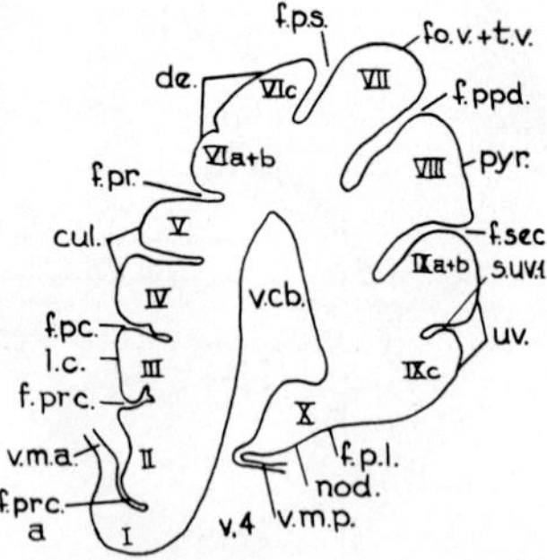

Figure 337D. Midsagittal section of cerebellum in a Duck embryo of 13 days (from LARSELL, 1967). f.ppd.: prepyramidal fissure; f.pl.: posterolateral fissure; f.pr.: fissura prima; f.pr.c.(a): precentral fissure and precentral fissure a; f.p.s.: posterior superior fissure; f.sec.: fissura secunda; nod.: nodulus; pl.ch.: choroid plexus of fourth ventricle; pyr.: pyramis; s. uv. 1: uvular sulcus 1; t: taenia of velum medullare posterius; v. cb.: cerebellar ventricle. The Roman numbers I to X designate the folia in LARSELL's nomenclature. Other abbreviations as in Figures 337A and 338D.

tions, namely *lobus anterior*, *lobus medius*, *lobus posterior*, and *lobus flocculo-nodularis*. The delimiting fissures would be *fissura prima*, *fissura secunda*, and *fissura posterolateralis*. This subdivision, also essentially adopted by BECCARI (1943) would remain valid regardless of the diverging views concerning fissura prima, (LARSELL, 1948, SAETERSDAL, 1956) and fissura secunda (ELLIOT SMITH, 1902; BRADLEY, 1903; BOLK, 1906; INGVAR, 1918; HALLER V. HALLERSTEIN, 1934; LARSELL, 1948, 1967).

Concerning the numerous variations of foliar development and topography in diverse Avian orders and species a detailed discussion can be found in the first volume of LARSELL's fundamental monograph (1967). A concise summary on this topic, with bibliographic references, is also contained in R. PEARSON's relevant treatise on the Avian brain (1972).

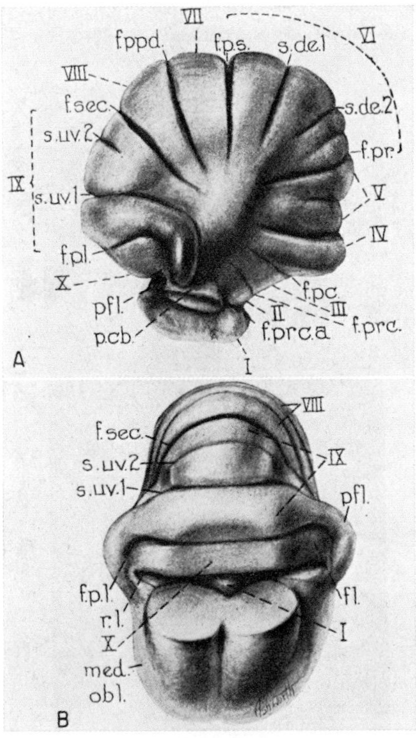

Figure 338 A, B. Cerebellum of adult Pigeon, Columba livia. A Lateral view. B Caudal view (from LARSELL, 1967). fl.: flocculus; p.cb.: 'cerebellar peduncle'; pfl.: paraflocculus; r.l.: recessus lateralis of fourth ventricle; s.de.1, 2: declival sulci 1 and 2. Other abbreviations as in preceding figures from LARSELL.

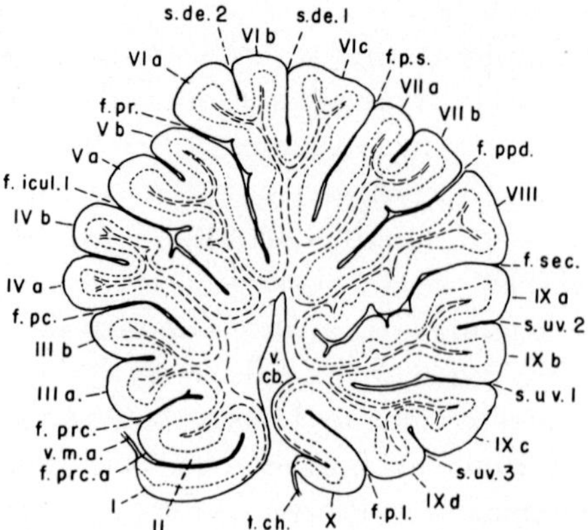

Figure 338C. Midsagittal section of cerebellum in the adult Pigeon (after LARSELL and WHITLOCK, 1952, from LARSELL, 1967). f. icul. 1: 'interculminar fissure 1'; t. ch.: tela chorioidea. Other abbreviations as in preceding figures.

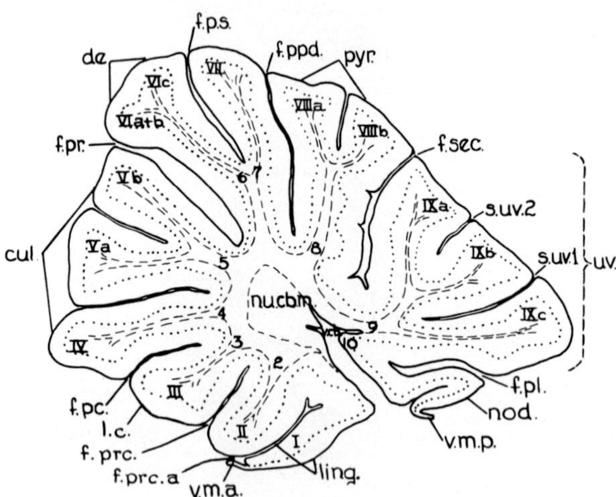

Figure 338D. Midsagittal section through the cerebellum of a young adult bantam Chicken (from LARSELL, 1967). Cul.: culmen; dec.: declive; ling.: lingula; nod.: nodulus; nuc. cb. m.: medial cerebellar nucleus; uv.: uvula. The numerals 2 to 10 indicate the so-called medullary rays. Other abbreviations as in preceding figures.

The histologic structure of the *cerebellar cortex* of Birds[76a] essentially corresponds to that of the Mammalian one, whose details have been dealt with in section 1 of this chapter, and require no further comments in this context. The Avian *cerebellar nuclei* are highly developed (Figs. 339A–D). Roughly speaking, two main grisea can be distinguished, namely, *nucleus medialis* (s. internus) and *nucleus lateralis cerebelli*. The latter may be considered homologous to the Mammalian dentate nucleus,[77] and the former to the other grisea, including the tectal or fastigial one. Connecting cellular groups, pointed out as 'gray bands' by KAPPERS *et al.* (1936) 'still betray their origin from subcerebellar vestibular regions' (cf. Fig. 337C).[78] The two main grisea are, to some extent, incompletely separated, and a more or less definable nucleus intermedius can be distinguished (cf. e.g. Fig. 339C).

The multiple variations in cellular packing and cytoarchitectural arrangement displayed by the Avian cerebellar nuclei have led to numerous discrepancies as well as differences in the nomenclature used by the investigators concerned with this topic, on which relevant critical reviews can be found in the the publications by LARSELL (1967) and by R. PEARSON (1972).

The fiber connections of the Avian cerebellum have been investigated with various techniques by many authors, including BOK (1915), CAJAL (1911, 1955), CRAIGIE (1928), FRENKEL (1909), FRIEDLÄNDER (1898), GROEBBELS (1928), INGVAR (1918), MÜNZER and WIENER (1898), SANDERS (1929), SHIMAZONO (1912), WALLENBERG (1900), and WHITLOCK (1952).

The *afferent connections* comprise spinocerebellar, vestibulocerebellar, cochleocerebellar, trigeminocerebellar, olivocerebellar, bulbo-

---

[76a] Recent speculations on the maturation and 'life cycle' of Avian cerebellar cortical neuronal elements can be found in the papers by BAFFONI (1963) and by MUGNAINI (1969). The latter author has particularly emphasized observations by means of electron microscopy.

[77] It also doubtless corresponds to the lateral cerebellar nucleus of Crocodilia. Despite the probable lack of true pontine nuclei and of a 'neocerebellum' in Reptilia, there can be little doubt that the Mammalian dentate nucleus originates from, and is topologically homologous to this lateral griseum. The differences in fiber connections and function can be entirely discounted, since we may here assume an instance of defective, respectively augmentative heteropractic and heterotypic kathohomology (cf. vol. 1, chapter III, section 3 on p. 200 and volume 3, part II, chapter VI, section 1A on p. 65).

[78] Despite differing interpretations expressed by other authors (cf. e.g. RÜDEBERG, 1961), I am inclined to agree with KAPPERS' view.

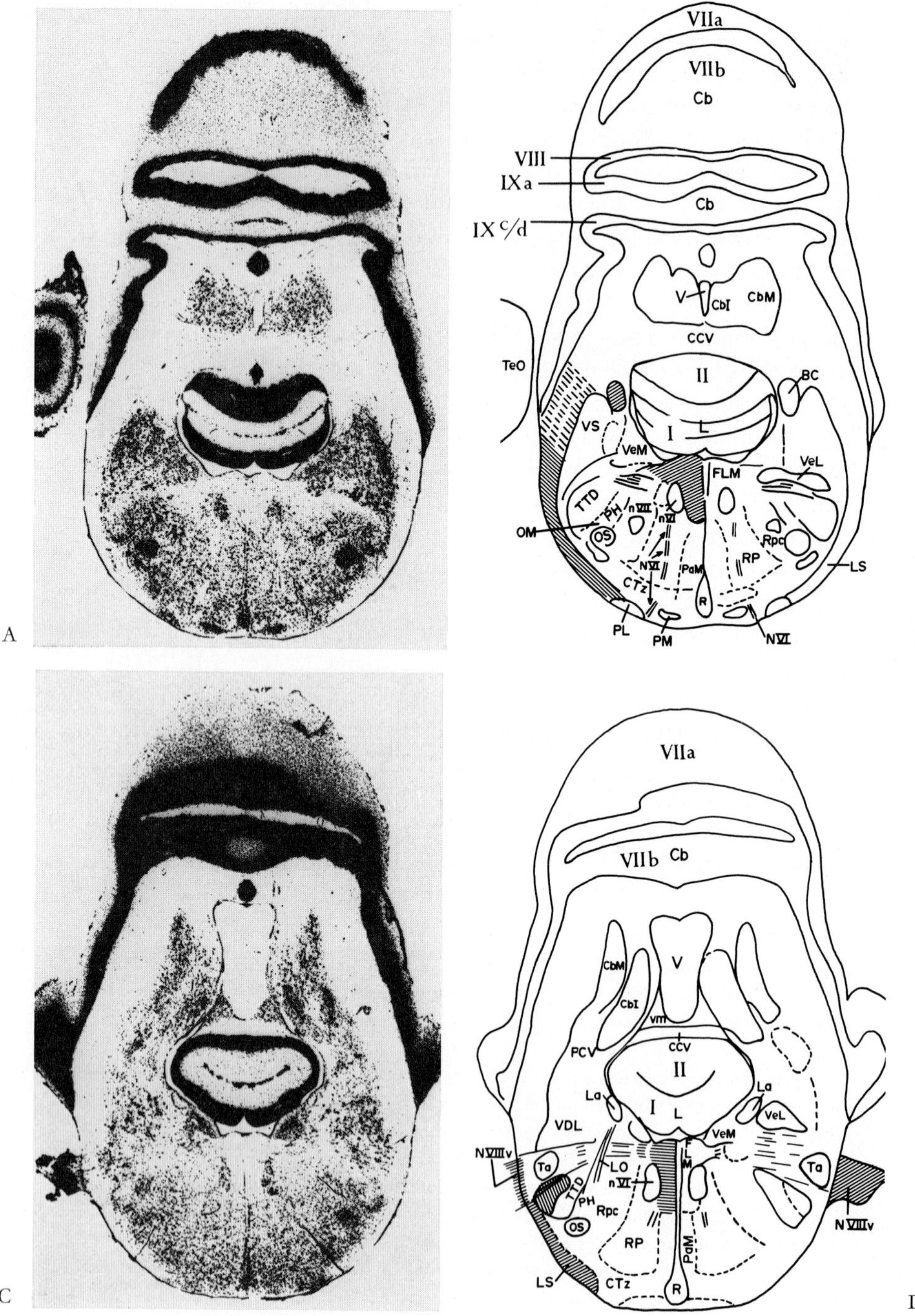
A
B
VIIa
VIIb
Cb
VIII
IXa
Cb
IX c/d
V
CbI
CbM
CCV
TeO
II
BC
VS
I
L
VeM
VeL
FLM
TTD
PH
nVII
nVI
OM
OS
Rpc
NVI
PaM
RP
LS
CTz
R
PL
PM
NVI
C
D
VIIa
VIIb
Cb
CbM
V
CbI
VM
CCV
PCV
II
La
I
L
La
VeL
VDL
VeM
NVIIIv
Ta
LO
L
M
Ta
TTD
nVI
PH
Rpc
NVIIIv
OS
RP
PaM
LS
CTz
R

cerebellar, tectocerebellar, and pontocerebellar channels. Whether additional systems, providing cerebellar input from hypothalamus or other rostral diencephalic, respectively diencephalo-mesencephalic grisea are present, remains uncertain.

The spinocerebellar system is relatively much more developed than in Reptiles and Amphibians. As KAPPERS *et al.* (1936) point out, the enlargement of the Avian corpus cerebelli may be substantially, but certainly not exclusively, related to the increase in spinocerebellar fibers which terminate mainly if not exclusively in this portion. Said fibers seem to arise from the entire length of the spinal cord.[79] A larger dorsal spinocerebellar tract and a somewhat smaller ventral one are generally distinguished. Figure 340 illustrates the distribution of spinocerebellar fibers as demonstrated by the *Marchi method.* It can be seen that folium and tuber vermis, pyramis, lingula, nodulus, and part of the declive appear to lack spinocerebellar input. Despite some uncertainties, it has been assumed that this channel does not effect direct connections with the cerebellar nuclei.

---

[79] Cf. chapter VIII, section 10, p. 174.

Figure 339A, B. Cross-section through cerebellum and oblongata (pons) of the Pigeon, Columba livia (from KARTEN and HODOS, 1967). A Photomicrograph *(Nissl stain).* B Explanatory drawing. BC: brachium conjunctivum; Cb.: cerebellum; Cbl.: 'nucleus cerebellaris internus'; CbM.: 'nucleus cerebellaris intermedius'; CCV.: 'Commissura cerebellaris ventralis' (anterior); CTz: 'corpus trapezoideum *(Papez)*'; FLM: fasciculus longitudinalis medialis; L: lingula; LS: 'lemniscus spinalis'; N VI: nervus abducens; n. VI: nucleus nervi abducentis; n. VII: nucleus nervi facialis; OM: tractus occipito-mesencephalicus; Os: 'nucleus olivaris superior'; PaM: 'nucleus paramedialis'; PH: '*plexus of Horsley*' (cf. explanation of fig. 255D); PL: nucleus pontis lateralis; PM: nucleus pontis medialis; R: nuclei of raphe; RP: 'nucleus reticularis pontis caudalis'; Rpc: 'nucleus reticularis parvocellularis'; TeO: tectum opticum; TTD: radix descendens trigemini and its nucleus; V: cerebellar ventricle; VeL: nucleus vestibularis lateralis; VeM: nucleus vestibularis medialis; VS: nucleus vestibularis superior. The designations I–IXc, indicating the cerebellar folia in LARSELL's terminology, have here been added in the reproduction of the original figure.

Figure 339C, D. Cross-section, somewhat caudal to the preceding level, through the cerebellum and oblongata (pons) of the Pigeon (from KARTEN and HODOS, 1967). C Photomicrograph *(Nissl stain).* D Explanatory drawing. PCV: 'nucleus cerebellaris lateralis'; La: nucleus laminaris; N VIIIv: nervus octavus, pars vestibularis; Rgc: 'nucleus reticularis gigantocellularis'; RP: 'nucleus reticularis pontis caudalis'; Rpc: 'nucleus reticularis parvocellularis'; Ta: nucleus tangentalis *(Cajal);* VDL: nucleus vestibularis dorsolateralis *(Sanders).* Other abbreviations as in Figure 339B.

The vestibulocerebellar system includes primary (root fiber) and secondary projections to the flocculonodular lobe, with an overlap to uvula (folium IX). *Connections with the cerebellar nuclei have been noted.* A cochleocerebellar channel originally reported by Bok (1915), and including fibers originating in nucleus laminaris, may be present, but has been doubted by others (e.g. Whitlock 1952).

The trigeminocerebellar channel seems to reach the region between folia V and VII. It originates in the main trigeminal nucleus, but is probably joined by components from the system of radix mesencephalica trigemini. The participation of trigeminal root fibers has been assumed by some authors and cannot be excluded, although the findings of other investigators were inconclusive or negative.

The bulbocerebellar system *sensu latiori* includes predominantly crossed olivocerebellar fibers apparently reaching all folia, reticulocerebellar connections and possibly fibers from other undefined bulbar grisea.

The tectocerebellar system connects tectum opticum proper as well as presumably other adjacent dorsal mesencephalic grisea with the cerebellum. This channel passes through anterior medullary velum and its

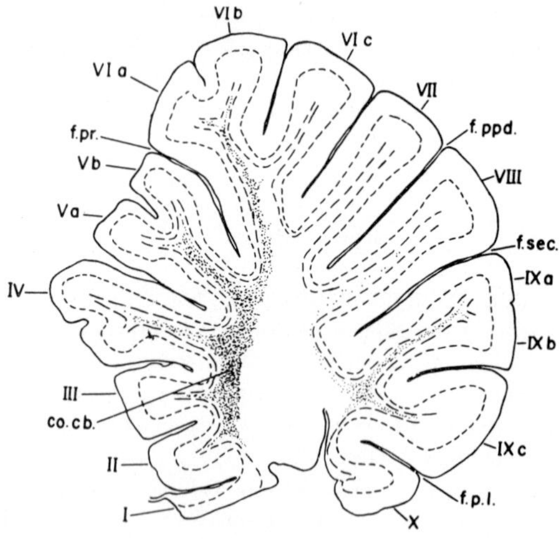

Figure 340. Midsagittal section through cerebellum of adult Pigeon, showing distribution of the degenerating spinocerebellar tracts, transected in the spinal cord, by means of the *Marchi technique* (after Whitlock, 1952, from Larsell, 1967). co.cb.: (anterior) cerebellar commissure. Designations of folia and fissures as in Figures 337–338.

lateral neighborhood as far as the so-called 'cerebellar peduncle', being perhaps predominantly homolateral.

The pontocerebellar system,[80] as demonstrated by BRODAL *et al.*, projects upon the rudimentary cerebellar 'hemisphere', the 'paraflocculus', and folia VI to IX. In the Penguin, LARSELL (1967) found an 'externally visible and relatively large brachium pontis'. It seems possible that the projections from the medial pontine grisea are contralateral, and those from the lateral ones homolateral.

Although the fiber systems terminating in the pontine nuclei have not been identified with reasonable certainty, it can be surmised that they are substantially provided by telencephalic output, perhaps in part identical with the dubious 'striocerebellar tract' of KAPPERS and some other authors. KAPPERS *et al.* (1936) also suggested that said tract might, in part at least, represent a cerebellostriate channel. In addition to the presumed telencephalic ('striatal') output, the pontine nuclei probably also receive terminals from additional undefined fiber systems.

The *efferent tracts* of the Avian cerebellum can be grouped as cerebellospinal, cerebellobulbar, and cerebellomesencephalic channels.[81] The fibers constituting these tracts arise chiefly from the cerebellar nuclei, but a part of the cerebellofugal fibers is of cortical origin, being represented by neurites of *Purkinje cells* (cf. e.g. KAPPERS *et al.*, 1936). The ratio of these so-called 'long corticofugal fibers' to the *Purkinje cell* neurites ending in the cerebellar nuclei (so-called 'short corticofugal fibers') remains uncertain and may differ *qua* diverse taxonomic forms.

A cerebellospinal and cerebellobulbar channel originates, according to SHIMAZONO (1912), in the 'nuclear' medial cerebellar grisea, with homolateral and contralateral fibers. These latter decussate within the cerebellum ventrally to the nuclei. The combined channel runs toward oblongata and spinal cord. This system gives off terminals to motor trigeminal, facial and perhaps other efferent cranial nerve nuclei, joining the lateral funiculus of the spinal cord and running as far as the cord's caudal levels.

---

[80] Cf. chapter IX, section 9, p. 503.

[81] LARSELL (1967) adopts the following grouping: fibers having cerebellotegmental, cerebellothalamic, and cerebellospinal connections, in addition to those that end in the vestibular nuclei. LARSELL, moreover, remarks that the interpretations concerning relations of most of the efferent cerebellar tracts with the subdivisions of the cerebellar nuclear mass, as well as their distribution, are in a state of confusion.

Other fibers, from undefined regions of the cerebellar nuclei, reach the reticular formation and may also, according to KAPPERS, join the fasciculus longitudinalis medialis. Cerebellovestibular fibers, originating in cortex of the flocculus and perhaps also in the cerebellar nuclei, terminate in the vestibular grisea.

The group of cerebellomesencephalic channels subsumes here all the rostrally directed efferent cerebellar systems. The brachium conjunctivum seems to originate in all cerebellar nuclear masses and decussates in the mesencephalic tegmentum, terminating in nucleus ruber. A component of that system appears to continue as cerebellodiencephalic tract to thalamic and perhaps other diencephalic grisea. Other efferent cerebellar fibers related to brachium conjunctivum *sensu latiori* join the fasciculus longitudinalis medialis and also reach the eye muscle nerve nuclei. Still other fibers of that system end in the mesencephalic reticular formation. It is likely that cerebellotectal fibers are included in the tectocerebellar channel. The possible presence of an insufficiently documented cerebellostriatal channel was mentioned above.

The *commissural system* of the Avian cerebellum includes a commissura cerebelli located rostrally to the cerebellar ventricular recess, and an essentially but not exclusively vestibular commissure in the region of flocculonodular lobe near the caudal margin of the cerebellum. Both commissures contain predominantly afferent fibers, which, in the commissura cerebelli are, to a large extent, the decussating components of the spinocerebellar channels (cf. Fig. 340). Primary and secondary trigeminal fibers as well as tectocerebellar ones may also cross in that commissure. The caudal commissure, in addition to primary and secondary vestibular fibers, likewise contains spinocerebellar as well as trigeminal components. The efferent fibers crossing in these commissures are cerebellobulbar and cerebellospinal within the rostral one, and presumably some cerebellovestibular fibers in the caudal commissure. The crossing fibers of the brachium conjunctivum system seem to decussate exclusively in the mesencephalic tegmentum. Whether some of the cerebellotectal and related fibers cross within the main (or rostral) cerebellar commissure remains an open question.

A concise review of *cerebellar functional aspects* has been presented in the treatise on the Avian brain by PEARSON (1972), which summarizes an extensive literature and also includes data on histochemistry. A brief reference to the following here relevant topics seems perhaps appropriate.

On the basis of experiments recording cerebellar activation by peripheral stimuli, WHITLOCK (1952) concluded that there obtains a 'somatotopic representation'. The caudal part of the body is said to project most rostrally, and the legs, wings, and facial areas following in sequence behind as far as folia VIc. The cochlea and the retina, respectively the optic tectum seem to project, with considerable overlap, upon folia VII and VIII, again in the here given sequence.

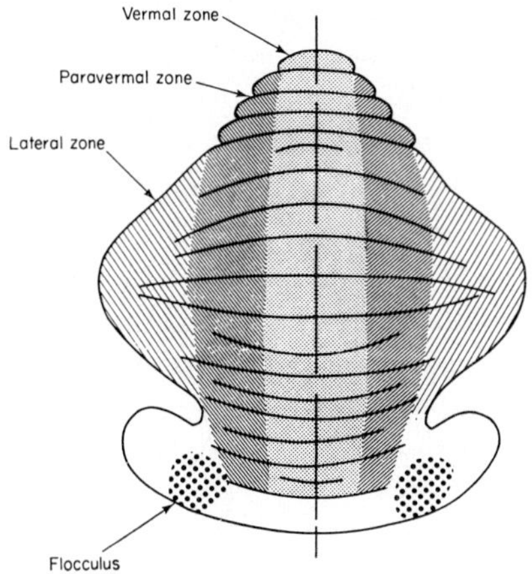

Figure 341A. Arrangement of longitudinal cerebellar zones corresponding to bands from which particular sorts of gross movements can be evoked. The longitudinal zones have been extended to reach those folia which were not reached by an electrode (after GOODMAN *et al.*, 1965, from PEARSON, 1972).

Folium	Face	Neck	Wing	Tail
III	Nictitating membrane	Extension		Down
IV		Extension		Erect
Va		Extension-retraction	Present	
Vb		Retraction		Mixed
VIa and b	Eyelids	Extension	Present	Mixed
VIc	Beak	Extension		
VII	Beak	Retraction	Present	Mixed
VIIIa	Beak and nicitating membrane	Extension-retraction	Present	Mixed
VIIIb	Beak and nictitating membrane	Extension-retraction	Present	Mixed-down
IXa		Extension		Down
IXb		Extension		Down

Figure 341B. Tabulation indicating a summary of responses which RAYMOND (1958) elicited by electrical stimulation of the cerebellar folia (from PEARSON, 1972).

Direct electrical stimulation of the cerebellar cortex, undertaken by various authors, particularly RAYMOND (1958) and GOODMAN *et al.* (1965) seem to indicate that Avian corpus cerebelli is divided into three longitudinal cortical zones, while the flocculus represents a fourth zone (Fig. 341A). This essentially agrees with a corticonuclear theory of cerebellar function proposed by CHAMBERS and SPRAGUE (1955) for Mammals. Stimulation of the median (or vermal) zone in Birds caused ipsilateral hindlimb flexion and protraction, coupled with contralateral hind limb extension and protraction. The head and neck rotated to the homolateral side. Opposite and complementary results were obtained from the 'paravermal zone'. Stimulation within the 'lateral zone' resulted in a response pattern somewhat similar to that obtained from the 'vermal zone'.[82] Figure 341B summarizes results recorded by RAYMOND (1958) with reference to the stimulated folia. A sceptical neurobiologist might finally add that experimental results are encumbered by considerable difficulties and ambiguities of interpretation, the observed movements, moreover, being doubtless influenced by a large number of poorly understood parameters. Thus, their evaluation should be accepted with an appropriate degree of caution.

9. Mammals (Including Man)

The Mammalian cerebellum constitutes, together with a macroscopically well defined pons, not present as such in other Vertebrates, the subdivision of the central neuraxis designated as metencephalon in standardized nomenclature (BNA, PNA).

In comparison with many 'submammalian' forms, the highly developed cerebellum of Mammals is remarkable by its substantial size[83] and its great relative width. As regards the *corpus cerebelli*, a lateral portion, the paired *cerebellar hemispheres*, can be roughly distinguished from an

[82] GOODMAN and SIMPSON (1960) interpret experiments which they performed on the Alligator as indicating that, in this Reptile, a functional equivalent of 'vermal' and 'paravermal' zones is present. In addition, an 'incipient cerebellar hemisphere' of the Alligator is believed, in the opinion of the cited authors, to be represented by a lateral segment of corpus cerebelli. The corpus cerebelli of the frog, however, is regarded as functionally equivalent to the 'vermal' zone.

[83] The substantial development of the cerebellum in Selachians, Teleosts, and Birds must nevertheless be kept in mind. The extreme development of the cerebellum in Mormyrids (cf. section 4) seems to represent a degree of cerebellar expansion unique in the entire Vertebrate series.

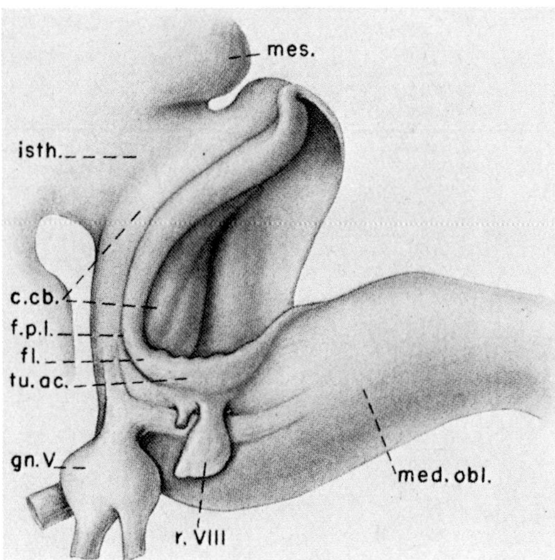

Figure 342. Early aspect of Mammalian cerebellar ontogenetic development: postero-lateral view of cerebellum in a 20 mm Pig embryo (from LARSELL, 1970). c.cb.: corpus cerebelli; fl.: flocculus; f.p.l.: fissura posterolateralis; gn.V: ganglion trigemini; isth.: isthmus rhombencephali; med.obl.: medulla oblongata; mes.: mesencephalon; r.VIII: root of nervus octavus; tu.ac.: 'acoustic tubercle'.

unpaired median subdivision, the vermis. In addition, as in other Vertebrates, an *auricular portion* remains connected with the *nodulus* which represents the caudal end of corpus *sensu latiori*. This auricular subdivision, however, develops in close spatial relationship with a laterally expanding parafloccular configuration, which may encroach upon the rostral parts of the auricular neighborhoods but can nevertheless be considered as pertaining to the hemisphere. The Mammalian *paraflocculus*, commonly displaying several subdivisions, ist essentially comparable respectively kathomologous with the paraflocculus of Birds.[84] Figures 342–344 illustrate relevant stages of Mammalian ontogenetic cerebellar development.

The morphology of the Mammalian cerebellum, based on an appropriate nomenclature for its numerous subdivisions, and allowing for a reasonably valid comparison with the subdivisions of the corpus cerebelli in Birds respectively Crocodilian Reptiles has presented considerable difficulties. The attempts at solving the problems pertaining

[84] Cf. the comments on the Avian paraflocculus in the preceding section 8.

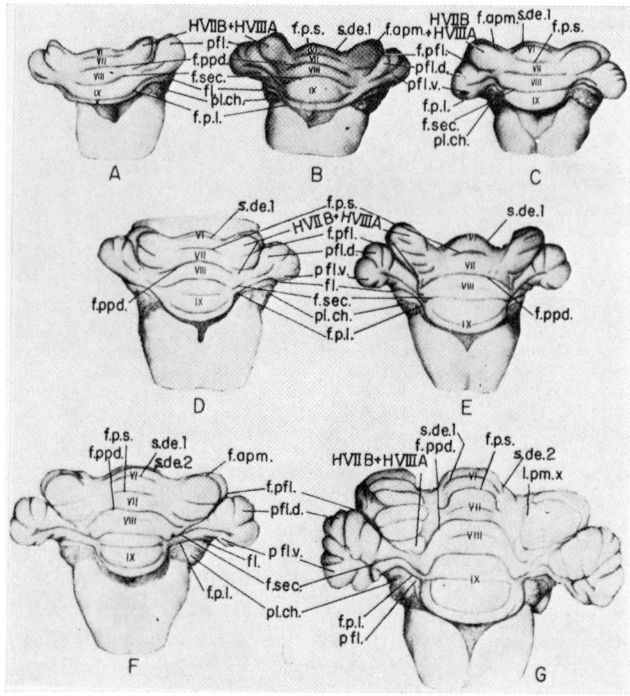

Figure 343 A–G. Dorsal view of the Rabbit's cerebellum from late fetal stages to newborn (from LARSELL, 1970). For abbreviations see legend to Figure 344A, B.

to this topic have, as it were, a fairly long history, culminating, at present, in the comprehensive formulation by LARSELL (1970, 1972).[85]

The studies by KUITHAN (1894, 1895) and STROUD (1895) initiated an embryologic approach particularly concerned with the significance of a '*fissura prima*' and of a '*floccular sulcus*'. ELLIOT SMITH (1902, 1903), by means of comparative studies, and BRADLEY (1903, 1904, 1905), who combined the embryologic and the comparative approach, added further data, which were taken into consideration in the schemes developed by BOLK (1906), INGVAR (1918), and HALLER V. HALLERSTEIN (1934).

A phylogenetic approach was undertaken by EDINGER (1910) and COMOLLI (1910) who attempted to delimit a *neocerebellum* from a *palaeo-*

[85] A more detailed account of these attempts can be found in volume 2 of LARSELL's posthumous monograph (1970).

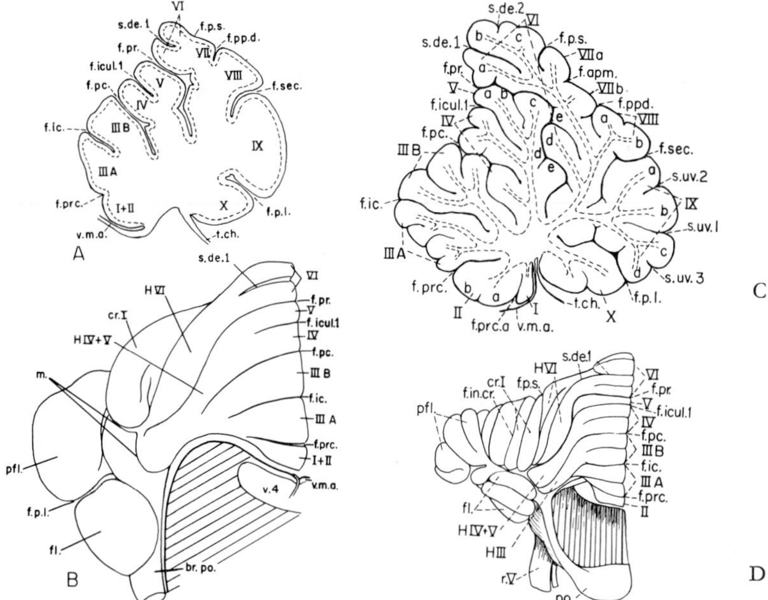

Figure 344A, B. Cerebellum of 28 days old rabbit fetus (from LARSELL, 1970). A Midsagittal section. B Anterior view. br. po.: brachium pontis; f. apm.: ansoparamedian fissure; f. ic.: intracentral fissure; f. icul. (1, 2): intraculminar fissures 1, 2; fl.: flocculus; f. pfl.: parafloccular fissure, f. p. l.: fissura posterolateralis; f. pp. d.: prepyramidal fissure; f. pr.: fissura prima; f. pr. c. (a): precentral fissure and prec. f. a; f. p. s.: posterior superior fissure; f. sec.: fissura secunda; H. II–X: hemispheral lobules II–X; l. pm. x: paramedian lobule, cut; m.: margin of cerebellar cortex; pfl. (d) (v): paraflocculus (dorsal, ventral); pl. ch.: choroid plexus; s. de. (1) (2): declival sulcus 1, 2; s. uv. (1) (2): uvular sulcus 1, 2; t. ch.: tela chorioidea; v. m. a.: velum medullare anterius; v. 4: fourth ventricle; I–X: cerebellar folia; a, b, c, etc., subdivisions of folia.

Figure 344C, D. Cerebellum of 6 days old Rabbit (from LARSELL, 1970). A Midsagittal section. B Anterior view. cr. I: so-called crus I; f. apm.: ansoparamedian fissure; f. in. cr.: intercrural fissure; po.: pons; r.: root of trigeminus. Other abbreviations as in preceding figures.

cerebellum. The investigations by LARSELL (1947, 1967, 1970, 1972) subsequently led to a now widely adopted concept of cerebellar morphology.[86]

---

[86] The minor discrepancies between the concepts of SAETERSDAL (1959) and LARSELL, mentioned in the preceding section 8, can here be ignored. Again, it should be understood that the short summary of the problem as dealt with in the present section 9 necessarily omits references to various additional contributions to this topic, which can be found in the treatises by KAPPERS *et al.*, 1936 and by LARSELL (1967, 1970, 1972).

Figures 345 to 346 illustrate the cited diverse schemes of Mammalian cerebellar morphology preceding that propounded by LARSELL, and shown in Figure 347A as adapted by JANSEN and BRODAL. It can be seen that in the latter scheme the cerebellum comprises an *anterior lobe* including folia I–V, and delimited from the *posterior lobe*, consisting of folia VI–IX, by the *fissura prima*. The posterior lobe, in turn, is separated from the *flocculonodular lobe* by the *posterolateral fissure*. The nodulus is represented by folium X which is laterally continuous with the auricular derivative forming the flocculi. While the nodulus could still be included into the corpus cerebelli *sensu latiori*, the entire flocculonodular

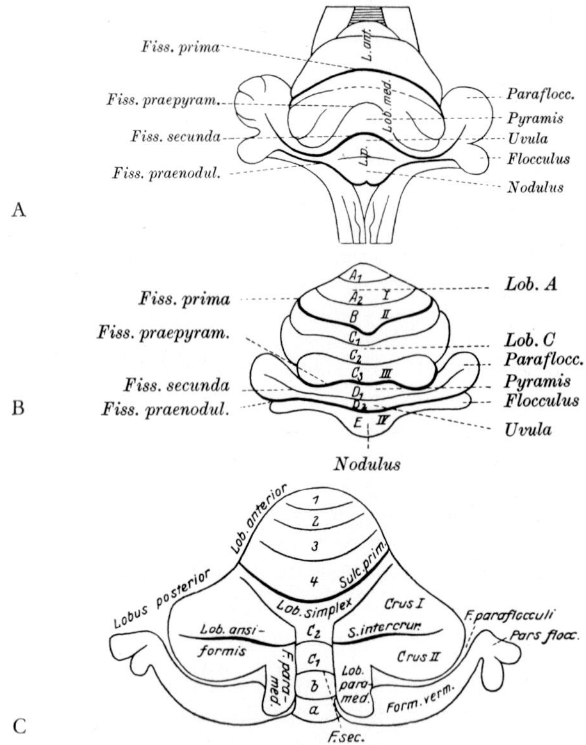

Figure 345A. Subdivisions of Mammalian cerebellum according to ELLIOT SMITH's scheme (from JAKOB, 1928).

Figure 345B. Subdivisions of Mammalian cerebellum according to BRADLEY's scheme (from JAKOB, 1928).

Figure 345C. Subdivisions of Mammalian cerebellum according to BOLK's scheme (from JAKOB, 1928).

lobe (or *archicerebellum* in obsolescent terminology) is usually not considered a part of corpus *sensu strictiori*.

The *posterior lobe* of corpus includes the rather complex configurations displayed by the *lobulus ansoparamedianus* consisting of ansiform sublobulus with two crura, and sublobulus paramedianus, which, in turn, is adjacent to the subdivisions of the *paraflocculus*. Compared with

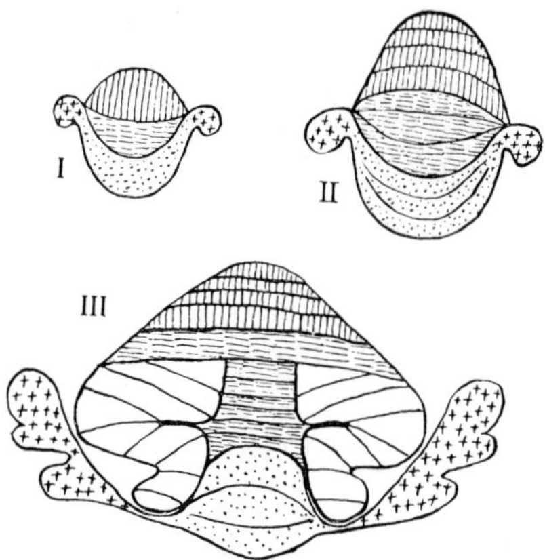

Figure 345D. Subdivisions of Crocodilian, Avian, and Mammalian cerebellum according to INGVAR's scheme (from K., 1927). I Crocodilian Reptile. II Bird. III Mammal. Vertical hatching: lobus anterior; horizontal hatching: parts of lobus medius; blank: lobulus ansoparamedianus; dots: lobus posterior; crosses: auricular lobe (flocculus, paraflocculus).

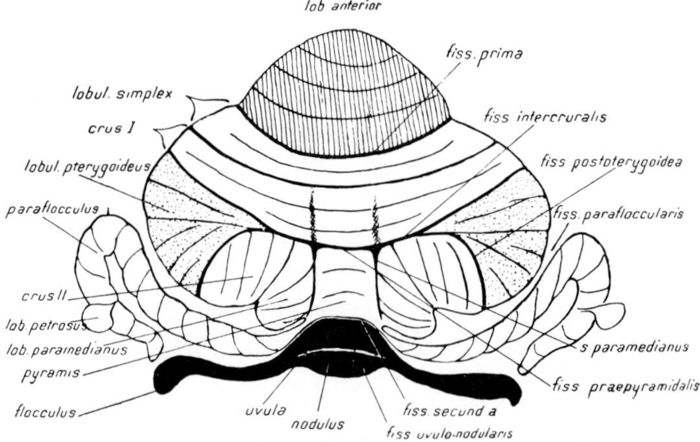

Figure 345E. Subdivisions of Mammalian cerebellum in the interpretation of *Graf* HALLER V. HALLERSTEIN (from BECCARI, 1943). Black: lobus posterior.

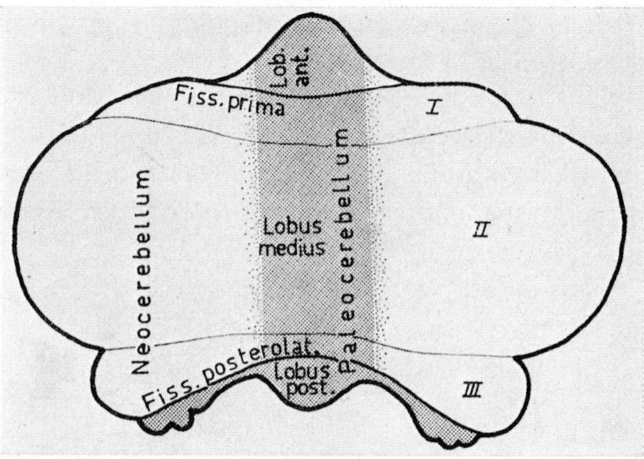

Figure 346. Subdivisions of Mammalian cerebellum according to EDINGER and COMOLLI (from VOOGD, 1967). I–III refer to subdivisions of neocerebellum.

the human paraflocculus ventralis, that of Cetacea appears particularly extensive (cf. Fig. 347B).

The scheme indicated by Figure 347A also indicates the *longitudinal* vermal, paravermal, and lateral zones mentioned above in section 8, and here indicated as vermis, pars intermedia, and pars lateralis of hemisphere. In a still more detailed subdivision, a pars intermediomedialis and a pars intermediolateralis of the paravermal zone have been distinguished. KORNELIUSSEN (1972) recently elaborated on these longitudinal zones with reference to corticogenesis.

As regards the morphology of the cerebellum in various Mammalian taxonomic forms, detailed data can be found in volume 2 (1970) of LARSELL's monograph. Comparative features, including fiber connections are also summarized in a contribution by VOOGD (1967). In the aspect here under consideration, it will be sufficient to illustrate, with two simplified sketches (Figs. 348A, B), the gross subdivisions of the human cerebellum. Two additional tabulations (Figs. 349A, B) show the detailed nomenclature of the BNA[87] with two now obsolete attempts at adaptation to the then current comparative anatomical concepts, as given in my 1927 '*Vorlesungen*'[88] and imitated, with slight

[87] Essentially retained, in a somewhat simplified manner, by the PNA.

[88] At the time, I rather prematurely and erroneously presumed that the indeed very valuable study by INGVAR '*eine endgültige Klärung der grobmorphologischen Verhältnisse des Reptilien-, Vogel- und Säugerkleinhirns gebracht haben dürfte*'.

modifications, by JAKOB (1928). The now widely accepted correlation of BNA respectively PNA terminology with that of comparative anatomy is shown in Figure 347. Figures 348C–E depict the gross configuration of an Ungulate cerebellum (Sheep) and Figures 263B–E of the preceding chapter IX illustrate some aspects of a 'lower' Mammalian cerebellum (Rat) in cross-sections.

The close correspondence between Avian and Mammalian cerebellar morphology, manifested by the orthohomologies in the differentiation of cerebellar lobes and folia, as demonstrated by LARSELL, is of particular interest. LARSELL's overall concept remains valid, regardless

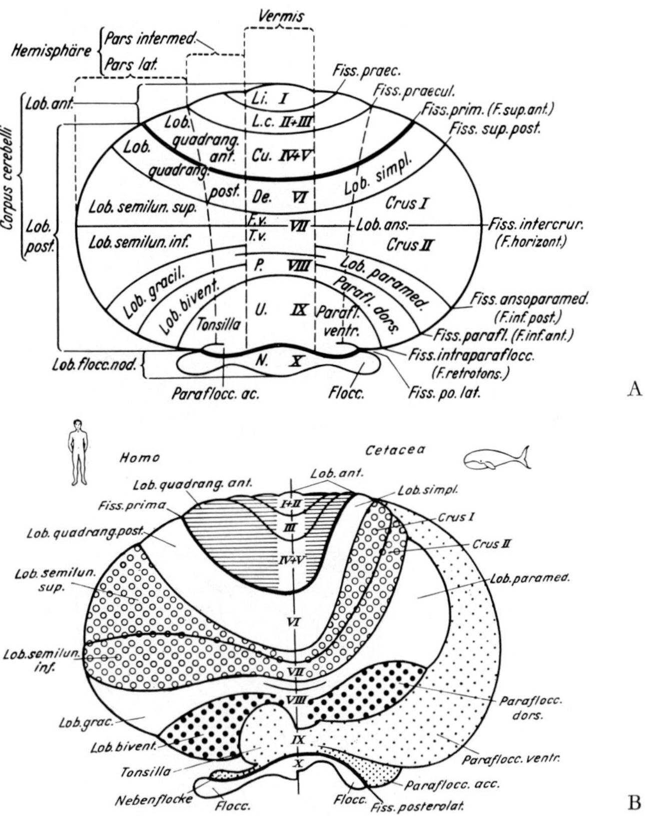

Figure 347A. Subdivisions of Mammalian cerebellum according to JANSEN and BRODAL, in adaptation to LARSELL's concepts (from VOOGD, 1967).

Figure 347B. Comparison of Human with Cetacean cerebellum (from JANSEN and BRODAL, 1958).

whether the modifications proposed by SAETERSDAL and mentioned in the preceding section 8, should be evaluated as valid or as non-valid.

In a comparison with Crocodilia, the Avian and Mammalian cerebellar subdivision is merely kathomologous, and only with regard to lobus anterior, lobus medius, and lobus flocculonodularis. With respect to other Reptiles and to all Anamnia, a kathomology merely obtains, at most, concerning (a) corpus cerebelli *sensu strictiori* and (b) flocculonodular lobe, or, at least, concerning (a) corpus cerebelli *sensu latiori* (cerebellar plate) and (b) auriculae.

It does not seem very likely that Crocodilia or closely related Reptilian forms represent a common ancestral form which led to Avian and Mammalian phylogenetic evolution. It appears perhaps much more probable that the close correspondence between details of Avian

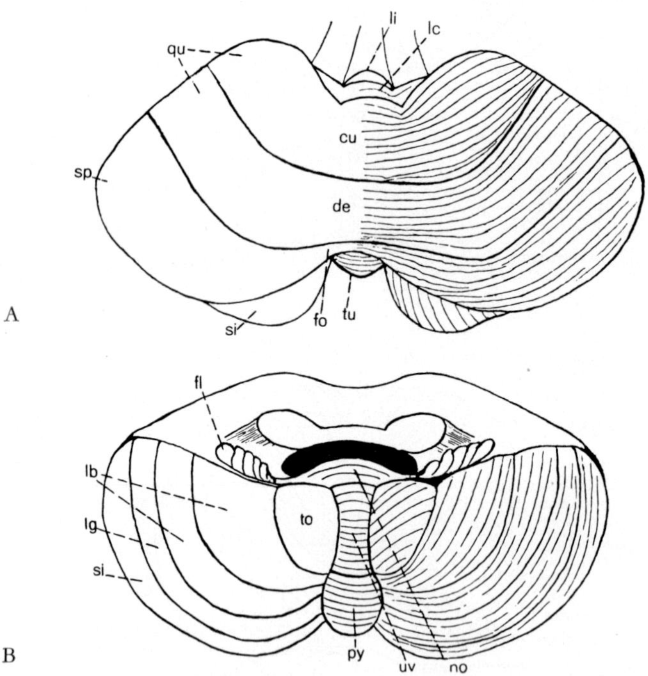

Figure 348 A, B. Simplified sketches of human cerebellum illustrating BNA nomenclature (from K., 1927). A Dorsal view. B Ventral view. cu: culmen monticuli; de: declive monticuli; fl: flocculus; fo: folium vermis; lb: lobulus biventer; lc: lobulus centralis; li: lingula; no: nodulus; py: pyramis; qu: lobulus quadrangularis; si: lobulus semilunaris inferior; sp: lobulus semilunaris superior; to: tonsilla; tu: tuber vermis; uv: uvula.

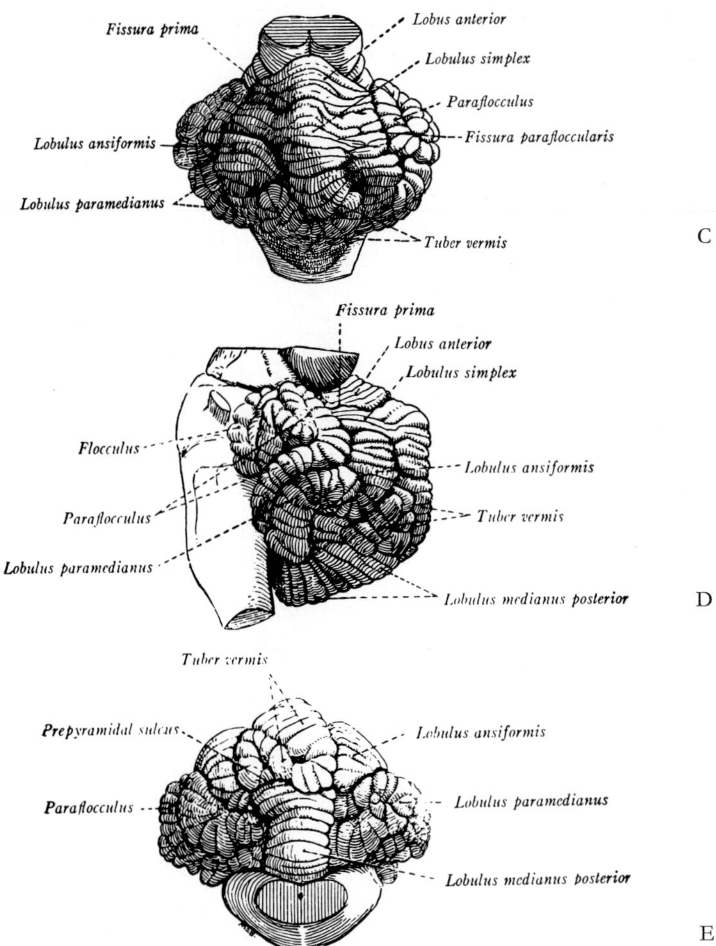

Figure 348C–E. Gross aspects of an Ungulate cerebellum (Sheep) in dorsal (C), lateral (D), and caudal (E) view. The right-left asymmetry of some vermian and paravermian portions will be noticed (from RANSON, 1943).

and Mammalian cerebellar morphology originated independently within diverging evolutionary radiations whose presumed common ancestral form, or 'greatest lower bound' [88a] remains unknown.

Another interesting feature of the remarkable Avian and Mammalian cerebellar homologies is the fact that they are essentially based on a specific pattern of transverse *sulci* or *fissures* developing in the

---

[88a] Cf. the comments on pp. 146–150 in chapter II of volume 1.

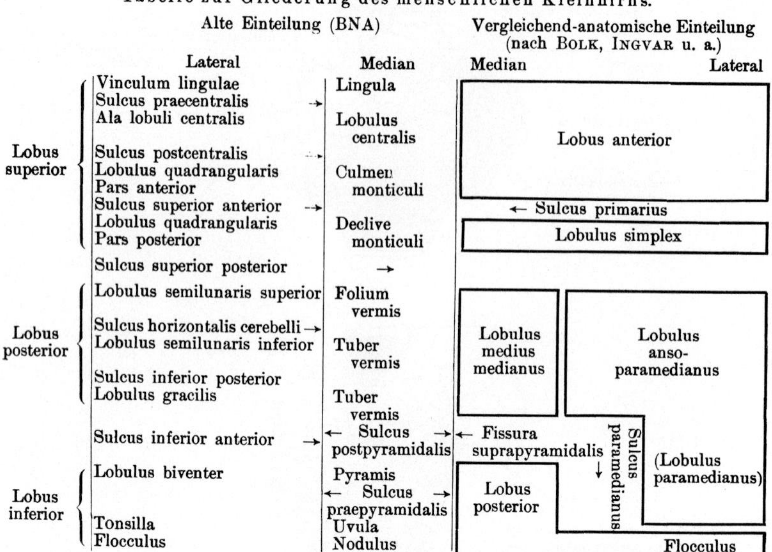

Figure 349A. Tabulation attempting to correlate BNA nomenclature with comparative concepts based on the interpretations of Elliot Smith, Bolk, Ingvar and previous authors (from K., 1927).

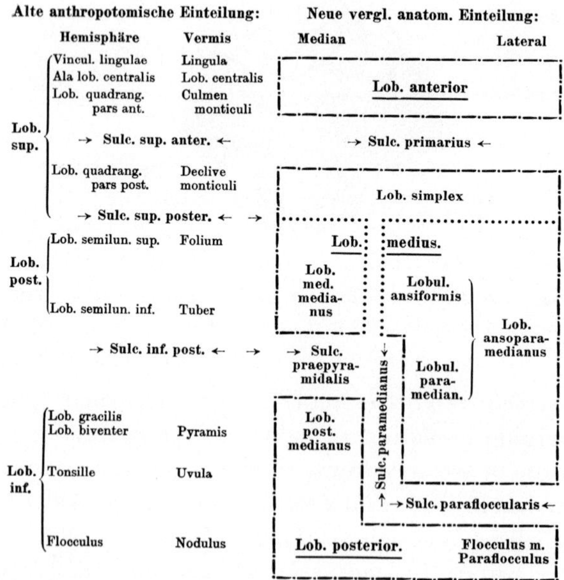

Figure 349B. Adoption of the preceding tabulation, with minor changes, by Jakob (from Jakob, 1928).

course of ontogenetic evolution. The fiber connections play here a more or less secondary role; the configuration of the cortical grisea, because of its uniformity, is, moreover, here of practically irrelevant significance. Actually obtaining minor differences, as e.g. stressed by KORNELIUSSEN and others, refer to a longitudinal zonal arrangement not directly relevant for the homologies based on the transverse, *i.e.* rostrocaudal morphologic pattern.

It is perhaps also of interest that the *telencephalic* morphologic pattern, dealt with in section 6, chapter VI of the preceding volume 3, part II, is to some extent related to the configuration of typical *sulci*, which, however, are *ventricular* as well as longitudinal. In the telencephalon, however, the longitudinal pattern is predominantly indicated by definable griseal zones. Strangely enough, while the cerebellar pattern, whose homologies are essentially indicated by *furrows*, has been widely accepted, the telencephalic pattern, primarily based on griseal zones but strongly supported by the arrangement of *sulci*, has been largely ignored. One may here recall the presumably sarcastic comment by KÄLLEN[88b] that I summarized my concept of telencephalic homologies in a number of 'Bauformeln' *(sic)* and then 'used the ventricular furrows as marks, after which the homologies were settled'. My studies, it is alleged, 'were chiefly carried out on adult material'.[88c]

With regard to ventricular *sulci*, and *mutatis mutandis*, much the same relationships obtain in the configuration of the Vertebrate *diencephalic* bauplan, discussed in section 5, chapter VI of the preceding volume and in chapter 2 of the treatise on the hypothalamus edited by HAYMAKER *et al.* (1969).

Omitting further comments on the Mammalian and human cerebellar cortex, which was dealt with in Section 1 of this chapter, the *cerebellar nuclei* shall briefly be considered. In *Prototheria* and *Metatheria*, respectively in 'Aplacentalia', BECCARI (1943) recognizes only two grisea, a nucleus medialis and a nucleus lateralis. The former is believed to represent the tectal group of Eutheria, and the latter to represent, or at

---

[88b] Lunds Universitets Arscrift, N.F. Avd. 2, Bd. 47, Nr. 5, 1951, p. 21.

[88c] That this last remark is entirely unfounded can easily be seen by consulting the relevant original publications on telencephalic morphology (Anat. Anz. 60, 1926; 67, 1929; J. comp. Neurol. 69, 1938) and on the diencephalic zones (Morph. Jb. 63, 1929; 66, 1931; 77, 1936; J. comp. Neurol. 66, 1937; 71, 1939) which are listed in the bibliography to chapt. VI, vol. 3, part II of this series. Cf. also the comments in footnote 222, p. 546 of vol. 3, part II.

least to include, the primordium of nucleus dentatus. Thus, ABBIE (1934) recognized a medial and a lateral nucleus in Echidna. HINES (1929) likewise describes four, i.e. two pairs of nuclei ('nn. cerebelli mediales' and 'nn. cerebelli laterales') in Ornithorhynchus, but states that the lateral nucleus consists of an anterior and a posterior portion. In the Metatherian Opossum, however, VORIS and HOERR (1932) recorded three cerebellar nuclei (n. fastigii, n. interpositus, and n. dentatus). LARSELL (1970) also noted three corresponding grisea. The medial griseum ('n. fastigii') appears as a well defined nucleus, while the lateral cell masses are rather indistinctly subdivided into 'nucleus interpositus' and 'nucleus dentatus' (cf. also FOLTZ and MATZKE, 1960). Thus, BECCARI's evaluation, as quoted above, could be essentially upheld.

In *Eutheria*, and depending partly on more or less distinct griseal fragmentation, and partly on the subjective criteria of parcellation adopted by the observers, three or four nuclei can be distinguished, namely, either (1) nucleus fastigii, (2) nucleus interpositus, and (3) nucleus dentatus, or, (1) n. fastigii, (2) n. globosus (better nn. globosi), (3) n. emboliformis, and (4) n. dentatus. KAPPERS *et al.* (1936) express the opinion that a typical or unequivocal differentiation of nucleus interpositus into nuclei globosi and nucleus emboliformis is found only in the 'Simian' suborder of the order Primates.

SPATZ (1922) demonstrated a peculiar extracellular and even partly intracellular iron reaction of the nucleus dentatus in Man and some higher Mammals. This reaction, characteristic for some grisea of the so-called extrapyramidal system, is also displayed by the nucleus ruber tegmenti, whose connections with nucleus dentatus are well established. Further comments on *Spatz's iron reaction* will be included in section 9 of chapter XI and in the relevant chapters of volume 5.

Figures 263C and D illustrate aspects of these nuclei in the Rat. It will be seen that CRAIGIE, ZEMAN, and INNES distinguish a globose, an emboliform and a dentate nucleus in the lateral griseum. In the Dog, however, BECCARI (1943) indicates merely nucleus tecti s. fastigii, n. interpositus, and n. cerebelli lateralis *('oliva cerebellare')* as shown in Figure 350A. Nevertheless, islands of gray interconnect nucleus fastigii with his nucleus interpositus. In many Eutherian Mammals, the nucleus dentatus becomes greatly expanded in correlation with the development of the cerebellar hemispheres and assumes the aspect of a folded griseal plate with a rostromedial hilus, and very similar to the main inferior olivary nucleus, being thus also occasionally designated

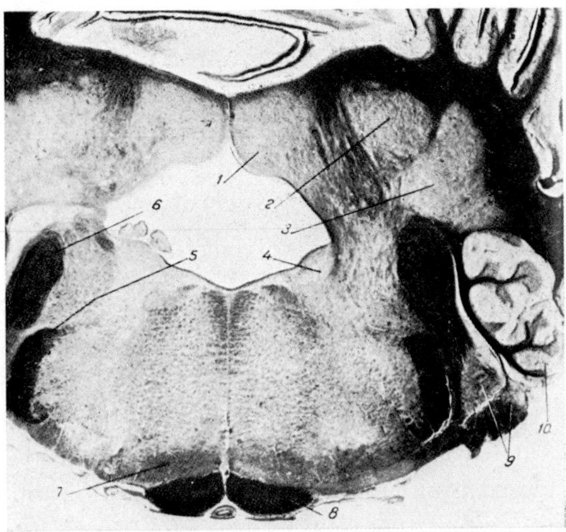

Figure 350A. Cross-section *(Weigert stain)* through cerebellum and oblongata of the Dog, showing cerebellar nuclei (from BECCARI, 1943). 1: medial cerebellar nucleus (n. tecti s. fastigii); 2: nucleus interpositus; 3: lateral cerebellar nucleus *('oliva cerebellare)*; 4: nucleus vestibularis medialis; 5: descending trigeminal root; 6: corpus restiforme (inferior cerebellar peduncle); 7: lemniscus medialis; 8: pyramid (tractus corticospinalis); 9: tuberculum acusticum; 10: flocculus.

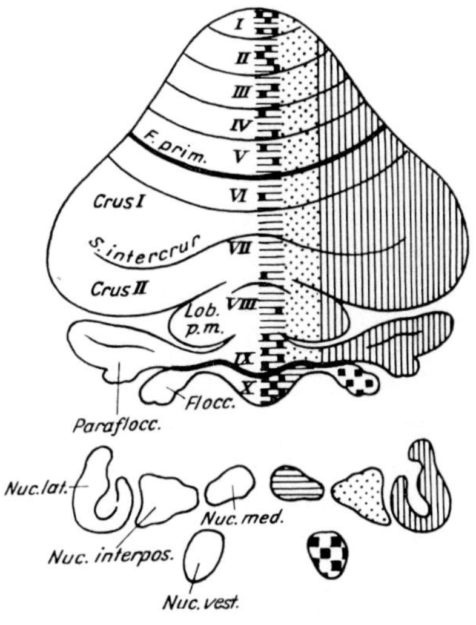

Figure 350B. Diagram showing projection from longitudinal zones of cortex cerebelli upon cerebellar nuclei and vestibular grisea (from JANSEN and BRODAL, 1958).

as 'cerebellar olive' (cf. the just mentioned '*oliva cerebellare*' of BECCARI). The fibers of the brachium conjunctivum originating in that griseum leave it through the hilus.

Recently, REIS *et al.* (1973) have claimed that, in the Cat, electrical stimulation at single sites in the rostral fastigial nucleus 'elicits hypertension, grooming, feeding, and attack behaviors'. The stimulus intensity and the 'availability of suitable goal objects determines the behavior', while bilateral lesions of the area 'fail to produce motor deficits'. According to the cited authors, 'the rostral fastigial nucleus may be a cerebellar area for behavioral and autonomic regulations'. Assuming that the reported results are reasonably accurate (cf. above, footnote 24 on p. 658), one could, however, interpret these findings as merely demonstrating that a disturbance, originating in the cerebellar fastigial nuclei and transmitted to the reticular formation and other grisea may, depending upon circumstances, trigger a variety of different responses by interfering with the multifactorial variable of a very large and undefined neuronal system.

As regards the cerebellar nuclei in the Primate brain, FIX (1967) has published an investigation of their development in different Prosimiae and in several Simian forms (Ceboidea, Cercopithecoidea, Pongidae, and Hominidae), with particular emphasis on nucleus lateralis s. dentatus.[89] Figures 303A–D illustrate some aspects of the human cerebellar nuclei, and Figure 350B indicates the cortical projections from corpus cerebelli and flocculonodular lobe upon cerebellar nuclei and vestibular grisea. It can be seen that this projection seems to be correlated with the three longitudinal zones (vermal, paravermal, and lateral, as discussed above).

Again, as already mentioned in section 10 of the preceding chapter IX, cells of nucleus loci caerulei and of nucleus radicis mesencephalicae trigemini may become dorsally displaced and can be found scattered in the periependymal region of the cerebellar medullary center. JAKOB (1928, Fig. 216 l.c.) depicts such large pigmented cells which were also noted by previous authors (MEYNERT, JACOBSOHN, ZIEHEN). As far as such elements seem identical with those of locus caeruleus, it seems possible to assume with JAKOB and others that these cells '*scheinen eine cerebellare Abteilung des Nuc. loci coerulei darzustellen*'.

---

[89] The cited author considers here also quantitative questions such as cell density, cell sizes, '*Grauzellkoeffizient*', ratio of glial to neuronal elements, and volumetric aspects. Concerning the general significance of these topics, cf. also section 8, chapter VI in volume 3, part II of this series.

Afferente Verbindungen des Kleinhirns

	Ursprung	Terminalgebiet	Funktion
Tr. spino-cerebellaris post.	Nuc. dorsalis	Lob. ant.: Vermis u. Pars intermed. Lob. post.: Pyramis u. Uvula	Propriozeptive Impulse
Tr. spino-cerebellaris ant.	?	Lob. ant.: Vermis u. Pars intermed. Lob. post.: Pyramis u. Uvula	Exterozeptive Impulse
Tr. cuneo-cerebellaris	Nuc. cuneatus accessorius	Lob. ant.: Vermis u. Pars intermed.	Propriozeptive und exterozeptive Impulse
Tr. olivo-cerebellaris	Nuc. olivaris	Ganze Kleinhirnrinde	Spinale und zerebrale Impulse
Tr. ponto-cerebellaris	Nuc. pontis	Ganze Kleinhirnrinde	Spinale und zerebrale Impulse
Tr. reticulo-cerebellaris	Nuc. reticularis lat. Nuc. reticularis tegm. pont. Nuc. reticularis paramedian.	Ganze Kleinhirnrinde Lob. ant.: Vermis u. Pars intermed. Lob. post.: Pyramis u. Uvula	Rückenmarks- und Hirnstammimpulse ?
Tr. perihypoglosso-cerebellaris	Nuc. perihypoglossi.	Lob. ant.: Vermis u. Pars intermed. Lob. post.: Pyramis u. Uvula	?
Tr. vestibulo-cerebellaris	Ganglion vestibulare Nuc. vestibularis med. et inf.	Lobus floc. nod. u. Nuc. fastigii	Vestibulare Impulse Vestibulo-spinale Impulse
Tr. trigemino-cerebellaris	Nuc. sens. n. V	Lob. post. (Declive, Folium, Tuber?)	Exterozeptive Impulse
Tr. tecto-cerebellaris	Tectum mesencephali	Lob. post. (Folium, Tuber, Pyramis?)	Auditive und visuelle Impulse
Tr. rubro-cerebellaris	Nuc. ruber	Nuc. dentatus	?

Figure 351 A. Tabulation of cerebellar input in JANSEN's interpretation (from JANSEN, 1961).

Concerning the extrinsic Mammalian cerebellar connections, already dealt with in section 1 of this chapter, only a few additional details need here be pointed out. With respect to the input channels, the tabulation of figure 351A summarizes the views of JANSEN (1961), based both on first hand experience and on an evaluation of the data recorded by numerous investigators. It will be noted that the dorsal spinocerebellar tract is considered to provide proprioceptive input, and the anterior (ventral) one conceived as an exteroceptive channel. The pontocerebellar connections, which are essentially, but not exclusively crossed, bring the Mammalian cerebellum, especially its so-called hemispheres, under the control of the cerebral cortex by way of the temporoparieto-occipitopontine tract and of the frontopontine tract.

The term 'perihypoglossal' nuclei refers to the grisea commonly designated as *nucleus of Roller*, *nucleus intercalatus Staderini*, and nucleus praepositus hypoglossi mentioned in the preceding chapter IX. In addition to the channels tabulated by JANSEN, one might also mention the possibility of a more direct cochlear input and the suspicion that interoceptive respectively 'visceral' input is also somehow reaching the Mammalian cerebellar cortex. Figures 351B and C illustrate projection areas of cerebellar input channels as mapped by various authors using experimental methods. The *olivocerebellar input channel*, although it may be considered 'phylogenetically old' assumes a substantial further development in Mammals The inferior olivary complex becomes here related to the central tegmental tract (cf section 9 of chapter XI). BRODAL (1972) emphasizes that a direct if 'modest' telencephalic cortical

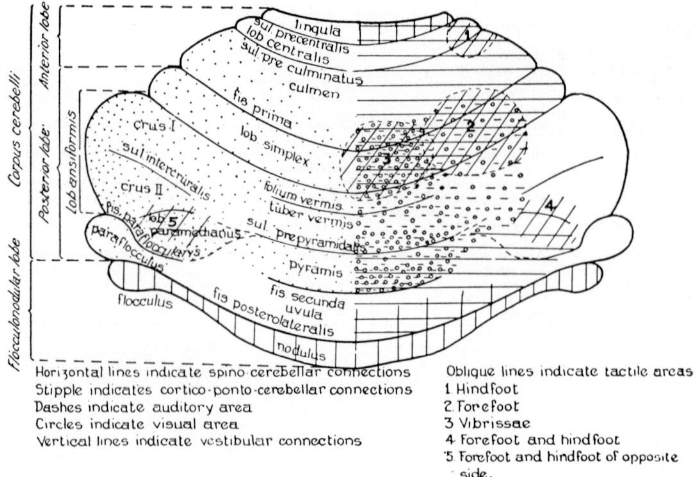

Figure 351B. Schema of Mammalian cerebellum showing areas of tactile, optic, and auditory input, adapted from various authors, and as interpreted by LARSELL (from LARSELL, 1942, 1951).

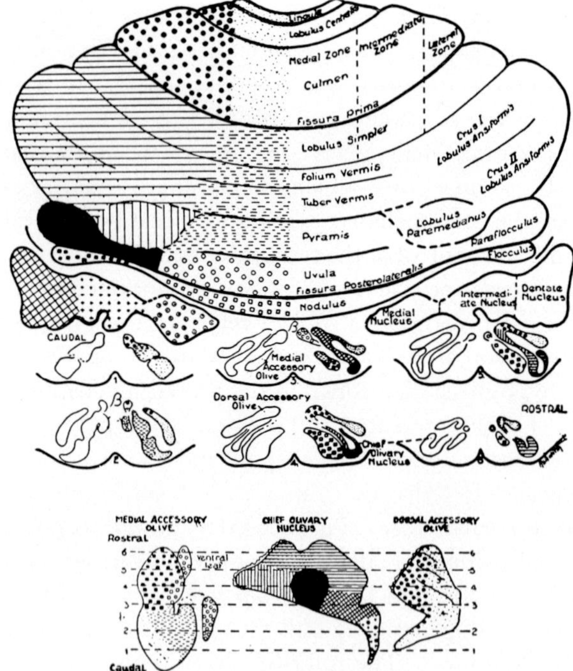

Figure 351C. Schema of Mammalian cerebellum, showing localization of olivocerebellar connections as recorded by BRODAL, 1940 (from LARSELL, 1942, 1951).

output reaches the inferior olivary complex. According to this author, moreover, the pontine nuclei receive fibers from all major parts of the cerebral cortex. In addition, the so-called nucleus reticularis tegmenti pontis (cf. Fig. 281B) is presumed to be an important 'precerebellar reticular nucleus' discharging into the cerebellum and receiving fibers from the cerebral cortex.[89a]

The pathways of the presumably indirect optic input to cerebellum are poorly understood, but a tectocerebellar bundle running through velum medullare anterius can be regarded as a channel transmitting optic impulses processed in tectum mesencephali.

The Mammalian *cerebellar output channels* can be roughly subdivided into three groups: (1) direct neurites of *Purkinje cells*, (2) fibers arising in nucleus fastigii, and (3) fibers arising in nucleus dentatus and nucleus interpositus (resp. globosus and emboliformis).

It seems likely that, in Mammals, direct neurites of *Purkinje cells* (1) reach only the vestibular grisea and that their origin is essentially if not exclusively restricted to the flocculonodular lobe. This output apparently mainly passes through the inner part (IAK) of the inferior cerebellar peduncle, but some fibers may run through brachium pontis (middle cerebellar peduncle).

Most of the output from nucleus fastigii (2) takes its course through the inferior cerebellar peduncle, but a substantial component of that channel, however, forming the co-called fasciculus uncinatus, crosses dorsally the brachium conjunctivum and then bends ventralward and caudad. It can thus be included as a part of brachium conjunctivum *sensu latiori* or superior cerebellar peduncle.[90] In addition, the tractus uncinatus is said to provide an ascending component of the brachium conjunctivum, the 'fasciculus uncinatus ascendens', which decussates within the cerebellum (cf. Fig. 352A).

Most or perhaps all output (3) arising from nucleus dentatus and nucleus interpositus[91] leaves the cerebellum through the upper pedun-

---

[89a] Also designated as the 'medial reticulo-tegmental pontile (or pontine) process'.

[90] In Man and at least a number of other Mammals, the tractus uncinatus crosses the brachium conjunctivum caudally to the transverse fibers of tractus spinocerebellaris ventralis, but partly overlapping with this latter.

[91] As regards the nucleus interpositus respectively the nucleus globosus, some ambiguity obtains in the literature, since it is occasionally questionable whether nondescript griseal neighborhoods between the medial and the lateral cerebellar nuclear aggregations should be considered lateral portions of nucleus fastigii or components of the 'interpositus' complex.

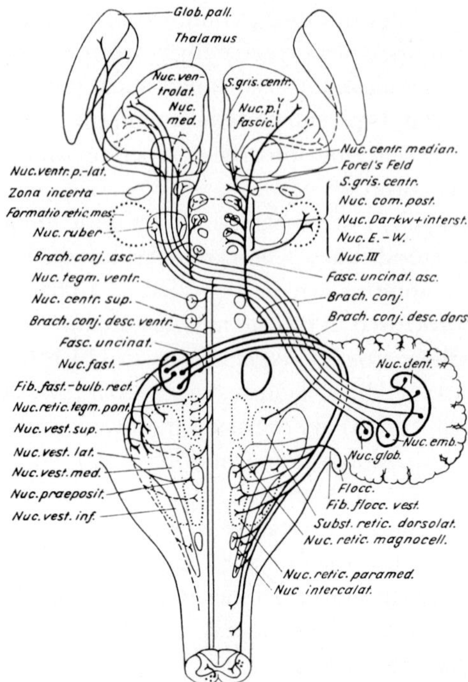

Figure 352 A. Detailed diagram of output from cerebellar nuclei, as interpreted by Jansen and Brodal (from Jansen and Brodal, 1958). Some of the more uncertain connections drawn as dotted lines.

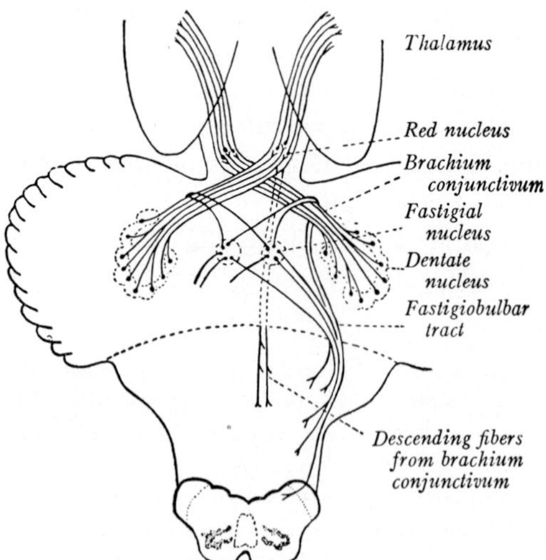

Figure 352 B. Simplified diagrams of output channels from cerebellar nuclei (from Ranson, 1943).

cle or brachium conjunctivum, whose fibers decussate in the mesencephalic tegmentum *(Werneking's decussation or commissure)*. Distally to that decussation, the large brachium conjunctivum ascendens component reaches the contralateral nucleus ruber and more rostral prosencephalic grisea. Through the mediation of thalamic nuclei, this output also reaches the cerebral cortex as well as pallidum and striatum. With regard to the premotor and motor cortex a feedback system thus exists, whose descending limb returns by way of the corticopontine channel and the brachium pontis to the cerebellum (cf. also K. 1954). This feedback, which presumably radiates collateral discharges, is evidently related to the close cooperation of cerebellum and of telencephalic grisea in motor and other (i.e. 'visceral') activities. A smaller brachium conjunctivum ventrale[92] component reaches grisea of pons and medulla oblongata. Because double decussations not infrequently obtain (e.g. those of brachium conjunctivum and of rubrospinal tract), each cerebellar hemisphere seems to be particularly linked with the homolateral side of the body. Figure 352A represents a detailed diagram of efferent cerebellar channels, and Figure 352B provides an appropriate simplified version as a useful first approximation.[93]

Experimental studies by means of either evoked potentials or by use of direct stimulation appear to have disclosed a so-called *somatotopic cerebellar localization*, at least within lobus anterior and lobulus paramedianus (cf. e.g. SNIDER, 1950; DOW and MORUZZI, 1958). Figures 353A–C illustrate concepts based on data of this type. It can be seen that the rostrocaudal sequence hind leg, foreleg, head, obtaining in lobus anterior, is reversed in the lobulus paramedianus. Comparable reversals of somatotopic or other topographic sequences also obtain in certain areas of the cerebral cortex (e.g. first and second sensorimotor area, first and second auditory area).

Reverting to the disturbances of cerebellar function briefly dealt with in the introductory section 1 of this chapter, a few additional comments on clinical aspects of cerebellar deficiency in Man may here

---

[92] The brachium conjunctivum descendens dorsale being a component of tractus s. fasciculus uncinatus.

[93] With regard to functional aspects, BRAIN (1969) formulates the following simplified version: 'the anterior lobe and the roof nuclei are concerned with the regulation of stretch reflexes and the anti-gravity posture. The flocculolobular lobe is an important equilibratory centre and lesions of this region cause swaying, staggering, and titubation. The neocerebellum regulates voluntary movement'.

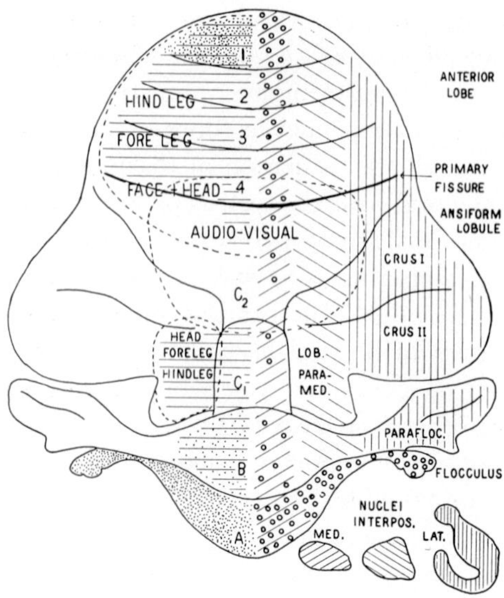

Figure 353A. Diagram indicating some aspects of somatotopic and functional localization in the cerebellum, based on the concepts of various authors (from RANSON-CLARK, 1961). Corticonuclear projections on right; circles indicate direct cortical projections to vestibular grisea. Input projections on left; dots indicate vestibular input, horizontal lines spinocerebellar input; blank space for main region of pontocerebellar input. Numbers and letters on midline structures refer to BOLK's notation (cf. Fig. 345C). 'Cortical' refers here, of course, to the *cerebellar* cortex.

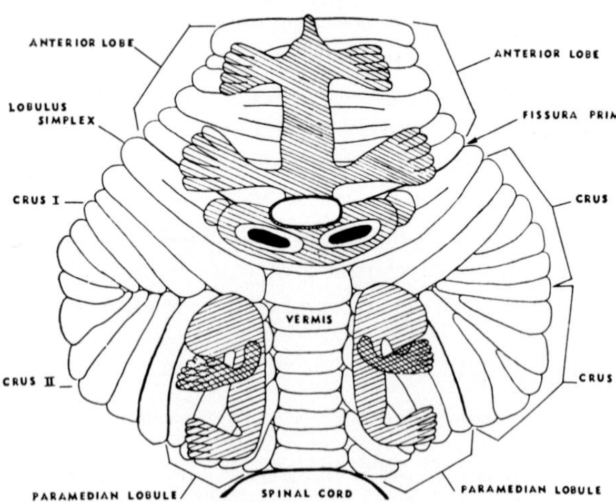

Figure 353B. Outline drawing purporting to show, in schematized form, the somatotopic tactile projections upon the Mammalian cerebellar cortex (after SNIDER, 1952, from CROSBY *et al.*, 1962).

be added. With a unilateral lesion of the cerebellar hemisphere, the shoulder on the affected side is often held at a lower level than the contralateral shoulder. In standing, the weight is thrown on the unaffected contralateral leg. The head tends to rotate so that the occiput is directed and flexed toward the shoulder on the side of the lesion. This, according to Brain (1969) may also be due to a coincident lesion of the vestibular tracts.

Cerebellar tremor can be static, i.e. manifested when the patient attempts to maintain a limb in a fixed posture, or upon the performance of movements. Inability to carry out rapidly alternating movements, e.g. pronation and supination of the hand, is another characteristic symptom, designated as *adiadochokinesis*, and can easily be detected

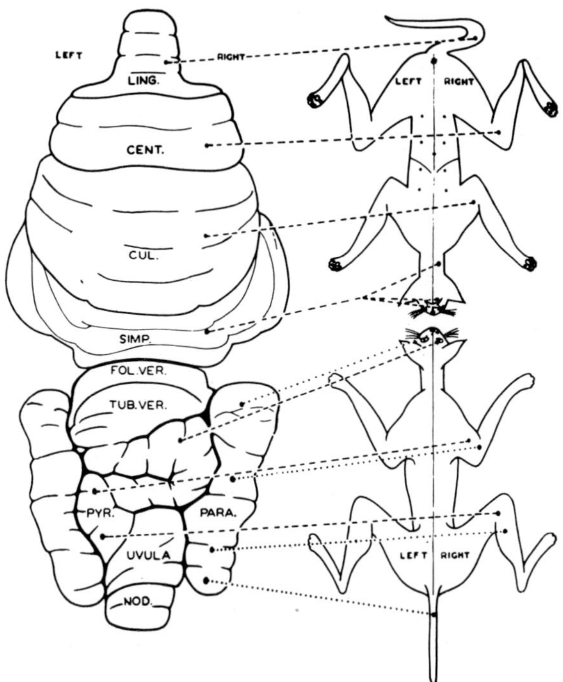

Figure 353C. Drawing purporting to show the double pattern of somatotopic localization in the cerebellum of the cat recorded by movements elicited upon electrical stimulation (after various authors, from Crosby *et al.*, 1962). At top of figure: movements elicited ipsilaterally; at bottom: movements elicited bilaterally (stimulation of the decerebrate animal); paraflocculus (PARA.) labeled on the right side.

upon standard neurological examination. Such test then results in uncoordinated, jerky, exaggerated associated movements. These latter, including vigorous grimaces, may also accompany speech or other intentional motor performances. The tremor concomitant with voluntary motions such as pointing or similar manipulations is thus also designated as *intention tremor*. Again, rapidly alternating opposite movements evidently require, for their smooth execution, properly timed and accurately gaged inhibition. Thus, the symptom of *adiadochokinesis* can be said to corroborate some aspects of contemporary concepts of cerebellar mechanisms stressing the *inhibitory* functions as well as the '*synchronizing*' respectively '*timing*' activities (cf. above, section 1, p. 665 and 666).

Deviations of movements, *overshooting of the mark*, or '*past-pointing*'[93a] are likewise common symptoms of the obtaining *dysmetria*, which, moreover, includes '*decomposition of movement*', whereby, instead of a simultaneous combined action, one joint is clumsily moved before another. The so-called '*rebound phenomenon*' is tested by requesting the patient to bend the elbow while his hand is restrained at the wrist by the examiner. Upon sudden release, a jerky uncontrolled flexion will then occur, and, if not properly prevented, the patient might strike himself strongly in the face.

Ocular disturbances in cerebellar lesions include skew deviation of the eyes and horizontal *nystagmus* on fixation in any direction, the slow phase being towards the position of rest and the quick phase towards the periphery.[94]

Cutaneous reflexes seem to be unaffected by cerebellar lesions but some tendon reflexes may display characteristic changes, such as the '*pendular knee-jerk*'. The elicited patellar reflex is here followed by a sequence of oscillations which are normally prevented by the after-shortening of the quadriceps.

[93a] Symptoms of this type are commonly detected by the well known finger-to-nose, finger-to-finger, and heel-to-knee tests.

[94] In *skew deviation* the eye on the affected side is deviated downwards and inwards, while the opposite eye is deviated outwards and upwards. *Nystagmus* is characterized by a more or less rhythmical oscillation of the eyebulb. If the motion is quicker in one direction than in the other, the quick phase is taken to indicate its direction. Nystagmus may be due to a number of different causes (ocular, labyrinthine, and central, involving, e.g. brain stem or cerebellar lesions).

Additional clinical manifestations of cerebellar lesions include (1) disorders of articulation and phonation, (2) disorders of gait, and (3) vertigo.[95]

As concisely stated by BRAIN (1969), the symptoms of a cerebellar lesion differ markedly in severity according to whether it develops rapidly or slowly. Many of the clinical descriptions of cerebellar deficiency are based on the examination of acute lesions. When the cerebellar involvement is slowly progressive, as e.g. in some neoplasms, symptoms of cerebellar deficiency are much less severe than in acute conditions. In these latter, considerable recovery from their effects can often be expected. These facts seem to imply that other parts of the central neuraxis can, to a large extent, compensate for loss of cerebellar function. This, of course, is in full agreement with the general comments on cerebellar mechanisms presented in section 1, and in particular with the there pointed out redundant aspect of these and other neural activities.

As regards specific 'diseases' in contradistinction to 'symptoms' affecting the cerebellum, the various sorts of congenital hereditary, and degenerative ataxias[96] might be mentioned in addition to lateral sclerosis and neoplasms. Secondary involvements of the cerebellum occur in vascular pathologic processes, in multiple sclerosis, various demyelinating diseases, and in congenital atresia of the *foramina of Luschka and of Magendie*. Concerning the clinical topics, which, however, are of general neurobiologic interest, the detailed chapter on diseases of the cerebellum by BROWN (1971) should be mentioned in addition to the more concise and generalized presentation by BRAIN (1969).

As regards age changes (cf. volume 3/I, pp. 720–723) various findings indicate that the *Purkinje cells* may pertain to those neuronal elements which are subject to (the somewhat controversial) *senile cell*

---

[95] It should be kept in mind that these three sorts of symptoms can, but must not be due to cerebellar disturbances, since they may be manifestations of damaging processes involving diverse other parts of the neuraxis. In other words, these symptoms are transforms which do not indicate a single operand (e.g. vestibular vertigo).

[96] *Friedreich's ataxia*, *Roussy-Lévy's dystasie aréflexique héréditaire*, *Refsum's heredopathia atactica polyneuritiformis*, olivopontocerebellar ataxia *(Dejerine-Thomas disease)*, spinocerebellar ataxia *(Marie's ataxia)*, primary lateral or posterolateral sclerosis. In several respects, a separation into diverse 'nosological entities' is arbitrary. With respect to other anomalies, including the *Arnold-Chiari malformation*, reference is made to pp. 274–277 of Section 1 C, Chapter VI in volume 3/II. Among additional cerebellar involvements, those related to numerous sorts of infection are of significance, and may be either localized (e.g. abscess, gumma, tuberculoma), or diffuse.

loss. Thus, according to BROWN (1971), 'the average person has lost about a third of the *Purkinje cells* by the age of 70'. This would be in essential agreement with my own observations concerning aged Human and Rat brains, in which I noticed, as a general rule, that the row of *Purkinje cells* displayed far more conspicuous and multiple lacunae than the occasional ones seen in young individuals.

10. References to Chapter X

ABBIE, A.A.: The brain-stem and cerebellum of Echidna aculeata. Philos. Trans. roy. Soc. London Ser. B. *224:* 1–74 (1934).

ADDISON, W.H.F.: A comparison of the cerebellar tracts in three Teleosts. J. comp. Neurol. *36:* 1–35 (1923).

ALTMAN, J.: Postnatal development of the cerebellar cortex in the Rat. I. The external germinal layer and the transitional molecular layer. J. comp. Neurol. *145:* 353–397 (1972).

ALTMAN, J.: Postnatal development of the cerebellar cortex in the Rat. II. Phases in the maturation of the Purkinje cells and of the molecular layer. III. Maturation of the components of the granular layer. J. comp. Neurol. *145:* 399–511 (1972).

ARONSON, L.R.: The central nervous system of sharks and bony fishes with special reference to sensory and integrative mechanisms; in GILBERT, Sharks and survival, pp.165–241 (Heath, Boston 1963).

BAFFONI, G.M.: Osservazioni sulla morfogenesi e sull'istogenesi del cerveletto di uccelli e sul ciclo vitale dei neuroni cerebellari. Riv. neurobiol. (Perugia) *9:* 453–580 (1963.)

BARTON, H.: Eine seltene Fehlbildung des Kleinhirns. Beitr. path. Anat. *93:* 219–237 (1934).

BECCARI, N.: Neurologia comparata anatomo-funzionale dei Vertebrati compreso l'uomo (Sansoni, Firenze 1943).

BENEDETTI, E.: Il cervello e i nervi cranici del Proteus anguineus Laur. Mem. Ist. ital. Speleologia, Ser. biol. III: 1–29 (1933).

BENNETT, M.L.V.: Electroreceptors in Mormyrids. Cold Spring Harb. Symp. quant. Biol. *30:* 245–262 (1965).

BENNETT, M.L.V.: Neuronal control of electric organs; in INGLE, The central nervous system and fish behavior, pp.147–169 (University of Chicago Press, Chicago 1968).

BENNETT, M.L.V. and GRUNDFEST, H.: Studies on the morphology and electrophysiology of electric organs. III. Electrophysiology of electric organs in Mormyrids; in CHAGAS and PAES DE CARVALHO, Bioelectrogenesis, pp.113–135 (Elsevier, Amsterdam 1961).

BERKELBACH VAN DER SPRENKEL, H.: The central relations of the cranial nerves in Silurus glanis and Mormyrus caschive. J. comp. Neurol. *25:* 5–63 (1915).

BIACH, P.: Zur normalen und pathologischen Anatomie der äusseren Körnerschicht des Kleinhirns. Arb. neurol. Inst. Wiener Univ. *18:* 13–30 (1910).

BIELSCHOWSKY, M. und SIMONS, A.: Über diffuse Hamartome (Ganglioneurome) des Kleinhirns und ihre Genese. J. Psychol. Neurol. *41:* 50–75 (1930).

Blinkov, S.M. and Glezer, I.I.: The human brain in figures and tables. A quantitative handbook (Plenum Press, New York 1968).

Bolk, L.: Das Cerebellum der Säugetiere. Eine vergleichende anatomische Untersuchung (Fischer, Jena 1906).

Bone, Qu.: The central nervous system; in Brodal and Fänge, The biology of Myxine, pp.50–91 (Universitetetsforlagat, Oslo 1963).

Bonin, G. von: Essay on the cerebral cortex (Thomas, Springfield 1950).

Bradley, O.C.: On the development and homology of the mammalian cerebellar fissures. J. Anat. *37:* 112–130; 221–240 (1903).

Bradley, O.C.: The mammalian cerebellum; its lobes and fissures. J. Anat. *38:* 448–475, *39:* 99–117 (1904–1905).

Brain, W.R.: Diseases of the nervous system. Posthumous 7th ed., revised by J.N. Walton (Oxford University Press, London 1969).

Braitenberg, V.: Is the cerebellum a biological clock in the millisecond range. Progr. Brain Res. *25:* 334–346 (1967).

Brodal, A.: Anatomical studies of cerebellar fibre connections with special reference to the problems of functional localization. Progr. Brain Res. *25:* 135–173 (1967).

Brodal, A.: Cerebrocerebellar pathways. Anatomical data and some functional implications. Acta neurol. scand. Suppl. *51:* 153–195 (1972).

Brodal, A.; Kristiansen, K., and Jansen, J.: Experimental demonstration of a pontine homologue in birds. J. comp. Neurol. *92:* 23–69 (1950).

Brouwer, B.: Über das Kleinhirn der Vögel nebst Bemerkungen über das Lokalisationsproblem im Kleinhirn. Fol. neurobiol. *7:* 349–377 (1913).

Brown, J.R.: Diseases of the cerebellum; in Baker's Clinical Neurology, 3rd ed., vol.2, chapt.29, pp.1–38 (Harper & Row, New York 1971).

Cajal, S.R.y: Los ganglios centrales del cerebelo de las aves. Trav. Rech. biol. Madrid *6:* 177–194 (1906).

Cajal, S.R.y: Histologie du système nerveux de l'homme et des vertébrés. 2 vols. (Maloine, Paris 1909, 1911; Instituto Ramon y Cajal, Madrid 1952, 1955).

Cajal, S.R.y: Studies on vertebrate neurogenesis (1929). Translated by Guth, L. (Thomas, Springfield 1960).

Cajal, S.R.y: Histology (Wood, Baltimore 1933).

Del Cerro, M.P. and Snider, R.S.: The Purkinje cell cilium. Anat. Rec. *163:* 127–140 (1969).

Chambers, W.W. and Sprague, J.M.: Functional localization in the cerebellum. I. Organization in longitudinal corticonuclear zones and their contribution to the control of posture, both extrapyramidal and pyramidal. J. comp. Neurol. *103:* 105–129 (1955a).

Chambers, W.W. and Sprague, J.M.: Functional localization in the cerebellum. II. Somatotopic organization in cortex and nuclei. Arch. Neurol. Psychiat. *74:* 653–680 (1955b).

Comolli, A.: Per una nuova divisione del cervelletto dei mammiferi. Arch. ital. Anat. Embriol. *9:* 247–273 (1910).

Craigie, E.H.: Observations on the brain of the humming bird (Chrysolampis mosquitus Linn., Chlorostilbon caribaeus Lawr.). J. comp. Neurol. *24:* 141–149 (1928).

Crosby, E.C.; Humphrey, T., and Lauer, E.W.: Correlative anatomy of the nervous system (Macmillan, New York 1962).

Dow, R. S. and Moruzzi, G.: The physiology and pathology of the cerebellum (University of Minnesota Press, Minneapolis 1958).

Eccles, J. C.; Ito, M., and Szentágothai, J.: The cerebellum as a neuronal machine (Springer, New York 1967).

Edinger, L.: Das Cerebellum von Scyllium canicula. Arch. mikr. Anat. *58:* 661–678 (1901).

Edinger, L.: Über die Einteilung des Cerebellums. Anat. Anz. *35:* 319–338 (1909–1910).

Edinger, L.: Vorlesungen über den Bau der nervösen Zentral-Organe des Menschen und der Tiere. I. Das Zentralnervensystem des Menschen und der Säugetiere. II. Vergleichende Anatomie des Gehirns (Vogel, Leipzig 1911, 1908).

Edinger, L.: Einführung in die Lehre vom Bau und den Verrichtungen des Nervensystems (Vogel, Leipzig 1912).

Fessard, A.: Les organes électriques; in Grassé, Traité de Zoologie, vol. 13, II, pp. 1143–1238 (Masson, Paris 1958).

Fessard, A.: La synchronisation des activités élémentaires dans les organes des poissons électriques; in Chagas and Paes de Carvalho, Bioelectrogenesis, pp. 202–211 (Elsevier, Amsterdam-New York 1961).

Fix, J. D.: Vergleichend-anatomische Untersuchungen an den Kernen des Primaten-Kleinhirns (Dissertation, naturwiss. Fak. d. Universität Tübingen, 1967).

Foerster, O. und Gagel, O.: Ein Fall von Gangliocytoma dysplasticum des Kleinhirns. Z. ges. Neurol. Psychiat. *146:* 792–803 (1933).

Foltz, F. and Matzke, H. A.: An experimental study on the origin, course and termination of the cerebellifugal fibers in the opossum. J. comp. Neurol. *114:* 107–125 (1960).

Fox, C. A.; Andrade, A., and Schwyn, R. C.: Climbing fiber branching in the granular layer; in Llinás, Neurobiology of cerebellar evolution and development, pp. 603–611 (American Medical Association, Chicago 1969).

Fox, C. A. and Barnard, J. W.: A quantitative study of the Purkinje cell dendrite branchlets and their relationship to afferent fibers. J. Anat. *91:* 299–313 (1957).

Fox, C. A.; Hillman, D. E.; Siegesmund, K. A., and Dutta, C. R.: The primate cerebellar cortex: A Golgi and electron microscopic study. Progr. Brain Res. *25:* 174–225 (1967).

Fox, C. A. and Snyder, R. S. (eds.): The cerebellum. Progr. in Brain Res. vol. 25 (Elsevier, New York 1967).

Franz, V.: Das Mormyridenhirn. Zool. Jb. Anat. *32:* 465–492 (1911–1912).

Franz, V.: Das Kleinhirn der Knochenfische. Zool. Jb. Anat. *32:* 401–464 (1911–1912).

Franz, V.: Faseranatomie des Mormyridengehirns. Anat. Anz. *45:* 271–279 (1913–1914).

Franz, V.: Zur mikroskopischen Anatomie der Mormyriden. Zool. Jb. Anat. *42:* 91–148 (1920–1921).

Freeman, J. A.: The cerebellum as a timing device: an experimental study in the frog; in Llinás, Neurobiology of cerebellar evolution and development, pp. 397–420 (American Medical Association, Chicago 1969).

Fujita, S.: Autoradiographic studies on histogenesis of the cerebellar cortex; in Llinás, Neurobiology of cerebellar evolution and development, pp. 743–747 (American Medical Association, Chicago 1969).

Fuse, G.: Die innere Abteilung des Kleinhirnstiels (Meynert, IAK) und der Deiterssche Kern. Arb. hirnanatom. Inst. Zürich *6:* 29–267 (1912).

Gerlach, J.: Über das Gehirn von Protopterus annectens. Anat. Anz. *75:* 305–448 (1933).

Gerlach, J.: Beiträge zur vergleichenden Morphologie des Selachierhirnes. Anat. Anz. *96:* 79–165 (1947).

GLEES, P.; PEARSON, C., and SMITH, A.G.: Synapses on the Purkinje cells of the frog. Quart. J. exp. Physiol. *43:* 52–60 (1958).

GONA, A.G.: Morphogenesis of the cerebellum in the frog tadpole during spontaneous metamorphosis. J. comp. Neurol. *146:* 133–142 (1972).

GOODMAN, D.C.; HOREL, J.A., and FREEMON, F.R.: Functional localization in the cerebellum of the bird and its bearing on cerebellar function. J. comp. Neurol. *123:* 45–53 (1965).

GOODMAN, D.G. and SIMPSON, J.T.: Cerebellar stimulation in the unrestrained and unanesthetized Alligator. J. comp. Neurol. *114:* 127–135 (1960).

GORONOWITSCH, N.: Das Gehirn und die Cranialnerven von Acipenser ruthenus. Morphol. Jb. *13:* 515–574 (1888).

GROEBBELS, F.: Die Lage und Bewegungs-Reflexe der Vögel IX. Die Wirkung von Kleinhirnläsionen und ihre anatomisch-physiologische Analyse. Pflügers Arch. *221:* 15–40 (1928).

HALLER V. HALLERSTEIN, GRAF, V.: Äussere Gliederung des Zentralnervensystems; in BOLK *et al.*, Handbuch der vergleichenden Anatomie der Wirbeltiere, Bd. II (Urban & Schwarzenberg, Berlin 1934).

HANAWAY, J.: Formation and differentiation of the external granular layer of the chick cerebellum. J. comp. Neurol. *131:* 1–14 (1967).

HAYASHI, T.: Autoradiographic studies on the histogenesis of the chick cerebellum. J. Kyoto Pref. Univ. Med. *75:* 1225–1239 (1966).

HAYLE, T.H.: A comparative study of spinocerebellar systems in three classes of poikilothermic vertebrates. J. comp. Neurol. *149:* 463-495 (1973).

HEIER, P.: Fundamental principles in the structure of the brain. A study of the brain of Petromyzon fluviatilis. Acta anat. (Suppl. VI): 1–213 (1948).

HEINLEIN, H. und FALKENBERG, K.: Beitrag zur Kasuistik der Ganglioneurome des Kleinhirns. Z. ges. Neurol. Psychiat. *166:* 128–135 (1939).

HERRICK, C.J.: The tactile centers in the spinal cord and brain of the sea robin, Prionotus carolinus. J. comp. Neurol. *17:* 307–327 (1907).

HERRICK, C.J.: The medulla oblongata of larval Amblystoma. J. comp. Neurol. *24:* 343–437 (1914).

HERRICK, C.J.: Origin and evolution of the cerebellum. Arch. Neurol. Psychiat. *11:* 621–652 (1924).

HERRICK, C.J.: An introduction to neurology, 5th ed. (Saunders, Philadelphia 1931).

HERRICK, C.L.: Contributions to the comparative morphology of the central nervous system. J. comp. Neurol. *1:* 5–37 (1891).

HESS, N.: De cerebelli gyrorum disquisitiones microscopicae (Dissert., Dorpat 1858).

HINES, M.: The brain of Ornithorhynchus anatinus. Philos. Trans. roy. Soc. London Ser. B. *217:* 155–287 (1929).

HOCHSTETTER, F.: Beiträge zur Entwicklungsgeschichte des menschlichen Gehirns. II (3). Die Entwicklung des Mittel- und Rautenhirns (Deuticke, Wien 1929).

HOEVELL, J.J.L.D. VAN: The phylogenetic development of the cerebellar nuclei. Proc. kon. ned. Akad. Wet. Sect. Sci. *18:* 1421–1434 (1916).

HOLMES, G.: The Croonian lectures on the clinical symptoms of cerebellar disease and their interpretation. Lancet *i:* 1177–1182; 1231–1237 (1922); *ii:* 59–65; 11–115 (1922).

HOLMGREN, N. and VAN DER HORST, C.J.: Contribution to the morphology of the brain of Ceratodus. Acta Zool. *6:* 59–165 (1925).

HOOGENBOOM, K.J.HOCKE: Das Gehirn von Polyodon folium Lacép. Z. mikr.-anat. Forsch. *18:* 311–392 (1929).

HORST, C.J. VAN DER: Das Kleinhirn der Crossopterygii. Bijdr. Dierk. *21:* 113–118 (1919).

HORST, C.J. VAN DER: The cerebellum of fishes. I. General morphology of the cerebellum. Proc. Kon. Akad. Wet. Amsterdam, Sect. Sci. *28:* 735–746 (1925).

HORST, C.J. VAN DER: The cerebellum of fishes II. The cerebellum of Megalops cyprinoides (Brouse) and its connections. Proc. Kon. Akad. Wet. Amsterdam, Sect. Sci. *29:* 44–53 (1926).

HUBER, G.C. and CROSBY, E.C.: On thalamic and tectal nuclei and fiber tracts in the brain of the American alligator. J. comp. Neurol. *40:* 97–227 (1926).

INGVAR, S.: Zur Phylo- und Ontogenese des Kleinhirns. Fol. neurobiol. *11:* 205–495 (1918).

JAKOB, A.: Das Kleinhirn; in v. MÖLLENDORFF, Handbuch der mikroskopischen Anatomie des Menschen, vol. IV/1, pp. 674–916 (Springer, Berlin 1928).

JANSEN, J.: Neure Ergebnisse der Kleinhirnforschung. Münch. med. Wochenschr. *86:* 488–499 (1961).

JANSEN, J. und BRODAL, A.: Das Kleinhirn. Ergänz. zu v. MÖLLENDORFF and BARGMANN, Handbuch der mikroskopischen Anatomie des Menschen, vol. IV/1 (Springer, Berlin-Heidelberg, 1958).

JOHNSTON, J.B.: The brain of Acipenser. Zool. Jb. Anat. Abt. *15:* 59–260 (1901).

JOHNSTON, J.B.: The brain of Petromyzon. J. comp. Neurol. *12:* 87–106 (1902).

JOHNSTON, J.B.: The nervous system of vertebrates (Blakiston, Philadelphia 1906).

KAPPERS, C.U.A.: Die vergleichende Anatomie des Nervensystems der Wirbeltiere und des Menschen, 2 vols. (Bohn, Haarlem 1920, 1921).

KAPPERS, C.U.A.: Anatomie comparée du système nerveux particulièrement de celui des mammifères et de l'homme. Avec la collaboration de E.H. STRASBURGER. (Masson, Paris 1947).

KAPPERS, C.U.A.; HUBER, G.C., and CROSBY, E.C.: The comparative anatomy of the nervous system of vertebrates including man, 2 vols. (Macmillan, New York 1936).

KARAMYAN, A.I.: Evolution of the function of the cerebellum and cerebral hemispheres. Gors. izd. med. lit. Leningrad (quoted after PEARSON, 1972).

KARTEN, W.J. and HODOS, W.: A stereotaxic atlas of the brain of the pigeon (Columba livia). (Johns Hopkins Press, Baltimore 1967).

KINGSBURY, B.F.: On the brain of Necturus maculatus. J. comp. Neurol. *5:* 140–198 (1895).

KINGSBURY, B.F.: The encephalic evaginations in Ganoids. J. comp. Neurol. *7:* 1–36 (1897).

KOELLIKER, A.: Handbuch der Gewebelehre des Menschen. Bd. II, 1. und 2. Abt. (Engelmann, Leipzig, 1893–1896).

KORNELIUSSEN, H.K.: Histogenesis of the cerebellar cortical zones; in LARSELL and JANSEN, The Human cerebellum, pp. 164–174 (University of Minnesota Press, Minneapolis 1972).

KRABBE, K.H.: Studies on the morphogenesis of the brain in some Urodeles and Gymnophions ('Amphibians') (Munksgaard, Copenhagen 1962).

KREHT, H.: Über die Faserzüge im Zentralnervensystem von Salamandra maculosa L. Z. mikr.-anat. Forsch. *23:* 239–320 (1930).

KREHT, H.: Über die Faserzüge im Zentralnervensystem von Proteus anguineus Laur. Z. mikr.-anat. Forsch. *25:* 381–425 (1931).

Kreht, H.: Die markhaltigen Faserzüge im Gehirn der Anuren und Urodelen und ihre Myelogenese. Z. mikr.-anat. Forsch. *48:* 192–285 (1940)

Kuhlenbeck, H.: Zur Morphologie des Gymnophionengehirns. Jen. Z. f. Naturw. *58:* 453–484 (1922).

Kuhlenbeck, H.: Vorlesungen über das Zentralnervensystem der Wirbeltiere (Fischer, Jena 1927).

Kuhlenbeck, H.: The transitory superficial granular layer of the cerebellar cortex. Its relationship to certain cerebellar neoplasms. J. Amer. med. Women's Assn. *5:* 349–351 (1950).

Kuhlenbeck, H.: The human diencephalon. A summary of development, structure, function, and pathology (Karger, Basel 1954, also suppl. ad. vol. 14, Confin. neurol., 1954).

Kuhlenbeck, H.: Brain and consciousness. Some prolegomena to an approach of the problem (Karger, Basel 1957).

Kuhlenbeck, H.: Mind and matter. An appraisal of their significance for neurologic theory (Karger, Basel 1961).

Kuhlenbeck, H. and Haymaker, W.: Neuroectodermal tumors containing neoplastic neuronal elements: ganglioneuroma, spongioneuroblastoma and glioneuroma. Milit. Surg. *99:* 273–292 (1946).

Kuhlenbeck, H. and Haymaker, W.: Neuroectodermal neoplasms of borderline dysplastic character. A survey of different types and their relationship to normal histogenesis (Abstract). Anat. Rec. *97:* 351 (1947).

Kuhlenbeck, H. and Niimi, K.: Further observations on the morphology of the brain in the Holocephalian Elasmobranchs Chimaera and Callorhynchus. J. Hirnforsch. *11:* 267–314 (1969).

Kuithan, W.: Die Entwicklung des Kleinhirns bei Säugetieren. Münch. med. Abhandl. *7:* 1–40 (1895).

Landau, E.: Über cytoarchitektonische Bauunterschiede in der Körnerschicht des Kleinhirns. Z. Anat. EntwGesch. *87:* 551–557 (1928).

Landis, D. M. D. and Reese, T. S.: Differencens in membrane structure between excitatory and inhibitory synapses in the cerebellar cortex. J. comp. Neurol. *155:* 93–125 (1974).

Larsell, O.: The cerebellum of reptiles: Chelonians and alligator. J. comp. Neurol. 56: 299–345 (1932).

Larsell, O.: Anatomy of the nervous system, 2nd ed. (Appleton, New York 1942, 1951).

Larsell, O.: The development of the cerebellum in man in relation to its comparative anatomy. J. comp. Neurol. *87:* 85–129 (1947).

Larsell, O.: The development and subdivisions of the cerebellum in birds. J. comp. Neurol. *89:* 123–189 (1948).

Larsell, O.: The comparative anatomy and histology of the cerebellum from Myxinoids through Birds. Edited by J. Jansen (University of Minnesota Press, Minneapolis 1967).

Larsell, O.: The comparative anatomy and histology of the cerebellum from Monotremes through Apes. Edited by J. Jansen (University of Minnesota Press, Minneapolis 1970).

Larsell, O. and Jansen, J.: The comparative anatomy and histology of the cerebellum. The human cerebellum, cerebellar connections, and cerebellar cortex (University of Minnesota Press, Minneapolis 1972).

LARSELL, O. and WHITLOCK, D.G.: Further observations on the cerebellum of birds. J. comp. Neurol. *97:* 545–566 (1952).

LISSMANN, H.W.: On the function and evolution of electric organs in fish. J. exp. Biol. *35:* 156–191 (1958).

LLINÁS, R. (ed.): Neurobiology of cerebellar evolution and development (American Medical Association, Chicago 1969).

LUCIANI, L.: Il cerveletto. Nuovi studi di fisiologia normale e patologica (Monnier, Firenze, 1891).

LUCIANI, L.: Das Kleinhirn (Besold, Leipzig 1893).

MACKAY, D.: Self-organization in the time domain; in YOVITS *et al.*, Self-organizing systems, pp. 37–48 (Spartan Books, Washington 1962).

MAGNUS, R.: Körperstellung (Springer, Berlin 1924).

MIES, H.: Kleinhirn; in LANDOIS-ROSEMANN, Lehrbuch der Physiologie des Menschen 28. Aufl., vol. 2, pp. 765–771 (Urban & Schwarzenberg, München 1962).

MUGNAINI, E.: Ultrastructural studies on the cerebellar histogenesis; in LLINÁS, Neurobiology of cerebellar evolution and development, pp. 749–782 (American Medical Association, Chicago 1969).

MUGNAINI, E.: The histology and cytology of the cerebellar cortex; in LARSELL and JANSEN, The comparative anatomy and histology of the cerebellum. The human cerebellum, cerebellar connections, and cerebellar cortex, pp. 201–262 (University of Minnesota Press, Minneapolis 1972).

MÜNZER, E. und WIENER, H.: Beiträge zur Anatomie und Physiologie des Centralnervensystems der Taube. Mschr. Psychiat. Neurol. *3:* 379–406 (1898).

NAWAR, G.: Observations on the brain of two members of the Nile Mormyridae. Nytt Magasin for Zool. *10:* 63–66 (1961).

NIEUWENHUYS, R.: Comparative anatomy of the cerebellum. Progr. Brain Res. *25:* 1–93 1967).

NIKLOWITZ, W.: Elektronenmikroskopische Untersuchungen zur Struktur der normalen und kollapsgeschädigten Purkinjezelle. Beitr. path. Anat. *127:* 424–449 (1962).

OBERSTEINER, H.: Der feinere Bau der Kleinhirnrinde beim Menschen und bei Tieren. Biol. Centralbl. *3:* 145–155 (1883).

OBERSTEINER, H.: Anleitung beim Studium des Baues der nervösen Zentralorgane im gesunden und kranken Zustande, 5. Aufl. (Deuticke, Leipzig 1912).

OLIVEIRA CASTRO, C. DE: Morphological data on the brain of Electrophorus electricas (L.); in CHAGAS and PAEZ DE CARVALHO, Bioelectrogenesis, pp. 171–184 (Elsevier, Amsterdam 1961).

PALMGREN, A.: Embryological and morphological studies on the midbrain and cerebellum of vertebrates. Acta zool. *2:* 1–74 (1921).

PEARSON, A.A.: The acustico-lateral centers and the cerebellum, with fiber connections, of fishes. J. comp. Neurol. *65:* 201–294 (1936).

PEARSON, R.: The Avian brain (Academic Press, London 1972).

RAAF, J. and KERNOHAN, J.W.: Study of external granular layer in cerebellum. Amer. J. Anat. *75:* 151–172 (1944).

RAAF, J. and KERNOHAN, J.W.: Relation of abnormal collections of cell in posterior medullary velum of cerebellum to origin of medulloblastoma. Arch. Neurol. Psychiat. *52:* 163–169 (1944).

RANSON, S.W.: The anatomy of the nervous system from the standpoint of development and function. 7th ed.; 10th ed. revised by S.L. Clark (Saunders, Philadelphia 1943, 1959, 1961).

Raymond, A.: Responses to electrical stimulation in the cerebellum of unanesthetised birds. J. comp. Neurol. *110:* 299–320 (1958).

Reis, D.J.; Doba, N., and Nathan, M.A.: Predatory attack, grooming, and consummatory behavior evoked by electrical stimulation of Cat cerebellar nuclei. Science *182:* 845–847 (1973).

Röthig, P.: Beiträge zum Studium des Zentralnervensystems der Wirbeltiere. 11. Über die Faserzüge im Mittelhirn, Kleinhirn und der Medulla oblongata der Urodelen und Anuren. Z. mikr.-anat. Forsch. *10:* 381–472 (1927).

Rüdeberg, S.I.: Morphogenetic studies on the cerebellar nuclei and their homologization in different vertebrates including man. Thesis (Ohlsson, Lund 1961).

Saccone, A. and Epstein, J.A.: Granuloblastoma, a primary neuroectodermal tumor of the cerebellum. J. Neuropath. exp. Neurol. *7:* 287–298 (1948).

Saetersdal, T.A.S.: On the ontogenesis of the avian cerebellum. III. Formation of fissures with a discussion of fissure homologies between the avian and mammalian cerebellum. Univ. Bergen Arb. Nat. *3:* 1–44 (1959).

Saito, T.: Über das Gehirn des japanischen Flussneunauges (Entosphenus japonicus Martens). Fol. anat. japon. *8:* 189–263 (1930).

Sanders, A.: Contribution to the anatomy of the central nervous system in vertebrate animals. On the brain of the Mormyridae. Phil. Trans. roy. Soc. London *173:* 927–959 (1882).

Sarnat, H.B.: Occurrence of fluorescent granules in the Purkinje cells of the cerebellum. Anat. Rec. *162:* 25–32 (1968).

Schaper, A.: Zur feineren Anatomie des Kleinhirns der Teleostier. Anat. Anz. *12:* 41–52 (1893).

Schaper, A.: Die morphologische und histologische Entwicklung des Kleinhirns der Teleostier. Morphol. Jb. *24:* 625–708 (1894).

Schaper, A.: Die frühesten Differenzierungsvorgänge im Centralnervensystem. Arch. Entw. Mech. *5:* 81–132 (1897).

Schaper, A.: The finer structure of the selachian cerebellum *(Mustelus vulgaris)* as shown by chrome silver preparations. J. comp. Neurol. *8:* 1–20 (1898).

Sherrington, Ch.: The integrative action of the nervous system (Yale University Press, New Haven 1906, 1947).

Shimazono, J.: Das Kleinhirn der Vögel. Arch. mikr. Anat. *80:* 397–449 (1912).

Smith, G. Elliot: The primary subdivision of the mammalian cerebellum. J. Anat. *36:* 381–385 (1902).

Smith, G. Elliot: Further observations on the natural mode of subdivision of the mammalian cerebellum. Anat. Anz. *23:* 368–384 (1903a).

Smith, G. Elliot: On the morphology of the brain in the mammalia. The cerebellum. Trans. Linnean Soc. Lond. Sec. Ser. *8:* 425–432 (1903b).

Smith, G. Elliot: The morphology of the human cerebellum. Rev. Neurol. Psychiat. *1:* 629–639 (1903c).

Snider, R.S.: Recent contributions to the anatomy and physiology of the cerebellum. Arch. Neurol. Psychiat. *64:* 196–219 (1950).

Sotelo, C.: Cerebellar neuroglia: morphological and histochemical aspects. Progr. Brain Res. *25:* 226–250 (1967).

Spatz, H.: Über den Eisennachweis im Gehirn, besonders in Zentren des extrapyramidal-motorischen Systems. Z. ges. Neurol. Psychiat. *77:* 261–390 (1922).

Spiegel, E.A.: Tonus; in Bethe *et al.*, Handbuch der normalen und pathologischen Physiologie, Bd. 9: Allgemeine Physiologie der Nerven und des Zentralnervensystems, pp. 711–740 (Springer, Berlin 1929).

Spiegel, E.A.: Comparative studies on the thalamic, cerebral, and cerebellar potentials. Amer. J. Physiol. *118:* 569–579 (1937).

Spiegel, E.A.: Book review: Eccles *et al.*, The cerebellum as a neuronal machine Confin. neurol. *30:* 272 (1968).

Steiner, J.: Die Functionen des Centralnervensystems und ihre Phylogenese. I. Untersuchungen über die Physiologie des Froschhirns. II. Die Fische. IV. Reptilien, Rückenmarksreflexe, Vermischtes (Vieweg, Braunschweig 1885, 1888, 1900).

Stendell, W.: Die Faseranatomie des Mormyridengehirns. Abh. Senckenberg. Ges. *36:* 3–40 (1914).

Sterzi, C.: Il sistema nervoso centrale dei vertebrati. Cyclostomi; Pesci, Selaci; Sviluppo (Draghi, Padova 1907, 1909, 1912).

Stevenson, L. and Echlin, F.: Nature and origin of some tumors of the cerebellum. Medulloblastoma. Arch. Neurol. Psychiat. *31:* 93–109 (1934).

Stroud, B.B.: The mammalian cerebellum. J. comp. Neurol. *5:* 71–118 (1895).

Szabo, T.: Les organes électriques des Mormyrides; in Chagas and Paes de Carvalho, Bioelectrogenesis, pp. 20–24 (Elsevier, Amsterdam 1961).

Szabo, T.: Anatomo-physiologie des centres nerveux spécifiques de quelques organes électriques; in Chagas and Paes de Carvalho, Bioelectrogenesis, pp. 185–201 (Elsevier, Amsterdam 1961).

Szabo, T.: Sense organs of the lateral line system in some electric fish of the Gymnotidae, Mormyridae and Gymnarchidae. J. Morphol. *117:* 229–249 (1965).

Szabo, T.: Activity of peripheral and central neurons involved in electroreception; in Cahn, Lateral line detectors, pp. 295–311 (Indiana University Press, Bloomington 1967).

Szabo, T. et Fessard, A.: Le fonctionnement des électrorécepteurs étudiés chez les Mormyres. J. Physiol. Paris *57:* 343–360 (1965).

Tretjakoff, D.: Das Nervensystem von Ammocoetes. II. Das Gehirn. Arch. mikr. Anat. *74:* 636–779 (1909).

Tuge (Tsuge), H.: Studies on cerebellar function in the teleost. I, II, III. J. comp. Neurol. *60:* 201–224; 225–236 (1934); *61:* 347–369 (1935).

Voogd, J.: Comparative aspects of the structure and fibre connections of the mammalian cerebellum. Progr. Brain Res. *25:* 94–134 (1967).

Voorhoeve, J.J.: Over den bouw van de kleine hersenen der Plagiostomen (Diss., Amsterdam 1917). Quoted after Kappers, 1921, and Gerlach, 1947.

Voris, H.C. and Hoerr, N.L.: The hindbrain of the opossum, Didelphis virginiana. J. comp. Neurol. *54:* 277–355 (1932).

Wallenberg, A.: Beiträger zur Kenntnis des Gehirns der Teleostier und Selachier. Anat. Anz. *31:* 369–399 (1907).

Weston, J.K.: The reptilian vestibular and cerebellar gray with fiber connections. J. comp. Neurol. *65:* 93–200 (1936).

Whitlock, D.G.: A neurohistological and neurophysiological study of the afferent fiber tracts and receptive areas of the avian cerebellum. J. comp. Neurol. *97:* 567–635 (1952).

Zeman, W. and Innes, J.R.M.: Craigie's neuroanatomy of the rat. Revised and expanded (Academic Press, New York 1963).

XI. The Mesencephalon

1. General Pattern and Basic Mechanisms; The Meaning of 'Tegmentum' and 'Tectum'

The mesencephalon or midbrain of Vertebrates represents the most rostral portion of the deuterencephalon. It includes thus, as is the case of spinal cord and rhombencephalon, grisea or 'centers' which are derivatives of both *basal and alar plates*.[1] It is likely, that, as claimed by KINGSBURY (1921–22; cf. also section 2 of vol. 3, part II), the *floor plate* ends in the neighborhood of fovea isthmi, and does not take part in the formation of the mesencephalic neuraxial tube. The *roof plate*, although extending rostralward into the prosencephalon (diencephalon and telencephalon) generally disappears in the mesencephalon by a process akin to absorption.[2] It is, nevertheless, commonly recognizable at ontogenetic stages, at least in form of a distinctive raphe.

The lumen of the mesencephalic neuraxial tube becomes a fairly large ventricle in most submammalian forms, particularly in Anamnia; in Mammals it is reduced to a comparatively narrow channel, the *aquaeductus Sylvii* (aqueductus cerebri, BNA, PNA), interconnecting the diencephalic third ventricle with the rhombencephalic or fourth ventricle.

The *sulcus limitans*, which indicates the boundary region between alar and basal plate, can be identified with reasonable certainty at ontogenetic stages of perhaps all Vertebrates, and also at the adult stage of numerous forms pertaining to all classes. As discussed in section 2, 4,

---

[1] It should here be recalled that the typical morphologic pattern or bauplan of the spinal cord, although, on one hand, definitely recognizable throughout the deuterencephalon as far as the mesencephalo-diencephalic boundary zone, becomes, on the other hand, progressively blurred in caudorostral direction by complex differentiations specific for the various deuterencephalic regions. As regards the alar plate, two such particularly conspicuous specific configurations are the cerebellum and the *tectum sensu latiori* mesencephali.

[2] In particular instances, however, it may persist and even form a choroid plexus (in Petromyzonts), or a distended lamina epithelialis (in many Osteichthyes whose protruding valvula cerebelli distends the mesencephalic ventricle).

and 5 of chapter VI in volume 3, part I, the rostral end of sulcus limitans, together with that of the basal plate, is assumed to be located within the mesencephalo-diencephalic boundary zone near, or in, the mammillary recess.

The *basal plate* component of the mesencephalon displays two griseal systems comparable to those of rhombencephalon, and of which they can be evaluated as the rostral extensions, namely, (1) motor respectively efferent cranial nerve nuclei, and (2) the nucleus s. formatio reticularis tegmenti with its derivatives.

As a first approximation, the *alar plate* components can be roughly enumerated as consisting of three components. (1) A retained primary afferent cranial nerve constituent or 'ingredient', namely, the nucleus radicis mesencephalicae trigemini, (2) a secondary afferent correlation neighborhood consisting of nucleus isthmi and nucleus visceralis (gustatorius) secundarius, particularly evident in Anamnia, and (3) an important correlation neighborhood of higher order, which, at least in part, becomes a typical 'suprasegmental griseum', the tectum mesencephali with its adjacent torus semicircularis of submammalian forms.

In accordance with this elementary configurational arrangement respectively subdivision of the alar and basal plates, the general bauplan of the Vertebrate mesencephalon could be described as follows.

The efferent cranial nerve nuclei, pertaining to nervus oculomotorius and nervus trochlearis, may be evaluated as the mesencephalic extensions of ventral and intermedioventral spinal and deuterencephalic longitudinal basal plate zones ('somatic efferent column' and 'visceral efferent column'). The more rostrally located oculomotor nucleus, whose neurites form the rostralmost deuterencephalic ventral root, includes a 'somatic motor' ventral zone component, and a preganglionic ('general visceral efferent' or cranial parasympathetic) intermedioventral zone portion. A *sulcus intermedioventralis*, however, approximately indicating the ventricular boundary of these two zones at late ontogenetic or at adult stages, does not seem to be commonly displayed. A transitory 'sulcus intermedioventralis' may, nevertheless, become manifest in some forms at early ontogenetic stages (cf. Fig. 161B, p. 379 of volume 3, part II), but, because of possible complex and perhaps secondary or 'cenogenetic' shifts or migrations of cell groups at these early ontogenetic periods, its significance cannot here be properly assessed.[3] The more caudally located nucleus trochlearis generally

[3] Migrations or shifts of that type, which still remain poorly elucidated, were briefly pointed out on pp. 386–388 in section 4, chapter VI of volume 3, part II.

appears as pertaining to the basal plate's ventral zone, but, particularly also with respect to the peculiar crossed and dorsal exit of the trochlear nerve, poses a number of poorly understood and unresolved morphologic respectively phylogenetic problems, pointed out in volume 3, part II (chapter VI, sections 3 and 4, chapter VII, sections 3 and 4).[4]

The mesencephalic *nucleus s. formatio reticularis tegmenti* represents the rostral extension of its rhombencephalic portion (or vice-versa), which was dealt with in chapter IX. Concomitantly with the development of particular fiber systems, several distinctive grisea become differentiated within the mesencephalic part of this basal plate component which has also been designated as nucleus motorius tegmenti.

Among such grisea, the *nucleus ruber tegmenti*[5] can be said essentially to derive from that neighborhood of reticular formation which receives its input from the cerebellum by way of the decussating brachium conjunctivum. The nucleus ruber is clearly distinguishable in Amniota (Reptiles, Birds, Mammals) but cannot be sufficiently well delimited within the diffusely arranged cell population of the Anamniote mesencephalic reticular griseum which, as it were, represents the 'matrix' of nucleus ruber. Various authors, however, not altogether without a limited justification, have described a 'nucleus ruber' in some Anamnia.

The Amniote nucleus ruber gives origin to a decussating rubrospinal tract[6] pertaining to the 'motor system' and taking its course independently of the fasciculus longitudinalis medialis (Fig. 354). In Mammals, moreover, the distinction between a magnocellular and a parvocellular component of nucleus ruber becomes conspicuous, the small celled constituent giving origin to rostralward directed channels.

---

[4] With respect to the mesencephalic nuclei of eye musculature, it might here be mentioned that the oculomotor nucleus and its nerve are absent in Myxinoid Cyclostomes and e.g. in the blind urodele Amphibian Proteus anguineus. This applies also to the trochlear nucleus and its nerve. A trochlear nucleus (and nerve) could not be identified in Gymnophione Amphibians. Some difficulties or doubts concerning the identification of the trochlear nucleus in Petromyzonts were mentioned in section 2 of the preceding chapter X and shall again be pointed out in section 2 of this chapter.

[5] The term nucleus ruber (BNA, PNA; also red nucleus) refers to its reddish appearance in fresh dissections of the human and of at least some mammalian brains.

[6] In Reptiles, however, where caudalward directed rubrobulbar fibers are doubtless present, their extension into the spinal cord as a possible rubrospinal tract does not seem to have been sufficiently well established. In Birds, a rubrospinal tract is generally presumed to obtain, although the available evidence is not altogether satisfactory.

A distinctive rostral differentiation of the reticular formation is the nucleus interstitialis of the fasciculus longitudinalis medialis (or *interstitial nucleus of Cajal)*, which seems to be present in all Vertebrates from Cyclostomes to Mammals. It gives origin to descending fibers of said fasciculus and is located in the region representing the mesence-

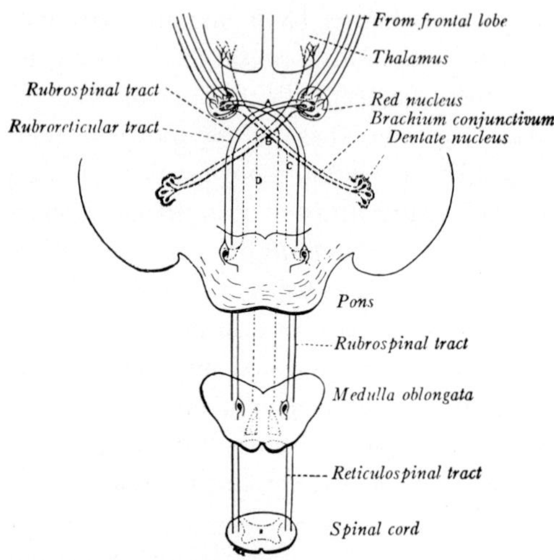

Figure 354. Simplified diagram indicating the main connections of red nucleus (nucleus ruber tegmenti), as presumed to obtain in the human respectively Mammalian brain (from RANSON, 1943). A: ventral tegmental decussation; B: decussation of brachium conjunctivum; C, D: descending homolateral and contralateral cerebellotegmental fibers from brachium conjunctivum.

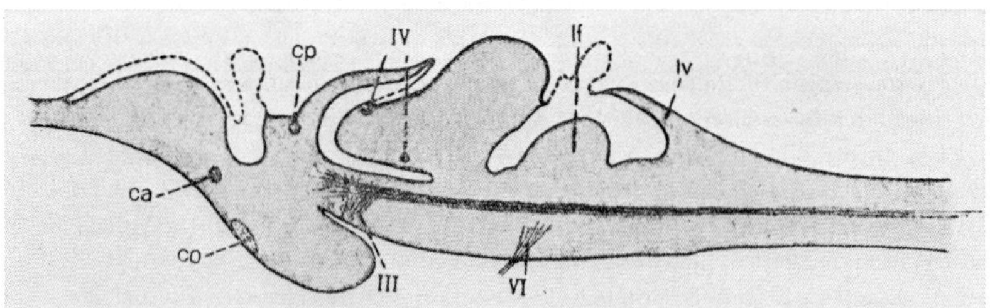

Figure 355. Simplified sketch showing the course of fasciculus longitudinalis medialis in a sagittal section through the brain of a Cyprinoid Teleost (modified after EDINGER, from K., 1927). ca: commissure anterior; co: optic chiasma and postoptic commissures; cp: commissure posterior; lf: lobus facialis; lv: lobus vagi; Roman numerals indicate cranial nerve roots.

phalo-diencephalic boundary zone, being located, within that topologic neighborhood, ventrally to the nucleus commissurae posterioris which, in turn, pertains to the caudal dorsal thalamic longitudinal zone,[7] and is a derivative of alar plate.

In Anamnia, the mesencephalic reticular formation including its *nucleus interstitialis of Cajal* represents the rostral part of STEINER's *'allgemeines Bewegungszentrum'*, whose efferent channel is the *fasciculus longitudinalis medialis*, as discussed in section 2 of chapter IX (cf. Figs. 355 and 381). In Amniota, concomitantly with the further differentiation of the relevant grisea, a further development of motor channels takes place. Accordingly, the tractus rubrospinalis becomes a central motor pathway of substantial importance. In Mammals, and particularly in man as well as in other 'higher' Primates, the corticospinal tract, an apparently exclusively Mammalian motor system, assumes a predominant functional significance. The fasciculus longitudinalis medialis of Anamnia thereby doubtless undergoes various changes in composition, connections, and performance. It retains, however, in particular its vestibular components and its close relationships with the eye muscle nerve nuclei and other primary efferent grisea.

Thus, in Man, the fasciculus longitudinalis medialis, especially conspicuous at mesencephalic levels, extends from the mesencephalo-diencephalic border zone caudalward into the spinal cord; it is subtantially concerned with reflectory control of correlated eye and head movements. A significant proportion of its fibers originate in the vestibular grisea, either bifurcating or ascending respectively descending on the contralateral or homolateral side. Synaptic junctions are effected in the nuclei of oculomotor, trochlear, abducens, and spinal accessory nerves as well as in the ventral horn of spinal cord. Other fibers are believed to connect the nn. III, IV, and VI with each other and also with the motor facial nerve nucleus. Still other fibers connect with the hypoglossal nucleus. Again, among the descending fibers, neurites from the (diencephalic) grisea of the posterior commissure seem to join the fasciculus while, as regards ascending fibers,[8] a component originating

[7] Additional complications, related to the differentiation of several 'subnuclei' of posterior commissure, and of a periventricular central gray including the *nucleus of Darkschewitsch* will be taken up in the subsequent sections dealing with the respective Vertebrate forms, with some further comments on the recorded or presumed fiber connections of said grisea, some of which are related to the *fasciculus longitudinalis dorsalis of Schütz*, mentioned in section 10 of chapter IX.

[8] Ascending fibers are, of course, also presumed to be present in the fasciculus longitudinalis medialis of Anamnia.

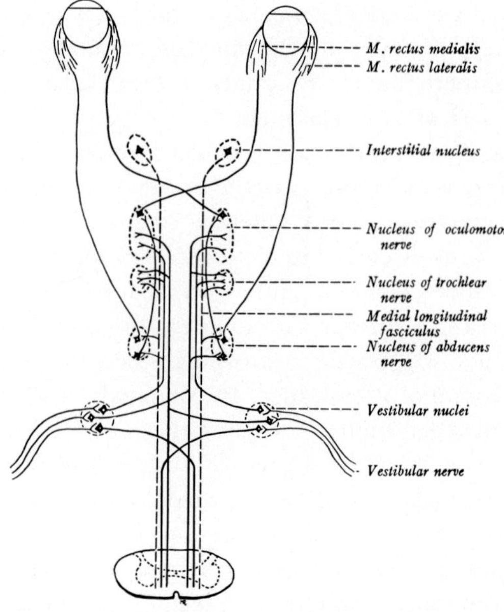

Figure 356A. Simplified diagram illustrating some connections of fasciculus longitudinalis medialis in the human respectively Mammalian neuraxis (modified after VILLIGER, 1920, from RANSON, 1943).

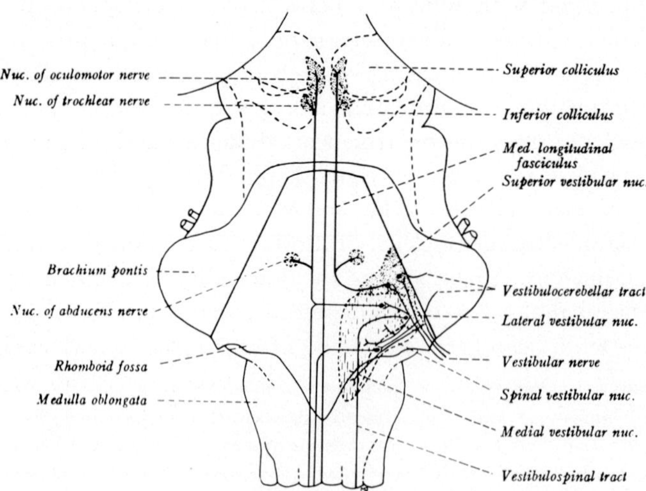

Figure 356B. Simplified diagram illustrating some relationships for the fasciculus longitudinalis medialis to the vestibular system in the human respectively Mammalian brain stem (from RANSON, 1943).

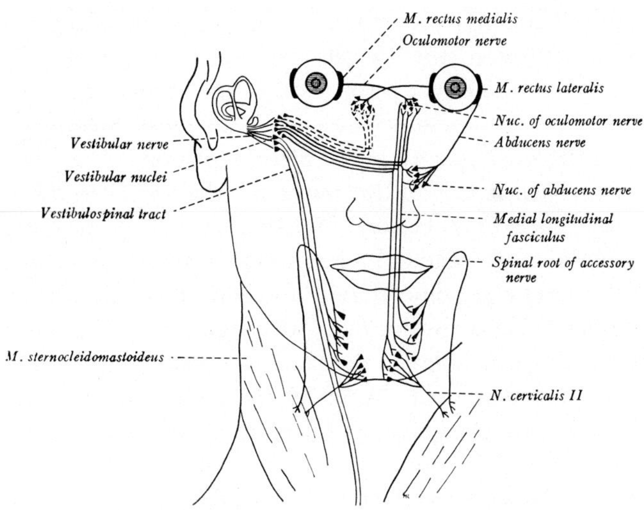

Figure 356C. Simplified diagram illustrating some vestibular reflex arcs related to combined eye and head movements and involving the participation of fasciculus longitudinalis medialis (slightly modified after EDINGER, 1912, from RANSON, 1943).

in the spinal cord is presumed to be included. Be that as it may, there is little doubt that not only in Mammals and Man, but quite generally in Amniote and Anamniote Vertebrates, the fasciculus longitudinalis medialis is a rather complex system, carrying a great variety of fibers.[9] Figures 356A–C illustrate reasonably well corroborated concepts concerning arrangement and connections of that multiplex channel as believed to obtain in the human brain.

Reverting to the *grisea* of the basal plate, an unpaired median configuration, the *nucleus interpeduncularis* (ganglion s. corpus interpedunculare) discovered by MEYNERT and first accurately described by FOREL (1877, 1907) in his 'Habilitationsschrift' of 1877,[10] is present in all Ver-

[9] Although generally agreeing in the essential overall aspects, the numerous experimental studies on the fasciculus longitudinalis medialis by a variety of authors using diverse techniques display many contradictory and controversial observations or conclusions. Relevant references to these investigations can be found in the treatises by BECCARI (1943), CROSBY *et al.* (1962), KAPPERS *et al.* (1936), and RANSON (1943, 1961).

[10] HERRICK (1948) states that it was discovereed by FOREL in 1872. There is, however, no mention of this griseum in FOREL's doctor-dissertation of 1872. In his '*Habilitationschrift*' of 1877, sponsored by v. GUDDEN, FOREL gives indeed the first specific description

tebrates from Petromyzon to Man. It represents one of the most conservative structures in the Vertebrate brain, being relatively much larger in some of the lower forms than in the higher ones (HERRICK, 1948). Thus, the interpeduncular nucleus of most Anamnia extends throughout the entire length of the mesencephalon.

The main *input channel* of this nucleus is the *habenulopeduncular tract* or *fasciculus retroflexus of Meynert* (Fig. 358 A) which, particularly in Anamnia, contains a substantial (higher order) olfactory component. The fibers of this tract end homolaterally and contralaterally within the interpeduncular nucleus. Additional, presumably mostly input connections are provided by fibers of the basal forebrain bundles, by a mammillo-interpeduncular tract, by fibers from the reticular formation, and, at least in some Anamnia, by fibers from the nucleus visceralis secundarius.

The *output channels* from the interpeduncular griseum are essentially directed toward the tegmentum (pedunculotegmental tract to dorsal tegmental nucleus of the central gray in Mammals, and dorsal portions of the reticular formation of all Vertebrates through fasciculus tegmentalis profundus). The functional significance of the interpeduncular griseum, discussed in detail by HERRICK (1948) remains obscure, but it must evidently exert a non-negligible influence upon the activities of the reticular formation. This influence, however, may greatly vary in accordance with the taxonomic status of the different Vertebrate classes.

The mesencephalic reticular formation *(sensu latiori)* of many, perhaps of all or at least most Vertebrates receives also direct optic input through the relatively thin tractus s. fasciculus opticus basalis, which branches off from the main optic tract centrally to its decussation and ends in the *nucleus opticus tegmenti*, reported in Amphibians by BELLONCI (1888). This griseum, also designated as nucleus opticus basalis (nucleus ectomammillaris, nucleus of tractus peduncularis transversus), can be evaluated as a derivative of the reticular formation[11]

---

of the ganglion interpedunculare, which he depicts in his Figure 27, pointing out that MEYNERT (1872) had indicated it in Figures 230 and 231 of STRICKER's handbook with the designation 'ganglion of lamina perforata posterior'. FOREL adds that the term 'ganglion interpedunculare' was suggested to him by v. GUDDEN who had previously noted this griseum but not yet published his relevant findings *('Alte, aber noch nicht veröffentlichte Benennung'* by v. GUDDEN, later reported in 1881).

[11] In those Anamnia who do not display a well defined basal optic nucleus, this griseum may be included within a diffuse region of the reticular formation.

(Figs. 358B, 387B). It has been identified in Reptiles and Birds (cf. BECCARI, 1943) and corresponds to the griseum in which the Mammalian basal optic root ends. This latter, also called tractus peduncularis transversus, was already described in Mammals more than hundred years ago by GUDDEN and was also dealt with by KÖLLIKER and by CAJAL (1911). Subsequently, the basal optic root has been particularly investigated by FREY (1937, 1938), who especially emphasized its hypothalamic connections. Generally speaking, the nucleus opticus teg-

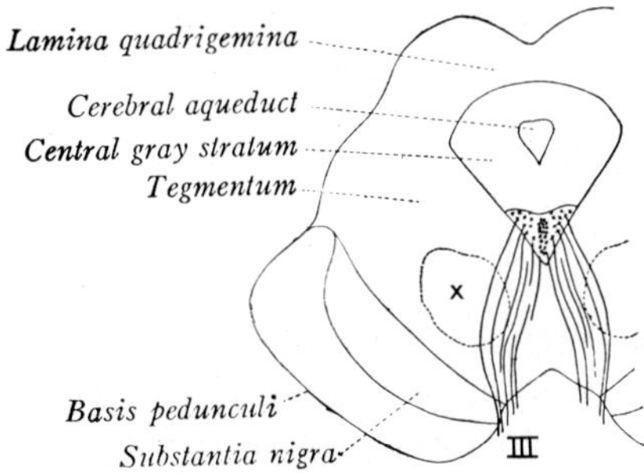

Figure 357. Diagrammatic cross-section through the human diencephalon at the level of oculomotor nuclei (from RANSON, 1943). The outline of nucleus ruber tegmenti is indicated by x.

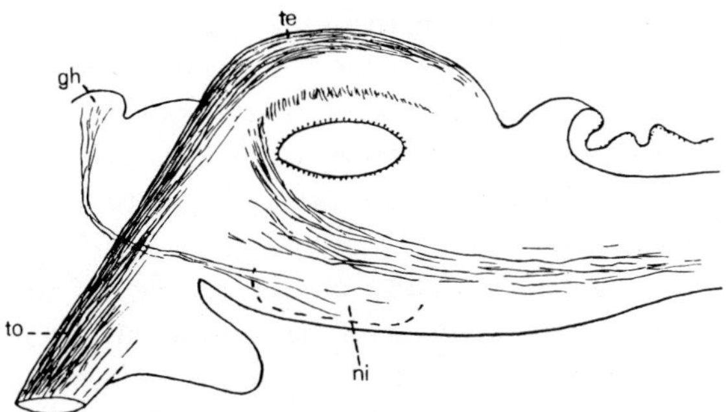

Figure 358A. Simplified diagram (sagittal section) showing the optic radiation to tectum mesencephali, afferent and efferent systems of tectal deep medulla, and the fasciculus retroflexus running from ganglion habenulae to nucleus interpeduncularis in the Amphibian brain (based on figures by EDINGER and on personal observations, from K., 1927). gh: ganglion (s. nucleus) habenulae; ni: nucleus interpeduncularis; te: tectum opticum; to: tractus opticus. The fiber channels related to the deep medulla (not labelled) include general lemniscus system and tectobulbar tracts.

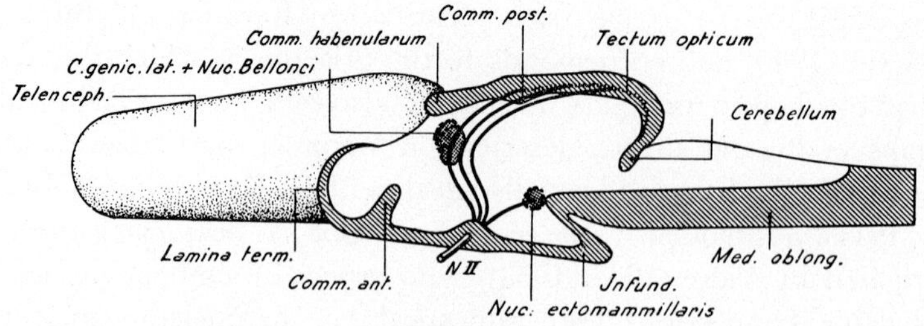

Figure 358B. Drawing based on a wax model and showing optic system in the brain of the urodele Amphibian Triturus taeniatus (from STRÖER, 1939). The nucleus ectomammillaris is the nucleus opticus tegmenti, a griseum of the basal optic tract. The lateral geniculate complex with its *nucleus of Bellonci* is a diencephalic optic griseum to be dealt with in vol. 5, chapter XII.

menti may be considered a griseum providing for the direct transmission of optic impulses upon the midbrain reticular system.[12]

Still another derivative of the mesencephalic basal plate is the *substantia nigra Soemmeringi*, characteristic for Mammals (cf. Figs. 357, 364F), and pertaining to the extrapyramidal motor system. The subdivisions of substantia nigra, including its connections with telencephalic basal ganglia and cortex, as well as with brain stem reticular formation, shall be considered further below in section 9 of this chapter. The substantia nigra separates the tegmentum from the pes or basis pedunculi, which is likewise a characteristic Mammalian configuration, formed by the bundles of pyramidal and corticopontine (frontopontine and temporo-parieto-occipitopontine) tracts. The so-called *pedunculus cerebri* (BNA, PNA) consists of *basis (pes)*, *substantia nigra*, and the 'covering' *tegmentum*.

Turning now to the derivatives of the *alar plate*, the nucleus radicis mesencephalicae trigemini, which can be considered a primary 'somatic afferent' component of that plate's dorsal zone, might here be men-

[12] Whether the presumably present centrifugal, retinopetal fibers of the optic nerve originate in the hypothalamic or mesencephalic grisea related to the basal optic root respectively in the tectum opticum remains a moot question. The findings of GRANIT (1955, p. 105), who recorded centrifugal effects upon the retina originating in the so-called 'mesencephalic reticular activating system' could perhaps be interpreted as suggesting the origin of at least some retinopetal fibers in the nucleus opticus tegmenti. Cf. also the short comment on centrifugal fibers of the VIIIth nerve in section 1 of chapter IX.

tioned first. The evaluation of its large 'pseudo-unipolar' vesicular cells as representing intracerebral primary afferent elements comparable to 'permanent' *Rohon-Beard cells* was given in section 3, chapter VII of volume 3, part II *et passim*. Their functional significance as first order proprioceptive neurons particularly but perhaps not exclusively related to the muscles of mastication seems reasonably well established. Figure 268A of chapter IX illustrates the general arrangement displayed by these neuronal elements as components of the trigeminal nerve. As regards their connections, CAJAL and others showed that (afferent) neurites of these cells join the mandibular nerve, while their relatively short intracerebral efferent neurites are connected with the motor nucleus of the trigeminus.[13] Nothing certain is known concerning synaptic junctions on the perikarya of these cells. If such synapses should be lacking then the 'nucleus mesencephalicus trigemini' would represent an intracerebral branchial cranial nerve 'ganglion' rather than a true 'nucleus'.[14] Nevertheless, HERRICK (1948) described the perikaraya of the nucleus radicis mesencephalica trigemini of Amblystoma tigrinum as 'imbedded in dense neuropil' which he interpreted as providing synaptic contacts. This, however, remains somewhat doubtful, since the relationship of the large vesicular cells to said neuropil, which is part of the lamination characterizing the tectum opticum, appear rather topographically incidental than functionally synaptic.

The cells of the trigeminal mesencephalic root are variously distributed throughout the tectum mesencephali, usually in the deepest layers or near the ependyma, and as far rostrad as the posterior commissure. They occur either as single cells or in small clusters, and may also extend into the base or body of the cerebellum.

---

[13] These findings, of course, do not exclude the possibility of other functionally similar peripheral and central connections.

[14] If the term 'nucleus' is restricted to central grisea including perikarya and neuropil with synaptic connections. A spinal or cranial nerve ganglion, on the other hand, is generally believed to contain 'pseudo-unipolar' or bipolar nerve cells without synaptic junctions on their perikarya. Nevertheless, the possibility that some synaptic structures or processes might obtain as 'special cases' in such ganglia cannot be altogether excluded (cf. the comments on this topic in section 1 of chapter VIII). Some synapse-like structures and aspects of their ontogenetic development were recently described on the basis of electron-microscopic pictures in the nucleus radicis mesencephalicae trigemini of the Hamster, a Rodent Mammalian (ALLEY, 1973).

Two secondary *correlating grisea* within the mesencephalic alar plate are *nucleus isthmi* and *nucleus visceralis secundarius*. The *nucleus isthmi* (s. 'ganglion isthmi'), representing a *ventral derivative of the alar plate's dorsal zone*, is located, in Anamnia, somewhat rostrally to the cerebellar nucleus, and receives mainly input from the VIIIth nerve system (Fig. 387G). Fibers originating in nucleus isthmi appear to reach the tectum opticum, the torus semicircularis, and perhaps unspecified diencephalic grisea. In some Reptiles (Fig. 391G) the nucleus isthmi is a rather prominent structure forming a dorsolateral protrusion on the external surface of the isthmus region.

In Birds, the nucleus isthmi assumes further development, and consists of at least two portions (pars principalis and nucleus isthmo-opticus, cf. further below in section 8). In Mammals, the nucleus isthmi presumably becomes the nucleus (dorsalis) lemnisci lateralis, including perhaps also other smaller grisea intercalated within that fiber system at mesencephalic levels (cf. K., 1935).

The *nucleus gustatorius secundarius* is a derivative *of the alar plate's intermediodorsal zone*, located slightly ventrally or ventrocaudally to the nucleus isthmi. It is particularly developed in various Teleosts (cf. Figs. 233A, B, 378B), but can also be identified in many Amphibians (cf. Fig. 387G). In other Anamnia this griseum appears to be present, although not very clearly demarcated. Concerning Amniota, the fragmentary available data on the secondary ascending gustatory tract, and the nondescript griseal differentiation in the mesencephalic or isthmic region corresponding to nucleus gustatorius s. visceralis secundarius do not provide adequate evidence for a definite identification of that nucleus. In Mammals, the ascending secondary gustatory tract doubtless reaches the diencephalon. It should be added that the term 'gustatory tract' refers to what is believed to be the major component of a channel which, however, presumably also includes 'general visceral afferent' fibers.

The most significant differentiations of the mesencephalic *alar plate's dorsal zone* are (1) *torus semicircularis* (respectively colliculus inferior of Mammals) and (2) *tectum mesencephali s. tectum opticum* (respectively colliculus superior of Mammals). Both (1) and (2) can be conceived as tectum mesencephali *sensu latiori*, while (2) represents the Anamniote tectum *sensu strictiori*. In the mesencephalo-diencephalic boundary zone, a number of mesencephalic and diencephalic grisea represent the nuclei of the pretectal region, or '*pretectal complex*', which comprises a dorsal thalamic group including interstitial nuclei of the posterior commissure, a mesencephalic tectal nuclear group, and a mesencephalic tegmental group, to which the *nucleus interstitialis of Cajal* pertains (K., 1939, 1956; K. and MILLER, 1942, 1949; KUMAMOTO-SHINTANI, 1959). The mesencephalic grisea of the 'pretectal complex' shall be pointed out in the various sections of the present chapter, and further comments on the pretectal region as a whole will be included in chapter XII of volume 5.

The *tectum mesencephali s. tectum opticum* in all Vertebrate classes receives a bundle of optic tract fibers (Figs. 358A, B). Particularly in Anamnia, this bundle essentially represents the main optic input.[15] In Amniota and particularly in Mammals, the connections with diencephalic grisea (corpus geniculatum laterale[16]) and the projections from these latter upon the telencephalon assume an increasing importance.

A useful approach to the first approximation of a proper understanding concerning the basic mechanisms of the Vertebrate tectum mesencephali *sensu latiori* may be provided by considering the very simple arrangement obtaining in the urodele Amphibian, Necturus, which was particularly investigated by HERRICK (1917, 1931). Although Necturus, being a tetrapod Anamnian, can hardly be evaluated as phylogenetically primitive with respect to the origin of the Vertebrate phylum, some of the neuronal arrangements in the central nervous system of this Urodele are doubtless among the very simplest encountered in the entire phylum.

A cross-section through the mesencephalon of Necturus (Fig. 359) shows that the optic nerve fibers reach the tectum's dorsal part, on both sides of the midline. More laterally there are terminals of second-

---

[15] A much smaller optic input channel is represented by the basal optic tract briefly mentioned further above. It should, moreover, be mentioned that the main or dorsal optic tract in many if not most Vertebrates, e.g. in Anurans (WLASSAK, 1893), can again be subdivided into a larger marginal and a smaller axial portion. The marginal portion, in turn, displays a medial (dorsal) and a lateral (ventral) bundle. However, the various subdivisions of the optic tract, as designated and classified by diverse authors, are not always sufficiently clearly recognizable (cf. e.g. HUBER and CROSBY, 1926, in the Alligator).

[16] A lateral geniculate griseal complex, consisting of dorsal and ventral lateral geniculate nucleus, is, of course, also present in Anamnia, and shall be dealt with in chapter XII of volume 5. It must also again be emphasized that, even in Anamnia, the tectum opticum, despite its preponderating significance for vision, is by no means the only griseum receiving direct optic input. Such input also reaches hypothalamus, ventral and dorsal thalamus (lateral geniculate complex), pretectal region (diencephalomesencephalic boundary region), and mesencephalic tegmentum. Secondary input to telencephalon is likewise doubtless obtaining in Anamnia.

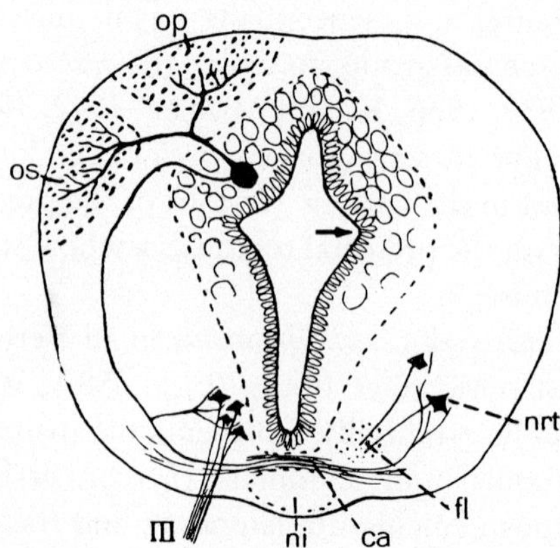

Figure 359. Simplified diagram depicting basic aspects of mesencephalic neuronal connections in the urodele Amphibian Necturus, as seen in a cross-section. Added arrow indicates sulcus limitans (based on figures by C. J. HERRICK combined with personal observations, from K., 1927). ca: commissura ansulata; fl: fasciculus longitudinalis medialis; ni: nucleus interpeduncularis; nrt: nucleus reticularis s. motorius tegmenti; op: endings of optic tract; os: ending of secondary octavus and general (bulbotectal and spinotectal) lemniscus systems; III: oculomotor nerve. Cf. also Figure 385B.

ary octavus fibers together with endings of ascending channels carrying 'sensory' input ('general body sensibility' including 'touch') from rhombencephalon and spinal cord. Particular neurons[17] of the tectum, here remaining in the periventricular griseum, extend one main dendrite into the optic terminals, and another one into those of afferent octavus and 'general sensory' channels. Such neurons can thus, either simultaneously or consecutively, register input from diverse sources (optic tract, eighth nerve, 'general body sensibility', possibly also impulses of 'visceral' origin). These neurons, presumably interacting with other local elements, may then not only register but to some extent also process their input and then, through their neurites, representing, as it were, a sort of 'common final path', discharge their output either directly upon primary motor nuclei (e.g. of n. III and IV), or into the important secondary motor center of nucleus motorius tegmenti, i.e. into STEINER's '*allgemeines Bewegungszentrum*'. Decussating neurites of said neurons of tectum mesencephali, crossing ventrally to the midbrain ventricle, form the commissura ansulata described in An-

[17] These cells represent the so-called '*Hauptzellen*' ('principal cells', 'chief cells') in the more highly differentiated tectum opticum of other Anamnia.

amnia by the classical pioneer investigators. It is likely that the neurites of the cells in question give off collaterals not only to other local elements but also to diencephalic grisea, by means of which further correlations may take place. Again, reciprocal connections with these grisea and also with other mesencephalic elements can be surmised.

In Anamnia with a more developed tectum mesencephali, the fibers of the octavus system tend to become connected with a griseum located ventrally to tectum opticum *sensu strictiori*. This griseum protrudes into the ventricular lumen, forming the *torus semicircularis* (Figs. 387E, F) separated from tectum opticum by the sulcus lateralis mesencephali (internus) of the ventricular wall. Since a sulcus intermedius dorsalis is not, as a rule, manifested, the sulcus limitans approximately represents the ventricular boundary between torus semicircularis and basal plate. In some instances, the bilateral-symmetric tori semicirculares can fuse in the midline, separating, at some cross-sectional levels, a dorsal and ventral portion of the mesencephalic ventricle from each other, the two portions, however, remaining rostrally respectively caudally in direct communication.

The griseum within the bulging torus semicircularis is the *nucleus mesencephali lateralis (s. ganglion mesencephali laterale)* essentially representing a correlation center of the nervus octavus, but doubtless also receiving some additional input channels. In some Reptiles, the tori semicirculares with their grisea show the tendency to be displaced in a dorsocaudal direction (Fig. 391A). In Mammals, the tori semicirculares have disappeared, and the arrangement of their grisea results in the formation of two external, dorsal protrusions, the *colliculi inferiores*, located caudally to the paired protrusions of tectum opticum. The dorsal surface of tectum mesencephali which, in submammalian forms essentially displays merely the two rounded prominences of tectum opticum,[18] also designated as the '*lobi optici*', becomes thus the *lamina quadrigemina* (BNA) of Mammals, the corpora quadrigemina consisting of the paired colliculi superiores and inferiores (BNA, PNA), of which the latter are homologous to tori semicirculares, and the former to tectum opticum of the submammalian neuraxis. In Mammals, moreover, the rostralward extension of the octavus input system becomes rather prominent, its diencephalic relay center being the medial geniculate complex (nucleus geniculatus medialis s. corpus geniculatum mediale,

[18] Discounting here the suggestion, in some Reptiles, of an 'inferior colliculus' (cf. Figs. 391A, G).

with dorsal and ventral thalamic components). This rostralward extension of the octavus channel, although in a much less clearly defined manner, is doubtless present even in Fishes and Amphibia.

The tectum opticum of Anamnia, respectively the superior colliculi of Mammals, however, are by no means an exclusively optic 'center', but a suprasegmental apparatus receiving input from the ascending 'general lemniscus system' (cf. Fig. 358A) as well as, directly or indirectly by way of the torus semicircularis or colliculus inferior, octavus ('lateral lemniscus') input.

As regards the *optic input channel*, the investigations of numerous authors[19] by means of different methods have demonstrated a point-to-point, or perhaps better a definitely region-to-region, i.e. topographic projection of the retina, respectively of the visual fields, upon the tectum mesencephali. This can be presumed to obtain, in essentially similar or 'identical' manner, for all Vertebrates, that is, for the tectum opticum of Anamnia, Reptiles, and Birds as well as for the superior colliculi of Mammals.[20] The general arrangement of this projection, roughly sketched in Figure 360A, is such that the ventral (inferior) half of the retina is represented on the medial part of the tectum, and the dorsal (superior) half on the lateral tectal region. The two rostromedial (respectively nasal) quadrants are represented caudally, and the two caudolateral (respectively temporal) quadrants rostrally. Figures 360B–D, which are self-explanatory, illustrate early concepts of CAJAL (1911, 1923), concerning the projection of the retinal image upon the Anamniote optic tectum, compared with an assumed projection based on the more recent data. Figure 360E depicts a concept of panoramic vision in Fishes elaborated by KAPPERS *et al.* (1936). With respect to this topic, however, it should be emphasized that no projection of an actual retinal *picture* upon tectum opticum respectively striate cortex (cf. chapter XV, vol. 5) takes place in the neural optic system. What is here 'projected' represents a *three-dimensional pattern of localized neural events*, characterized by regional neighborhoods, which merely corresponds to some topologic features of the essentially two-dimensional

---

[19] E.g. APTER (1945), BODIAN (1937), BROUWER *et al.* (1923), LASHLEY (1934), STONE (1930), STRÖER (1939, 1940). Cf. also the conference on subcortical visual systems edited by INGLE and SCHNEIDER (1970).

[20] Such topographic projection of the optic channel as obtaining for the lateral geniculate griseum and, in Mammals, for the telencephalic optic cortex ('area striata') shall be dealt with in volume 5 of this series.

picture projected by the refractive media upon the retina. It can, moreover, be seen that this topologic mapping of the total visual panorama (cf. the arrows in Fig. 360D) is transposed from right to left, folded upon itself, and, as it were, bisected (cut up) by a process akin to *topologic 'tearing'* (cf. vol. 1, chapter III, section 2, p. 178).

In Anamnia, respectively in Sauropsida, that is, in forms with almost complete decussating of the optic nerves in the chiasma opticum, the tectal optic projections are, of course, mainly contralateral. In those Vertebrates whose chiasmatic decussation tends to be incomplete, however, the temporal retinal quadrants provide a variable proportion of uncrossed (homolateral) optic tract fibers, which project (perhaps to some extent overlapping with the contralateral fibers) on the corresponding rostral (lateral and medial) tectal areas.

Experiments e.g. by Sperry (1944, 1945) and Stone (1947, 1953) concerning cut, rotated or exchanged, and replanted Amphibian eyes have shown that the regenerating optic nerve fibers commonly re-establish proper connections with the optic tectum according to their re-

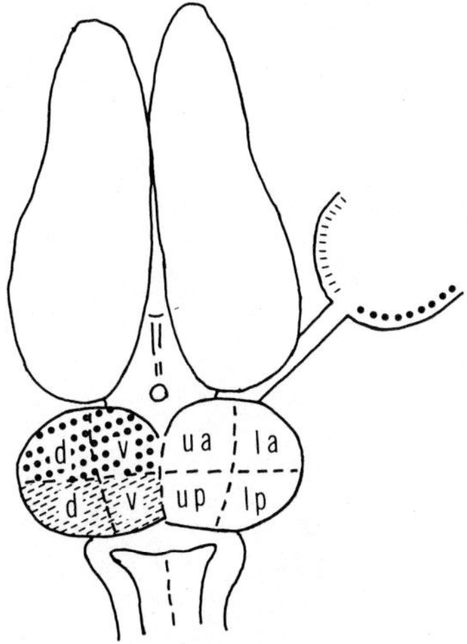

Figure 360A. Diagrammatic sketch roughly illustrating projection of contralateral retinal quadrants upon the tectum opticum in a 'generalized Amphibian'. Coarse dots: posterior (temporal) retinal quadrants; hatching: anterior (nasal) retinal quadrants; d: dorsal retinal quadrants; v: ventral retinal quadrants. On the right tectum opticum, the corresponding visual field quadrants of the left eye are indicated. la: ventral (lower) anterior visual field quadrant; lp: lower (ventral) posterior visual field quadrant; ua: upper (dorsal anterior visual field quadrant; up: upper (dorsal) posterior visual field quadrant.

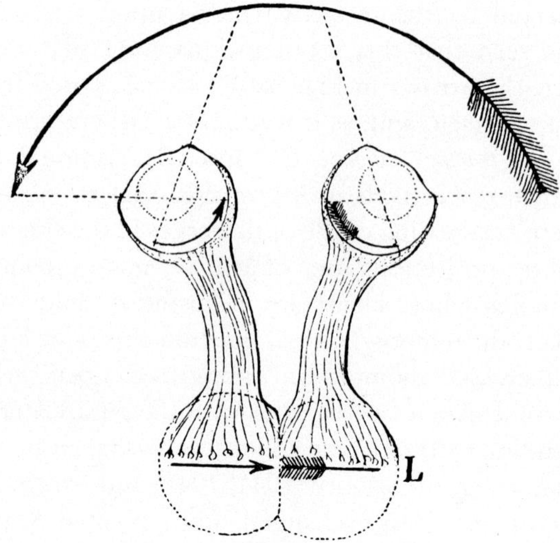

Figure 360 B

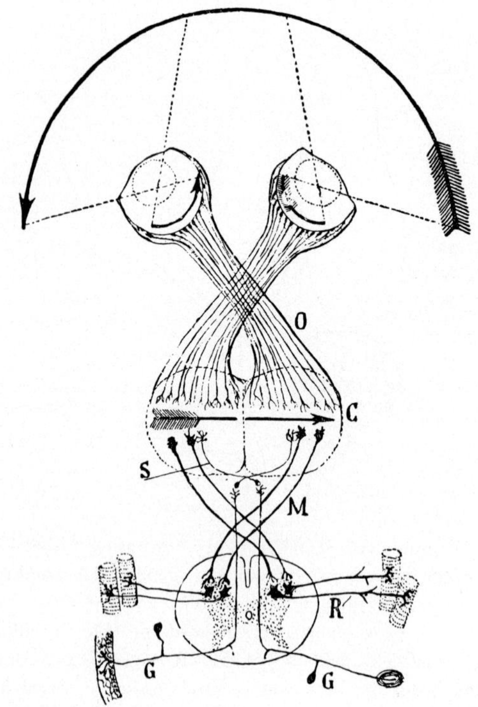

Figure 360 C

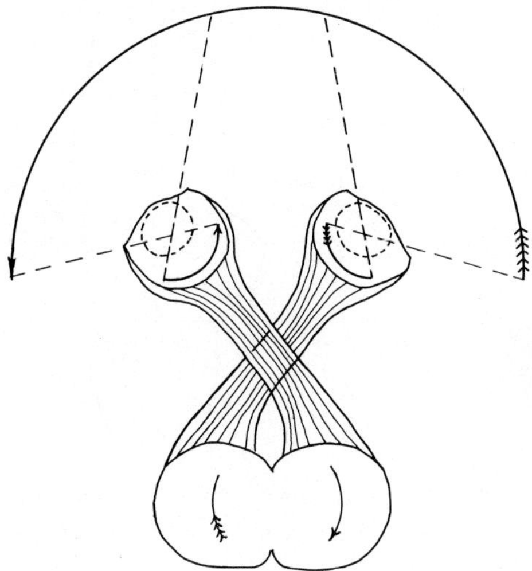

Figure 360D. Partial modification of CAJAL's scheme in preceding Figure C as required by the subsequently recorded data (cf. Fig. A).

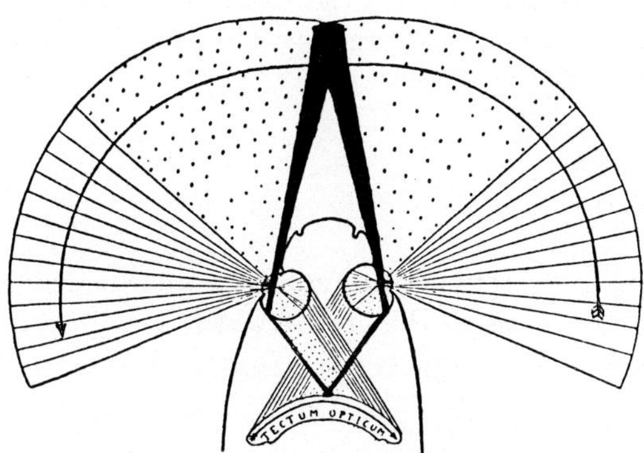

Figure 360E. KAPPERS' concept of mesencephalic (tectal) panoramic vision in a Fish with total optic tract decussation. A slight overlap of the adjoining nasal visual fields was here assumed to obtain because of convergence of optic tract fibers in the optic tectum (from KAPPERS *et al.*, 1936).

Figure 360 B, C. Diagrams illustrating CAJAL's concept concerning the significance of the optic chiasma (from CAJAL, 1911). Figure B shows the (topological) 'disparity' of the retinal image projection upon optic tectum in a lower Vertebrate whose optic tract would not undergo the chiasmatic decussation. Figure C indicates the 'usefulness' *(utilité)* of optic chiasma and compensatory decussations of motor as well as sensory channels in a lower Vertebrate. C: optic center of mesencephalic tectum; G: spinal ganglion and afferent root; L: tectum opticum; M: decussating motor pathway; O: decussating optic pathway; R: motor roots of spinal cord; S: decussation of central sensory pathway.

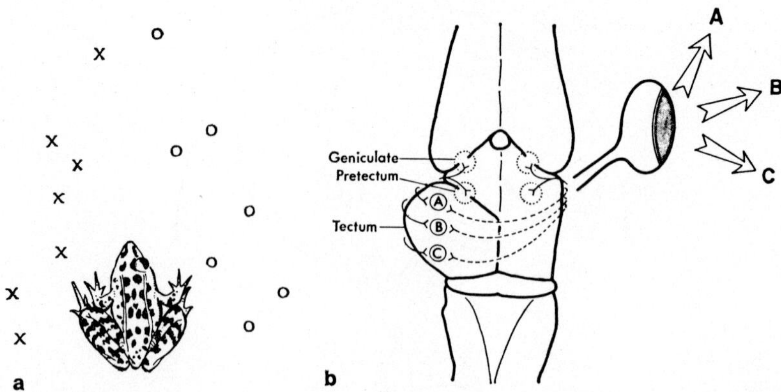

Figure 360F. Sketches illustration results of unilateral tectum opticum extirpation in the Frog (from INGLE, 1973). a: illustration of the mirror-symmetrical relation between a prey stimulus (head of a moving worm, indicated by circles) and the frog's corresponding snap (center of the tongue print, indicated by crosses), b: schematic illustration of optic tract regeneration (solid lines, but exact route of fiber regrowth unknown) to homolateral tectum (the normal contralateral fibers are indicated by dotted lines. The arrows A, B, C indicate direction of visual response upon homolateral input to tectal loci A, B, C, as if received via the contralateral eye. The optic input to diencephalic grisea is supposed to have remained intact on the lesion's side. Further explanations in text. It should be noted that, in a, the left tectum opticum has been extirpated.

spective origin in the retinal quadrants as primarily determined before the operation. By rotating the eye 180°, the resulting changed functional quadrants of the retina become thus, as it were, connected in reverse.[21]

Still more recent experiments with Frogs (INGLE, 1973b) indicate that, after unilateral removal of the tectum opticum in adult specimens, the thereby cut contralateral optic tract regenerates by growing into the remaining homolateral tectum. These fibers remap here a topographic projection of the retina, which is superposed on, or overlaps with, the normal and undamaged contralateral projection. If the eye with normal projection was sewn shut, the frogs reacted to moving optic stimuli (moving worms or threats) by mirror-symmetrically di-

[21] Some comments on the significance of these experiments can be found on p. 140 and 196 of the monograph 'Brain and Consciousness' (K., 1957; p. 175 and 243/244 of the 1973 German edition, translated by Prof. GERLACH and Dr. PROTZER). Cf. also p. 97 of the monograph 'The Human Diencephalon' (K., 1954).

rected responses (Fig. 360F). On the other hand, the frogs correctly localized stationary objects such as barriers. The cited author assumes that, since diencephalic optic input (to geniculate and pretectal grisea) remained undisturbed by the operation, 'thalamic and tectal visual mechanisms can operate independently'.[21a] In this connection, INGLE cites investigations by SCHNEIDER (1969), based upon experiments performed with Hamsters, where lesions of the telencephalic visual cortex produced pattern discrimination deficits, while tectal ablations eliminated turning of the head toward objects such as food. It was thus concluded that the Mammalian telencephalic visual cortex and the superior colliculus might have 'distinctive behavioral functions' within the overall optic system.

In most Anamnia, in Reptiles, and in Birds, the tectum opticum displays a multilaminated cortical structure with a complex pattern of synaptic connections. In the superior colliculus of Mammals, the cortical aspect of its griseum is less conspicuous, but the laminar arrangement remains, nevertheless, clearly recognizable. The '*cortex lobi optici*' has doubtless a more intricate structure than the cerebellar cortex, but shares with this latter an essentially uniform arrangement, devoid of obviously distinctive cytoarchitectonic areal differentiations. CAJAL (1911), reviewing and expanding previous studies by his brother PEDRO RAMÓN, presents an elaborate account of structure and lamination of the lobus opticus in 'lower Vertebrates'. In accordance with PEDRO RAMÓN, 15 layers are described in Birds, Reptiles, and in the Amphibian frog,[22] this number being reduced to about 10 or less layers in Fishes. BELLONCI (1888), BECCARI (1943) and others have adopted slightly different classifications respectively enumerations of the tectal cortical layers.

A group of investigators concerned with the interpretation of neural circuit processes in terms of communication theory and engi-

---

[21a] The cited author does not refer to the basal optic channel reaching the nucleus opticus tegmenti, respectively the reticular formation. This pathway might likewise be concerned with responses to optic input.

[22] In his standard work, CAJAL (1911, 1955) enumerates and describes these layers of the Bird's tectum opticum beginning with the optic tract fibers (1), and ending at the ependymal lining (15). In Reptiles and in the frog, however, he retains the *inverse* sequence as there depicted by PEDRO RAMÓN, and beginning with the ependyma (1). He merely states in the legend to his Figure 140 (Lacerta): '*Le numérotage des couches est fait ici en sens inverse de celui adopté pour les oiseaux.*'

Figure 361 A. Transverse section ('coupe frontale') through the tectum opticum of the frog, as seen in a Golgi impregnation (after PEDRO RAMÓN, from CAJAL, 1911). A, B, C, D, E: various sorts of granule cells (cells A, B, C, however, located in the granular layer, are interpreted by CAJAL as '*cellules à crosses*', i.e. as chief cells, although not located in the chief cell layers 10, 9, 8); F: ependymal cell; a, b, c: strata of optic fiber arborizations. The fibers marked by a small *a* are presumed to be neurites. The neurite of cell D is supposed to be a centrifugal fiber ending in the retina. 1–14: layers in P. RAMÓN's concept of stratification.

neering has published two reports referring to the Frog's retinal mechanisms, and including some comments on the pertinent input reaching the tectum opticum (LETTVIN *et al.*, 1959; MATURANA *et al.*, 1960).[23] In the present section, dealing with general pattern and basic mecha-

[23] One of these papers (LETTVIN *et al.*, 1959) is published in a journal of radio engineering and entitled: 'What the frog's eye tells the frog's brain'. The other, 'Anatomy and physiology of vision in the frog (Rana pipiens)' by MATURANA *et al.*, 1960, appeared in the Journal of general physiology. Both papers are by the same group of authors, including MCCULLOCH and PITTS, whose adaptation, to neuronal networks, of SHANNON's circuit algebra (based on *Boolean algebra)* is discussed in the monograph 'Brain and Consciousness' (K., 1957, 1973).

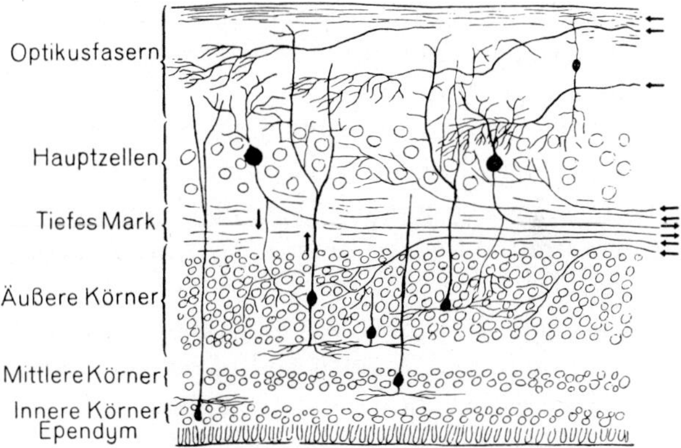

Figure 361 B. Semidiagrammatic sketch showing layers of tectal cortex in the frog, partly based on P. RAMÓN's findings, but interpreted on the basis of personal observations (from K., 1927).

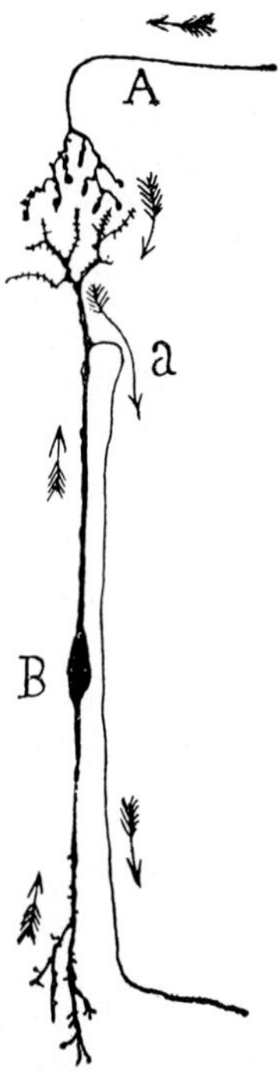

Figure 361 C. Semidiagrammatic sketch of a *crozier cell (cellule à crosse, celula de cayado)* in the tectum opticum of 'lower' Vertebrates *(peces, batracios y reptiles)* showing characteristically bent origin of neurite from the peripheral dendrite and the presumed course *('la marcha de las corrientes')* of the impulse transmission (from CAJAL, 1923). A: terminals of optic tract; B: cell body at level of nucleus; a: origin of neurite.

nisms, it is therefore perhaps appropriate to include a short discussion of the structural arrangements of the Frog's tectum mesencephali as a useful paradigm.

Figure 361A depicts the cortex lobi optici as interpreted by PEDRO RAMÓN and displaying 15 layers, while figure 361B shows my own interpretation of 1927 which adopted a simplified enumeration of layers, and particularly stressed the *deep medulla (tiefes Mark)*, corresponding to RAMÓN's layer 7.

The optic layer (15–10 of RAMÓN, Optikusfasern, K., 1927) consists of the peripheral optic tract bundles and of several strata formed by their terminal arborizations. RAMÓN showed three such strata, which I likewise depicted. According to the postulates by LETTVIN *et al.* (1959), to be discussed further below, one should expect four. This may indeed be the case, since Figure 132 of CAJAL (1911, 1955), reproduced as Figure 403D in section 8, actually shows four such strata to obtain in the Bird's tectum. Because of their intrinsic vagaries, the *Golgi preparations* interpreted by RAMÓN as well as by myself may have failed clearly to display a pattern of four strata in the Frog.

Internally to, and partly overlapping with, the optic strata, the layer of chief cells *(Hauptzellen)* is located. At least two sorts of chief cells may be distinguished, namely the rod-like elements designated as *'crozier cells' (cellules à crosse, células de cayado)* of CAJAL and PEDRO RAMÓN and the multipolar large *'cellules ganglionaires'* of the just cited authors. The crozier cells evidently correspond to the simpler type of cell described by HERRICK in Necturus, and depicted above in figure 359. The peripherally directed dendrite of these elements effects, through its various ramifications, connections with the optic input. The centrally directed dendrite and its ramifications receive input from the deep medullary layer, which adjoins, respectively overlaps with, the inner boundary of the chief cell stratum. In contradistinction to the rod-like *'crozier cells'* depicted in Figure 361C[24] the multipolar *chief cells*

---

[24] The rod-like *crozier cells* display diverse variations of their typical overall configuration, depending partly on the taxonomic status of the particular species, and partly on the location of the perikaryon within deeper or more superficial levels of the chief cell layer, whose outer and inner boundaries may be ill-defined. The neurite frequently arises from the peripherally directed dendrite, but can also originate from the centralward directed one. Two rather typical *crozier cells* were depicted in Figure 53, p. 76, of volume 3, part I.

(Fig. 403C) display dendrites extending 'horizontally' and peripherally toward the optic fiber layer. The neurite usually originates here from the cell body. While some of these multipolar chief cells seem to connect mainly (or exclusively ?) with optic tract end-arborization, others have evident connections with end-arborizations of input fibers from the deep medulla. Their synaptic relationships correspond thus to those of the rod-like cells and of the more primitive neuronal elements described in the tectum of Necturus. Additional connections with other nerve cells of the tectal cortex, such e.g. as *Golgi II cells* and similar neurons *(Schaltzellen)* doubtless obtain for both types of chief cells, whose neurites provide the output of tectum opticum through tecto-bulbar and tectospinal tracts, with perhaps some additional collateral connections. It will be seen that in RAMÓN's *Golgi picture* (Fig. 361A) the chief cells were not impregnated, being merely indicated as vague outlines in layer 8. My own sketch, on the other hand, although depicting chief cells, merely schematically indicates one type, namely the '*cellules ganglionaires*'.

The *deep medulla (tiefes Mark)* consists of both output and input channels. The former are the just mentioned neurites of chief cells, the latter are represented by the general spinal and bulbar lemniscus (tractus spino- et bulbo-tectalis). As regards octavus input, of which the lateral lemniscus is a main ascending channel, it seems likely that fibers either by-passing the torus semicircularis, or neurites originating in this griseum, or, again lateral lemniscus collaterals are included within the deep medulla. Thalamotectal, hypothalamotectal and cerebellotectal connections likewise seem to obtain (cf. e.g. BECCARI, 1943). The presumptive 'visceral input' of the tectum may be mediated either by the hypothalamotectal fibers or by the general lemniscus system.

Internally to this layer, three strata of 'granule cells' can be noted (layers 6, 4, and 2 of RAMÓN) which were designated as *äussere Körner*, *mittlere Körner*, and *innere Körner*. The outer granular layer is the widest one, and the inner granular stratum directly adjoins the ependyma. Inner and outer granular layer are each separated by a 'molecular' or 'plexiform' layer (5 and 3 of RAMÓN) from the intermediate granular stratum *(mittlere Körner)*.

The three granular layers of the tectal cortex contain a number of different neuronal elements predominantly representing nerve cells of *Golgi II type*, perhaps even including some elements comparable to the amacrine cells of the retina. Small neurons of horizontal or stellate type, with complex and poorly understood presumably 'intracortical' connections[25] also occur in the optic and in the chief cell layers (cf. Figs. 377, 379B, C, 403).

Summarizing and reviewing my own concept of tectal cortical lamination, as expressed in the '*Vorlesungen*' of 1927, I believe that, in a useful simplified approximation, and in essential agreement with the views of HOUSER (1901) and STERZI (1909) four main layers could be distinguished, namely (1) optic layer, (2) layer of chief cells, (3) deep medulla, and (4) layer of granule cells. However, the layer of chief cells, the deep medulla, and the layer of granule cells display numerous variations in the diverse taxonomic forms. Tectal and non-optic input fibers may, moreover, proceed through layers 5 and 3 of RAMÓN, while non-optic input channels can also run through an ill-defined fiber layer, intermingled with chief cells, between layer (1) and layer (2) of my simplified classification. BECCARI (1943) and HUBER and CROSBY (1933) have used a notation distinguishing five respectively six layers as shown below in Figures 379C (Teleost) and 394A–H (Reptiles). The useful layer concept of SENN (1968b), illustrated by Figure 395 shall be discussed in section 7 of the present chapter.

As regards the optic input into the tectum mesencephali, the above-mentioned studies by LETTVIN *et al.* (1959) and by MATURANA *et al.* (1960) are of considerable interest. These authors emphasize (1959) that their experimental work was done on the frog and that their 'interpretation applies only to the frog'. It seems nevertheless likely that, *mutatis mutandis*, comparable even if somewhat modified mechanisms obtain in other submammalian Vertebrates, whose optic system does not include a projection to the neocortex. In higher Mammals, particularly Primates and Man, the neocortical optic area appears to play a predominant role for vision.

The cited authors reach the conclusion that the retina performs several complex analytical operations on the physical optic image projected upon its area. It is here significant that the ganglion cells which give origin to the optic tract fibers are connected (by way of the bipo-

---

[25] Thus, the tectal cortex contains, in addition to the chief cells, '*zahlreiche kurze Schaltneurone, Zellen vom II. Golgischen Typ oder unpolarisierte Körnerzellen, welche eine weitgehende Verteilung der Impulse besorgen*' (K., 1927). The details of these complex connections remain unknown but are evidently related to the function of the mesencephalic tectum as a 'suprasegmental' cortical structure of prime importance for the relevant processing of sensory input and for the discharge of output signals into nucleus motorius tegmenti as well as other grisea, including diencephalic ones.

lar elements) not to single rods or cones, but to a group of such receptors representing a neighborhood (responsive receptive field, RRF).

As regards such neighborhoods, it is generally not the light intensity itself, but rather the pattern of local variation of intensity which provides the 'exciting factor'. Five 'natural classes' of retinal ganglion cells are presumed to obtain. Four of them react upon the optic image by the performance of complex analytical operations that remain invariant under changes of the general illumination and changes of the general configuration of the optic picture. The fifth class registers the light intensity. These classes of ganglion cells and their operations are described as follows. (1) Sustained edge detection, with non-erasable holding; (2) convex edge detection, with erasable holding; (3) changing contrast detection; (4) dimming detection; (5) darkness detection.

The cells of classes 1 to 4 respond to moving objects, those of class 1 registering only a sharp edge in their receptive field, emitting burst of activity while the edge moves across that field, and a sustained response if the edge stops in the middle of it. This response is not erased by a transient darkening of the field. Cells of class 2 respond in a similar manner only to convex dark edges, but the response to such edge stopped in their receptive field is erased by a transient darkening of the field. Cells of class 3 respond (with small bursts) only to changing pattern. Cells of class 4 respond to any adequate darkening of the receptive field with a prolonged off discharge; they also respond to a moving object in proportion to the relative darkening that the object produces during motion.

The five classes of ganglion cells form five populations uniformly distributed over the 8th (or ganglionic) layer of the retina with great overlapping of receptive fields. In the tectum mesencephali the neurites of each cell class end in a separate stratum of the optic layer, the terminals of class 5, however, ending in the stratum formed by the terminals of class 3. Thus, as mentioned above on p. 800, four strata of terminals can be assumed on the basis of the experimental electric recordings. It seems likely that merely because of the vagaries inherent in the *Golgi method*, Pedro Ramón depicted only three strata of terminals (cf. Fig. 361A). The most superficial strata are those of the contrast detectors (the 'slowest' fibers), beneath them but not so distinctly separate, are the convexity detectors. Deeper, and rather well separated, are the moving edge detectors, admixed with the ill-defined axons of actual darkness level detectors. Deepest lie the terminals of 'dimming detectors'. It seems that the depths at which the various optic nerve fibers end with terminals 'is directly related to their speed of conduction'.

The retina[26] transforms the physical optic picture from a mosaic of luminous points to a system of overlapping '*qualitative contexts*' in which any neighborhood is related 'to what is around it'. These 'qualitative contexts' are (a) standing edges, (b) curvatures, (c) changing contrast, (d) local lessening of light intensity, and (e) general measure of illumination. Such transformation probably arises from intraretinal summing and differencing of the receptor activity, whereby local continuity and curvature of boundary are taken, giving rise at the same time to the coding of local extension (MATURANA *et al.*, 1960). In still more generalized terms, neuronal processing activities of this sort evidently represent a coding of uniform types of events or, in other words an '*abstraction of invariants*'.

The point-to-point (or neighborhood-to-neighborhood) projection of the retina upon the tectum mesencephali, discussed above on p. 792, is accordingly complicated by a stratification of the retinal map in depth; this stratification corresponds to the diverse processing operations performed by the retina.[27] The results recorded and interpreted by the cited authors are doubtless of considerable interest. Still, one might wonder, since the frog's retina includes rods as well as cones, these latter being generally considered as recording frequency (i.e. 'color') of light waves, whether the frog's vision does or does not involve some particular sort of 'color vision' mechanisms, for which no proviso is made in the model under discussion.

[26] The significance of the retina as a neural griseum with complex processing functions is now well established and generally recognized. Since the retina represents an evagination (optic vesicle) with secondary invagination (optic cup) of the prosencephalon, remaining connected, as a peripherally displaced central griseum, with the hypothalamic subdivision of the diencephalon by means of the optic 'nerve', the eye as an organ of sense shall be dealt with in chapter XII of volume 5. Further discussion of the optic pathway will likewise be included in that and other chapters of volume 5. Comments on the optic system, and on the recent investigations by HUBEL and WIESEL as well as other authors can also be found in the recent German edition of '*Brain and Consciousness*' (K., 1973). It should also be recalled that functional analoga of the Vertebrate retinal grisea are represented by the large 'optic lobes' of Invertebrates (e.g. Insects and Cephalopoda), these lobes remaining parts of the brain rather than of the eye proper (cf. vol. 2 of this series, pp. 233–236, 267–277 *et passim*).

[27] The well known regeneration of the transected Amphibian optic nerve with re-establishment of the original areal projections upon the optic tectum (cf. above p. 793) has been confirmed by LETTVIN *et al.* (1959) and MATURANA *et al.* (1960) with respect to the original order in depth 'with no mistakes'. 'If there is any randomness in the connections of this system, it must be at a very fine level indeed' (LETTVIN *et al.*, 1959).

Color *(sensu strictiori)* of course, is just as 'light' and 'darkness', a mental (i.e. conscious) experience, namely a P-event. This color experience, in turn, is related to a number of events in the postulated physical world, namely (a) frequencies of 'light waves', (b) transduction processes by retinal receptors, (c) further processing of the receptor signals by the retinal grisea, and (d) an additional series of processing operations by grisea of the central neuraxis. The events under (a) represent R-events, and those under (b) to (d) N-events, of which the hierarchically 'highest' are Np-events, 'functionally' correlated with P-events.

Quite evidently, nothing whatsoever can be ascertained about the conscious experiences of a frog, but, from a behavioristic viewpoint, it is possible to ascertain whether the central nervous system of a frog or, for that matter of any Vertebrate respectively Invertebrate animal registers patterns of light frequencies, that is, what, in terms of our consciousness, we call 'color'.

As regards Man, whose conscious experiences are known by 'introspection' or can be ascertained by verbal communications in correlation with behavioristic (objective) tests, numerous well-known 'classical' color theories have been elaborated, none of which is entirely satisfactory, particularly because most of these theories overemphasize the processes classified above under (b) and (c). A more recent theory, based on interesting experiments,[28] and elaborated by LAND (1961), seems to indicate that experienced colors are not specifically correlated with particular light wave lengths (frequencies), but rather with the interplay of longer and shorter wavelengths 'over the entire scene'. Accordingly, 'classical color theories' may model only 'special cases' in a complex and still poorly understood series of neural processes by means of which, as P-events, human vision 'shows the world in all its variegated hues'. It is thus quite possible that not only cone signals but also rod signals could play a role in color vision. The general topic of vision, including color perception, shall again be considered in the pertinent chapters of volume 5.

---

[28] LAND (1961) has shown that human vision 'can build colored worlds of its own out of informative materials that have always been supposed to be inherently drab and colorless'. LAND's experiments demonstrated the production of full-color pictures by projecting two black-and-white photo-transparencies together on a screen, these transparencies being taken through different 'color filters'. The only requirement is that the long-wave-length photographs should be illuminated by the longer band, and the 'short record' by the shorter band. Indeed, one of the bands may be as wide as the entire visible spectrum, i.e. it may be 'white light'.

Reverting to some general structural features of the tectum mesencephali, a commonly obtaining *commissura tecti* or lamina commissuralis tecti, interconnecting the two antimeric alar plate portions, usually at the level of the deep medulla (stratum medullary profundum, *Tiefes Mark)* remains to be mentioned.[29] Rostrally the commissura tecti merges or overlaps with the *commissura posterior* which represents the mesencephalo-diencephalic boundary region. The commissura posterior includes decussating fibers of diencephalic cell groups (nucleus or nuclei commissurae posterioris) as well as of mesencephalic tegmental and tectal grisea. Within a fairly constant overall pattern, the composition of the commissura posterior may differ in the diverse Vertebrate forms.

An important reflex, experimentally investigated particularly in Mammals, and clinically studied in Man, is the *pupillary reflex on light*, which is presumably mediated by (diencephalic) pretectal grisea and seems to involve fibers passing through the commissura posterior.

The width of the pupil is controlled by two mutually antagonistic 'smooth' muscles, muscular sphincter pupillae and musculus dilatator pupillae. The latter is innervated by postganglionic sympathetic fibers originating in the superior cervical ganglion. The corresponding preganglionic neurons are located in the intermediolateral cell column of the spinal cord at about segments C VIII to Th II *('centrum ciliospinale')* and their neurites reach the sympathetic trunk by way of the so-called 'white' communicating rami. The musculus sphincter pupillae is innervated by cranial parasympathetic fibers originating in the ciliary ganglion. The preganglionic fibers of this channel pertain to the oculomotor nerve, being neurites of nerve cells located in the 'general visceral efferent' oculomotor nucleus *(Edinger-Westphal nucleus)*, to be dealt with again in section 9 of this chapter.

If one eye is exposed to light, a constriction of both pupils normally occurs, that of the pupil in the stimulated eye being the direct reaction, and that in the contralateral eye the consensual reaction. Constriction of the pupil is also associated with accommodation, which is performed by the 'smooth' musculus ciliaris, innervated by the oculomotor nerve. The preganglionic fibers of this channel presumably origi-

---

[29] The commissura tecti is, of course, missing in those parts of the tectum whose alar plate components become separated by the presence of a lamina epithelialis derived from the roof plate, as e.g. in Petromyzont Cyclostomes and some Osteichthyes with extensive valvula cerebelli.

nate in the above-mentioned *Edinger-Westphal nucleus*, and the postganglionic fibers are neurites of cells located in the ciliary ganglion. Accommodation is commonly combined with convergence, and it seems likely that, in Mammals and man, convergence of the eyebulbs, through undefined neuronal connections, may by itself cause or contribute to, pupillary constriction.

Dilation of the pupil in darkness or dimmed light seems to be mediated by poorly understood neural channels, some of which could be included in the tectobulbar and tectospinal system. It can be assumed that, upon illumination of the retina, these connections exert an inhibiting action upon the centrum ciliospinale.

Again, upon strong stimulation of diverse sensory system (particularly but not necessarily involving pain) a dilation of the pupil occurs, e.g. in the so-called *pupillary-skin reflex*. Several pathways may mediate this reaction. One of these channels seems to include bulbar and spinal lemniscus fibers reaching the tectum opticum and either suppressing inhibitory effects of tectospinal connections to ciliospinal center, or by way of stimulating bulbospinal respectively tectospinal systems, activating said center. Dilation of the pupil also occurs in states of emotion which imply cortical, thalamic, and hypothalamic grisea. This neural mechanism might, *inter alia*, be provided by a corticofugal path reaching the hypothalamus, from which 'iridodilator fibers' proceed to midbrain, bulbar, and spinal grisea (cf. BRAIN-WALTON, 1969). Figure 362 illustrates, in a simplified manner, some of the pathways mediating pupillary reactions.[30]

In summary, the midbrain with its suprasegmental tectal cortex and its grisea of 'nucleus motorius tegmenti' may be evaluated as the dominating subdivision of the brain in most Anamnia, particularly in those characterized by a well developed optic system. The tectum mesencephali correlates optic input, and all additional sensory input, that of nervus octavus being, to a large extent, transmitted to the tectum proper by way of the tori semicirculares. The processed input is then transmitted, through tectotegmental respectively tectobulbar and tectospinal channels, to the neuraxial output grisea. In contradistinction to the suprasegmental cerebellum, which is connected in parallel *(im Nebenschluss)* to the afferent and efferent systems and seems to have, figuratively speaking, an essentially 'regulating' or 'smoothening'

---

[30] Various clinical implications, including iridoplegia and the *Argyll-Robertson (or Romberg) pupil* shall be considered in section 9 of this chapter.

function as an 'auxiliary mechanism', the tectum mesencephali, together with STEINER's '*allgemeines Bewegungszentrum*', appears to represent a 'directing' center *(bestimmendes Zentrum*, K., 1927). It is thus easily understandable that decerebrate frogs, deprived of prosencephalon, are capable of catching flies (in some instances even more readily than normal frogs), provided their optic channel to the tectum has not been damaged by the operation. Likewise, such decerebration in Fishes does

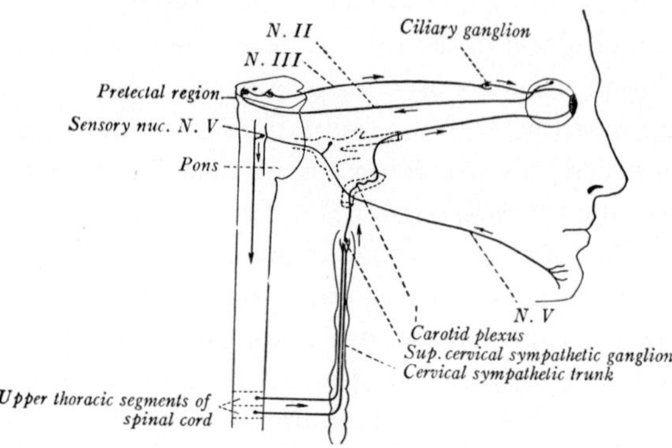

Figure 362. Simplified diagram illustrating some aspects of the pupillary reflex arcs in Man (from RANSON, 1943).

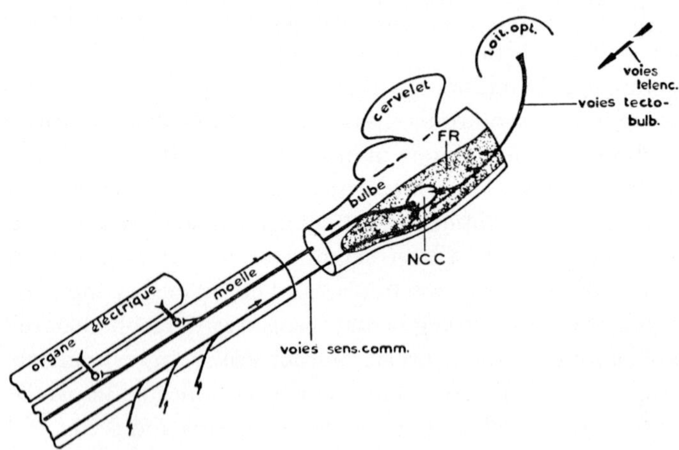

Figure 363. Diagram illustrating the general arrangement of the neuronal apparatus related to the discharge of electric organs, and '*commun à tous les poissons électriques*'. The diagram includes the relevant tectobulbar channel (from SZABO, 1961). FR: reticular formation; NCC: *noyau de la commande centrale.*

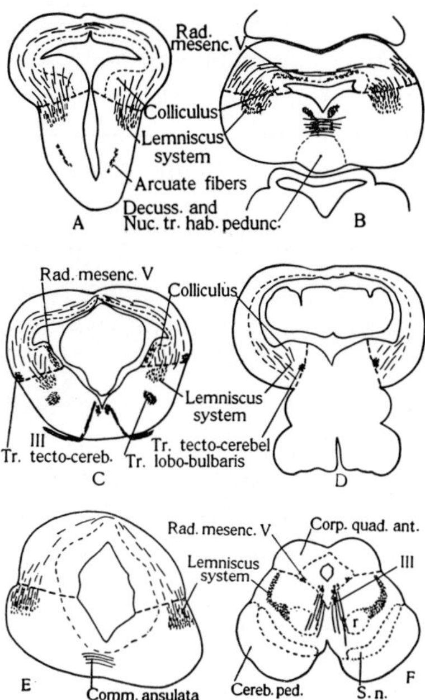

Figure 364. Outline drawings of cross-sections through the mesencephalon of various Vertebrates to illustrate diverse changes of form of the tectal region (from JOHNSTON, 1906). A: Petromyzont Cyclostome; B: Selachian; C: Ganoid (Acipenser); D: Teleost (note valvula cerebelli in mesencephalic ventricle); E: (Urodele) Amphibian; F: Man. 'Colliculus' in A–D stands here for torus semicircularis; Cereb. ped.: pes pedunculi; Nuc. tr. hab. pedunc.: nucleus interpeduncularis; r: nucleus ruber tegmenti; S. n.: substantia nigra.

not seem to be followed by easily detectable behavioral changes.[31] Again, in electric fishes, the tegmental '*noyau de commande centrale*' seems to be, in turn, controlled by the tectum mesencephali by way of the tectobulbar tract (Figs. 363, 380), although descending channels from the forebrain (telencephalon and diencephalon) might likewise have some sort of effect on the functioning of the electric discharge system.

---

[31] This, of course, does not exclude various behavioral changes or deficiencies detectable by more refined analysis. Evidently, the olfactory system and various other processing events performed by telencephalic and diencephalic grisea can be presumed, and have been shown, to influence behavioral activities. Moreover, the commonly obtaining predominance of olfactory input to the Anamniote telencephalon in no way precludes the presumption that other sensory input of all categories may reach telencephalic

In *Amniota* (such as Sauropsida), an increasing significance of the prosencephalic grisea becomes manifest,[32] representing a rostrad shifting *(Verschiebung nach vorn)* of functionally relevant neural mechanisms respectively connections. Complex reactions (e.g. 'instincts') become here far more noticeably correlated with the activities of telencephalic grisea. In Mammals, a most remarkable shift toward 'telencephalization' has occurred concomitantly with the progressive development of a '*neocortex*' accompanied by further differentiations of parahippocampal and hippocampal cortices. Generally speaking, in the course of Amniote evolution, the overall behavioristic significance of the mesencephalon tends more and more toward reduction to a, nevertheless still highly important, subordinated auxiliary respectively reflectory mechanism as well as region of passage for numerous channels. Figure 364 illustrates, as interpreted by JOHNSTON (1906), general aspects of mesencephalic configuration and fiber systems in some representative Vertebrates.

2. Cyclostomes

The *Petromyzont* mesencephalon is approximately delimited from the diencephalon by the external sulcus praetectalis of SAITO (1930) on the lateral surface, and by the region of commissura posterior on the dorsal surface. Its boundary against the rhombencephalon is provided by a conspicuous constriction at the isthmus region. Dorsally, this constriction becomes the groove between cerebellum and tectum. Between the caudal edge of posterior commissure, which also includes here a rostral component of commissura tecti, and the more caudal parts of tectum opticum, the mesencephalic roof is formed by an evaginated and folded lamina chorioidea derived from the here persistent roof plate (Figs. 365A, B; cf. also Fig. 305A). Figures 366A–F illus-

---

grisea of the lower Vertebrates by way of the basal forebrain bundles. Such input from ascending channels could be directly relayed through the diencephalon or, in part, indirectly by way of intercalated mesencephalic grisea. Thus, the shift toward 'telencephalization' displayed in the Vertebrate taxonomic series should be understood, figuratively speaking, as representing merely an *increase in emphasis* involving further differentiation of '*phylogenetically old*' channels.

[32] Yet, to some extents, as the experiments by STEINER and others have shown, decerebration is compatible with a 'fairly normal' behavior in Reptiles and, to a lesser degree, even in Birds.

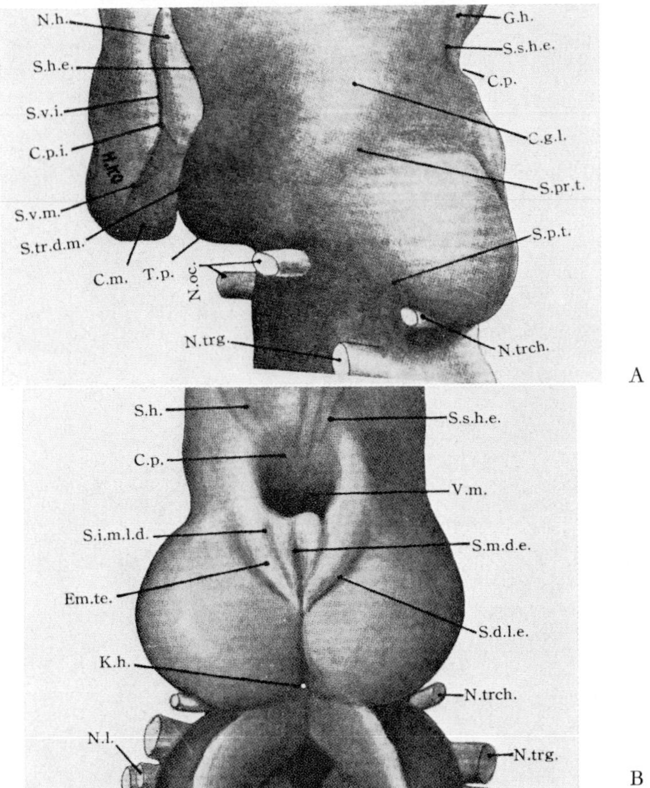

Figure 365A, B. External aspects of the mesencephalon in the Petromyzont Entosphenus in lateral (A) and in dorsal (B) view, as displayed by a wax model (from Saito, 1930). C.g.l.: prominence of lateral geniculate body; C.p.: commissura posterior; C.p.i.: commissura postinfundibularis; Em.te.: eminentia tecti (dorsalis); G.h.: Ganglion habenulae; K.h.: cerebellar plate (corpus cerebelli); N.l.: nervus lateralis; N.oc.: nervus oculomotorius; N.trig.: nervus trigeminus; N.troch.: nervus trochlearis; S.d.l.e.: 'sulcus dorsolateralis externus'; S.h.: sulcus habenularis; S.h.e.: 'sulcus hypothalamicus externus'; S.i.m.l.d.: 'sulcus intermediolateralis dorsalis'; S.m.d.e.: 'sulcus medianus dorsalis externus'; S.p.t.: 'sulcus posttectalis'; S.pr.t.: 'sulcus praetectalis'; S.s.h.e.: 'sulcus subhabenularis externus'; S.tr.d.m.: 'sulcus transversalis dorsalis mammillaris'; S.v.i.: 'sulcus ventralis infundibuli'; S.v.m.: 'sulcus ventralis mammillaris; V.m.: ventriculus mesencephalicus.

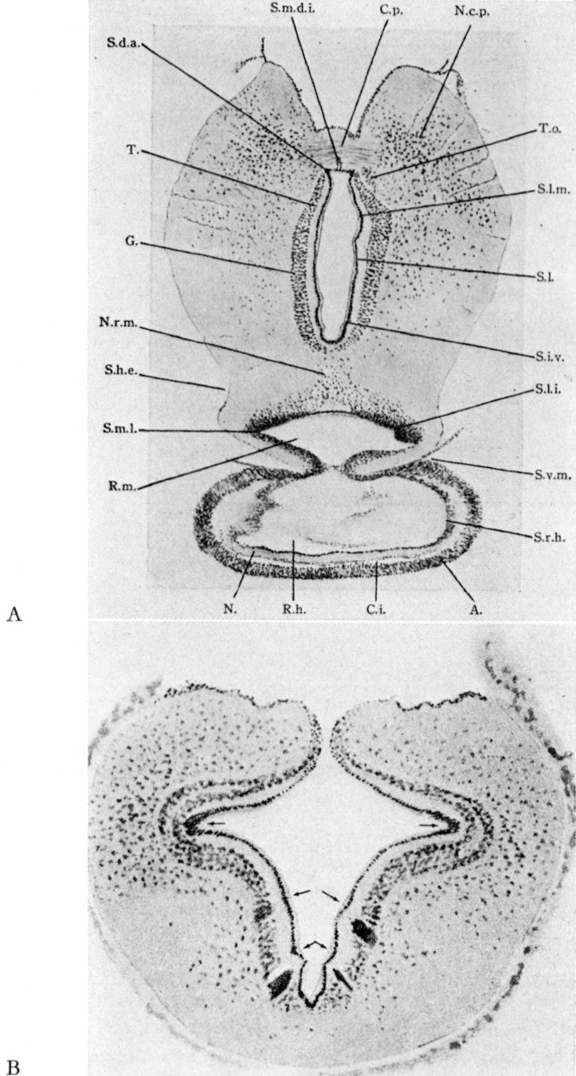

Figures 366 A–D. Cross sections *(Nissl stain)* through the mesencephalon of the Petromyzont Entosphenus at successive levels from commissura posterior to that of oculomotor root exit. For the trochlear nerve and for a sagittal section through the caudal tectum, Figures 305A and C should be consulted (B and C from Saito, 1928, A and D from Saito, 1930). The dorsal and ventral cells of the first '*Doppelpaar*' of *Müller cells* are shown in B, and the double pair can be recognized in C. Levels B and C correspond to the 'open' portion of ventriculus mesencephali, whose epithelial roof plate has been removed. Arrows indicate here relevant sulci. Abbr. for A.: A: 'adenohypophysis';

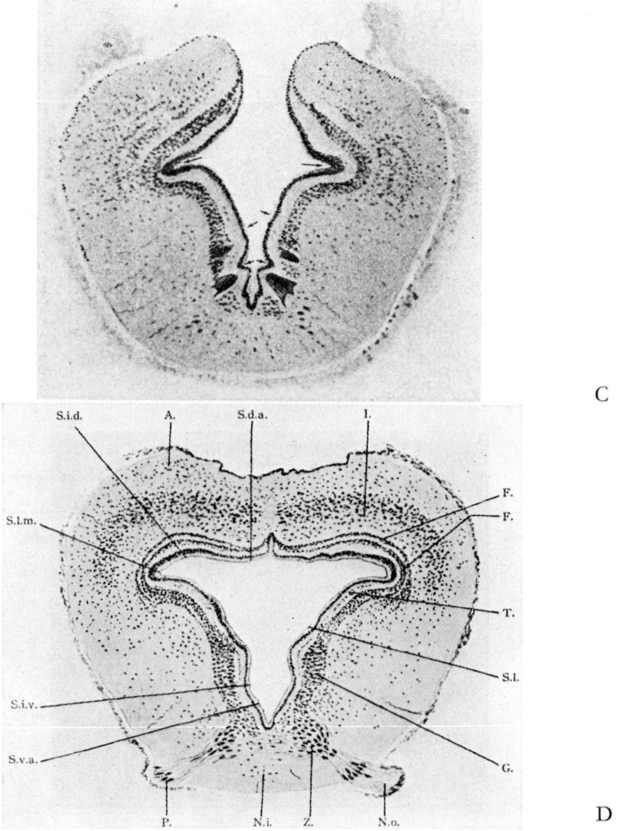

C

D

Ci.: 'interhypophysial cleft'; C.p.: commissura posterior; G.: rostral basal plate of mesencephalon (tegmental cell plate); N.: neurohypophysis; N.c.p.: nucleus commissurae posterioris; N.r.m.: cell group of tuberculum posterius (probably rostral end of deuterencephalic basal plate); R.h.: 'hypophysial recess' ('infundibular recess'); R.m.: 'mammillary recess'; S.d.a.: probably accessory lateral sulcus of posterior commissure; S.h.e.: 'sulcus hypothalamicus externus; S.i.v.: sulcus intermedius ventralis; S.l.: sulcus limitans; S.l.i.: 'sulcus lateralis infundibuli'; S.l.m.: sulcus lateralis mesencephali near fusion with sulcus synencephalicus; S.m.d.i.: 'sulcus medianus dorsalis internus'; S.m.l.: 'sulcus mammillaris lateralis' (probably part of S.l.i.); S.r.h.: 'sulcus recessus hypophysei; S.v.m.: 'sulcus ventralis mammillaris' (externus); T., T.o.: transition neighborhood of posterior thalamus dorsalis and mesencephalic alar plate. For nucleus lentiformis mesencephali cf. Fig. 189, p. 416 of volume 3, part II, depicting a section at a slightly different angle and in another Petromyzont species. Added arrows in Figures B and C indicate sulcus lateralis mesencephali (dorsally), sulcus limitans, and sulcus intermedioventralis (ventrally). Abbreviations for D.: A: stratum opticum; F.: granular layers of tectum opticum; G: dorsal ('intermediodorsal') portion of basal plate; I: layer of chief cells

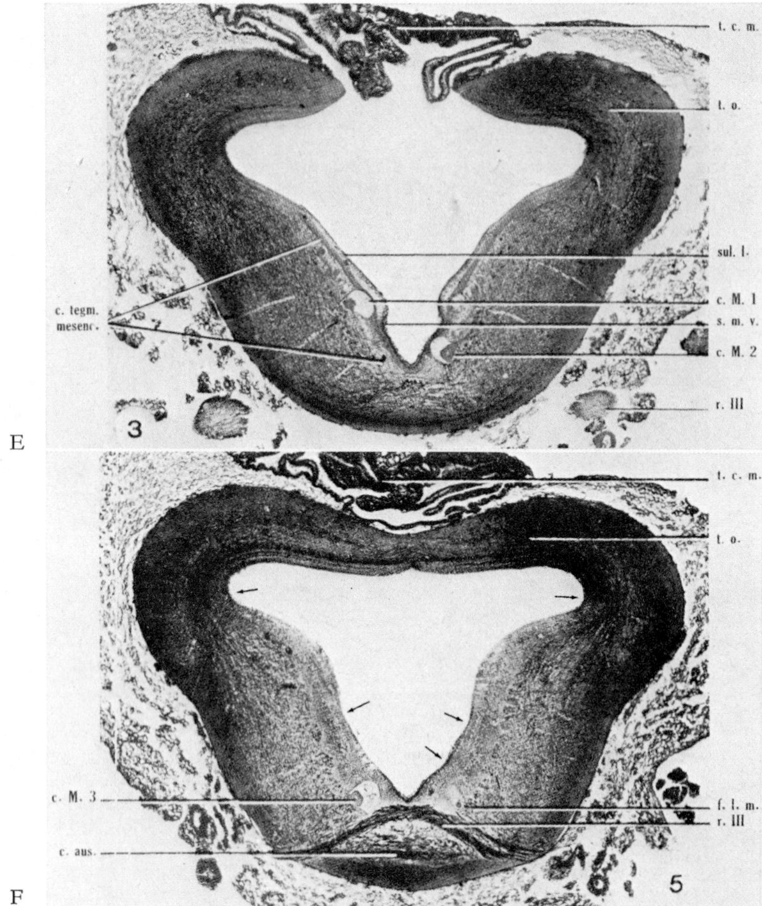

Figure 366E, F. Cross-sections through the mesencephalon of Petromyzon marinus *(Cajal silver impregnation)* at the level of 'open' mesencephalic ventricle (E) and at that of oculomotor root (F). Section E displays two *Müllerian cells* (c. M. 1, c. M. 2) of SAITO's first '*Doppelpaar*', and section F a *Müllerian cell* (c. M. 3) of SAITO's second pair (from STEFANELLI, 1934). c. ans.: commissura ansulata; c. tegm. mesenc.: '*centro tegmentale mesencefalico;* c. M. 1, 2, 3: *Müller cells* in STEFANELLI's enumeration; f. l. m.: fasciculus longitudinalis medialis; r. III: oculomotor root; s. m. v.: '*solco medioventrale*' (s. intermedius ventralis of Saito); sul. l.: sulcus limitans; t. c. m.: tela chorioidea mesencephali (respectively roof plate); t. o.: tectum opticum. Added arrows as in C.

(partly within this layer and partly between it and the outer granular layer lies the deep 'medullary' layer); N. i.: Nucleus interpeduncularis; N. o.: oculomotor root; P.: peripheral oculomotor nucleus; S. d. a.: 'sulcus dorsalis accessorius'; S. i. d.: sulcus intermedioventralis; S. l.: sulcus limitans; S. l. m.: sulcus lateralis mesencephali; S. v. a.: 'sulcus ventralis accessorius'; T: torus semicircularis; Z.: central oculomotor nucleus.

trate relevant aspects of the Petromyzont mesencephalon as seen in cross-sections.[33]

The midbrain ventricle displays dorsally a prominent sulcus lateralis mesencephali (internus), approximately delimiting the *tectum opticum* from the *torus semicircularis*. The torus, in turn, is roughly delimited from the basal plate by the sulcus limitans.[34] A sulcus intermediodorsalis does not seem to be displayed. Within the basal plate, a sulcus intermedioventralis can usually be recognized, and a sulcus ventralis accessorius is frequently present.

The dorsal portion of the alar plate's dorsal zone is provided by the *tectum opticum* in which the main layers dealt with in the preceding section (cf. Fig. 361B) are discernible. An external stratum opticum with nondescript substrata is formed by optic tract fibers and their terminal arborizations. It also contains scattered nerve cells. The layer of chief cells is rather thick. Some of its elements are oriented with their long axes at a right angle to the surface and presumably represent crozier cells, but, as far as I know, sufficiently complete and convincing Golgi impregnations have not been obtained. The stratum opticum overlaps with this layer, which Heier (1948) designates as 'stratum opticum', the outer portion of the optic layer being Heier's 'stratum album et griseum superficiale'. The deep medulla with its efferent and afferent channels also contains scattered neuronal (and perhaps glial) elements. Internally to the deep medulla, one or two closely packed layers of granule cells are adjacent to the ependyma. Narrow cell-free 'molecular' or 'plexiform' layers are commonly intercalated between the two granular layers, respectively the inner one and the ependyma. How-

---

[33] The various authors whose investigations have made significant contributions to the anatomy of the Petromyzont brain were cited in section 3 of chapter IX and listed in that chapter's bibliography. In order to avoid undue repetitions, additional references to these pioneer workers shall be made in the present section only as regards specific topics. *Mutatis mutandis*, this restriction applies to the other sections of this chapter, for which, again, the bibliography to chapter IX may be consulted.

[34] In contradistinction to Saito and myself, a recent author (Nieuwenhuys, J. comp. Neurol. 145, 1972, p. 176) states that he could not trace the sulcus limitans beyond the isthmus rhombencephali. Yet, the sulcus limitans was clearly depicted by Stefanelli (1933, cf. Figs. 366E, F of the present text), by Heier (1948), and by Schober (1964, cf. the photomicrograph Fig. 25 of that author's paper). It should be added, however, that in some instances, as likewise documented by Schober's photomicrographs, the mesencephalic portion of sulcus limitans may become flattened out. Nevertheless, as Schober's detailed illustrations clearly show, a definite cytoarchitectural distinction between alar and basal plate cell groups remains quite conspicuous.

ever, these various layers may be rather indistinct, with considerable overlap and, *qua* granular layers, some degree of fusion (cf. Figs. 366A–F). Even in one and the same specimen, the distinctiveness of the tectal layer frequently displays variability at the diverse cross-sectional levels. Rostralward, the layer of chief cells and the neuronal elements of the adjacent strata merge into an ill-defined nucleus lentiformis mesencephali pertaining to the pretectal cell group, and located laterally to the diencephalic nucleus commissurae posterioris (cf. Fig. 189 of chapter V, p. 416 of vol. 3, part II). This is the 'primordial pretectal nucleus' of KAPPERS *et al.* (1936). It is evident that, at the level of commissura posterior, that is, in the mesencephalo-diencephalic boundary zone, parts of the deuterencephalic and of the diencephalic zonal systems become, as it were, 'telescoped' into each other.

Along a portion of the ventricular surface of the posterior commissure a peculiar differentiation of the epithelial roof forms the *subcommissural organ*, characterized by a columnar epithelium differing from that of the general ependymal lining.[35] Although the 'organ' extends caudalward to what may be considered rostral mesencephalic levels, it seems to derive from the synencephalic roof plate of the caudalmost diencephalic neighborhood and can be evaluated as a *grenzgebiets* structure. In Petromyzonts, this organ is commonly paired, each antimere again consisting of a prominent lateral and a medial less prominent ridge (*'Leiste'*, ADAM, 1956), interconnected, in the midline, by lower roof plate epithelium (cf. Fig. 189 of chapter V, vol. 3, part II).

A nucleus of the mesencephalic trigeminal root has not been recorded in the tectum mesencephali of Petromyzonts, or for that matter, of Cyclostomes in general. The lack of this nucleus in these jawless Vertebrates (Agnatha) may be correlated with the lack of the corresponding musculature, whose proprioceptive innervation is provided by the radix mesencephalica trigemini.[36]

[35] The subcommissural organ, pertaining to the group of 'paraependymal and ependymal circumventricular organs' respectively structures, was dealt with in section 5, chapter V of volume 3, part I, pp. 354–367 *et passim*. Further comments can be found on p. 139 and 404 of volume 3, part II, and, *qua Reissner's fiber*, in chapter VIII, section 3, p. 66 of the present volume.

[36] KAPPERS (1947) justly comments: *'Il est important de noter que la racine mésencéphalique du trijumeau, racine proprioceptive par excellence, présente chez tous les autres Vertébrés, manque chez les Cyclostomes. Ce fait s'explique probablement par la fonction tout-à-fait différente de la musculature trigéminale de ces animaux'*. It should here also be recalled that, at least in some Gnathostome Vertebrates, the trigeminal mesencephalic root might also be related to proprioceptive 'sensibility' of non-trigeminal musculature innervated by other cranial nerves.

The *torus semicircularis*, slightly bulging into the ventricle, displays a dense inner layer, dorsally continuous with the granular layers of the tectum, and a nondescript external layer containing more scattered nerve cells intermingled with the afferent, efferent and intrinsic fiber connections. Rostralward, the cell groups of torus semicircularis and of tectum opticum tend to merge, and become caudally adjacent to the above-mentioned nucleus lentiformis mesencephali as well as to the nucleus commissurae posterioris. A typical nucleus isthmi and a nucleus visceralis secundarius cannot be clearly recognized, but neuronal groups representing these nuclei in Gnathostome Anamnia may be included in caudobasal portions of the torus semicircularis.

The basal plate of the mesencephalon contains in its ventral zone *the oculomotor nucleus* which is subdivided into a central and into a peripheral portion (SAITO, 1930, cf. Fig. 366D).[37] The *nucleus of the nervus trochlearis*, located in the isthmus region, lies in close apposition to cerebellar structures, and its proper identification presents considerable difficulties. According to KAPPERS and other authors, it may have been 'displaced' into the alar plate, but according to SAITO (1930), with whom I am inclined to agree, this nucleus seems to be represented by a small cell group within the intermedioventral zone of the isthmic region.[38] Even as regards the oculomotor nerve, some poorly elucidated questions remain. The innervation of the m. rectus inferior is said to be provided by a rhombencephalic cell group located rostrally to the motor trigeminal group. According to KAPPERS (1947) who also quotes ADDENS and STEFANELLI, '*ces fibres radiculaires s'associent à celles de l'abducens qui naît d'un noyau situé au niveau posterieur du noyau moteur trijumeau et dont les fibres radiculaires sortent au côté medial de celles du trijumeau*'.

The mesencephalic *basal plate* in general contains the rostral portion of the nucleus motorius s. reticularis tegmenti with three *Müller cells* ('*erstes Doppelpaar*' and '*zweites Paar*' of SAITO, 1928). The first cell is located between sulcus limitans and sulcus intermedioventralis, the two others (1′ and 2) ventrally to that sulcus, within the ventral zone.[39] Whether an actual, functionally 'visceral' griseum obtains within the

---

[37] HEIER (1948) distinguishes, in addition, a nucleus intermedius of the oculomotor complex. Root fibers of the central portion are said to decussate and to emerge on the contralateral side (chiasma oculomotorii, HEIER, 1948).

[38] This, of course, might also indicate a 'dorsal migration'. Cf. the comments on p. 669 and footnotes 36, 37, and 38 of chapter X. The trochlearis problem is also discussed by HEIER (1948) on pp. 178–183 loc. cit.

[39] The *Müller cells* were dealt with in section 3 of chapter IX.

intermedioventral zone remains an open question. The rostral end of the mesencephalic nucleus reticularis tegmenti is located as a 'tegmental cell plate or cell cord' in the mesencephalo-diencephalic boundary zone, protruding, as a neighborhood of tuberculum posterius, into the caudal portion of the third ventricle. A nucleus opticus tegmenti has not been identified in Petromyzon, but it is not impossible that some small cell group within the 'reticular' tegmentum may represent, as it were, a 'forerunner' of this nucleus. Much the same could be said about the nucleus interstitialis of CAJAL. Its 'precursor' seem to be neuronal elements in the rostral tegmentum, whose neurites presumably join the fasciculus longitudinalis medialis.

An additional nucleus of the basal plate is the unpaired *ganglion interpedunculare (s. nucleus interpeduncularis)* located in the basal midline. Its configuration is frequently most conspicuous in the intermediate planes between rostralmost and caudalmost mesencephalic levels, but the nucleus is still clearly recognizable in the isthmic region toward the mesencephalo-rhombencephalic boundary neighborhood. It consists of scattered, rather small stellate or bipolar nerve cells imbedded in a neuropil. The terminal fibers of the habenulo-interpeduncular tract (fasciculus retroflexus) which represents its main input channel apparently run through the nucleus in spirals crisscrossing the midline, as described by HERRICK (1948) in Amblystoma and depicted in Figure 386 E (cf. also HEIER, 1948).

The fiber systems of the Vertebrate mesencephalon in general, and thus that of the Petromyzont one in particular, could be classified as *(1) input channels*, *(2) output channels*, *(3) intrinsic connections*, and *(4) channels of passage*. In Petromyzonts, the input or afferent systems (1) include (a) optic tract fibers to tectum and possibly a few basal ones to interpeduncular nucleus or to nucleus motorius tegmenti; (b) the predominantly but presumably not entirely 'somatic' general spinal and bulbar lemniscus, likewise reaching the tectum, and (c) the vestibulolateral lemniscus, mainly reaching the torus semicircularis. The so-called tractus octomotorii of HEIER (1948) can be considered part of the vestibulolateral lemniscus, with collaterals to the nucleus reticularis tegmenti. All these systems are essentially contralateral, having crossed in the various decussations. Other input channels are (d) habenulotectal, as well as possibly thalamotectal and hypothalamotectal connections, (e) the fasciculus retroflexus as mentioned above, moreover (f) presumed cerebellotectal fibers, and poorly identified descending input (g) such as thalamic and hypothalamic crossed and uncrossed fibers reaching the tegmentum. A mammillotegmental tract, in part connecting with the interpeduncular nucleus may also be present. It is doubtful but not impossible that some telencephalic output, by way of the nondescript basal forebrain bundle, may end in the mesencephalic tegmentum.

In addition, the mesencephalic tegmentum presumably also receives collaterals from the lemniscus channels, moreover ascending components of the fasciculus longitudinalis medialis system, interconnecting bulbar tegmental or vestibular cell groups with mesencephalic tegmental ones.

As regards (2) the mesencephalic output channels, the efferent systems of the tectum (a), running through the 'deep medullary stratum'[40] comprise several subsystems such as tectobulbar and tectospinal ones, which may even, in part, share the same neurites giving off bulbar collaterals and then extending into the spinal cord. This channel seems predominantly to cross in the *commissura ansulata* (tractus tectobulbaris etc. cruciatus) but may also include homolateral fibers (tractus tectobulbaris etc. rectus). To this channel pertain also the tectotegmental fibers,[41] of which those ending in the mesencephalic tegmentum can be subsumed under the intrinsic connections (3). Nothing certain is known about inferred or suspected tectocerebellar channels which, in the wider sense of 'tectum', may include fibers from torus semicircularis. Other suspected tectal output systems are tectohabenular, tectothalamic, and tectohypothalamic respectively 'tectolobar' and tectomammillary connections.

Among (b) the efferent channels of the tegmentum, the fasciculus longitudinalis medialis represents the main central motor tract, which includes neurites of the *Müller cells*. Besides these giant fibers, discussed in the pertinent sections of chapters VIII and IX, crossed or uncrossed neurites from smaller mesencephalic tegmental cells, perhaps also from the diencephalic nucleus commissurae posterioris, seem to

---

[40] It should be recalled that medullated nerve fibers in the generally accepted sense do not seem to occur in Cyclostomes.

[41] One could, strictly speaking, define tectotegmental fibers as e.g. connecting with nucleus reticularis tegmenti, and tectobulbar respectively tectospinal fibers as those reaching primary efferent cell groups including preganglionic ones (which, in a rigorous terminology, are, although giving origin to root fibers, not 'primary' since they do not, with exception of those innervating true chromaffine cells, terminate on the effectors). Yet, one and the same tectofugal fiber might have collateral endings in reticular formation and in primary efferent 'nuclei'.

join the fasciculus longitudinalis medialis. In addition to this fasciculus, crossed and uncrossed tegmentothalamic and tegmentohypothalamic connections may be present.

Except for some of the short comments given above on the intrinsic connections (3) no reliable data concerning their specific or significant arrangement in Cyclostomes are extant. This remark also applies (4) to the channels of passage, which play a substantial role e.g. in the Mammalian brain. At most, it could be said that those fibers of the general ascending lemniscus which may reach diencephalic structures might be subsumed under this category. Conversely, descending systems of the prosencephalon, if reaching rhombencephalic levels, could also here be included.[42]

As regards the mesencephalic *commissural systems*, the commissura tecti interconnects the antimeric portions of tectum *sensu strictiori* and of torus. It consists of a caudal portion (commissura posttectalis of HEIER) and of a rostral portion, which adjoins the commissura posterior. Both parts of the tectal commissure are separated by the protruding lamina epithelialis of the mesencephalic ventricle. The components of this commissure are insufficiently identified, but may include both true antimeric interconnections, and decussations of ascending or descending systems. Much the same may be said about the posterior commissure at the mesencephalo-diencephalic boundary, whose fibers are predominantly related to diencephalic cell groups but include fibers related to the mesencephalon. Some neurites originating in the nucleus commissurae posterioris join that commissure, while other neurites from said nucleus decussate in the basal commissura ansulata which represents the ventral or tegmental commissural system whose general composition respectively fiber tracts were repeatedly dealt with above in the discussion of the main mesencephalic communication channels.

Turning now to the *Myxinoid* mesencephalon, the peculiar configurational characteristics of the brain displayed in that Cyclostome order should be recalled (cf. section 3, chapter IX). Figures 193A, B, and 194 will indicate the overall mesencephalic boundaries, some of

---

[42] As regards the mesencephalic fiber systems of Cyclostomes, both KAPPERS *et al.* (1936) and HEIER (1948) give rather detailed summaries based on a different classification. Yet, because of the rather flimsy available evidence, whose unreliability is intrinsic to the limitations of the applicable technical methods, the descriptions given by these authors, with whom I essentially agree, remain, of necessity, no less generalized and vague than my own summary.

which are difficult to define because many of the typical landmarks are blurred respectively distorted. Nevertheless, the rostral boundary zone is indicated by a vaguely defined region caudally to the diencephalic habenular bodies, and corresponding to the commissura posterior, while the caudal boundary of the midbrain is marked by the deep isthmic fissure. The dorsocaudal extremity of mesencephalic tectum, however, includes the rudimentary corpus cerebelli which is reduced to a 'cerebellar commissure'.[43]

Figures 367A–D illustrate cross-sections through the mesencephalon of Myxine, displaying the narrow (aqueduct-like) mesencephalic ventricle. Figures 195A and B of chapter IX depict still more caudal levels, corresponding to isthmus region respectively rhombomesencephalic boundary zone.

Although rudimentary eyes and optic nerves are present in Myxinoids (HOLMGREN, 1919; JANSEN, 1930) the optic input channel can be regarded as structurally and functionally negligible. Thus, the tectum mesencephali does not here represent a typical tectum opticum. The alar plate contains merely a rather diffuse, nondescript cell mass constituting a tectum *sensu strictiori* fused with the torus semicircularis and associated structures. The input of this cell population is provided by an ill-defined ascending lemniscus system as well as by descending channels from the prosencephalic region. Many coarse fibers decussate in what JANSEN (1930) and others interpret as commissura posterior. It is, however, not unlikely that the cited author's commissura posterior includes fibers of the tectal commissure, the two commissures being 'telescoped' into each other.[44] The commissura tecti mesencephali of JANSEN (Fig. 195A) may be regarded as the caudal portion of the entire tectal commissural system, which is, in turn, continuous with the 'cerebellar commissure'.

Again, the rostral cell populations of the tectum mesencephali cannot be sufficiently well delimited from the caudal dorsal thalamic (posthabenular) cell masses. This ill-defined neighborhood may represent a modified pretectal region. JANSEN, moreover, delimited a vaguely outlined nucleus commissurae posterioris lateral to the ventricular lumen, and related to the commissural fibers (Fig. 367D).

---

[43] In this particular respect, however, similar conditions obtain in Gymnophiona and even in Urodeles such as Proteus (cf. chapter X).

[44] The ependymal lining of the narrow mesencephalic ventricle at these levels represents the subcommissural organ (ADAM, 1956). This, of course, tends to support JANSEN's interpretation of commissura posterior.

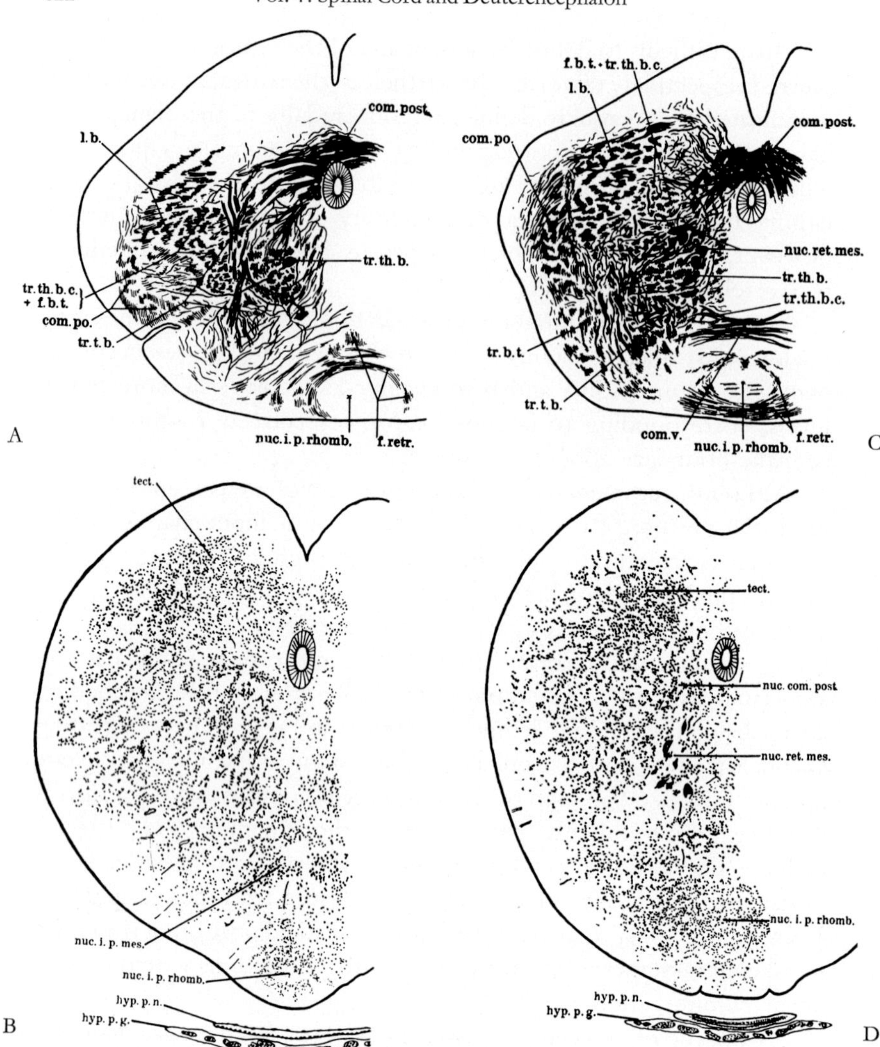

Figures 367 A–D. Cross-sections of two levels, in rostro-caudal sequence, through the mesencephalon of Myxine glutinosa. Level A approximately corresponds to that of B, and C to that of D. A and C *Bielschowsky silver impregnations*, B and D modified *Nissl stain* (from JANSEN, 1930). com. po.: fibers of commissura postoptica; com. post.: commissura posterior; com. v.: commissura ventralis (sive ansulata); f. b. t.: 'fasciculus basalis telencephali' (fibers of basal forebrain bundle); f. retr.: fasciculus retroflexus; hyp. p. g.: adenohypophysis; hyp. p. n.: neurohypophysis; l. b.: lemniscus bulbaris; nuc. com. post.: nucleus commissurae posterioris; n. i. p.: nucleus interpeduncularis (mes.: 'pars mesencephalica, rhomb.: 'pars rhombencephalica'); tect.: tectum mesencephali; tr. b. t.: tractus bulbotectalis; tr. t. b.: tractus tectobulbaris; tr. th. b.: tractus thalamobulbaris (c.: cruciatus).

In the basal plate, the nuclei of the eye muscle nerves III and IV are not present, but within the diffuse tegmental cell masses, EDINGER (1906) and JANSEN (1930) depicted large cells of the nucleus reticularis tegmenti mesencephali.[45] In addition, a well recognizable nucleus interpeduncularis is present. JANSEN distinguished a dorsal pars mesencephalica from a ventral and caudalward more extensive pars rhombencephalica. It seems, however, more likely that both parts are mesencephalic.

The analysis of the fiber tracts in the Myxinoid midbrain has presented considerable difficulties. The various input and output channels are here inextricably intermingled with the intrinsic mesencephalic fiber systems and commissures.

Afferent channels are represented by a general spino-bulbo-tectal lemniscus containing fibers of the diverse functional categories and presumably to some extent also rostralward proceeding into the diencephalon. It is likely that these ascending systems, either through collaterals or through separate channels, also provide input for the mesencephalic tegmentum. Descending input is probably included in the uncrossed and crossed thalamobulbar and hypothalamobulbar tracts which may give off collaterals to midbrain cell groups. Direct hypothalamotectal and thalamotectal fibers, these latter passing through the pretectal region, may also be present. The nucleus interpeduncularis receives crossed and perhaps also uncrossed fibers from the fasciculus retroflexus, which may also effect some connections with other adjacent or more caudal regions.

The *mesencephalic tectal output* seems represented by channels which JANSEN (1930) described as tractus tectobulbaris et spinalis, and tractus tectothalamicus, whose fibers join the rostralward proceeding portion of the general lemniscus. The fasciculus longitudinalis medialis provides the main *tegmental output channel*,[46] and extends throughout the entire length of the spinal cord. Ascending connections of both tegmentum and tectum with hypothalamus as well as other diencephalic cell masses can be presumed to obtain.

The mesencephalic commissural systems include the basal commissura ansulata (commissura tuberculi posterioris of JANSEN, 1930), and

---

[45] The disputed question whether these cells should or should not be evaluated as *Müller cells* was briefly dealt with in section 3 of chapter IX.

[46] It should again be recalled that the fasciculus longitudinalis medialis, despite its significance as a highly important descending channel, also contains ascending components.

the dorsal commissura posterior together with the commissura tecti mesencephali, as already pointed out above on p. 824.

With respect to the different views on dubious question concerning the interpretation of the peculiar configurational aspects displayed by the Myxinoid brain, including the delimitation of its main subdivisions, further comments and references can be found in various chapters of a treatise on the biology of Myxine (BRODAL and FÄNGE, 1963) and in volume 2 of the handbook by KAPPERS *et al.* (1936).

3. Selachians

The mesencephalon of Plagiostomes, although manifesting a general configurational pattern essentially identical with that obtaining in Petromyzont Cyclostomes, does *not* display a thin roof plate with folded laminal chorioidea within a median portion of the tectum, which thus remains here entirely massive. In various Plagiostomes, the corpus cerebelli may completely cover the tectum mesencephali,[47] which is commonly divided by a midsagittal groove of various depth into antimeric 'corpora bigemina', as is also to some extent the case in Cyclostomes. A sulcus lateralis mesencephali externus may approximately delimit the tectum proper from the more basal midbrain portions. It was designated as sulcus tecto-tegmentalis by KAPPERS *et al.* (1936). However, this somewhat shallow and variable surface groove, depending on the development of tectum opticum, may either approximately correspond to a boundary between tectum proper and lateral aspects of the torus semicircularis zone, or to a boundary between this latter and the true tegmentum, i.e. between alar and basal plate zones. Only in this latter case would it represent a true sulcus tecto-tegmentalis.

Figure 368 illustrates some aspects of the Selachian midbrain in a sagittal section, and Figures 369 to 371A depict transverse sections at different levels. It will be seen that the approximate boundary of alar plate and basal plate (tegmentum) is more or less distinctly indicated by the sulcus limitans. The torus semicircularis, slightly protruding toward the ventricular lumen, is separated from the tectum mesencephali *sensu strictiori* (or tectum opticum) by the sulcus lateralis mesencephali (internus). Where the ventricular surface of the tectum opticum reaches its farthest lateral extension, an accessory sulcus related to the fold-

---

[47] Cf. Figures 306A, 306D, 308A, and Figure 246 on p.482 of volume 3, part II.

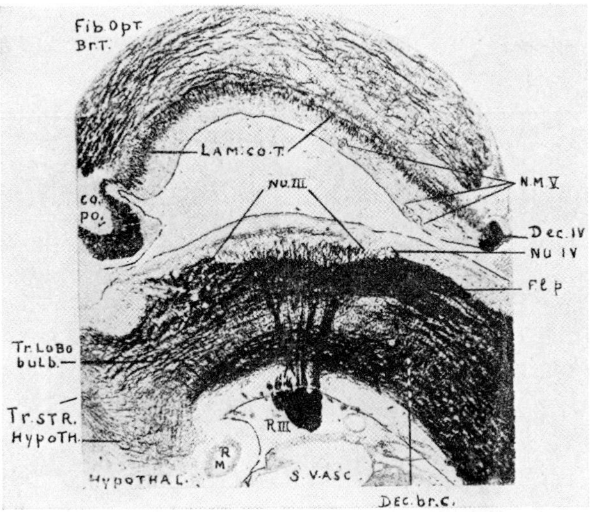

Figure 368. Sagittal section (myelin stain) through the mesencephalon of the Plagiostome Scyllium (from Kappers, 1947). Co. po.: posterior commissure; Dec. br. c.: decussation of superior cerebellar peduncle (tractus cerebellotegmentalis et cerebellolobaris); Dec. IV: decussation of trochlear root; Fib. opt. Br. T: optic tract ('brachium tecti'); F. l. p.: fasciculus longitudinalis medialis; Lam. Co. T.: commissura tecti; N. M. V.: nucleus radicis mesencephalicae trigemini; Nuc. III, IV: oculomotor and trochlear nuclei; R. III: oculomotor root; R. M.: 'mammillary recess'; S. vasc.: saccus vasculosus; T. lobobulb.: 'tractus lobobulbaris'; tr. Str. Hypoth.: 'tractus strio-hypothalamicus' (hypothalamic component of basal forebrain bundle.

ing of the tectum *(Ventrikelumschlagsfurche)* may be displayed and represents a sulcus lateralis mesencephali (internus) accessorius. Although pertaining to the 'system' of sulcus lateralis mesencephali, this accessory groove does not possess a morphologic significance comparable with that of the main sulcus, which suggests a relevant boundary between *formbestandteile* of the *bauplan*, namely between tectum opticum and torus.

With respect to the cross-sectional levels, the mesencephalo-diencephalic boundary zone, a topologically open neighborhood, is located in the region of commissura posterior,[48] where mesencephalic and diencephalic configurations overlap and may even be conspicuous-

[48] This boundary region includes the *subcommissural organ* repeatedly dealt with elsewhere in this treatise.

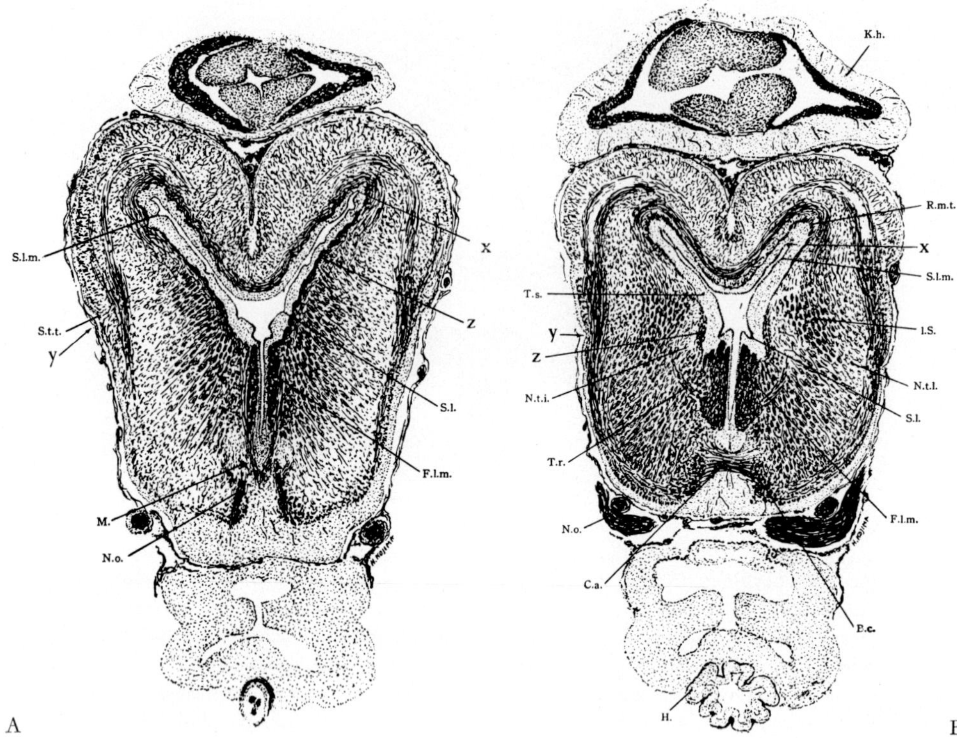

Figure 369 A, B. Cross-sections (myelin stain) through the mesencephalon of the Plagiostome Acanthias mitsukurii, A being somewhat rostral to B (from SAITO, 1930). B.c.: 'brachium conjunctivum; C.a.: commissura ansulata; F.l.m.: fasciculus longitudinalis medialis; H.: saccus vasculosus; K.h.: cerebellum; l.s.: lateral lemniscus (fasciculus longitudinalis lateralis, lemniscus octavolateralis, tractus octavomesencephalicus); M: large cell of formatio reticularis mesencephali; N.o.: root fibers of oculomotor nerve; N.t.i.: nucleus tegmentalis medialis; N.t.l.: nucleus tegmentalis lateralis; R.m.t.: radix mesencephalica trigemini; S.l.: sulcus limitans; S.l.m.: sulcus lateralis mesencephali (internus); T.r.: fiber system joining fasciculus longitudinalis medialis; T.s.: torus semicircularis; tt: tectotegmental, tectobulbar, and bulbotectal fiber systems; x: sulcus lateralis mesencephali (internus) accessorius; y: sulcus lateralis mesencephali externus; z: deep tectotegmental fibers. Nucleus interpeduncularis (ventral to commissura ansulata), and the basally adjacent lobi inferiores hypothalami with part of hypophysis (in A) remain unlabelled but can easily be identified.

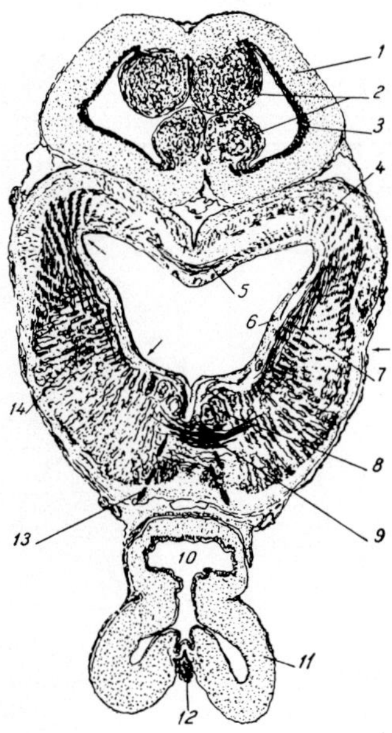

Figure 369C. Cross-section (myelin stain) through the mesencephalon of Acanthias vulgaris at the level of oculomotor root exit (modified after KAPPERS, from BECCARI, 1943). 1, 2, 3: molecular layer, granular eminences, and *Purkinje cell* layer of cerebellum; 4: tectum opticum; 5: commissura tecti; 6: torus semicircularis; 7: tractus tectobulbaris; 8: fasciculus longitudinalis medialis; 9: dorsal portion of commissura ansulata; 10: recessus posterior ('mammillaris') of hypothalamus; 11: lobi inferiores hypothalami; 12: hypophysis; 13: oculomotor root; 14: 'tractus spinotectalis et mesencephalicus' (includes here the diverse bulbo-mesencephalic channels). Arrows point to sulcus lateralis mesencephali internus et externus, and to sulcus limitans.

ly telescoped into each other, as e.g. in Chimaeroids (K. and NIIMI, 1969). A comparable transitional zone, representing the caudal boundary of the mesencephalon, obtains at the isthmus rhombencephali (cf. Fig. 309D).

The tectum opticum, a derivative of the alar plate's dorsal zone, displays a stratification essentially corresponding to that found in other Vertebrates with a well developed mesencephalic roof. Figures 371A to D illustrate this arrangement as interpreted, with slight differences in emphasis, by GERLACH (1947), KAPPERS (1947) and myself (1927). In a simplified manner, it seems appropriate to distinguish an outer layer of

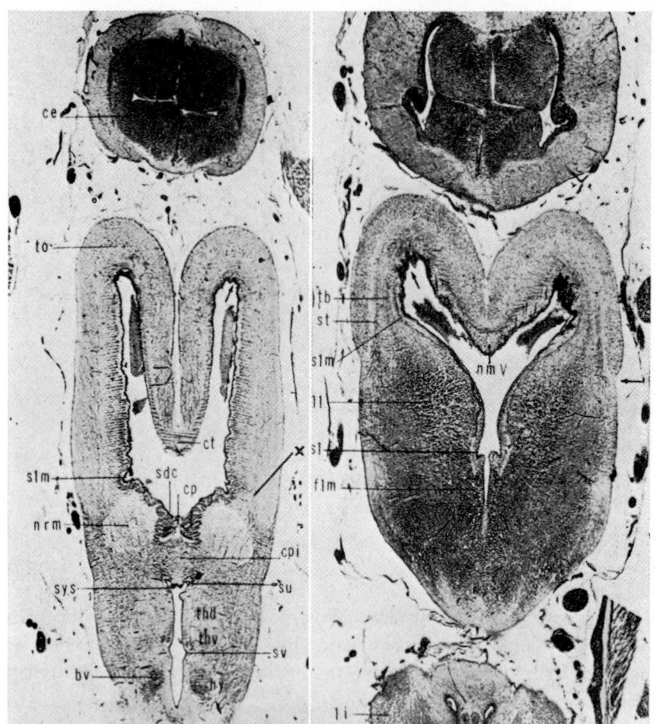
ce
to
tb
st
slm
nmV
ll
ct
sl
x
slm
sdc
cp
flm
nrm
cpi
sys
su
thd
thv
sv
bv
hy
li

A B

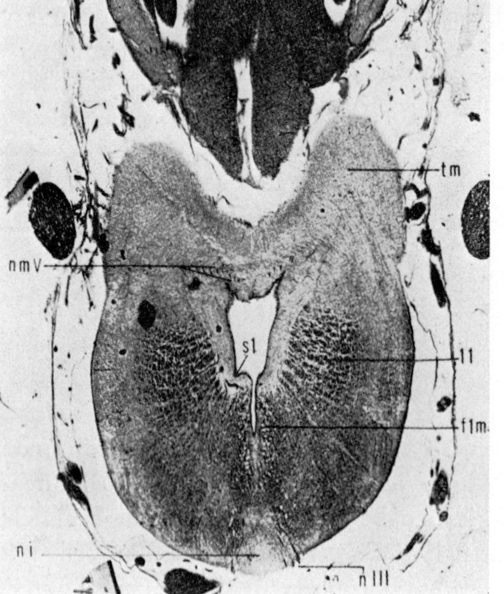
tm
nmV
sl
ll
flm
ni
nIII

C

optic fibers[49] with predominantly sagittal course (cf. Fig. 368), a layer of chief cells, a lamina medullaris interna with essentially transverse course,[50] and an internal cellular layer adjacent to the ventricular lining.[51]

The *nucleus radicis mesencephalicae trigemini* (nucleus magnocellularis tecti of GERLACH, 1947) is commonly constituted by a medial single row of large 'vesicular' cells[52] near the ventricular lumen on both sides of the midline, and extending from the isthmus level as far rostrad as the region of commissura posterior. Depending on taxonomic forms, and even individual variations in one and the same species, the nucleus radicis mesencephalicae trigemini may display, *qua* distribution of its cells, a number of minor differences within the given overall pattern. The fibers of the mesencephalic trigeminal root usually gather in sizeable bundles ventrally to the trochlear decussation.

---

[49] The *stratum opticum*, whose outer sublayer may appear as a 'molecular layer', usually contains various sorts of small neuronal elements. Its inner sublayer, with the terminals of optic tract fibers, merges into, or overlaps, the layer of chief cells.

[50] The *deep medulla*, containing efferent and afferent fibers of tectum opticum, may split into variable sublayers differing in the diverse taxonomic forms, and even from level to level in a given specimen (cf. e.g. Figs. 369A–C, 370A, B).

[51] A 'molecular' or 'plexiform' layer may separate the internal cellular (or 'granular') layer from the ependyma (cf. Fig. 371C).

[52] In older specimens, these cells may display paraphytes (cf. vol. 3, part I, pp. 79–88). KAPPERS (1947) refers to apparent fusions between such paraphytes pertaining to adjacent cells, as recorded by GLEES.

Figures 370A–C. Cross-sections (hematoxylin-eosin stain), in rostrocaudal sequence, C being at an isthmus level, through the mesencephalon of the Holocephalian Plagiostome Chimaera Colliei (from K. and NIIMI, 1969). bv: ventral part of basal forebrain bundle; ce: cerebellum; cp: commissura posterior; cpi: inferior subdivision of commissura posterior; ct: commissura tecti; flm: fasciculus longitudinalis medialis; hy: hypothalamus; li: lobi inferiores hypothalami; ll: lemniscus lateralis; ni: nucleus interpeduncularis; nmV: nucleus radicis mesencephalicae trigemini; nrm: nucleus rotundus mesencephali; n. III: root fibers of oculomotor; sdc: sulcus dorsalis commissurae posterioris (a Chimaeroid peculiarity); sl: sulcus limitans; slm: sulcus lateralis mesencephali (internus); st: tractus spino- et bulbotectalis; su: subcommissural organ; sys: synencephalic sulcus system; tb: tractus tectobulbaris (et tectospinalis); thd: thalamus dorsalis; thv: thalamus ventralis; tm: caudal end of tectum mesencephali. Added x in A: nucleus lentiformis mesencephali; lower arrow in B: sulcus lateralis mesencephali externus; other arrow: system of sulcus lateralis mesencephali (internus) accessorius. Compare with Figure 309D for lower isthmus level.

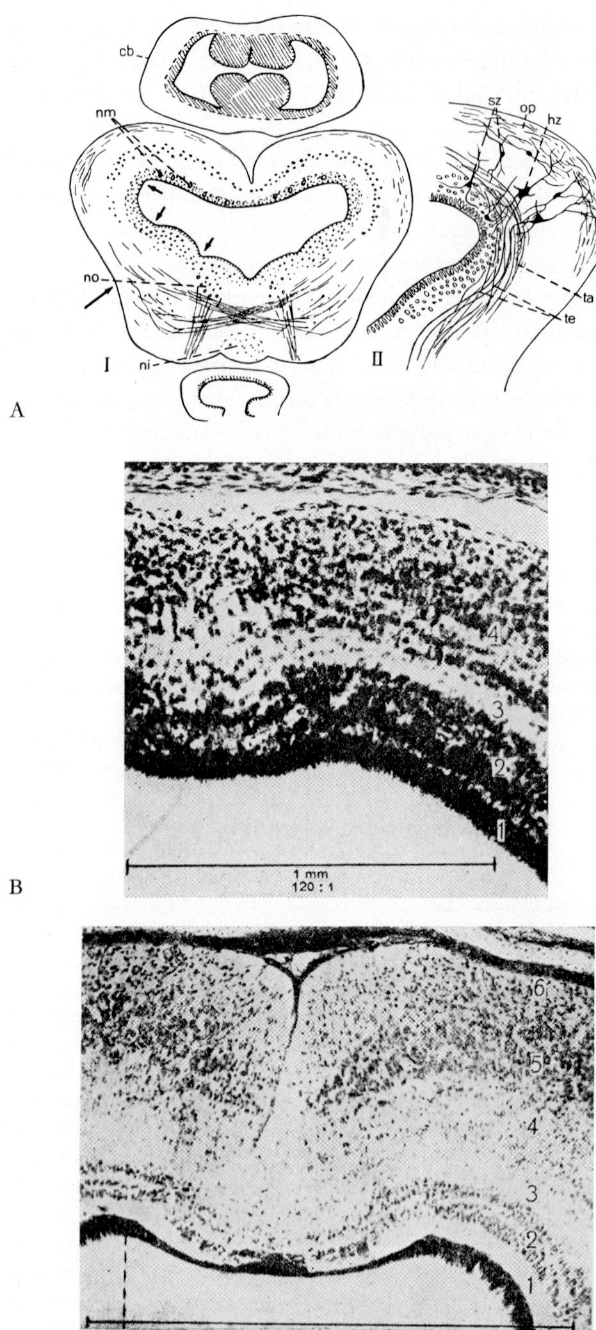
cb
nm
no
ni
I
sz
op
hz
ta
te
II
A
4
3
2
1
1 mm
120 : 1
B
6
5
4
3
2
1
1 mm
80 : 1
Ependym
C

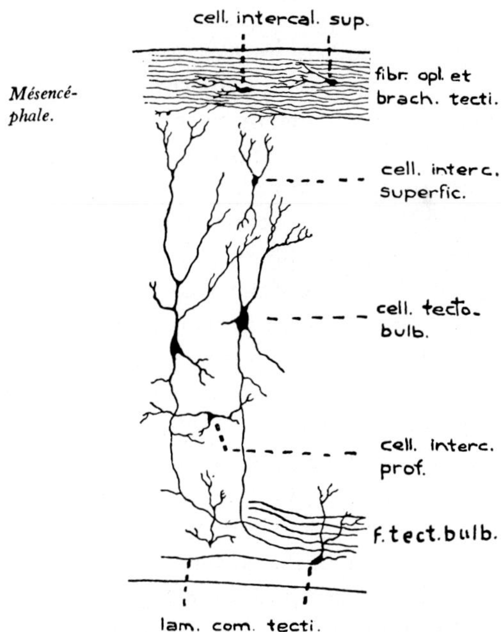

Figure 371 D. Some structural details of the Plagiostome tectum opticum (partly based on observations by HOUSER and STERZI, from KAPPERS, 1921).

Figure 371 A. Simplified sketches illustrating aspects of mesencephalon of Squalus acanthias in cross-sections (from K., 1927). I Overall features. II Details of tectum opticum (partly adapted from KAPPERS, 1921). cb: cerebellum; hz: chief cell; ni: nucleus interpeduncularis; nm: nucleus radicis mesencephalicae trigemini; no: nucleus nervi oculomotorii; op: layer of optic tract fibers; sz: correlating neurons *(Schaltzellen)*; ta: afferent components of deep medulla; te: efferent fibers of deep medulla; arrows in I: upper: sulcus lateralis mesencephali (internus) accessorius; middle: sulcus lateralis mesencephali (internus); lower: sulcus limitans; the lead to no passes through sulcus lateralis mesencephali externus (outer arrow).

Figure 371 B, C. Two ontogenetic stages of tectum opticum in Scyllium canicula *(Nissl stain)*, B at 4.2 cm length and C at 9.5 cm (from GERLACH, 1947). 1: subependymal layer; 2: granular layer; 3: deep medulla; 4, 5: layer of chief cells (4 in B oberlapping with deep medulla); 6: optic tract layer. The scale in B should read: 0.5 mm.

At levels of the mesencephalo-diencephalic boundary, where some of the optic tract fibers radiate into tectum mesencephali within the rostral wall of the tectum, a *nucleus lentiformis mesencephali* can be recognized. It is commonly represented by a vaguely band-like cell group located rostrobasally to the tectal formation, and roughly dorsolaterally to the commissura posterior respectively to its nucleus (cf. Fig. 192, p. 419 of vol. 3, part II). In Chimaera, the nucleus lentiformis mesencephali, traversed by optic tract fibers, presumably corresponds to the structure described by KAPPERS and CARPENTER (1911) as ganglion geniculatum laterale and depicted in their Figure 14 (loc.cit.). An ill-defined portion of the nucleus lentiformis mesencephali is indicated in the present Figure 370A. The two antimeric halves of the tectum are joined by a median commissura tecti which begins at the posterior commissure and extends caudalward as far as the isthmus region, there blending in various ways with the isthmic decussations, including that of the trochlear nerve.[53] Lateralward, the commissura tecti tends to merge with the fiber system of the deep medulla.

Basally to the tectum opticum the diffuse cellular substratum of the *torus semicircularis* represents the *nucleus lateralis mesencephali*. In Chimaera, a rostral portion of this griseum becomes the nucleus rotundus mesencephali (Fig. 370A), interpreted by KAPPERS and CARPENTER (1911) as nucleus dorsalis thalami, and presumably also corresponding to the nucleus lentiformis depicted by KAPPERS *et al.* (1936, Fig. 435, p. 894) in Acanthias.[54] The fiber system of lemniscus lateralis may indistinctly separate the griseum of torus into a paraventricular and a lateral portion. Because of the diffuse cellular arrangements, the boundary between torus grisea and tegmental grisea remains rather indistinct.[55]

---

[53] Cf. e.g. the particular relationships described in Chimaera by K. and NIIMI (1969). The peculiar 'pattern distortion' obtaining in Chimaera involves here a number of displacements within a topologically invariant bauplan.

[54] Again, the same griseum of Acanthias is labelled nucleus lateralis tegmenti by KAPPERS (1947) in his Figure 58 loc. cit.

[55] Thus, what KAPPERS *et al.* 1936 in their figure 434 designate as nucleus lateralis tegmenti is, in my opinion, a griseum of the torus. Again, in the same figure, their upper lead to a nucleus medialis tegmenti designates, in my interpretation, a part of the torus, while their lower lead doubtless points to a genuine tegmental griseum. In Figure 431 loc. cit., at a level close to the isthmus, the neighborhood designated as nucleus profundus mesencephali seems likewise tegmental, its dorsal extension, as depicted in the myelin stained section, being, however, a griseum of the lateral lemniscus, and most probably an alar plate derivative. It may represent the nucleus isthmi, as also indicated by KAPPERS (1947) in his Figure 62 loc. cit.

A caudal extension of the torus cell masses seems to represent the *nucleus isthmi* of Plagiostomes (Fig. 372). This griseum is less developed in Selachians than in Osteichthyes, Amphibians and Amniota. We could not unequivocally delimit a nucleus isthmi in our Chimaeroid material (K. and Niimi, 1969) but it was identified in other Plagiostomes by Gerlach (1947) and Kappers (1947). A *nucleus visceralis (s. gustatorius) secundarius* is not clearly recognizable in Selachians, but it may presumably be present as a caudal differentiation of the intermediodorsal zone (submerged within basimedial parts of the torus), and closely adjacent to the griseum of nucleus isthmi.

The mesencephalic *tegmentum* (or *basal plate*) is characterized by diffusely distributed cell masses between the numerous longitudinal fiber bundles traversed by transverse arcuate fibers. It contains the cell population representing the mesencephalic portion of *nucleus reticularis tegmenti* investigated by Saito (1930b). Gerlach (1947) has designated an indistinct lateral condensation as nucleus tegmentalis lateralis, and other authors describe a nucleus profundus mesencephali.

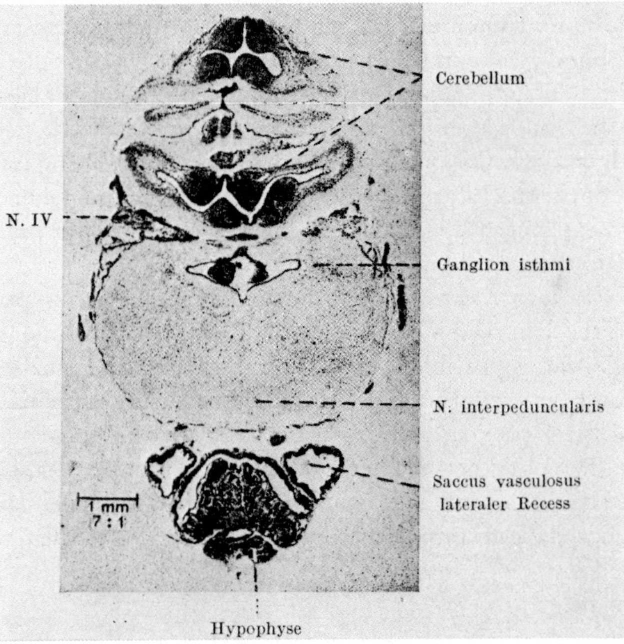

Figure 372. Cross-section *(Nissl stain)* through the mesencephalo-rhombencephalic isthmus region of the Plagiostome Galeus canis (from Gerlach, 1947).

Rostral tegmental cell groups, which cannot be clearly demarcated, may correspond to the nucleus of the fasciculus longitudinalis medialis (*nucleus interstitialis of Cajal* in more differentiated Vertebrate forms). While it is of course possible to consider scattered tegmental cells related to efferent cerebellar fibers as a primordium kathomologous to nucleus ruber tegmenti, it does not seem justified to claim the actual presence of this nucleus as a circumscribed griseum in Plagiostomes. At the base of the tegmentum, the *nucleus interpeduncularis* is easily recognizable as a conspicuous griseum in the midline (Figs. 369, 370, 371A, 372). A nucleus opticus tegmenti cannot be identified with any certainty, but it may be presumed that such a cell group is included within the nondescript tegmental cell masses.

As regards the subdivisions of the *basal plate*, a shallow sulcus intermedioventralis ventrally to sulcus limitans is occasionally discernible (cf. Fig. 369A, unlabelled.) The *nucleus nervi oculomotorii* (Fig. 371A) consists of loosely spaced cells between the bundles of the tegmentum, laterally to fasciculus longitudinalis medialis. It pertains to the basal plate's ventral zone, but its morphologically nondescript preganglionic component[55a] presumably derives from the intermedioventral zone. The oculomotor root fibers leave the basal surface of the tegmentum in scattered bundles, the most caudal of which may occasionally be seen at the level of isthmus rhombencephali. Caudally and rather closely adjacent to the oculomotor griseum lies the smaller *nucleus nervi trochlearis*, apparently pertaining to the basal plate's ventral zone. The intracerebral root of the trochlear nerve decussates in the velum medullare anterius at the isthmus (Fig. 309D) and emerges here from the brain rather closely to the dorsal midline.

The *input and output channels* of the mesencephalon are, in principle, similar to those in the preceding section 2 for Petromyzonts. In order to avoid, as far as possible, unnecessary repetition, they shall here be grouped as (1) rostral, (2) caudal, and (3) commissural systems.

The rostral connections, *qua* input, include the following main channels. The optic nerve, after a practically complete decussation is the optic chiasma, forms a fairly massive *optic tract* which radiates into the mesencephalic tectum with some, perhaps mostly collateral end-

---

[55a] In Plagiostomes, the autonomic fibers of the oculomotor seem to innervate the musculus dilator pupillae (cf. vol. 3, part II, chapter VII, section 6, p. 928).

ings in the pretectal nucleus lentiformis mesencephali.[56] GERLACH (1947) distinguishes a pars lateralis and a pars medialis of the main optic tract, pars lateralis being the larger subdivision, which reaches the more caudal regions of the tectum, while pars medialis essentially reaches the rostral tectal area. It seems likely that a small basal optic tract, related to hypothalamus and mesencephalic tegmentum, is present, although I was unable to identify this subsystem with sufficient certainty and did not find any pertinent conclusive data on this topic in the available literature.

The *tractus habenulo-peduncularis* or *fasciculus retroflexus* consists essentially of descending fibers providing input for the nucleus interpeduncularis, and partly decussates, displaying complex sinuous windings, within the mesencephalic tegmentum. The right tract, to a large extent non-medullated, is conspicuously thinner than the left one, which contains a much greater proportion of medullated fibers.[57]

Fiber bundles connecting thalamus and hypothalamus with mesencephalic tectum and particularly tegmentum form a rather *diffuse diencephalo-mesencephalic system* presumably containing reciprocal (descending and ascending channels) which include the various 'tracts' such e.g. as the tractus lobo-peduncularis, described by KAPPERS *et al.* (1936) and numerous other authors. It is not impossible that descending telencephalic components, contained in the basal forebrain bundles, or in the tractus pallii, directly or indirectly reach the mesencephalic tegmentum. Again, some components of the just mentioned fiber systems might convey ascending impulses from the tegmentum. At approximately a right angle to these longitudinal channels a likewise diffuse system of ventrodorsal (respectively dorsoventral) fibers runs between diencephalon and mesencephalon as well as subtectal grisea.

With respect to *mesencephalic output*, the rostral connections seem to include fibers from lateral tectum *sensu latiori* and from tegmentum which decussate in the diencephalic ventral supraoptic commissure (s.

---

[56] The main optic tract effects, of course, also some connections (collateral or direct or possibly both) with poorly outlined lateral cell groups of dorsal and ventral thalamus (dorsal and ventral 'nucleus geniculatus lateralis'), as well as with caudal (posthabenular) dorsal thalamic grisea, including nucleus commissurae posterioris. These grisea represent the diencephalic components of the pretectal complex. A few uncrossed optic tract fibers to tectum may be present.

[57] The left habenular 'ganglion' of Plagiostomes is noticeably larger than its antimere. This asymmetry is reversed with respect to that obtaining in Cyclostomes (Petromyzonts and Myxinoids) whose right 'ganglion' is much larger than the left one.

commissura transversa). Scattered decussating fibers, some of which become gathered into a thin so-called suprainfundibular commissure, are located in the caudal roof of the hypothalamus, namely in the midline dorsal to the hypothalamic portion of third ventricle expanding into the lateral recesses and can be found as far caudal as the roof of lobi inferiores basal to the midbrain (cf. e.g. Fig. 370B).

Again, the rostral connections of the mesencephalon include some channels originating in more caudal brain regions and passing rostralward through the midbrain. Thus, a fairly distinct medullated bundle entering the posterior hypothalamus seems to be traceable through the mesencephalic tegmentum as far as the cerebellum and represents the *tractus cerebellohypothalamicus* or *cerebellolobaris* autorum.[58] It may contain crossed and uncrossed, direct and indirect cerebellar connection. Moreover, a perhaps minor extension of the bulbar (and even spinal ?) lemniscus systems into diencephalic grisea does not seem entirely improbable. Descending connections passing through the mesencephalon may include components of the tractus lobopeduncularis extending into the oblongata (*'tractus lobobulbaris'* of KAPPERS, 1947) as depicted in Figure 368.

The *caudal fiber connections* of the mesencephalon are those with cerebellum, oblongata, and spinal cord. Among the *caudal input systems*, the *lemniscus lateralis*,[59] particularly reaching the torus semicircularis, represents a complex and large bundle, which comprises ascending fibers from the vestibulo-lateralis area and may contain some other ascending as well as descending fibers. The *tractus spino- et bulbotectalis* carries ascending fibers from various other sensory systems into tectum mesencephali.[60]

A compound *tractus cerebellotegmentalis*, (or *brachium conjunctivum)*, essentially decussating in the tegmentum, can be followed as far as the diencephalon and includes the *tractus cerebellolobaris* mentioned above.

The *caudal output* of the mesencephalon includes lateral and medial, crossed or uncrossed *tectobulbar*, perhaps even *tectospinal fibers*. Vaguely circumscribed bundles located between tractus tectobulbaris and lem-

---

[58] *'Faisceau cérébello-thalamique'* in KAPPERS' (1947) Figure 59 illustrating a myelin-stained cross section through the mesencephalon of Acanthias.

[59] The nucleus isthmi can be regarded as essentially a relay- or 'interstitial' griseum of that tract.

[60] Ascending 'general visceral afferent' fibers may be connected with an undefined 'nucleus visceralis secundarius' located medially or medioventrally to nucleus isthmi.

niscus lateralis seem to connect with the cerebellum and presumably correspond to the *fasciculus tectocerebellaris* of Kappers (1947, Fig. 61, loc. cit.). A slender tectocerebellar bundle may also run through the velum medullare anterius.

The *fasciculus longitudinalis medialis* is a large and compact fiber system located in the ventricular floor of the tegmentum on both sides of the midline (Figs. 369, 370). It begins rostrally in the mesencephalo-diencephalic boundary zone at about the level of tuberculum posterius and runs caudalward throughout the entire mesencephalic and rhombencephalic tegmentum into the spinal cord. In Selachians, as in other Vertebrates, the numerous different and to some extent variable components of this complex fiber system are ascending and descending, as well as both crossed and uncrossed. Generally speaking, however, it can be considered the main central motor tract of the Anamniote brain, representing, *in the spinal cord*, an essentially and perhaps exclusively *descending channel*. As far as the Selachian mesencephalon is concerned, the fasciculus longitudinalis medialis includes output channels of the nucleus reticularis tegmenti, with the inclusion of perhaps some of the tectofugal fibers, moreover possibly also some fibers from the essentially diencephalic nucleus commissurae posterioris. On the other hand, the fasciculus longitudinalis includes input to mesencephalic tegmentum from rhombencephalic reticular and vestibular cell groups, as discussed, from a more generalized viewpoint, in section 1 of the present chapter.

As regards the various *commissural systems* related to the Selachian mesencephalon, it may be sufficient to summarize these decussations as follows. A mesencephalic component is included in the diencephalic supraoptic and postoptic commissures as pointed out above. In addition to *commissura posterior*, shared by diencephalon[61] and mesencephalon, and to *commissura tecti*, which were both mentioned in connection with the tectum mesencephali, there is a conspicuous tegmental commissural system, which consists of several, in part poorly demarcated components. The *commissura tegmenti ventralis* or *commissura ansulata* begins rostrally at the level of tuberculum posterius, and extends as far as the isthmic region. Most of its fibers cross dorsally to nucleus interpeduncularis. In addition, there are a number of crossing fibers running

[61] The significance of the commissura posterior, especially related to the pretectal grisea, shall again be dealt with in chapter XII of volume 5.

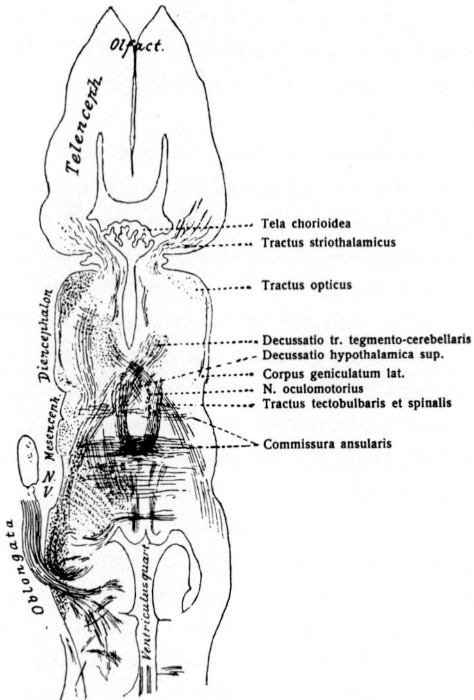

Figure 373. Horizontal section through the brain of the Plagiostome Scyllium canicula (myelin stain), displaying decussations in the mesencephalic tegmentum and some other fiber system (from Edinger, 1908). Commissura ansularis = c. ansulata; tractus striothalamicus = lateral forebrain bundle. The 'tractus tegmento-cerebellaris' is here probably the tractus lobocerebellaris, and the bundle designated as 'tractus tectobularis' rather seems to be the fasciculus retroflexus.

through other, dorsal or ventral levels of the tegmentum.[62] These ill-defined commissures seem to include the crossed fibers of tectobulbar systems, cerebellotegmental systems, and fasciculus longitudinalis medialis. Decussating *cerebellar connections* are particularly numerous and compact ventrally to fasciculus longitudinalis medialis at caudal mesencephalic tegmental levels and include the '*decussation du pédoncule cérébelleux supérior*' depicted by Kappers (1947, Fig. 62 loc. cit.). Figure 373 illustrates, in an 'horizontal' section, the various fiber systems

[62] Gerlach (1947) distinguished 4 components of the commissura ansulata: (1) rostrodorsal and (2) caudodorsal, both including decussating fibers of tractus tectobulbares dorsales, moreover (3) and (4) two caudoventral components, with crossing fibers of tractus tectobulbares ventrales and of tractus cerebello-meso-diencephalici.

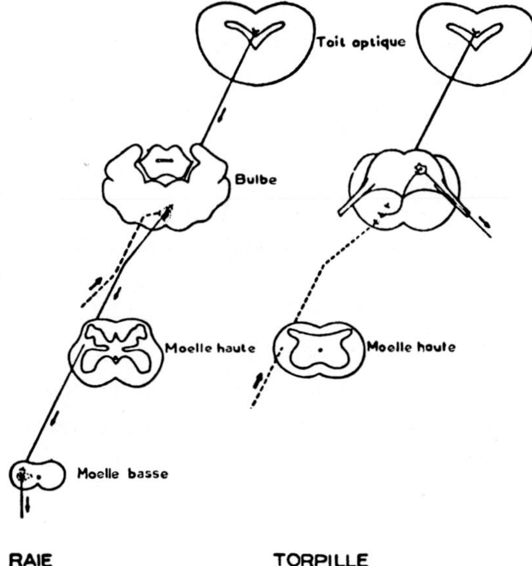

Figure 374. Diagrams indicating assumed efferent mesencephalic tectal channels assumed to trigger neural discharge mechanisms of electric organs in the Plagiostomes Raja and Torpedo (from SZABO, 1961).

(whose details still remain very poorly elucidated) and some of their decussations in the ventral region of the neuraxis, and as interpreted by EDINGER (1908). The diagrams of Figure 374 show, in SZABO's (1961) interpretation, the tectobulbar channel presumed to exert an influence on the bulbar '*noyau de la commande centrale*' controlling the discharge of the electric organs in Raja and Torpedo.

With respect to the visual functions of the Plagiostome tectum opticum, a recent communication (GRAEBER *et al.*, 1973) claims that, after 'complete removal' of the tectum, Nurse Sharks *(Ginglymostoma cirratum)* 'can learn to discriminate black versus white and horizontal versus vertical stripes'. Assuming that the interpretation of the discrimination test as reported by the cited authors is essentially correct,[63] their

[63] Considerable scepticism with regard to the interpretations, descriptions, and classifications of behavioral activities is evidently justified (cf. chapter X, p. 658, footnote 24).

far-reaching further elaborations on these findings can be evaluated as greatly exaggerated. The obtained results are said to contradict 'the traditional belief of exclusive tectal control over visuomotor behavior in lower vertebrates' and to suggest 'a role for the telencephalon in the vision of these primitive animals'. Additional observations by the quoted authors, said to reveal 'severe visual dysfunction' following lesions of the 'posterior telencephalon', and to have displayed 'short latency, visually evoked potentials in the same area' are believed to 'suggest that portions of the nonlaminated posterior telencephalon of sharks are remarkably similar to the laminated visual cortex of mammals. Such a view, if correct, would necessitate a revision of our notions on the evolution of the brain, especially the view that the primitive telencephalon is dominated by olfaction' (GRAEBER *et al.*, 1973). The cited authors refer here particularly to the views expressed by HERRICK.

One could, however, reply that all competent comparative neurobiologists, including the late C. J. HERRICK, were quite familiar with the optic input to diencephalon (lateral geniculate nuclei) and with non-olfactory, presumably including optic, input from diencephalon to telencephalon, regardless of the here predominating olfactory input. There is little doubt that the (lateral and medial) forebrain bundles are two-way channels through which a variety of 'somatic' and 'visceral' input can reach the Plagiostome and quite generally the Anamniote telencephalon. Thus, in my early studies more than fifty years ago, I assumed, somewhat overenthusiastically, the presence of a true 'neopallium' and of a 'corpus callosum' in urodele Amphibians, and that the 'neopallium' was already recognizable in Dipnoan fishes. It is hardly necessary to add that, *qua* morphologic relationships, I no longer uphold this interpretation. While non-olfactory 'sensory' input into the Plagiostome telencephalon may nevertheless be regarded as reasonably certain, the detailed tracing of tracts by means of the newer 'terminal degeneration techniques' as reported by EBBESON (1972) and others should be regarded with considerable scepticism. Because of highly ambiguous 'degeneration' pictures characterized by nondescript detritus displayed at various 'stages' of said methods (including the *Fink-Heimer technique*), it seems possible to claim some sort of 'evidence' for almost any particular postulated type of fiber connections. The basal optic tract, with input to hypothalamus and mesencephalic tegmentum, was likewise a matter of common knowledge, although this accessory optic tract, present in Anamnia, has not been unequivocally identified in Selachians. Thus, the allegedly 'traditional belief of exclusive tectal control over visuomotor behavior' was certainly never upheld by any competent comparative neurologist.

While the possibility that telencephalic[64] optic connections played some role in the asserted 'visual discrimination' of Nurse Sharks with destroyed tectum opticum[65] cannot be dogmatically excluded, it appears more likely that diencephalic mechanisms via the lateral geniculate nuclei, and tegmental connection by way of the presumed basal optic tract were here involved. Even in Anamnia, the neuronal networks of the neuraxis doubtless display considerable redundancy enabling the execution of performances through a number of alternative pathways.[66]

4. Ganoids and Teleosts; Latimeria

The configuration of the Ganoid and Teleostean mesencephalon displays more (taxonomically related) variations than that of Elasmobranchs. Particularly in Teleosts, wide differences *qua* external morphology and in details of internal structural arrangements are manifested.[67] In some Ganoids, however, the overall configuration, except for greater or lesser protrusion of the valvula cerebelli into the mesencephalic ventricle, is quite similar to that obtaining in Selachians (cf.

---

[64] STEINER (1888) reported behavioral changes (reduction or loss of 'spontaneous activity') after removal of telencephalon and even merely of olfactory bulbs in Sharks. It is thus quite possible that the loss of telencephalic impulses might affect the 'visuomotor behavior' of Plagiostomes. As regards this latter, STEINER (loc. cit.) already showed that, in general, Sharks have a very poor 'daylight vision', their optic mechanisms being essentially scotopic, i.e. depending on the 'dark-adapted' condition. While it is, of course, possible that some species of Sharks may have photopic ('daylight') vision, the retina of Nurse Sharks and even of other Plagiostomes being now said to contain not only rods but also (rather nondescript) cones, the interpretation of the 'visual discrimination tests' described by GRAEBER and EBBESSON (in Comp. Biochem. Physiol. 42, A: 131–139, 1972) can be evaluated as rather unconvincing.

[65] The 'total tectal ablation' depicted by GRAEBER *et al.* (1973, Fig. 1 loc. cit.), although fairly complete, is not entirely convincing, since some ventrolateral remnants of tectum appear identifiable.

[66] Cf. vol. 1, pp. 20–21, and Brain and Consciousness (K., 1957) pp. 138–139.

[67] Because of the polymorph variations displayed by the multitudinous Teleosts, KAPPERS *et al.* (1936), in their very detailed standard treatise, were compelled to restrict their account to only brief mention 'of the most outstanding features' of the mesencephalic and diencephalic regions in Osteichthyes, referring, for further details, to the original reports by investigators of these topics (p. 903 loc. cit.).

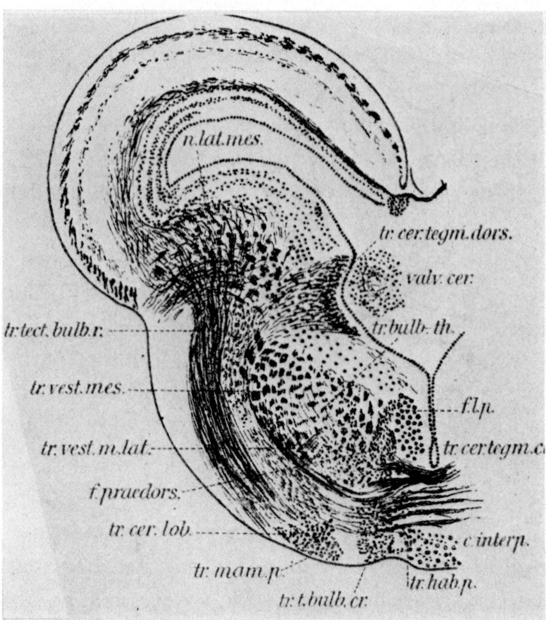

Figure 375. Cross-section through the mesencephalon of the Ganoid Amia calva (from KAPPERS, 1907). c. interp.: nucleus interpeduncularis; f.l.p.: fasciculus longitudinalis medialis; f. praedors.: 'fasciculus praedorsalis' (tecto-bulbar system); n. lat. mes.: nucleus lateralis mesencephali; tr. bulb. th.: 'tractus bulbo-thalamicus'; tr. cer. lob.: 'tractus cerebello-lobaris; tr. cer. tegm. (cruc., dors.): tractus cerebello-tegmentalis (cruciatus, dorsalis); tr. hab. o.: tractus habenulo-interpeduncularis; tr. mam. p.: 'tractus mammilo-peduncularis'; tr. t. bulb. cr.: tractus tectobulbaris cruciatus; t. tect. bulb. r.: tractus tectobulbaris rectus; tr. vest. mes.: 'tractus vestibulo-mesencephalicus'; tr. vest. m. lat.: 'tractus vestibulo-mesencephalicus lateralis'; valv. cer.: rostral end of valvula cerebelli.

Fig. 375). In many Teleosts, on the other hand, the particular development of various grisea and tracts leads to considerable pattern distortions which, nevertheless, preserve the topologic relationships within (or under) the obtaining transformation. These distortions, however, often present substantial difficulties of interpretation.[68] Mesencephalon and diencephalon, as well as mesencephalon and rhombencephalon, tend to be 'telescoped' into each other, but tendencies of this sort are

[68] Many of these difficulties result from the fact that, in general, relatively few data on the relevant key stages displaying the ontogenetic development of mesencephalic (and diencephalic) centers in Teleosts are available. Extrapolations and interpolations become thereby necessary.

also conspicuously manifested by other Anamnia, e.g. by the Cyclostome Myxinoids and the Chimaeroid Plagiostomes.

Figures 375, 376A–D, and 378A–I illustrate relevant aspects of the mesencephalon of Ganoids and Teleosts as seen in cross-sections.[69] At the adult stage, *sulcus limitans* and *sulcus lateralis mesencephali (internus)* remain generally identifiable, on the whole more readily in Ganoids than in Teleosts. In the former, the *sulcus intermedius ventralis* is occasionally displayed.

As in Plagiostomes, the *alar plate* comprises the optic tectum, the torus semicircularis, the nucleus isthmi, and the nucleus visceralis s. gustatorius. This latter, especially in many Teleosts, is much better developed than in Plagiostomes, where its identification generally remains rather tentative. In correlation with the valvula cerebelli, the nucleus lateralis valvulae, dealt with in section 4 of the preceding chapter X, represents a griseum not displayed by other Anamnia.

The deep layer of the tectum opticum contains, with various minor differences of cellular distributions, the large elements of the *nucleus radicis mesencephalicae trigemini*, frequently in the vicinity of the commissura posterior (Figs. 376A, 378G).

A peculiar bilateral median formation of the Ganoid[70] and particularly of the Teleostean mesencephalic tectum is the *torus longitudinalis*, a more or less rounded rostrocaudal ridge protruding into the ventricular space. The torus longitudinalis contains relatively small cells and receives fibers, through the tectal commissure, from the deep tectal layers as well as, through a longitudinal tractus cerebellotoralis, from the corpus cerebelli. Output of the torus is said to reach the molecular layer of the tectum and may also descend to the corpus cerebelli. Additional rostral connections with diencephalic grisea can likewise be assumed. Kudo (1923, 1924) who investigated the torus longitudinalis and refers to the studies by previous authors, does not believe that it has direct relations to the optic system, as suggested in some of the earlier studies which he quotes. According toKudo (1923), the torus longitudinalis seems to play a role related to the 'static functions' of the tectum opticum. The torus commonly displays its greatest expansion at rostral levels and caudalward gradually decreases in size (cf. Figs. 376A–D). According to Kappers (1947), the cerebellotoral tract joins the torus at

[69] Cf. also Figures 316A and 322A/B.

[70] The Ganoid Acipenser, however, does not display a recognizable torus longitudinalis (cf. also Kudo, 1923 and Fig. 364C; in Fig. 364D representing Johnston's diagram of a Teleost mesencephalon, the torus is included, but unlabelled and barely suggested).

levels of the commissura posterior. The cited author likewise denies a relationship of optic system to torus longitudinalis and points out that this structure is well developed in the blind Teleosts Troglichthys as recorded by CHARLTON (1933).

With respect to the *tectum opticum*, Figure 377 depicts some details of its structural arrangement in the Ganoid Acipenser according to the investigations of JOHNSTON (1906). The main four layers of the simplified enumeration dealt with above in section 1 are roughly displayed.

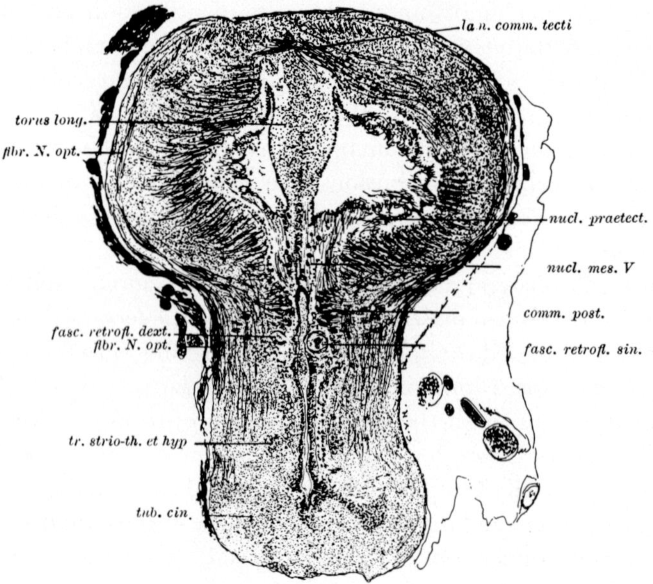

Figure 376A. Cross-section through the mesencephalon of the ganoid Polyodon *(Weigert stain)* in the mesencephalo-diencephalic boundary region (from HOOGENBOOM, 1929). The 'nucl.praetect.' is presumably the nucleus lentiformis mesencephali.

Figure 376B. Cross-section through the mesencephalon of Polyodon, somewhat caudal to the preceding level (from HOOGENBOOM, 1929). Added arrows indicate sulcus lateralis mesencephali (upper) and sulcus limitans (lower); 'nucl. entop. p. dors.' probably lateral tegmental griseum; 'n. entop. p. ventr.' is presumably a posterior hypothalamic griseum; 'nucl. comm. post.' is apparently rostral end of tegmental cell plate.

Figure 376C. Cross-section through the mesencephalon of Polyodon, caudal to the preceding levels and including the rostral tips of the cerebellar auriculae (from HOOGENBOOM, 1929).

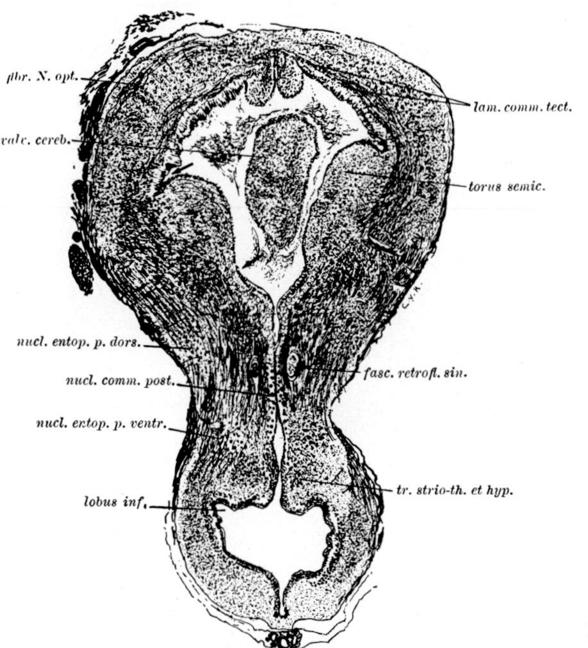
fibr. N. opt.
valv. cereb.
lam. comm. tect.
torus semic.
nucl. entop. p. dors.
nucl. comm. post.
nucl. entop. p. ventr.
lobus inf.
fasc. retrofl. sin.
tr. strio-th. et hyp.

376 B

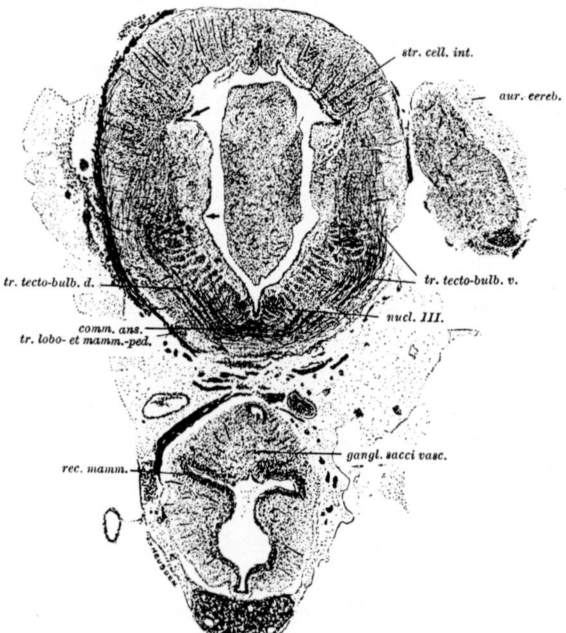
str. cell. int.
aur. cereb.
tr. tecto-bulb. d.
tr. tecto-bulb. v.
nucl. III.
comm. ans.
tr. lobo- et mamm.-ped.
gangl. sacci vasc.
rec. mamm.

376 C

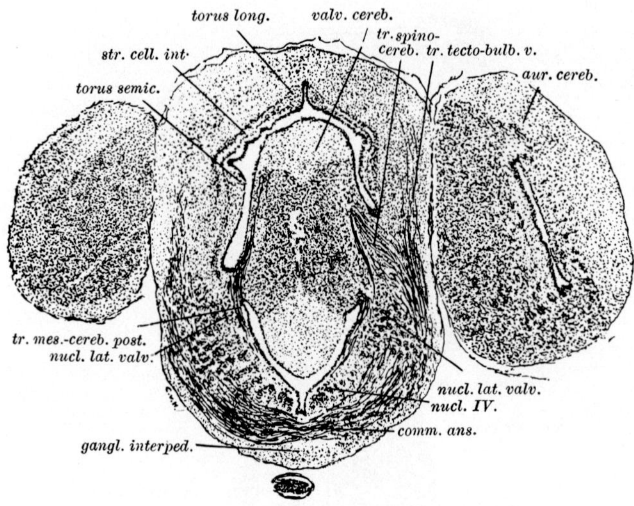

Figure 376D. Cross-section through a caudal portion of the mesencephalon in Polyodon (from HOOGENBOOM, 1929).

In presumably most Teleosts, the neuronal structure of the tectum opticum, as studied by various authors quoted in the treatise of KAPPERS *et al.* (1936), seems to be more complex than in Plagiostomes and Ganoids, and is depicted by Figures 379A–E. It will be seen that LEGHISSA (1955) has suggested a somewhat differing and rather complex, but not altogether convincing scheme of stratification. The cited author also described the relationship of the general and lateral lemniscus to the tectum opticum as shown in Figure 379E. Figure 369F indicates topographic features of the optic tract's projection upon the tectum according to STRÖER (1939). At the mesencephalodiencephalic boundary, a *nucleus lentiformis mesencephali*, formed by a more or less distinct rostrobasal derivative of the tectum, represents a mesencephalic alar plate component of the *pretectal complex* (cf. Fig. 376A) and also doubtless receives optic input. It includes the cell group of the so-called nucleus corticalis mentioned by KAPPERS *et al.* (1936). The diencephalic, dorsal thalamic components extending into the mesencephalodiencephalic boundary zone include a pretectal nuclear group and the nucleus of the posterior commissure.[70a] On the ventricular surface of

[70a] Figures 193 to 197, p. 420–421 of volume 3/II illustrate the configuration of the pretectal region in Osteichthyes.

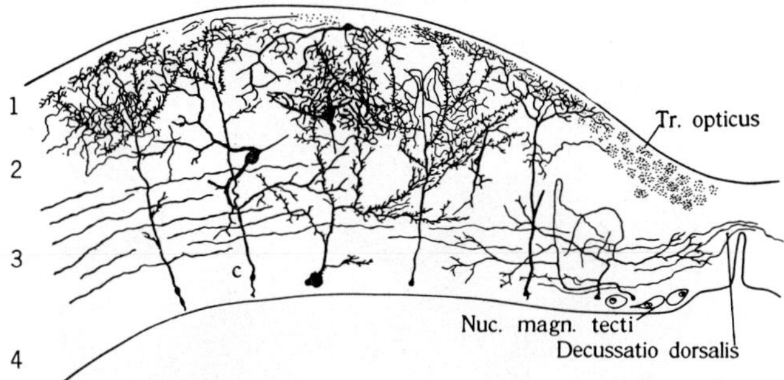

Figure 377. Part of a cross-section *(Golgi impregnation)* through the tectum opticum of the Ganoid Acipenser (from Johnston, 1906). c is either an ependymal cell or a neuronal granular element. Decussatio dorsalis is the decussatio tecti. Nuc. magn. tecti is the nucleus of the mesencephalic trigeminal root. Added numbers 1–4 indicate layers according to the simplified enumeration adopted in the present chapter.

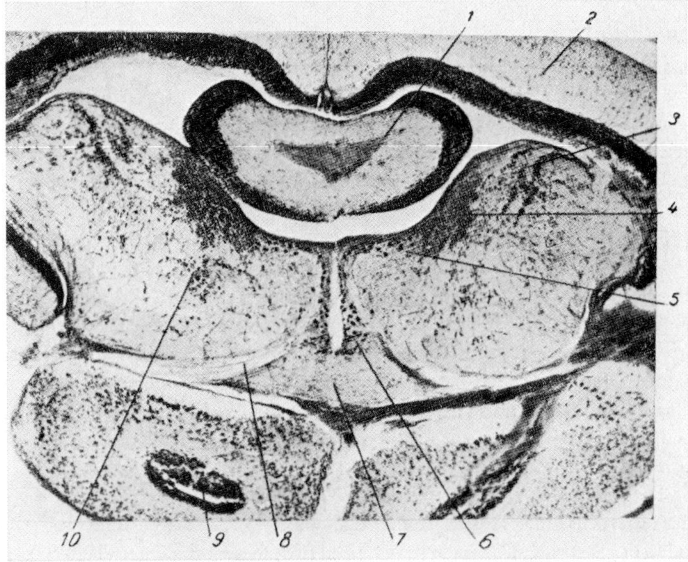

Figure 378 A. Cross-section *(Giemsa stain)* through the mesencephalon of the Teleost Gasterosteus osseus at the level of the oculomotor root (from Beccari, 1943). 1: valvula cerebelli; 2: tectum opticum; 3: torus semicircularis; 4: nucleus lateralis valvulae; 5, 6: dorsal and ventral cell groups of oculomotor nucleus; 7: commissura ansulata; 8: oculomotor root; 9: lobi inferiores hypothalami; 10: nucleus lateralis profundus (tegmentalis, reticularis) mesencephali.

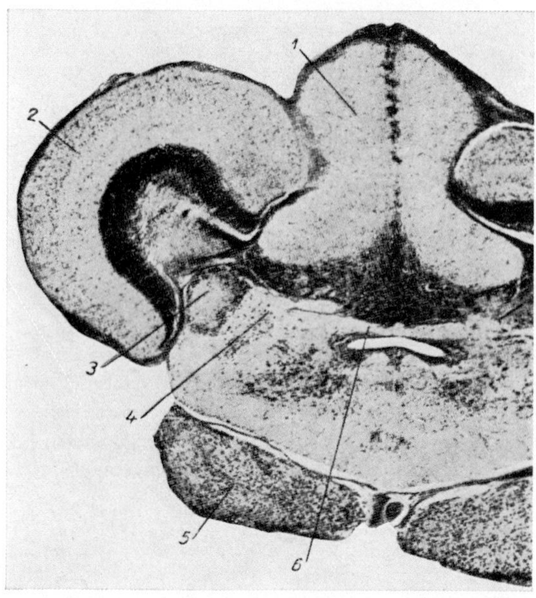

Figure 378B. Cross-section *(Nissl stain)* through the mesencephalon of Gasterosteus aculeatus close to the isthmus (from BECCARI, 1943). 1: cerebellum (transition of corpus to valvula); 2: tectum opticum; 3: nucleus isthmi; 4: nucleus gustatorius secundarius; 5: lobi inferiores hypothalami; 6: 'commissura gustatoria'. It can be seen that the two antimeric halves of tectum mesencephali become separated by the protruding cerebellum (cf. also Fig. 316A). Nucleus interpeduncularis, not labelled, can be identified.

Figure 378C. Cross-section (modified *Cajal impregnation)* through a caudal portion of the mesencephalon in the Teleost Gambusia (from BECCARI, 1943). 1: torus longitudinalis; 2: 'commissura gustatoria'; 3: tectum opticum; 4: crossed trochlear nerve root; 5: torus semicircularis; 6: nucleus gustatorius secundarius; 7: nucleus isthmi; 8: fasciculus longitudinalis lateralis (octavolateral lemniscus); 9: lobi inferiores hypothalami; 10: fasciculus longitudinalis medialis; 11: trochlear root before decussation; 12: 'brachium conjunctivum'; 13, 14: anterior and posterior 'mesencephalo-cerebellar bundle; 15: valvula cerebelli.

Figure 378D. Cross-section (myelin stain) through the mesencephalon of the Teleost Idus idus (after VAN DER HORST, from BECCARI, 1943). 1: commissura tecti; 2: torus longitudinalis; 3: valvula cerebelli; 4: posterior mesencephalo-cerebellar fasciculus; 5, 6: dorsolateral and ventromedial cell groups of oculomotor nucleus; 7: anterior mesencephalo-cerebellar fasciculus (tectocerebellar tract; 8: fasciculus longitudinalis medialis; 9: commissura ansulata (fasciculus tectobulbaris cruciatus); 10: lobi inferiores hypothalami; 11: oculomotor root; 12: 'ventral tectobulbar fasciculus'; 13: commissura transversa (commissura postoptica ventralis, decussating in hypothalamus); 14: fasciculus longitudinalis lateralis (octavolateral lemniscus); 15: torus semicircularis; 16: tectum opticum.

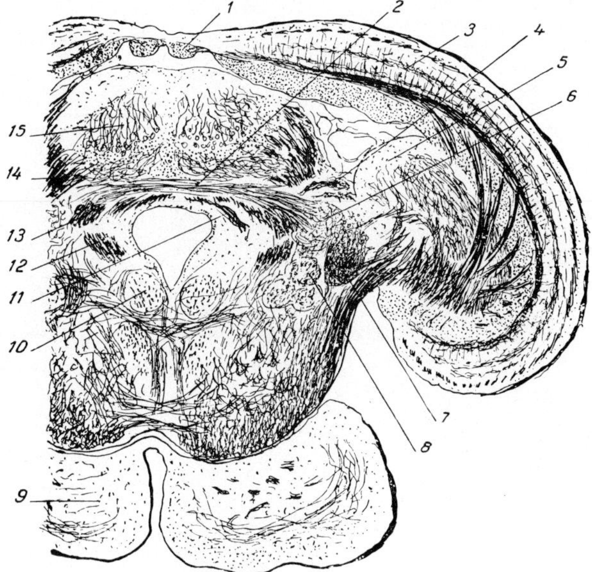

378 C

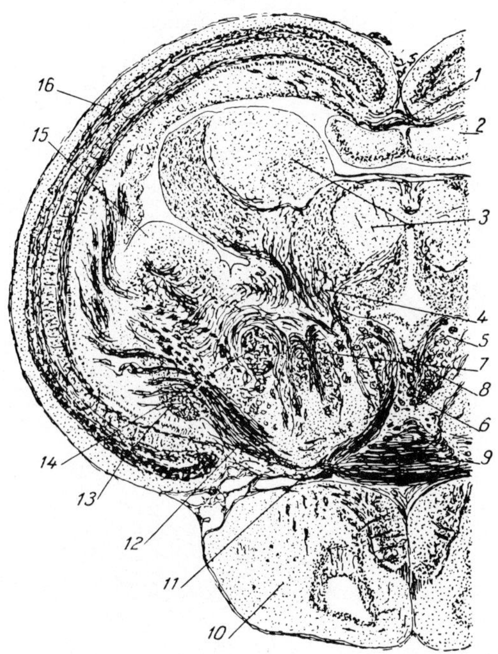

378 D

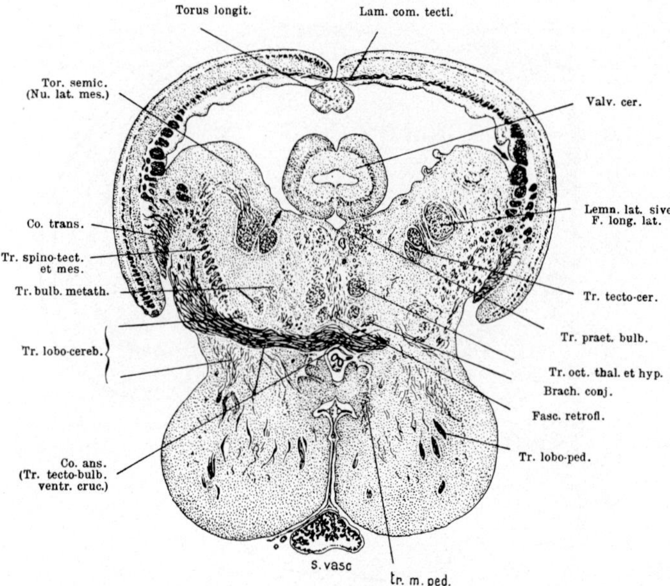

Figure 378E. Cross-section (myelin stain) through the mesencephalon of the Teleost Gadus morrhua, showing commissura ansulata (from KAPPERS, 1921). Co. ans.: commissura ansulata; co. trans.: commissura transversa (cf. preceding figure); tr. bulb. metath.: rostral extensions of general lemniscus reaching diencephalon; tr. praet. bulb. and tr. oct. thal. et hyp.: probably parts of fasciculus longitudinalis medialis. Other abbr. self-explanatory.

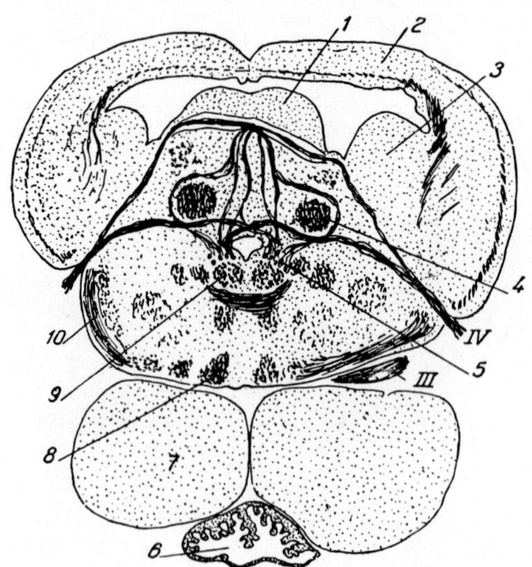

Figure 378 F

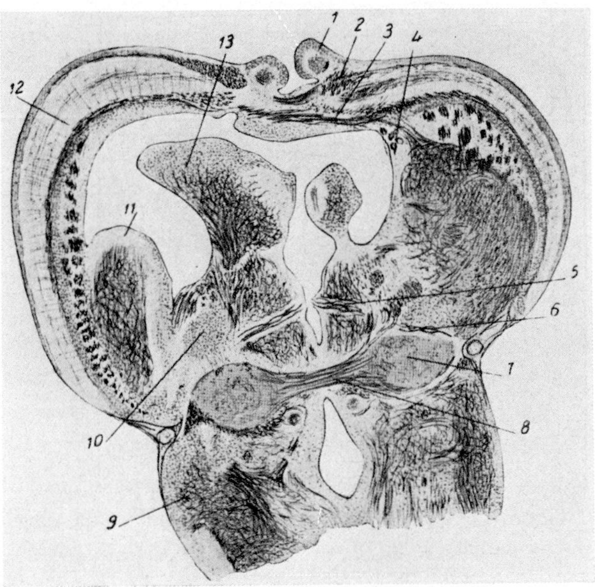

Figure 378G. Oblique rostrodorsal-ventrocaudal section *(Cajal impregnation)* passing through the mesencephalon of Carassius auratus (from BECCARI, 1943). 1: ganglion habenulae; 2: medial fasciculus of optic tract; 3: posterior commissure; 4: nucleus radicis mesencephalicae trigemini; 5, 6: trochlear nerve root before and after decussation; 7, 8: nucleus gustatorius secundarius with decussation; 9: eminentia granularis of cerebellum; 10: nucleus lateralis valvulae; 11: torus semicircularis; 12: tectum opticum; 13: valvula cerebelli.

Figure 378F. Cross-section (combined from several myelin stained sections) through the mesencephalon of Gadus morrhua at the level of the trochlearis decussations (after VAN DER HORST, from BECCARI, 1943). 1: valvula cerebelli; 2: tectum opticum; 3: torus semicircularis; 4: posterior mesencephalo-cerebellar fasciculus; 5: nucleus of trochlear nerve; 6: saccus vasculosus; 7: lobi inferiores hypothalami; 8: fasciculus tectobulbaris cruciatus; 9: fasciculus longitudinalis medialis; 10: fasciculus tectobulbaris rectus; III, IV: roots of oculomotor and trochlear nerves. The decussation below 9 is presumably part of the commissura ansulata system.

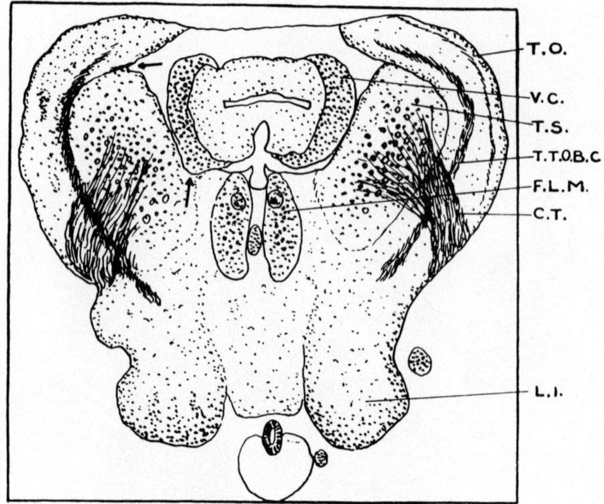

Figure 378H. Cross-section through the mesencephalon of the deep see Teleost Bathypterois, showing relationship of commissura transversa with torus semicircularis (from SHANKLIN, 1935). C.T.: commissura transversa; F.L.M.: fasciculus longitudinalis medialis; L.I.: lobi inferiores hypothalami; T.O.: tectum opticum; T.S.: torus semicircularis; T.T.OB.C.: tractus tectobulbaris cruciatus; V.C.: valvula cerebelli. Note the lamina epithelialis of root plate separating the antimeric tecta. Added arrows: sulcus lateralis mesencephali (above) and sulcus limitans (below).

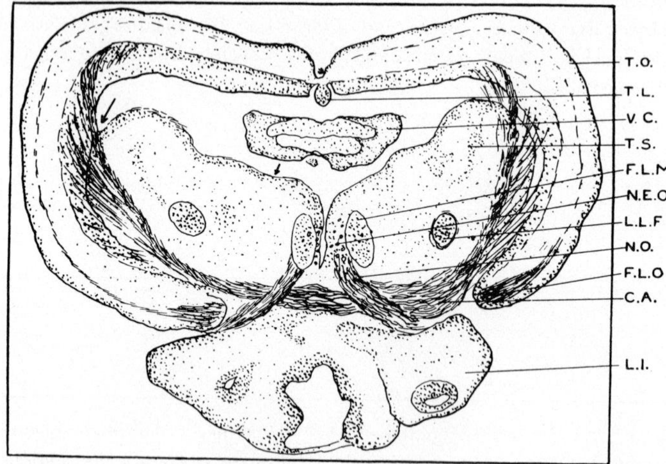

Figure 378 I. Cross-section through the mesencephalon of the deep sea Teleost Saurida at level of oculomotor root (from SHANKLIN, 1935). C.A.: commissura ansulata; F.L.O.: 'fasciculus lateralis tractus optici'; L.L.F.: fasciculus longitudinalis lateralis (lemniscus lateralis); N.E.O.: nucleus oculomotorii; N.O.: radix oculomotorii; T.L.: torus longitudinalis. Other abbreviations as in preceding Figure A. Added arrows: sulcus lateralis mesencephali (left), sulcus limitans (right).

this latter commissure, a subcommissural organ is present. From the pretectal region in general, an ill-defined pretecto-bulbar fiber system seems to arise, which may, however also contain ascending, i.e. reciprocal connections.

CHARLTON (1933) has investigated the optic tectum and its related fiber systems in the blind Teleosts (cave fishes) Troglichthys and Typhlichthys. He also compared his findings with those by KAPPERS (1921) and JANSEN (1929) in specimens of Teleosts with one atrophic eye. In the blind cave fishes, the deep medulla of the tectum predominates, but the outer layer (optic layer in other Teleosts) is still clearly recognizable. Its fibers are provided by an apparently expanded 'commissura superficialis tecti', 'which follows almost exactly the divisions of the normal optic tract' (CHARLTON, 1934).

Interesting data on the regenerative capacities of the Teleostean tectum opticum following experimental lesions have been recorded by W. KIRSCHE (1960), K. KIRSCHE and W. KIRSCHE (1961), and RICHTER (1965). This regeneration seems to take place by the migration of proliferating indifferent (undifferentiated) cellular elements from dorsal matrix zones of the midbrain.

Degeneration of optic fibers in the Teleost tectum, as recorded by electron microscopy, was recently investigated by LAUFER and VANEGAS (1974). These authors (VANEGAS *et al.*, 1974) also report on diverse secondary structural details concerning the tectum opticum, purporting to 'provide an adequate framework for the interpretation of electrophysiological data', but it cannot be claimed that, so far, this attempt has met with success.

As regards other alar plate grisea, the *torus semicircularis* with its nucleus lateralis mesencephali,[71] the nucleus lateralis valvulae, the nucleus isthmi, and the nucleus gustatorius secundarius, as already enumerated above, constitute the reasonably well recognizable and definable components. The torus semicircularis and its grisea represent an im-

---

[71] KAPPERS *et al.* (1936) describe it as a 'subtectal segment of the regio subtectalis', containing the 'nucleus tegmentalis medialis'. There is, however, in my opinion little doubt about the appertainance *(Zugehörigkeit)* of torus semicircularis and its cell masses to the alar plate or tectum *sensu latiori*. Since these cell masses, located dorsally to the sulcus limitans, extend laterad, the term nucleus lateralis mesencephali may be appropriate (K., 1927). Nevertheless, in various instances, a medial component (torus semicircularis *sensu strictiori* with a 'nucleus mesencephali medialis'), separated from a lateral component (the nucleus lateralis *sensu strictiori)* by fiber masses, might perhaps be distinguished.

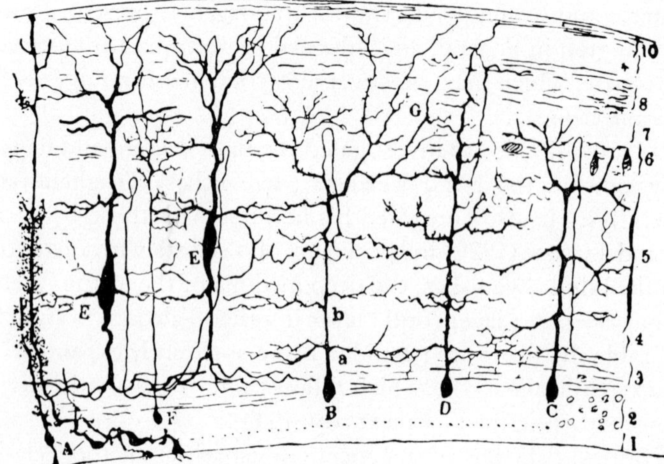

Figure 379A. Cross-section *(Golgi method)* through the tectum opticum of the Teleost Barbus fluviatilis (after PEDRO RAMÓN, from CAJAL, 1911). 1: ependyma; 2: granular layer; 3: deep medulla; 4: *'substance grise moyenne'*; 5: *'grande couche plexiforme'*; 6: *'couche des cellules fusiformes et des fibres optiques profondes'*; 7–10: additional optic layers. A: ependymal cells; b: *'cellule à cylindre-axe en anse'*; C: crozier cell; D: cell with centrifugal axon, supposedly reaching the retina; E: large crozier cells of 5th layer; G, a, b: various unidentified cell processes presumed to be dendrites; F: *'cellule sans cylindre-axe'* (amacrine cell).

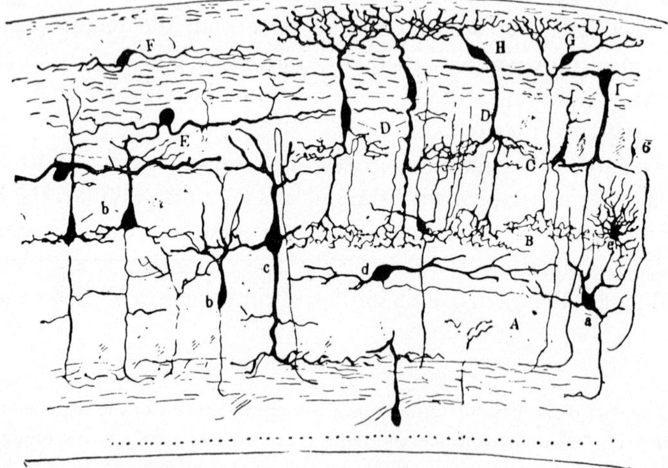

Figure 379B. Cross-section *(Golgi method)* through the tectum opticum of Barbus, showing additional details (after P. RAMÓN, from CAJAL, 1911). A: lower stratum of 5th layer; B, C: middle and upper stratum of 5th layer; D, E, F, G, H, I: diverse unidentified neuronal elements; a: nerve cells of 5th layer; b: *'cellules pyramidales à cylindre axe central'*; c: crozier cells; d: 'transversal cells' of 5th layer; e: small stellate cells.

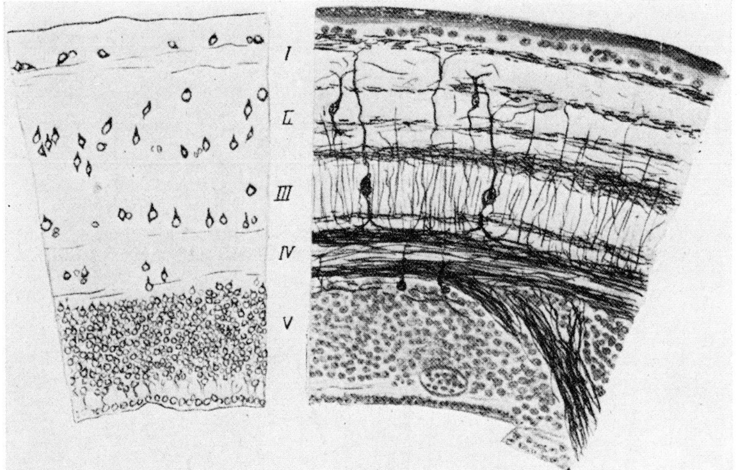

Figure 379C. Cross-sections through the tectum opticum of the Teleost Gambusia, showing *Nissl picture* at left, and a modified *Cajal impregnation* at right (from Beccari, 1943). I–V sequence of layers in the enumeration by Huber and Crosby. I and II are optic layers, III is the main layer of chief cells, IV is the deep medulla, and V is the granular layer.

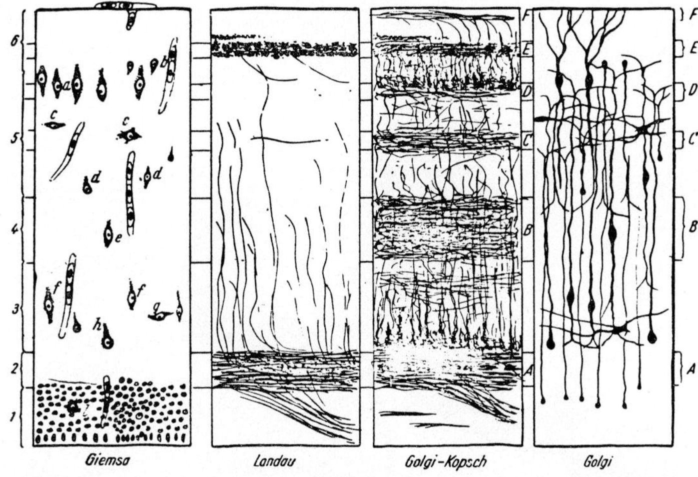

Figure 379D. Transverse sections through the cortex tecti mesencephali in the Teleost Carassius auratus (from Leghissa, 1955). *Giemsa stain* for cytoarchitecture, *Landau stain* for myelin, *Golgi-Kopsch impregnation* for medullated and non-medullated fibers, *Golgi impregnation* for neuronal relationships. 1: periventricular gray; 2: deep medulla; 3: internal gray layer; 4: internal plexiform layer; 5: external gray layer; 6: plexiform retinal fibers and '*strato fibroso marginale*'; A: 'centrifugal fibers'; B: 'plexiform layer'; C: 'plexiform layer' of external griseal stratum; D: 'external or retinal plexiform layer'; E: 'afferent optic and olfactory fibers'; F: 'marginal fiber plexus'; a: 'pyramidal neuron'; b: 'marginal neuron'; c: 'horizontal associative neuron'; d: 'small pyramidal neuron'; e, f, g: 'pyramidal neurons' of respective layers; g: 'multipolar associative neurons'; i: 'piriform periventricular neurons'.

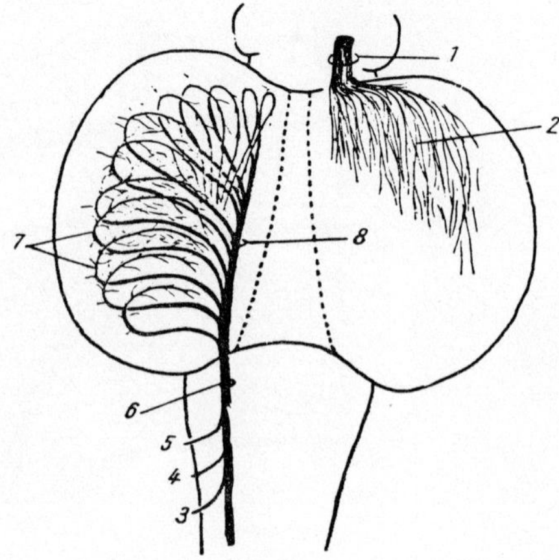

Figure 379E. Diagram purporting to show lemniscal and olfactory input into tectum opticum of Carassius (from LEGHISSA, 1955). 1, 2: 'fasciculus striotectalis' and its tectal projection (conceived as essentially 'olfactory' by the cited author); 3: octavolateral lemniscus components; 4: lateralis components of facial nerve; 5: trigeminal lemniscus components; 6: general bulbar lemniscus; 7: tectal projections of lemniscus systems; 8: common lemniscus (general lemniscus and fasciculus longitudinalis lateralis, i.e. octavolateral lemniscus).

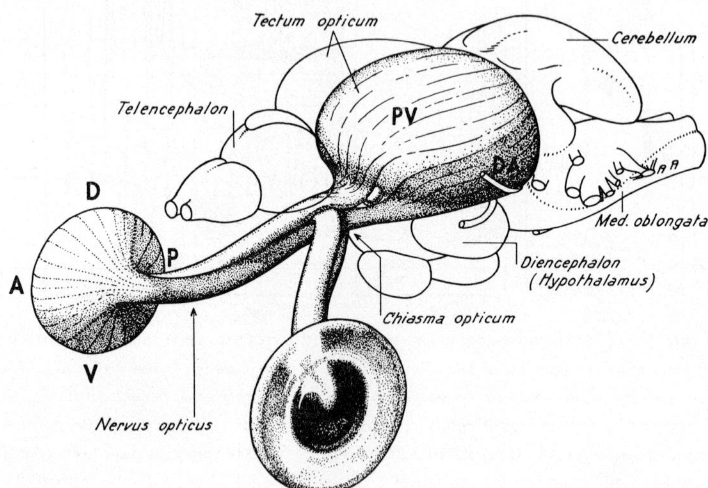

Figure 379F. Sketch showing the course of the optic pathway from retina to tectum opticum in the Teleost Salmo, and illustrating some features of the projection of retinal quadrants (from STRÖER, 1939). A, D, P, V: rostral, dorsal, caudal, and ventral retinal quadrants (cf. Fig. 360A).

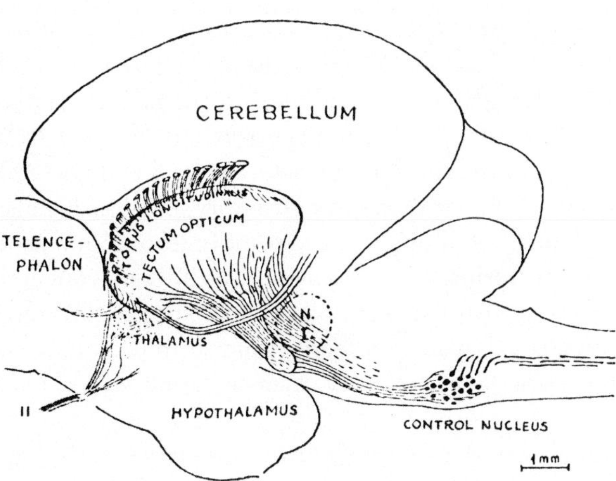

Figure 380. Semidiagrammatic sketch illustrating some connections of the tectum opticum and torus longitudinalis, projected upon a midsagittal plane, in the electric fish Gymnotus (Electrophorus) electricus (from DE OLIVEIRA CASTRO, 1961). II: optic nerve (note its thinness); N.I.: nucleus isthmi; control nucleus is the reticular griseum *(noyau de commande of Fessard)* supposed to control the electric discharge activities.

portant terminal 'center' of the vestibulolateral lemniscus (lemniscus lateralis, fasciculus longitudinalis lateralis). The nucleus lateralis valvulae *(Übergangsganglion)*, located medially to the nucleus lateralis mesencephali, is closely associated with this latter. Both grisea have connections with the tectum opticum by way of the deep medulla. Whether lemniscus lateralis fibers proceed into the tectum as in Plagiostomes, or whether the vestibulolateral input reaching tectum opticum is entirely relayed by the nucleus lateralis mesencephali of Osteiichthyes remains a moot question.

The *nucleus isthmi*, in the caudal region of the alar plate, but still representing a derivative of its dorsal zone, is likewise a griseum of the vestibulolateral lemniscus and also receives other, insufficiently understood ascending bulbar tracts. The nucleus isthmi is interconnected with tectum and pretectal region by means of reciprocal isthmotectal and tecto-isthmal, respectively isthmopretectal etc. tracts.[72] According to KAPPERS *et al.* (1936) the Osteichthyan nucleus isthmi 'forms a correlation between photostatic and gravistatic impulses'.

[72] E.g. the tractus isthmo-opticus of FRANZ. This fascicle, however, presumably only traverses the bundles of the optic tract.

The *nucleus gustatorius seu visceralis secundarius ('Rindenknoten')* can be evaluated as a derivative of the mesencephalic alar plate's intermediodorsal zone. This griseum, depending on the pattern distortions displayed by Osteichthyes, is located ventromedially, medially, dorsomedially, or caudally to the nucleus isthmi (cf. e.g. Figs. 223A, 223B, 233A, B, 378B, C, G). It correlates gustatory, presumably also 'general visceral', and olfactory impulses, and some of its output reaches the lobi inferiores s. posteriores hypothalami. Additional connections with tectum (tractus gustotectalis), torus semicircularis, cerebellum via nucleus lateralis valvulae, and tegmentum seem to obtain, including a descending secondary gustatory tract as discussed in section 5 of chapter IX.

Turning now to the components of the *basal plate*, the efferent *nuclei of oculomotor and trochlear nerve* may be considered first. The *nucleus oculomotorius* (III) is located either between, or dorsally to, the bundles of the fasciculus longitudinalis medialis. The oculomotor root fibers emerge basally, and are both uncrossed (homolateral) and crossed (contralateral). A distinct nuclear subdivision, containing the preganglionic elements of this nerve, cannot be generally recognized, although, in some forms, a dorsal and a ventromedial oculomotor cell group may be seen (cf. Fig. 378A). The dorsal group might perhaps include the preganglionic cells, and, at least in part, represent a derivative of the basal plate's intermedioventral zone.[73] The *nucleus of the trochlear nerve* (IV) is located at a short distance caudally to that of the oculomotor, commonly in line with this latter (cf. Fig. 381A) or somewhat more dorsally (cf. Fig. 381B). The bundles of fasciculus longitudinalis medialis generally run ventrally to the trochlear nucleus all or most of whose fibers decussate in the isthmus region respectively in velum medullare anterius. The trochlear decussation may manifest various displacements, distortions, or splittings correlated with the pattern distortions obtaining in many Teleosts (cf. Fig. 378F). Although seemingly pertaining to the basal plate's ventral zone, the nucleus of this morphologically rather peculiar nerve has also been interpreted, from

[73] It will be recalled that, at least in the Teleosts investigated with respect to autonomic effects on intrinsic eye musculature, the oculomotor innervates the m. dilator pupillae (cf. p. 930, chapter VII, section 6 of vol. 3, part II). The transformation of extrinsic eye musculature into the elctric organ of the Teleost Astroscopus, and the innervation of this organ by the oculomotor, may also again be pointed out (cf. p. 891, chapter VII, section 5, vol. 3, part II).

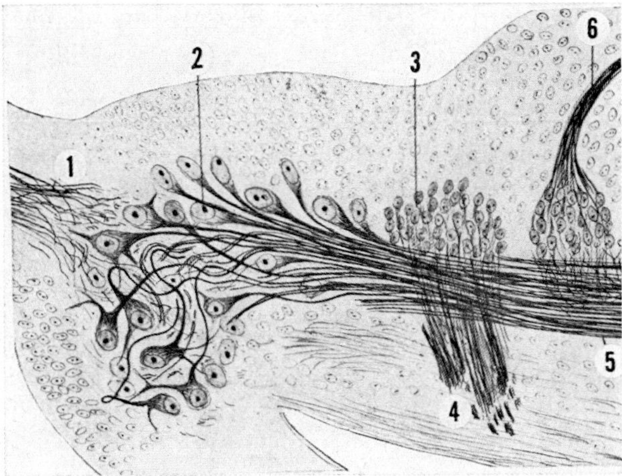

Figure 381 A. Paramedian sagittal section through the mesencephalic tegmentum in an advanced larva of the Teleost Trutta fario (trout), showing origin of fasciculus longitudinalis fibers from *Cajal's nucleus interstitialis* at the rostral end of the mesencephalic tegmentum, as displayed by means of that author's silver impregnation technique (from CAJAL, 1923). 1: fibers of posterior commissure reaching nucleus interstitialis; 2: other elements of mesencephalic tegmentum contributing to the fasciculus *('nucleus of van Gehuchten');* 3: oculomotor nucleus and its root fibers (4); 5: fasciculus longitudinalis medialis with collaterals to nucleus trochlearis; 6: root and nucleus of trochlear nerve.

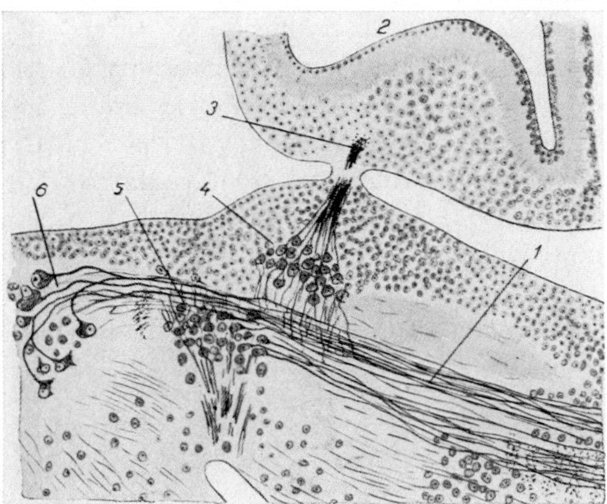

Figure 381 B. Paramedian section *(Cajal method)* through mesencephalic tegmentum of an advanced larva of the Teleost Salmo irideus (from BECCARI, 1943). 1: fasciculus longitudinalis medialis; 2: valvula cerebelli; 3, 4: decussation and nucleus of trochlear nerve; 5: nucleus of oculomotor nerve; 6: *Cajal's interstitial nucleus* of fasciculus longitudinalis medialis.

a phylogenetic viewpoint, as a derivative of the intermedioventral zone.[74]

The mesencephalic *nucleus s. formatio reticularis tegmenti* includes a number of more or less diffusely arranged cell groups which display considerable variations in the different Teleostean groups. Figure 378A shows a fairly compact '*nucleo laterale profundo del mesencefalo*' pointed out by BECCARI (1943). Although the mesencephalic tegmentum receives decussating fibers of the rostral cerebellotegmental tract ('brachium conjunctivum') a typical 'nucleus ruber tegmenti' cannot be delimited. At the mesencephalo-diencephalic boundary, however, a well developed mesencephalic tegmental nucleus of the fasciculus longitudinalis medialis *(interstitial nucleus of Cajal)* is present (Figs. 381A, B). In the basal midline of the tegmentum, a *nucleus interpeduncularis* with its neuropil has essentially similar features in all examined Ganoids and Teleosts (Figs. 376D, 378B). Although a typical nucleus opticus tegmenti has not been recorded, there is little doubt that direct *optic input* reaches the mesencephalic tegmentum. JANSEN (1929) described in Teleosts a small bundle of optic fibers, the fasciculus dorsomedialis, which terminates in the tegmentum. It could be considered analogous (rather than 'homologous') to components of the basal optic tract of other Vertebrates.[75] A similar system can be presumed to obtain in Ganoids. It was, as JANSEN points out, probably already noted by WALLENBERG (1913).

The fiber tracts related to the mesencephalon of Ganoids and Teleosts can be classified in accordance with the criteria adopted in the preceding sections of this chapter either as (1) afferent, (2) efferent, (3) intrinsic, and (4) transit channels, or, with respect to (1) and (2), as (a) rostral input, (b) rostral output, (c) caudal input, and (d) caudal output pathways, moreover (5) as commissural systems. It should be kept in mind that, except for some of the main features displayed by the most conspicuous channels, such as optic tract, fasciculus retroflexus, bulbospinal lemniscus, lateral lemniscus (octavolateral), cerebelloteg-

---

[74] Cf. the discussion by BECCARI (1943, p. 276) and the comments on the 'trochlearis problem' in chapter VII, section 3, p. 844 of vol. 3, part II).

[75] JANSEN (1929) distinguishes 5 components of the Teleostean optic tract: (1) fasciculus dorsalis, (2) fasciculus lateralis, (3) fasciculus medialis, (4) fasciculus dorsomedialis, (5) tractus preoptico-opticus. JANSEN believes that this latter bundle is, at least in part, efferent, i.e. providing input to the eye bulb. Nevertheless, it might be essentially afferent and likewise represent a component of the basal optic tract, which JANSEN assumes to be lacking in Teleosts.

mental tract, tectocerebellar connections, mesencephalodiencephalic systems, tectobulbar system, and fasciculus longitudinalis medialis and related tegmental longitudinalis channels, the available data concerning detailed interrelations remain rather ambiguous and flimsy.[76] Many of these channels, e.g. the fasciculus longitudinalis medialis and several others, are doubtless reciprocal, i.e. ascending and descending, or mediating both input and output for various grisea.

Generally speaking, the mesencephalic fiber system in Ganoids and Teleosts are essentially similar to those described in section 3 for Plagiostomes. Significant differences related to peculiarities displayed by Osteichthyes were considered above, and include the connections of *torus longitudinalis*, of *nucleus lateralis valvulae*, as well as of the cerebellum, with its tectocerebellar and cerebellotegmental fasciculi. Again, the intrinsic mesencephalic and the extrinsic connections of the well developed *nucleus gustatorius secundarius* of many Teleosts, and of the *nucleus isthmi* manifest diverse and variable characteristics differing from those fiber channels in Plagiostomes. Figure 380 illustrates, with respect to the peculiarities obtaining in some Teleosts, connections of torus longitudinalis, tectum opticum, and of bulbar 'control nucleus' for the electric organ of the Gymnotid Electrophorus electricus.

The *dorsal mesencephalic commissural systems* in Ganoids and Teleosts include the partly diencephalic commissura posterior, the commissura tecti, and the decussation of the trochlear nerve. In addition, a substantial '*commissura gustativa*', described by BECCARI (1943) and others, may interconnect the antimeric grisea of nucleus gustatorius secundarius and take its course dorsally to the ventricle (cf. Fig. 378C).

Within the *basal plate*, the decussation of the cerebellotegmental channel and of the *commissura ansulata*, which shows diverse components and variations, are the most conspicuous.[77] Decussating fibers related to mesencephalic grisea are also included in the complex postoptic respectively supraoptic commissures (e.g. commissura transversa) located in the diencephalon and to be discussed in chapter XII of volume 5.

---

[76] This comment applies, of course, not only to Osteichthyes but also to the other Vertebrates. It is related both to 'psychologic' factors in the interpretation by the observers and to the inherent shortcoming of all technical methods, be they normal fiber stains respectively impregnations, the diverse degeneration techniques or the various recordings obtained by physiologic experimentation including procedures of electrophysiology.

[77] Decussating fibers of the oculomotor root should here also again be mentioned.

Few investigations have been undertaken on the brain of deep sea Teleosts, which live in a rather unusual environment. Data on the midbrain (and diencephalon) in a few of these forms, with additional references, were recorded by SHANKLIN (1935). In some species, an exceedingly massive *commissura transversa* was noted. In the nearly blind *Bathypteroid*, a torus longitudinalis is not present, and the antimeric optic tecta are separated by the lamina epithelialis of a persistent roof plate (Fig. 378H). In *Saurida*, on the other hand, where the optic system is better developed, the torus longitudinalis is present, but the torus semicircularis is smaller, and the commissura transversa is reduced, the commissura ansulata, however, related to efferent tectal fibers, being quite conspicuous (Fig. 378I).

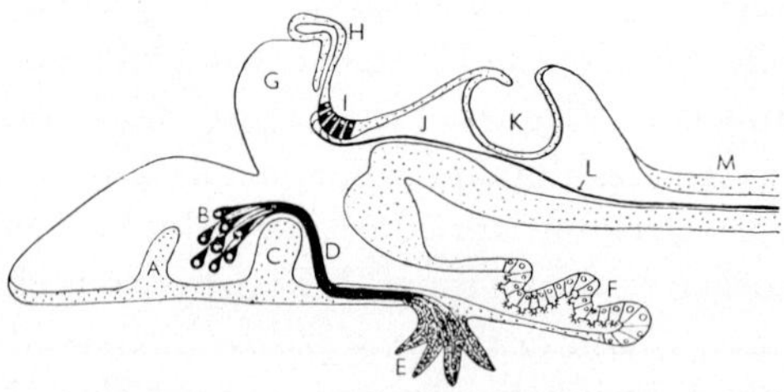

Figure 381C. Diagrammatic midsagittal section through the brain of the Ganoid Polypterus showing origin of *Reissner's fiber* at subcommissural organ (from OLSSON and WINGSTRAND, 1954). A: anterior commissure; B: preoptic nucleus; C: chiasmatic ridge; D: tractus praeoptico-hypophyseus; E: neurohypophysis; F: saccus vasculosus with primary sensory cells; G: saccus dorsalis (roof plate); H: epiphysis; I: subcommissural organ; J: mesencephalic ventricle; K: cerebellum; L: Reissner's fiber in fourth ventricle; M: approximate rostral end of spinal cord.

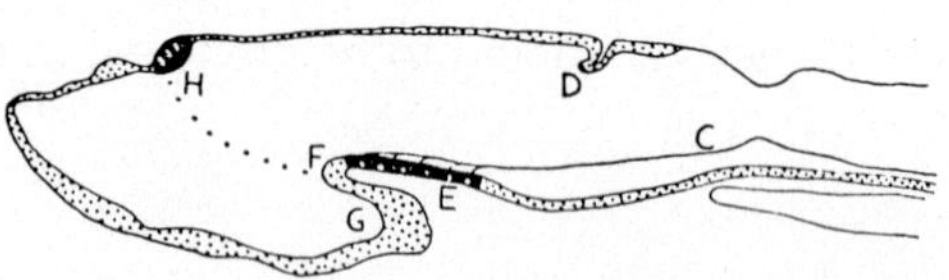

Figure 381D. Diagrammatic midsagittal section through the brain in a 14 days old (5.6 mm) embryo of the Teleost Salmo, showing origin of Reissner's fiber at the mesencephalic 'flexural organ' (from OLSSON, 1956). C: *Reissner's fiber;* D: plica rhombomesencephalica; E: 'flexural organ'; F: 'plica ventralis' (tuberculum posterius); G: 'infundibulum' (lobi inferiores hypothalami); H: subcommissural organ.

Reissner's fiber, dealt with in chapter VIII, seems to arise from the subcommissural organ in practically all Vertebrates (Fig. 381C), but from the so-called infundibular organ of Amphioxus (cf. Fig. 25B). In this respect, it is of interest that at embryonic stages of the Teleosts Salmo and Esox (and perhaps also at comparable stages of the Cyclostome Petromyzon and the anuran Amphibian Xenopus), OLSSON (1956) found *Reissner's fiber* to arise not at the subcommissural organ, as in the adult animals, but at a secretory ependyma of the mesencephalic floor (Fig. 381D), which the cited author designates at the '*flexural organ*'.

The mesencephalon of the *Crosspterygian Latimeria*, as described by MILLOT and ANTHONY (1965), is a tubular configuration of relatively substantial length '*ovalaire en coupe transversale, plus haut que large sur la majeure part de son étendue*' (p. 18, loc. cit.). At the isthmus region and at the level of commissura posterior, the transverse (right-left) diameter becomes longer than the dorsoventral one (Figs. 382A–C).

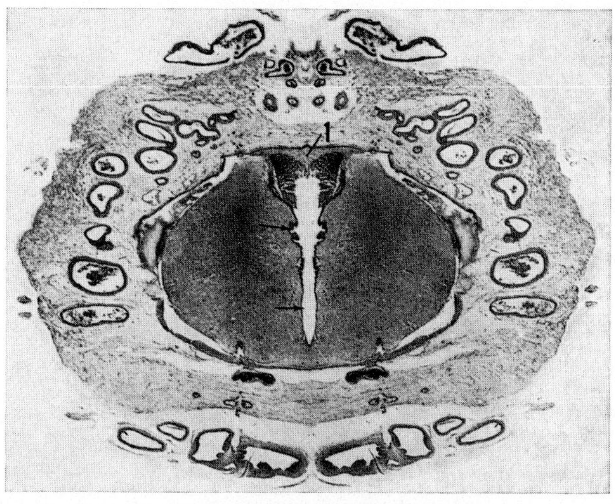

Figure 382A. Cross-section (protargol impregnation) through the rostral part of the mesencephalon in the Crossopterygian Latimeria (from MILLOT and ANTHONY, 1965). Added legends: 1: commissura posterior with adjacent subcommissural organ; upper arrow: sulcus lateralis mesencephali system, perhaps overlapping with caudal end of sulcus synencephalicus; lower arrow: approximate location of sulcus limitans. This and the following figure is a 'montage' compounded from a section through a single (antimeric) half (hemisection).

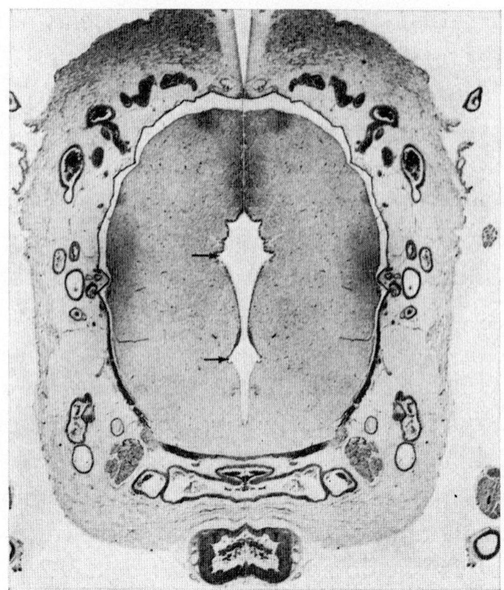

Figure 382B. Cross-section through an intermediate level of the mesencephalon in Latimeria (from MILLOT and ANTHONY, 1965). Added arrows: sulcus lateralis mesencephali (internus) above, and sulcus limitans, below.

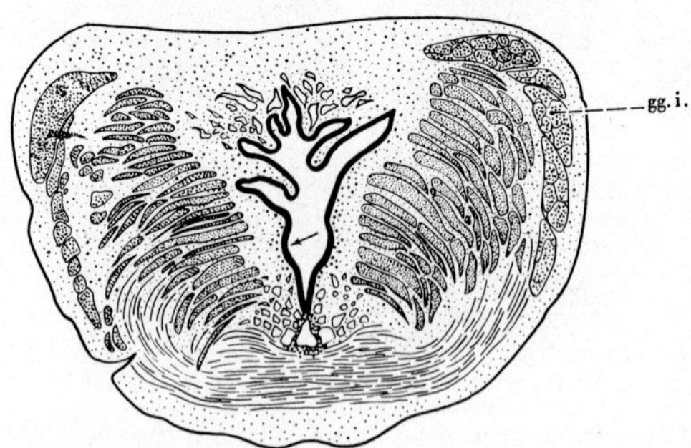

Figure 382C. Sketch illustrating a cross-section through caudal portion of mesencephalon in Latimeria, closely rostral to isthmus (from MILLOT and ANTHONY, 1965). gg.i.: nucleus isthmi; added arrow: sulcus limitans; dorsally a system of accessory sulci, perhaps corresponding to caudal end of sulcus lateralis mesencephali (internus); the large basal commissure corresponds to the commissura ansulata system.

Millot and Anthony (1965) justly remark that '*la forme générale du mésencéphale, d'une grande simplicité, s'écarte de celle que l'on observe chez la plupart des Vertébrés inférieurs*'. This is particularly the case with regard to the narrow mesencephalic ventricle which, however, displays the sulcus limitans, and, at intermediate levels of its length, a sulcus lateralis mesencephali (internus)[78] with some more dorsally located accessory sulci (Fig. 382B). It is thus possible to distinguish a tectum opticum *sensu strictiori*, a torus semicircularis, and a tegmentum. A torus longitudinalis does not seem to be present.[79]

The *tectum* is rostrally bounded by the posterior commissure with its rather large subcommissural organ (Fig. 382A). An ill-defined griseum, representing the dorsal thalamic, pretectal nucleus of that commissure, blends with the tectal cell masses. A typical nucleus lentiformis mesencephali, although presumably present, cannot be delimited on the available photomicrographs, and is not mentioned by the cited authors. The caudally adjacent commissura tecti seems to be inconspicuous and is likewise not mentioned by Millot and Anthony who point out a dorsal '*raphé névroglique*' separating the antimeric halves of the alar plate (Fig. 382B). The tectum displays an external layer of longitudinal optic fibers internally to which a stratum of interwoven fibers and scattered neuronal elements is located.[80] Still more internally follows a 'stratum griseum' or '*zone feuilletée*', formed by a nondescript sequence of about 3 to 9 overlapping indistinct cellular laminae vaguely separated by fiber zones, and bordering on the ependyma. In addition, Millot and Anthony identified the radix mesencephalica trigemini which takes its course between 'stratum album' and 'stratum griseum'. The large cells of the nucleus radicis mesencephalicae trigemini are located near the dorsal midline and found only in the caudal two thirds of the tectum.

The *torus semicircularis*, although not very prominent, nevertheless protrudes into the ventricle and presumably receives the vestibulola-

---

[78] Millot and Anthony (1965) who mention the sulcus limitans, do not refer to the sulcus lateralis mesencephali clearly shown in their photomicrographs.

[79] The cited authors state: '*On n'observe d'autre part rien qui ressemble, de près ou de loin, aux tores longitudinaux des Téléostéens*'. One might, nevertheless, wonder whether this structure is not vaguely suggested by the slight protrusion shown on both sides of the dorsal ventricular midline between the median sulcus and the laterally adjacent dorsalmost accessory sulcus in Figure 381B).

[80] This might perhaps in part correspond to the layer of chief cells, the deep medulla being possibly represented by the fiber lamellae internally to that stratum.

teral lemniscus, not specifically mentioned by the quoted authors, who describe, however, a *nucleus isthmi* (Fig. 382C) which they consider part of the tegmentum. The two antimeres of this griseum, which is in my opinion a component of the *alar plate*, are interconnected by a commissure passing through the velum medullare anterius. A nucleus gustatorius seu visceralis secundarius was not identified, but might be represented by cell groups located medially to ganglion isthmi.

As regards the *basal plate* or *true tegmentum*, the nuclei of oculomotor and trochlear nerves are present as recognizable cell groups, the trochlear nucleus being separated by a short distance from that of the third nerve, and extending caudad as far as the level of decussation of its root fibers in the velum medullare anterius. The reticular formation consists of diffusely distributed neuronal elements, large multipolar cells being particularly found in the rostral tegmentum and corresponding in part to the nucleus interstitialis of the fasciculus longitudinalis medialis. The *tuberculum posterius* at the rostral end of the tegmentum is highly flattened and rather indistinct. Instead of being telescoped into each other, as in some other Anamnia, mesencephalon and diencephalon are, figuratively speaking stretched out, since the hypothalamus is '*étiré en avant*' without a conspicuous '*courbure du plancher vers l'arrière*'. A *nucleus interpeduncularis*, not mentioned by the quoted authors, is doubtless present and can be vaguely recognized in some of their illustrations. The question whether a tegmental optic griseum is or is not present, must be left unanswered. Generally speaking, the available photomicrographs seem to indicate that, as e.g. in various Selachians (including Chimaeroids, cf. K. and Niimi, 1969) the grisea in the brain of Latimeria display a very diffuse arrangement apparently precluding a well-defined delimitation of cytoarchitectural units. Likewise, most fiber tracts do not seem to be very clearly outlined.

As regards the fiber systems of the mesencephalon, Millot and Anthony (1965) mention, in addition to the radix mesencephalica trigemini, the optic tract, ending in the tectum with a dorsal and a ventral fascicle, Edinger's tractus spinomesencephalicus (the general spinobulbar lemniscus), and the descending tectobulbar and tectospinal system. A doubtless present lateral (octavolateral) lemniscus is not specifically pointed out. With respect to the tegmentum, the fasciculus retroflexus, the fasciculus longitudinalis medialis, a thalamotegmental tract, a hypothalamotegmental tract, and components of the basal forebrain bundle reaching the tegmentum (tractus striotegmentalis and striopeduncularis) were reported by the cited authors.

Concerning the *commissural systems*, posterior commissure, commissura tecti, commissure of the nucleus isthmi, and decussation of the trochlear nerve were mentioned above, the two latter systems of crossing fibers passing through the velum medullare anterius. In addition, a vaguely outlined system of tegmental commissures is present in the basal plate, and can be subsumed under the concept of commissura ansulata *sensu latiori* (cf. Fig. 382C). This tegmental commissural system presumably includes decussating fibers of the fasciculus retroflexus.

5. Dipnoans

Data on the mesencephalon of *Lungfishes* are summarized for *Ceratodus* in the studies by HOLMGREN and VAN DER HORST (1925) and for *Protopterus* by GERLACH (1933). Both papers contain a bibliography referring to the findings of previous authors. Generally speaking, the

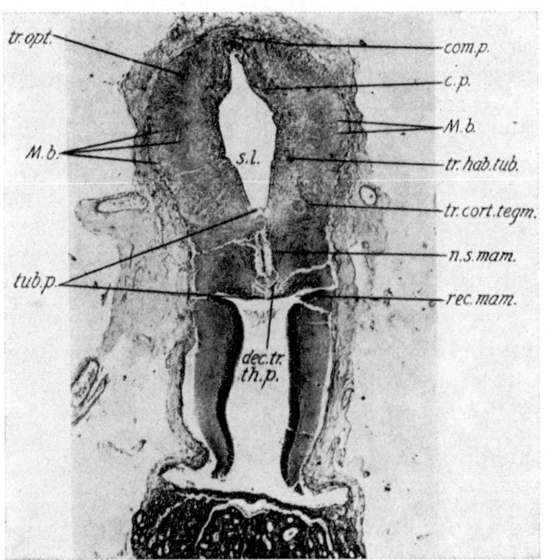

Figure 383 A. Cross-section (silver impregnation) through the rostral portion of the mesencephalon in the Dipnoan Ceratodus (from HOLMGREN and VAN DER HORST, 1925). com.p.: posterior commissure; c.p.: nucleus of posterior commissure; dec.tr.th.p.: 'decussatio tractus thalamopeduncularis'; M.b.: fasciculus retroflexus; n.s.mam.: 'nucleus supramammillaris' (probably rostral end of tegmenta cell plate); s.l.: sulcus limitans; tr.cort.tegm.: 'tractus cortico-tegmentalis complex' (probably mesencephalic extension of basal forebrain bundle system); tr.hab.ped.: 'tractus habenulotubercularis'; tub.post.: tuberculum posterius; tr.opt. and rec.mam. self-explanatory.

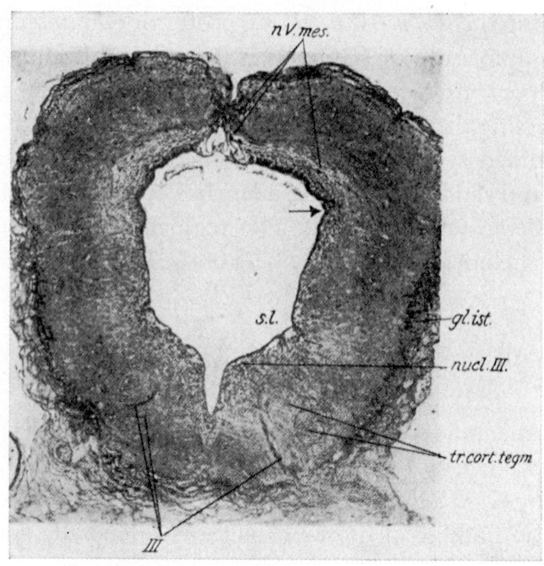

Figure 383B. Cross-section (silver impregnation) through the mesencephalon of Ceratodus at the level of oculomotor nucleus and root (from Holmgren and van der Horst, 1925). gl.ist.: nucleus isthmi; n.V mes.: nucleus of mesencephalic trigeminal root; s.l., tr.cort.tegm.cf.preceding figure. Added arrow: sulcus lateralis mesencephali. The nucleus interpeduncularis, not labelled, can be faintly recognized in the basal midline region.

Dipnoan mesencephalon appears less well developed than that of Plagiostomes and of most Osteichthyes. Again, the mesencephalon of Ceratodus manifests a higher degree of differentiation than that of Protopterus. In this latter form, the tectum displays a conspicuous rostral and caudal recess (cf. Figs. 325A, B, D, E, 384A, B), not present in Ceratodus, where, at most, a caudal recess is barely suggested (cf. Fig. 324A).[81]

Figures 383A, B, illustrate cross-sections through the mesencephalon of Ceratodus, while Figure 324B, at the level of corpus cerebelli, still includes caudal, respectively isthmic mesencephalic structures, namely nucleus isthmi and nucleus gustatorius secundarius of the alar plate, and nucleus interpeduncularis within the tegmentum.

---

[81] Cf. also the sagittal sections through the brain of Ceratodus in Figure 72, p.197 of volume 3, part II, and the plate Figures 1 and 2, illustrating a model of that brain, in the paper by Holmgren and van der Horst (1925).

It can be seen that the *sulcus limitans* and a *sulcus lateralis mesencephali (internus)* are recognizable. A sulcus lateralis mesencephali externus may be faintly suggested. Sulcus intermedius dorsalis and sulcus intermedius ventralis cannot be properly identified in the mesencephalon of Ceratodus (nor in that of Protopterus). The cell populations retain an essentially periventricular arrangement.[82]

As regards the *alar plate*, the tectum opticum has its rostral end at the level of commissura posterior with its subcommissural organ and its nucleus (Fig. 383A). An ill-defined *nucleus lentiformis mesencephali* is presumably present and may, at least in part, correspond to what HOLMGREN and VAN DER HORST have described as their 'primordium of the pretectalis nucleus'. The *tectum opticum* displays some nondescript relatively thick and dense periventricular layers, and more peripheral cells diffusely distributed throughout the whole stratum album, without forming distinct layers as e.g. in many Teleosts and in anuran Amphibians. The *nucleus of the mesencephalic trigeminal root* consists of large elements more or less distributed over the whole tectum. A median and a more lateral, as well as a superficial and a ventricular row can roughly be distinguished.

The *torus semicircularis* contains a periventricular cell aggregation, laterally to which a group of scattered neuronal elements form a more superficial griseum. Near the isthmus, a *nucleus isthmi* and a *nucleus visceralis secundarius* are present (Figs. 383B, 324B).

The *tegmentum (i.e. the basal plate)* has its rostral end at the *tuberculum posterius*, which in Ceratodus displays some pecularities characterized by what could be called an intertubercular recess (Fig. 383A) which HOLMGREN and VAN DER HORST describe as a slight depression between two eminences. A 'nucleus supramammillaris' located about this recess may, in my interpretation, represent the rostral end of the tegmental cell plate.

---

[82] In describing the cellular arrangements, HOLMGREN and VAN DER HORST (1925) refer to a concept of four columns elaborated by PALMGREN (1921), which they accept with some modifications. Their 'dorsal column' seems to be the dorsal zone of the alar plate, their 'lateral column' may correspond to the intermediodorsal zone which, however, is barely suggested as such in the Dipnoan mesencephalon. Their 'ventral column' (PALMGREN's medial), pertaining to the basal plate, includes apparently both the intermedioventral zone and the greater part of the ventral zone. Their 'medial column' (PALMGREN's ventral) seems to represent a raphé component of the ventral zone and perhaps derivatives of the floor plate. Again, the intermedioventral zone is barely identifiable, if at all, in the Dipnoan mesencephalon.

The cell groups described by the cited authors as nucleus tuberis posterioris, nucleus fasciculi longitudinalis medialis, and 'nucleus ruber' can be considered indistinct differentiations of the *nucleus reticularis tegmenti*. In addition, an *interpeduncular nucleus* is present. The neuronal elements of the formatio reticularis s. nucleus motorius tegmenti seem, on the whole, smaller than the *motoneurons of the oculomotor nerve* (Fig. 383B). Some smaller elements in the rostral part of the nucleus of this nerve might be preganglionics. The root fibers of the oculomotor which include a decussating component, emerge at caudal levels of the nucleus. The *nucleus of the trochlear nerve* is separated by a short but noticeable distance from the oculomotor nucleus and, like this latter, is clearly located in the ventral zone of the basal plate, even slightly more medially. The motoneurons of the trochlear nucleus, of medium size, are not as large as those of the oculomotor. The scattered trochlear root fibers[83] seem to undergo a complete dorsal decussation in the velum medullare anterius and closely adjacent rostral parts of corpus cerebelli.

Figures 384A–D illustrate representative sections through the mesencephalon of Protopterus. The lesser degree of differentiation in comparison with the midbrain of Ceratodus is here quite evident, particularly also with respect to the dense periventricular arrangement of the cell masses.[84] *Sulcus limitans* and faintly suggested *sulcus lateralis mesencephali (internus)* are recognizable,[85] but there is relatively little cytoarchitectural distinction between the fundamental longitudinal zones respectively subdivisions of alar and basal plate. Nevertheless, in the *alar plate*, the location of *nucleus commissurae posterioris* and *nucleus lentiformis mesencephali* within the mesencephalo-diencephalic boundary neighborhood can be identified (cf. Figs. 384B). *Tectum opticum* and *torus semicircularis* are likewise distinguishable (cf. Figs. 384C, 325F). In the rostral mesencephalic recess (Fig. 384A) only the optic tectum *sensu strictiori* seems to be represented, while the caudal recess (fig. 325D) might include a portion of torus semicircularis. The tectum opticum, which in-

---

[83] Holmgren and van der Horst attempted to count the number of motoneurons and root fibers of the nervus trochlearis in Ceratodus. Roughly, a bilateral total of about 200 cells and fibers (i.e. 100 for each antimeric portion) seems to obtain.

[84] It should, however, be pointed out that the depicted sections pertain to a rather young specimen, which, however, *qua* overall morphologic features, may be considered fully developed.

[85] A faint sulcus lateralis mesencephali externus is suggested at some of the transverse levels.

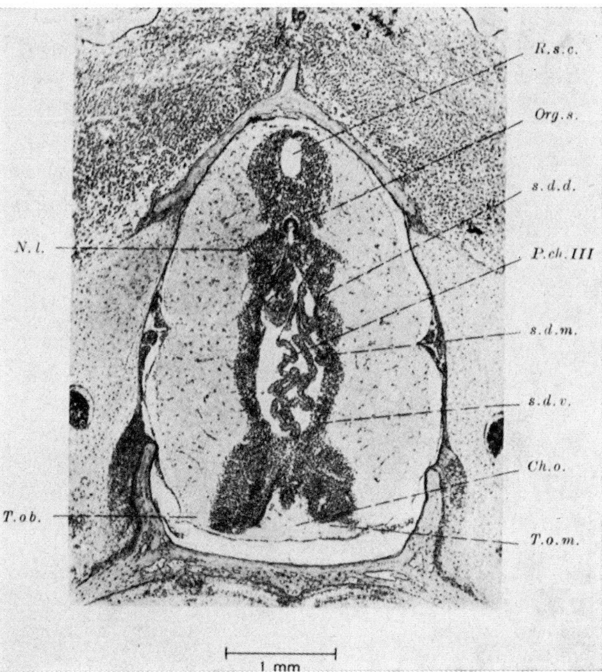

Figure 384A. Cross-section through the mesencephalo-diencephalic boundary in the Dipnoan Protopterus (from GERLACH, 1933). Ch. o.: optic chiasma and postoptic commissures; N. l.: nucleus lentiformis mesencephali; Org. s.: subcommissural organ; P. ch. III.: choroid plexus of third ventricle; R. s. c.: rostral respectively supracommissural recess of mesencephalic ventricle; S. d. d.: sulcus synencephalicus; s. d. n.: sulcus diencephalicus medius; s. d. v.: sulcus diencephalicus ventralis; T. ob.: Tractus opticus basalis; T. o. m.: Tractus opticus marginalis.

cludes the large cells representing the *nucleus radicis mesencephalicae trigemini*,[86] does not display a recognizable stratification and is, in that respect, comparable to the tectum of Necturus (Fig. 359). *Nucleus isthmi* and *nucleus visceralis secundarius*, whose approximate location within the periventricular cell masses at isthmic levels can be inferred, remain quite indistinct.

Within the *basal plate*, whose cell population includes the *nucleus reticularis tegmenti*, *oculomotor* and *trochlear nerve nuclei* are likewise present

[86] The nucleus has a median location in the thin mesencephalic roof which gives the appearance of a persistent roof plate (cf. Fig. 325F).

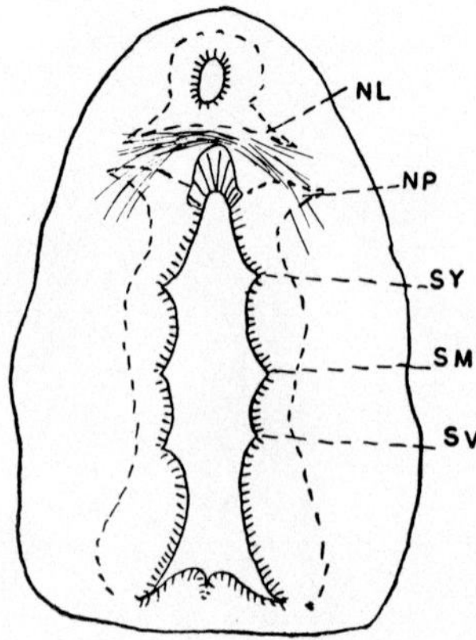

Figure 384B. Sketch of a cross-section through the mesencephalo-diencephalic boundary zone of Protopterus, approximately corresponding to the level illustrated in the preceding figure. The voluminous intraventricular choroid plexus has been omitted (from K., 1956). NL: nucleus lentiformis mesencephali; NP: nucleus commissurae posterioris with fibers of commissure; SM, SV: sulcus diencephalicus medius respectively ventralis; SY: sulcus synencephalicus.

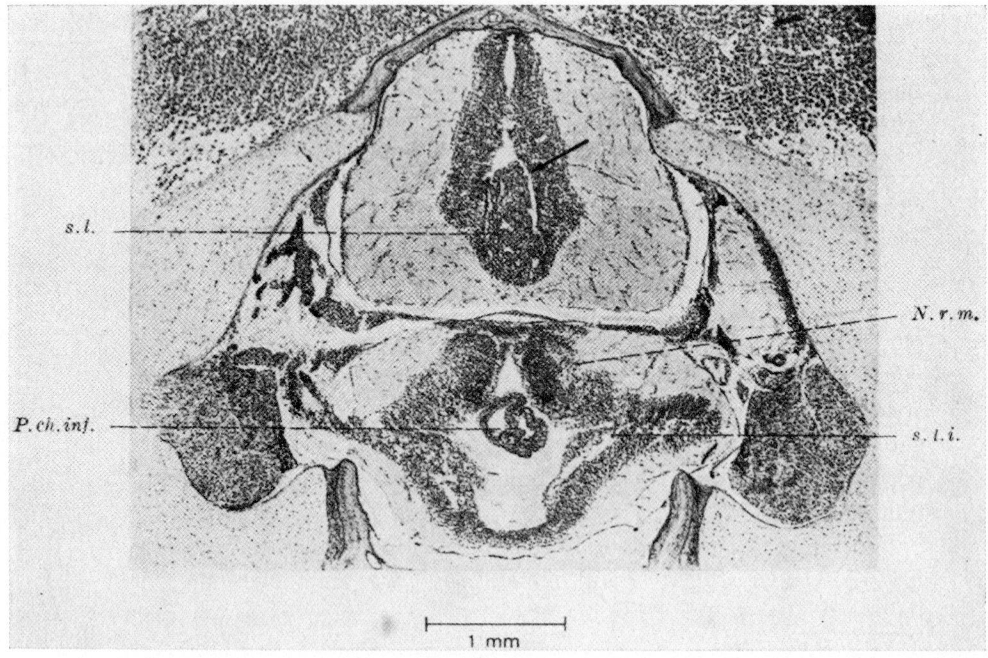

Figure 384 C

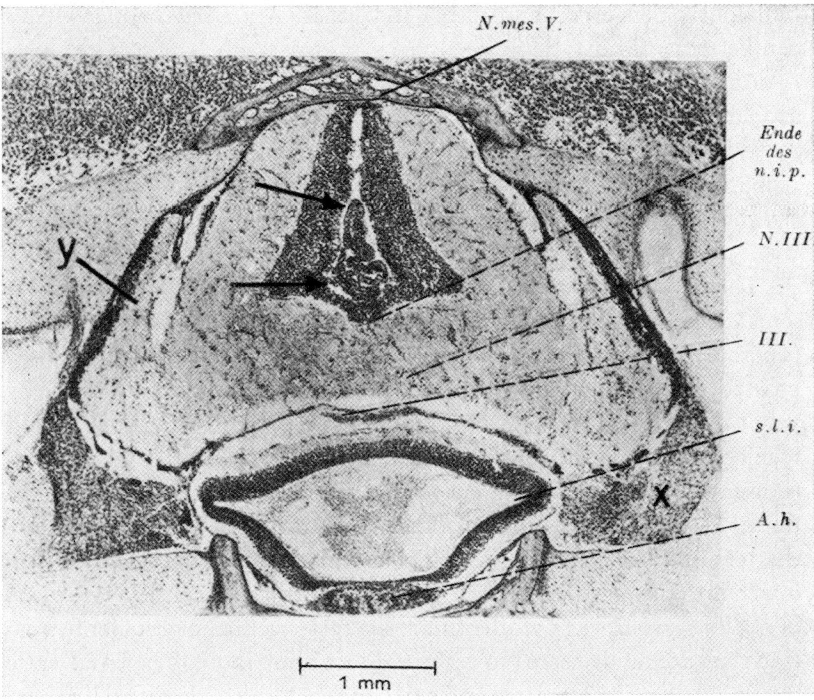

Figure 384D. Cross-section through a still more caudal level of the mesencephalon in Protopterus, showing oculomotor nucleus and nerve, as well as rostral tips of cerebellar auricles with an external layer of granule cells (from GERLACH, 1933). A.h.: adenohypophysis; *Ende des* n.i.p.: caudal portion of interpeduncular nucleus; N.mes.V.: nucleus of the mesencephalic trigeminal root: N.III, III nucleus respectively fibers of the oculomotor nerve; S.l.i.: sulcus lateralis hypothalami inferioris (sulcus lateralis infundibuli). Added designations: x: part of trigeminal ganglion; y: rostral tip of cerebellar auriculae (cf. Fig. 325F). Added arrows: sulcus lateralis mesencephali (above); sulcus limitans (below).

Figure 384C. Cross-section through the mesencephalon of Protopterus slightly caudal to tuberculum posterius (from GERLACH, 1933). N.r.m.: nucleus recessus mammillaris; P.ch.i: 'Plexus chorioideus infundibuli'; s.l.: sulcus limitans; s.l.i.: 'sulcus lateralis infundibuli'. Added arrow indicates approximate location of sulcus lateralis mesencephali. It can be seen that at this and at the next level (D) the choroid plexus of third ventricle extends into the lumen of the mesencephalic ventricle.

but poorly developed and rather indistinct. An *interpeduncular nucleus* can be recognized.

As regards the *fiber connections* of the Dipnoan mesencephalon, the available data for Protopterus are rather sketchy and uncertain. In Ceratodus, however, HOLMGREN and VAN DER HORST (1925) were able to recognize some general features of relevant fiber tracts. *Mutatis mutandis*, this overall arrangement may essentially also obtain for Protopterus.

The rostral connections in Ceratodus include the *habenulo-interpeduncular tract* together with a habenulo-tubercular subdivision to tuberculum posterius.[87] *Thalamo-* and *hypothalamo-peduncular pathways*, crossed as well as uncrossed, are likewise present and may even, by way of the basal forebrain bundles, contain connections interrelating telencephalon and mesencephalon. The *optic pathway*, after its decussation in the chiasma, divides into the marginal or main optic tract, the dorsal axial optic tract, and the basal optic tract. The two former reach the tectum, while the latter, which is relatively thick, could be followed into the mesencephalic tegmentum. Although a distinctive *nucleus opticus tegmenti* was not identified, the mesencephalic reticular formation thus definitely seems to receive direct optic input. The basal optic tract as well as the marginal and axial ones were also identified by GERLACH (1933) in Protopterus.

Other recorded mesencephalic fiber connections include the tectotegmental and tectobulbar systems, a tectocerebellar pathway, a rudimentary 'brachium conjunctivum', a tractus tecto-isthmicus, the octavolateral and the general spinobulbar lemniscus, and the fasciculus longitudinalis medialis. The relevant *commissures* include *commissura posterior*, *commissura tecti* and the system of *commissura ansulata*.[88] This system may also involve decussating fibers (*'commissura transversa'*) in the diencephalic supraoptic commissures.

---

[87] The left tractus habenulo-interpeduncularis contains medullated as well as nonmedullated fibers, the right tractus only nonmedullated ones. Again, the unusually lateral position of this tract in Ceratodus was pointed out by the cited authors (cf. also Fig. 383A).

[88] Commissura posterior, a faint commissura tecti, the commissura ansulata, and the fasciculus longitudinalis medialis were also clearly identified and described by GERLACH (1933) in Protopterus.

6. Amphibians

The mesencephalon of *Amphibians* displays a high degree of development in many Anurans, but remains essentially at the stage of a more or less differentiated periventricular cell layer in most Urodeles. The mesencephalon of Gymnophiones appears still less well developed than that of Urodeles.

As regards its external dorsal aspect, the midbrain roof of *Anurans* with well developed eyes is characterized by two conspicuous rounded antimeric lobes, separated by a pronounced midsagittal sulcus. In Urodeles, whose eyes are usually smaller, the midbrain roof is less expanded and may not show externally well demarcated antimeric lobes. This is also the case in Gymnophiona which are characterized by reduced eyes. Yet, both in some Urodeles and some Gymnophiona (cf. e.g. Figs. 385D, 386C, 389B), a faint midsagittal sulcus may occur, particularly at caudal roof levels. Again, in the small-eyed Anuran toad Pipa, the optic lobes are confluent, as in most Urodeles.

Figures 385A to 386D illustrate representative cross-sections through the Urodele midbrain which has been extensively studied by HERRICK (1948).[89] Relevant data on the medullated fiber tracts were also recorded by KREHT (1930, 1931, 1940) and by RÖTHIG (1927). As regards the ventricular sulci, *sulcus lateralis mesencephali* and *sulcus limitans* are commonly present or at least suggested, and approximately indicate ventricular boundaries between tectum *sensu strictiori*, torus semicircularis, and basal plate (tegmentum).

Rostrodorsally, the boundary neighborhood between mesencephalon and diencephalon is indicated by the *commissura posterior*, as in practically all Vertebrates. A rostral tectal recess, such as e.g. displayed by the Dipnoan Protopterus, is not present. A poorly outlined (pretectal) *nucleus lentiformis mesencephali* is faintly indicated dorsolaterally to the periventricular *nucleus commissurae posterioris*. Although the cell masses of *tectum opticum* proper likewise retain a periventricular arrangement, an ill-defined and frequently variable stratification can be noted. HERRICK (1942) attempted to distinguish 8 layers or 'concentric zones'

---

[89] HERRICK's monograph of 1948 contains an extensive bibliography of his own numerous preceding papers as well as relevant references to investigations by other authors. Further references can be found in the publications by KREHT (1930, 1931), and in the treatises by KAPPERS (1947), KAPPERS *et al.* (1936), and by BECCARI (1943).

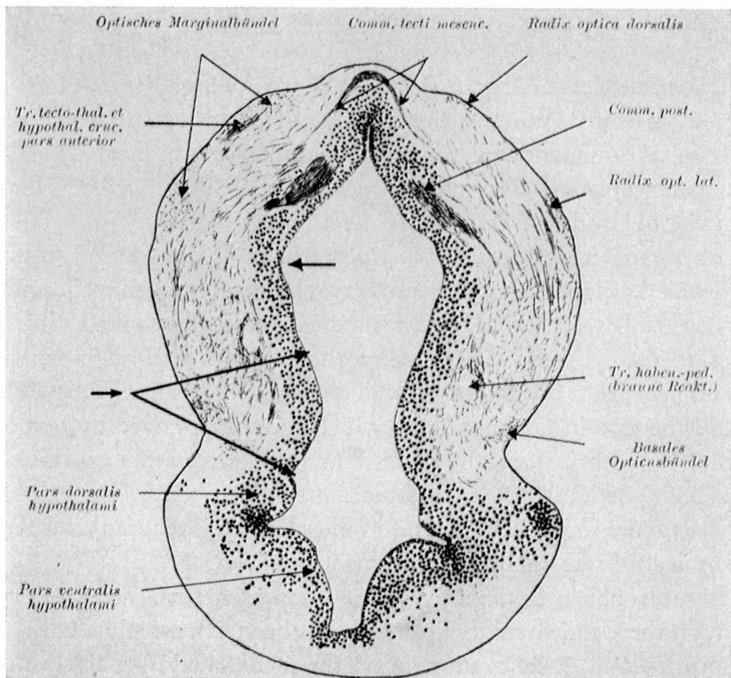

Figure 385 A. Cross-section through a rostral level of the mesencephalon in the Urodele Salamandra maculosa (from KREHT, 1930). Added arrows: sulcus lateralis mesencephali overlapping with sulcus synencephalicus (upper); curved end of sulcus limitans including tegmental cell plate (lower).

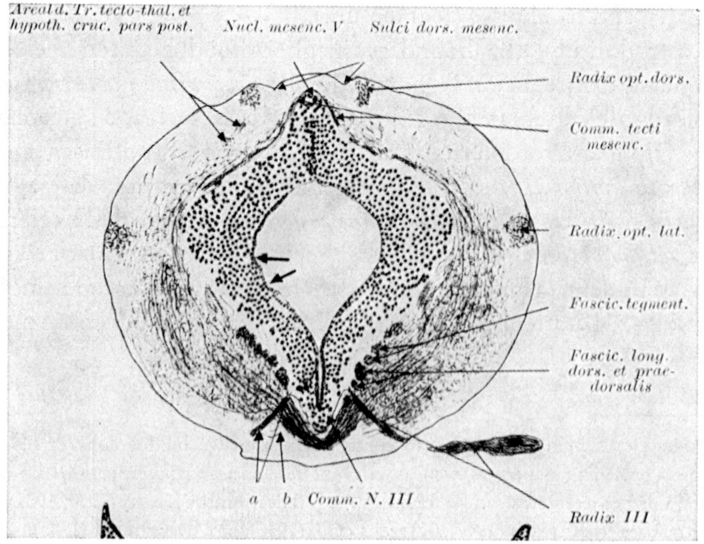

385 B

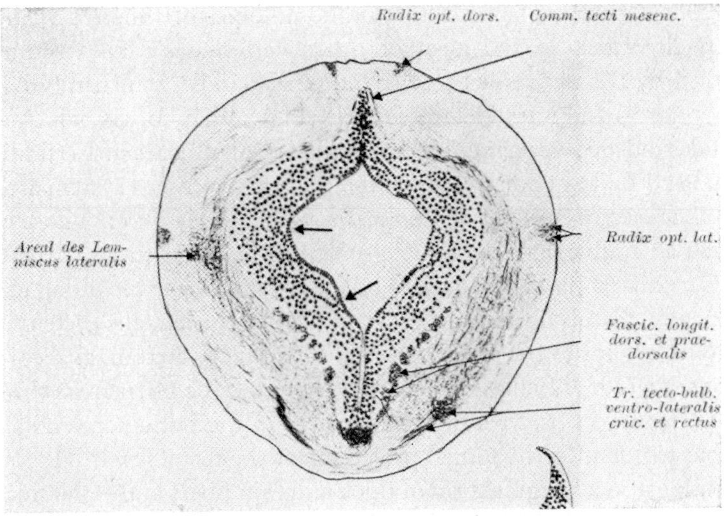

Figure 385C. Cross-section through the mesencephalon of Salamandra showing slight bulge of torus semicircularis (from KREHT, 1930). Added arrows as in preceding figure.

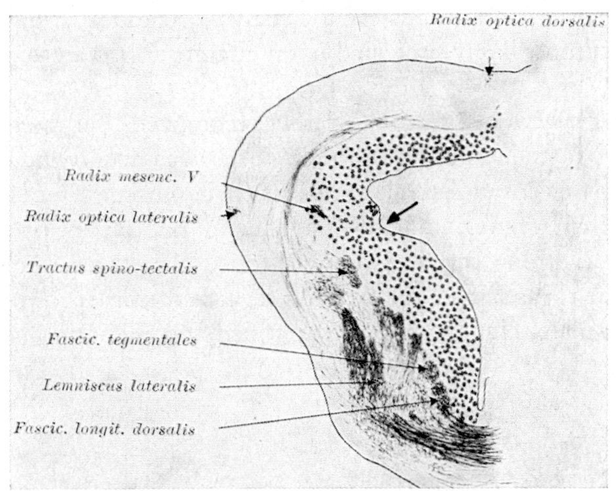

Figure 385D. Cross-section through the mesencephalon of Salamandra, showing 'fasciculi tegmentales' and commissura ansulata (from KREHT, 1930). Added arrow: sulcus lateralis mesencephali, above which an accessory sulcus can be seen.

Figure 385B. Cross-section through the mesencephalon of Salamandra at level of oculomotor root (from KREHT, 1930). Comm. N. III: commissura ansulata; the fasciculus longitudinalis medialis is here designated as fascic. long. dors.; added arrows indicate sulcus lateralis mesencephali and approximate location of sulcus limitans.

(Fig. 386C) but emphasized 'that no homologies are implied with the layers described in the tecta of other vertebrates'. The tectum is reached by the optic tract's main marginal bundle, which subdivides into a dorsal and into a lateral fascicle (Figs. 385A, B, C), which also includes most components of the so-called axial optic tract. In addition, spinal and bulbar components of the general lemniscus system ascend into the tectum. A *nucleus of the mesencephalic trigeminal root*, whose cells display a variable distribution (Figs. 385B, 386A), is present. *Tectal output* is mediated through a variety of *fasciculi tegmentales* including tectotegmental, tectobulbar, tectospinal, and tectodiencephalic systems.[90]

With regard to the retinal projection upon the tectum, as recorded by STRÖER (1939, 1940), HERRICK (1948) found the fibers from the different quadrants 'inextrically intertwined' in the optic nerve and was unable to identify the retinal projection at the tectal levels. He found in Amblystoma a segregation of thick and thin fibers at the chiasma. In the tectum, the fibers of both kinds seemed to mingle and to be distributed nearly uniformly to all parts of the tectum. This, of course, does not contradict the results obtained by STRÖER, and HERRICK justly adds that since Amblystoma can localize objects in the visual field, the retinal loci most likely projected upon circumscribed areas of the tectum.

A *torus semicircularis* can be clearly recognized, but does not very markedly protrude into the ventricle. It receives mainly the vestibular respectively vestibulolateral lemniscus or lemniscus lateralis (Figure 385C), which HERRICK (1948) includes in his tractus bulbotectalis lateralis.[91] Nothing certain is known concerning the further connections of the torus semicircularis.[92] As regards the other form-elements of the alar plate, HERRICK (1925) identified a nondescript *nucleus visce-*

---

[90] HERRICK (1948) elaborates a complex notation enumerating up to 10 different tegmental fasciculi, but admits that this classification is arbitrary, most of the bundles being 'mixtures of fibers of diverse sorts from unexpectedly widely separated sources'.

[91] HERRICK (1948) uses the term 'lemniscus systems', which subsumes (1) the spinal lemniscus complex, (2) the general bulbar lemniscus, (3) the tractus bulbotectalis lateralis, and (4) the ascending secondary visceral-gustatory tract.

[92] HERRICK (1948) designates the torus semicircularis as 'dorsal tegmentum' (cf. Fig. 386D). The torus, however, is an alar plate derivative, and it seems preferable to use the term tegmentum exclusively for derivatives of the basal plate.

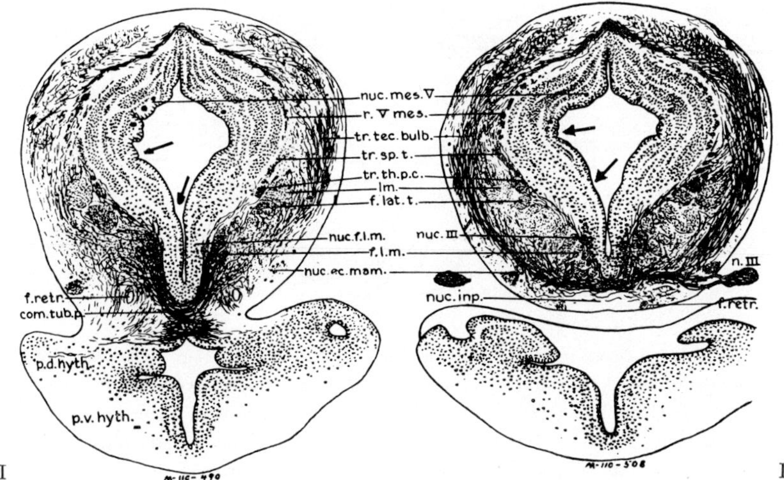

Figure 386A. Cross-sections (myelin stain) through the mesencephalon of the Urodele Amblystoma (I) at level of tuberculum posterius, and (II) of oculomotor nerve exit (from HERRICK, 1925). com.tub.p.: commissura tuberculi posterioris; f.lat.t.: 'fasciculus lateralis telencephali' (basal forebrain bundle); f.l.m.: fasciculus longitudinalis medialis; f.retr.: fasciculus retroflexus (habenulo-interpeduncularis); f.v.t.: 'fasciculus ventralis tegmenti'; lm.: lemniscus acusticolateralis; nuc.ec.mam.: 'nucleus ectomammillaris' (nucleus opticus tegmenti); nuc.inp.: nucleus interpeduncularis; nuc.mes.V: nucleus radicis mesencephalicae trigemini; nuc.p.t.: 'nucleus posterior tecti' (caudal end of optic tectum); nuc.vis.s.: nucleus visceralis secundarius; nuc. and n.III: nucleus respectively root of oculomotor; p.d. and p.v.hyth.: pars dorsalis respectively ventralis hypothalami (corresponding to lobi inferiores hypothalami posterioris in Fishes); r.V.mes.: radix mesencephalica trigemini; tr.sp.t.: tractus spinotectalis; tr.tec.bulb.: tractus tectobulbaris; tr.th.p.c.: 'tractus thalamopeduncularis cruciatus. Added arrows indicate sulcus lateralis mesencephali and sulcus limitans.

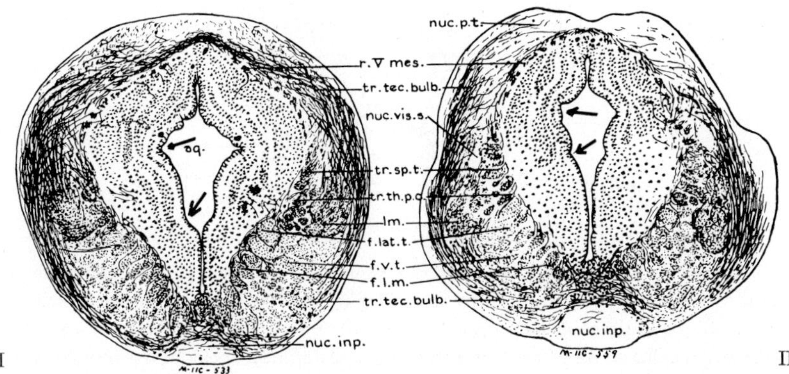

Figure 386B. Cross-section (myelin stain) through (I) the midregion of tectum mesencephali, and (II) the caudal portion of mesencephalon, close to isthmus, of Amblystoma (from HERRICK, 1925). Abbreviations as in preceding figure.

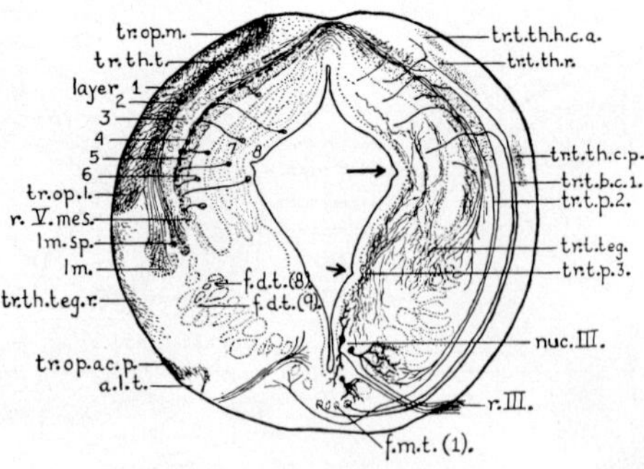

Figure 386C. Cross-section, based on various *Weigert* and *Golgi preparations*, of the mesencephalon in Amblystoma, indicating layers and tracts according to Herrick's interpretation (from Herrick, 1942). a.l.t.: 'area lateralis tegmenti' (neuropil of nucleus opticus tegmenti); f.d.t.: 'fasciculi dorsalis tegmenti'; f.m.t.: 'fasciculus medianus tegmenti; lm.sp.: 'spinal lemniscus'; r.III.: oculomotor root; tr.op.ac.p.: 'tractus opticus accessorius posterior' (tegmental portion of basal optic root); tr.op.lat.: tractus opticus lateralis; tr.op.m.: tractus opticus medialis; tr.t.b.c.1: tractus tectobulbaris cruciatus; tr.t.p.2, 3: tractus tectopeduncularis; tr.th.t.: 'tractus thalamotectalis'; tr.th.teg.r.: tractus thalamotegmentalis rectus; tr.t.teg.: tractus tecto-tegmentalis; tr.t.th.h.c.a., c.p.: tractus tectothalamicus et hypothalamicus cruciatus anterior (a) and posterior (p.). Other abbr. and added arrows as in preceding figure. The fasciculus longitudinalis medialis, not labelled, is represented by the facicles located dorsally and ventrally to the oculomotor root.

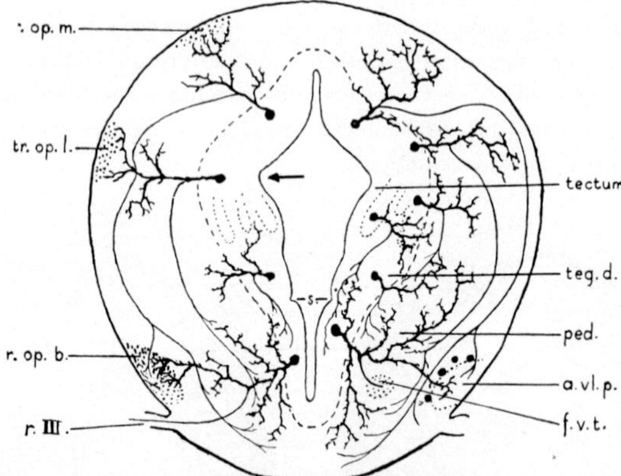

Figure 386D. Cross-section, based on various *Golgi preparations*, through the middle of the mesencephalon in Amblystoma (from Herrick, 1948). a.vl.p.: 'area ventrolateralis pedunculi' (neuropil of nucleus opticus tegmenti); f.v.t.: 'fasciculi ventrales tegmenti'; ped.: 'pedunculus cerebri' (true tegmentum, basal plate); r.III: oculomotor root; s: sulcus limitans; teg.d.: 'tegmentum dorsale mesencephali' (torus semicircularis, alar plate derivative); tr.op.b., l., m.: basal, lateral and medial optic tracts. Added arrow: sulcus lateralis mesencephali. The unlabelled sulcus dorsal to sulcus limitans is probably sulcus intermedius dorsalis.

ralis seu gustatorius secundarius as a neuropil which receives visceral afferent fibers from the bulbar lemniscus system (Fig. 386B). The *nucleus isthmi* is poorly developed, being probably represented by periventricular neuronal elements located at the caudal end of torus semicircularis slightly dorsally to the cells whose dendrites form the neuropil of 'nucleus visceralis secundarius'. The dendrites of 'nucleus isthmi' presumably receive input from the lateral lemniscus system.

The components of the *basal plate* which provides the true tegmentum or the so-called 'motor zone' of HERRICK (1948) have their rostral beginning at the *tuberculum posterius* with its tegmental cell plate. Relatively large cells in this region represent the nucleus of the fasciculus longitudinalis medialis *(interstitial nucleus of Cajal)*. The *nucleus motorius tegmenti* or *reticular formation* is provided by periventricular cells whose dendrites extend into peripheral neuropil condensations. HERRICK (1948) attempted, in a not altogether convincing manner, to distinguish various areae. His generalized summary, mainly referring to Amblystoma, can, however, be accepted as essentially valid and applying to most if not all Urodeles: the reticular formation 'receives fibers from practically all parts of the brain above the isthmus, and its primary function is control of mass movements of the trunk and limbs and conjugate movements of the eyes. Control of local reflexes is effected elsewhere. It is intimately connected with the hypothalamus, and various visceral adjustments are made here, though of these little is known' (HERRICK, 1948). This, of course, agrees on the whole with many of the views expressed by STEINER (1885–1900). A '*brachium conjunctivum*' is included in HERRICK's 'fasciculus tegmentalis profundus', described by this author as a 'mixed tract'. It should, moreover, be added that presumably reciprocal connections with other rhombencephalic, diencephalic and perhaps also telencephalic grisea obtain.

The *nucleus opticus tegmenti* of Urodeles is represented by periventricular cells whose dendrites extend toward the periphery and form, together with the endings of the basal optic tract, a region of neuropil in HERRICK's 'area lateralis tegmenti'.

The *interpeduncular nucleus* of Urodeles is well developed and its details have been recorded by HERRICK (1948), who described its synapses formed by 'glomerular dendrites', its neuropil, and the spiralling course of the tractus habenulopeduncularis (Figs. 386E, F). Other connections of the interpeduncular nucleus are provided by tractus mammillotegmentalis, basal forebrain bundle, and additional fiber systems within the tegmentum.

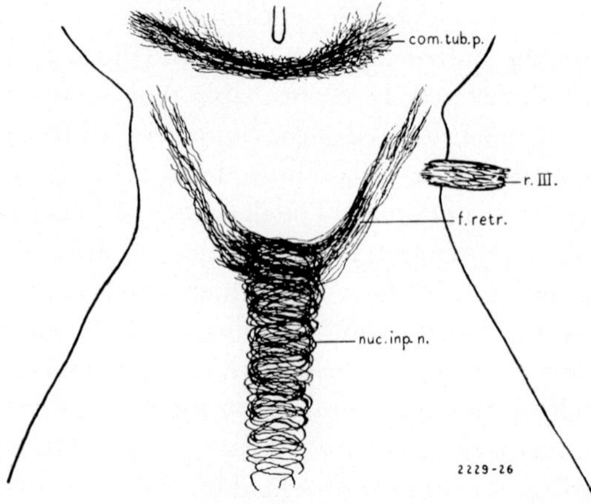

Figure 386E. Horizontal section through mesencephalic tegmentum of Amblystoma showing decussation and spiral endings of fasciculus retroflexus in nucleus interpeduncularis, as reconstructed from several *Golgi preparations* (from HERRICK, 1948). com. tub. p.: 'commissura tuberculi posterioris' (rostral extremity of commissura ansulata system); f. retr.: fasciculus retroflexus sive tractus habenulopeduncularis; nuc. inp. n.: neuropil of nucleus interpeduncularis; r.III: radix oculomotorii. Note caudal extension of n. interpeduncularis into rhombencephalon.

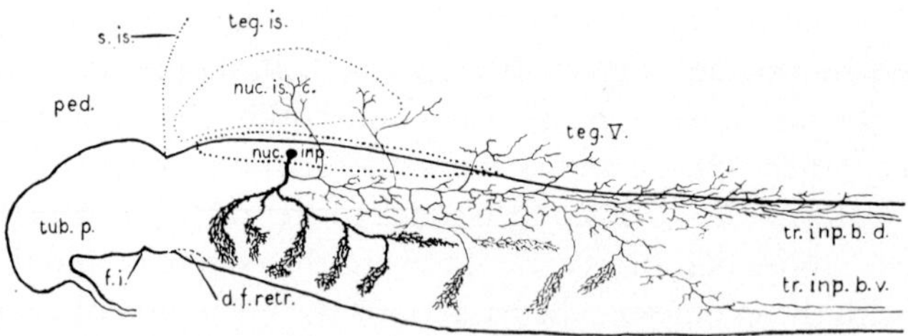

Figure 386F. Semidiagrammatic sagittal section through the deuterencephalic tegmentum of Amblystoma, reconstructed from a variety of *Golgi preparations* (from HERRICK, 1948). d. f. retr.: decussation of fasciculus retroflexus; f. i.: 'fovea isthmi' (external isthmic sulcus); nuc. inp.: nucleus interpeduncularis; nuc. is. c.: 'nucleus centralis isthmi'; ped.: region of 'pedunculus cerebri'; s. is.: 'sulcus isthmi' (internus, basally ending in true, ventricular, fovea isthmi); teg. is.: 'tegmentum isthmi, pars magnocellularis' (an alar plate region); teg. V.: 'tegmentum trigemini'; tr. inp. b. d., v.: tractus interpedunculobulbaris dorsalis and ventralis.

The tegmentum also contains the *oculomotor* and *trochlear nuclei*.[93] The nucleus oculomotorii is usually better distinguishable than that of the trochlear which nevertheless can be clearly recognized in the rostral isthmic tegmentum at an appreciable distance from the oculomotor nucleus, and like this latter, appears as pertaining to the basal plate's ventral longitudinal zone. The trochlear root fibers, mostly myelinated, 'ascend along the outer border of the gray to decussate in the anterior medullary velum in the usual way' (HERRICK, 1948). Preganglionic fibers are presumed to be present in the oculomotor nerve, but specific and definite data on this topic apparently have not been recorded. Further comments on the mesencephalic fiber systems and commissures will be found further below, following a short description of the Anuran and Caecilian midbrain.

Figures 387A to 387G illustrate representative cross-sections through the mesencephalon in Salientia. It can easily be seen that, as compared with conditions prevailing in Urodeles, a higher degree of mesencephalic development and differentiation obtains in Anurans. The *tectum opticum* manifests a considerable lateral expansion. *Sulcus lateralis mesencephali internus* and *externus* are prominent. The *torus semicircularis* is large and commonly displays an extensive secondary fusion in the midline. The mesencephalic ventricle extends into two rostral tectal diverticula (Fig. 387B) opening into a common ventricular space (Figs. 387C, D) which, immediately caudalward, becomes again subdivided into a lateralward flaring common dorsal tectal ventricle and a

---

[93] According to BENEDETTI (1933) oculomotor and trochlear nerves with their nuclei are completely absent in the blind Urodele Proteus anguineus. KREHT (1931), however, noted a variable small number of oculomotor root fibers and a nondescript oculomotor nucleus. This author did not find trochlear root fibers, but believed to have seen cells corresponding to a trochlear nucleus (*'dagegen liessen sich die bei anderen Urodelen'... 'als mutmasslicher Kern des Trochlearis gedeuteten grossen Zellen im Isthmusgebiet auch bei Proteus feststellen'*). Nevertheless, BENEDETTI, referring to KREHT's findings, states: '*Per conto nostro no possiamo che riconfirmare quanto fu esposto piu sopra, e cioè che in nessuno dei cervelli da noi esaminati si trova alcuna traccia ne di fibre ne di cellule dell'oculomotore*'. In this respect, it seems likely that both authors are right, and that, in the case of a regressively rudimentary structure as displayed by the ocular system of the blind Proteus, appreciable variations with respect to different populations of that species may obtain. KREHT's tentative identification of a 'trochlear nucleus', however, appears unconvincing. Quite evidently, the locus of said nucleus must be present in Proteus and also include cellular elements, but their designation as 'trochlear nucleus' remains a matter of arbitrary semantics and may be questioned from a functional viewpoint. At most, this locus could be evaluated as representing a 'defective kathomologon' of trochlear nucleus.

subtoral ventricular space (Figs. 387E, F). The *sulcus limitans* is commonly well recognizable. A *sulcus intermedioventralis* can occasionally be noted.

At rostral levels, a mesencephalic pretectal *nucleus lentiformis mesencephali*, receiving, *inter alia*, optic input, is fairly conspicuous. Dorsally to the diencephalic pretectal *nucleus commissurae posterioris*, a derivative of

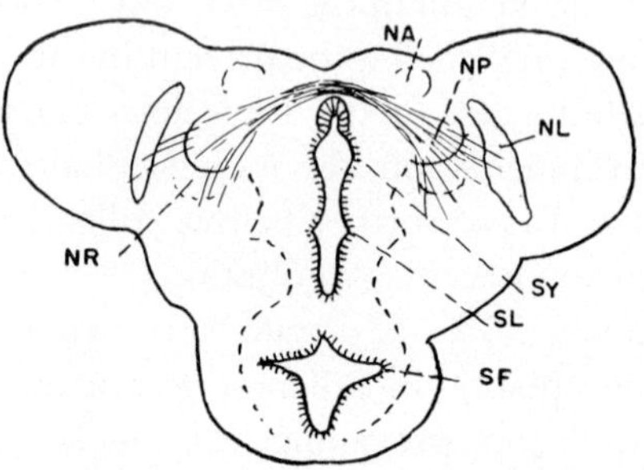

Figure 387A. Semidiagrammatic cross-section through the rostral end of the mesencephalon in the Anuran Rana (from K., 1956). NA: area praetectalis; NL: nucleus lentiformis mesencephali; NP: nucleus commissurae posterioris; NR: primordium of nucleus praetectalis; SF: 'sulcus lateralis infundibuli'; SL: sulcus limitans; SY: sulcus synencephalicus. Below the posterior commissure, the subcommissural organ (not labelled) is indicated by the columnar ependyma.

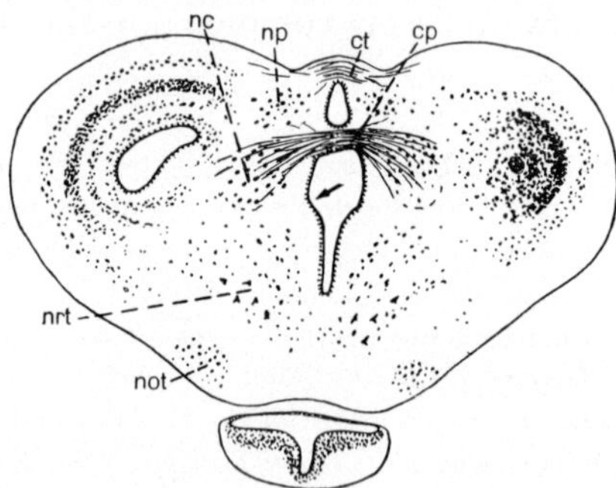

Figure 387B. Cross-section through the mesencephalon of the Frog at the level of the posterior commissure's caudal end (from K., 1927). cp: commissura posterior; ct: commissura tecti; nc: nucleus commissurae posterioris; not: nucleus opticus tegmenti; np.: area praetectalis; nrt: nucleus reticularis tegmenti; arrow: sulcus limitans. The slightly oblique section, from which the drawing was made, shows at left the rostral recess of the tectal ventricle and, above the posterior commissure a rostral median recess of the mesencephalic ventricle (a fairly common individual variation occurring in diverse Anurans).

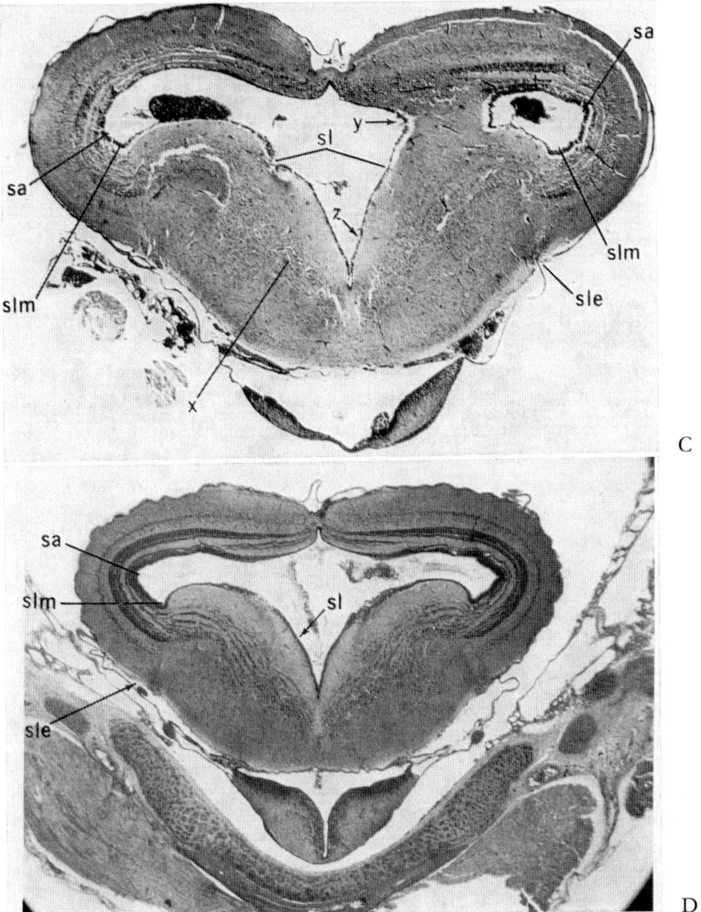

Figure 387C, D. Two cross-sections through the mesencephalon of the Frog, showing (C) communication of left rostral tectal diverticulum with common mesencephalic ventricle, and (D) the common ventricle rostral to midline fusion of tori semicirculares (hematoxylin eosin stain; × ca. 30, red. $^1/_2$). sa: sulcus lateralis mesencephali (internus) accessorius; sl: sulcus limitans; sle: sulcus lateralis mesencephali externus; slm: sulcus lateralis mesencephali (internus); x: caudal end of nucleus interstitialis fasciculi longitudinalis medialis; y: accessory sulcus; z: sulcus intermedius ventralis.

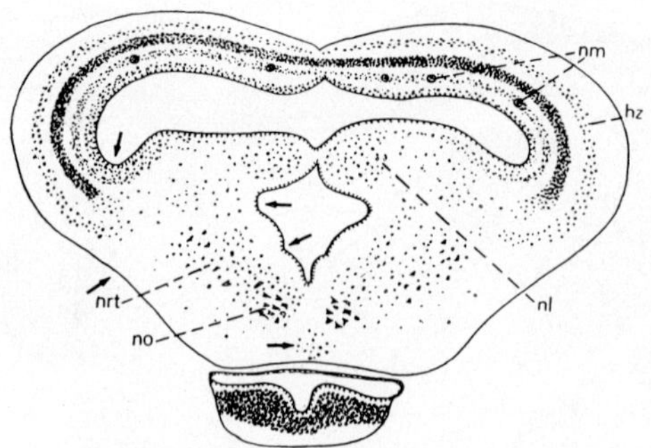

Figure 387E. Cross-section through the mesencephalon of the Frog at the level of the oculomotor nucleus (from K., 1927). hz: chief cell layer of tectum opticum; nl: 'nucleus mesencephali lateralis' within torus semicircularis; nm: cells of nucleus radicis mesencephalicae trigemini; no: nucleus oculomotorii; nrt: nucleus reticularis tegmenti; arrows in sequence from above: sulcus lateralis mesencephali (internus), sulcus limitans, sulcus intermedioventralis. The lead to nl passes approximately through the sulcus lateralis mesencephali externus. Lowest arrow: nucleus interpeduncularis.

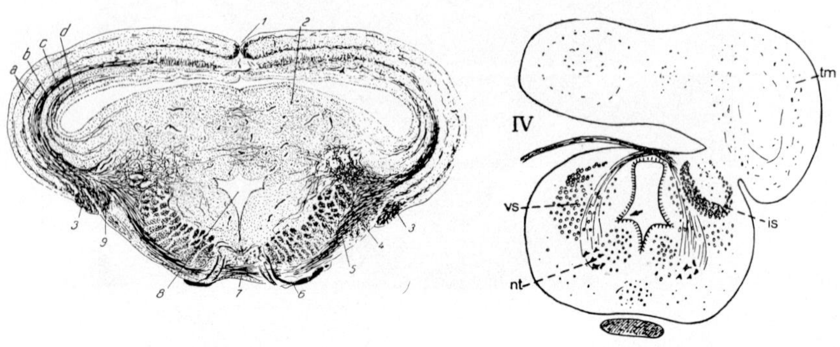

Figure 387 F

Figure 387 G

Figure 387F. Cross-section (myelin stain) through the mesencephalon of Rana catesbyana at the level of the oculomotor exit (after Kappers and Hammer, 1918, from Beccari, 1943). 1: fasciculus opticus medialis; 2: torus semicircularis; 3: fasciculus opticus lateralis; 4: general lemniscus system; 5: tractus tectobulbaris ventralis; 6: oculomotor root; 7: commissura ansularis; 8: ventral part of mesencephalic ventricle; 9: fibers of commissura transversa system; a: superficial optic layers *('stratto fibroso e grigio superficiale')* of optic tectum; b: deep medulla; c, d: external and internal (subependymal) granular layers. The inner lead from a indicates the layer of chief cells.

Figure 387G. Cross-section through the isthmus rhombencephali of Rana, showing the trochlear nerve decussation (from K., 1927). is: nucleus isthmi; nt: nucleus trochlearis; tm: caudal end of tectum mesencephali; vs: nucleus visceralis secundarius. In the basal midline the nucleus interpeduncularis, not labelled; arrow: sulcus limitans.

the dorsal thalamic regio posthabenularis is differentiated as a neuropil with scattered cells, corresponding to the *area praetectalis* of Amniota.[94]

The *tectum opticum* contains the large neuronal elements of the *nucleus radicis mesencephalicae trigemini* (Fig. 387E) whose distribution displays individual as well as species variations. Since the lamination of the Anuran tectum opticum was discussed and depicted in section 1 of chapter (Figs. 361A, B) no further comments on this topic are required in the present context.

However, with respect to the retinotectal projection it might here be added that, after dorsoventral and anteroposterior surgical rotation of an eye cup at early larval stages of the Anuran Xenopus, normal retinal projection upon the tectum was observed. At a somewhat later stage, rotation of the eye cup resulted in anteroposterior inversion of the projection, and at still later stages anteroposterior and dorsoventral inversion took place. It is therefore assumed that the 'spatial specification' of the retinal ganglion cells does not occur before a certain ontogenetic stage, and subsequently takes place first in the anteroposterior axis of the eye cup and next in the dorsoventral axis.[95]

The large *torus semicircularis* contains more or less diffusely arranged cell groups. Although in a subtectal position with respect to the tectum opticum (cf. Figs. 329C, 388), it evidently corresponds to the posterior (or inferior) colliculi of Mammals, the topologically invariant transformation, clearly indicated in several Reptiles, merely requiring a

---

[94] Since the pretectal region comprises a number of different grisea which, from a morphologic viewpoint, include derivatives of both diencephalon and mesencephalon, the relevant details of its composition shall be discussed in chapter XII of volume 5. It might here be mentioned, however, that recent experimental studies have claimed an 'inhibiting' effect of 'pretectal' grisea. A 'disinhibited feeding' by frogs after 'pretectal lesions' is said to be 'paralleled by the consistent failure of tectal neurons to ignore' *(sic)* 'moving buglike stimuli' (INGLE, 1973).

[95] It has also been shown, according to KEATING and others (as quoted in a report by INGLE and SCHNEIDER entitled Brain Mechanisms and Vision: Subcortical Systems. Science *168:* 1493–1494, 1970) that, although in the African frog Xenopus retino-tectal connections are determined during early development, 'the pattern of tectotectal connections appears to be determined by visual experience about the time of metamorphosis. The tectotectal system develops a reversed topographic order when the tadpole larva is reared with one eye rotated surgically by 180°. Thus, at early developmental stages, plasticity seems here to obtain, and it was suggested that in such cases distant populations of neurons which are 'simultaneously activitated by the same visual stimulus (via opposite eyes) can selectively grow, or retain, synaptic associations'.

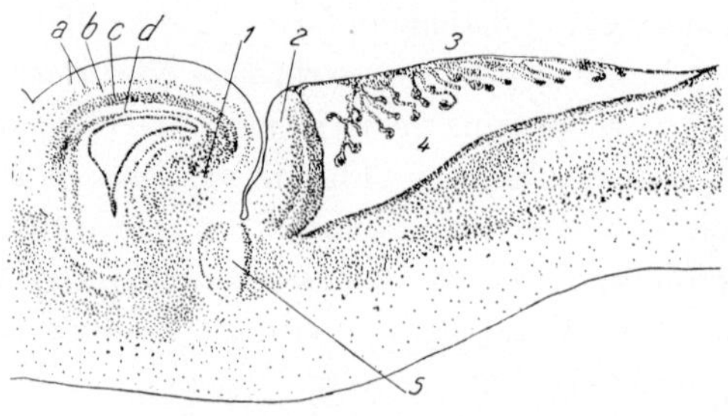

Figure 388. Paramedian sagittal section *(Nissl stain)* through the deuterencephalon of Rana (from Beccari, 1943). 1: torus semicircularis *('corpo posteriore')* 2: cerebellum; 3: roof plate, with choroid plexus, of fourth ventricle; 4: fourth ventricle; 5: nucleus isthmi; a: stratum opticum and chief cell layer of tectum opticum; b: deep medulla; c, d: granular layers of tectum. This figure should be compared with Figures 329C and 330A.

caudal evagination of the torus (cf. Figs. 391A, G). The conspicuous development of this region in Anurans is doubtless related to the progressive differentiation of the otic apparatus in this Amphibian order as an 'organ of hearing'. The grisea of the torus semicircularis receive a substantial input from the *lemniscus lateralis system (fasciculus longitudinalis lateralis)* which includes here vestibular as well as 'cochlear' components.[96] Nothing definite is known about the numerous additional connections of the torus, which is also related to the diencephalic *postoptic commissura transversa.*

In the caudal region of the mesencephalic *alar plate* a large, oval *nucleus isthmi* is characterized by a central fiber mass, containing only few and scattered cells (Figs. 387G, 388). Its main connections are with the *lemniscus lateralis*, but the termination of other bulbar, particularly trigeminal fibers have been recorded. Crossed and uncrossed connections with tectum opticum and with the commissura transversa system likewise obtain. Medially to the nucleus isthmi lies a relatively small nucleus visceralis secundarius.

The *basal plate* contains the *nucleus reticularis tegmenti* with a rostral *nucleus interstitialis of the fasciculus longitudinalis medialis* and additional scattered, ill-defined cell groups, some of which display rather large multipolar elements. They include a forerunner of the Amniote nucle-

---

[96] A typical 'cochlea' is, of course, not present in Amphibians respectively Anurans, but structures with presumably analogous functions obtain (cf. section 1 of chapter IX).

us ruber, but this griseum does not seem sufficiently well circumscribed to warrant the designation applying to said Amniote griseum.[97]

Crossed fibers of the medullated basal optic tract can be followed to a lateral tegmental cell group, the *nucleus opticus tegmenti* or '*nucleus ectomammillaris*' located at the level of the oculomotorius nucleus or slightly rostrally to this latter (Fig. 387B). There is, in addition, an *axial optic bundle* (HERRICK, 1925, KREHT, 1930) which besides proceeding toward the tectum opticum seems to connect with hypothalamic grisea and apparently also joins the basal forebrain bundle.[97a] Through this pathway, optic fibers may reach other tegmental structures, pertaining to the reticular formation. The basal midline of the tegmentum contains the *nucleus interpeduncularis*, whose main input channel is the *fasciculus retroflexus* (Figs. 387E, G). The *nuclei of the oculomotor and of the trochlear nerves*, in close relation to the fasciculus longitudinalis medialis, are commonly better demarcated than in Urodeles but display essentially similar relationships (Figs. 387E, F, G), with a variable distance between the two nuclei.

The mesencephalon of *Gymnophiones* is poorly developed and represents a narrow tube, whose configuration is complicated by the exaggerated ventral concavity of the mesencephalic flexure (Figs. 330A–G, 389A, B). Most of its cell masses are arranged as a more or less compact periventricular layer. In some instances, depending on species or individual variations, and on the cross-sectional planes, a *sulcus limitans* (Fig. 389A) or even a *sulcus lateralis mesencephali internus* labelled and easily identifiable in Fig. 330D) can be recognized.

The *tectum opticum* includes variously distributed cells of the *nucleus radicis mesencephalicae trigemini* and occasionally displays a nondescript lamination reminiscent to that described by HERRICK (1948) for certain Urodeles. In my report of 1922, primarily concerned with the telencephalon, and only incidentally recording features pertaining to the other regions of the rarely investigated Gymnophione brain, I could dis-

---

[97] This, of course, being merely an arbitrary question of terminology, remains a matter of opinion.

[97a] The axial optic bundle (also called 'axial optic tract', cf. also p. 878) consists of coarser optic fibers segregated from the finer ones at the chiasma, and taking a deeper (i.e. more medial) course in the optic tract before this latter separates into medial and lateral branches to the tectum (respectively pretectal grisea. These branches then contain both thick and thin fibers (cf. HERRICK, 1948). The fibers of the axial optic bundle presumably reaching the basal grisea mentioned in the text are, at best, only a rather small component of said bundle.

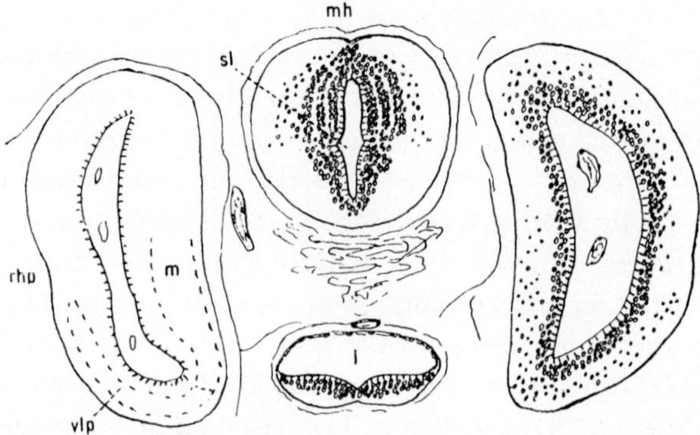

Figure 389A. Cross-section through the midbrain of the Gymnophione Siphonops annulatus at a level corresponding to the dorsal convexity of the midbrain flexure (from K., 1922). i: 'infundibular' recess of hypothalamus; m: area medialis pallii (D_3) of telencephalon; mh: mesencephalon; rhp: sulcus endorhinalis posterior; sl: sulcus limitans; vlp: area ventrolateralis posterior (basal cortex, 'cortical amygdaloid nucleus'). This and the following Figure 389B should be compared with Figures 330A (sagittal view) and 330B–G (transverse sections).

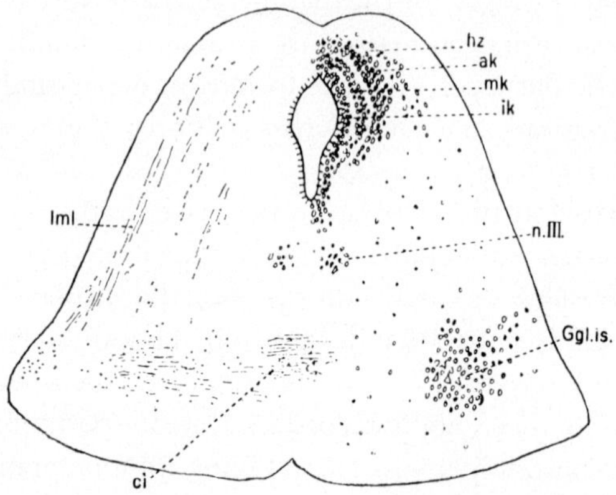

Figure 389B. Cross-section through the midbrain of Siphonops, passing through the caudal limb of the mesencephalic flexure (from K., 1922). ak: 'external granular layer'; ci: 'commissura intertrigemina'? (it is possible that these fibers pertain to the trigeminal portion of the 'auricular complex', cf. Fig. 330B); Ggl. is.: nucleus isthmi (probably; its basal, but lateral or alar plate location being related to the topography of the mesencephalic flexure's caudal limb); hz: 'chief cell layer'; ik: 'inner granular layer'; lml: fibers of lemniscus system (perhaps lemniscus lateralis or general lemniscus or both); mk: 'intermediate granular layer'; n.III: nucleus oculomotorius (its dorsal but medial position with respect to nucleus isthmi is related to the bend of the mesencephalic flexure's caudal limb).

tinguish four cell layers which I designated as subependymal inner granular layer, intermediate granular layer, outer granular layer, and layer of chief cells.[97b] *Tectum proper* and *torus semicircularis* are not clearly differentiated from each other; the combined tectum mesencephali *sensu latiori* receives the lateral and general bulbospinal lemniscus. A well developed *commissura posterior* with associated periventricular cells can be seen in the mesencephalo-diencephalic boundary zone. Depending on species and individual variations, a practically negligible amount of the reduced optic tract's fibers may also reach the tectum proper.

The rudimentary degree of mesencephalic development in Gymnophiones is doubtless related to the various stages of *eye-degeneration* displayed by this Amphibian order. It will be recalled that three species of Urodeles, namely Proteus, Typhlotriton, and Typhlomolge, all inhabitants of caves, are also blind at the adult stage. The eye of Proteus has been regarded as essentially a case of arrested development. The American Typhlomolge exhibits further degenerative changes, the eye muscles and the lens have here vanished. Noble (1931, 1954) assumes that in Typhlomolge, as in Typhlotriton and Proteus, the eyes develop normally until a certain stage when growth is checked, differentiation ceases, and degenerative changes arise. It seems, of course, in several respects difficult to define an exact distinction between arrested development and degeneration. Typhlotriton is the only blind Urodele which metamorphoses, the blind Typhlomolge and the likewise blind Proteus anguineus being permanent larvae. In Gymnophiones the typical eye musculature[98] has been modified by the degeneration of some muscles and the transfer of others to adjacent regions where they have different functions. The retractor bulbi is transformed into a retractor tentaculi; the rectus internus, into a retractor of the tentacular sheath, and the levator bulbi, into compressor and dilator muscles of the orbital glands (Norris, 1917, Noble, 1931, 1954). According to Nishi (1938), however, the levator bulbi of Urodeles and Anurans, and the compressor glandulae orbitalis of Gymnophiones represent derivatives of the trigeminus musculature.

[97b] At that time, I failed to recognize the not very prominent nucleus radicis mesencephalicae trigemini in the available material (cf. footnote 131, p. 455 of chapter IX).

[98] In addition to the usual six eye muscles of Vertebrates, the Amphibia possess a special retractor bulbi which pulls the eyeball within the orbit, and a levator bulbi, which raises the eyeball again (Noble, 1931, 1954). The retractor bulbi is said to be innervated by the abducens in Rana, but apparently by the oculomotor in Salamandra (cf. Nishi, 1938).

Within the caudal limb of the mesencephalic flexure, just rostrally to the mainly trigeminal auricular complex, a cell group can perhaps be interpreted as *nucleus isthmi* (Fig. 389B). It is probably medialward continuous with a nondescript *nucleus visceralis secundarius*.

The mesencephalic *basal plate* does not show any conspicuous differentiation. Its periventricular layer, together with some scattered more peripheral nerve cells presumably includes the equivalent of a rudimentary nucleus reticularis tegmenti and a likewise indistinct rostral nucleus interstitialis of the fasciculus longitudinalis medialis, whose mesencephalic portion appears rather slender, becoming more prominent in the rhombencephalon, concomitantly with the appearance of the nucleus centralis. Cells and neuropil in the caudal mesencephalic midline neighborhood of the basal plate probably correspond to an indistinct *interpeduncular nucleus*, since a fairly well outlined fasciculus retroflexus can be traced toward that region. A nucleus of the *nervus oculomotorius*, slightly separated from the periventricular cell population, was tentatively identified (Fig. 389B). A *nucleus trochlearis* and its nerve could not be recognized. Reliable data on mesencephalic fiber tracts and commissura ansulata do not seem to be available.

Generally speaking the *fiber systems of the Amphibian mesencephalon* are essentially similar to those dealt with, and subsumed under diverse categories of classification, in the preceding sections concerning the other Anamnia.

The *optic input* to tectum and tegmentum, the *fasciculus retroflexus*, and the various, still poorly understood communication channels connecting with diencephalic[99] as well as presumably with telencephalic grisea, repeatedly discussed above, require here no further comments.

The variously developed *secondary ascending tracts* from medulla oblongata and spinal cord, believed to be essentially or predominantly crossed, include *the general, somatic and visceral lemniscus*, and the *vestibulolateral*, respectively, in Anurans, the *vestibulo-'cochlear' lemniscus*. The visceral component of the general lemniscus has, moreover, connections with a more or less developed '*nucleus visceralis s. gustatorius secundarius*'. A comparable 'relay station' of the lateral lemniscus is the *nucleus isthmi*.[100] Said lemniscus, especially in Anurans, is particularly related

[99] Among the mesencephalo-diencephalic channels are included tecto-habenular (and presumably habenulo-tectal) bundles providing a tectal connection with epithalamus.

[100] In Anurans with sharply circumscribed nucleus isthmi, a small-celled and a large-celled portion of that griseum has been described. A commissura isthmi interconnects the two antimeric nuclei by way of the velum medullare anterius.

to the *torus semicircularis* and its cell groups (*'nucleus lateralis mesencephali'*, *'nucleus profundus mesencephali'*). Fibers of the 'general lemniscus' enter the *tectum proper* through the deep medulla in forms where this layer is displayed, and a rostral extension of this system seems to reach diencephalic (thalamic) grisea. The *'brachium conjunctivum'* carries cerebellar output to the tegmentum. Figure 390 illustrates an oblique section through the isthmus region of Necturus, showing some fiber pathways in JOHNSTON's (1906) interpretation.

The *descending tectal pathways*, to a large extent, also run through the deep medulla if this latter is present. A dorsal and a ventral system of *tectobulbar channels* has been distinguished, which include intrinsic mesencephalic tectotegmental fibers or collaterals, as well as presumably tectospinal fibers. The descending tegmental pathways again, to a large extent, are included in the complex channel of the *fasciculus longitudinalis medialis*. More or less diffuse and presumably reciprocal pathways seem to connect tectum and torus with the cerebellum.

The *dorsal commissural system* comprises, in addition to the largely diencephalic pretectal *commissura posterior*, the *commissura tecti* and the *isthmic commissure*. The *ventral commissural system* is represented by the variably developed *commissura ansulata sensu latiori*, which displays diverse components and contains the crossed components of the descending tegmental pathway. Said commissure is partly intermingled

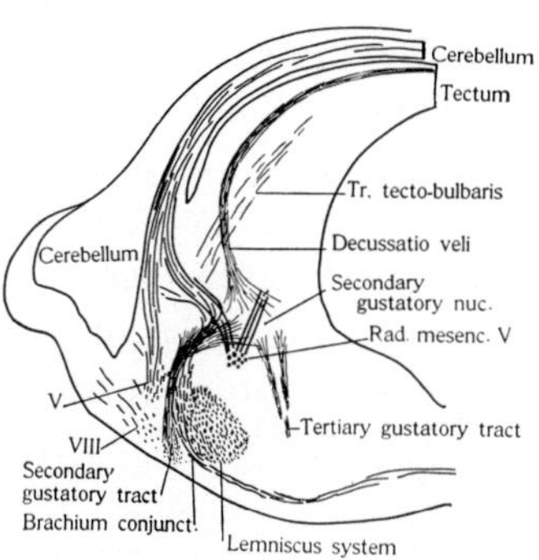

Figure 390. Oblique transverse section through cerebellum and caudal mesencephalon of the Urodele Necturus, showing some fiber systems in JOHNSTON's interpretation (from JOHNSTON, 1906). The legend cerebellum at left is placed in the lumen of the auricular recess. Decussatio veli: isthmic commissure. Trochlearis decuss. not shown.

with decussating habenulo-interpeduncular fibers. Tectal and tegmental areas are also associated with the diencephalic supraoptic commissural system, particularly through the variously described *commissura transversa*. A more detailed discussion of the major mesencephalic fiber connections, with reference to the rather vague specific data recorded in the literature, can be found in the treatises by KAPPERS *et al.* (1936) and by HERRICK (1948). It could be said that no subsequent significant new observations clarifying these details have become extant.

7. Reptiles

The Reptilian mesencephalon, in contradistinction to the cerebellum of this Vertebrate class (cf. chapter X, section 7), is rather well developed and in various respects similar to that found in diverse Anuran Amphibians. It displays, moreover, as KAPPERS *et al.* (1936) and other authors have pointed out, a few structural differentiations suggestive of those characteristic for Mammals. Although both order and species differences in degree of differentiation and general configuration obtain, these differences are on the whole much less pronounced than in the class of Amphibia.[101] Figures 391 A–H, 392 A–D, and 393 A, B illustrate representative sections through the mesencephalon of Squamata, Chelonia, and Crocodilia, respectively. Figures 394 to 397 depict some further relevant structural details. With regard to the Squamate 'suborder' Ophidia, a detailed study of nuclear topography and fiber tracts in the mesencephalon (and diencephalon) of the Rattlesnake Crotalus has been undertaken by WARNER (1946).

Externally, the *tectum mesencephali* displays the typical paired 'optic lobes', which may be relatively small, as in Snakes, or rather large and similar to, but somewhat more pronounced than those in Frogs, as e.g. seen in the Crocodile. In certain forms, such as diverse Ophidia and some Lacertilia, the caudal enlargement of the torus semicircularis, generally remaining covered by tectum opticum, may here bulge out

[101] Nevertheless, with respect to such differences, the mesencephalic (and diencephalic) grisea in the Ophidian Typhlopidae, burrowing snakes with a reduction of the optic system, are of interest. SENN (1969), who studied diencephalon and mesencephalon in two representatives of that family (Anomalepis and Liotyphlops), reports a rather simple degree of cytoarchitectural differentiation, which can be interpreted as intermediate between a Saurian and Ophidian pattern.

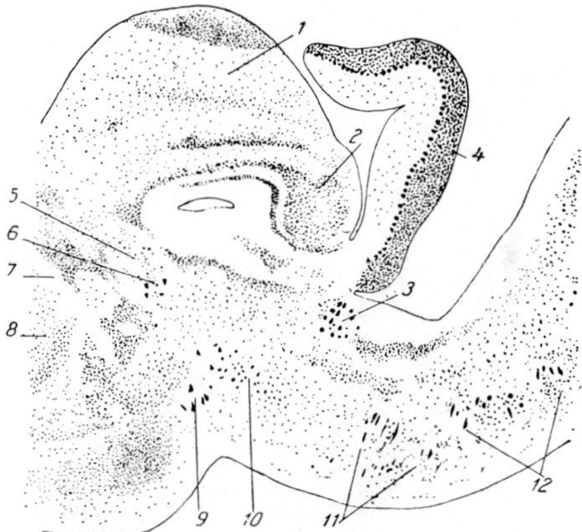

Figure 391A. Fairly lateral parasagittal section *(Nissl stain)* through mesencephalon and rhombencephalon of the Squamate Lacerta viridis (from BECCARI, 1943). 1: tectum opticum; 2: caudal torus semicircularis ('corpo posteriore', 'colliculus inferior'); 3: nucleus isthmi; 4: cerebellum; 5, 6: dorsal parvocellular and ventral magnocellular nucleus of posterior commissure; 7: fasciculus retroflexus; 8: nucleus rotundus thalami; 9: nucleus of fasciculus longitudinalis *(interstitial nucleus of Cajal)*; 10: nucleus ruber tegmenti; 11, 12: anterior and posterior cell groups of thombencephalic nucleus reticularis tegmenti.

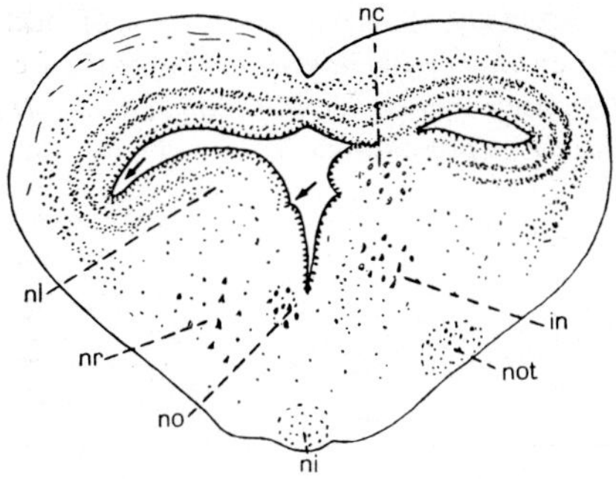

Figure 391B. Fairly rostral cross-section through the mesencephalon of Lacerta (from K., 1927). in: *nucleus interstitialis of Cajal;* nc: nucleus commissurae posterioris; ni: nucleus interpeduncularis; nl: griseum of torus semicircularis ('nucleus mesencephali lateralis'); no: nucleus oculomotorii; not: nucleus opticus tegmenti; nr: nucleus ruber tegmenti. Added arrows: sulcus lateralis mesencephali (internus), above, and sulcus limitans, below. The more rostral right half of the oblique section shows the rostral tectal ventricular recess and the caudal end of the large-celled nucleus commissurae posterioris.

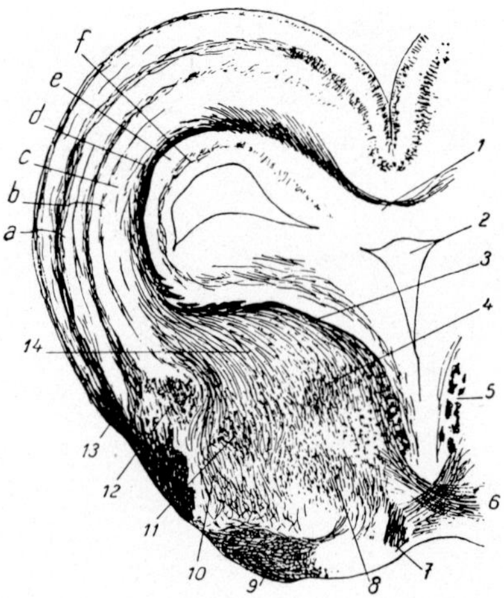

Figure 391C. Cross-section *(Weigert stain)* through the mesencephalon of Lacerta muralis at the level of rostral tectal ventricular recess (from BECCARI, 1943). 1: commissura tecti; 2: median part of mesencephalic ventricle; 3: fasciculus tectobulbaris dorsalis; 4: fasciculus praetectalis descendens; 5: fasciculus longitudinalis medialis; 6: dorsal tegmental commissure; 7: fasciculus retroflexus; 8: fasciculus geniculatus descendens; 9: nucleus opticus tegmenti *('nucleo ottico basale')*; 10: fasciculus tectobulbaris ventralis; 11: fasciculus bulbotectalis et thalamicus (general bulbar, especially trigeminal lemniscus); 12: fasciculus spinotectalis et thalamicus (spinal lemniscus); 13: fasciculus opticus lateralis of marginal optic tract; 14: fasciculus tectobulbaris intermedius; a: layer of optic fibers; b: *'strato fibroso e grigio superficiale'* (probably mostly layer of chief cells including some afferent deep medulla'); c: *'strato grigio centrale'* (probably chief cells intermingled with granule cells); d: *'strato bianco centrale'* (deep medulla); e: *'strato grigio periventriculare'* (granule layer); f: *'strato fibroso periventriculare'*.

Figure 391D. Cross-section *(Nissl stain)* through the mesencephalon of Lacerta vivipara (from FREDERIKSE, 1931). 1: nucleus reticularis tegmenti dorsalis (dorsally and laterally to this nucleus lies the griseum of torus semicircularis); 2: nucleus reticularis tegmenti ventralis; 3: nucleus basalis nervi optici (nucleus opticus tegmenti); 4: nucleus mammillaris; 5: recessus mammillaris.

Figure 391E. Cross-section (myelin stain) through the mesencephalon of Lacerta vivipara (from FREDERIKSE, 1931). Com. post.: caudobasally directed fiber traced to commissura posterior; Com. postinf.: probably rostral and basal part of tegmental commissural system (pertaining to commissura ansulata of Anamnia). Other abbreviations self-explanatory.

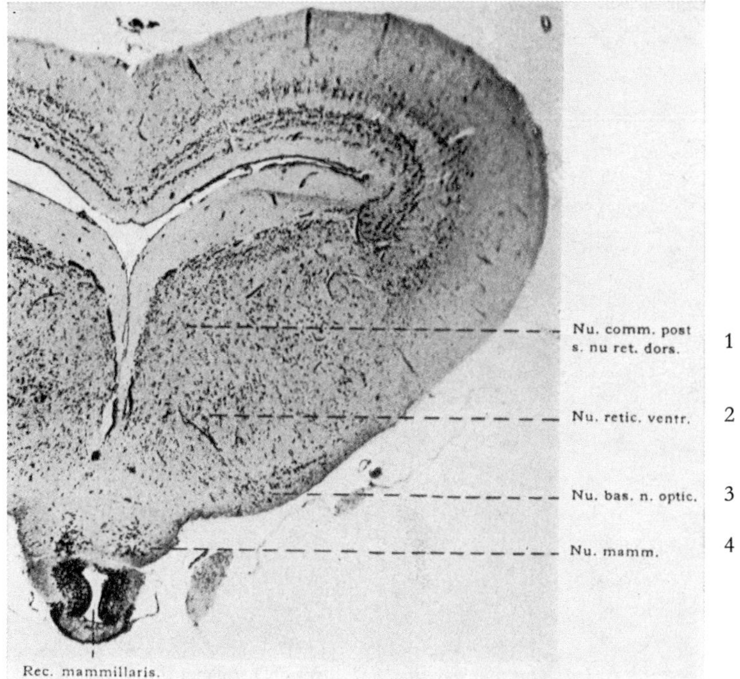

Figure 391 D

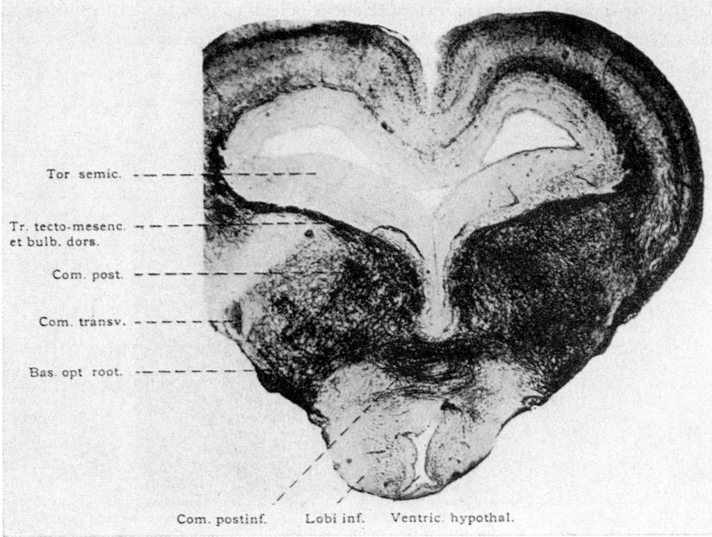

Figure 391 E

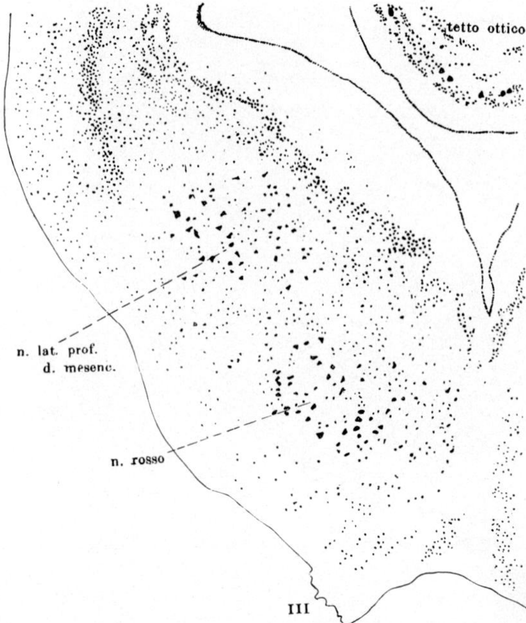

Figure 391F. Cross-section *(Nissl stain)* through torus semicircularis, tegmentum, and part of the tectum in adult Lacerta muralis (from BECCARI, 1943). It is uncertain whether the n. lat. prof. mesenc. of BECCARI represents a dorsolaterally located component of nucleus reticularis tegmenti dorsalis (basal plate) or a portion of the torus (alar plate). The ventral tegmentum shows here the 'nucleus ruber'. In the tectum opticum, some large cells of n. radicis mesencephalicae trigemini can be recognized.

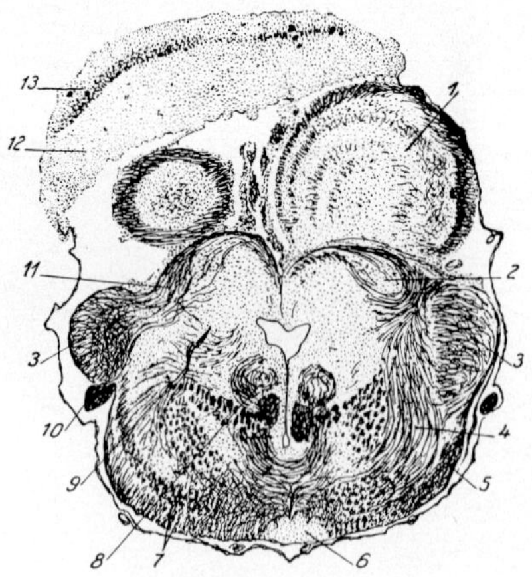

Figure 391 G

dorsalward, forming a slight protuberance, comparable to the *colliculus inferior (corpora quadrigemina posteriora)* of mammals.

As regards the *mesencephalic ventricle*, a rostral tectal recess commonly occurs (Figs. 391B, C). The *sulcus lateralis mesencephali (internus)* indicating an approximate boundary between tectum opticum and torus semicircularis, corresponds to that generally seen in Amphibia. The

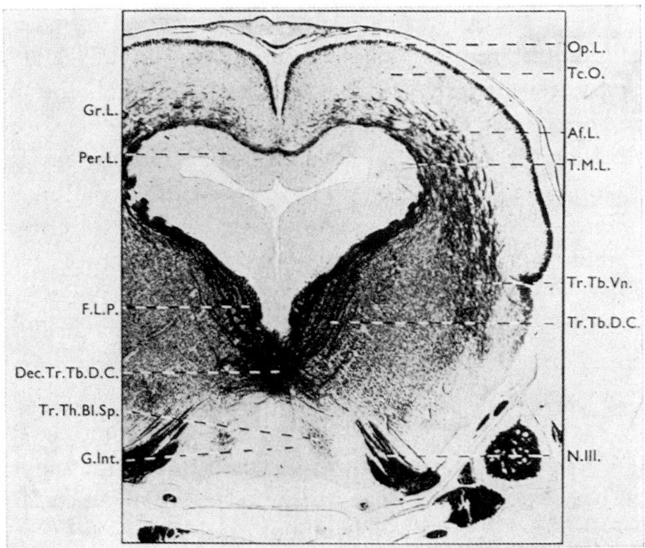

Figure 391 H. Cross-section (myelin stain) through the mesencephalon of the Ophidian Crotalus adamanteus (American Rattlesnake) at the level of oculomotor root (from WARNER, 1946). Af.L.: afferent layer of tectum (stratum lemnisci); Dec. Tr. Tb. D.C.: dorsal tectobulbar decussation; F.L.P.: fasciculus longitudinalis medialis; G.Int.: nucleus interpeduncularis; Gr.L.: stratum griseum periventriculare; N.III: oculomotor root; Op.L.: Stratum opticum tecti; Per. L.: periventricular fiber system; Tc.O.: tectum opticum (chief cell layer); T.M.L.: deep medulla of tectum; Tr.Tb.D.C.: tractus tectobulbaris dorsalis cruciatus; Tr.Tb.Vn.: tractus tectobulbaris ventralis; Tr.Th.Bl.Sp.: 'tractus thalamobulbaris'.

Figure 391 G. Cross-section (myelin stain) through the caudal part of the mesencephalon in the Squamate Chamaeleon (after DELANGE and KAPPERS, from BECCARI, 1943). 1: tectum opticum; 2: caudal end of torus semicircularis ('colliculus inferior'); 3: nucleus isthmi; 4: lemniscus lateralis; 5: tractus spino-mesencephalicus (general lemniscus system); 6: nucleus interpeduncularis; 7: tractus tectobulbaris; 8: nucleus trochlearis; 9: radix mesencephalica trigemini; 10: trochlear nerve; 11: 'fasciculus tecto-isthmicus'; 12, 13: molecular and granular layer of cerebellum.

sulcus limitans (Figs. 391B, 392D, 393A) may or may not remain recognizable in adult forms. This sulcus tends to be smoothened out by the growth pattern of the torus. *Sulcus intermediodorsalis* and *intermedioventralis* are generally absent, but a bare suggestion of the latter one may occasionally be noted.

The *alar plate grisea* at the mesencephalo-diencephalic boundary zone include dorsally the diencephalic *nucleus commissurae posterioris* with several subdivisons, and a *nucleus lentiformis mesencephali* with a magnocellular and a parvocellular component (Fig. 392B). Various diencephalic pretectal grisea such as *area praetectalis* and other pretectal grisea, to be dealt with in chapter XII of volume 5, are likewise included in that open topologic neighborhood. The stratification of the tectum opticum, which also contains in its paraventricular layers the large cells of *nucleus radicis mesencephalicae trigemini*, has been described in detail by Huber and Crosby (1933), who distinguished a sequence of six layers (Fig. 394A–H), and by Senn (1966, 1968a, b, 1969), who emphasized three main sequences or *Schichtengruppen* of lamination (Fig. 395). The layer of optic fibers is provided by the medial, lateral and corresponding axial bundles of the optic tract. The deep medulla is formed by the input channels of the general lemniscus system and the tectotegmental, tectobulbar and tectospinal output channels.

It can be seen that the stratification in the Squamate Anolis essentially corresponds to that found in anuran Amphibians but that numerous variations are displayed by other forms and particularly concern the so-called granular layer of the simplified notation discussed in section 1. Thus, the non-optic input and the tectal output channels, subsumed under the term '*deep medulla*' (layer d or 'stratum album centrale' of Huber and Crosby) may be partly distributed upon superficial strata of the chief cell layer, and partly upon its deep stratum, adjacent to the 'granular layer' (cf. also Fig. 391C).

The stratification proposed by Senn (1968a, b) is of particular interest (Fig. 395). The superficial set of laminae, corresponding to the optic layer (1) of my simplified scheme of stratification (cf. Fig. 361B) contains an external and an internal stratum opticum. The 'central' set of layers includes layer of chief cells (2) and deep medulla (3). The periventricular set of laminae comprises the layer of granule cells (4) with its subdivisions. Senn's classification is based upon a study of the ontogenetic development displayed by the optic tectum's cell population in accordance with that author's concept of stratification discussed in chapter VI, section 2 et passim, of volume 3, part II.

At the rostroventral extremity of the tectal formation, close to the ventral edge of the nucleus lentiformis mesencephali, Curwen and Miller (1939) noted, in the turtle Pseudemys, a small but distinctive cell group which they designated as x. It is related to the optic system and, on the basis of its topologic relationships, may be considered homologous with the nucleus parageniculatus tecti optici of Birds (cf. section 8).

The *torus semicircularis*, which mainly receives input from lemniscus lateralis, protrudes to a variable degree into the ventricle. There is no extensive fusion between the antimeric tori comparable to that obtaining in Frogs, but a midline fusion of the caudalmost torus portions can not uncommonly occur, particularly in those forms, mentioned above, which display an 'inferior colliculus'.

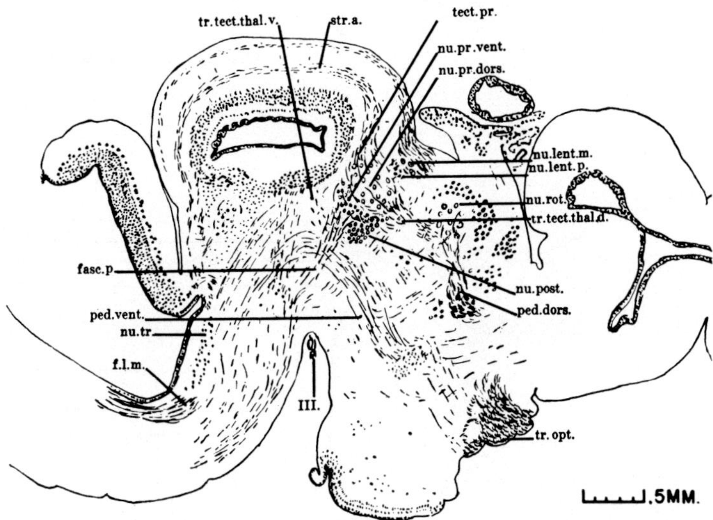

Figure 392A. Parasagittal section *(Weil stain)* through the brain of the Chelonian Pseudemys, showing some mesencephalic relationship (from Curwen and Miller, 1939). fasc.p.: fasciculus praetectalis descendens; f.l.m.: fasciculus longitudinalis medialis; nu.lent.m., p.: nucleus lentiformis mesencephali, pars magnocellularis (m), pars parvocellularis (p); nu.post.: nucleus posterodorsalis (thalami dorsalis); nu.pr. dors., vent.: nucleus praetectalis, pars dorsomedialis respectively ventrolateralis; nu.rot.: nucleus rotundus (thalami dorsalis); nn.tr.: nucleus trochlearis; ped.dors., vent.: pedunculus ventralis respectively dorsalis (lateral and medial basal forebrain bundle); str.a.: deep medulla; tect.pr.: tractus tecto-praetectalis; tr.opt.: tractus opticus; tr.tect.thal.d., v.: tractus tectothalamicus dorsalis respectively ventralis; III: exit of oculomotor nerve.

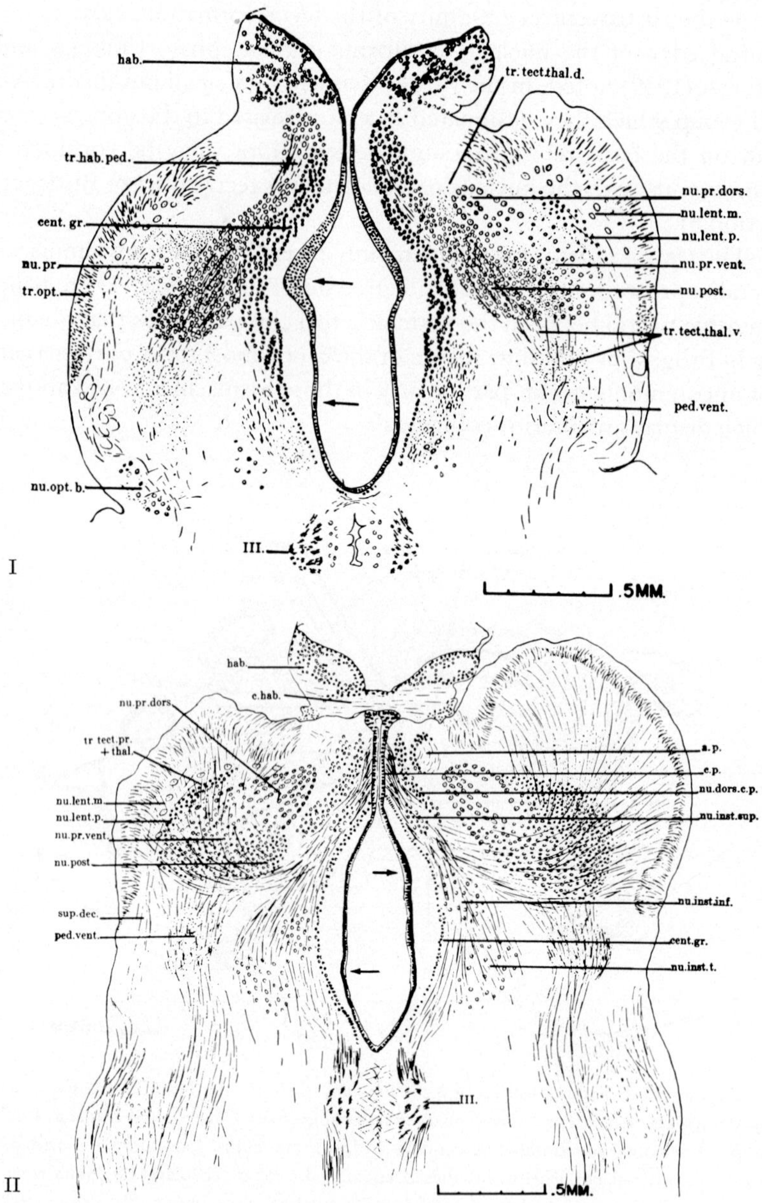

Figure 392B. Two slightly oblique cross-section *(Weil stain)* through the mesencephalo-diencephalic boundary regions of the turtle Pseudemys (from CURWEN and MILLER, 1939). The plane of I passes through synencephalic recess, that of II is slightly more

Generally speaking, the cell masses of the torus appear as ventral extension of the tectum opticum's granular layers, whose abventricular portion commonly differentiates into a 'nucleus', while the periventricular portion usually remains as a cell band. The development of the torus in the Lizard, as studied by SENN (1970) is shown in Figure 396. As regards the primordia of the colliculi inferiores, SENN (1969) found that in Saurians (Lacertilians) these 'colliculi' have the structure of torus semicircularis, being formed by the deep group of layers, while in

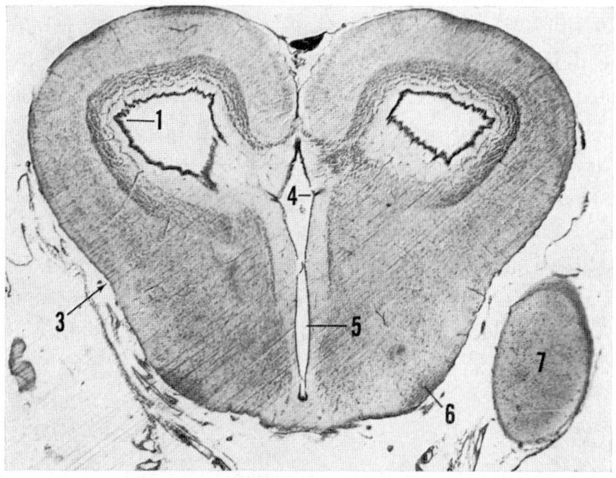

Figure 392C. Cross-section through the mesencephalon of the turtle Chrysemys at the level of rostral tectal diverticulum or recess (hematoxylin eosin, × ca. 40; red $^1/_2$). 1: sulcus lateralis mesencephali (internus) accessorius; 2: sulcus lateralis mesencephali accessorius; 3: sulcus lateralis mesencephali externus; 4 sulcus accessorius; 5: remnant of sulcus limitans; 6: nucleus opticus tegmenti; 7: caudal tip of hemisphere.

---

caudal, but still displaying sulcus synencephalicus. Also, the right side is slightly more caudal than the left. a.p.: area praetectalis; cent. gr.: paracentral gray; c.hab.: commissura habenulae; c.p.: commissura posterior; hab.: nucleus sive ganglion habenulae; nu.dors. c.p.: nucleus dorsalis commissurae posterioris; nuc. inst.inf., sup.: nucleus interstitialis inferior respectively superior of posterior commissure; nuc. inst. t.: nucleus interstitialis tegmentalis commissure posterioris; nuc. opt. b.: nucleus opticus basalis seu tegmenti; nuc.post.: nucleus posterodorsalis (thalami dorsalis); nu.pr.: nucleus praetectalis; sup.dec.: supraoptic decussation; tr.hab.ped.: tractus habenulo-(inter)peduncularis seu fasciculus retroflexus; tr.opt.: tractus opticus; tr.tect.pr. thal.: tractus tectopraetectalis and tectothalamicus; III: nucleus of oculomotor. Other abbreviations as in Figure 392A. Added arrows indicate synencephalic recess, synencephalic sulcus (in II) and sulcus limitans.

Ophidia the 'inferior collicular prominences' consist of a dorsal part (*'paratorus'*) containing an extension of the central group of layers, and of a ventromedial part (torus), which includes the caudal extension of the periventricular layer group. The 'torus' of Lacertilia and Ophidia is said to be homologous, while the paratorus represents an additional configuration. Thus, the *'colliculus inferior'* of Saurians, displayed by the 'torus' and that of Ophidians, provided by the 'paratorus', are interpreted as 'analogous' but not as 'homologous'. While there is some justification for this evaluation, which is supported by certain differences in detailed aspects of fiber connections, one could also consider both types of colliculi as kathomologous. With regard to their relevant input channels, both are evidently related to the lemniscus lateralis system. There is, moreover, some evidence that, even in Lacerta (cf. Fig. 391A) an extension of the tectum's central group of layers likewise participates to some degree in the composition of the overall torus griseum.

The *nucleus isthmi* (Fig. 391A) is located caudally and ventrolaterally to the torus semicircularis with whom it has relevant functional relationships as a relay griseum of lemniscus lateralis. In various Reptiles, the nucleus isthmi forms a protrusion at the level of velum medullare

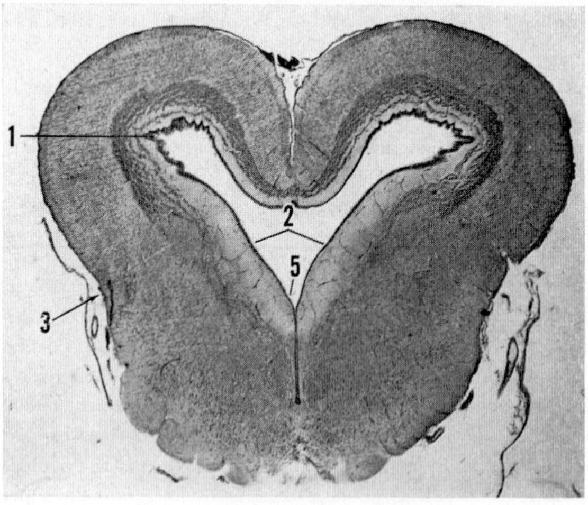

Figure 392D. Cross-section through the mesencephalon of the turtle Chrysemys caudal to level of preceding figure, and showing common mesencephalic ventricle with its tectal expansion. Designations and magnification as in preceding figure.

anterius, directly rostrally to the exit of trochlear nerve (Figs. 391G, 393B). The nucleus is larger in Crocodilia than in Chelonia, and unusally large in the Squamate Chamaeleon. In some other Squamates, including Ophidia, it is relatively small (KAPPERS *et al.*, 1936). A conspicuous *nucleus visceralis secundarius* cannot be recognized, but may be represented by cell groups located within the periventricular gray medially to nucleus isthmi.

The *basal plate* or true tegmental components of the Reptilian mesencephalon comprise the grisea related to the formatio reticularis tegmenti, and the nuclei of oculomotor and trochlear nerves.

The *reticular formation* is represented by various grisea which include the rostrally located *nucleus interstitialis of the fasciculus longitudinalis medialis* (Figs. 397A, B). Slightly distinguishable from rostral portions of this griseum, a (mesencephalic) tegmental interstitial nucleus of the posterior commissure can be recognized, at least in the turtle Pseudemys (CURWEN and MILLER, 1939). In addition, the tegmentum contains diffusely arranged cell groups which, nevertheless, display some more or less distinctive condensations. Among these latter, a magnocellular group can be identified as corresponding to the Mammalian nucleus ruber (DELANGE, 1912; BECCARI, 1923, 1943; SENN, 1968; and other authors). It receives fibers of the crossed brachium conjunctivum or superior cerebellar peduncle, and gives off an essentially crossed descending tract presumably reaching the spinal cord.

Other tegmental cell groups have been designated as nucleus reticularis ventralis and dorsalis. It is difficult to delimit this latter from the ventral and ventrolateral cell population of the torus semicircularis.[102] The nucleus lateralis profundus mesencephali of SENN (1968) is presumably identical with a portion of the nucleus reticularis dorsalis.

Dorsomedially to the nucleus interstitialis, a paraventricular cell group presumably corresponds to the Mammalian *nucleus of Darkschewitsch* (Fig. 397A).[103] The griseum designated as *nucleus zeta* by BEC-

[102] Cf. the '*nucleo laterale profondo del mesencefalo*' of BECCARI, depicted in Figure 391F. This nucleus is not identical with a definitely tegmental '*nucleus profundus mesencephali*' located laterally to nucleus ruber in the Chelonian Testudo and depicted by KAPPERS *et al.* (1936, Fig. 480 loc. cit.) following DELANGE, and interpreted as a relay nucleus of the lateral lemniscus.

[103] BECCARI (1943) included it with the nuclei of the posterior commissure. This is doubtless justified, but the position of this cell group nevertheless clearly suggest that it represents, in contradistinction to alar plate grisea of posterior commissure, a paraventricular tegmental derivative homologous to the *nucleus of Darkschewitsch*. KAPPERS likewise applies this term to said nucleus (cf. KAPPERS *et al.*, 1936, p. 984).

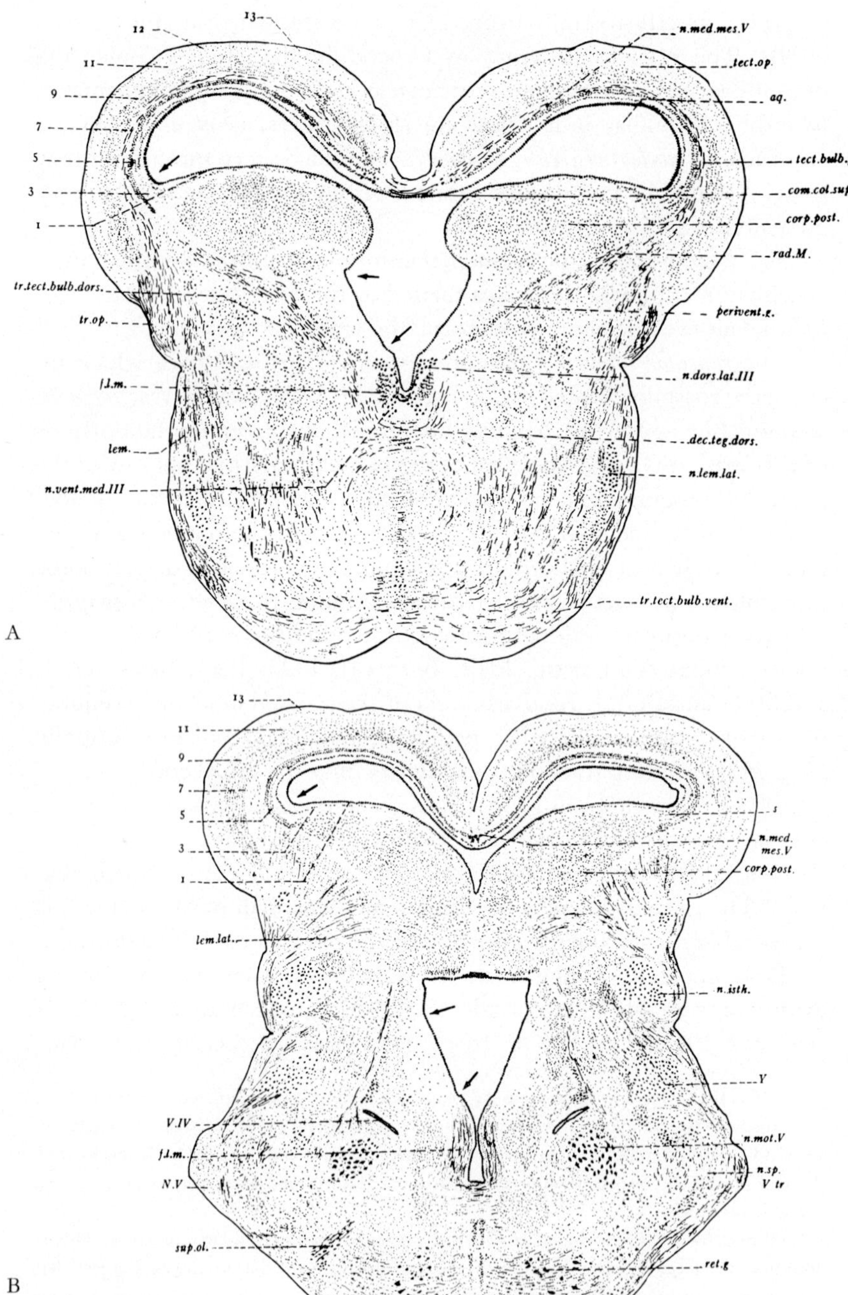
13
12
11
9
7
5
3
1
tr.tect.bulb.dors.
tr.op.
f.l.m.
lem.
n.vent.med.III
n.med.mes.V
tect.op.
aq.
tect.bulb.f.
com.col.sup.
corp.post.
rad.M.
perivent.g.
n.dors.lat.III
dec.teg.dors.
n.lem.lat.
tr.tect.bulb.vent.
A
13
11
9
7
5
3
1
lem.lat.
V.IV
f.l.m.
N.V
sup.ol.
s
n.med.
mes.V
corp.post.
n.isth.
Y
n.mot.V
n.sp.
V tr
ret.g
B

Figure 393

CARI (1943) most likely represents a 'forerunner of the Mammalian substantia nigra', as suggested by the cited author (Fig. 397B). Other, well identifiable tegmental grisea are the lateral nucleus opticus tegmenti and the unpaired, median interpeduncular nucleus (Fig. 397D).

TUGE (1932) who has investigated structural aspects of somatic motor mechanisms in the midbrain and medulla oblongata of the Chelonian Chrysemys elegans, propounds, in his otherwise competent paper, a highly unconvincing distinction between 'nucleus motorius tegmenti' and 'nucleus reticularis tegmenti'. The former is said to send most of its axons into the fasciculus longitudinalis medialis, while most axons of the 'reticular nucleus' are presumed to be intrinsic, i.e. 'associated with the reticular formation'. According to TUGE (TSUGE)[104] 'the nucleus reticularis is an apparatus of correlation having only indirect, or at best only questionable connection with motor centers'.

[104] The author, whose name is pronounced TSUGE and would also be spelled TSUGE, in the generally accepted *Hepburn-rômaji* system of transcription for the Japanese language, uses a peculiar system of transcription, introduced by Japanese purists, which is only intelligible to those already familiar with the Japanese syllabary *(go-ju-on)* in which the sounds *ti, tu, si* do not exist, being modified to *chi, tsu, shi*. Since the purpose of a transcription could be said to spell out words from a foreign script in a manner intelligible to the unacquainted outsider, it is evident that a system in which the actual sounds, *chi, tsu, shi* are written *ti, tu, si* (e.g. *Tiba* for *Chiba*, *niti* for *nichi*, *Tuge* for *Tsuge*, *Nisi* for *Nishi)* does not seem very appropriate.

Figure 393 A, B. Cross-sections through optic tectum (A) at the level of oculomotor nucleus and (B) close isthmus region, through the midbrain of Alligator mississippiensis (from HUBER and CROSBY, 1926). The drawings are based on both *Nissl preparations* and preparations processed by the *Cajal silver impregnation method.* aq: ventriculus mesencephali; com.col.sup.: commissura tecti; corp. post.: torus semicircularis; dec.teg.dors.: dorsal tegmental decussation; f.l.m.: fasciculus longitudinalis medialis; lem., lem.lat.: lemniscus system, especially lemniscus lateralis; n.dors.lat.III: dorsolateral oculomotor nucleus; n.isth.: nucleus isthmi; n.lem.lat.: 'nucleus of lemniscus lateralis'; n.med.mes.V.: nucleus medialis radicis mesencephalicae trigemini; n.mot.V.: nucleus motorius trigemini; n.sp.V.tr.: nucleus radicis descendentis ('spinalis') trigemini; n.vent.med.III.: ventromedial oculomotor nucleus; N.V.: trigeminal root; perivent.g.: periventricular gray; rad.M.: *'radiation of Meynert'* (tectotegmental respectively tectobulbar system); ret.g.: formatio reticularis tegmenti; s: extension of tectal granular layers into torus; sup.ol.: superior olivary griseum; tect.bulb.f.: tectobulbar fibers; tec.op.: tectum opticum; tr.op.: tractus opticus; tr.tect.bulb.dors.,vent.: dorsal and ventral tectobulbar tracts; V IV: obliquely cut recessus of fourth ventricle; Y: medial (small celled) group of nucleus isthmi (the bend of the lead lies in what presumably is the main sensory trigeminal nucleus); alternate numbers 1–13 represent tectal layers in P. RAMÓN's notation. Added arrows: sulcus lateralis mesencephali (internus), sulcus limitans, and sulcus intermedius ventralis. Nucl. of lemn. lat. probably part of n. isthmi.

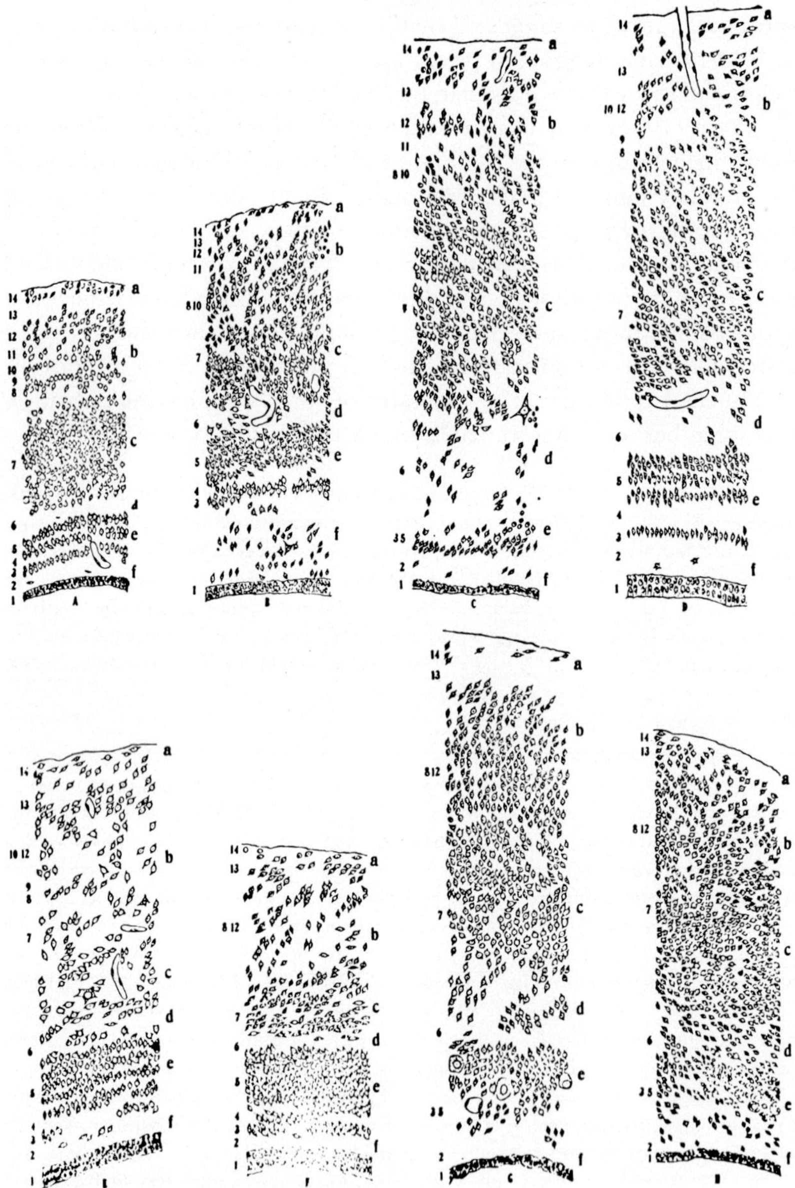

Figure 394 A–H. Cytoarchitectural aspects (modified *Nissl stain*) of the tectum opticum in various Reptiles (after HUBER and CROSBY, 1933, from KAPPERS *et al.*, 1936). A: Anolis carolinensis; B: Heloderma suspectum; C: Varanus griseus; D: Alligator mississippiensis; E: Chelhydra serpentina, old adult; F: Chelhydra serpentina, young

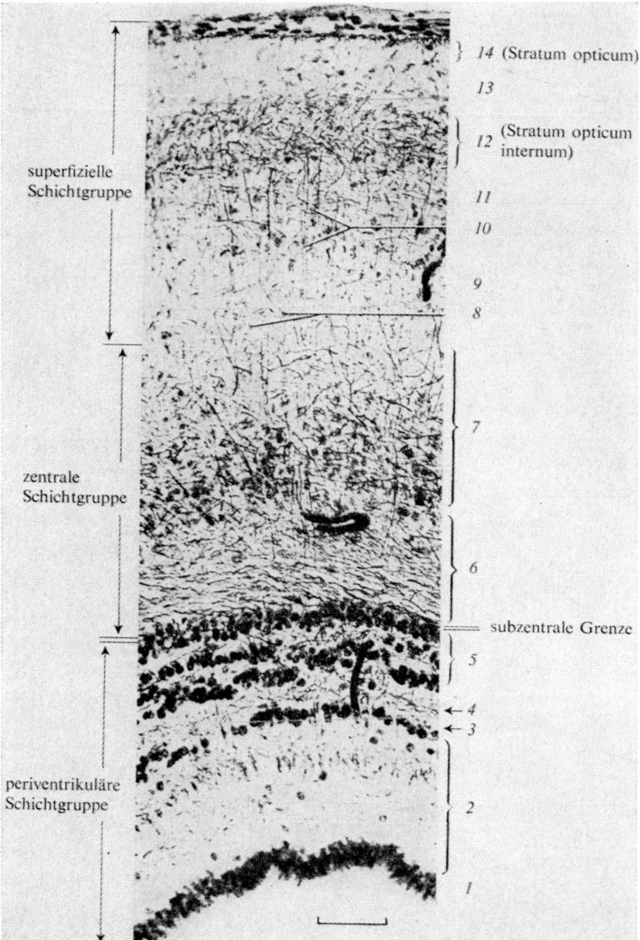

Figure 395. Fundamental laminar structure of the Reptilian tectum opticum as conceived by Senn, and shown in Lacerta sicula (from Senn, 1968a). The scale indicates 50μ, the preparation is processed by a combined silver proteinate and *Nissl technique.*

---

specimen; G: Natrix; H: Thamnophis sirtalis (an Ophidian). The numbers represent P. Ramón's notation. Huber's and Crosby's notation are indicated by the letters. a: stratum opticum; b: stratum fibrosum et griseum superficiale; c: stratum griseum centrale; d: stratum album centrale; e: stratum griseum periventriculare; f: stratum griseum periventriculare.

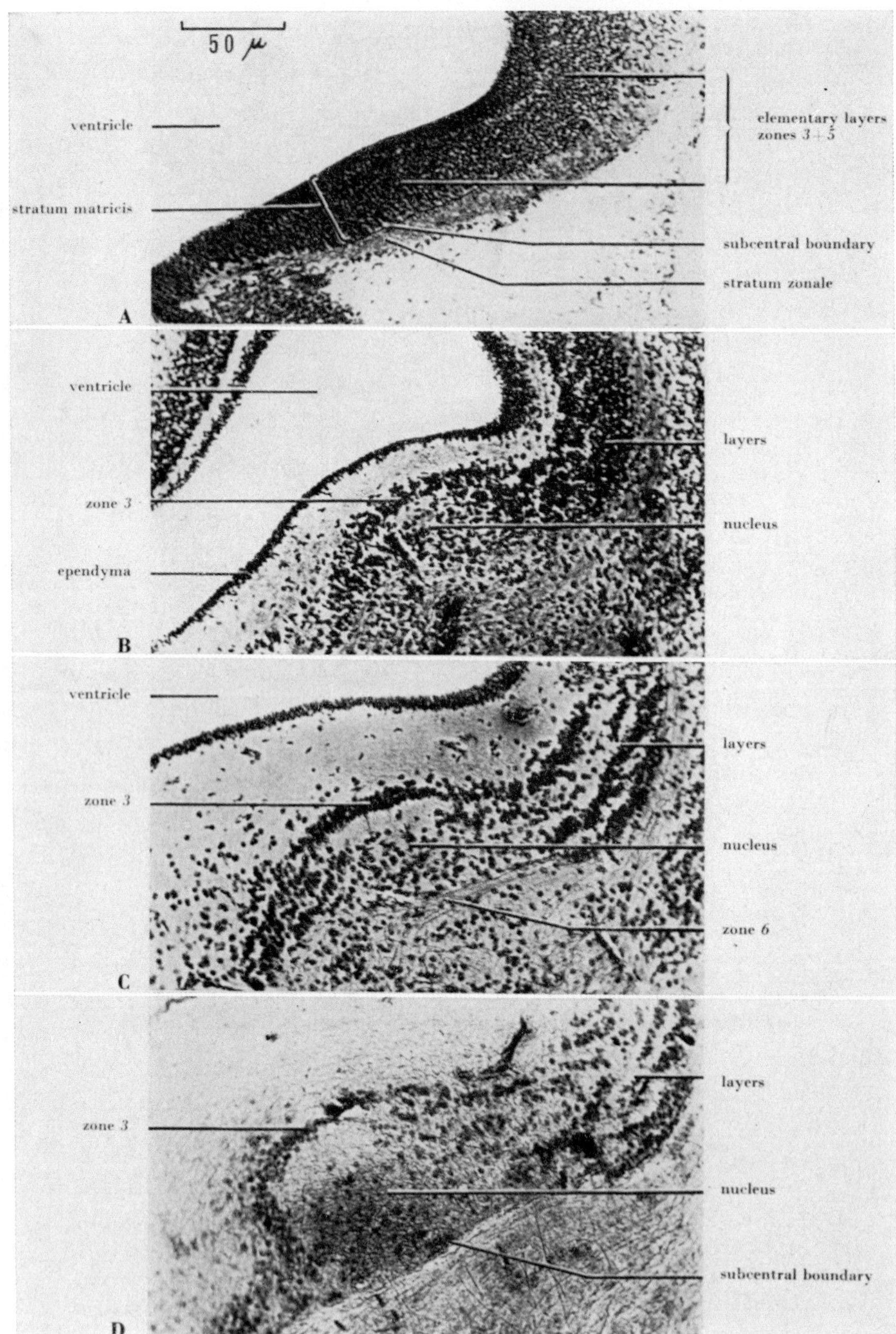
50 μ
ventricle
stratum matricis
elementary layers
zones 3+5
subcentral boundary
stratum zonale
A
ventricle
zone 3
ependyma
layers
nucleus
B
ventricle
zone 3
layers
nucleus
zone 6
C
zone 3
layers
nucleus
subcentral boundary
D

As regards the *nuclei of the oculomotor and trochlear nerves*, diverse stages of differentiation obtain in Reptiles, and the mutual relations of these two grisea likewise show numerous variations. In Caiman, Chelone and Boa, these nuclei are separated by a distinct gap, being, on the other hand, closely adjacent in Varanus. Detailed data on the eye muscle nerve nuclei in the brain of squamate Reptilia have been recorded by SENN (1966, 1968a).

The simplest condition of the oculomotor nucleus is displayed by some Chelonia. In other Reptiles, a dorsolateral and a ventromedial cell group is conspicuous. KAPPERS and ADDENS, moreover, have recorded a dorsal accessory nucleus in Varanus as well as in some other

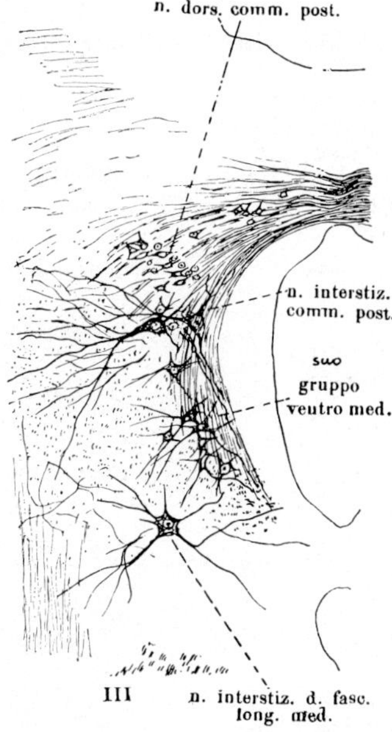

Figure 397A: Cross-section *(Cajal's silver impregnation)* showing subdivisions of nucleus commissurae posterioris and a large cell of *Cajal's nucleus interstitialis* in an advanced embryo of Lacerta muralis (from BECCARI, 1943). It is not improbable that the '*gruppo ventro med.*' pertains to the basal plate and represents the *nucleus Darkschewitschi* of Mammals.

Figure 396A–D. Cross-section through torus semicircularis at various stages of the Lizard's ontogenic development (from SENN, 1970). A: Lizard embryo of 10 days; B: embryo of 16 days, showing layered and nuclear differentiation; C: embryo about to hatch, fiber systems appear between layers and within nucleus; D: adult stage.

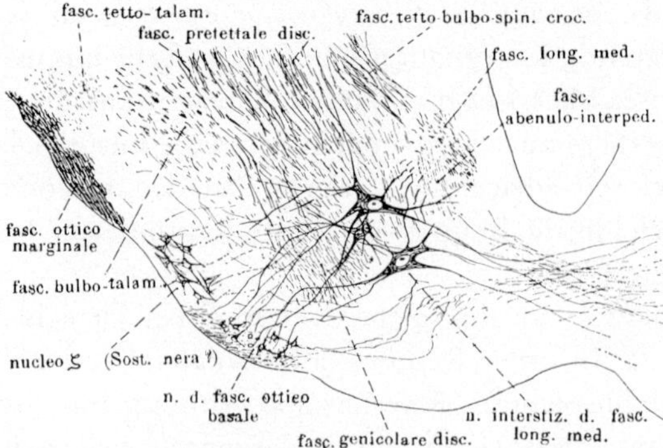

Figure 397B. Cross-section *(Cajal's silver impregnation)* through the rostral tegmentum of an advanced embryo of Lacerta muralis, somewhat caudal to level of preceding figure (from Beccari, 1943). The nucleus Zeta may indeed represent a 'forerunner' of the Mammalian substantia nigra. Basomedially to this cell group lies the nucleus opticus tegmenti (n.d.fasc.ottico basale). The large multipolar cells are those of *Cajal's nucleus interstitialis.*

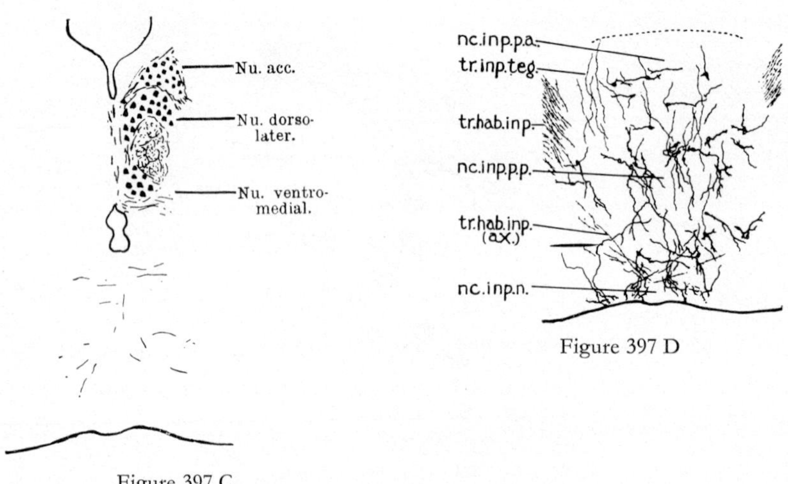

Figure 397 C

Figure 397 D

Figure 397C. Cross-section through the mesencephalic tegmentum in Varanus, showing subdivisions of oculomotor nucleus (from Kappers, 1920). Nu.acc.: nucleus accessorius, presumably representing the *Edinger-Westphal nucleus.*

Figure 397D. Cross-section through the interpeduncular nucleus of the Chelonian Chrysemys, combined from several *Golgi sections* (from Tuge, 1932). nc.inp.n.: neuropil of nucleus interpeduncularis; nc.inp.p.a.: nucleus interpeduncularis, pars anterior; nc.inp. p.p.: nucleus interpeduncularis, pars posterior; tr.inp.teg.: tractus interpedunculo-tegmentalis; tr.hab.inp.: tractus habenulo-interpeduncularis, respectively (ax.) single axons of that tract.

forms (Fig. 397C). It is likely that this cell group represents the preganglionic nucleus of the oculomotor, comparable to the *Edinger-Westphal nucleus of Mammals* (KAPPERS *et al.*, 1936).

Summarizing the relevant configurational aspects of Reptilian mesencephalic grisea, suggesting a 'phylogenetically advanced' Amniote characteristics, the following features could be enumerated. (1) a further differentiation of pretectal grisea at the mesencephalo-diencephalic boundary, including diencephalic nuclei of posterior commissure, nucleus lentiformis mesencephali, and mesencephalic tegmental *nucleus of Darkschewitsch*. (2) Tendency toward formation of inferior colliculi comparable to those of the Mammalian corpora quadrigemina. (3) Differentiation of a tegmental nucleus ruber and presence of a primordial 'substantia nigra'. (4) Tendency toward segregation of a distinct preganglionic or *Edinger-Westphal oculomotor nucleus*.

With respect to the mesencephalic fiber systems the overall composition and distribution of communication channels is essentially comparable to those found in anuran Amphibia, or, for that matter generally obtaining in Anamnia. It will here be sufficient to mention the major and fundamental, but *qua* specific details quite variable or diversified fiber systems such as optic tract and its subdivisions, basal forebrain bundles with their mesencephalic connections, fasciculus retroflexus, fasciculus longitudinalis medialis, general spinal and bulbar lemniscus system, lateral lemniscus, a number of tectotegmental systems with tectobulbar and tectospinal extensions, and the several channels interconnecting cerebellum with tectum or tegmentum mesencephali. The *commissures* include commissura posterior with its subcommissural organ, commissura tecti, commissura transversa, and the various decussations in the tegmentum such as those of brachium conjunctivum and those of efferent tectal systems.[105] HOLLÄNDER (1917), who investigated the descending components of the fasciculus longitudinalis medialis in the Sauropsidan mesencephalon, recorded relevant relations of that bundle to the nuclei of the posterior commissure and other grisea. Yet, despite detailed descriptions by numerous other authors (e.g. HUBER and CROSBY, 1926; KAPPERS *et al.*, 1936), the available data on hodology, remain, *qua* specific synaptology and relevant griseal interconnection, very poorly elucidated, their validity being re-

[105] The generalized term *commissura ansulata*, subsuming the decussating channels in the mesencephalic tegmentum of Anamnia is not usually employed in descriptions of the corresponding system in the brain of Amniota.

stricted to generalized statements.[106] Still less well known are the relevant intrinsic connections.

As regards the cochlear component of the lateral lemniscus, however, it has been recently claimed that, in Crocodilians, a 'tonotopically organized auditory region' is present in the 'central nucleus' of torus semicircularis, from which an ascending channel originates and reaches thalamic grisea (MANLEY, 1971; PRITZ, 1974).

In comparison with Anamnia, the reciprocal connections with the diencephalon are doubtless more differentiated, including, e.g. a tractus tecto-rotundus, and the ascending lemniscus systems may have a greater proportion of thalamic endings.[107] Among the rostral channels, tractus tectothalamicus dorsalis and ventralis, tractus tectopretectalis, fasciculus geniculotectalis, geniculotegmentalis *(fascio genicolare discendente* of BECCARI), fasciculus praetectalis descendens, investigated by BECCARI (1923, 1943) and by CURWEN and MILLER (1939) seem to be particularly well recognizable in Reptilia. It is uncertain whether the tractus thalamobulbaris, noted by DELANGE (1913), HUBER and CROSBY (1933), and WARNER (1946, cf. Fig. 391H), represents predominantly an ascending or a descending tract. This fiber system may overlap with BECCARI's fasciculus geniculatus descendens, fasciculus praetectalis descendens, and with parts of the habenulopeduncular tract (cf. Fig. 391C). Also, rostral extensions of the brachium conjunctivum might be expected to be present within the tegmental bundles reaching diencephalic grisea (of thalamus and hypothalamus). Connections of hypothalamus with basal and dorsal mesencephalic neighborhoods are suggested by the course of fibers displayed in suitably prepared sagittal sections. However, the relevant details concerning origin, termination, and functional significance of said fiber systems have not been satisfactorily ascertained and thus remain obscure.

In addition, a mesencephalic *periventricular system* has been described in the Alligator by HUBER and CROSBY (1926) and was also depicted in Lacerta by BECCARI (1943). It may be related to the tectum

[106] As regards '*intrinsic connections*', those within the cerebellar cortex, due to its relatively simple and uniform structure, are slightly better elucidated than those of most neuraxial grisea.

[107] In many fishes, however, the connections of the posterior hypothalamic centers (lobi inferiores) are relatively more massive, the lobi inferiores being relatively reduced as well as presumably functionally quite modified in Amniota.

(cf. Fig. 391C) and could also, in a general way, represent a forerunner of the Mammalian fiber bundles of the central gray which include the *fasciculus longitudinalis dorsalis of Schütz*.

Again, among the caudal channels, the presence of a crossed, but at the Reptilian stage still rather small rubrospinal tract (KAPPERS, 1947) should be mentioned. According to BECCARI, however, this fiber system can only be followed into the oblongata, '*ma non è stata vista una via rubro-spinale*'.

8. Birds

The adult Avian mesencephalon displays, with respect to its external configuration, conspicuous differences in comparison with the midbrain of Reptiles. The large *tectum opticum* is displaced lateralward and even ventralward. The *torus semicircularis* remains completely covered or 'buried' by the tectum, into whose lateral displacement it becomes involved. In contradistinction to Reptiles, the tendency toward the formation of posterior (or inferior) colliculi is not manifested. A fairly large part of mesencephalic tectum is covered by the cerebellum and by the caudal extensions of the massive telencephalic hemispheres (cf. Fig. 336).

At early and intermediate embryonic stages the mesencephalon of Birds is, nevertheless, closely similar to that of Reptiles. *Sulcus lateralis mesencephali (internus)* and *sulcus limitans* are clearly recognizable, and a sulcus intermedioventralis may be faintly indicated. Sulcus lateralis mesencephali remains conspicuous at the adult stage in which the sulcus limitans, although occasionally identifiable, tends to be flattened out. The peculiar configuration of the Avian mesencephalon and diencephalon is presumably correlated with the optic system's high degree of development and with the substantial differentiation of the epibasal (D_1) and lateral basal (B_1 and B_2) grisea as well as with the general expansion of the telencephalon. The mesencephalic ventricle comprises paired rostral and caudal *tectal recesses*, which jointly communicate, at a restricted, rostrocaudally short transverse level, with the ventricle's more or less flattened and rather narrow median portion, whose greatest but here slit-like dorsoventral extension is found at the mesencephalo-diencephalic boundary. Figures 398A–C, 399A–D, 400A–C, and 401 illustrate representative cross-sections through the Avian midbrain. As regards some significant details of taxonomic differences, relevant data, including bibliographic references, can be found in the publications by CRAIGIE (1928, 1930), HUBER and CROSBY (1929), JUNGHERR (1945), PEARSON (1972), PORTMANN and STINGELIN (1961), and SANDERS (1929). Stereotaxic atlases have been prepared for the Pigeon's brain by KARTEN and HODOS (1967) and for the Chick's brain (telencephalon, diencephalon and mesencephalon) by VAN TIENHOVEN and JUHASZ (1962). Among older reports including data on the Avian

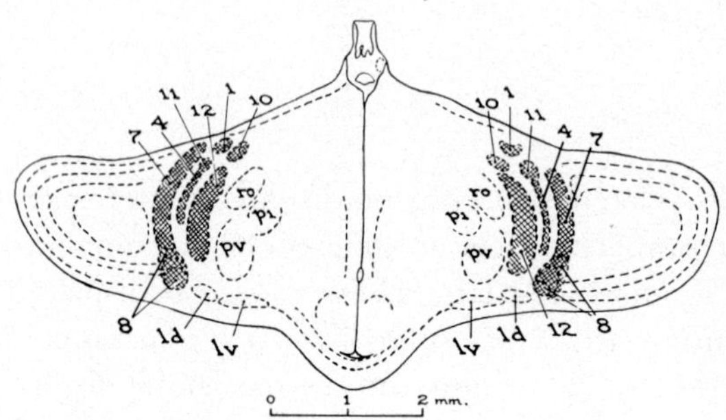

Figure 398A. Outline of cross-section through pretectal region and rostral end of tectum opticum in the newly hatched chick (from K., 1939). 1: area praetectalis; 2, 2′: subdivisions of nucleus diffusus commissurae posterioris; 4: nucleus interstitialis tractus praetecto-subpraetectalis; 7: nucleus lentiformis mesencephali (pars parvocellularis); 8: nucleus parageniculatus tractus optici; 9: nucleus praetectalis lateralis; 10: nucleus praetectalis medialis; 11: nucleus praetectalis principalis; 12: nucleus principalis praecommissuralis; 13: nucleus spiriformis dorsomedialis; 14: nucleus spiriformis ventromedialis; 15: nucleus subpraetectalis; ld, lv: corpus geniculatum laterale (dorsale, ventrale); ot: nucleus opticus tegmenti; pi: nucleus posterointermedius; pv: nucleus posteroventralis; ro: nucleus rotundus (thalami dorsalis).

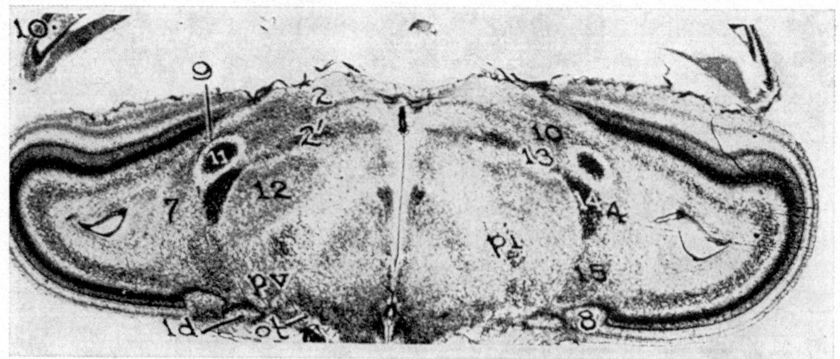

Figure 398B. Cross-section *(Nissl stain)* through pretectal region and rostral portion of tectum opticum in the newly hatched chick (from K., 1939, × 12, red. $^{4}/_{5}$). For abbreviations cf. preceding Figure A.

mesencephalon, those by EDINGER and WALLENBERG (1899) and by WALLENBERG (1898, 1904) deserve particular mention. Details on mesencephalic centers and connections, together with a review of Avian brain literature can also be found in the paper by HUBER and CROSBY (1929).

In the *alar plate*, at the levels of the mesencephalo-diencephalic boundary zone, the *dorsal thalamic pretectal grisea* and *nuclei of the posterior commissure*, as well as the *mesencephalic group* of *pretectal nuclei*, derived from the tectum opticum, are particularly well differentiated (Figs. 398A, B). The boundary between diencephalon and mesencephalon, a region of open topologic neighborhoods, becomes blurred and difficult to establish in the adult. Likewise, the terminology applied to various grisea by different authors has led to some degree of ambiguity. On the basis of embryologic observations, however, the identifications and provenance of the diencephalic and mesencephalic pretectal grisea can be reasonably well clarified (K., 1939).

As regard the dorsal thalamic pretectal nuclei caudally protruding into said boundary region, the following grisea can be noted: area praetectalis, nucleus principalis praecommissuralis, nucleus principalis

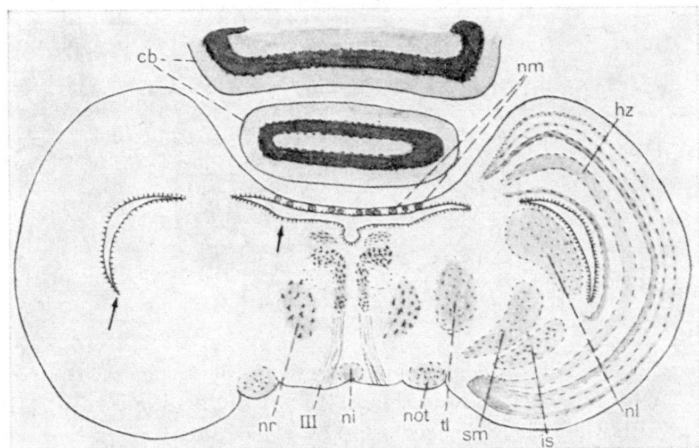

Figure 398C. Semidiagrammatic cross-section through the mesencephalon of the Pigeon at level of oculomotor (from K., 1927). cb: cerebellum; hz: layer of chief cells; is: nucleus isthmi, pars magnocellularis; ni: nucleus interpeduncularis; nl: nucleus mesencephali lateralis (torus semicircularis); nm: nucleus radicis mesencephalicae trigemini; not: nucleus opticus tegmenti; nr: nucleus ruber; sm: nucleus isthmi, pars parvocellularis (nucleus semilunaris); tl: nucleus lateralis tegmenti; III: oculomotor root. Added arrows: sulcus lateralis mesencephali (left), sulcus limitans (right).

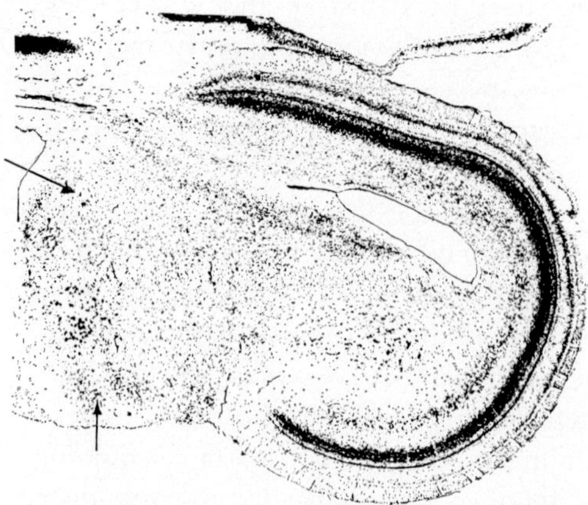

Figure 399A. Cross-section *(Nissl stain)* through the mesencephalon of the Pigeon at the level of nucleus ruber (from KARTEN and HODOS, 1967). Added arrows indicate nucleus interstitialis tegmentalis commissurae posterioris (above) and 'substantia nigra' (below).

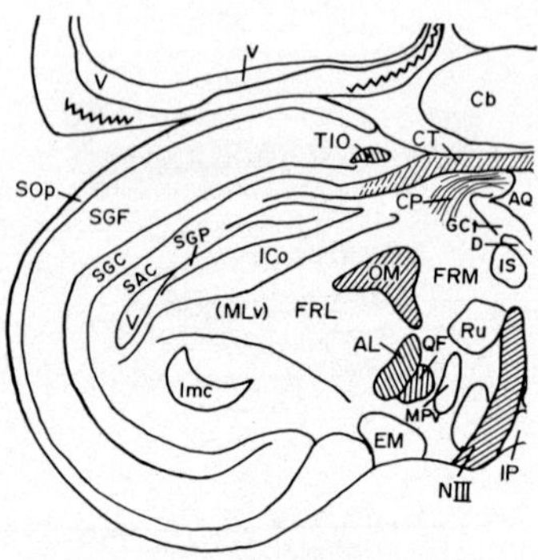

Figure 399B. Outline drawing of preceding figure, illustrating the interpretation by the cited authors (from KARTEN and HODOS, 1967). AL: 'ansa lenticularis'; AQ: 'aquaeductus cerebri'; BC: brachium conjunctivum; BCA: 'brachium conjunctivum ascendens'; BCS: 'brachium colliculi superioris'; Cb: cerebellum; CP: commissura posterior; CT: commissura tecti; D: '*nucleus of Darkschewitsch*'; EM: nucleus ectomammillaris (nucleus opticus tegmenti); EW: *Edinger-Westphal nucleus;* FLM: fasciculus longitudinalis medialis; FRL: formation reticularis lateralis mesencephali; FRM: formatio reticularis medialis mesencephali; GCt: substantia grisea centralis; ICo: 'nucleus intercollicularis'; Imc: nucleus isthmi pars magnocellularis; Ip: nucleus interpeduncularis; MLd: nucleus interpeduncularis; MLd: nucleus mesencephalicus lateralis, pars dorsalis; MLv: nucleus mesencephalicus lateralis, pars ventralis; MNV: nucleus radicis mesencephalicae tri-

praetectalis, nucleus praetectalis medialis and lateralis, nucleus interstitialis tractus praetecto-subpraetectalis, and nucleus subpraetectalis. Of these grisea, the nucleus praetectalis principalis is commonly most conspicuous as a crowded and rounded cellular aggregate surrounded by a ring-like fiber mass including neuropil (cf. Fig. 398B).

The likewise dorsal thalamic nuclei of the posterior commissure are represented by the magnocellular nuclei spiriformes (dorsomedialis and ventrolateralis), moreover by the nucleus diffusus parvocellularis commissurae posterioris.[108]

The *mesencephalic tectal group of pretectal grisea* comprises nucleus lentiformis mesencephali pars magnocellularis and pars parvocellularis, moreover a rather basally located nucleus parageniculatus tecti optici, receiving fibers from the marginal optic tract. This nucleus can be considered homologous with the cell group x recorded by CURWEN and MILLER (1939) in the turtle Pseudemys, and with FUSE's nucleus olivaris colliculi superioris in Mammals (cf. section 7 and 9 of the present chapter).

One can agree with the statement by KAPPERS *et al.* (1936) that the 'pretectal' and 'subpretectal' areas in Birds are extremely well developed, probably more so than in other Vertebrates. Despite lack of sufficiently documented details *qua* connections and synaptology, these grisea can be evaluated 'as way stations to and from the tectum', 'intermediaries between diencephalic and tectal centers', including also connections with the telencephalon and with the deuterencephalic tegmentum.

The lateroventral displacement of the tectum opticum is correlated with a rather thin and, as it were, stretched median portion of the alar

[108] A *nucleus interstitialis tegmentalis commissurae posterioris*, pertaining to the basal plate, is briefly dealt with further below. Further details concerning the pretectal region as a whole shall be included in chapter XII of volume 5.

---

gemini; MPv: nucleus mesencephalicus profundus, pars ventralis; N III: oculomotor root OM: tractus occipitomesencephalicus; OMd: nucleus oculomotorii, pars dorsalis; OMv: nucleus oculomotorii, pars ventralis; Pap: 'nucleus papillioformis'; QF: tractus quintofrontalis; Ru: nucleus ruber; SAC: stratum album centrale; SGC: stratum griseum centrale; SGF: stratum griseum et fibrosum superficiale; SGP: substantia grisea et fibrosa periventricularis; SOp: stratum opticum; TIO: tractus isthmo-opticus; TPc: 'nucleus tegmenti pedunculo-pontinus, pars compacta'; TVM: 'tractus vestibulo-mesencephalicus'; V: ventricular lumen.

plate, which contains the *commissura tecti* and the *nucleus of the radix mesencephalicae trigemini* which, in some Birds, is mainly restricted to this location[109] (Figs. 398C, 399C, D, 404A).

The *optic tectum* displays the fundamental layers present in Reptiles, but with a more pronounced cellular lamination of the sublayers of SENN's terminology (1968b). At least five to six cellular strata can generally be distinguished, which are separated by fiber layers. Three or four strata, of which the innermost is particularly dense and massive, pertain to SENN's '*superfizielle Schichtgruppe*'; a fairly wide but less dense layer, which I interpreted as *Hauptzellenschicht*, lies in SENN's '*zentrale Schichtgruppe*', while the cells of the cited authors '*periventrikuläre Schichtgruppe*' is relatively inconspicuous (Figs. 395, 398B, C, 399A). At the rostroventral extremity of the superficial and central layer system, in close apposition to the dorsal lateral geniculate griseum, the nucleus parageniculatus tecti optici, already mentioned above, forms a distinct griseum, whose rostralmost portion is contiguous with nucleus lentiformis mesencephali, pars parvocellularis (Figs. 398A, B).

Structural arrangements in the Avian optic tectum, as disclosed by the *Golgi method*, are shown in Figures 403A–C. Figure 403D discloses at least 3 if not 4 different levels of optic nerve end arborizations. These different levels are of interest with respect to the theory propounded by LETTVIN *et al.* (1959) and discussed in section 1 of the present chapter. Figure 403E illustrates the pattern of electric potential gradients recorded in the Chick's optic tectum by CRAGG *et al.* (1954) in an attempt to correlate electrical responses with the obtaining histologic structures. Attempts of this sort, however, have not, until now, disclosed data useful for a relevant clarification of the complex tectal mechanisms.

---

[109] In some instances, however, cells of this nucleus may be found in the deeper layers of the tectum opticus proper (cf. KAPPERS, 1947). Thus, in the Chick, ROGERS and COWAN (1973) distinguish a larger medial division within the tectal commissure and a smaller lateral division within the stratum periventriculare of the tectum opticum. During ontogenetic development, the cited authors recorded a 'dramatic loss of cells' from a maximum population of about 4000 at the 9th day of incubation to a final population of slightly over 1000 after the 13th day. This is explained as resulting from the failure to make contact with the developing muscle spindles at the critical period. The observations of these authors also suggest that the Chick jaw musculature is bilaterally innervated by the proprioceptive mesencephalic trigeminal nucleus. As regards the cell loss which may occur as a relevant factor in the course of normal ontogenetic development, cf. the comments concerning overproduction and subsequent cell disintegration on p. 45, section 1, chapter V, vol. 3/I, and p. 218, section 1 C, chapter VI, vol. 3/II.

In order to ascertain some aspects of optic input reaching the Avian tectum mesencephali and telencephalon, we performed experiments with evoked potentials upon unilateral optic stimulation in the Chick (K. and SZEKELY, 1963). Direct monopolar as well as bipolar electroencephalographic recordings were taken while the eye was stimulated by single light flashes and by various flicker frequencies. Identifiable homo- and contralateral evoked responses, including synchronization, were obtained from both the tectum opticum and the explored dorso-

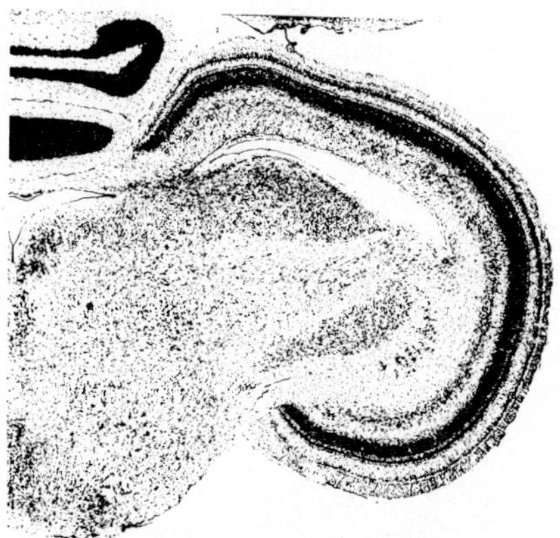

Figure 399C. Cross-section *(Nissl stain)* through the mesencephalon of the Pigeon at the level of the oculomotor nuclei (from KARTEN and HODOS, 1967).

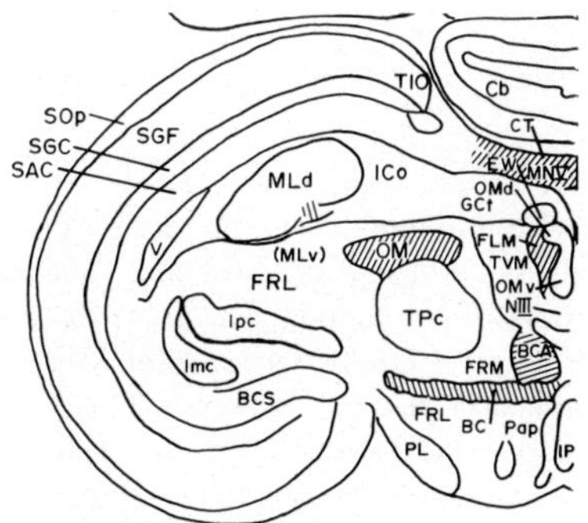

Figure 399D. Outline drawing of preceding figure illustrating the interpretation by the cited authors (from KARTEN and HODOS, 1967). For abbreviations cf. legend to Figure 399B. Cf. also Figure 398C. The sectional plane of this latter figure is at a dorsocaudal-ventrorostral inclination to the planes of Figures 399A–D.

lateral part of the telencephalon. These responses were predominantly very definite, although, without ascertainable regularity, at times weak or missing. Again, the contralateral responses were more pronounced, while the corresponding homolateral evoked potentials were at times of rather small amplitude, and, moreover, the tectal evoked potentials were less pronounced than some of the telencephalic ones. Occasionally, a repeated flicker stimulation would result in conspicuous inhibition of 'spontaneous' tectum opticum activity. This inhibitory effect

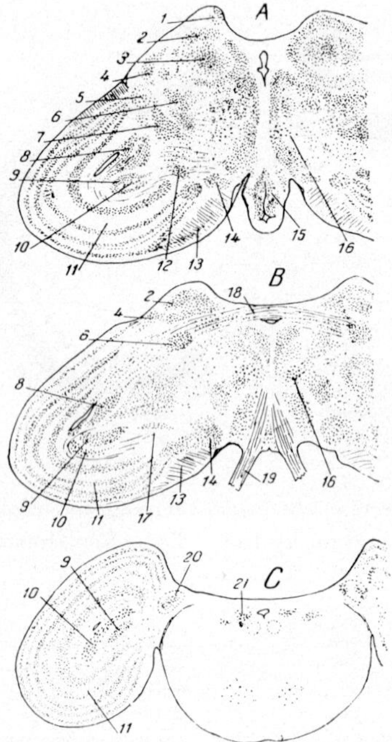

Figure 400. Cross-sections through three levels of the mesencephalon in the sparrow Passer domesticus (slightly modified and simplified after Huber and Crosby, 1929, from Beccari, 1943). The sections, based on modified *Nissl stain* preparations, pass through rostral (A), intermediate (B), and caudal (C) levels of tectum opticum. 1: habenular nuclei; 2: 'nucleus dorsolateralis' thalami; 3: 'anterior nucleus dorsolateralis' thalami; 4: 'nucleus superficialis parvocellularis' thalami; 5: 'nucleus praetectalis'; 6, 7: medial and lateral subdivisions of nucleus spiriformis; 8: dorsal part of nucleus lateralis mesencephali (torus); 9, 10: parvocellular and magnocellular subdivisions of nucleus isthmi (principalis); 11: tectum opticum; 12: 'nucleus subpraetectalis'; 13: tractus opticus; 14: nucleus ectomammillaris (nucleus opticus tegmenti); 15: nucleus mammillaris dorsalis; 16: nucleus ruber tegmenti; 17: nucleus semilunaris; 18: commissura posterior; 19: radix oculomotorii; 20: nucleus isthmo-opticus; 21: nucleus nervi trochlearis.

was characterized by depressed amplitude, considerably reduced frequency, and lack of synchronization. In the telencephalon, the responses were obtained from the caudal two-thirds. Generally speaking, responses from the surface were here of smaller amplitude than those obtained from various depths.

The projection of the retinal quadrants upon the tectum opticum, investigated by various recent authors (e.g. HAMDI and WHITTERIDGE, 1954; DELONG and COULOMBRE, 1965) essentially corresponds to the

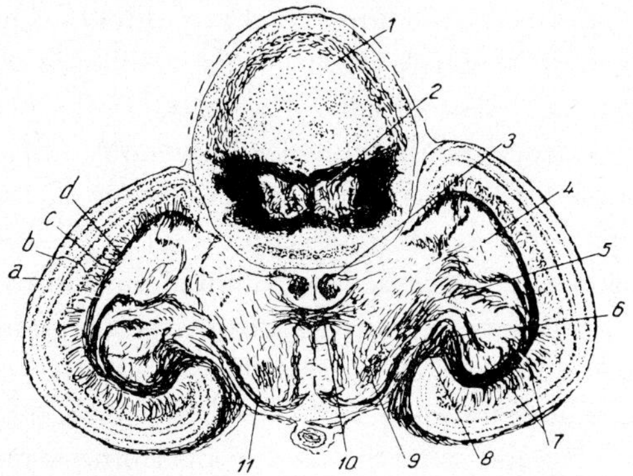

Figure 401. Cross-section (myelin stain) through the brain of Pratincola rubicola at the level of decussatio brachii conjunctivi and nuclei isthmi (after KAPPERS, 1921, from BECCARI, 1943). 1: cerebellum; 2: medial nucleus of cerebellum; 3: root of trochlear nerve before decussation; 4: nucleus lateralis mesencephali (torus); 5: lemniscus lateralis; 6: 'fasciculus tecto-thalamicus'; 7: nucleus isthmi (principalis); 8: nucleus semilunaris; 9: 'fasciculus bulbothalamicus et hypothalamicus'; 10: decussatio brachii conjunctivi; 11: tractus tectobulbaris ventralis; a: stratum opticum tecti; b: stratum fibrosum et griseum superficiale; c: stratum griseum centrale; d: stratum album centrale.

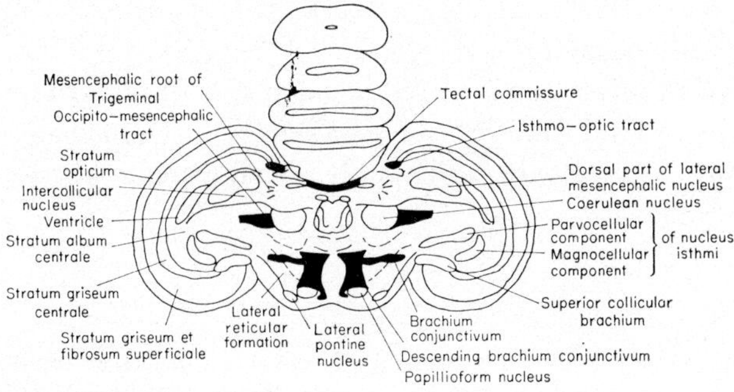

Figure 402. Diagram showing the generalized organization of the Avian mesencephalon (from PEARSON, 1972).

general optic projection pattern upon the Vertebrate tectum as dealt with in section 1 of the present chapter. Although the projection from the retina can be considered predominantly contralateral, our results (K. and SZEKELY, 1963) clearly demonstrated an involvement of homolateral tectal activities, mediated through a small homolateral optic tract component as suggested by KNOWLTON (1964), or through the commissural systems (commissura posterior, commissura tecti), or through both sorts of connections.[110]

The *torus semicircularis* of Birds, located ventrally respectively ventrolaterally to the tectal portions of the midbrain ventricle, is also known as the ganglion laterale (WALLENBERG, 1898) or nucleus mesencephalicus lateralis, pars dorsalis (KAPPERS, 1921; KAPPERS *et al.*, 1936). It consists of several ill-defined subdivisions, distinguishable by populations of medium-sized and of larger cellular elements. This griseum, which appears to be an important center of the cochlear respectively vestibular system, receives a substantial input through the cochleo-vestibular lemniscus lateralis and from the isthmomesencephalic bundle. Other connections seem to be those with the trigeminal portion of the general lemniscus, with the diencephalic commissura supraoptica ventralis, and with the tectum opticum. A medial subdivision of the torus grisea, extending toward the median (alar plate) roof of the mesencephalic ventricle is the so-called *nucleus intercollicularis* (cf. Figs. 399A–D). According to ZIGMOND *et al.* (1973), androgen-concentrating cells can be detected by autoradiography, using tritiated testosterone, in tectum, torus and tegmentum of the Chaffinch Fringilla coelebs, but most of these cells were localized in the 'nucleus intercollicularis'. Since it is claimed that 'vocalizations' result from electrical stimulation of the intercollicular area in Birds, the cited authors 'suggest that the nucleus intercollicularis is a site in the action of androgens on avian vocal behavior'.

The Avian *isthmic grisea* represent a complex of cell groups formed by the ventrolateral *nucleus isthmi magnocellularis*, the medially adjacent *nucleus isthmi parvocellularis*, and a more dorsal griseum, the *nucleus isth-*

[110] In this respect, it should be recalled that the lateral geniculate complex and the diencephalic components of the pretectal grisea have intricate fiber connections with the optic tectum. The pretectal grisea, moreover, are closely related to, or interconnected by, the posterior commissure and perhaps components of the supraoptic commissural system.

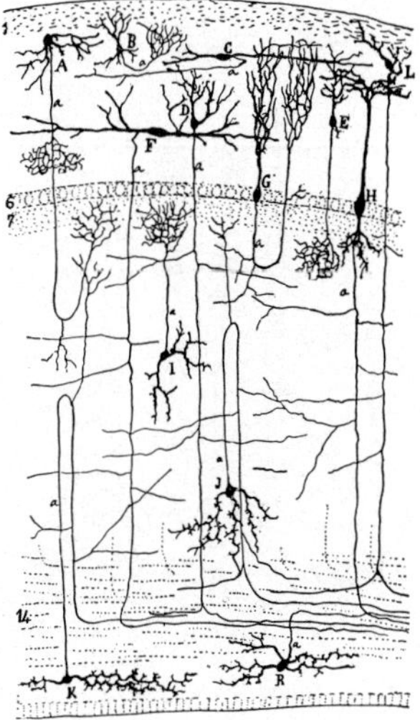

Figure 403A. Composite drawing of a sagittal section through the tectum opticum of a Sparrow based on diverse *Golgi preparations* (partly after P. Ramón, from Cajal, 1911). A, B, C, E, F, G, H, L: diverse cell types, described in detail by Cajal, and located in Senn's superficial laminar group. I: cell presumably in upper layer of Senn's central laminar group; J: modified crozier cell (chief cell) in central laminar group; K, R: cells in Senn's periventricular laminar group; a: presumed neurite; u: deep medulla; 1, 6, 7 refer to a layer notation adopted by P. Ramón and by Cajal for the Avian tectum. The layers are here enumerated in opposite direction from that being followed in Lizards and Frogs, as e.g. depicted by Figure 361A (*'numérotage des couches en sens inverse de celui adopté pour les oiseaux'*).

mo-opticus, located medially to the tectum opticum[111] (Figs. 398C, 399A–D, 400A–C, 401). The nucleus isthmi parvocellularis was also described as *nucleus semilunaris;* some authors restrict this term to a cell group which may appear as a separate medial extension of the parvocellular isthmic nucleus, but seems to be merely a frequently directly continuous component of this latter. It may represent an interstitial nucleus of the lateral lemniscus. The isthmo-optic nucleus receives optic tract fibers and is also linked to the tectum, the oculomotor and trochlear nuclei, the reticular formation, and presumably also to the other isthmic grisea. A conspicuous fiber tract connected with the isthmo-

[111] Depending on the cross-sectional levels in one and the same species, as well as on the different taxonomic forms, the isthmic grisea display diverse minor variations of their mutual topographic relationships within said overall configurational pattern (cf. also Pearson, 1972).

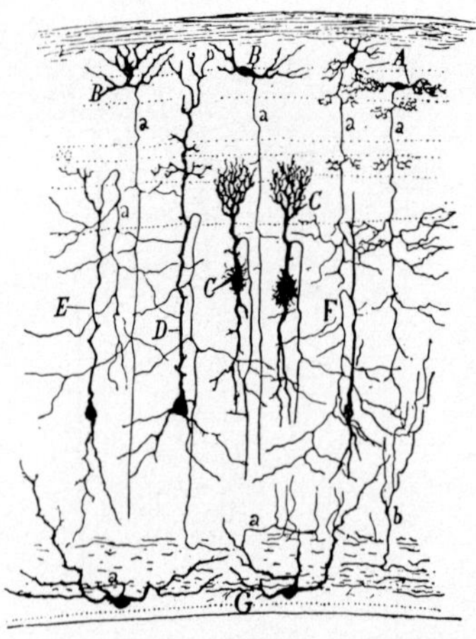

Figure 403B. Sagittal section *(Golgi impregnation)* through tectum opticum in a Sparrow a few days after hatching (from CAJAL, 1911). A: small stellate cells in superficial laminar group; B: cells with long axon in superficial laminar group; C: chief cell with short crozier; D: chief cell with long crozier; E: cell with looped neurite; F: crozier cell with short neurite; G: cells in central laminar group; a: neurites; b: branches of fiber pertaining to an afferent system ending in tectum.

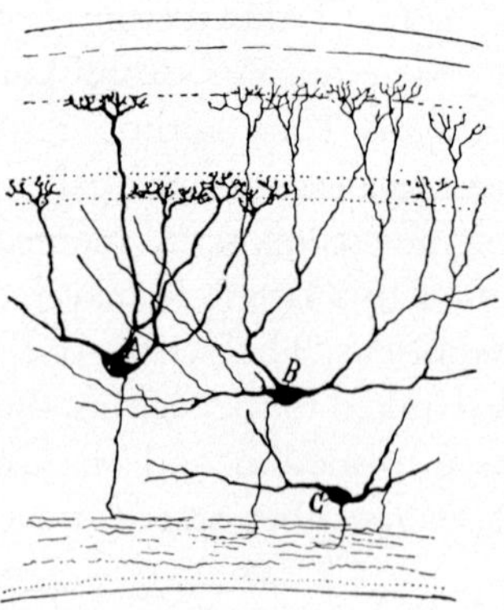

Figure 403C. Cross-section *(Golgi impregnation)* through the tectum opticum in a Sparrow a few days after hatching (after P. RAMÓN, from CAJAL, 1911). A: chief cell with thick dendrites; B: chief cell with thin dendrites; C: nerve cell adjacent to deep medulla.

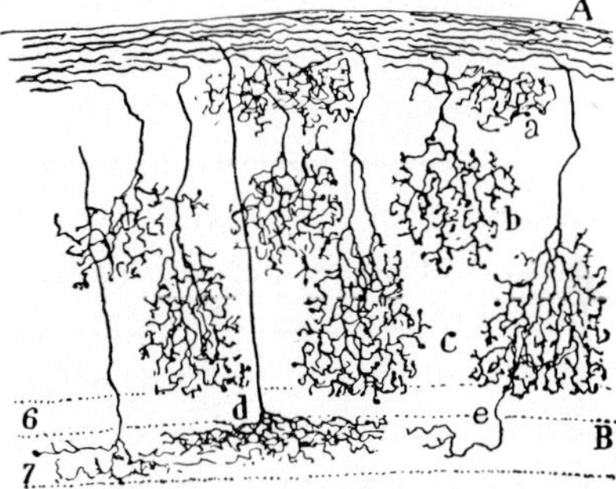

Figure 403D. Optic tract fibers ending in four different sublayers of the superficial laminar group a young sparrow, Passer domesticus, as shown by *Golgi impregnations* (from CAJAL, 1911). A: superficial optic fibers; B: 'seventh optic tectum layer'; a, b, c, d: optic arborizations at first, second, third, and fourth level; 6, 7: layers of optic tectum in RAMON's notation.

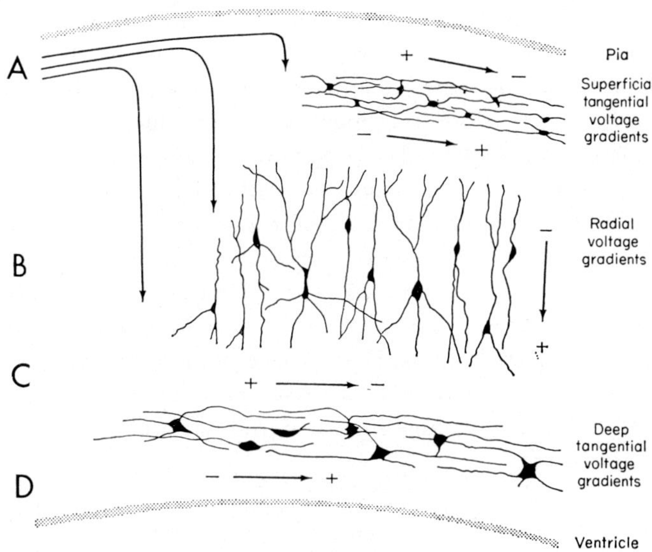

Figure 403E. Diagram illustrating the spread of impulses through the three main layers of the Avian tectum opticum (after CRAGG *et al.*, 1954, from PEARSON, 1972). The following designations, indicating my interpretation, have been added: A: superficial laminar group of SENN; B: layer of chief cells; C: deep medulla; D: periventricular laminar group of SENN.

optic nucleus, and apparently including several components, is the isthmo-optic tract. Moreover, centrifugal fibers are believed to originate in nucleus isthmo-opticus and to reach the contralateral retina. Centrifugal fibers of this sort may also originate in the deeper layers of the tectum opticum as described by CAJAL (1911). The main or principal isthmic nuclei (magnocellularis, parvocellularis respectively semilunaris) are mainly linked to the lateral lemniscus, of which they presumably represent relay stations, and to torus semicircularis. Cerebellar connections also seem to be involved.[112] A griseum comparable to the *nucleus visceralis secundarius* of Anamnia, derived from the intermediodorsal zone of the alar plate, could not be recognized in the material at my disposal, nor was it identified, as far as I know, by other authors. It will be recalled that the presence of this nucleus in Reptilians remains rather doubtful (cf. section 7, p. 905).

Turning now to the components of the *basal plate*, the reticular grisea and their derivatives shall be considered first. The nucleus of the fasciculus longitudinalis medialis *(nucleus interstitialis of Cajal)* is well developed. Medially and dorsomedially to it, in the central gray, the *nucleus of Darkschewitsch* is present. Laterally to the nucleus of fasciculus longitudinalis medialis, a rather diffuse but nevertheless distinctive tegmental interstitial *nucleus of the posterior commissure* can be identified (K., 1939). The *nucleus ruber* is commonly located laterally to the oculomotor root, and in close neighborhood to the more diffuse medial cell populations of the reticular formation. Although better developed than in Reptiles, it does not include a small-celled component as characteristic for the Mammalian nucleus ruber. One of its main input channels is the brachium conjunctivum, and its output seems to be represented by an essentially crossed rubrobulbar and rubrospinal tract. Other clearly identifiable grisea are *nucleus opticus tegmenti (nucleus ectomammillaris)* and the unpaired median *nucleus interpeduncularis*. Their connections, whose details remain poorly elucidated, seem to correspond to those generally obtaining for Vertebrate and dealt with in the preceding sections.

---

[112] Numerous recent experimental studies with evoked potentials or direct stimulation, and with degeneration techniques are reviewed by PEARSON (1972). However, the data provided by these studies remain ambiguous and not always convincing. I do not believe that they have substantially clarified, as regards dubious details, the analysis of fiber tracts based on the investigations by the older authors and summarized by BECCARI (1943), KAPPERS (1947), and KAPPERS *et al.* (1936).

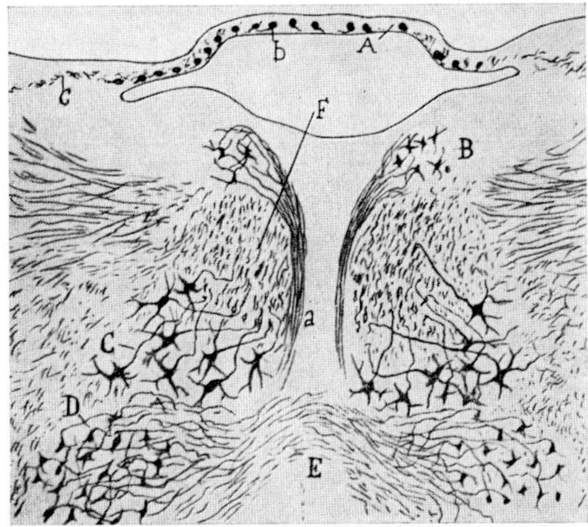

Figure 404 A. Semidiagrammatic cross-section *(Cajal's silver impregnation)* through a portion of the mesencephalon in the Magpie (Pica) at the level of nucleus ruber (from CAJAL, 1911). A: thin median roof of mesencephalon *('valvule de Vieussens');* B: probably *Edinger-Westphal nucleus* or perhaps nucleus medialis anterior; C: nucleus interstitialis fasciculi longitudinalis medialis; D: nucleus ruber; E: decussation of rubrobulbar (and spinal?) tract; F: fasciculus longitudinalis medialis; a: fibers originating in nucleus B; b: nucleus radicis mesencephalicae trigemini (interpreted by CAJAL as motor: *'noyau moteur descendant du trijumeau');* c: fibers of radix mesencephalica quinti.

The more diffusely arranged components of the *reticular formation* defy a precisely defined parcellation. Nevertheless, a nucleus reticularis medialis and a nucleus reticularis lateralis can be roughly distinguished. In addition, nondescript cell groups have been recorded, such as a nucleus mesencephalicus profundus, a nucleus tegmenti pedunculopontinus and a caudomedial nucleus papillioformis (lateral to interpeduncular nucleus). Near the isthmus region, a tegmental 'nucleus of the locus caeruleus' is depicted by KARTEN and HODOS (1967) and was briefly referred to in section 9 of chapter IX.

CRAIGIE (1928) noted and depicted a tegmental griseum in Humming Bird which might correspond to the Mammalian *substantia nigra*, located just caudally to the posterior end of nucleus ruber and dorsomedially to basal optic nucleus, moreover close to the end of the caudalward extending hypothalamic grisea. CRAIGIE (1928) very cautiously suggests the Mammalian homology, which does not seem improbable, particularly since this Avian 'substantia nigra' might also correspond to BECCARI's (1943) *nucleus zeta* mentioned in the preceding sec-

Figure 404B. Cross-section *(Cajal's silver impregnation)* through the oculomotor complex in an embryo of the Sparrow Passer italiae near the hatching stage (from BECCARI, 1943). 1: dorsomedial oculomotor nucleus; 2: *Edinger-Westphal nucleus (nucleo accessorio o superiore);* 3: dorsolateral oculomotor nucleus; 4: fasciculus longitudinalis medialis; 5: ventral oculomotor nucleus; 6: radix oculomotorii.

tion 7. While in Reptiles and Mammals the nucleus opticus tegmenti is ventromedial to substantia nigra, the ventrolateral location of said nucleus to the substantia nigra of Birds could easily be understood as related to the ventrolateral expansion of the tectum opticum. KARTEN and HODOS (1967) do not record a substantia nigra, but this griseum might be represented by the cell group which I have indicated in Figure 399A.

The prerubral tegmentum, derived from the embryonic tegmental cell plate, seems to display a considerable rostral expansion into diencephalic neighborhoods, as shown in Figure 398B (not labelled, but easily identifiable medially to 'pi'). It is likely that the grisea designated by KARTEN and HODOS (1967) as '*campi Foreli*', 'stratum cellulare internum' and 'statum cellulare externum' pertain to this mesencephalic tegmental formation.

Concerning the efferent nuclei of the two mesencephalic nerves III and IV, the *oculomotor nuclear complex* generally displays four distinctive cell groups (Fig. 404B), namely a dorsomedial, a dorsolateral, a ventromedial, and an accessory nucleus.[113] This latter presumably repre-

[113] The problem concerning proprioceptive components of the Avian eye muscle nerves has been investigated by ROGERS (1957) and is also considered by PEARSON (1972). With respect to Vertebrates in general, this topic is also briefly discussed in chapter VII of volume 3, part II.

sents the preganglionic (general visceral efferent) component, comparable to the *Edinger-Westphal nucleus* of Man and other Mammals. All the somatic motor nuclei (or subnuclei), particularly the ventromedial one, are closely adjacent to the fasciculus longitudinalis medialis. Some of the root fibers, perhaps mainly those of the ventromedial nucleus, immediately decussate, thus providing a contralateral component of the oculomotor nerve. A rostral extension of the accessory nucleus, within the periventricular gray, protrudes into the mesencephalo-diencephalic boundary zone within the rostral end of the tegmentum corresponding to the tegmental cell cord. It may be a separate griseum comparable with the nucleus medianus anterior of Mammals.

The avian *nucleus trochlearis* (Fig. 400C) is commonly in close apposition to the caudal end of oculomotor nucleus, and is found dorsally to the fasciculus longitudinalis medialis. It does not display a subdivision into separate cell groups. Caudal portions of the oculomotor nuclei may extend medially between the antimeric trochlear nuclei, thus displaying some degree of overlap.[114]

Summarizing the main *communication channels* of the Avian mesencephalon, the massive optic input to tectum, moreover the isthmo-optic tract, the additional optic input to tegmentum, and the assumed centrifugal fibers reaching the retina may be mentioned first, but require no further comments since data concerning these systems were pointed out above.

As regards other *rostral channels*, telencephalic connections provided by lateral and medial forebrain bundles appear to be of substantial importance. The Avian forebrain bundles are far more differentiated than those of Anamnia and Reptiles. Fascicles pertaining to the lateral forebrain bundle and reaching the mesencephalon are contained in a composite so-called *striomesencephalic* and *striocerebellar tract*,[115] which may include some reciprocal, i.e. 'ascending' connections, and, with respect to the mesencephalon, reaches not only tegmental grisea but also the tectum opticum and probably with some fibers the torus (nucleus mesencephalicus lateralis) and adjacent cell

---

[114] At ontogenetic stages, oculomotor and trochlear nuclei are separated by an interval, which subsequently becomes reduced respectively completely diappears (MESDAG, 1909, and other authors).

[115] A ventral and caudal part of this system is also designated as 'ansa lenticularis' (e.g. KARTEN and HODOS, 1967; cf. also Figs. 399A, B of the present chapter). Caudal extensions of this channel may reach basal bulbar grisea including the 'nuclei pontis'.

groups. Again, the so-called *tractus occipito-mesencephalicus (et bulbaris)*, connecting caudolateral telencephalic regions with diencephalic, mesencephalic, and (rhombencephalic) bulbar grisea, can be considered a special subdivision of the lateral forebrain bundle. At the mesencephalo-diencephalic boundary zone, it is particularly related to the 'spiriform' nuclei (nn. of posterior commissure).

The complex Avian *medial forebrain bundle* is represented by the so-called *septo-mesencephalic tract (Scheidewandbündel* of early investigators such as EDINGER and WALLENBERG). In the telencephalon, it passes through the paraterminal ('septal') region and contains a number of different components, some of which are analogous to the Mammalian fornix.[116] The descending channels of this tract originate in dorsal, dorsomedial and medial telencephalic grisea. Upon reaching the diencephalon at the hemispheric stalk, the main portion of said channels swings lateralward, ventrally to the lateral forebrain bundle, runs along the lateral and dorsal surface of the diencephalon, finally connecting with pretectal and medial tectal grisea.

Other rostral channels or 'tracts' are provided by *tectothalamic* and *pretectothalamic fibers*, by the *fasciculus retroflexus (habenulopeduncularis)*, and by additional rather inconspicuous and thus poorly recognizable fiber systems interrelating mesencephalon with diencephalon. The *quinto-frontal tract*, presumed to connect afferent trigeminal grisea with diencephalon and telencephalon, seems to run through the mesencephalic tegmentum ventrally to the lateral forebrain bundle which is also called '*striobulbar*' or '*striotegmental tract*', or '*ansa lenticularis*' (cf. Fig. 399B). From the torus semicircularis homolateral and contralateral fibers (crossing in the ventral supraoptic decussation) reach nucleus posterior decussationis supraopticae ventralis and nucleus ovoidalis of the diencephalon.

The *caudal channels* comprise the *fasciculus longitudinalis medialis*, the *rubrobulbar (and spinal?) tract*, other ascending and descending fibers of the reticular formation, the *brachium conjunctivum*, the *general (spino-bulbo-mesencephalic and thalamic lemniscus*, and the *lateral lemniscus*, repeatedly dealt with above.

---

[116] Further details concerning the systems of medial and lateral forebrain bundles in relation to diencephalic and telencephalic grisea shall be dealt with in chapter XII and XIII of volume 5.

The complex *efferent systems of the tectum opticum* include an essentially crossed dorsal *tectobulbar* and presumably *tectospinal tract*, whose fibers may or may not join the fasciculus longitudinalis medialis, and a crossed as well as uncrossed ventral *tectobulbar (and spinal?)* tract with further subdivisions. A *tectocerebellar connection* seems likewise present.[117]

The conspicuous *pretecto-subpretectal tract* appears, at least in part, to represent an intrinsic system of the mesencephalo-diencephalic boundary zone, but a *fasciculus praetectalis descendens (faisceau prétecto-bulbaire*, KAPPERS, 1947) has been noted, which may join the multiplex channel of fasciculus longitudinalis medialis.

In addition to the tegmental decussations mentioned in connection with efferent tectal fibers, with brachium conjunctivum, some oculomotor root fibers, and other systems, as well as to the dorsally decussating trochlear root fibers, the main *commissures* related to mesencephalon are commissura posterior, commissura tecti, and the dorsal and ventral supraoptic commissures located in the hypothalamic diencephalon (cf. HUBER and CROSBY, 1929). There are doubtless close, but still poorly elucidated relations between posterior commissure and fasciculus longitudinalis medialis, such that grisea interconnected by that commissure may contribute contralateral or homolateral fibers to said fasciculus.

With respect to mesencephalic grisea, the dorsal supraoptic commissure seems to include fibers of some tegmental cell groups, while the ventral supraoptic decussation appears to contain tectal as well as some tegmental fibers. A (dorsal) part of that commissure may be analogous to the so-called commissura transversa of Reptiles and various Anamnia. The relationship to torus semicircularis was pointed out above.

---

[117] KARTEN (1965), also quoted by PEARSON (1972) brings a brief but detailed enumeration of diverse efferent tectal fibers as inferred by experiment with the *Nauta-Gygax technique* in the Pigeon. Homolateral fibers were traced to lateral pontine nucleus, lateral reticular formation, nucleus intercollicularis, nucleus isthmo-opticus, and central gray. Via the brachium of superior colliculus (cf. Figs. 399C, D) fibers were found to reach subpretectal nucleus and diencephalic grisea, particularly nucleus rotundus. Additional 'extrabrachial' fibers seem to reach pretectal and diencephalic nuclei. Crossed connections to reticular grisea via the decussation of tractus tectobulbaris, and contralateral commissural fibers by way of commissura tecti to antimeric tectal cortex were likewise traced. PEARSON (1972) justly emphasizes the widespread tectal connections.

9. Mammals (Including Man)

The general morphology of the Mammalian midbrain manifests overall features easily comparable with those obtaining in Reptiles. The *tectum mesencephali* displays two paired prominences, the *superior* and the *inferior colliculi*. The former correspond to the tectum opticum, and the latter to the torus semicircularis of lower Vertebrates. In Mammals, however, the torus has been caudally displaced by a morphogenetic process of externation respectively promination.[118] This process, *qua* torus semicircularis, is not manifested in Birds, but very definitely in various Reptiles, some of which display a rudimentary colliculus inferior, as dealt with above in section 7. The *brachium quadrigeminum superius* connects, along the external brain surface, the superior colliculus with the diencephalon, and gives passage to optic tract fibers together with other fiber systems. The inferior colliculus is connected with the diencephalic corpus geniculatum mediale through the *brachium quadrigeminum inferius*, essentially pertaining to the lemniscus lateralis system. While the tectum mesencephali of Reptiles is uncovered in dorsal view, and that of Birds is dorsally covered by the cerebellum but protrudes lateralward, the tectum *sensu latiori* (corpora quadrigemina) of Mammals tends to be still more extensively covered by the cerebellum or by the occipital lobes of the telencephalon or by both. In higher Mammals, such as Primates including Man, the cerebellum, in turn, becomes covered by the occipital lobes of the hemispheres, and the mesencephalic roof tends to be covered by the splenium of corpus callosum.[119]

The *basal portion* of the Mammalian mesencephalon is characterized by a specific component, not present as such in Anamnia nor in submammalian Amniota, namely the *pes pedunculi*, containing the pyramidal and the corticopontine tracts, separated from the tegmentum by the substantia nigra.[120] The pes (or basis) pedunculi, relatively small in

---

[118] The phylogenetic and ontogenetic events leading to *externation* (REMANE) respectively *internation* or *promination* respectively *introversion* (SPATZ) were pointed out in volume 3, part II of this series (p. 172 *et passim*).

[119] Cf. Figures 67, 78E, and 145 of volume 3, part II.

[120] The topologic neighborhood represented by pes pedunculi, ventrally to substantia nigra, is of course included in the basal tegmentum of Sauropsida and even of Anamnia. Although pyramidal and corticopontine tracts are most likely lacking in Sauropsida, a telencephalopontine ('striatopontine') system is presumably present at least in Birds.

lower Mammals, is considerably expanded in higher forms, particularly in Man. The more or less protruding bulge of pes pedunculi is laterally demarcated by a (secondary) *sulcus mesencephali lateralis externus*, and medially by the *sulcus mesencephali medialis seu oculomotorii*, along which root fibers of oculomotor nerve emerge. The space between the two medial sulci becomes the *fossa interpeduncularis* essentially corresponding to the location of nucleus interpeduncularis.

The mesencephalic ventricle of adult Mammals is the *aquaeductus cerebri (Sylvii)*, a relatively quite narrow, single median tubular passage between third and fourth ventricle. It may display a fairly prominent lateral groove, representing, as it were, a fused sulcus lateralis mesencephali internus and sulcus limitans system. In addition, a variable number of nondescript minor accessory grooves or rugae can be present. At isthmus levels, a remnant of sulcus limitans is occasionally suggested. The *subcommissural organ*, located in the dorsal midline, ventrally to the fibers of the posterior commissure, pertains to the mesencephalodiencephalic boundary zone. A '*mid-aqueductal ependymal organ*' (QUAY, 1971) was recently described in the dorsal wall of the aqueduct of the Hyrax (Procavia capensis) which can be classified as pertaining to the Paenungulata. During a period of ontogenetic development, however, the mesencephalic ventricle is, as in other Vertebrates, relatively quite wide, sulcus lateralis mesencephali internus and sulcus limitans being definitely identifiable at certain stages.

Figures 405A to 410C illustrate cross-sectional levels through the midbrain of some Prototherian, Metatherian, and Eutherian Mammals. The Primate human mesencephalon shall be considered separately further below. As regards the brain of the Prototherian Ornithorhynchus anatinus, investigated by HINES (1929) and others, a peculiar pattern distortion related to the displacement of neighborhoods pertaining to rhombencephalon and prosencephalon as well as to ventrally concave 'cephalic flexure' is evident at various mesencephalic levels (cf. e.g. Fig. 406). Illustrations and detailed descriptions of the nuclear pattern in the Mammalian midbrain and isthmus region, with numerous bibliographic references, are included in the series of papers published by HUBER *et al.* (1943). Cross-sectional levels in a variety of

---

Thus, depending on an arbitrarily assumed viewpoint, the pes pedunculi can be evaluated as a ne-encephalic configuration or as the further development of a phylogenetically 'old' topologic neighborhood.

Mammalian groups are also depicted in the recent treatise by VERHAART (1970) which contains a discussion of diverse recorded data, and further bibliographic references.

The *alar plate derivatives* of the mesencephalo-diencephalic boundary zone provide an array of various pretectal grisea of diencephalic and mesencephalic origin (K., 1954; K. and MILLER, 1942, 1949; KUMAMOTO-SHINTANI, 1959).

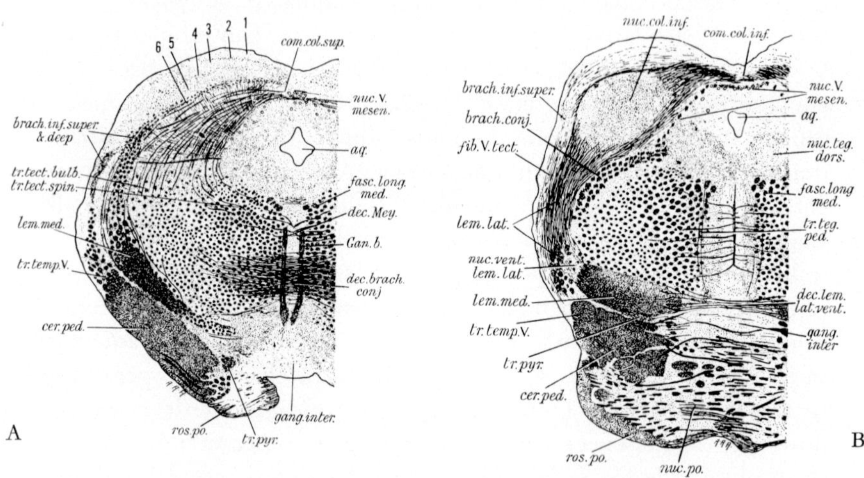

Figure 405A. Cross-section (myelin stain) through the mesencephalon of the Prototherian Mammal Echidna at the level of superior colliculus (from ABBIE, 1934). brach. inf.sup. & deep: superficial and deep portion of brachium quadrigeminum inferius; cer. ped.: pes pedunculi; dec. Mey.: *dorsal tegmental decussation of Meynert;* Gan.b.: *Ganser's bundle* (interpedunculo-tegmental tract); gang. inter.: interpeduncular nucleus; ros. po: 'rostrum pontis'; tr. pyr.: pyramidal tract 'beginning to leave the medial border of the cerebral peduncle'; tr. temp.V: 'tractus temporo-trigeminalis'; 1: stratum griseum superficiale; 2: stratum opticum; 3: stratum griseum medium; 4: stratum medullare medium; 5: stratum griseum profundum; 6: stratum medullare profundum (1–6 have been relabelled in accordance with the terminology adopted in the present treatise). Other abreviations self-explanatory. In Echidna, the nuclei of oculomotor and trochlear nerve are separated by a substantial interval The planes of both Figures 405A and B pass through that interval. The substantia nigra (not labelled) is poorly developed, but can easily be recognized in both Figures A and B as a thin band of gray dorsomedially respectively dorsolaterally to pes pedunculi.

Figure 405B. Cross-section through mesencephalon (and pons) of Echidna at level of colliculus inferior (from ABBIE, 1934). fib.V tect.: 'trigeminal fibres' passing to tectum; nuc.po.: nuclei pontis; nuc.teg.d.: nucleus dorsalis tegmenti; nuc.vent.lem.lat.: 'ventral nucleus' of lateral lemniscus. Other abbreviations as in preceding Figure A or self-explanatory.

The following enumeration of the pretectal nuclei pertaining to the alar plate[121] can be given. The numbers in parenthesis refer to a notation adopted in our investigations and applicable to all Mammalian pretectal grisea.

I. *Diencephalic pretectal cell masses*
1. (3) Nucleus praetectalis (principalis)
2. (4) Nucleus posterior (thalami dorsalis)
3. (8) Area praetectalis *(Papez)*
4. (9b) Nucleus interstitialis principalis commissurae posterioris
5. (9c) Nucleus centralis subcommissuralis commissurae posterioris
6. (9d) Nucleus interstitialis supracommissuralis commissurae posterioris
7. (9e) Nucleus interstitialis magnocellularis commissurae posterioris
8. (9f) Nucleus medianus infracommissuralis commissurae posterioris

II. *Mesencephalic tectal group*
1. (5) Nucleus lentiformis mesencephali, pars magnocellularis
2. (6) Nucleus lentiformis mesencephali, pars parvocellularis
3. (12) Nucleus sublentiformis
4. (7) Nucleus olivaris colliculi superioris *(Fuse)*

The functional significance and the exact fiber connections of this complex nuclear group still remain poorly understood. Taking into account the available data, it can be assumed that these grisea comprise a number of essentially '*interstitial centers*' (i.e. 'centers' intercalated within fiber systems) correlating in diverse ways the tectum opticum, thalamic nuclei, hypothalami nuclei, tegmental grisea, and to a certain extent telencephalic cortical areas. Direct optic input by way of the brachium quadrigeminum superius into all pretectal nuclei of the mesencephalic group appears conspicuous and presumably also obtains for nucleus praetectalis principalis, area praetectalis, at least some of the nuclei of posterior commissure, and perhaps also nucleus posterior thalami. Direct or indirect connections with corpus striatum cannot be excluded.

Experimental studies in cats by SPIEGEL (1933), SCALA and SPIEGEL (1936), and a few subsequent authors seem to indicate that the *pupillary*

[121] The pretectal nuclei pertaining to the basal plate are considered further below in dealing with the tegmental grisea *(nucleus interstitialis of Cajal, nucleus of Darkschewitsch,* nucleus interstitialis tegmentalis commissurae posterioris, tegmental cell cord, and perhaps nucleus medialis anterior).

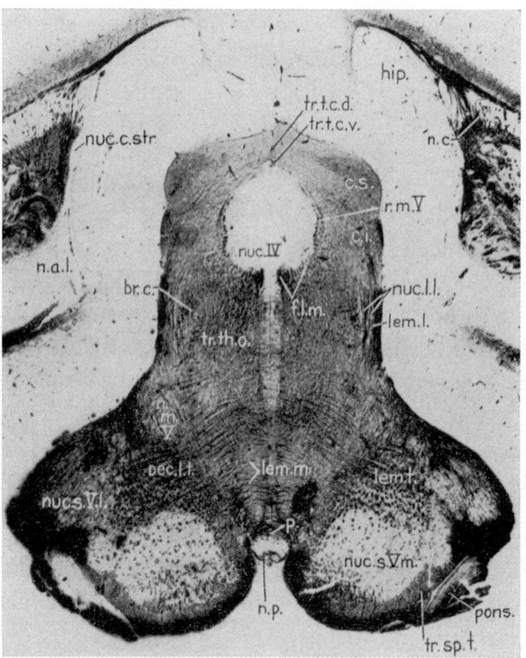

Figure 406. Cross-section *(Weigert stain)* through mesencephalon and adjacent regions in the brain of the Prototherian Ornithorhynchus anatinus (from HINES, 1929). br.c.: brachium conjunctivum; c.i.: colliculus inferior; c.s.: colliculus superior; Dec.l.t.: decussatio lemnisci trigemini; f.l.m.: fasciculus longitudinalis medialis; hip.: hippocampus; lem.n.: lemniscus medialis; lem.t.: lemniscus trigemini; n.a.l.: nucleus amygdalae; n.m.V.: nucleus motorius trigemini; n.p.: nuclei pontis; nuc.c.str.: 'nucleus corporis striati'; nuc.l.l.: nuclei lemnisci lateralis; nuc.s. V l., m.: nucleus sensibilis trigemini (lateralis respectively medialis); nuc. IV: nucleus nervi trochlearis; P.: pons; r.m.V: radix mesencephalica trigemini; tr. sp. t.: 'tractus spinalis trigemini'; tr.t.c.d., v.: tecto-cerebellaris (dorsalis respectively ventralis); tr.th.o.: 'tractus thalamo-olivaris' (central tegmental tract). The rostralward protrusion of the sensory trigeminal nuclei in the posterior limb of the cephalic flexure leads to considerable pattern distortion, such that the motor tegmental trigeminal nucleus appears topographically dorsal to the sensory ones. A similar, but somewhat less pronounced pattern distortion is displayed in Figure 393B (Alligator) which should be compared with Figure 406.

Figure 408 A. Slightly oblique cross-section (myelin stain) through the mesencephalon of the Eutherian Mus musculus at the level of nucleus colliculi inferioris (from CAJAL, 1911). A: nucleus colliculi inferioris; B: lemniscus lateralis; C: intercollicular griseum *('partie postérieure de l'écorce du tubercule antérieur')*; D: caudal portion of oculomotor nucleus; E: pes pedunculi; F: dorsal tegmental decussation ventrally to fasciculus longitudinalis medialis *('voie optico-acoustique réflexe descendante')*; G: interpeduncular nucleus; H: central

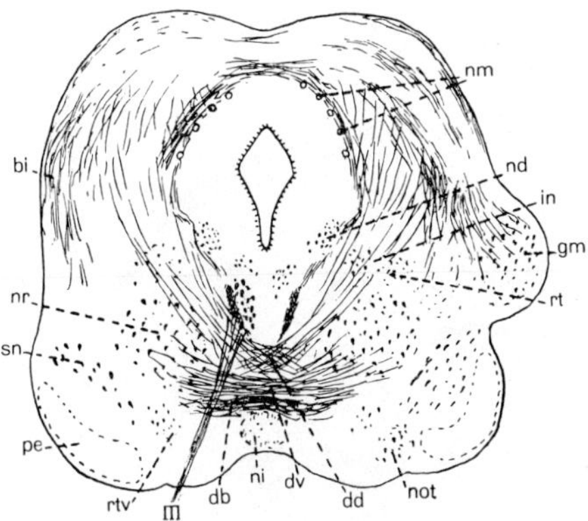

Figure 407. Simplified sketch of a cross-section through mesencephalon of the Metatherian Didelphys virginiana at the level of superior colliculus (from K., 1927). bi: brachium quadrigeminum inferius; db: decussation of brachium conjunctivum; dd: dorsal tegmental decussation; dv: ventral tegmental decussation; gm: medial geniculate body; in: nucleus interstitialis; nd: *nucleus of Darkschewitsch;* ni: nucleus interpeduncularis; nm: nucleus radicis mesencephalicae trigemini; not: nucleus opticus tegmenti; nr: nucleus ruber; pe: pes sive basis pedunculi; rt: nucleus reticularis tegmenti (nucleus tegmentalis profundus mesencephali; rtv: nucleus reticularis ventralis; sn: substantia nigra.

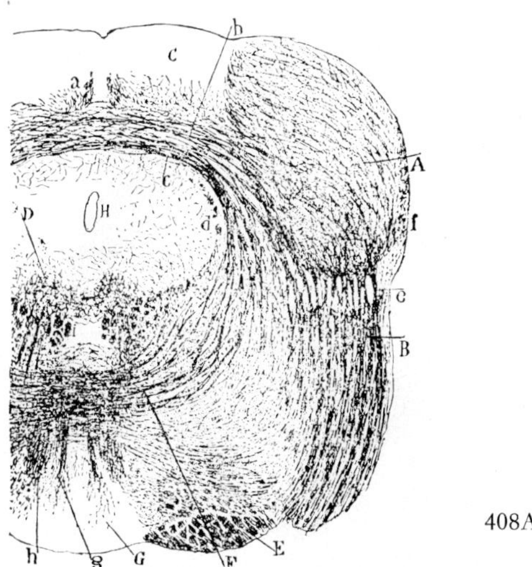

408A

gray; a: fibers of tractus opticus; b, c: two levels of commissura tecti; d: radix mesencephalica trigemini; e: nuclei of lateral lemniscus; f: brachium quadrigeminum inferius; g: interpedunculo-tegmental *tract of Ganser;* h: tegmental fibers (fasciculus mammillo-tegmentalis ?).

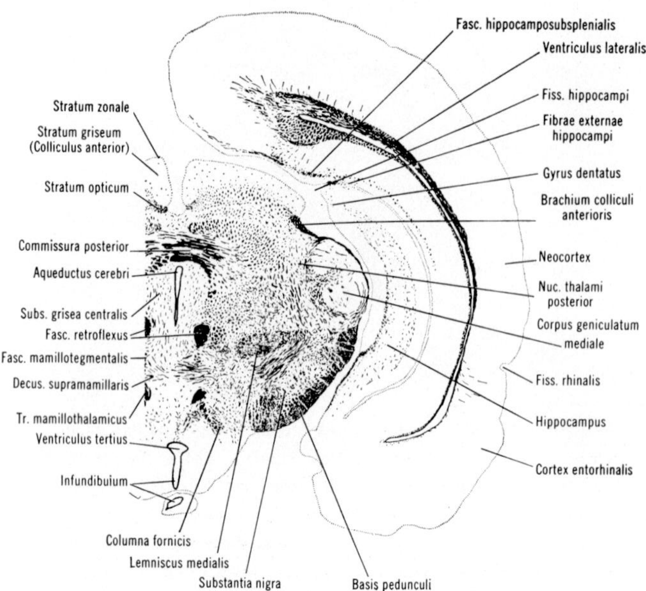

Figure 408B. Cross-section (myelin stain) through the mesencephalon of the Rat at level of commissura posterior (after Craigie, 1925, from Zeman and Innes, 1963).

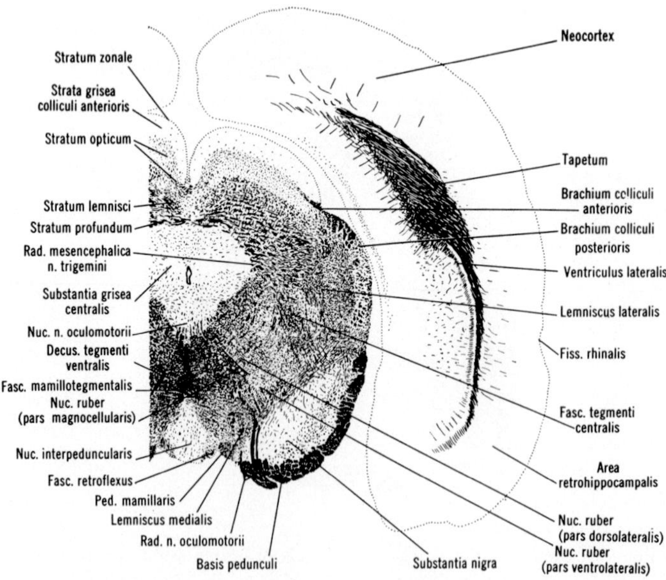

Figure 408C. Cross-section (myelin stain) through the mesencephalon of the Rat at the level of oculomotor nucleus (after Craigie, 1925, from Zeman and Innes, 1963).

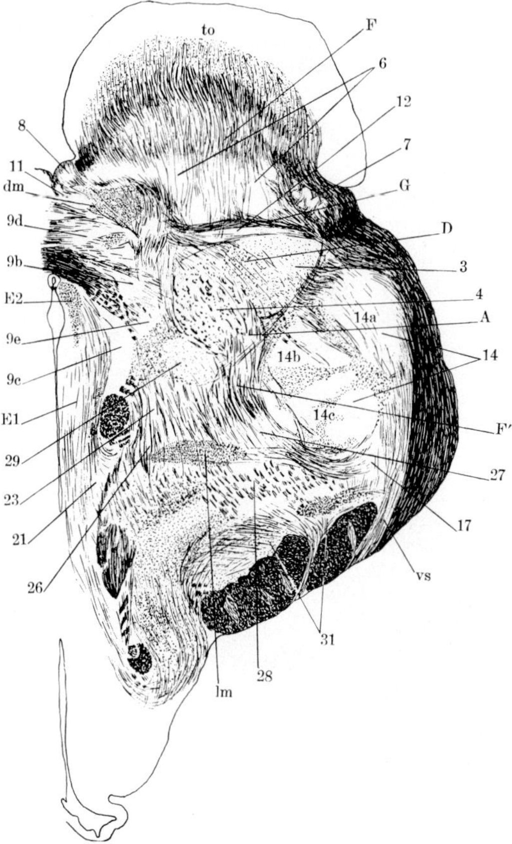

Figure 408D. Cross-section (myelin stain) through rostral portion of tectum mesencephali at level of commissura posterior in the Rabbit (from K. and MILLER, 1942). A: ventromedial arciform fiber system; D: dorsolateral arciform system; E: periventricular system, vertical (1), dorsal longitudinal (2), ventrolateral (3); F: perpendicular system; F′: tractus praetectalis descendens; G: 'tractus sublentiformis'; K: 'tractus tectoreuniens'; dm: dorsomedial tectothalamic fibers; fr: fasciculus retroflexus; lm: lemniscus medialis; ms: supramammillary commissure; to: tectum opticum; vs: ventral supraoptic commissure; 3: nucleus praetectalis principalis; 4: nucleus posterior (thalami); 6: nucleus lentiformis mesencephali (pars parvocellularis); 7: nucleus olivaris colliculi superioris; 8: area praetectalis (propria); 9: nuclei of posterior commissure (b: n. interstitialis principalis; c: n. centralis subcommissuralis; d: n. supra commissuralis interstitialis; e: n. interstitialis magnocellularis; t: n. interstitialis tegmentalis); 10: nucleus suprageniculatus; 11: area praetectalis, pars supracommissuralis; 12: nucleus sublentiformis; 14 (a, b, c): nucleus geniculatus medialis dorsalis with subdivisions; 15: nucleus geniculatus medialis ventralis; 17: nucleus geniculatus lateralis ventralis; 21: tegmental cell cord or 'plate'; 23: nucleus parafascicularis (thalami); 26: nucleus ventromedialis thalami; 27: nucleus ventrolateralis thalami; 28: nucleus reticularis thalami ventralis (zona incerta); 30: caudal end of n. ventralis thalami; 31, 32: substantia nigra.

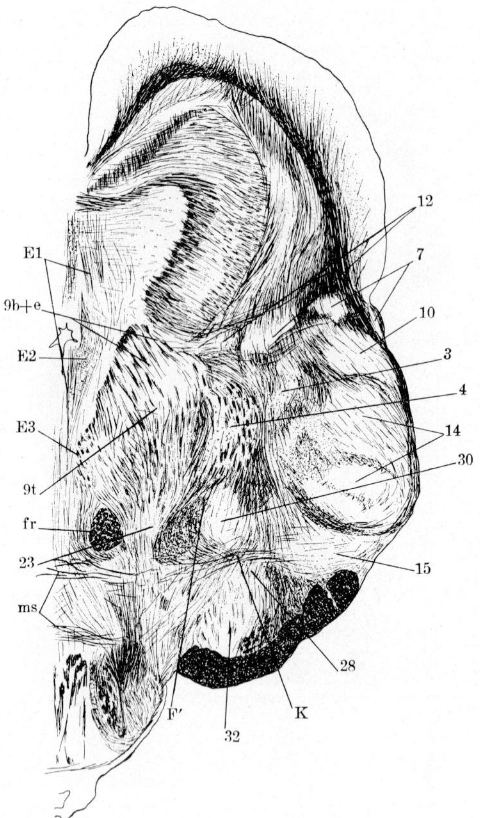

Figure 408E. Cross-section (myelin stain) through rostral part of tectum opticum in the Rabbit, about 900 μ caudal to preceding figure (from K. and MILLER, 1942). Abbreviations as in Figure 408D.

light reflex is one of the specific functions of the pretectal region.[122] Nucleus lentiformis mesencephali, area praetectalis, nucleus principalis praetectalis, and nucleus olivaris colliculi superioris may here be involved, together with commissura posterior including some of its alar plate nuclei.

Since impulses from all the main sensory systems appear to reach the pretectal grisea, it is likely that, in addition to the correlating and reflex activities mentioned above, complex discharge patterns can

---

[122] Our clinicopathologic observations, confirming SPIEGEL's experimental results, shall be pointed out further below in dealing with the human mesencephalon.

A

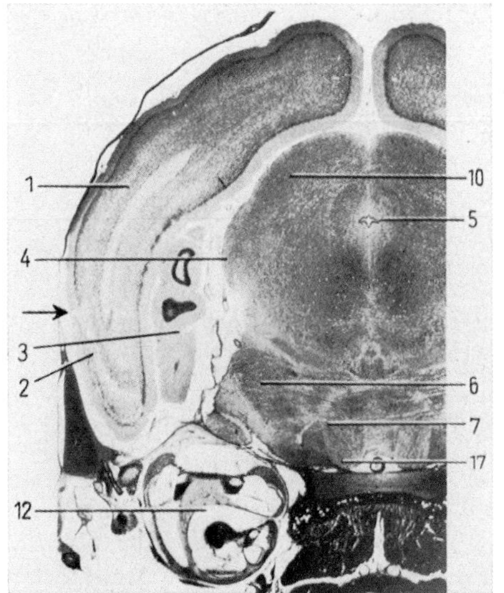

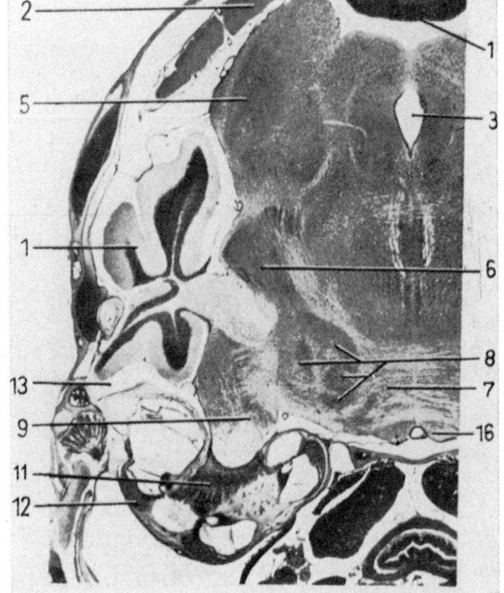

B

Figure 409 A. Cross-section through the brain of the Chiropteran Mammal Pteronotus suapurensis at the level of superior colliculi (from SCHNEIDER, 1957). 1: telencephalic neopallium (arrow indicates its basal boundary); 2: telencephalic posterior piriform lobe; 3: portion of hippocampal formation; 4: corpus geniculatum mediale; 5: cerebral aqueduct; 6: superior olivary complex; 7: trapezoid body; 10: colliculus superior; 12: internal ear (labyrinth); 17: pyramidal tract.

Figure 409B. Cross-section through the brain of the Chiropteran Pteronotus suaparensis at level of inferior colliculi (from SCHNEIDER, 1957). 1: cerebellum; 2: caudal pole of telencephalon; 3: cerebral aqueduct; 5: inferior colliculus; 6: superior olivary complex; 7: trapezoid body; 8: griseum of trapezoid body; 9: tuberculum acusticum; 11: ganglion spirale of cochlea; 12: osseous capsule of labyrinth; 13: 'fossa subarcuata'; 16: pyramidal tract. The extensive development of the inferior colliculi and of other structures pertaining to the octavus system should be noted.

again be channelled into the various efferent pathways which include the fasciculus longitudinalis medialis, the central tegmental tract, the more diffusely distributed bundles of the reticular formation, the *fasciculus longitudinalis dorsalis of Schütz* in the central gray, and other systems.[123]

---

[123] A distinctive fasciculus praetectalis descendens as noted in Reptiles (cf. section 7) could not be identified in Man, but is presumably included in the above mentioned complex channels.

The Mammalian *superior colliculus* or tectum opticum displays *prima facie*, as seen in routine preparations, a less evident lamination respectively 'cortical' structure than in submammalian forms and appears of relatively smaller size. This apparent 'reduction' has been interpreted as related to the progressive development and functional significance of diencephalic and telencephalic grisea concerned with the processing of optic and other sensory input handled by the submammalian tectum opticum.[124] According to CRAIGIE (1925), the lamination in Mammals is less complex than in lower forms, the reduction being chiefly in the outer layers. Again, these latter are said to be more largely developed in lower Mammals than in Man.

Be that as it may, the Mammalian tectum opticum is nevertheless very definitely stratified, but certainly differs from that of submammalian forms by a modified arrangement of the optic input channel. The optic tract fibers, which form a superficial layer in the tectum of Anamnia and Sauropsida, represent a much deeper layer in the Mammalian tectum (cf. Figs. 411A–D).

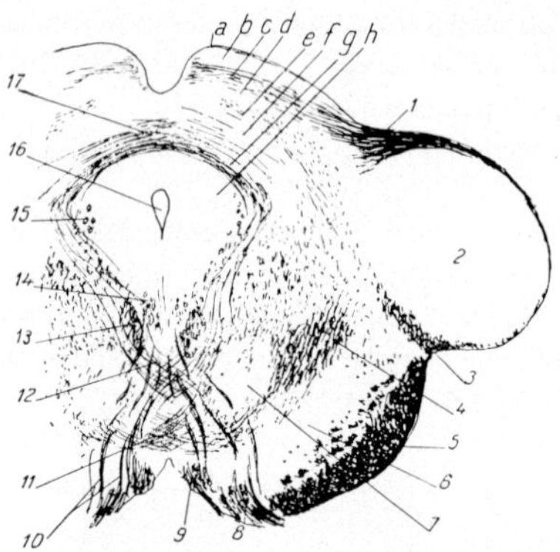

Figure 410A. Cross-section *(Weigert stain)* through the mesencephalon of the Carnivore Dog at level of superior colliculus and oculomotor root (from BECCARI, 1943). 1: tractus opticus; 2: corpus geniculatum mediale; 3: brachium quadrigemium inferius; 4: lemniscus medialis; 5: pes pedunculi; 6: substantia nigra; 7: nucleus ruber; 8: 'mammillary peduncle' (tegmento-mammillar tract); 9: fasciculus retroflexus; 10: radix nervi oculomotorii; 11: *ventral tegmental decussation of Forel;* 12: dorsal tegmental decussation of Meynert; 13: fasciculus longitudinalis medialis; 14: oculomotor nucleus; 15: nucleus radicis mesencephalicae trigemini; 16: aquaeductus cerebri; 17: commissura tecti; a: stratum zonale; b: stratum griseum superficiale; c: stratum opticum; d: stratum griseum intermedium; e: stratum medullare (s. album) intermedium; f: stratum griseum profundum; g: stratum medullare profundum; h: central gray.

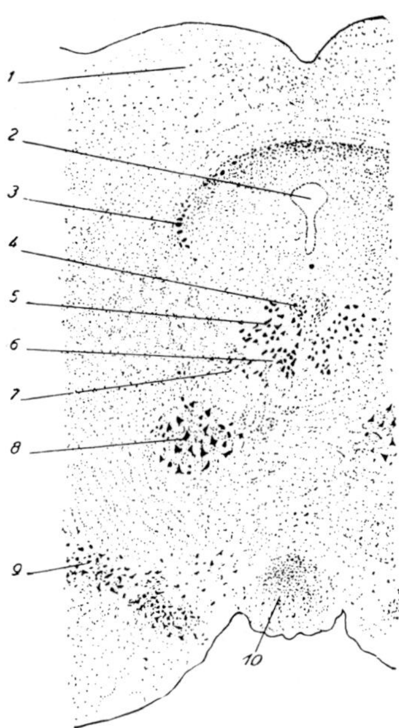

Figure 410B. Cross-section *(Nissl stain)* through the mesencephalon of the Dog at level of superior colliculus and caudal portion of nucleus ruber (from BECCARI, 1943). 1: tectum opticum; 2: cerebral aqueduct; 3: nucleus radicis mesencephalicae trigemini; 4: *Edinger-Westphal nucleus;* 5; 6: dorsolateral and ventromedial magnocellular oculomotor nucleus; 7: tegmental cell group labelled '*n. del Roller*' by BECCARI; 8: caudal portion of nucleus ruber; 9: substantia nigra; 10: interpeduncular nucleus.

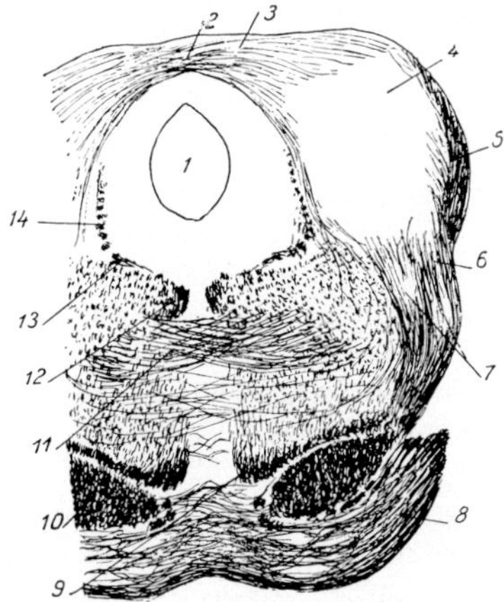

Figure 410C. Cross-section *(Weigert stain)* through the mesencephalon of the Dog at level of inferior colliculus (from BECCARI, 1943). 1: cerebral aqueduct; 2: commissura tecti (pars posterior); 3: nucleus intercollicularis; 4: nucleus colliculi posterioris; 5: brachium quadrigeminum inferius; 6: lemniscus lateralis; 7: grisea lemnisci lateralis; 8: rostral extremity of pons; 9: lemniscus medialis; 10: decussation of brachium conjunctivum; 12: fasciculus longitudinalis medialis; 13: radix nervi trochlearis; 14: radix mesencephalica trigemini.

The *optic tract fibers*[125] run here as an essentially sagittally directed flattened bundle, covered by at least two griseal layers.[125a] This bundle, which displays an external and an internal subdivision *(courant superficiel des fibres optiques et courant profond*, CAJAL, 1911) gives off terminals to the superficial layers and to the internal ones. According to CAJAL (1911) only the external subdivision consists of optic tract fibers.[126]

Concerning the *stratification* of the Mammalian tectum opticum, a variety of different notations have been used (TARTUFERI, 1885; CAJAL, 1911; and other authors). With some modifications, the simplified enumeration adopted, on the basis of personal investigations, in my '*Vorlesungen*' (K., 1927) essentially corresponds to the data recorded by the earlier investigators and to that elaborated in the Opossum by TSAI (1925). The following layers might be distinguished:

(1) *Stratum zonale* consisting of fine medullated and non-medullated fibers interspersed with small neuronal elements.

(2) *Stratum griseum superficiale*, containing somewhat larger nerve cells and many terminal arborizations of the optic tract.

(3) *Stratum opticum*, consisting of optic tract fibers and other descending input channels for tectum opticum (external and internal sublayers as mentioned above.

(4) *Stratum griseum medium (s. intermedium)*, containing 'association cells' *(Schaltneurone)* and efferent chief cells.

(5) *Stratum album medium (s. intermedium)*, mainly containing input fibers from lemniscus medialis and also from lemniscus lateralis.

---

[124] Thus, KAPPERS *et al.* (1936) state that 'the reduction of the superior colliculus phylogenetically is to a considerable extent in inverse proportion to the development of the dorsal part of the lateral geniculate nucleus, since this latter center assumes some of the functions formerly carried out by the superior colliculus'.

[125] Further details on optic chiasma, crossed and uncrossed optic tract fibers, as well as other details concerning the optic system will be found in volume 5 (diencephalon and telencephalon).

[125a] '*Chez les mammifères les fibres optiques cheminent toutes très profondement dans le tubercule pour monter ensuite aux couches grises où leurs ramifications doivent se répandre. Il n'en est pas de même chez les vertébrés inférieurs, chez qui ces fibres restent d'habitude à la surface du lobe optique*' (CAJAL, 1911).

[126] In *Marchi degenerations* obtained after enucleation of one eye in the Rabbit, CAJAL (1911, Fig. 121 loc. cit.) noted and depicted degeneration essentially restricted to the external subdivision of optic fibers. He assumed that the internal subdivision might be provided by fibers from prosencephalic grisea *('probablement originaires du cerveau')*. However, it is not impossible that some homolateral optic tract fibers might be included in that layer.

(6) *Stratum griseum profundum*, containing various sorts of neuronal elements, including chief cells, and efferent neurites of chief cells *(fibers of Meynert, radiatio Meynerti)*.

(7) *Stratum medullare profundum (tiefes Mark)*, containing *Meynert's fibers* (output channels), commissural tectal fibers, and fibers as well as large vesicular cells of the radix mesencephalica trigemini.

(8) *Stratum griseum centrale seu periventriculare* with various small-celled neuronal groups and fibers of the *fasciculus longitudinalis dorsalis of Schütz*.

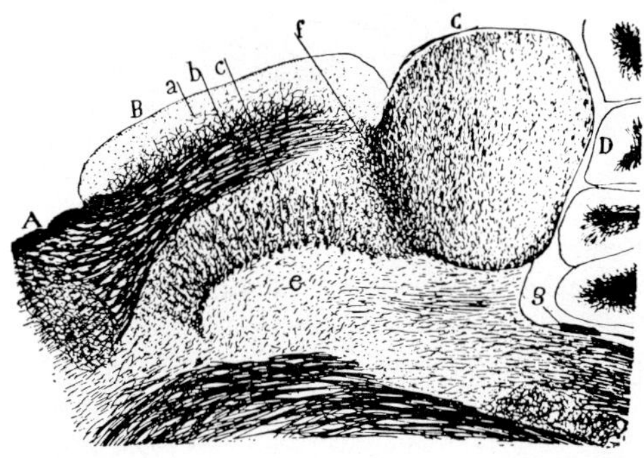

Figure 411 A. Parasagittal section *(Weigert stain)* through superior and inferior colliculi in the Rodent Mammal Cavia (from CAJAL, 1911). A: tractus opticus; B: colliculus superior; C: colliculus inferior; D: cerebellum; a: stratum griseum superficiale; b: stratum opticum; c: transverse fibers of medullary strata; e: central gray; f: commissura (posterior) tecti; g: decussation of trochlear root.

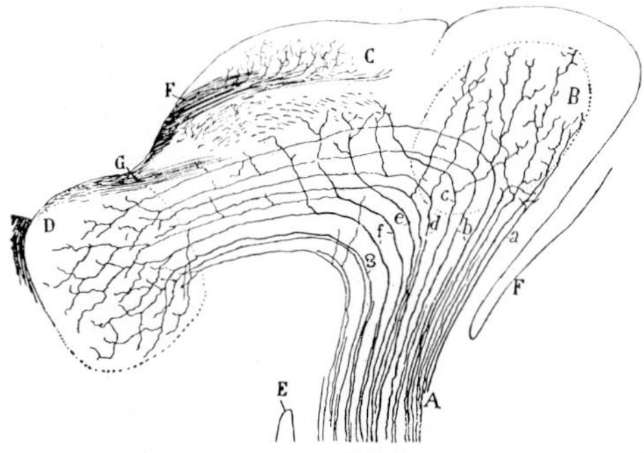

Figure 411 B. Semidiagrammatic parasagittal section *(Golgi impregnation)* through the mesencephalon of the Mouse (from CAJAL, 1911). A: lemniscus lateralis; B: nucleus colliculi inferioris; C: colliculus superior; D: corpus geniculatum mediale; E: interpeduncular fossa; F: tractus opticus; G: corticotectal fibers (?); a–g: diverse fibers of lateral lemniscus. F at right indicates rhombencephalon.

Figure 411D illustrates structural aspects of the superior colliculus as investigated by CAJAL. It can be seen that, generally speaking, the cells in stratum zonale tend to be transversely oriented parallel to the surface, and that many cells in the deeper layers are 'radially' oriented. Some crozier cells, comparable to those of submammalian Vertebrates, can be found in the internal portion of superficial gray layer and in the stratum griseum medium.

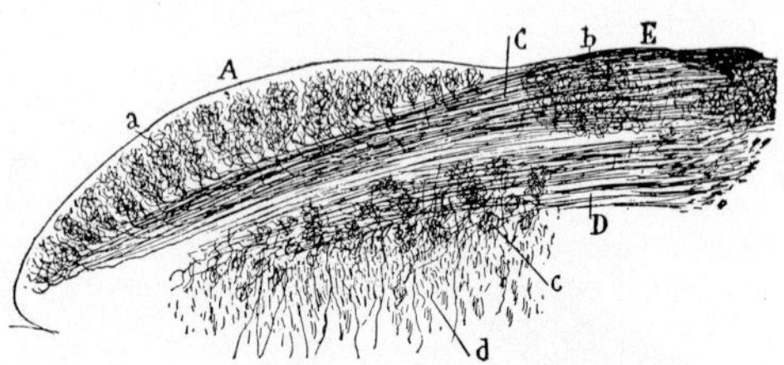

Figure 411C. Parasagittal section *(Golgi impregnation)* showing optic tract fibers entering superior colliculus in the newborn Mouse (from CAJAL, 1911). A: colliculus superior; C: optic tract *('courant superficiel des fibres optiques')*; D: deep sagittal fiber stratum *('courant profond'*; it is doubtful to which extent these fibers represent optic tract or other components, such, e.g. as pretectotectal or corticotectal channels); E: *'région postérieure du corps genouillé externe'* in CAJAL's interpretation (more likely pretectal region); a: optic tract terminals in stratum griseum superficiale; b: *'foyer où se terminent des collatérales des fibres optiques'* (doubtless a pretectal griseum); c: *'nids péricellulaires formés par les fibres optiques'*; d: transverse fibers of tectum.

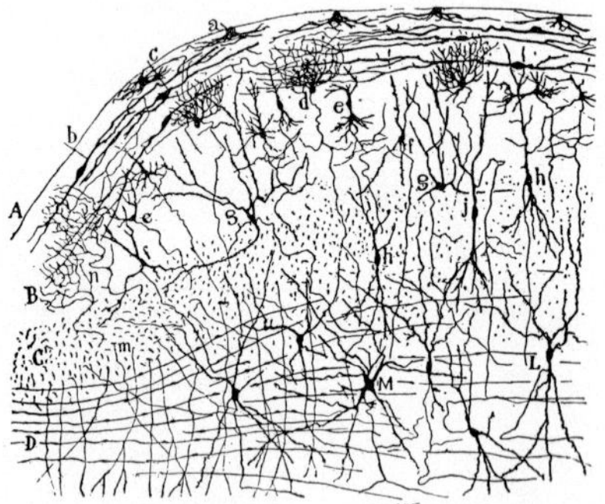

Figure 411D. Cross-section *(Golgi impregnation)* through tectum opticum in a young Rabbit (from CAJAL, 1911). A: neighborhood of dorsal midline; B: stratum griseum superficiale; C: stratum opticum, showing optic tract fibers in cross-section; D: stratum griseum (inter)medium; L, M: chief cells; a, b, c: marginal cells in stratum zonale; d, e: cells in superficial layer of stratum griseum superficiale; f, g: cells in deeper layer of said stratum; h, j: fusiform (radially oriented) cells in tratum opticum; m: optic tract collaterals to deeper layers of tectum; n: terminal arborizations of optic tract fibers.

Although, in the Mammalian taxonomic series, increasingly complex activities of visual discrimination are doubtless performed by cortex cerebri, apparently culminating in the striate cortex and adjacent cortical regions of Man, the tectum mesencephali, particularly in subprimate forms, seems to retain significant functional capabilities. Thus, superficial lesions of the tectum in the tree shrew Tupaia are said to produce deficits in form discrimination, while deeper lesions caused 'inability to track objects' (CASSAGRANDE *et al.*, 1972). This latter effect, however, might be essentially related to disturbances of the 'nucleus motorius tegmenti' by interfering with tectobulbar output required for integrated motor performances.

Removal of the occipital cortex spares both brightness and pattern discrimination in Rats, Rabbits, and Tree Shrews, while ablation of the occipital lobes in Primates seems to abolish all vision other than brightness registration (cf. GIOLLI and POPE, 1971, including relevant references listed by these authors).

Reverting to the nucleus radicis mesencephalicae trigemini mentioned above, the connections of its central neurites with the motor trigeminal nucleus, as seen by CAJAL, were mentioned in section 10 of chapter IX and depicted in Figure 268A. In addition, some such fibers

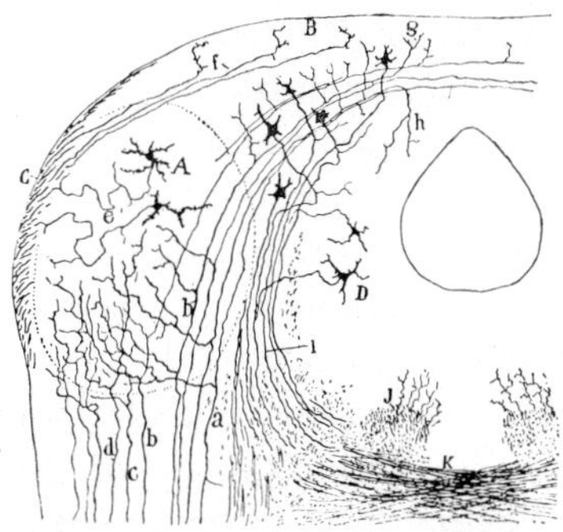

Figure 412A. Semidiagrammatic cross-section *(Golgi impregnation)* through the inferior colliculus in a young Mouse (from CAJAL, 1911). A: nucleus colliculi inferioris; B: intercollicular griseum; C: brachium quadrigeminum inferius; D: central gray; I: efferent fibers of intercollicular gray; J: fasciculus longitudinalis medialis; K: decussation of brachium conjunctivum; a, b, c, d: fibers of lemniscus lateralis; e: fibers originating in nucleus colliculi inferioris and joining brachium quadrigeminum inferius; f: fibers from brachium ending in griseum intercolliculare; g, h: lateral lemniscus fibers ending in griseum intercolliculare respectively centrale.

have been described as forming a small bundle, the *tract of Probst*, which runs caudad, ventrally to the *nucleus vestibularis of Deiters*, and reaching, with a few fibers, the general region of nucleus dorsalis vagi. *Deiters' nucleus* and additional bulbar grisea are believed to receive collaterals or terminals of that tract. Likewise, collaterals to the cerebellum have been assumed (cf. OBERSTEINER, 1912; PROBST, 1902).

The *inferior colliculus* (Figs. 411A, B, 412A), whose structure appears less complex than that of tectum opticum, contains laterally an ovoid griseum, nucleus colliculi posterioris, which receives collaterals as well as straight, i.e. non-bifurcated terminals of the lateral lemniscus. The brachium quadrigeminum inferius contains the fibers of lemniscus lateralis proceeding to the diencephalic medial geniculate body and additional poorly elucidated connections of inferior colliculus. Within the medial portion of that colliculus lies a griseum representing the caudal extension of the cortex colliculi superioris, but with a less complex stratification and apparently without optic tract fibers. This griseum doubtless pertains to the functional system of colliculus inferior and gives off efferent fibers included in *Meynert's radiation*.

The *nucleus isthmi* of lower Vertebrates seems to be represented by an interstitial griseum of the lateral lemniscus (cf. Fig. 408A). In addition to the main portion of this griseum, designated as nucleus dorsalis lemnisci lateralis, several variable and nondescript additional cell groups (sagulum, n. cuneiformis) associated with lemniscus lateralis may be evaluated as Mammalian differentiations of the submammalian nucleus isthmi. Fibers from or passing through the nucleus dorsalis lemnisci lateralis turn in a medioventral direction, traverse the brachium conjunctivum as 'fibrae perforantes' and cross the tegmental midline dorsally to decussatio of brachium conjunctivum, forming the *commissure of Probst* (Fig. 418A), which presumably pertains to the system of lemniscus lateralis.

Recent experimental studies (SUGA and SHIMOZAWA, 1974) seem to indicate that, in *Chiroptera*, the nucleus of the lateral lemniscus is involved in a central neural mechanism synchronized with vocalization, which attenuates, by about 15 decibels, responses to self-vocalized orientation sounds, and to nonorientation sounds. Said mechanism appears to represent an 'adaptation' to the emitted intense orientation sounds for *echolocation* (cf. vol. 2, p. 186–187). If such sounds would directly and strongly stimulate the hearing mechanism of the Bat, the detection of echoes from short distances would evidently be impaired.

A *nucleus visceralis secundarius*, characteristic for most or many Anamnia, cannot be recognized as a definable griseum in Mammals. It is, however, not improbable that a kathomologon of said nucleus is included either within the intermediodorsal portion of the Mammalian periaqueductal gray, perhaps within caudal parts of nucleus lateralis grisei centralis, or within the region of the so-called nucleus cuneiformis.[126a] This griseum is located between lemniscus lateralis and central gray at levels of the inferior colliculus. The lateral part of the nondescript subcollicular 'nucleus cuneiformis' might also in part be related to lemniscus lateralis.

The *central gray* surrounding the cerebral aqueduct is, as a rule, particularly well delimited from the more peripheral grisea and fiber tracts. In this respect, it seems to represent a characteristic Mammalian feature, not manifested, in a comparable manner, by submammalian forms. This central gray (Figs. 412B, 413) includes grisea pertaining to alar and to basal plate, as well as a dorsal and ventral system of mostly longitudinal, nonmedullated or poorly medullated fibers. The innermost gray stratum, adjacent to the ependymal lining, may or may not display a distinct subependymal cell plate.

Alar plate components of the periaquaeductal gray are central subcommissural and median infracommissural nuclei of posterior commissure rostrally, and more caudally the nucleus lateralis grisei centralis. *Basal plate components* are the nucleus medialis anterior (a rostral extension of *the Edinger-Westphal nucleus*), *the nucleus of Darkschewitsch*,[127] and, more caudally, the nucleus dorsalis tegmenti, mentioned in section 10 of chapter IX, and extending toward the isthmus region. As regards the dorsal and ventral periventricular fiber systems, the ventral bundle pertains to *the fasciculus longitudinalis dorsalis of Schütz*. The mes-

---

[126a] A recent experimental study of ascending gustatory pathways in the albino rat (Norgren and Leonard, 1973) seems to corroborate the conclusion that a Mammalian 'secondary gustatory nucleus' is present within this ill-defined region.

[127] The *nucleus of Darkschewitsch* and perhaps also the nucleus medialis anterior are present in Sauropsida. In these forms, periventricular fibers corresponding to the *fasciculus longitudinalis dorsalis of Schütz* might be included in the fasciculus longitudinalis medialis. Thus, the characteristic aspect of the central gray in Mammals could be evaluated as a further differentiation of structural arrangements 'already' present in Sauropsida. In Mammals, however, at least in Primates (Macaca), 'projections' from the medial occipital (parastriate) cortex to the *nucleus of Darkschewitsch* have been claimed on the basis of experiments using the *Fink-Heimer technique* (Wright *et al.*, 1974).

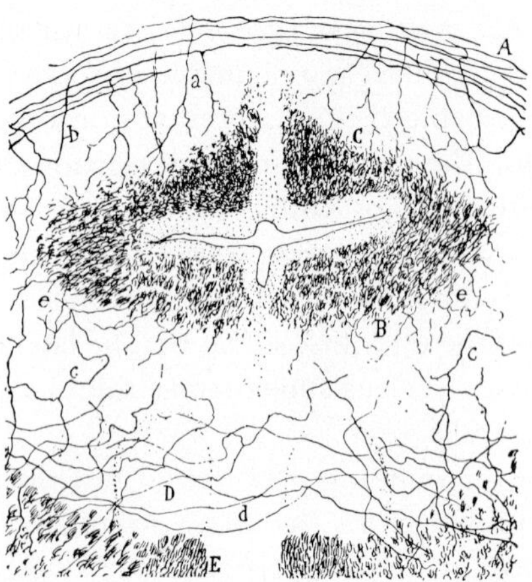

Figure 412B. Transverse section *(Golgi impregnation)* through the mesencephalic central gray in a young Mouse (from CAJAL, 1911). A: transverse fibers (commissura tecti) of stratum medullare profundum; B: fasciculus longitudinalis dorsalis of *Schütz;* C: longitudinal bundles of dorsal central gray; D: region of oculomotor nucleus; E: fasciculus longitudinalis medialis; a, b: collaterals of transverse fibers (A); c, d, e: undefined tegmental fibers, partly related to ventral central gray.

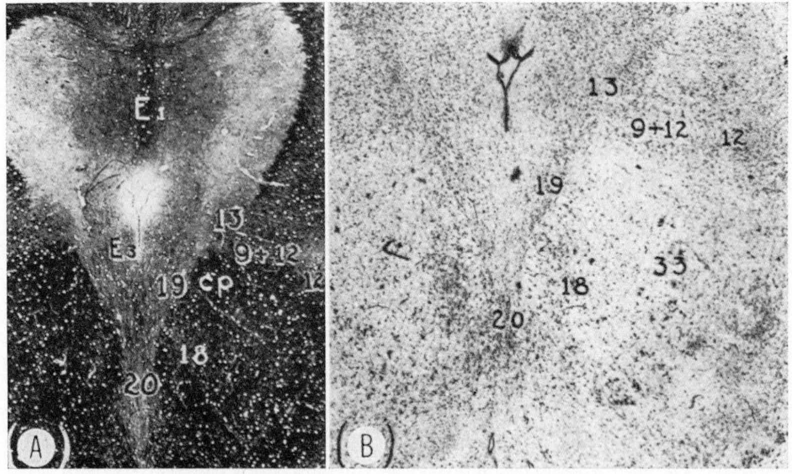

Figure 413. Rostral region of central gray in the mesencephalon of the Rabbit, as seen in two adjacent levels of a cross-sectional series alternatingly stained (A) by a myelin and (B) by a modified *Nissl technique* (from K. and MILLER, 1942). E 1, 3: dorsal vertical and ventral periventricular fibers of central gray; cp: fibers pertaining to caudal limb of commissura posterior; 9+12: overlapping grisea of posterior commissure and nucleus sublentiformis (12); 13: nucleus lateralis grisei centralis; 18: nucleus interstitialis fasciculi longitudinalis medialis; 19: *nucleus of Darkschewitsch;* 20: nucleus medialis anterior; 33: nucleus ruber tegmenti.

encephalic periaqueductal gray with its fiber systems is doubtless in part related to the vegetative or 'visceral' nervous system. It seems also to participate in the regulation of the sleep-wakefulness rhythm. Moreover, lesions of the mesencephalic central gray appear to produce states of 'stupor' or 'unconsciousness'. It is thus possible to include this region into the larger group of so-called 'activating centers' or '*centrencephalic grisea*' whose activities may have a significant facilitating or inhibiting effect, or both, on corticothalamic events. Additional comments concerning the 'catecholamine fiber pathway' included in the periventricular fiber system will be found below on p. 987.

A peculiar griseum of the prerubral tegmentum, the *nucleus ellipticus*, was described by HATSCHEK and SCHLESINGER (1902) in the brain of the Dolphin and subsequently recorded by various authors in other Cetacea. It was also identified in the Elephant (DEXLER, 1907; PŘCECHTĚL, 1925). This nucleus, rounded or ovoid in cross-sections, attains a substantial size, being commonly delimited or almost surrounded by a ring of medullated fibers ('*Markring*', OGAWA, 1935) and protrudes rostralward into the thalamus. It seems to be a lateral derivative or expansion of the *nucleus of Darkschewitsch*, although diverse other interpretations (relation to nucleus medialis anterior or even to nucleus ruber) have been propounded, as discussed by VERHAART (1970). A so-called medial tegmental tract, located basally to fasciculus longitudinalis medialis, and losing itself caudally in the tegmentum, is said to originate from that griseum, respectively from the *nucleus of Darkschewitsch* in forms lacking a nucleus ellipticus. VERHAART (1970) believes that a hypertrophic inferior olive is related to medial tegmental tract and nucleus ellipticus. VERHAART then adds 'that one could be inclined to conclude to a certain relationship between the elephant and the cetaceans, which otherwise is not evidenced. The similarity in this respect is the more remarkable, because the difference in the mode of life between the two orders makes convergence phenomena very unlikely. Moreover in the Sirenia, the hypertrophic olive and the other characters mentioned are not found. Why they developed in the two orders therefore as yet is not readily comprehensible'.

The grisea in the main portion of the *basal plate* externally to the periaqueductal gray comprise, as in Sauropsida, *the reticular formation* with associated tegmental structures (nucleus ruber, substantia nigra, nucleus opticus tegmenti, nucleus interpeduncularis, etc.) and the two mesencephalic eye muscle nerve nuclei (III, IV).

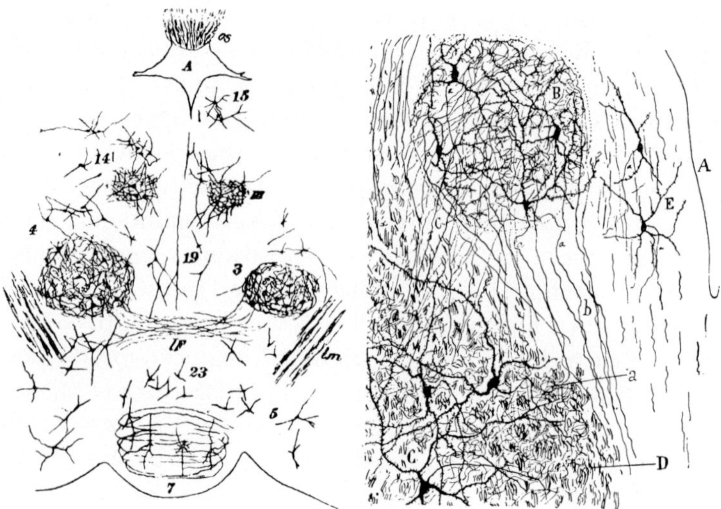

Figure 414A. Cross-section *(Golgi-Cox impregnation)* through the mesencephalic tegmentum in an 88 mm long fetus of Cavia cobaya (from CASTALDI, 1923). A: cerebral aqueduct; F: *Forel's ventral tegmental decussation;* lm: lemniscus medialis; os: subcommissural organ; III: oculomotor nucleus; 3: nucleus ruber; 4: nucleus lateralis profundus (tegmenti) mesencephali; 5: substantia nigra; 7: nucleus interpeduncularis; 14: nucleus ventralis grisei centralis; 15: nucleus ventrolateralis grisei centralis; 19: 'nucleus linearis rostralis'; 23: supra-interpeduncular cell group of reticular formation.

Figure 414B. Nucleus of Darkschewitsch and nucleus interstitialis *(Golgi impregnation)* as displayed in a cross section through the mesencephalon in the Cat a few days after birth (from CAJAL, 1911). A: cerebral aqueduct; B: *nucleus of Darkschewitsch;* C: nucleus interstitialis; D: fasciculus longitudinalis medialis; E: neuronal elements of central gray; a: neuropil of nucleus interstitialis; b: neurites from *nucleus of Darkschewitsch;* c: collaterals ending in preceding nucleus. Small cursive *a* designates axons.

The Mammalian reticular formation and its ontogenetic differentiation have been described in great detail, and with numerous bibliographic references, in the painstaking study by CASTALDI (1923–1928), based upon the development of the mesencephalon in Cavia cobaya (cf. Fig. 414A). The diffusely arranged grisea of the reticular formation include the nucleus (tegmentalis) lateralis profundus, located laterally and dorsally to nucleus ruber. It extends into the mesencephalo-diencephalic boundary as the *prerubral tegmentum* of PAPEZ. A medial portion of this cell population, dorsal to nucleus ruber, is the *nucleus interstitialis tegmentalis commissurae posterioris.*[128] The rostral part of the

[128] Nucleus *9* t of our notation (K. and MILLER, 1942, 1949; KUMAMOTO-SHINTANI, 1959).

prerubral tegmentum, together with nucleus medialis anterior and adjacent periventricular gray, is a derivative of the Mammalian embryonic *tegmental cell plate*. More caudally, and ventrally to nucleus dorsalis tegmenti of the central gray as well as to fasciculus longitudinalis medialis, the reticular formation is represented by the nucleus ventralis tegmenti, which, in turn, overlaps or merges with *Castaldi's median 'nucleo lineare'*. A still more ventral part of this griseum, between *Forel's decussation*, to be discussed further below, and the interpeduncular nucleus, is *Castaldi's 'foyer pediculaire'*.[129] A rostral area of mesencephalic reticular formation, within the mesencephalodiencephalic boundary zone, and included in the nondescript so-called ventral tegmental area of TSAI (1925) and of CROSBY *et al.* (1962), represents the 'nucleus of the mammillary peduncle', perhaps providing input for the hypothalamic mammillary grisea to be dealt with in chapter XII.

The *nucleus interstitialis of Cajal*, a well defined tegmental derivative present in the entire Vertebrate series, is found at its typical location in the mesencephalo-diencephalic boundary zone (Figs. 413B, 414B). The Mammalian *nucleus ruber* is, generally speaking, a larger and more conspicuous tegmental structure than in Sauropsida. It displays a *magnocellular* and a *parvocellular* component which are commonly more or less intermingled (Fig. 415A, B). The magnocellular component gives origin to the rubrobulbar and rubrospinal tract which crosses in the *ventral tegmental decussation of Forel*. From the parvocellular component arise crossed and uncrossed fibers reaching the thalamus and other (e.g. tegmental) grisea.

The parvocellular component appears more extensive in 'higher' than in 'lower' Mammalian forms. Likewise, there are, in this respect, some variations concerning the caudo-rostral expansion of the entire nucleus. An important input channel of the nucleus ruber is the brachium conjunctivum which decussates ventrally to fasciculus longitudinalis at caudal mesencephalic levels (cf. Fig. 410C). Besides this cerebellar connection, dealt with in chapter X, the nucleus ruber receives input from prosencephalic centers, to be discussed in volume 5. Additional collateral input from the lemniscus medialis has been recorded and depicted by CAJAL (cf. Fig. 415C). In Man and some other but not all Mammals, the nucleus ruber manifests the *iron reaction of Spatz* referred to further below.

---

[129] Nucleus δ of OLZEWSKI and BAXTER (1954) in the human brain.

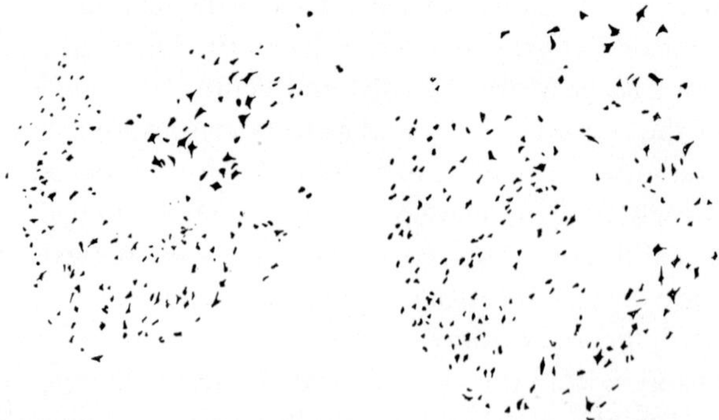

Figure 415A. Cytoarchitecture of nucleus ruber *(Nissl stain)* in the Dog, showing distribution of magnocellular and parvocellular elements as seen in cross-sections (from BECCARI, 1943). Rostral level at left, intermediate level at right. Lateral side in both sections at left.

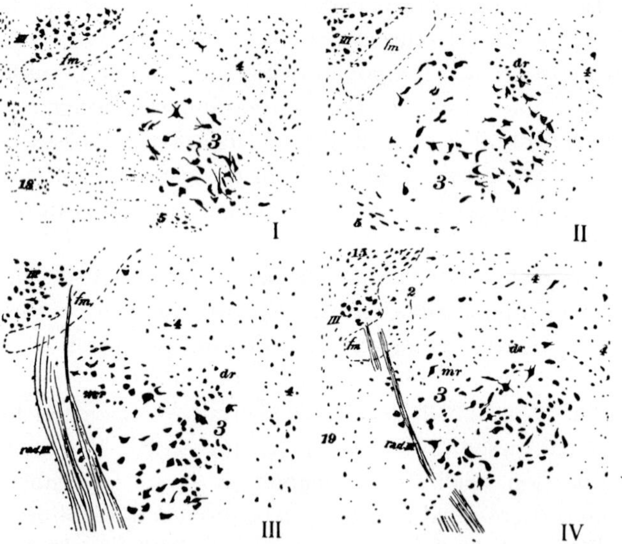

Figure 415B. Four cross-sectional levels *(Nissl stain)* of the nucleus ruber in Cavia cobaya (from CASTALDI, 1928). I: near caudal pole; II: caudal third; III: intermedial third; IV: rostral third; dr: dorsolateral reticular nucleus; fm: fasciculus longitudinalis medialis; mr: mediodorsal reticular nucleus; rad. III: oculomotor root; III: oculomotor nucleus; 2: nucleus interstitialis; 3: nucleus ruber; 4: *nucleus profundus (tegmentalis) mesencephali of Edinger-Castaldi;* 5: substantia nigra; 15: *nucleus lateroventralis grisei centralis of Castaldi;* 19: *nucleus linearis rostralis of Cajal-Castaldi.*

The *substantia nigra*, a band- or plate-like griseum which separates the tegmentum from the pes pedunculi, can be evaluated as a phylogenetic derivative of nucleus motorius tegmenti, and as an ontogenetic derivative of the matrix from which the reticular formation originates. The designation substantia nigra refers to its appearance in Man,

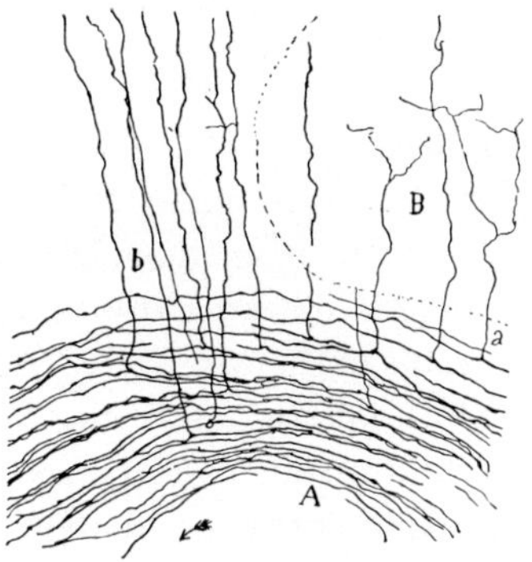

Figure 415C. Sagittal section *(Golgi impregnation)* through mesencephalic tegmentum in a young Mouse, showing collaterals of medial lemniscus, some of which end in nucleus ruber (from CAJAL, 1911). A: lemniscus medialis; B: nucleus ruber; a: collaterals to nucleus ruber; b: prerubral collaterals, perhaps to pretectal grisea *('collatérales plus nombreuses passant en avant du noyau rouge et allant se perdre dans le noyau postérieur de la couche optique')*. Rostral side at left.

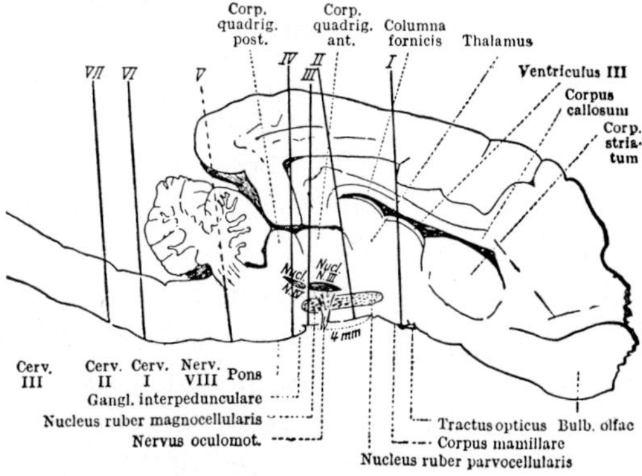

Figure 415D. Semidiagrammatic sagittal section through the brain of a Cat, showing various levels of transections performed by MAGNUS and RADEMAKER in studies of decerebrate rigidity (from FULTON, 1949). Section I: Thalamus animal with heat regulation and normal posture; II: midbrain animal with normal posture but abolished thermal regulation; III: decerebrate rigidity; IV: if unilateral gives homolateral rigidity; V: just rostral to vestibular grisea, rigidity persists; VI: rigidity abolished; VII: neck reflexes abolished, 'true spinal animal'.

where the dorsal portion *(zona compacta)* of said griseum is characterized by neuronal elements containing a considerable amount of melanin pigment, recognizable upon macroscopic inspection of sections through the mesencephalon. The melanin, usually not displayed in the newborn, becomes conspicuous in the third year and seems to reach its adult concentration about the tenth year. In diverse subprimate forms, the melanin does not seem to be present at all, but smaller amounts than in Man have been recorded in several Mammals such as Dog, Horse, and others (cf. KAPPERS *et al.*, 1936; BROWN, 1944).[130] The ventral portion or *zona reticulata* does not contain melanin but gives, in Man and some other Mammals (Cat, Dog, large Ungulates, Primates) the *iron reaction of Spatz* (1922). This reaction is faint in the Rabbit, and apparently not demonstrable in the Mouse.[131]

Although cell groups homologous to the Mammalian substantia nigra seem to be present in Sauropsidans (cf. section 7 and 8), this griseum reaches its full development in the Mammalian series, particularly in the higher forms of that class. Again, in lower Mammalian forms, the distinction between zona compacta and zona reticulata is less pronounced than in the higher ones. Both zones consist of fairly large multipolar nerve cells, those of zona compacta being commonly larger than those of zona reticulata. Rostral connections of the substantia nigra are striatonigral as well as nigrostriatal fibers, similar relations with globus pallidus, subthalamic body, and corticonigral fibers from diverse portion of the telencephalon. In addition, some fiber connections with the superior colliculus have been suspected. The caudal connections seem to be represented by poorly elucidated crossed and uncrossed nigrotegmental and tegmentonigral fibers, mainly providing interrelations of substantia nigra with the reticular formation of the brain stem and with the nucleus ruber (cf. Fig. 416C).

At the dorsolateral edge of the rostral portion of substantia nigra within the mesencephalo-diencephalic boundary zone, and overlying the dorsolateral surface of the pes pedunculi, a *nucleus peripeduncularis* has been described as perhaps derived from the midbrain reticular gray. The results of my own investigations of this griseum in the brain of Man, Macaca mulatta, Cat, Rabbit, Rat, and Mouse have shown,

---

[130] The melanin of the human substantia nigra is said to be also present in albinos (cf. CROSBY *et al.*, 1962).

[131] Cf. the comments in chapter V, section 2, pp. 135–136 of volume 3, part I.

however, that it consists of several components, all of which pertain to caudal portions of the thalamus ventralis (K., 1962). In order to avoid confusion, it seems advisable to use the term *peripeduncular region* which subsumes three different ventral thalamic grisea, namely pars ventralis of corpus geniculatum laterale, pars ventralis of corpus geniculatum mediale, and an ill-defined peripeduncular portion of zona incerta. The conspicuous peripeduncular region of the human brain is shown in Figure 421A.

The *nucleus opticus tegmenti* (or *nucleus ectomammillaris*) is closely contiguous to the rostromedial portion of the substantia nigra (cf. Fig. 407) and receives fibers from the basal optic tract.[132] In some Mammals this tract (or 'basal optic root') may form or participate in, a macroscopically visible bundle, the tractus peduncularis transversus, which runs externally dorsad over the pes pedunculi (cf. Fig. 417B).

The *nucleus interpeduncularis*, receiving the fasciculus retroflexus (tractus habenulopeduncularis), is present in all investigated Mammals. It appears to be relatively larger in macrosmatic than in microsmatic form. The *tegmental bundle of Ganser*, particularly conspicuous in macrosmatic Mammals, connects the interpeduncular nucleus with nucleus dorsalis tegmenti and also with nucleus ventralis tegmenti. Said bundle is predominantly an output channel of nucleus interpeduncularis but may also include reciprocal (i.e. input) fibers. Structural details of this nucleus, whose further connections and functional significance (correlation of olfactory impulses with activities of other systems) can merely be inferred in very vague generalized terms, are depicted in Figures 416A and B.

Concerning the Mammalian mesencephalic *eye muscle nerve nuclei*, three different arrangements seem to obtain (Kappers *et al.*, 1936). (1) Oculomotor and trochlear nerve nuclei may be completely separated, as e.g. in Echidna, a relation comparable to conditions found in some reptiles. (2) The two nuclei, although more or less separated from each other by nondescript fiber systems, are in fairly close contiguity. (3)

---

[132] Details concerning the basal optic root are contained in the publications by Frey (1935, 1937, 1938), who was particularly concerned with its possible diencephalic (hypothalamic connections), as well as by Castaldi (1923), Gillilan (1941), Hayhow *et al.* (1960), Tsai (1925), and others. The relationship of the crossed to uncrossed optic tract fibers in the basal optic root remains controversial, but crossed fibers may predominate. Likewise, hypothalamic connections of the basal optic root have been questioned by some authors.

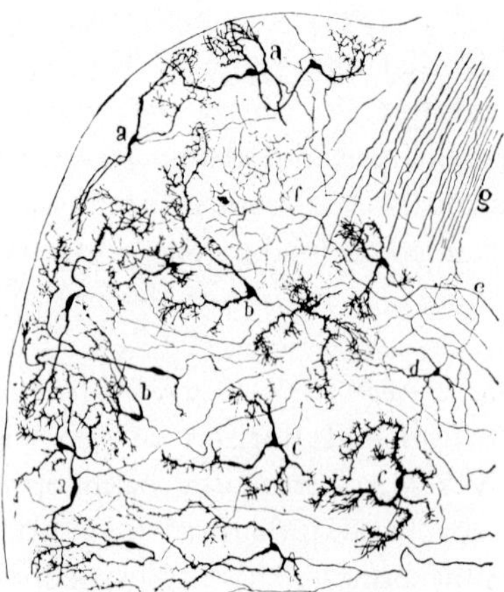

Figure 416 A. Sagittal section *(Golgi impregnation)* through the interpeduncular nucleus of a young Rabbit (from CAJAL, 1911). a: 'transversal peripheral nerve cells'; b: 'elongated cells'; c: 'stellate cells'; d: *Golgi II cell;* e, f: fiber from pontine levels *('venue de la protubérance')* and its terminal arborization; g: entrance of fasciculus retroflexus into interpeduncular ganglion. Rostral side of section at top, ventral side at left. Compare with Figure 386F.

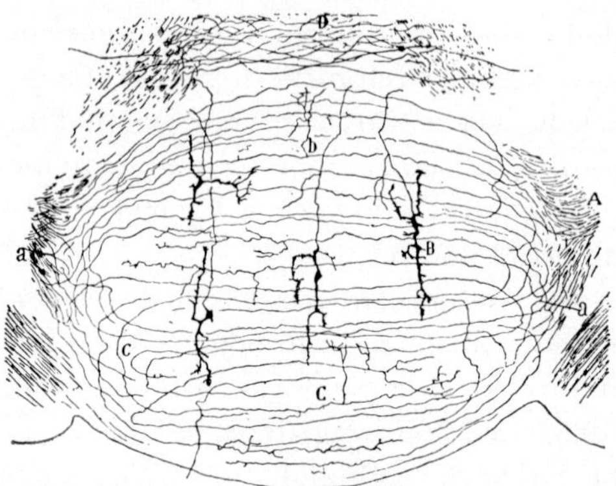

Figure 416B. Slightly oblique cross-section *(Golgi impregnation)* through the interpeduncular nucleus of a young Mouse (from CAJAL, 1911). A: terminal fibers of *Meynert's fasciculus retroflexus* before entrance into ganglion; B: neuronal element of ganglion as seen in cross-section *('vue de profil'*, of sagittal aspect in preceding figure); C: terminal arborizations of fasciculus retroflexus; D: *ventral tegmental decussation of Forel;* a: bifurcating fiber of fasciculus retroflexus fiber on the contralateral side *('crochets onduleux de chaque fibre dans le côté opposé'* cf. Fig. 386E). The unlabelled ventral bilateral fiber bundle on both sides of ganglion interpedunculare is presumably the 'mammillary peduncle', considered to be an essentially afferent channel for the mammillary grisea from the brain stem tegmentum.

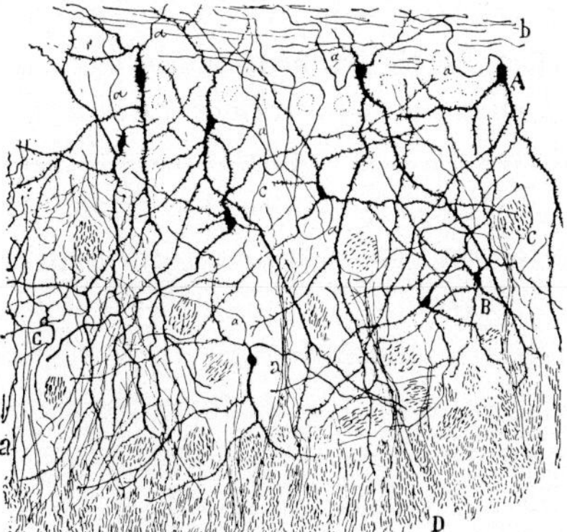

Figure 416C. Cross-section *(Golgi impregnation)* through the substantia nigra in a young Cat (from CAJAL, 1911). A: zone compacta *('cellules supérieures')*; B: zona reticulata *('cellules inférieures')*; C: *Golgi II cells?* (CAJAL's question mark); D: pes pedunculi; a: collaterals from pes pedunculi ramifying in substantia nigra; b: transverse fiber bundle at dorsal margin of substantia nigra.

Oculomotor and trochlear nucleus form a continuous nuclear column, as e.g. in the Opossum and in Man.[133]

The Mammalian *oculomotor nucleus*, generally located dorsally to fasciculus longitudinalis medialis, includes the preganglionic, small-celled *nucleus of Edinger-Westphal*, a derivative of the basal plate's intermedioventral zone, and various large-celled groups.[134] These latter can be subdivided into a dorsolateral, and a ventromedial main group. An additional unpaired median group, the *nucleus of Perlia*, is present in Man and some other Mammals, but not in other forms such e.g., as Rodentia, Marsupialia, Cetacea. In lower forms, the large-celled oculomotor nucleus is certainly less differentiated than in most Birds, and even the dorsolateral and ventromedial group may represent a more or less unified griseum. There is little doubt that a variable proportion of oculomotor root fibers decussates and joins the contralateral oculomotor nerve root.

---

[133] KAPPERS *et al.* (1936) emphasize that the primitive Vertebrate trochlear nucleus is quite distinct from that of the oculomotor. The gradual rostrally directed shift of the trochlear nucleus and its close approximation to the oculomotor grisea is evaluated as 'a secondarily acquired relation' occurring in the course of Vertebrate phylogenetic development.

[134] *Nucleus of Darkschewitsch* respectively nucleus ellipticus, dealt with further above, are in close topographic relation to the oculomotor complex.

The *nucleus of the trochlear nerve*, whose root undergoes an essentially complete decussation in the velum medullare anterius, appears, if directly continuous with the oculomotor nucleus, as a caudal extension of the dorsolateral large-celled oculomotor subgroup. In some forms, e.g. the Horse and perhaps a few other Ungulata, the trochlear nucleus is located ventrally to the fasciculus longitudinalis medialis. But even here, said nucleus appears continuous with the dorsolateral rather than with the ventromedial oculomotor subgroup (KAPPERS *et al.*, 1936). In a large number of Mammals, the trochlear nucleus lies partly within bundles of the fasciculus longitudinalis medialis.

Many authors have attempted to establish a '*somatotopic' localization*, within the large-celled subdivisions of the oculomotor nucleus, of the muscles innervated by the third nerve, but the obtained results have remained controversial (cf. CROSBY *et al.*, 1962). This question shall again be taken up further below in dealing with the Human mesencephalon.

Various fiber connections of the Mammalian midbrain were briefly pointed out in dealing with the diverse grisea. For an overall summary of these channels, *rostral*, *caudal*, and *intrinsic connections*, moreover fibers predominantly of *passage*, and *commissural systems* might be distinguished. Again, ascending and descending connections, moreover fibers arising from or ending in mesencephalic grisea could be enumerated. None of these classifications is sufficiently satisfactory, quite apart from the fact that the recorded data on fiber systems are, to a significant extent, rather vague and in part controversial. Most statements concerning precise details of fiber tracts, regardless of the techniques by which these tracts were studied, must be evaluated with the proper degree of caution. Moreover, it must be kept in mind that many or most central tracts are not only composite bundles comprising fibers of diverse origin and diverse destination, but may also include channels conducting in opposite directions, e.g. some ascending, and some descending fibers. In many instances, such two-way connections may be reciprocal, that is, providing a mutual interrelation between two particular grisea.

Among the *rostral channels*, the *optic input* to pretectal grisea and superior colliculus through the subdivisions of the main optic tract is of substantial importance. Additional optic input reaches the nucleus opticus tegmenti through the basal optic root, and indirectly or even directly, the reticular formation.

As regards further rostral channels, the pretectal grisea doubtless play an important role mediating between tectum and prosencephalic grisea, to be dealt with in volume 5. Concerning the tectum proper, a tectohabenular tract, a dorsomedial and a dorsolateral tectothalamic tract have been recorded (cf. e.g. K. and MILLER, 1942, 1949). With respect to the telencephalon, corticotectal connections are presumed to obtain. The brachium quadrigeminum superius and the brachium quadrigeminum inferius provide a pathway for some of the various channels interconnecting colliculus superior respectively inferior with prosencephalic grisea.

As regard the *tegmentum*, corticotegmental and hypothalamotegmental fibers are mentioned by CROSBY *et al.* (1962) whose treatise contains a fairly recent critical review of the relevant fiber systems. The rostral connections of substantia nigra, and of nucleus ruber were mentioned above in dealing with these grisea. Among the hypothalamotegmental channels, the mammillotegmental tract connecting with dorsal and ventral tegmental nucleus, the mammillo-interpeduncular tract, and the perhaps predominantingly ascending mammillary peduncle are generally recognized.

Recent investigations by means of the glyoxylic acid fluorescence method have disclosed an ascending *catecholamine fiber pathway* in the mesencephalic periventricular system. In addition to fibers of the here ventrally located *fasciculus longitudinalis dorsalis of Schütz* (dorsalis with respect to the tegmentum), said pathway also includes a dorsal circumventricular component (LINDVALL *et al.*, 1974, with additional relevant bibliography). This system is said to project upon pretectal grisea, epithalamus, and medial as well as midline thalamic nuclei.

Another channel, the *locus caeruleus system* (cf. chapter IX, section 10, p. 545) is described as ascending through the dorsal tegmental fasciculi, reaching the diencephalic neighborhood of zona incerta, with extensive endings in many thalamic grisea (including lateral and medial geniculate complex as well as pretectal region).

Important *caudal descending channels* are represented by (1) the fibrae efferentes tecti, (2) components of fasciculus longitudinalis medialis, (3) tractus rubro-bulbospinalis, (4) components of central tegmental tract, (5) components of *fasciculus longitudinalis dorsalis of Schütz*, (6) tectocerebellar connections, and (7) *the tract of Probst.*

(1) The *fibrae efferentes tecti*, essentially provided by the neurites of chief cells, run mostly through the stratum album (or medullare) profundum of superior colliculus. Some such fibers also originate from cells in the intercollicular griseum of inferior colliculus. This system of efferent tectal fibers *(Meynert's fibers)* provides the medial and the later-

al tectobulbospinal tracts as well as intrinsic mesencephalic connections (tectotegmental, tectonigral, tectorubral, tecto-oculomotor fibers).

The medial tectobulbospinal tract decussates in *Meynert's dorsal tegmental decussation ('fontänenartige Haubenkreuzung')*, running caudad as the fasciculus praedorsalis ventrally to fasciculus longitudinalis medialis. The lateral tectobulbospinal tract seems to be essentially uncrossed and runs caudad through the lateral tegmentum.

(2) *Descending components* of the fasciculus longitudinalis medialis originate in the *nucleus interstitialis of Cajal.* Connections and significance of this important and complex channel, repeatedly discussed in chapters VIII and IX, as well as in section 1 of the present chapter (Figs. 356A–C), do not require further comments in this context.

(3) The *tractus rubrospinalis or rubrobulbospinalis* originates from the magnocellular component of nucleus ruber and essentially crosses, at levels of that griseum, in *Forel's ventral tegmental decussation.*

(4) The *central tegmental or so-called thalamo-olivary tract* is still poorly understood. Noticed by some earlier authors, it was first carefully studied by BECHTEREW (1885) who believed that said bundle represented a connection between the cerebrum *(Grosshirn)* and the inferior olivary nucleus. BECHTEREW (1889) subsequently introduced the term central tegmental tract while other authors used the term *tractus thalamo-olivaris.* Generally speaking, this channel runs laterally to fasciculus longitudinalis medialis, assuming a central position in the pontine tegmentum and shifts to the lateral side of the main inferior olivary nucleus in the medulla oblongata. Although, as displayed by myelin stains, said tract commonly forms a very distinct entity within the reticular formation, the determination of its rostral origin and the analysis of its different components have proven to be a difficult problem. Our own observations, based on normal myelin stained material, led us to assume that this tract, although mainly connecting with the inferior olive, is a complex system originating from various centers, and possibly varying from level to level. In addition to a pretectal (thalamic) component and contributions from the periaquaeductal gray (the *anulo-olivary component* stressed by METTLER, 1944), tegmental and diverse prosencephalic components do not seem improbable (K. and MILLER, 1942, 1949; K. 1956a). Further critical comments on the widely diverging views concerning the central tegmental tract, together with relevant bibliographic references, can be found in the publications by JANSEN and BRODAL (1958) and by LARSELL and JANSEN (1972).

As regards additional *descending channels*, it will here be sufficient merely to enumerate the *fasciculus longitudinalis dorsalis of Schütz* (5), mentioned above in connection with the central gray, the tectocerebellar tract (6) dealt with in chapter X, and the *tract of Probst* (7) pointed out above in reference to the nucleus radicis mesencephalicae trigemini.

Important *caudal ascending channels* are provided by the general spinobulbar lemniscus medialis system and by that of the lateral lemniscus. Spinotectal and bulbotectal fibers or collaterals of the general lemniscus reach the superior colliculus particularly by way of the stratum album medium, while a substantial portion of this lemniscus system[135] represents a channel of passage reaching the thalamus.

The *lemniscus lateralis* has substantial connections with the inferior colliculus[136] (cf. Fig. 411B) and proceeds, via the brachium quadrigeminum inferius, to the diencephalon (corpus geniculatum mediale), being joined by fibers originating in the nucleus colliculi inferioris.

The *brachium conjunctivum*, decussating in the tegmentum, ascending components of fasciculus longitudinalis medialis and those within the *fasciculus longitudinalis dorsalis of Schütz* are likewise ascending caudal input channels for mesencephalic and also diencephalic grisea, as repeatedly mentioned above.

In addition to the ascending fibers of passage represented by the diencephalic (thalamic) input components of the two main lemniscus systems, *descending fibers* of passage, taking their course through the pes pedunculi, are the corticopontine (frontopontine, temporoparieto-occipitopontine) and the pyramidal tracts.[137]

The *intrinsic fiber connections* of the mesencephalon still defy attempts at a convenient classification for purposes of enumeration; some of these connections, however, were mentioned above in dealing with the diverse grisea.

Discounting the crossed components of the oculomotor root, the decussation of the trochlear nerve and basal decussations within the

---

[135] In addition to the ascending secondary trigeminal fibers joining the general (or medial) lemniscus system, the existence of a separate dorsal ascending secondary trigeminal tract is generally assumed (cf. section 10 of chapter IX). The relationship of this tract to mesencephalic grisea remains uncertain.

[136] Some collaterals also reach the superior colliculus (cf. Fig. 411B).

[137] It remains uncertain to which extent collaterals of these systems are or are not given off to mesencephalic grisea. In particular, there remain many open questions concerning the course of cortical fibers presumed to innervate the mesencephalic eye muscle nuclei.

fasciculus retroflexus, the main *commissural systems*[138] related to the mesencephalon are (1) commissura posterior, (2) commissura tecti, (3) components of supraoptic commissures, (4) supramammillary commissure, (5) dorsal tegmental decussation, (6) ventral tegmental decussation, (7) decussation of brachium conjunctivum *(Werneking's decussation)*, and (8) *commissure of Probst.*

(1) *The commissura posterior* with its heavily medullated fibers is a landmark of the mesencephalo-diencephalic boundary zone and can be evaluated as pertaining to both brain subdivisions. It seems to be in part a commissure *sensu strictiori* interconnecting antimeric nuclei, and in part a decussation of fibers interrelating differing grisea. A variety of mostly pretectal diencephalic and mesencephalic nuclei related to that commissure were pointed out further above. Crosby *et al.* (1962) justly state that the specific interrelations of the posterior commissure's fibers 'are not well understood for any mammal' (cf. also K., 1954). The *subcommissural organ*, located ventrally to said commissure, was likewise repeatedly dealt with in the preceding sections.[139] The caudally adjacent commissura tecti (2) can again be subdivided into the more extensive commissura colliculi superioris and the less extensive commissura colliculi inferioris. Caudally to this latter, some crossing fibers of uncertain significance, not pertaining to the decussation of the trochlear nerve, have been described as 'commissure of the anterior velum'.

(3) The *supraoptic commissures or decussations* are formed by fibers which, although not belonging to the optic tract, cross the midline in the chiasmatic ridge of the hypothalamus. A dorsal and ventral supraoptic decussation can be distinguished. As far as fiber systems related to the midbrain are concerned, undefined fibers related to mesencephalic grisea may pass through the dorsal supraoptic decussation, and likewise undefined tectopretectothalamic respectively tectopretecto-hypothalamic fibers may pass through the ventral supraoptic decussation (cf. K., 1954; Crosby *et al.*, 1962).[140]

---

[138] It should here be reiterated that, unless specified, the term 'commissure' is used *sensu latiori*, that is designating any transverse system of fibers crossing the midline, and subsuming both commissural fibers *sensu strictiori* and merely decussating ones, as again mentioned below in the text.

[139] This organ pertains to the group of ependymal and paraependymal organs or structures dealt with on pp. 354–367, section 5, chapter V, of volume 3, part I.

[140] Crosby *et al.* (1962) remark that the supraoptic decussations are 'among the least well-understood connections'.

(4) The *supramammillary or posterior hypothalamic decussation* was apparently first described by FOREL (1872, 1907). Many of its fibers arise in the *subthalamic nucleus of Luys* and terminate in contralateral tegmental centers. The caudal portion of the supramammillary decussation belongs very definitely to the (mesencephalic) tegmentum and might even be set apart as postmammillary or anterior tegmental decussation. The entire supramammillary commissural system is also occasionally referred to as *Forel's decussation* but should not be confused with *Forel's ventral tegmental decussation* (6). This latter, and the *dorsal tegmental decussation* (5) *of Meynert's fibers*, the *decussation of the brachium conjunctivum* (7) and the *commissure of Probst* (8) were already dealt with above in connection with their relevant grisea.

In attempting to summarize the functional significance of the Mammalian mesencephalon the following appraisal of its main griseal subdivisions seems perhaps justified. The diencephalo-mesencephalic *pretectal region*, besides being a 'relay-center' between mesencephalon and prosencephalon, appears independently to mediate some reflexes, such, e.g. as the pupillary constriction reflex upon illumination.

The *superior colliculus*, rostrally connected with pretectal grisea, with thalamic pulvinar and lateral geniculate body, remains, as in submammalian Vertebrates, a suprasegmental 'cortex' of substantial significance, capable of 'directing' numerous behavioral activities in addition to the performance of more simple, particularly optic reflexes.[141] It could, nevertheless be inferred that the functional significance of the superior colliculi decreases in higher Mammalian forms (e.g. Primates), becoming inversely related to a progressively increasing dominance of the pallial cortex telencephali. The *inferior colliculi*, containing a 'relay griseum' of the lateral lemniscus, are generally evaluated as an 'auditory reflex center'.

The *reticular formation, together with the periventricular gray*,[142] seems to be in part a component of the so-called 'centrencephalic' or 'activat-

[141] Pupillary dilatation upon diminished light is presumably a multifactorial event. It seems possible that, with increasing illumination, an inhibiting effect of tectospinal fibers on the centrum ciliospinale (chapter VIII, section 11, p. 195) becomes reduced respectively abolished (cf. Fig. 362).

[142] The *fasciculus longitudinalis of Schütz* (Fig. 428B) related to the central gray, can be evaluated as a channel predominantly pertaining to the vegetative nervous system. This latter, in the wider sense, may be taken as synonymous with the 'visceral nervous system'. *Sensu strictiori*, of course, the vegetative or autonomic nervous system comprises exclusively the efferent components as defined in chapter VII, section 6, of volume 3, part II.

ing' respectively 'inactivating' systems influencing corticothalamic activities. As a mesencephalic portion of the overall brain stem reticular formation, the mesencephalic tegmental reticular formation presumably participates in the manifold further functions attributed to this extensive griseum dealt with in section 2 of chapter IX. These functions, *inter alia*, involve the so-called 'extrapyramidal motor system' to which *qua* additional tegmental mesencephalic grisea, *nucleus ruber* and *substantia nigra* pertain.

In lower Mammals, transection of the brain at diencephalic levels, or even at diencephalo-mesencephalic levels, leaving the brain stem intact up to the rostral level of superior colliculi, is compatible with normal distribution of tonus and essentially normal reflexes.[143] Primates appear somewhat more affected by this procedure.

In contradistinction, if a transection of the brain stem is performed *between superior colliculi and rostral level of vestibular nuclei*, *decerebrate rigidity* occurs, which is characterized by a continuous spasm of the extensor musculature in all four limbs (SHERRINGTON, 1898). The most hypertonic muscles are those which normally counteract the effect of gravity, and this state was therefore also compared to a caricature of normal standing position. Transection of the brain stem caudally to the bulk of the vestibular nuclei abolishes decerebrate rigidity (cf. Fig. 415D).

The levels of transection causing decerebrate rigidity extend from superior colliculi to the rostral pons. Removal of the cerebellum does not abolish the rigidity. However, this condition disappears after the destruction of the vestibular nuclei and adjoining parts of the reticular formation, but not after lesions restricted exclusively to the vestibular nuclei (BERNIS and SPIEGEL, 1925, and others). Removal of the labyrinth alone, on the other hand, modifies but does not completely abolish decerebrate rigidity. As was already demonstrated by SHERRINGTON, intact dorsal (afferent) spinal nerve roots are necessary for the occurrence of typical decerebrate rigidity in the corresponding extremities. Nevertheless, deafferentiation does not seem to abolish all phases or manifestations of decerebrate rigidity in the corresponding limbs.

The *nucleus ruber* is believed to be an important relay center of the extrapyramidal system and appears to be involved in postural adjustments. RADEMACHER (1926) assumed that normal distribution of tonus might be one of the main functions of the red nucleus and that decerebrate rigidity is caused by the destruction of that griseum. According

[143] Discounting those upon optic input, if both optic tracts are severed.

to RADEMACHER, said rigidity does not occur when the nucleus ruber remains intact. Subsequent investigators, however, failed to observe decerebrate rigidity after isolated destruction of the nucleus ruber and its descending output, if the central nervous system remained otherwise undamaged (KELLER and HARE, 1934, and others). Rigidity experimentally produced by ligation of carotids and basilar arteries caused a much more pronounced decerebrate rigidity than that after brain stem sections. After unilateral transections just caudally to nucleus ruber (Fig. 415D), the rigidity develops on the homolateral side, suggesting that the here already crossed rubrobulbospinal tract is thereby involved.

On the basis of all available observations it seems reasonable to interpret Mammalian decerebrate rigidity as a multifactorial '*release phenomenon*', presumably caused by the release of vestibular and closely associated bulbar reticular grisea from higher 'extrapyramidal motor' control. The descending output of the vestibular and reticular centers involved in producing the rigidity is possibly transmitted through vestibulospinal and reticulospinal channels.

The above-mentioned apparently contradictory statements by RADEMACHER and KELLER and HARE can be evaluated as not mutually exclusive. Following RADEMACHER, it is still indicated to assume that the nucleus ruber may be responsible for a large share of the extrapyramidal preventive control over the neural discharge circuits causing decerebrate rigidity. This latter condition, however, seems to depend on a number of variables, many of which have an additive effect. Besides the nucleus ruber, neighboring reticular cell groups and diencephalic as well as telencephalic centers doubtless exert an influence similar to that of the red nucleus. In accordance with the number of variables, manifestations of decerebrate rigidity differ in details, depending to a certain extent on the various species of Mammals used for experimentation. Caution is thus advisable in drawing conclusions applying to clinical neurology[144] from results obtained in Mammals, especially in lower forms. SPIEGEL (1929) justly points out that, besides the nucleus ruber, other (e.g. diencephalic and telencephalic) centers of the human neuraxis doubtless play an important role for the distribution of muscular tonus *(normale Tonusverteilung)*. Decerebrate rigidity is here

---

[144] Cf. our own comments on decerebrate rigidity in Man (K. and MAHER, 1957; K. *et al.*, 1957; K. *et al.*, 1959).

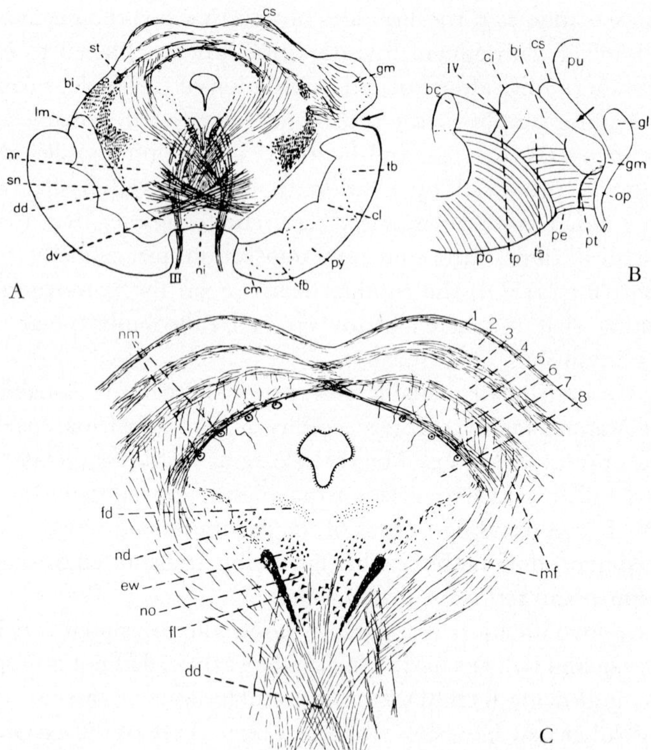

Figure 417A. Simplified sketch of a slightly oblique cross-section through the human mesencephalon at the level of superior colliculus. The right side, showing caudal end of corpus geniculatum mediale, is slightly more rostral than the left (from K., 1927). bi: brachium quadrigeminum inferius; cl: tractus corticobulbaris lateralis; cm: tractus corticospinalis medialis; cs: colliculus superior; dd: decussatio tegmenti dorsalis *('Meynertsche fontänenartige Haubenkreuzung');* dv: decussatio tegmenti ventralis (*Forel's decussation);* fb: frontopontine tract; gm: corpus geniculatum mediale; lm: lemniscus medialis; ni: nucleus interpeduncularis; nr: nucleus ruber tegmenti; py: corticospinal tract; sn: substantia nigra; st: tractus spino- et bulbotectalis; tb: temporopontine tract; added arrow: sulcus lateralis mesencephali externus.

Figure 417B. Simplified sketch of human mesencephalon as seen in lateral view (from K., 1927). bc: brachium conjunctivum; bi: brachium quadrigemium inferius; ci: colliculus inferior; cs: colliculus superior; gl: corpus geniculatum laterale; gm: corpus geniculatum mediale; op: optic tract; pe: basis sive pes pedunculi; po: *pons Varol(i)i;* pt: tractus peduncularis transversus; pu: pulvinar thalami; ta: pars anterior trigoni lemnisci; tp: pars posterior trigoni lemnisci; added arrow: brachium quadrigeminum superius.

Figure 417C. Simplified sketch of a cross-section through the human tectum opticum and adjacent subtectal tegmentum (from K., 1927). 1: stratum zonale; 2: stratum griseum superficiale; 3: stratum opticum (their transverse course is here a diagrammatic exaggeration: although the optic tract fibers run indeed slightly transversely near their entrance at

known to result from disturbances of these higher centers without involvement of the nucleus ruber and its descending output channel.

Decerebrate rigidity may be interpreted, with reference to the concepts propounded by W. R. ASHBY, discussed in chapter I (vol. 1), as a disturbance in the circuits of stable or even ultrastable systems, resulting in loss of stability. Since the stability depends on different combinations of discharging grisea, that is, on a diversity of fluctuating impulse patterns, serious disturbances may be set up by severe damage to various sets of significant nodal centers.[145]

The *substantia nigra* (Figs. 407, 408B, C, 410A, B, 416C) is generally regarded as a griseum of the 'extrapyramidal motor system' functionally related to telencephalic basal grisea and to 'extrapyramidal' cortical areas. Its significance with respect to the *Parkinsonian syndrome* will be discussed further below in dealing with the Human mesencephalon. A recent report (ROUTTENBERG and HOLZMAN, 1973) claims that electrical stimulation of the substantia nigra, zona compacta (but not zona reticularis), during training experiments with Rats in learning a simple task 'disrupted retention of that task after original learning'. Stimulation after the trials 'also disrupted retention performance'. The cited authors consider here the zona compacta 'as part of a system important for the memory storage of processed inputs and executed responses'. Assuming that the reported data might be trustworthy, such 'memory disruption' could perhaps be attributed to an influence of the substantia nigra's zona compacta upon unspecified 'activating' or 'centrencephalic' centers operative for the performance of cortical storage mechanisms. It should, moreover, be recalled that stimulation of other brain regions (e.g. amygdala, nucleus caudatus, hippocampus, etc.) seem to have an effect on the performance of memory mechanisms. While, with regard to these latter, some very general concepts concerning their

[145] A detailed monograph on the neuronal mechanisms of postural regulation, based on extensive experimental work, and including the relevant bibliography up to that time, was published by MAGNUS (1924).

---

rostral tectal levels, their further course is predominantly sagittal); 4: stratum griseum medium; 5: stratum lemnisci (stratum album s. medullare medium); 6: stratum griseum profundum; 7: stratum album s. medullare profundum *(tiefes Mark)*; 8: stratum griseum centrale s. periventriculare; dd: *Meynert's dorsal tegmental decussation;* ew: *nucleus of Edinger-Westphal;* fd: *fasciculus longitudinal dorsalis of Schütz;* fl: fasciculus longitudinalis medialis; nm: nucleus radicis mesencephalicae trigemini; mf: fibrae efferentes tecti *(Meynertsche Fasern)*; nd: *nucleus of Darkschewitsch;* no: nucleus oculomotorii *(grosszelliger Lateralkern)*.

overall nature seem justified (cf. K., 1927), present-day speculations about their specific details still remain very hazy and should be evaluated with considerable scepticism.

Turning now to various aspects of the *Human mesencephalon*, the simplified sketch of Figure 417B illustrates some external landmarks as seen in lateral view (cf. also Figs. 273A and 274A). The superficial *trigonum lemnisci*, between the inferior colliculus, the brachium quadrigeminum inferius, and the sulcus lateralis mesencephali externus (which delimits the pes s. basis pedunculi) consists of an anterior and of a posterior subdivision. In the fairly triangular field of the anterior (or rostral) subdivision, fibers pertaining to the general lemniscus system, particularly to those of lateral spinothalamic tract, run close to the surface, covered by a thin, presumably essentially glial gray layer. Below the surface of the posterior or caudal portion of the trigone run fibers of the lemniscus lateralis, intermingled with those of lateral spinothalamic tract, and covered by the griseum of the so-called sagulum. The part of the lemniscus lateralis which does not terminate in the inferior colliculus and is augmented by fibers originating in this latter, proceeds rostrad through the brachium quadrigeminum inferius toward the diencephalic *medial geniculate body*. The sketch likewise indicates the relationship of brachium quadrigeminum superius to lateral geniculate body and to optic tract; the sketch also shows a not infrequently macroscopically recognizable *tractus peduncularis transversus*, mentioned above in connection with the nucleus opticus tegmenti. Said tract, however, displays numerous variations.

As regards the standardized nomenclature of Human anatomy (BNA, PNA), *the brachium conjunctivum* (BNA) is also termed pedunculus cerebellaris superior (PNA). The *pedunculus cerebri* comprises the tegmentum, the substantia nigra, and the '*crus cerebri*' (PNA) or '*basis pedunculi*' (BNA). The BNA distinguish a sulcus lateralis and a sulcus nervi oculomotorii delimiting the basis pedunculi, moreover a *fossa interpeduncularis (Tarini)* with recessus anterior and posterior, as well as a *substantia perforata posterior* forming the floor (or rather roof) of the fos-

---

Figure 418A. Cross-section *(Weigert stain)* through the human isthmus mesencephali at level of trochlear root and decussation (×7; red. $^3/_4$). x: part of trochlear root in cross-section; y: *Probst's commissure* (consisting of scattered fibers).

Figure 418B. Explanatory outline corresponding to preceding figure (from HAYMAKER-BING, 1956). Original sections Figures 418A–421A from former Dept. of Anat., Woman's Medical College of Pa.

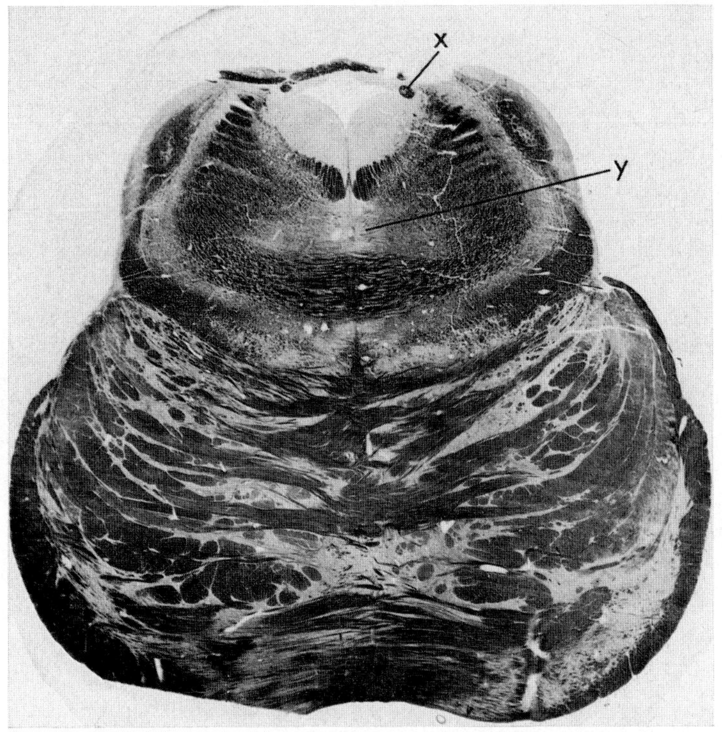
x
y

418 A

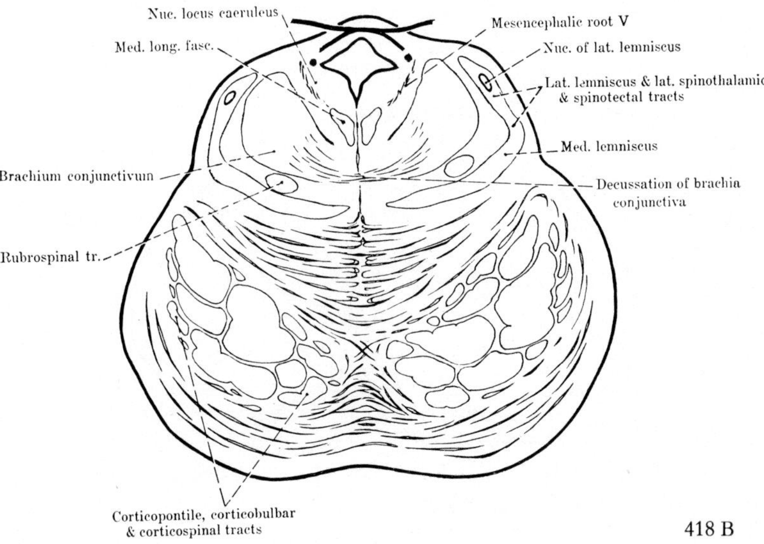
Nuc. locus caeruleus
Med. long. fasc.
Mesencephalic root V
Nuc. of lat. lemniscus
Lat. lemniscus & lat. spinothalamic & spinotectal tracts
Med. lemniscus
Brachium conjunctivum
Decussation of brachia conjunctiva
Rubrospinal tr.
Corticopontile, corticobulbar & corticospinal tracts

418 B

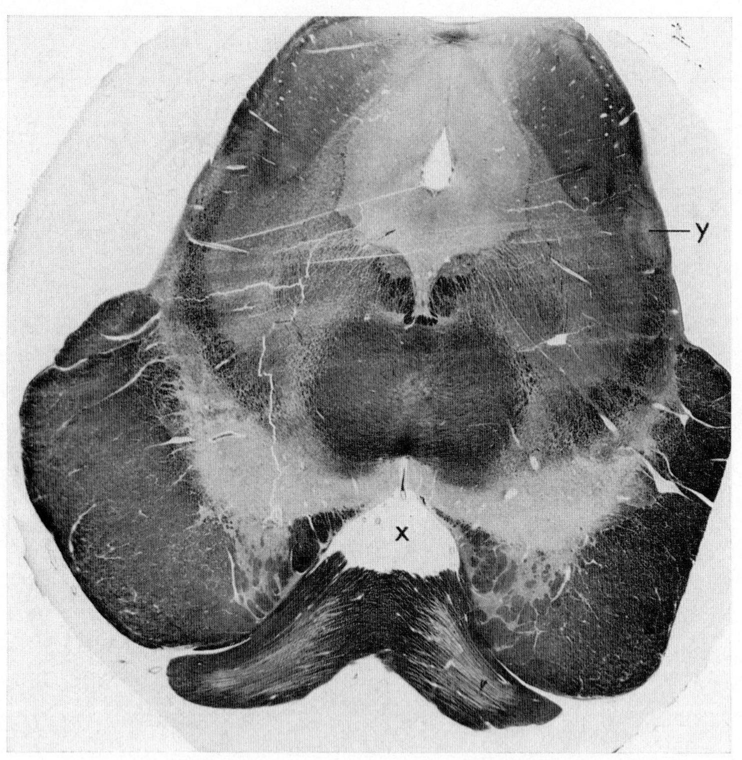
y
x

419 A

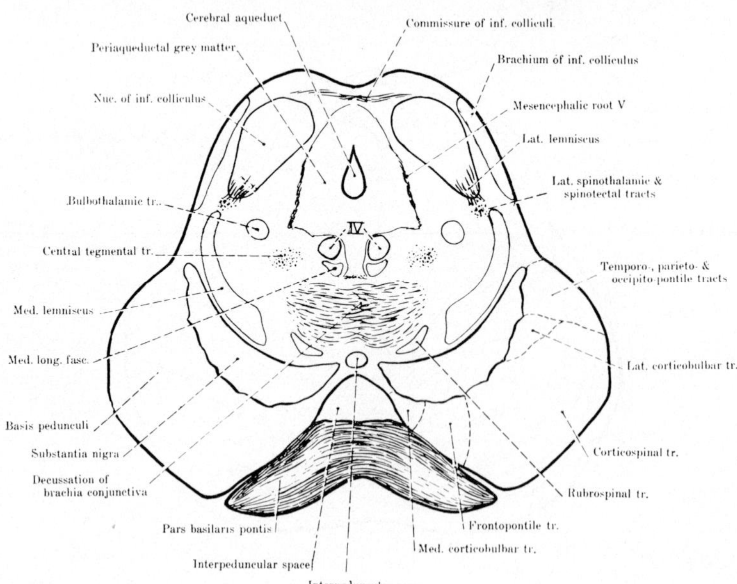
Cerebral aqueduct
Commissure of inf. colliculi
Periaqueductal grey matter
Brachium of inf. colliculus
Nuc. of inf. colliculus
Mesencephalic root V
Lat. lemniscus
Lat. spinothalamic & spinotectal tracts
Bulbothalamic tr.
IV
Central tegmental tr.
Temporo-, parieto- & occipito-pontile tracts
Med. lemniscus
Med. long. fasc.
Lat. corticobulbar tr.
Basis pedunculi
Substantia nigra
Corticospinal tr.
Decussation of brachia conjunctiva
Rubrospinal tr.
Pars basilaris pontis
Frontopontile tr.
Med. corticobulbar tr.
Interpeduncular space
Interpeduncular nuc.

419 B

sa. The PNA merely list crus cerebri, sulcus medialis cruris cerebri, and substantia perforata posterior. The *lamina quadrigemina* (BNA) becomes the lamina tecti (PNA). *Brachium quadrigeminum superius and inferius* (BNA) become brachium colliculi superioris and inferioris (PNA).

The simplified sketches of Figures 417A and C depict overall features of Human mesencephalon and superior colliculus as seen in cross-sections, while Figures 418 to 426 illustrate, in caudorostral sequence, details concerning grisea and fiber tracts at representative cross-sectional levels.

At the level of Figure 418, *the lateral spinothalamic tract*, a channel for signals related to pain and temperature, is more or less randomly intermingled with the system of lateral lemniscus. RASMUSSEN and PEYTON (1941, 1948) have outlined the course and the endings of the human lateral spinothalamic tract and medial lemniscus by means of the *Marchi method* in cases of cordotomy and tractotomy.[146]

At the level of Figure 419, the lateral spinothalamic tract tends to separate itself from the lateral lemniscus system, being completely disengaged at the level of superior colliculus (Fig. 420). Along its course to the mesencephalon, the joint lemniscus medialis system (spino- and bulbothalamic tracts) gives off fibers, perhaps mainly collaterals, to the superior colliculus, and possibly to the reticular formation.

With respect to the various grisea and tracts indicated in these figures, it will here be sufficient to recall the dubious interpretations of the so-called *bulbothalamic tract*, already dealt with in chapter IX, section 10, p. 587, and evaluated, *inter alia*, as a 'dorsal secondary ascending trigeminal tract'.[147] The unqualified designation 'bulbothalamic

[146] Further data on the results of operative procedures eliminating the pain channels can be found in the work of SPIEGEL and WYCIS (1952, 1962) dealing with Stereoencephalotomy, a surgical approach introduced into clinical neurology by the cited authors.

[147] The BNA list a lemniscus medialis and a lemniscus lateralis, while the PNA enumerate four lemnisci at midbrain levels: lemniscus spinalis, trigeminalis, medialis, and lateralis. No explanation is given for the distinction between lemniscus spinalis and medialis which remains unclarified, unless the former is meant to represent the lateral spinothalamic channel. The PNA 'lemniscus trigeminalis' is vaguely defined as including 'all the fibres which arise from the sensory nuclei of the trigeminus and ascend to reach the thalamus'.

Figure 419A. Cross-section *(Weigert stain)* through human mesencephalon at level of inferior colliculus (×7; red. $^{3}/_{4}$). x: posterior recess of fossa interpeduncularis *(Tarini)*; y: so-called sagulum, caudally blending with grisea of lateral lemniscus.

Figure 419B. Explanatory outline corresponding to preceding figure (from HAYMAKER-BING, 1956).

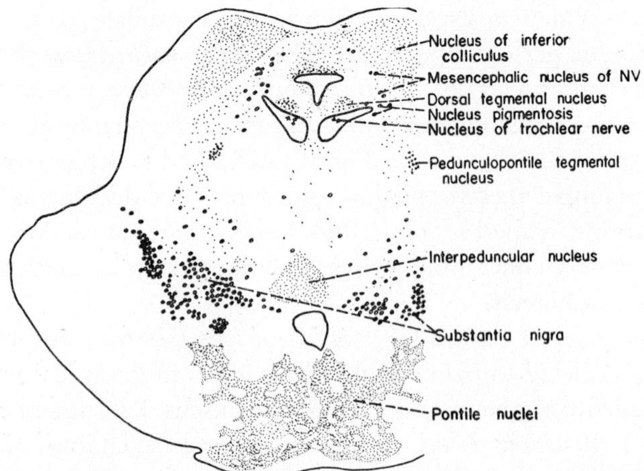

Figure 419C. Cross-section at level of inferior colliculus, approximately corresponding to preceding figures and indicating configuration of cell groups (slightly modified after JACOBSOHN, 1909, from KUNTZ, 1950). The term 'pedunculopontile tegmental nucleus' of JACOBSOHN ('nucleus tegmentalis pedunculopontinus' of OLZEWSKI-BAXTER, 1954) refers to a nondescript group of the reticular formation.

tract' does not seem particularly appropriate, since a substantial part of the general lemniscus medialis represents a well documented and generally recognized bulbothalamic tract.

Figures 420A–C depict an intermediate level of superior colliculus, and Figures 421A, B show a somewhat more rostral level, at which mesencephalic and diencephalic configurations are 'telescoped' into each other. Figure 422 illustrates the relevant differences between superior colliculus and *pretectal region.* Cross-sections through this latter are not infrequently labelled 'tectum opticum' in atlas[148] and textbook

[148] E.g. 'nucleus corporis quadrigemini anterioris' in MARBURG's (1927) atlas, and 'colliculus superior' in the detailed atlas by RILEY (1943, Section T9–480, loc. cit.). RILEY, nevertheless, depicts a nondescript 'area praetectalis' in his section T9–550, loc. cit. OLZEWSKI and BAXTER (1954) recognize part of the regio praetectalis but, in contradistinction to their detailed parcellation of brain stem grisea, do not indicate distinguishable nuclei in that region (plates XLI and XLII, loc. cit.). CLARA's textbook (1959) includes a relevant description of the pretectal region and its grisea.

Figure 420A. Cross-section *(Weigert stain)* through human mesencephalon at an intermediate level of superior colliculus (× 7; red. $^3/_4$). 1: stratum opticum; 2: stratum lemnisci; 3: stratum medullare profundum; x: striatonigral channels; y: probably nigrotegmental (caudal) channels; z: oculomotor root fibers.

Figure 420B. Explanatory outline corresponding to preceding figure (from HAYMAKER-BING, 1956).

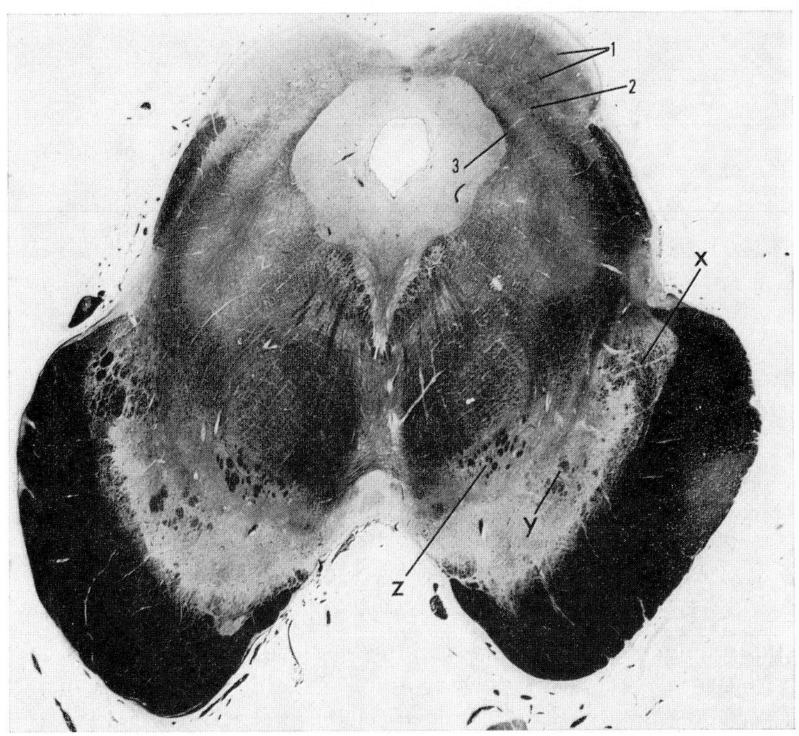
1
2
3
X
Y
Z

420 A

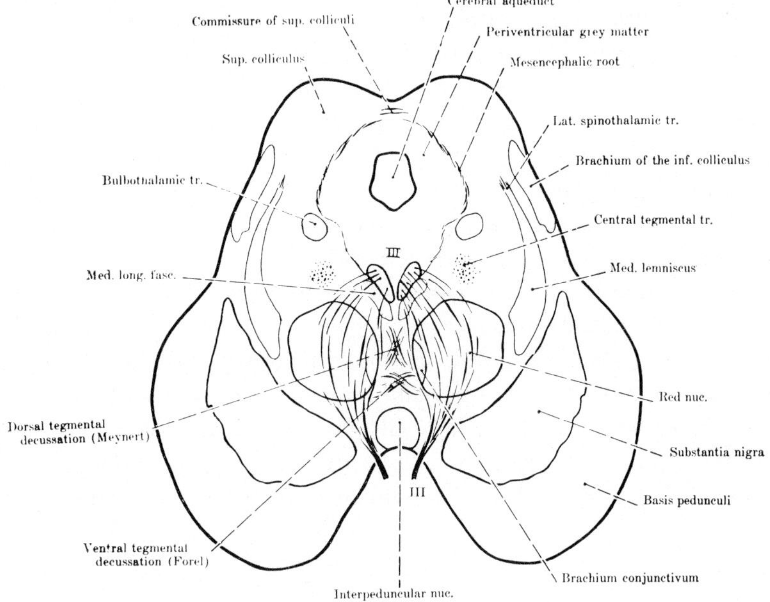
Cerebral aqueduct
Commissure of sup. colliculi
Periventricular grey matter
Sup. colliculus
Mesencephalic root
Lat. spinothalamic tr.
Brachium of the inf. colliculus
Bulbothalamic tr.
Central tegmental tr.
III
Med. long. fasc.
Med. lemniscus
Red nuc.
Dorsal tegmental decussation (Meynert)
Substantia nigra
III
Basis pedunculi
Ventral tegmental decussation (Forel)
Brachium conjunctivum
Interpeduncular nuc.

420 B

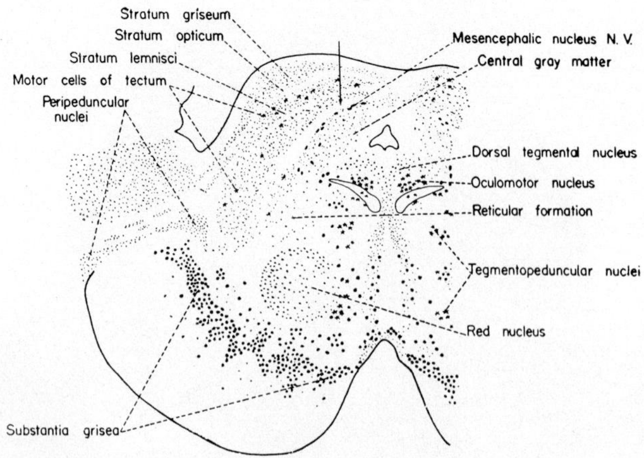

Figure 420C. Cross-section through human mesencephalon, somewhat rostral to level of preceding two figures, and indicating configuration of cell groups (slightly modified after JACOBSOHN, 1909, from KUNTZ, 1950). The so-called 'motor cells of tectum' presumably represent the 'chief cells'. Unlabelled dorsal lead: stratum medullare profundum.

figures, but a comparison between the two levels A and B will clearly show the relevant distinguishing features. Additional details of the pretectal region, to be again dealt with in chapter XII of volume 5, are illustrated in Figures 423–425.

Being rather sceptical about excessive parcellation into arbitrary cytoarchitectural subdivisions, I felt very reluctantly compelled, in the course of our own studies of the Mammalian pretectal region, to distinguish relatively numerous 'nuclei' with nondescript architectural characteristics and ill defined boundaries. Some of these 'nuclei' might indeed be questioned as another instance of excessive parcellation without intrinsic significance. There is, nevertheless, in two important respects, a considerable difference between the complexity of the pretectal region and the parcellation of neocortex or of many other grisea.[149] First, the various pretectal nuclei are rather clearly outlined in some submammalian Amniota, particularly in Birds, just as the cortical lamination of the tectum opticum is much more prominent in many lower Vertebrates than in Mammals. Again, in lower Mammals, such e.g. as the Rabbit, the pretectal nuclei are frequently more distinct than in Man. Since it is possible to recognize in the human pretectal region

[149] Cf. K., 1954, p.62f.

all the component neighborhoods present in lower forms, their identification seems of significance for comparative anatomy. Second, while excessive cortical, thalamic, and other nuclear parcellation represents an arbitrary drawing of boundaries within a 'continuum', even the ill-defined human pretectal 'nuclei' have very definite topographic rela-

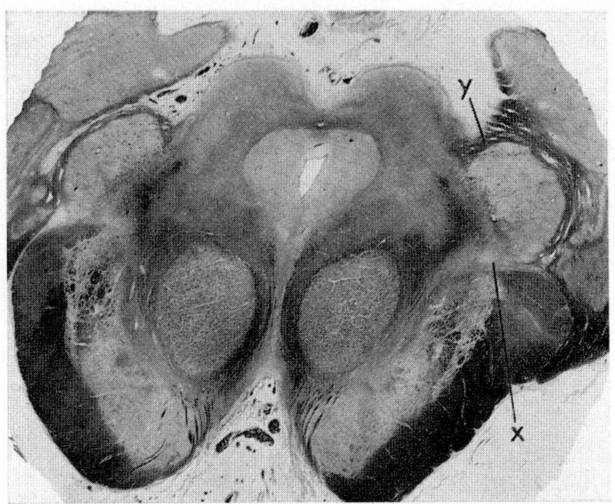

Figure 421 A. Cross-section *(Weigert stain)* through human mesencephalon at a rostral level of superior colliculus (× 6; red. $^1/_2$). x: 'peripeduncular' grisea; y: brachium colliculi superioris, with optic tract fibers, entering superior colliculus.

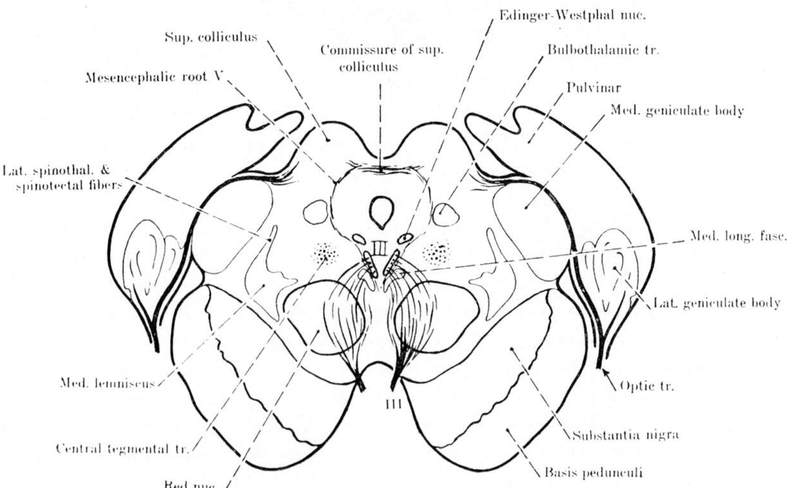

Figure 421 B. Explanatory outline corresponding to preceding figure (from HAYMAKER-BING, 1956).

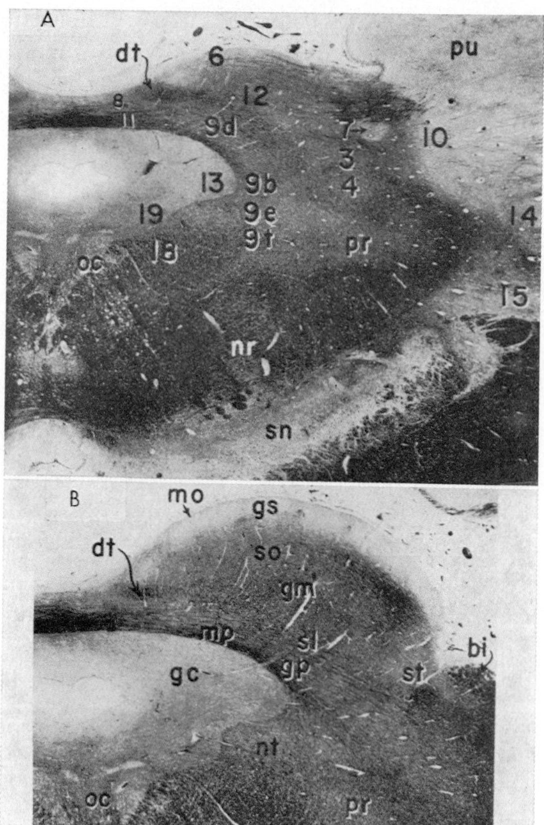

Figure 422. Cross-sections *(Weil's myelin stain)* through pretectal region (A) and superior colliculus (B) of the adult human brain, illustrating the distinguishing features of these levels (from K. and MILLER, 1949). bi: brachium quadrigeminum inferius; dt: tractus tectothalamicus dorsomedialis; gc: griseum centrale; gm: stratum griseum medium; gp: stratum griseum profundum; gs: stratum griseum superficiale; mo: stratum moleculare sive zonale; mp: stratum medullare profundum; nr: nucleus ruber; nt: tegmental griseum apparently related to commissura tecti ('nucleus tegmentalis commissurae tecti'); oc: oculomotor grisea; pr: nucleus profundus (tegmentalis) mesencephali; pu: pulvinar; sl: stratum lemnisci; sn: substantia nigra; so: stratum opticum; st: tractus spinothalamicus (lateralis); 3: nucleus praetectalis (principalis); 4: nucleus posterior (thalami); 6: nucleus lentiformis mesencephali (pars parvocellularis); 7: nucleus olivaris colliculi superioris; 8: area praetectalis (propria); 9b: nucleus interstitialis (principalis) commissurae posterioris; 9d: nucleus centralis subcommissuralis com. post.; 9e: nucleus interstitialis magnocellularis com. post.; 9t: nucleus interstitialis tegmentalis com. post.; 10: nucleus suprageniculatus; 11: area praetectalis (pars supra- and intracommissuralis); 12: nucleus sublentiformis; 13: nucleus lateralis grisei centralis; 14: nucleus geniculatus medialis (dorsalis); 15: nucleus geniculatus medialis (ventralis); 18: nucleus interstitialis fasciculi longitudinalis medialis; 19: *nucleus of Darkschewitsch.*

tionships to distinguishable components of the very complex fiber systems in the pretectal region. These fiber masses, although at least up to now inextricable[150] *qua* specific details, are: *posterior commissure*, *pretectal* and *tegmental fields*, *tractus tectothalamicus dorsomedialis*, and *brachium quadrigeminum superius* with associated radiations.

As regards the *oculomotor grisea*, a subdivision into three cell groups is generally accepted. The rostral and dorsomedial *Edinger-Westphal nucleus* lies very close to the central gray. Said nucleus represents the general visceral efferent, parasympathetic, preganglionic component of the oculomotor complex, and is presumed to mediate pupillary constriction as well as accommodation. A rostral, an intermediate, and a more caudal portion have been described.[151] Individual variations, such as midline fusions of rostral and caudal parts seem to occur. There remains some doubt as to the delimitation of the rostral portion from the 'nucleus medialis anterior'.

The somatic efferent *lateral large celled oculomotor nucleus* is the most extensive subdivision. An additional, rather caudal, and individually variable large-celled subdivision is the median *nucleus of Perlia*[152] whose significance as a nucleus of convergence remains controversial.[153]

Numerous authors have attempted, with widely conflicting results, to describe a localization pattern in the Mammalian and Human oculomotor nucleus. Figure 426A depicts a once widely accepted and possibly still valid interpretation by BERNHEIMER (1897) supported by BROUWER (1918), while Figure 426B shows a more recent concept elaborated by WARWICK (1964) and essentially based on investigations in the Rhesus monkey. Further details on this problem, with bibliographic references, can be found in the publications by OLZEWSKI and BAXTER (1954) and by CROSBY *et al.* (1962). These latter authors comment as follows on the obtaining discrepancies: 'It is surprising that there should be such a variation of pattern in mammals if the differ-

---

150 Inextricable regardless of the hitherto applied techniques: normal myelin stains in serial sections, experimental procedures with tigrolysis, *Marchi*, *Glees*, *Nauta-Gygax* or *Fink-Heimer techniques*, biochemical fluorescence studies, stimulations respectively evoked potentials.

151 OLZEWSKI and BAXTER (1954), who use a slightly different terminology.

152 Described by this author in Bd. 35, Abt. 4, p. 287 of *Graefe's Arch. f. Ophthalm.*, 1889 (quoted after KAPPERS *et al.*, 1936).

153 *Perlia's median large-celled nucleus* may occasionally be easily confused with a more rostral median variant of the *Edinger-Westphal small-celled nucleus*. These unpaired midline nuclei are also referred to as the 'central oculomotor nuclei'.

ences are due only to the animals studied. It is of course possible, although not likely, that there is no fixed pattern, and that such pattern as can be demonstrated in an animal is due to chance association of the peripheral oculomotor roots with a muscle. Actually, until several independent observers have demonstrated a common pattern, a final decision as to which is the correct one cannot be made.'

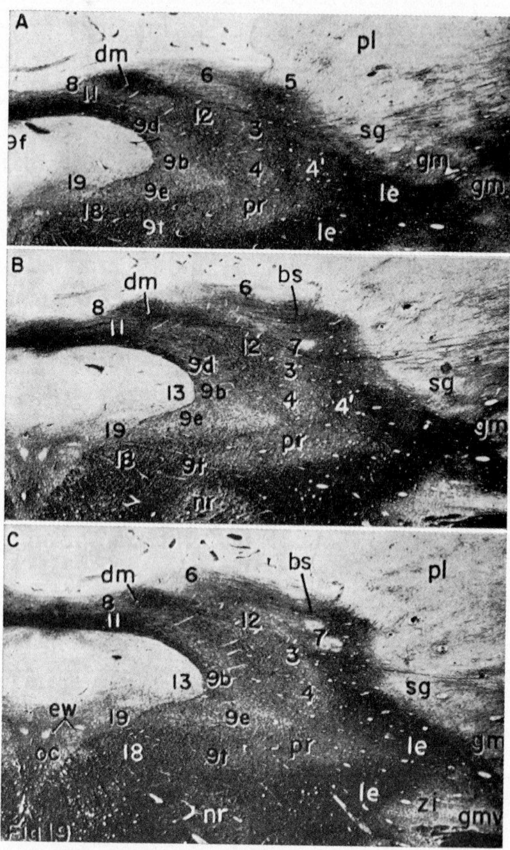

Figure 423. Cross-sections (myelin stain) through caudal part of adult human pretectal region, showing unfolding of nucleus olivaris colliculi superioris (× 9; red. $^1/_2$; from K., 1954). bs: brachium quadrigeminum superius; dm: tractus tectothalamicus dorsomedialis; ew: *Edinger-Westphal nucleus;* gm: corpus geniculatum mediale (dorsale); gmv: corpus geniculatum mediale (ventrale); le: lemniscus medialis and associated systems; pl: pulvinar; pr: nucleus profundus (tegmentalis) mesencephali; zi: zona incerta; 4′: nucleus posterior (thalami), pars lateralis; 5: nucleus lentiformis mesencephali, pars magnocellularis; 7: nucleus olivaris colliculi superioris; 9f: nucleus medianus infracommissuralis commissurae posterioris. Other notations as in Figure 422. A–C in rostrocaudal sequence at about 120 μ distance.

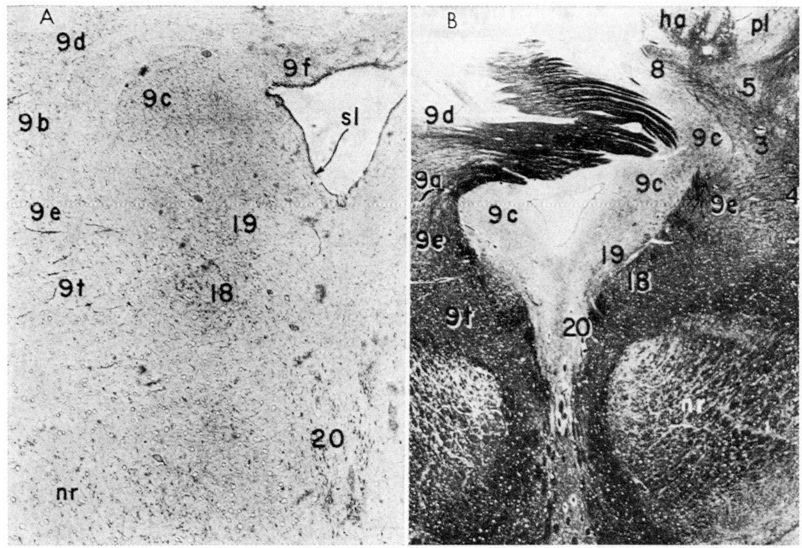

Figure 424. Two sections *(Nissl-* and *Weil stains)* through rostral part of adult human pretectal region, showing central gray and nuclei of posterior commissure (from K., 1954). ha: ganglion habenulae; sl: recognizable sulcus limitans; 9a: nucleus interstitialis commissurae posterioris, pars anterior; 9c: nucleus centralis subcommissuralis commissurae posterioris; 20: nucleus medianus anterior. Other notations as in Figures 422 and 423. Section B is rostral to section A and includes caudal end of habenular griseum at right.

In this respect it seems of interest that experiments by MARINA (1915) have demonstrated the recovery of normal conjugate movements of the eye bulbs in Monkeys whose external and internal recti muscles were severed and re-attached in an interchanged position, thus reversing their effect. Subsequent authors performed additional experiments with similar results which confirmed the here obtaining *ultrastability* of at least some Mammalian central neural mechanisms.[154]

Reverting to the course of certain fiber systems through the human diencephalon, the lateral position of the temporo-(occipito)-pontile tract and the medial location of the pontopontile tract in the pes pedunculi, as shown in Figure 427A, should be pointed out. The corticospinal and corticobulbar (pyramidal) system is located between the just mentioned tracts, but a medial corticobulbar component tends to shift toward the medial aspect of the frontopontile tract.

---

[154] Ultrastability in ASHBY's formulation, and including a reference to MARINA's experiments, is discussed in K., 1957, pp. 138–142.

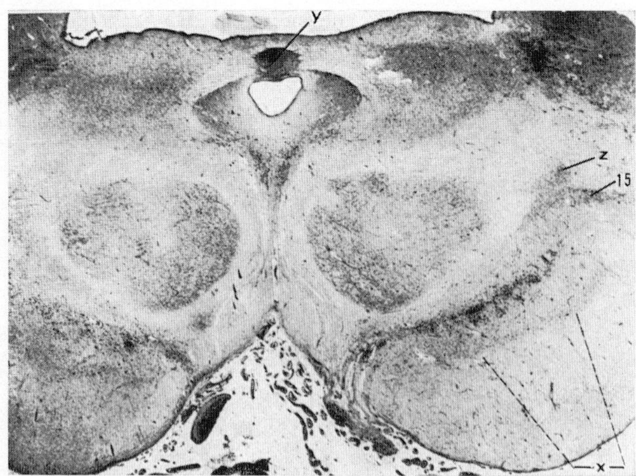

Figure 425A. Cross-section *(Nissl stain)* through adult human pretectal region and tegmentum (courtesy of Prof. Hugo Spatz). 15: nucleus geniculatus medialis (thalami) ventralis; x: ventral border of zona reticulata substantiae nigrae; y: subcommissural organ; z: probably rostral end of lateral prerubral tegmentum (15, y, z have been added to Spatz' original figure).

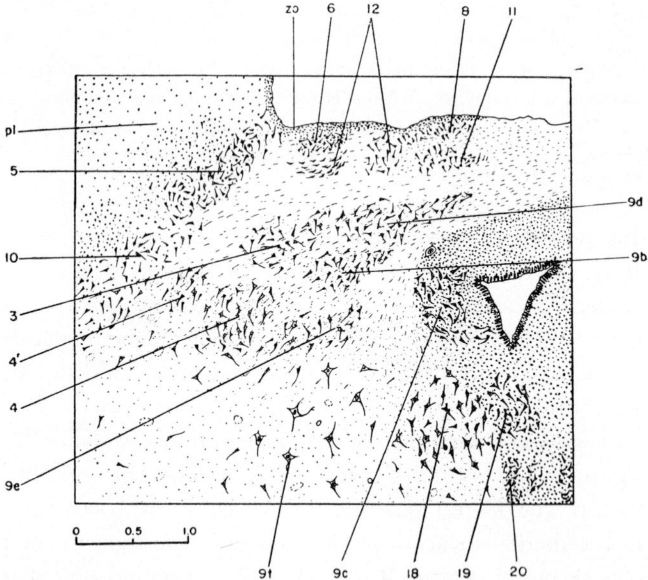

Figure 425B. Semidiagrammatic drawing of cytoarchitecture in adult human pretectal region (from K., 1954). zo: superficial intrazonal cell plate. Other notations as in preceding Figures 422, 423, 424. The cytoarchitectural grouping of pretectal cell masses can be recognized with sufficient clarity only in thick sections of 40–60 and more micra. In this orientation sketch, drawn by Dr. Ruth Miller in 1949 and subsequently published in 1954, the zones of separation between some of the pretectal groups are overemphasized for greater diagrammatic distinctiveness.

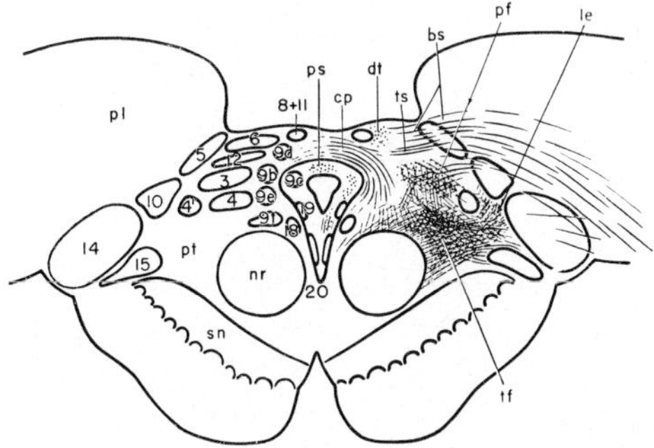

Figure 425C. Simplified diagram of the human pretectal region (schematic cross-section). The left side illustrates topography of cell masses, and right side shows topography of pretectal fiber systems (slightly modified after K. and MILLER, 1949, from K., 1954). bs: brachium quadrigeminum superius; cp: commissura posterior; dt: tractus tectothalamicus dorsomedialis; pf: pretectal field; ps: periventricular fibers; pt: prerubral tegmentum; tf: tegmental fiber field lateral to nucleus ruber; ts: tractus sublentiformis. Other notations as in preceding Figures 422–424.

In the *pes pedunculi*, the corticospinal fibers for the lower extremity are believed to be most laterally located, while those for trunk seem to run medially adjacent, and the still more medial channel for upper extremity becomes laterally adjacent to the medial corticobulbar respectively to the frontopontile tract.

According to DEJERINE (1895–1901; 1914), one component of the medial corticobulbar tract accompanies the corticospinal channel as far as the pars basilaris pontis and the pyramids of medulla oblongata. Another component, forming the system of aberrant pyramidal fibers, leaves the main channel at successive levels of the brain stem and enters the tegmentum, running caudad through the region occupied by portions of the medial lemniscus system (cf. Fig. 427B). These aberrant fibers seem to undergo an incomplete decussation in the raphe before reaching the efferent cranial nerve nuclei.[155] The aberrant fibers appear to be grouped in small bundles, the decussating ones crossing over at the level of their terminal destination. Considerable individual variations of this 'aberrant' system seem to obtain, and additional variations may be displayed by the main grouping of fibers into lateral and medial corticobulbar tracts and by the assemblage within the corticospinal system.

[155] This involves dubious questions of homolateral, contralateral, or bilateral 'supranuclear innervation' pointed out in chapter IX, p. 542 *et passim*.

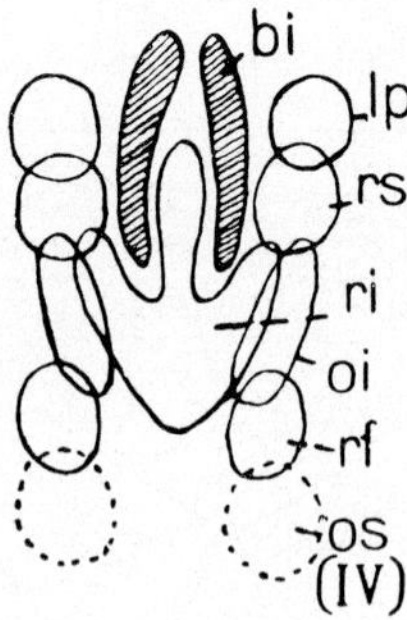

Figure 426A. Diagram of functional localization in the human oculomotor complex according to Brouwer (simplified after Brouwer, 1918, from K., 1927). bi: smooth musculature of eyebulb (Binnenmuskeln), Edinger-Westphal nucleus; lp: levator palpebrae superioris; oi: obliquus inferior; os: obliquus superior (nucleus nervi trochlearis); rf: rectus inferior; ri: rectus internus (s. medialis); rs: rectus superior.

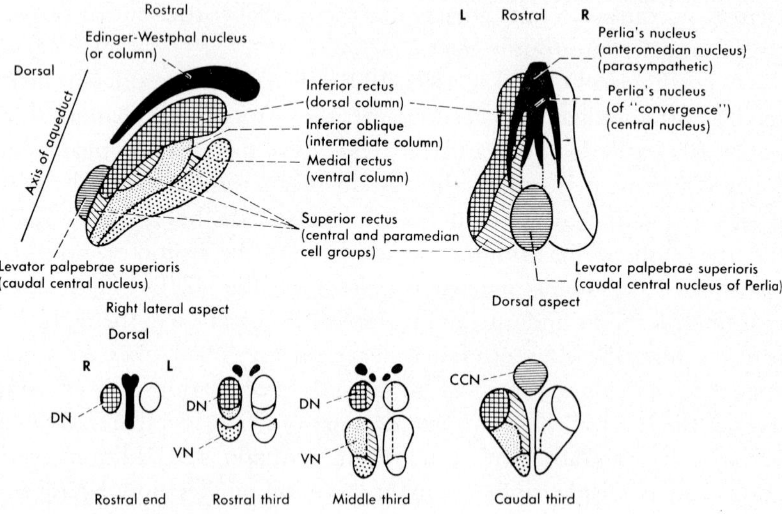

Figure 426B. Diagrams of functional localization in Primate oculomotor complex according to Warwick (after Warwick, 1964, from Haymaker-Bing, 1969). CCN: caudal central nucleus *(of Perlia)*; DN: dorsal magnocellular subdivision; VN: ventral magnocellular subdivision.

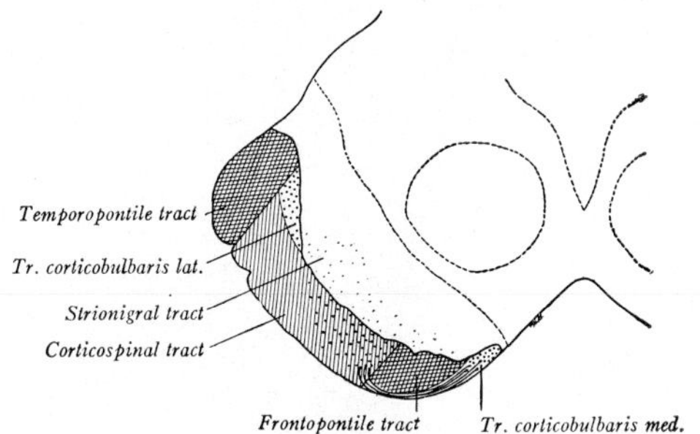

Figure 427A. Diagrammatic sketch of cross-section through pes pedunculi; showing location of the relevant fiber tracts (from RANSON, 1943).

Figure 428A shows a widely accepted interpretation of the human *lateral lemniscus system* ending in the medial geniculate body, from which neural signals are relayed to the telencephalic cortex of the temporal lobe. Although the mammalian and human lemniscus lateralis is doubtless a predominantly cochlear channel, the presence, within that system, of some vestibular fibers reaching the medial geniculate grisea cannot be excluded. The vestibular channels ascending to prosencephalic grisea are poorly understood. Thus a pathway running through the reticular formation and perhaps identical with the so-called 'bulbothalamic tract' ('dorsal ascending secondary trigeminal tract') has been suggested by HASSLER (1959).

Figure 428B illustrates tentatively assumed overall relationships of the *fasciculus longitudinalis dorsalis of Schütz*, another poorly understood channel which runs through the central gray and presumably pertains, entirely or predominantly, to the central subdivision of the vegetative (or autonomic) nervous system, probably containing ascending as well as descending fibers of various length and destination. At present, substantial evidence indicates that the fasciculus longitudinalis dorsalis includes components of the ascending 'catecholamine pathway' mentioned above on p. 953, and again to be dealt with in chapter XII of volume 5.

Anatomical features of the human mesencephalon related to *clinical symptoms* are illustrated in Figures 429 and 430. Lesions causing *hemiplegia alterna oculomotoria*, comparable to the 'crossed paralyses' dealt

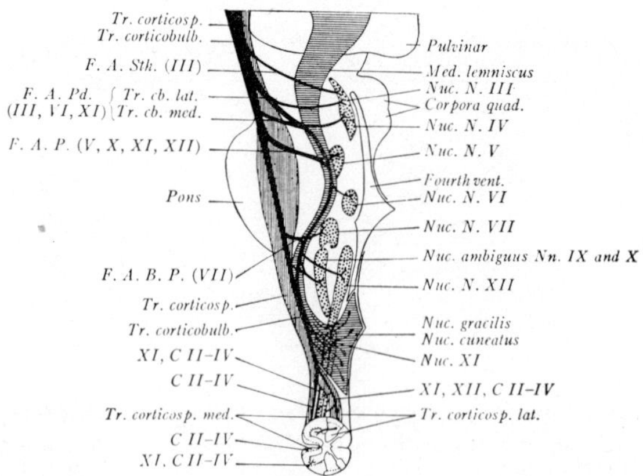

Figure 427B. Diagrammatic sketch of brain stem with adjacent diencephalic and spinal neighborhoods in longitudinal aspect, showing the course of corticobulbar tract components (redrawn after Dejerine, from Ranson, 1943). F.A.B.P.: bulbopontile aberrant fibers; F.A.P.: aberrant fibers in pons; F.A.Pd.: aberrant fibers in pes pedunculi; F.A.Sth.: 'subthalamic aberrant fibers'; Tr.cb.lat.: tractus corticobulbaris lateralis; Te.cb.med.: tractus corticobulbaris medialis; solid black: corticobulbar tract with some (cervical) corticospinal components; vertical hatching: corticospinal tract; horizontal hatching: medial lemniscus. Roman numerals indicate nuclei of cranial nerves and cervical spinal nerves supplied by the various bundles. Other abbreviations self-explanatory.

with in section 10 of chapter IX, are depicted by Figures 429 and 430D. In both instances, a left lower neuron oculomotor paralysis is combined with a right spastic hemiplegia (upper neuron paralysis).

In the case of Figure 429, the contralateral paresis involved the lower subdivision of the *facial nerve*[156] and a weakness, without atrophy, of tongue and palate musculature. On the homolateral side, the ptosis is due to paralysis of musculus levator palpebrae superioris. If the lid is lifted by the examiner, the eyebulb is seen turned outward and slightly downward, indicating a paralysis of all extrinsic eye muscles except rectus lateralis and obliquus superior.[157] The pupil shows dila-

---

[156] The sparing of the upper facial subdivision (rami temporales et zygomatici) is presumably related to bilateral 'supranuclear innervation' (cf. preceding footnote 155).

[157] Deviations of the eye bulb which the patient cannot overcome are designated as *strabismus* (squint) or *tropias*, such e.g. as esotropia (convergent strabismus). Strabismus, particularly at its onset, is commonly accompanied by *diplopia* (double vision). This subjective symptom can easily be experienced by gently displacing the position of one eyebulb with the fingers. It also occurs for the blurred distant objects if one accommodates for a near fixation point.

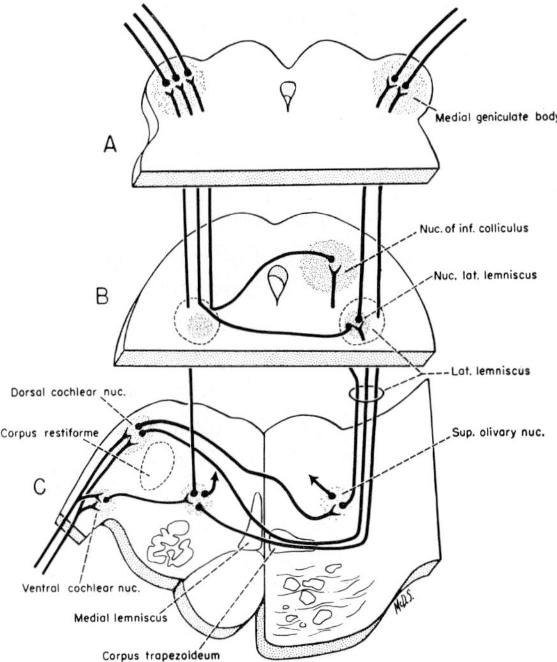

Figure 428 A. Diagrammatic sketch showing some widely accepted features of the ascending cochlear channel respectively lateral lemniscus system (from HAYMAKER-BING, 1956). A: level of superior colliculus and medial geniculate body; B: level of inferior colliculus; C: bulbopontine level.

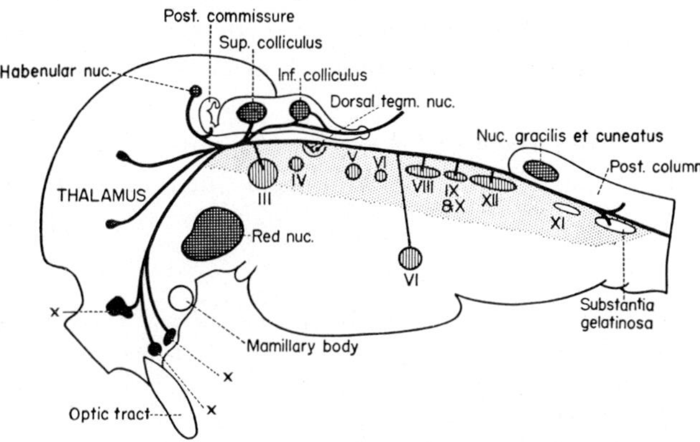

Figure 428 B. Diagrammatic sketch showing a tentative concept concerning overall features of the *fasciculus longitudinalis dorsalis of Schütz* (from HAYMAKER and K., 1955). X: paraventricular hypothalamic grisea. It can be assumed that the *bundle of Schütz* is a complex channel with fibers of diverse length as well as descending and ascending components.

tion *(mydriasis)*, and accommodation is abolished. The symptoms caused by the lesions depicted by Figures 430A–D are listed in the tabulation Figure 430E.

With regard to the *pupillary reflexes*, briefly dealt with in section 1 of the present chapter, it is evident that the musculus constrictor pupillae may be paralyzed by a lesion involving the relevant parasympathetic outflow at any neighborhood between its termination in the eye bulb and its origin in the *Edinger-Westphal nucleus*. Much the same applies to accommodation performed by the musculus ciliaris, whose preganglionics likewise seem to originate in caudal portions of the just mentioned nucleus.[158]

Paralysis of the musculus dilator pupillae *(miosis)*[159] caused by involvement of the sympathetic preganglionic channel was briefly dealt with in footnote 166 of chapter VIII (spinal cord). Miosis, however, may also result from damage to central 'iridodilator' fibers originating in the midbrain and, to some extent, generally occurs as part of the *Argyll Robertson symptom* discussed further below. It can also be noted in some lesions of pons and oblongata ('pin-point pupils' in pontine hemorrhage and miosis in thrombosis of posterior inferior cerebellar artery).

The mediation of the *pupillary light reflex* through the *pretectal grisea* and the participation of the posterior commissure, as established by Spiegel and others in experimental animals was mentioned above on p. 806 and 937. With regard to the human brain, these results were corroborated by our own findings in an analysis of clinicopathologic aspects of fatal missile-caused cerebral injuries (K. and MILLER, 1949; K., 1954).[160]

The term '*reflex iridoplegia*' designates a failure of the pupil to react to light. BRAIN and WALTON (1969) point out that the term '*Argyll*

---

[158] This portion may be unpaired median, and may be confused with *Perlia's large celled nucleus*, which is still somewhat more caudally located (cf. above footnote 153).

[159] *Miosis* from μείων, compar. of μικρός; also μειοῦν, to reduce, make smaller. In German, it should be spelled *Meiosis*.

[160] A condensed and partial report of this analysis was published in volume I, Neurosurgery, of the official history of the U.S. Army Medical Department in World War II (CAMPBELL, KUHLENBECK *et al.*, 1958) which merely contains a passing reference to pretectal structures on p. 300. However, details of our findings concerning damage to the pretectal region were included in K. and MILLER, 1949. Limitations of space precluded their discussion in the cited official report.

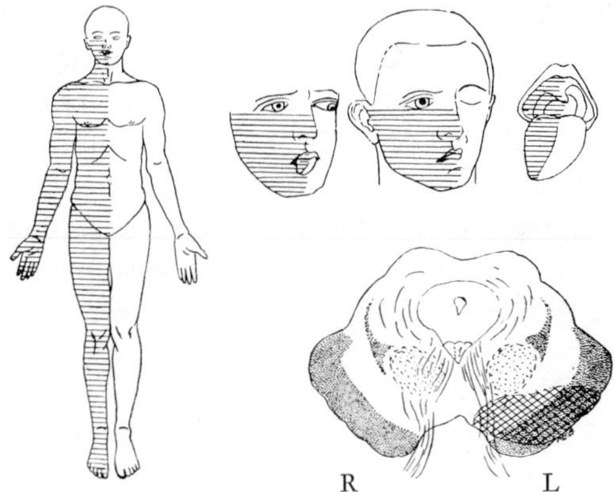

Figure 429. Diagrams illustrating site of lesion and extent of paralysis respectively paresis in a case of hemiplegia alterna oculomotoria *(Weber's syndrome)* involving the left oculomotor root. The external deviation of eye-bulb and the ptosis on left side should be noted (from RANSON, 1943).

Robertson pupil' should be restricted to a special form of iridoplegia[161] in which the pupil is small, constant in size, and unaltered by light or shade; it contracts promptly and fully on convergence and dilates again promptly when the effort to converge is relaxed; it dilates slowly and imperfectly to mydriatics. Very rarely the *Argyll Robertson pupil* reacts paradoxically to light by a slight dilatation. Most of these features had been described many years before ROBERTSON by ROMBERG (cf. BRAIN-WALTON, 1969).

Additional pupillary reactions, namely that on *accommodation* and *convergence*, as well as the *pupillary-skin reflex*, were pointed out in section 1 of this chapter.

As regards the *substantia nigra*, whose relationship to the so-called 'extrapyramidal' motor system is well established, and which is also substantially involved in the disturbances designated as *parkinsonism*, some comments on that condition appear appropriate in the present context.

[161] The *Argyll Robertson pupil* commonly occurs in luetic affections of the central nervous system such as tabes dorsalis and general paresis. In some of these instances, however, this symptoms may barely or not at all be present.

The *Parkinsonian syndrome* is characterized by disturbances of motor function involving a particular type of tremor, muscular rigidity, and slowing or enfeeblement of emotional and voluntary movements. It is named after JAMES PARKINSON (1755–1824) who first described paralysis agitans *(Parkinsons's disease)* in 1817. The tremor, characterized by rhythmic alternating movements of opposing muscular groups commonly displays peculiar finger motions with participation of the thumb in the typical so-called 'pill-rolling' pattern. A stooped, shuff-

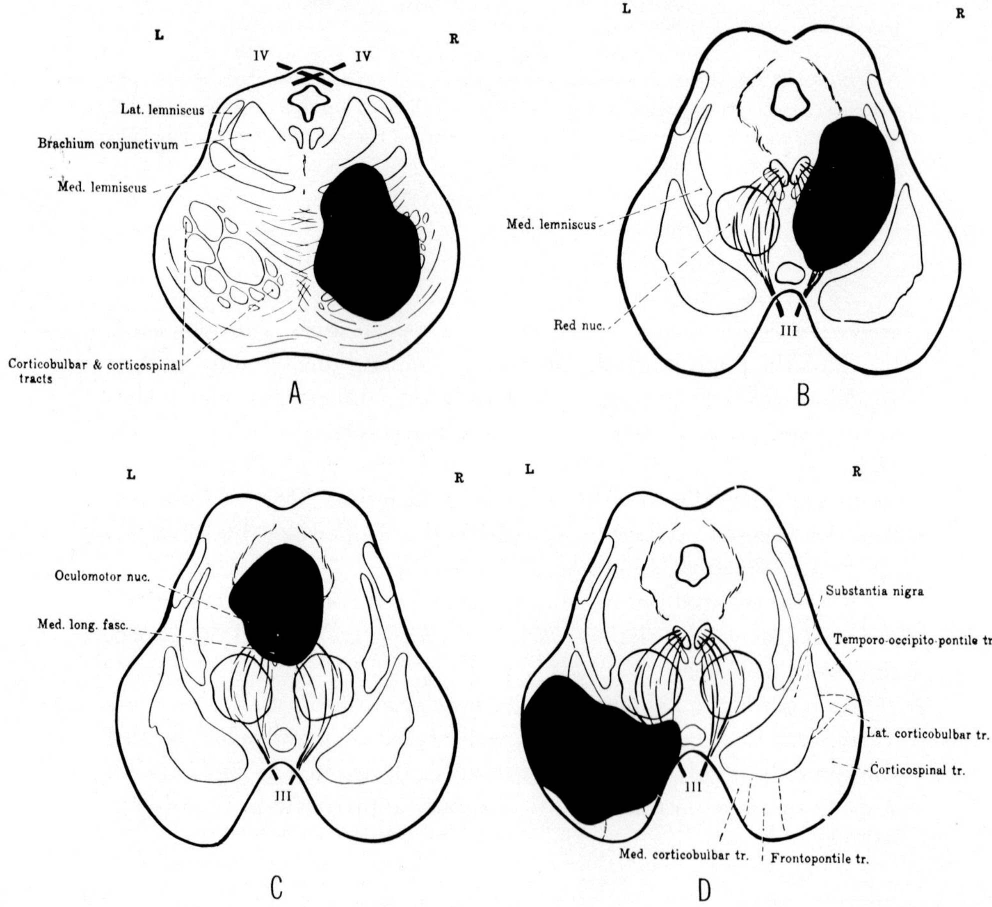

Figure 430 A–D. Diagrams of lesions involving isthmic and mesencephalic structures (from HAYMAKER-BING, 1956).

		Left	Right
A	Head	Weakness but no atrophy of face and tongue	
	Trunk and Limbs	Spastic hemiplegia, reduction in deep sensibility	
B	Head	Trigeminal hypesthesia or anesthesia	Oculomotor nerve palsy
	Trunk and Limbs	Hemianesthesia, choreoathetotic movements	
C	Head	Bilateral ophthalmoplegia	
	Trunk and Limbs		
D	Head	Oculomotor nerve palsy Weakness but no atrophy of face and tongue	
	Trunk and Limbs		Spastic hemiplegia

Figure 430E. Tabulation listing clinical signs resulting from the lesions illustrated by Figures 430A–D (from HAYMAKER-BING, 1956).

ling gait with festination (involuntary tendency to increase speed in walking) is likewise a common symptom[162]. An instant diagnosis *(Augenblicksdiagnose)* upon first sight of the patient is thus frequently possible if the condition is conspicuously developed.

Mental changes are not necessarily present in *parkinsonism*, but do occur in various instances of this condition. Emotional experience may be quite unimpaired behind a 'mask-like' facial expression (cf. also BRAIN-WALTON, 1969).

Parkinsonism is one of the most frequent 'extrapyramidal' disorders and may be caused by a number of different pathologic states producing lesions in corpus striatum, globus pallidus and particularly in substantia nigra. The condition is chronic and usually progressive.

Moreover, in addition to the 'extrapyramidal disturbances',

---

[162] Cf. also the biographic sketch of PARKINSON in HAYMAKER's 2nd edition of 'Founders of Neurology' (1970).

various symptoms related to disturbances of the vegetative nervous system are commonly displayed in *parkinsonism*. These symptoms include excessive salivation as well as flushing of the skin, and, rather rarely, cyanosis and edema. There is frequently also an unusual tolerance of, and indifference to, cold, while heat is less well tolerated. This 'vegetative involvement' suggests close interrelationships between 'extrapyramidal' and 'vegetative' 'centers' in full accordance with our own views concerning a suitable subdivision of the human hypothalamus (K. and HAYMAKER, 1949; K., 1954). It will be recalled that we distinguish here (1) a dorsal and entopeduncular griseal group, (2) an anterior, (3) a middle, and (4) a posterior griseal group, the three latter ones being mainly related to the autonomic (or vegetative) nervous system. The dorsal and entopeduncular group, consisting of globus pallidus, nucleus entopeduncularis, nucleus ansae lenticularis, nucleus of the inferior thalamic peduncle, and *nucleus subthalamicus of Luys*, was interpreted to be largely associated with the 'extrapyramidal motor system' including the mesencephalic substantia nigra, without, nevertheless, precluding close interrelations, pointed out in our reports, with the predominantly 'vegetative centers' of the three griseal groups commonly conceived as hypothalamus *sensu strictiori*.

Since about 15 years ago, relevant findings concerning the *neurochemistry* of *parkinsonism* were made by a number of investigators (cf. the recent report by LLOYD and HORNYKIEWICZ, 1973). It became established that *dopamine* (3,4-dixydroxyphenylethylamine) has a characteristic distribution within the Mammalian brain. Appromixately 80% of the total dopamine in the brain was found in the subcortical 'extrapyramidal' grisea, notably in nucleus caudatus, globus pallidus, and substantia nigra. In patients with *Parkinson's disease* the concentration of dopamine and its major metabolite homovanillic acid becomes greatly decreased, presumably as a result of degeneration in the 'nigro-striatal' dopaminergic neuronal system. CARLSSON (1959) noted that, in Rabbits, reserpine depletes the 'corpus striatum' of its principal amine (dopamine), and may produce 'extrapyramidal' symptoms similar to those of parkinsonism. This effect could be reversed by administering l-dopa (the amino acid levodihydroxyphenylanaline) which is the immediate 'precursor' of dopamine. Subsequently, the administration of l-dopa was introduced for the treatment of patients with parkinsomism and proved to be at least partially successful. This therapy seems more effective particularly *qua* relief of rigidity and bradykinesia than any other medication, but cannot be considered entirely reliable, quite apart from not infrequent undesirable side-effects.[163] The treatment seems less effective in the relief of tremor, which is frequently completely abolished by the surgical procedures of stereoencephalotomy introduced by SPIEGEL and WYCIS (1952, 1962). These methods, however, have their own drawbacks, particularly if undertaken by eager and uncritical neurosurgeons.[164]

The microscopically detectable changes in substantia nigra and other grisea involved in *parkinsonism* include 'ballooning' of nerve cells with presence of lipochrome granules, neurofibrillar tangles of *Alzheimer type*,[165] and '*Lewy-type*' *inclusion bodies*.[166]

As regards the various forms of *parkinsonism*, the so-called 'idiopathic' primary degeneration of *Parkinson's disease sensu stricto* (shaking palsy, paralysis agitans) generally has its onset in late middle life. *Postencephalitic parkinsonism* is a not uncommon sequela of encephalitis lethargica. Other instances seem to represent *arteriosclerotic parkinsonism*, usually complicated by other features of the vascular lesions. Severe head injuries, carbon monoxide and manganese poisoning may also produce *parkinsonian symptoms*.

The *substantia nigra*, like other grisea of the neuraxis, displays thus a '*selective vulnerability*, or '*pathoclisis*' (C. and O. VOGT, 1929). In this respect, it is of interest that the Californian star thistle is not poisonous to Sheep, but is toxic for Horses, whose substantia nigra and globus pallidus become selectively affected (INNES and SAUNDERS, 1962; ZEMAN and INNES, 1963).[167] There is little doubt that morphologically re-

---

[163] These, more or less unpredictable, i.e. apparently randomly occurring side effects include hematologic disturbances (i.e. granulocytopenia), nausea, anorexia, and vomiting, cardiac irregularities, orthostatic hypertension, moreover involuntary choreiform, dystonic and other adventitious movements. Psychiatric side-effects can be manifested by agitation, anxiety, depression with suicidal tendencies, confusion, hallucinations, nightmares, insomnia, paranoid episodes, and increased libido with 'bizarre sexual behavior'. A synergistic effect on the clinical symptoms was occasionally noted upon simultaneous use of l-dopa and anticholinergic drugs.

[164] The cautious and critical attitude displayed by the cited inaugurators of stereoencephalotomy deserves here special emphasis.

[165] Cf. volume 3, part I, chapter V, section 9, p. 722.

[166] Cf. volume 3, part I, chapter V, section 8, p. 665.

[167] It is of interest that the recent studies in neurochemistry seem to indicate substantial differences in the occurrence respectively distribution of catecholamines and related substances within the neuraxis of various animals. Thus, according to CARLSSON (1959) no dopamines but only adrenaline and noradrenaline was detected in the brain of Amphibians. It should be added that 'pathoclisis' seems to be much more conspicuously displayed by some grisea than by others.

spectively histologically definable grisea represent not only morphologic but also biologic units (*'topistic units'* in the VOGTS' terminology) with functional characteristics (including pathoclisis) correlated with their ontogenetic and phylogenetic origin resulting from morphogenetic field properties (cf. also K., 1954, p. 63–65).

10. References to Chapter XI

ABBIE, A.A.: The brain stem and cerebellum of Echidna aculeata. Philos. Trans. roy. Soc. London Ser. B. vol. *224:* 1–74 (1934).

ADAM, H.: Der III. Ventrikel und die mikroskopische Struktur seiner Wände bei Lampetra (Petromyzon) fluviatilis und Myxine glutinosa L. nebst einigen Bemerkungen über das Infundibularorgan von Branchiostoma (Amphioxus) lanceolatum Pall; in KAPPERS Progress in Neurobiology, pp. 146–158 (Elsevier, Amsterdam 1956).

ALLEY, K. E.: Quantitative analysis of the synaptogenic period in the trigeminal mesencephalic nucleus. Anat. Rec. *177:* 49–60 (1973).

APTER, J.T.: Projection of the retina on superior colliculus of cats. J. Neurophysiol. *8:* 123–134 (1945).

BECCARI, N.: Neurologia comparata anatomo-funzionale dei Vertebrati, compreso l'uomo (Sansoni, Firenze 1943).

BECHTEREW, W. VON: Über eine bisher unbekannte Verbindung der grossen Oliven mit dem Grosshirn. Neurol. Centralbl. *4:* 194–196 (1885).

BECHTEREW, W. VON: Die Leitungsbahnen im Gehirn und Rückenmark (Georgi, Leipzig 1889).

BELLONCI, J.: Über die centrale Endigung des Nervus opticus bei den Vertebraten. Z. wiss. Zool. *47:* 1–46 (1888).

BENEDETTI, E.: Il cervello e i nervi cranici del Proteus anguineus Laur. (Mem. Ist. ital. Speleolog. Ser. biol., III, 77 pp., 1933).

BERNHEIMER, S.: Experimentelle Studien zur Kenntnis der Innervation der inneren und äusseren, von Oculomotorius versorgten Muskeln des Auges. Graefes Arch. Ophthalm. *44:* 481–525 (1897).

BERNIS, W.J. und SPIEGEL, E.A.: Zentren der statischen Innervation und ihre Beeinflussung durch Gross- und Kleinhirn. Arb. neurol. Inst. Univ. Wien *27:* 197–224 (1925).

BODIAN, D.: An experimental study of the optic tracts and retinal projection of the Virginia opossum. J. comp. Neurol. *66:* 113–144 (1937).

BRAIN, W.R. and WALTON, J.W.: *Brain's* diseases of the nervous system, 7th ed. (Oxford University Press, London 1967).

BROUWER, B.: Klinisch-anatomische Untersuchungen über den Oculomotoriuskern. Z. ges. Neurol. Psychiat. *40:* 152–193 (1918).

BROUWER, B.; ZEEMAN, W.P.C., und HOUWER, A.W.: Experimentell-anatomische Untersuchungen über die Projektion der Retina auf die primären Opticuszentren. Schweiz. Arch. Neurol. Psychiat. *13:* 118–138 (1923).

BROWN, J.A.: Pigmentation of certain mesencephalic tegmental nuclei in the dog and cat. J. comp. Neurol. *81:* 249–257 (1944).

CAJAL, S. R. Y: Histologie du système nerveux de l'homme et des vertébrés, 2 vols. (Maloine, Paris 1909, 1911; Instituto Ramon y Cajal, Madrid 1952, 1955).

CAMPBELL, E.; KUHLENBECK, H.; CAVENAUGH, R. I., and NIELSEN, A. E.: Clinicopathological aspects of fatal missile-caused craniocerebral injuries. Chapt. XV, pp. 335–399, Surgery in World War II, Neurosurgery vol. 1 (Dept. of the Army, Washington 1958).

CARLSSON, A.: The occurrence, distribution, and physiological role of catecholamines in the nervous system. Pharmacol. Rev. *11:* 490–493 (1959).

CASSAGRANDE, V.A.; HARTING, J.K.; HALL, W.C., and DIAMOND, I.T.: Superior colliculus of the tree shrews: A structural and functional subdivision into superficial and deep layers. Science *177:* 444–447 (1972).

CASTALDI, L.: Studi sulla struttura e sullo sviluppo del mesencefalo. Ricerche in Cavia Cobaya. I, II, III, IV. Arch. ital. Anat. Embriol. *20:* 23–225 (1923); *21:* 172–263 (1924); *23:* 481–609 (1926); *25:* 157–306 (1928).

CATOIS, E. M.: Recherches sur l'histologie et l'anatomie microscopiques de l'encéphale chez les poissons. Bull. scient. France et Belgique *36:* 1–166 (1901).

CHARLTON, H. H.: The optic tectum and its related fiber tracts in blind fishes. A. Troglichthys Rosae and Typhlichthys Eigenmanni. J. comp. Neurol. *57:* 285–325 (1933).

CLARA, M.: Das Nervensystem des Menschen, 3. Aufl. (Barth, Leipzig 1959).

CRAGG, B.G.; EVANS, D.H.L., and HAMLYN, L.H.: The optic tectum of Gallus domesticus: a correlation of the electrical responses with the histological structure. J. Anat. *88:* 292–307 (1954).

CRAIGIE, E. H.: An introduction to the finer anatomy of the central nervous system based upon that of the albino rat (Blakiston, Philadelphia 1925).

CRAIGIE, E. H.: Observations on the brain of the Humming Bird (Chrysolampis mosquitus Linn., and Chlorostilbon Caribaeus Lawr.) J. comp. Neurol. *45:* 377–481 (1928).

CRAIGIE, E. H.: Studies on the brain of the Kiwi (Apteryx australis). J. comp. Neurol. *49:* 223–357 (1930).

CRAIGIE, E. H.: The cell masses in the diencephalon of the Humming Bird. Proc. Kon. Akad. Wetensch. Amsterdam *34:* 1038–1050 (1931).

CROSBY, E.C.; HUMPHREY, T., and LAUER, E.W.: Correlative anatomy of the nervous system (Macmillan, New York 1962).

CURWEN, A.O. and MILLER, R. N.: The pretectal region of the turtle, Pseudemys scripta troostii. J. comp. Neurol. *71:* 99–120 (1939).

DEJERINE, J.: Anatomie des centres nerveux, 2 vols. (Rueff, Paris 1895, 1901).

DEJERINE, J.: Sémiologie des affections du système nerveux (Masson, Paris 1914).

DEXLER, H.: Zur Anatomie des Zentralnervensystems von Elephas indicus. Arb. neurol. Inst. Wiener Univers. *15:* 137–281 (1907).

EBBESSON, S.O.E.: New insights into the organization of the shark brain. Comp. Biochem. Physiol. *42A:* 121–129 (1972).

EDINGER, L.: Über das Gehirn von Myxine glutinosa. Phys. Abh. Kgl. preuss. Akad. Wiss. 1906.

EDINGER, L.: Vorlesungen über den Bau der nervösen Zentral-Organe des Menschen und der Tiere, Bd. I: Das Zentralnervensystem der Menschen und der Säugetiere, 8. Aufl.; Bd. II: Vergleichende Anatomie des Gehirns, 7. Aufl. (Vogel, Leipzig 1911, 1908).

EDINGER, L. und WALLENBERG, A.: Untersuchungen über das Gehirn der Tauben. Anat. Anz. *15:* 245–271 (1899).

FOREL, A.: Untersuchungen über die Haubenregion und ihre oberen Verknüpfungen im Gehirn des Menschen und einiger Säugethiere, mit Beiträgen zu den Methoden der Gehirnuntersuchung. Arch. Psychiat. *7:* 393–495 (1877).

FOREL, A.: Gesammelte hirnanatomische Abhandlungen (Reinhardt, München 1907).

FREY, E.: Die basale optische Wurzel des Meerschweinchens. Proc. Kon. Akad. Wetensch. Amsterdam *38:* 775–783 (1935).

FREY, E.: Vergleichend-anatomische Untersuchungen über die basale optische Wurzel, die Commissura transversa Gudden und über eine Verbindung der Netzhaut mit dem vegetativen Gebiet im Hypothalamus durch eine dorsale hypothalamische Wurzel des Nervus opticus bei Amnioten (Orell Füssli, Zürich 1937; also Schweiz. Arch. Neurol. Psychiat. *39* and *40*, 1937).

FREY, E.: Studien über die hypothalamische Optikuswurzel der Amphibien. I. Rana mugiens, Rana esculenta, Bombinator pachypus und Pipa pipa. II. Proteus anguineus und die phylogenetische Bedeutung der hypothalamischen Optikuswurzel. Proc. Kon. Akad. Wetensch. Amsterdam *41:* 1003–1021 (1938).

FULTON, J. F.: Physiology of the nervous system, 3rd ed. (Oxford University Press, New York 1949).

GERLACH, J.: Über das Gehirn von Protopterus annectens. Ein Beitrag zur Morphologie des Dipnoerhirnes. Anat. Anz. *75:* 310–406 (1933).

GERLACH, J.: Beiträge zur vergleichenden Morphologie des Selachierhirnes. Anat. Anz. *96:* 79–165 (1947).

GILLILAN, L.A.: The connections of the basal optic root (posterior accessory optic tract) and its nucleus in various mammals. J. comp. Neurol. *74:* 367–408 (1941).

GIOLLI, R.A. and POPE, J.E.: The anatomical organization of the visual system in the rabbit. Documenta ophthalmol. *30:* 9–31 (1971).

GRAEBER, R.C. and EBBESSON, S.O.E.: Visual discrimination learning in normal and tectal-ablated Nurse Sharks (Ginglymostoma cirratum). Comp. Biochem. Physiol. *42A:* 131–139 (1972).

GRAEBER, R.C.; EBBESSON, S.O.E., and JANE, J.A.: Visual discrimination in sharks without optic tectum. Science *180:* 413–415 (1973).

GRANIT, R.: Receptors and sensory perception (Yale University Press, New Haven 1955).

GUDDEN, B. v.: Mitteilung über das Ganglion interpedunculare. Arch. Psychiat. *11:* 424–427 (1881).

HAMDI, F.A., and WHITTERIDGE, D.: The representation of the retina on the optic tectum of the pigeon. J. exper. Physiol. *39:* 111–118 (1954).

HASSLER, R.: Anatomy of the thalamus; in SCHALTENBRAND and BAILEY Introduction to stereotaxis with an atlas of the human brain, pp. 230–290 (Thieme, Stuttgart 1959).

HATSCHEK, R. und SCHLESINGER, H.: Der Hirnstamm des Delphins (Delphinus delphis). Arb. neurol. Inst. Wiener Univ. *9:* 1–117 (1902).

HAYHOW, W. R.; WEBB, C., and JERVIE, A.: The accessory optic fiber system in the rat. J. comp. Neurol. *115:* 187–215 (1960).

HAYMAKER, W. and BING, R.: *Bing's* local diagnosis in neurological diseases, 14th ed., 15th ed. (Mosby, St. Louis, 1956, 1969).

HAYMAKER, W. and KUHLENBECK, H.: Diseases of the brain stem and its cranial nerves. Chapter 23, vol 2, pp. 1260–1324, 1st ed.; chapter 29, vol. 3, pp. 1456–1526, 2nd ed.; chapter 30, vol. 3, pp. 1–82, 3rd ed. in BAKER Clinical Neurology (Hoeber, Harper & Row, New York 1955, 1962, 1971).

HEIER, P.: Fundamental principles in the structure of the brain. A study of the brain of Petromyzon fluviatilis (Håkan, Lund 1948; also Acta anat. Suppl. VI).

HERRICK, C.J.: The amphibian forebrain. III. The optic tracts and centers in Ambystoma and the frog. J. comp. Neurol. *39:* 433–489 (1925).

HERRICK, C.J.: An introduction to neurology (Saunders, Philadelphia 1931).

HERRICK, C.J.: Optic and postoptic systems in the brain of Ambystoma tigrinum. J. comp. Neurol. *77:* 191–353 (1942).

HERRICK, C.J.: The brain of the tiger salamander Ambystoma tigrinum (University of Chicago Press, Chicago 1948).

HINES, M.: The brain of Ornithorhynchus anatinus. Philos. Trans. roy. Soc. London Ser. B., vol. *217:* 155–287 (1929).

HOLLÄNDER, P. P.: Über den Ursprung der aus dem Mittelhirn im dorsalen Längsbündel absteigenden Nervenfasern bei Sauropsiden. Jen. Z. Naturwiss. *55:* 203–220 (1917).

HOLMGREN, N.: Zur Anatomie des Gehirns von Myxine. Kungl. svensk. vet. Akad. Handl. 60, part. 7: 1–96 (1919).

HOLMGREN, N. and HORST, C.J. VAN DER: Contributions to the morphology of the brain in Ceratodus. Acta Zool. *6:* 59–165 (1925).

HOOGENBOOM, K.J. HOCKE: Das Gehirn von Polyodon folium Lacép. Z. mikr.-anat. Forsch. *18:* 311–392 (1929).

HORNYKIEWICZ, O.: Dopamine (3-hydroxytyramine) and brain function. Pharmacol. Rev. 18*:* 925–964 (1966).

HOUSER, G.L.: The neurones and supporting elements of the brain of a selachian. J. comp. Neurol. *11: 6–175 (1901).*

HUBER, G.C. and CROSBY, E.C.: On thalamic and tectal nuclei and fiber paths in the brain of the American Alligator. J. comp. Neurol. *40:* 97–227 (1926).

HUBER, G.C. and CROSBY, E.C.: The nuclei and fiber paths of the avian diencephalon, with consideration of some telencephalic and certain mesencephalic connections. J. comp. Neurol. *48:* 1–225 (1929).

HUBER, G.C. and CROSBY, E.C.: The reptilian optic tectum. J. comp. Neurol. *57:* 57–163 (1933).

HUBER, G.C.; CROSBY, E.C.; WOODBURNE, R.T.; GILLILAN, L.A.; BROWN, J.O., and TAMTHAI, B.: The mammalian midbrain and isthmus regions. J. comp. Neurol. *78:* 129–534 (1943).

INGLE, D.: Disinhibition of tectal neurons by pretectal lexions in the frog. Science *180:* 422–424 (1973a).

INGLE, D.: Two visual systems in the Frog. Science *181:* 1053–1055 (1973b).

INGLE, D. and SCHNEIDER, G. E. (eds): Subcortical visual systems. Brain, Behav. Evol. *3:* 2–352 (1970).

INNES, J. R. M. and SAUNDERS, L. Z.: Comparative Neuropathology (Academic Press, New York 1962).

JACOBSOHN, L.: Über die Kerne des menschlichen Hirnstamms (Medulla oblongata, Pons, und Pedunculus cerebri). Abh. d. k. preuss. Akad. wiss. phys.-math. Kl., Anhang (1909).

JACOBSON, M.: Retinal ganglion cells: specification of central connections in larval Xenopus laevis. Science *155:* 1106–1108 (1967).

JANSEN, J.: A note on the optic tract in teleosts. Proc. kon. Akad. Wetensch. Amsterdam *32:* 1105–1117 (1929).

Jansen, J.: The brain of Myxine glutinosa. J. comp. Neurol. *49:* 359–507 (1930).

Jansen, J. und Brodal, A.: Das Kleinhirn; in Möllendorff und Bargmann Handbuch der mikroskopischen Anatomie des Menschen, vol. IV, 8 (Springer, Berlin 1958).

Johnston, J. B.: The nervous system of vertebrates (Blakiston, Philadelphia 1906).

Jungherr, E.: Certain nuclear groups of the avian mesencephalon. J. comp. Neurol. *82:* 55–175 (1945).

Kappers, C. U. A.: Die vergleichende Anatomie des Nervensystems der Wirbeltiere und des Menschen, 2 vols. (Bohn, Haarlem 1920, 1921).

Kappers, C. U. A.: Anatomie comparée du système nerveux particulièrement de celui des mammifères et de l'homme. Avec la collaboration de E. H. Strasburger (Masson, Paris 1947).

Kappers, C. U. A. und Carpenter, F. W.: Das Gehirn von Chimaera monstrosa. Fol. neurobiol. *5:* 127–160 (1911).

Kappers, C. U. A. und Hammer, E.: Das Zentralnervensystem des Ochsenfrosches (Rana catesbyana). Psych. en neurol. Bl. 1918, Feestbundel *Winkler:* 368–415 (1918).

Kappers, C. U. A.; Huber, G. C., and Crosby, E. C.: The comparative anatomy of the nervous system of vertebrates, including man., 2 vols. (Macmillan, New York 1936).

Karten, H.: Projections of the optic tectum of the pigeon (Columba livia). (Abstract). Anat. Rec. *151:* 369 (1965).

Karten, H. J. and Hodos, W.: A stereotaxic atlas of the brain of the pigeon (Columba livia). (The Johns Hopkins Press, Baltimore 1967).

Keller, A. D. and Hare, W. K.: The rubrospinal tract in the monkey. Effects of experimental section. Arch. Neurol. Psychiat. *32:* 1253–1272 (1934).

Kirsche, K. und Kirsche W.: Experimentelle Untersuchungen zur Frage der Regeneration und Funktion des Tectum opticum von Carassius carassius L. Z. mikr.-anat. Forsch. *67:* 140–182 (1961).

Kirsche, W.: Zur Frage der Regeneration des Mittelhirns der Teleostei. Verh. anat. Ges. 56. Vers., Erg. H. Anat. Anz. *106/107:* 259–270 (1960).

Knowlton, V. Y.: Abnormal differentiation of embryonic avian brain centers associated with unilateral anophthalmia. Acta anat. *58:* 222–254 (1964).

Kosaka, K. und Hiraiwa, K.: Zur Anatomie der Sehnervenbahnen und ihrer Zentren. Fol. neurobiol. *9:* 367–389 (1915).

Kreht, H.: Über die Faserzüge im Zentralnervensystem von Salamandra maculosa L. Z. mikr.-anat. Forsch. *23:* 239–320 (1930).

Kreht, H.: Über die Faserzüge im Zentralnervensystem von Proteus anguineus L. Z. mikr.-anat. Forsch. *25:* 376–427 (1931). *48:* 192–285 (1940).

Kudo, K.: Über den Torus longitudinalis der Knochenfische. Anat. Anz. *56:* 359–367 (1923).

Kudo, K.: Beiträge zur Anatomie des Zwischen- und Mittelhirns der Knochenfische. III. Eine frontale Verbindung des Torus longitudinalis. Anat. Anz. *57:* 271–275 (1924).

Kuhlenbeck, H.: Zur Morphologie des Gymnophionengehirns. Jen. Z. Naturw. *58:* 453–484 (1922).

Kuhlenbeck, H.: Vorlesungen über das Zentralnervensystem der Wirbeltiere (Fischer, Jena 1927).

Kuhlenbeck, H.: Über die morphologische Stellung des Corpus geniculatum mediale. Anat. Anz. *81:* 28–37 (1935).

Kuhlenbeck, H.: The development and structure of the pretectal cell masses in the chick. J. comp. Neurol. *71:* 361–387 (1939).

Kuhlenbeck, H.: The human diencephalon. A summary of development, structure, function, and pathology. Suppl. ad vol. 14, Confin. neurol. (Karger, Basel 1954).

Kuhlenbeck, H.: Die Formbestandteile der Regio praetectalis des Anamnier-Gehirns und ihre Beziehungen zum Hirnbauplan. Fol. anat. japon. (*Nishi* Festschrift) *28:* 23–44 (1956).

Kuhlenbeck, H.: Brain and consciousness. Some prolegomena to an approach of the problem (Karger, Basel 1957).

Kuhlenbeck, H.: Morphologic significance of the so-called peripeduncular nucleus in the mammalian brain (Abstract). Anat. Rec. *142:* 314 (1962).

Kuhlenbeck, H.: Gehirn und Bewusstsein (Translated by Prof. J. Gerlach and Dr. U. Protzer). Erfahrung und Denken. Schriften zur Förderung der Beziehungen zwischen Philosophie und Einzelwissenschaften, vol. 39 (Duncker & Humblot, Berlin 1973).

Kuhlenbeck, H.; Hafkesbring, R., and Ross, M.: Further observations on a living 'decorticate' (hydranencephalic) child. J. amer. med. Women's Ass. *14:* 216–225 (1959).

Kuhlenbeck, H. and Maher, I.: Decerebrate rigidity in man (Abstract). Anat. Rec. *127:* 427 (1957).

Kuhlenbeck, H.; Maher, I.; Ross, M., and Eastwood, R.: Hydranencephaly with univentricular telencephalic malformation. General comments, with clinical observations on three cases. Confin. neurol. *17:* 100–118 (1957).

Kuhlenbeck, H. and Miller, R. N.: The pretectal region of the rabbit's brain. J. comp. Neurol. *76:* 323–365 (1942).

Kuhlenbeck, H. and Miller, R. N.: The pretectal region of the human brain. J. comp. Neurol. *91:* 369–407 (1949).

Kuhlenbeck, H. and Niimi, K.: Further observations on the morphology of the brain in the Holocephalian Elasmobranchs Chimaera and Callorhynchus. J. Hirnforsch. *11:* 267–314 (1969).

Kuhlenbeck, H. and Szekely, E. G.: Evoked potentials from tectum mesencephali and telencephalon of the chicken after unilateral optic stimulation (Abstract). Anat. Rec. *145:* 332 (1963).

Kuntz, A.: A text-book of neuroanatomy, 5th ed. (Lea & Febiger, Philadelphia 1950).

Land, E. H.: Experiments in color vision; in Teevan and Birney, pp. 162–183, Color vision (Van Nostrand, New York 1961).

De Lange, S. J.: Das Zwischenhirn und das Mittelhirn der Reptilien. Fol. neurobiol. *7:* 67–138 (1913).

Larsell, O. and Jansen, J.: The comparative anatomy and histology of the cerebellum. The human cerebellum, cerebellar connections, and cerebellar cortex (University of Minnesota Press, Minneapolis 1972).

Lashley, K. S.: The mechanism of vision. VII. The projection of the retina upon the primary centers in the rat. J. comp. Neurol. *59:* 341–373 (1934).

Laufer, M. and Vanegas, H.: The optic tectum of a perciform Teleost. II. Fine structure. J. comp. Neurol. *154:* 61–95 (1974).

Laufer, M. and Vanegas, H.: The optic tectum of a perciform Teleost. III. Electron microscopy of degenerating retino-tectal afferents. J. comp. Neurol. *154:* 97–115 (1974).

LEGHISSA, S.: La struttura microscopica e la citoarchitettonica del tetto ottico dei pesci teleostei. Z. Anat. Entw. Gesch. *118:* 427–463 (1955).

LETTVIN, J.Y.; MATURANA, H. R.; MCCULLOCH, W. S., and PITTS, W. H.: What the frog eye tells the frog's brain. Proc. Inst. Radio Engin (IRE) *47:* 1940–1957 (1959).

LINDVALL, O.; BJORKLUND, A.; NOBIN, A., and STENEVI, U.: The adrenergic innervation of the rat thalamus as revealed by the glyoxylic acid fluorescence method. J. comp. Neurol. *154:* 317–347 (1974).

LLOYD, K. and HORNYKIEWICZ, O.: Parkinson's disease: activity of 1-dopa decarboxylase in discrete brain regions. Science *170:* 1212–1213 (1970).

DE LONG, G.R. and COULOMBRE, A.J.: Development of the retino-tectal topographic projection in the chick embryo. Exper. Neurol. *13:* 351–363 (1965).

MAGNUS, R.: Körperstellung (Springer, Berlin 1924).

MANLEY, J.A.: Single unit studies in the midbrain auditory area of Caiman. Z. vergl. Physiol. *71:* 255–261 (1971).

MARBURG, O.: Mikroskopisch-topographischer Atlas des menschlichen Zentralnervensystems, 3. Aufl. (Deuticke, Leipzig 1927).

MARINA, A.: Die Relationen des Palaenceephalon (Edinger) sind nicht fix. Neurol. Centralbl. *34:* 338–345 (1915).

MATURANA, H. R.; LETTVIN, J.Y.; MCCULLOCH, W. S., and PITTS, W. H.: Anatomy and Physiology of vision in the frog (Rana pipiens). J. gen. Physiol. *43*, Suppl. 2: 129–175 (1960).

MESDAG, T. M.: Bijdrag tot de ontwikkelings-geschiednis van de structuur der hersenen bij het kip (Dissertation, Groningen 1909).

METTLER, F.A.: The tegmento-olivary and central tegmental fasciculi. J. comp. Neurol. *80:* 149–175 (1944).

MEYNERT, TH.: Vom Gehirne der Säugethiere; in STRICKER Handbuch der Lehre von den Geweben des Menschen und der Thiere, Bd. 2, Cap. 31, pp. 694–808 (Engelmann, Leipzig 1872).

NISHI, S.: Muskeln des Rumpfes. Muskeln des Kopfes. Parietale Muskulatur; in BOLK *et al.* Handbuch der vergleichenden Anatomie, vol. 5, pp. 351–406 (Urban & Schwarzenberg, Berlin 1938).

NOBLE, G. K.: The biology of the Amphibia (McGraw-Hill, New York 1931; Dover, New York 1954).

NORGREN, R. and LEONARD, CH.M.: Ascending gustatory pathways. J. comp. Neurol. *150:* 217–237 (1973).

NORRIS, H.W.: The eyeball and associated structures in the blindworms. Proc. Iow. Acad. *24:* 299–300 (1917).

OBERSTEINER, H.: Anleitung beim Studium des Baues der nervösen Zentralorgane im gesunden und kranken Zustande (Deuticke, Leipzig 1912).

OGAWA, T.: Über den nucleus ellipticus und den nucleus ruber beim Delphin. Arb. anat. Inst. kaiserl. japan. Univers. Sendai *17:* 55–61 (1935).

OLIVEIRA CASTRO, G. DE: Morphological data on the brain of Electrophorus electricus (L.); in CHAGAS and PAES DE CARVALHO Bioelectrogenesis, pp. 171–184 (Elsevier, Amsterdam 1961).

OLSSON, R.: The development of Reissner's fibre in the brain of the Salmon. Acta zool. *37:* 235–250 (1956).

OLSSON, R. and WINGSTRAND, K.C. (1954) cf. references to chapter VIII.

Olzewski, J. and Baxter, D.: Cytoarchitecture of the human brain stem (Karger, Basel 1954).

Palmgren, A.: Embryological and morphological studies on the midbrain and cerebellum of vertebrates. Acta zool. *2:* 1–94 (1921).

Parkinson, J.: Essay on the shaking palsy (Sherwood, Neely, and Jones, London 1817).

Pearson, R.: The Avian Brain (Academic Press, London-New York 1972).

Pollock, L.J. and Davis, L. E.: The reflex activities of a decerebrate animal. J. comp. Neurol. *50:* 377–411 (1930).

Portmann, A. and Stingelin, W.: The central nervous system; in Marshall The biology and comparative physiology of birds, vol. 2, pp. 1–36 (Academic Press, New York 1961).

Přcechtěl, A.: Some notes upon the finer anatomy of the brain stem and basal ganglia of Elephas indicus. Proc. kon. Akad. Wetensch. Amsterdam *28:* 81–93 (1925).

Pritz, M.B.: Ascending connections of a midbrain auditory area in a Crocodile, Caiman crocodilus. J. comp. Neurol. *153:* 179–197 (1974).

Probst, M.: Experimentelle Untersuchungen über die Anatomie und Physiologie der Leitungsbahnen des Gehirnstammes. Arch. Anat. Physiol. Anat. Abtg. 1902, Suppl.: 147–254 (1902).

Quay, W. B.: A mid-aqueductal ependymal organ in the brain of the Hyrax (Procavia capensis). J. comp. Neurol. *142:* 249–256 (1971).

Rademacher, G.G.J.: Die Bedeutung der roten Kerne und des übrigen Mittelhirns für Muskeltonus, Körperstellung und Labyrinthreflexe. Übers. von E. Le Blanc (Springer, Berlin 1926).

Ramon, P.: (quoted by Cajal, 1911, 1955).

Ranson, S.W.: The anatomy of the nervous system from the standpoint of development and function, 7th ed., 9th and 10th ed. revised by S. L. Clark (Saunders, Philadelphia 1943, 1953, 1961).

Rasmussen, A.T. and Peyton, W.T.: The location of the lateral spinothalamic tract in the brain stem of man. Surgery *10:* 699–710 (1941).

Rasmussen, A.T. and Peyton, W.T.: Course and termination of the medial lemniscus in man. J. comp. Neurol. *88:* 411–424 (1948).

Richter, W.: Regeneration im Tectum opticum bei Leucaspius delineatus (Heckel 1843). Z. mikr.-anat. Forsch. *74:* 46–68 (1965).

Riley, H.A.: An atlas of the basal ganglia, brain stem and spinal cord. Based on myelin-stained material (Williams & Wilkins, Baltimore 1943).

Rogers, K.T.: Ocular muscle proprioceptive neurons in the developing chick. J. comp. Neurol. *107:* 427–435 (1957).

Rogers, L.A. and Cowan, W.M.: The development of the mesencephalic nucleus of the trigeminal nerve in the Chick. J. comp. Neurol. *147:* 291–319 (1973).

Romanoff, A. L.: The avian embryo. Structural and functional development (Macmillan, New York 1960).

Röthig, P.: Beiträge zum Studium des Zentralnervensystems der Wirbeltiere. Über die Faserzüge im Mittelhirn, Kleinhirn und der Medulla oblongata der Urodelen und Anuren. Z. mikr.-anat. Forsch. *10:* 381–472 (1927).

Routtenberg, A. and Holzman, N.: Memory disruption by electrical stimulation of substantia nigra, pars compacta. Science *181:* 83–85 (1973).

Saito, T.: Über die Müllerschen Zellen im Gehirn des japanischen Flussneunauges (Entosphenus japonicus Martens). Fol. anat. japon. *6:* 457–473 (1928).

SAITO, T.: Über das Gehirn des japanischen Flussneunauges (Entosphenus japonicus Martens) Fol. anat. japon. *8:* 189–263 (1930a).

SAITO, T: Über die retikulären Zellen im Gehirn des japanischen Dornhaies (Acanthias mitsukurii Jordan et Fowler). Fol. anat. japon. *8:* 323–343 (1930b).

SANDERS, E. B.: A consideration of certain bulbar, midbrain, and cerebellar centers and fiber tracts in birds. J. comp. Neurol. *45:* 155–221 (1929).

SCALA, N. P. and SPIEGEL, E. A.: The pupillary reaction in combined lesions of the posterior commissure and of the pupillodilator tracts. A contribution to the pathogenesis of the Argyll Robertson pupil. Arch. Ophth. *15:* 195–216 (1936).

SCHNEIDER, G. E.: Two visual systems: brain mechanisms for localization and discrimination are dissociated by tectal and cortical lesions. Science: *163:* 895–902 (1969).

SCHNEIDER, R.: Morphologische Untersuchungen am Gehirn der Chiroptera (Mammalia). Abh. Senkenberg. naturf. Ges. *495:* 1–92 (1957).

SCHOBER, W.: Vergleichend-anatomische Untersuchungen am Gehirn der Larven und adulten Tiere von Lampetra fluviatilis (LINNÉ, 1784) und Lampetra planeri (BLOCH, 1784). Journ. f. Hirnforsch. *7:* 107–209 (1964).

SENN, D. G.: Über das optische System im Gehirn squamater Reptilien. Acta anat., Suppl. 1 ad vol. *65:* 1–87 (1966).

SENN, D. G.: Bau und Ontogenese von Zwischen- und Mittelhirn bei Lacerta sicula (Rafinesque). Acta anat., Suppl. 1 ad vol. *71:* 1–150 (1968a).

SENN, D. G.: Der Bau des Reptiliengehirns im Licht neuer Ergebnisse. Verh. naturf. Ges. Basel *79:* 25–43 (1968b).

SENN, D. G.: The saurian and ophidian colliculi posteriores of the midbrain. Acta anat. *74:* 114–120 (1969a).

SENN, D. G.: Über das Zwischen- und Mittelhirn von zwei typhloiden Schlangen, Anomalepis aspinosus und Liotyphlops albirostris. Verh. naturf. Ges. Basel *80:* 32–48 (1969b).

SENN, D. G.: Über die Bedeutung der Hirnmorphologie für die Systematik. Eine Untersuchung mit Beispielen der Reptilien und Vögel. Verh. naturf. Ges. Basel *80:* 49–55 (1969c).

SENN, D. G.: The stratification in the reptilian central nervous system. Acta anat. *75:* 521–552 (1970).

SHANKLIN, W. M.: On diencephalic and mesencephalic nuclei and fibre paths in the brain of deep sea fish. Phil. Trans. roy. Soc., Ser. B, vol. *224:* 361–419 (1935).

SHERRINGTON, C. S.: Decerebrate rigidity, and reflex coordination of movements. J. Physiol. *22:* 319–332 (1898).

SHINTANI-KUMAMOTO, Y.: The nuclei of the pretectal region of the mouse brain. J. comp. Neurol. *113:* 43–60 (1959).

SPATZ, H.: Über den Eisennachweis im Gehirn, besonders in Zentrent des extrapyramidal-motorischen Systems. Z. ges. Neurol. Psychiat. *77:* 261–390 (1922).

SPERRY, R. W.: Optic nerve regeneration with return of vision in anurans. J. Neurophysiol. *7:* 57–69 (1944).

SPERRY, R. W.: Restoration of vision after crossing of optic nerves and after contralateral transplantation of eye. J. Neurophysiol. *8:* 15–28 (1947).

SPIEGEL, E. A.: Further experiments on the localization of the Argyll Robertson phenomenon (injuries to the posterior commissure). Vol. jubil. en l'honneur du Prof. G. MARINESCO, Bucharest, pp. 625–633 (1933).

SPIEGEL, E.A. and WYCIS, H.T.: Stereoencephalotomy (thalamotomy and related procedures). Part I. Methods and stereotaxic atlas of the human brain. Part II. Clinical and physiologic applications (Grune & Stratton, New York 1952, 1962).

STEFANELLI, A.: I centri tegmentali dell'encefalo dei Petromyzonti. Arch. zool. ital. *20:* 117–202 (1934).

STEINER, J.: Die Functionen des Centralnervensystems und ihre Phylogenese. I–IV. (Vieweg, Braunschweig 1885, 1888, 1898, 1900).

STERZI, G.: Il sistema nervoso centrale dei vertebrati. Vol. 1. Ciclostomi. Vol. 2. Pesci, Selaci. Vol. 3. Sviluppo (Draghi, Padova 1907, 1909, 1912).

STONE, L. S.: Heteroplastic transplantation of eyes between the larvae of 2 species of Amblystoma. J. exper. zool. *55:* 193–261 (1930).

STONE, L. S.: Return of vision and functional polarization in the retinae of transplanted eyes. Trans. ophthal. Soc. U. Kingd. *67:* 349–367 (1947).

STONE, L. S.: Normal and reversed vision in transplanted eyes. Arch. Ophthal. *49:* 28–35 (1953).

STRÖER, W. F. H.: Zur vergleichenden Anatomie des primären optischen Systems bei Wirbeltieren. Z. Anat. EntwGesch. *110:* 301–321 (1939a).

STRÖER, W. F. H.: Über den Faserverlauf in den optischen Bahnen bei Amphibien. Proc. kon. nederl. Akad. Wetensch. *42:* 649–656 (1939b).

STRÖER, W. F. H.: Das optische System beim Wassermolch (Triturus taeniatus). Acta neerl. Morphol. *3:* 178–195 (1940).

SUGA, N. and SHIMOZAWA, T.: Site of neural attenuation responses to self-vocalized sounds in echolocating Bats. Science *183:* 1211–1212 (1974).

SZABO, TH.: Anatomo-physiologie des centres nerveux spécifiques de quelques organes électriques; in CHAGAS and PAEZ DE CARVALHO Bioelectrogenesis, pp. 185–201 (Elsevier, Amsterdam-New York 1961).

TARTUFERI, F.: Sull'anatomia minuta dell'eminenze bigemine anteriori dell'uomo, Arch. ital. Mal. nerv. *22:* 3–37 (1885).

TELLO, F.: Contribución al conocimiento del encephalo de los teleósteos. I. Los nucleos bulbares. Trab. Lab. Invest. biol. Univers. Madrid *7:* 1–29 (1909).

TIENHOVEN, A. VAN and JUHASZ, L. P.: The chicken telencephalon, diencephalon and mesencephalon in stereotaxic coordinates. J. comp. Neurol. *118:* 185–198 (1962).

TSAI, CH.: The optic tracts and centers of the opossum, Didelphys virginiana. J. comp. Neurol. *39:* 173–216 (1925).

TUGE (TSUGE), H.: Somatic motor mechanisms in the midbrain and medulla oblongata of Chrysemys elegans (Wied). J. comp. Neurol. *55:* 185–271 (1932).

VANEGAS, H.; LAUFER, M., and AMAT, J.: The optic tectum of a perciform Teleost. I. General configuration and cytoarchitecture. J. comp. Neurol. *154:* 43–60 (1974).

VERHAART, W. J. C.: Comparative anatomical aspects of the mammalian brain stem and the cord, 2 vols. (Van Gorkum, Assen 1970).

VOGT, C. und VOGT, O.: Über die Neuheit und den Wert des Pathoklisenbegriffes. J. Psychol. Neurol. *38:* 147–154 (1929).

WALLENBERG, A.: Die secundäre Acusticusbahn der Taube. Anat. Anz. *14:* 357–369 (1898).

WALLENBERG, A.: Neue Untersuchungen über den Hirnstamm der Taube. Anat. Anz. *24:* 357–369 (1904).

WALLENBERG, A.: Beitrag zur Kenntnis der Sehbahnen der Knochenfische. Le Névraxe *14:* 251–275 (1913).

Warner, F. J.: The diencephalon and midbrain of the American rattlesnake (Crotalus adamanteus). Proc. Zool. Soc. *116:* 531–550 (1946).

Warwick, R.: Oculomotor organization; in Bender The oculomotor system, pp. 173–204 (Harper & Row, New York 1964).

Wlassak, R.: Die optischen Leitungsbahnen des Frosches. Arch. Anat. Physiol. Physiol. Abt., Suppl. Bd. 1: 1–28 (1893).

Wright, S. Jr.; Hilsz, J., and Locke, S.: Medial occipital projections to the nucleus of Darkschewitsch of the Rhesus monkey. Anat. Rec. *178:* 667–670 (1974).

Zeman, W. and Innes, J. R. M.: *Craigie's* neuroanatomy of the rat. Revised and expanded (Academic Press, New York 1963).

Zigmond, R. E.; Nottebohm, F., and Pfaff, D. W.: Androgen-concentrating cells in the midbrain of a songbird. Science *179:* 1005–1007 (1973).

Corrigenda to Volume 3/I

P. 257, line 6 from bottom, instead of: Schmidt-Lantermann, read: Schmidt-Lanterman.

P. 258, line 5 from bottom, likewise read Schmidt-Lanterman incisures (Lanterman, A. J.: Über den feineren Bau der markhaltigen Nervenfasern. Arch. mikr. Anat. *13:* 1–8, 1877).

P. 273, line 17 from bottom: instead of: efferent (sensory) center, read: afferent (sensory) center.

Corrigenda to Volume 3/II

P. 16, in Figure 2, on the scale at left, below 1, instead of: 10, read: 10^{-4}.

P. 64, line 1 from top, instead of: homorphic, read homomorphic.

P. 241, first line of footnote 92a, read: *contra Paganos* instead of: *contra Pagamos*.

P. 294, line 15 from top, read: or *Endleiste* instead of: on *Endleiste*.

P. 410, in legend to Figure 184 read: sf: sulcus lateralis hypothalami posterioris instead of: thalami posterioris.

P. 462, in legend to Figure 237A read: 16: lateral ganglionic hill (D_1 of telencephalon); 17: medial ganglionic hill (B_{1+2}); instead of: lateral ganglionic hill (B_1 of telencephalon), and medial ganglionic hill (B_2).

P. 919, line 11 from bottom, read: display the structure instead of: displays.